Periodic table of the elements[†]

Symbol — Cl 17 — Atomic number
Atomic mass[‡] — 35.453
$3p^5$ — Electron configuration

Group I	Group II	Transition elements										Group III	Group IV	Group V	Group VI	Group VII	Group 0
H 1 1.0080 $1s^1$																	He 2 4.0026 $1s^2$
Li 3 6.94 $2s^1$	Be 4 9.012 $2s^2$											B 5 10.81 $2p^1$	C 6 12.011 $2p^2$	N 7 14.007 $2p^3$	O 8 15.999 $2p^4$	F 9 18.998 $2p^5$	Ne 10 20.18 $2p^6$
Na 11 22.99 $3s^1$	Mg 12 24.31 $3s^2$											Al 13 26.98 $3p^1$	Si 14 28.09 $3p^2$	P 15 30.97 $3p^3$	S 16 32.06 $3p^4$	Cl 17 35.453 $3p^5$	Ar 18 39.948 $3p^6$
K 19 39.102 $4s^1$	Ca 20 40.08 $4s^2$	Sc 21 44.96 $3d^14s^2$	Ti 22 47.90 $3d^24s^2$	V 23 50.94 $3d^34s^2$	Cr 24 51.996 $3d^54s^1$	Mn 25 54.94 $3d^54s^2$	Fe 26 55.85 $3d^64s^2$	Co 27 58.93 $3d^74s^2$	Ni 28 58.71 $3d^84s^2$	Cu 29 63.54 $3d^{10}4s^1$	Zn 30 65.37 $3d^{10}4s^2$	Ga 31 69.72 $4p^1$	Ge 32 72.59 $4p^2$	As 33 74.92 $4p^3$	Se 34 78.96 $4p^4$	Br 35 79.91 $4p^5$	Kr 36 83.80 $4p^6$
Rb 37 85.47 $5s^1$	Sr 38 87.62 $5s^2$	Y 39 88.906 $4d^15s^2$	Zr 40 91.22 $4d^25s^2$	Nb 41 92.91 $4d^45s^1$	Mo 42 95.94 $4d^55s^1$	Tc 43 (99) $4d^55s^2$	Ru 44 101.1 $4d^75s^1$	Rh 45 102.91 $4d^85s^1$	Pd 46 106.4 $4d^{10}5s^0$	Ag 47 107.87 $4d^{10}5s^1$	Cd 48 112.40 $4d^{10}5s^2$	In 49 114.82 $5p^1$	Sn 50 118.69 $5p^2$	Sb 51 121.75 $5p^3$	Te 52 127.60 $5p^4$	I 53 126.90 $5p^5$	Xe 54 131.30 $5p^6$
Cs 55 132.91 $6s^1$	Ba 56 137.34 $6s^2$	57– 71[§]	Hf 72 178.49 $5d^26s^2$	Ta 73 180.95 $5d^36s^2$	W 74 183.85 $5d^46s^2$	Re 75 186.2 $5d^56s^2$	Os 76 190.2 $5d^66s^2$	Ir 77 192.2 $5d^76s^2$	Pt 78 195.09 $5d^96s^1$	Au 79 196.97 $5d^{10}6s^1$	Hg 80 200.59 $5d^{10}6s^2$	Tl 81 204.37 $6p^1$	Pb 82 207.2 $6p^2$	Bi 83 208.98 $6p^3$	Po 84 (210) $6p^4$	At 85 (218) $6p^5$	Rn 86 (222) $6p^6$
Fr 87 (223) $7s^1$	Ra 88 (226) $7s^2$	89–103[‖]	Rf 104 (261) $6d^27s^2$	Ha 105 (262) $6d^37s^2$	106 (263)	107 (261)											

§Lanthanide series	La 57 138.91 $5d^16s^2$	Ce 58 140.12 $5d^14f^16s^2$	Pr 59 140.91 $4f^36s^2$	Nd 60 144.24 $4f^46s^2$	Pm 61 (147) $4f^56s^2$	Sm 62 150.4 $4f^66s^2$	Eu 63 152.0 $4f^76s^2$	Gd 64 157.25 $5d^14f^76s^2$	Tb 65 158.92 $5d^14f^86s^2$	Dy 66 162.50 $4f^{10}6s^2$	Ho 67 164.93 $4f^{11}6s^2$	Er 68 167.26 $4f^{12}6s^2$	Tm 69 168.93 $4f^{13}6s^2$	Yb 70 173.04 $4f^{14}6s^2$	Lu 71 174.97 $5d^14f$
‖Actinide series	Ac 89 (227) $6d^17s^2$	Th 90 (232) $6d^27s^2$	Pa 91 (231) $5f^26d^17s^2$	U 92 (238) $5f^36d^17s^2$	Np 93 (239) $5f^56d^07s^2$	Pu 94 (239) $5f^66d^07s^2$	Am 95 (243) $5f^76d^07s^2$	Cm 96 (245) $5f^76d^17s^2$	Bk 97 (247) $5f^86d^17s^2$	Cf 98 (249) $5f^{10}6d^07s^2$	Es 99 (254) $5f^{11}6d^07s^2$	Fm 100 (253) $5f^{12}6d^07s^2$	Md 101 (255) $5f^{13}6d^07s^2$	No 102 (255) $6d^07s^2$	Lr (257 $6d^17s$

[†] Atomic mass values given are averaged over isotopes in percentages they occur in nature.
[‡] For an unstable element, mass number of the most stable known isotope is given in parentheses.

Physics

COVER PHOTOGRAPH

Laser beams (see Chapter 27) illuminate the evening sky over Kassel, Germany, in celebration of a modern-art exhibition. (Courtesy of Horst H. Baumann and Spectra-Physics, Inc.)

Physics
Principles with applications

Douglas C. Giancoli
California State Polytechnic University, Pomona

PRENTICE-HALL, INC.
Englewood Cliffs, New Jersey 07632

Douglas C. Giancoli: *Physics: Principles with Applications*

Printed in the United States of America

10 9 8 7 6 5

Library of Congress Cataloging in Publication data

Giancoli, Douglas C
Physics, principles with applications.

Includes bibliographies and index.
1. Physics. I. Title.
QC23.G399 530 79-14164
ISBN 0-13-672600-3

Prentice-Hall International, Inc., London
Prentice-Hall of Australia Pty. Limited, Sydney
Prentice-Hall of Canada, Ltd., Toronto
Prentice-Hall of India Private Limited, New Delhi
Prentice-Hall of Japan, Inc., Tokyo
Prentice-Hall of Southeast Asia Pte. Ltd., Singapore
Whitehall Books Limited, Wellington, New Zealand

For Andrea, Francesca, and Mary Teresa

Summary contents

	Complete contents	*ix*
	Preface	*xxi*
	Introduction	1
1	The description of motion: kinematics	7
2	Motion and force: dynamics	36
3	Circular and rotational motion; gravitation	60
4	Bodies in equilibrium	93
5	Momentum and energy	116
6	Fluids	147
7	Temperature and the kinetic theory	181
8	Applications of kinetic theory	208
9	Heat	226
10	The first and second laws of thermodynamics	250
11	Vibrations and waves	274
12	Sound	308
13	Electric charge and electric field	341
14	Electric potential and electric energy	364
15	Electric currents	384
16	Dc circuits and instruments	408
17	Magnetism	436
18	Electromagnetic induction and power generation	463
19	Ac circuits and electronics	489
20	Electromagnetic waves	520
21	Light: geometrical optics	537
22	The wave nature of light	563
23	Optical instruments	592
24	Relativity	618
25	Early quantum theory	648
26	Models of the atom; quantum mechanics	667
27	Quantum theory of atoms and molecules	688
28	The nucleus and radioactivity	710
29	Nuclear reactions and nuclear energy	737
30	Elementary particles	750

Appendix A Mathematical review 769
Appendix B Order of magnitude: rapid estimating 780
Appendix C Dimensional analysis 781
Appendix D Conversion factors 784
Appendix E Selected isotopes 785

Answers to odd-numbered Problems *788*
Index *794*

Complete contents

▮ Summary contents, vii

▮ Preface, xxi

▮ Introduction 1

▮ The practice of science, 1

▮ Physics and its relation to other fields, 3

▮ Models, theories, and laws, 4

▮ Mathematics, 5

For additional reading, 6

1 ▮ The description of motion: kinematics 7

1-1 ▮ Speed and velocity, 7

1-2 ▮ Standards and units, 10

1-3 ▮ Acceleration, 12

1-4 ▮ Uniformly accelerated motion, 14

1-5 ▮ Falling bodies, 17

1-6 ▮ Addition of vectors: relative velocity, 21

***1-7** ▮ Graphical analysis of linear motion, 27

Summary, 30 Questions, 31 Problems, 32

For additional reading, 35

2 ▮ Motion and force: dynamics 36

2-1 ▮ Force, 36

2-2 ▮ Galileo vs. Aristotle on horizontal motion, 37

2-3 ▮ Newton's laws of motion, 38

2-4 ▮ Units and systems of measurement, 42

2-5 ▌ Weight: the force of gravity, 44

2-6 ▌ Applications of Newton's laws: vector forces, 45

2-7 ▌ Projectile motion, 51

2-8 ▌ Notes on problem solving, 55

Summary, 55 Questions, 56 Problems, 57
For additional reading, 59

3 ▌ Circular and rotational motion; gravitation 60

3-1 ▌ Angular quantities, 60

3-2 ▌ Circular motion of a particle, 64

3-3 ▌ Gravitation, 71

***3-4** ▌ Kepler's laws and Newton's synthesis, 76

3-5 ▌ Dynamics of rotation, 79

3-6 ▌ Center of gravity, 83

3-7 ▌ General motion, 86

Summary, 86 Questions, 87 Problems, 88
For additional reading, 92

4 ▌ Bodies in equilibrium 93

4-1 ▌ Statics: the study of forces in equilibrium, 93

4-2 ▌ The conditions for equilibrium, 94

***4-3** ▌ Applications to muscles and joints, 98

4-4 ▌ Stability and balance, 100

4-5 ▌ Elasticity: stress, strain, and fracture, 101

***4-6** ▌ Spanning a space: arches and domes, 107

Summary, 110 Questions, 111 Problems, 111
For additional reading, 115

5 ▌ Momentum and energy 116

5-1 ▌ Momentum and impulse, 116

5-2 ▌ Conservation of momentum, 119

5-3 ▌ Angular momentum, 121

5-4 ▌ Work, 123

5-5 ▌ Energy: types and transformations, 127

5-6 ▌ The conservation of energy, 133

5-7 ▮ Power, 139

Summary, 141 Questions, 142 Problems, 142
For additional reading, 146

6 ▮ Fluids 147

6-1 ▮ Density and specific gravity, 147
6-2 ▮ Pressure and its measurement, 149
6-3 ▮ Buoyancy and Archimedes' principle, 156
*6-4 ▮ Surface tension and capillarity, 159
*6-5 ▮ Negative pressure and the cohesion of water; the rise of fluids in trees, 163
6-6 ▮ Fluids in motion: Bernoulli's principle, 164
*6-7 ▮ Viscosity, 170
*6-8 ▮ Flow in tubes—laminar and turbulent, 170
*6-9 ▮ Object moving in a fluid; sedimentation and drag, 173

Summary, 176 Questions, 177 Problems, 177
For additional reading, 180

7 ▮ Temperature and the kinetic theory 181

7-1 ▮ Atoms, 181
7-2 ▮ Temperature, 183
*7-3 ▮ The thermostat: an example of feedback, 186
7-4 ▮ Thermal expansion, 187
7-5 ▮ The ideal gas law, 191
7-6 ▮ Partial pressure, 197
7-7 ▮ Kinetic theory and the molecular interpretation of temperature, 197
7-8 ▮ Distribution of molecular speeds, 201

Summary, 204 Questions, 205 Problems, 205
For additional reading, 207

8 ▮ Applications of kinetic theory 208

*8-1 ▮ Vapor pressure and humidity, 208
*8-2 ▮ Real gases and changes of phase, 212

*8-3 ▌ Diffusion, 216

*8-4 ▌ Osmosis, 221

Summary, 223 Questions, 224 Problems, 225
For additional reading, 225

9 ▌ Heat 226

9-1 ▌ Heat as energy transfer, 226

9-2 ▌ Heat capacity and calorimetry, 229

9-3 ▌ Latent heat, 233

9-4 ▌ Heat transfer: conduction, convection, and radiation, 236

*9-5 ▌ Space heating and solar energy, 243

Summary, 245 Questions, 245 Problems 246
For additional reading, 249

10 ▌ The first and second laws of thermodynamics 250

10-1 ▌ The first law of thermodynamics, 250

10-2 ▌ Applications of the first law of thermodynamics, 251

*10-3 ▌ Human metabolism and the first law, 253

*10-4 ▌ Heat capacities of gases and the equipartition of energy, 255

10-5 ▌ The second law of thermodynamics; heat engines, 259

10-6 ▌ Entropy and probability, 264

Summary, 270 Questions, 270 Problems, 271
For additional reading, 273

11 ▌ Vibrations and waves 274

11-1 ▌ Simple harmonic motion, 274

11-2 ▌ The pendulum, 280

11-3 ▌ Nonharmonic motion, 283

11-4 ▌ Forced vibrations; resonance, 284

11-5 ▌ Wave motion, 286

11-6 ▌ Types of waves, 289

*11-7 Energy transmitted by waves, 292

11-8 Behavior of waves: reflection, refraction, interference, and diffraction, 293

11-9 Standing waves, 300

Summary, 302 Questions, 303 Problems, 304
For additional reading, 307

12 Sound 308

12-1 Characteristics of sound, 308

12-2 Sound intensity, 310

12-3 Sound detection; the ear, 313

12-4 Sources of sound; musical instruments, 316

12-5 Quality; music and noise, 321

12-6 Musical scales and harmony, 324

12-7 Interferences of sound waves; beats, 326

*12-8 Acoustic and medical applications of sound and ultrasound, 329

12-9 Doppler effect, 332

*12-10 Shock waves and the sonic boom, 334

Summary, 336 Questions, 337 Problems, 337
For additional reading, 340

13 Electric charge and electric field 341

13-1 Electric charges and forces, 341

13-2 Coulomb's law, 345

13-3 Electric field, 348

*13-4 Electric forces in molecular biology, 353

*13-5 Muscle action, 358

Summary, 360 Questions, 361 Problems, 362
For additional reading, 363

14 Electric potential and electric energy 364

14-1 Electric potential and potential difference, 364

14-2 Capacitors, 369

14-3 Electric energy storage, 371

14-4 Electric potential of single point charges, 373

*14-5 ▮ Electric dipoles, 374

*14-6 ▮ Interaction energy and weak bonds in molecular biology, 376

*14-7 ▮ The electrocardiogram, 378

Summary, 380 Questions, 381 Problems, 381
For additional reading, 383

15 ▮ Electric currents 384

15-1 ▮ The electric battery, 384

15-2 ▮ Electric current and Ohm's law, 387

15-3 ▮ Resistivity and superconductivity, 390

15-4 ▮ Electric power, 393

15-5 ▮ Alternating current, 395

*15-6 ▮ Electrolysis and electrophoresis, 397

*15-7 ▮ The nervous system and nerve conduction, 398

Summary, 403 Questions, 404 Problems, 405
For additional reading, 407

16 ▮ Dc circuits and instruments 408

16-1 ▮ Circuits involving resistors; internal resistance, 408

*16-2 ▮ Kirchhoff's rules, 412

16-3 ▮ Ammeters and voltmeters, 416

*16-4 ▮ The potentiometer, 420

*16-5 ▮ The Wheatstone bridge, 421

*16-6 ▮ Transducers and the thermocouple, 422

16-7 ▮ Circuits containing capacitors, 424

16-8 ▮ Effects of current on the human body, 427

Summary, 430 Questions, 431 Problems, 431
For additional reading, 435

17 ▮ Magnetism 436

17-1 ▮ Magnets and magnetism, 436

17-2 ▮ Electric currents produce magnetism, 439

17-3 ▮ Magnetic force on currents and moving charges, 441

17-4 Discovery and properties of the electron; the mass spectrograph, 445

17-5 Applications—meters, motors, and loudspeakers, 449

17-6 Determination of magnetic field strengths; Ampère's law, 451

***17-7** Operational definition of the ampere and coulomb; force between two parallel conductors, 455

***17-8** Hysteresis, 456

Summary, 458 Questions, 459 Problems, 460
For additional reading, 462

18 Electromagnetic induction and power generation 463

18-1 Induced emf and Faraday's law, 463

18-2 The electric generator, 469

18-3 Counter emf and torque; eddy currents, 471

18-4 Transformers; transmission of power, 473

***18-5** Inductance, 476

***18-6** Power plants and resources, 479

Summary, 484 Questions, 485 Problems, 486
For additional reading, 488

19 Ac circuits and electronics 489

***19-1** Ac circuits and impedance, 489

***19-2** *LRC* series circuit, 494

***19-3** Resonance in ac circuits, 497

***19-4** Thermionic emission; vacuum tubes, 498

***19-5** Cathode ray tube; television and oscilloscope, 500

***19-6** Semiconductors; diodes and transistors, 501

***19-7** Semiconductor diodes and rectification, 504

***19-8** Transistors; integrated circuits, 506

***19-9** Instrumentation; amplifiers, 509

***19-10** Feedback and control, 511

***19-11** ▮ Impedance matching, 513

Summary, 514 Questions, 516 Problems, 516
For additional reading, 519

20 ▮ Electromagnetic waves 520

20-1 ▮ Changing electric fields produce magnetic fields, 520

20-2 ▮ Production of electromagnetic waves, 523

20-3 ▮ Speed of EM waves and the electromagnetic spectrum, 525

***20-4** ▮ Energy in EM waves, 530

***20-5** ▮ Radio and television, 532

Summary, 534 Questions, 535 Problems, 535
For additional reading, 536

21 ▮ Light: geometrical optics 537

21-1 ▮ Characteristics of light; the ray model, 537

21-2 ▮ Reflection; the plane mirror, 539

***21-3** ▮ Formation of images by spherical mirrors, 541

21-4 ▮ Refraction, 547

21-5 ▮ Total internal reflection; fiber optics, 548

21-6 ▮ Thin lenses; the lens equation, 550

***21-7** ▮ The lens-maker's equation, 556

Summary, 558 Questions, 559 Problems, 560
For additional reading, 562

22 ▮ The wave nature of light 563

22-1 ▮ Waves versus particles; Huygens' principle, 563

22-2 ▮ Interference—Young's double-slit experiment, 566

22-3 ▮ Visible spectrum and dispersion, 568

22-4 ▮ Diffraction by a single slit or disk, 569

22-5 ▮ Diffraction grating, 572

22-6 ▮ The spectroscope and spectroscopy, 574

22-7 ▮ Interference by thin films, 576

*22-8 ▌ Michelson interferometer, 579

22-9 ▌ Polarization, 580

*22-10 ▌ Optical activity, 583

*22-11 ▌ Double refraction: birefringence and dichroism, 585

*22-12 ▌ Scattering of light by the atmosphere, 586

Summary, 587 Questions, 588 Problems, 589
For additional reading, 591

23 ▌ Optical instruments 592

23-1 ▌ The camera, 592

23-2 ▌ The human eye, 595

23-3 ▌ The magnifying glass, 598

23-4 ▌ Telescopes, 600

23-5 ▌ Compound microscope, 602

23-6 ▌ Lens aberrations and compound lenses, 604

23-7 ▌ Limits of resolution, 607

*23-8 ▌ Specialty microscopes and contrast, 610

Summary, 614 Questions, 615 Problems, 615
For additional reading, 617

24 ▌ Relativity 618

24-1 ▌ Galilean-Newtonian relativity, 618

*24-2 ▌ The Michelson-Morley experiment, 622

24-3 ▌ Einstein and the special theory of relativity, 626

24-4 ▌ Simultaneity, 627

24-5 ▌ Time dilation and the twin paradox, 629

24-6 ▌ Length contraction, 633

24-7 ▌ Four-dimensional space-time, 635

24-8 ▌ Mass increase, 636

24-9 ▌ The ultimate speed, 636

24-10 ▌ $E = mc^2$; mass and energy, 637

24-11 ▌ Relativistic addition of velocities, 639

***24-12** ▮ The Galilean and Lorentz transformations, 640

Summary, 644 Questions, 645 Problems, 646
For additional reading, 647

25 ▮ Early quantum theory 648

25-1 ▮ Planck's quantum hypothesis, 648

25-2 ▮ Photon theory of light and the photoelectric effect, 650

25-3 ▮ Photon interactions; pair production, 654

25-4 ▮ X rays and x-ray diffraction, 655

25-5 ▮ Wave-particle duality; the principle of complementarity, 658

25-6 ▮ Wave nature of matter, 659

***25-7** ▮ Electron microscopes, 661

Summary, 664 Questions, 664 Problems, 665
For additional reading, 666

26 ▮ Models of the atom; quantum mechanics 667

26-1 ▮ Early models of the atom, 667

26-2 ▮ Atomic spectra: key to the structure of the atom, 668

26-3 ▮ The Bohr model, 670

26-4 ▮ De Broglie's hypothesis, 675

26-5 ▮ Quantum mechanics, 676

26-6 ▮ Quantum mechanical view of atoms, 679

26-7 ▮ The Heisenberg uncertainty principle and its implications, 680

Summary, 684 Questions, 685 Problems, 686
For additional reading, 687

27 ▮ Quantum theory of atoms and molecules 688

27-1 ▮ Quantum mechanics of the hydrogen atom; quantum numbers, 688

27-2 ▮ Complex atoms and the exclusion principle, 690

27-3 ▮ The periodic table of elements, 692

27-4 ▮ X rays and atomic number, 693

27-5 ▮ Bonding in molecules and solids, 694

***27-6** ▮ Potential-energy diagrams, 698

***27-7** ▮ Activation energy, enzymes, and ATP, 699

***27-8** ▮ Molecular spectra, 700

***27-9** ▮ Fluorescence and phosphorescence, 701

***27-10** ▮ Lasers and holography, 702

Summary, 707 Questions, 708 Problems, 708
For additional reading, 709

28 ▮ The nucleus and radioactivity 710

28-1 ▮ Structure of the nucleus, 710

28-2 ▮ Binding energy and nuclear forces, 712

28-3 ▮ Discovery of radioactivity, 714

28-4 ▮ Radioactive decay modes: α, β, and γ, 715

28-5 ▮ Half-life and rate of decay, 720

28-6 ▮ Radioactive dating, 724

***28-7** ▮ Biological uses and dangers of radioactivity, 726

28-8 ▮ Measurement of radiation; dosimetry, 729

28-9 ▮ Detection of radiation, 731

Summary, 733 Questions, 734 Problems, 734
For additional reading, 736

29 ▮ Nuclear reactions and nuclear energy 737

29-1 ▮ Nuclear reactions and the transmutation of elements, 737

29-2 ▮ Nuclear fission, 739

29-3 ▮ Nuclear reactors, 744

29-4 ▮ Fusion, 745

Summary, 747 Questions, 748 Problems, 748
For additional reading, 749

30 ▮ Elementary particles 750

30-1 ▮ Particle accelerators, 750

30-2 ▮ Beginnings of elementary particle physics —the Yukawa particle, 755

30-3 ▮ Particles and antiparticles, 758

30-4 ▮ Particle interactions and conservation laws, 758

30-5 ▮ Particle classification, 760

30-6 ▮ Particle stability and resonances, 761

30-7 ▮ Strange particles, 762

30-8 ▮ Quarks: how physics became charming and found truth and beauty, 763

Summary, 766 Questions, 767 Problems, 767
For additional reading, 768

Appendixes

A ▮ Mathematical review 769

A-1 ▮ Relationships, proportionality, and equations, 769

A-2 ▮ Accuracy and significant figures, 770

A-3 ▮ Exponents, 771

A-4 ▮ Powers of 10, or exponential notation, 772

A-5 ▮ Algebra, 773

A-6 ▮ Plane geometry, 776

A-7 ▮ Logarithms, 777

B ▮ Order of magnitude: rapid estimating 780

C ▮ Dimensional analysis 781

D ▮ Conversion factors 784

E ▮ Selected isotopes 785

▮ Answers to odd-numbered Problems, 788

▮ Index, 794

▮ Color plate (Figure 22-22), facing 582

Preface

Two intentions motivated the writing of this introductory physics textbook: to give students a thorough understanding of the basic concepts of physics and, by means of many interesting applications, to prepare them to use physics in their own lives and professions. The book is thus appropriate for the introductory one-year physics courses commonly taken by students majoring in the life sciences, premedicine, architecture, technology, environmental sciences, and other disciplines. It may also be suitable for use in colleges that offer only one introductory physics course. It makes use of algebra and elementary trigonometry, but not calculus.

What this book offers, above all, is an in-depth presentation of physics. In developing the concepts, I have tried to avoid the dogmatic approach of giving general principles and then deriving conclusions—an approach that makes physics seem like a collection of facts. Instead, I have attempted to motivate students by opening a topic with observations (including ones that they can make themselves) and then going on to generalizations and/or development of physical theories. This procedure is closer to the way science is practiced and thus gives students a more honest picture of science; it also gives them, I believe, a better understanding of the fundamentals because the students can relate them to their own experience. In addition, where possible, I have endeavored to present the basic concepts in their historical and philosophical context. I have also attempted to bring each topic to a natural conclusion by discussing why it is important; topics are not left hanging, with students asking, "Why did we study that?" (One example is statics: this subject has practical importance because real materials are elastic and can fracture; therefore elasticity and fracture are included in Chapter 4 on statics.)

The second important feature of this book is the inclusion of a wide range of examples from, and applications to, other fields: biology, medicine, architecture, technology, earth sciences, the environment, and daily living. These serve not only to enliven the book but also to show how physics is important in our lives and in other disciplines and professions. Some applications serve only as examples of physical principles. Many other applications, however, are treated in depth, with whole sections or subsections devoted to them (among these are the study of the nervous system, constructing arches and domes, effects of radiation, and feedback). Yet the applications do not dominate the text (that is, after all, a physics book!). They have been carefully chosen and integrated into the text so as not to interfere with the development of the physics but rather to illuminate it.

A great deal of attention is given to problem solving. Methods of attacking problems are discussed in several places (for example, Section 2-8). Some 250 Examples are fully worked out in the text; these help students to fix ideas in their minds, to demonstrate interesting applications, and to help students develop problem-solving skills. Many examples are taken from everyday life and aim at being realistic. There are over 2000 end-of-chapter exercises, including no fewer than 500 Questions that require verbal answers based on an understanding of the concepts and more than 1500 Problems involving mathematical calculation. The wide range of Problems relates directly to the physics as well as the applications. They are arranged by sections, and they are graded according to difficulty: (I) simple problems requiring an understanding of basic definitions and concepts and typically the use of a single equation; (II) problems requiring more thought and often the combination of two different concepts; (III) the most difficult, often requiring synthesis of three or more concepts or perhaps dealing with more advanced material. I suggest that instructors assign a significant number of the type I and type II Problems and only a small number of type III; although most type I may seem easy, they help to build self-confidence—an important part of learning, especially in physics.

Throughout the text *Système international* (SI) units are used. Other metric and British units are defined for informational purposes.

Topics are presented in more or less traditional order, beginning with mechanics and ending with modern physics; nearly all topics customarily taught in introductory physics courses are included here (including those listed in the *New MCAT Student Manual*). The tradition of beginning with mechanics is sensible, I believe, since it was developed first, historically, and since so much else in physics depends on it. Within mechanics there is a variety of ways to order the topics. Although this book allows for flexibility, my reasons for the order as it stands may be of interest. I have placed statics (Chapter 4) after dynamics partly because many students seem to have trouble with the concept of force. (They tend to associate force with motion, and it seems to help if they understand the nature of this connection before dealing with forces without motion.) This order also allows full development of the concept of torque before it is used in statics. Moreover, statics is a special case of dynamics—we study statics so that we can prevent structures from becoming dynamic (falling down)—and that sense of being at the limit of dynamics is intuitively helpful. Also in mechanics I have treated rotational motion before energy because angular variables and torque follow the pattern of kinematics and Newton's laws already discussed for translational motion. The conservation laws—energy, momentum, and angular momentum—offer an entirely different approach to mechanics. It therefore seems easier for students if we treat these two approaches separately in a unified way. The chapters are written, nonetheless, so that most of Chapter 5, on momentum and energy, can be treated right after Chapter 2, on Newton's laws, with the appropriate sections on rotational kinetic energy and angular momentum following Chapter 3. Another option is whether light should follow electricity and magnetism or instead precede it by coming directly after waves and sound. The former is more traditional, and it is the order I have used. The material on light (Chapters 21 to 23), however, is so treated that it can be taught immediately after sound (Chapter 12).

The book contains more material than can be covered in most one-year courses. This was done to give instructors flexibility in choice of topics. The wide range of subjects also means that students can learn about many topics even though there is not class time for them. This aspect makes the book more valuable to students as a resource and as a reference book. Many sections and subsections are marked with a star (asterisk) to indicate they are *optional.* These sections contain slightly more advanced physics material (often material not usually covered in typical courses) and/or interesting applications. They contain no material needed in later chapters (except, perhaps, in later optional sections). This is not to imply that all nonstarred sections must be covered; there still remains considerable flexibility in the choice of material to suit the needs of students and instructors.

As is only too well known, mathematics can be an obstacle to student understanding. To avoid frightening students with an initial chapter on mathematics, I have instead used an appendix for review of algebra, geometry, accuracy and significant figures, exponents, powers of ten, proportions, and equations. Other appendixes cover order-of-magnitude estimating and dimensional analysis. Other important mathematical tools, such as addition of vectors and trigonometry, are dealt with in text where first needed.

Difficult language too can hinder understanding; and to put students at their ease, I have tried to write in a relaxed, colloquial style, avoiding jargon. New or unusual terms are carefully defined when first used.

An extremely important aspect of the text is the more than 800 line drawings and photographs, which illustrate the physical principles as well as the applications and exercises. They are an integral part of the material, and it may be helpful to point out to students how valuable they are for understanding physics.

I wish to thank the Rutherford High Energy Laboratory, Didcot, England, and the University of California at Berkeley, who were so hospitable to me during parts of the writing process, and to express my gratitude to the Physics Department at California State Polytechnic University, Pomona, for continual support and encouragement.

I also wish to thank the many people who contributed in various ways to making this a better book. The professors who read part or all of the manuscript and offered many excellent comments include Isaac Bass, Paul A. Bender, Joseph Boyle, Peter Brancazio, William Klink, William Eidson, David J. Ernst, Laurent Hodges, Gordon Jones, Robert Messina, Kwangjai Park, D. Lee Rutledge, William Riley, Paul Urone, Jearl Walker, Gareth Williams, and Peter Zimmerman. Special thanks go to Logan Campbell, physics editor at Prentice-Hall, who coordinated the project in an effective and humanistic manner, to Keith Brown, who worked out all the solutions to the Problems (and made me delete a few that could not be done), to Isaac Bass, who allowed me to use some of his interesting problems, to Bob Messina, who checked the examples for correctness, and to Donna Botash, who typed nearly all the manuscript. The responsibility for errors, of course, lies with me, and I welcome corrections and comments.

D. C. G.

▮ Notes to students and instructors on the format

1 Sections and subsections marked with a star (*) are considered optional (see the Preface).

2 The customary conventions are used: Symbols for quantities are italicized (such as *m* for mass), whereas units are not italicized (m for meter); boldface (**F**) is used for vectors, and this is discussed in the text.

3 Important terms are italicized where they are introduced, and the most important are in boldface (such as *coefficient of friction* and **acceleration**).

4 The most basic laws and equations are set off and preceded by a colored rectangle (▮).

5 Worked-out Examples and their Solutions in the text are set off with a rectangle (▮) at the beginning and two rectangles (▮▮) at the end.

6 Each chapter ends with a Summary, giving a brief review of important concepts and terms (the most important ones are italicized here).

7 Following the Summary in each chapter are sets of Questions that students should attempt to answer (to themselves at least) and Problems arranged according to section and difficulty (see the Preface). Questions and Problems that relate to optional sections are starred.

8 Each chapter contains a bibliography for further reading.

9 The appendixes contain useful tables and other valuable material, such as a mathematic review and discussions of order-of-magnitude estimating and dimensional analysis. Frequently used tables are located inside the front and back covers.

10 The extensive Index can be a useful tool. For example, it can be used to look up concepts or words whose meanings have been forgotten.

Physics

Introduction

Physics is the most basic of the sciences. It deals with the behavior and structure of matter. The field of physics is usually divided into the areas of motion, fluids, heat, sound, light, electricity and magnetism, and the modern topics of relativity, atomic structure, and nuclear physics. We will cover all these topics in this book, beginning with motion (or mechanics, as it is sometimes called). But before we begin on the physics itself, let us take a brief look at how this activity called "science," including physics, is actually practiced.

The practice of science

Unfortunately, many people think science is a mechanical process of collecting facts and deriving theories. This, however, is not the case. Science is a creative activity that ranks with the other creative achievements of the human mind.

The principal aim of science is generally considered to be the search for order in the world around us. A simplified view of how this aim is accomplished is called the "scientific method" and involves three steps: the *observation* of the pertinent facts; the invention of *hypotheses* or theories to order or explain these observations; and the *testing* of the hypotheses to see if their predictions are actually observed. It is a mistake, however, to think that scientists rigidly work through these three steps.

Instead, a great deal of creativity and inspiration is involved. Even something as seemingly simple as "observing the pertinent facts" requires imagination, for scientists can never include everything in a description of what they observe. Hence, scientists must make judgments about what is relevant in their observations. As an example, let us consider how two great minds, Aristotle (384–322 B.C.) and Galileo (1564–1642), interpreted motion along a horizontal surface. Aristotle saw that any object that is given an initial push along the ground (or on a table top) always slows down and comes to rest. He concluded that the natural state of a body is at rest. Galileo, in his reexamination of horizontal motion, judged something else to be relevant—that different objects slow down at different rates, depending on the smoothness of their surfaces. In fact, Galileo imagined that, if friction could be eliminated, an object given an initial push along a horizontal surface would continue to move indefinitely without stopping. He concluded that for an object to be in motion was just as natural as to be at rest. By seeing something new in the same "facts," Galileo founded our modern view of motion (more details in Chapter 2). This seeing of something new was surely inspired thinking.

The act of creation is most obvious in the putting forth of scientific hypotheses or theories. Theories are never derived from observations—they are *created* to explain observations. They are inspirations that come from the minds of human beings. For example, the idea that matter is made up of atoms (the atomic theory) was certainly not arrived at because someone observed atoms (they would be much too small to be seen even under a microscope!). Rather, the idea sprang from a creative mind. The theory of relativity, the electromagnetic theory of light, and the quantum theory were likewise the result of inspiration.

The great theories of science may be compared, as creative achievements, with great works of art or literature. But how does science differ from other creative activities? One important difference is that science requires *testing* of its ideas or theories to see if predictions are borne out by experiment.

Although the testing of theories can be considered to distinguish science from other creative fields, it should not be assumed that a theory is "proved" by testing. First of all, no measuring instrument is perfect; so exact confirmation cannot be possible. Furthermore, it is not possible to test a theory in every single possible circumstance (after all, we don't live forever). Hence a theory can never be absolutely verified. In fact theories themselves are generally not perfect—a theory rarely agrees exactly, within experimental error, in every single case in which it is tested. Indeed, the history of science tells us that theories come and go; that long-held theories are replaced by new ones. The process of one theory replacing another is an important subject in the philosophy of science today; we can discuss it here only briefly.

A new theory is accepted by scientists in some cases because its predictions are quantitatively in much better agreement with experiment than the older theory. But in many cases, a new theory is accepted only if it explains a greater range of phenomena than the older one. Copernicus's sun-centered theory of the universe, for example, was no more accurate than Ptolemy's earth-centered theory for predicting the motion of heavenly bodies. But Copernicus's theory did give a simple explanation for the calendar year and the seasons, which Ptolemy's theory did not. A simpler theory, one which unifies and explains a greater variety of phenomena, is more useful and beautiful to a scientist. And this aspect, as well as quantitative agreement, plays a major role in the acceptance of a theory.

An important aspect of any theory is how well it can quantitatively predict phenomena; and from this point of view, a new theory may often seem to be only a minor advance over the old one. For example, Einstein's theory of relativity arrives at predictions that differ very little from the older theories of Galileo and Newton in nearly all everyday situations; its predictions are better mainly in the extreme case of very high speeds, those close to the speed of light. From this point of view, the theory of relativity might be considered as mere "fine tuning" of the older theory. But quantitative prediction is not the only important outcome of a theory. Our view of the world is affected as well. As a result of Einstein's theory of relativity, for example, our concepts of space and time have been completely changed; and we have come to see mass and energy as a single entity (via the famous equation $E = mc^2$). Indeed, our view of the world underwent a major change when relativity theory came to be accepted.

Science is truly a creative human activity. In this book, we will explore the subject matter of physics. We will apply the principles of physics to other fields such as biology and architecture. And we shall spend a little time, at least, pointing out the humanity and creativity of science.

Physics and its relation to other fields

For a long time science was more or less a united whole known as natural philosophy. Not until the last century or two did the distinctions between physics and chemistry and even the life sciences become prominent. Indeed, the sharp distinctions we now see between the arts and the sciences is itself but a few centuries old. It is no wonder then that the development of physics has both influenced, and been influenced by, other fields. For example, Leonardo da Vinci's notebooks contain the first references to the forces present within a structure, a subject we consider as physics today; but Leonardo was interested, at least in part, because of the relevance to architecture and building. The discovery of electric current—certainly an important one in physics—is generally credited to an eighteenth-century physiologist, Luigi Galvani (1737–1798); he noticed the twitching of frogs legs in response to an electric spark and later that the muscles twitched when in contact with two dissimilar metals (see Chapter 15). At first this phenomenon was known as "animal electricity," but it shortly became clear that current electricity itself could exist in the absence of an animal. More recently, in the 1930s and 1940s, a number of scientists trained as physicists became interested in applying the ideas and techniques of physics to problems in microbiology. Among the most prominent were Max Delbrück (1906–) and Erwin Schrödinger (1887–1961). They hoped, among other things, that by studying biological organisms, some new unsuspected laws of physics might be uncovered. Alas, this hope has not been realized; but their efforts gave rise to the field we now call "molecular biology" and have resulted in a dramatic increase in our understanding of the genetics and structure of living beings.

One does not have to be a research scientist in, say, medicine or molecular biology to be able to use physics in his or her work. A zoologist, for example, may find it useful to know how prairie dogs and other animals can live underground without suffocating. A botanist may wonder how water can get to the tops of trees. A physical therapist will do a more effective job if aware of the principles of lever action, center of gravity, and the action of forces within the human body. A knowledge of the operating principles of optical and electronic equipment is helpful in a variety of fields. Life scientists and architects alike will be interested in the nature of heat loss and gain in human beings and the resulting comfort or discomfort. Architects themselves may never have to calculate, for example, the dimensions of the pipes in a heating system or the forces involved in a given structure to determine if it is stable; but they must know the principles behind these systems in order to make realistic designs and to communicate effectively with engineering consultants and other specialists. From the aesthetic or psychological point of view, too, architects must be aware of the forces involved in a structure—for instability, even if only apparent, can be discomforting to those who must live or work in the structure. Indeed, many of the features we admire in the architecture of the past three millennia were introduced not for their decorative effect but rather for practical purposes. For example, the cornices and pediments of windows protected the facades of buildings from the erosive effects of water. The base and capital of a column acted to diffuse the load from the column onto the masonry above and below. The development of the arch as a means to span a space and at the same time support a heavy load will be discussed in Chapter 4. There we will see that the pointed, or Gothic, arch was not originally a decorative device but a technological development of considerable importance.

The list of how physics relates to other fields is extensive. In the chapters that follow we will discuss many such applications in detail, as we carry out our principal aim of explaining basic physics.

Models, theories, and laws

When scientists are trying to understand a particular set of phenomena, they often make use of a *model.* A model, in the scientists' sense, is a kind of analogy or mental image of the phenomena in terms of something we are familiar with. One example is the wave model of light. We cannot see waves of light as we can water waves; but it is valuable to think of light as if it were made up of waves because experiments on light indicate that it behaves as water waves do in many respects.

The purpose of a model is to give us a mental or visual picture—something to hold onto—when we cannot see what actually is happening. Models often give us a deeper understanding; for the analogy to a known system (for instance, water waves in the above example) can suggest new experiments to perform and can provide ideas about what other related phenomena might occur.

The creation of models is useful in many fields besides physics. For example, molecular biologists, through many indirect experiments, have created a model for protein synthesis involving several types of ribonucleic acid (RNA), enzymes, and so on. No one has seen this intricate mechanical model in action, but it is consistent with experiments. Nonetheless, one must be aware of the limitations of a model.

No model is ever perfect, and scientists are constantly trying to refine their models or to think up new ones when old ones do not seem adequate. The atomic model of matter has gone through many refinements. At one time or another, atoms were imagined as tiny spheres with hooks on them (to explain chemical bonding) or tiny billiard balls continually bouncing against each other. More recently, the "planetary model" of the atom visualizes the atom as a nucleus with electrons revolving around it, just as the planets revolve about the sun.

You may by now wonder what the difference is between a theory and a model. Sometimes the words are used interchangeably. Usually, however, a model is fairly simple and provides a structural similarity to the phenomena being studied, whereas a theory is broader and more detailed. Often, as a model is developed and modified and corresponds more closely to experiment over a wide range of phenomena, it may come to be referred to as a theory. The atomic theory is an example, as is the wave theory of light.

Models can be very helpful, and they often lead to important theories; but it is important not to confuse a model, or a theory, with the real system or phenomena themselves.

Scientists give the title *law* to certain concise but general statements about how nature behaves (that momentum is conserved, for example); sometimes the statement takes the form of a relationship or equation between quantities (such as Newton's second law of motion, $F = ma$).

To be called a law, a statement must be found experimentally valid over a wide range of observed phenomena; in a sense, the law brings a unity to many observations. For less general statements, the term *principle* is often used (such as Archimedes' principle). Where to draw the line between laws and principles is, of course, arbitrary, and there is not always complete consistency.

Scientific laws are different from political laws in that the latter are *prescriptive*: they tell us how we must behave. Scientific laws are *descriptive*: they do not say how nature *must* behave, but rather describe how nature *does* behave. As with theories, laws cannot be tested in the infinite variety of cases possible. So we cannot be sure that any law is absolutely true. We use the term law when its validity has been tested over a wide range of cases, and when any limitations and the range of validity are clearly understood. Even then, as new information comes in, certain laws may have to be modified or discarded.

Scientists normally do their work as if the accepted laws and theories were true; but they are obliged to keep an open mind in case new experiments should alter the validity of any given law or theory.

Mathematics

The appendixes that follow the last chapter contain a brief summary of simple mathematical techniques that will be used in this book. You may find it useful to examine them now to review old topics or learn any new ones. You may also want to reread them later when you need those concepts. Other mathematical techniques, such as vectors and trigonometric functions, are treated in the text itself.

For additional reading

Bronowski, J., *Science and Human Values*, rev. ed., Harper & Row, Publishers, Inc., New York, 1965.

Kuhn, T., *The Structure of Scientific Revolutions*, 2nd ed., University of Chicago Press, Chicago, 1970.

Schroeer, D., *Physics and the Fifth Dimension: Society*, Addison-Wesley Publishing Co., Inc., Reading, Mass., 1972.

Snow, C. P., *Two Cultures and the Scientific Revolution*, Cambridge University Press, Cambridge, 1959.

Watson, J. D., *The Double Helix*, Atheneum Publishers, New York, 1968.

The references given at the end of each chapter offer interesting historical background, applications, or expanded explanations of certain topics. The reader may also find it helpful to consult other standard textbooks (perhaps at the library) to get a different point of view of the basic material of each chapter.

1 ▌The description of motion: kinematics

Animals, baseballs, cars, and almost everything else in our environment are in motion, or can be in motion, even the sun and the moon. The motion of objects was undoubtedly the first aspect of the physical world to be thoroughly studied, and this study can be traced back to the ancient civilizations in Asia Minor.

In their studies of motion the ancients acquired significant insights, but not until the Renaissance were the important concepts established. Many contributed to this understanding, but two individuals stand out above the rest: Galileo Galilei (1564–1642) and Isaac Newton (1642–1727). We will presently see the important role these two great thinkers played.

The study of motion is customarily divided into two parts: *kinematics*, which is the description of motion, and *dynamics*, the reasons why a particular body moves as it does. This chapter deals with kinematics. To describe the motion of an object we will need, in addition to distance and time, two concepts: velocity (or speed) and acceleration.

In this chapter and the next we will discuss objects that move without rotating. Such motion is called *translational motion.* We will also restrict ourselves, in the first five sections of this chapter, mainly to *linear motion*—that is, to motion along a straight line.

1-1 ▌Speed and velocity

Speed

Speed refers to how far an object moves in a given time interval. If a car is driven 160 kilometers (km) in two hours, we say its average speed was 80 kilometers per hour (we can abbreviate this as 80 km/h).† In general, the **average speed** of an object is defined as the distance traveled divided by the time it takes to travel this distance:

$$\text{Average speed} = \frac{\text{distance traveled}}{\text{time elapsed}}.$$

† Units of measurement are discussed in Section 1–2.

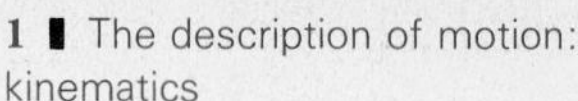

This definition can be written briefly as:

$$\bar{v} = \frac{d}{t}, \tag{1-1}$$

where d stands for distance, t for the elapsed time, and v for speed (v is the abbreviation for the word velocity). The bar ($\bar{}$) over the v is the standard symbol meaning "average."

It is unlikely that the car mentioned above was moving at 80 km/h during every instant of its two-hour trip. To deal with this situation we need the concept of **instantaneous speed**. By this we mean the speed an object has at any instant of time (this is what a speedometer is supposed to indicate). However, this is not a very precise definition of instantaneous speed because we haven't specified what we mean by an "instant." Instantaneous speed can be defined more precisely as the average speed over a very short time interval—an interval so short that the speed can be considered not to change during that short time. We can state this in shorthand (or algebraic) terms by making use of the Greek letter Δ (delta) to mean "small change in" or "small amount of." Then:

$$v = \frac{\Delta d}{\Delta t} \quad [\Delta t \text{ very small, approaching zero}]. \tag{1-2}$$

Here v represents the instantaneous speed, and Δd represents the small distance traveled during the very short time interval Δt. In what follows, we will often drop the word "instantaneous" when referring to instantaneous speed; but when we are referring to the average speed, we will always specify that it is the average.

If an object moves at a uniform (or constant) speed over a particular time interval, then its instantaneous speed at any instant is the same as its average speed (see Figure 1-1*a*). But in most situations this will not be the case. For example, a car may start from rest, speed up to 50 km/h, remain at that speed for a time, then slow down to 20 km/h in a traffic jam, and finally stop at its destination after traveling a total of 15 km in a half hour. This trip is plotted on the graph of Figure 1-1*b*. Also shown on the graph is the average speed (dashed line), which is $\bar{v} = d/t = 15\text{ km}/0.5\text{ h} = 30\text{ km/h}$.

If the average speed of an object is known, it is an easy matter to calculate how far it will travel in a given time. This can be done by solving for distance in the equation that defines average speed, $\bar{v} = d/t$. This is done by multiplying both sides of this equation by t, which does not change the equality, and we obtain

$$d = \bar{v}t. \tag{1-3}$$

For example, if an airplane flies at an average speed of 900 km/h, in four hours it will have gone 3600 km.

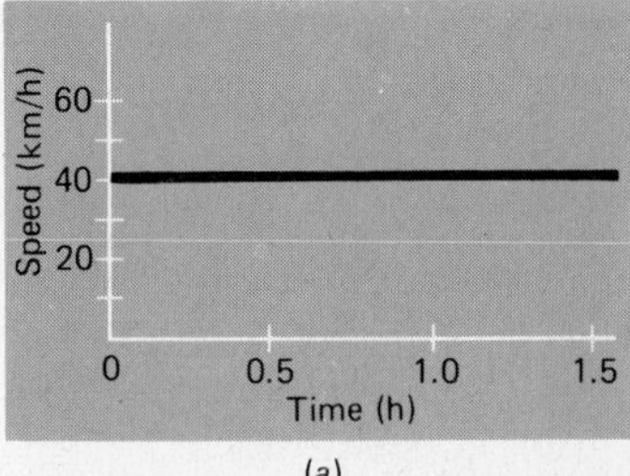

(*a*)

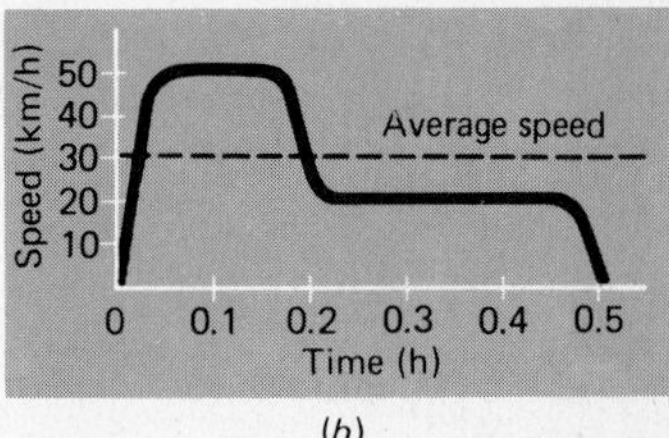

(*b*)

FIGURE 1-1
Speed of an object as a function of time: (*a*) at constant speed, (*b*) with varying speed.

Reference frames and coordinate systems

When riding in a train you may observe a bird flying by overhead and remark that it looks as if it is moving at a speed of 30 km/h. But do you mean it is traveling 30 km/h with respect to the train, or with respect to the ground?

Every measurement must be made with respect to a **frame of reference**. For example, while on a train traveling at 80 km/h, you might notice a person walk past you toward the front of the train at a speed, of, say, 5 km/h. Of course this is the person's speed with respect to the train. With respect to the ground that person is moving at 85 km/h. It is always important to specify the frame of reference when stating a speed. We almost always mean "with respect to the earth" without even thinking about it, but the reference frame should be specified whenever there might be confusion.

The values of other physical quantities also depend on the frame of reference. For example, there is no point in telling you that Yosemite National Park is 300 km away unless I specify 300 km from where. Distances are always measured in some frame of reference. Furthermore, when specifying the motion of an object, it is important to specify not only the speed but also the direction of motion. For example, if a friend leaves New York on a jet plane that travels at a speed of 1000 km/h, you would like to know in what direction this person is going—toward Washington, San Francisco, Paris, or wherever. Often we can specify a direction by using the cardinal points, north, east, south, and west, and by "up" and "down." This is not always convenient; so in physics we often draw a set of coordinate axes as shown in Figure 1-2, to represent a frame of reference. Objects positioned to the right of the origin of coordinates (O) on the x axis are usually said to have an x coordinate with a positive value, if to the left a negative value. The position along the y axis is usually considered positive when above and negative when below O, although the reverse convention can be used if convenient. Any point on the plane can then be specified by giving its x and y coordinates. In three dimensions, a z axis perpendicular to the x and y axes is also used.

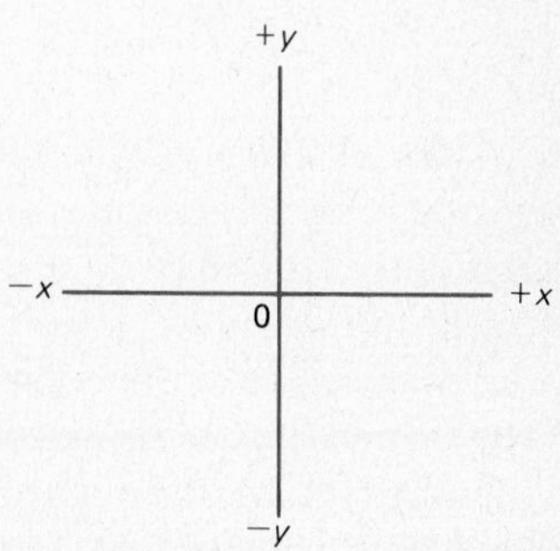

FIGURE 1-2
Standard (Cartesian) set of coordinate axes.

Although most measurements are made in reference frames fixed on the earth, it is important to recognize that reference frames other than the earth are perfectly legitimate. For example, scientific measurements are often made on moving ships and even on the moon.

Velocity

The terms velocity and speed are used interchangeably in ordinary language. However, in physics we make a distinction between the two. Most importantly, the term **velocity** is used to signify both the *magnitude* (numerical value) of how fast an object is moving and its *direction.* Speed, on the other hand, is a magnitude only. There is a second difference between speed and velocity: namely, the magnitude of the average velocity is defined in terms of "displacement," rather than total distance traveled; that is

$$\text{average velocity} = \frac{\text{displacement}}{\text{time elapsed}},$$

where **displacement** means the net distance of the object from the starting point after a given elapsed time. To see the distinction between distance (which is used in calculating speed) and displacement (used for velocity), imagine a person walking 50 meters (m) to the east and then turning around and walking back (west) a distance of 10 m. The total distance traveled is 60 m but the displacement is only 40 m since the person is now only 40 m from the starting point. If this walk took 40 seconds (s), the average speed was (60 m)/(40 s) = 1.5 m/s, but the average velocity in this case was (40 m)/(40 s) = 1.0 m/s. This discrepancy between the magnitude of velocity and the speed occurs in some cases, but only for the *average* values and we rarely need to be concerned with it. The magnitude of the instantaneous velocity and the instantaneous speed are always the same. The **instantaneous velocity** is defined, like instantaneous speed, as the average velocity over a very short time interval—that is, as the very small displacement divided by the very short time interval Δt it takes to make this tiny displacement.

As another illustration, consider an object that moves along the x axis of Figure 1-3 from the point $x = +50$ m to $x = +30$ m. Since the distance from the origin of the coordinates *decreases,* this is considered a negative displacement. That is, the displacement is -20 m. If this motion requires 10 s, the average velocity is -2.0 m/s. The minus sign tells us that the object is moving in the direction of decreasing x and thus gives us directional information in this one-dimensional case.

Equations 1-1, 1-2, and 1-3 which apply to speed, also are valid for velocity; but in this case, d must stand for the displacement.

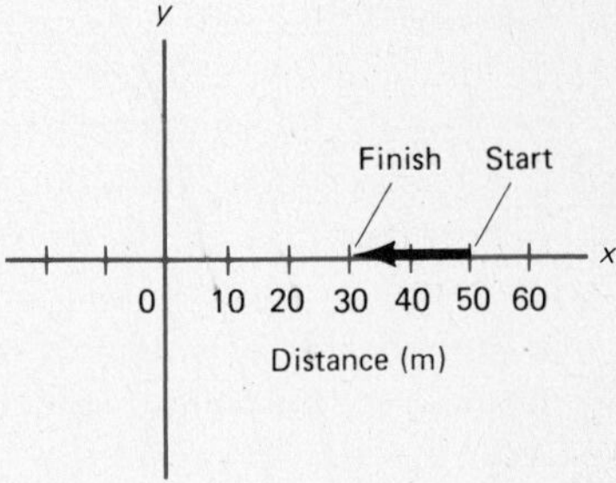

FIGURE 1-3
Object moves from $x = 50$ m to $x = 30$ m. The displacement is -20 m.

Vectors

A quantity such as velocity, which has direction as well as a magnitude, is called a **vector**. Other quantities that are also vectors are displacement, force, and momentum. However, many quantities such as time, temperature, and energy, have no direction associated with them; they only have a magnitude; that is, they are specified completely by giving a number (and units, if any). Such quantities are called **scalars**.

Drawing a diagram of a particular physical situation is very helpful in physics, and this is especially true when dealing with vectors. On a diagram, each vector is represented by an arrow. The arrow is always drawn so that it points in the direction of the vector it represents; the length of the arrow is drawn proportional to the magnitude of the vector. For example, in Figure 1-4, arrows have been drawn representing the velocity of a car at various places as it rounds the curve. The magnitude of the velocity at each point can be read off this figure by measuring the length of the corresponding arrow.

When we write the symbol for a vector, we will always use boldface type.† Thus for velocity we write **v**. If we are only concerned with the magnitude we will simply write v. We will deal with vectors more fully in Section 1-6. For now, we deal mainly with motion along a line (linear motion); vectors which point in one direction along this line will be assigned a positive value, and those pointing in the opposite direction will have a negative sign.

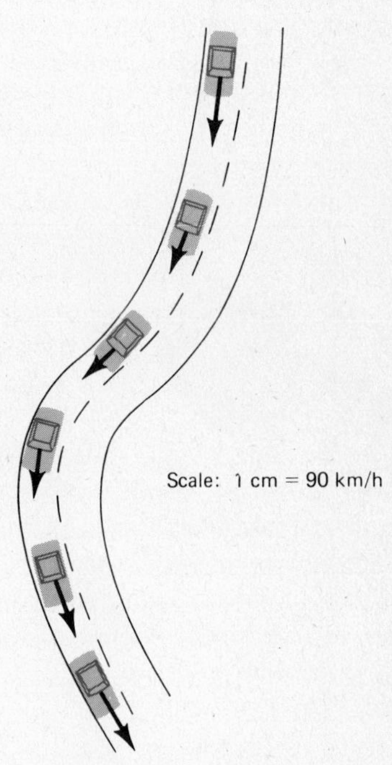

FIGURE 1-4
Car traveling on a road. The arrows represent the velocity vector at each position.

1-2 ▮ Standards and units

Standards

All measurements are made relative to particular standards or units, and these units must be specified along with the numerical value of any

† The symbol for a vector is usually indicated in handwritten work with an arrow over it, as $\vec{v}$ for velocity.

quantity. For example, length can be measured in units of inches, feet, or miles or in the metric system in centimeters, meters, or kilometers. Until about 200 years ago, the units of measurement were not standardized, and that made scientific communication difficult. Different people used different units: cubits, leagues, hands; and even the length of the foot varied from place to place.

The first real internationl standard was the establishment of the standard *meter* by the French Academy of Sciences in 1791.† The meter was defined as the distance between two finely engraved marks on a particular bar of platinum–iridium alloy kept near Paris at the International Bureau of Weights and Measures. Accurate copies of the standard meter were sent to laboratories around the world. In 1960, a new, more accurate and reproducible standard was chosen. The meter is now defined as 1,650,763.73 wavelengths of a particular orange light emitted by the gas krypton 86. British units (inch, foot, mile) are defined in terms of the meter. The inch (in) is defined as precisely 2.54 centimeters (cm; 1 cm = 0.01 m). Other conversion factors between metric and British units of length and volume are given in Table 1-1; additional conversion factors are given in Appendix D.

TABLE 1-1

Conversion factors for length and volume

1 in	= 2.540 cm = 25.40 mm
1 ft	= 12 in = 30.48 cm
1 mi	= 5280 ft = 1.61 km
	= 1610 m
1 cm	= 0.394 in
1 m	= 39.37 in = 3.28 ft
1 km	= 1000 m = 3280 ft
	= 0.621 mi
1 l	= 1000 cm^3
	= 1.06 qt (U.S.)
1 qt (U.S.)	= 946 cm^3

In the metric system, the larger and smaller units are defined in multiples of 10 from the standard unit, and this makes calculation particularly easy. Thus one centimeter is 1/100 m, one kilometer (km) is 1000 m, and so on. The prefixes "centi-," "kilo-," and others are listed in the inside back cover for ready reference (along with abbreviations for all units used in this book) and can be applied not only to units of length, but units of volume, mass, or any other metric units. For example a centiliter (cl) is 1/100 of a liter (l), and a kilogram (kg) is 1000 grams (g). The conversion factors between the various units of the British system (for example, 12 inches in a foot) make it unwieldy for calculation. This is one reason why scientists and nearly all countries of the world have adopted the metric system for practical purposes. Indeed, the United States has also decided to convert to metric units and Britain already has done so. For this reason, metric units will be used throughout this book.

The standard unit of *time* is the *second* (s). For many years, the second was defined as 1/86,400 of a mean solar day. The standard second is now defined more precisely in terms of the vibrations within a cesium atom. Specifically, one second is defined as the time required for 9,192,631,770 vibrations of cesium atoms when they are vibrating in a specific manner. There are, of course, 60 seconds in one minute (min) and 60 minutes in an hour (h).

The definitions of other standard units for other quantities will be defined as we encounter them in later chapters.

Changing units

It is often necessary or useful to change from one set of units to another. For example, it is often preferable to specify the speed of a car in m/s rather than in km/h when measuring braking distances or quick

† In a spirit of rationality, the standard meter was originally chosen to be one ten-millionth of the distance from the earth's equator to either pole. Modern measurement of the earth's circumference reveal that the intended length is off by about one-fiftieth of one percent. This is of little practical importance, however.

accelerations since the time and distances involved are usually seconds and meters rather than hours and kilometers. Also, one sometimes must change units from metric to British units, or vice versa.

To determine what a speed of 80 km/h is in m/s, we proceed as follows. There are 1000 m in 1 km and 60 min, each with 60 s, for a total of 3600 s in 1 h. Thus:

$$80 \text{ km/h} = \left(\frac{80 \not{\text{km}}}{1 \not{\text{h}}}\right)\left(\frac{1000 \text{ m}}{1 \not{\text{km}}}\right)\left(\frac{1 \not{\text{h}}}{3600 \text{ s}}\right)$$

$$= (80)\left(\frac{1000}{3600}\right)\frac{\text{m}}{\text{s}}$$

$$= (80)(0.278) \text{ m/s} = 22 \text{ m/s}.$$

Notice that in the first line we have multiplied our original number (80 km/h) by the two conversion factors (1000 m/1 km and 1 h/3600 s) which in essence are equal to one and thus do not change the equation. The units of hours and of kilometers cancel out so we obtain m/s. (Note also that we kept only two significant figures in the final answer, 22 m/s, since the least significant number in the product contained two significant figures. See Appendix A for a discussion of significant figures.)

When changing units it is often a problem to figure out whether the conversion factor is to go in the numerator or in the denominator. The easiest way to be sure is to check if the units cancel out as they did above. We would have gotten a wrong result if we had written (80 km/h) (1000 m/1 km)(3600 s/1 h) and this would have been obvious since the hour units do not cancel.

In the last line in the calculation above, we see the factor 0.278; this is the conversion factor between m/s and km/h. This can more quickly be obtained by noting that 1 km/h is the same as 1000 m in 3600 s. That is, 1 km/h = 1000 m/3600 s = 0.278 m/s. Thus, any speed given in km/h can be changed to m/s by multiplying by the factor of 0.278.

Changes between metric and British units can be done in a similar way, and we leave these calculations as an exercise. Table 1-2 shows a comparison between several units for a number of different speeds.

TABLE 1-2

Equivalent speeds in different units

km/h	m/s	mi/h	ft/s
40 km/h =	**11 m/s =**	**25 mi/h =**	**37 ft/s**
80 km/h =	**22 m/s =**	**50 mi/h =**	**74 ft/s**
120 km/h =	**33 m/s =**	**75 mi/h =**	**111 ft/s**

1-3 ▮ Acceleration

An object whose velocity is changing is said to be accelerating. An automobile increasing in speed from rest to 80 km/h is accelerating. If one car can accomplish this change in speed in less time than another, it is said to undergo a greater acceleration. **Average acceleration** is defined as the change in velocity divided by the time taken to make this change:

$$\text{Average acceleration} = \frac{\text{change of velocity}}{\text{time elapsed}}.$$

In symbols we define the average acceleration, $\bar{\mathbf{a}}$, as:

$$\bar{\mathbf{a}} = \frac{\mathbf{v} - \mathbf{v}_0}{t}, \qquad (1\text{-}4)$$

where $\mathbf{v}_0$ is the "initial velocity," the velocity at the beginning of the time interval we are considering; and $\mathbf{v}$ is the "final velocity," the velocity after a time t has elapsed. (We can choose the initial point as we please; it is not necessarily when the motion actually started.)

Equation 1-4 is a vector equation. For linear motion a, v, v_0, and t can be treated as scalars so we need only use numerical values. (In Section 1-6 we will see how to add and subtract velocity vectors.)

Instantaneous acceleration can be defined in analogy to instantaneous velocity:

$$\mathbf{a} = \frac{\Delta \mathbf{v}}{\Delta t} \quad [\Delta t \text{ very small, approaching zero}],$$

where $\Delta \mathbf{v}$ represents the small change in velocity during the very short time interval Δt.

The concept of acceleration is often confused with velocity. To help make the distinction, let us take an example; suppose a car accelerates from rest to 60 km/h in 10 s; the magnitude of its average acceleration was then

$$\bar{a} = \frac{60 \text{ km/h} - 0 \text{ km/h}}{10 \text{ s}} = \frac{60 \text{ km/h}}{10 \text{ s}} = 6.0 \text{ km/h/s}.$$

This is read as "six kilometers per hour per second" and means that on the average the speed changed by 6 km/h during each second. In other words, assuming the acceleration was uniform, during the first second the car's speed increased from zero to 6 km/h. During the next second its speed increased by another 6 km/h, to 12 km/h, and so on. (Of course, if the acceleration was not precisely uniform, the instantaneous acceleration at any moment may have varied from the calculated average.) Acceleration is the rate at which the velocity changes whereas velocity or speed is the rate at which position changes.

In this calculation we used two different time units, hours and seconds. We usually prefer to use only seconds, so we change the km/h to m/s by multiplying by 0.278 as we saw in the previous section. Then

$$\bar{a} = \frac{17 \text{ m/s} - 0 \text{ m/s}}{10 \text{ s}} = 1.7 \text{ m/s}^2,$$

where instead of m/s/s we have written m/s^2 for short. According to this calculation, the speed changes on the average by 1.7 m/s during each second, for a total change of 17 m/s over the 10 s.

As a second illustration, consider an object which is moving to the right along the x axis (in some chosen coordinate system) and is decreasing in speed, say an automobile braking. Suppose the initial speed is 15.0 m/s and it takes 5.0 s to reach the final speed of 5.0 m/s. Then

$$\bar{a} = \frac{5.0 \text{ m/s} - 15.0 \text{ m/s}}{5.0 \text{ s}} = -2.0 \text{ m/s}^2.$$

The negative sign in this last example appears because the final velocity is less than the initial velocity and thus indicates that the speed is decreasing. This situation is sometimes called deceleration; but, as can be seen here, deceleration can be considered as a negative acceleration.

The definition of acceleration involves the change in *velocity*. Thus an acceleration results not only when the *magnitude* of the velocity changes but also if the *direction* changes. For example, a person riding in a car traveling at constant speed around a curve or a child riding on a merry-go-round, will both experience an acceleration because of a change in direction of the velocity. Such accelerations due to change in direction of velocity will be discussed later.

1-4 ▮ Uniformly accelerated motion

Many practical situations occur in which the acceleration is constant in magnitude and direction. In many other situations the variation in acceleration is sufficiently small that we are justified in assuming it is constant. We now treat this situation of uniformly accelerated motion: that is, when the magnitude of the acceleration is constant and the motion is in a straight line. In this case, the instantaneous and average accelerations are equal.

A common problem is to determine the velocity of an object after a certain time given its acceleration. To solve such problems we need only solve Equation 1-4 for the final velocity v:

$$v = v_0 + at. \tag{1-5}$$

For example, it may be known that the acceleration of a particular motorcycle is 4.0 m/s^2 and we wish to determine how fast it will be going after, say, 6.0 s. Assuming it starts from rest ($v_0 = 0$), after 6.0 s the velocity will be $v = at = (4.0\ \text{m/s}^2)(6.0\ \text{s}) = 24\ \text{m/s}$.

Next, let us see how to calculate the distance traveled by an object when undergoing uniform acceleration. We make use of Equation 1-3, $d = \bar{v}t$, but first we have to figure out the average velocity, $\bar{v}$. Because the velocity increases at a uniform rate, the average velocity will be halfway between the initial and the final velocities; that is:

$$\bar{v} = \frac{v + v_0}{2}. \tag{1-6}$$

In the motorcycle example above, $\bar{v} = (24\ \text{m/s} + 0\ \text{m/s})/2 = 12\ \text{m/s}$. The distance traveled is then $d = \bar{v}t = (12\ \text{m/s})(6.0\ \text{s}) = 72\text{m}$.

To calculate this distance, we used three different equations, Equations 1-3, 1-5, and 1-6. If we combine these three equations algebraically, we obtain one equation which will make future calculations of distance much simpler. We start with Equation 1-3:

$$d = \bar{v}t;$$

now substitute Equation 1-6 for $\bar{v}$:

$$d = \left(\frac{v + v_0}{2}\right)t$$

and then put in Equation 1-5 for v:

$$d = \left(\frac{v_0 + at + v_0}{2}\right)t = \frac{2v_o t + at^2}{2},$$

$$d = v_0 t + \tfrac{1}{2}at^2. \tag{1-7}$$

Using this new equation, we can calculate the distance traveled by the motorcycle more quickly: $d = 0 + \frac{1}{2}(4.0\ \text{m/s}^2)(6.0\ \text{s})^2 = 72\ \text{m}$, as before.

There is one more equation that is useful in kinematics, and we will derive it now. It is useful in a situation, for instance, when the acceleration, distance, and initial velocity are known, and the final velocity is desired, but the time t is not known. To obtain the final velocity v in terms of v_0, a, and d, we begin as above with Equations 1-3 and 1-6:

$$d = \bar{v}t = \left(\frac{v + v_0}{2}\right)t.$$

Next we solve Equation 1-5 for t, and obtain $t = (v - v_0)/a$, and substitute this in the above equation to obtain

$$d = \left(\frac{v + v_0}{2}\right)\left(\frac{v - v_0}{a}\right) = \frac{v^2 - v_0^2}{2a}.$$

Solving for v^2 we obtain the desired equation:

$$v^2 = v_0^2 + 2ad. \qquad (1\text{-}8)$$

We now have four equations relating the various quantities important in uniformly accelerated motion. We collect these Equations 1-5, 1-6, 1-7, and 1-8, in one place for further reference:

$$v = v_0 + at \qquad [a = \text{constant}] \qquad (1\text{-}9a)$$

$$d = v_0 t + \tfrac{1}{2}at^2 \qquad [a = \text{constant}] \qquad (1\text{-}9b)$$

$$v^2 = v_0^2 + 2ad \qquad [a = \text{constant}] \qquad (1\text{-}9c)$$

$$\bar{v} = \frac{v + v_0}{2}. \qquad [a = \text{constant}] \qquad (1\text{-}9d)$$

These equations will allow us to solve problems of uniformly accelerated motion simply and quickly.

▮ EXAMPLE 1-1 Suppose a planner is working on the preliminary design of a new community and wonders if there is sufficient space for an airport for small planes. One kind of airplane that might use this airfield must reach a speed before takeoff of 200 km/h (55.6 m/s) and can accelerate at $12.0\ \text{m/s}^2$. If the runway is 100 m long, can this airplane reach the proper speed to take off?

SOLUTION We use Equation 1-9c with $v_0 = 0$, $d = 100$ m and $a = 12.0\ \text{m/s}^2$; then

$$v^2 = 0 + 2\,(12.0\ \text{m/s}^2)(100\ \text{m}) = 2400\ \text{m}^2/\text{s}^2$$

$$v = \sqrt{2400\ \text{m}^2/\text{s}^2} = 49\ \text{m/s}.$$

Unfortunately, this length runway is not sufficient. By solving Equation 1-9c for d, you can determine how long a runway is needed for this plane. ▮▮

One of the difficulties in doing kinematics problems (and other problems, too, for that matter) is knowing which equation to use. The best thing to do when you're not sure is to use the following procedure: (1) write down what *is* "known," or "given," and what you *want* to know: (2)

search out an equation that involves only known quantities and one desired unknown, but contains no other unknowns; (3) if the desired unknown is buried within the equation—for example, you may want d from Equation 1-9c—solve the equation for the desired unknown. In some instances a combination of equations may be needed. An important aspect of doing problems is keeping track of units. Note that an equals sign implies the units on each side must be the same just as the numbers must. By treating units carefully, mistakes can often be avoided in calculations.

▌ EXAMPLE 1-2 How long does it take a car to travel 30 m if it accelerates from rest at a rate of 2.0 m/s^2?

SOLUTION. First we make a table:

Known	Wanted
$d = 30$ m	t
$a = 2.0$ m/s^2	
$v_0 = 0$	

Equation 1-9a is not helpful in this case since it contains v, an unknown, as well as the desired unknown t. Equation 1-9c is worse—it contains v but not t. Equation 1-9b is perfect since the only unknown quantity is t. Before we solve for t in this equation, we simplify it by setting $v_0 = 0$ first. Then

$$d = \tfrac{1}{2}at^2$$

$$t^2 = \frac{2d}{a} = \frac{2\,(30\text{ m})}{2.0\text{ m/s}^2} = 30\text{ s}^2$$

$$t = \sqrt{30\text{ s}^2} = 5.5\text{ s.}$$ ▌▌

▌ EXAMPLE 1-3 We now consider the stopping distances for a car which are important not only for traffic safety but are needed for traffic design problems. This problem is best dealt with in two parts. First we consider the distance traveled between the time the driver decides to apply the brakes and the time the brakes are actually applied. (This is the "reaction time.") In the second part we calculate the distance traveled during the deceleration due to braking. The stopping distance depends on the reaction time of the driver, the initial speed of the car (the final speed is zero), and the deceleration rate of the car. For a dry road, good brakes can decelerate a car at a rate of about 5 m/s^2 to 8 m/s^2. We make the calculation for an initial speed of 90 km/h (25 m/s) and assume the deceleration rate of the car is -6.0 m/s^2. Reaction time for normal drivers varies from perhaps 0.3 s to about 1.0 s; we take it to be 0.50 s. For the first part of the problem, the car travels at a constant speed of 25 m/s during the time the driver is reacting (0.50 s). Thus:

Known	Wanted
$t = 0.50$ s	d
$v_0 = 25$ m/s	
$v = 25$ m/s	
$a = 0$	

We can use either Equation 1-9*b* or 1-9*c*; Equation 1-9*b* looks easier:

$$d = v_0 t + 0 = (25 \text{ m/s})(0.50 \text{ s}) = 12 \text{ m}.$$

Now for the second part during which the brakes are applied and the car brought to rest:

Known	Wanted
$v_0 = 25$ m/s	d
$v = 0$	
$a = -6.0$ m/s^2	

Equation (1-9*a*) doesn't contain *d*; Equation 1-9*b* contains *d* but also the unknown *t*. Equation 1-9*c* is what we want; we solve for *d*:

$$v^2 - v_0^2 = 2ad$$

$$d = \frac{v^2 - v_0^2}{2a}$$

$$= \frac{0 - (25 \text{ m/s})^2}{2(-6.0 \text{ m/s}^2)} = \frac{-625 \text{ m}^2/\text{s}^2}{-12 \text{ m/s}^2} = 52 \text{ m}.$$

The car traveled 12 m while the driver was reacting and another 52 m during the braking period before coming to a stop; the total distance traveled was then 64 m.

Under wet or icy conditions, the value of *a* may be only one-third the value for a dry road since the brakes cannot be applied as hard without skidding, and hence stopping distances are much greater. ▮▮

▮ EXAMPLE 1-4 A baseball pitcher throws a fastball with a speed of 40 m/s. Calculate approximately the acceleration of the ball during the throwing motion. It is observed that in throwing a baseball, a pitcher accelerates the ball through a total distance of about 3.5 m from behind the ear to the point where it is released (Figure 1-5).

SOLUTION. We want to find the acceleration *a* given that $d = 3.5$ m, $v_0 = 0$, and $v = 40$ m/s. We use Equation 1-9*c* and solve for *a*:

$$a = \frac{v^2 - v_0^2}{2d}$$

$$= \frac{(40 \text{ m/s})^2 - (0 \text{ m/s})^2}{2(3.5 \text{ m})} = \frac{1600 \text{ m}^2/\text{s}^2}{7.0 \text{ m}} = 230 \text{ m/s}^2.$$ ▮▮

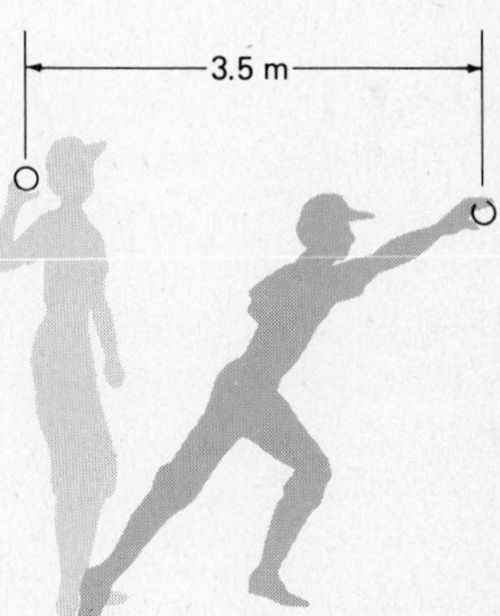

FIGURE 1-5
A baseball pitcher accelerates the ball over a distance of about 3.5 m.

The analysis of motion we have been discussing in this chapter is basically algebraic. It is sometimes helpful to use a graphical interpretation as well and this is done in Section 1-7 (optional).

1-5 ▮ Falling bodies

One of the commonest examples of uniformly accelerated motion is that of an object allowed to fall vertically to the earth. That a falling body is accelerating may not be obvious at first. Indeed, it was not considered so by the ancients. Aristotle believed that a falling object acquired a particular speed immediately upon release and that it maintained this constant speed until it reached the ground. Furthermore Artistotle was convinced that heavier bodies fall faster than light bodies, and that the speed of fall is proportional to how heavy the object is.

Although these notions about falling bodies were occasionally challenged during the Middle Ages, it was Galileo who first gave an accurate quantitative description of free fall. Galileo questioned not only Aristotle's theories of motion, but he attacked the very methods by which the ancients came to their conclusions. Their method had relied on logical reasoning and "common sense" in order to arrive at an understanding of nature. Galileo however insisted that no theory or hypothesis about nature was tenable unless it had been thoroughly tested by experiment. In his analysis, Galileo used an important technique: that of imagining what would happen in ideal situations, such as in the absence of friction or air resistance; and on this basis he was able to establish a successful theory of motion. Because of his insistence on testing, as well as for his highly imaginative analysis of motion, Galileo is generally regarded as the father of modern science.

FIGURE 1-6
(a) A ball and a light piece of paper are dropped at the same time. (b) Repeated, with the paper wadded up.

Galileo performed many simple experiments† and came to the conclusion that objects fall with constantly increasing speed, i.e. with a uniform acceleration. To support this claim, Galileo made use of the following argument: a heavy stone dropped from a height of 2 m will drive a wooden stake into the ground much further than if the same stone is dropped from a height of only 10 cm; the stone must clearly be moving faster in the former case. Secondly, Galileo claimed that all objects, light or heavy, fall with the *same* acceleration, at least in the absence of air. Now common sense may say that Aristotle is closer to the truth; for if you hold a piece of paper horizontally in one hand and a heavier object, say, an eraser or a baseball, in the other hand and release them at the same time (see Figure 1-6*a*), surely the heavier object will reach the ground first. Repeat the experiment but this time crumple up the paper into a small wad (see Figure 1-6*b*). You will find that the two objects reach the floor at very nearly the same time. Galileo believed that air acts like a sort of friction on very light objects that have a large surface area; but for most ordinary objects this air resistance is negligible. In a chamber from which the air has been removed even light objects like a feather or a horizontally held piece of paper will fall with the same acceleration as any other object (see Figure 1-7).

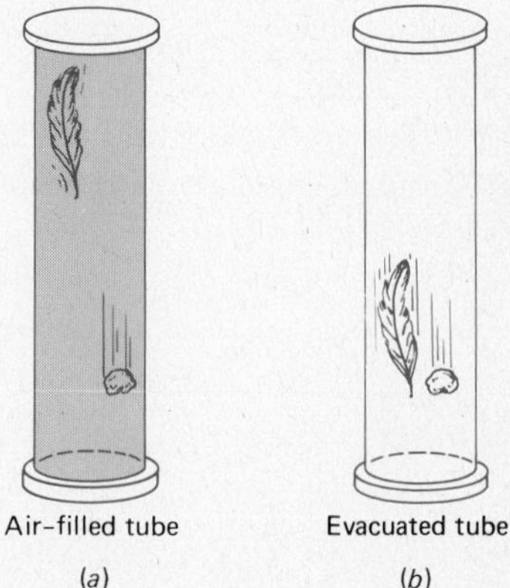

FIGURE 1-7
A rock and a feather are dropped simultaneously (a) in air, (b) in a vacuum.

Galileo's contribution to our understanding of the motion of falling objects can be summarized as follows:

▮ At a given location on the earth and in the absence of air resistance, all objects fall with the same uniform acceleration.

We call this acceleration the *acceleration due to gravity* on the earth, and we give it the symbol g. Its value is approximately:

$$g = 9.80 \text{ m/s}^2.$$

TABLE 1-3

Acceleration due to gravity at various locations

Location	Elevation (ft)	g (m/s^2)
New York	**0**	**9.803**
San Francisco	**350**	**9.800**
Denver	**5400**	**9.796**
Pikes Peak	**14,100**	**9.789**
Equator	**0**	**9.780**
North Pole (calculated)	**0**	**9.832**

† Although Galileo is often said to have dropped objects of varying weight from the Leaning Tower of Pisa, we have no record that he actually did so.

In British units g is about 32 ft/s^2. Actually, g varies slightly according to latitude and elevation (see Table 1-3) but these variations are so small that we will ignore them for most purposes. The effects of air resistance are usually small and we will neglect them for the most part. However, air resistance will be noticeable even on a reasonably heavy object if the distance of fall is very large.†

When dealing with freely falling objects we can make use of Equations 1-9 where for a we use the numerical value of g given above.

▮ EXAMPLE 1-5 Suppose that a ball is dropped from a tower 70 m high. How far will it have fallen after 1.0, 2.0, and 3.0 s?

SOLUTION. Equation 1-9b is the appropriate one, with $v_0 = 0$. Thus, after 1.0 s,

$$d = \tfrac{1}{2}at^2 = (\tfrac{1}{2})(9.80\ \text{m/s}^2)(1.0\ \text{s})^2 = 4.9\ \text{m}.$$

After 2.0 s

$$d = (\tfrac{1}{2})(9.80\ \text{m/s}^2)(2.0\ \text{s})^2 = 20\ \text{m}.$$

After 3.0 s

$$d = (\tfrac{1}{2})(9.80\ \text{m/s}^2)(3.0\ \text{s})^2 = 44\ \text{m}.$$

See Figure 1-8. ▮▮

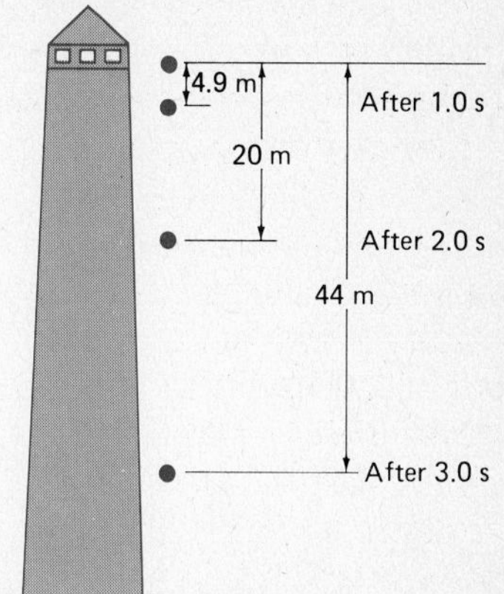

FIGURE 1-8
When an object is dropped from the top of a tower, it falls with progressively greater speed and covers greater distance with every successive second.

▮ EXAMPLE 1-6 A person throws an object *upward* into the air with an initial velocity of 15.0 m/s. We wish to find out (a) how high it goes, and (b) how long it takes before it comes back to his hand. We are not concerned here with the throwing action, but only with the motion of the ball *after* it leaves the thrower's hand, Figure 1-9.

SOLUTION. This is a somewhat more complicated situation than when an object is merely dropped: the initial velocity is not zero, and there is motion both upwards and downwards. Since velocity and acceleration are vectors, we must keep track of whether they are pointing up or down; to do this, we use either a plus or a minus sign for "up," and the opposite sign for "down." It is arbitrary which direction we choose to be positive; it doesn't matter as long as we are consistent throughout a problem.

Let us choose up as positive and down as negative. Then the acceleration will have a negative sign (since it changes the velocity in a downward direction), $a = -g = -9.80\ \text{m/s}^2$. Now, to calculate ($a$) the maximum height we calculate the position of the ball when its velocity equals zero ($v = 0$ at the highest point). We know $v_0 = 15.0$ m/s, $v = 0$, $a = -9.80$ m/s and we wish to find d. We use Equation 1-9c and solve for d:

$$v^2 = v_0^2 + 2ad$$

$$d = \frac{v^2 - v_0^2}{2a} = \frac{0-(15.0\ \text{m/s})^2}{2(-9.80\ \text{m/s}^2)} = \frac{-225\ \text{m}^2/\text{s}^2}{-19.6\ \text{m/s}^2} = 11.5\ \text{m}.$$

So the object reaches a maximum height of 11.5 m.

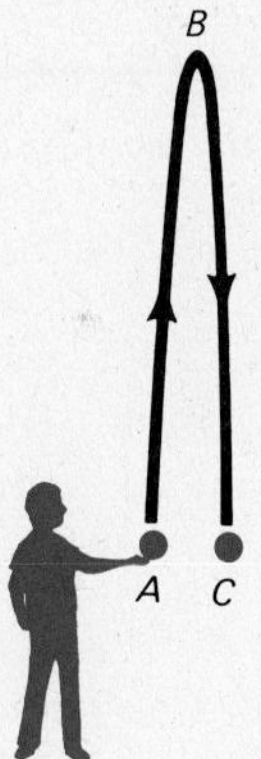

FIGURE 1-9
Object thrown into air leaves thrower's hand at *A*, reaches its maximum height at *B*, and returns to original height at *C*. (See Example 1-6.)

† The speed of an object falling in air (or other fluid) does not increase indefinitely; if it falls far enough, it will reach a maximum velocity called the *terminal velocity*. The maximum velocity is reached when the force of air resistance (which increases with speed) balances the force of gravity.

(*b*) We divide this part into two calculations. First we calculate the time required for the ball to reach its highest point, and then calculate the time it takes to fall back down. To go up from A to B in Figure 1-9, we have $v_0 = 15.0$ m/s, $v = 0$, $d = 11.5$ m and we use Equation 1-9*a* to find t:

$$t = \frac{v - v_0}{a} = \frac{0 - 15.0\ \text{m/s}}{-9.80\ \text{m/s}^2} = 1.53\ \text{s}.$$

To calculate the time to fall from B to C in Figure 1-9, we now take a, v, d as positive since they are all in the same direction (downward); for this part of the problem $v_0 = 0$ (the initial point is now B where the ball has just stopped and is starting its downward journey), $a = 9.80\ \text{m/s}^2$ and $d = 11.5$ m. We use Equation 1-9*b* with $v_0 = 0$. Thus $d = \frac{1}{2}at^2$ and solving for t:

$$t = \sqrt{\frac{2d}{a}} = \sqrt{\frac{2 \times 11.5\ \text{m}}{9.80\ \text{m/s}^2}} = 1.53\ \text{s}.$$

It takes the same time to come down as to go up. Use of Equation 1-9*a* will show that the velocity at C is the same as at A, although in the opposite direction. The up and down motions are symmetrical, which is a general rule when air resistance can be ignored. The total time is 3.06 s. Part (*b*) can be done another way that is quicker and involves only one step. We consider the motion from A to B to C in one step by using Equation 1-9*b*, where v_0 represents the initial velocity ($v_0 = 15.0$ m/s) and d represents the displacement, which in this case is $d = 0$ since the object returns to its original point. Thus, at both A and C, $d = 0$. We use Equation 1-9*b* with $a = -9.80\ \text{m/s}^2$ and find:

$$d = v_0t + \tfrac{1}{2}at^2$$

$$0 = (15.0\ \text{m/s})t + (\tfrac{1}{2})(-9.80\ \text{m/s}^2)t^2.$$

There are two solutions: $t = 0$ and $t = (2)(15.0\ \text{m/s}^2)/(9.80\ \text{m/s}^2) = 3.06$ s. The first solution corresponds to the initial point, A, and the second to the return, C. Thus we obtain the same result as before, 3.06 s for the total trip from A to B to C. ▌▌

▌ EXAMPLE 1-7 The jump of a flea can be analyzed using slow-motion photographs. The motion can be separated into two parts. The first is the "push off," lasting about 10^{-3} s, during which the flea's legs push against the ground and accelerate it to a speed of about 1.0 m/s. The second part is the flight of the flea into the air subject only to gravity (Figure 1-10). Calculate: (*a*) the acceleration of the flea during push off expressed as a multiple of g, the acceleration of gravity; (*b*) the distance above the ground the flea reaches during push off; and (*c*) the height the flea should reach during the second period when its acceleration is that of gravity. Assume the jump is vertical.

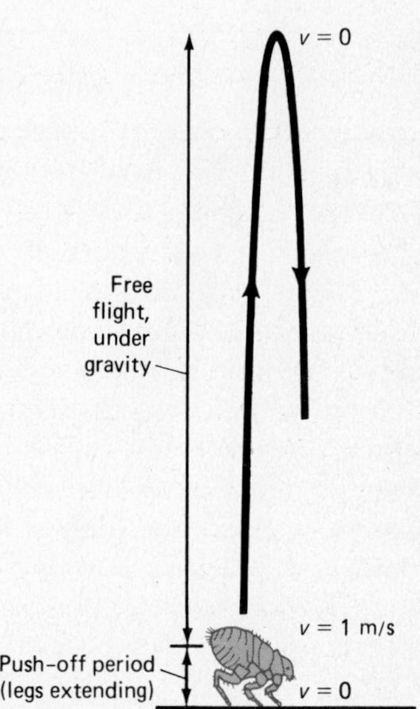

FIGURE 1-10
The jump of a flea (Example 1-7).

SOLUTION. (*a*) During the push off

$$a = \frac{v - v_0}{t} = \frac{1.0\ \text{m/s} - 0}{10^{-3}\ \text{s}} = 1000\ \text{m/s}^2.$$

The calculated acceleration is $(1000\ \text{m/s}^2)/(9.80\ \text{m/s}^2) \approx 100$ times the acceleration of gravity. Thus $a \approx 100\ g$.

(*b*) We use Equation 1-9*b*

$$d = \tfrac{1}{2}at^2 = 0.5(10^3\ \mathrm{m/s^2})(10^{-3}\ \mathrm{s})^2$$
$$= 0.5 \times 10^{-3}\ \mathrm{m} = 0.05\ \mathrm{cm}.$$

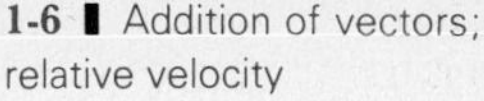

(*c*) Now the acceleration is $a = -g = -9.80\ \mathrm{m/s^2}$ and the initial velocity for this second period is 1.0 m/s. We use Equation 1-9*c*, with $v = 0$ (we are considering the highest point):

$$d = \frac{v^2 - v_0^2}{2a} = \frac{0 - (1.0\ \mathrm{m/s})^2}{2(-9.80\ \mathrm{m/s^2})}$$
$$= 0.05\ \mathrm{m} = 5\ \mathrm{cm}.$$

The film indicates that the flea only jumps about two-thirds this high. Can you guess why our calculation is off? ▌▌

1-6 ▌ Addition of vectors; relative velocity

Until now we have been dealing with motion along a straight line. In order to deal with motion that is not in a straight line, we must discuss vectors more fully, and in particular see how vectors are added. Because vectors are quantities that have direction as well as magnitude, they must be added in a special way. In this section we will mainly deal with displacement (**d**) and velocity (**v**) vectors. But the results will apply for other vectors as well, such as force and momentum.

Addition of vectors: graphical methods

We use simple arithmetic for adding scalars, such as time. For example, if a TV is run for 4 h one day and 2 h the next, it has been on for a total of 4 h + 2 h = 6 h. Simple arithmetic can also be used for adding vectors if they are in the *same* direction. For example, if a person walks 8 km east one day, and 6 km east the next day, the person will be 8 km + 6 km = 14 km from the point of origin. We say that the **net** or **resultant** displacement is 14 km. If, on the other hand, the person walks 8 km east on the first day, and 6 km *west* (in the reverse direction) on the second day, then he will be 2 km from the origin after the two days so his resultant displacement is 2 km to the east. In this case, the resultant displacement is obtained by subtraction: 8 km − 6 km = 2 km.

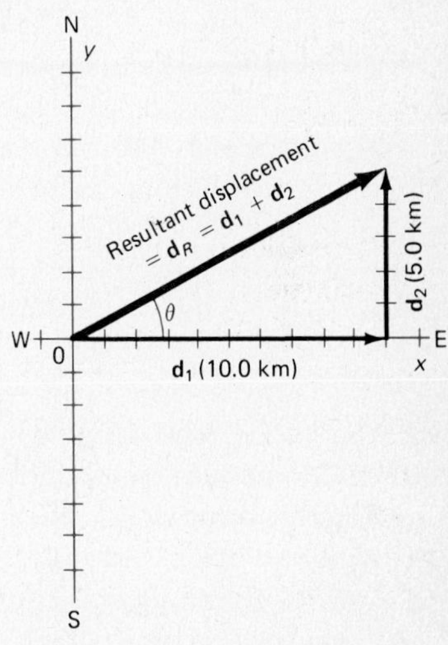

FIGURE 1-11
A person walks 10.0 km east and then 5.0 km north. These two displacements are represented by the vectors $\mathbf{d}_1$ and $\mathbf{d}_2$, which are shown as arrows. The resultant displacement vector, $\mathbf{d}_R$, which is the sum of $\mathbf{d}_1$ and $\mathbf{d}_2$, is also shown. Measurement on the graph with ruler and protractor shows that $\mathbf{d}_R$ has magnitude 11.2 km and it is at an angle $\theta = 27°$ north of east.

But simple arithmetic cannot be used if the two vectors are *not* along the same line. For example, suppose a person walks 10.0 km east and then walks 5.0 km north. This motion can be represented on a graph in which the positive *y* axis points north and the positive *x* axis points east, Figure 1-11. On this graph, we draw an arrow, labeled $\mathbf{d}_1$ to represent the displacement vector of the 10.0 km displacement to the east; and a second arrow, $\mathbf{d}_2$ to represent the 5.0 km displacement to the north. Both vectors are drawn to scale.

After taking this walk, the person is now 10.0 km east and 5.0 km north of the point of origin. The *resultant displacement* is represented by the arrow labeled $\mathbf{d}_R$ on the diagram. If you use a ruler and a protractor, you will measure on this diagram that the person is 11.2 km from the origin at an angle of 27° north of east. In other words, the resultant displacement vector has a magnitude of 11.2 km and makes an angle θ with the x axis of 27°. The magnitude (length) of $\mathbf{d}_R$ can also be obtained using the theorem of Pythagoras† since d_1, d_2, and d_R form a right triangle with d_R as the hypotenuse. Consequently, $d_R = \sqrt{d_1^2 + d_2^2} = \sqrt{(10.0\text{ km})^2 + (5.0\text{ km})^2} = \sqrt{125\text{ km}^2} = 11.2$ km. You can use this theorem, of course, only when the vectors are *perpendicular* to each other.

The resultant displacement vector, $\mathbf{d}_R$, is defined as the *sum* of the vectors $\mathbf{d}_1$ and $\mathbf{d}_2$; that is, $\mathbf{d}_R = \mathbf{d}_1 + \mathbf{d}_2$. This is a *vector* equation. Notice an important feature of adding two vectors: the magnitude of the **resultant vector** (sometimes referred to as the *sum vector* or, in this case, as the *net displacement* or *total displacement*) is *not* equal to the sum of the magnitudes of the two separate vectors.

Figure 1-11 illustrates the general rules for adding two vectors together, no matter what angles they make, to get their sum. Specifically, the rules are (1) on a diagram, draw one of the vectors—call it $\mathbf{V}_1$—to scale; (2) next draw the second vector $\mathbf{V}_2$, to scale, placing its tail at the tip of the first vector and being sure its direction is correct; (3) the arrow drawn from the tail of the first vector to the tip of the second represents the *sum*, or *resultant*, of the two vectors. The length of the resultant can be measured and compared to the scale. Angles can be measured with a protractor. This method is known as the *tail-to-tip method of adding vectors.*

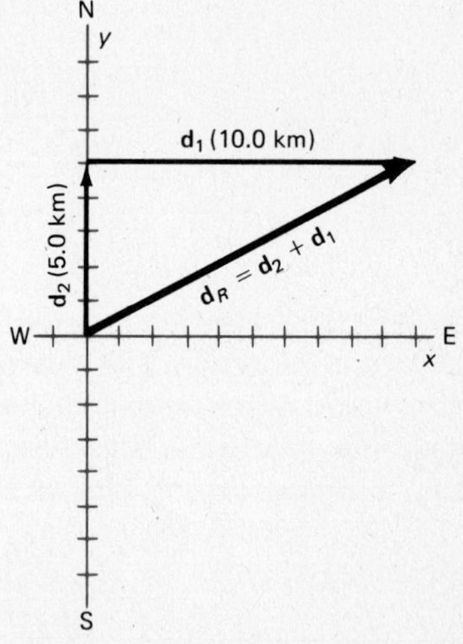

FIGURE 1-12
If the order of the vectors is reversed, the resultant is the same; see Figure 1-11.

Note that it is not important in which order the vectors are taken. For example, a displacement of 5.0 km north, to which is added a displacement of 10.0 km east, yields a resultant of 11.2 km and angle $\theta = 27°$ (Figure 1-12)—the same as when they were added in reverse order (Figure 1-11); that is $\mathbf{d}_1 + \mathbf{d}_2 = \mathbf{d}_2 + \mathbf{d}_1$.

The tail-to-tip method can be extended to three or more vectors. An example is shown in Figure 1-13; the three vectors could represent displacements (northeast, south, west) or perhaps three forces. Check for yourself that you get the same resultant no matter in which order you add the three vectors.

A second way to add two vectors is the *parallelogram method*: it is equivalent to the tail-to-tip method. In this method, the two vectors are drawn from a common origin and a parallelogram is constructed using these two vectors as adjacent sides. The resultant is the diagonal drawn from the common origin. An example is shown in Figure 1.14*b*. In Figure 1-14*a* the tail-to-tip method is shown, and it is clear that both methods yield the same result. It is a common error to draw the sum vector as the diagonal running between the tips of the two vectors, as in Figure 1-14*c*; this is incorrect.

† See Appendix A-6.

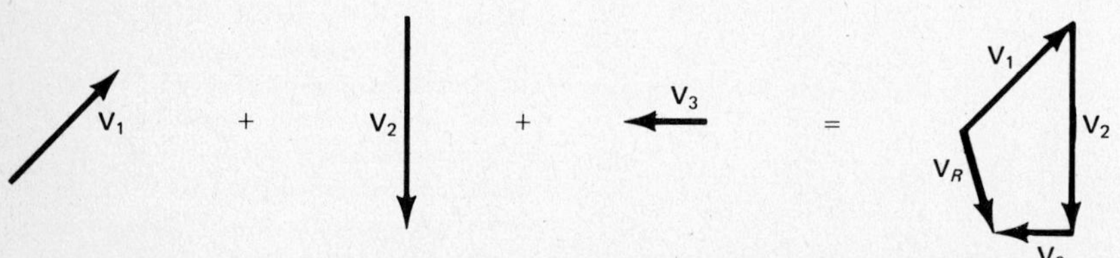

FIGURE 1-13
The resultant of three vectors, $\mathbf{V}_R = \mathbf{V}_1 + \mathbf{V}_2 + \mathbf{V}_3$.

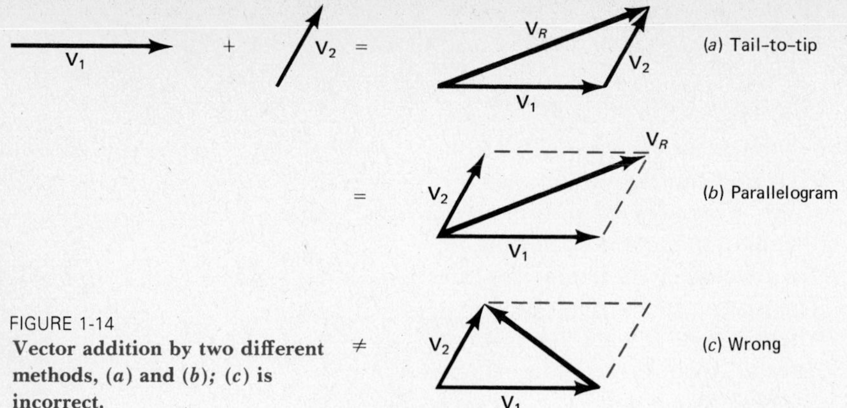

FIGURE 1-14
Vector addition by two different methods, (*a*) and (*b*); (*c*) is incorrect.

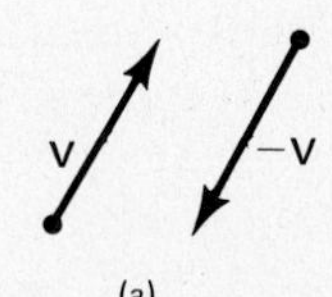

Subtraction of vectors

The subtraction of one vector from another is readily done by noting that the negative of any vector is a vector of the same magnitude but that points in the opposite direction as shown in Figure 1-15*a*. We can write the difference of two vectors $\mathbf{A}-\mathbf{B}$ as $\mathbf{A}-\mathbf{B}=\mathbf{A}+(-\mathbf{B})$ so that the differences of two vectors is equal to the sum of the first plus the negative of the second. Thus our rules for addition of vectors can be applied as shown in Figure 1-15*b* using the tail-to-tip method.

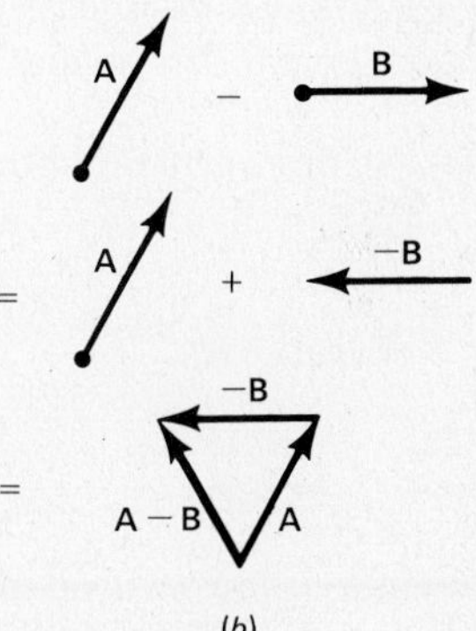

FIGURE 1-15
(*a*) The negative of a vector is a vector with the same length but opposite direction;
(*b*) subtracting two vectors: A − B.

Relative velocity

Adding velocity vectors is a more interesting situation than adding displacement vectors. This leads us to the concept of *relative velocity.* For example, if two trains are approaching one another, each with a speed of 80 km/h with respect to earth, the speed of one train relative to the other is 160 km/h. That is, to an observer on one train, the other train seems to be approaching at 160 km/h. Similarly, when one car traveling 90 km/h passes a second car traveling 75 km/h, the first car has a speed relative to the second car of 90 km/h − 75 km/h = 15 km/h.

When the velocities are along the same line, simple addition or subtraction is sufficient to obtain the relative velocity. But if they are not along the same line, we must make use of vector addition as the following examples show. But we note, as mentioned in Section 1-1, that when specifying a velocity, it is important to specify what the reference frame is.

▌ EXAMPLE 1-8 A boat can travel 1.20 m/s in still water. (*a*) If the boat heads directly across a stream whose current is 0.75 m/s, what is the velocity (magnitude and direction) of the boat relative to the shore? (*b*) What will be the position of the boat, relative to its point of origin, after 3.0 s?

SOLUTION (*a*) As shown in Figure 1-16, the boat is pulled downstream by the current. The boat's velocity with respect to the shore, $\mathbf{v}_{BS}$, is the sum of its velocity with respect to the water, $\mathbf{v}_{BW}$, plus the velocity of the water with respect to the shore, $\mathbf{v}_{WS}$:

$$\mathbf{v}_{BS}=\mathbf{v}_{BW}+\mathbf{v}_{WS}.$$

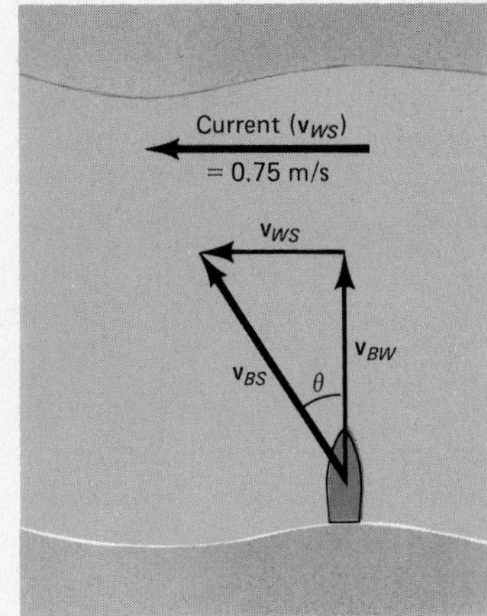

FIGURE 1-16
A boat heading directly across a river whose current moves at 0.75 m/s. See Example 1-8.

(Note the use of subscripts: the first refers to the object—B for boat, W for water, S for shore—and the second is what the velocity is with respect to.) We can measure θ and v_{BS} (the magnitude of $\mathbf{v}_{BS}$) from the diagram, and obtain $v_{BS} = 1.4$ m/s, and $\theta = 32°$. Or, since v_{BW} is perpendicular to v_{WS}, we can get v_{BS} using the theorem of Pythagoras, $v_{BS} = \sqrt{v_{BW}^2 + v_{WS}^2} = \sqrt{(1.20 \text{ m/s})^2 + (0.75 \text{ m/s})^2} = 1.4$ m/s. (*b*) To find the position after 3.0 s, we determine the displacement vector, **d**. The vector equivalent of Equation 1-3 (or Equation 1-9*b* with $a = 0$) tells us that $\mathbf{d} = \mathbf{v}t$, where t is the elapsed time. Since t is a scalar, it does not affect directions; so the direction of **d** is the same as that for **v**. The magnitude of **d** in our case is $d = v_{BS}t = (1.4 \text{ m/s})(3.0 \text{ s}) = 4.2$ m. So the boat, after 3.0 s, is 4.2 m from its point of origin in a direction 32° downstream from directly across the river.

▮ EXAMPLE 1-9 An airplane whose air speed is 200 km/h heads due north. But a 100-km/h northeast wind suddenly begins to blow. What is the resulting velocity of the plane with respect to the ground?

SOLUTION The vector diagram is shown in Figure 1-17. We cannot use the Pythagorean theorem in this case because the vectors do not form a right triangle. Measurement on a carefully drawn diagram gives the magnitude of $\mathbf{v}_{PG}$ as 150 km/h at an angle θ of 29° W of N. ▮▮

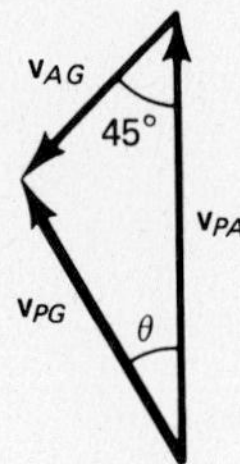

FIGURE 1-17
Example 1-9; the speed of the airplane with respect to the air (v_{PA}) is 200 km/h; speed of the air (coming from the northeast at 45°) with respect to the ground (v_{AG}) is 100 km/h; $\mathbf{v}_{PG}$ is measured to have magnitude 150 km/h and direction $\theta = 29°$ west of north.

Resolution of vectors

We have discussed how two or more vectors can be added together to obtain a single vector, their sum, whose effect is the same as the original vectors acting together. It is useful to do the reverse: to express a given single vector in a plane as the sum of two other vectors, called the **components** of the original vector. The components are usually chosen to be along two perpendicular directions. The process of finding the components is known as *resolving the vector into components.* For example, in Figure 1-18*a*, we see the velocity vector for an airplane traveling 500 km/h at a 30° angle north of east; we wish to resolve this vector into its x and y (east and north) components. We first draw in the x and y axes. Then from the tip, A, of our given vector we draw a line AB perpendicular to the x axis, and a line AC perpendicular to the y axis. Then the lines OB and OC represent the x and y components of **v** respectively. We call these components $\mathbf{v}_x$ and $\mathbf{v}_y$, and, as can be seen in Figure 1-18*b*, they add up to give **v** by the parallelogram method. Using a ruler, you will find that the components of **v** in Figure 1-18 have magnitudes of about $v_x = 430$ km/h and $v_y = 250$ km/h.

Space is made up of three dimensions, and sometimes it is necessary to resolve a vector into components along three mutually perpendicular directions. The components are then called v_x, v_y, and v_z. Although resolution of a vector in three dimensions is merely an extension of the above technique, we will mainly be concerned with situations in which the vectors are in a plane and two components are all that are necessary.

Adding and resolving vectors by measuring the length of arrows on graphs is time consuming and not very accurate. Quicker and much more precise results can be obtained using trigonometric functions.

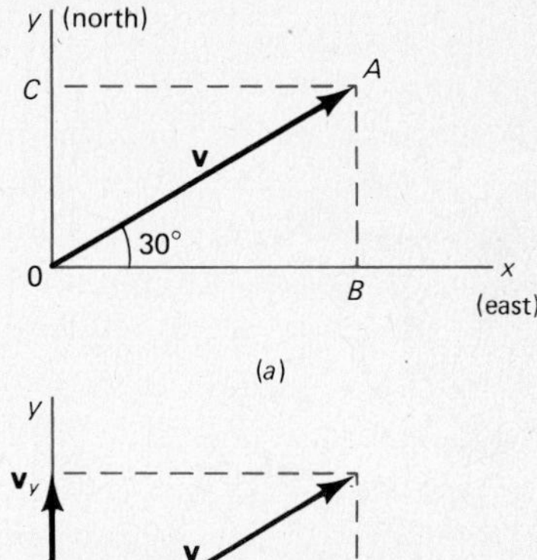

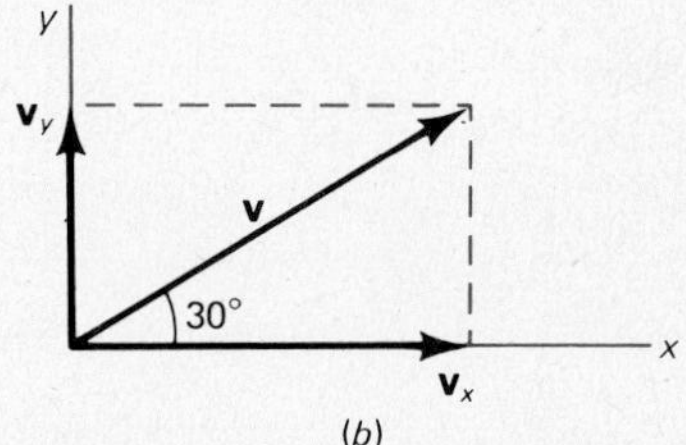

FIGURE 1-18
Resolving a vector v into its components along an arbitrarily chosen set of x and y axes.

Trigonometric functions

Given any angle, θ, as in Figure 1-19*a*, a right triangle can be constructed by drawing a line perpendicular to either of its sides, as in Figure 1-19*b*. The longest side of a right triangle, opposite the right angle is called the hypotenuse, which we label *c*. The side opposite the angle θ is labeled *o*, and the side adjacent is labeled *a*; we let *c*, *o*, and *a* represent the lengths of these sides, respectively. We now define the three trigonometric functions, sine, cosine, and tangent (abbreviated sin, cos, tan) in terms of the right triangle, as follows:

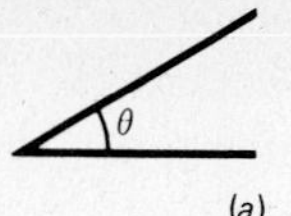

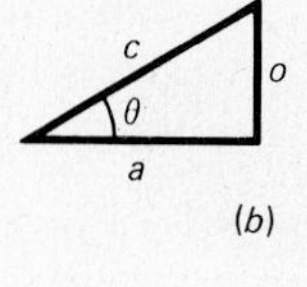

$$\sin\theta = \frac{\text{side opposite}}{\text{hypotenuse}} = \frac{o}{c}$$

$$\cos\theta = \frac{\text{side adjacent}}{\text{hypotenuse}} = \frac{a}{c} \qquad (1\text{-}10)$$

$$\tan\theta = \frac{\text{side opposite}}{\text{side adjacent}} = \frac{o}{a}.$$

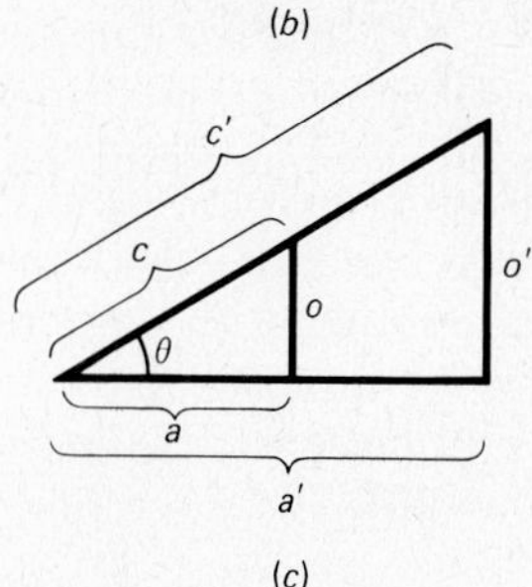

FIGURE 1-19
Starting with an angle θ (*a*), we can construct right triangles of different sizes, (*b*) and (*c*), but the ratio of the lengths of the sides does not depend on the size of the triangle.

Now it is an interesting fact that if we make the triangle bigger, but keep the same angles, then the ratio of the length of one side to the other, or of one side to the hypotenuse, remains the same; that is, in Figure 1-19*c* $a/c = a'/c'$; $o/c = o'/c'$; and $o/a = o'/a'$. Thus the values of sin, cos, and tan do not depend on how big the triangle is. They depend only on the size of the angle. The values of sin, cos, and tan for different angles have been computed precisely, and a table of them is located on the inside back cover of this book.

The use of trigonometric functions for finding the components of a vector is illustrated in Figure 1-20 where it is seen that a vector and its two components can be thought of as making up a right triangle. If we multiply the definition of $\sin\theta = V_y/V$ by V on both sides, we get

$$V_y = V\sin\theta; \qquad (1\text{-}11a)$$

similarly we obtain

$$V_x = V\cos\theta. \qquad (1\text{-}11b)$$

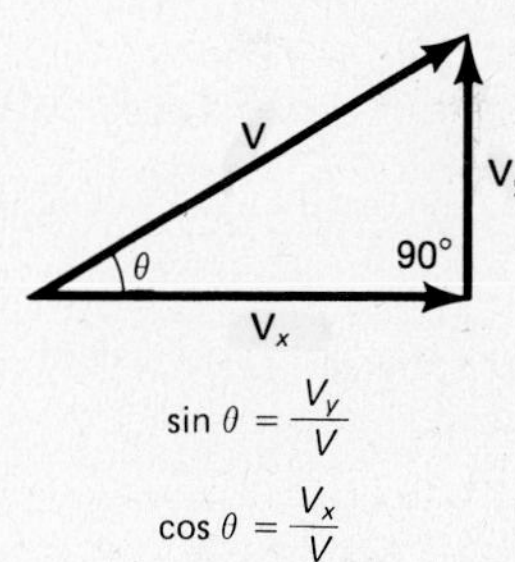

$$\sin\theta = \frac{V_y}{V}$$

$$\cos\theta = \frac{V_x}{V}$$

$$\tan\theta = \frac{V_y}{V_x}$$

$$V^2 = V_x^2 + V_y^2$$

FIGURE 1-20
Finding the components of a vector using trigonometric functions.

These two equations, along with the definition of sin, cos, and tan, and the theorem of Pythagoras:

$$V^2 = V_x^2 + V_y^2,$$

are the principal trigonometric relations we will need.†

Using these trigonometric relations, we can calculate v_x and v_y for the example illustrated in Figure 1-18, in which $v = 500$ km/h. From the tables, $\sin 30° = 0.500$ and $\cos\theta = 0.866$. Thus $v_x = v\cos\theta = (500\text{ km/h})(0.866) = 433$ km/h and $v_y = v\sin\theta = (500\text{ km/h})(0.500) = 250$ km/h.

Addition of vectors using components

We can now make use of a powerful and accurate method of adding vectors based on resolving vectors into their components. The technique

† In three dimensions, the theorem of Pythagoras becomes $V^2 = V_x^2 + V_y^2 + V_z^2$, where V_z is the component along the third or *z* axis.

is as follows: (1) Resolve each vector into its x and y components. (2) Add the x components of all the vectors together; this gives the x component of the resultant, v_{Rx}; similarly add the y components together to obtain v_{Ry}. (3) Then use the theorem of Pythagoras to obtain the magnitude of the resultant vector, $v=\sqrt{v_{Rx}^2+v_{Ry}^2}$; the angle (relative to the x-axis) at which it acts can be obtained from any of the trigonometric functions, such as $\tan\theta = v_{Ry}/v_{Rx}$. It is necessary to specify components along the x and y axes as positive or negative according to the usual convention, Figure 1-2.

▌ EXAMPLE 1-10 Use the component method of adding vectors to find the resultant velocity of the airplane in Example 1-9, Figure 1-17. Assume the numbers are accurate to three significant figures.

SOLUTION The two velocity vectors are shown resolved in Figure 1-21a. They are drawn with a common origin for convenience, and have been relabeled to avoid confusion from too many subscripts: $\mathbf{v}_1 = \mathbf{v}_{PA}$, $\mathbf{v}_2 = \mathbf{v}_{AG}$.

Since $\mathbf{v}_1$ is along the y axis, it only has a y component:

$$v_{1x} = 0 \text{ km/h},$$

$$v_{1y} = v_1 = 200 \text{ km/h}.$$

The components of $\mathbf{v}_2$ are

$$v_{2x} = v_2 \cos 45° = -(100 \text{ km/h})(0.707) = -70.7 \text{ km/h}$$

$$v_{2y} = v_2 \sin 45° = -(100 \text{ km/h})(0.707) = -70.7 \text{ km/h}.$$

Both v_{2x} and v_{2y} are negative because they are, respectively, along the negative x and y axes. The components of the resultant velocity are (we use $\mathbf{v}_R = \mathbf{v}_{PG}$, the velocity of the plane with respect to the ground):

$$v_{Rx} = 0 \text{ km/h} - 70.7 \text{ km/h} = -70.7 \text{ km/h},$$

$$v_{Ry} = 200 \text{ km/h} - 70.7 \text{ km/h} = 129 \text{ km/h}.$$

We find the magnitude of the resultant velocity using the Pythagorean theorem:

$$v_R = \sqrt{v_{Rx}^2 + v_{Ry}^2} = \sqrt{(70.7 \text{ km/h})^2 + (129 \text{ km/h})^2} = 147 \text{ km/h}.$$

To find the angle θ that $\mathbf{v}_R$ makes with the x axis (Figure 1-21b), we use

$$\tan\theta = \frac{v_{Ry}}{v_{Rx}} = \frac{129 \text{ km/h}}{70.7 \text{ km/h}} = 1.82;$$

From the trigonometry tables, we see that $\tan 61° = 1.804$ and $\tan 62° = 1.881$. So our angle θ is about 61°. (Notice that this angle θ is the complement, $90° - \theta$, to the angle used in Figure 1-17, but is fully consistent with the result obtained there.) ▌▌

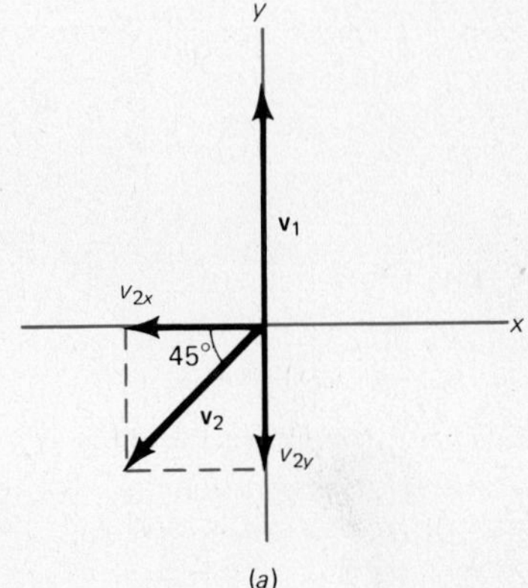

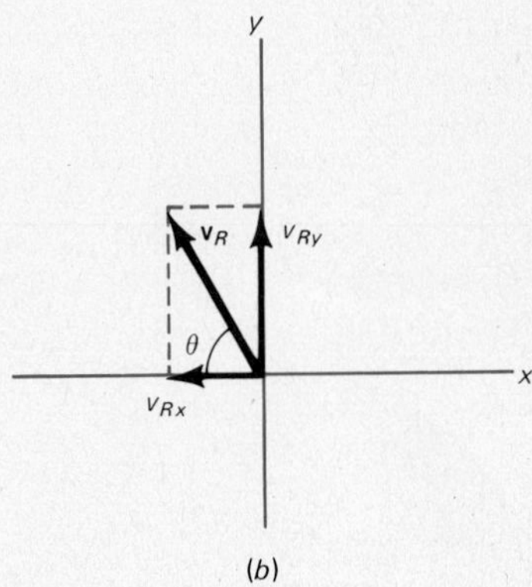

FIGURE 1-21
Example 1-10: same as Example 1-9 but using components, and with $\mathbf{v}_1 = \mathbf{v}_{PA}$ and $\mathbf{v}_2 = \mathbf{v}_{AG}$ (see Figure 1-17). (a) Resolving $\mathbf{v}_1$ and $\mathbf{v}_2$ into components. (b) The resultant velocity and its components.

▌ EXAMPLE 1-11 If a boat is to cross a river directly to the opposite side when the current is 12 km/h, at what angle upstream must the boat head? The boat's speed in still water, v_{BW}, is 20 km/h.

SOLUTION Let $\mathbf{v}_{WS}$ be the velocity of the water with respect to the shore. Then the velocity of the boat with respect to the shore, $\mathbf{v}_{BS}$, is

$$\mathbf{v}_{BS} = \mathbf{v}_{BW} + \mathbf{v}_{WS}.$$

(Note the arrangement of subscripts: the W's are in the middle; putting them in the correct order helps in setting up the equation correctly.) What we wish to find is the direction, θ, of $\mathbf{v}_{BW}$. So we draw the rough diagram of Figure 1-22; note that it is $\mathbf{v}_{BS}$ that must be straight across the river. Since we have a right triangle and know the magnitudes v_{WS} and v_{BW}, we can write

$$\sin\theta = \frac{v_{WS}}{v_{BW}} = \frac{12\ \text{km/h}}{20\ \text{km/h}} = 0.600,$$

and from the trigonometry tables, $\theta = 37°$. So the boat must head upstream at a 37° angle. ▮▮

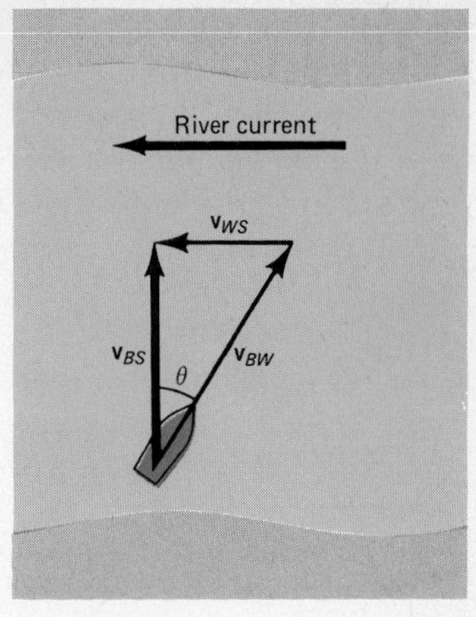

FIGURE 1-22
Boat must head upstream at an angle θ if it is to move directly across the river.

*1-7 ▮ Graphical analysis of linear motion†

Figure 1-1 shows the graph of the velocity of a car versus time for two cases of linear motion: (*a*) constant velocity, and (*b*) a particular case where the magnitude of the velocity varied. It is also useful to graph (or "plot") the displacement as a function of time. The time t is considered the independent variable and is measured along the horizontal axis (called the abscissa); and the displacement d, the dependent variable, is measured along the vertical axis (ordinate). We assume that at $t = 0$ the displacement $d = 0$. First we consider a car moving at a constant speed of 40 km/h, which is equivalent to 11 m/s (see Section 1-2). From Equation 1-3, $d = vt$, we see that d increases by 11 m every second. Thus, the displacement increases linearly in time, so the graph of d vs. t is a straight line as shown in Figure 1-23. Each point on this straight line tells us the car's displacement at a

† Certain sections and subsections of this book, such as this one, are considered *optional* at the discretion of the instructor. See the Preface for more details.

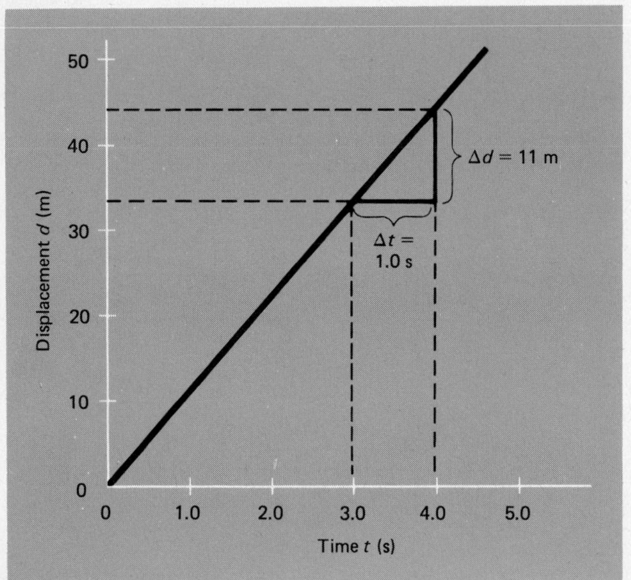

FIGURE 1-23
Graph of displacement vs. time for object moving at a uniform speed of 11 m/s.

particular time. For example, at $t = 3.0$ s, the displacement is 33 m, and at $t = 4.0$ s, $d = 44$ m; this is indicated by the dashed lines. The small triangle on the graph indicates the *slope* of the straight line which is defined as the change in the dependent variable (Δd) divided by the corresponding change in the independent variable (Δt):

$$\text{slope} = \frac{\Delta d}{\Delta t}.$$

We see, using the definition of speed (Equation 1-2), that the slope of the d vs. t graph is equal to the magnitude of the velocity. And, as can be seen from the little triangle on the graph, $\Delta d/\Delta t = (11\text{ m})/(1.0\text{ s}) = 11$ m/s, which is the given velocity.

The slope of the d vs. t graph is everywhere the same if the velocity is constant. But if the velocity changes, the slope of the d vs. t graph also varies. Consider, for example, a car that accelerates uniformly from rest to 15 m/s in 10 s, after which it remains at a constant speed of 15 m/s for the next 10 s, decelerates to 5.0 m/s during the following 5.0 s, and then remains at this constant speed; the velocity as a function of time is shown in the graph of Figure 1-24*a*. Now, to construct the d vs. t graph, we can use Equation 1-9*b* with constant acceleration for the intervals $t = 0$ to $t = 10$ s and $t = 20$ s to $t = 25$ s, and with constant speed for the periods $t = 10$ s to $t = 20$ s and after $t = 25$ s. The result is the d vs. t graph of Figure 1-24*b*.

From the origin to point A on the plot, the d vs. t graph is not a straight line, but a curve. The *slope* of the curve at any point is defined as the *slope of the tangent to the curve at that point*. For example, the tangent to the curve at the time $t = 5.0$ s is drawn on the graph (it is labeled "tangent"). A triangle is drawn with $\Delta t = 4.0$ s; Δd can be measured off the graph and is found to be 30 m. Thus the slope of the curve at $t = 5.0$ s, which equals the instantaneous velocity at that instant, is $v = \Delta d/\Delta t = 30\text{ m}/4.0\text{ s} = 7.5$ m/s. (This is the same result we get if we apply $v = at$ with $t = 5.0$ s and $a = [15\text{ m/s} - 0\text{ m/s}]/[10\text{ s}] = 1.5\text{ m/s}^2$.) In the region between A and B (Figure 1-24) the d vs. t graph is a straight line and the slope can be measured using the triangle shown between $t = 15$ s and $t = 20$ s: $\Delta d/\Delta t = 75\text{ m}/5.0\text{ s} = 15$ m/s.

Suppose we were given the d vs. t graph of Figure 1-24*b*; we could measure the slopes at a number of points and plot these slopes as a function of time. Since the slope equals the speed, we would thus have

FIGURE 1-24
(*a*) Velocity vs. time and (*b*) displacement vs. time for an object with variable speed. See text.

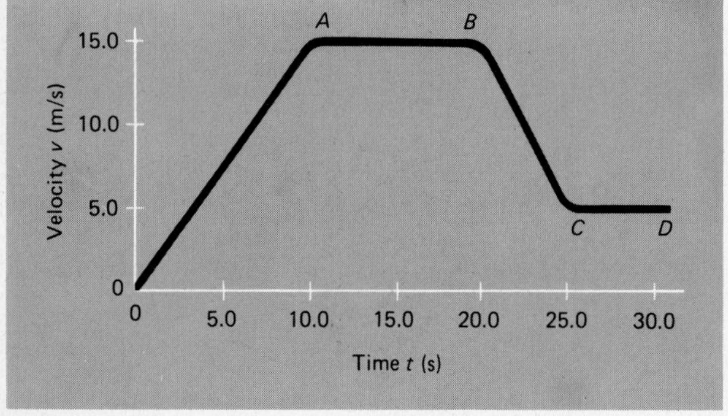

(*a*)

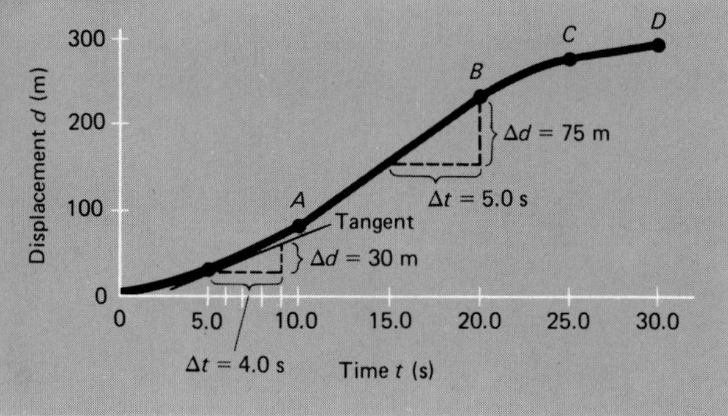

(*b*)

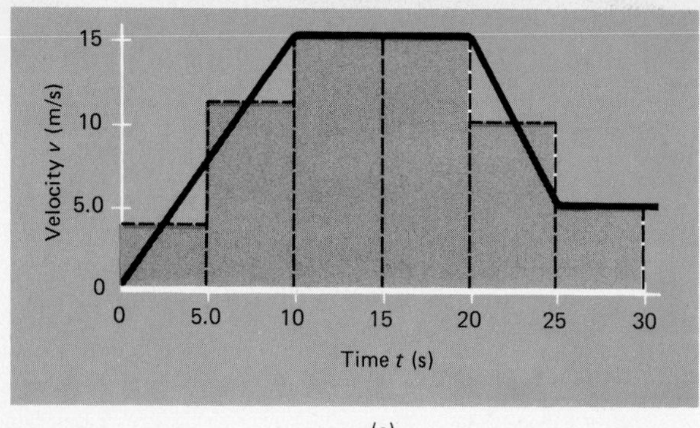

(a)

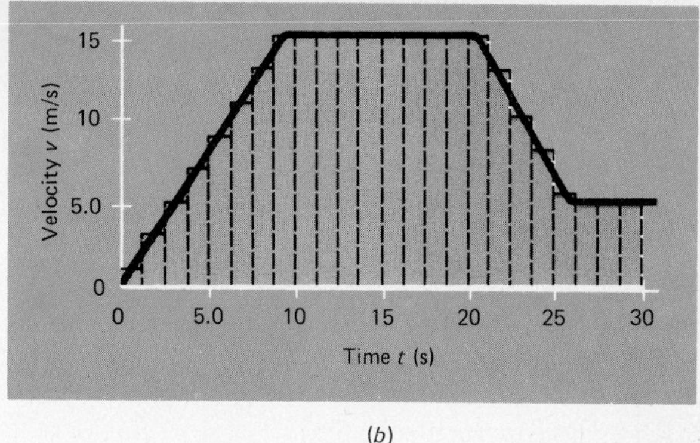

(b)

FIGURE 1-25
Determining displacement from v vs. t graph is done by calculating areas.

reconstructed the v vs. t graph! In other words, if we were given the graph of d vs. t, we could determine the speed as a function of time using graphical methods, instead of using equations. This technique is particularly useful when the acceleration is not constant, for then Equations 1-9 cannot be used.

The reverse process is also posssible. If we are given the v vs. t graph, we can determine the displacement d as a function of time. To do so, we use the following procedure which we apply to the v vs. t graph of Figure 1-25*a* (which is the same as Figure 1-24*a*). We first divide the time axis into many subintervals (in Figure 1-25*a*, only six for simplicity) which are indicated in the figure by the dashed vertical lines. In each interval, a *horizontal* dashed line is drawn to indicate the average speed during that time interval. For example, in the first interval, the speed increases at a constant rate from zero to 7.5 m/s so $\bar{v} = 3.75$ m/s; and in the third interval the speed is a constant 15 m/s, so $\bar{v} = 15$ m/s (no horizontal dashed line is shown since it coincides with the curve itself). The displacement during any subinterval is, from the definition of average speed, $\Delta d = \bar{v}\Delta t$. Thus the displacement during each subinterval equals the product of $\bar{v}$ and Δt, and this is just the *area of the rectangle*, shown in gray, for that interval. The total displacement after, say, 25 s, will be the sum of the first five rectangles.

If the speed varies a great deal, it may be difficult to estimate $\bar{v}$ from the graph. To reduce this difficulty, more—but narrower—subintervals are used; that is, we make each Δt smaller, as in Figure 1-25*b*. Ideally, we could let Δt approach zero; this leads to the subject of integral calculus, which we don't discuss here. The result, in any case, is that *the total displacement between any two times is equal to the area under the v vs. t graph between these two times.*

EXAMPLE 1-12 A space vehicle accelerates uniformly from 50 m/s at $t = 0$ to 150 m/s at $t = 10$ s. How far did it move between $t = 2.0$ s and $t = 6.0$ s?

SOLUTION The graph of v vs. t is shown in Figure 1-26. We simply need to calculate the gray area shown, which is a trapezoid. The area will be the average of the heights (in units of velocity) times the width (which is 4.0 s). At $t = 2.0$ s, $v = 70$ m/s; and at $t = 6.0$ s, $v = 110$ m/s. Thus

$$d = \left(\frac{70 \text{ m/s} + 110 \text{ m/s}}{2}\right)(4.0 \text{ s}) = 360 \text{ m}.$$

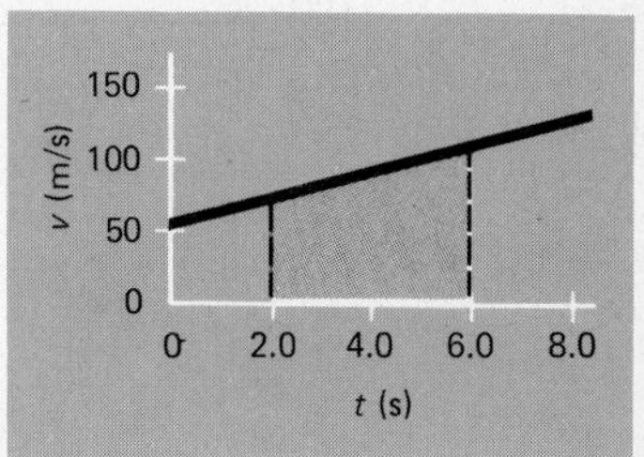

FIGURE 1-26
The gray area represents the displacement during the time interval $t = 2.0$ s to $t = 6.0$ s.

In cases where the acceleration is not constant, the area can be obtained by counting squares on graph paper (see the problems).

▮ Summary

The motion of a body is described by giving its position (or displacement), its velocity, and its acceleration. The *average speed* is defined as the distance traveled divided by the elapsed time. The *average velocity* is the displacement vector divided by the elapsed time. *Displacement* is the vector representing the position of an object relative to its position at some chosen earlier time (point of origin). Whereas speed is a *scalar* (a magnitude only), velocity is a *vector* (magnitude and direction). The *instantaneous velocity*, whose magnitude is the same as the *instantaneous speed*, is the average velocity taken over an indefinitely short time. Velocity (as well as other qualities describing motion) are always measured with respect to some *reference frame*.

Acceleration is the rate of change of velocity: the change in velocity divided by the elapsed time; it is a vector. If an object moves in a straight line with constant acceleration, the velocity, v, and acceleration, a, are related to the initial velocity, v_0, displacement, d, and time, t, by the equations:

$$v = v_0 + at, \qquad d = v_0 t + \tfrac{1}{2}at^2, \qquad v^2 = v_0^2 + 2ad, \qquad \bar{v} = \frac{v + v_0}{2}.$$

Objects allowed to fall freely, without air resistance, all fall with the same constant acceleration, $g = 9.8\ \text{m/s}^2$.

Addition of vectors can be done graphically by placing the tail of each successive arrow (representing each vector) at the tip of the previous one. The sum or *resultant vector* is the arrow drawn from the tail of the first to the tip of the last. Two vectors can also be added using the parallelogram method. Vectors can be added more accurately using the analytical method of adding their *components* along chosen axes with the aid of trigonometric functions. The velocity of an object relative to one frame of reference can be found by vector addition if its velocity relative to a second frame of reference, and the relative velocity of the two reference frames, are known.

Questions

1 One car travels due east at 40 km/h, and a second car travels north at 40 km/h. Are their velocities equal? Explain.

2 Can the velocity of an object be zero at the same instant its acceleration is not zero? Give an example.

3 If an object has a greater speed, does it necessarily have a greater acceleration? Explain, using examples.

4 Compare the acceleration of a motorcycle that accelerates from 80 km/h to 90 km/h with a bicycle that accelerates from rest to 10 km/h in the same time.

5 Can you conclude that a car is not accelerating if its speedometer indicates a steady 60 km/h?

6 A car rounds a curve at a steady 50 km/h. If it rounds the same curve at a steady 70 km/h, will its acceleration be any different? Explain.

7 Will the acceleration of a car be the same if it travels around a sharp curve at 60 km/h as when it travels around a gentle curve at the same speed? Explain.

8 Does the odometer of a car measure a scalar or a vector quantity? What about the speedometer?

9 Two rowers who can row at the same speed set off across a river at the same time. One heads straight across and is pulled downstream somewhat by the current. The other one heads upstream at an angle so as to arrive at a point opposite the starting point. Which rower reaches the opposite side first?

*10 Describe in words the motion plotted in Figure 1-27.

*11 Describe in words the motion of the object graphed in Figure 1-28.

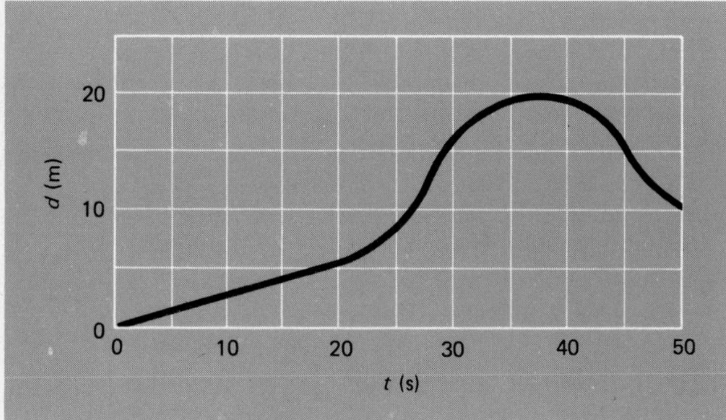

FIGURE 1-27

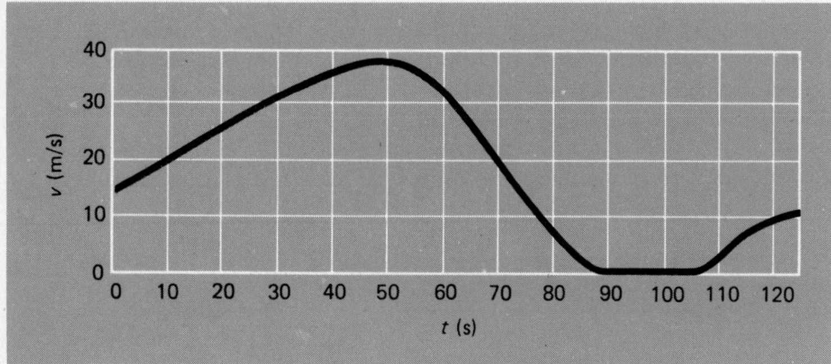

FIGURE 1-28

Problems

[*The problems at the end of each chapter are arranged by section and marked according to difficulty*: (I) *is easiest*, (III) *the most difficult, as discussed in the Preface.*]

SECTIONS 1-1 AND 1-2

1 (I) What must your average speed be in order to travel 680 km in 8.0 h?

2 (I) At an average speed of 18 km/h, how far will a bicyclist travel in $3\frac{1}{2}$ h?

3 (I) A bird can fly 30 km/h. How long does it take to fly 235 km?

4 (I) If you are driving 90 km/h and you look to the side for 2.0 s, how far do you travel during this inattentive period?

5 (I)Use a ruler to determine the speed of the car in Figure 1-4 at each point shown.

6 (I) Calculate the conversion factor between mi/h and km/h. (Do not use Table 1-2 except as a check.)

7 (I) A rabbit travels 4.0 km in 3.5 h. What is its average speed in m/s?

8 (II) A rock thrown horizontally at a large bell 50 m away is heard to hit the bell 4.5 s later. If the speed of sound is 330 m/s, what was the speed of the rock? (Disregard the effect of gravity.)

9 (II) An airplane travels 1800 km at a speed of 1000 km/h. It then encounters a headwind that slows it to 850 km/h for the next 2300 km. What was the average speed of the plane for this trip?

10 (II) A race car driver must average 180 km/h for four laps to qualify for a race. Because of engine trouble, the car averages only 150 km/h over the first two laps. What average speed must be maintained for the last two laps?

11 (II) Calculate the carrying capacity (number of cars passing a given point per hour) on a freeway with three lanes (in one direction) using the following assumptions: the average speed is 90 km/h, the average length of a car is 6.0 m, and the average distance between cars should be (see Example 1-3) 65 m.

SECTION 1-3

12 (I) A dolphin accelerates from 1.0 m/s to 7.6 m/s in 5.5 s. What was its acceleration?

13 (I) A car accelerates from rest to 100 km/h in 7.0 s. What is its acceleration in m/s^2?

14 (I) At high speeds, a particular automobile is capable of an acceleration of about 0.50 m/s^2. At this rate how long does it take to accelerate from 90 km/h to 100 km/h?

SECTION 1-4

15 (I) The principal kinematic equations, 1-9*a* through 1-9*d*, become particularly simple if the initial speed is zero. Write down the equations for this special case.

16 (I) A car accelerates from 40 km/h up to 80 km/h in 5.0 s. What was its acceleration in m/s^2, and how far did it travel in this time?

17 (I) A car decelerates from a speed of 30 m/s to rest in 6.0 s. How far did it travel in that time?

18 (I) An eagle accelerates from 15 m/s to 20 m/s over a distance of 50 m. What was its acceleration?

19 (II) A linebacker starts from rest and accelerates at a constant rate for 2.0 s to make a tackle 7.0 m away. What was the acceleration and the speed when making the tackle?

20 (II) A 100-m long train accelerates uniformly from rest. If the front of the train passes a railway worker 150 m down the track at a speed of 25 m/s, what will be the speed of the last car as it passes the worker?

21 (II) A car traveling 80 km/h decelerates at 1.5 m/s^2. Calculate (*a*) the distance it goes before it stops, (*b*) the time it takes to stop, and (*c*) the distance it travels during the first and third seconds.

22 (II) Make up a table of stopping distances for an automobile with an initial speed of 80 km/h and human reaction time of 1.0 s: (*a*) for a deceleration $a = -4.0\ m/s^2$; (*b*) for $a = -8.0\ m/s^2$.

23 (II) Repeat Problem 22 using a reaction time of 0.40 s.

24 (III) Show that the equation for the stopping distance of a car is $d_s = v_0 t_R - v_0^2/(2a)$ where v_0 is the initial speed of the car, t_R is the driver's reaction time, and a is the rate of deceleration (and is negative).

25 (III) In designing traffic signals, it is necessary to allow the yellow light to remain on long enough so that a driver can either stop or pass completely through the intersection. Thus, if a driver is less than the stopping distance, d_s (calculated in Problem 24 above) from the intersection, then the light must remain on long enough for him to travel this distance plus the width of the intersection, d_I. (*a*) Show that the light should remain on for a time $t = t_R - v_0/(2a) + d_I/v_0$ where v_0 is a typical expected speed of a car approaching the intersection, and a and t_R are defined in Problem 24. (*b*) A traffic planner expects cars to approach a 12-m wide intersection at speeds between 30 and 50 km/h. To be safe, he calculates the time for both speeds, assuming $t_R = 0.50$ s, $a = -4.0\ m/s^2$, and chooses the longest time to be safe. What is his result?

26 (III) In the design of a rapid transit system, it is necessary to balance out the average speed of a train against the distance between stops. The more stops there are, the slower the train's average speed. To get an idea of this problem, calculate the time it takes a train to make a 30-km trip in two situations: (*a*) The stations at which the trains must stop are 0.80 km apart; (*b*) the stations are 3.0 km apart. Assume that at each station the train accelerates at a rate of $1.5\ m/s^2$ until it reaches 80 km/h, then stays at this speed until it puts on its brakes to arrive at the next station at which time it decelerates at $-3.0\ m/s^2$. Assume it stops at each station for 20 s.

27 (III) For the design of a rapid transit system as discussed in the previous problem, derive a general formula for the average speed of a train. Specify the symbols used for all quantities involved such as the acceleration, deceleration, maximum velocity, distance between stations, and time stopped at each station.

28 (III) A runner hopes to complete the 5000-m run in less than 13.0 min. After exactly 11.0 min, there are still 800 m to go. The runner must accelerate at $0.20\ m/s^2$ for how many seconds in order to achieve the desired time?

SECTION 1-5

29 (I) Calculate the acceleration of the baseball in Example 1-4 in "*g*'s."

30 (I) A stone is dropped from the top of a cliff. It is seen to hit the ground below after 5.5 s. How high is the cliff?

31 (I) A baseball is thrown vertically into the air with a speed of 24 m/s. (*a*) How high does it go? (*b*) How long does it take to return to the ground?

32 (I) (*a*) How long does it take a brick to reach the ground if dropped from a height of 80 m? (*b*) What will be its velocity just before it reaches the ground?

33 (I) A trout jumps a waterfall 2.5 m high. With what minimum speed did it leave the water below to reach the top?

34 (I) The flea in Example 1-7 actually reached a height of only 3.5 cm because of air resistance. Using this height, and the initial speed of 1.0 m/s, calculate the actual deceleration rate of the flea, assuming it to be constant.

35 (II) A stone is dropped from the roof of a high building. A second stone is dropped 1.0 s later. How far apart are the stones when the second one has reached a speed of 23 m/s?

36 (II) A kangeroo jumps to a vertical height of 2.8 m. How long was it in the air?

37 (II) A helicopter is ascending vertically with a speed of 8.0 m/s; at a height of 120 m above the earth, a package is dropped from a window. How much time does it take for the package to reach the ground?

38 (II) A person who is properly constrained by a shoulder harness has a good chance of surviving a car collision if the deceleration does not exceed 30 *g*'s. Assuming uniform deceleration at this rate, calculate the distance over which the front end of the car must be designed to collapse if a crash occurs at 100 km/h.

39 (II) Draw a graph of (*a*) the speed and (*b*) the distance fallen, as a function of time, for a body falling under the influence of gravity for $t=0$ to $t=5$ s.

40 (II) If air resistance is neglected, show that a ball thrown vertically upward with a speed v_0 will have the same speed, v_0, when it comes back down to the starting point.

41 (II) A falling stone takes 0.30 s to pass a window 2.1 m high. From what height did the stone fall?

42 (III) A stone is thrown vertically upward with a speed of 19.0 m/s. (*a*) How fast is it moving when it reaches a height of 15.0 m? (*b*) How long is required to reach this height? (*c*) Why are there two answers to (*b*)?

43 (III) A rock is dropped from a seacliff and the sound of it striking the ocean is heard 3.0 s later. If the speed of sound is 330 m/s, how high is the cliff?

44 (III) Pelicans tuck their wings and free fall straight down when diving for fish. Suppose a pelican starts its dive from a height of 20 m and cannot change its path once committed. If it takes a fish 0.10 s to perform evasive action, at what minimum height must it spot the pelican to escape? Assume the fish is at the surface of the water.

45 (III) Suppose you adjust your garden hose nozzle for a hard stream of water. You point the nozzle vertically upward at a height of 1.5 m above the ground. When you quickly move the nozzle away from the vertical, you hear the water striking the ground next to you for 2.0 s. What is the water speed as it leaves the nozzle?

SECTION 1-6

46 (I) A dog walks at a speed of 1.8 m/s along the deck toward the front of a boat which is traveling at 7.6 m/s with respect to the water. What is the velocity of the dog with respect to the water? What if the dog were walking toward the back of the boat?

47 (I) The three vectors in Figure 1-13 can be added in six different orders ($\mathbf{V}_1+\mathbf{V}_2+\mathbf{V}_3$, $\mathbf{V}_1+\mathbf{V}_3+\mathbf{V}_2$, etc.). Show on a diagram that the same resultant is obtained no matter what the order.

48 (I) A car is driven 30 km west and then 30 km southwest. What is the displacement of the car from the point of origin (magnitude and direction)?

49 (I) A delivery truck travels 8 blocks north, 6 blocks east, and 10 blocks south. What is its final displacement from the origin? Assume the blocks are equal length.

50 (I) A vacationer walks 4.0 km/h directly across a cruise ship whose speed relative to the earth is 12 km/h. What is the speed of the vacationer with respect to the earth?

51 (I) Show that the vector labeled "wrong" in Figure 1-14 is actually the difference of the two vectors, $\mathbf{V}_2 - \mathbf{V}_1$.

52 (I) An airplane is traveling 1000 km/h in a direction 37° east of north. (*a*) Find the components of the velocity vector in the northerly and easterly directions. (*b*) How far north and how far east has the plane traveled after 2.0 h?

53 (II) Two trains approach each other on parallel tracks. Each has a speed of 80 km/h with respect to the earth. If they are initially 10 km apart, how long will it be before they pass each other?

54 (II) (*a*) A skier is accelerating down a 30° hill at 3.6 m/s^2. What is the vertical component of her acceleration? (*b*) How long will it take her to reach the bottom of the hill, assuming she starts from rest and accelerates uniformly, if the elevation change is 150 m?

55 (II) An airplane is heading due north at a speed of 300 km/h. If a wind begins blowing from the southwest at a speed of 50 km/h (average), calculate: (*a*) the velocity (magnitude and direction) of the plane, and (*b*) how far off course it will be after 30 min.

56 (II) A helicopter heads due south with an air speed of 50 km/h. The pilot observes, however, that they have covered 30 km in the previous 45 min in a southwesterly direction. What is the wind speed and direction?

57 (II) A swimmer is capable of swimming 1.80 m/s in still water. (*a*) If she swims directly across a 200-m wide river whose current is 0.80 m/s, how far downstream (from a point opposite her starting point) will she land? (*b*) At what upstream angle must the swimmer aim if she is to arrive at a point directly across the stream?

58 (II) A motorboat whose speed in still water is 9.5 km/h must aim at a 55° angle upstream in order to travel directly across the stream. (*a*) What is the speed of the current? (*b*) What is the resultant speed of the boat with respect to the shore?

59 (II) Determine the speed of the boat with respect to the shore in Example 1-11.

60 (II) An airplane, whose air speed is 300 km/h, is supposed to fly in a straight path 30° N of E. But a steady 100 km/h wind is blowing from the north. In what direction should the plane head?

61 (III) An automobile traveling 90 km/h overtakes a 1.0-km long train traveling in the same direction on a track parallel to the road. If the train's speed is 60 km/h, how long does it take the car to pass it and how far will it have traveled in this time? What are the results if the car and train are traveling in opposite directions?

62 (III) A motorcycle traveling 90.0 km/h approaches a car traveling in the same direction at 80.0 km/h. When the motorcycle is 50 m behind the car, the rider pushes down on the accelerator and passes the car 10.0 s later. What was the acceleration of the motorcycle?

63 (III) An unmarked police car traveling a constant 80 km/h is passed by a speeder traveling 100 km/h. Precisely 1.0 s after the speeder passes, the policeman steps on the accelerator; if the police car's acceleration is 2.0 m/s^2, how much time passes before the police car overtakes the speeder (assumed moving at constant speed)?

64 (III) Assume in the previous problem that the speeder's speed is not known. If the police car accelerates uniformly as given above for 6.0 s, what was the speeder's speed?

*SECTION 1-7

*65 (I) The position of a rabbit along a straight tunnel as a function of time is plotted in Figure 1-27. What is its instantaneous velocity (*a*) at $t = 10$ s, and (*b*) at $t = 30$ s? What is its average velocity (*c*) between $t = 0$ and $t = 5$ s, (*d*) between $t = 25$ s and $t = 30$ s, and (*e*) between $t = 40$ s and $t = 50$ s?

*66 (I) In Figure 1-27, (*a*) during what time periods, if any, is the rabbit's velocity constant? (*b*) At what time is its velocity the greatest? (*c*) At what time, if any, is the velocity zero? (*d*) Does the rabbit run in one direction or in both along its tunnel during the time shown?

*67 (I) Figure 1-28 shows the velocity of a train as a function of time. (*a*) At what time was its velocity greatest? (*b*) During what periods, if any, was the velocity constant? (*c*) During what periods, if any, was the acceleration constant? (*d*) When was the magnitude of the acceleration greatest?

*68 (II) In Figure 1-28, estimate the distance the train traveled (*a*) during the first minute, (*b*) during the second minute.

*69 (II) Construct the v vs. t graph for the object whose displacement as a function of time is given by Figure 1-27.

*70 (II) Construct the d vs. t graph for the object whose velocity as a function of time is given by Figure 1-28.

For additional reading

Casper, B. M., and R. J. Noer, *Revolution in Physics*, W. W. Norton & Co., Inc., New York, 1972.

Galilei, Galileo, *Dialogues Concerning Two New Sciences*, Dover Publications, Inc., New York, 1914.

Hay, J. G., *The Biomechanics of Sports Techniques*, Prentice-Hall, Inc., Englewood Cliffs, N.J., 1973.

Magie, W. F., *Source Book in Physics*, Harvard University Press, Cambridge, Mass., 1963. Contains selections from the writings of many important physicists throughout history, including those from Galileo on acceleration and falling bodies.

Shonle, J. I., *Environmental Applications of General Physics*, Addison-Wesley Publishing Co., Inc., Reading, Mass., 1974.

2 ▌ Motion and force: dynamics

We have discussed how motion is described in terms of velocity and acceleration. Now we deal with the question of *why* objects move as they do: what makes an object at rest begin to move? What causes a body to accelerate or decelerate? What is involved when an object moves in a circle? We might answer that in each case a force is required. In this chapter we will investigate the connection between force and motion. Before we delve into this subject of *dynamics*, we must first examine what we mean by the concept of force.

2-1 ▌ Force

Intuitively, we can define **force** as any kind of a push or a pull. When you push a grocery cart, you are exerting a force on it. When children pull a wagon, they are exerting a force on the wagon. When a motor lifts an elevator, or a hammer hits a nail, or the wind blows the leaves of a tree, a force is being exerted; we say that an object falls because of the *force of gravity*. Forces do not always give rise to motion. For example, you may push very hard on a heavy desk or sofa and it won't move.

Whether or not an object moves when a force is exerted on it, the object does change shape. This is obvious when you squeeze a balloon, or push on a mattress. You can also see the slight deformation of the metal when you push on the side of a refrigerator or the fender of a car. There is always some deformation when a force is exerted, although it may take delicate instruments to detect it for a very rigid object like a heavy steel plate.

One way to quantitatively measure the magnitude (or strength) of a force is to make use of a spring scale (Figure 2-1). Normally such a spring scale is used to find the weight of an object; by weight we mean the force of gravity acting on the body. The spring scale, once calibrated, can be used to measure other kinds of forces as well, such as the pulling force shown in Figure 2-1.

A force has direction as well as magnitude and is therefore a vector. We can represent any force on a diagram by an arrow, just as we did with

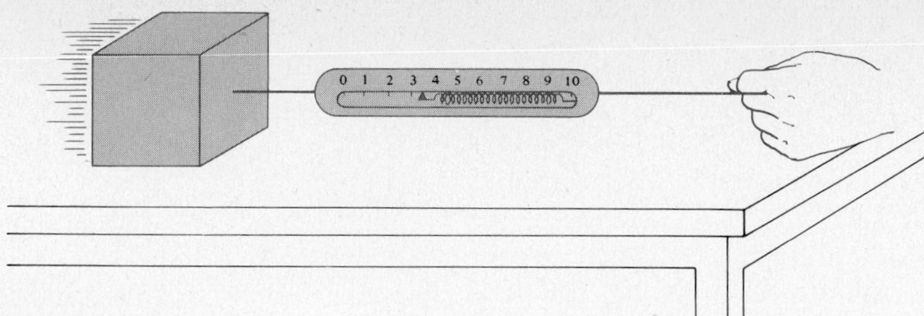

FIGURE 2-1
A spring scale used to measure a force.

velocity. The direction of the arrow is of course in the direction of the push or pull, and its length is drawn proportional to the strength or magnitude of the force. Although the definition of force as a push or a pull is adequate for the moment, we will give a more precise definition in Section 2-3.

2-2 ▌ Galileo vs. Aristotle on horizontal motion

What is the exact connection between force and motion? Aristotle believed that a force was required to keep an object moving along a horizontal plane. He would argue that to make a book move across the table, you would have to exert a force on it continuously. To Aristotle, the natural state of a body was at rest, and a force was believed necessary to keep a body in motion.

Some 2000 years later, Galileo, skeptical about these Aristotelian views just as he was of those on falling bodies, performed his own experiments and came to a radically different conclusion. Galileo claimed that it is just as natural for an object to be in horizontal motion with a constant speed as it is to be at rest. To understand Galileo's idea, consider the following observations involving motion along a horizontal plane (where the effects of gravity do not enter). It will take a certain amount of force to push an object with a rough surface along a table top at constant speed. To push an equally heavy object with a very smooth surface across the table at the same speed will require less force. Finally, if a layer of oil or other lubricant is placed between the surface of the object and the table, then almost no force is required to move the object. (These observations may well be obvious to you, but if not, you should do these simple experiments for yourself). Notice that in each step the force required was less and less. As the next step we can extend the data to a situation in which the object does not rub against the table at all—or there is a perfect lubricant between them—and theorize that once started, the object would move across the table at constant speed with *no* force applied. A steel ball bearing rolling on a hard horizontal surface approaches this situation closely.

This is the kind of reasoning Galileo used to reach his remarkable conclusion that if no force is applied to a moving object, it will continue to move with constant speed in a straight line. An object slows down only if a force is exerted on it. Galileo thus interpreted friction as a force akin to ordinary pushes and pulls.

To push an object across a table at constant speed requires a force from your hand only to balance out the force of friction; the pushing force is equal in magnitude to the friction force but they are in opposite directions, so the *net* force on the object is zero, Figure 2-2. This is consistent with Galileo's viewpoint, for the object moves with constant speed when no net force is exerted on it.

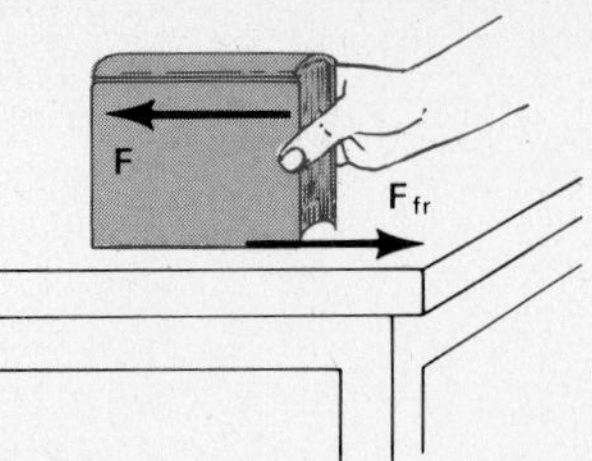

FIGURE 2-2
F represents the force applied by the person and $\mathbf{F}_{fr}$ represents the force of friction.

The difference between Aristotle's view and Galileo's is not simply one of right or wrong. Aristotle's view was not really wrong, for our everyday experience indicates that moving objects do tend to come to a stop if not continually pushed. The real difference lies in the fact that Aristotle's view about the "natural state" of a body was essentially a final statement—no further development was possible. Galileo's analysis, on the other hand, could be extended to explain a great many more phenomena. By making the creative leap of imagining the experimentally unattainable situation of no friction and by interpreting friction as a force, Galileo was able to reach his conclusion that an object will continue moving with constant velocity if no force acts to change this motion. Upon this foundation, Newton built his great theory of motion that we now discuss.

2-3 ▮ Newton's laws of motion

Newton's first law: the law of inertia

Newton's analysis of motion is summarized in his famous "three laws of motion." In his great work, the *Principia* (which contains nearly all his work on motion), Newton readily acknowledged his debt to Galileo. In fact, **Newton's first law of motion** is essentially a restatement of Galileo's results:

> ▮ **Every body continues in its state of rest or of uniform speed in a straight line unless it is compelled to change that state by forces acting on it.**

The tendency of a body to maintain its state of rest or of uniform motion in a straight line is called **inertia**. As a result, Newton's first law is often called the *law of inertia*.

Inertia and mass

Newton's second law makes use of the concept of *mass*. Newton himself used the term mass as a synonym for *quantity of matter*. This intuitive notion of the mass of a body is not very precise because the concept "quantity of matter" is itself not well defined. A more precise definition is that **mass** *is a measure of the inertia of a body*. The more inertia a body has, the harder it is to change its state of motion; it is harder to start it moving from rest, or to stop it when it is moving, or to change its motion sideways out of a straight line path. A piano or a truck has much more inertia than a pencil or a paper clip, and therefore much more mass. The definition of mass as a measure of the inertia of a body is completely compatible with the notion of mass as "quantity of matter."

43 (III) A bicyclist can coast down a 4° hill at 6.0 km/h. If the force of friction is proportional to the speed v so that $F_{fr} = cv$, calculate: (*a*) the value of the constant c, and (*b*) the average force that must be applied in order to descend the hill at 20 km/h. The mass of the cyclist plus bicycle is 80 kg.

44 (III) A motorcyclist is coasting with the engine off at a steady speed of 12 m/s but enters a sandy stretch where the coefficient of friction is 0.80. Will the cyclist emerge from the sandy stretch without having to start the engine if the sand lasts for 15 m? If so, what will be the speed upon emerging?

45 (III) A city planner is working on the redesign of a hilly portion of a city. An important consideration is how steep the roads can be so that even small cars can get up the hills without slowing down. It is given that a particular small car, with a mass of 1000 kg, can accelerate on the level from rest to 14 m/s (50 km/h) in 8.0 s. Using this figure, calculate the maximum steepness of a hill.

SECTION 2-7

46 (I) Calculate the height of the locust's jump in Example 2-13, ignoring air resistance.

47 (I) A diver running 3.5 m/s dives from the edge of a vertical cliff and reaches the water below 2.0 s later. How high was the cliff and how far from its base did the diver hit the water?

48 (I) A tiger leaps horizontally from a 15-m high rock with a speed of 7.0 m/s. How far from the base of the rock will it land?

49 (II) A ball thrown horizontally at 18 m/s from the roof of a building lands 24 m from the base of the building. How high is the building?

50 (II) Show that the speed with which a projectile leaves the ground is equal to its speed just before it strikes the ground at the end of its journey, assuming the firing level equals the landing level.

51 (II) An airplane traveling 150 km/h wants to drop supplies to flood victims isolated on a patch of land 200 m below. The supplies should be dropped how many seconds before the plane is directly overhead?

52 (II) A football is kicked with a speed of 21.0 m/s at an angle of 37° to the horizontal. How much later does it hit the ground? Ignore air resistance.

53 (II) A hunter aims directly at a target (on the same level) 220 m away. If the bullet leaves the gun at a speed of 550 m/s, by how much will it miss the target?

54 (II) An athlete throws the shotput (mass = 7.3 kg) with an intitial speed of 14 m/s at a 40° angle to the horizontal. Calculate the distance traveled. The shot leaves the shotputter's hand at a height of 2.2 m above the ground.

55 (III) An athlete executing a long jump leaves the ground at a 30° angle and travels 8.90 m. What was the takeoff speed?

56 (III) Determine how much further a person can jump on the moon as compared to the earth if the takeoff speed and angle are the same. The acceleration due to gravity on the moon is one-sixth what it is on earth.

57 (III) (*a*) Show that the range R of a projectile, which is defined as the horizontal distance traveled when the final point is at the same level as the initial point, is given by the equation $R = (v_0^2 \sin 2\theta)/g$. [*Hint*: Use the trigonometry identity $\sin 2\theta = 2 \sin \theta \cos \theta$.] Ignore air resistance. (*b*) Show that the maximum range is obtained when $\theta = 45°$. (*c*) When performing the long jump, athletes gain speed by running before making the jump. Give a practical reason why the longest jumps are made at angles somewhat less than 45°.

58 (III) Prove that, in the absence of air resistance, the path of an object projected horizontally is a parabola.

59 (III) Police agents flying 200 km/h in a low-flying airplane wish to drop an explosive onto a master criminal's automobile traveling 130 km/h on a level highway 80 m below. At what angle (with the horizontal) should the car be in their sights when the bomb is released?

60 (III) A fire hose shoots water at a speed of 15 m/s. At what angle(s) should the nozzle point in order that the water land 18 m away? Why are there two different angles?

For additional reading

Andrade, E. N., *Sir Isaac Newton: His Life and Work*, Doubleday & Co., New York, 1958.

Cooper, L. N., *An Introduction to the Meaning and Structure of Physics*, Harper & Row, Publishers, Inc., New York, 1970.

Drake, S., and J. MacLachlan, "Galileo's Discovery of the Parabolic Trajectory," *Sci. American*, **232**:3 (Mar., 1975), p. 102. Projectile motion.

Hay, J. G., *The Biomechanics of Sports Techniques*, Prentice-Hall, Inc., Englewood Cliffs, N.J., 1973.

Magie, W. F., *Source Book in Physics*, Harvard University Press, Cambridge, Mass., 1963. Excerpts from Newton on the laws of motion and Galileo on projectile motion.

Newton, I., *Mathematical Principles of Natural Philosophy . . . (Principia)*, tr. A. Motte, ed. F. Cajori, University of California Press, Berkeley, 1947.

3 ▮ Circular and rotational motion; gravitation

An object moves in a straight line if the net force on it acts in the direction of motion, or is zero. If the net force acts at an angle to the direction of motion at any moment, then the object moves in a curve. We have already discussed one example of this in projectile motion. Another important situation is that of an object moving in a circle, such as a ball at the end of a string revolving around one's head, the motion of a wheel or a centrifuge, or the nearly circular motion of the moon about the earth. These are examples of circular or rotational motion.

3-1 ▮ Angular quantities

Radian measure for angles

To describe circular or rotational motion, we make use of angular quantities, such as angular velocity and angular acceleration. These are defined in analogy to the corresponding quantities in linear motion.

Consider a small particle revolving in a circle of radius r as shown in Figure 3-1. To indicate the position of the particle, or how far it has revolved, we specify the angle θ. The particle moves through an angle θ when it travels the distance s measured along the circumference of the circle. Angles are commonly measured in degrees, but the mathematics of circular motion is much simpler if we use the radian as the unit for angular measure: one *radian* (rad) is defined as the angle subtended by an arc whose length is equal to the radius. Thus, in Figure 3-1, if $s = r$, then θ is exactly equal to 1 rad. In general, any angle θ is given by

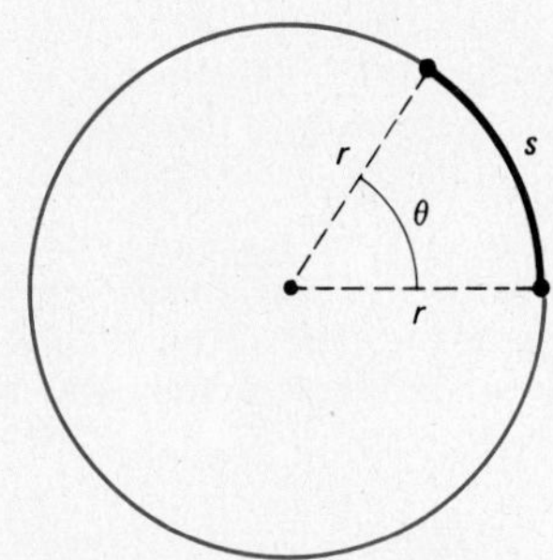

FIGURE 3-1
A particle moves in a circular path of radius r; here it is shown to have moved through an angle θ and a distance s along the circle.

$$\theta = \frac{s}{r}, \tag{3-1}$$

where r is the radius of the circle and s is the arc length subtended by the angle θ; θ itself will then be in radians. Radians can be related to degrees in the following way. In a complete circle there are 360° and this of course must correspond to an arc length equal to the circumference of the circle, $s = 2\pi r$; thus $\theta = s/r = 2\pi r/r = 2\pi$ rad in a complete circle, so an angle of $360° = 2\pi$ rad. One radian is then $360°/2\pi = 360°/6.28 = 57.3°$. The following example illustrates the usefulness of radian measure.

▮ EXAMPLE 3-1 A particular bird's eye can just distinguish objects that subtend an angle no smaller than about 3×10^{-4} rad (how many degrees is this?). How small an object can the bird just distinguish when flying at a height of 100 m (Figure 3-2a)?

SOLUTION We solve Equation 3-1 for s:

$$s = r\theta.$$

Strictly speaking, s is the arc length, but for small angles the linear distance is approximately the same (Figure 3-2b).† Since $r = 100$ m and $\theta = 3 \times 10^{-4}$ rad we find that

$$s = (100\text{ m})(3 \times 10^{-4}\text{ rad}) = 3 \times 10^{-2}\text{ m} = 3\text{ cm}.$$

Had the angle been given in degrees, we would first have had to change it to radians to make this calculation. ▮▮

Note that in this example we used the fact that because the radian is the ratio of two lengths, it is dimensionless (has no units) (see Equation 3-1).

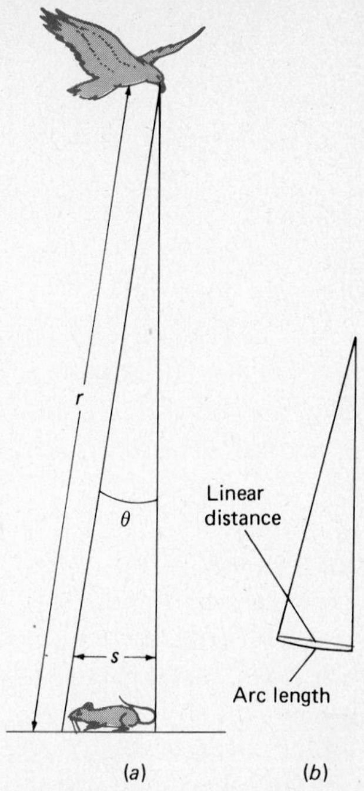

FIGURE 3-2
(a) Example 3-1; (b) for small angles, arc length s and the linear distance (the chord) are nearly equal.

Angular velocity and acceleration

Angular velocity is defined in analogy with ordinary linear velocity. Instead of distance traveled, we use the angular distance θ, Figure 3-1. Thus the **average angular velocity** (denoted by ω, the Greek lowercase letter omega) is defined as

$$\bar{\omega} = \frac{\theta}{t}, \tag{3-2}$$

where θ is the angle through which the particle has rotated in the time t. We can also define the **instantaneous angular velocity** as the small angle, $\Delta\theta$, through which the particle turns in the very short time interval Δt:

$$\omega = \frac{\Delta\theta}{\Delta t} \quad [\Delta t \text{ very small, approaching zero}].$$

Angular velocity is generally measured in radians per second. We can relate the angular velocity of a particle to its linear speed v and its distance r from the axis of rotation: since v = distance/time, we have, using Equation 3-1:

$$v = \frac{r\theta}{t},$$

or, since $\theta/t = \omega$:

$$v = r\omega. \tag{3-3}$$

This equation is valid both instantaneously and on the average.

Angular acceleration, in analogy to ordinary linear acceleration, is defined as the change in angular velocity divided by the time required to make this change:

$$\bar{\alpha} = \frac{\omega - \omega_0}{t}. \tag{3-4}$$

† Even for an angle as large as 15°, the error in making this estimate is only 1 percent, but for larger angles the error increases rapidly.

This is the *average* angular acceleration (denoted by α, the Greek lowercase letter alpha) over the time interval t; ω_0 is the angular velocity initially and ω is the angular velocity after the time t has passed. Instantaneous angular acceleration can be defined in the usual way:

$$\alpha = \frac{\Delta\omega}{\Delta t}.$$

However, we will mainly be considering cases in which the angular acceleration is constant and therefore equal to the average. When ω is measured in rad/s and t in seconds, α will be expressed as rad/s/s or rad/s^2.

When radian measure is used, α is related to the linear acceleration of a particle by

$$a_T = r\alpha, \qquad (3\text{-}5)$$

a fact that can be arrived at in the same way as that in which we obtained Equation 3-3. In this equation r is the radius of the circle in which the particle is moving, and the subscript T in a_T, which stands for "tangential," is added because the acceleration here is along the circle (i.e., tangential to it).

The definitions and equations that we have discussed so far apply for a particle moving in a circle of radius r. They can also be applied to an extended object that is rotating, such as a wheel. We can, in fact, consider an extended object as being made up of a great many particles (or atoms) situated at different distances from the axis of rotation.

When a wheel rotates through an angle θ (Figure 3-3), each point of the wheel moves through the same angle θ. However, each point does not move the same "linear" distance. The farther the point is from the center, the greater distance it travels for the same rotation angle θ. That is, $s = r\theta$, where s is the distance moved and r the distance of the particle from the axis of rotation. If the wheel turns with an angular velocity ω, each of its particles follows a circular path with the same angular velocity ω, but each has its own particular radius. Referring to Equation 3-3, we see that particles near the edge have a higher *linear* velocity than those closer to the center even though their angular velocity is the same. The same considerations apply for acceleration. The defining equations for ω (Equation 3-2) and for α (Equation 3-4) apply equally well to rotating objects as to particles moving in a circle. But Equations 3-3 and 3-5 apply *only* to a particle, although they can apply to a particle of the wheel, in which case r is the distance of that particle from the axis of rotation.

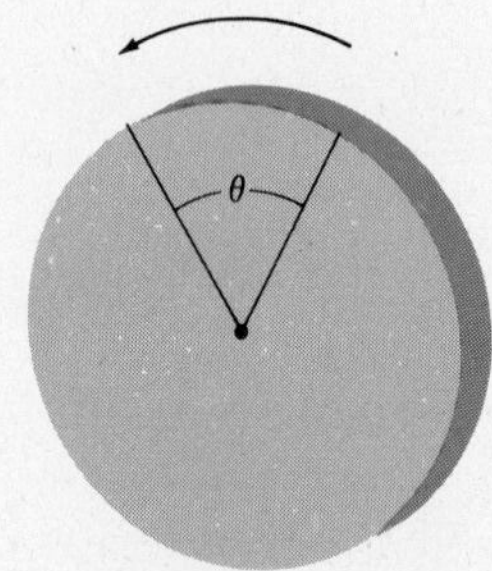

FIGURE 3-3
A wheel rotated through an angle θ.

The **frequency** of rotation, f, is defined as the number of complete revolutions (rev) made per second. Since one revolution (of, say, a wheel) corresponds to an angle of 2π radians, and thus 1 rev/s $= 2\pi$ rad/s, then in general the frequency f is related to angular velocity ω by

$$f = \frac{\omega}{2\pi}$$

or

$$\omega = 2\pi f.$$

▮ EXAMPLE 3-2 What is the linear speed of a point on the edge of a 33-rpm (revolutions per minute) phonograph record whose diameter is 30 cm?

SOLUTION First we find the angular velocity in radians per second: the frequency $f = 33\text{ rpm} = 33\text{ rev}/60\text{ s} = 0.55\text{ rev/s}$; then $\omega = 2\pi f = 3.5\text{ rad/s}$. The radius r is 0.15 m, so the speed v at the edge is

$$v = r\omega = (0.15\text{ m})(3.5\text{ rad/s}) = 0.52\text{ m/s}.$$ ▮▮

▮ EXAMPLE 3-3 A centrifuge rotor is accelerated from rest to 20,000 rpm in 5.0 min. What is its angular acceleration?

SOLUTION

$$\omega = (20{,}000\text{ rpm})\left(\frac{2\pi\text{ rad/rev}}{60\text{ s/min}}\right) = 2100\text{ rad/s};$$

then, since $\omega_0 = 0$,

$$\alpha = \frac{(2100\text{ rad/s}) - 0}{300\text{ s}} = 7.0\text{ rad/s}^2.$$ ▮▮

Angular velocity and acceleration, like their linear counterparts, are actually vectors and thus have a direction as well as a magnitude. For example, the angular veloctiy ω for a rotating body can be represented by a vector that points along the axis of rotation. We will not usually have to deal with the directions of angular quantities, however, except briefly in Chapter 5.

Kinematic equations for uniformly accelerated rotational motion

In Chapter 1 we derived the important equations (1-9*a* through 1-9*d*) that relate acceleration, velocity, and distance for the situation of uniform acceleration. Those equations were derived from the definitions of linear velocity and acceleration assuming constant acceleration. The definitions of angular velocity and angular acceleration are the same as for their linear counterparts, except that θ has replaced the distance d, ω has replaced v, and α has replaced a. Therefore, the angular equations for constant angular acceleration will be analogous to Equations 1-9 with d replaced by θ, v by ω, and a by α. We summarize them here, opposite their linear equivalents:

Angular	Linear	
$\omega = \omega_0 + \alpha t$	$v = v_0 + at$	(3-6*a*)
$\theta = \omega_0 t + \frac{1}{2}\alpha t^2$	$d = v_0 t + \frac{1}{2}at^2$	(3-6*b*)
$\omega^2 = \omega_0^2 + 2\alpha\theta$	$v^2 = v_0^2 + 2ad$	(3-6*c*)
$\bar{\omega} = \dfrac{\omega + \omega_0}{2}$	$\bar{v} = \dfrac{v + v_0}{2}$	(3-6*d*)

Now let us apply these equations in an example.

▮ EXAMPLE 3-4 Through how many turns has the centrifuge rotor of Example 3-3 turned during its acceleration period?

SOLUTION We know that $\omega_0 = 0$, $\omega = 2100$ rad/s, $\alpha = 7.0$ rad/s^2, and $t = 300$ s. We could use either Equation 3-6*b* or 3-6*c*; the former gives

$$\theta = 0 + \tfrac{1}{2}(7.0 \text{ rad/s}^2)(300 \text{ s})^2 = 3.2 \times 10^5 \text{ rad}.$$

To find the total number of revolutions we divide by 2π and obtain 5.0×10^4 revolutions. (To decide whether to multiply or divide by 2π, it helps to remember that there are more radians than revolutions since 2π rad = 1 rev.) ▌▌

3-2 ▌ Circular motion of a particle

Centripetal acceleration and uniform circular motion

In this section we consider the important special case of a particle moving in a circle of radius r with constant angular velocity ω. This is called "uniform circular motion". Since the angular acceleration α is zero, there is nothing to learn by applying Equations 3-6. However, although the particle has zero *angular* acceleration, its *linear* acceleration is not zero. True, the magnitude of its linear velocity, v, is constant and equal to $r\omega$; but the *direction* of its velocity is continually changing as it moves around the circle (Figure 3-4). As we saw in Chapter 1, acceleration is defined as the rate of change of velocity, and a change in direction of velocity constitutes an acceleration just as much as a change in magnitude. We now investigate this acceleration quantitatively.

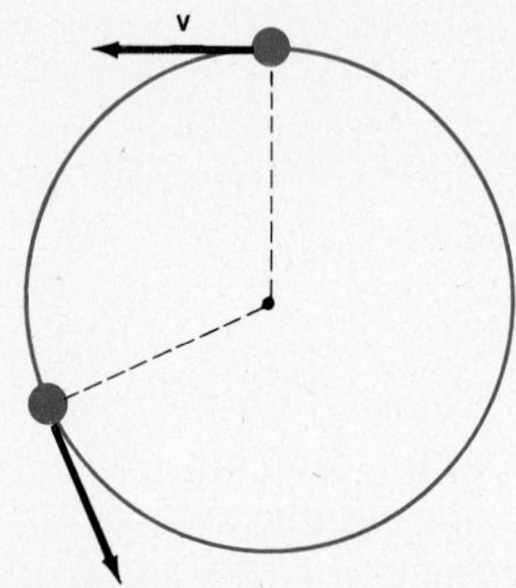

FIGURE 3-4
A particle moving in a circle, showing how the velocity changes. Note that at each point, the instantaneous velocity is in a direction tangent to the circular path.

Acceleration is defined as

$$\mathbf{a} = \frac{\mathbf{v} - \mathbf{v}_0}{\Delta t} = \frac{\Delta \mathbf{v}}{\Delta t},$$

where $\Delta\mathbf{v}$ is the change in velocity during the short time interval Δt. We will eventually consider the situation when Δt approaches zero and thus obtain the instantaneous acceleration; but for purposes of making a clear drawing, Figure 3-5, we consider a nonzero time interval. During the time Δt the particle in Figure 3-5*a* moves from point A to point B, covering a

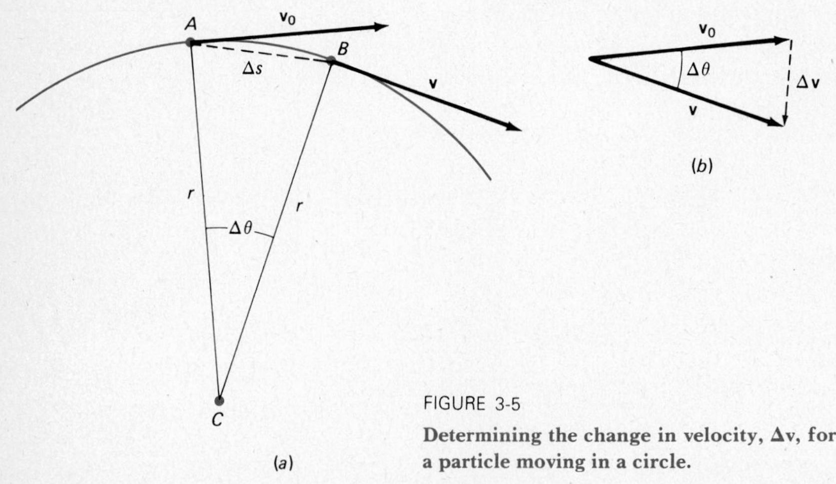

FIGURE 3-5
Determining the change in velocity, $\Delta\mathbf{v}$, for a particle moving in a circle.

small distance Δs which subtends a small angle $\Delta\theta$. The change in the velocity vector is $\mathbf{v}-\mathbf{v}_0=\Delta\mathbf{v}$. If we transfer $\mathbf{v}_0$ to the right side of this equation, we obtain $\mathbf{v}=\mathbf{v}_0+\Delta\mathbf{v}$. Thus $\Delta\mathbf{v}$ added to $\mathbf{v}_0$ is equal to $\mathbf{v}$, so $\Delta\mathbf{v}$ must be the vector shown dashed in Figure 3-5*b*. In this diagram we notice that when Δt is very small (approaching zero), and therefore Δs and $\Delta\theta$ are also very small, $\mathbf{v}$ will be almost parallel to $\mathbf{v}_0$ and $\Delta\mathbf{v}$ will be essentially perpendicular to them. Thus $\Delta\mathbf{v}$ points toward the center of the circle. Since $\mathbf{a}$, by its definition above, is in the same direction as $\Delta\mathbf{v}$, it too must point toward the center of the circle. Therefore, this acceleration is called **centripetal acceleration** ("center-seeking" acceleration), and henceforth we denote it by $\mathbf{a}_c$.

Now that we have determined the direction of the acceleration, we next determine the magnitude of the centripetal acceleration, a_c. The vectors $\mathbf{v}$, $\mathbf{v}_0$ and $\Delta\mathbf{v}$ in Figure 3-5*b* form a triangle that is geometrically similar† to triangle ABC in Figure 3-5*b*. This relies on the fact that the angle between $\mathbf{v}_0$ and $\mathbf{v}$ is equal to $\Delta\theta$, defined as the angle between CA and CB in Figure 3-5*a*; this is true because CB is perpendicular to $\mathbf{v}$, and CA is perpendicular to $\mathbf{v}_0$. Thus we can write

$$\frac{\Delta v}{v}\approx\frac{\Delta s}{r}.$$

This is an exact equality when Δt approaches zero, for then the arc length Δs equals the chord length AB. Since we want to find the instantaneous acceleration, which is the case for which Δt approaches zero, we write this as an equality and solve for Δv:

$$\Delta v=\frac{v}{r}\Delta s.$$

To get the centripetal acceleration, a_c, we divide Δv by Δt:

$$a_c=\frac{\Delta v}{\Delta t}=\frac{v}{r}\frac{\Delta s}{\Delta t},$$

and since $\Delta s/\Delta t$ is the linear speed, v, of the object:

$$a_c=\frac{v^2}{r}. \tag{3-7a}$$

To summarize, a particle moving in a circle of radius r with constant speed v (and therefore constant angular velocity ω) has an acceleration whose direction is toward the center of the circle and whose magnitude is $a_c=v^2/r$. It is not surprising that this acceleration depends on v and r. For the greater the speed v, the faster the velocity changes direction; and the larger the radius, the less rapidly the velocity changes direction.

We can write the centripital acceleration in terms of the angular velocity, ω, by substituting Equation 3-3, $v=r\omega$. Then

$$a_c=\omega^2 r. \tag{3-7b}$$

Either of Equations 3-7 will give us the magnitude of the acceleration of any body moving in a circle of radius r at constant speed.

† Appendix A contains a review of geometry.

EXAMPLE 3-5 What is the magnitude of the acceleration of a speck of dust on the edge of a 33-rpm phonograph record whose diameter is 30 cm?

SOLUTION In Example 3-2 we calculated the angular velocity to be 3.5 rad/s and the linear velocity v at the edge to be 0.52 m/s. Since the radius is 0.15 m, we use Equation 3-7*b* to obtain

$$a_c = \omega^2 r = (3.5\ \text{rad/s})^2(0.15\ \text{m}) = 1.8\ \text{m/s}^2.$$

Using Equation 3-7*a*, we get the same result:

$$a_c = \frac{v^2}{r} = \frac{(0.52\ \text{m/s})^2}{0.15\ \text{m}} = 1.8\ \text{m/s}^2.$$

An object moving in a circle, such as a ball on the end of a string, must have a force applied to it to keep it in that circle; that is, a force is necessary to give it a centripetal acceleration. The magnitude of the required force can be calculated using Newton's second law, $F = ma$, where we use the value of the centripetal acceleration, $a_c = v^2/r$, and F must be the total (or net) force:

$$F = ma_c = m\frac{v^2}{r}.$$

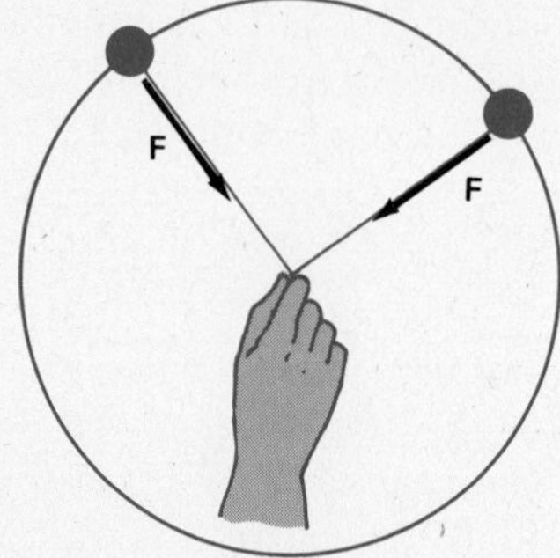

FIGURE 3-6
A force is required to keep an object moving in a circle. If the speed is constant, the force is directed toward the center of the circle.

Since a_c is directed toward the center of the circle, the force too must be directed toward the center of the circle. That a force is necessary can be seen intuitively from the fact that if no force were exerted on the object, it would not move in a circle but in a straight line, as Newton's first law tells us. To pull an object out of its "natural" straight-line path, a force to the side is necessary; for uniform circular motion, this sideways force must act toward the circle's center (see Figure 3-6). This force is sometimes called a centripetal force. But be aware that "centripetal force" does not indicate some new kind of force. The term merely means that the force is directed toward the circle's center. The force must be applied by some object. For example, when a person swings a ball on the end of string in a circle, the person (via the string) exerts the force on the ball.

There is a common misconception that an object moving in a circle has an outward force acting on it, a so-called centrifugal ("center-fleeing") force. Consider, for example, a person swinging a ball on the end of a string around his head (Figure 3-7). If you have ever done this yourself, you know that you feel a force pulling outward on your hand. The misconception arises when this pull is interpreted as an outward "centrifugal" force pulling on the ball which is transmitted along the string to the hand. But this is not what is happening at all. To keep the ball moving in a circle, the person pulls inwardly on the ball. The ball, then, exerts an equal and opposite force on the hand (Newton's third law), and *this* is the force your hand feels. The force *on the ball* is the one exerted *inwardly* on it by the person. For even more convincing evidence that a centrifugal force does not act on the ball, consider what happens when you let go of the string. If a centrifugal force were acting, the ball would fly outward, as shown in Figure 3-8*a*. But it does not; it flies off tangentially (Figure 3-8*b*), in the direction of the velocity it had at the moment it was released since the inward force no longer acts.

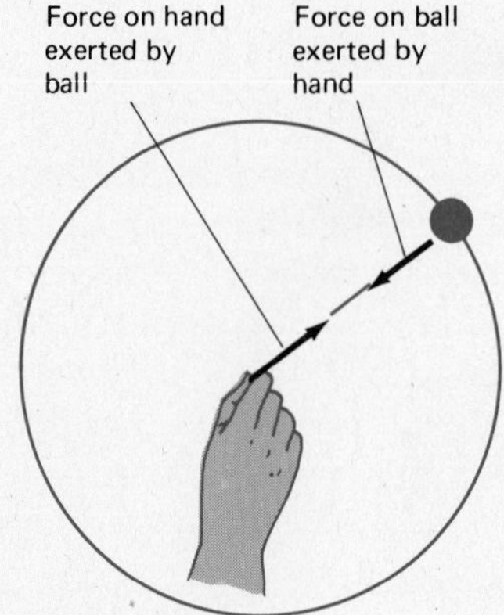

FIGURE 3-7
Swinging a ball on the end of a string. Your hand pulls in on the ball, and the ball pulls outward on your hand (Newton's third law).

An example of centripetal acceleration occurs when a fast-moving automobile rounds a curve. In such a situation you may feel that you are thrust outward. This is not some mysterious centrifugal force pulling on you. What is happening is that you tend to move in a straight line whereas the car curves "in front of you," so to speak. To make you go in the curved path, the back of the seat or the door of the car exerts a force on you (Figure 3-9). The car itself must have an inward force on it if it is to move in a curve. On a flat road this force is supplied by friction between the tires and the pavement. If the friction force is not great enough, as under icy conditions, sufficient force cannot be applied and the car will skid out of a circular path into a more nearly straight path.

▮ EXAMPLE 3-6 A 1000-kg car rounds a curve on a flat road of radius 50 m at a speed of 50 km/h (14 m/s). Will the car make the turn if (*a*) the pavement is dry and the coefficient of static friction is 0.60, (*b*) the pavement is icy and $\mu = 0.20$?

SOLUTION First we calculate the net force F required to accelerate the car around the curve:

$$F = m\frac{v^2}{r} = \frac{(1000\ \text{kg})(14\ \text{m/s})^2}{(50\ \text{m})} = 3900\ \text{N}.$$

The normal force, F_N, on the car is equal to the weight since the road is flat: $F_N = mg = (1000\ \text{kg})(9.8\ \text{m/s}^2) = 9800\ \text{N}$. For (*a*), $\mu = 0.60$ and the maximum friction force attainable is

$$F_{fr} = \mu F_N = (0.60)(9800\ \text{N}) = 5900\ \text{N}.$$

Since a force of only 3900 N is needed, and that is how much will in fact be exerted, the car can make the turn. But in (*b*) the maximum friction force possible is

$$F_{fr} = \mu F_N = (0.20)(9800\ \text{N}) = 2000\ \text{N}.$$

The car will skid. ▮▮

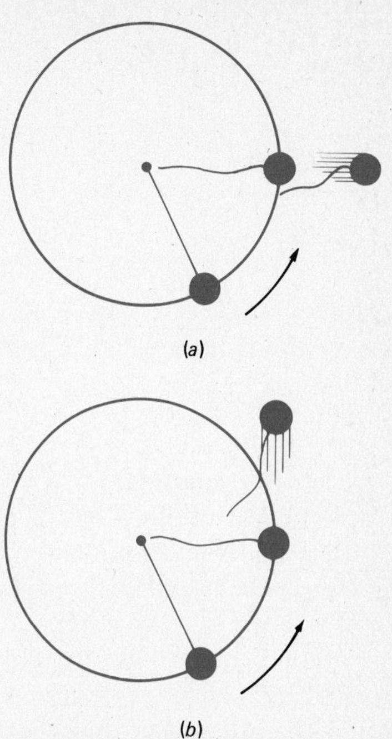

FIGURE 3-8
If centrifugal force existed, the ball would fly off as in (*a*) when released. In fact, it flies off as in (*b*).

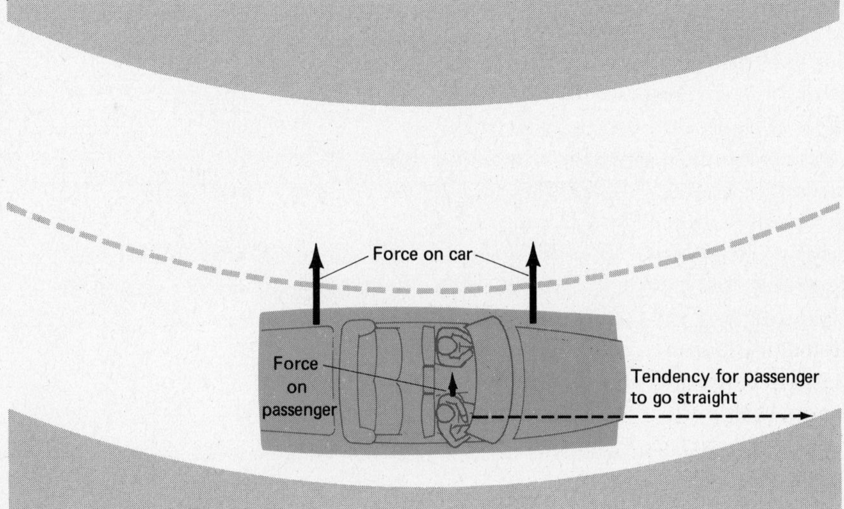

FIGURE 3-9
The road exerts an inward force on a car to make it move in a circle; and the car exerts an inward force on the passenger.

The banking of curves can reduce the chance of skidding because the normal force of the road will have a component toward the center of the circle (Figure 3-10), thus reducing the need for friction. For a given angle of banking, there will be one speed for which no friction at all is required. This will be the case when the horizontal component of the normal force, $F_N \sin\theta$, is just equal to the force required to give a vehicle its centripetal acceleration; that is,

$$F_N \sin\theta = m\frac{v^2}{r}.$$

The banking angle of a road, θ, is chosen so that this condition holds for a particular speed, called the "design speed." An example is given in the problems.

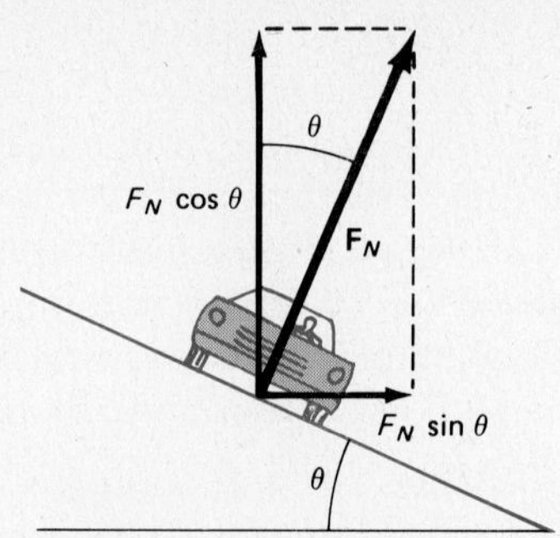

FIGURE 3-10
Normal force on a car (rounding a banked curve), resolved into its horizontal and vertical components.

*Centrifugation

The centrifuge is an important tool used in biology and other fields to sediment materials quickly or to separate materials with slightly different characteristics. The main component of a centrifuge or ultracentrifuge is a rotor that turns at high angular speeds about its central axis. (In an ultracentrifuge, which is designed to turn at very high angular velocity, the rotor turns in a vacuum to reduce heating by friction, and the rotor is constructed to withstand the greater forces involved.) Two common types of rotor are the fixed angle rotor, which contains a set of evenly spaced cylindrical holes into which test tubes can be placed (Figure 3-11*a*), and the swinging bucket rotor, which has individual test-tube holders which hang vertically when at rest but swing up smoothly to a horizontal position as the rotor gains speed (Figure 3-11*b*). Figure 3-12 shows a top view of a rotating rotor. Only one test tube is shown, and it is shown in two different positions as the rotor turns. The small circle represents a small particle, perhaps a cell organelle or a macromolecule, in the liquid-filled test tube. When the tube is at position A and the rotor is turning, the particle has a tendency to move in a straight line in the direction of the dashed arrow. The liquid resists the motion of the particles, thus exerting the centripetal force necessary to keep the particle moving in a circle.

Usually, the resistance of the liquid does not quite equal the required centripetal force and the particles eventually reach the bottom of the tube. As the particles collect, they form a pellet at that point on the base of the tube that is farthest from the axis, and the bottom of the tube exerts the force necessary to keep the particles moving in a circle. In fact, the bottom of the tube must exert a force on the whole tube of liquid, keeping it moving in a circle. If the tube is not strong enough to exert this force, it will break. In certain applications of centrifugation, namely "equilibrium density ultracentrifugation," the resisting force of the liquid increases as the particles move down the tube and at some point equals the required centripetal force. At this point, the particles stop moving down the tube. Particles with different densities can be separated thereby. The details of these forces and how they operate in different techniques of centrifugation will be discussed in the chapter on fluids.

The kinds of materials placed in a centrifuge are those that do not

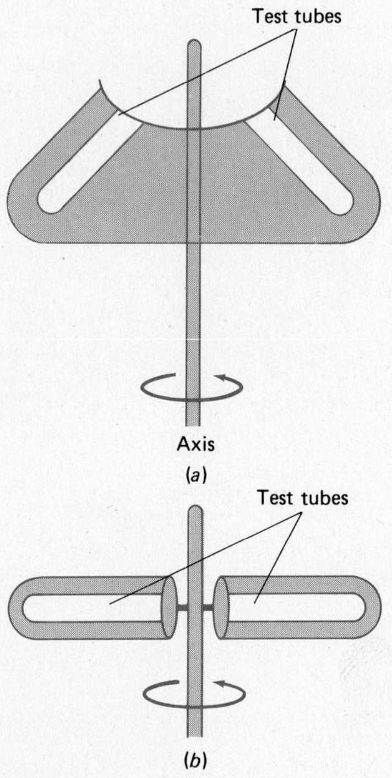

FIGURE 3-11
Centrifuge rotors: (*a*) fixed-angle and (*b*) swinging-bucket types.

sediment rapidly under the action of gravity. The point of a centrifuge is that the effects of gravity are multiplied because of the high rotational speeds, and the particles move down the tube more quickly.

▮ EXAMPLE 3-7 An ultracentrifuge rotor rotates at 50,000 r.p.m. The top of a tube is 6.0 cm, and the bottom is 10.0 cm, from the axis of rotation. Calculate the centripetal acceleration in "g's."

SOLUTION First, we calculate the angular velocity ω. Since there are 2π rad in 1 rev, $\omega = 6.28 \times 50{,}000 = 3.14 \times 10^5$ rad/min $= 5200$ rad/s. From Equation 3-7*b*, $a_c = \omega^2 r$, we see that the acceleration increases with distance along the tube. We will calculate a_c for the top and the bottom of the tube. For the top, $r = 0.060$ m, so

$$a_c = \omega^2 r = (5200 \text{ rad/s})^2(0.060 \text{ m}) = 1.6 \times 10^6 \text{ m/s}^2.$$

We divide by $g = 9.8 \text{ m/s}^2$ and find $a_c = 1.6 \times 10^5$ *g*'s. At the bottom of the tube

$$a_c = (5200 \text{ rad/s})^2(0.10 \text{ m}) = 2.7 \times 10^6 \text{ m/s}^2,$$

which is equal to 2.6×10^5 *g*'s. ▮▮

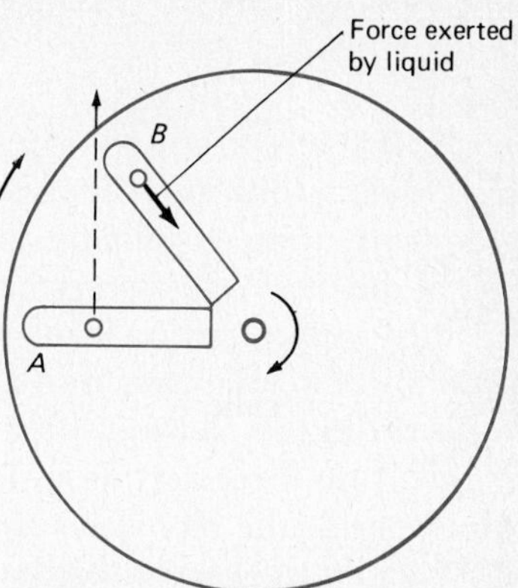

FIGURE 3-12
Rotating test tube in a centrifuge.

▮ EXAMPLE 3-8 The contents of the tube in Example 3-7 has a total mass of 12 grams. What force must the bottom of the tube withstand?

SOLUTION Since the acceleration varies with distance from the axis, we will make the calculation as if the mass were all at the average radius, $r = 0.080$ m. Then the required force is

$$\begin{aligned} F_c &= ma_c = m\omega^2 r \\ &= (0.012 \text{ kg})(5200 \text{ rad/s})^2(0.080 \text{ m}) \\ &= 26{,}000 \text{ N}. \end{aligned}$$

This is equivalent to a mass of $(26{,}000 \text{ N})/(9.8 \text{ m/s}^2) \approx 2600$ kg! ▮▮

* Nonuniform circular motion

Circular motion at constant speed occurs when the force on an object is exerted toward the center of the circle. If the force is not directed toward the center but is at an angle as shown in Figure 3-13, the force has two components. The component directed toward the center of the circle, F_c, gives rise to the centripital acceleration a_c and keeps the object moving in a circle. The component F_T, tangent to the circle, acts to increase (or decrease) the speed. Thus there will be a component of the acceleration tangent to the circle, a_T. It is a_T that is related to the angular acceleration α by Equation 3-5, for when an F_T acts, the speed of the object is changed.

In summary, a_T is related to the change in the *magnitude* of the velocity of the object, whereas a_c is related to the change in *direction* of the velocity.

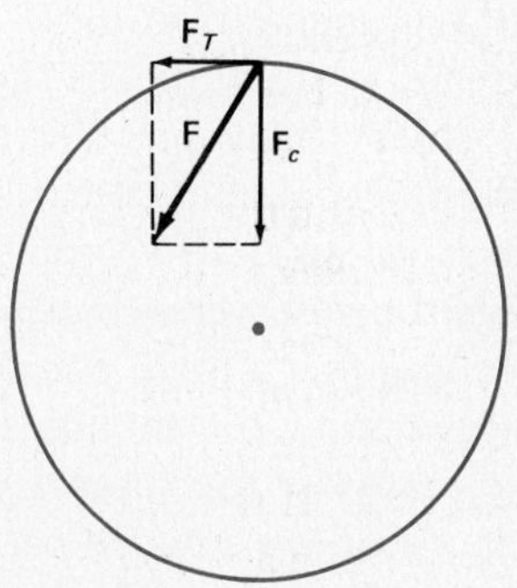

FIGURE 3-13
The speed of a particle moving in a circle changes if the force on it has a tangential component.

When you first start revolving a ball on the end of string around your head, you must give it tangential acceleration. You do this by pulling on the string with your hand displaced from the center of the circle. In athletics, a hammer thrower accelerates the hammer tangentially in a similar way so that it reaches a high speed before release.

*Rotating frames of reference

We have so far examined rotational motion from the outside, as an observer fixed on the earth. Sometimes it is convenient to place ourselves (in theory, if not physically) into the rotating system. Let us, for example, examine the motion of objects from the point of view, or frame of reference, of a person seated on a rotating platform such as a merry-go-round. It looks to him as if the rest of the world is going around *him*. But let us focus attention on what he observes when he places a tennis ball on the floor of the merry-go-round that we assume is frictionless. If he puts it down gently, without giving it any push, he will observe that it accelerates from rest and moves outward in a path similar to that shown in Figure 3-14*a*. According to Newton's first law, an object initially at rest should stay at rest if no force acts on it. But, according to the observer on the merry-go-round, the ball starts moving even though there is no force applied to it. To an observer on the earth, this is all very clear; for to him, the ball has an initial velocity when it is released (because the platform is moving), and it simply continues moving in a straight-line path as shown in Figure 3-14*b*, in accordance with Newton's first law.

But what shall we do about the frame of reference of the person on the merry-go-round? Clearly, Newton's first law, the law of inertia, does not hold in this rotating frame of reference. (For this reason, such a frame is sometimes called a *noninertial* reference frame. An *inertial* reference frame is one in which the law of inertia—Newton's first law—does hold, and so do Newton's second and third laws.) Newton's second law does not hold either in a rotating frame of reference. For instance in the situation described above, there is no force on the ball; yet, with respect to the merry-go-round, it accelerates.

Because Newton's laws do not hold when observations are made with respect to a rotating frame of reference, calculation of motion can be difficult. However, we can still apply Newton's laws in such a reference frame if we make use of a trick. We write down the equation $F = ma$ as if a force equal to mv^2/r (or $m\omega^2 r$) were acting on the object† in addition to any other forces that may be acting. This extra force, which can be designated as "centrifugal force" since it acts outwards, is called a "pseudoforce" ("pseudo" means "false"). It is a pseudoforce because there is no other object that exerts this force; besides, when viewed from the earth or other inertial reference frames, the force disappears entirely. We have made it up so that we can make calculations using the relation $F = ma$. Thus the observer in Figure 3-14*a* can calculate the motion of the ball by assuming that a force equal to mv^2/r acts on it.

We can examine the motion of a particle in a centrifuge from the frame of reference of the rotor. In this frame of reference, the particles move in a more-or-less straight path down the tube. (From the reference frame of the earth, the particles go round and round.) The acceleration of a particle with respect to the rotor can then be calculated using $F = ma$ if we include a pseudoforce, "F," equal to $m\omega^2 r$ (or mv^2/r) in addition to the resisting force F_R exerted by the liquid on the particle (Figure 3-15).

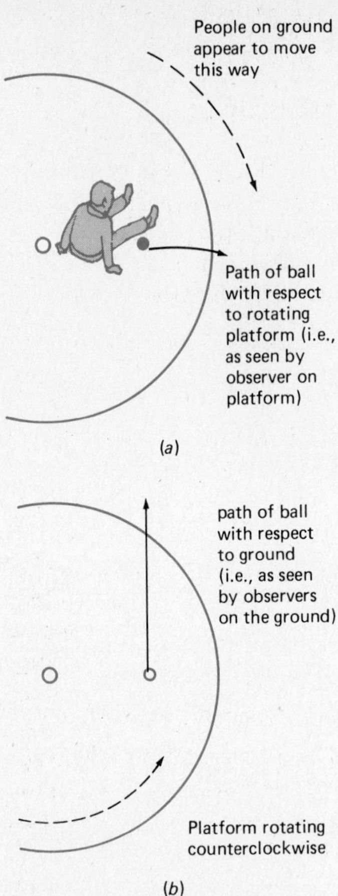

FIGURE 3-14
Path of a ball released on a rotating merry-go-round (*a*) in the reference frame of the merry-go-round and (*b*) in the reference frame of the earth.

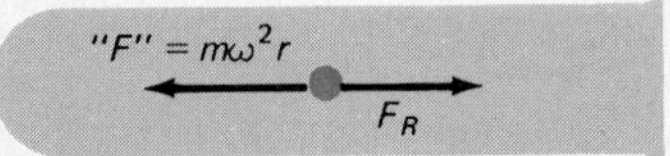

FIGURE 3-15
The forces on a particle in a rotating centrifuge, in the reference frame of the centrifuge.

† This is slightly simplified. Careful analysis shows that other pseudoforces may also be present as a result of the rotation; one such is the Coriolis force. These are, however, often small enough to be ignored.

The earth itself is rotating on its axis. Thus, strictly speaking, Newton's laws are not valid on the earth. However, the effect is usually so small that it can be ignored, although it does affect the movement of large air masses and ocean currents.

Because of the earth's rotation, over eons of time the material of the earth has tended to concentrate slightly at the equator. The earth is thus not a perfect sphere but is slightly fatter at the equator than it is at the poles.

3-3 ▌ Gravitation

Newton's law of universal gravitation

At about the time when Newton was developing his three laws of motion, he was also examining the motion of the heavenly bodies—the planets and the moon. In particular, he wondered about the nature of the force that kept the moon in its nearly circular orbit around the earth. For it was quite clear to Newton that some kind of centripetal force, directed toward the earth, must act to keep the moon in its circular path.

He was also thinking about the apparently unrelated problem of gravity. Since falling bodies accelerate, Newton had concluded that they must have a force exerted on them, a force we call the force of gravity. But, he asked, what *exerts* this force of gravity—for, as we have seen, he had earlier concluded that whenever a body has a force exerted *on* it, that force is exerted *by* some other body. Every object on the surface of the earth feels this force of gravity, and no matter where the object is, the force is directed toward the center of the earth. Newton concluded that it must be the earth itself that exerts the gravitational force on objects at its surface.

One day, according to an early biographer, Newton was sitting in his garden and saw an apple fall from a tree. At that moment an inspiration came to him: if the force of gravity acts at the tops of trees, and even at the tops of mountains, then perhaps it acts all the way to the moon! From this brilliant insight, Newton developed his great theory of gravitation, based on the idea that it is gravity that keeps the moon in its orbit.

Newton set about determining the magnitude of the gravitational force that the earth exerts on the moon as compared to the force on objects at the earth's surface. At the surface of the earth, the force of gravity accelerates objects at 9.8 m/s^2. But what, Newton asked himself, is the centripetal acceleration of the moon? Since the moon moves with nearly uniform circular motion, the acceleration can be calculated from $a_c = \omega^2 r$, where $r = 380{,}000$ km is the distance of the moon from the earth and ω is the angular velocity of the moon around the earth. Since it takes the moon 27.3 days to make one revolution (2π rad) about the earth, and 27.3 days equals $27.3 \times 24 \times 60 \times 60$ seconds $= 2.36 \times 10^6$ s, then

$$a_c = \omega^2 r = \left(\frac{2\pi \text{ rad}}{2.36 \times 10^6 \text{ s}}\right)^2 (3.8 \times 10^7 \text{ m})$$

$$= 0.0027 \text{ m/s}^2.$$

In terms of the acceleration of gravity at the earth's surface, g:

$$a_c \approx \frac{1}{3600} g.$$

The acceleration of the moon toward the earth is $\frac{1}{3600}$ as great as the acceleration of objects at the earth's surface. Now the moon is 380,000 km from the earth, which is 60 times the earth's radius of 6400 km. That is, the moon is 60 times farther from the earth's center than are objects at the earth's surface. But $60 \times 60 = 3600$—again that number 3600! Newton concluded that the gravitational force exerted by the earth on any object decreases with the square of its distance, r, from the earth's center†:

$$\text{force of gravity} \propto \frac{1}{r^2}.$$

The moon, being 60 earth radii away, feels a gravitational force only $\frac{1}{60^2} = \frac{1}{3600}$ times as strong as it would if it were at the earth's surface. Any object placed 380,000 km from the earth would experience the same acceleration due to the earth's gravity as the moon: $0.0027\ \text{m/s}^2$.

Newton realized that the force of gravity on an object depends not only on distance but also on the object's mass. In fact, it is directly proportional to its mass, as we have seen. According to Newton's third law, when the earth exerts its gravitational force on an object, such as the moon, that object exerts an equal and opposite force on the earth (Figure 3-16). Because of this symmetry, Newton reasoned, the force of gravity must be proportional to *both* the masses. Thus

$$F \propto \frac{m_e m_0}{r^2}$$

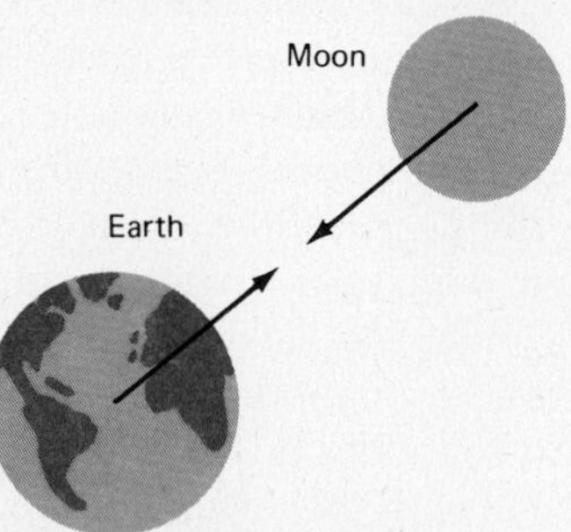

FIGURE 3-16
The earth exerts a (gravitational) force on the moon; the moon must, according to Newton's third law, exert an equal but opposite (gravitational) force on the earth.

where m_e is the mass of the earth, m_0 the mass of the other object, and r the distance from the earth's center to the center of the object.

Newton went one step further in his analysis of gravity. He had determined that the force required to hold the different planets in their orbits around the sun diminished as the inverse square of their distance from the sun. This led him to believe that it was also the gravitational force that acted between the sun and each of the planets to keep them in their orbits. And if gravity acts between these objects, why not between all objects? Thus he proposed his famous **law of universal gravitation**:

> **▍Every body in the universe attracts every other body with a force that is proportional to the product of their masses and inversely proportional to the square of the distance between them. This force acts along the line joining the two bodies.**

In symbols, this law can be written

$$F \propto \frac{mm'}{r^2},$$

† Actually, when Newton first conceived the idea that the gravitational force might decrease as the square of the distance, the accepted value for the radius of the earth was slightly inaccurate, and the calculations did not accurately support his theory. Some 20 years later an accurate determination of the earth's radius was made; when he used this new value, Newton found excellent agreement, as calculated above. It is interesting that Newton did not publish his theory until after these new data were available to support his theory (in the 1680s).

where m and m' are the masses of the two objects and r is the distance between their centers. To make this into an equation, we insert a constant of proportionality, G:

$$F = G\frac{mm'}{r^2}. \tag{3-8}$$

The value of G must be very small since we are not aware of any force existing between ordinary-sized objects—say, between two baseballs. The force between two ordinary objects was first measured over 100 years later, in 1798, by Henry Cavendish. Using an apparatus like the one illustrated in Figure 3-17, Cavendish not only confirmed Newton's hypothesis that any two bodies attract one another and that Equation 3-8 accurately describes this force, but he measured the value of the constant G as well. The accepted value today is

$$G = 6.67 \times 10^{-11}\ \text{N} \cdot \text{m}^2/\text{kg}^2.$$

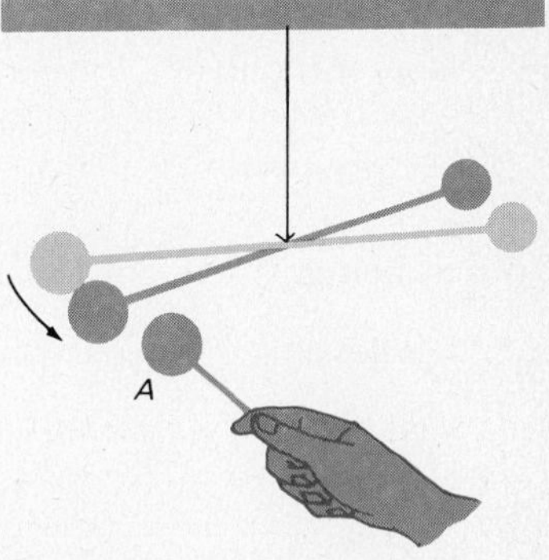

FIGURE 3-17
Schematic diagram of Cavendish's apparatus. Two spheres are attached to a light horizontal rod, which is suspended at its center by a thin fiber. When the sphere labeled *A* is brought close to one of the suspended spheres, the gravitational force causes the latter to move, and this twists the fiber slightly. Cavendish had earlier determined what force twisted the fiber a given amount. He was thus able to determine the magnitude of the gravitational force between two objects whose masses were known.

When Equation 3-8 is applied to the gravitational force between the earth and an object at its surface, m' becomes the mass of the earth m_e and r becomes the distance from the earth's center,† namely the radius of the earth, r_e; this force of gravity due to the earth is the weight of the object, which we have been writing as mg. Thus

$$mg = G\frac{mm_e}{r_e^2};$$

hence

$$g = G\frac{m_e}{r_e^2}. \tag{3-9}$$

Thus the acceleration of gravity at the surface of the earth, g, is determined by m_e and r_e. In Chapter 1 we saw that the value of g varied slightly from place to place. From Equation 3-9 we would expect g to be slightly less on the tops of mountains, for example, than at sea level, since r_e, the distance from the earth's center, is slightly greater there. Indeed, this is what is found experimentally (Table 1-3).

When dealing with the weight of objects at the surface of the earth, we can continue to use simply mg. If we wish to calculate the force of gravity on an object some distance from the earth, or the force due to some other heavenly body such as the moon, we can calculate the effective value of g from Equation 3-9, replacing r_e (and m_e) by the appropriate distance (and mass); or we can use Equation 3-8 directly.

Until G was measured, the mass of the earth was not known. But once G was measured, Equation 3-9 could be used to calculate the earth's mass, and Cavendish was first to do so. Since $g = 9.80\ \text{m/s}^2$ and the radius of the earth $r_e = 6400\ \text{km} = 6.4 \times 10^6\ \text{m}$, then from Equation 3-9 we obtain

$$m_e = \frac{gr_e^2}{G} = \frac{(9.8\ \text{m/s}^2)(6.4 \times 10^6\ \text{m})^2}{6.67 \times 10^{-11}\ \text{N} \cdot \text{m}^2/\text{kg}^2} = 6.0 \times 10^{24}\ \text{kg}$$

for the mass of the earth.

† That the distance is measured from the earth's center does not imply that the force of gravity somehow emanates from that one point. Rather, all parts of the earth attract gravitationally, but the net effect is a force acting toward the earth's center.

▮ EXAMPLE 3-9 What is the force of gravity acting on a spacecraft when it is 6400 km above the earth's surface?

SOLUTION It is twice as far from the earth's center as when at the surface of the earth; therefore, since the force of gravity decreases as the square of the distance (and $1/2^2 = 1/4$), the force of gravity on it will be only one-fourth as great. ▮▮

▮ EXAMPLE 3-10 What is the effective value of g at a height of 1000 km above the earth's surface? That is, what is the acceleration due to gravity of objects allowed to fall freely at this altitude?

SOLUTION We use Equation 3-9, with r_e replaced by $r = 7400\text{ km} = 7.4 \times 10^6\text{ m}$:

$$g = \frac{(6.67 \times 10^{-11}\text{ N}\cdot\text{m}^2/\text{kg}^2)(6.0 \times 10^{24}\text{ kg})}{(7.4 \times 10^6\text{ m})^2}$$

$$= 7.3\text{ m/s}^2.$$ ▮▮

The law of universal gravitation should not be confused with Newton's second law of motion, $F = ma$. The former describes a particular force, gravity, and how its strength varies with the distance and masses involved. Newton's second law, on the other hand, relates the force on a body—it can be any force—to the mass and acceleration of that body.

*Gravitational versus inertial mass

We have dealt with two aspects of mass as if they were the same. First, in Chapter 2 we defined mass as a measure of the inertia of a body. Newton's second law relates the force acting on a body to its acceleration and its *inertial mass*, as we call it. Then in this chapter we have dealt with mass as a property relating to the gravitational force—that is, mass as a quantity that determines the strength of the gravitational force between two bodies. This we call the *gravitational mass*.

Now, it is not at all obvious that the inertial mass of a body should be equal to its gravitational mass. (The force of gravity might have depended on a completely different property of a body, just as the electrical force depends on a property called electric charge, as we shall see later.) Newton's and Cavendish's experiments indicated that these two types of mass are precisely equal for any body, and modern experiments confirm it to an accuracy of about 1 part in 10^{11}.

Satellites and weightlessness

Artificial satellites circling the earth are now commonplace. A satellite is put into orbit by accelerating it to a sufficiently high tangential speed with the use of rockets, as shown in Figure 3-18. If the speed is too high, the spacecraft will not be confined by the earth's gravity and will escape, never to return. If too low, it will fall back to earth. Satellites are usually put into circular, or nearly circular, orbits. This requires the least takeoff speed. It is sometimes asked: "What keeps a satellite up?" If a satellite stopped moving, it would of course, fall back to earth. But at the very high speed a satellite has, it would quickly fly out into space if it were not for the

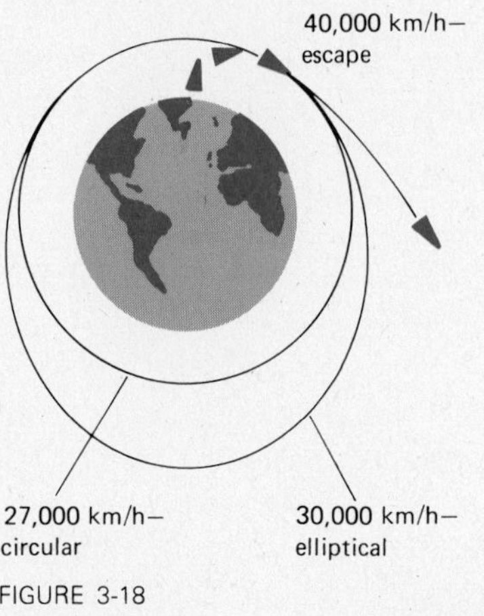

FIGURE 3-18 **Artificial satellites.**

gravitational force of the earth pulling it into orbit. So it is the satellite's speed, or inertia, that keeps it up.

For satellites that move (at least approximately) with uniform circular motion, the acceleration is v^2/r. The force that gives a satellite this acceleration is the force of gravity, and since a satellite may be at a considerable distance from the earth, we must use Equation 3-8 for the force acting on it. When we apply Newton's second law, $F = ma$, we find that

$$G\frac{mm_e}{r^2} = m\frac{v^2}{r}. \tag{3-10}$$

This equation relates the distance of the satellite from the earth's center, r, to its speed, v. Note that only one force, gravity, is acting on the satellite.

▮ EXAMPLE 3-11 Calculate the velocity required for a satellite moving in a circular orbit 200 km above the earth's surface.

SOLUTION We solve Equation 3-10 for v:

$$v = \sqrt{\frac{Gm_e}{r}}.$$

The value of r is 6400 km + 200 km = 6600 km, which is the same as 6.6×10^6 m. Thus,

$$v = \sqrt{\frac{(6.67 \times 10^{-11}\ \mathrm{N \cdot m^2/kg^2})(6.0 \times 10^{24}\ \mathrm{kg})}{6.6 \times 10^6\ \mathrm{m}}} = 7.8 \times 10^3\ \mathrm{m/s}.$$

This is approximately 27,000 km/h. Note that the velocity required will decrease with increasing height. ▮▮

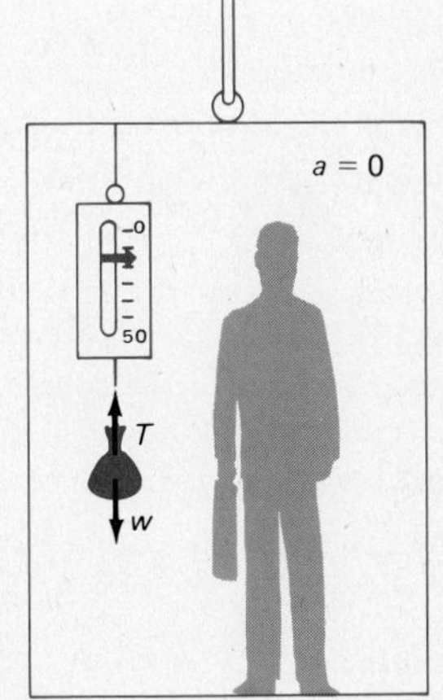

(a)

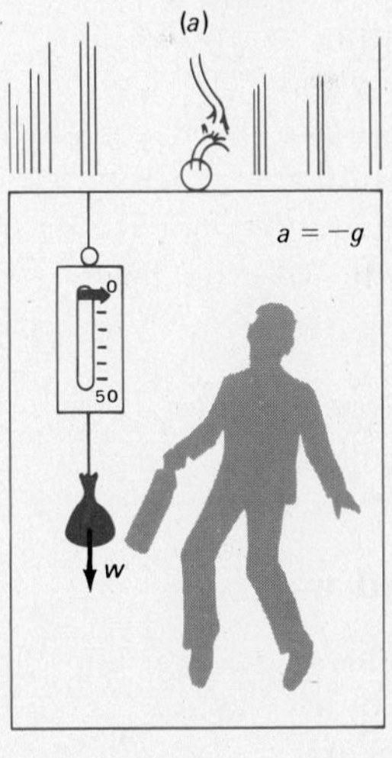

(b)

FIGURE 3-19
An object in an elevator at rest exerts a force on a spring scale equal to its weight (*a*); but in a freely falling elevator (*b*), it experiences "weightlessness."

People and other objects in a satellite circling the earth are said to experience "weightlessness." Before tackling the case of a satellite, however, let us first look at the simpler case of a falling elevator. In Figure 3-19*a* we see an elevator at rest. Let us consider an object such as a bag hanging from a spring scale as shown. The scale indicates the downward force exerted on it by the bag. This force, exerted *on* the scale, is just equal and opposite to the force exerted *by* the scale upward on the bag. We call this force T. Since the mass, m, is not accelerating, we apply $F = ma$ to the bag and obtain

$$T - w = 0,$$

where w is the weight of the bag; thus, $T = w$, and since the scale indicates the force T exerted on it by the bag, it registers a force equal to the weight of the bag, as we expect. Now, if the elevator has an acceleration a, when we apply $F = ma$ to the object, we obtain $T - w = ma$. We then solve for T and obtain

$$T = w + ma.$$

We have chosen the positive direction up. Thus, if the acceleration is up, a is positive; and the scale, which measures T, will read more than w. If the elevator accelerates downward, a will be negative and T will be less than w. For example, if the elevator's acceleration is $-\frac{1}{2}g$, then using the fact that $w = mg$ we find $T = mg - \frac{1}{2}mg = \frac{1}{2}mg$. That is, the scale reads one-half the actual weight. If the elevator is in *free fall*—for example, if the cables break—then $a = -g$ and $T = mg - mg = 0$. The scale reads zero! See Figure 3-19*b*. The object seems weightless. If the person in the elevator dropped a pencil, say, it would not fall to the floor. True, the pencil would fall with acceleration g. But so does the floor of the elevator, and the person. The pencil would hover right in front of the person. This is called *apparent weightlessness* because, in fact, gravity is still acting on the object. The objects seem weightless only because the elevator is accelerating at $-g$.

The weightlessness experienced by people in a satellite orbiting close to the earth is the same apparent weightlessness experienced in a freely falling elevator. It may seem strange, at first, to think of a satellite as freely falling. But a satellite is indeed falling toward the earth, as shown in Figure 3-20; the force of gravity causes it to "fall" out of its natural straight-line path. The acceleration of the satellite must be the acceleration due to gravity, since the only force acting on it is gravity. (We made use of this fact in obtaining Equation 3-10; the fact that we wrote the acceleration as v^2/r does not alter this fact.) Thus, although the force of gravity acts on objects within the satellite, the objects experience an apparent weightlessness because they, and the satellite, are accelerating as in free fall.

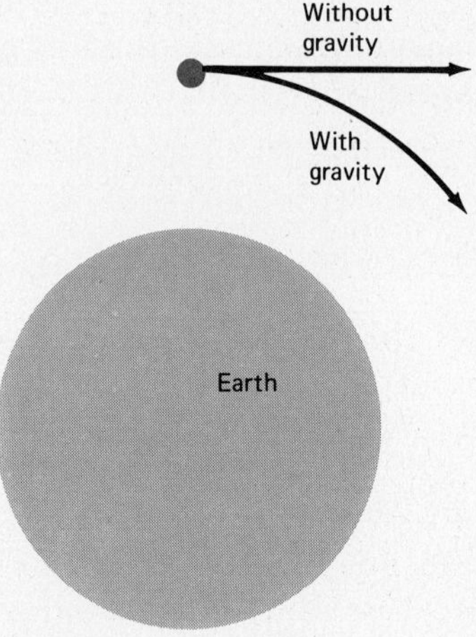

FIGURE 3-20
A moving satellite "falls," out of a straight-line path, toward the earth.

A completely different situation occurs when a spacecraft is out in space far from the earth and other attracting bodies such as the moon. The force of gravity due to the earth and other heavenly bodies will then be quite small because the distances are large, and persons in such a spacecraft will experience real weightlessness.

The effects on human beings of weightlessness (whether real or apparent makes no difference) is interesting. In ordinary circumstances, for example, a person can become quite tired holding out his arms horizontally. But for a person experiencing weightlessness, no effort is needed. The arms will just "float" there, since there is no sensation of weight. This effect has many applications in athletics; during a jump or a dive, while on a trampoline, and even between strides while running, a person is experiencing apparent weightlessness or free fall, although only for a short time. During these brief periods, limbs can be moved much more easily, since inertia only, not gravity, needs to be overcome. The loss of control because of lack of contact with the ground is compensated for by the increased mobility.

Prolonged weightlessness in space, however, can have deleterious effects on health. Red blood cells diminish, blood collects in the thorax, bones lose calcium and become brittle, and muscles lose their tone. These effects are at present being carefully studied.

*3-4 ▮ Kepler's laws and Newton's synthesis

A half century before Newton proposed his three laws of motion and his law of universal gravitation, the German astronomer Johannes Kepler (1571–1630) had worked out a detailed description of the motion of the planets about the sun. Kepler spent years examining data collected by

Tycho Brahe (1546–1601) on the positions of the planets in their motion through the heavens. Based on these detailed experimental observations, Kepler was able to summarize the motion of the planets in what are now called *Kepler's laws of planetary motion.* These are summarized as follows, with additional explanation in Figure 3-21.

> *Kepler's first law*: The path of each planet about the sun is an ellipse with the sun at one focus (Figure 3-21*a*).
>
> *Kepler's second law*: Each planet moves so that an imaginary line drawn from the sun to the planet sweeps out equal areas in equal periods of time (Figure 3-21*b*).
>
> *Kepler's third law*: The ratio of the squares of the periods of any two planets revolving about the sun is equal to the ratio of the cubes of their average distances from the sun. That is, if T_1 and T_2 represent the periods (time needed for one revolution about the sun, and r_1 and r_2 represent their average distances from the sun, then
>
> $$\left(\frac{T_1}{T_2}\right)^2=\left(\frac{r_1}{r_2}\right)^3.$$

Kepler formulated these laws based on experimental observation. Newton, a half century later, was able to *derive* Kepler's law from his law of universal gravitation and his laws of motion; and he showed that for any reasonable form for the gravitational force law, only one that depends on the inverse square of the distance is fully consistent with all three of Kepler's laws. He thus used Kepler's laws as evidence in favor of his law of universal gravitation, Equation 3-8.

Kepler's third law is the easiest to derive, and we do it here for the special case of a circular orbit. (Most of the planetary orbits are fairly close to a circle, which is a special case of an ellipse.) First, we write down Newton's second law of motion $F=ma$; then for F we substitute the law of universal gravitation, Equation 3-8, and for a the centripetal acceleration, v^2/r:

$$F=ma,$$

$$G\frac{m_1M}{r_1^2}=m_1\frac{v_1^2}{r_1}.$$

Here m_1 is the mass of a particular planet, r_1 its average distance from the sun, and v_1 its speed in orbit; M is the mass of the sun, since it is the gravitational attraction of the sun that keeps each planet in its orbit. Now the period T_1 of the planet is the time required for one complete orbit, a distance equal to $2\pi r_1$; thus

$$v_1=\frac{2\pi r_1}{T_1}.$$

We substitute this formula for v_1 into the equation above,

$$G\frac{m_1M}{r_1^2}=m_1\frac{4\pi^2r_1}{T_1^2},$$

and rearrange this to get

$$\frac{T_1^2}{r_1^3}=\frac{4\pi^2}{GM}.$$

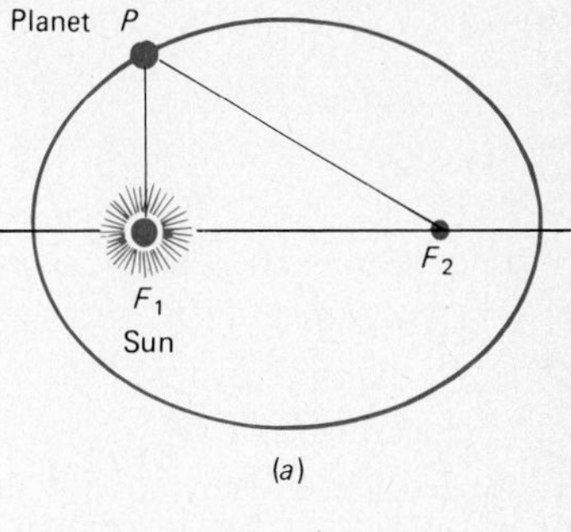

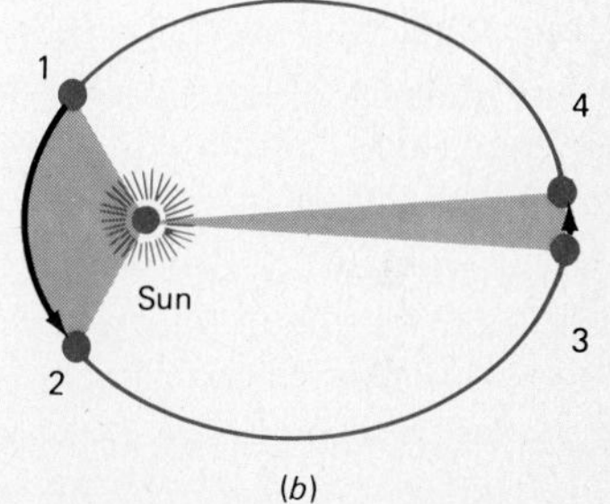

FIGURE 3-21
(*a*) Kepler's first law. An ellipse is a closed curve such that the sum of the distances from any point *P* on the curve to two fixed points (called the foci, F_1 and F_2) remains constant. That is, the sum of the distances F_1P+F_2P is the same for all points *P* on the curve. A circle is a special case of an ellipse when the two foci coincide, at the center of the circle. (*b*) Kepler's second law. The two shaded regions have equal areas. It takes the planet the same time to move from point 1 to point 2 as it takes it to move from point 3 to point 4. Planets move fastest in that part of their orbit where they are closest to the sun.

We derived this for planet 1 (say Mars); the same derivation would apply for a second planet (say Saturn):

$$\frac{T_2^2}{r_2^3}=\frac{4\pi^2}{GM},$$

where T_2 and r_2 are the period and orbit radius for the second planet. Since the right sides of these two equations are equal, we have $T_1^2/r_1^3 = T_2^2/r_2^3$ or, rearranging,

$$\frac{T_1^2}{T_2^2}=\frac{r_1^3}{r_2^3},$$

which is Kepler's third law.

Accurate measurements on the orbits of the planets indicated later that they do not follow perfectly elliptical orbits. Actually, this is to be expected from Newton's law of universal gravitation ("every body in the universe attracts every other body . . .") because each planet exerts a gravitational force on the other planets. Since the mass of the sun is much greater than that of any planet, the force on one planet due to any other planet will be small in comparison to the force on it due to the sun. (The derivation of perfectly elliptical orbits ignores the forces due to other planets.) But because of this small force, each planetary orbit should depart from a perfect ellipse, especially when a second planet is close to it. Such deviations, or *perturbations*, as they are called, from perfect ellipses are indeed observed. In fact, they led to the discovery of Neptune and Pluto. Deviations in the orbit of Uranus, for example, could not be accounted for by perturbations due to the other known planets; careful calculation in the nineteenth century indicated that these deviations could be accounted for if there were another planet farther out in the solar system. The position of this planet was predicted from the deviations in Uranus's orbit, and telescopes focused on that region of the sky quickly found it; the new planet was called Neptune. Similar, but much smaller perturbations of Neptune's orbit led to the discovery of Pluto in 1930.

The development by Newton of the law of universal gravitation and the three laws of motion was a major achievement in Western thought. For with these laws, Newton was able to describe the motion of objects on earth and how this motion was caused. Furthermore, the motion of the planets about the sun, and the motion of the moon about the earth, were explained on the same basis. The motions of heavenly bodies and bodies on earth were seen to follow the same laws (something not previously recognized generally, although Galileo had argued strongly in its favor). For this reason, we sometimes speak of Newton's "synthesis."

Newton's work was so encompassing that it constituted a theory of the universe, and influenced philosophy and other fields. The laws formulated by Newton are referred to as *causal laws.* By *causality* we mean the idea that one occurrence can cause another. We have repeatedly observed, for example, that when a rock strikes a window, a moment later the window breaks. We infer that the rock *caused* the window to break. This idea of "cause and effect" took on more forceful meaning with Newton's laws. For the motion—or rather the acceleration—of any object was seen to be *caused* by the net force acting on it. As a result, the universe came to be pictured by many scientists and philosophers as a big machine whose parts move in a predictable and predetermined way—according to natural laws. However, this *deterministic* view of the universe had to be rejected by scientists in the twentieth century, as we shall see in Chapters 25 and 26.

3-5 ▌ Dynamics of rotation

Torque

We return now to the general case of the rotation of a body about an axis, such as a wheel on an axle or a door on its hinges. In Section 3-1 we discussed the description of rotational motion in terms of angle, angular velocity, and angular acceleration. Now we discuss the dynamics, or causes, of rotational motion. Just as we found analogies between linear and rotational motion when *describing* motion, so rotational equivalents for dynamics exist as well. For example, the rotational equivalent of Newton's first law states that a freely rotating body will continue to rotate with constant angular velocity as long as no forces (or as we shall see shortly, no torques) act to change that motion. More difficult is the question of a rotational equivalent for Newton's second law, that is, the problem of what gives rise to angular acceleration. To make an object start rotating about an axis clearly requires a force; but where this force is applied is also important. Take, for example, an ordinary situation such as the door in Figure 3-22. If you apply a force $\mathbf{F}_1$ to the door as shown, you will find that the greater the magnitude of $\mathbf{F}_1$, the more quickly the door opens. But now if you apply the same magnitude force at a point closer to the hinge, say $\mathbf{F}_2$ in Figure 3-22, you will find that the door will not open so quickly. The effect of the force is less.

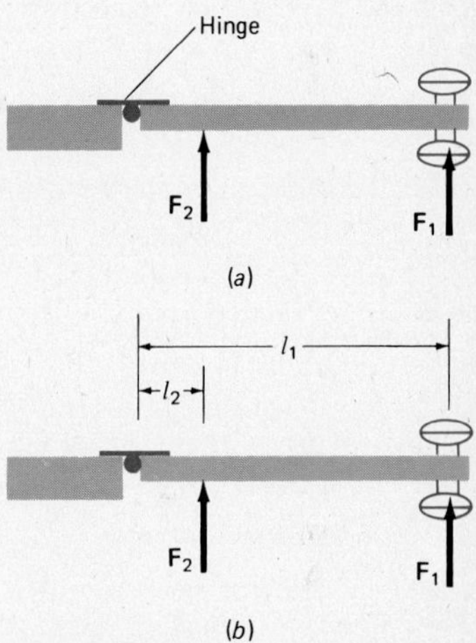

FIGURE 3-22
Applying the same force with different lever arms, *l*.

Careful observation shows that the angular acceleration of the door is proportional not only to the magnitude of the force, but is also proportional to the perpendicular distance from the axis of rotation to the line along which the force acts. Thus, if the distance l_1 in Figure 3-22*b* is three times larger than l_2, the angular acceleration of the door is found to be three times as great, assuming of course that the magnitudes of the forces are the same. To say it another way, if $l_1 = 3l_2$, then F_2 must be three times as large as F_1 to give the same angular acceleration. The distances l_1 and l_2 are called the *lever arms* of the respective forces. The angular acceleration, then, will be proportional to the product of the force times the lever arm; this product is called the *moment* of the force, or **torque**, and is abbreviated τ (Greek lowercase letter tau): $\tau = lF$. The angular acceleration α of an object is found to be directly proportional to the applied torque τ:

$$\alpha \propto \tau. \tag{3-11}$$

Thus, we see that it is torque that gives rise to angular acceleration.

We define the lever arm as the *perpendicular* distance of the axis of rotation from the line of action of the force (by which we mean an imaginary line drawn along the direction of the force); we do this to take into account the effect of forces acting at an angle. It is clear that a force

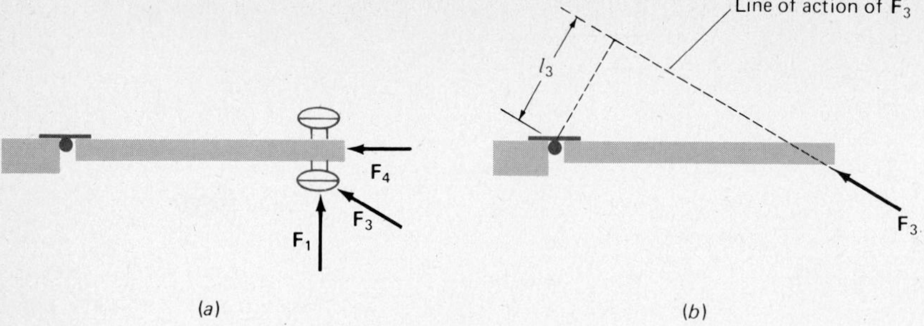

FIGURE 3-23
(*a*) Forces acting at different angles. (*b*) The lever arm is defined as the perpendicular distance from the axis of rotation to the line of action of the force.

applied at an angle, such as F_3 in Figure 3-23, will be less effective than a force of the same magnitude applied straight on, such as F_1 (Figure 3-23*a*). And if you push on the end of the door so that the force is directed at the hinge (the axis of rotation) as indicated by $\mathbf{F}_4$, the door will not move at all!

The lever arm for a force such as $\mathbf{F}_3$ is found by drawing a line along the direction of $\mathbf{F}_3$ (the line of action of $\mathbf{F}_3$), and then drawing another line from the axis of rotation perpendicular to the first line. The length of this second line is the lever arm for F_3 and is labeled l_3 in Figure 3-23*b*.

The torque associated with F_3 is then l_3F_3. This short lever arm and the corresponding smaller torque associated with F_3 is consistent with the observed fact that $\mathbf{F}_3$ is less effective in opening the door than is $\mathbf{F}_1$. When the lever arm is defined in this way, experiment shows that relation 3-11 is valid in general. Notice in Figure 3-23*a* that the line action of the force $\mathbf{F}_4$ passes through the hinge and hence its lever arm is zero. Consequently, F_4 produces zero torque and thus gives rise to no angular acceleration.

In general, then, we must write the torque as

$$\tau = l_{\perp}F, \tag{3-12a}$$

where $l_{\perp}$ is the lever arm and we put the perpendicular symbol ($\perp$) as a subscript to remind us that we must use the perpendicular distance from the axis of rotation to the line of action of the force (see Figure 3-24*a*). An alternative but equivalent way of determining the torque associated with a force is to resolve the force into components parallel and perpendicular to a line joining the point of application of the force to the axis as shown in Fig. 3-24*b*. Then the torque will be equal to $F_{\perp}$ times the distance l from the axis to the point of application of the force:

$$\tau = lF_{\perp}. \tag{3-12b}$$

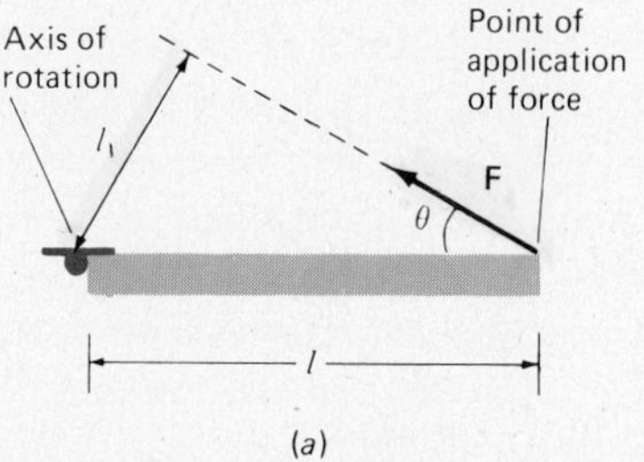

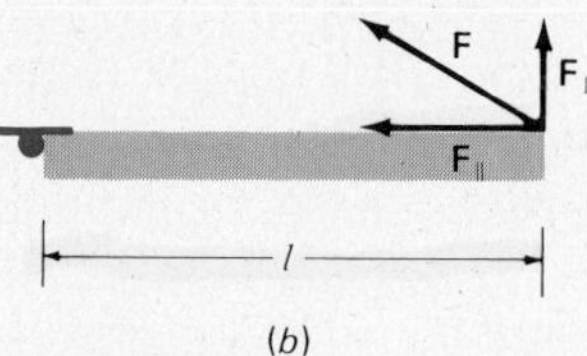

FIGURE 3-24
Torque $= l_{\perp}F = lF_{\perp}$.

That this gives the same result as Equation 3-12*a* can be seen from the fact that $F_{\perp} = F \sin\theta$ and $l_{\perp} = l \sin\theta$; so $\tau = Fl \sin\theta$ in either case. We can use either Equation 3-12*a* or *b* to calculate the torque, whichever is easier.

Since torque is a force times a distance, it is measured in units of N · m in SI units, dyne · cm in the cgs system, and lb · ft in the English system.

▮ EXAMPLE 3-12 What is the torque applied by the biceps on the lower arm in Figure 3-25*a* and *b*? The axis of rotation is the elbow joint and the muscle is inserted 5.0 cm from the joint.

SOLUTION (*a*) $F = 250$ N, $l_{\perp} = 0.050$ m; so $\tau = (0.050 \text{ m})(250 \text{ N}) = 12.5$ N · m. (*b*) $F = 250$ N, $l_{\perp} = (0.050 \text{ m})(\sin 45°)$; so $\tau = (0.050 \text{ m})(0.71)(250 \text{ N}) = 8.9$ N · m. ▮▮

Rotational inertia

We have seen that the angular acceleration α of a rotating body is proportional to the torque τ applied to it:

$$\alpha \propto \tau.$$

This corresponds to Newton's second law for translational motion, $a \propto F$, where torque has taken the place of force and correspondingly the angular acceleration α takes the place of the linear acceleration a. In the linear case, the acceleration is not only proportional to the net force but is also inversely proportional to the inertia of the body, which we call its mass m; thus we could write $a = F/m$. What then plays the role of mass for the rotational case?

Clearly, a heavy millstone is harder to start rotating than a roller-skate wheel; once it is rotating, it is also harder to stop. The millstone is said to have more *rotational inertia* than the skate wheel. The rotational inertia of an object depends not only on its mass, however, but also on how that mass is distributed. For example, a large-diameter cylinder will have greater rotational inertia than one of equal mass but smaller diameter (and therefore longer); see Figure 3-26*a*. The former will be harder to start rotating, and harder to stop. When the mass is concentrated farther from the axis of rotation, the rotational inertia is greater. To see exactly what the relation is, let us consider a simple case, that of a small object of mass m rotating in a circle of radius r at the end of a string whose mass we can ignore (Figure 3-26*b*). The torque that gives rise to its angular acceleration is $\tau = rF$. If we make use of Newton's second law for linear quantities, $F = ma$, and Equation 3-5, relating the angular acceleration to the tangential linear acceleration, $a = r\alpha$, we have

$$F = ma = mr\alpha.$$

When we multiply both sides by r, we find that the torque $\tau = rF$ is given by

$$\tau = mr^2\alpha. \tag{3-13a}$$

Here, at last, we have a direct relation between the angular acceleration and the applied torque τ. The quantity mr^2 represents the rotational inertia of the particle and is called its *moment of inertia.*

An ordinary object such as a wheel rotating on its axis, a door turning on its hinges, or an arm rotating about the shoulder joint can be thought of as consisting of many particles located at various distances from the axis of rotation. We can apply the relation 3-13*a* to each particle and sum over all the particles. The sum of the various torques is just the total torque, which we now call τ, so we obtain

$$\tau = \left(\sum mr^2\right)\alpha. \tag{3-13b}$$

The $\sum$ (Greek capital letter sigma) means "the sum of." The sum, $\sum mr^2$, represents the sum of the masses of each particle in the body multiplied by the square of the distance of that particle from the axis of rotation. If we give each particle a number $(1, 2, 3, \ldots)$, then $\sum mr^2 = m_1r_1^2 + m_2r_2^2 + m_3r_3^2 + \cdots$. This quantity is called the **moment of inertia** of the body, I:

$$I = \sum mr^2. \tag{3-14}$$

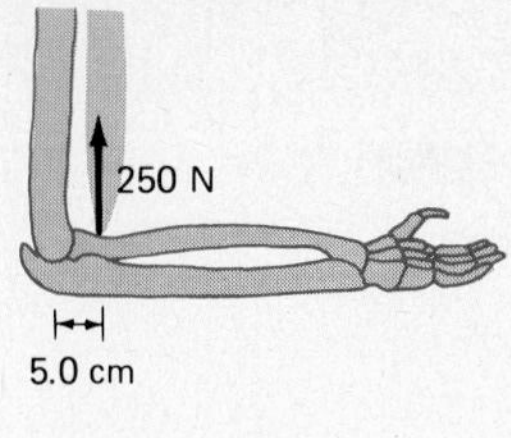

(*a*)

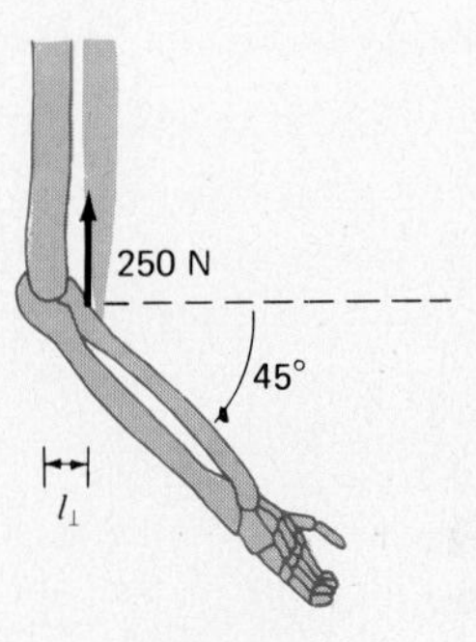

(*b*)

FIGURE 3-25
Example 3-12.

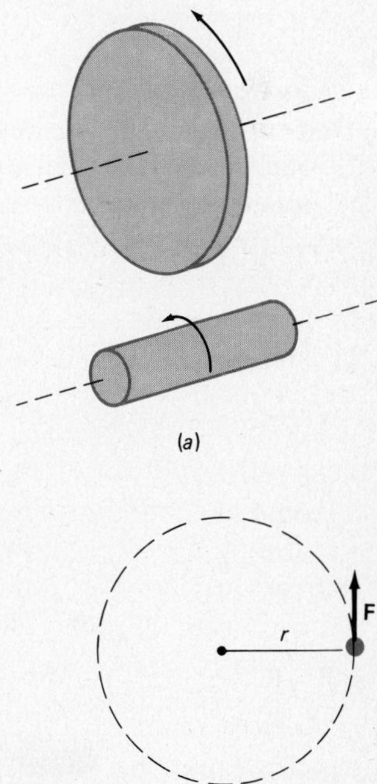

(*b*)

FIGURE 3-26
Rotational inertia.

Combining Equations 3-13b and 3-14, we can write

$$\tau = I\alpha. \qquad (3\text{-}15)$$

This is the rotational equivalent of Newton's second law. We see that the quantity we defined as the moment of inertia is the actual measure of the rotational inertia of a body. (Torque, like angular velocity and acceleration, can be represented by a vector; however, we will mainly be concerned with rotation in a plane with the axis of rotation remaining fixed, so that we need deal only with the magnitude of the torque.)

Whenever Equation 3-15 is used, it must be remembered to give α in rad/s^2, and the other terms in a consistent set of units. The moment of inertia, I, has units of kg · m^2 in SI units.

▌ EXAMPLE 3-13 Two objects, of mass 5.0 kg and 7.0 kg, are mounted 4.0 m apart on a light rod (whose mass can be ignored) as shown in Figure 3-27. Calculate the moment of inertia of the system (a) when rotated about an axis passing halfway between the masses, Figure 3-27a, and (b) when the system rotates about an axis located 0.50 m to the left of the 5.0-kg mass (Figure 3-27b).

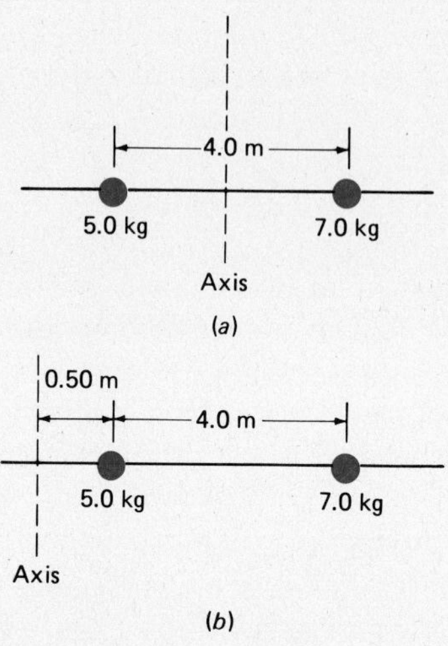

Calculating moment of inertia (Example 3-13).

SOLUTION (a) Both particles are the same distance, 2.0 m, from the axis of rotation. Thus,

$$I = \sum mr^2 = (5.0 \text{ kg})(2.0 \text{ m})^2 + (7.0 \text{ kg})(2.0 \text{ m})^2 = 48 \text{ kg} \cdot \text{m}^2.$$

(b) The 5.0-kg mass is now 0.50 m from the axis and the 7.0-kg mass is 4.50 m from the axis. Then

$$\begin{aligned} I = \sum mr^2 &= (5.0 \text{ kg})(0.50 \text{ m})^2 + (7.0 \text{ kg})(4.50 \text{ m})^2 \\ &= 1.3 \text{ kg} \cdot \text{m}^2 + 142 \text{ kg} \cdot \text{m}^2 \\ &= 143 \text{ kg} \cdot \text{m}^2. \end{aligned}$$

▌▌

Example 3-13 illustrates two important points. First, the moment of inertia of a given system is different for different axes of rotation. Second, part (b) illustrates the fact that mass close to the axis of rotation contributes very little to the total moment of inertia; in this example, the 5.0-kg object contributed less than 1 percent of the total.

For most ordinary bodies, the mass is distributed continuously and calculation of the moment of inertia, $\sum mr^2$, can be difficult. Expressions can be worked out, however, for the moments of inertia of regularly shaped bodies in terms of their dimensions. Table 3-1 gives these expressions for a number of solids rotated about the axes specified. The only one for which the result is obvious is that for the thin ring rotated about an axis passing through its center perpendicular to the plane of the ring. For this object, all the mass is concentrated at the same distance from the axis, R. Thus $\sum mr^2 = (\sum m)R^2 = MR^2$, where M is the total mass of the ring.

When discussing moments of inertia, it is often convenient to work with the *radius of gyration, k*, which is a sort of average radius. In particular, the radius of gyration of an object is defined so that if all the mass of the object were concentrated at this distance from the axis, it would have the same moment of inertia as the original object. For example (see Table 3-1), the radius of gyration of a cylinder is $(1/\sqrt{2})R \approx 0.71R$; this means that the cylinder of radius 100 cm has the same moment of inertia as an equal-mass thin ring of radius 71 cm. The moment of inertia of any object

TABLE 3-1

Moments of inertia for various solids

Object†	Location of axis	Moment of inertia	Radius of gyration
Thin ring of radius *R*	**Through center**	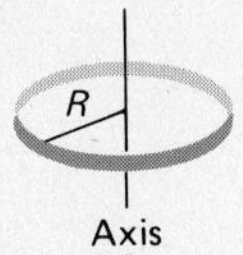 MR^2	R
Cylinder of radius *R*	**Through center**	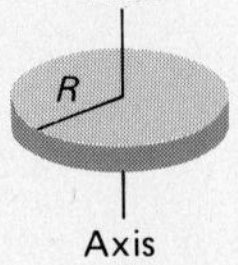 $\frac{1}{2}MR^2$	$\frac{R}{\sqrt{2}}$
Sphere of radius *R*	**Through center**	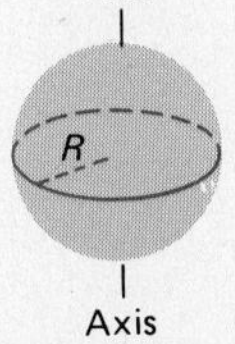 $\frac{2}{5}MR^2$	$\sqrt{\frac{2}{5}}R$
Long rod of length *L*	**Through center**	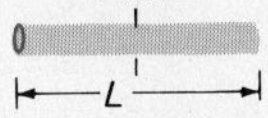 $\frac{1}{12}ML^2$	$\frac{L}{\sqrt{12}}$
	Through end	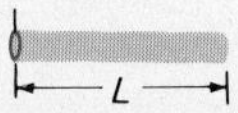 $\frac{1}{3}ML^2$	$\frac{L}{\sqrt{3}}$

† All objects are assumed to be of uniform composition.

can be written in terms of its radius of gyration as

$$I = Mk^2.$$

For unusual or irregularly shaped objects it is common to specify the radius of gyration.

▌ EXAMPLE 3-14 A 15-N force is applied to a cord wrapped around a 4.0-kg wheel of radius 33 cm, and radius of gyration 30 cm, Figure 3-28. If there is a frictional torque of 1.1 N · m at the hub, what is the acceleration of the wheel?

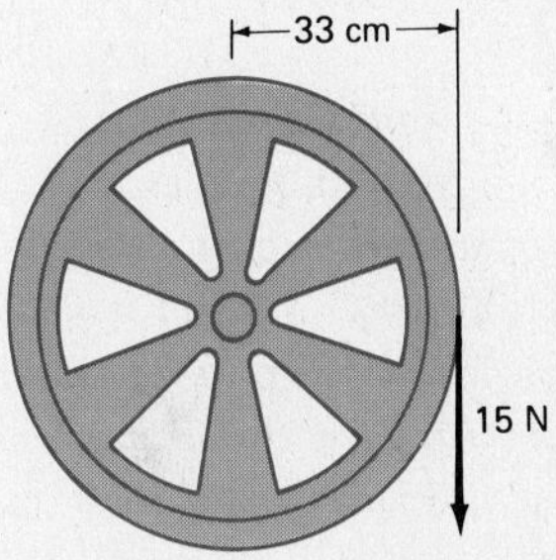

SOLUTION First, we calculate I:

$$I = Mk^2 = (4.0\text{ kg})(0.30\text{ m})^2 = 0.36\text{ kg}\cdot\text{m}^2.$$

There are two torques involved, that due to the 15-N force equal to (0.33 m)(15 N) = 5.0 N · m, and the opposing friction torque of 1.1 N · m. Thus we use Equation 3-15 and obtain

$$\alpha = \frac{\tau}{I} = \frac{(5.0\text{ N}\cdot\text{m}) - (1.1\text{ N}\cdot\text{m})}{0.36\text{ kg}\cdot\text{m}^2} = 11\text{ rad/s}^2.$$ ▌▌

33 cm

15 N

FIGURE 3-28
Example 3-14.

3-6 ▌ Center of gravity

Observations of the behavior of solid bodies under the action of gravity indicate that there is a single point in a body, called the **center of gravity** (cg), at which the force of gravity can be considered to act. Of course, the force of gravity actually acts on all the different parts or particles of a body as shown in Figure 3-29*a*. But for purposes of determining the motion of a body as a whole, we can assume that the entire weight of the body w (which is the sum of the weights of all its parts) acts at the cg (Figure 3-29*b*). For regularly shaped bodies of uniform composition, such as cylinders, spheres, and rectangular solids, the cg is located at the geometric center of the body. The cg for other objects can be found by making use of a technique based on Equation 3-15, $\tau = I\alpha$. To see how this works, let us consider a body whose cg is known, say, a rectangle. In Figure 3-30 the rectangle is shown pivoted from a point near its edge. In the position shown, there is a net torque acting on the object: the force of gravity w times the lever arm. Therefore, the object will swing to the left; eventually it will come to rest with the cg directly below the pivot point because in this position the lever arm will be zero and thus the torque will be zero. To determine the cg of any odd-shaped object, we make use of this fact—that when it is at rest, its cg will lie on a vertical line directly below whatever point it is suspended from. If the object is two-dimensional or has a plane of symmetry, it need only be hung from two different pivot points and the respective vertical lines drawn; then the cg will be at the intersection of the two lines (Figure 3-31). If the object is suspended from a third point, a vertical line will also pass through the cg. If the object does not have a plane of symmetry, the cg with respect to the third dimension is found by suspending the object from at least three points.

A related concept is that of *center of mass* (cm), which is defined as that point in a body at which its *mass* can be considered to be concentrated. Strictly speaking, there is a conceptual difference between the center of gravity and the center of mass, but for practical purposes they are generally the same point.†

Sometimes one must deal with a system made up of parts whose individual centers of gravity are known and the cg of the whole system is required. This can be calculated using torque. Suppose, for example, that we wish to find the cg of a series of masses located along a massless rod as shown in Figure 3-32. Let us assume that the cg is a distance x_{cg} from the end of the rod. If the various masses are at known distances x_1, x_2, and x_3 from the end, then the unknown distance x_{cg} can be calculated as follows. If we calculate the torque exerted by gravity on this system about the end‡ of the rod, the total torque is

$$\tau = m_1gx_1 + m_2gx_2 + m_3gx_3$$

since the force on each mass is just the force of gravity. By definition of the cg, the torque must be the same if we consider all the mass $(m_1 + m_2 + m_3)$ located at the cg. Thus

$$\tau = (m_1 + m_2 + m_3)gx_{cg}.$$

When we combine these two equations and solve for x_{cg} we find that

† There would be a difference between the two only if a body were large enough so that the acceleration due to gravity would be different at different parts of the body.

‡ This is arbitrary. The final result will be the same if we calculate the torque about any other point.

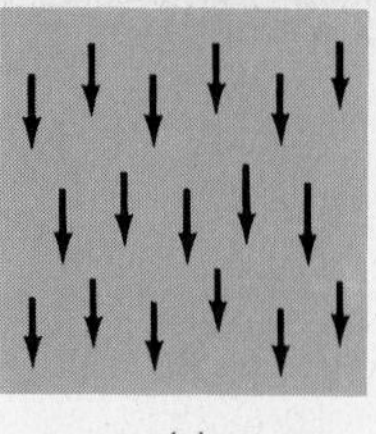

(*a*)

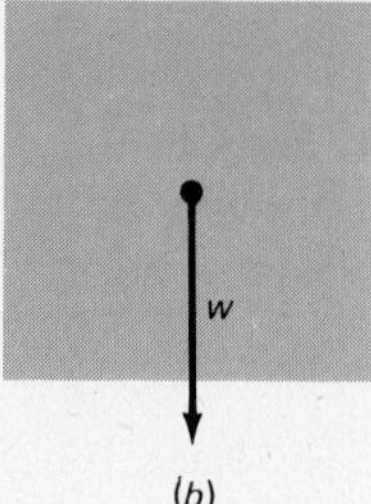

(*b*)

FIGURE 3-29
Force of gravity acts on all parts of a body (*a*), but can be considered to act at one point (*b*), the center of gravity.

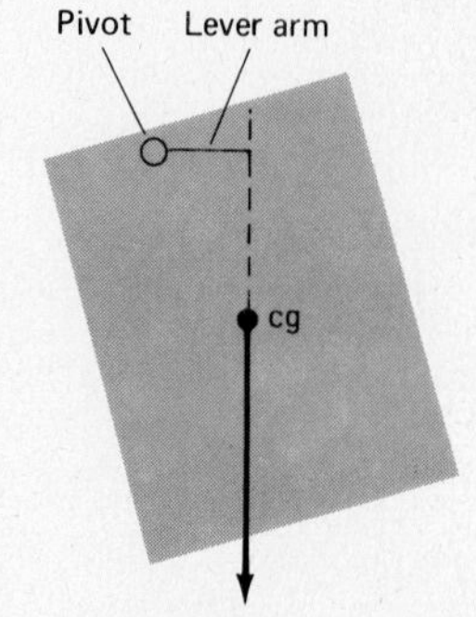

FIGURE 3-30
Torque will cause the object to rotate unless the cg is below the point of support; only there is the lever arm zero.

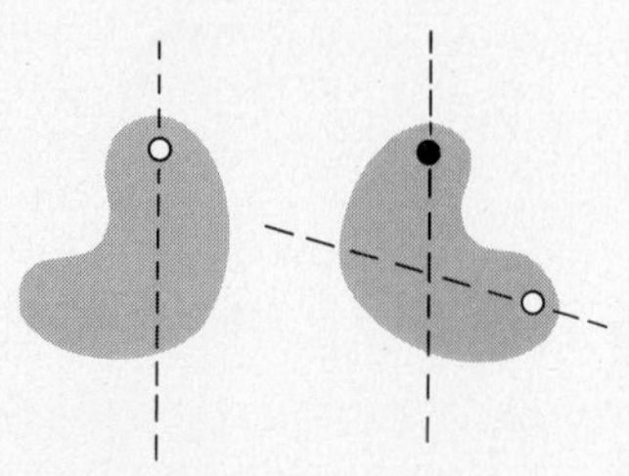

FIGURE 3-31
Finding the cg of an object.

$$x_{cg} = \frac{m_1x_1 + m_2x_2 + m_3x_3}{m_1 + m_2 + m_3}. \tag{3-16}$$

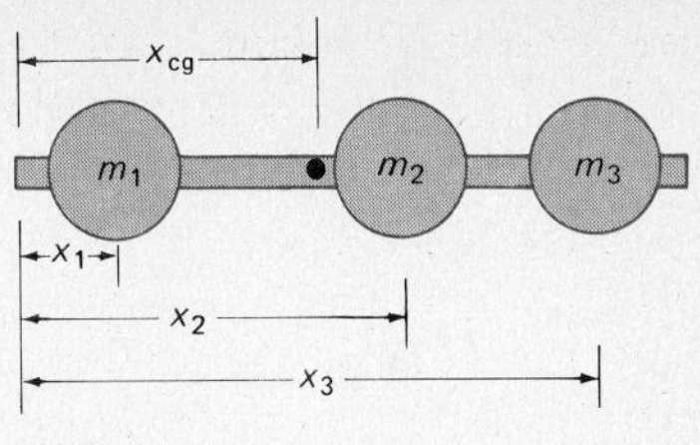

FIGURE 3-32
Finding the cg by calculation.

Thus, if $x_1 = 1.0$ m, $x_2 = 5.0$ m, and $x_3 = 6.0$ m and the masses each have a mass of 2.0 kg,

$$x_{cg} = \frac{(2.0 \text{ kg})(1.0 \text{ m}) + (2.0 \text{ kg})(5.0 \text{ m}) + (2.0 \text{ kg})(6.0 \text{ m})}{6.0 \text{ kg}} = 4.0 \text{ m}.$$

Equation 3-16 can be extended to any number of bodies that lie in a line. If the bodies are spread out over a plane, or three dimensions, Equation 3-16 is applied to the second (and third) dimension as well. To see how this is done, we consider an example involving the human body. In Table 3-2 the cg and hinge points (joints) for the different components of a "representative" person are shown. Of course, there are wide variations among people, so these data represent only a rough average. Note that the numbers represent a percentage of the total height, which is regarded as 100 units.

EXAMPLE 3-15 Calculate the cg of a whole leg when (*a*) stretched out and (*b*) bent at 90°, as shown in Figure 3-33.

SOLUTION (*a*) We will use the percentage units; that is, the person has a mass of 100 units and a height of 100 units. We measure the distance from the hip joint using Table 3-2 and obtain the numbers shown on the diagram. Using Equation 3-16, we obtain

$$x_{cg} = \frac{(21.5)(9.6) + (9.6)(33.9) + (3.4)(50.3)}{21.5 + 9.6 + 3.4}$$

$$= 20.4 \text{ units}.$$

Thus, the center of mass of the leg and foot is 20.4 units below the hip joint or $52.1 - 20.4 = 31.7$ units above the base of the foot. (*Note*: If the person is 170 cm tall, this is $31.7 \times 170 \text{ cm}/100 = 54$ cm.) (*b*) This is a

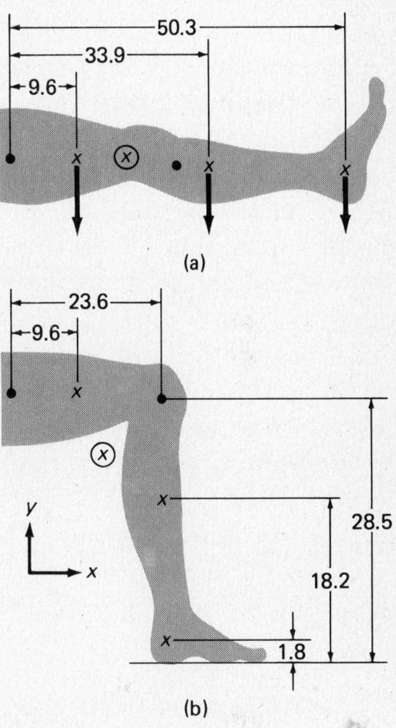

FIGURE 3-33
Example 3-15; finding the cg of a leg in two different positions (⊗ represents calculated cg).

TABLE 3-2

Center of gravity of parts of the male human body†

Distance of hinge points	Hinge points (·)(joints)	Center of gravity (×) (% height above floor)		Percent mass
91.2	**Base of skull on spine**	**Head**	93.5	6.9
81.2	**Shoulder joint**	**Trunk and neck**	71.1	46.1
	elbow 62.2	**Upper arms**	71.7	6.6
	wrist 46.2	**Lower arms**	55.3	4.2
52.1	**Hip**	**Hands**	43.1	1.7
		Upper legs (thighs)	42.5	21.5
28.5	**Knee**	**Lower legs**	18.2	9.6
4.0	**Ankle**	**Feet**	1.8	3.4
			58.0	100.0

† From *NASA LIFE SCIENCE DATA BOOK*, Washington, D.C.

two-dimensional problem. We use an xy coordinate system as shown in Figure 3-33b. First we calculate how far to the right of the hip joint the cg lies:

$$x_{cg} = \frac{(21.5)(9.6)+(9.6)(23.6)+(3.4)(23.6)}{21.5+9.6+3.4} = 14.9 \text{ units.}$$

Next, we calculate the distance, y_{cm}, of the cm above the floor:

$$y = \frac{(3.4)(1.8)+(9.6)(18.2)+(21.5)(28.5)}{21.5+9.6+3.4} = 23.1 \text{ units.}$$

Thus the cg is located 23.1 units above the floor and 14.9 units to the right of the hip joint. ▮▮

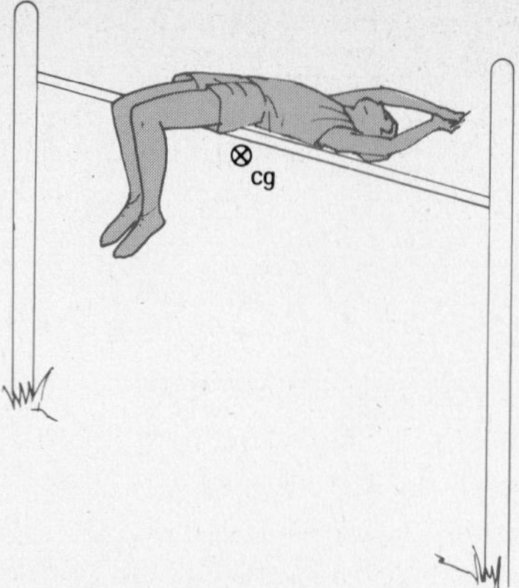

FIGURE 3-34
The cg of a high jumper may actually pass beneath the bar.

This example illustrates the fact that the cg can lie outside the body itself. If this seems implausible, notice that the cg of a doughnut is at the center of the hole.

Knowing the cg of the body when it is in various positions is of great use in studying body mechanics. One simple example from athletics is shown in Figure 3-34. If high jumpers can get in the position shown, their cg would actually pass below the bar, which means that for a particular takeoff speed, they could clear a higher bar. This is indeed what they try to do.

3-7 ▮ General motion

We devote this brief section to a single theorem. This theorem is a summation of what we have studied so far, and its validity will hopefully be evident to you.

> ▮ **The motion of any solid body can be analyzed as the sum of two separate motions: translational motion of the center of mass, plus rotational motion of the object about its center of mass.**

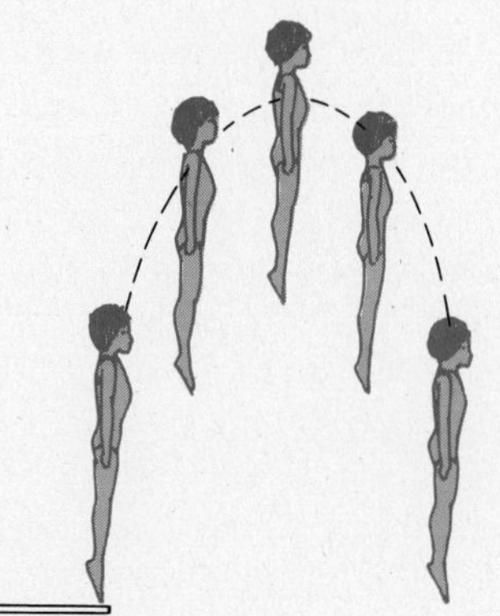

FIGURE 3-35
Only translational motion occurs here.

By "translational motion" we mean the path through space taken by a point object. In the case of an ordinary extended body, it is said to undergo translational motion if it maintains the same orientation with respect to some fixed set of reference axes, that is, if it does not rotate. For example, the person who jumps off a diving board and remains vertical throughout is undergoing translational motion only (Figure 3-35). The diver shown in Figure 3-36 is undergoing both translational and rotational motion, and this illustrates our theorem. Her cg follows a parabolic path, as we expect for a projectile, and in addition her body rotates about its cg. General motion will be discussed more in Chapter 5.

FIGURE 3-36
The motion of the diver is translational (in a parabola as for any projectile) of the cg, plus rotation about the cg.

▮ Summary

Angular quantities are defined in analogy with linear quantities. Angles can be measured in degrees, revolutions, or radians: 2π rad = 1 rev = 360°. *Angular velocity*, ω, is the rate at which angle changes: $\omega = \theta/t$. Angular velocity is related to *frequency* f by $\omega = 2\pi f$. *Angular acceleration*, α, is the rate at which angular velocity changes. The linear velocity v of a point fixed at a distance r from the axis of rotation is $v = r\omega$, and its tangential linear acceleration is $a_T = r\alpha$. The equations describing uniformly accelerated rotational motion (α = constant) have the same form as for uniformly accelerated linear motion:

$$\omega = \omega_0 + \alpha t; \quad \theta = \omega_0 t + \tfrac{1}{2}\alpha t^2; \quad \omega^2 = \omega_0^2 + 2\alpha\theta; \quad \bar{\omega} = \frac{\omega + \omega_0}{2}.$$

A particle moving with constant speed v in a circle of radius r has a linear *centripetal* (toward the center) *acceleration* $a_c = v^2/r = \omega^2 r$, because the velocity vector is continually changing in direction. A force acting toward the center is thus needed to keep a particle revolving in a circle. If the particle is revolving in a circle with nonuniform speed, it will have both centripetal and tangential linear acceleration.

Newton's *law of universal gravitation* states that every body in the universe attracts every other body with a force proportional to the product of their two masses and inversely proportional to the square of the distance between them. It is this force of gravity that keeps the moon in its orbit around the earth and the planets in their orbits around the sun.

The dynamics of rotation is analogous to the dynamics of linear motion; force is replaced by *torque*, τ, which is defined as the product of force times lever arm (perpendicular distance from the line of action of the force to the axis of rotation); mass is replaced by *moment of inertia*, I, which depends not only on the mass of the body but also on how the mass is distributed about the axis of rotation; and linear acceleration is replaced by angular acceleration. The rotational equivalent of Newton's second law is then $\tau = I\alpha$.

The *center of gravity* of a body is that point at which the force of gravity can be considered to act for purposes of determining the motion of the body as a whole. The complete motion of a body can be described as the translational motion of its center of gravity plus rotation about its center of gravity.

Questions

1 Sometimes it is said that water is removed from clothes in a spin dryer by centrifugal force throwing the water outward. Is this correct? Analyze in terms of the discussion of centripetal acceleration in Section 3-2.

2 Astronauts who spend time in outer space are affected by weightlessness. One way to simulate gravity is to shape the spaceship like a bicycle wheel which rotates about an axis just like a wheel, and the astronauts walk on the inside of the "tire." Explain how this simulates gravity.

*3 Technical reports often specify only the rpm for centrifuge experiments. Why is this inadequate? (*Hint*: See Equation 3-7*b*.)

4 Does an apple exert a gravitational force on the earth? If so, how large a force?

5 People sometimes ask, "What keeps a satellite up in its orbit around the earth?" How would you respond?

6 How can the distance from the earth to the moon be measured? From this measurement, how is the diameter of the moon determined?

7 If the earth's mass were suddenly to double, in what ways would the moon's orbit change?

8 Explain why a running person experiences free fall or "weightlessness" between steps.

9 The earth moves faster in its orbit around the sun in winter than in summer. Is it closer to the sun in summer or in winter?

10 Why is it more difficult to do situps when your knees are bent than when your legs are stretched out?

11 Expert bicyclists use very lightweight "sew-up" tires. They claim that reducing the mass of the tires is far more significant than an equal reduction in mass elsewhere on the bicycle. Explain why this is true.

12 Mammals that depend on being able to run fast have slender lower legs with flesh and muscle concentrated high, near the shoulder. On the basis of rotational dynamics, explain why this distribution of mass is advantageous.

13 Make an approximate guess for the radius of gyration of the insect wing shown in Figure 3-37, assuming it is thin and of uniform thickness.

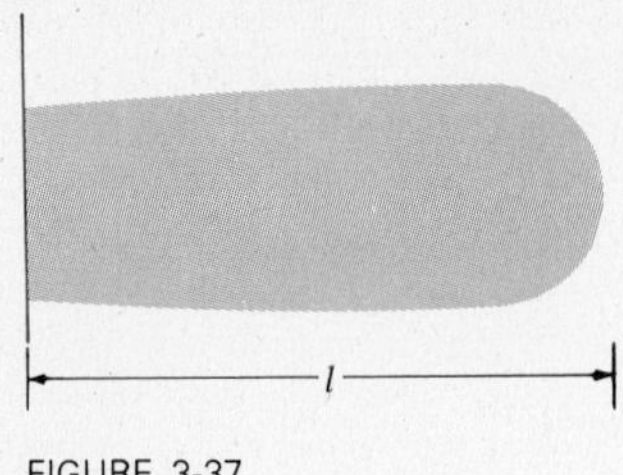

FIGURE 3-37

14 A sphere and a cylinder of equal radius and mass roll down an inclined plane. Which one do you expect will reach the bottom first? (*Hint*: See Table 3-1.)

15 A spool of thread rests on the floor with the thread emerging from the underside. If this thread is gently pulled in a direction parallel to the floor, in which direction will the spool move? (Try it and see.) Explain. What happens if the thread is pulled at an angle of 60° with the floor? Why? (*Hint*: Draw a diagram showing the forces on the spool. The thread is wound on a cylinder whose radius is less than that of the edges of the spool in contact with the floor.)

16 Why is the cg of a 1-m length of pipe at its midpoint whereas this is not true for your arm or leg?

17 Show on a diagram how your cg shifts when you change from a lying position to a sitting position.

18 A cyclist rides over the top of a hill. Is the bicycle's motion rotational, translational, or a combination?

Problems

SECTION 3-1

1 (I) What are the following angles in radians: (*a*) 30°, (*b*) 90°, and (*c*) 420°?

2 (I) The Eiffel Tower is 300 m tall. When you are standing at a certain place in Paris, it subtends an angle of 5°. How far are you, then, from the Eiffel Tower?

3 (I) A laser beam is directed at the moon, 380,000 km from earth. The beam diverges at an angle of 1.8×10^{-5} rad. How large a spot will it make on the moon?

4 (I) A bicycle with 68-cm-diameter tires travels 2.0 km. How many revolutions do the wheels make?

5 (I) A 20-cm-diameter grinding wheel rotates at 2000 rpm. Calculate its angular velocity in rad/s.

6 (I) What is the linear speed of a point on the edge of the grinding wheel in problem 5?

7 (I) A 70-cm-diameter wheel rotating at 1200 rpm is brought to rest in 15 s. Calculate its angular acceleration.

8 (I) A 33-rpm phonograph record reaches its rated speed 2.8 s after turning it on. What was the angular acceleration?

9 (II) Estimate the angle subtended by the moon using a ruler and your finger or other object to just blot out the moon. Describe your measurement and the result obtained and then use it to estimate the diameter of the moon. The moon is about 380,000 km from the earth.

10 (II) A centrifuge accelerates from rest to 10,000 rpm in 420 s. Through how many revolutions did it turn in this time?

11 (II) Calculate the angular velocity of the earth (*a*) in its orbit around the sun, and (*b*) about its axis.

12 (II) Starting from the definitions of ω and α, derive Equations 3-6*a*, 3-6*b*, and 3-6*c*, assuming constant angular acceleration.

13 (II) A cyclist accelerates from rest at a rate of 0.80 m/s^2. How fast will a point on the rim of the tire (diameter = 68 cm) at the top be moving after 4.0 s? (*Hint*: At any moment, the lowest point on the tire is in contact with the ground and hence is at rest.)

SECTION 3-2

14 (I) A child moves with a speed of 1.80 m/s when 12.4 m from the center of a merry-go-round. Calculate (*a*) the centripetal acceleration of the child, and (*b*) the net force exerted on the child (mass = 25 kg).

15 (I) A jet plane traveling 1800 km/h (500 m/s) pulls out of a dive by moving in an arc of radius 3.0 km. What is the plane's acceleration in *g*'s?

16 (I) How fast (in rpm) must a centrifuge rotate if a particle 10.0 cm from the axis of rotation is to experience an acceleration of 100,000 *g*'s?

17 (I) Calculate the centripetal acceleration of the earth in its orbit around the sun and the net force exerted on it. What exerts this force on the earth? Assume that the earth's orbit is a circle of radius 1.5×10^{11} m.

18 (II) Tarzan plans to cross a gorge by swinging in an arc from a hanging vine. If his arms are capable of exerting a force of 1500 N on the rope, what is the maximum speed he can tolerate at the lowest point of his swing? His mass is 85 kg; the vine is 4.0 m long.

19 (II) A ball on the end of a string is revolving at a uniform rate in a vertical circle of radius 1.10 m, as shown in Figure 3-38. If its speed is 3.75 m/s and its mass is 0.355 kg, calculate the tension in the string when the ball is (*a*) at the top of its path, and (*b*) at the bottom of its path.

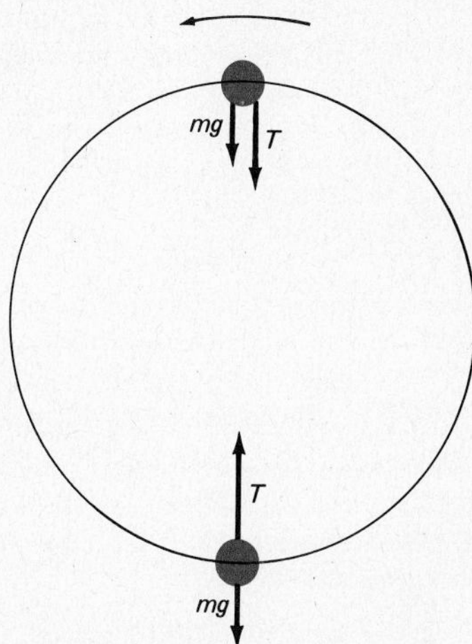

FIGURE 3-38

20 (II) A coin is placed 15 cm from the axis of a rotating turntable of variable speed. When the angular speed of the turntable is slowly increased, the coin remains fixed on the turntable until a rate of 66 rpm is reached, at which point the coin slides off. What is the coefficient of static friction between the coin and the turntable?

21 (II) In a "Rotor-ride" at a carnival, riders are pressed against the inside of a vertical cylinder 2.8 m in radius rotating at a speed of 3.2 rad/s when the floor drops out. What minimum coefficient of friction is needed so a person will not slip down?

22 (II) What is the maximum speed with which a 1000-kg car can round a turn of radius 100 m on a flat road if the coefficient of friction between tires and road is 0.50? Is this result independent of the mass of the car?

23 (II) If the static coefficient of friction between tires and road is 0.70, what is the minimum turning radius on a level road for a car traveling at (*a*) 30 km/h, and (*b*) 60 km/h?

24 (II) What minimum speed must a roller coaster be traveling when upside down at the top of a circle (Figure 3-39) if the passengers are not to fall out? Assume a radius of curvature of 8.0 m.

25 (III) A car rounds a curve of radius 50 m banked at an angle of 16°. If the car is traveling at 70 km/h, will a friction force be required? If so, how much and in what direction? The car's mass is 1200 kg.

26 (III) (*a*) Referring to Figure 3-10, show that the angle θ at which a road should be banked so that the required centripetal force will be supplied entirely by the horizontal component of the normal force, F_N, is given by $\tan\theta = v^2/rg$. (*b*) Calculate this angle assuming a design speed of $v = 50$ km/h and a radius $r = 50$ m. (*c*) If a car traveling 90 km/h rounds this curve banked at the angle calculated above, what minimum value must the coefficient of friction have if the car is not to skid?

27 (III) If a plant is allowed to grow from seed on a rotating platform, it will grow at an angle, pointing inward. Calculate what this angle will be (put yourself in the rotating frame) in terms of g, r, and ω.

28 (III) In a rotating frame of reference, Newton's first and second laws remain valid if we assume that a pseudoforce equal to $m\omega^2 r$ is acting. What effect does this assumption have on the validity of Newton's third law?

SECTION 3-3

29 (I) Calculate the force of gravity between two bowling balls each of which has a mass of 7.0 kg, when they are 0.50 m apart (center to center).

30 (I) Calculate the force of gravity on a spacecraft 12,800 km above the earth's surface if its mass is 700 kg.

31 (I) Calculate the acceleration due to gravity on the moon. The moon's radius is about 1.7×10^6 m and its mass is 7.4×10^{22} kg.

32 (I) Calculate the effective value of g, the acceleration of gravity, 6400 km above the earth's surface.

FIGURE 3-39

The "Great American Revolution" in an amusement park in California. (D. C. Giancoli.)

33 (I) Calculate the velocity of a satellite moving in a stable circular orbit about the earth at a height of 3200 km.

34 (II) How far above the earth's surface will the acceleration of gravity be half what it is on the surface?

35 (II) Compare the force of gravity due to the sun on the moon with the force exerted by the earth on the moon. The sun's mass is about 3×10^5 that of the earth and is about 1.5×10^8 km from the moon whereas the earth is only 3.8×10^5 km away.

36 (II) (*a*) Calculate the force of the sun ($m = 2 \times 10^{30}$ kg) and that of the moon ($m = 7.4 \times 10^{22}$ kg) on the earth. (*b*) Explain why the moon has the greater influence on the earth's tides. The sun is 1.5×10^{11} m from earth.

37 (II) At what height above the earth's surface must a satellite be placed if it is to remain over the same geographical point on the equator of the earth?

38 (II) Derive a formula for the mass of a planet in terms of its radius, r, the acceleration due to gravity at its surface, g_p, and the gravitational constant, G.

39 (II) At what distance from the earth will a spacecraft on the way to the moon experience zero net force because the earth and moon pull with equal and opposite forces?

40 (II) A 60-kg person carries a 125-N suitcase onto an elevator. As the elevator descends, the weight of the suitcase felt by the person is only 100 N. Calculate (*a*) the acceleration of the elevator, (*b*) the force exerted by the floor on the person, and (*c*) the net force on the person.

41 (III) How long would a day be if the earth were rotating so fast that objects at the equator would be weightless?

42 (III) (*a*) Show that the mass of the sun is given by the formula $M_s = (4\pi^2 r^3)/(GT^2)$, where T is the period of any planet (the time to make one revolution about the sun), r the radius of its orbit (assumed to be circular), and G the gravitational constant. (*b*) Calculate the mass of the sun using the earth as the planet. The earth is about 1.49×10^{11} m from the sun.

43 (III) Two stars maintain a constant distance apart of 8.0×10^{10} m and rotate about a point midway between them at an angular speed of one revolution every 12.6 years. (*a*) Why don't the two stars crash into one another as a result of the gravitational force between them? (*b*) What must be the mass of each star? (Take the two masses to be the same.)

*SECTION 3-4

*44 (I) The asteroid Icarus, although only a few hundred meters across, orbits the sun like the other planets. Its period is about 410 days. What is its average distance from the sun?

*45 (I) Venus is an average distance of 108,000,000 km from the sun. How long is the year on Venus? (The earth is 149,000,000 km from the sun on the average.)

*46 (II) At what height above the earth must an artificial satellite orbit if it is to remain at the same place relative to earth? [*Hint*: Apply Kepler's third law to bodies orbiting the earth (one of which includes the moon).]

*47 (III) Use Kepler's second law to show that the ratio of the speeds of a planet at its near and far points from the sun is equal to the inverse ratio of the near and far distances: $v_N/v_F = d_F/d_N$.

SECTION 3-5

48 (I) A person exerts a force of 18 N on the end of a door 84 cm wide. What is the magnitude of the torque if the force is exerted (*a*) perpendicular to the door, and (*b*) at a 60° angle to the face of the door?

49 (I) What is the maximum torque exerted by a 60-kg person riding a bike if the person puts all the weight on each pedal when climbing a hill? The pedals rotate in a circle of radius 18 cm.

50 (I) Calculate the moment of inertia of a 10-kg sphere of radius 0.20 m when the axis of rotation is through its center.

51 (I) Calculate the moment of inertia of a 67-cm-diameter bicycle wheel. The rim and tire have a combined mass of 1.3 kg. The mass of the hub can be ignored. (Why?)

52 (II) A small 2.4-kg ball on the end of a light rod is rotated in a horizontal circle of radius 1.2 m. Calculate: (*a*) the moment of inertia of the ball, and (*b*) the torque needed to keep the ball rotating at constant angular velocity if air resistance exerts a force of 0.020 N on the ball.

53 (II) A grinding wheel is a uniform cylinder of radius 8.2 cm and mass 0.88 kg. Calculate: (*a*) its moment of inertia, and (*b*) the torque needed to accelerate it from rest to 1200 rpm in 4.0 s if a frictional torque of 0.014 N · m is also acting.

54 (II) What is the radius of gyration of a 13.6-kg wheel that accelerates from rest to 600 rpm in 10 s when a net torque of 3.2 N · m acts on it?

55 (II) A 0.84-m-diameter sphere can be rotated about an axis through its center by a torque of 12.3 N · m which accelerates it uniformly from rest through a total of 180 rev in 15.0 s. What is the mass of the sphere?

56 (II) A merry-go-round accelerates from rest to 3.0 rad/s in 34 s. Assuming that the merry-go-round is a uniform disc of radius 8.0 m and mass 31,000 kg, calculate the net torque required to accelerate it.

57 (II) What is the acceleration of the 0.50-kg block shown in Figure 3-40? The uniform cylinder of radius 0.20 m has a mass of 0.80 kg. Ignore friction. (*Hint*: The force exerted by the rope on the edge of the cylinder, T, is *not* equal to the weight of the block. Why not?)

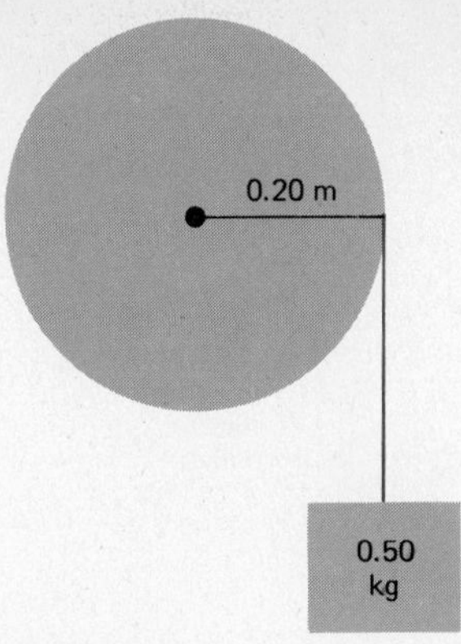

FIGURE 3-40

58 (II) A centrifuge rotor rotating at 10,000 rpm is shut off and is eventually brought to rest by a frictional torque of 0.20 N · m. If the mass of the rotor is 4.3 kg and its radius of gyration is 0.070 m, through how many revolutions will the rotor turn before coming to rest and how long will it take?

59 (II) The forearm in Figure 3-41 accelerates a 3.5-kg ball at 8.0 m/s^2 by means of the triceps muscle as shown. Calculate: (*a*) the torque needed, and (*b*) the force that must be exerted by the triceps muscle. Ignore the mass of the arm.

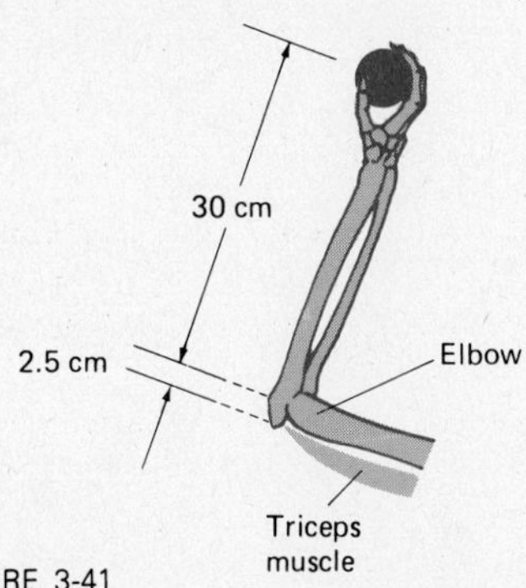

FIGURE 3-41

60 (III) Assume that a 1.0-kg ball is thrown solely by the action of the forearm, which rotates about the elbow joint under the action of the triceps muscle, Figure 3-41. The ball is accelerated from rest to 10.0 m/s in 0.22 s, at which point it is released. Calculate: (*a*) the angular acceleration of the arm, and (*b*) the force required of the triceps muscle. Assume that the forearm has a mass of 3.2 kg and rotates like a uniform rod about an axis at its end.

61 (III) A hammer thrower accelerates the hammer (mass = 7.3 kg) from rest within four full turns (revolutions) and releases it at a speed of 27.2 m/s. Assuming a uniform rate of increase in angular velocity and a radius of 2.0 m, calculate: (*a*) the angular acceleration, (*b*) the (linear) tangential acceleration, (*c*) the centripetal acceleration just before release, (*d*) the net force being exerted on the hammer by the athlete just before release, and (*e*) the angle of this force with respect to the radius of the circular motion.

SECTION 3-6

62 (I) The masses of the earth and moon are 6.0×10^{24} kg and 7.35×10^{22} kg, respectively, and they are separated by about 3.8×10^8 m. Where, approximately, is the cg of this system located?

63 (I) Determine the cg of an outstretched arm using Table 3-2.

64 (I) Assume that your proportions are the same as those in Table 3-2, and calculate the mass of one of your legs.

65 (I) An empty 1500-kg car has its cg 3.10 m from the front of the car. How far from the front of the car will the cg be if two people sit in the front seat 2.60 m from the front of the car and three in the back seat 3.85 m from the front? Assume that each person has a mass of 65 kg.

66 (I) From the data in Table 3-2, determine the height above the ground of the cg of the entire body for a 175-cm-tall person.

67 (II) Calculate how far below the torso's median line the cg will be when a gymnast's arms and legs are hanging vertically and the trunk and head are horizontal. Will this be outside the body?

68 (II) Use Table 3-2 to calculate the position of the cg of an arm bent at a right angle. Assume that the person is 170-cm tall.

69 (II) You are standing on a train accelerating at 0.50g. At what angle must your body lean to avoid falling? (*Hint*: The torque about your cg must be zero.)

70 (III) A square uniform raft, 30 m by 30 m of mass 7400 kg, is used as a ferryboat carrying cars northward across a river. (*a*) If three cars, each of mass 1000 kg, occupy the NE, SE, and SW corners, determine the cg of the loaded ferryboat. (*b*) If the car in the SW corner accelerates northward at 1.80 m/s^2 relative to the raft, where will the cg be after 4.0 s?

71 (III) A projectile is fired at a 45° angle above level ground with an initial speed of 250 m/s. The projectile explodes into two equal pieces just as it reaches the highest point of its path. One fragment falls vertically to the ground. How far from the firing point does the second fragment land? (*Hint*: The cg of the projectile follows the normal projectile path.)

For additional reading

Andrade, E. N., *Sir Isaac Newton: His Life and Work,* Doubleday & Co., Inc., New York, 1958.

Bronowski, J., *The Ascent of Man,* Little, Brown & Company, Boston, 1974. Chapters 6 and 7.

Casper, B. M., and R. J. Noer, *Revolutions in Physics,* W. W. Norton & Co., Inc., New York, 1972.

Hay, J. G., *The Biomechanics of Sports Techniques,* Prentice-Hall, Inc., Englewood Cliffs, N.J., 1973.

Magie, W. F., *Source Book in Physics,* Harvard University Press, Cambridge, Mass., 1963. Reprint of Cavendish's paper on measurement of *G.*

Newton, I., *Mathematical Principles of Natural Philosophy . . . (Principia),* tr. by A. Motte, ed. by F. Cajori, University of California Press, Berkeley, 1947.

Wilson, C., "How Did Kepler Discover His First Two Laws?" *Sci. American,* **226**:3 (Mar., 1972), p. 92.

4 ▌Bodies in equilibrium

4-1 ▌Statics: The study of forces in equilibrium

In this chapter we will study a special case of motion—when the acceleration of the object is zero. We will mainly be concerned with objects at rest, although the same considerations apply to a body moving at constant velocity. Now you may think that the study of objects at rest is not very interesting since the body has neither velocity nor acceleration; and from Newton's second law the net force on such an object is zero. But it is just this last aspect that makes this subject of *statics* interesting. True, the *net* force on an object at rest is zero, but this does not imply that no forces act on it. In fact it is virtually impossible to find a body on which no forces act at all. Objects within our experience have at least one force acting on them (gravity) and if they are at rest then there must be other forces acting on them as well. Figure 4-1 shows a commonplace situation we have already met, an object at rest on a table. Its weight, the force of gravity, acts downward on it; and the table exerts a force upward on it which, since the net force is zero, must be equal to the force of gravity acting downward. (Do not confuse this with the equal and opposite forces of Newton's third law which act on different bodies; here both forces act on the same body). Such a body is said to be in **equilibrium** (Latin for "equal forces") under the action of these two forces. A body in equilibrium at rest is said to be in *static equilibrium* whereas a body moving uniformly at constant velocity is in *dynamic* equilibrium; we will consider mainly the former.

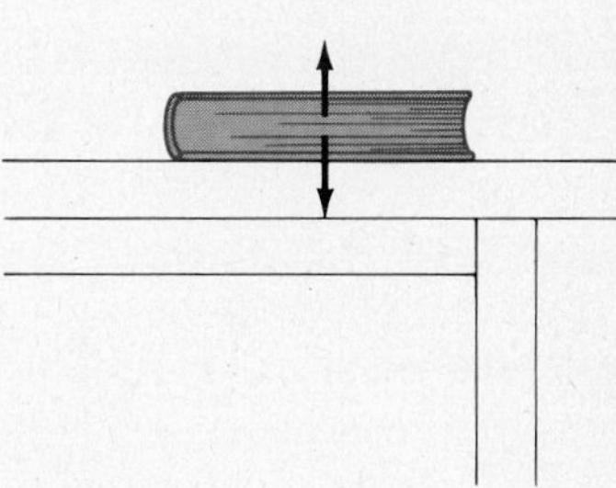

FIGURE 4-1
The book is in equilibrium; the net force on it is zero.

The subject of statics is concerned with the calculation of the forces acting on bodies which are in equilibrium. The techniques for doing this find application in a wide range of fields. A knowledge of the forces in muscles and joints is of great value in medicine and physical therapy, and is also valuable in the study of athletic activity. Architects and engineers must be able to calculate the forces on the structural members of a building, for any material will break or buckle if too much force is applied.

Let us take a simple example of the addition of forces applied to orthodonture.

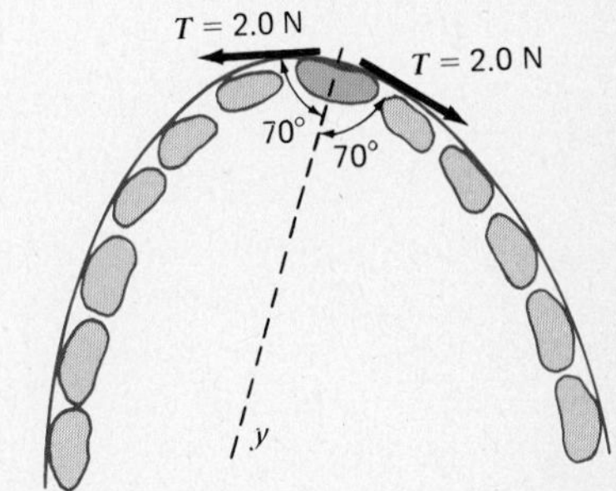

FIGURE 4-2
Forces on a tooth, Example 4-1.

▌EXAMPLE 4-1 The wire band shown in Figure 4-2 has a tension T of 2.0 N along it. It therefore exerts a force of 2.0 N on the tooth in the two directions shown. Calculate the resultant force on the tooth due to the wire.

SOLUTION Since the two forces are equal, their sum will be directed along the line that bisects the angle between them, which we have labeled the y axis. The x components of the two forces add up to zero. The y component of each force is $(2.0\text{ N})(\cos 70°) = 0.68\text{ N}$ and adding the two together we get a total force of 1.36 N. Note that if the wire is firmly attached to the tooth, the tension to the right, say, can be made larger than to the left and the net force would correspondingly be directed more toward the right. ▌▌

Although it is intended that the tooth in this example should move, it moves very slowly and is essentially in equilibrium. The net force on it is therefore zero. The gums, then, must exert a force of 1.36 N on the tooth in a direction opposite to that exerted by the wire. In a similar way the force exerted on the leg by the traction apparatus shown in Figure 4-3 can be calculated to be 320 N. Notice that if the pulleys are essentially frictionless, there is a tension of $20\text{ kg }(9.8\text{ m/s}^2) = 200\text{ N}$ all along the cord; thus there are two 200-N forces acting at 37° angles on the central pulley and on the leg and the resultant is 320 N acting to the right. Now the leg is in equilibrium so there must be another 320-N force acting on the leg to keep it at rest. What exerts this force?

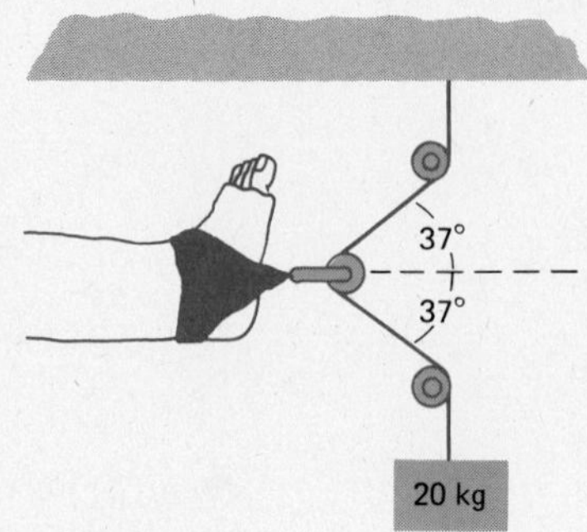

FIGURE 4-3
Traction apparatus exerts force on a leg.

4-2 ▌ The conditions for equilibrium

For a body to be at rest, the sum of the forces acting on it must add up to zero. Since force is a vector, the components of the net force must each be zero. Hence, if the forces on the object act in a plane, a condition for equilibrium is that

$$\sum F_x = 0, \qquad \sum F_y = 0. \tag{4-1}$$

If the forces act in three dimensions (a situation we will not consider) then the third component must be zero as well, $\sum F_z = 0$. We must remember that if a particular force component points along the negative x or y axis, it must have a negative sign. Equation 4-1 is called the *first condition* for equilibrium.

▌ EXAMPLE 4-2 Calculate the tension in the two cords used to support the 200-kg chandelier shown in Figure 4-4.

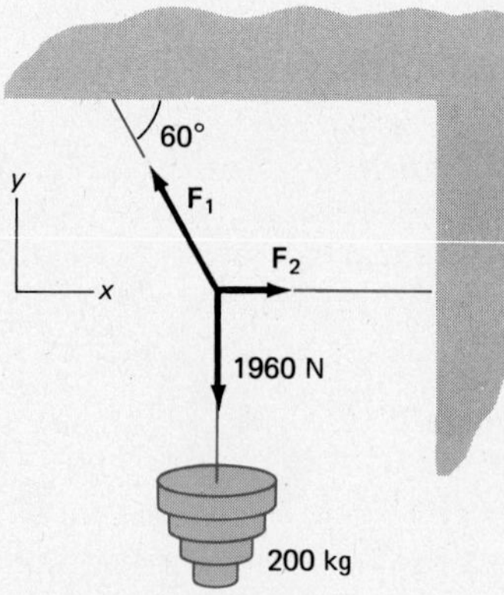

SOLUTION The three forces, $\mathbf{F}_1$, $\mathbf{F}_2$, and the weight of the 200-kg chandelier, act at a common point. Their vector sum must be zero. We first resolve $\mathbf{F}_1$ into its horizontal (x) and vertical (y) components. F_1 and F_2 are the unknowns and we can write $F_{1x} = F_1 \cos 60°$ and $F_{1y} = F_1 \sin 60°$. $\mathbf{F}_2$ has only an x component. In the vertical direction we have only the weight of the chandelier $= (200\text{ kg})(g)$ acting downward and the vertical component of $\mathbf{F}_1$ upwards. Since $\sum F_y = 0$, we have:

$$\sum F_y = F_1 \sin 60° - (200\text{ kg})(g) = 0$$

so

$$F_1 = \frac{(200\text{ kg})(g)}{\sin 60°} = (230\text{ kg})(g) = 2260\text{ N}.$$

FIGURE 4-4
Example 4-2.

In the horizontal direction,

$$\sum F_x = F_2 - F_1 \cos 60° = 0.$$

Thus

$$F_2 = F_1 \cos 60° = (230\text{ kg})(g)(0.500) = (115\text{ kg})(g) = 1130\text{ N}.$$

The magnitudes of $\mathbf{F}_1$ and $\mathbf{F}_2$ determine the strength of cord or wire that must be used. In this case, the wire must be able to hold at least 230 kg. ▌▌

Note in this example that we didn't insert the value of g, the acceleration due to gravity, until the end; in this way we found the magnitude of the force in terms of the weight of an equivalent number of kilograms (which may be a more familiar quantity than newtons).

Although Equation 4-1 must be true if an object is to be in equilibrium, it is not a sufficient condition. Figure 4-5 shows an object on which the net force is zero. Although the two forces† labeled **F** add up to give zero net force on the object, they do give rise to a torque which will rotate the object. Referring to Equation 3-15, we see that if a body is to remain at rest, the net torque applied to it must be zero. Thus we have the *second condition* for equilibrium: that the sum of the torques acting on a body must be zero:

$$\sum \tau = 0. \tag{4-2}$$

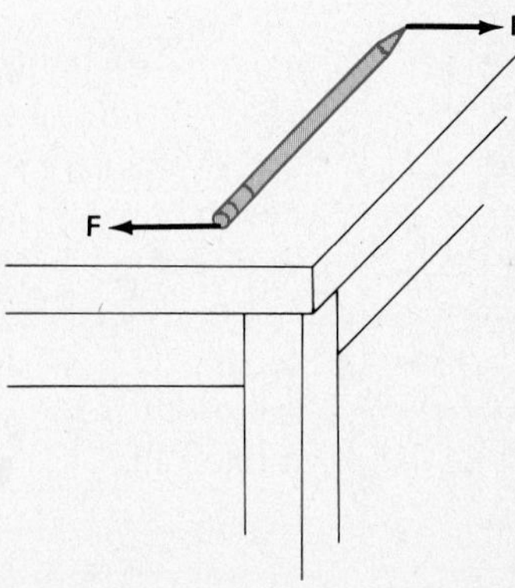

FIGURE 4-5
Although the net force on it is zero, the pencil will move (rotate).

Equations 4-1 and 4-2 are all that is required for a body to be in equilibrium.

If we know some of the forces acting on an object in equilibrium, we can use Equations 4-1 and 4-2 to calculate the other forces acting. There is no single technique for attacking such statics problems, but the following procedure is helpful. (1) Choose one body at a time for consideration, and show on a diagram all the forces acting on the body and the points at which these forces act (this is called a "free body diagram"). (2) Choose a convenient coordinate system and resolve the forces into their components using sines and cosines. (3) Using letters to represent unknowns, write down equations for $\sum F_x = 0$, $\sum F_y = 0$, and $\sum \tau = 0$, and then solve these equations.‡ Note that if there are three equations, you can solve for no more than three unknowns. Undoubtedly the hardest step is (1): *all* the forces *on* the body must be included, but the forces exerted *by* this body on other objects must *not* be included.

It is useful to refer back to Example 4-2 to see how this procedure was followed. Notice that we did not bother to consider the forces on the chandelier (which are simply gravity downward and the rope pulling upward with an equal force). Since we wanted to know the tension in the two cords, we chose the junction point (or knot) of the cords as our object. Since the object is essentially a point, there would be no torques, so the torque equation was not used. Of course, one does not have to use all three equations if they are not needed.

† A pair of equal forces acting in opposite directions but at different points on a body as in Figure 4-5 is referred to as a *couple*.

‡ If the forces are not in a plane, the situation is more complicated; there will be three force equations plus three torque equations (torque calculated about three mutually perpendicular axes). We won't usually be concerned with such complicated situations.

In many situations we must make use of the torque equation 4-2. Just as the components of a force are positive or negative depending on their direction, so too are torques. If a torque that tends to rotate the object counterclockwise is considered positive then a torque that tends to rotate it clockwise must be considered negative.

Another way of stating the equillibrium conditions is that the sum of all clockwise torques is equal to the sum of all counterclockwise torques; and for the forces: the sum of the upward forces is equal to the sum of the downward forces, and the sum of horizontal forces to the left is equal to the sum of horizontal forces to the right.

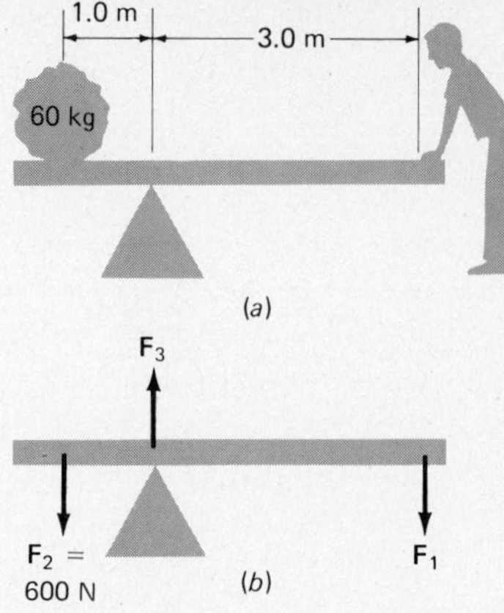

FIGURE 4-6
Example 4-3.

▌ EXAMPLE 4-3 How much force must the person exert to balance the 60-kg stone shown in Figure 4-6*a*? Ignore the weight of the lever.

SOLUTION We will consider the forces acting on the lever and these are shown in Figure 4-6*b*. Besides the (60 kg)(g) force exerted by the stone and the force $\mathbf{F}_1$ exerted by the man, the pivot support exerts an unknown force $\mathbf{F}_3$ upwards. Since all the forces are vertical there is no $\sum F_x = 0$ equation. The $\sum F_y = 0$ equation gives us

$$F_3 - F_1 - (60\text{ kg})(g) = 0.$$

For the torques, we notice that the stone gives rise to a counterclockwise torque about the pivot equal to $(1.0\text{ m}) \times (60\text{ kg})(g)$. The push exerted by the man gives rise to a clockwise torque of magnitude $(3.0\text{ m}) \times (F_1)$. Thus $\sum \tau = 0$ yields

$$(1.0\text{ m})(60\text{ kg})(g) - (3.0\text{ m})F_1 = 0.$$

From this equation $F_1 = (20\text{ kg})(g)$ or 200 N. (We rounded off g to 10 m/s^2). To get our desired answer we only needed the torque equation. However, we can use the other equation to find that $F_3 = (80\text{ kg})(g) =$ 800 N. ▌▌

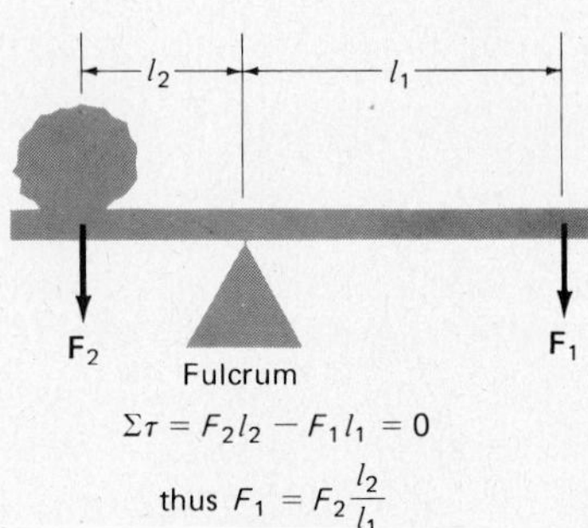

FIGURE 4-7
Law of the lever.

This example illustrates the well-known law of the lever: if l_2 is the distance of the load F_2 from the pivot point (or *fulcrum*) and l_1 is the distance to the applied force F_1, then† $F_1 = (l_2/l_1)F_2$. See Figure 4-7. The ratio $(F_2/F_1) = (l_1/l_2)$ is called the *mechanical advantage* of the lever. The greater this ratio, the greater load a given force can lift.

We next consider a common problem in architecture and engineering: calculating the forces on a beam. The beam could be supporting a floor or a roof. Such examples are also useful in the life sciences for the limbs of an animal resemble a beam.

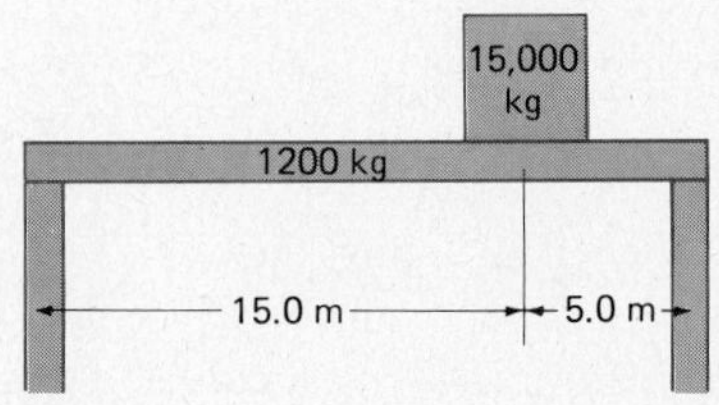

FIGURE 4-8
Beam supporting a 15,000-kg load. Example 4-4.

Figure 4-8 shows a 1200-kg beam which supports 15,000 kg of machinery as shown. We want to calculate the force on each of the supports. We will have to make use of the torque equation and the question is, from which point do we measure the distance for calculating torque? The answer is that we can use whatever point we like! The reason for this is that we can consider any point of a body as a potential axis of rotation; that is, if we drilled a tiny hole and inserted a tiny axle, nothing would be changed, and since the body is in equilibrium the net torque calculated about the arbitrarily chosen axis must be zero. Since it doesn't

† This relation is not quite true if we take into account the weight of the lever itself—see the problems at the end of the chapter. We will discuss levers more fully in Chapter 5.

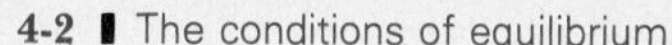

matter which point we choose as the axis for writing the torque equation, we can choose one that is convenient. This is a general rule that can be used for any problem in statics.

▮ EXAMPLE 4-4 Calculate the force on each of the supports in Figure 4-8.

SOLUTION We analyze the forces on the beam, since the force it exerts on the supports is equal and opposite to the forces exerted by the supports on the beam. We call the latter $\mathbf{F}_1$ and $\mathbf{F}_2$ in Figure 4-9. The weight of the beam itself acts at its center of gravity, 10 m from either end. If we calculate the torques about the point of application of $\mathbf{F}_1$, then $\mathbf{F}_1$ will not enter (its lever arm will be zero) and we will have an equation in only one unknown, F_2. Thus $\sum \tau = 0$ gives

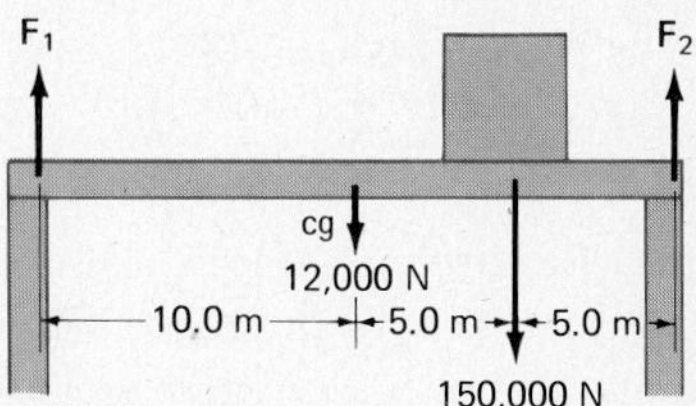

FIGURE 4-9
Forces acting on the beam; Example 4-4 shows $F_1 = 42{,}000$ N and $F_2 = 120{,}000$ N.

$$-(10\text{ m})(1200\text{ kg})(g) - (15\text{ m})(15{,}000\text{ kg})(g) + (20\text{ m})F_2 = 0.$$

Solving for F_2 we find $F_2 = (12{,}000\text{ kg})(g)$ or, with g rounded off to 10 m/s^2, $F_2 = 120{,}000$ N. To find F_1 we use $\sum F_y = 0$:

$$\sum F_y = F_1 - (1200\text{ kg})(g) - (15{,}000\text{ kg})(g) + F_2 = 0.$$

Putting in $F_2 = (12{,}000\text{ kg})(g)$ we find $F_1 = (4200\text{ kg})(g)$ or 42,000 N. ▮▮

Figure 4-10 shows a beam which extends beyond its support. Such a beam is called a *cantilever*. The forces acting on the beam in this figure are those due to the supports, $\mathbf{F}_1$ and $\mathbf{F}_2$, and the force of gravity which acts at the cg, 5.0 m to the right of the right-hand support. If you follow the procedure of the last example to calculate F_1 and F_2 assuming they point upward as shown in the figure (see the problems) you will find that F_1 comes out negative; if the beam has a mass of 1200 kg, then $\mathbf{F}_2 = 15{,}000$ N and $\mathbf{F}_1 = -3000$ N. Whenever an unknown force comes out negative, it merely means that the force actually points in the opposite direction from what you assumed. Thus in Figure 4-10 F_1 actually points downwards. With a little reflection it should become obvious that the left-hand support must indeed pull downward on the beam if it is to be in equilibrium.

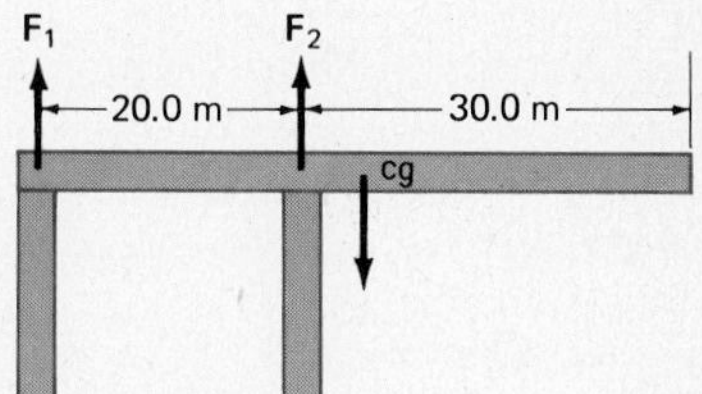

FIGURE 4-10
A cantilever.

As another example, we calculate the stability of a building under the action of wind. It is obviously an important consideration in the design of high buildings to determine if the wind will knock them over. In Figure 4-11, the forces on the building are shown. The weight $\mathbf{w}$ acts downward at the cg (assumed to be at the center of the building). $\mathbf{F}_a$ represents the force of the air (wind), which, if we assume it acts uniformly over the whole face of the building, can be considered as a single force acting at the center of that face of the building. Finally, $\mathbf{F}_E$ represents the force of the earth on the building which counterbalances the first two forces to keep the building in equilibrium. Normally the upward and sideways force of the earth ($\mathbf{F}_E$) acts all along the base of the building. But we are interested in the worst possible situation, when the building has started to tip slightly and $\mathbf{F}_E$ acts at the last point of support, the corner of the building. The components of $\mathbf{F}_E$ can be calculated from the force equations, but what we are really interested in is whether the building will reach this rather dire situation of starting to tip over. To determine this, we must calculate torques about the potential pivot point.

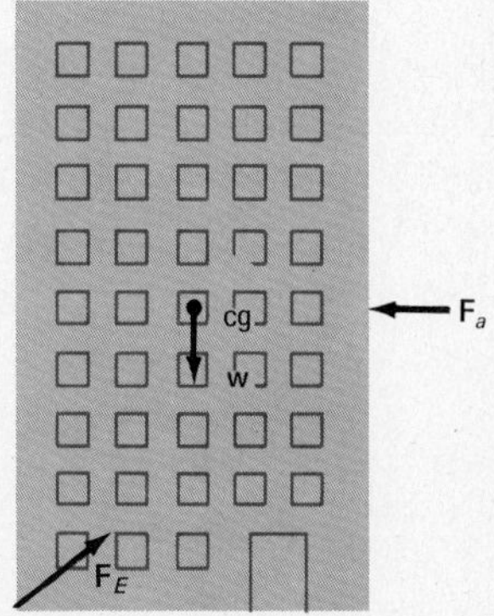

FIGURE 4-11
Forces on a building subjected to wind ($\mathbf{F}_a$) and gravity ($\mathbf{w}$); $\mathbf{F}_E$ is the force on the building due to the earth in the situation when the building is just about to tip.

▌ EXAMPLE 4-5 A 50-story building is being planned. It is to be 200 m high with a base 40 m by 70 m. Its total mass will be about 1.5×10^7 kg and its weight therefore about 1.5×10^8 N. Will this building tip over in a 200-km/h wind† which exerts a force of 1450 N/m² over the 70-m-wide face?

SOLUTION To get F_a we must multiply the force per area by the total area: $F_a = (1450\ \text{N/m}^2)(70\ \text{m})(200\ \text{m}) = 2.0 \times 10^7$ N. We calculate the torque about the potential pivot point, the lower rear edge of the building (where F_E acts in Figure 4-11). The wind, as we saw above, can be considered to act at the midpoint of the face, which is 100 m above the ground. The torque it produces, tending to overturn the building, is equal to $(100\ \text{m})(2.0 \times 10^7\ \text{N}) = 2.0 \times 10^9\ \text{N} \cdot \text{m}$. The torque tending to keep the building upright is the force of gravity acting through a lever arm equal to half the width of the building, or 20 m; this torque equals $(20\ \text{m})(1.5 \times 10^8\ \text{N}) = 3.0 \times 10^9\ \text{N} \cdot \text{m}$. The building will not tip over, but it is close; a safety factor of two or more would be better. ▌▌

*4-3 ▌ Applications to muscles and joints

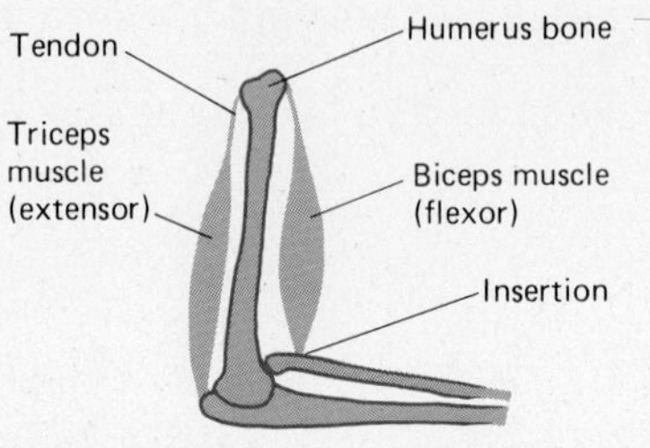

FIGURE 4-12
Diagram showing the biceps (flexor) and triceps (extensor) muscles in human arm.

The techniques for calculating forces on bodies in equilibrium can readily be applied to the human (or animal) body. This is of great use in studying the forces on muscles, bones, and joints for organisms in motion or at rest. When studying the physics of limb movement, one becomes aware of the wonder of the human body and all the muscles that allow the different forms and directions of movement which we take for granted.

Generally a muscle is attached, via tendons, to two different bones (see Figure 4-12). The points of attachment are called *insertions.* The two bones are flexibly connected at a *joint,* such as those at the elbow, knee, and ankle. A muscle exerts a pull when its fibers contract under stimulation by a nerve, but it cannot exert a push. Thus there are two main kinds of muscles associated with limb motion: *flexor* muscles that bring the two limbs closer together such as the biceps muscle in the upper arm; and *extensor* muscles that are used to extend the limb outward such as the triceps muscles, Figure 4-12. The flexor muscle in the upper arm is used when lifting an object in your hand; the extensor muscle is used when throwing a baseball.

▌ EXAMPLE 4-6 How much force must the biceps muscle exert when a 5.0-kg mass is held in the hand with the arm outstretched as in Figure 4-13*a*, and when the arm is at a 45° angle as in Figure 4-13*b*? Assume the mass of forearm and hand together is 2.0 kg and their cg is as shown.

SOLUTION (*a*) The forces acting on the forearm are shown in part (*a*) of the figure and include the upward force F_M exerted by the muscle and a force F_J exerted at the joint by the bone in the upper arm. We wish to find F_M, which is done most easily by using the torque equation, calculated about the joint so that F_J does not enter:

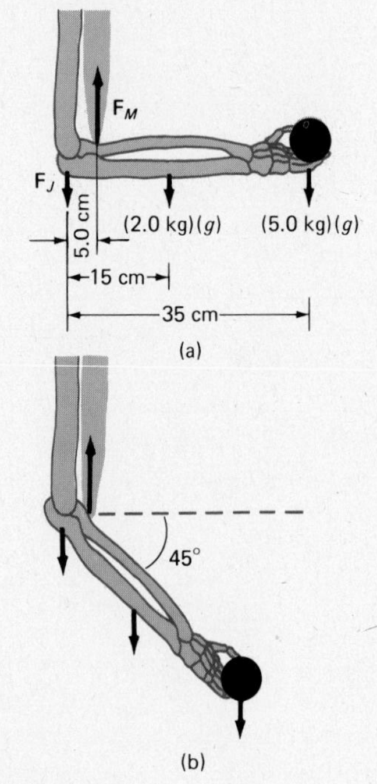

FIGURE 4-13
Example 4-6.

† The force (in N) exerted per square meter on a flat surface by a wind of speed v (in km/h) is given roughly by $0.035\,v^2$, assuming the wind blows perpendicular to the surface.

$$(0.050\text{ m})(F_M)-(0.15\text{ m})(2.0\text{ kg})(g)-(0.35\text{ m})(5.0\text{ kg})(g)=0.$$

We solve this for F_M and find $F_M=(41\text{ kg})(g)=400$ N. (*b*) The lever arm for the muscle is reduced to (0.050 m) sin 45° = 0.035 m. But the lever arms for the downward forces are reduced by the same ratio; so the same result will be obtained, $F_M=400$ N. ▮▮

Note in the example that the force required of the muscle is quite large compared to the weight of the object lifted. Indeed, the muscles and joints of the body are generally subjected to quite large forces.

The point of insertion of a muscle varies from person to person. A slight increase in the point of insertion of the biceps muscle from 5.0 cm to 5.5 cm can be a considerable advantage for lifting and throwing. Indeed, athletes are often found to have muscle insertions farther from the joint than the average person, and if this applies to one muscle it usually applies to all.

As another example of the large forces acting within the human body we consider the muscles used to support the trunk when bending forward, Figure 4-14*a*. The lowest vertebra on the spinal column (fifth lumbar vertebra), acts as a fulcrum for this bending position. The "erector spinae" muscles that support the trunk act at an effective angle of about 12° to the axis of the spine. Figure 4.14*b* is a simplified schematic drawing showing the forces on the upper body; we assume the trunk makes an angle of 30° with the horizontal. The force exerted by the muscles is represented by $\mathbf{F}_M$ and the force exerted on the base of the spine at the lowest vertebra is $\mathbf{F}_V$. $\mathbf{w}_1$, $\mathbf{w}_2$, $\mathbf{w}_3$ represent the weights of the head, freely hanging arms, and trunk respectively. The values shown are approximations taken from Table 3-2; the distances (in cm) refer to a person 180 cm tall, but are approximately in the same ratio of 1 : 2 : 4 for an average person of any height and the answer below is then independent of the height of the person.

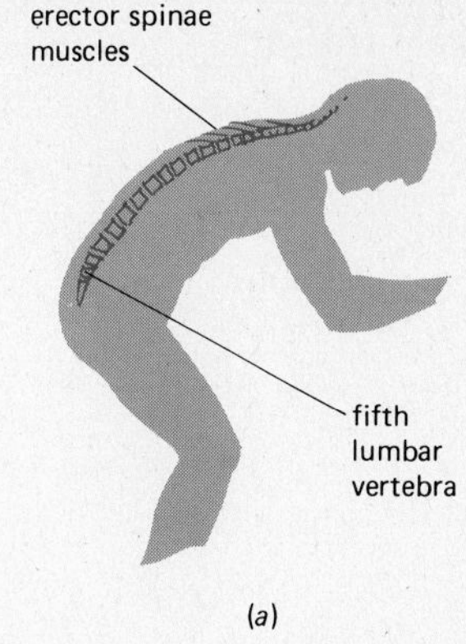

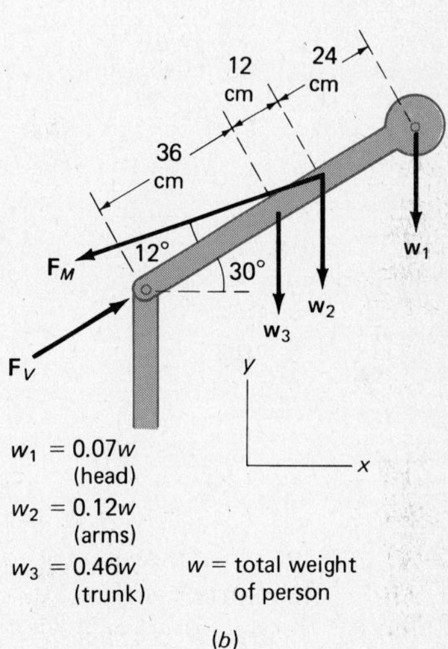

FIGURE 4-14
(*a*) Person bending over. (*b*) Forces required of back muscles ($\mathbf{F}_M$) and in the vertebrae ($\mathbf{F}_V$) when a person bends over.

▮ EXAMPLE 4-7 Calculate the magnitude and direction of the force $\mathbf{F}_V$ acting on the fifth lumbar vertebra for the example shown in Figure 4-14*b*.

SOLUTION First we calculate F_M using the torque equation about the base of the spine:

$$(0.48\text{ m})(\sin 12°)(F_M)-(0.72\text{ m})(\cos 30°)(w_1)$$
$$-(0.48\text{ m})(\cos 30°)(w_2)-(0.36\text{ m})(\cos 30°)(w_3)=0.$$

Putting in the values for w_1, w_2, w_3 given in the figure, we find $F_M=2.2w$, where w is the total weight of the body. To get the components of F_V we use the x and y components of the force equation:

$$\sum F_y=F_{Vy}-F_M\sin 18°-w_1-w_2-w_3=0$$
$$F_{Vy}=1.3w.$$
$$\sum F_x=F_{Vx}-F_M\cos 18°=0$$
$$F_{Vx}=2.1w.$$

Then $F_V=\sqrt{F_{Vx}^2+F_{Vy}^2}=2.5w$. The angle θ that F_V makes with the horizontal is given by $\tan\theta=F_{Vy}/F_{Vx}=0.62$, so $\theta=32°$. ▮▮

The force on the lowest vertebra is thus $2\frac{1}{2}$ times the body weight! This force is transmitted from the "sacral" bone at the base of the spine, through the fluid-filled and somewhat flexible *intervertebral disc*. The discs at the base of the spine are being compressed under very large forces.

If the person in Figure 4-14 has a mass of 90 kg, and is holding 20 kg in his hands (this increases w_2 to $0.33w$) then F_V is increased to nearly five times the person's weight ($5w$)! (For this 200-lb person, the force on the disc would be 1000 lb!) With such strong forces acting, it is little wonder that so many people suffer from lower back pain at one time or another in their lives.

When gravity acts on the upper body through a long lever arm, it necessitates large forces when bending forward. To reduce the stress, ordinary bending and lifting tasks are best done from a squatting position, or in any case by reducing the angle of bend which reduces the lever arm.

4-4 ▌ Stability and balance

A body in static equilibrium, if left undisturbed, will undergo no translational or rotational acceleration since the sum of all the forces and the sum of all the torques acting on it are zero. However, if the object is displaced slightly, three different outcomes are possible: (1) the object may return to its original position, in which case it is said to be in **stable equilibrium**; (2) the object will move even further from its original position, in which case it is said to be in **unstable equilibrium**; or (3) it remains in its new position, in which case it is said to have **neutral equilibrium**. Consider the following examples. A ball suspended freely from a string is in stable equilibrium for if it is displaced to one side it will quickly return to its original position (Figure 4-15*a*). On the other hand, a pencil standing on its point is in unstable equilibrium. If its cg is directly over its tip (Figure 4-15*b*), the net force and net torque on it will be zero. But if it is displaced ever so slightly—say by a slight vibration or tiny air current—there will be a torque on it and it will continue to fall in the direction of the original displacement. Finally, an example of an object in neutral equilibrium is a sphere resting on a horizontal table top. If it is placed slightly to one side, it will remain in its new position.

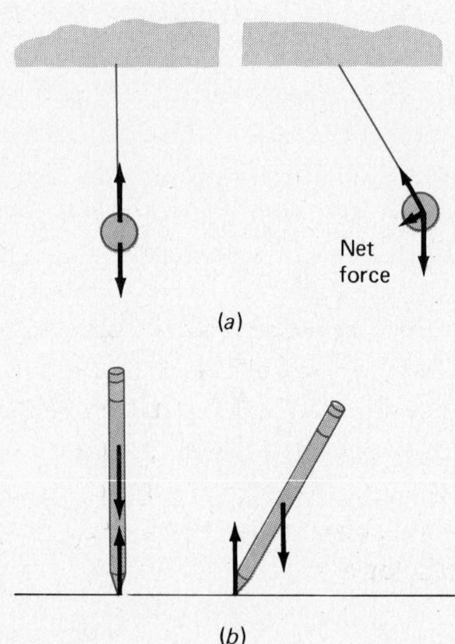

FIGURE 4-15

(*a*) Stable equilibrium; (*b*) unstable equilibrium.

In most situations, such as in the design of structures and in working with the human body, we are interested in maintaining stable equilibrium or *balance*, as we sometimes say. In general, an object whose cg is below its point of support, such as a ball on a string, will be in stable equilibrium. If the cg is above the base of support we have a more complicated situation. Consider a block standing on its end (Figure 4-16*a*). If it is tipped slightly, it will return to its original position due to the torque on it as shown in Figure 4-16*b*. But if it is tipped too far, Figure 4-16*c*, it will fall over. The critical point is reached when the cg is no longer above the base of support. In general, *a body whose cg is above its base of support will be stable if a vertical line projected downward from its cg falls within the base of support.* This is a result of the fact that the upward force on the object (which balances out gravity) can only be exerted within the area of contact, so that if the force of gravity acts beyond this area, a net torque will act to topple the

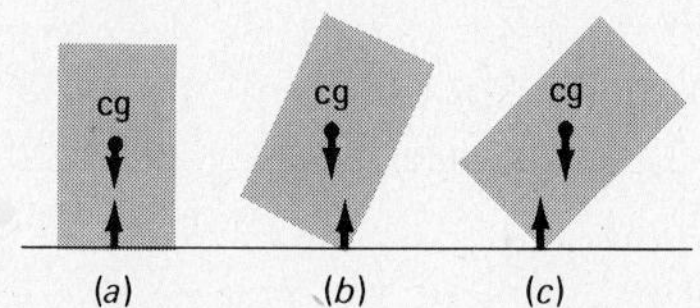

FIGURE 4-16

Equilibrium of a block resting on a surface.

object. Stability in this case is relative. A brick lying on its widest face is more stable than a brick standing on its end, for it will take more of an effort to tip it over. In the extreme case of the pencil in Figure 4-15*b*, the base is practically a point and the slightest disturbance will topple it. In general, the larger the base and the lower the cg, the more stable the object.

In this sense, humans are much less stable than four-legged mammals, which not only have a much larger base of support because of their four legs, but have a lower cg. The human species has had to develop special apparatus, such as certain very strong muscles, in order to deal with the problem of keeping a person upright in a not terribly stable position. Because of this upright position, humans suffer from numerous ailments such as low back pain due to the large forces involved as we saw in Example 4-7. When walking and performing other kinds of movement a person continually shifts the body so its cg is over the feet, although in the ordinary adult this requires no conscious thought. Even as simple a movement as bending over requires moving the hips backward so that the cg remains over the feet, and this repositioning is done without thinking about it. To illustrate this, position yourself with your heels and back to a wall and try to touch your toes. Persons carrying heavy loads automatically adjust their posture so that the cg of the total mass is over their feet, as shown in Figure 4-17. However, to people who suffer from muscle weakness in the trunk or legs, or who have suffered damage to the spinal cord, even walking can be difficult. A conscious effort is required to balance the body over the feet and the slightest miscalculation can lead to a fall.

FIGURE 4-17

Humans adjust their posture when carrying loads to achieve stability.

4-5 ▮ Elasticity: stress, strain, and fracture

In the first part of this chapter we studied how to calculate the forces on objects in equilibrium. In this section we study the effects of these forces, for any object changes shape under the action of applied forces; and if the forces are great enough, the object will break or *fracture*. To know how much force it can take before it breaks is of obvious importance in a variety of situations from designing a building to the action of forces on human bones.

Hooke's law

If a force is exerted on an object, such as the suspended metal bar shown in Figure 4-18, the object stretches. If the amount of elongation, ΔL, is small compared to the length of the object, experiment shows that ΔL is proportional to the weight or force exerted on the object; this was first noted by Robert Hooke (1635–1703). This proportionality can be written as an equation and is known as **Hooke's law**:

$$F = k\,\Delta L \tag{4-3}$$

where F represents the force (or weight) pulling on the object, ΔL is the increase in length, and k is a proportionality constant. Hooke's law is found to be valid for almost any solid material from iron to bone, but it is

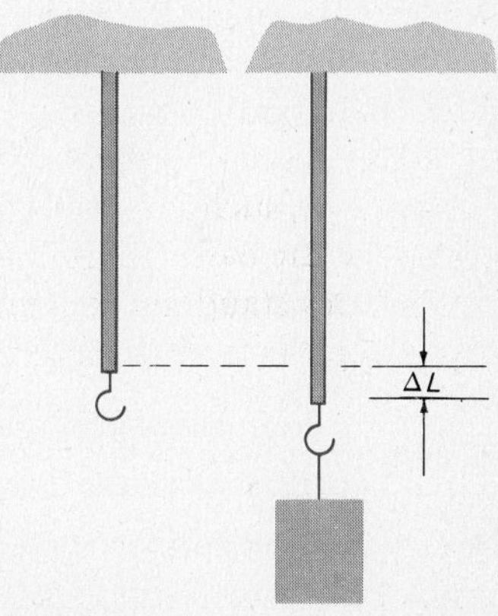

FIGURE 4-18

Hooke's Law: $\Delta L \propto$ applied force.

valid only up to a point. For if the force is too great, the object stretches excessively and eventually breaks. Figure 4-19 shows a typical graph of elongation versus applied force. Up to a point called the *elastic limit*, the object will return to its original length if the applied force is removed. This is called the *elastic region*. If the object is stretched beyond the elastic limit, it will be permanently deformed. For most common materials Hooke's law is a good approximation almost up to the elastic limit and the graph is a straight line. Beyond this point, the graph deviates from a straight line and no simple relationship exists between F and ΔL. If the object is stretched much beyond the elastic limit, it will break. The maximum force that can be applied without breaking is called the *ultimate strength* of the material.

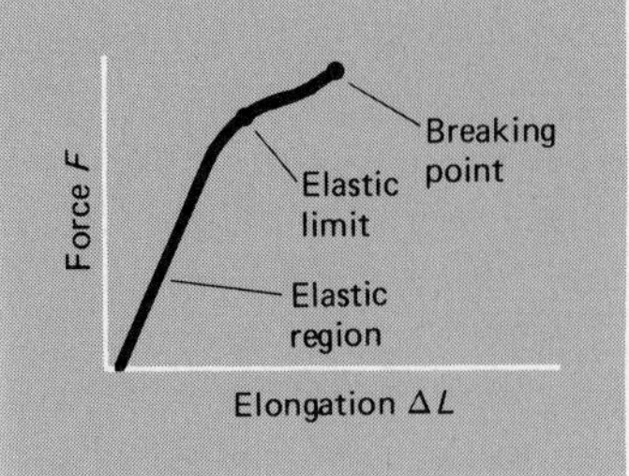

FIGURE 4-19
Applied force vs. elongation for a typical solid.

Young's modulus: stress and strain

The amount of elongation of an object, such as the bar shown in Figure 4-18, depends not only on the force applied to it, but also on the material from which it is made and on its dimensions; that is, the constant k in Equation 4-3 can be written in terms of these factors. If we compare bars made of the same material but of different lengths and cross-sectional areas, it is found that for the same applied force the amount of stretch (again assumed small compared to the total length) is proportional to the original length and inversely proportional to the cross-sectional area. That is, the longer the object, the more it elongates for a given force; and the fatter it is, the less it elongates. These experimental facts can be combined with Hooke's law to yield the relation:

$$\Delta L = \frac{1}{E}\frac{F}{A}L_0. \tag{4-4}$$

L_0 is the original length of the object, A is the cross-sectional area, and ΔL is the change in length due to the applied force F; E is a constant of proportionality† known as the *elastic modulus*, or Young's modulus, and its value depends only on the material. The value of Young's modulus for various materials is given in Table 4-1. Because E is a property only of the material and is independent of the object size or shape, Equation 4-4 is far more useful for practical calculation than Hooke's law itself. From Equation 4-4 we see that the change in length of an object is directly proportional to the product of the object's length L_0 and the force per unit area F/A applied to it. It is general practice to define the force per unit area as the **stress**:

$$\text{stress} = \frac{\text{force}}{\text{area}} = \frac{F}{A}.$$

Also, the **strain** is defined to be the ratio of the change in length to the original length:

$$\text{strain} = \frac{\text{change in length}}{\text{original length}} = \frac{\Delta L}{L_0}.$$

† The fact that E is in the denominator so that $1/E$ is the actual proportionality constant is merely a convention.

Strain is thus the fractional change in length of the object and is a measure of how much the bar has been deformed. Equation 4-4 can be rewritten as

$$\frac{F}{A} = E\frac{\Delta L}{L_0} \tag{4-5}$$

or

$$E = \frac{F/A}{\Delta L/L_0} = \frac{\text{stress}}{\text{strain}}.$$

Thus we see that the strain is directly proportional to the stress.

▌ EXAMPLE 4-8 A 1.60-m long steel piano wire has a diameter of 0.20 cm. How great is the tension in the wire if it stretches 0.30 cm when tightened?

SOLUTION We solve for F in Equation 4-4 or 4-5 and note that the area $A = \pi r^2 = (3.14)(0.0010\text{ m})^2 = 3.1 \times 10^{-6}\text{ m}^2$. Then

$$F = E\frac{\Delta L}{L_0}A$$

$$= (2.0 \times 10^{11}\text{ N/m}^2)\left(\frac{0.0030\text{ m}}{1.60\text{ m}}\right)(3.1 \times 10^{-6}\text{ m}^2) = 1200\text{ N}$$

where we obtained the value for E from Table 4-1. ▌▌

TABLE 4-1

Elastic moduli

Material	Elastic modulus E (N/m²)	Shear modulus G (N/m²)	Bulk modulus B (N/m²)
Solids			
Iron, cast	100×10^9	40×10^9	90×10^9
Steel	200×10^9	80×10^9	140×10^9
Brass	100×10^9	35×10^9	80×10^9
Aluminum	70×10^9	25×10^9	70×10^9
Concrete	20×10^9		
Brick	14×10^9		
Marble	50×10^9		70×10^9
Granite	45×10^9		45×10^9
Wood (pine)			
(parallel to grain)	10×10^9		
(perpendicular to grain)	1×10^9		
Nylon	5×10^9		
Bone (limb)	15×10^9	80×10^9	
Liquids			
Water			2.0×10^9
Alcohol (ethyl)			1.0×10^9
Mercury			2.5×10^9

Tension, compression and shear

The bar shown in Figure 4-18 is said to be under **tension** or **tensile stress**. For not only is there a force pulling down on the bar at its lower end, but since the bar is in equilibrium we know that the support is exerting an equal upward force† on the bar at its upper end (see Figure 4-20*a*). In fact, this tensile stress exists throughout the material. Consider for example the lower half of a suspended bar as shown in Figure 4-20*b*. This lower half is in equilibrium so there must be an upward force on it to balance the downward force at its lower end. What exerts this upward force? It must be the upper part of the bar. Thus we see that external forces applied to an object give rise to internal forces, or stress, within the material itself.

Strain or deformation due to tensile stress is but one type of stress to which materials can be subjected. There are two other common types of stress: compression and shear. **Compression** or **compressive stress** is the exact oppposite of tensile stress. Instead of being stretched, the material is compressed: the forces act inwardly on the body. Any sort of column that supports a weight, such as the columns of a Greek temple (Figure 4-21), is subjected to a compressive stress. Equations 4-4 and 4-5 apply equally well to compression and tension, and the values for E are usually the same.

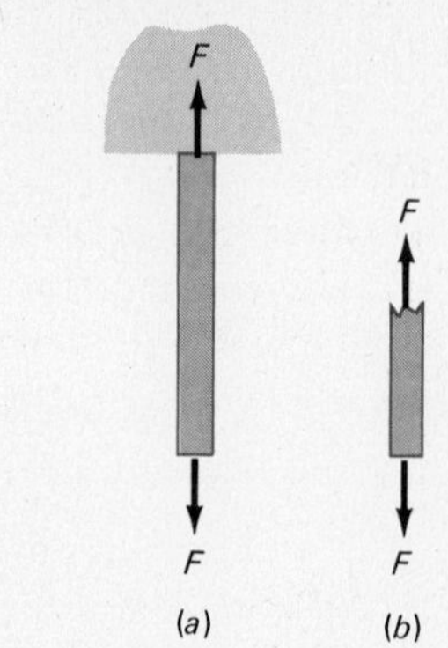

FIGURE 4-20
Stress exists *within* the material.

Figure 4-22 compares tensile and compressive stresses as well as the third type, shear stress. An object under **shear stress** has equal and opposite forces applied *across* its opposite faces. An example is a book or brick firmly attached to a table top with a force exerted parallel to the top surface; the table exerts an equal and opposite force along the bottom surface. Although the dimensions of the object do not change significantly, the shape of the object does change as shown in the figure. An equation similar to 4-4 can be applied to calculate shear strain:

$$\Delta L = \frac{1}{G}\frac{F}{A}L_0 \tag{4-6}$$

but ΔL, L_0, and A must be reinterpreted as indicated in Figure 4-22*c*. Note that A is the area of the surface *parallel* to the applied force (and not perpendicular as for tension and compression), and ΔL is *perpendicular* to

FIGURE 4-21
A Greek temple (this one is in Paestum, Italy) (D. C. Giancoli.)

† They are precisely equal only if we ignore the weight of the bar.

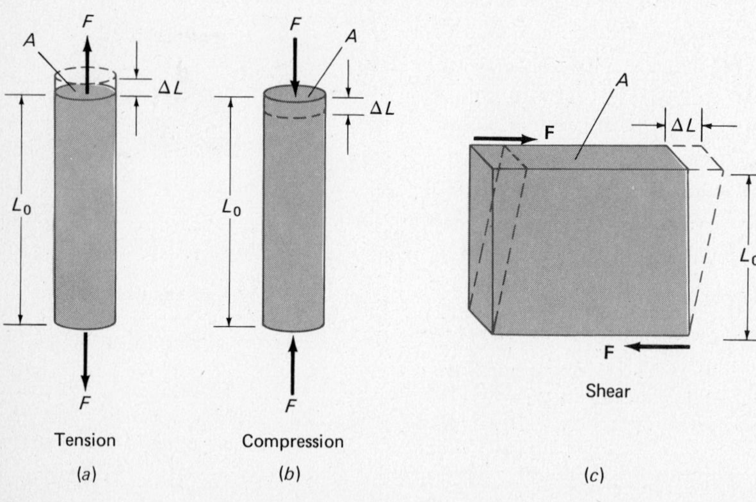

FIGURE 4-22
The three types of stress.

L_0. The constant of proportionality, G, is called the *shear modulus* and is generally one-half to one-third the value of the elastic modulus, E (see Table 4-1).

The rectangular object undergoing shear in Figure 4-22c is not actually in equilibrium under the forces shown, for a net torque exists. If the object is in fact in equilibrium, there must be two more forces acting on it which balance out this torque. One acts vertically upward on the right, and the other acts vertically downward on the left as shown in Figure 4-23. This is generally true of shear forces. If the object is a brick or book lying on a table, these two additional forces can be exerted by the table.

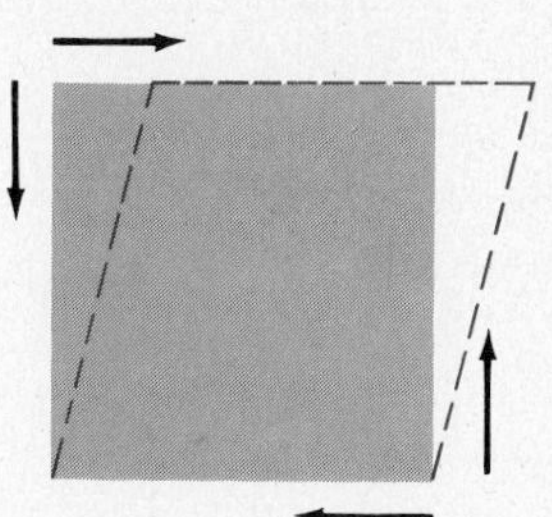

FIGURE 4-23
Balance of forces and torques for shear stress.

Bulk modulus

If an object is subjected to a pressure on all sides, its volume will be compressed. A common situation is a body submerged in a fluid; for in this case, the fluid exerts a pressure on the object in all directions, as we shall see in Chapter 6. Pressure is defined as force per area and thus is the equivalent of "stress." For this situation the change in volume ΔV is found to be proportional to the original volume V_0 and to the increase in the pressure ΔP. We thus obtain a relation of the same form as Equations 4-4 and 4-5 but with a proportionality constant called the *bulk modulus* B:

$$\frac{\Delta V}{V_0} = -\frac{1}{B}\Delta P \qquad (4\text{-}7)$$

or

$$B = -\frac{\Delta P}{\Delta V/V_0}.$$

The minus sign is included to indicate that the volume *decreases* with an increase in pressure. Values for the bulk modulus are given in Table 4-1. Since liquids and gases do not have a fixed shape, only the bulk modulus applies to them.

Fracture

If the stress on a solid object is too great, the object fractures or breaks, Figure 4-24. In Table 4-2 are listed the ultimate tensile strength, compressive strength, and shear strength for a variety of materials. These give the maximum force per unit area that an object can withstand under each of these three types of stress. They are, however, representative values only and the actual value for a given specimen can differ considerably. It is therefore necessary to maintain a safety factor of from three to perhaps ten or more—that is, the actual stresses on a structure should not exceed one-tenth to one-third of the values given in the table. You may encounter tables of the "allowable stresses" in which an appropriate safety factor has already been included.

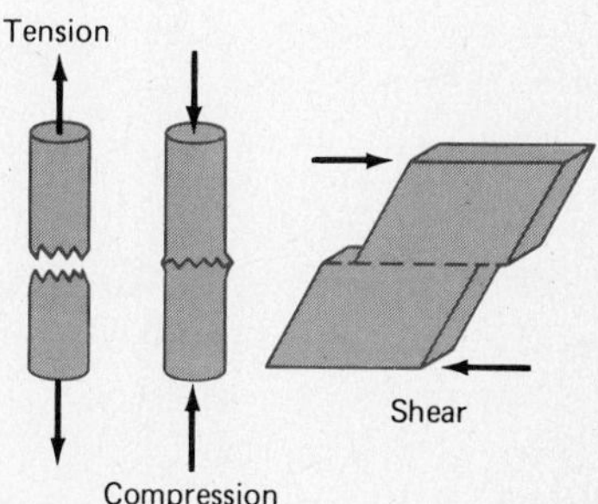

FIGURE 4-24
Fracture as a result of the three types of stress.

▮ EXAMPLE 4-9 (a) What minimum cross-sectional area should the supports have to support the beam of Example 4-4 (Figures 4-8 and 4-9), assuming the supports are made of concrete and a safety factor of 6 is required? (b) How much will the chosen supports compress under the given load?

TABLE 4-2

Ultimate strengths of materials (force/area)

Material	Tensile strength (N/m^2)	Compressive strength (N/m^2)	Shear strength (N/m^2)
Iron, cast	170×10^6	550×10^6	170×10^6
Steel	500×10^6	500×10^6	250×10^6
Brass	250×10^6	250×10^6	200×10^6
Aluminum	200×10^6	200×10^6	200×10^6
Concrete	2×10^6	20×10^6	2×10^6
Brick		35×10^6	
Marble		80×10^6	
Granite		170×10^6	
Wood (pine)			
(parallel to grain)	40×10^6	35×10^6	5×10^6
(perpendicular to grain)		10×10^6	
Nylon	500×10^6		
Bone (limb)	130×10^6	170×10^6	

SOLUTION (*a*) The right-hand support (see Example 4-4) receives the larger force, 1.2×10^5 N. It is clearly under compression and from Table 4-2 we see that the ultimate compressive strength of concrete is $2.0\times10^7\ \mathrm{N/m^2}$. Using a safety factor of six, the maximum allowable stress is $3.3\times10^6\ \mathrm{N/m^2}$. Since $F/A = 3.3\times10^6\ \mathrm{N/m^2}$ and $F = 1.2\times10^5$ N we can solve for A and find

$$A = \frac{1.2\times10^5\ \mathrm{N}}{3.3\times10^6\ \mathrm{N/m^2}} = 3.6\times10^{-2}\ \mathrm{m^2} \text{ or } 360\ \mathrm{cm.}^2$$

A support of 18 cm × 20 cm will be adequate.

(*b*) We solve for

$$\frac{\Delta L}{L_0} = \frac{1}{E}\frac{F}{A} = \left(\frac{1}{2.0\times10^{10}\ \mathrm{N/m^2}}\right)(3.3\times10^6\ \mathrm{N/m^2}) = 1.7\times10^{-4}.$$

Thus, if the support has a length $L_0 = 5.0$ m, $\Delta L = 0.85\times10^{-3}$ m, or about 1 mm. This calculation was for the right-hand support. If the left-hand support is made of the same cross-sectional area, it will compress less and this should be taken into account. ▌▌

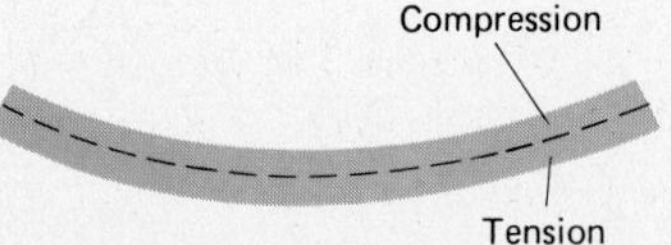

FIGURE 4-25
A beam sags, at least a little (but is exaggerated here), even under its own weight. The beam thus changes shape so that the upper portion is compressed, and the lower portion is elongated (under tension). Shearing stress also occurs within the beam.

*Prestressed concrete

As can be seen in Table 4-2, concrete (like stone and brick) is reasonably strong under compression but extremely weak under tension. Thus concrete can be used as vertical columns placed under compression but is of little value as a beam since it cannot withstand the tensile forces that arise—see Figure 4-25. Reinforced concrete, in which iron rods are embedded in the concrete, is much stronger in terms of stability; but the

FIGURE 4-26
Round arches in the Roman Forum. The one in the background is the Arch of Titus. (D. C. Giancoli.)

concrete on the lower edge of a loaded beam still tends to crack. This problem is solved with prestressed concrete which also contains iron rods or a wire mesh, but during the pouring of the concrete, the rods or wire are held under tension. After the concrete dries, the tension on the iron is released and this puts the concrete under compression. The amount of compressive stress is carefully predetermined so that when the design loads are applied to the beam, they reduce the compression on the lower edge but never put it into tension.

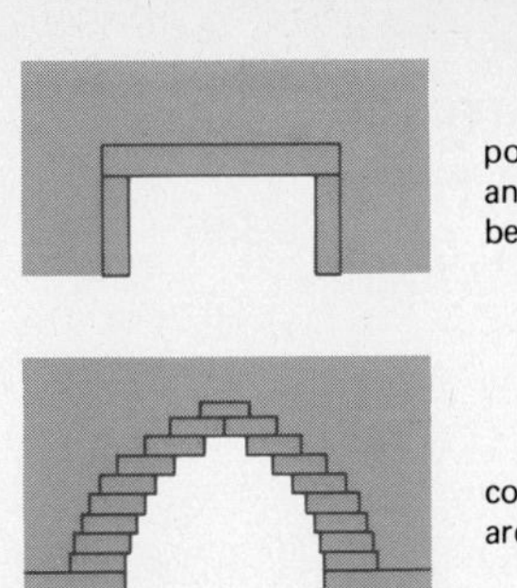

*4-6 ▌ Spanning a space: arches and domes

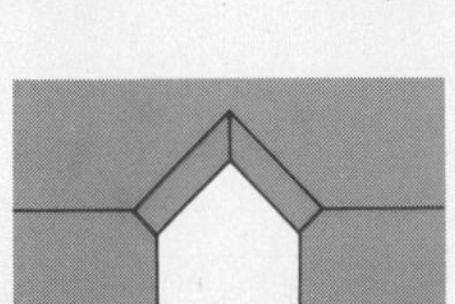
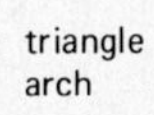

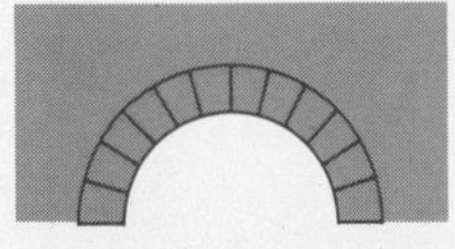

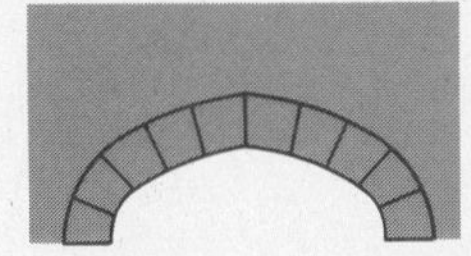

FIGURE 4-27
Various means to span a space.

There are a great many areas where the arts and humanities overlap the sciences and this is especially clear in architecture. Many of the features we admire in the architecture of the past were introduced not just for their decorative effect but often for practical purposes. One example of this that relates to the subject matter of this chapter is the development of methods to span a space, from the simple beam to arches and domes.

It might be said that the first important architectural invention was the so-called post-and-beam (or post-and-lintel) construction in which two upright posts support a horizontal beam. It is still used today; but until steel was introduced in the nineteenth century, the length of a beam was quite limited because the strongest building materials were then stone and brick. Hence the width of a span was limited by the size of available stones. But equally important, all of these materials—though strong under compression—are very weak under tension and shear; all of these stresses occur in a beam as shown in Figure 4-25. The minimal space that could be spanned using stone is shown by the closely spaced columns of the great Greek temples (Figure 4-21).

The introduction of the semicircular arch by the Romans (Figure 4-26) aside from its aesthetic appeal, was a tremendous technological innovation. It had been preceded by the so-called "triangle arch" and the "corbeled arch," but these were only slight improvements over the post-and-beam (see Figure 4-27). The advantage of the "true" or semicircular arch is that if well designed, its wedgeshaped stones experience stress which is mainly compressive even when supporting a large load such as the wall and roof of a cathedral. Because the stones are forced to squeeze against each other, they are thus mainly under compression (see Figure 4-28). Note however, that the arch transfers horizontal as well as vertical forces to the supports. A round arch consisting of many well-shaped stones, could span a very wide space. However, considerable buttressing on the sides was needed to support the horizontal components of the forces. The buttressing was frequently provided by heavy walls leading off to the side.

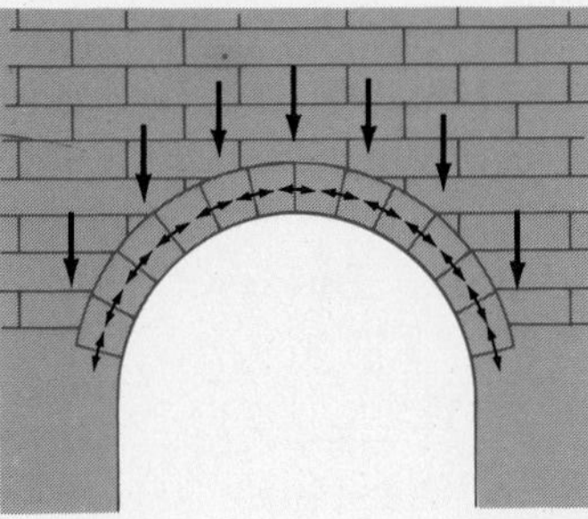

FIGURE 4-28
Stones in a round (or "true") arch are mainly under compression.

The pointed arch came into use in Europe about A.D. 1100 and soon became the hallmark of the great Gothic cathedrals. It too was an important technical innovation, and it is likely that this was at least part of the reason it was originally introduced: it was first used to support heavy loads such as the tower of a cathedral and as the central arch across the nave; the lesser arches in these "transitional" buildings often remained round. Apparently the builders realized that because of the steepness of the pointed arch, the forces due to the weight above could be brought down more nearly vertically so less horizontal buttressing would be needed. The pointed arch reduced the load on the walls, so there could be more openness and light. The smaller amount of buttressing needed was provided on the outside by graceful flying buttresses (Figure 4-29).

The technical innovation of the pointed arch was achieved not through calculation but through experience and intuition. The beginnings of the science of statics and building appeared many hundreds of years later, in the notebooks of Leonardo da Vinci (1452–1519) and the writings of Galileo, and it was not until somewhat later that detailed calculations, such as those presented earlier in this chapter, came into use. To make an accurate analysis of a stone arch is quite difficult in practice. But if we make some simplifying assumptions, we can show why the horizontal component of the force at the base is less for a pointed arch than for a round one. Figure 4-30 shows a round arch and a pointed arch, each with an 8.0 m span. The height of the round arch is thus 4.0 m whereas that of the pointed arch is larger and has been chosen to be 8.0 m. Each arch supports a weight of 12.0×10^4 N (= 12,000 kg × g) which for simplicity we have divided into two parts (each 6.0×10^4 N) acting on the two halves of each arch as shown. To be in equilibrium, each of the supports must exert an upward force of 6.0×10^4 N; they also exert a horizontal force, F_H, at the base of the arch, and it is this we want to calculate. We focus only on the right half of each arch and will set equal to zero the total torque calculated about the apex of the arch due to the forces exerted on that half arch, as if there were a hinge at the apex. For the round arch, the torque equation is:

$$(4.0\text{ m})(6.0 \times 10^4\text{ N}) - (2.0\text{ m})(6.0 \times 10^4\text{ N}) - (4.0\text{ m})(F_H) = 0.$$

Thus $F_H = 3.0 \times 10^4$ N. For the pointed arch the torque equation is

$$(4.0\text{ m})(6.0 \times 10^4\text{ N}) - (2.0\text{ m})(6.0 \times 10^4\text{ N}) - (8.0\text{ m})(F_H) = 0.$$

Solving, we find $F_H = 1.5 \times 10^4$ N—only half as much! From this calculation we can see that the horizontal buttressing force required for a pointed arch is less because the arch is higher and there is therefore a longer lever arm for this force. Indeed, the steeper the arch, the less the horizontal component of the force needs to be, and hence the more nearly vertical is the force exerted at the base of the arch.

The further development of the arch was one of a decline; for the subsequent flattened arches, such as the Tudor arch (Figure 4-27), were structurally weaker than the simple pointed arch. However, with the coming of advanced methods of calculation in the nineteenth and twentieth centuries it became possible to calculate the best shape of arch for a given load condition. For example, if the load is uniform across the span, it can be shown that the stresses within the arch will be purely compressive if the arch has a parabolic shape.

Whereas an arch spans a two-dimensional space, a dome—which is basically an arch rotated about a vertical axis—spans a three-dimensional space. The Romans built the first large domes; their shape was hemispherical and some still stand, such as that of the Pantheon in Rome

FIGURE 4-29
A flying buttress (on the cathedral of Notre Dame, in Paris). (D. C. Giancoli.)

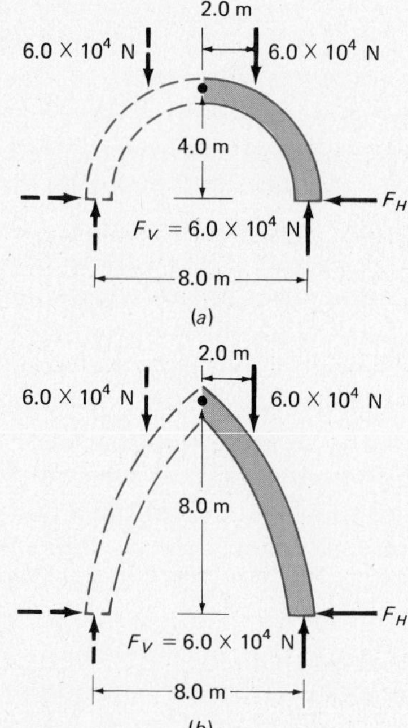

FIGURE 4-30
Forces in a round arch (*a*) compared with those in a pointed arch (*b*).

(Figure 4-31). By the time of the Renaissance, the technique for constructing large domes seems to have been lost. Indeed, the dome of the Pantheon was a source of wonder to Renaissance architects. (Even today we do not know precisely how it was built.) The problem came to the fore in fifteenth century Florence with the designing of a new cathedral which was to have a dome 43 m in diameter to rival that of the Pantheon. Now a dome, like an arch, is not stable until all the stones are in place. It had been the custom to support a dome during a construction with a wooden framework. But no trees big enough or strong enough could be found to support the 43 m space required for the cathedral in Florence. In 1418, after the cathedral was finished except for the dome, a competition for the design of the dome was held and was won by Brunelleschi. One problem that had to be dealt with was that the dome was to rest on a drum that had been completed with no external abutments; and there was no place to put any. Hence the dome must exert the minimum of horizontal force. Brunelleschi solved this by designing a pointed dome (Figure 4-32), since a pointed dome, like a pointed arch, exerts a smaller side thrust against its base.

FIGURE 4-31
The Pantheon in Rome, built in the first century. (Italian Government Travel Office.)

The other major problem was how to support the dome during construction. Instead of using a wooden framework, Brunelleschi built the dome in horizontal layers. Each layer was bonded to the previous one which held it in place until the last stone of the circle was placed. Each closed ring was then strong enough to support the next layer. It was an amazing feat.

FIGURE 4-32
Brunelleschi's dome for the cathedral of Florence, Italy. (Italian Government Travel Office.)

To end this section we will consider the forces necessary to support a modern dome, that of the Little Sports Palace in Rome, Figure 4-33. A dome, like an arch, is statically most stable when under compression. The 36 buttresses which support the 1.2×10^6 kg dome are positioned at a 38° angle and connect smoothly with the dome.

EXAMPLE 4-10 Calculate the components of the force, F_V and F_H, which each buttress exerts on the dome so that the force acts compressively, that is, at a 38° angle (Figure 4-34).

SOLUTION The vertical load on *each* buttress is 1/36 of the total weight. Thus

$$F_V = \frac{(1.2 \times 10^6 \text{ kg})(9.8 \text{ m/s}^2)}{36} = 3.4 \times 10^5 \text{ N}.$$

FIGURE 4-33
The dome of the Little Sports Palace in Rome, built for the 1960 Olympics. (Italian Government Travel Office.)

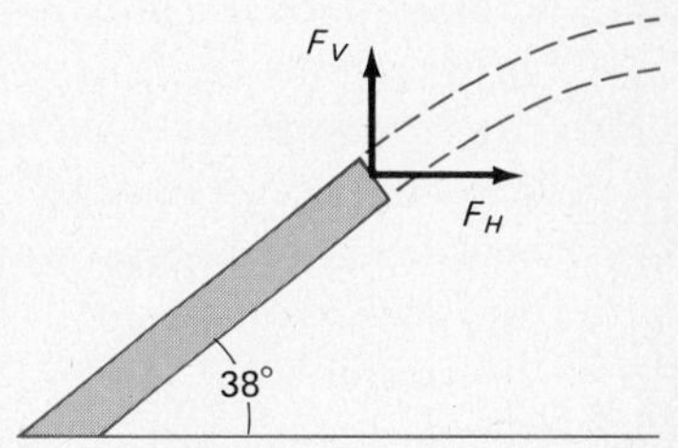

FIGURE 4-34
Example 4-10.

The force must act at a 38° angle at the base of the dome in order to be purely compressive. Thus

$$\tan 38° = \frac{F_V}{F_H} = \frac{340{,}000 \text{ N}}{F_H}$$

$$F_H = \frac{340{,}000 \text{ N}}{\tan 38°} = 430{,}000 \text{ N}.$$

In order that each of the buttresses be able to exert this 430,000-N horizontal force, a prestressed concrete tension ring surrounds the base of the buttresses beneath the ground. ▮▮

▮ Summary

A body at rest, or one in uniform motion at constant velocity, is said to be in *equilibrium*. The determination of the forces within a structure at rest is the field called statics. The two necessary conditions for a body to be in equilibrium are (1) the vector sum of all the forces on it must be zero, and (2) the sum of all the torques (calculated about any arbitrary point as axis) must also be zero. It is important when doing statics problems to apply the equilibrium conditions to only one body at a time.

A body in static equilibrium is said to be in (*a*) *stable*, (*b*) *unstable*, or (*c*) *neutral equilibrium*, depending on whether a slight displacement leads to (*a*) a return to the original position, (*b*) further movement, or (*c*) rest in the new position. An object in stable equilibrium is also said to be in *balance*.

Hooke's law applies to many elastic solids and states that the change in length of an object is proportional to the applied force. If the force is too great, the object will exceed its elastic limit, which means it will no longer return to its original shape when the distorting force is removed. If the force is even greater, the ultimate strength of the material can be exceeded and the object fractures. The force per unit area acting on a body is called the *stress*, and the resulting fractional change in length is called the *strain*. The stress on a body is present within the body and can be of three types: *compression*, *tension*, and *shear*. The ratio of stress to strain is called the elastic modulus of the material; Young's modulus applies for compression and tension, and the shear modulus for shear; bulk modulus applies to an object whose volume changes as a result of pressure on all sides. All three moduli are constants for a given material when distorted within its elastic region.

The subject of statics is especially useful for calculating forces within muscles and bones, and in structures such as buildings and bridges; many architectural innovations—for example, the round and pointed arches—resulted from a consideration of static forces and the need to avoid fracture.

Questions

1 Why is it more difficult to do a situp with your hands behind your head than when they are outstretched in front of you? A diagram may help you to answer this.

2 Explain why touching the toes while seated on the floor with outstretched legs produces less stress on the lower spinal column than when touching the toes from a standing position. Use a diagram.

3 An earth retaining wall is shown in Figure 4-35*a*. The earth, particularly when wet, can exert a significant force F on the wall. (*a*) What force produces the torque to keep the wall upright? (*b*) Explain why the retaining wall in Figure 4-35*b* would be much less likely to overturn.

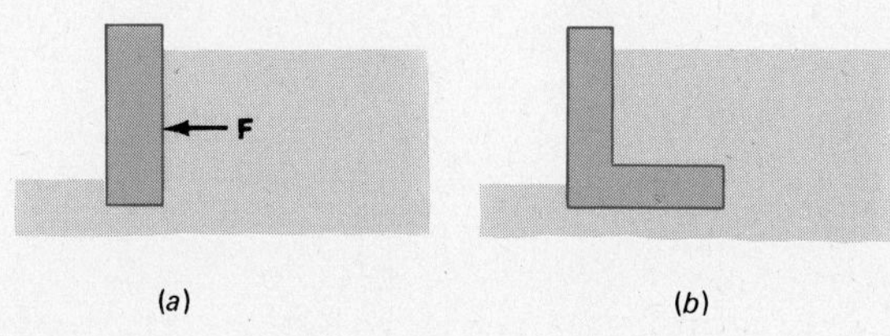

FIGURE 4-35

4 Discuss why surgical removal of the *patella*, a small bone above the knee (Figure 4-36), requires therapy to strengthen the quadriceps muscle above.

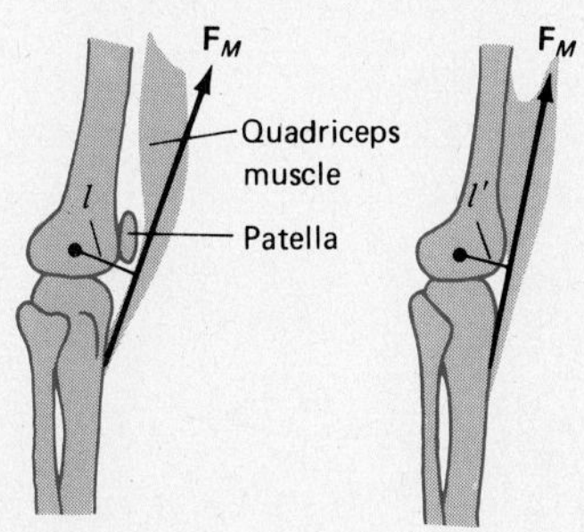

FIGURE 4-36

5 Name the type of equilibrium for each position of the ball in Figure 4-37.

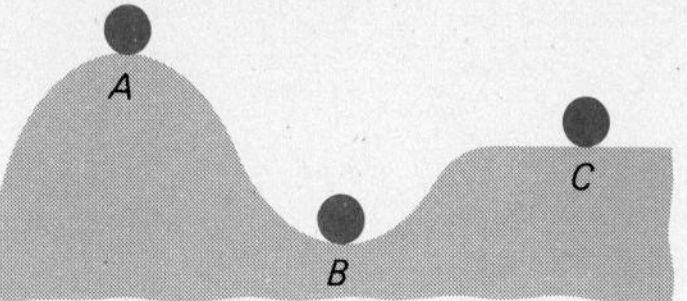

FIGURE 4-37

6 In what state of equilibrium is a cube (*a*) when resting on its face, (*b*) when on its edge?

7 Why do you tend to lean backwards when carrying a heavy load in you arms?

8 What purpose does a walking stick serve when you are hiking in rough country? Be specific.

9 Place yourself facing the edge of an open door. Position your feet astride the door with your nose and abdomen touching the door's edge. Try to rise on your tiptoes. Why can't this be done?

10 Why is it not possible to sit upright in a chair and rise to one's feet without first leaning forward?

11 Explain why a rectangular brick can be placed so that slightly less than half its length can be suspended over the edge of a table, but no more.

12 Which of the configurations of brick, (*a*) or (*b*) of Figure 4-38, is the more likely to be stable? Why?

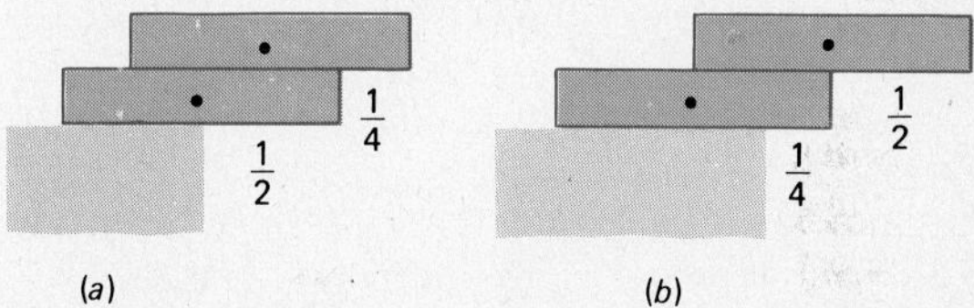

FIGURE 4-38

The dots indicate the cg of each brick. The fractions $\frac{1}{4}$ and $\frac{1}{2}$ indicate what portion of the brick is hanging beyond its support.

13 Materials such as ordinary concrete and stone are very weak under tension or shear. Would it be wise to use such a material for either of the supports of the cantilever shown in Figure 4-10? If so, which one(s)?

14 Examine how a pair of scissors or shears cuts through a piece of cardboard. Is the name "shears" justified?

Problems

SECTION 4-1

1 (I) what should be the tension in the wire if the net force exerted on the tooth in Figure 4-2 is to be 0.50 N? Assume that the angle between the two forces is 160° rather than the 140° in the figure.

2 (II) Assume in Example 4-1 that the net force as calculated is 10° to the left of where it should point, if the tooth is to move correctly. If the tension to the left is 2.0 N, what should the tension to the right be to make the net force act in the correct direction.

SECTION 4-2

3 (I) Calculate the torque about the front support of a diving board exerted by a 70-kg person 3.0 m from that support.

4 (I) Calculate the mass m needed in order to suspend the leg shown in Figure 4-39. Assume the leg has a mass of 11 kg, and its cg is 35 cm from the hip joint; the sling is 75 cm from the hip joint.

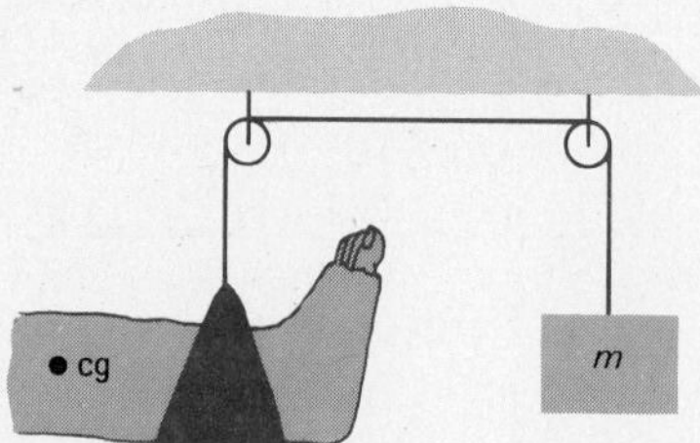

FIGURE 4-39

5 (I) Two cords support a chandelier in the manner shown in Figure 4-4 except the upper wire makes an angle of 45° with the ceiling. If the cords can sustain a force of 1000 N without breaking, what is the maximum weight chandelier than can be supported?

6 (II) Find the tension in the two wires shown in Figure 4-40.

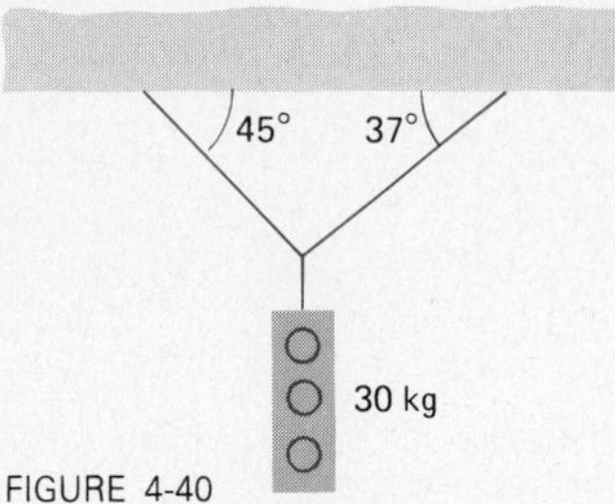

FIGURE 4-40

7 (II) Calculate the tension F_T in the wire which supports the 20-kg beam shown in Figure 4-41 and the force $\mathbf{F}_w$ exerted by the wall on the beam (give magnitude and direction).

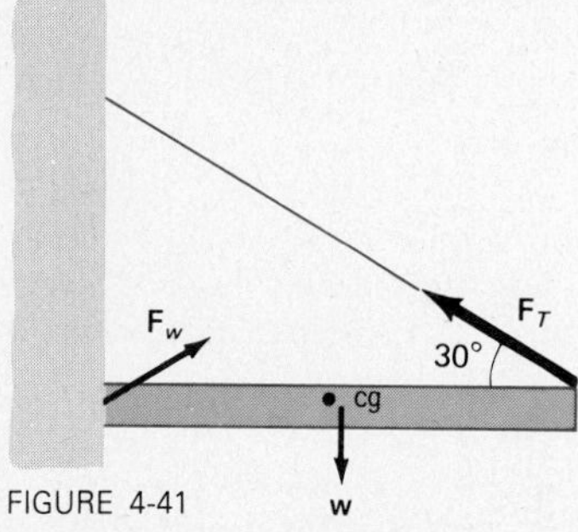

FIGURE 4-41

8 (II) Calculate the force required, F_1, in Figure 4-6 taking into account the lever's mass of 12 kg. Assume the cg of the lever is at its center.

9 (II) Calculate F_1 and F_2 for the uniform cantilever shown in Figure 4-10, whose mass is 1200 kg.

10 (II) Calculate the net force required to topple a 60-kg rectangular retaining wall 3.0-m high and 1.2-m thick. Assume the toppling force acts 1.5 m above the base, and that the force exerted by the ground to resist toppling acts at the corner of the base (as in Figure 4-11).

11 (II) Calculate the vertical force of traction on the neck for the traction device shown in Figure 4-42.

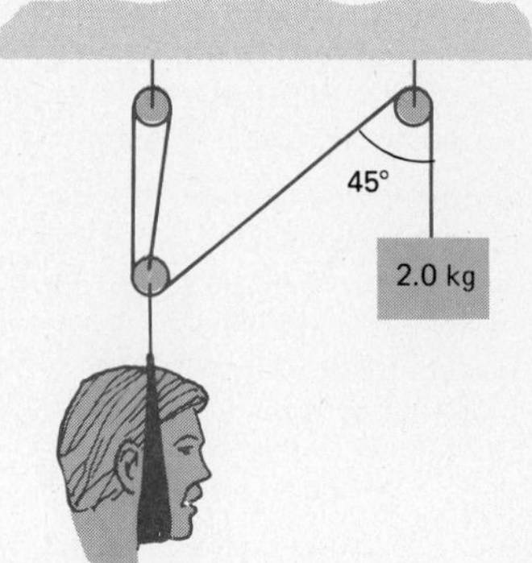

FIGURE 4-42

12 (II) Calculate the forces F_1 and F_2 that the supports exert on the diving board of Figure 4-43 when a 50-kg person stands at its tip. Ignore the weight of the board.

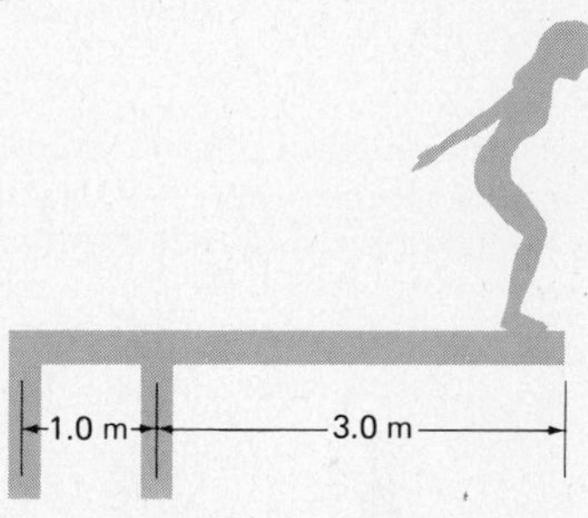

FIGURE 4-43

13 (II) Repeat the problem above taking into account the board's mass of 40 kg. Assume the board's cg is at its center.

14 (II) Calculate the mass m required in Figure 4-39 to support the leg using the result of example 3-15 and the values given in Table 3-2 assuming an 85-kg person 170 cm tall. The leg pivots about the hip joint and the support acts at the ankle joint.

15 (II) Calculate $\mathbf{F}_1$ and $\mathbf{F}_2$ for the beam shown in Figure 4-44.

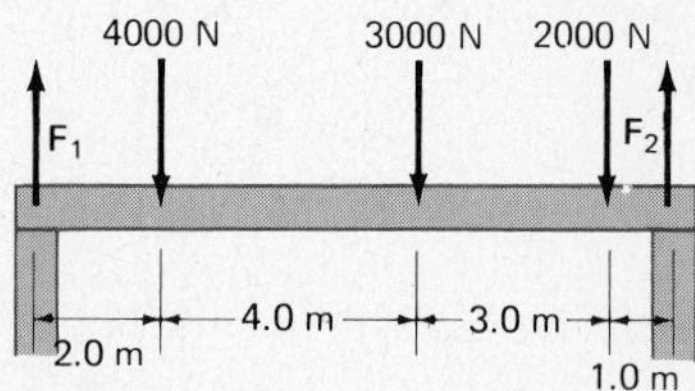

FIGURE 4-44

16 (II) A large 60-kg board leans at a 45° angle against the edge of a barn door, 2.4 m from the hinge. How great a horizontal force must a person behind the door exert (at the edge) in order to open it? Assume there is negligible friction between the door and the board, but that the board is firmly set against the ground.

17 (II) Repeat the above problem assuming the coefficient of friction between the board and the door is 0.40.

18 (II) A 150-cm-tall person lies on a light (massless) board which is supported by two scales, one under the feet and one beneath the top of the head. The two scales read, respectively, 32.0 kg and 29.4 kg. Where is the center of gravity of this person?

19 (II) A tightly stretched "high wire" is 50 m long and sags 3.5 m when a 60-kg tightrope walker stands at its center. What is the tension in the wire? Is it possible to increase the tension in the wire so that there is no sag?

20 (II) A 7.5-m-long ladder of mass 15.0 kg leans steeply (60°) against a smooth wall (so the force exerted by the wall, F_w, is perpendicular to the wall). The ground is rough. Calculate the components of force exerted by the ground on the ladder at its base, and determine what the coefficient of friction at the base of the ladder must be if the ladder is not to slip when a 70-kg person is three-fourths of the way up the ladder.

21 (II) If the coefficient of friction between the ladder and the ground in the situation described in Problem 20 is 0.45, how far up the ladder can the person climb before the ladder starts to slip?

22 (III) A person wants to push a lamp (mass 10 kg) across the floor. Assuming the person pushes at a height of 60 cm above the ground and the coefficient of friction is 0.20, determine whether the lamp will slide or tip over (Figure 4-45).

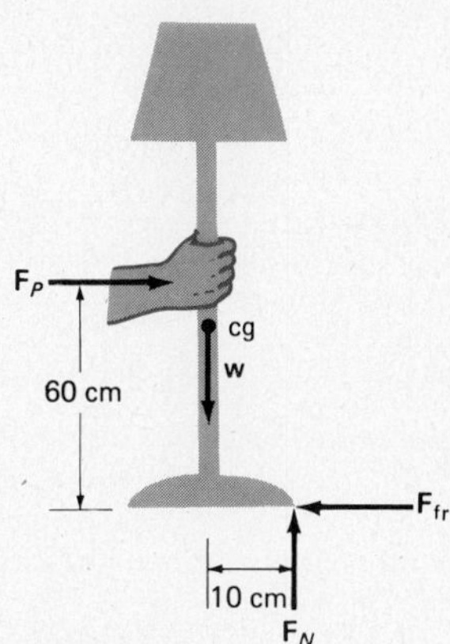

FIGURE 4-45

23 (III) In problem 22, calculate the maximum height above the floor at which the person can push the lamp so its slides rather than tips.

24 (III) A 30-kg round table is supported by three legs placed equal distances apart on the edge. What minimum mass, placed on the edge, will cause the table to overturn?

25 (III) Two guy wires run from the top of a pole 2.4 m tall that supports a volleyball net. The two wires are anchored to the ground 2.0 m apart and each 2.0 m from the pole. The tension in each wire is 65 N. What is the tension in the net, assumed horizontal and attached at the top of the pole?

26 (III) A home mechanic wants to raise the 200-kg engine out of a car. The plan is to stretch a rope from the engine to a branch of a tree 30 m above. The rope passes over a firmly placed pulley just above the engine so that when the mechanic climbs the tree and pulls horizontally on the rope at its midpoint, the engine rises out of the car. How much force must the mechanic exert to raise the engine 0.50 m?

*SECTION 4-3

*27 (I) If the biceps muscle inserts 4.5 cm from the joint, what force F_M would be required to hold the 5.0-kg mass in Figure 4-13*a*?

*28 (I) If the point of insertion of the biceps muscle into the lower arm shown in Figure 4-13*a* were 6.0 cm, how much mass could the person hold with a muscle exertion of 400 N?

*29 (I) Approximately what force, F_M, must the extensor muscle in the upper arm exert on the lower arm to hold a 7.3 kg shotput (Figure 4-46)? Assume the lower arm has a mass of 2.8 kg and its cg is 12 cm from the pivot point.

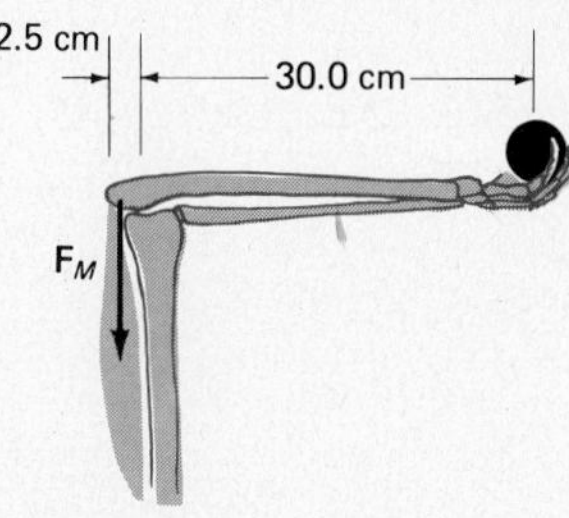

FIGURE 4-46

*30 (II) Calculate the force required of the "deltoid" muscle, F_M, to hold up the outstretched arm shown in Figure 4-47. The total mass of the arm is 2.8 kg.

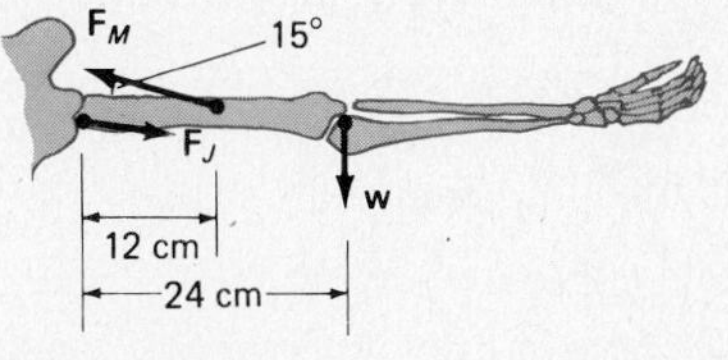

FIGURE 4-47

*31 (II) Suppose the hand in the above problem holds a 10-kg mass. What force, F_M, is required of the deltoid muscle assuming the mass is 50 cm from the shoulder joint?

*32 (II) Calculate the magnitude of the force F_J exerted by the shoulder on the upper arm at the joint for Problem 30.

*33 (II) Calculate the force F_J exerted by the shoulder on the upper arm at the joint for Problem 31.

*34 (II) A person is standing on one foot with the heel raised, Figure 4-48. $\mathbf{F}_T$ is the force exerted by the muscles of the lower leg on the foot acting through the Achilles tendon and $\mathbf{F}_J$ is the force with which the rest of the person's body acts on the foot at the ankle joint. (*a*) Is there another force acting on the foot? If so, where does it act and what is its magnitude? (*b*) Determine F_T and F_J in terms of the weight w of the person assuming they act in the vertical direction and that $\mathbf{F}_T$ acts at a point one and one half ($1\frac{1}{2}$) times further from the pivot point than $\mathbf{F}_J$.

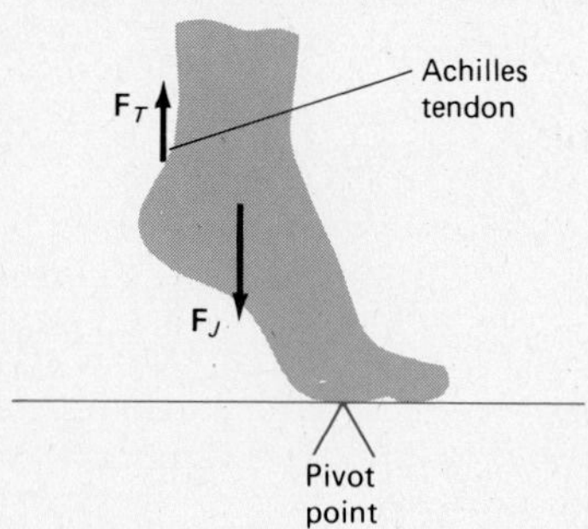

FIGURE 4-48

*35 (II) A person standing on one foot, as in Figure 4-48, decides to jump; assuming the ground exerts a force upward at the toes equal to $2.0w$ just before the foot starts moving (where w is the person's weight), what is the force F_J exerted on the ankle joint? Assume F_T acts $1\frac{1}{2}$ times further from the toes than does F_J.

*36 (II) if the tension in the biceps muscle in Figure 4-13*a* is 520 N when a 7.0-kg mass is held in the hand, is the forearm extending, flexing, or in equilibrium?

*37 (III) Calculate the magnitude of the force at the base of the spine, F_V in Figure 4-14*b*, if a mass of 20 kg is held in the person's hands. Assume the person has a mass of 70 kg.

*38 (III) When walking, a person momentarily puts all the weight on one foot. The cg of the body lies over the supporting foot. Figure 4-49 shows the supporting leg and the forces on it. Calculate the force exerted by the "hip abductor" muscles, F_M, and the x and y components of the force $\mathbf{F}_J$ acting on the joint. Take the whole leg as the object under discussion. The person's mass is 70 kg.

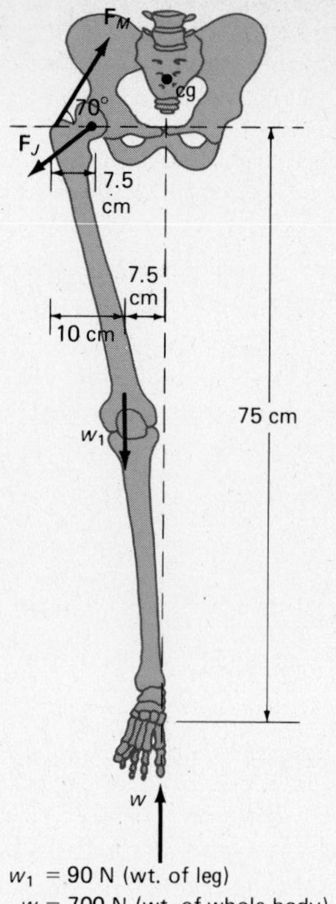

w_1 = 90 N (wt. of leg)
w = 700 N (wt. of whole body)

FIGURE 4-49

*39 (III) Calculate F_M and F_J as in the above problem assuming the person carries a 20-kg suitcase in each hand. [*Hint*: $w = (110 \text{ kg})(g)$.]

*40 (III) Calculate F_M and F_J as in the previous problem assuming the person carries a suitcase only in the hand on the side opposite the supporting leg. [*Hint*: first calculate the common cg of the person plus suitcase; this point will be above the foot, thus shifting the horizontal measurements from those shown in Figure 4-49.] Assume the suitcase is at the edge of the hip.

SECTION 4-4

41 (II) The Leaning Tower of Pisa is 55 m tall and about 7.0 m in diameter. The top is 4.5 m off-center. Is the Tower in stable equilibrium? If so, how much further can it lean before it becomes unstable? Assume the Tower is of uniform composition.

42 (III) Bricks are stacked as shown in Figure 4-38*b*. (*a*) Show that if a four-brick stack is to be stable, the bricks must extend no more than (starting at the top) one-half, one-quarter, one-sixth, one-eighth of their length beyond the one below. (*b*) Is the top brick completely beyond the base? (*c*) Determine a general formula for the maximum total distance spanned by N bricks if they are to remain stable. (*d*) A builder wants to construct a corbeled arch (Figure 4-27) based on the principle of stability discussed in (*a*) and

(*c*) above. What minimum number of bricks, each 0.30 m long, is needed if the arch is to span 1.0 m?

SECTION 4-5

43 (I)A marble column of cross-sectional area 2.0 m^2 supports a mass of 20,000 kg. (*a*) What is the stress within the column? (*b*) What is the strain?

44 (I) By how much is the column in the previous problem shortened if it is 10-m high?

45 (I) If the supports in Example 4-9 were 10 m high, how much would they shorten?

46 (II) How much force is required to stretch a steel piano wire of 0.050 cm diameter by 0.01 percent?

47 (II) If the two wires in Figure 4-40 are made of steel wire 1.0 mm in diameter, how much does each stretch because of the load?

48 (II) A 15-cm-long animal tendon was found to stretch 3.1 mm by a force of 12.4 N. The tendon was approximately round with an average diameter of 8.5 mm. Calculate the elastic modulus of this tendon.

49 (II) How much pressure is needed to compress the volume of an iron block by 0.10 percent? Express answer in N/m^2, and compare it to atmospheric pressure (1.0×10^5 N/m^2).

50 (II) One liter of alcohol (1000 cm^3) in a flexible container is carried to the bottom of the sea where the pressure is 2.4×10^6 N/m^2. What will its volume be now?

51 (II) If a compressive force of 3.0×10^4 N is exerted on the end of a 20-cm-long bone of cross-sectional area 3.6 cm^2, (*a*) will the bone break, and (*b*) if not, by how much does it shorten?

52 (II) What is the minimum cross-sectional area required of a vertical steel wire from which is suspended a 250-kg chandelier? Assume a safety factor of 4.

53 (II) The femur bone in the leg has an average effective cross section of about 3.0 cm^2 ($=3.0 \times 10^{-4}$ m^2). How much compressive force can it withstand before breaking?

54 (II) Assume the supports of the cantilever shown in Figure 4-10 (mass = 1200 kg) are made of wood. Calculate the minimum cross-sectional area required of each, assuming a safety factor of 8.

55 (II) An iron bolt is used to connect two iron plates together. The bolt must withstand shear forces up to about 2000 N. Calculate the minimum diameter for the bolt, based on a safety factor of 4.

56 (III) A pole projects horizontally from the front wall of a shop. A 4.0-kg sign hangs from the pole at a point 2.0 m from the wall. (*a*) What is the torque due to this sign calculated about the point where the pole meets the wall? (*b*) If the pole is not to fall off, there must be another torque exerted to balance it. What exerts this torque? Use a diagram to show how this torque must act. (*c*) Discuss whether compression, tension, and/or shear play a role in (*b*)

57 (III) A steel cable is to support an elevator whose total (loaded) mass is not to exceed 2200 kg. If the maximum acceleration of the elevator is 1.0 m/s^2, calculate the diameter of cable required. Assume a safety factor of 5.

*SECTION 4-6

*58 (II) How high must a pointed arch be if it is to span a space 8.0 m wide and exert one-third the horizontal force at its base that a round arch would?

*59 (II) The subterranean tension ring that exerts the balancing horizontal force on the abutments in Figure 4-33 is 36 sided so each segment makes a 10° angle with the adjacent one, Figure 4-50. Calculate the tension *F* that must exist in each segment so that the required force of 4.3×10^5 N can be exerted at each corner (Example 4-10).

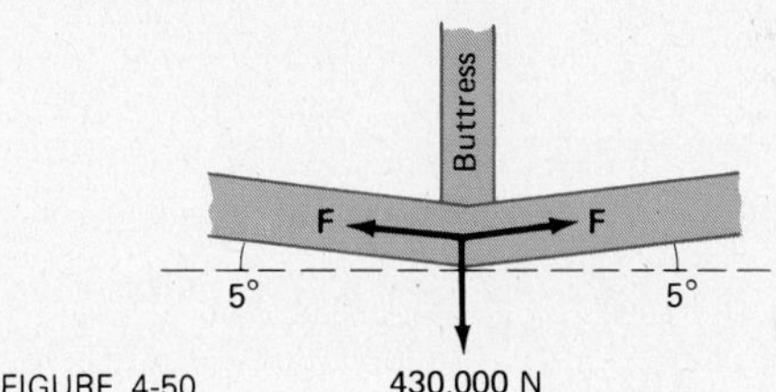

FIGURE 4-50

For additional reading

Alexander, R. M., *Animal Mechanics*, University of Washington Press, Seattle, 1968. Interesting applications of forces within animals, center of gravity, elasticity.

Ambrose, J. E., *Building Structures Primer*, John Wiley & Sons, Inc., New York, 1967.

Bronowski, J., *The Ascent of Man*, Little, Brown & Company, Boston, 1974. Chapter 3.

Davidovits, P., *Physics in Biology and Medicine*, Prentice-Hall, Inc., Englewood Cliffs, N.J., 1975.

Mark, R., "Structural Analysis of Gothic Cathedrals," *Sci. American*, **227**: 15 (Nov., 1972), p. 90.

Salvadori, M. G., and R. Heller, *Structure in Architecture: The Building of Buildings*, Prentice-Hall, Inc., Englewood Cliffs, N.J., 1975.

Williams, M., and H. R. Lissner, *Biomechanics of Human Motion*, W. B. Saunders Company, Philadelphia, 1962. Statics applied to the human body.

5 ▌ Momentum and energy

Until now we have been studying the mechanics of bodies at rest and in motion in terms of Newton's three laws of motion. In our analysis, *force* has played a central role as the quantity determining the motion. In the present chapter we discuss an alternative analysis of motion in terms of the quantities *momentum* and *energy*. The importance of these two quantities is that they are *conserved*; that is, in quite general circumstances they remain constant. The fact that such quantities are conserved gives us not only a deeper insight into the nature of the world but also gives us another way to attack practical problems.

The conservation laws of momentum and energy are especially valuable in dealing with systems of many objects in which a detailed consideration of the forces involved would be difficult.

5-1 ▌ Momentum and impulse

Momentum and its relation to force

The **momentum** of a body is defined as the product of its mass and its velocity. Momentum is usually represented by the symbol **p**. If we let m represent the mass of a body and **v** represent its velocity, then its momentum, **p**, is

$$\mathbf{p} = m\mathbf{v}. \tag{5-1}$$

Since velocity is a vector, momentum is a vector. The direction of the momentum is the direction of the velocity; and the magnitude of the momentum, p, at any moment is equal to the numerical product of the mass m times the instantaneous speed v. The units of momentum are just those of mass × velocity, which in SI units are kg · m/s.

Everyday usage of the term momentum is in accord with the definition above. For according to Equation 5-1, a fast-moving car has more momentum than a slow-moving car of the same mass; and a heavy truck has more momentum than a small car moving with the same speed. The more momentum an object has, the harder it is to stop it, and the greater effect it will have if it is brought to rest by impact or collision. A football player is more likely to be stunned if tackled by a heavy opponent running at top speed than by a lighter or slower-moving tackler. A heavy fast-moving truck can do more damage than a slow-moving small car.

A force is required to change the momentum of an object whether it is to increase the momentum, decrease it (such as to bring a moving object to rest), or to change its direction. Newton in fact originally stated his second law in terms of momentum (although he called the product mv the "quantity of motion"). Newton's statement of the second law, translated into modern language, is as follows: "the rate of change of momentum of a body is proportional to the net force applied to it." For a constant force we can write this as:

$$\mathbf{F}=\frac{\Delta \mathbf{p}}{\Delta t}, \tag{5-2}$$

where $\mathbf{F}$ is the net force applied to the object and $\Delta\mathbf{p}$ is the resulting momentum change that occurs during the time interval Δt. The familiar form of the second law, $\mathbf{F}=m\mathbf{a}$, can readily be derived from Equation 5-2, for the case of constant acceleration, using Equation 1-9*a* now written in vector form:

$$\mathbf{v}=\mathbf{v}_0+\mathbf{a}t;$$

to prove this, we assume an object of mass m is accelerated uniformly (although the result is valid even if it is not uniform) from a velocity $\mathbf{v}_0$ to velocity $\mathbf{v}$ in a time interval Δt. We start with Equation 5-2, with t replacing Δt:

$$\mathbf{F}=\frac{\Delta \mathbf{p}}{t}=\frac{m\mathbf{v}-m\mathbf{v}_0}{t}=\frac{m}{t}(\mathbf{v}-\mathbf{v}_0).$$

We then substitute in Equation 1-9*a* and obtain the desired result:

$$\mathbf{F}=\frac{m}{t}(\mathbf{a}t)$$

$$\mathbf{F}=m\mathbf{a}.$$

Newton's statement, Equation 5-2, is actually more general than the more familiar one since it includes the situation in which the mass may change. This is important in certain situations such as rockets which lose mass as they burn fuel. Another interesting example is the following.

EXAMPLE 5-1 Water leaves a hose at a rate of 5.0 kg/s with a speed of 50 m/s. It strikes a wall, which stops it. What is the force exerted by the water on the wall?

SOLUTION In each second, water with a momentum of (5.0 kg) (50 m/s) = 250 kg·m/s is brought to rest. The magnitude of the force required to change the momentum by this amount is

$$F=\frac{\Delta p}{\Delta t}=\frac{0-250\ \mathrm{kg\cdot m/s}}{1.0\ \mathrm{s}}=-250\ \mathrm{N}.$$

The minus sign indicates that the force on the water is opposite to its original velocity. The wall exerts a force of 250 N to stop the water, so by Newton's third law, the water exerts a force of 250 N on the wall.

Impulse

If we multiply both sides of Equation 5-2 by the time interval Δt we obtain

$$\mathbf{F}\,\Delta t = \Delta \mathbf{p}. \tag{5-3}$$

The quantity on the left, the product of the force **F** times the time Δt over which the force acts, is called the **impulse**. We see that the total change in momentum is equal to the impulse. The same change in momentum can be achieved by a large force acting over a short time, or by a small force acting over a long time. The concept of impulse is of most help when dealing with forces that act over a short time, such as when a bat hits a baseball. The force is generally not constant and often its variation in time is like that graphed in Figure 5-1. It is often sufficient to approximate this by an average force $\bar{F}$ acting over a time Δt, as indicated by the dashed lines in Figure 5-1.

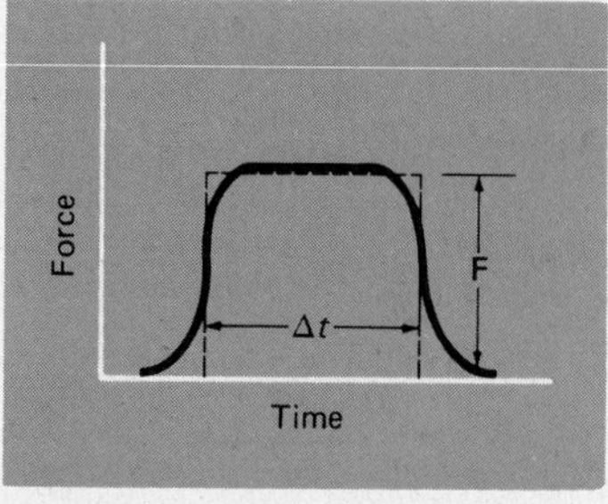

FIGURE 5-1
Impulse. The solid line represents a typical impulsive force as a function of time; the dashed line represents an average force, for this case, acting over the time Δt.

▮ EXAMPLE 5-2 (*a*) Calculate the impulse suffered when a 70-kg person lands on firm ground after jumping from a height of 5.0 m; (*b*) determine if the person will suffer a fractured tibia (lower-leg bone; cross-sectional area about 3.0 cm^2).

SOLUTION (*a*) After a fall of 5.0 m, the velocity will be (using Equation 1-9*c*, with $a = g = 9.8\ \text{m/s}^2$ and initial velocity $v_0 = 0$):

$$v = \sqrt{2ad} = \sqrt{2(9.8\ \text{m/s}^2)(5.0\ \text{m})} = 9.9\ \text{m/s}.$$

After striking the ground, the momentum is quickly brought to zero, Figure 5-2. The impulse is then

$$F\,\Delta t = \Delta p = p - p_0 = 0 - (70\ \text{kg})(9.9\ \text{m/s}) = -690\ \text{N}\cdot\text{s}.$$

The minus sign tells us the force must be opposed to the original momentum—that is, upward.

(*b*) According to Table 4-2, the ultimate compressive strength of bone is about $170 \times 10^6\ \text{N/m}^2$. To calculate the average force due to the impulse we need to know the time *t* during which the force acts. If the person lands stiff-legged, the impulse lasts† only about 0.002 s so the force will be large. If, however, the legs are bent upon impact, the impulse is stretched over a longer time,† about 0.050 s so the force will be much less. We will do the calculation for the latter case and leave the former for the problems. Since $F\,\Delta t = 690\ \text{N}\cdot\text{s}$ and $t = 0.050$ s,

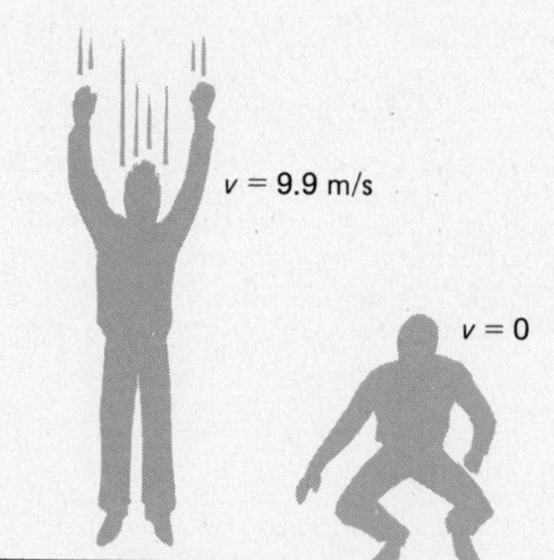

FIGURE 5-2
Period during which impulse acts (Example 5-2).

$$F = \frac{690\ \text{N}\cdot\text{s}}{0.050\ \text{s}} = 1.4 \times 10^4\ \text{N}.$$

The force *F* is the *net* force on the person, which is the sum of the upwards force F_{gr} exerted by the ground, and the downward force of gravity, mg ($= 690$ N); thus $F_{gr} = F + mg = 1.4 \times 10^4\ \text{N} + 700\ \text{N} = 1.5 \times 10^4$ N. The stress, then, is $F_{gr}/A = (1.5 \times 10^4\ \text{N})/(3.0 \times 10^{-4}\ \text{m}^2) = 50 \times 10^6\ \text{N/m}^2$, which is less than the ultimate compressive strength, so the bone doesn't break. ▮▮

† These times can be estimated directly from motion picture film, or by observing the distance moved by the person's cg during impact (about 1 cm when stiff-legged and perhaps 50 cm when the knees are bent) and then using Equations 1–9 to obtain the time.

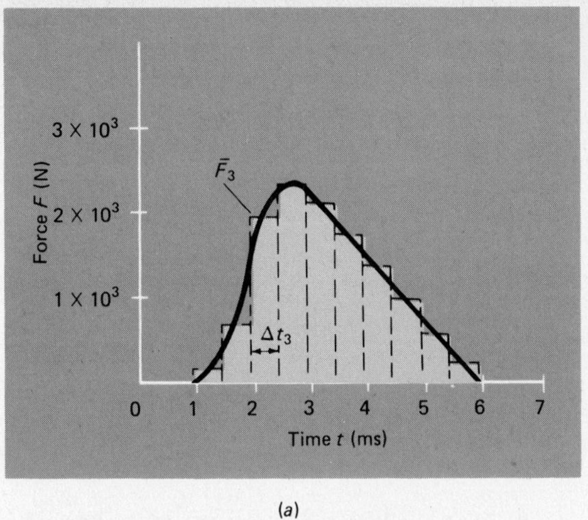

(a)

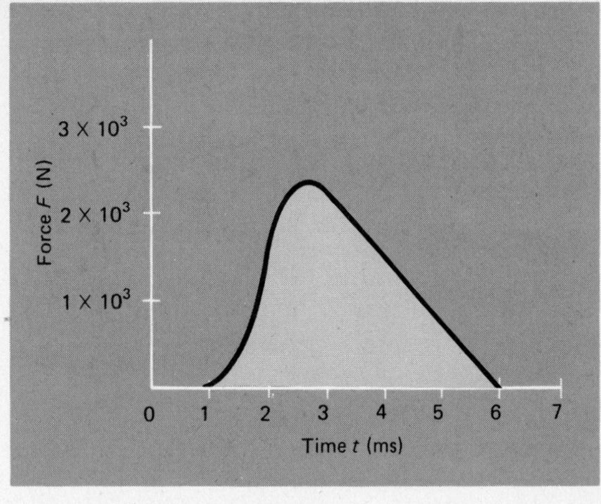

(b)

FIGURE 5-3
(a) Impulse for a variable force, graphed over a number of time subintervals. (b) as Δt approaches zero, the impulse approaches the area under the curve.

*Impulse created by a variable force

When a force varies in time, the impulse can be calculated graphically; the technique is the same as that used in Section 1-7 for calculating displacement from the v vs. t graph.

If a variable force acts over a certain time interval, we divide this interval into many subintervals as shown in Figure 5-3*a*. If the width of, say, the third subinterval is Δt_3, then the impulse due only to the third subinterval is $F_3 \Delta t_3$, where $\bar{F}_3$ is the average force acting during the time Δt_3; this is just the area of the rectangle of height $\bar{F}_3$ and width Δt_3. The total impulse, and therefore the total momentum change, equals the sum of the areas of all the rectangles. In many cases, $\bar{F}$ can be estimated for each subinterval, and a reasonable estimate of the total impulse can be readily obtained by adding up these areas (see the Problems). Ideally, we can subdivide the time interval into a greater and greater number of subintervals so that Δt for each subinterval becomes very small. In the limit as Δt approaches zero, the area of all the rectangles approaches the area under the curve of F vs. t (Figure 5-3*b*). Thus, in general, the total change in momentum equals the area under the F vs. t curve.

5-2 Conservation of momentum

The real importance of the concept of momentum lies in the fact that, under certain circumstances, it is a conserved quantity. In the midseventeenth century, shortly before Newton's time, it had been observed that the sum of the momenta of two colliding objects remains constant. Consider, for example, the head-on collision of two billiard balls shown in Figure 5-4. Although the momentum of each of the two balls changes as a result of the collision, the *sum* of their momenta is found to be the same before as after the collision. If $m_1\mathbf{v}_1$ is the momentum of ball number one and $m_2\mathbf{v}_2$ the momentum of ball two, both measured before the collision, then the total momentum of the two balls before the collision is $m_1\mathbf{v}_1 + m_2\mathbf{v}_2$. After the collision, the balls each have a different velocity and

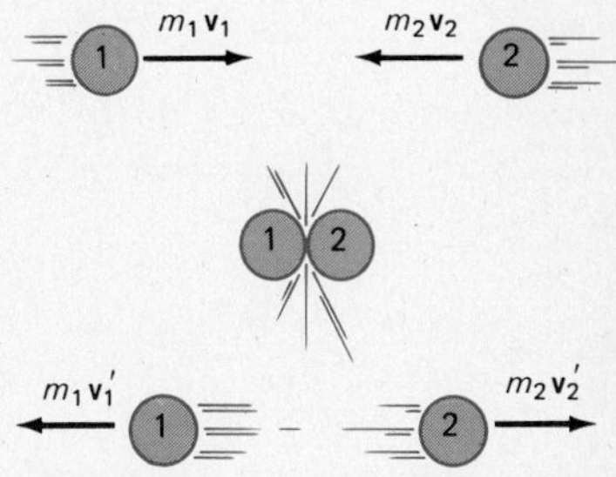

FIGURE 5-4
Momentum is conserved in collision of two balls.

momentum which we will designate by a "prime" on the velocity: $m_1\mathbf{v}_1'$ and $m_2\mathbf{v}_2'$. The total momentum after the collison is $m_1\mathbf{v}_1' + m_2\mathbf{v}_2'$. No matter what the velocities and masses involved are, it is found that the total momentum before the collision is the same as afterwards whether the collision is head-on or not:

$$m_1\mathbf{v}_1 + m_2\mathbf{v}_2 = m_1\mathbf{v}_1' + m_2\mathbf{v}_2'. \tag{5-4}$$

That is, the total momenta of the two balls is conserved: it stays constant.

Although the conservation of momentum principle was discovered experimentally, it is closely connected to Newton's laws of motion and can in fact be derived from them. We will do the derivation for the one-dimensional case illustrated in Figure 5-4. According to Newton's second law as expressed in Equation 5-3, the impulse $F\,\Delta t$ applied to, say, ball one is equal to its change in momentum,

$$F\,\Delta t = m_1v_1' - m_1v_1$$

where v_1 represents the velocity initially and v_1' the velocity after the collision. During the collision, the force F on ball one is exerted by ball two; and according to Newton's third law, ball one exerts an equal and opposite force $(-F)$ on ball two. Thus the change in momentum of ball two equals $-F\,\Delta t$, which is exactly equal and opposite to the change in momentum of ball one:

$$m_1v_1' - m_1v_1 = -(m_2v_2' - m_2v_2).$$

Any momentum lost by one ball is gained by the other. Thus the total momentum remains constant. This equation is readily rearranged to give Equation 5-4.

This derivation can be extended to include any number of interacting bodies. To show this in a simple way, we let $\mathbf{p}$ in Equation 5-2 represent the total momentum of a system (that is, the vector sum of the momenta of all objects in the system). Then, if the net force $\mathbf{F}$ on the system is zero, $\Delta\mathbf{p} = 0$ so the total momentum doesn't change. Thus the general statement of the **law of conservation of momentum** is:

▌ The total momentum of an isolated system of bodies remains constant.

By a *system* we simply mean a set of objects that interact with each other. An *isolated* system is one in which the only forces present are those between the objects of the system; that is, there is no net external force. If external forces do act—that is, forces exerted by objects outside the system—then the momentum may not be conserved. However, if the system is redefined so as to include the other objects, then the conservation of momentum principle will apply. For example, if we take as our system a falling rock, it does not conserve momentum since an external force, the force of gravity exerted by the earth, is acting on it. However, if we include the earth in the system, the total momentum of rock plus earth is conserved. (This of course means that the earth comes up to meet the ball; naturally you don't notice this motion of the earth since its mass is so great that the required velocity is very tiny).

▎**EXAMPLE 5-3** A 10,000-kg railroad car traveling at a speed of 20 m/s strikes an identical car at rest. If the cars lock together as a result of the collision, what is their common speed afterwards?

SOLUTION The initial total momentum is

$$m_1 v_1 + m_2 v_2 = (10{,}000\text{ kg})(20\text{ m/s}) + (10{,}000\text{ kg})(0\text{ m/s})$$
$$= 2.0 \times 10^5\text{ kg}\cdot\text{m/s}.$$

After the collision, the total momentum will be the same but it will be shared by both cars; since the two cars become attached, they will have the same velocity, call it v'. Then:

$$(m_1 + m_2)v' = 2.0 \times 10^5\text{ kg}\cdot\text{m/s}$$
$$v' = \frac{2.0 \times 10^5\text{ kg}\cdot\text{m/s}}{2.0 \times 10^4\text{ kg}} = 10\text{ m/s}.$$ ▮▮

The law of conservation of momentum is particularly useful when dealing with fairly simple systems such as collisions and certain types of explosions. For example, rocket propulsion, which we saw in Chapter 2 can be understood on the basis of action and reaction, can also be explained on the basis of the conservation of momentum. Before a rocket is fired, the total momentum of rocket plus fuel is zero. As the fuel burns, the total momentum remains unchanged: the backward momentum of the expelled gases is just balanced by the forward momentum of the rocket itself. Similar examples are the recoil of a gun and the throwing of a package from a boat.

▎**EXAMPLE 5-4** Calculate the recoil velocity of a 4.0 kg rifle which shoots a 0.050-kg bullet at a speed of 280 m/s.

SOLUTION We use Equation 5-4, and let the subscripts B represent the bullet and R the rifle; the final velocities are indicated by primes:

$$m_B v_B + m_R v_R = m_B v'_B + m_R v'_R$$
$$0 + 0 = (0.050\text{ kg})(280\text{ m/s}) + (4.0\text{ kg})(v'_R)$$
$$v'_R = -3.5\text{ m/s}.$$

Since the rifle has a much larger mass, its velocity is much less than that of the bullet. The minus sign indicates that the velocity (and momentum) of the rifle is in the opposite direction to that of the bullet. Notice that it is the *vector sum* of the momenta that is conserved. ▮▮

The last two examples illustrate the conservation of momentum applied to one-dimensional situations. For two- or three-dimensional situations, Equation 5-4 (or its equivalent) is applied to the x, y (and z) components separately (see Problems).

5-3 ▎Angular momentum

We saw in Chapter 3 that by using the appropriate angular variables, the kinematic and dynamic equations for rotational motion are analogous to those for ordinary linear motion. In like manner, the ordinary or *linear momentum* we have been discussing so far, mv, has a rotational analogue. It is called **angular momentum,** L, and is defined as

$$\text{angular momentum} = L = I\omega, \tag{5-5}$$

where I is the moment of inertia of a rotating object and ω is its angular velocity. Angular momentum, like linear momentum, is a conserved quantity under certain conditions. The **law of conservation of angular momentum** can be stated as

▮ **The total angular momentum of an isolated object or system of objects remains constant.**

Again we say "isolated," for if there are torques due to objects external to the system, the law does not apply. The law of conservation of angular momentum follows from the rotational equivalent of Newton's second law, Equation 3-15: $\tau = I\alpha$. For, by the definition of angular acceleration (Section 3-1), $\alpha = (\omega - \omega_0)/t$; then $\tau = I\alpha = I(\omega - \omega_0)/t = (L - L_0)/t$ where L_0 is the initial angular momentum of an object or system of objects and L is the angular momentum after a time t; if the net torque on that object or system is zero, then $(L - L_0) = 0$ and the angular momentum doesn't change, which is what we claimed above.

Many interesting everyday phenomena can be understood on the basis of conservation of angular momentum. For example, a skater doing a spin on ice rotates at a relatively low speed when her arms are outstretched; but when she brings her arms in close to her body she suddenly spins much faster. Remembering the definition of moment of inertia as $I = \sum mr^2$, it is clear that when she pulls her arms in closer to the axis of rotation her moment of inertia is reduced. Since the angular momentum $I\omega$ remains constant (we ignore the small torque due to friction), then if I decreases the angular velocity ω must increase. If she reduces her moment of inertia by a factor of 2 then she will rotate with twice the angular velocity.

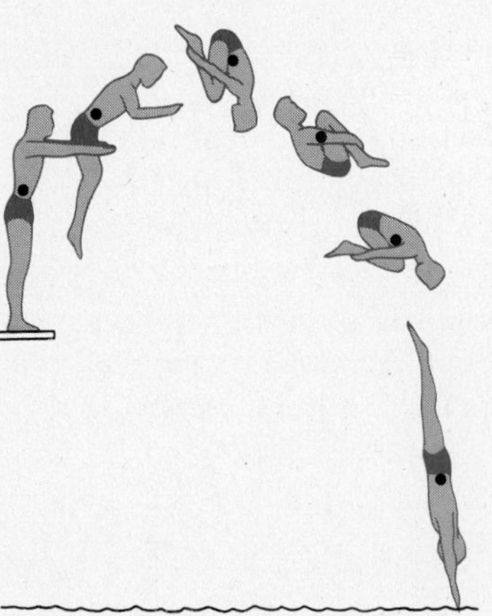

FIGURE 5-5
Diver rotates faster when arms and legs are tucked in than when outstretched. Conservation of angular momentum.

A similar example is the diver shown in Figure 5-5. The push as he leaves the board gives him an initial angular momentum about his cg and he rotates rapidly when his body is tucked in. At the end he stretches out, which increases his moment of inertia and so reduces the angular velocity to a small value, and then enters the water. The change in moment of inertia from the straight position to the tuck position can be as much as a factor of $3\frac{1}{2}$.

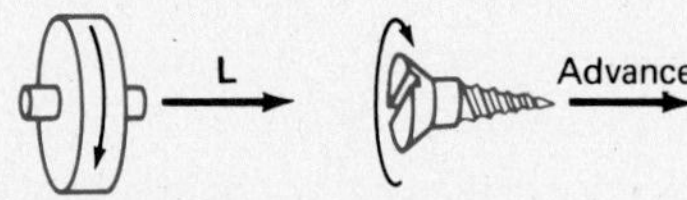

FIGURE 5-6
Direction of angular momentum is that in which a right-hand screw moves when turned.

Angular momentum, like linear momentum, is a vector quantity. It can be represented by a vector that points along the axis of rotation in the direction that a right-handed screw would advance, Figure 5-6. A child's top illustrates the conservation of angular momentum as a vector. If set on its tip, a top immediately falls over. But when it is spinning it remains upright for some time. If the spinning top were to fall, the axis of rotation would have changed drastically, and such a fall would correspond to a drastic change in the *direction* of the angular momentum. But the external torque on a spinning top is quite small so the angular momentum remains nearly constant and the top does not fall over; the stability of a bicycle or motorcycle depends partly on the same principle. A bike at rest tips over; but when it is moving, the angular momentum of the spinning wheels resists any tendency to change direction.

The gyroscope used by mariners also depends on the conservation of angular momentum. The rapidly spinning wheel is mounted on a complicated set of bearings so that even when the mount moves no net torque acts to change the direction of the angular momentum. Thus the axis of the wheel remains in the same direction in space (say, pointing at the north star), which is a great help for determining one's bearings at sea.

5-4 ▮ Work

The rest of this chapter is devoted to the very important concept of *energy* and the closely related concepts of *work* and *power*. All of these quantities are scalars and have no direction associated with them. Energy derives its importance from two sources. First, it is a conserved quantity; and secondly, energy is a concept that is useful not only in the study of motion, but in all areas of physics and other sciences as well. But before discussing energy† itself, we first examine the concept of work.

The concept of work

The word *work* has a variety of meanings in everyday language. But in physics, work is given a very specific meaning to describe what is accomplished by the action of a force. Specifically, the **work** done by a constant force on an object is defined to be *the product of the magnitude of the applied force times the distance through which the force acts.*

For example, if a person pushes a loaded grocery cart a distance of 50 m by exerting a horizontal force of 30 N on the cart, the person does 30 N × 50 m = 1500 N · m of work on the cart.

As this example shows, in SI units work is measured in newton-meters; for convenience, however, a special name is given to this unit, the *joule* (J): 1 J = 1 N · m. In the cgs system the unit of work is called the erg and is defined as: 1 erg = 1 dyne-centimeter. In British units, work is measured in ft · lb.‡

A force can be exerted on an object and yet do no work. For example, if you hold a heavy bag of groceries in your hands at rest, you do no work on it; you may become tired (and indeed energy is being expended by your muscles), but because the bag is not moved through a distance, the product of force × distance = 0. You also do no work if you carry the bag of groceries as you walk horizontally across the floor as shown in Figure 5-7. No horizontal force is required to move the package at a constant velocity. However, you do exert an upward force **F** on the package equal to its weight. But this upward force has nothing to do with the horizontal motion of the package; hence the upward force is doing no work.

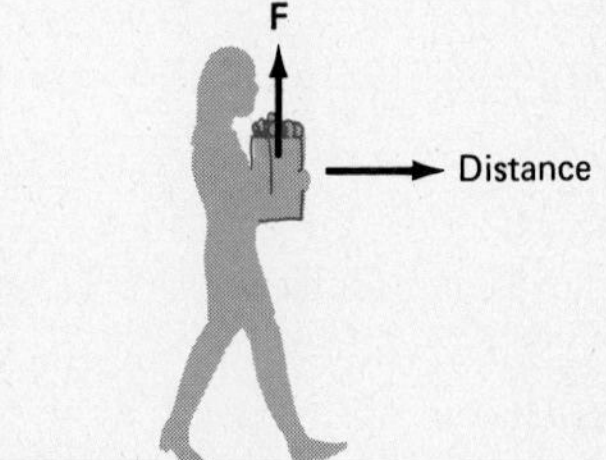

FIGURE 5-7
Work = 0 in this case since F is perpendicular to distance.

† Energy sources for practical use in homes and industry are discussed in Chapter 18, Section 18-6.

‡ It is easy to show that 1 J = 10^7 erg = 0.74 ft · lb. Note also that N · m and ft · lb serve both for work and for torque (Chapter 3). But the two quantities are very different; for one thing torque is a vector and work is a scalar. The special unit name, *joule*, is not used for torque; it is used only for work (and for energy, as we shall see).

Whenever the force is perpendicular to the motion, no work is done. This is what was meant when work was defined as the force times the distance *through which the force acts.*

In general terms, if a constant force acts at an angle to the direction of motion, the work can be calculated in one of two ways. Either we multiply the distance, d, times the component of the force parallel to the direction of motion (which we call $F_{\parallel}$); or, we multiply the force F times the component of distance parallel to the force, $d_{\parallel}$. In symbols, the work W is given by

$$W = F_{\parallel}d \tag{5-6a}$$

or

$$W = Fd_{\parallel}. \tag{5-6b}$$

These two relations are equivalent and we can use whichever one is easier to apply. To illustrate this, let us consider a boy pulling a wagon by exerting a force **F** at an angle θ to the horizontal as shown in Figure 5-8. If he pulls it a distance d along the ground, the work done is readily found using Equation 5-6*a*:

$$W = F_{\parallel}d$$

$$= (F\cos\theta)d = Fd\cos\theta. \tag{5-6c}$$

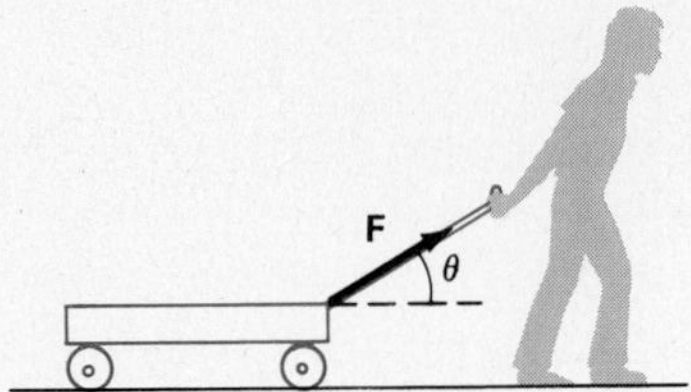

FIGURE 5-8
Work done by a force F acting at an angle θ to the ground is $Fd\cos\theta$.

If the force is 20 N, the angle 30°, and the wagon is pulled 100 m, the total work done is (20 N)(100 m)(0.866) = 1700 J. If we had used Equation 5-6*b* we would have $W = F(d\cos\theta)$, which is exactly equivalent. Equation 5-6*c* is valid in general where θ is the angle between the force **F** and the path taken.

As a second example, let us calculate the work done against gravity to carry a backpack up a hill as shown in Figure 5-9. Neglecting any acceleration that might occur, the person exerts an upward force F on the pack equal to its weight. Since the path followed is curved, Equation 5-6*a* would be difficult to use. Therefore we use Equation 5-6*b*. The distance parallel to the direction of the force, $d_{\parallel}$, is simply the height of the hill, h. Thus, if the pack weighs 150 N (15 kg) and the vertical height of the hill is 10 m, then the work done is 1500 J. The work done depends only on the change in elevation and not on the steepness of the hill or the path taken. To find the total work done by the person against gravity to climb the hill, we would proceed in the same way using the weight of the person plus the pack.

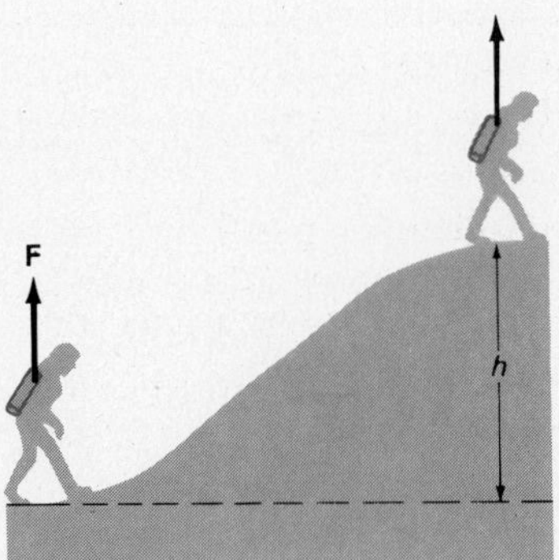

FIGURE 5-9
Work done in climbing a hill is found using Equation 5-6*b*.

When dealing with work, as with force, it is necessary to specify whether you are talking about work done *by* a specific object, or done *on* a specific object. It is also important to specify if the work done is due to a particular force, or due to the total net force on the object. For example, when a person gently lifts a 5-kg package from the ground up to a table 1.0-m high, there are two forces exerted on the package: the person pulling upward and gravity (its weight = 50 N) acting downward. If the package is lifted gently at constant speed, the upward force exerted by the person is equal to the weight and the work done *by the person* is $W =$ (50 N)(1.0 m) = 50 J. If we take the upward direction as positive, then the work done *by gravity* on the package will be $W = (-50\text{ N})(1.0\text{ m}) = -50\text{ J}$. (Note that, when the force is in the direction opposite to the change of position, the work is negative.) But the *net work* done on the package is

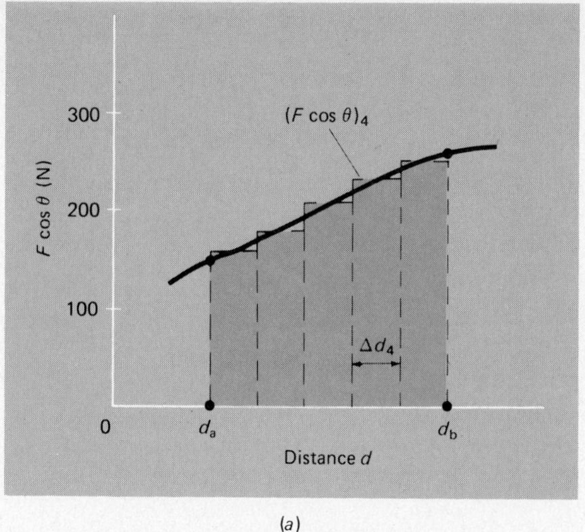

(a)

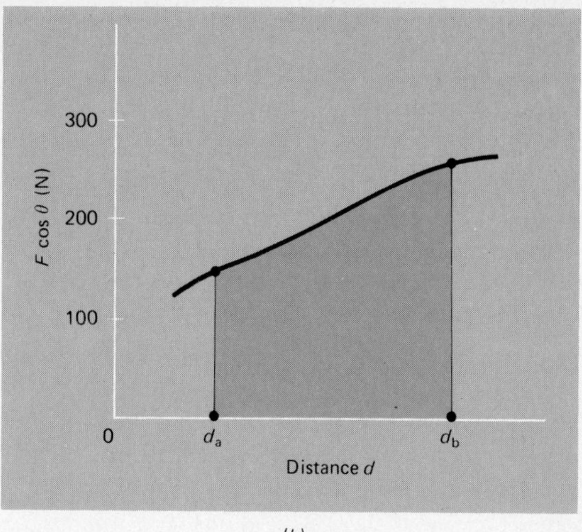

(b)

FIGURE 5-10
Work done by a force F can be calculated (*a*) by taking sum of areas of the rectangles; (*b*) area under the curve of $F \cos \theta$ vs. d.

zero (50 J − 50 J = 0 J), which is consistent with the fact that the net force on it is zero.

*Work done by a varying force

If the force acting on an object is constant, the work done by that force can be calculated using any of Equations 5-6. But in many cases the force varies in magnitude or direction during a process. For example, as a rocket moves away from earth, work is done against the force of gravity which varies as the inverse square of the distance from the earth's center; and the force exerted by a spring increases with the amount of stretch.

The work done by a varying force can be determined graphically. The procedure is like that for determining displacement when the velocity is known as a function of time (Section 1-7), and for determining the impulse delivered by a varying force (Section 5-1). To determine the work done by a variable force, we plot $F \cos \theta$ as a function of distance d, as in Figure 5-10*a*. We divide the distance into small segments Δd. For each segment, we indicate the average of $F \cos \theta$ by a horizontal dashed line. Then the work done during each segment is $\Delta W = (F \cos \theta)(\Delta d)$, which is the area of a rectangle (Δd) wide and $(F \cos \theta)$ high. The total work done to move the object a total distance $d = d_b - d_a$ is the sum of the areas of the rectangles (5 in the case shown in Figure 5-10*a*). Usually the average value of $(F \cos \theta)$ for each segment must be estimated, and a reasonable approximation of the work done can then be made. If we subdivide the distance into many more segments, Δd can be made smaller and our estimate more accurate. In the limit as Δd approaches zero, the total area of the many narrow rectangles approaches the area under the curve, Figures 5-10*b*. That is, *the work done by a variable force in moving an object between two points is equal to the area under the* $(F \cos \theta)$ *vs d curve between those two points.*

*Machines

Man has devised a great variety of machines to aid him in doing work. We are all familiar with many kinds of simple machines such as levers, pulleys, gears, and the wheel and axle. Although machines are often designed to reduce the force needed to lift a heavy load, they do not in fact save us work. Consider, for example, the lever shown in Figure 5-11. If the stone is lifted slowly, it is essentially in equilibrium, and using the torque rule ($\sum \tau = 0$; see Chapter 4) we see that a force of only $(50 \text{ kg})(g) = 500$ N is needed to lift the stone whose weight is $(200 \text{ kg})(g) = 2000$ N. In a sense, the force has been multiplied by a factor of 4, and we say the lever gives us a *mechanical advantage* of 4. However, from the geometry of the situation as shown in Figure 5-11*b*, the 500-N force must be exerted through a distance of 2.0 m to raise the stone 0.50 m. Thus the work done by the person is (500 N)(2.0 m) = 1000 J, and the work done on the stone is (2000 N)(0.50 m) = 1000 J, also. In other words, the input work equals the output work. Although the force has been multiplied, the work done has not been increased. This does not lessen the usefulness of the lever, for the rock couldn't be lifted at all without it.

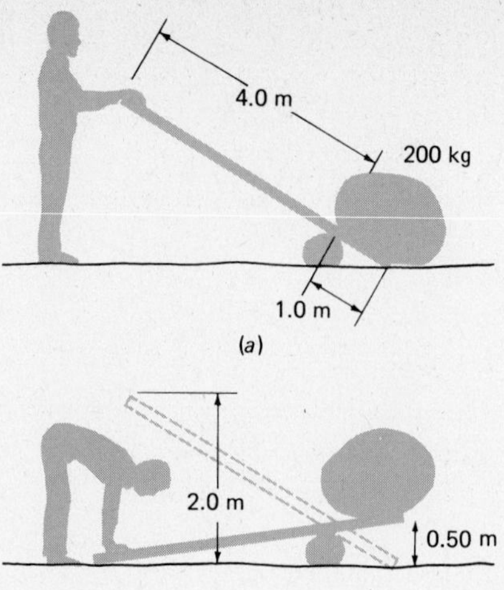

FIGURE 5-11
A simple lever.

This example has been idealized, for in practice it would be found that the force necessary to lift the 2000-N stone would be different than 500 N. This is due to two factors common to all machines: friction, and the weight of the machine itself (for example, in Figure 5-11 we would need to include the weight of the lever in calculating the torques). We therefore define an "ideal" and an "actual" mechanical advantage. The *ideal mechanical advantage* (IMA) is defined as the (output force)/(input force) for the ideal situation, which means the output work = input work, or $F'_i d_i = F'_o d_o$; the subscript "*o*" means output and "*i*" means input and the primes remind us we are talking of the force in the ideal case. From this relation we have $F'_o/F'_i = d_i/d_o$. Thus the ideal mechanical advantage is also the ratio of the distance d_i moved by the input force to the distance d_o moved by the load (the output):

$$\text{IMA} = \frac{d_i}{d_o}.$$

The *actual mechanical advantage* (AMA) is the ratio of the actual output force to actual input force for the particular machine

$$\text{AMA} = \frac{F_o}{F_i}.$$

Finally, the efficiency of any machine, be it simple or complex, is defined as (work output)/(work input). This can be written in terms of the IMA and AMA:

$$\text{efficiency} = \frac{F_o d_o}{F_i d_i}$$

$$= \left(\frac{F_o}{F_i}\right) \Big/ \left(\frac{d_i}{d_o}\right) = \frac{\text{AMA}}{\text{IMA}}.$$

For example, suppose it actually took a 550-N force to lift the 2000 N stone in Figure 5-11. The IMA = 2.0 m/0.50 m = 4.0, whereas the AMA = 2000 N/550 N = 3.6. The efficiency is then (3.6)/(4.0) = 0.90 or 90 percent.

Pulleys are often used to mechanical advantage. The pulley system shown in Figure 5-12 has an IMA of 2.0. This is readily seen by noting that

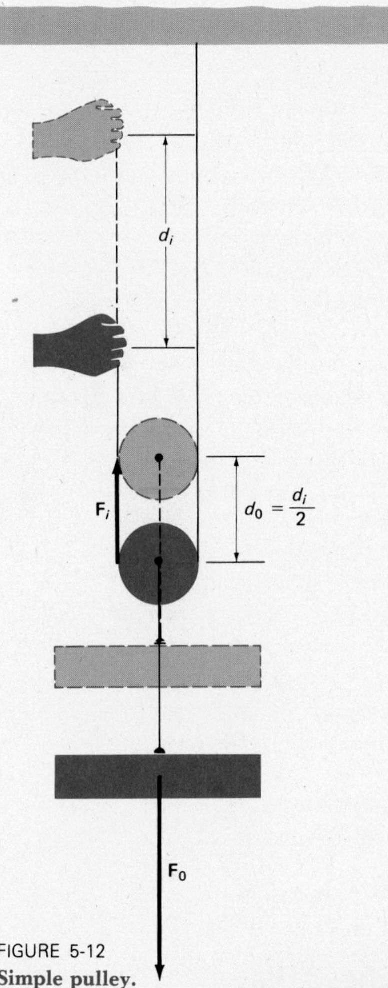

FIGURE 5-12
Simple pulley.

$d_o = d_i/2$. This can also be seen from the fact that there are two ropes supporting the weight and in the ideal case each need carry only half the weight. For more complicated systems, such as those shown in Figure 5-13, similar considerations will reveal that the IMA in general is equal to the number of ropes pulling up on the load. For pulleys, as with any type of machine, the AMA will always be less than the IMA. Hence the work output will be less than the work input.

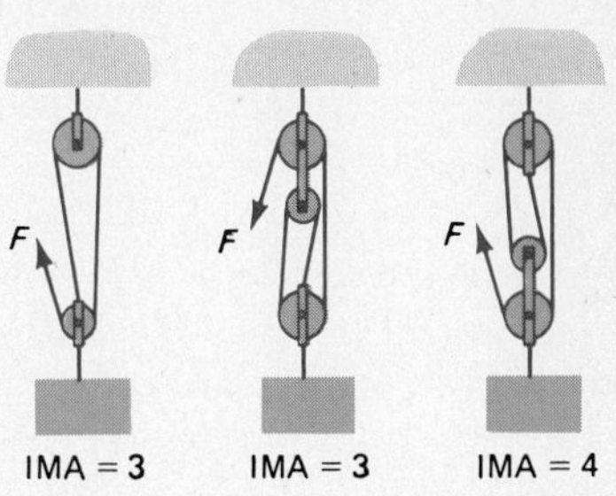

FIGURE 5-13
Various pulley systems.

5-5 ▌ Energy: types and transformations

Energy is defined as the *ability to do work.* Energy takes on many forms and we now discuss the two most basic, kinetic energy and potential energy.

Kinetic energy

A moving object can do work on another object it strikes. A flying cannonball does work on a brick wall it knocks down; a moving hammer does work on a nail it strikes. In either case, a moving object exerts a force on a second object and moves it through a distance. An object in motion has the ability to do work and thus can be said to have energy. We call the energy of motion **kinetic energy**, from the Greek word "kinetic," meaning motion.

In order to obtain a quantitative formula for kinetic energy, let us calculate how much work a moving object actually can do in a particular case. We consider, for the moment, only translational motion. Suppose the moving object has mass m and velocity v, and it strikes a second object on which it does work. To be concrete, we might think of a hammer striking a nail. Upon striking the nail, the hammer exerts an average force F on the nail over a distance d, Figure 5-14, and thus does an amount of work on the nail, $W = Fd$. By Newton's third law, the nail exerts a force equal to $-F$ on the hammer, and it is this force that accelerates (negatively) the hammer from its initial speed v to rest. Thus, $-F = ma$. The acceleration a (assumed constant) can be related to the initial speed v and the distance d using Equation 1-9c; since the final speed is zero, $v^2 = -2ad$. Combining these equations, we find the work done is

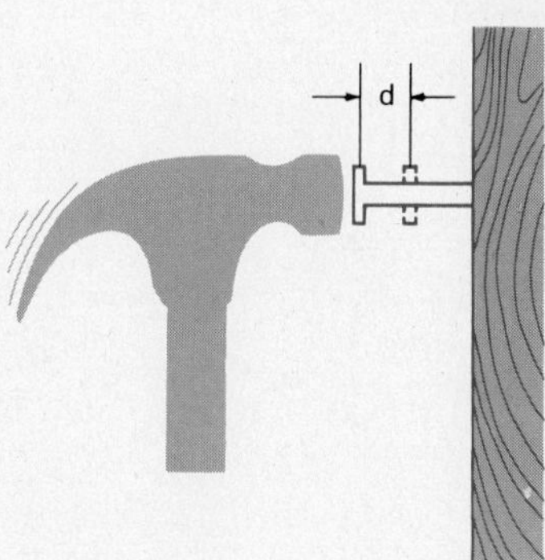

FIGURE 5-14
A hammer does work when striking a nail.

$$
\begin{aligned}
W &= Fd \\
&= -mad \\
&= m\left(\frac{v^2}{2d}\right)d \\
&= \tfrac{1}{2}mv^2.
\end{aligned}
$$

Thus, a body of mass m moving with a velocity v can do an amount of work equal to $\frac{1}{2}mv^2$ when it is brought to rest. We therefore *define* the quantity $\frac{1}{2}mv^2$ as the *translational kinetic energy* (KE) of the body:

$$\text{KE} = \tfrac{1}{2}mv^2. \tag{5-7}$$

Note that whereas the translational KE is directly proportional to the mass of the object, it is proportional to the *square* of the velocity. Thus, if the mass is doubled, the KE is doubled; but if the velocity is doubled, the object has four times as much KE and is therefore capable of doing four times as much work.

We have seen that a moving object can do work. The opposite is true as well: work must be done on an object to give it KE. To find the precise relationship, we reverse the above argument. Suppose an object of mass m is moving with an initial speed v_0, and to accelerate it (uniformly) to a speed v a net force F is exerted on it parallel to its motion over a distance d; then the net work done on it is $W=Fd$. Using Newton's second law, $F=ma$, and Equation 1-9c, $v^2=v_0^2+2ad$, we find

$$\begin{aligned} W &= Fd \\ &= mad \\ &= m\left(\frac{v^2-v_0^2}{2d}\right)d \\ W &= \tfrac{1}{2}mv^2-\tfrac{1}{2}mv_0^2. \end{aligned}$$

That is

▮ **The net work done on an object is equal to its change in kinetic energy.**

This is known as the **work-energy theorem**. Notice, however, that since we made use of Newton's second law, $F=ma$, F must be the *net* force—that is, the sum of all forces acting on the object.

The connection between work and kinetic energy thus operates in both directions. First, if work is done on an object, its kinetic energy increases. Second, if an object has kinetic energy, it can do work on something else; and if it does, its own KE decreases.

Because of the direct connection between work and kinetic energy, energy must be measured in the same units as work: joules in SI units, ergs in the cgs, and ft · lbs in the British system.

▮ EXAMPLE 5-5 An 80-gram bird flies at a speed of 10 m/s (36 km/h). (*a*) What is its KE? (*b*) How much work was done to reach this speed starting from rest?

SOLUTION (*a*) $\text{KE}=\tfrac{1}{2}mv^2=\tfrac{1}{2}(0.080\text{ kg})(10\text{ m/s})^2=4.0\text{ J}$. (*b*) Since the initial KE was zero the work done is just equal to the final KE, 4.0 J. ▮▮

▮ EXAMPLE 5-6 How much work is required to accelerate a 1000-kg car from 20 m/s to 30 m/s?

SOLUTION The work needed is equal to the increase in kinetic energy:

$$\begin{aligned} W &= \tfrac{1}{2}mv^2-\tfrac{1}{2}mv_0^2 \\ &= \tfrac{1}{2}(1000\text{ kg})(30\text{ m/s})^2-\tfrac{1}{2}(1000\text{ kg})(20\text{ m/s})^2 \\ &= 2.5\times10^5\text{ J}. \end{aligned}$$

▮▮

Rotational kinetic energy

The quantity $\tfrac{1}{2}mv^2$ is the kinetic energy of a body undergoing translational motion. A body rotating about an axis is said to have *rotational kinetic energy* and by analogy we would expect this to be given by the formula $\frac{1}{2}I\omega^2$ where I is the moment of inertia of the body and ω is its angular velocity. We can indeed prove that this is true. Consider any rotating object as made up of many tiny particles, each of mass m. If we let r represent the distance of any one particle from the axis of rotation, then its linear velocity is $v = r\omega$. The total kinetic energy of the whole body will be the sum of the KEs of all its particles:

$$\begin{aligned} \text{KE} &= \sum(\tfrac{1}{2}mv^2) \\ &= \sum(\tfrac{1}{2}mr^2\omega^2) \\ &= \tfrac{1}{2}(\sum mr^2)\omega^2 \end{aligned}$$

where we have factored out the $\frac{1}{2}$ and the ω^2 since they are the same for every particle. Since $\sum mr^2 = I$, the moment of inertia, we see that the kinetic energy of a rotating object, as expected, is:

$$\text{rotational KE} = \tfrac{1}{2}I\omega^2. \qquad (5\text{-}8)$$

An object that rotates while its cg undergoes translational motion will have both translational and rotational KE. (Such "general motion" was previously mentioned in Section 3-7. A worked-out example is given in Section 5-6).

Potential energy

An object can have energy not only by virtue of its motion, but also by virtue of its shape or position. This is called **potential energy** (PE).

A wound up watch spring, for example, has PE. For, as it unwinds, it does work to move the watch hands around. The watch spring acquired its potential energy because work was done *on* it by the person winding the watch.

Perhaps the most common example of PE is *gravitational potential energy*. A heavy brick held high in the air has potential energy because of its position. It has the ability to do work, for if it is released it will fall to the ground and can do work on (say) a stake, driving it into the ground. Let us determine quantitatively the gravitational PE of an object. In order to lift an object of mass m, a force at least equal to its weight, mg, must be exerted on it, say, by a person's hand. In order to lift it to a height h above the ground, Figure 5-15, the work done on it by the person will be the product of the force, mg, and the distance h; that is, $W = mgh$. If we instead allow the object to fall freely under the action of gravity and drive a stake into the ground, it will do an amount of work equal to mgh on the stake, a fact which can be verified by using the kinematical equations as we did a few pages back for kinetic energy. To summarize, an amount of work equal to mgh is needed to raise an object of mass m to a height h; and if allowed to fall a height h, an object can do an amount of work equal to mgh. Therefore we define the *gravitational potential energy* of a body as the product of its weight mg and its height h:

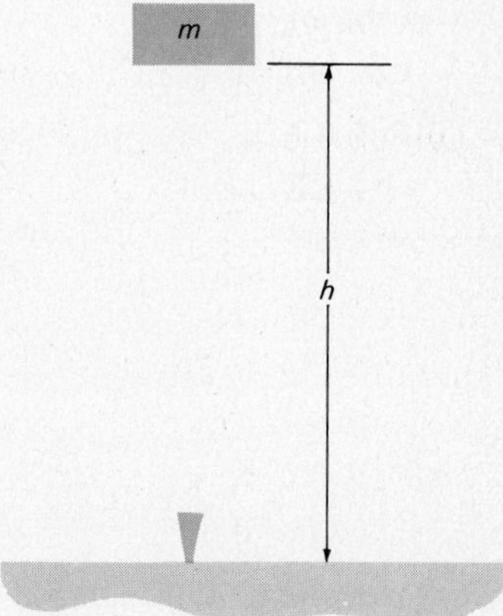

FIGURE 5-15
An object of mass m at height h above the ground can do an amount of work $W = mgh$ when it falls.

$$\text{gravitational PE} = mgh. \qquad (5.9a)$$

Notice that the gravitational PE depends only on the *vertical height* of the object above some reference level, in this case the ground. As we saw earlier in this chapter, the work required to lift an object a height h does not depend on the path taken—that is, on whether one lifts it vertically or goes up a hill as in Figure 5-9. Similarly, the work it can do when descending does not depend on whether the object falls vertically or by some other path (as down a frictionless incline), but depends only on the vertical height h.

In some situations you may wonder from what point to measure the height h. For example, if you hold a book above a table, should you measure h from the top of the table or from the floor? The answer is that it doesn't matter as long as you use the same level throughout a given problem or situation. This is true because it is only *changes* in PE that are important and changes do not depend on the reference level, as the following example shows.

▌ EXAMPLE 5-7 A 1000-kg car moves from point A, Figure 5-16, to point B and then point C. (a) What is its PE at B and C relative to point A? (b) What is its change in potential energy when it goes from B to C?

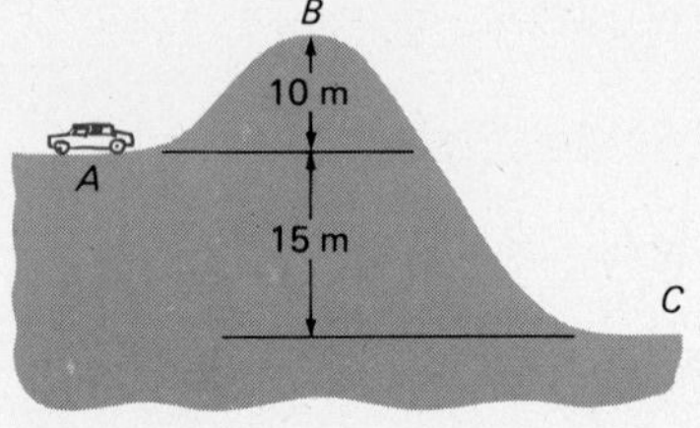

FIGURE 5-16
Example 5-7.

SOLUTION (a) Let us measure the heights from point A, so initially the car has zero PE. At point B, where $h = 10$ m,

$$PE_B = mgh = (1000\ \text{kg})(9.8\ \text{m/s}^2)(10\ \text{m}) = 1.0 \times 10^5\ \text{J}.$$

At point C, $h = -15$ m since C is below A. (We have arbitrarily chosen the upward direction as positive.) Therefore,

$$PE_C = mgh = (1000\ \text{kg})(9.8\ \text{m/s}^2)(-15\ \text{m}) = -1.5 \times 10^5\ \text{J}.$$

(b) In going from B to C, the potential energy change is:

$$PE_C - PE_B = (-1.5 \times 10^5\ \text{J}) - (1.0 \times 10^5\ \text{J}) = -2.5 \times 10^5\ \text{J}.$$

That is, the car loses 2.5×10^5 J of potential energy. ▌▌

We could have chosen the reference point in this last example to be, say, point C. In this case, the car would have had a PE initially equal to $(1000\ \text{kg})(9.8\ \text{m/s}^2)(15\ \text{m}) = 1.5 \times 10^5$ J since $h = 15$ m in this case. At B, its PE would have been 2.5×10^5 J, and at C it would have been zero. But the change in PE going from B to C (or from A to B) gives the same result as before. In general, if the object is initially a height h_0 above the reference point, then the change in potential energy when it moves to a different height h is:

$$\text{change in PE} = mgh - mgh_0. \tag{5-9b}$$

There are other kinds of PE besides gravitational. In general, we define the *change in PE of an object when it is moved between two points (or shapes) as equal to the work needed to take it from one point to the other.* We consider now one other type of PE, that associated with elastic materials. This includes a great variety of practical applications. To take a simple example, consider a spring such as that in a watch, or better yet the simple coil spring shown in Figure 5-17. The spring has potential energy when compressed (or stretched); for when it is released it can do work on a ball as shown. Like other elastic materials, a spring is accurately described by Hooke's law: the force F required to compress (or stretch) it a distance x from its normal length is proportional to this distance x. That is, $F = kx$

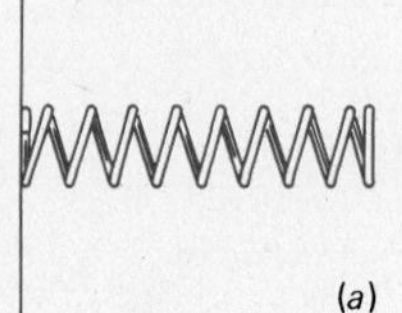

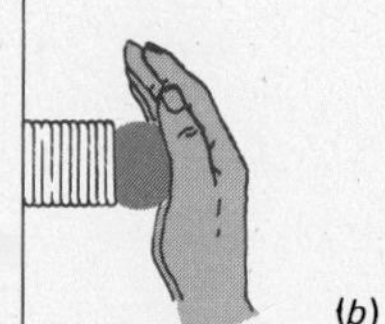

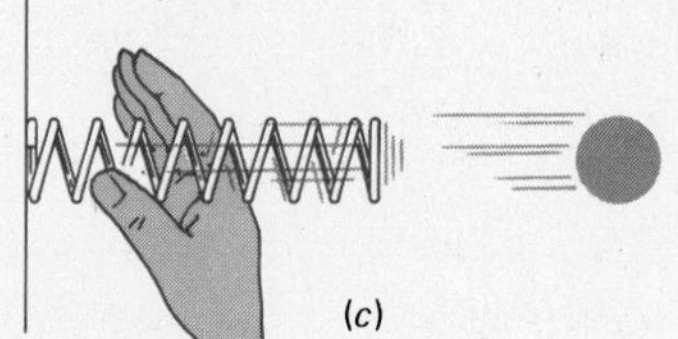

FIGURE 5-17
A spring (a) can store energy (elastic PE) when compressed (b), which can be used to do work when released (c).

where k is a proportionality constant often called the "spring stiffness constant." In order to calculate the potential energy of a compressed spring, we merely need to calculate the work required to compress it (Figure 5-17*b*), or the work it does when released (Figure 5-17*c*). In either case, the work done is $W = Fx$ where x is the amount it is compressed from its normal length. The force F varies over this distance, becoming greater the more the spring is compressed. We want the average force, and since it varies from zero at the uncompressed position to kx when fully compressed, the average force must be $\frac{1}{2}kx$. The work done is then $W = (\frac{1}{2}kx)(x) = \frac{1}{2}kx^2$. Hence the *elastic potential energy* is proportional to the square of the amount of compression:

$$\text{elastic PE} = \tfrac{1}{2}kx^2. \tag{5-10}$$

If a spring is *stretched* a distance x beyond its normal length, it too has potential energy given by this equation. Thus x can be either the amount compressed or amount stretched from the normal position.

In each of the above examples of potential energy—from a brick held at a height h to a compressed spring—an object has the capacity or *potential* to do work even though it is not yet actually doing it. That is why we use the term "potential" energy. From these examples we can also see that energy can be *stored*, for later use, in the form of potential energy.

*Conservative forces and the work-energy theorem

The work done against gravity in moving an object from one point to another does not depend on the path taken. For example, it takes the same work ($=mgh$) to lift an object of mass m vertically a height h as to carry it up a hill of height h as in Figure 5-9. (See discussion after Equations 5-6 in Section 5-4.) Forces such as gravity for which the work done does not depend on the path taken but depends only on the initial and final positions are called *conservative forces.*† Friction, on the other hand, is not a conservative force since the work done in, say, pushing a crate across a floor from one point to another depends on whether the path taken is straight, or is curved or zigzag. (In the two latter cases more work is required since the distance is greater and, unlike the gravitational force, the friction force is always directed precisely opposite to the direction of motion—so the work done depends on the particular path taken.)

An important fact is that *potential energy can be defined only for a conservative force.* Thus, although potential energy is always associated with a force, we can't write PE for just any force—such as friction which is a nonconservative force.

We can now state the **work-energy theorem** (discussed earlier in the subsection on kinetic energy) to include potential energy:

$$\text{Work done by a force} = \Delta\text{KE} + \Delta\text{PE}.$$

That is, *the work done by a force on a body is equal to the total change in kinetic and potential energy of that body.* This is true, however, only if any *other* forces acting (like gravity) are included in the PE term on the right. We can, however, bring the PE term over to the left so that all forces are treated as

† You should be able to convince yourself that the elastic force of, say, a spring (or other elastic material) in which $F = kx$ is also a conservative force.

doing work. (Indeed, it is arbitrary whether a given force is considered as doing work, or as involved in a change in PE—except that nonconservative forces such as friction must be included in the work term.) In this case the work done is the *net work* since it includes the sum of all the forces. Hence we have

$$\text{net work done} = \Delta\text{KE}$$

which is the original form of the work–energy theorem discussed earlier.

Other forms of energy

Besides the KE and PE of ordinary objects, other forms of energy exist as well. These include electric energy, nuclear energy, thermal energy, and the chemical energy stored in food and fuels. With the advent of the atomic theory, these other forms of energy have come to be considered as kinetic or potential energy at the atomic or molecular level. For example, according to the atomic theory, thermal energy is interpreted as the kinetic energy of rapidly moving molecules—when an object is heated, the molecules that make up the object move faster. On the other hand, the energy stored in food and fuel such as gasoline can be regarded as potential energy stored by virtue of the relative positions of the atoms within a molecule. For this energy to be used to do work, it must be released, usually through a chemical reaction. This is something like a compressed spring which, when released, can do work. Enzymes in our bodies allow the release of energy stored in food molecules. The violent spark of a spark plug in an automobile allows the mixture of gas and air to react chemically, releasing the stored energy and doing work against the piston to propel the car forward. Electric, magnetic, and nuclear energy can also be considered examples of kinetic and potential (or stored) energy. We will deal with these other forms of energy in detail in later chapters.

Transformation of energy

Energy can be transformed from one form to another, and we have already encountered several examples of this. A stone held high in the air has potential energy; as it falls, it loses potential energy since its height above the ground decreases (Equation 5-9); at the same time it gains in kinetic energy since its velocity is increasing. Potential energy is being transformed into kinetic energy.

Often the transformation of energy involves a transfer of energy from one body to another. The PE stored in the spring of Figure 5-17*b* is transformed into KE of the ball, Figure 5-17*c*. Water at the top of a dam has potential energy which is transformed into kinetic energy as the water falls; at the base of the dam, the kinetic energy of the water can be transferred to turbine blades and further transformed to electric energy as we shall see in a later chapter. The potential energy stored in a bent bow can be transformed into kinetic energy of the arrow.

In each of these examples, the transfer of energy is accompanied by the performance of work. The spring of Figure 5-17 does work on the ball; water does work on the turbines blades; the bow does work on the arrow. This observation gives us a further insight into the relation between work and energy: *Work is done whenever energy is transferred from one object to another.* Another example is when a person throws a ball or pushes a grocery cart. The work done is a manifestation of energy being transferred from the person (ultimately derived from the chemical energy of food) to the ball or cart.

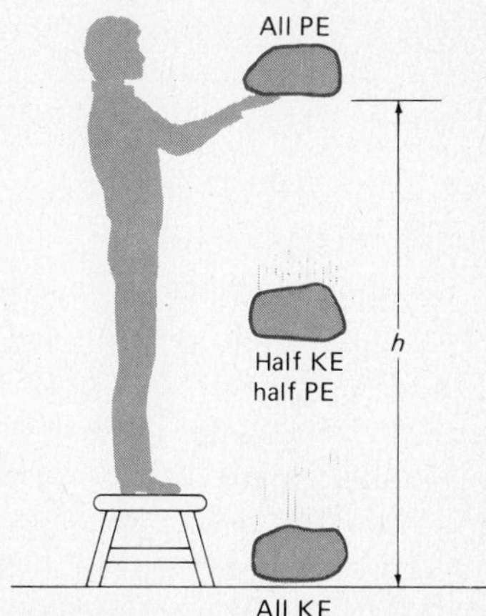

FIGURE 5-18
The stone's PE changes to KE as it falls.

5-6 ▮ The conservation of energy

The law of conservation of energy

Whenever energy is transformed, it is found that no energy is gained or lost in the process. As an example, let us consider a stone that is allowed to fall toward the ground (Figure 5-18). Before it is dropped, it has potential energy equal to mgh (we take the ground as the reference level). As it falls, its PE decreases but its KE increases. Just before hitting the ground, it has only kinetic energy. In fact, its KE at the bottom is exactly equal to the PE it had at the top. We can prove this by using Equation 1-9c; since $v_0 = 0$ and $a = g$, we have $v^2 = 2gh$ where h is the distance fallen. Thus, the KE after falling a distance h is

$$\begin{aligned}\tfrac{1}{2}mv^2 &= \tfrac{1}{2}m(2gh)\\ &= mgh.\end{aligned}$$

In other words, the original PE, equal to mgh, has been transformed entirely into KE, with the total amount neither increasing nor decreasing.

This example is an illustration of one of the most important principles in physics, the **law of conservation of energy**:

> **▮ The total energy is neither increased nor decreased in any process. Energy can be transformed from one kind to another, but the total amount remains constant.**

For mechanical systems, this law can be derived from Newton's laws and thus is equivalent to them. But in its full generality, the validity of the law of conservation of energy rests on experimental observation. Even though Newton's laws are found to fail in the submicroscopic world of the atom, the law of conservation of energy (and of momentum and angular momentum as well) has been found to hold in every experimental situation so far tested.

We need a brief aside here to distinguish the use of the word "conservation" in everyday life from its specific use in physics. In everyday usage, conservation means "saving" or "using wisely"—such as when it is said that we must "conserve energy" (what is really meant is that we must conserve *fuel*). In physics the word conservation refers to a quantity that remains strictly *constant.* To be clear, we can say that energy *is* conserved (we don't have to try to conserve it), but we should try to conserve our fuel resources so that we don't use them up too quickly.

Also, it is important to recognize that energy is not a material entity. It is not substance or matter of any kind. Energy is a concept, an abstraction. The concept of energy was invented by human minds. Some physicists would say that energy is a mathematical function. This, of course, doesn't lessen its importance. That we can define a quantity which can take various forms (such as $\frac{1}{2}mv^2$ or mgh) but which always remains constant is very remarkable—and very useful, as we shall see.

Let us now consider some simple illustrations of the law of conservation of energy. For the moment we consider mechanical systems in which friction can be neglected. In this case the law of conservation of energy can be stated in the following way:

$$\text{KE} + \text{PE} = \text{constant.} \tag{5-11}$$

That is, the sum of the kinetic plus potential energies of an object or system of objects, which is called the *total mechanical energy*, remains constant. For example, in the case of the rock falling under gravity, Figure 5-18, the rock initially has only potential energy. As it falls, its PE decreases, but its KE increases to compensate so that the sum of the two remains constant. At any point along the path, the total mechanical energy is given by $\frac{1}{2}mv^2 + mgh$ where h is its height above the ground at that point and v is its velocity at that point. If we let the subscript 1 represent the rock at one point along its path (e.g., the initial point), and 2 represent it at some other point, then according to Equation 5-11, we have:

$$\tfrac{1}{2}mv_1^2 + mgh_1 = \tfrac{1}{2}mv_2^2 + mgh_2. \tag{5-12}$$

To see the practical value of this relation, suppose the original height of the stone in Figure 5-18 is $h_1 = 3.0$ m, and we wish to calculate its speed when it has fallen to 1.0 m above the ground. Then, since $v_1 = 0$ (the moment of release), $h_2 = 1.0$ m, and $g = 9.8\ \text{m/s}^2$, Equation 5-12 tells us

$$0 + (m)(9.8\ \text{m/s}^2)(3.0\ \text{m}) = \tfrac{1}{2}mv_2^2 + (m)(9.8\ \text{m/s}^2)(1.0\ \text{m}).$$

The m's cancel out, and solving for v_2^2, we find $v_2^2 = 2[(9.8\ \text{m/s}^2)(3.0\ \text{m}) - (9.8\ \text{m/s}^2)(1.0\ \text{m})] = 39.2\ \text{m}^2/\text{s}^2$, and $v_2 = \sqrt{39.2}\ \text{m/s} = 6.3\ \text{m/s}$.

Equation 5-12 can be applied to any object moving without friction under the action of gravity. For example, Figure 5-19 shows a car starting from rest at the top of a hill, and coasting without friction to the bottom and up the hill on the other side. Initially, the car has only PE. As it coasts down the hill, it loses PE and gains in KE, but the sum of the two remains constant. At the bottom of the hill, it has its maximum KE and as it climbs up the other side, the KE changes back to PE. When the car comes to rest again, all of its energy will be PE. Since PE is proportional to height and because energy is conserved, the car comes to rest at a height equal to its original height. If the two hills are the same height, the car just barely reaches the top of the second hill when it stops. If the second hill is lower than the first, not all of the car's KE will be transformed to PE and the car continues over the top and down the other side. If the second hill is higher, the car will only reach a height on it equal to its original height on the first hill. This is true no matter how steep the hill is, since PE depends only on the vertical height.

FIGURE 5-19
Car coasting down a hill illustrates conservation of energy.

▌ EXAMPLE 5-8 Assuming the height of the hill in Figure 5-19 is 40 m, calculate (a) the velocity of the car at the bottom of the hill and (b) at what height it will have half this speed.

SOLUTION (a) We use Equation 5-12 with $v_1 = 0$, $h_1 = 40$ m, and $h_2 = 0$. Then

$$0+(m)(9.8\ \mathrm{m/s^2})(40\ \mathrm{m})=\tfrac{1}{2}mv_2^2+0.$$

The m's cancel out and we find $v_2=28$ m/s. (*b*) We use the same equation but now $v_2=14$ m/s and h_2 is unknown:

$$0+(m)(9.8\ \mathrm{m/s^2})(40\ \mathrm{m})=\tfrac{1}{2}(m)(14\ \mathrm{m/s})^2+(m)(9.8\ \mathrm{m/s^2})(h_2).$$

we cancel the m's and solve for h_2 and find $h_2=30$ m. That is, the car has a speed of 14 m/s when it is 30 *vertical* meters above the lowest point, both when descending the left-hand hill and when ascending the right-hand hill. ▌▌

There are many interesting examples of the conservation of energy in athletics, one of which is the pole vault illustrated in Figure 5-20. In terms of energy, the sequence of events is as follows: KE of the running athlete is transformed (as he leaves the ground) into gravitational PE plus elastic PE of the bending pole; and then as he reaches the top and the pole straightens out again, it has all been transformed to gravitational PE (plus a tiny amount of KE). The pole does not supply any energy, but it acts as a very convenient device to *store* energy and thus aid in the transformation of KE into gravitational PE, which is the net result. The energy required to pass over the bar depends on how high the cg of the vaulter must be raised. By bending their bodies pole vaulters can keep their cg so low that it actually passes slightly beneath the bar, thus enabling them to cross over a higher bar than would otherwise be possible.

▌ EXAMPLE 5-9 Calculate the kinetic energy and the velocity required for a 70-kg pole vaulter to pass over a 5.0-m high bar. Assume the vaulter's cg is initially 0.90 m off the ground and reaches its maximum height at the level of the bar itself.

SOLUTION We equate the total energy just before the vaulter leaves the ground with his total energy as he passes over the bar (we ignore the small amount of KE at this point). Thus

$$\tfrac{1}{2}mv^2+0=0+mgh$$

and

$$\mathrm{KE}=\tfrac{1}{2}mv^2=mgh$$
$$=(70\ \mathrm{kg})(9.8\ \mathrm{m/s^2})(4.1\ \mathrm{m})=2.8\times10^3\ \mathrm{J}.$$

The velocity $v=\sqrt{\dfrac{2\mathrm{KE}}{m}}=\sqrt{\dfrac{2(2800\ \mathrm{J})}{70\ \mathrm{kg}}}=8.9\ \mathrm{m/s}.$ ▌▌

FIGURE 5-20
Transformation of energy during a pole vault.

Next we consider an example involving rotation: a sphere rolling down a hill as in Figure 5-21. The kinetic energy consists of two parts: the "translational" kinetic energy of the cg of the sphere ($\frac{1}{2}mv^2$) plus the rotational KE of the sphere about its cm ($\frac{1}{2}I\omega^2$).

▌ EXAMPLE 5-10 What will be the speed of a sphere of mass M and radius R when it reaches the bottom of an incline if it starts rolling from rest at a vertical height H?

FIGURE 5-21
Sphere rolling down a hill has both translational and rotational kinetic energy, Example 5-10.

SOLUTION The total energy at any point a distance h above the ground is

$$\tfrac{1}{2}Mv^2 + \tfrac{1}{2}I\omega^2 + Mgh.$$

We equate the total energy at the top ($h = H$ and $v = \omega = 0$) to the total energy at the bottom ($h = 0$):

$$0 + 0 + MgH = \tfrac{1}{2}Mv^2 + \tfrac{1}{2}I\omega^2 + 0.$$

Referring back to Chapter 3, we see† that $\omega = v/R$ and $I = \frac{2}{5}mr^2$. Hence

$$\tfrac{1}{2}Mv^2 + \tfrac{1}{2}(\tfrac{2}{5}MR^2)\left(\frac{v^2}{R^2}\right) = MgH.$$

Canceling the Ms and Rs, we obtain

$$(\tfrac{1}{2} + \tfrac{1}{5})v^2 = gH$$

or

$$v = \sqrt{\tfrac{10}{7}gH}.$$

This can be compared to the velocity of an object that slides the same distance without rotating, in which case, $v = \sqrt{2gH}$, which is greater. ▌▌

We have neglected friction up to now, but in many situations it cannot be ignored. In a real situation the car in Figure 5-19, for example, will not in fact reach the same height on the second hill as it had on the first hill because of friction. In this and in other natural processes the sum of the kinetic and potential energies does not remain constant but decreases. Because of this, frictional forces are called *dissipative forces.* Historically this fact hindered the formulation of a comprehensive conservation of energy law until well into the nineteenth century. It was not until then that heat, which is always produced when friction is present (try rubbing your hands together), was recognized as energy. Quantitative studies by nineteenth century scientists (this is discussed in Chapters 9 and 10) demonstrated that if heat is interpreted as energy (properly called *thermal energy*) then the total energy is conserved in any process. For example, if the car in Figure 5-19 is subject to frictional forces, then the initial total energy of the car will be equal to the kinetic plus potential energy of the car at any subsequent point along its path plus the amount of thermal energy produced in the process. The thermal energy produced by a

† In the equation $v = \omega R$, v usually refers to the speed of a point on the edge of the rotating object with respect to the center or axis. In the present case, v is the speed of the center with respect to the point of contact which at any instant is momentarily at rest. From the symmetry of the situation, it can be seen that these two speeds are the same, which justifies our use of $v = \omega R$.

friction force F_{fr} is equal to the work done against this force: $W = F_{fr}s$, where s is the distance over which the force acts. Equation 5-12 can then be rewritten to take into account friction:

$$\tfrac{1}{2}mv_1^2 + mgh_1 = \tfrac{1}{2}mv_2^2 + mgh_2 + F_{fr}s$$

where s is the actual distance along the path traveled by the object in going from point 1 to point 2.

▌ EXAMPLE 5-11 The car in Example 5-8 is found to reach a vertical height of only 25 m on the second hill before coming to a stop. It traveled a total distance of 400 m. Calculate the average friction force on the car. Assume the car has a mass of 1000 kg.

SOLUTION We take point one to be the instant when the car started coasting and point two when it stopped. Then $v_1 = 0$, $h_1 = 40$ m, $v_2 = 0$, $h_2 = 25$ m, and $s = 400$ m. Thus

$$0 + (1000\text{ kg})(9.8\text{ m/s}^2)(40\text{ m}) = 0 + (1000\text{ kg})(9.8\text{ m/s}^2)(25\text{ m}) + F_{fr}(400\text{ m}).$$

We solve this for F_{fr} to find:

$$F_{fr} = 370\text{ N}.$$

▌▌

Another example of the transformation of kinetic energy into thermal energy occurs when an object, such as the rock of Figure 5-18, strikes the ground. Both the rock and the ground will be slightly warmer as a result of their collision. A more apparent example of this transformation of kinetic energy into heat can be observed by vigorously striking a nail several times with a hammer and then gently touching the nail with your finger.

When other forms of energy are involved, such as chemical or electrical energy, the total amount of energy is always found to be conserved. Hence the law of conservation of energy is believed to be universally valid.

Energy and momentum conservation in collisions

We saw in Section 5-2 that in the collision of two objects such as billiard balls the total momentum is conserved. If the two objects are very hard and elastic and no heat is produced in the collision, then kinetic energy is conserved as well. By this we mean that the sum of the kinetic energies of the two objects is the same after the collision as before. Of course for the brief moment during which the two objects are in contact, some (or all) of the energy is stored momentarily in the form of elastic potential energy; but if we compare the total kinetic energy before the collision with the total after the collision, they are found to be the same. Such a collision is said to be *perfectly elastic*. If we use the subscripts 1 and 2 to represent the two objects, we can write this as

$$\tfrac{1}{2}m_1v_1^2 + \tfrac{1}{2}m_2v_2^2 = \tfrac{1}{2}m_1v_1'^2 + \tfrac{1}{2}m_2v_2'^2 \quad \begin{pmatrix}\text{elastic}\\ \text{collision}\end{pmatrix} \tag{5-13}$$

where primed quantities (′) mean after the collision and unprimed mean before the collision.

Although at the atomic level the collisions of atoms and molecules are often elastic, in the "macroscopic" world of ordinary objects an elastic collision is an ideal that is never quite reached since at least a little thermal energy (and perhaps sound and other forms of energy) is always produced during a collision. The collision of two hard elastic balls, such as billiard balls, however, is very close to perfectly elastic and we often treat it as such. Even when the KE is not conserved, the *total* energy is, of course, conserved.

Collisions in which kinetic energy is not conserved are said to be *inelastic*. The kinetic energy that is lost is changed into other forms of energy, usually thermal energy, so that the total energy is conserved. In this case we can write that $KE_1 + KE_2 = KE_1' + KE_2' +$ thermal and other forms of energy.

If the collision is elastic, we can make use of the conservation of momentum (Equation 5-4) and of kinetic energy (Equation 5-13) to determine the speeds of the objects after a collision, if we know their speeds beforehand as well as their directions. But if the collision is inelastic we cannot do this unless somehow the amount of heat generated can be measured (which is not often easy to do). However, if we know the initial and final velocities we can calculate the heat energy produced as result of the collision.

❚ EXAMPLE 5-12 A billiard ball at rest is struck head on by a second billiard ball moving at a speed of 2.0 m/s, Figure 5-22. The balls each have a mass of 0.10 kg. Calculate the speeds of the two balls after the collision. Ignore spin and frictional effects and assume the collision is perfectly elastic and that all the motion is along a straight line.

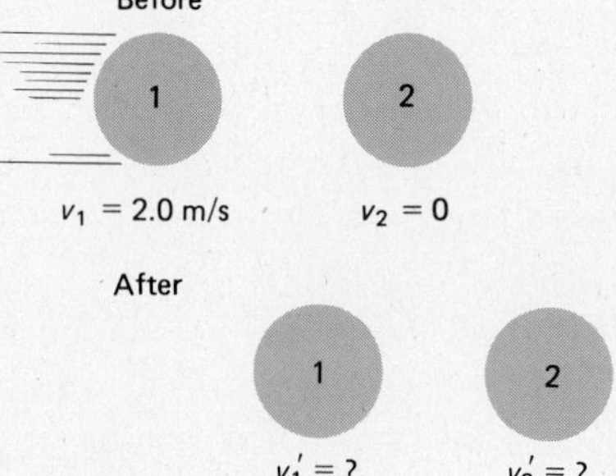

FIGURE 5-22
Collision of two billiard balls, Example 5-12.

SOLUTION If the ball initially at rest is called ball two, then $v_2 = 0$ and $v_1 = 2.0$ m/s. We wish to find v_1' and v_2', the speeds after the collision. From Equation 5-4 we have:

$$(0.10\text{ kg})(2.0\text{ m/s}) + 0 = (0.10\text{ kg})v_1' + (0.10\text{ kg})v_2'$$

and from Equation 5-13

$$\tfrac{1}{2}(0.10\text{ kg})(2.0\text{ m/s})^2 + 0 = \tfrac{1}{2}(0.10\text{ kg})v_1'^2 + \tfrac{1}{2}(0.10\text{ kg})v_2'^2.$$

Since we have two equations in two unknowns, the problem can be solved. From the first equation we find

$$v_1' = (2.0\text{ m/s}) - v_2'$$

and substituting this into the second equation we find

$$v_2'^2 - (2.0\text{ m/s})v_2' = 0.$$

Solving,† we find $v_2' = 2.0$ m/s and $v_1' = 2.0\text{ m/s} - v_2' = 0$. Hence, ball number one is brought to rest and ball number two, initially at rest, acquires the original speed of ball one, a fact often observed by billiard players. ❚❚

❚ EXAMPLE 5-13 For the inelastic collision we considered in Example 5-3, calculate how much heat is produced.

† There is a second solution to this equation, $v_2' = 0$, and $v_1' = 2.0$ m/s; but this merely corresponds to the two balls missing each other.

SOLUTION Initially, the total kinetic energy is $\frac{1}{2}m_1v_1^2 = \frac{1}{2}(10{,}000\ \text{kg})(20\ \text{m/s})^2 = 2.0\times10^6\ \text{J}$. After the collision, the total kinetic energy is $\frac{1}{2}(20{,}000\ \text{kg})(10\ \text{m/s})^2 = 1.0\times10^6\ \text{J}$. Hence the heat energy produced is $2.0\times10^6 - 1.0\times10^6\ \text{J} = 1.0\times10^6\ \text{J}$.

5-7 ▌ Power

Power is defined as the rate at which work is done or the rate at which energy is transformed:

$$\text{power} = \frac{\text{work}}{\text{time}} = \frac{\text{energy transformed}}{\text{time}}.$$

The power of a horse refers to how much work it can do per unit time; the power rating of an engine refers to how much chemical or electrical energy can be transformed into mechanical energy per unit time. In SI units, power is measured in joules per second and this is given a special name, the *watt* (W): 1 W = 1 J/s. We are most familiar with the watt for measuring the rate at which an electric light bulb or heater changes electric energy into light or heat energy, but it is used for other types of energy transformation as well. In the British system, the unit of work is the ft · lb/s. For practical purposes a larger unit is often used, the *horsepower*. One (British) horsepower† (hp) is defined as 550 ft · lb/s, which equals 746 W. In the metric system, 1 hp is defined as 750 W. The metric and British horsepower units differ by less than one percent.

To see the distinction between energy and power, consider the following example. A person is limited in the work he can do not only by the total energy required, but also by the rate this energy is used; that is, by power. For example a person may be able to walk a long distance or climb many flights of stairs before having to stop because he has used up so much energy. On the other hand, if a person runs very quickly upstairs, he may fall exhausted after only a flight or two. He is limited in this case by power, the rate at which his body can transform chemical energy into mechanical energy.

▌ EXAMPLE 5-14 A 70-kg man runs up a long flight of stairs in 4.0 s. The vertical height of the stairs is 4.5 m. Calculate his power output in watts and horsepower.

SOLUTION The work done is against gravity. Hence the power P is

$$P = \frac{mgh}{t} = \frac{(70\ \text{kg})(9.8\ \text{m/s}^2)(4.5\ \text{m})}{4.0\ \text{s}} = 770\ \text{W}.$$

Since there are 750 W in 1 hp, the man is doing work at a rate of just over 1 hp. It is worth noting that a human cannot do work at this rate for very long. ▌▌

† The unit was first chosen by James Watt (1736–1819) who needed a way to specify the power of his newly developed steam engines. He found by experiment that a good horse can work all day at an average rate of 360 ft · lb/s. So as not to be accused of exaggeration in the sale of his steam engines, he multiplied this by 1.5 when he defined the hp.

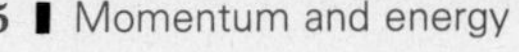

Automobiles do work to overcome the force of friction (and air resistance), to climb hills, and to accelerate. A car is limited by the rate it can do work, which is why automobile engines are rated in horsepower. A car needs power most when it is climbing hills and when accelerating (as for passing). In the next example we will calculate how much power is needed in these situations for a car of reasonable size. Even when a car travels on the level, it needs some power for it must do work to overcome friction and air resistance; these forces depend on the conditions but are usually in the region of 400–1000 N.

It is often convenient to write the power in terms of the net force F applied to an object and its speed v. This is readily done since $P = W/t$ and $W = Fd$ where d is the distance traveled. Then

$$P = \frac{W}{t} = \frac{Fd}{t}$$

$$P = F\bar{v} \qquad (5\text{-}14)$$

where $\bar{v} = d/t$ is average speed of the object during the interval under discussion.

▌ EXAMPLE 5-15 Calculate the power required of a 1400-kg car under the following circumstances: (*a*) the car climbs a 10° hill at a steady 80 km/h and (*b*) the car accelerates from 90 to 110 km/h in 6.0 s to pass another car. Assume the force of friction on the car is 700 N. See Figure 5-23.

SOLUTION (*a*) To move at a steady speed up the hill, the car must exert a force equal to the sum of the friction force, 700 N, and the component of gravity parallel to the hill, $mg \sin 10° = (1400\text{ kg})(9.8\text{ m/s}^2)(0.174) = 2400\text{ N}$. Since $\bar{v} = 80\text{ km/h} = 22\text{ m/s}$,

$$P = F\bar{v}$$

$$= (2400\text{ N} + 700\text{ N})(22\text{ m/s}) = 6.8 \times 10^4\text{ W}$$

$$= 91\text{ hp}.$$

(*b*) The car accelerates from 25.0 m/s to 30.6 m/s (90 to 110 km/h). Thus the car must exert a force that overcomes the 700 N friction force plus that required to give it the acceleration $a = (30.6\text{ m/s} - 25.0\text{ m/s})/6.0\text{ s} = 0.93\text{ m/s}^2$. Since the mass of the car is 1400 kg, the force required for acceleration is $F = ma = (1400\text{ kg})(0.93\text{ m/s}^2) = 1300\text{ N}$, and the total force required is then 2000 N. The average speed is $\bar{v} = 27.8\text{ m/s}$. Thus, using Equation 5-14, the power needed is

$$P = (2000\text{ N})(27.8\text{ m/s})$$

$$= 5.6 \times 10^4\text{ W}$$

$$= 75\text{ hp}.$$

▌▌

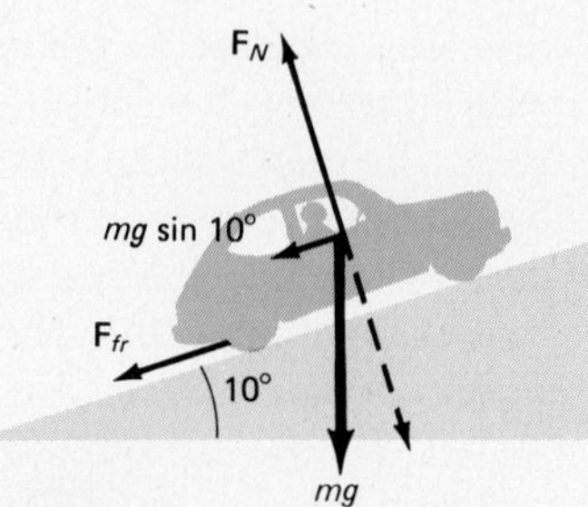

FIGURE 5-23
Example 5-15, calculation of power needed for car to climb a hill.

Even taking into account the fact that only 60 to 80 percent of the engine's power reaches the wheels, it is clear from these calculations that an engine of 100 to 150 hp is quite adequate from a practical point of view, even in a reasonably heavy car. In view of fuel requirements and environmental damage, larger engines are indeed an extravagance that society has already begun to question.

Summary

The *momentum*, $\mathbf{p}$, of a body is defined as its mass times its velocity: $\mathbf{p} = m\mathbf{v}$. In terms of momentum, Newton's second law can be written $\mathbf{F} = \Delta\mathbf{p}/\Delta t$: the rate of change of momentum equals the net applied force; or we can say the *impulse*, defined as the force times the time over which it acts, equals the change in momentum.

Momentum is a conserved quantity: the law of *conservation of momentum* states that the total momentum of an isolated system of objects remains constant. An isolated system is one on which the net external force is zero.

The *angular momentum* of a rotating rigid body is defined as $L = I\omega$. The total angular momentum of an isolated body or system of bodies remains constant in magnitude and direction. This is the law of *conservation of angular momentum.*

Work is done on an object by a force when the force moves the object though a distance, d. If the direction of the force makes an angle θ with the direction of motion, the work done by this force is $W = Fd \cos\theta$.

Energy is defined as the ability to do work. Both work and energy are measured in joules ($1\ \text{J} = 1\ \text{N} \cdot \text{m}$) in SI units. *Kinetic energy* (KE) is energy of motion; a body of mass m and speed v has translational KE equal to $\frac{1}{2}mv^2$; if the object is rotating, it also has rotational KE given by $\frac{1}{2}I\omega^2$. An object can have *potential energy* (PE) by virtue of its position or shape. Examples are gravitational PE (equal to mgh where h is the height of the object of mass m above an arbitrary reference point), elastic PE (such as in a compressed spring), as well as chemical, electrical, and nuclear energy. The change in potential energy of an object when it changes position is defined as the work needed to take it from one position to the other. The *work–energy theorem* states that the *net* work done on a body (by the *net* force) equals the change in kinetic energy of that body.

The law of *conservation of energy* states that energy can be transformed from one type to another, but the total energy remains constant. It is valid even when friction is present since the heat generated by friction can be considered a form of energy. Momentum is conserved in any collision between objects; energy is conserved too, but kinetic energy is conserved only in so-called *elastic* collisions in which other forms of energy do not change.

Power is defined as the rate at which work is done, or the rate at which energy is transformed. The SI unit of power is the watt ($1\ \text{W} = 1\ \text{J/s}$).

Questions

1 Explain, on the basis of conservation of momentum, how a fish propels itself forward by swishing its tail back and forth.

2 Why, when you release an inflated balloon, does it fly across the room?

3 It is said that in ancient times a rich man with a bag of gold coins was frozen to death stranded on the surface of a frozen lake. Because the ice was frictionless, he could not push himself to shore. What could he have done to save himself had he not been so miserly?

4 According to Equation 5-3, the shorter the impact time of a given impulse, the greater the force must be and hence the greater the deformation of the object on which the force acts. Explain on this basis the value of "air bags" which are intended to inflate during an automobile collision and reduce the possibility of fracture or death.

5 In a collision between two cars, which would you expect to be more damaging to the occupants: if the cars collide and remain together, or if the two rebound backwards? Explain.

6 On the basis of the law of conservation of angular momentum, discuss why a helicopter must have more than one rotor (or propeller). Discuss one or more ways the second propeller can operate in order to keep the body stable.

7 A quarterback leaps into the air to throw a forward pass. As he throws the ball, the upper part of his body rotates; if you look quickly you will notice that his hips and legs rotate in the opposite direction. Explain.

8 When a motorcyclist leaves the ground on a jump, if the throttle is left on (so the rear wheel spins) why does the front of the cycle rise up?

* 9 Some car jacks work on the principle of a screw. Determine the IMA of a jack in terms of the pitch (distance between threads) P, radius r_1 of the screw, and the turning radius r_2 of the jack handle. Ignore any other effects, such as that due to gears or levers in the jack.

10 What does a mechanical advantage of less than one signify?

11 A light body and a heavy body have the same kinetic energy. Which has the greater momentum?

12 A sphere, a cylinder, and a hoop, all of the same diameter, roll down an inclined plane. Which arrives at the bottom first? Which arrives last?

13 Describe the energy transformations that take place when a skier starts skiing down a hill, but after a time is brought to rest by striking a snowdrift.

14 Describe the energy transformations when a child hops about on a pogo stick.

15 When an experienced hiker comes to a log across his path, he prefers to step over it rather than surmounting it and jumping down on the other side. Why?

16 Does the following statement make sense? "Fiberglass poles led to higher pole vaults because the additional potential energy of bending was converted into gravitational potential energy." Explain.

17 If a falling ball were to make a perfectly elastic collision with the floor, would it rebound to its original height? Explain.

18 A rock thrown with a certain speed from the top of a cliff will enter the water below with the same speed whether the rock is thrown horizontally or at any angle. Explain.

19 Analyze the motion of a simple swinging pendulum in terms of energy: (*a*) ignoring friction, and (*b*) taking it into account. Explain why a grandfather clock has to be wound up.

Problems

SECTIONS 5-1 AND 5-2

1 (I) What is the momentum of an 8.0-g sparrow flying with a speed of 15 m/s.

2 (I) The tennis ball may leave the racket of a top player on the serve with a speed of 65 m/s. If the ball's mass is 60 g and is in contact with the racket for 0.030 s, what is the average force on the ball?

3 (I) A 90-kg fullback is traveling 5.0 m/s and is stopped by a tackler in one second. Calculate (*a*) the original momentum of the fullback, (*b*) the impulse imparted to the tackler, and (*c*) the average force exerted on the tackler.

4 (I) The air in a 200 km/h wind strikes a $30\text{ m} \times 20\text{ m}$ face of a building at a rate of 5.4×10^4 kg/s. Calculate the net force on the building, assuming the air is brought to rest on impact.

5 (I) Calculate the force exerted on a rocket given that the propelling gases are expelled at a rate of 1000 kg/s with a speed of 60,000 m/s.

6 (I) A child throws a 5.0-kg package horizontally from a boat with a speed of 10.0 m/s. Calculate the resulting velocity of the boat assuming it was initially at rest. The mass of the child is 20 kg and that of the boat is 80 kg.

7 (II) A 15,000-kg railroad car travels alone on a level frictionless track with a constant speed of 30 m/s. A 5000-kg additional load is dropped onto the car. What then will be its speed?

8 (II) A 140-kg tackler moving at 3.0 m/s meets head on (and tackles) a 90-kg halfback moving at 7.5 m/s. What will be their mutual speed immediately after the collision?

9 (II) In Example 5-2, assume the person lands stiff-legged so that $t \approx 0.0020$ s. (*a*) Calculate the stress in the tibia (area $= 3.0 \times 10^{-4}$ m^2) and (*b*) determine if the bone will break.

10 (II) A 44-g bullet strikes a 1.54-kg block of wood placed on a horizontal surface just in front of the gun. If the coefficient of kinetic friction between the block and the surface is 0.28, and the impact drives the block a distance of 18.0 m before it comes to rest, what was the muzzle speed of the bullet?

11 (II) A gun is fired vertically into a 3.40-kg block of wood directly above it. If the bullet has a mass of 62.0 g and a speed of 340 m/s, how high will the block rise into the air?

12 (II) An atomic nucleus at rest decays radioactively into an alpha particle and a smaller nucleus. What will be the speed of this recoiling nucleus if the speed of the alpha particle is 6.2×10^5 m/s? Assume the nucleus has a mass 57 times greater than that of the alpha particle.

13 (II) A 150-kg astronaut (including space suit) acquires a speed of 2.5 m/s by pushing off with his legs from a 2200-kg space capsule. (*a*) What is the change in speed of the space capsule? (*b*) If the push lasts 0.20 s, what was the average force exerted by each on the other? As the reference frame, use the position of the capsule before the push.

14 (III) A billiard ball of mass $M_A = 0.40$ kg moving with a speed $v_A = 2.00$ m/s strikes a second ball, initially at rest, of mass $m_B = 0.40$ kg. As a result of the collision, the first ball is deflected off at an angle of 30° with a speed $v'_A = 1.20$ m/s as shown in Figure 5-24. (*a*) Taking the *x* axis to be the original direction of motion of ball *A*, write down the equations expressing the conservation of momentum for the components in the *x* and *y* directions separately. (*b*) Solve these equations for the speed, v'_B, and angle θ, of ball *B*.

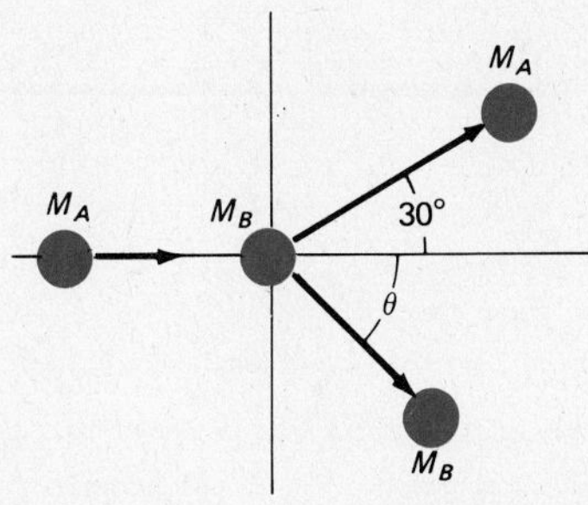

FIGURE 5-24

15 (III) A 60-kg parachutist is falling with a speed of 10.2 m/s (so-called "terminal velocity"). To avoid breaking a leg upon striking the ground, the person must bend the knees so that the cg of the whole body comes to rest gradually. Calculate the minimum distance over which the person's cg must move in order to avoid breaking the tibia which has a cross-sectional area of 2.8 cm^2.

16 (III) From what maximum height can a 60-kg person jump without breaking the lower leg bone ($A = 3.0 \times 10^{-4}$ m^2). Ignore air resistance and assume the cg of the person moves a distance of 0.60 m from the standing to the seated position (i.e., in breaking the fall).

17 (III) With what speed must a 1.2-kg rectangular brick 15 cm × 6.0 cm × 4.0 cm be thrown against a steel wall in order to break? Assume the brick strikes the wall directly on its largest face, and that the compression of the brick is much greater than that of the steel (ignore compression of the steel). State other simplifying assumptions that may be necessary.

*18 (III) Suppose the force acting on a tennis ball (mass 0.060 kg) as a function of time is given by the graph of Figure 5-3*a*. Use graphical methods to (*a*) estimate the total impulse given the ball; (*b*) estimate the speed of the ball after being struck assuming the ball is being served so it is nearly at rest initially.

*19 (III) The force on a bullet is given by the formula $F = 480 - 1.6 \times 10^5 t$ over the time interval $t = 0$ to $t = 3.0 \times 10^{-3}$ s; in this formula t is in seconds and F is in newtons. (*a*) Plot a graph of F vs. t for $t = 0$ to $t = 3.0$ ms. (*b*) Estimate, using graphical methods, the impulse given the bullet. (*c*) If the bullet achieves a speed of 320 m/s as a result of this impulse, given to it in the barrel of a gun, what must its mass be?

20 (III) A 4200-kg rocket is traveling in outer space with a velocity of 150 m/s toward the sun. It wishes to alter its course by 30°, and can do this by shooting its rockets briefly in a direction perpendicular to its motion. If the rocket gases are expelled at a speed of 2700 m/s, what mass of gas must be expelled?

SECTION 5-3

21 (I) What is the angular momentum of a 200-g ball rotating on the end of a string in a circle of radius 1.0 m at an angular speed of 9.4 rad/s?

22 (I) What is the angular momentum of a 2.1-kg uniform cylindrical grinding wheel rotating at 1500 rpm? Its radius is 27.4 cm.

23 (I) A person stands, hands at the side, on a platform that is rotating at a rate of 1.20 rev/s. If the person now raises both arms, the speed of rotation decreases to 0.80 rev/s. (*a*) Why does this occur? (*b*) By what factor has the moment of inertia of the person changed?

24 (II) How much torque is required to stop the grinding wheel of Problem 22 in 6.0 s?

25 (II) A diver such as that shown in Figure 5-5 can reduce his moment of inertia by a factor of about 3.5 when changing from the straight position to the tuck position. If he makes two rotations in 1.5 s when in the tuck position, what is his angular speed (rev/s) when in the straight position?

26 (II) A mass attached to a string revolves in a circle on a frictionless tabletop. The other end of the string passes through a hole in the table. Initially the ball is rotating with an angular velocity of 8.2 rad/s in a circle of radius 48.0 cm, but the string is being pulled through the hole so the radius is reduced to 32.0 cm. What will be the angular velocity of the mass now?

27 (II) A 4.0-m diameter merry-go-round is rotating with an angular velocity of 0.80 rad/s; its total moment of inertia is $1400\ \text{kg}\cdot\text{m}^2$. Four people standing on the ground, each of 65-kg mass, suddenly jump on at the edge. What will be the angular velocity of the merry-go-round now? What if they were on it initially and then jumped off?

28 (III) Suppose a 60-kg person stands at the edge of a 6.0-m diameter merry-go-round turntable which is mounted on frictionless bearings and has a moment of inertia of $1800\ \text{kg}\cdot\text{m}^2$. The turntable is at rest initially, but when the person begins running at a speed of 4.2 m/s (with respect to the turntable) around its edge, the turntable begins to rotate in the opposite direction. Calculate the angular velocity of the turntable.

SECTION 5–4

29 (I) A 50-kg woman climbs a flight of stairs 5.0-m high. How much work is required?

30 (I) A 550-N crate rests on the floor. How much work is required to move it at constant speed (*a*) 2.0-m along the floor against a friction force of 150 N, and (*b*) 2.0-m vertically?

31 (I) How much work did a horse do that pulled a 200-kg wagon 80 km along a level road if the effective coefficient of friction was 0.060?

32 (I) How far must a 200-kg pile driver fall if it is to do 13,000 J of work?

33 (I) One kind of car jack is essentially a lever with a ratchet to hold the lifted object in place while the lever is returned for a new thrust. If the lever is pivoted 2.0 cm from the ratchet and the handle is 35.0 cm long, what is the IMA?

34 (II) How much work is needed to push a 1000-kg car 45 m up a 3° incline? (*a*) Ignore friction. (*b*) Assume the coefficient of friction is 0.30.

35 (II) Eight bricks, each 6.0 cm thick with mass 1.5 kg, lie flat on a table. How much work is required to stack them one on top of another?

36 (II) In pedaling a bicycle, a particular cyclist exerts a downward force of 90 N during each stroke. If the diameter of the circle traced by each pedal is 36 cm, calculate how much work is done in each stroke.

*37 (II) Design a pulley system which can lift 80 kg with a force of ($18\ \text{kg}\times g$) assuming an efficiency of 75 percent.

*38 (II) A 38-cm long crank is attached to a spindle with a 2.5-cm radius. Assuming an efficiency of 80 percent, how much weight can be lifted on a cord wrapped around the spindle if the person can exert a steady 50-N force on the crank handle?

*39 (II) An inclined plane (such as used by the Egyptians to build the pyramids) can be considered to be a simple machine since a heavy weight can be raised by the application of a smaller force. (*a*) Calculate the IMA of a plane inclined at an angle of 15° to the horizontal. (*b*) If the coefficient of friction for a particular object on this plane is 0.40, what is the AMA?

40 (III) Assume a cyclist of weight w can exert a force on the pedals equal to $0.90w$ on the average. If the pedals rotate in a circle of radius 18 cm, the wheels have a radius of 34 cm, and the front and back sprockets on which the chain runs have 52 and 17 points respectively, determine the maximum steepness of hill the cyclist can climb. Assume the mass of the bike is 12 kg and of the rider is 60 kg. Ignore friction.

*41 (III) In Figure 5-10*a*, assume the distance axis is linear and that $d_a = 5.0$ m and $d_b = 32.0$ m. Estimate the work done by this force in moving a 20-kg object from d_a to d_b.

*42 (III) A 1000-kg space vehicle falls vertically from a height of 2000 km above the earth's surface. Determine how much work is done by the force of gravity in bringing the vehicle to the earth's surface. (First construct an F vs. r graph, where r is the distance from the earth's center; then determine the work graphically.)

SECTION 5-5

43 (I) What is the initial KE of a 0.50-mg flea that leaves the ground at a speed of 100 cm/s? ($1\ \text{mg} = 10^{-3}$ gram.)

44 (I) A carbon atom of mass 2.0×10^{-26} kg has 3.8×10^{-19} J of kinetic energy. How fast is it moving?

45 (I) How much work must be done to stop a 1000-kg car traveling at 100 km/h?

46 (I) A 5.0-kg monkey swings from one branch to another 2.0 m higher. What is the change in its potential energy?

47 (II) (*a*) A spring of spring constant k is initially compressed a distance x_0 from its normal length. What is the change in potential energy if it is now compressed by an amount x from equilibrium? (*b*) The spring is now *stretched* a distance x_0 from the normal length. What is the change in potential energy as compared to when it is compressed by an amount x_0 from normal length?

48 (II) A centrifuge rotor has a moment of inertia of 4.0×10^{-2} kg · m^2. How much energy is required to bring it from rest to 10,000 rpm?

49 (II) A merry-go-round has a mass of 1300 kg and a radius of gyration of 13 m. How much work is required to accelerate it from rest to a rotation rate of one revolution in 10 s?

50 (III) A scallop forces open its shell with an elastic material called abductin whose elastic modulus (see Chapter 4) is about 2.0×10^6 N/m^2. If this piece of abductin is 3.0-mm thick and has a cross-sectional area of 0.50 cm^2, how much potential energy does it store when compressed 1.0 mm?

SECTION 5-6

51 (I) Tarzan is running at top speed (8.0 m/s) and grabs a vine hanging vertically from a tall tree in the jungle. How high can he swing upwards? Does the length of the vine (or rope) affect your answer?

52 (I) An object slides down a frictionless 30° incline 10 cm high. How fast is it going when it reaches the bottom?

53 (II) In the high jump, the KE of an athlete is transformed into gravitational potential energy without the aid of a pole. With what minimum speed must the athlete leave the ground in order to lift his cg 2.20 m and cross the bar with a speed of 0.80 m/s?

54 (II) A projectile is fired at an upward angle of 45° from the top of a 300-m cliff with a speed of 500 m/s. What will be its speed when it strikes the ground at the base of the cliff? (Use conservation of energy.)

55 (II) Calculate the translational speed of a solid cylinder when it reaches the foot of an incline 18 m high. Assume it starts from rest and rolls without slipping.

56 (II) A spring whose spring constant is 850 N/m is compressed 0.40 m. What speed can it give to a 0.30-kg ball when released?

57 (II) A mass m is attached to the end of a spring of spring constant k, Figure 5-25. The mass is given an initial displacement x_0, after which it oscillates back and forth. Write down a formula for the total mechanical energy (ignore friction) in terms of position x and speed v.

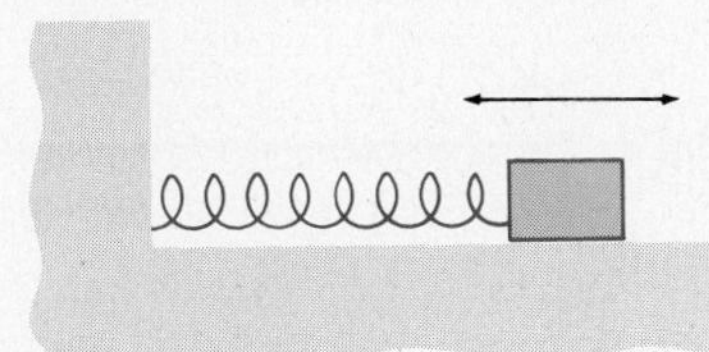

FIGURE 5-25

58 (II) A 24-kg child descends a slide 5.0-m high and reaches the bottom with a speed of 2.9 m/s. How much heat was generated in this process?

59 (II) Two railroad cars, each of mass 4200 kg and traveling 80 km/h, collide head-on and come to rest. How much thermal energy is produced in this collision?

60 (II) According to the data of Example 1-7, a flea should jump to a height of 5.0 cm but because of air resistance it actually reaches a height of only 3.5 cm. What fraction of its energy is lost to air resistance?

61 (II) A 0.25-kg pine cone falls from a branch 20 m above the ground. (*a*) With what speed would it hit the ground if air resistance could be ignored? (*b*) If it actually hits the ground with a speed of 9.0 m/s, what was the average force of air resistance exerted on it?

62 (II) A ball of mass 0.40 kg moving with a speed of 8.0 m/s collides head-on with a 0.20-kg ball at rest. If the collision is perfectly elastic, what will be the speeds and directions of the two balls after the collision?

63 (II) Two equal mass billiard balls undergo a perfectly elastic head-on collision. The speed of one ball was initially 2.0 m/s, and of the other 3.0 m/s in the opposite direction. What will be their speeds after the collision?

64 (II) A roller coaster is shown in Figure 5-26. Assuming no friction, calculate the speed at points *B, C, D*, assuming it has a speed of 2.0 m/s at point *A*.

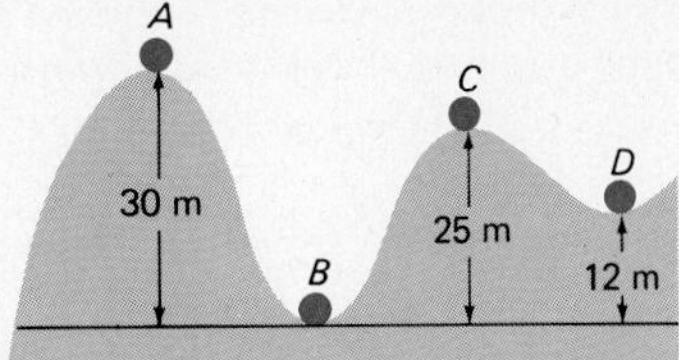

FIGURE 5-26

65 (II) The roller coaster in Figure 5-26 passes point *A* with a speed of 1.0 m/s. If the average force of friction is equal to 1/5 of its weight, with what speed will it reach point B? The distance traveled is 70 m.

66 (III) A 5.0-m long pole is balanced vertically on its tip. What will be the speed of the tip of the pole just before it hits the ground? Assume the lower end of the pole does not slip.

67 (III) A cyclist intends to cycle up a 10° hill 100 m high. Assuming the mass of bicycle plus person is 80 kg, (*a*) calculate how much work must be done against gravity; (*b*) if each complete revolution of the pedals moves the bike 5.1 m along its path, calculate the average force that must be exerted on the pedals during each stroke. Neglect friction and other losses. The pedals turn in a circle of diameter 36 cm.

68 (III) A small mass m slides without friction along the looped apparatus shown in Figure 5-27. If the object is to remain on the track, even at the top of the circle (whose radius is r), from what minimum height h must it be released?

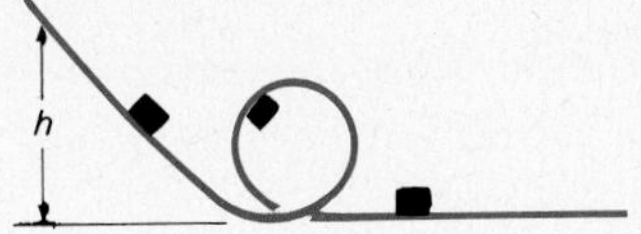

FIGURE 5-27

69 (III) The speed of a projectile, such as a bullet, can be measured using a so-called ballistic pendulum, Figure 5-28. The projectile is fired into the pendulum bob which then holds it (such as a bullet fired into a block of wood). The pendulum bob swings upward to a maximum height h. Using the conservation of momentum and energy laws where they are appropriate, show that the velocity of the projectile v is given by $v = \sqrt{2gh}(m + M)/m$, where m is the mass of the projectile, M is the mass of the pendulum bob, and g is the acceleration due to gravity.

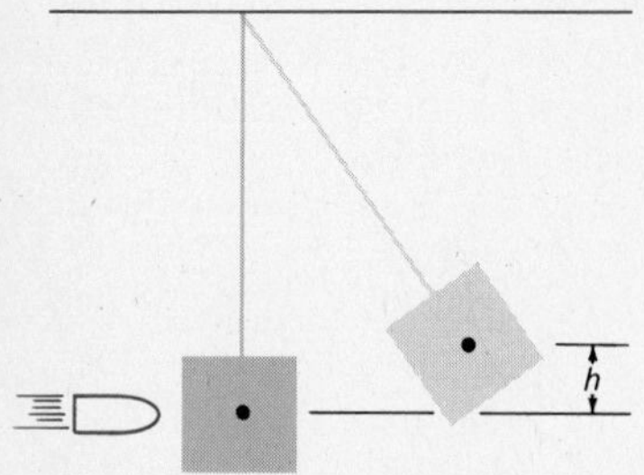

FIGURE 5-28 **A ballistic pendulum.**

70 (III) A 200-g wood block is firmly attached to a horizontal spring, Figure 5-25. The block can slide along a table where the coefficient of friction is 0.40. A force of 10 N compresses the spring 18 cm. If the spring is released from this position, how far beyond its equilibrium position will it stretch on its first swing?

SECTION 5-7

71 (I) If a car generates 15 hp when traveling at a steady 80 km/h, what must be the average force exerted on the car due to friction and air resistance?

72 (I) How long will it take a 2.0-hp motor to lift a 400-kg piano to a sixth-story window 20 m above?

73 (I) Show that a British horsepower is equal to 746 W. What is the percent difference between this unit and the metric hp?

74 (I) Electric energy is often stated in kilowatt-hours. Show that the kilowatt-hour (kwh) is a unit of energy and is equal to 3.6×10^6 J.

75 (II) A shotputter accelerates a 7.3-kg shot from rest to 15 m/s. If this motion takes 2.0 s, what average power was developed?

76 (II) How much work can a 2.0-hp motor do in 1.0 h?

77 (II) A 70-kg hiker climbs to the top of a 4200-m high mountain. The climb is made in 4.0 h starting at an elevation of 3100 m. Calculate (*a*) the work done against gravity, (*b*) the average power output in watts and in horsepower, and (*c*) assuming the body is 15 percent efficient, how much energy input was required.

78 (II) How fast must a cyclist climb a 10° hill to maintain a power output of 0.25 hp? Ignore friction and assume the mass of cyclist plus bicycle is 85 kg.

79 (II) A pump is to lift 4.0 kg of water per minute through a height of 2.6 m. What output rating should the pump motor have?

80 (II) A bicyclist coasts down a 6° hill at a steady speed of 7.0 km/h. Assuming a total mass of 75-kg (bicycle plus rider), what must be the cyclist's power output to climb the same hill at the same speed?

81 (II) A 1000-kg car has a maximum power output of 100 hp. How steep a hill can it climb at a constant speed of 60 km/h if the frictional forces add up to 500 N?

82 (III) Water flows over a dam at the rate of 800 kg/s and falls vertically 150 m before striking the turbine blades. Calculate (*a*) the velocity of the water just before striking the turbine blades, and (*b*) the rate at which mechanical energy is transferred to the turbine blades. Assume the water loses 80 percent of its speed when striking the blades, and that 12 percent of the initial energy is transformed to heat.

83 (III) A bicyclist of mass 80 kg (including the bicycle) can coast down a 4° hill at a steady speed of 6.0 km/h. Pumping hard, the cyclist can descend the hill at a speed of 30 km/h. Using the same power, at what speed can the cyclist climb the same hill? Assume the force of friction is directly proportional to the speed v; that is, $F_{fr} = bv$, where b is a constant.

For additional reading

Einstein, A., and L. Infeld, *Evolution of Physics*, Simon & Schuster, Inc., New York, 1938.

Hay, J. G., *The Biomechanics of Sports Techniques*, Prentice-Hall, Inc., Englewood Cliffs, N.J., 1973.

Magie, W. R., *Source Book in Physics*, Harvard University Press, Cambridge, Mass., 1963. Various selections on energy and its conservation.

6 ▌ Fluids

The three common states, or *phases*, of matter are solid, liquid, and gas. We can distinguish these three phases as follows. A *solid* maintains a fixed shape and a fixed size; even if a large force is applied to a solid, it does not readily change its shape or volume. A *liquid* does not maintain a fixed shape—it takes on the shape of its container—but like a solid it is not readily compressible and its volume can be changed significantly only by a very large force. A *gas* has neither a fixed shape nor a fixed volume—it will expand to fill its container. For example when air is pumped into an automobile tire, the air does not all run to the bottom of the tire as a liquid would; it fills the whole volume of the tire. Since liquids and gases do not maintain a fixed shape, they each have the ability to flow; they are thus often referred to collectively as **fluids**.

The division of matter into three states is not always simple. How, for example, should butter be classified? And what about glass—have you ever noticed that a very old pane of window glass is thicker at the bottom than at the top? Glass flows, although very slowly, and is thus often considered to be a liquid. We should also mention that a fourth state of matter can be distinguished. This state is called the *plasma*, and it occurs only at very high temperatures, as we will discuss in Chapter 29. Some scientists believe that so-called colloids (suspensions of tiny particles in a liquid) should also be considered a separate state of matter. However, for our present purposes we will mainly be interested in the three ordinary states of matter. In Chapter 4 we discussed some of the properties of solid materials. In this chapter we will discuss the properties of fluids.

6-1 ▌ Density and specific gravity

Density

It is sometimes said that iron is "heavier" than wood. This cannot really be true since a large log clearly weighs more than an iron nail. What we should say is that iron is more *dense* than wood.

TABLE 6-1

Densities of substances

Substance†	Mass density ρ (kg/m^3)	Substance†	Mass density ρ (kg/m^3)
Solid		**Liquids**	
Aluminum	**2.70 $\times 10^3$**	**Water (4°C)**	**1.00 $\times 10^3$**
Iron and steel	**7.8 $\times 10^3$**	**Blood, plasma**	**1.03 $\times 10^3$**
Copper	**8.9 $\times 10^3$**	**Blood, whole**	**1.05 $\times 10^3$**
Lead	**11.3 $\times 10^3$**	**Seawater**	**1.025 $\times 10^3$**
Gold	**19.3 $\times 10^3$**	**Mercury**	**13.6 $\times 10^3$**
Concrete	**2.3 $\times 10^3$**	**Alcohol, ethyl**	**0.79 $\times 10^3$**
Granite	**2.7 $\times 10^3$**	**Gasoline**	**0.68 $\times 10^3$**
Wood	**0.3–0.9 $\times 10^3$**	**Gases†**	
Glass	**2.6 $\times 10^3$**	**Air**	**1.29**
Ice	**0.917 $\times 10^3$**	**Helium**	**0.179**
Bone	**1.7 $\times 10^3$**	**Carbon dioxide**	**1.98**
		Water (steam) (100°C)	**0.60**

† Densities are given at 0°C and 1 atm pressure unless otherwise specified.

The **density**, ρ, of a substance is defined as its mass per unit volume:

$$\rho = \frac{m}{V}, \tag{6-1}$$

where m is the mass of an amount of the substance whose volume is V. Density is a characteristic property of a substance; objects made of a given substance, say pure iron, can have any size or mass, but the density will be the same for each.

The SI unit for density is kg/m^3. Sometimes densities are given in g/cm^3. Note that since 1 kg/m^3 = 1000 g/(100 cm)^3 = 10^{-3} g/cm^3, a density given in g/cm^3 must be multiplied by 1000 to give the result in kg/m^3. Thus the density of aluminum is $\rho = 2.70$ g/cm^3, which is equal to 2700 kg/m^3. The densities of a variety of materials are listed in Table 6-1. Temperature affects the density of substances (although the effect is slight for liquids and solids), and hence the temperature is specified in Table 6-1.

▮ EXAMPLE 6-1 Calculate the mass of a lead sphere of radius 0.50 m.

SOLUTION The volume of the sphere is

$$V = \tfrac{4}{3}\pi r^3 = (\tfrac{4}{3})(3.14)(0.50\text{ m})^3 = 0.52\text{ m}^3$$

From Table 6-1, the density of lead is $\rho = 11{,}300$ kg/m^3, so we have from Equation 6-1

$$m = \rho V = (11{,}300\text{ kg/m}^3)(0.52\text{ m}^3) = 5800\text{ kg}.$$ ▮▮

Specific gravity

The **specific gravity** of a substance is defined as the ratio of the density of that substance to the density of water at 4°C. Specific gravity (abbreviated SG) is a pure number, without dimensions or units. Since the density of water is $1.00\ \mathrm{g/cm^3} = 1.00 \times 10^3\ \mathrm{kg/m^3}$, the specific gravity of any substance will be precisely equal numerically to its density specified in $\mathrm{g/cm^3}$, or 10^{-3} times its density specified in $\mathrm{kg/m^3}$. For example (see Table 6-1), the SG of lead is 11.3 and of alcohol 0.79.

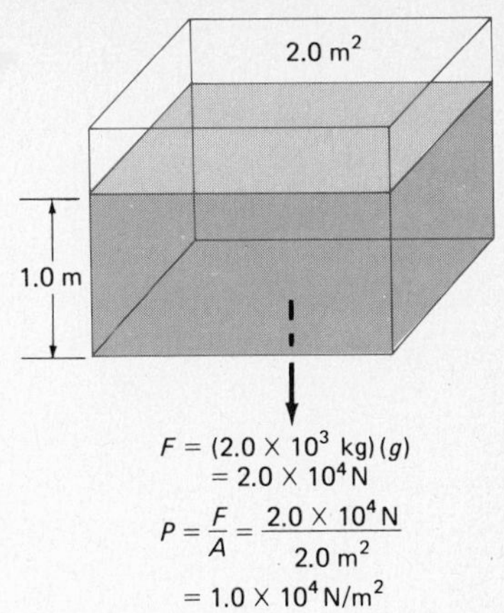

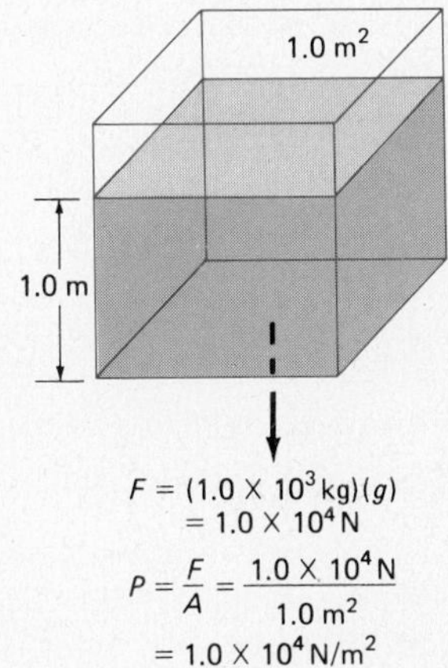

FIGURE 6-1
Pressure in a liquid depends only on depth.

6-2 ▌ Pressure and its measurement

Pressure is defined as force per unit area, where the force F is understood to be acting perpendicular to the surface area A:

$$\text{Pressure} = P = \frac{F}{A}. \qquad (6\text{-}2)$$

The SI unit of pressure is $\mathrm{N/m^2}$. This unit has the official name pascal (Pa): $1\ \mathrm{Pa} = 1\ \mathrm{N/m^2}$; however, we will generally use $\mathrm{N/m^2}$ for clarity. Other units sometimes used are $\mathrm{dynes/cm^2}$, $\mathrm{lb/in^2}$ (sometimes abbreviated "psi"), and $\mathrm{kg/cm^2}$ (as if kilograms were a force: that is, $1\ \mathrm{kg/cm^2} = 9.8\ \mathrm{N/cm^2} = 9.8 \times 10^4\ \mathrm{N/m^2}$). The last two are often used on tire gauges. We will meet several other units shortly.

As an example of equation 6-2, a 60-kg person whose two feet cover an area of $500\ \mathrm{cm^2}$ exerts a pressure of $F/A = mg/A = (60\ \mathrm{kg})(9.8\ \mathrm{m/s^2})/(0.050\ \mathrm{m^2}) = 12 \times 10^3\ \mathrm{N/m^2}$ on the ground. If the person stands on one foot, the force is the same but the area will be half; so the pressure will be twice as much: $24 \times 10^3\ \mathrm{N/m^2}$.

Pressure in fluids

The concept of pressure is particularly useful in dealing with fluids, which is why we introduced it in this chapter. We will first deal with fluids at rest. To help understand pressure in a fluid imagine two containers, one whose bottom area is twice that of the other. Each is filled with a liquid, say water, to the same height, Figure 6-1. The larger container holds twice the weight of liquid as the smaller one and therefore the liquid exerts twice the force on the bottom. But the pressure on the bottom due to the weight of the liquid is the same for both since the container with twice the area has twice the force exerted on it. This is a general result: *the pressure at equal depths in a liquid is the same.* At different depths, of course, the pressure is different. As any diver knows, the pressure under water increases with depth. We will look at this more closely in a moment.

It is also an experimental fact that *a fluid exerts a pressure in all directions.* This is well known to swimmers and divers who feel the water pressure on all parts of their bodies. At a particular depth in a fluid at rest, the pressure is the same in all directions. This is illustrated in Figure 6-2, where we consider a tiny cube of the fluid which is so small that we can

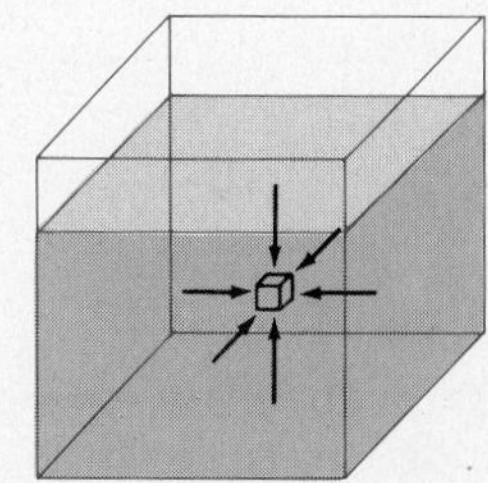

FIGURE 6-2
Pressure is the same in every direction in a fluid at a given depth; if it weren't, the fluid would be in motion.

ignore the force of gravity upon it. Then the pressure on one side of it must equal the pressure on the opposite side. If this weren't true, the net force on this cube would not be zero, and it would move until the pressure did become equal; if the fluid is at rest (which is the case we are considering now) then the pressures must be equal.

Another important property of a fluid at rest is that the force due to fluid pressure always acts *perpendicularly* to any surface it is in contact with. If there were a component of the force parallel to the surface as shown in Figure 6-3, then according to Newton's third law, the surface would exert a force back on the fluid that also would have a component parallel to the surface; this component would cause the fluid to flow, in contradiction to our assumption that the fluid is at rest.

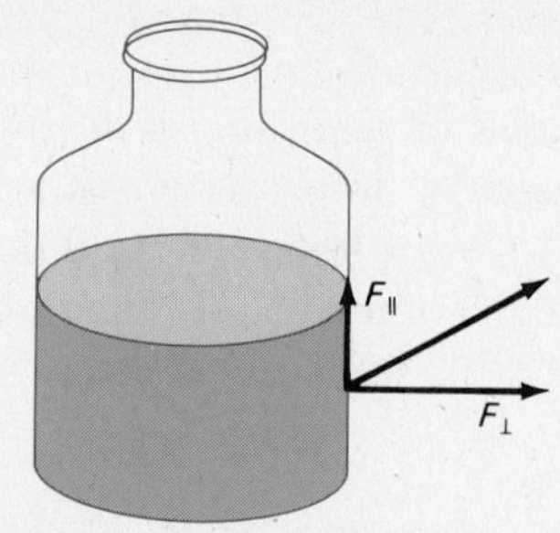

FIGURE 6-3
The force a fluid exerts on a solid surface is perpendicular to the surface; that is, $F_\parallel = 0$.

Let us now calculate quantitatively how the pressure in a liquid varies with depth. The liquid pressure at a depth h in a liquid, as shown in Figure 6-4, is due to the weight of the column of liquid above it. Thus the force acting on the area is $F = mg = \rho Ahg$, where Ah is the volume of the column, ρ is the density of the liquid (assumed to be constant), and g is the acceleration of gravity, and we have used Equation 6-1, $m = \rho V = \rho Ah$. The pressure P is then

$$P = \frac{F}{A} = \frac{\rho Ahg}{A}$$

$$P = \rho gh. \tag{6-3a}$$

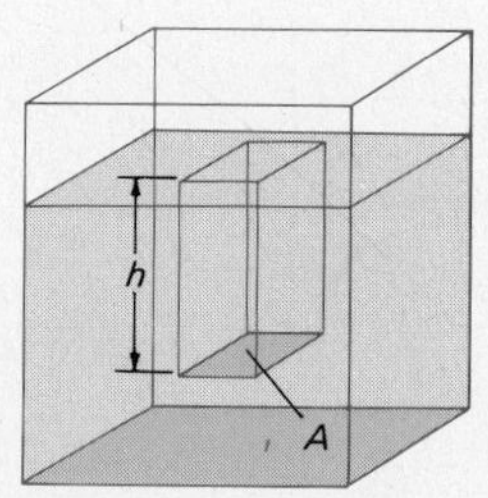

FIGURE 6-4
Calculating the pressure at a depth h in a liquid.

Thus the pressure is directly proportional to the depth in the liquid; it is also proportional to the density of the liquid.

▮ EXAMPLE 6-2 The surface of the water in a storage tank is 30 m above a water faucet in the kitchen of a house. Calculate the water pressure at the faucet:

SOLUTION We use Equation 6-3a:

$$\begin{aligned} P &= \rho gh \\ &= (1.0 \times 10^3\ \text{kg/m}^3)(9.8\ \text{m/s}^2)(30\ \text{m}) \\ &= 2.9 \times 10^5\ \text{N/m}^2, \end{aligned}$$

where $\rho = 1.0 \times 10^3\ \text{kg/m}^3$ is the density of water (Table 6-1). ▮▮

The height h is sometimes called the *pressure head.* In this example, the head of the water is 30 m.

Equation 6-3a is extremely useful. It is valid for fluids whose density is constant and does not change with depth—that is, if the fluid is *incompressible.* This is usually a very good approximation for liquids (although at great depths in the ocean the density of water does increase substantially by compression due to the great weight of water above). Gases, on the other hand, are very compressible and density can vary significantly with depth. If the density varies only slightly, Equation 6-3 can be used to determine the difference in pressure ΔP at different heights with ρ being the average density:

$$\Delta P = \rho g\ \Delta h. \tag{6-3b}$$

Atmospheric pressure and gauge pressure

The pressure of the earth's atmosphere, as in any fluid, decreases with decreased depth (or increased height). But the earth's atmosphere is somewhat complicated because not only does the density of air vary greatly with altitude, but there is no distinct top surface to the atmosphere from which h (in Equation 6-3a) can be measured. We can, however, calculate the approximate difference in pressure between two altitudes using Equation 6-3b.

The pressure of the air at a given place varies slightly according to the weather. At sea level, the pressure of the atmosphere on the average is $1.013 \times 10^5\ \mathrm{N/m^2}$ (or $14.7\ \mathrm{lb/in^2}$). This value is used to define a new unit of pressure in common use, the *atmosphere* (abbreviated atm):

$$1\ \mathrm{atm} = 1.013 \times 10^5\ \mathrm{N/m^2}.$$

Another unit of pressure sometimes used (in meteorology and on weather maps) is the *bar*, which is defined as $1\ \mathrm{bar} = 1.00 \times 10^5\ \mathrm{N/m^2}$; thus standard atmospheric pressure is slightly more than one bar.

The pressure due to the weight of the atmosphere is exerted on all objects immersed in this great sea of air, including our bodies. How does a human body withstand the enormous pressure on its surface? The answer lies in the fact that living cells maintain an internal pressure that just balances the external pressure. Similarly, the pressure inside a balloon balances the outside pressure of the atmosphere. An automobile tire, because of its rigidity, can maintain pressures much greater than the external pressure.

One must be careful, however, when determining the pressure in a tire, for tire gauges, and most other pressure gauges as well, register the pressure over and above atmospheric pressure. This is called **gauge pressure**. Thus, to get the absolute pressure P, one must add the atmospheric pressure, P_a, to the gauge pressure, P_G:

$$P = P_a + P_G.$$

For example if a tire gauge registers $28.5\ \mathrm{lb/in^2}$, the actual pressure within the tire is $14.7\ \mathrm{lb/in^2} + 28.5\ \mathrm{lb/in^2} = 43.2\ \mathrm{lb/in^2}$. In SI units, the absolute pressure would be $(43.2\ \mathrm{lb/in^2}/14.7\ \mathrm{lb/in^2})(1.013 \times 10^5\ \mathrm{N/m^2}) = 2.98 \times 10^5\ \mathrm{N/m^2}$.

Pascal's principle

The earth's atmosphere exerts a pressure on all objects with which it is in contact, including other fluids. Atmospheric pressure acting on a fluid is transmitted throughout that fluid. For instance, according to Equation 6-3a, the water pressure at a depth of 100 m below the surface of a lake is $P = \rho g h = (1000\ \mathrm{kg/m^3})(9.8\ \mathrm{m/s^2})(100\ \mathrm{m}) = 9.8 \times 10^5\ \mathrm{N/m^2}$ or 9.7 atm. However, the total pressure at this point is due to the pressure of water plus the pressure of the air above it. Hence the total pressure would be $9.7\ \mathrm{atm} + 1.0\ \mathrm{atm} = 10.7\ \mathrm{atm}$. This is just one example of a general principle attributed to the French philosopher and scientist, Blaise Pascal (1623–1662). **Pascal's principle** states that *pressure applied to a confined fluid is transmitted throughout the fluid and acts in all directions.*

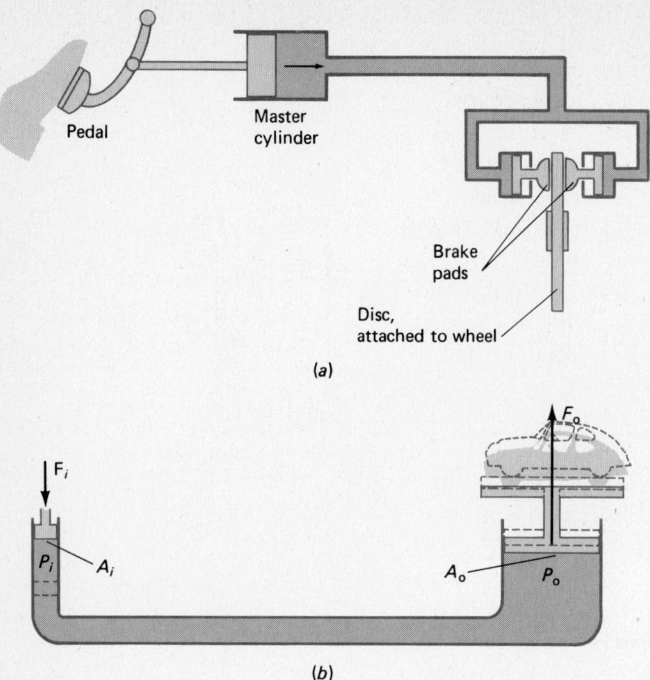

FIGURE 6-5
Applications of Pascal's principle: *(a)* hydraulic brakes; *(b)* hydraulic lift.

A number of practical devices make use of Pascal's principle. Two examples, hydraulic brakes in an automobile and the hydraulic lift, are illustrated in Figure 6-5. In the case of a hydraulic lift, a small force can be used to exert a large force by making the area of one piston (the output) larger than the area of the other (the input). This is another example of a simple machine (Chapter 5) and relies on the fact that the pressures on the input and output cylinders are the same at equal heights. (If the difference in height of the two pistons is not too great this result is not much altered.) Thus, if the input quantities are represented by the subscript "i" and the output by "o," we have

$$P_o = P_i$$

$$\frac{F_o}{A_o} = \frac{F_i}{A_i},$$

or, finally

$$\frac{F_o}{F_i} = \frac{A_o}{A_i}.$$

F_o/F_i is the ideal mechanical advantage of the hydraulic press, and is equal to the ratio of the areas. For example, if the area of the output piston is 20 times that of the input cylinder, the force is multiplied by a factor of 20; thus a force of 200 lb could lift a 4000-lb car.

A number of living organisms make use of hydraulic pressure. A sea anemone can achieve a variety of shapes by the action of muscles on its seawater-filled body cavity. Its body is sometimes referred to as a "hydrostatic skeleton." Earthworms move forward by successive contractions of circular muscles along the body axis acting on the hydrostatic skeleton. The legs of spiders have flexor muscles but possess no extensor muscles; their legs are extended by fluid driven into them under pressure.

Measurement of pressure

Many devices have been invented to measure pressure, some of which are shown in Figure 6-6. The simplest is the open-tube *manometer* (Figure 6-6*a*) which is a U-shaped tube partially filled with a liquid, usually mercury or water. The pressure P being measured is related to the difference in height of the two levels of the liquid by the relation (see Equation 6-3*b*)

$$P = P_0 + \rho g h,$$

where P_0 is atmospheric pressure (which is acting on the top of the fluid in the left-hand tube) and ρ is the density of the liquid. Note that the quantity $\rho g h$ is the "gauge pressure"—the amount by which P exceeds atmospheric pressure. If the liquid in the left-hand column were lower than in the right-hand column, this would indicate that P was less than atmospheric pressure (and h would be negative).

Instead of calculating out the product $\rho g h$, it is common to merely specify the height h. In fact pressures are sometimes specified as so many "millimeters of mercury" (mm Hg), and sometimes as so many "mm of water." The unit 1 mm Hg is equivalent to a pressure of 133 N/m^2, since 1 mm $= 1 \times 10^{-3}$ m and

$$\rho g h = (13.6 \times 10^3\ \text{kg/m}^3)(9.8\ \text{m/s}^2)(1.00 \times 10^{-3}\ \text{m})$$
$$= 1.33 \times 10^2\ \text{N/m}^2.$$

The unit mm Hg is also called the "torr" in honor of Evangelista Torricelli (1608–1647), who invented the barometer (see below). Conversion factors among the various units of pressure (an incredible nuisance!) are given in Table 6-2. It should be added that only N/m^2 = Pa, the proper SI unit, is to be used in calculations involving other quantities specified in SI units.

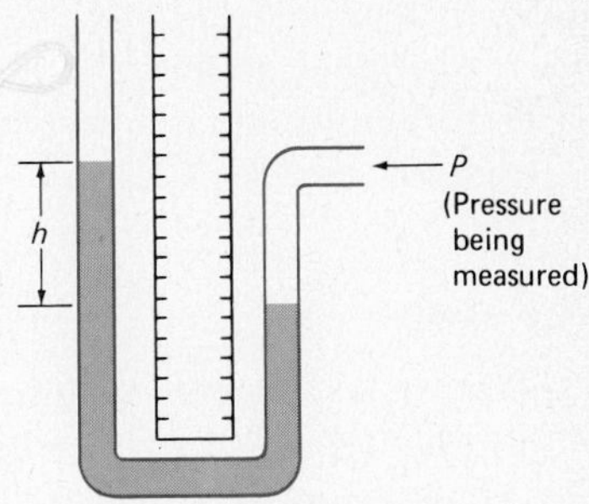

(*a*) Open-tube manometer

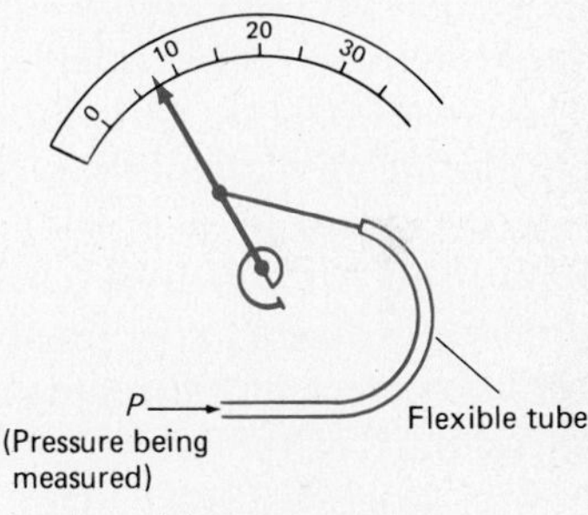

(*b*) Bourdon gauge

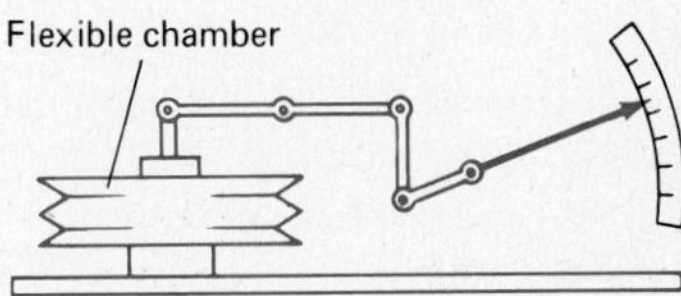

(*c*) Aneroid gauge (mainly used for air pressure and then is called an aneroid barometer)

FIGURE 6-6
Pressure gauges: (*a*) open-tube manometer; (*b*) Bourdon gauge; (*c*) aneroid gauge.

TABLE 6-2

Conversion factors between different units of pressure

In terms of Pa=N/m^2		Related to 1 atm
1 atm	**= 1.013 × 10^5 N/m^2**	**1 atm = 1.013 × 10^5 N/m^2**
1 bar	**= 1.000 × 10^5 N/m^2**	**1 atm = 1.013 bar**
1 dyne/cm^2	**= 0.1 N/m^2**	**1 atm = 1.013 × 10^6 dyne/cm^2**
1 kg/cm^2	**= 9.85 × 10^4 N/m^2**	**1 atm = 1.03 kg/cm^2**
1 lb/in^2	**= 6.90 × 10^3 N/m^2**	**1 atm = 14.7 lb/in^2**
1 lb/ft^2	**= 47.9 N/m^2**	**1 atm = 2.12 × 10^3 lb/ft^2**
1 cm Hg	**= 1.33 × 10^3 N/m^2**	**1 atm = 76 cm Hg**
1 mm Hg	**= 133 N/m^2**	**1 atm = 760 mm Hg**
1 torr	**= 133 N/m^2**	**1 atm = 760 torr**
1 mm H_2O(4°C)	**= 9.81 N/m^2**	**1 atm = 1.03 × 10^4 mm H_2O (4°C)**

Other types of pressure gauge are the Bourdon gauge (Figure 6-6*b*) in which increased pressure tends to straighten out the delicate tube that is attached to a pointer, and the aneroid gauge (Figure 6-6*c*) in which the pointer is connected to the flexible ends of an evacuated thin metal chamber. In a more sophisticated type of gauge, the pressure to be measured is applied to a thin metal diaphragm and the resulting distortion of the diaphragm is detected electrically with the use of a so-called pressure transducer (see Chapter 16).

Blood pressure

Blood pressure is measured using either a mercury-filled manometer or one of the other types of gauge mentioned above; it is usually calibrated in mm Hg. The gauge is attached to a closed air-filled jacket which is wrapped around the upper arm at the level of the heart, Figure 6-7. Two values of blood pressure are measured: the maximum pressure when the heart is pumping, which is called *systolic pressure*; and the pressure when the heart is in the resting part of the cycle, which is called *diastolic pressure*. Initially, the air pressure in the jacket is increased high above the systolic pressure by means of a hand pump, and this compresses the main (brachial) artery in the arm and cuts off the flow of blood. The air pressure is then slowly reduced to the point where blood again begins to flow into the arm; it is detected by listening with a stethoscope to the characteristic tapping sound† of the blood returning to the forearm. At this point, systolic pressure is just equal to the air pressure in the jacket which can be read off the gauge. The air pressure is subsequently reduced further and the tapping sound disappears when blood at low pressure can enter the artery. At this point, the gauge indicates the diastolic pressure. Normal systolic pressure is around 120 mm Hg whereas normal diastolic pressure is around 80 mm Hg.

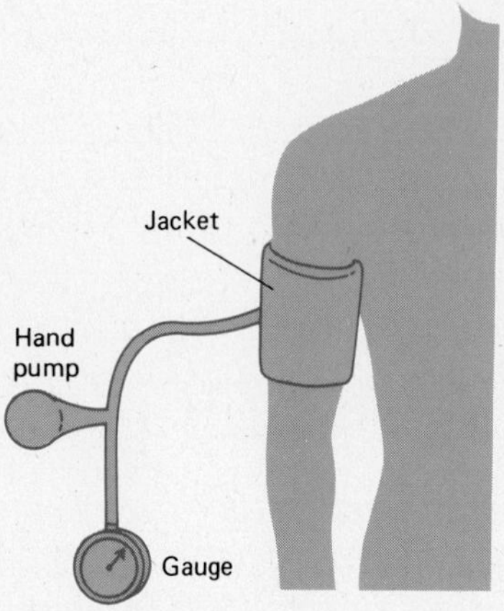

Jacket

Hand pump

Gauge

FIGURE 6-7
Device for measuring blood pressure.

The barometer

Atmospheric pressure is measured by a modified kind of mercury manometer with the end closed. This is called a mercury barometer, Figure 6-8. The glass tube is filled with mercury and then inverted into the bowl of mercury. If the tube is long enough, the level of the mercury will drop, leaving a vacuum at the top of the tube, since atmospheric pressure can only support a column of mercury 76 cm high. That is, a column of mercury 76 cm high exerts the same pressure as the whole atmosphere. Note that the formula $P = \rho g h$, with $\rho = 13.6 \times 10^3$ kg/m^3 for mercury and $h = 76.0$ cm gives us

$$P = (13.6 \times 10^3 \text{ kg/m}^3)(9.80 \text{ m/s}^2)(0.760 \text{ m})$$
$$= 1.013 \times 10^5 \text{ N/m}^3$$
$$= 1.00 \text{ atm.}$$

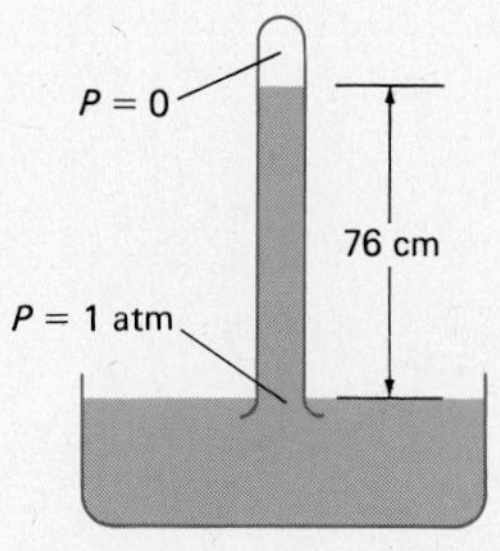

FIGURE 6-8
A mercury barometer, when the air pressure is 76 cm Hg.

Household barometers are usually of the aneroid type, Figure 6-6*c*.

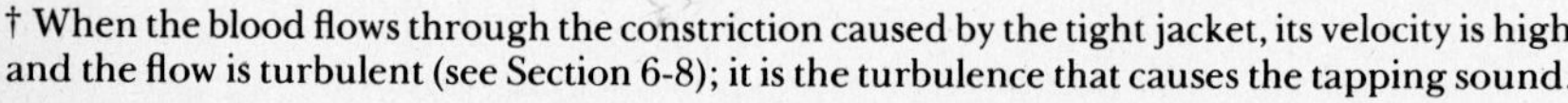
† When the blood flows through the constriction caused by the tight jacket, its velocity is high and the flow is turbulent (see Section 6-8); it is the turbulence that causes the tapping sound.

A calculation similar to that above will show that atmospheric pressure can maintain a column of water 10.3 m high in a tube whose top is under vacuum. It was a source of wonder and frustration not so long ago that no matter how good a vacuum pump was, it could not lift water more than about 10 m. (It was a great practical difficulty in trying to pump water out off deep mine shafts.) The point is that a pump does not really suck water up a tube. It is air pressure that pushes the water up the tube if the other end is under vacuum, just as it is air pressure that pushes the mercury 76 cm high in a barometer.

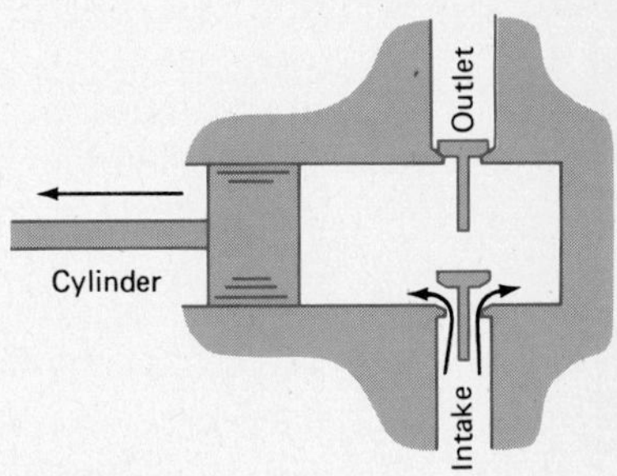

FIGURE 6-9
An example of one kind of pump. As the diagram indicates, the intake valve opens and air (or fluid that is being pumped) fills the empty space when the cylinder moves to the left. When the cylinder moves to the right (not shown), the outlet valve opens and fluid is forced out.

Pumps: the heart

Pumps can be classified into categories according to their function. A *vacuum pump* is designed to reduce the pressure (usually of air) in a given vessel. A *force pump*, on the other hand, is a pump which is intended to increase the pressure—for example, to lift a liquid (such as water from a well) or to push a fluid through a pipe. Figure 6-9 illustrates the principle behind a simple "reciprocating" pump. It could be a vacuum pump, in which case the "intake" is connected to the vessel to be evacuated; a similar mechanism is used in some force pumps, and in this case the fluid is forced under increased pressure through the outlet. Other kinds of pumps are illustrated in Figure 6-10. The centrifugal pump, or any force

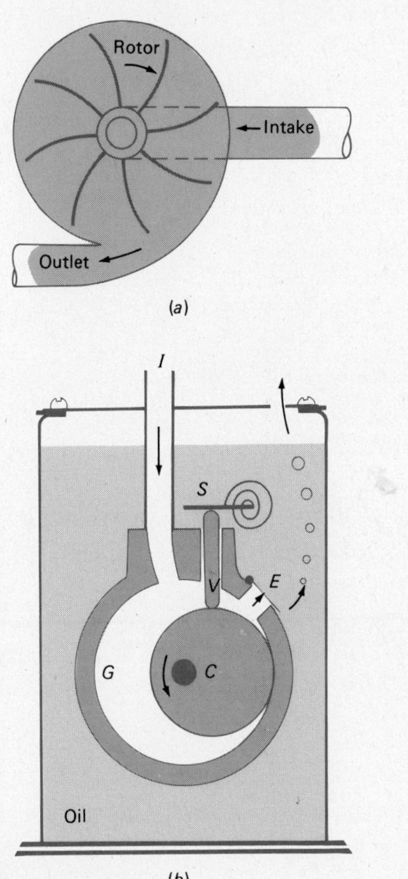

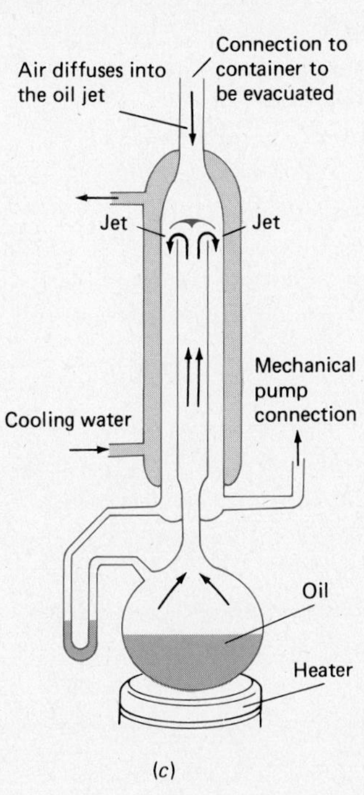

FIGURE 6-10
(*a*) Centrifugal pump: the rotating blades force fluids through the outlet pipe; this kind of pump is used in vacuum cleaners and as a water pump in automobiles. (*b*) Rotary oil pump; used to obtain vacuum as low as 10^{-4} mm Hg: gas (usually air) from the vessel to be evacuated diffuses into the space *G* via the intake pipe *I*; the rotating off-center cylinder *C* traps the gas in *G* and pushes it out the exhaust valve *E*, in the meantime allowing more gas to diffuse into *G* for the next cycle. The sliding valve *V* is kept in contact with *C* by a spring *S* and this prevents the exhaust gas from returning to *G*. (*c*) Diffusion pump, used to obtain vacuums as low as 10^{-8} mm Hg: air molecules from the vessel to be evacuated diffuse into the jet, where the rapidly moving jet of oil sweeps the molecules away. The "forepump" is a mechanical pump, such as the rotary type (*b*), and acts as a first stage in reducing the pressure.

FIGURE 6-11
(*a*) In the diastole phase, the heart relaxes between beats. Blood moves into the heart; both atria are filled rapidly. (*b*) When the atria contract, the systole or pumping phase begins. The contraction pushes the blood through the mitral and tricuspid valves into the ventricles. (*c*) The contraction of the ventricles forces the blood through the semilunar valves into the pulmonary artery (which leads to the lungs) and to the aorta (the body's largest artery), which leads to the arteries serving all the body (Figure 6-26). (*d*) When the heart relaxes, the semilunar valves close; blood fills the atria, beginning the cycle again.

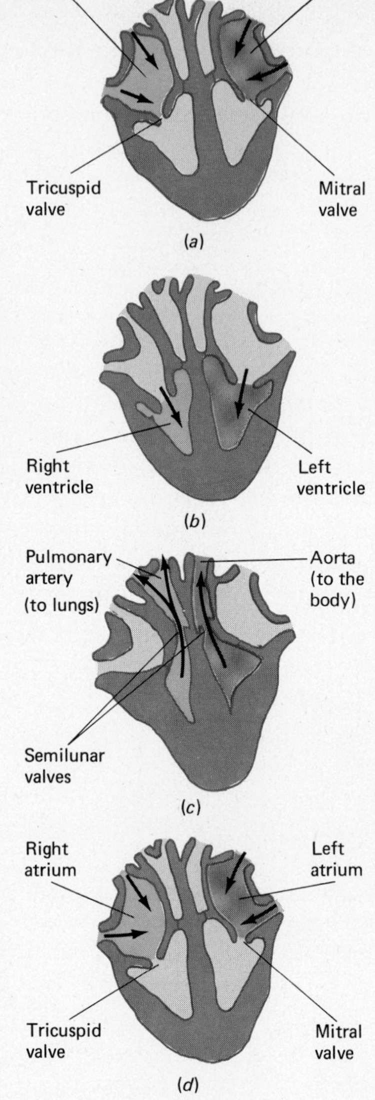

pump, can be used as a *circulating pump*—that is, to circulate a fluid around a closed path, such as the cooling water or lubricating oil in an automobile.

The heart of a human (and of other animals as well) is essentially a circulating pump. The action of a human heart is shown in Figure 6-11. There are actually two separate paths for blood flow. The longer path takes blood to the parts of the body, via the arteries, bringing oxygen to body tissues and picking up carbon dioxide which it carries back to the heart via veins. This blood is then pumped into the lungs (the second path) where the carbon dioxide is released and oxygen is taken up. The oxygen-laden blood is returned to the heart where it is again pumped to the tissues of the body.

6-3 Buoyancy and Archimedes' principle

Objects submerged in a fluid appear to weigh less. For example, a large rock that you would have difficulty lifting off the ground can often be easily lifted from the bottom of a stream; when the rock breaks through the surface of the water it suddenly seems to be much heavier. Many objects, such as wood, float on the surface of the water. These are two examples of buoyancy. In each example, the force of gravity is acting downward; but in addition, an upward buoyant force is exerted by the liquid.

The buoyant force arises from the fact that the pressure in a fluid increases with depth. Thus the upward pressure on the bottom surface of a submerged object is greater than the downward pressure on its top surface. To see the effect of this, consider a cylinder of height h whose top and bottom ends have an area A and which is completely submerged in a fluid of density ρ_f as shown in Figure 6-12. The fluid exerts a pressure

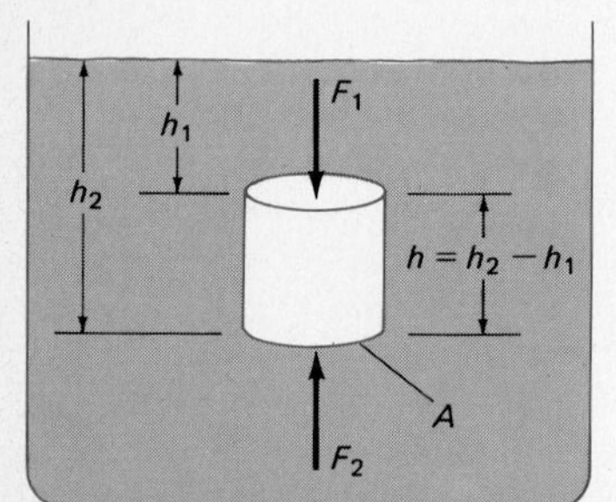

FIGURE 6-12
Determination of buoyant force.

$P_1 = \rho_f g h_1$ against the top surface of the cylinder; the force due to this pressure on top of the cylinder is $F_1 = P_1 A = \rho_f g h_1 A$, and it is directed downward. Similarly, the fluid exerts an upward force on the bottom of the cylinder equal to $F_2 = p_2 A = \rho_f g h_2 A$. The net force due to the fluid pressure, which is the *buoyant force* F_B, acts upward and has the magnitude

$$\begin{aligned} F_B &= F_2 - F_1 \\ &= \rho_f g A (h_2 - h_1) \\ &= \rho_f A h g \\ &= \rho_f V g, \end{aligned}$$

where $V = Ah$ is the volume of the cylinder. Since ρ_f is the density of the fluid, the product $\rho_f V g = m_f g$ is the weight of the fluid which takes up a volume equal to the volume of the cylinder. Thus the buoyant force on the cylinder is equal to the weight of the fluid displaced by the cylinder. This result is valid no matter what the shape of the object. It was first discovered by Archimedes (287?–212 B.C.) and hence is called **Archimedes' principle:** *The buoyant force on a body immersed in a fluid is equal to the weight of the fluid displaced† by that object.*

We can see that Archimedes' principle is true in general by the following simple but elegant argument. The irregularly shaped object D shown in Figure 6-13*a*, is acted on by its weight, w, and the buoyant force, F_B. (If there is no other force acting on the object, such as a hand pulling up, the object shown is in the process of moving downward since $w > F_B$.) We wish to determine F_B and to do so we next consider a body of fluid D' of the same shape and size as the original object, and located at the same depth (Figure 6-13*b*); you might think of this body of fluid as being separated from the rest of the fluid by an imaginary transparent membrane. The buoyant force F_B on this body of fluid will be exactly the same as that on the original object since the surrounding fluid, which exerts F_B, is in exactly the same configuration. Now the body of fluid D' is in equilibrium (the fluid as a whole is at rest) and hence $F_B = w'$, which is the weight of the body of fluid. Hence F_B is equal to the weight of the body of fluid whose volume is equal to the volume of the original submerged object, which is Archimedes' principle.

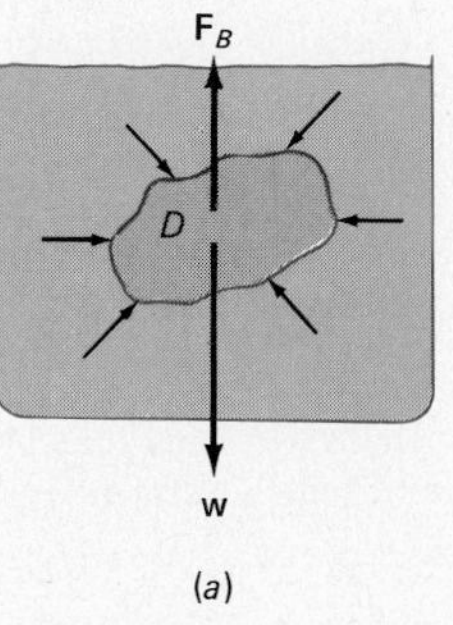

(a)

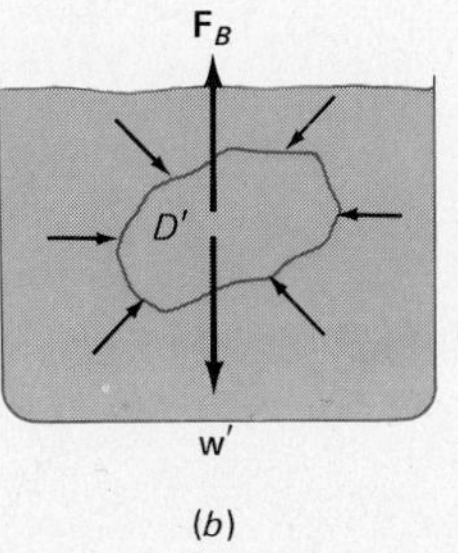

(b)

FIGURE 6-13
Archimedes' principle.

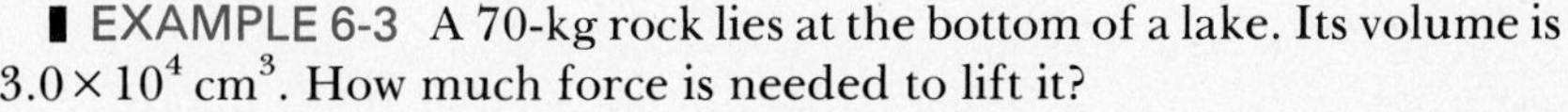

▌ EXAMPLE 6-3 A 70-kg rock lies at the bottom of a lake. Its volume is $3.0 \times 10^4\ \mathrm{cm}^3$. How much force is needed to lift it?

SOLUTION The buoyant force on the rock due to the water is equal to the weight of $3.0 \times 10^{-2}\ \mathrm{m}^3$ of water:

$$\begin{aligned} F_B &= m_{H_2O} g = \rho_{H_2O} V g \\ &= (1.0 \times 10^3\ \mathrm{kg/m^3})(3.0 \times 10^{-2}\ \mathrm{m^3})(9.8\ \mathrm{m/s^2}) \\ &= 2.9 \times 10^2\ \mathrm{N}. \end{aligned}$$

† By "the fluid displaced" we mean a volume of fluid equal to the volume of the object, or that part of the object submerged if it floats. If the object is placed in a glass initially filled to the brim with water, the water that flows over the top represents the water displaced by the object.

The weight of the rock is $mg = (70\ \text{kg})(9.8\ \text{m/s}^2) = 6.9 \times 10^2\ \text{N}$. Hence the force needed to lift it is $690\ \text{N} - 290\ \text{N} = 400\ \text{N}$. It is as if the rock weighed only $(400\ \text{N})/(9.8\ \text{m/s}^2) = 41\ \text{kg}$. ▌▌

Archimedes is said to have discovered his principle in his bath while thinking how he might determine whether the king's new crown was pure gold or a fake. Gold has a specific gravity of 19.3, somewhat higher than most metals, but a determination of specific gravity or density was not readily done directly since the volume of an irregularly shaped object (a crown) was not easily determined. However, if the object is weighed in air ($= w$) and also "weighed" while it is submerged in the fluid ($= w_s$), the density can be determined using Archimedes' principle since the apparent weight of the submerged object, w_s, equals its actual weight w minus the buoyant force F_B; that is:

$$w_s = w - F_B = \rho V g - \rho_f V g$$

where V is the volume of the object, ρ its density, and ρ_f the density of the fluid (water). Then, using $w = \rho V g$ and the above equation, we can write:

$$\frac{w}{w - w_s} = \frac{\rho V g}{\rho_f V g} = \frac{\rho}{\rho_f}. \tag{6.4}$$

Thus $w/(w - w_s)$ is equal to the specific gravity of the object if the fluid is water ($\rho_f = 1.00$).

▌EXAMPLE 6-4 A 14.7-kg crown has an effective mass of 13.4 kg when submerged in water. Is it gold?

SOLUTION

$$\frac{\rho}{\rho_{H_2O}} = \frac{w}{w - w_s} = \frac{14.7\ \text{kg}}{1.3\ \text{kg}} = 11.3.$$

This corresponds to a density of $11{,}300\ \text{kg/m}^3$. The crown seems to be made of lead! ▌▌

Archimedes' principle applies equally well to objects that float such as wood. In general, *an object floats on a fluid if its density is less than that of the fluid.* For example, a log whose specific gravity is 0.60 and whose volume is $2.0\ \text{m}^3$ will have a mass of 1200 kg. If the log is fully submerged, it will displace a mass of water $m = \rho V = (1000\ \text{kg/m}^3)(2.0\ \text{m}^3) = 2000\ \text{kg}$. Hence the buoyant force on it will be greater than its weight and it will float to the top. It will come to equilibrium when it displaces 1200 kg of water, which means that $1.2\ \text{m}^3$ or 0.60 of its volume will be submerged. In general, that the fraction of the object submerged is given by the ratio of the object's density to that of the fluid.

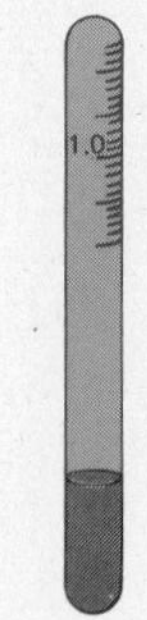

FIGURE 6-14
A hydrometer.

▌EXAMPLE 6-5 A hydrometer is a simple instrument used to measure specific gravity. A particular hydrometer (Figure 6-14) consists of a glass tube, weighted at the bottom, which is 25.0 cm long, $2.0\ \text{cm}^2$ in cross-sectional area, and has a mass of 45 g. How far from the end should the 1.000 mark be placed?

SOLUTION The hydrometer has a density

$$\rho = \frac{m}{V} = \frac{45\ \text{g}}{(2.0\ \text{cm}^2)(25\ \text{cm})} = 0.90\ \text{g/cm}^3.$$

Thus, when placed in water, it will come to equilibrium when 0.90 of its volume is submerged. It is of uniform cross section, so (0.90)(25.0 cm) = 22.5 cm of its length will be submerged. Since the specific gravity of water is defined to be 1.000, the mark should be placed 22.5 cm from the end. ▮▮

When the density of a submerged object is equal to the density of the fluid surrounding it, the buoyant force is equal to the weight of the object; the object is in equilibrium and will neither sink deeper nor rise to the surface. This is the situation with fish and other sealife. Many fish possess a gas-filled sack known as a *swim bladder* which allows them to remain suspended in water. The swim bladder reduces the specific gravity of a fish to near 1.00 whereas without it the SG would be about 1.07. Fish exert some control over their overall density by secreting gas into or out of the swim bladder, thus allowing them to rise or descend. Cuttlefish maintain a low density by means of the cuttlebone, which contains many tiny chambers partly filled with gas. Similarly, the chambers of the shell of the nautilus are filled with gas, to balance out the large density of its thick-walled shell so that it can hover in the water.

Air is a fluid and so it too exerts a buoyant force. Ordinary objects weigh less in air than they do when weighed in a vacuum. Because the density of air is so small, the effect for ordinary solids is slight. There are objects, however, that float in air—helium balloons for example.

▮ EXAMPLE 6-6 What volume of helium is needed if a balloon is to lift a load of 800 kg (including the weight of the empty balloon)?

SOLUTION The buoyant force on the helium, F_B, which is equal to the weight of displaced air, must at least be equal to the weight of the helium plus the load:

$$F_B = (m_{\text{He}} + 800\text{ kg})\,g$$

where g is the acceleration due to gravity. This equation can be written in terms of density:

$$\rho_{\text{air}} Vg = (\rho_{\text{He}} V + 800\text{ kg})\,g.$$

We solve now for V and find:

$$V = \frac{800\text{ kg}}{\rho_{\text{air}} - \rho_{\text{He}}} = \frac{800\text{ kg}}{(1.29\text{ kg/m}^3 - 0.18\text{ kg/m}^3)}$$

$$= 720\text{ m}^3.$$

▮▮

*6-4 ▮ Surface tension and capillarity

*Surface tension

Up to now in this chapter we have been studying what happens beneath the surface of a liquid (or gas). But the actual surface of a liquid also behaves in an interesting way. A number of common observations suggest that the surface of a liquid acts like a stretched membrane under tension. For example a drop of water on the end of a dripping faucet, or hanging from a thin branch in the early morning dew, forms into a nearly spherical shape as if it were a tiny balloon filled with water; and a steel needle can be made to float on the surface of water even though it is denser than the water. The surface of a liquid acts as if it is under tension, and this tension, acting parallel to the surface, arises from the attractive forces between the molecules. This effect is called *surface tension*; more specifically a quantity called the *surface tension*, γ (the Greek letter "gamma"), is defined as the force F per unit length L that acts across any line in a surface, tending to pull the surface open:

$$\gamma = \frac{F}{L}. \tag{6-5}$$

To understand this, consider the U-shaped apparatus shown in Figure 6-15; because of surface tension, a force F is required to pull the movable wire (of length l) and thus increase the surface area of the liquid. The liquid contained by the wire apparatus is a thin film having both a top and a bottom surface; hence the length of the surface being increased is $2l$. Thus the surface tension $\gamma = F/2l$. A delicate apparatus of this type can be used to measure the surface tension of various liquids. The surface tension of water is 0.072 N/m at 20°C. Table 6-3 gives the values for other liquids; note that the temperature considerably affects surface tension.

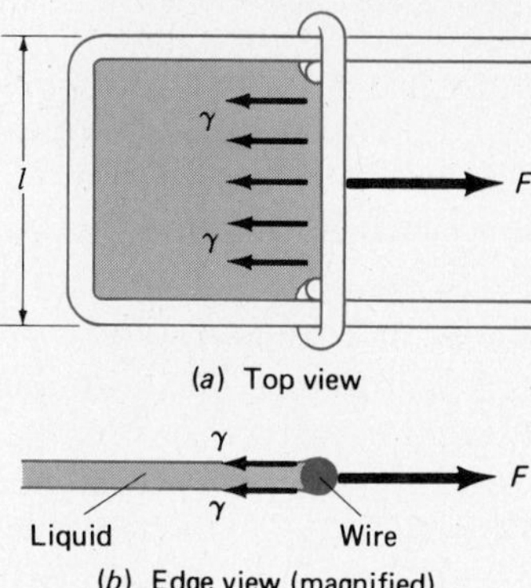

FIGURE 6-15
U-shaped wire apparatus holding film of liquid to measure surface tension.

We can see how surface tension arises by examining the process from the molecular point of view. The molecules of a liquid exert attractive forces on each other; these attractive forces due to other molecules are shown acting, in Figure 6-16, on a molecule at the surface and on a second molecule deep below the surface. We see that a molecule deep below the surface has no net force exerted on it since it is surrounded by other molecules exerting attractive forces on it in all directions. The attractive forces on a molecule at the surface, however, are exerted only by molecules below it. Because of this downward force, the surface layer is slightly compressed—but only to the point where this downward force is balanced by an upward (repulsive) force due to collisions with the molecules below.† This compression of the surface means that the liquid, in essence, tries to minimize its surface area. This is why water tends to form spherical droplets, for the sphere represents the minimum surface area for a given volume.

TABLE 6-3

Surface tension (N/m) of some substances

Substance	γ(N/m)
Mercury (20°C)	**0.44**
Blood, whole (37°C)	**0.058**
Blood, plasma (37°C)	**0.073**
Alcohol, ethyl (20°C)	**0.023**
Water (0°C)	**0.076**
(20°C)	**0.072**
(100°C)	**0.059**

In order to increase the surface area of a liquid, a force is required and work must be done to bring molecules from the interior to the surface (Figure 6-17). This work increases the potential energy of the molecules and is sometimes called *surface energy*. The greater the surface area, the greater the surface energy.

The amount of work needed to increase the surface area by an amount ΔA can be calculated from Figure 6-15 and Equation 6-5: if the movable wire is pulled a distance Δx, the increase in area is $\Delta A = l\,\Delta x$, so

$$\begin{aligned} W &= F\,\Delta x \\ &= \gamma l\,\Delta x \\ &= \gamma\,\Delta A. \end{aligned}$$

We can therefore write

$$\gamma = \frac{W}{\Delta A}.$$

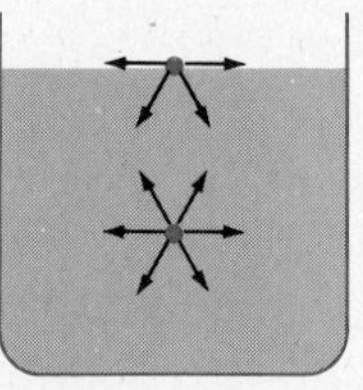

FIGURE 6-16
Molecular theory of surface tension.

† The air molecules above exert a force, but this is a small effect since the air molecules are so far apart. The surface tension therefore does depend on the material above the surface but is generally small if the latter is a dilute gas. But because of this small effect, surface tensions are specified for the boundary between two materials. If the second material is not mentioned, it is assumed to be air.

Thus, the surface tension γ is not only equal to the force per unit length; it is also equal to the work done per unit increase in surface area. Consequently, γ can be specified in units of N/m or J/m^2.

Because of surface tension, insects can walk on water; and objects more dense than water, such as a steel needle or a duck, can actually float on the surface. Figure 6-18*a* shows how the surface tension can support the weight w of an object. Actually, w is the "effective weight" of the object—its true weight less the buoyant force—since the object sinks slightly into the fluid. If the object is spherical in shape, which is close for the base of an insect's leg (Figure 6-18*b*), the surface tension acts at all points along the circumference $2\pi r$ of a circle of radius r. Only the vertical component, $\gamma \cos \theta$, acts to balance w. Hence the net upward force due to surface tension is $2\pi r\gamma \cos \theta$.

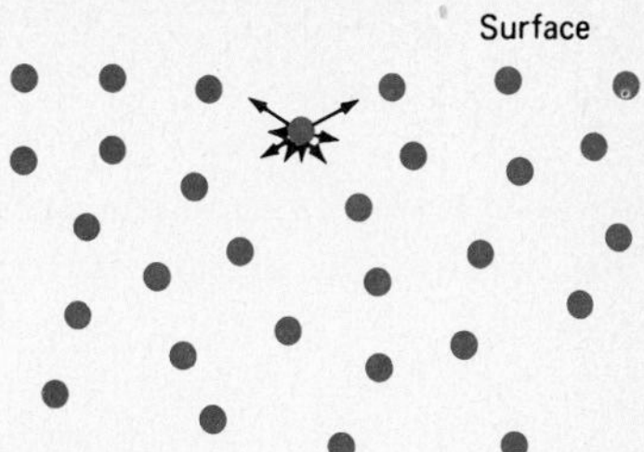

FIGURE 6-17
Forces on a molecule being brought to the surface because of an increase in the surface area. The molecules at the surface must exert a strong force on the molecule being pulled up, and they in turn feel a strong force back (Newton's third law). This is the surface-tension force.

▌ EXAMPLE 6-7 The base of an insect's leg is approximately spherical in shape with a radius of about 2.0×10^{-5} m. The 3.0×10^{-6} kg mass of the insect is supported equally by the six legs. Calculate the angle θ (see Figure 6-18). Assume the water temperature is 20°C.

SOLUTION Since

$$2\pi r\gamma \cos \theta = w$$

where w is one-sixth the weight of the insect (since it has six legs), we have

$$(6.28)(2.0 \times 10^{-5}\ \text{m})(0.072\ \text{N/m}) \cos \theta = \tfrac{1}{6}(3.0 \times 10^{-6}\ \text{kg})(9.8\ \text{m/s}^2)$$

$$\cos \theta = \frac{0.49}{0.90} = 0.54$$

$$\theta = 57°.$$

Notice that if $\cos \theta$ were greater than one, this would indicate that the surface tension would not be great enough to support the weight of the insect. ▌▌

Calculations such as that above are not always accurate since the radius r of the surface depression is not precisely equal to the radius of the object. Approximate estimates can nonetheless be useful.

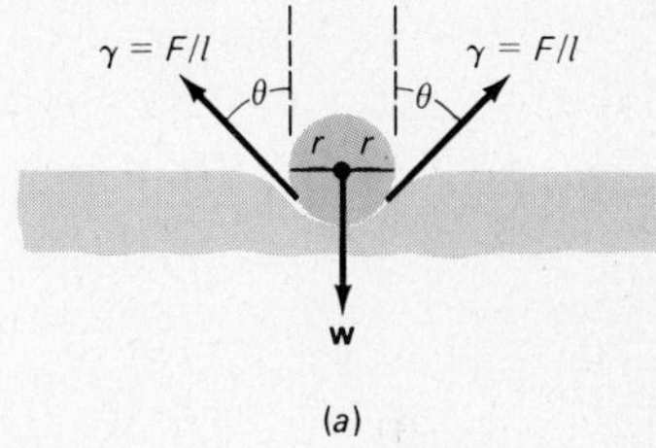

(*a*)

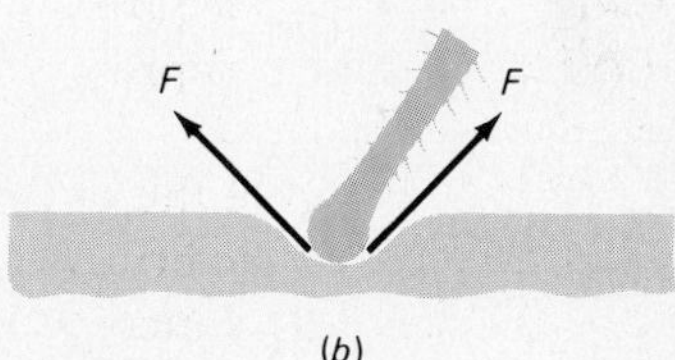

(*b*)

FIGURE 6-18
Surface tension acting on (*a*) a sphere, (*b*) an insect leg.

Soaps and detergents have the effect of lowering the surface tension of water. This is desirable for washing and cleaning since the high surface tension of pure water prevents it from penetrating between the fibers of material and into tiny crevices. Substances which reduce the surface tension of a liquid are called *surfactants*.

Surface tension and its reduction play an important role in the human body; one example is in the lungs. In order to facilitate the interchange of oxygen and CO_2 between the air and the blood capillaries in the lungs, the surface of the interior of the lungs is very large, the size of a tennis court! To fit this large area into a human, the surface of the lungs is very convoluted, containing myriads of tiny sacs called alveoli. The alveoli are lined with mucous fluid tissue which has a normal surface tension of about 0.050 N/m. This rather high surface tension would resist the expansion of the lungs (and consequent increase in surface area) during inhalation. However, in normal persons the lungs secrete a surfactant that reduces the effort needed to expand the lungs by reducing the surface tension to as low as 0.002 N/m.

*Capillarity

It is a common observation that water in a glass container rises up slightly where it touches the glass, Figure 6-19*a*. The water is said to "wet" the glass. Mercury, on the other hand, is depressed when it touches the glass, Figure 6-19*b*; the mercury does not wet the glass. Whether or not a liquid wets a solid surface is determined by the relative strengths of the cohesive forces between the molecules of the liquid compared to the adhesive forces between the molecules of the liquid and those of the solid. (*Cohesion* refers to the force between like molecules and *adhesion* to the force between unlike molecules.) Water wets glass because the water molecules are more strongly attracted to the glass molecules than they are to other water molecules. The opposite is true for mercury: the cohesive forces are stronger than the adhesive forces. The angle that the tangent to the liquid surface makes with the solid surface is called the *angle of contact* ϕ. Its value depends on the relative strengths of the cohesive and adhesive forces (see Figure 6-19 and Table 6-4). When ϕ is less than 90°, the liquid wets the solid; if ϕ is greater than 90°, the liquid does not wet the solid.

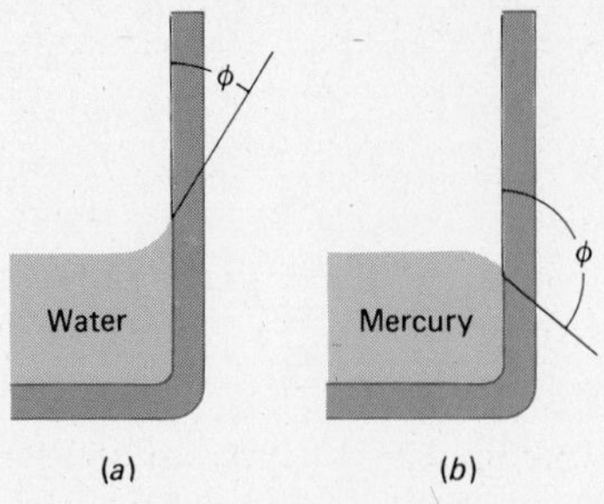

FIGURE 6-19
Water (*a*) "wets" the surface of glass, whereas mercury (*b*) does not.

TABLE 6-4

Contact angle of various substances

Substance	Contact angle (degrees)
Water–glass	**0**
Organic liquids (most)–glass	**0**
Mercury–glass	**140**
Water–paraffin	**107**
Kerosene–glass	**26**

In very small diameter tubes, liquids are observed to rise or fall relative to the level of the surrounding liquid. This phenomenon is called *capillarity*, and such thin tubes are called *capillaries*. Whether the liquid rises or falls depends on the relative strengths of the adhesive and cohesive forces, Figure 6-20. Thus water rises in a glass tube whereas mercury falls. The actual amount of rise or fall depends on the surface tension—since it is this which keeps the liquid surface from breaking apart—as well as on the contact angle ϕ and the radius r of the tube. To calculate h, the height of rise, we refer to Figure 6-21. The surface tension γ acts at an angle ϕ all around a circle of radius r. The magnitude of the vertical force F due to surface tension is, by Equation 6-5, $F = (\gamma \cos \phi)(L)$. Since $L = 2\pi r$, we have $F = 2\pi r\gamma \cos \phi$. This force is just balanced by the weight of the liquid below, which is approximately a cylinder of height h and volume $V = \pi r^2 h$. Hence

$$\begin{aligned} 2\pi r\gamma \cos \phi &= mg \\ &= \rho V g \\ &= \rho \pi r^2 h g, \end{aligned}$$

where ρ is the density of the liquid. Solving for h we find

$$h = \frac{2\gamma \cos \phi}{\rho g r}. \qquad (6\text{-}6)$$

For many liquids, such as water in glass, ϕ is close to zero, and since $\cos 0° = 1$, Equation 6-6 reduces to the simpler form $h = 2\gamma/\rho g r$.

Equation 6-6 is also valid when the liquid drops, as in the case of mercury in a glass tube. In this situation, the contact angle ϕ is greater than 90°, and $\cos \phi$ will be negative; this makes h negative, which corresponds to a decrease in level. Notice from Equation 6-6 that the thinner the tube, the greater the rise (or fall) of the liquid.

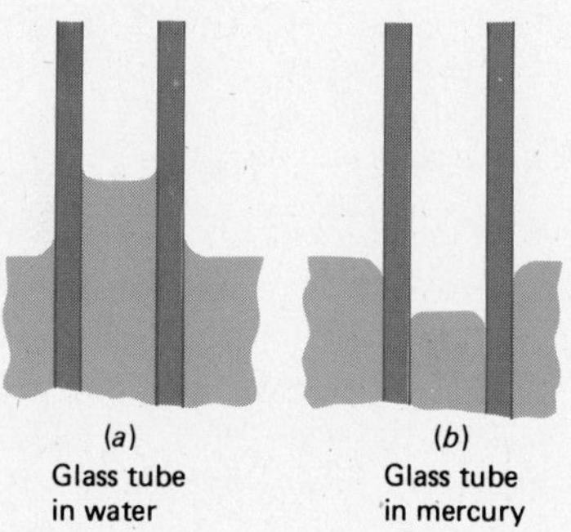

FIGURE 6-20
Capillarity.

EXAMPLE 6-8 If the xylem (tiny tubes which carry nutrients upwards) of a plant has a radius of 0.0010 cm, calculate how high the surface tension can be expected to pull a column of water. Assume $\phi = 0°$.

SOLUTION Using $\gamma = 0.072$ N/m (Table 6-3) in Equation 6-6 we find

$$h = \frac{(2)(0.072\ \text{N/m})(1.0)}{(1.0\times 10^{3}\ \text{kg/m}^3)(9.8\ \text{m/s}^2)(1.0\times 10^{-5}\ \text{m})}$$

$$= 1.5\ \text{m}.$$

One example of how capillary action is important in practical situations is in regard to buildings. Water on the exterior (say, from rain) can be brought to the interior by capillary action along tiny cracks or joints. To prevent this, whenever materials are joined together such that a narrow channel exists—say, at a window—it is general practice to widen the channel at one point as shown in Figure 6-22.

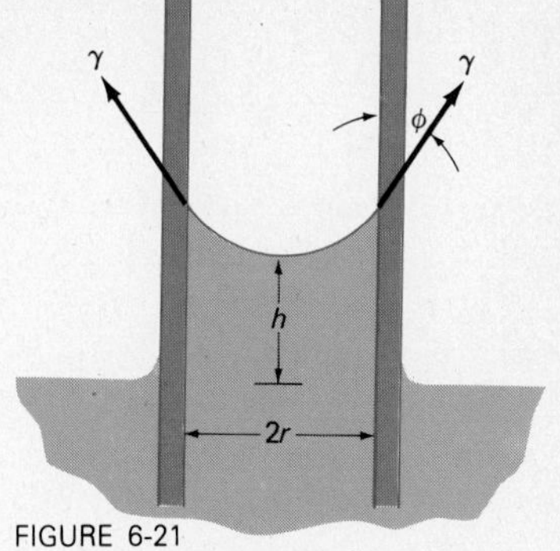

FIGURE 6-21
A liquid rises a height h in a capillary tube given by $h = 2\gamma \cos\phi/\rho g r$ (see text).

*6-5 Negative pressure and the cohesion of water; the rise of fluids in trees

Normally, a fluid exerts an outward pressure on its container and the container exerts an inward pressure on the fluid. Pressures can range from zero up to very large positive values. But can fluid pressure be negative? The answer, strange as it may seem, is yes; under certain circumstances a negative pressure can exist in a liquid, but not in a gas.

Negative pressures have been obtained, although it is difficult, using the apparatus shown in Figure 6-23. The closed tube is initially filled with liquid and then the reservoir on the right is pumped down with a vacuum pump. According to Equation 6-3 the pressure difference between points A and B is $\rho g h$ where ρ is the density of the liquid:

$$P_B - P_A = \rho g h.$$

When the pump has reduced the pressure to near zero, $P_B = 0$ since point B is at the same level as the surface of the liquid in the reservoir. Hence the pressure at A must be negative:

$$P_A = -\rho g h.$$

Normally we would expect the liquid to fall out of the tube when the pump reduces the pressure. But if the tube is scrupulously clean and the liquid is free from impurities, it may remain in the tube. Negative pressures as low as -270 atm have been achieved using this technique. The liquid under negative pressure is unstable, however, and the smallest disturbance will cause the liquid to shatter into droplets, and it falls.

A liquid under negative pressure actually pulls inward on its container; the liquid acts as if it were all under tension (not just its surface) much like a rope pulled at the two ends. How can this occur? It is due to the cohesive forces between the liquid molecules as well as adhesive forces between the liquid and the container. The cohesive forces between water molecules are quite strong—the tensile strength of pure water is about 30×10^6 N/m^2. These are the same forces that hold the water molecules

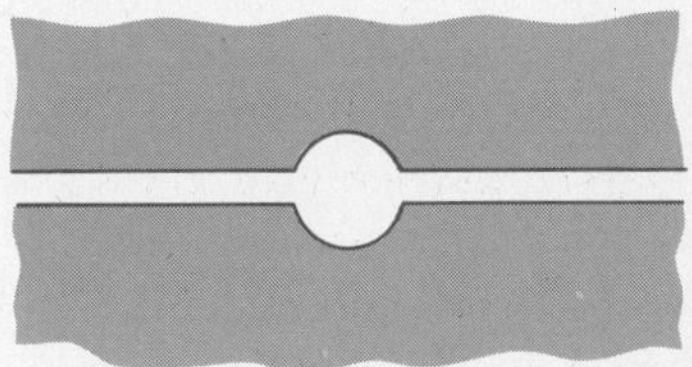
FIGURE 6-22
A window frame and/or the sill have a widened channel to prevent inward seepage of moisture due to capillarity.

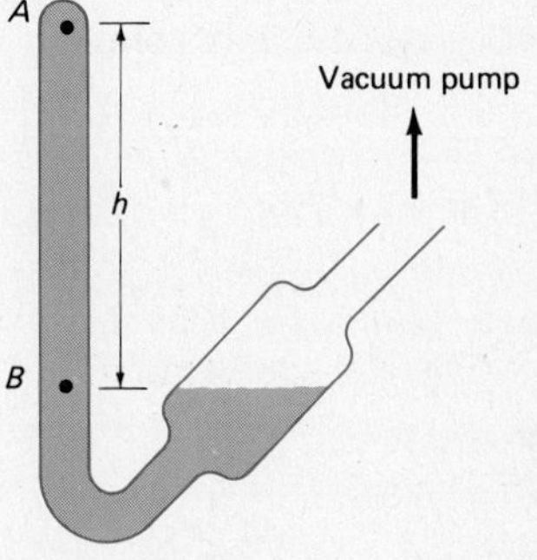

FIGURE 6-23
Producing a negative pressure.

together as ice at low temperatures. The difference is that in the liquid state the molecules have considerable kinetic energy and are free to move over one another. Liquid water does not normally sustain a tension force as a solid does because the slightest intrusion of an impurity (such as air) allows the water to flow and take on another shape; it can then fall under gravity.

It has long been wondered how water can rise to the top of tall trees such as redwoods that reach heights of over 100 m. Water, containing mineral nutrients from the soil, is carried upward by tiny capillaries called the xylem. The radius of these capillaries varies from about 0.01 mm to 0.3 mm. As we saw in Example 6-8, even the smallest-sized capillaries can raise water by only about 1.5 m. On the other hand, atmospheric pressure can raise water only about 10 m even when the top end is under high vacuum (Section 6-2). Neither of these effects account for the rise of water in tall trees. Neither can the process of osmosis (see Chapter 8). Today it is generally accepted that, although the above effects may help, it is the cohesive forces between water molecules acting under tension at negative pressure that are responsible, particularly for very tall trees. This theory was first proposed in the last century, but until recently there was much opposition to it. Within the last few years actual measurements of the pressure within the xylem of trees have been made; they were indeed found to be negative, often as low as −25 atm at the top. When water evaporates from leaves, other water molecules move in to take their place. Through cohesive forces, the water below is pulled upward.

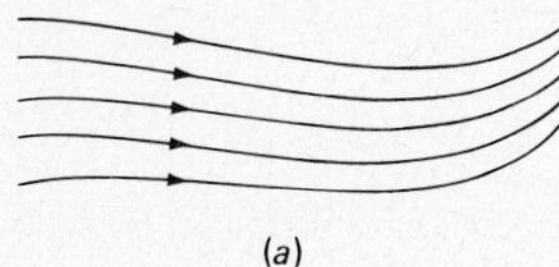

(*a*)

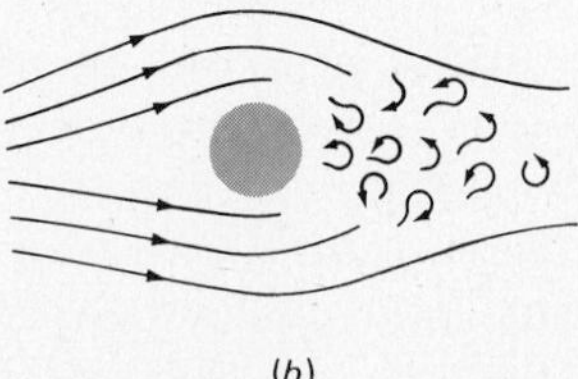

(*b*)

FIGURE 6-24
(*a*) Streamline or laminar flow; (*b*) turbulent flow.

6-6 ▮ Fluids in motion: Bernoulli's principle

We now turn from the study of fluids at rest to the more complex subject of fluids in motion. Many aspects of fluid motion are still not completely understood; nonetheless, with certain simplifying assumptions, a good comprehension of this subject can be obtained.

To begin with, we usually distinguish two types of fluid flow. If the flow is smooth such that neighboring layers of the fluid slide by each other smoothly, the flow is said to be **laminar** or **streamline** flow. This kind of flow is characterized by the fact that each particle of the fluid follows a smooth path, called a streamline, and these paths do not cross over one another, Figure 6-24*a*. Above a certain speed, which depends on a number of factors as we shall see later, the flow becomes turbulent. **Turbulent** flow is characterized by flow in small whirlpoollike circles called eddy currents or eddies, Figure 6-24*b*. Eddies absorb a great deal of energy and although a certain amount of internal friction, called **viscosity**, is present during streamline flow, it is much greater when the flow is turbulent. A few tiny drops of ink or food coloring dropped into a moving liquid can quickly reveal whether the flow is streamline or turbulent.

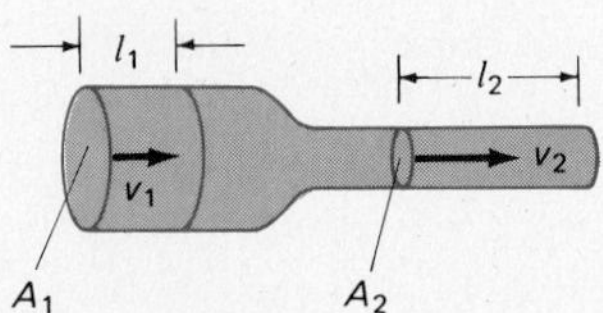

FIGURE 6-25
Fluid flow through a pipe of varying diameter.

Flow rate and the equation of continuity

Let us consider the steady laminar flow of a fluid through an enclosed tube or pipe as shown in Figure 6-25. First we determine how the speed of the fluid changes with the size of the tube. The flow rate is defined as the mass m of fluid that passes a given point per unit time t: m/t. In Figure 6-25, the volume of fluid passing point 1 in a time t is just $A_1 l_1$, where l_1 is the distance the fluid moves in time t and A_1 is the cross-sectional area of the tube at that point. Since the velocity† of fluid passing point 1 is $v_1 = l_1/t$, the flow rate m/t past point 1 is (where V refers to volume)

$$\frac{m}{t} = \frac{\rho_1 V_1}{t} = \frac{\rho_1 A_1 l_1}{t} = \rho_1 A_1 v_1.$$

Similarly, at point 2, the flow rate is $\rho_2 A_2 v_2$. Since no fluid flows in or out the sides, the flow rate at points 1 and 2 must be equal. Thus

$$\rho_1 A_1 v_1 = \rho_2 A_2 v_2.$$

This is called the *equation of continuity*. If the fluid is incompressible, which is an excellent approximation for liquids under most circumstances (and often for gases as well), then $\rho_1 = \rho_2$ and the equation of continuity becomes

$$A_1 v_1 = A_2 v_2. \tag{6-7}$$

Notice that the product Av represents the *volume* rate of flow (volume of fluid passing a given point per second), since $V/t = Al/t = Av$. Equation 6-7 tells us that where the cross-sectional area is large the velocity is small, and where the area is small the velocity is high. That this makes sense can be seen by looking at a river; a river flows slowly and languidly through a meadow where it is broad, but speeds up to torrential speed when passing through a narrow gorge.

Equation 6-7 can be applied to the flow of blood in the body. Blood flows from the heart into the aorta, from which it passes into the major arteries; these branch into the small arteries (arterioles), which in turn branch into myriads of tiny capillaries; the blood returns to the heart via the veins, Figure 6-26.

† If there were no viscosity, the velocity would be the same across a cross section of the tube. Real fluids have viscosity and this internal friction causes different layers of the fluid to flow at different speeds. In this case v_1 and v_2 represent the average speeds at each cross section.

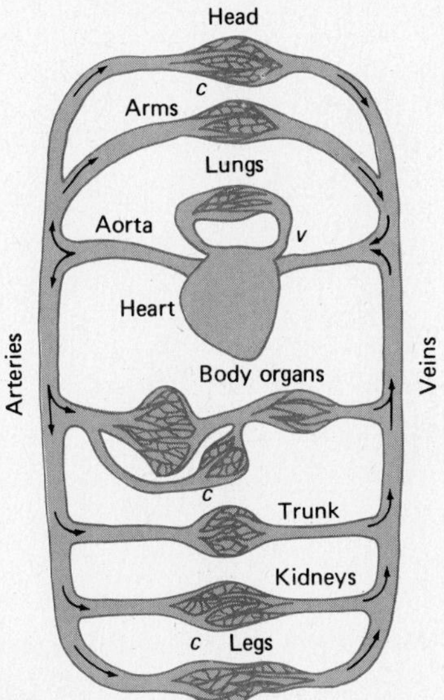

FIGURE 6-26
Human circulatory system.

v—valves
c—capillaries

▮ EXAMPLE 6-9 The radius of the aorta is about 1.0 cm and the blood through it has a speed of about 30 cm/s. Calculate the average speed of the blood in the capillaries using the fact that although each capillary has a diameter of about 8×10^{-4} cm, there are literally billions of them so that their total cross section is about 2000 cm^2.

SOLUTION The speed of blood in the capillaries is

$$v_2 = \frac{v_1 A_1}{A_2} = \frac{(0.30 \text{ m/s})(3.14)(0.010 \text{ m})^2}{(2 \times 10^{-1} \text{ m}^2)}$$

$$= 5 \times 10^{-4} \text{ m/s}$$

or 0.5 mm/s. ▮▮

Another example that makes use of Equation 6-7 and the argument leading up to it is the following.

▮ EXAMPLE 6-10 How large must a heating duct be if air moving 3.0 m/s along it can replenish the air in a room of 300 m^3 volume every 15 min?

SOLUTION We consider the room as a large section of duct for purposes of applying Equation 6-7 (call it point 2). Reasoning in the same way we did to obtain Equation 6-7, we see that $A_2 v_2 = A_2 l_2 / t = V_2 / t$ where V_2 is the volume of the room. Then $A_1 v_1 = A_2 v_2 = V_2 / t$ and

$$A_1 = \frac{V_2}{v_1 t} = \frac{300 \text{ m}^3}{(3.0 \text{ m/s})(900 \text{ s})} = 0.11 \text{ m}^3.$$

Since $A = \pi r^2$ we find that the radius must be 0.19 m or 19 cm. ▮▮

Bernoulli's equation

Have you ever wondered how air can circulate in a prairie dog's burrow, why smoke goes up a chimney, or how a sailboat can move against the wind? These are examples of a principle worked out by Daniel Bernoulli (1700–1782) in the early eighteenth century. In essence, *Bernoulli's principle* states that, *where the velocity of a fluid is high, the pressure is low, and where the velocity is low, the pressure is high.* For example, if the pressures at points 1 and 2 in Figure 6-25 are measured, it will be found that the pressure is lower at point 2 (where the velocity is higher) than it is at point 1. At first glance, this might seem strange; you might expect that the higher speed at point 2 would imply a greater pressure. But this cannot be the case; for if the pressure at point 2 were higher than at 1, this higher pressure would slow the fluid down, whereas in fact it has speeded up. Thus the pressure at 2 must be less than at 1, which will allow the fluid to speed up.

Bernoulli developed an equation that expresses this principle quantitatively. To derive Bernoulli's equation, we assume the flow is laminar, the fluid is incompressible, and the viscosity is small enough to be ignored. To be general, we assume the fluid is flowing in a tube with nonuniform cross section that varies in height above some reference level, Figure 6-27. We will consider the amount of fluid shown in color, and calculate the work done to move it from the position shown in (*a*) to that shown in (*b*).

In this process fluid at point 1 flows a distance l_1 and forces the fluid at point 2 to move a distance l_2. The fluid to the left of point 1 exerts a pressure P_1 on the fluid and does an amount of work $W_1 = F_1 l_1 = P_1 A_1 l_1$. At point 2, the work done is $W_2 = -P_2 A_2 l_2$; the negative sign is present because the force exerted on the fluid is opposite to the motion (thus the fluid shown in color does work on the fluid to the right of point 2). Work is also done on the fluid by the force of gravity; since the net effect of the process shown in Figure 6-27 is to move a mass m of volume $A_1 l_1 (= A_2 l_2)$ from point 1 to point 2, the work done by gravity is $W_3 = -mg(h_2 - h_1)$. Notice that this term is negative since the motion is uphill against the force of gravity. The net work W done on the fluid is thus:

$$W = W_1 + W_2 + W_3$$

$$W = P_1 A_1 l_1 - P_2 A_2 l_2 - mgh_2 + mgh_1.$$

According to the work-energy theorem (Section 5-5), the net work done on a system is equal to its change in kinetic energy. Thus

$$\tfrac{1}{2}mv_2^2 - \tfrac{1}{2}mv_1^2 = P_1 A_1 l_1 - P_2 A_2 l_2 - mgh_2 + mgh_1.$$

The mass m has volume $A_1 l_1 = A_2 l_2$ and thus we can substitute $m = \rho A_1 l_1 = \rho A_2 l_2$ and obtain (after dividing through by $A_2 l_2 = A_2 l_2$ and rearranging):

$$P_1 + \tfrac{1}{2}\rho v_1^2 + \rho g h_1 = P_2 + \tfrac{1}{2}\rho v_2^2 + \rho g h_2. \tag{6-8}$$

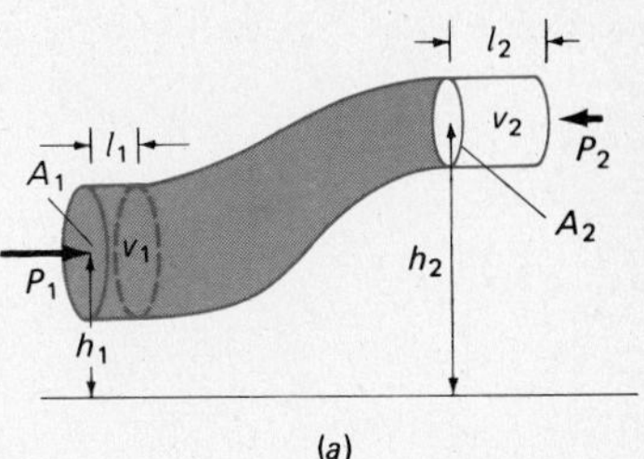

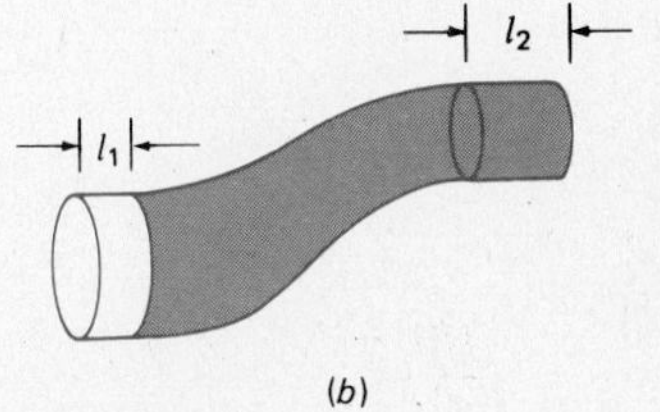

FIGURE 6-27
Fluid flow: for derivation of Bernoulli's equation.

This is **Bernoulli's equation**. It states that the quantity $P + \frac{1}{2}\rho v^2 + \rho g h$ is the same at every point in the fluid.

Bernoulli's equation can be applied to a great many situations. One example is to calculate the velocity, v_1, of a liquid out of a spigot at the bottom of a reservoir, Figure 6-28. We choose point 2 in Equation 6-8 to be the top surface of the liquid, and assuming the diameter of the reservoir is large compared to that of the spigot, v_2 will be almost zero. The pressure at points 1 (the spigot) and 2 are each equal to atmospheric pressure, so $P_1 = P_2$. Then Bernoulli's equation becomes

$$\tfrac{1}{2}\rho v_1^2 + \rho g h_1 = \rho g h_2,$$

or

$$v_1 = \sqrt{2g\,(h_2 - h_1)}. \tag{6-9}$$

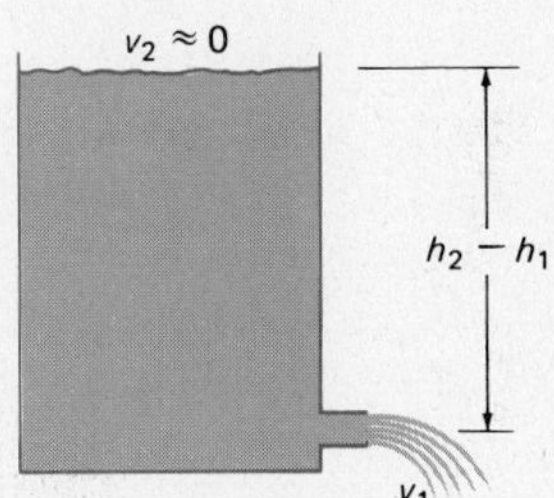

FIGURE 6-28
Torricelli's theorem.

This result is called *Torricelli's theorem*. Although it is seen to be a special case of Bernoulli's equation, it was discovered a century before Bernoulli by Evangelista Torricelli; hence its name. Notice that the liquid leaves the spigot with the same speed that a freely falling object would attain falling the same height. This should not be too surprising for the derivation of Bernoulli's equation relies on the conservation of energy.

Another special case of Bernoulli's equation arises when the fluid is flowing but there is no appreciable change in height; that is, $h_1 = h_2$. Then Equation 6-8 becomes

$$P_1 + \tfrac{1}{2}\rho v_1^2 = P_2 + \tfrac{1}{2}\rho v_2^2. \tag{6-10}$$

This tells us quantitatively that where the speed v is high the pressure P

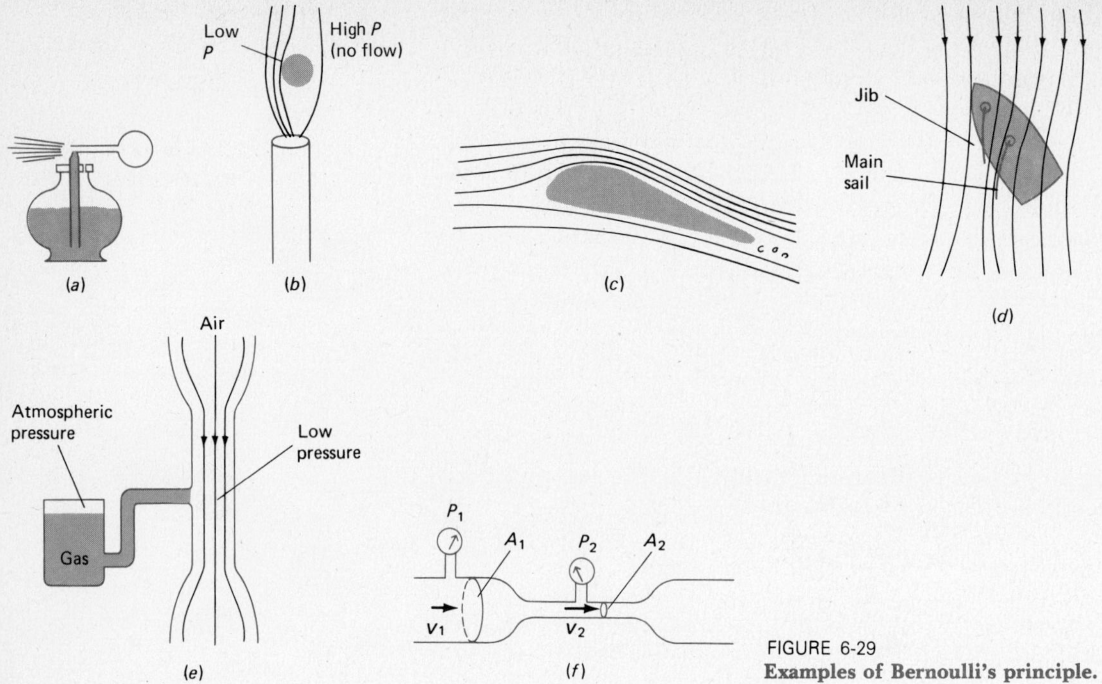

FIGURE 6-29
Examples of Bernoulli's principle.

must be low, and vice versa. It explains many everyday phenomena, some of which are illustrated in Figure 6-29. The pressure in the air blown at high speed across the top of the vertical tube on a perfume atomizer (Figure 6-29*a*) is less than the normal air pressure acting on the surface of the liquid in the bowl; thus perfume is pushed up the tube because of the reduced pressure at the top. A Ping-Pong ball can be made to float above a blowing jet of air (some vacuum cleaners can blow air), Figure 6-29*b*; if the ball begins to leave the jet of air, the higher pressure outside the jet pushes the ball back in.

Airplane wings and other airfoils are designed to deflect the air so although streamline flow is largely maintained, the streamlines are crowded together above the wing, Figure 6-29*c*. Just as the flow lines are crowded together in a pipe constriction where the velocity is high, so the crowded streamlines above the wing indicate the air speed is greater there than it is below the wing. Hence the air pressure above the wing is less than that below and there is thus a net upward force; this is called *lift*. Actually, the flight of a wing is rather complicated and involves other effects, such as turbulence, in addition to Bernoulli's principle.

A sailboat can move against the wind, Figure 6-29*d*, if its sails are arranged so the air velocity increases in the narrow constriction between them. The normal pressure behind the main sail is larger than the reduced pressure in front of it and this pushes the boat forward.

A *Venturi tube* is essentially a pipe with a narrow constriction (the throat). One example of a Venturi tube is the barrel of a carburetor in a car, Figure 6-29*e*. The flowing air speeds up as it passes this constriction (Equation 6-10), so the pressure is lower there. Because of the reduced pressure, gasoline under atmospheric pressure in the carburetor reser-

FIGURE 6-30
Bernoulli's principle is responsible for air flow in underground burrows.

voir is forced into the air stream and mixes with the air before entering the cylinders. The Venturi tube is also the basis of the Venturi meter, which is used to measure the flow speed of fluids, Figure 6-29*f*. Venturi meters can be used to measure the flow velocities of gases and liquids and have even been designed to measure blood velocity in arteries. (See the Problems.)

Why does smoke go up the chimney? It's partly because hot air rises (Section 9-4). But Bernoulli's principle also plays a role. Because wind blows across the top of a chimney, the pressure is less there than inside the house. Hence, air and smoke are pushed up the chimney. Even on an apparently still night there is usually enough ambient air flow at the top of a chimney to allow upward flow of smoke.

Gophers, prairie dogs, rabbits, and moles are among the animals that live underground. To avoid suffocation, the air must circulate in their burrows. These animals always have at least two entrances to their burrows (Figure 6-30). The speed of air flow across different holes will usually be slightly different. This results in a slight pressure difference which forces a flow of air through the burrow a la Bernoulli. The flow of air is enhanced if one hole is higher than the other (and this is often done by animals) since wind speed tends to increase with height.

▌EXAMPLE 6-11 Water circulates throughout a house in a hot water heating system. If the water is pumped at a speed of 0.50 m/s through a 10-cm-diameter pipe in the basement under a pressure of 3.0 atm, what will be the flow speed and pressure in a 6.0-cm-diameter pipe on the second floor 5.0 m above?

SOLUTION We first calculate the velocity v_2 using the equation of continuity, Equation 6-7:

$$v_2 = v_1\frac{A_1}{A_2} = (0.50\text{ m/s})\frac{(\pi)(0.050\text{ m})^2}{(\pi)(0.030\text{ m})^2} = 1.4\text{ m/s}.$$

To find the pressure we use Bernoulli's equation:

$$\begin{aligned} P_2 &= P_1 + \rho g(h_1 - h_2) + \tfrac{1}{2}\rho(v_1^2 - v_2^2) \\ &= (3.0\times10^5\text{ N/m}^2) + (1.0\times10^3\text{ kg/m}^3)(9.8\text{ m/s}^2)(-5.0\text{ m}) \\ &\quad + \tfrac{1}{2}(1.0\times10^3\text{ kg/m}^3)[(0.50\text{ m/s})^2 - (1.4\text{ m/s})^2] \\ &= 3.0\times10^5\text{ N/m}^2 - 4.9\times10^4\text{ N/m}^2 - 8.5\times10^2\text{ N/m}^2 \\ &= 2.5\times10^5\text{ N/m}^2. \end{aligned}$$

Notice that the velocity term contributes very little in this case. ▌▌

Bernoulli's equation ignores the effects of friction within a fluid. The energy that is transformed to thermal energy by this friction can be taken into account by adding a term to the right side of Equation 6-8. This term is difficult to calculate theoretically and is normally done empirically. We will not pursue it here, but merely note that it does not significantly alter the explanations for the phenomena described above.

*6-7 ▮ Viscosity

As already mentioned, real fluids have a certain amount of internal friction which is called viscosity. It exists in both liquids and gases, and is essentially a frictional force between different layers of fluid as they move past one another. In liquids it is due to the cohesive forces between the molecules; in gases it arises from collisions between the molecules.

Different fluids possess different amounts of viscosity: syrup is more viscous than water; grease is more viscous than engine oil; liquids in general are much more viscous than gases. The viscosity of different fluids can be expressed quantitatively by a *coefficient of viscosity*, η (the Greek eta), which is defined in the following way. A thin layer of fluid is placed between two flat plates. One plate is stationary and the other is made to move, Figure 6-31. The fluid directly in contact with either plate is held to the surface by the adhesive force between the molecules. Thus the upper surface of the fluid moves with the same speed v as the upper plate, whereas the fluid in contact with the stationary plate remains stationary. The stationary layer of fluid retards the flow of the layer just above it; this layer retards the flow of the next layer and so on. Thus the velocity varies smoothly from 0 to v as shown. The increase in velocity divided by the distance over which this change is made—equal to v/l—is called the *velocity gradient*. To move the upper plate requires a force, a fact you can verify by moving a flat plate across a puddle of syrup on a table. For a given fluid, it is found that the force F required is proportional to the area of either plate A and to the speed v, and is inversely proportional to the separation l of the plates: $F \propto vA/l$. For different fluids, the more viscous the fluid, the greater is the required force. Hence the proportionality constant for this equation is defined as the *coefficient of viscosity* η:

$$F = \eta \frac{vA}{l}. \qquad (6\text{-}11)$$

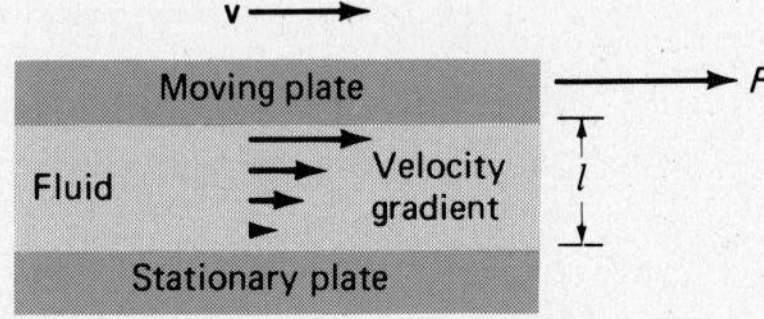

FIGURE 6-31
Determination of viscosity.

Solving for η, we find $\eta = Fl/vA$. Thus the SI units of η are $\text{N} \cdot \text{s/m}^2 = \text{Pa} \cdot \text{s}$ (pascal · second); in the cgs system the unit is $\text{dyne} \cdot \text{s/cm}^2$ and this unit is called a *poise* (P). Viscosities are often stated in *centipoise* (cP). (Note that $1\,\text{cP} = 10^{-2}\,\text{P} = 10^{-3}\,\text{Pa} \cdot \text{s}$.) Table 6-5 lists the coefficient of viscosity for various fluids; the temperature is also specified, since it has a strong effect—the viscosity of liquids such as motor oil, for example, decreases rapidly as temperature increases.

*6-8 ▮ Flow in tubes—laminar and turbulent

*Laminar flow: Poiseuille's law

If a fluid had no viscosity, it could flow through a level pipe without a force being applied. Because of viscosity, a pressure difference between the ends of a tube is necessary for the steady flow of any real fluid, be it water or oil in a pipe, or blood in the circulatory system of a human.

The rate of flow of a fluid in a tube depends on the viscosity of the fluid, the pressure difference, and the dimensions of the tube. A French scientist, J. L. Poiseuille (1799–1869), who was interested in the physics of blood circulation (and after whom the "poise" is named), determined how

TABLE 6-5

Coefficient of viscosity for various fluids

Fluid	Temperature (°C)	Coefficient of viscosity, η (Pa · s)†
Water	**0**	**1.8×10^{-3}**
	20	**1.0×10^{-3}**
	100	**0.3×10^{-3}**
Whole blood	**37**	**$\approx 4 \times 10^{-3}$**
Blood plasma	**37**	**$\approx 1.5 \times 10^{-3}$**
Ethyl alcohol	**20**	**1.2×10^{-3}**
Engine oil (SAE 10)	**30**	**200×10^{-3}**
Glycerine	**20**	**1500×10^{-3}**
Air	**20**	**0.018×10^{-3}**
Hydrogen	**0**	**0.009×10^{-3}**
Water vapor	**100**	**0.013×10^{-3}**

† $1\ \text{Pa} \cdot \text{s} = 10\ \text{P} = 10^3\ \text{cP}$

these variables affect the flow rate of an incompressible fluid undergoing laminar flow in a cylindrical tube. His result, known as *Poiseuille's law*, is given by the equation

$$\frac{V}{t} = \frac{\pi r^4 (P_1 - P_2)}{8\eta L}, \tag{6-12}$$

where r is the radius of the tube, L is its length, $P_1 - P_2$ is the pressure difference between the ends, η is the viscosity coefficient, and V/t is the volume rate of flow (volume of fluid flowing past a given point per unit time). The derivation of Equation 6-12, which is long and requires the use of calculus, can be found in more advanced textbooks.

▌ EXAMPLE 6-12 Water is being pumped through a horizontal pipe of diameter 1.20 m. Pumps, placed at intervals, can exert a pressure of 1.0 atm over atmospheric pressure. How far apart must the pumps be placed to maintain a flow rate of $2.0 \times 10^3\ \text{m}^3/\text{s}$ at 20°C?

SOLUTION We must solve for L in Equation 6-12 and put all terms in SI units:

$$L = \frac{(P_1 - P_2)\pi r^4}{8\eta (V/t)}$$

$$L = \frac{(1.01 \times 10^5\ \text{N/m}^2)(3.14)(0.60\ \text{m})^4}{(8)(1.0 \times 10^{-3}\ \text{N} \cdot \text{s/m}^2)(2.0 \times 10^3\ \text{m}^3/\text{s})}$$

$$= 2.6 \times 10^3\ \text{m},$$

or 2.6 km apart. ▌▌

Poiseuille's law tells us that the flow rate is directly proportional to the pressure drop per unit length, $(P_1 - P_2)/L$, which is called the *pressure gradient*; and it is inversely proportional to the viscosity of the fluid. This is just what we might expect. It may be surprising, however, that V/t also depends on the *fourth* power of the tube's radius. This means that for the same pressure gradient, if the tube radius is doubled the flow rate is increased by a factor of sixteen! Thus the rate of flow, or alternately the pressure required to maintain a given flow rate, is greatly affected by only a small change in tube radius.

*Blood flow

This last fact has important ramifications for the flow of blood in the human body. We must note, however, that Poiseuille's law is valid only for the streamline flow of an incompressible fluid, and when the viscosity η is constant. These requirements are not fully met by blood—the flow is not completely without turbulence, the blood contains corpuscles (whose diameter is almost equal to that of a capillary), and η depends to a certain extent on the flow speed v. Nonetheless, Poiseuille's law does give a good first approximation to the flow of blood in the body. The body controls the flow of blood by means of tiny bands of muscle surrounding the arteries. Contraction of these muscles reduces the diameter of an artery and, as we noted earlier, the flow rate is greatly reduced for only a small change in radius (due to r^4 in Equation 6-12). Very small actions by these muscles can thus control precisely the flow of blood to different parts of the body.

The radius of arteries is reduced as a result of arteriosclerosis (hardening of the arteries) and by cholesterol buildup; when this happens, the pressure gradient must be increased to maintain the same flow rate. If the radius is reduced by half, the heart would have to increase the pressure by a factor of about 16 in order to maintain the same blood flow rate. The heart must work much harder under these conditions but usually cannot maintain the original flow rate. Thus, high blood pressure is an indication that the heart is working harder and that blood flow rate is reduced, both of which are not desirable.

Another aspect of blood flow can be gleaned from Equation 6-12. Since no blood is lost from the body, the flow rate through the aorta is the same as the flow rate through all of the arteries combined and in turn is equal to the total flow rate through all the capillaries. Equation 6-12 then tells us that where the radius is large, the pressure difference is small; and where the radius is small, the pressure difference is large. Indeed, a pressure difference of only 3 mm Hg is needed to maintain normal blood flow through the aorta whereas the pressure drop across the major arteries is about 15 mm Hg. The major drop in pressure occurs across the small arteries and capillaries, as shown in Figure 6-32.

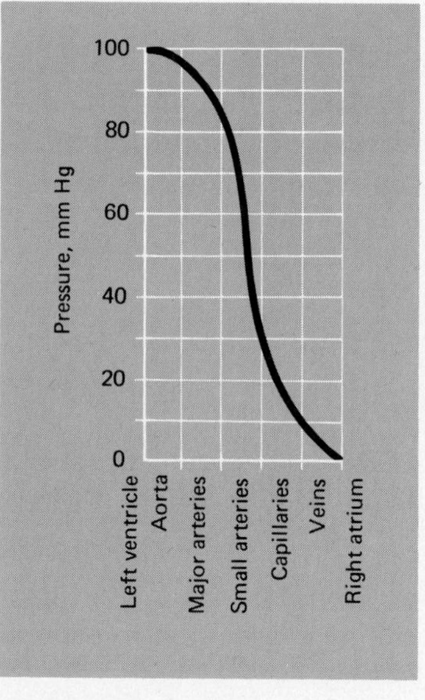

FIGURE 6-32
Blood pressure at various locations in the circulatory system.

*Turbulent flow: Reynolds number

If the flow velocity is large, the flow through a tube will become turbulent and Poiseuille's law will no longer hold. When the flow is turbulent, the flow rate for a given pressure difference will be less then for laminar flow as given in Equation 6-12. This is a result of the fact that friction forces are much greater when turbulence is present.

The onset of turbulence is often abrupt and is characterized by the so-called *Reynolds number* (*Re*):

$$Re = \frac{2\bar{v}r\rho}{\eta}$$

where $\bar{v}$ is the average speed of the fluid, ρ is its density, η is its viscosity, and r is the radius of the tube in which the fluid is flowing. Experiments show that the flow is laminar if *Re* has a value less than about 2000 but is turbulent if *Re* exceeds this value.

▮ EXAMPLE 6-13 The average speed of blood in the aorta ($r = 1.0$ cm) during the resting part of the heart's cycle is about 30 cm/s. Is the flow laminar or turbulent?

SOLUTION To answer this we calculate the Reynolds number using the values of ρ and η from Tables 6-1 and 6-5:

$$Re = \frac{(2)(0.30 \text{ m/s})(0.010 \text{ m})(1.05 \times 10^3 \text{ kg/m}^3)}{(4.0 \times 10^{-3} \text{ N} \cdot \text{s/m}^2)} = 1500.$$

The flow will be laminar. ▮▮

Notice in this example that since $1 \text{ N} = 1 \text{ kg} \cdot \text{m/s}^2$, *Re* has no units; hence Reynolds number is said to be a dimensionless quantity, and it does not matter in what set of units it is calculated as long as they are consistent.

*6-9 ▮ Object moving in a fluid; sedimentation and drag

In the previous section we saw how viscosity (and other factors) affect the flow of a fluid through a tube. In this section we will examine a slightly different situation, that of an object moving relative to a fluid. It could be an obstacle that obstructs the flow of a fluid such as a large rock in a river, or it could be an object moving in a fluid such as a glider or car moving through air, a fish in water, or a molecule sedimenting in an ultracentrifuge.

When an object moves relative to a fluid, the fluid exerts a force on the object. This force, which is referred to as a *drag force*, is due to the viscosity of the fluid and also, at high speeds, to turbulence behind the object.

*Reynolds number

To characterize the motion of an object relative to a fluid, it is useful to define a Reynolds number $Re = vL\rho/\eta$ where L is the length of the object, v is its velocity relative to the fluid, and ρ and η are the density and viscosity of the fluid, respectively. This Reynolds number must be clearly distinguished from the one used for fluid flow in a tube (although the form is similar) since the phenomena are quite different. When *Re* is less than about one,† the flow around an object is essentially laminar and it is found experimentally that the viscous force F_v is directly proportional to the speed v of the object:

$$F_v = kv. \tag{6-13}$$

† An object 1 mm long moving at a speed of 1 mm/s through water has a *Re* equal to one. So does an object 2 mm long traveling 7 mm/s in air. This situation then applies mainly to fairly small objects such as rain drops, pollen grains, and molecules in a centrifuge.

The magnitude of k depends on the size and shape of the object and on the viscosity of the fluid. For a sphere of radius r it has been calculated to be

$$k = 6\pi r\eta.$$

Thus the viscous force on a small sphere, when the flow is laminar, is given by an equation known as *Stokes' law*:

$$F_v = 6\pi r\eta v \quad \text{(for a sphere).}$$

For larger *Re*, there will be turbulence behind the body known as the wake (see Figure 6-24*b*), and the drag force will be much larger. For more streamlined objects, however, there will be less turbulence and hence less drag. When turbulence is present, experiment shows that the drag force increases as the square of the speed, $F_v \propto v^2$. The increase with speed is thus much more rapid than in the case of strictly laminar flow. When the Reynolds number reaches a value of about 10^6, the drag force increases abruptly. For above this value, turbulence exists not only behind the object but also in the layer of fluid lying next to the body (called the boundary layer) all along its sides.

*Sedimentation

Sedimentation refers to small objects falling in a fluid—examples are tiny particles of rock or minerals sedimenting under the sea, and red blood cells sedimenting in the fluid plasma in a laboratory.

An object of mass m falling through a fluid under the action of gravity has several forces on it as shown in Figure 6-33: the force of gravity, $w = mg$; F_B, the buoyant force of the fluid; and F_v the viscous force. By Newton's second law, the net force is equal to the mass times the acceleration of the object:

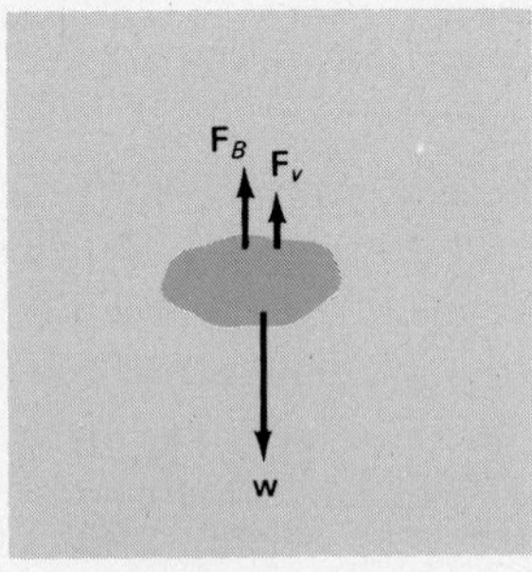

FIGURE 6-33
Forces on a small object falling through a fluid.

$$w - F_B - F_v = ma.$$

The buoyant force F_B is equal to the weight of fluid displaced; that is, $F_B = \rho_f Vg$, where ρ_f is the density of the fluid, V is the volume of the object (and hence the volume displaced), and g is the acceleration due to gravity. We can also write $w = mg = \rho_o Vg$ where ρ_o is the density of the object. Using Equation 6-13, we can write the above equation as:

$$(\rho_o - \rho_f)Vg - kv = ma.$$

The first term is just the weight of the object in the fluid and $(\rho_o - \rho_f)V$ is sometimes called the "effective mass." As the object increases in speed, the viscous force increases until it just balances the effective weight of the object. At this point the acceleration is zero and the speed increases no further. This maximum speed, v_T, is called the *terminal velocity* or *sedimentation velocity* and is obtained from the previous equation by setting $a = 0$:

$$v_T = \frac{(\rho_o - \rho_f)Vg}{k}.$$

The sedimentation velocity of tiny objects such as macromolecules and other constituents of cells is extremely small. This can be remedied using a centrifuge or ultracentrifuge. As we saw in Chapter 3, the effect of a centrifuge is to exert a force on a particle as if the acceleration of gravity were increased to a value of $\omega^2 r$; here, ω is the angular velocity of the rotor and r is the distance of the object from the axis of rotation. Thus, we can use the equation above for a centrifuge if we substitute the effective acceleration of gravity, equal to $\omega^2 r$, for g:

$$v_T = \frac{(\rho_o - \rho_f) V\omega^2 r}{k}. \tag{6.14}$$

Ultracentrifugation is often used to separate similar but slightly different particles or macromolecules (such as two types of nucleic acid) as well as to obtain valuable information on size and mass. Two principal techniques are used. The first, known as *sedimentation–velocity centrifugation*, depends on the fact that the particles quickly reach the terminal velocity v_T at which they travel down the tube. Particles of different densities and sizes will travel at different rates, as we saw in Equation 6-14. If the particles are carefully placed on the top of the tube, v_T is readily determined as the distance traveled down the tube divided by the time the centrifuge was run. If we divide v_T (Equation 6-14) by the acceleration $\omega^2 r$, we obtain a quantity called the sedimentation constant s,

$$s = v_T/\omega^2 r, \tag{6-15}$$

which depends only on molecular properties. The units of s are $(\text{m/s})/(\text{m/s}^2) = \text{s}$. A value of $s = 10^{-13}$ s is called a svedberg (S). Sedimenting particles are often characterized by their sedimentation constant (in svedbergs) in a standard fluid.† For example, ribosomes (particles in cells on which protein synthesis takes place) in the bacterium *Escherichia coli* have a sedimentation constant of 70 S. The ribosome actually consists of two subunits whose names, the 30 S and 50 S subunits, come from their (approximate) sedimentation rates.

▮ EXAMPLE 6-14 Calculate approximately how long a centrifuge should run at 50,000 rpm if the 30 S ribosome subunit is to reach the midpoint of a tube that extends from 6 to 10 cm from the rotation axis.

SOLUTION We assume the solution is close to the standard fluid so that $s = 30\,\text{S}$. The particles move a distance l of 2.0 cm and their average distance from the axis of rotation is 7.0 cm. Thus since $\omega = 2\pi f = (6.28)(50{,}000\,\text{rev/min})/(60\,\text{s/min}) = 5200\,\text{rad/s}$, we have, using Equation 6-15,

$$t = \frac{l}{v_T} = \frac{l}{s\omega^2 r} = \frac{(2.0\times 10^{-2}\,\text{m})}{(30\times 10^{-13}\,\text{s})(5.2\times 10^3\,\text{s}^{-1})^2(7.0\times 10^{-2}\,\text{m})}$$

$$= 3.5\times 10^3\,\text{s},$$

or a little under an hour. ▮▮

The second technique, called *density gradient equilibrium centrifugation*, relies on a fluid—usually a solution of cesium chloride (CsCl)—whose density increases from the top to the bottom of the tube. If the

† Usually water at 20°C. Note that from Equation 6-14 v_T depends on the density of the fluid.

sedimenting particles have a density intermediate between these extremes they will come to rest in the fluid at a point where their density is equal to that of the fluid (where $F_B = w$, Figure 6-33). This technique was used, for example, by M. S. Meselson and F. W. Stahl in 1958 to show that DNA replicates in the so-called semiconservative mode as predicted by James D. Watson and Francis H. C. Crick five years before. Their experiment involved growing bacteria in heavy nitrogen ^{15}N (which is an isotope of nitrogen that is slightly more dense than ordinary nitrogen, ^{14}N; see Chapter 28) and then suddenly transferring the bacteria to ^{14}N. Sections of deoxyribonucleic acid (DNA) containing these two different isotopes have slightly different densities and appear as two separate bands in an ultracentrifuge tube. The actual details of this classic experiment can be found in almost any book on molecular biology.†

▮ Summary

The three common phases of matter are solid, liquid, and gas. Liquids and gases are collectively called *fluids*, meaning they have the ability to flow. The *density* of a material is defined as its mass per unit volume. *Specific gravity* is the ratio of the density of the material to the density of water.

Pressure is defined as force per unit area. The pressure at a depth h in a liquid is given by ρgh where ρ is the density of the liquid and g is the acceleration due to gravity. In addition, if an external pressure is applied to a confined fluid this pressure is transmitted throughout the fluid; this is known as *Pascal's principle*. Pressure is measured using a manometer or other type of gauge. A barometer is used to measure atmospheric pressure. Standard atmospheric pressure (average at sea level) is 1.013×10^5 N/m^2. *Gauge pressure* is the total pressure less atmospheric pressure. *Archimedes' principle* states that an object submerged wholly or partially in a fluid is buoyed up by a force equal to the weight of fluid it displaces. This principle is used in a method to determine specific gravity, and explains why objects whose density is less than that of a liquid will float in that liquid.

Fluid flow rate is the mass or volume of fluid that passes a given point per unit time. The equation of continuity states that for an incompressible fluid flowing in an enclosed tube, the product of the velocity of flow and the cross-sectional area of the tube remains constant: $Av =$ constant. *Bernoulli's equation* tells us that where the velocity of a fluid is high the pressure in it is low, and where the velocity is low the pressure is high. Bernoulli's principle explains many common phenomena (Figure 6-29).

Fluid flow can be characterized either as *streamline* (sometimes called *laminar*), in which the layers of fluid move smoothly and regularly along paths called streamlines, or as *turbulent*, in which case the flow is not smooth and regular but is characterized by irregularly shaped whirlpools. *Viscosity* refers to friction within a fluid that prevents the fluid from flowing freely and is essentially a frictional force between different layers of fluid as they move past one another.

† For example, J. D. Watson, *Molecular Biology of the Gene*, W. A. Benjamin, New York, 1976.

Questions

1 Which of the following are fluids at room temperature: gold, mercury, air, glass, alcohol, carbon dioxide?

2 What is the specific gravity of (*a*) gold, (*b*) ice, (*c*) air?

3 If one material has a higher density than another, does this mean the molecules of the first must be heavier than those of the second? Explain.

4 A small amount of water is boiled in a one gallon gasoline can. The can is removed from the heat and the lid put on. Shortly thereafter, the can collapses. Explain.

5 It is often said that "water seeks its own level." Explain.

6 Is there a limit to the depth at which a skin diver can use a snorkel beneath the sea? Explain.

7 When measuring blood pressure, why must the jacket be held at the level of the heart?

8 Consider what happens when you push both a pin and a stick against your skin with the same force. Decide what determines whether your skin suffers a cut—the net force applied to it or the pressure.

9 Explain how the tube in Figure 6-34, known as a siphon, can transfer liquid from one container to a lower one even though the liquid must flow uphill for part of its journey. (Note that the tube must be filled with liquid to start with.) Why doesn't the liquid in each side of the tube flow back into its container?

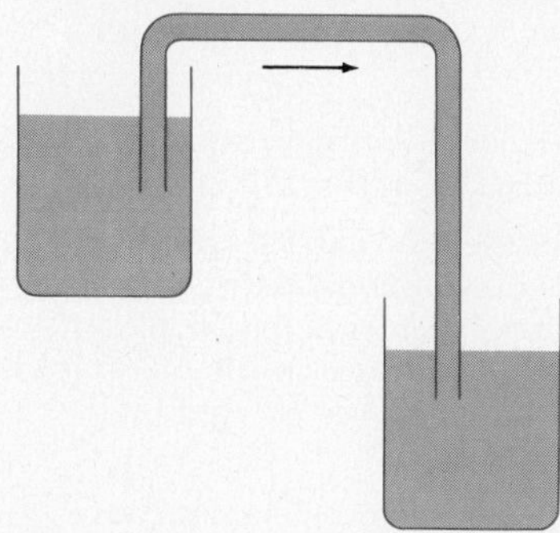

FIGURE 6-34
A siphon.

10 Does the buoyant force on a diving bell deep beneath the ocean have precisely the same value as when the bell is just beneath the surface? Explain.

11 Will a helium balloon rise indefinitely in the air? Explain.

12 A barge loaded with rocks approaches a low bridge over the river and cannot quite pass under it. Should rocks be added to or removed from the barge?

13 It is harder to pull the plug out of the drain of a bathtub when the tub is full of water than when it is empty. Is this a contradiction of Archimedes' principle? Explain.

14 An ice cube floats in a glass of water filled to the brim. As the ice melts, will the glass overflow?

*15 A duck can float in water because it preens its feathers to apply a layer of grease. Explain how the increased surface tension allows the duck to float.

16 Roofs of houses are sometimes pushed off during a tornado or hurricane. Explain, using Bernoulli's principle.

17 Why does the canvas top of a convertible bulge out when the car is traveling at high speed?

18 Two ships moving in parallel paths close to one another risk colliding. Why?

19 Blood corpuscles tend to flow in the center of blood vessels. Explain.

20 Children are told to avoid standing too close to a rapidly moving train because they might get sucked under it. Is this possible? Explain.

21 Hummingbirds expend 20 times as much energy to hover in front of a flower as they do in normal flight. Explain.

22 Explain why the speed of wind increases with increased height above the earth's surface. (*Hint*: See Figure 6-31.) Explain how this affects a mole that builds its burrow with one entrance higher than others (Figure 6-30).

*23 Identical steel ball bearings are dropped into tubs of water at 10°C and 40°C. In which tub will the ball bearing reach the bottom more quickly?

Problems

SECTION 6-1

1 (I) What is the volume of a whale bone whose mass is 12.8 kg?

2 (I) The approximate volume of the granite monolith known as El Capitan in Yosemite National Park is about 10^8 m^3. What is its approximate mass?

3 (I) What is the approximate mass of air in a living room 6.5 m × 4.4 m × 2.7 m?

4 (I) If 5.0 l of antifreeze solution (SG = 0.80) is added to 4.0 l of water to make a 9.0 l mixture, what is the SG of the mixture?

5 (II) A bottle has a mass of 31.20 g when empty and 98.44 g when filled with water. When filled with another fluid the mass is 88.78 g. What is the SG of the fluid?

SECTION 6-2

6 (I) The arm of a record player exerts a force of (1.0 g) $\times g$ on a record. If the diameter of the stylus is 0.0013 cm ($=0.5$ mil $= 0.5 \times 10^{-3}$ in), calculate the pressure on the record groove in N/m^2 and in atmospheres.

7 (I) A typical value for systolic blood pressure is 120 mm Hg. Convert this to (*a*) torr, (*b*) N/m^2, (*c*) atm, (*d*) lb/in^2.

8 (I) What is the difference in blood pressure between the top of the head and bottom of the feet of a 1.60-m-tall person standing vertically?

9 (I) What is the total force and the absolute pressure on the bottom of a swimming pool 8.0 m by 15.0 m whose uniform depth is 2.0 m? What will be the pressure against the *side* of the pool near the bottom?

10 (I) (*a*) Calculate the total force of the atmosphere acting on the top of a table which measures 2.0 m $\times$ 1.2 m. (*b*) What is the total force acting upward on the underside of the table?

11 (II) Intravenous infusions are often made under gravity as shown in Figure 6-35. Assuming the fluid has a density of 1.00 g/cm^3, at what height h should the bottle be placed so the liquid pressure is (*a*) 60 mm Hg, (*b*) 600 mm H_2O? (*c*) If the blood pressure is 18 mm Hg above atmospheric pressure, how high should the bottle be placed so that the fluid just barely enters the vein?

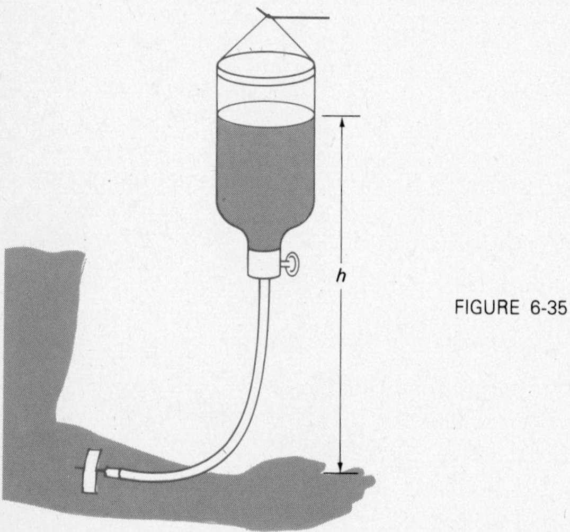

FIGURE 6-35

12 (II) The gauge pressure in each of the four tires of an 1800-kg automobile is 2.0 kg/cm^2. How much area of each tire is in contact with the ground?

13 (II) A 3.0 N force is applied to the plunger of a hypodermic needle. If the diameter of the plunger is 1.0 cm and that of the needle 0.20 mm, (*a*) with what force does the fluid leave the needle? (*b*) What force on the plunger would be needed to push fluid into a vein where the gauge pressure is 18 mm Hg?

14 (II) What is the approximate difference in air pressure between the top and the bottom of the World Trade Center Building in New York City which is 410 m tall and is located at sea level?

15 (II) When you drive up into the mountains, or descend rapidly from the mountains, your ears "pop," which means that the pressure behind the ear drum is being equalized to that outside. If this did not happen, what would be the approximate force on an eardrum of area 0.50 cm^2 if a change in altitude of 1000 m takes place?

16 (II) What is the total force on a rectangular-shaped dam 60 m high and 110 m wide if the water is filled to the top?

17 (II) One arm of a U-shaped tube (open at both ends) contains water and the other alcohol. If the two fluids meet exactly at the bottom of the U, and the alcohol is at a height of 16.0 cm, at what height will the water be?

18 (II) At best, a person can reduce the pressure in the lungs to -80 mm Hg gauge pressure. How high can water then be sucked up a straw?

19 (II) The maximum gauge pressure in a hydraulic lift is 18 atm. What is the largest size vehicle (kg) it can lift if the diameter of the output line is 20 cm?

20 (II) Determine the minimum gauge pressure needed in the water pipe leading into a building if water is to come out of a faucet on the twelfth floor 30 m above.

21 (II) Show that the work done when a pressure P acts to move a volume of fluid ΔV is $W = P\,\Delta V$.

22 (II) The contraction of the left ventricle (chamber) of the heart pumps blood to the body. Assuming that the inner surface of the left ventricle has an area of 85 cm^2 and the maximum pressure in the blood is 120 mm Hg, calculate the net force exerted by the ventricle at maximum pressure.

23 (III) During each heartbeat, approximately 70 cm^3 of blood is pushed from the heart at an average pressure of 105 mm Hg. Calculate the power output of the heart, in watts, assuming 60 beats per minute.

24 (III) In working out his principle, Pascal showed dramatically how force can be multiplied with fluid pressure. He placed a long tube of 0.30 cm radius vertically into a 20-cm-radius wine barrel. He found that, when the barrel was filled with water and the tube filled to a height of 12 m, the barrel burst. Calculate (*a*) the mass of fluid in the tube, (*b*) the net force on the lid of the barrel.

25 (III) The bulk modulus of water (see Chapter 4) is 2.0×10^9 N/m^2. Use this value to calculate the density of pure water at a depth of 1.0 km.

SECTION 6-3

26 (I) Determine the specific gravity of a fish weighing 320 g in air and 21 g in pure water.

27 (I) The hydrometer of Example 6-5 sinks to a depth of 22.3 cm when placed in a fermenting vat. What is the density of the brewing liquid?

28 (I) A geologist finds that a moon rock whose mass is 7.20 kg has an apparent of mass of 5.88 kg when submerged in water. What is the density of the rock?

29 (I) What fraction of a piece of iron will be submerged when it floats in mercury?

30 (II) A 0.40-kg piece of wood floats in water but is found to sink in alcohol (SG = 0.79) in which it has an apparent mass of 0.020 kg. What is the SG of the wood?

31 (II) A 142.0-g cuttlefish is found to have an effective mass of only 0.5 g when submerged in pure water. (*a*) Calculate its density. (*b*) How much would it weigh in seawater (SG = 1.025)?

32 (II) A 75-kg person has an effective mass of 54 kg (because of buoyancy) when standing in water that comes up to the hips. Calculate the mass of each leg. Assume the body has SG = 1.00.

33 (II) A freighter has a horizontal cross-sectional area of 2800 m^2 at the water line. When loaded, the ship lowers 6.5 m. What is the mass of the load?

34 (II) A small animal is found to remain suspended in a mixture of 18 percent (by weight) alcohol and 82 percent water. What is the density of the animal?

35 (II) A row boat has a volume of 1.5 m^3 and a mass of 35 kg. How many people, of mass 70 kg each, can the boat hold without sinking?

36 (II) Archimedes' principle can be used not only to determine the SG of a solid using a known liquid (Equation 6-4); the reverse can be done as well. (*a*) As an example, a 2.00-kg aluminium ball has an apparent mass of 1.40 kg when submerged in a particular liquid; calculate the density of the liquid. (*b*) Derive a simple formula for determining the density of a liquid using this procedure.

37 (II) Calculate the true mass (in vacuum) of a piece of aluminum whose apparent mass is 1.0000 kg when weighed in air.

38 (II) The specific gravity of ice is 0.92, whereas that for seawater is 1.03. What fraction of an iceberg is above the surface of the water?

39 (III) A 1.24-kg piece of wood (SG = 0.50) floats on water. What minimum mass of lead, hung from it by a string, will cause it to sink?

40 (III) If an object floats in water, its density can be determined by tying a "sinker" on it so that both the object and weight sink. Show that the specific gravity is given by $w/(w_1 - w_2)$ where w is the weight of the object alone in air, w_1 is the apparent weight when a "sinker" is tied to it and the sinker only is submerged, and w_2 is the apparent weight when both the object and the sinker are submerged.

*SECTION 6-4

***41** (I) If the force F needed to move the wire in Figure 6-15 is 4.1×10^{-3} N, calculate the surface tension γ of the enclosed fluid. Assume $l = 0.060$ m.

***42** (I) How high will water rise in a glass tube 0.10 mm in radius?

***43** (II) If the base of an insect's leg has a radius of about 3.0×10^{-5} m and its mass is 0.015 g, would you expect the six-legged insect to remain on top of the water?

***44** (II) The surface tension of a liquid can be determined by measuring the force F needed to just lift a circular platinum ring of radius r from the surface of the liquid. (*a*) Find a formula for γ in terms of F and r. (*b*) At 30°C, if $F = 7.40 \times 10^{-3}$ N and $r = 2.5$ cm calculate γ for the tested liquid.

***45** (II) A pencil 1.0 cm in diameter is held vertically in a glass of water. The water wets the pencil so that the contact angle is 0°. Calculate the magnitude and direction of the net force on the pencil due to surface tension.

***46** (II) A glass tube 1.0 mm in diameter is placed in mercury. Where will the level of the mercury be in the tube relative to the rest of the liquid?

***47** (II) A small pool of water on a table is broken into 100 droplets. By what factor does the surface energy change, assuming the original pool is flat and of depth h and the droplets are hemispheres of radius h.

***48** (III) How high will water rise due to capillarity between two flat plates of glass 0.10 mm apart when placed vertically in water?

***49** (III) Show that inside a soap bubble there must be a pressure ΔP in excess of that outside equal to $\Delta P = 4\gamma/r$ where r is the radius of the bubble and γ is the surface tension. (*Hint*: Think of the bubble as two hemispheres in contact with each other; and remember that there are two surfaces to the bubble. Note that this result applies to any kind of membrane where $T = 2\gamma$ is the tension per unit length in that membrane.)

SECTION 6-6

50 (I) Using the data of Example 6-9 calculate the average speed of blood flow in the major arteries of the body which have a total cross-sectional area of about 2.0 cm^2.

51 (I) If wind blows at 25 m/s over your house, what is the net force on the roof if its area is 250 m^2?

52 (I) How high should the pressure head be if water is to come from a faucet at a speed of 8.0 m/s? Ignore viscosity.

53 (I) A 15-cm-radius air duct is used to replenish the air of a room 10 m × 6.0 m × 3.0 m every 10 min. How fast does the air flow in the duct?

54 (I) What gauge pressure in the water mains is necessary if a firehose is to spray water to a height of 25 m?

55 (II) What is volume rate of flow of water from a 1.0 cm diameter faucet if the pressure head is 10 m?

56 (II) What is the lift (in newtons) due to Bernoulli's principle on a wing of area 50 m^2 if the air passes over the top and bottom surfaces at speeds of 320 m/s and 290 m/s respectively?

57 (II) Water at a pressure of 3.3 atm at street level flows into an office building at a speed of 0.50 m/s through a pipe 5.0 cm in diameter. The pipes taper down to 2.5 cm diameter by the top floor, 25 m above. Calculate the flow velocity and the pressure in such a pipe on the top floor. Ignore viscosity.

58 (II) A 4-cylinder automobile has a displacement of 2000 cm^3; this is approximately how much air is brought into the four cylinders per revolution. If the engine runs at a speed of 1500 rpm, and the venturi of the carburetor has a radius of 2.5 cm, (*a*) what is the speed of air flow through the venturi; (*b*) what is the pressure (in atmospheres) in the venturi?

59 (III) (*a*) Show that the flow velocity measured by a Venturi meter is given by the relation $v_1 = A_2\sqrt{2(P_1 - P_2)/\rho(A_1^2 - A_2^2)}$. See Figure 6-29*f*. (*b*) A Venturi tube is measuring the flow of water; it has a main diameter of 3.0 cm tapering down to a throat diameter of 1.0 cm; if the pressure difference is measured to be 18 mm Hg, what is the velocity of the water?

*SECTION 6-8

*60 (II) Blood from an animal is placed in a bottle 1.50 m above a 3.8-cm-long needle of inside diameter 0.40 mm from which it flows at a rate of 4.1 cm^3 per minute. What is the viscosity of this blood?

*61 (II) What must be the pressure difference between the two ends of a 2.0-km section of pipe 40 cm in diameter if it is to transport oil ($\rho = 950\,kg/m^3$, $\eta = 0.20$ Pa · s) at a rate of 400 cm^3/s?

*62 (II) What diameter must a 30-m-long air duct have if the pressure of a ventilation and heating system is to replenish the air in a room 10 m × 18 m × 4.0 m every 10 min? Assume the pump can exert a gauge pressure of 0.64×10^{-3} atm.

*63 (II) A gardener feels it is taking him too long to water a garden with a $\frac{3}{8}$-in-diameter hose. By what factor will his time be cut if he uses a $\frac{5}{8}$-in-diameter hose? Assume nothing else is changed.

*64 (II) During heavy exercise, the flow speed of blood increases by perhaps a factor of two. Referring to Example 6-13, would you expect the flow in the aorta to be laminar? Why or why not?

*65 (II) Calculate the pressure drop per cm along the aorta using the data of Example 6-9 and Table 6-5.

*66 (III) What is the (approximate) maximum flow rate, Q, of water in a 10-cm diameter pipe if turbulence is to be avoided?

*67 (III) A patient is to be given a blood transfusion. The blood is to flow through a tube from a raised bottle to a needle inserted in the vein (Figure 6-35). The inside diameter of the 4.0-cm-long needle is 0.30 mm and the required flow rate is 4.0 cm^3 of blood per minute. How high should the bottle be placed above the needle? Obtain ρ and η from the tables. Assume the blood pressure is 20 torr above atmospheric pressure.

*SECTION 6-9

*68 (II) Calculate the magnitude and direction of the terminal velocity of a 1.0 mm radius air bubble rising in oil of viscosity 0.20 Pa · s and SG 0.90.

*69 (III) Referring to Example 6-14, how far down the tube would you expect to find the 50 S subunit of the ribosomes?

*70 (III) If an object requires 30 min to sediment in an ultracentrifuge rotating at 30,000 rpm at an average distance from the axis of 8.0 cm, how long would it take to sediment under gravity in the same tube standing vertically in the lab?

*71 (III) (*a*) Show that the terminal velocity of a small sphere of density ρ_0 falling through a fluid of density ρ_F and viscosity η is

$$v_T = \frac{2}{9}\frac{(\rho_0 - \rho_F)r^2 g}{\eta}.$$

(*b*) What is the terminal velocity of a spherical raindrop of radius $r = 0.020$ cm falling in air?

For additional reading

Alexander, R. M., *Animal Mechanics*, University of Washington Press, Seattle, 1968. Interesting application to biology of fluids at rest and in motion.

Almgren, F. J., and J. E. Taylor, "Geometry of Soap Films and Soap Bubbles," *Sci. American*, **235**:1 (Jul., 1976), p. 82.

Davidovits, P., *Physics in Biology and Medicine*, Prentice-Hall, Inc., Englewood Cliffs, N.J., 1975.

Hay, J. G., *The Biomechanics of Sports Techniques*, Prentice-Hall, Inc., Englewood Cliffs, N.J., 1973.

Stanford, A. L., Jr., *Foundations of Biophysics*, Academic Press, Inc., New York, 1975. Sedimentation.

Swinney, H. L., and J. P. Gallub, " Transition to Turbulence," *Physics Today*, **31**:8 (Aug., 1978), p. 41.

7 ▮ Temperature and the kinetic theory

Much of this chapter will be devoted to an investigation of the theory that matter is made up of atoms and that these atoms are in continuous random motion. This theory is called the *kinetic theory*. ("Kinetic," you may recall, is Greek for "moving.")

Before investigating this atomic view of matter, we must first discuss the concept of temperature as well as the experimentally measured properties of gases; for these will serve as a foundation for testing the kinetic theory.

7-1 ▮ Atoms

The idea that matter is made up of atoms dates back to the ancient Greeks. According to the Greek philosopher Democritus, if a given substance—say a piece of iron—were cut into smaller and smaller bits, eventually a smallest piece of that substance would be obtained which could not be divided further. This smallest piece was called an **atom**, which in Greek means "indivisible." The only real alternative to the atomic theory of matter was the idea that matter is continuous and can be subdivided indefinitely.

Today the atomic theory is generally accepted by scientists. The experimental evidence in its favor, however, came mainly in the eighteenth and nineteenth centuries, and much of it was obtained from the analysis of chemical reactions. This analysis led to a distinction being made between elements and compounds, and between atoms and molecules. Some substances, such as iron, copper, and gold, could not be broken down into simpler materials by any chemical means. These materials were considered to be the basic elements of matter and hence were called **elements**. The smallest piece of an element, according to atomic theory, is an atom. Other substances, such as salt, carbon dioxide, and water, were found to be made up of combinations of elements and were called **compounds**. Water, for example, is a compound of the elements hydrogen and oxygen. The tiniest piece of a compound is called a **molecule**, and a molecule is itself made up of atoms. A molecule of water, for example, contains two atoms of hydrogen and one of oxygen;

the chemical formula for water is then written as H_2O. To complete this discussion, we should distinguish compounds from mixtures. A compound is a wholly new substance, completely different from the substances from which it is made; it is also uniform. A *mixture*, on the other hand, is not uniform, and the separate materials can be recognized, although this may require careful observation and experimentation. A mixture of the two gases hydrogen and oxygen is clearly different from the compound of hydrogen and oxygen that we call water.

A crucial piece of evidence in favor of the atomic theory was the *law of definite proportions*, a summation of experimental results collected during the half century prior to 1800; it states that, when two or more elements combine to form a compound, they always do so in the same proportions by weight. For example, table salt is always formed from 23 parts (by weight) of sodium and 35 parts of chlorine; and water is formed from 1 part hydrogen and 8 parts oxygen. A continuous theory of matter could hardly account for the law of definite proportions but, as John Dalton (1766–1844) pointed out, the atomic theory could. Dalton reasoned that the weight proportions of the elements forming a compound correspond to the relative weights of the combining atoms; thus, he argued, one atom of sodium (Na) combines with one atom of chlorine (Cl) to form one molecule of salt (NaCl), and one atom of sodium has a mass 23/35 times as large as one of chlorine. By measuring the relative amounts of each element needed to form a large variety of compounds, the relative weights of atoms was established. Hydrogen, the lightest atom, was arbitrarily assigned the relative weight of 1; on this scale, carbon was about 12, oxygen 16, sodium 23, and so on.

Today, we speak of the relative masses of atoms and molecules—what we call the **atomic mass** or **molecular mass**, respectively†—and these are based on assigning an ordinary carbon atom, ^{12}C, the value of exactly 12.000 atomic mass units (u). The atomic mass of hydrogen is then precisely 1.008 u, and the values for other atoms are listed in Appendix E as well as on the periodic table inside the cover of this book.

Another important piece of evidence for the atomic theory is the so-called *Brownian movement*, named after botanist Robert Brown, who is credited with its discovery in 1827. Brown noticed under his microscope that tiny pollen grains suspended in water moved about in tortuous paths even though the water appeared to be perfectly still. The atomic theory easily explains Brownian movement if the further reasonable assumption is made that the atoms of any substance are continually in motion. Thus Brown's tiny pollen grains are jostled about by the vigorous barrage of rapidly moving molecules of water.

In 1905, Albert Einstein examined Brownian movement from a theoretical point of view and was able to calculate from the experimental data the approximate size and mass of atoms and molecules. His calculations showed that the diameter of a typical atom is about 10^{-10} m.

It is interesting that almost 2000 years before Brown, in the first century B.C., the Roman poet Lucretius noted that, when a ray of sunlight shines into an otherwise darkened room, one can see tiny particles of dust

† The terms *atomic weight* and *molecular weight* are popularly used for these quantities, but properly speaking we are comparing masses.

bouncing around. And Lucretius pointed to this phenomenon as evidence for the atomic theory of matter! A man before his time?

At the start of Chapter 6, we distinguished the three states of matter—solid, liquid, gas—based on *macroscopic*, or "larger-scale," properties. Now let us see how the three phases of matter differ from the atomic, or *microscopic*, point of view. (Perhaps "submicroscopic" would be a better term since atoms are too small to be seen under an ordinary microscope.) Atoms and molecules must clearly exert attractive forces on each other. For how else could a brick or a piece of aluminum stay together in one piece? These forces are of an electrical nature (we will discuss this in later chapters). On the other hand, if the molecules come too close together, the force becomes repulsive (due to the electrical repulsion between their outer electrons); thus they maintain a minimum distance from each other. In a solid material, the attractive forces are strong enough that the atoms or molecules are held in more or less fixed positions, usually in an array known as a crystal lattice, as shown in Figure 7-1*a*. The atoms or molecules in a solid are in motion—they vibrate about their nearly fixed positions. In a liquid, the atoms or molecules are moving more rapidly, or the forces between them are weaker, so that they are sufficiently free to roll over one another as in Figure 7-1*b*. In a gas, the forces are so weak, or the speeds so high, that the molecules do not even stay close together. They move rapidly every which way (Figure 7-1*c*), filling any container and occasionally colliding with one another. On the average, the speeds are sufficiently high in a gas that when two molecules collide, the force of attraction is not strong enough to keep them close together and they fly off in new directions.

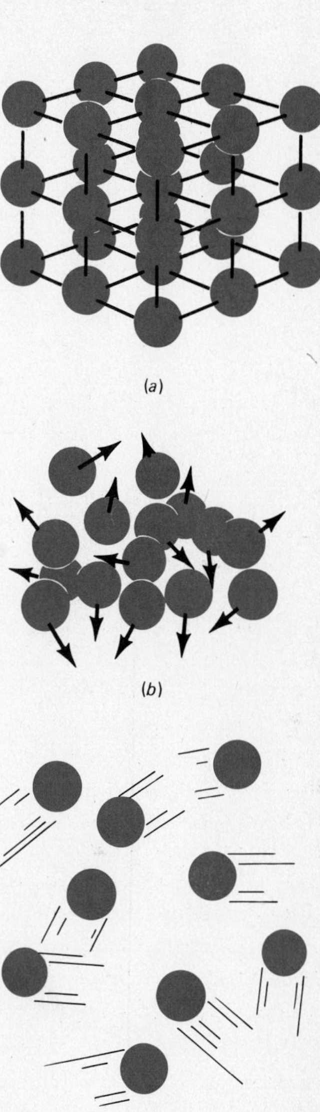

FIGURE 7-1
How atoms are arranged in (*a*) a crystalline solid, (*b*) a liquid, (*c*) a gas.

7-2 ▮ Temperature

Temperature refers to how hot or cold an object is. A hot oven is said to have a high temperature whereas a cold tray of ice is said to have a low temperature.

Many properties of matter change with temperature. For example, most materials expand when heated. An iron beam is longer when hot than when cold; concrete roads and sidewalks expand and contract slightly according to temperature, which is why compressible spacers are placed at regular intervals. The electrical resistance of matter changes with temperature (Chapter 15). And so too does the color of objects, at least at high temperatures: you may have noticed that the heating element of an electric stove glows with a red color when hot; at higher temperatures, solids such as iron turn orange and then white; and the white light from an ordinary incandescent light bulb comes from an extremely hot tungsten wire.

Thermometers

An instrument designed to measure temperature is called a **thermometer**. There are many kinds of thermometers, but they all have in common the fact that they depend on some property of matter that changes with temperature. Most common thermometers rely on the expansion of a material with an increase in temperature. The first thermometer, invented by Galileo, made use of the expansion of a gas. An early clinical thermometer (around 1660) is illustrated in Figure 7-2.

Common thermometers today consist of a hollow glass tube filled with mercury or alcohol colored with a red dye (Figure 7-3*a*). The liquid expands more than the glass when the temperature is increased, so the liquid level rises in the tube. Although metals also expand with temperature, the change in length of, say, a metal rod is generally too small to measure accurately for ordinary changes in temperature. However, a useful thermometer can be made by bonding together two dissimilar metals whose rate of expansion is different (Figure 7-3*b*). When the temperature is increased, the different amounts of expansion cause the bimetallic strip to bend. Often the bimetallic strip is in the form of a coil, one end of which is fixed and the other attached to a pointer. This kind of thermometer (Figure 7-3*c*) is used as ordinary air thermometers, as oven thermometers, and in automobiles as the automatic choke.

FIGURE 7-2
Model of a seventeenth-century clinical thermometer. The original, now in the Museum of the History of Science, Florence, Italy, was used to measure a person's temperature. The glass, blown in the shape of a frog, contains several small spheres of slightly different density floating in alcohol; straps attached to the frog's legs secured the instrument to the patient's arm. With an increase in temperature the density of the alcohol would decrease, and one or more of the spheres would sink; thus the number of spheres that sank would represent how high the patient's fever was. (Burndy Library, Norwalk, Conn.)

Temperature scales

In order to measure temperature quantitatively, some sort of numerical scale must be defined. The most common scale today is the *Celsius* scale, sometimes called the *centigrade* scale. In the U.S. the *Fahrenheit* scale is also common. Another scale that is important in scientific work is the absolute, or Kelvin, scale, and it will be discussed later in the chapter.

In order to define a temperature scale, two readily reproducible temperatures are assigned arbitrary values. For both the Celsius and Fahrenheit scales these two points are the freezing point and the boiling point† of water, both taken at atmospheric pressure. On the Celsius scale, the freezing point of water is chosen to be 0°C ("zero degrees Celsius") and the boiling point 100°C. On the Fahrenheit scale, the freezing point is defined as 32°F and the boiling point 212°F.

A practical thermometer is calibrated by first placing it in a container of water mixed with ice. A mark is made where the level of mercury, or the pointer, comes to rest. The thermometer is then placed in a container of boiling water and a second mark is made. The first mark represents 0°C and the second 100°C, assuming the calibration is done at 1 atm pressure. The distance between the two marks is then divided into 100 equal intervals separated by small marks representing each degree on the Celsius scale between 0 and 100 (hence the name "centigrade scale" meaning "100 steps"). For a Fahrenheit scale, the two points are labeled 32°F and 212°F, and the distance between them is divided into 180 equal intervals. For temperatures below the freezing point of water and above

FIGURE 7-3
(*a*) Mercury or alcohol in glass thermometer; (*b*) bimetallic strip; (*c*) bimetallic-strip thermometer.

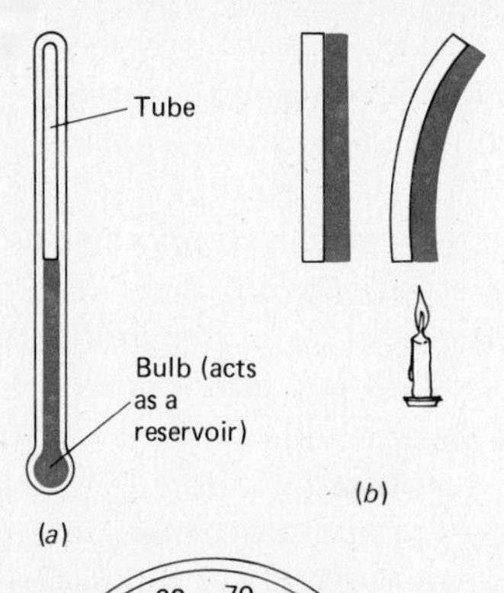

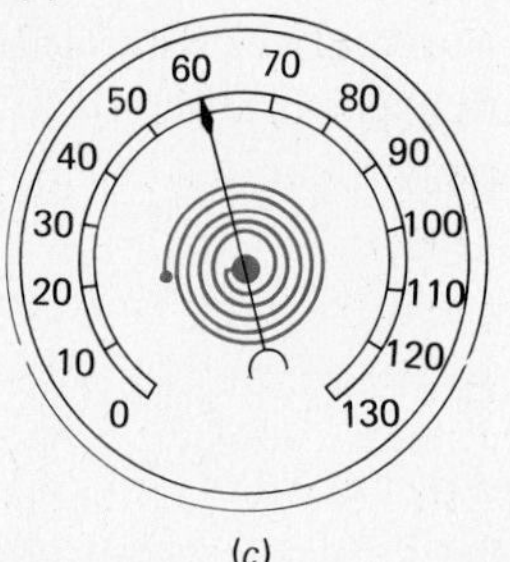

† The freezing point of a substance is defined as the temperature at which the solid and liquid phases coexist in equilibrium—that is, without the liquid changing into the solid or vice versa. Experimentally, this is found to occur at only one definite temperature, for a given pressure. Similarly, the boiling point is defined as that temperature at which the liquid and gas coexist in equilibrium. Since these points vary with pressure, the pressure must be specified (usually it is 1 atm).

the boiling point of water the scales can be extended using the same equally spaced intervals. However, ordinary thermometers can be used only over a limited temperature range because of their own limitations—for example, the mercury or alcohol in the thermometer of Figure 7-3*a* solidifies at some point, below which the thermometer will be useless; it is also rendered useless above temperatures where the fluid vaporizes. For very low or very high temperatures, specialized thermometers are required, some of which we will mention later.

Every temperature on the Celsius scale corresponds to a particular temperature on the Fahrenheit scale (Figure 7-4). It is easy to convert from one to the other if you remember that 0°C corresponds to 32°F and that a range of 100°C on the Celsius scale corresponds to a range of 180°F on the Fahrenheit scale. Thus, one Fahrenheit degree (1 F°) corresponds to $100/180 = 5/9$ of a Celsius degree; that is, 1 F° = 5/9 C°. (Notice that when we refer to a specific temperature, we say "degrees Celsius," as in 20°C; but when we merely refer to a change in temperature or a temperature interval, we say "Celsius degrees," as in "1 C°.")

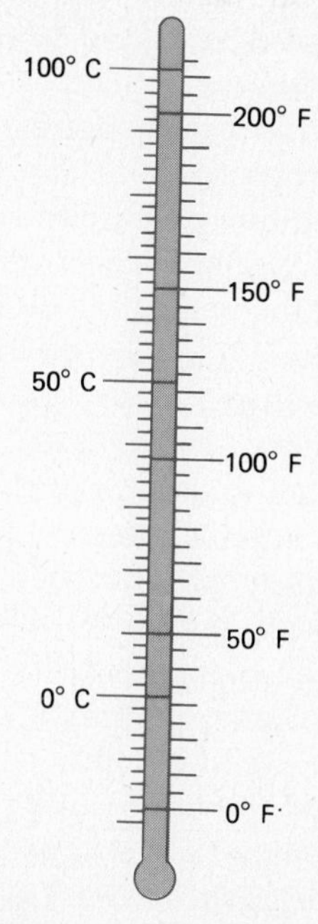

FIGURE 7-4
Celsius and Fahrenheit scales compared.

▮ EXAMPLE 7-1 Normal body temperature is 98.6°F. What is this on the Celsius scale?

SOLUTION First we note that 98.6°F is $98.6 - 32.0 = 66.6$ F° above the freezing point of water. Since each F° is equal to 5/9 C°, this corresponds to $66.6 \times 5/9 = 37.0$ C° above the freezing point; as the freezing point is 0°C, the temperature is 37.0°C ▮▮

The constant volume gas thermometer: a standard

Different materials do not expand in quite the same way over a wide temperature range. Consequently, if we calibrate different kinds of thermometers exactly as described above, they will not always agree precisely. Because of how we calibrated them, they will certainly agree at 0°C and at 100°C. But because of slightly different expansion properties, they may not agree at intermediate temperatures (remember we arbitrarily divided the thermometer scale into 100 equal divisions between 0°C and 100°C). Thus a carefully calibrated mercury in glass thermometer might register a temperature of 52.0°C whereas a carefully calibrated alcohol in glass thermometer might read 52.6°C.

Because of this discrepancy, some standard kind of thermometer must be chosen so that these intermediate temperatures can be precisely defined. For this purpose, scientists have chosen the so-called *constant volume gas thermometer*. As shown schematically in Figure 7-5, this thermometer consists of a bulb filled with a dilute gas connected by a thin tube to a mercury manometer. The volume of the gas is kept constant by raising or lowering the right-hand tube of the manometer so that the mercury in the left tube coincides with the reference mark. An increase in temperature causes an increased pressure in the bulb; so the tube must be lifted higher to keep the gas volume constant. Thus, the height of the mercury in the right-hand column is a measure of the temperature. This thermometer can be calibrated, and the resulting scale is defined as the standard temperature scale.

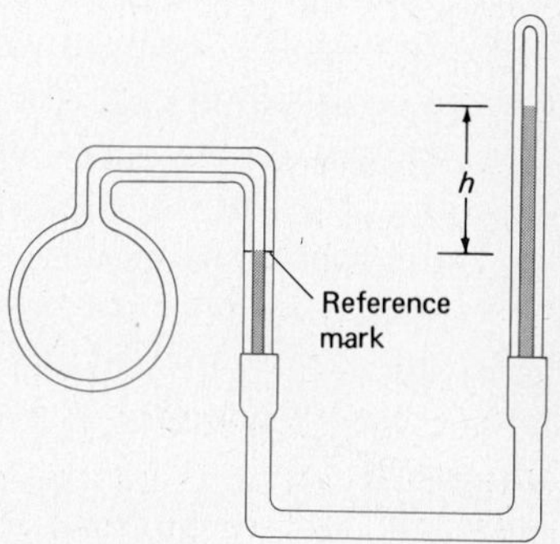

FIGURE 7-5
A constant volume gas thermometer.

*7-3 ▮ The thermostat: an example of feedback

A *thermostat* is a device used in conjunction with a heating or cooling system to maintain a relatively constant temperature. A typical thermostat for use with a furnace is shown schematically in Figure 7-6. When the temperature drops, the bimetallic strip bends downward and touches the contact. This completes an electric circuit that starts the heater. As the room heats up, the bimetallic strip will bend upward again; the circuit will be broken and the heater will be shut off. The position of the bimetallic strip can be altered so that the heater can be made to turn on at any desired temperature.

The thermostat-heater system is an example of feedback. *Feedback* is said to exist when a portion of the output of a system is returned ("fed back") as input into the system.† In the case of the thermostat, heat affects the temperature which is fed back to the thermostat to "tell" the furnace to turn off (or on).

Because the concept of feedback is important today, we discuss it briefly now. Another example of feedback concerns the size of the opening of the eye (the pupil); when a large amount of light reaches the retina, the brain reacts by causing muscles in the eye to reduce the size of the pupil. When the light is dim, the size of the pupil is increased allowing more light to reach the retina. The thermostat and the pupil of the eye are examples of *negative feedback*; that is, an increased input causes a decrease in output (and vice versa). When negative feedback occurs, the system remains stable. In the reverse case, called *positive feedback*, an increase in input causes a further increase in output. A thermostat hooked up in reverse is an example; in this case the thermostat would tell the heater to go on when the temperature was high and off when the temperature was low. Thus, if the heater were on initially, the temperature would increase indefinitely. (If it were off initially, it would never turn on.) Positive feedback often results in a system going out of control. The population explosion is an example of positive feedback, for the more people there are, the more people they produce. However, other factors also play a role; a finite food supply, for example, acts as negative feedback. The automatic control of machines depends on negative feedback. One example is the centrifugal governor on a steam engine (Figure 7-7).

The human body maintains its internal stability and well being by means of negative feedback systems. This is known as "homeostasis." One example is the eye mentioned above. Two of the many other common

† The output of a system is the result of the system's activities (e.g., heat); the input refers to the "instructions" given to the system.

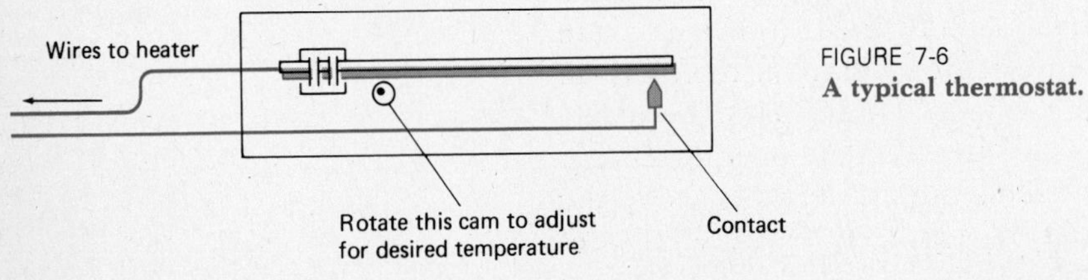

FIGURE 7-6
A typical thermostat.

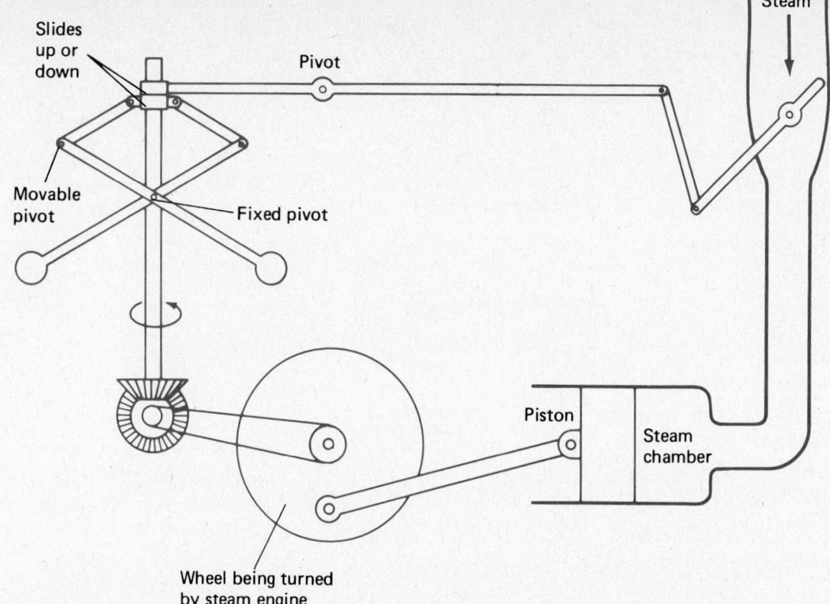

FIGURE 7-7
If the amount of steam fed to the engine is too little, the engine stalls; if too much, the engine runs too fast and perhaps damages itself. The amount of steam depends on the throttle opening and also on the pressure in the boiler, which can vary. In order to maintain the proper engine speed, a governor is used that detects changes in the engine's speed as shown. When the engine runs too fast, the centrifugal pendulum balls swing outward and through the linkage close the throttle, which reduces the flow of steam from the boiler. When the engine runs too slowly, the throttle valve is opened wider. This is an example of negative feedback used to control a machine.

examples involve body temperature regulation and hunger. Sensors in the body send temperature information to a portion of the brain known as the hypothalamus ("the body's thermostat") which controls the various mechanisms for heating and cooling the body, such as constriction of the blood vessels under the skin and production of moisture by the sweat glands. The sensation of hunger is triggered by sensors which detect the amount of glucose in the blood; thus a feeling of hunger occurs when the level of glucose falls below that needed for healthy functioning of the body. Clearly, control is intimately linked to negative feedback.

The concept of feedback was first developed with regard to electronic amplifiers, and we will mention it again in that context (Chapter 19).

7-4 Thermal expansion

Linear expansion

Most substances expand when heated and contract when cooled. However, the amount of expansion or contraction varies, depending on the material.

Experiments indicate that the change in length ΔL of almost all solids is, to a very good approximation, directly proportional to the change in temperature ΔT. As might be expected, the change in length is also proportional to the original length of the object, L_0. That is, for the same temperature change, a 4-m-long iron rod will increase twice as much in length as a 2-m-long iron rod. We can write this proportionality as an equation:

$$\Delta L = \alpha L_0 \Delta T, \tag{7-1}$$

where α, the proportionality constant, is called the *coefficient of linear expansion* for the particular material and has units of $(\mathrm{C}°)^{-1}$. The values of α for various materials† at 20°C are listed in Table 7-1.

It should be noted that α does vary slightly with temperature (which is why thermometers made of different materials do not agree precisely). However, if the temperature range is not too great, the variation can usually be ignored.

▌ EXAMPLE 7-2 A steel girder is 200 m long at 20°C. If the extremes of temperature to which it might be exposed are −30°C to +40°C, how much will it contract and expand?

SOLUTION From Table 7-1, we find that $\alpha = 12 \times 10^{-6} (\mathrm{C}°)^{-1}$. The increase in length when it is at 40°C will be

$$\Delta L = (12 \times 10^{-6}/\mathrm{C}°)(200\ \mathrm{m})(40°\mathrm{C} - 20°\mathrm{C}),$$
$$= 4.8 \times 10^{-2}\ \mathrm{m},$$

or 4.8 cm. When the temperature decreases to −30°C, $\Delta T = -50\,\mathrm{C}°$; so the change in length is

$$\Delta L = (12 \times 10^{-6}/\mathrm{C}°)(200\ \mathrm{m})(-50\ \mathrm{C}°) = -12.0 \times 10^{-2}\ \mathrm{m},$$

or a decrease in length of 12 cm. ▌▌

† For certain crystalline materials, α may be different for the three spatial directions. We won't be concerned with this.

TABLE 7-1

Coefficients of expansion at 20°C

Material	Coefficient of linear expansion, α $(\mathrm{C}°)^{-1}$	Coefficient of volume expansion, β $(\mathrm{C}°)^{-1}$
Solid		
Aluminum	25×10^{-6}	75×10^{-6}
Brass	19×10^{-6}	56×10^{-6}
Iron or steel	12×10^{-6}	35×10^{-6}
Lead	29×10^{-6}	87×10^{-6}
Glass (Pyrex)	3×10^{-6}	9×10^{-6}
Glass (ordinary)	9×10^{-6}	27×10^{-6}
Quartz	0.4×10^{-6}	1×10^{-6}
Concrete and brick	$\approx 12 \times 10^{-6}$	$\approx 36 \times 10^{-6}$
Marble	$1.4–3.5 \times 10^{-6}$	$4–10 \times 10^{-6}$
Liquid		
Gasoline		950×10^{-6}
Mercury		180×10^{-6}
Ethyl alcohol		1100×10^{-6}
Glycerin		500×10^{-6}
Water		210×10^{-6}
Gas		
Air (and most other gases at atmospheric pressure)		3400×10^{-6}

Thermal expansion must be taken into account when designing any large structure. Various kinds of expansion joints are used to prevent buckling, rupture, or damage to a building. Long steel girders, for example, often have the bolts placed in slotted holes at the ends. It is also important to be aware of possible unequal expansion under some circumstances. For example, on a fireplace, the mortar should have the same coefficient of expansion as the bricks, or cracks may quickly develop; similar considerations apply to tile and the bonding materials used to attach it to a wall. We are told to add water to an automobile engine slowly and with the engine running— this is necessary to avoid "cracking the block" due to different amounts of expansion of different parts of the engine. If one part of a glass container is heated or cooled more rapidly than adjacent parts, the glass may crack. Special heat-resistant glasses such as Pyrex have been developed to counteract this common mishap. The principal virtue of such glass is that the expansion coefficient is far smaller than that for ordinary glass (see Table 7-1).

Thermal stresses

In some situations, the ends of a rod or slab of material are rigidly fixed, which prevents expansion or contraction. If the temperature should change, large compressive or tensile stresses will be set up; these are sometimes referred to as *thermal stresses*. The magnitude of these stresses can be calculated using the concept of elastic modulus developed in Chapter 4. To calculate the internal stress, we can think of this process as occurring in two steps. The rod expands (or contracts) by an amount ΔL given by Equation 7-1, and then a force is applied to compress (or expand) the material back to its original length. The force F required is given by Equation 4-4:

$$\Delta L = \frac{1}{E}\frac{F}{A}L_0,$$

where E is Young's modulus for the material. To calculate the internal stress, F/A, we then set ΔL in Equation 7-1 equal to ΔL in Equation 4-4 above and find

$$\alpha L_0 \Delta T = \frac{1}{E}\frac{F}{A}L_0. \qquad (7\text{-}2)$$

▌ EXAMPLE 7-3 Blocks of concrete 10 m long are placed end to end with no space in between them to allow for expansion. If the blocks were placed at a temperature of 10°C, what will be the force of compression when the temperature reaches 40°C? The contact area between each block is 0.20 cm^2. Will fracture occur?

SOLUTION We solve for F in Equation 7-2 and use the value of E found in Table 4-1:

$$F = \alpha\,\Delta T\,EA$$

$$= (12\times10^{-6}/\text{C°})(30\ \text{C°})(20\times10^{9}\ \text{N/m}^2)(0.20\ \text{m}^2)$$

$$= 1.4\times10^{6}\ \text{N}.$$

The stress, F/A is $(1.4\times10^6\ \text{N})/(0.20\ \text{m}^2)=7.0\times10^6\ \text{N/m}^2$. This is not far from the ultimate strength of concrete under compression (Table 4-2) and exceeds it for tension and shear. Hence, assuming the concrete is not perfectly aligned, part of the force will act in shear and fracture is likely. ▮▮

Volume expansion

The change in volume of a material which undergoes a temperature change is given by a relation which is similar to Equation 7-1, namely

$$\Delta V=\beta V_0\,\Delta T, \tag{7-3}$$

where ΔT is the change in temperature, V_0 is the original volume, ΔV is the change in volume, and β is the *coefficient of volume expansion*. The units of β are $(\text{C}°)^{-1}$. The values of β for a number of materials are given in Table 7-1. Notice that for solids β is approximately equal to 3α. (This is not true however for solids that are not isotropic—that is, not the same in all directions.) Notice also that linear expansion has no meaning for liquids and gases since they do not have fixed shapes. For a liquid, however, if it is contained so it can expand mainly in one direction (as mercury or alcohol in a thermometer), α can be taken as approximately one-third β.

▮ EXAMPLE 7-4 The 70-liter tank of a car is filled to the top with gasoline at 20°C. The car is then left to sit in the sun and the tank reaches a temperature of 50°C. How much gasoline do you expect to overflow from the tank?

SOLUTION We ignore the expansion of the tank itself since its coefficient of volume expansion is much less than that of the gasoline. Then

$$\begin{aligned}\Delta V&=\beta V\,\Delta T\\&=(950\times10^{-6}/\text{C}°)(70\ \text{l})(30\ \text{C}°)\\&=2.0\ \text{liters}.\end{aligned}$$

▮▮

Equations 7-1 and 7-3 are accurate only if ΔL or ΔV is small compared to L_0 or V_0. This is of particular concern for liquids and even more so for gases because of the large values of β. Furthermore, the variation of β with temperature is also quite large for gases. Therefore a more convenient way of dealing with gases is needed and this is discussed in Section 7-5.

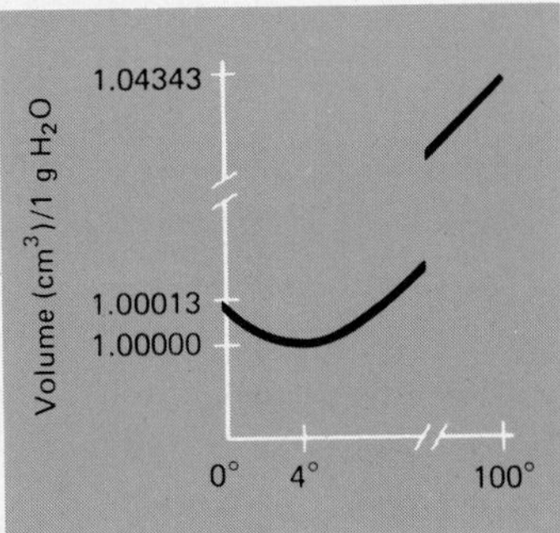

FIGURE 7-8
Volume (in cm³) of 1.0 g of water as a function of temperature.

Water exhibits unusual behavior

Most substances expand more or less uniformly with an increase in temperature (as long as no phase change occurs). Water, however, does not follow the usual pattern. If water at 0°C is heated, it actually *decreases* in volume until it reaches 4°C; above 4°C water behaves normally and expands as the temperature is increased, Figure 7-8. Water thus has its greatest density at 4°C. This fact is of considerable importance for the survival of aquatic life during cold winters. As the water in a lake or river cools the water at the surface sinks because of its greater density and it is replaced by warmer water from below. This mixing continues until the temperature reaches 4°C. As the surface water cools further, it remains on the surface because it is less dense than the 4°C water below. Water then freezes first at the surface. The water at the bottom remains at 4°C until almost the whole body of water is frozen. If water were like most substances, becoming more dense as it cools, the water at the bottom would be frozen first; lakes would freeze solid more easily since circulation would bring the warmer water to the surface to be efficiently cooled. The complete freezing of a lake would cause severe damage to its plant and animal life. Because of the unusual behavior of water below 4°C, it is rare for any large body of water to freeze completely.

7-5 ▌ The ideal gas law

The use of Equation 7-3 to describe the expansion of a gas is not very useful, partly because the expansion can be so great, and partly because gases generally expand to fill whatever container they are in. Indeed, Equation 7-3 is meaningful only if the pressure is kept constant. The volume of a gas depends very much on the pressure as well as on the temperature. It is therefore valuable to determine a relation between the volume, the pressure, the temperature, and the mass of a gas. Such a relation is called an *equation of state.*† (By the world *state* we mean the physical condition of the system.) If the state of a system is changed we assume that we have waited until the pressure and temperature have reached the same values throughout. We thus consider only equilibrium states of a system. Such a state is specified by giving the temperature, pressure, and volume. We shall now investigate the behavior of gases in detail, and will eventually arrive at an equation of state.

The gas laws and absolute temperature

For a given quantity of gas it is found experimentally that to a good approximation *the volume of a gas is inversely proportional to the pressure applied to it when the temperature is kept constant.* That is,

$$V \propto \frac{1}{P} \qquad [\text{at constant } T].$$

For example, if the pressure on a gas is doubled, the volume is reduced to half its original volume. This relation is known as *Boyle's law,* after Robert Boyle (1627–1691), who first stated it on the basis of his own experiments. Boyle's law can also be written:

$$PV = \text{constant} \qquad [\text{at constant } T].$$

That is, at constant temperature, if either the pressure or volume of the gas is allowed to vary, the other changes too so that the product PV remains constant.

Temperature also affects the volume of a gas, but a quantitative relationship between V and T was not found until more than a century

† An equation of state can also be sought for solids and liquids since for them too the volume depends on the mass, temperature, and external pressure, although the temperature and pressure have much less effect than on gases. However, the situation with solids and liquids, because of their complexity, is much more complicated.

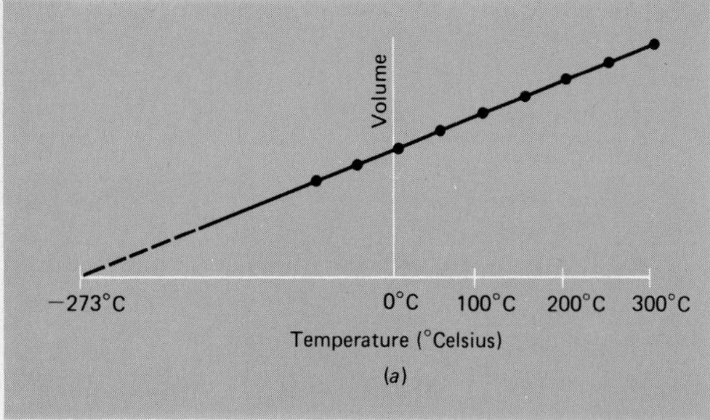

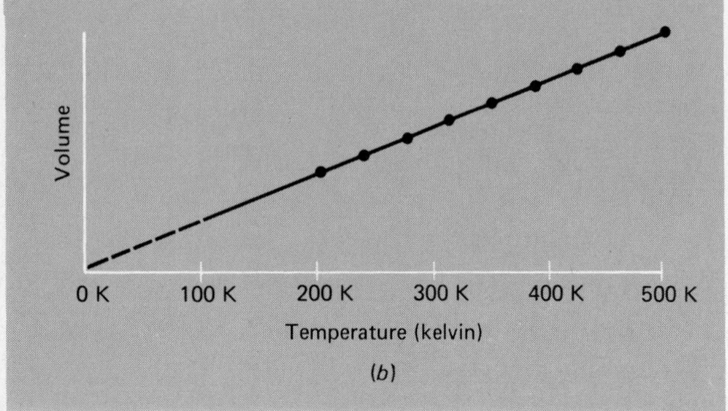

FIGURE 7-9
Volume of a gas as a function of (*a*) Celsius temperature, (*b*) Kelvin temperature.

after Boyle's work. The Frenchman Jacques Charles (1746–1823) found that when the pressure is kept constant, the volume of a gas increases with temperature at a constant rate as shown in Figure 7-9*a*. However, all gases liquefy at low temperatures (for example, oxygen liquefies at −183°C) and so the graph cannot be extended below the liquefaction point. Nonetheless, the graph is a straight line and if projected to lower temperatures as shown by the dashed line, it crosses the axis at −273°C.

This kind of graph can be drawn for any gas, and the straight line always projects back to −273°C at zero volume. This seems to imply that if a gas could be cooled to −273°C it would have zero volume, and at lower temperatures a negative volume—which makes no sense, of course. This led to the suggestion that perhaps −273°C was the lowest temperature possible, and many other more recent experiments indicate that this is so. This temperature is referred to as the *absolute zero* of temperature. Its precise value has been measured to be −273.15°C; however, we will often round it off to −273°C.

Absolute zero forms the basis of a temperature scale known as the *absolute* or *Kelvin scale*, and it is used extensively in scientific work. On this scale the temperature is specified as degrees Kelvin or simply as kelvins (K) without the degree sign. The intervals are the same as for the Celsius scale and the zero on this scale (0 K) is chosen as absolute zero itself. Thus the freezing point of water (0°C) is 273.15 K and the boiling point 373.15 K. Indeed, any temperature on the Celsius scale can be changed to kelvins by adding 273.15 to it.

Now let us look at Figure 7-9*b* where we see that the graph of the volume of a gas versus absolute temperature is essentially a straight line that passes through the origin. Thus, to a good approximation, *the volume* *of a given amount of gas is directly proportional to the absolute temperature when the pressure is kept constant.* This is known as *Charles' law*, and can be written

$$V \propto T \qquad \text{[at constant } P\text{]}.$$

A third gas law, known as *Gay-Lussac's law* [after Joseph Gay-Lussac (1778–1850)], states that *at constant volume, the pressure of a gas is directly proportional to the absolute temperature*:

$$P \propto T \qquad \text{[at constant } V\text{]}.$$

A familiar example of this is the fact that a closed jar or aerosol can thrown into a fire will explode due to the increase in air pressure inside. This relation is also the basis for the constant volume gas thermometer (Section 7-2).

Ideal gas law

The gas laws of Boyle, Charles, and Gay-Lussac were obtained by using a technique that is very useful in science: namely, to hold one or more variables constant in order to see clearly the effects of changing only one of the variables. These laws can now be combined into a single more general law that relates the pressure, volume, and temperature of a fixed quantity of gas:

$$PV \propto T.$$

This relation indicates how any of the quantities P, V, or T will vary when the other two quantities change. This relation reduces to Boyle's, Charles', or Gay-Lussac's law when either the temperature, the pressure, or the volume, respectively, is held constant.

Finally, we must incorporate the effect of the quantity of the gas—that is, the mass of gas present. Anyone who has blown up a balloon knows that the more air forced into the balloon, the bigger it gets. Indeed, careful experiment shows that at constant temperature and pressure, the volume V of an enclosed gas increases in direct proportion to the mass m of gas present. Hence we can write

$$PV \propto mT.$$

This proportion relates the significant variables for gases. It can be made into an equation by inserting a constant of proportionality. Experiment shows that this constant has a different value for different gases. However, the constant of proportionality turns out to be the same for all gases if instead of the mass m we use the number of moles; one **mole** (abbreviated mol) is defined as the amount of substance which contains as many elementary entities (atoms or molecules) as there are in 0.012 kg of carbon 12 (whose atomic mass is exactly 12 u); a simpler but equivalent definition is: 1 mole is that number of grams numerically equal to the molecular mass of the substance. For example, the molecular mass of hydrogen gas (H_2) is 2.0 u (since each molecule contains two atoms of hydrogen and each atom has an atomic mass of 1.0 u); thus 1 mol of H_2 has a mass of 2.0 grams or 0.0020 kg. Similarly, one mol of neon gas has a mass of 0.020 kg, and 1 mol of CO_2 has a mass of $(12+2\times 16)\times 10^{-3}$ kg $=0.044$ kg. The mol is the official unit in the SI system. Sometimes the kilogram-mole (kg · mol) is used, which is that number of kilograms numerically equal to the molecular mass of the substance; 1 kg · mol = 10^3 mol.

Now we write the above proportion as an equation:

$$PV = nRT, \tag{7-4}$$

where n represents the number of moles and R is the constant of proportionality. R is called the *universal gas constant* because its value is found experimentally to be the same for all gases; its value, in several sets of units (the first is the SI value), is

$$R = 8.314\ \text{J/(mol}\cdot\text{K)} \qquad \text{[SI units]}$$

$$= 0.082(\text{liter}\cdot\text{atm})/(\text{mol}\cdot\text{K})$$

$$= 1.99\ \text{calories/(mol}\cdot\text{K)}.$$

Equation 7-4 is called the **ideal gas law**, or the *equation of state for an ideal gas*. We use the term "ideal" because real gases do not follow Equation 7-4 precisely, particularly at high pressures or when the gas is near the liquefaction point. However, at pressures on the order of 1 atm or less, and when T is not close to the liquefaction point of the gas, Equation 7-4 is quite accurate.

The ideal gas law is an extremely useful tool, and we now consider some examples. We will often refer to "standard conditions" or "standard temperature and pressure" (STP) which means $T = 273$ K (0°C) and $P = 1.0$ atm $= 1.01 \times 10^5$ N/m^2.

▮ EXAMPLE 7-5 Determine the volume of 1 mol of any gas at STP.

SOLUTION We solve for V in Equation 7-4:

$$V = \frac{nRT}{P} = \frac{(1.00\ \text{mol})(8.31\ \text{J/mol}\cdot\text{K})(273\ \text{K})}{(1.01\times 10^5\ \text{N/m}^2)}$$

$$= 22.4 \times 10^{-3}\ \text{m}^3.$$

Since 1 liter is 1000 cm^3 = 10^{-3} m^3, 1 mol of any gas has a volume of 22.4 liters at STP. ▮▮

The value for the volume of 1 mol of gas at STP (22.4 liters) is worth remembering for it sometimes makes calculation simpler as the following example shows.

▮ EXAMPLE 7-6 A flexible container of oxygen (molecular mass = 32 u) at STP has a volume of 10.0 m^3. What is the mass of gas enclosed?

SOLUTION Since 1 mol occupies a volume of 22.4×10^{-3} m^3 the 10.0 m^3 of oxygen corresponds to $(10.0)/(22.4 \times 10^{-3}) = 446$ mol. Since 1 mol has a mass of 0.032 kg, the mass of oxygen is (446 mol) × (0.032 kg/mol) = 14.3 kg. ▮▮

It is frequently the case that volume is specified in liters and pressure in atmospheres. Rather than convert these to SI units, we can instead use the value of R given above as 0.082 liter · atm/(mol · K).

▮ EXAMPLE 7-7 What is the pressure (in atm) in a 1000-liter tank containing 42 kg of nitrogen gas (N_2) at 20°C?

SOLUTION Since N_2 has a molecular mass of $2 \times 14.0u = 28.0u$, we have $(42)/(0.028) = 1500$ mol. From Equation 7-4

$$P = \frac{nRT}{V} = \frac{(1500 \text{ mol})(0.082 \text{ liters} \cdot \text{atm})/\text{mol} \cdot \text{K})(293 \text{ K})}{(1000 \text{ liters})}$$

$$= 36 \text{ atm.}$$ ▮▮

In many situations it is not necessary to use the values of R at all. For example, many problems involve a change in the pressure, temperature and volume of a fixed amount of gas. In this case, $PV/T = nR =$ constant, since n and R remain constant. If we now let P_1, V_1, and T_1 represent the appropriate variables initially, and P_2, V_2, T_2 represent the variables after the change is made, then we can write

$$\frac{P_1 V_1}{T_1} = \frac{P_2 V_2}{T_2}. \tag{7-5}$$

If we know any five of the quantities in this equation, we can solve for the sixth.

▮ EXAMPLE 7-8 5.00 m^3 of a gas initially at STP is placed under a pressure of 4.00 atm. The final temperature is measured to be 25°C. What is the final volume?

SOLUTION Using Equation 7-5 we find

$$V_2 = \left(\frac{P_1 V_1}{T_1}\right)\left(\frac{T_2}{P_2}\right) = \frac{(1.00 \text{ atm})(5.00 \text{ m}^3)(298 \text{ K})}{(273 \text{ K})(4.00 \text{ atm})}$$

$$= 1.35 \text{ m}^3.$$ ▮▮

▮ EXAMPLE 7-9 An automobile tire is filled to a pressure of 200 k Pa at 10°C. After driving 100 km the temperature within the tire rises to 40°C. What is the pressure within the tire now?

SOLUTION Since the volume remains essentially constant, $V_1 = V_2$ and therefore

$$\frac{P_1}{T_1} = \frac{P_2}{T_2}.$$

This is, incidentally, a statement of Gay-Lussac's law. Since the pressure given is the gauge pressure, we must add atmospheric pressure = 103 kPa to get the absolute pressure $P_1 =$ 200 kPa + 100 k Pa = 300 k Pa. Then

$$P_2 = \frac{P_1}{T_1} T_2 = \frac{(300 \text{ k Pa})(313 \text{ K})}{(283 \text{ K})} = 330 \text{ kPa}.$$

Substracting out atmospheric pressure we find the resulting gauge pressure to be 230 k Pa. This example illustrates why car manuals suggest checking the pressure in tires when they are cold. Note also in this example that the units are not SI units. It really doesn't matter what units are used in a ratio equation like Equation 7-5 as long as they are the same on both sides of the equation. ▮▮

Ideal gas law in terms of molecules: Avogadro's number

The fact that the gas constant, R, has the same value for all gases is a remarkable reflection of simplicity in nature. This fact was first recognized, although in a slightly different form, by the Italian scientist Amedeo Avogadro (1776–1856). Avogadro stated that *equal volumes of gas at the same pressure and temperature contain equal numbers of molecules.* This is sometimes called *Avogadro's hypothesis.* That this is consistent with R being the same for all gases can be seen as follows. First of all from Equation 7-4 we see that for the same number of moles n, and the same pressure and temperature, the volume will be the same for all gases as long as R is the same. And secondly the number of molecules in 1 mol is the same for all gases; this follows directly from the definition of a mole.† Thus Avogadro's hypothesis is equivalent to R being the same for all gases.

The number of molecules in a mole is known as **Avogadro's number**, N_A. Although Avogadro conceived the notion, he was not able to actually determine the value of N_A. Indeed, precise measurements had to await the twentieth century. A number of methods have been devised to measure N_A and the accepted value is

$$N_A = 6.02 \times 10^{23} \qquad \text{[molecules/mol]}.$$

Since the total number of molecules N in a gas is equal to the number per mole times the number of moles ($N = nN_A$), the ideal gas law, Equation 7-4, can be written in terms of the number of molecules present:

$$PV = nRT = \frac{N}{N_A} RT,$$

or

$$PV = NkT, \tag{7-6}$$

where $k = R/N_A$ is called *Boltzmann's constant* and has the value $k = (8.314\,\text{J/mol}\cdot\text{K})/(6.02 \times 10^{23}/\text{mol}) = 1.38 \times 10^{-23}\,\text{J/K}$.

▮ EXAMPLE 7-10 Use Avogadro's number to determine the mass of a hydrogen atom.

SOLUTION One mole of hydrogen (atomic mass = 1.008) has a mass of 1.008×10^{-3} kg and contains 6.02×10^{23} atoms. Thus one atom has a mass

$$m = \frac{1.008 \times 10^{-3}\,\text{kg}}{6.02 \times 10^{23}} = 1.67 \times 10^{-27}\,\text{kg}.$$

Historically, the reverse process was one method used to obtain N_A—that is, from the measured mass of the hydrogen atom. ▮▮

† For example, the molecular mass of H_2 gas is 2.0 while that of O_2 gas is 32.0. Thus 1 mol of H_2 has a mass of 0.0020 kg and of O_2 gas 0.032 kg. Now the number of molecules in 1 mol is equal to the total mass M of 1 mol divided by the mass m of one molecule; since this ratio (M/m) is the same for all gases by definition of the mole, 1 mol of any gas must contain the same number of molecules.

7-6 ▌ Partial pressure

When two or more gases occupy the same volume, the total pressure can be written as the sum of the partial pressures of the individual gases. The *partial pressure* of a gas is defined as the pressure each gas would exert if it alone occupied the whole volume. An experimental law due to Dalton, and known as *Dalton's law of partial pressures*, states that *each gas in a mixture exerts a partial pressure proportional to its molecular concentration.* These ideas are consistent with the ideal gas law, Equation 7-6. For suppose we have a mixture of three gases with N_1, N_2, and N_3 molecules respectively. The total pressure is

$$P = \frac{NkT}{V},$$

where N is the total number of molecules and $N = N_1 + N_2 + N_3$. Then we can write

$$\begin{aligned} P &= \frac{N_1kT}{V} + \frac{N_2kT}{V} + \frac{N_3kT}{V} \\ &= P_1 + P_2 + P_3, \end{aligned}$$

where P_1, P_2 and P_3 are the partial pressures of each of the three gases, and N_1/V, N_2/V, and N_3/V represent their concentrations. Thus the total pressure is equal to the sum of the partial pressures and each of the latter is proportional to its molecular concentration.

For example, 78 percent (by volume) of air molecules are nitrogen and 21 percent oxygen, with much smaller amounts of water vapor, argon, and other gases. At an air pressure of 1 atm, oxygen exerts a partial pressure of 0.21 atm and nitrogen 0.78 atm.

Dalton's law is important in a variety of biological situations. For example, in the lungs, the exchange of gases with blood depends on the partial pressure of each gas, not on the total pressure: Oxygen flows from the air into the blood because the partial pressure of oxygen in the lungs is greater than in the blood; and CO_2 flows out of the blood into the air because its partial pressure is higher in the blood. This process is called diffusion and is discussed in more detail in the next chapter.

7-7 ▌ Kinetic theory and the molecular interpretation of temperature

The concept that matter is made up of atoms which are in continual random motion is called the *kinetic theory.* We now investigate the properties of a gas from the point of view of kinetic theory and in particular we will calculate the pressure in a gas in terms of its molecular properties. We will also arrive at an important relation between the average kinetic energy of molecules in a gas and the absolute temperature.

We make the following assumptions about the molecules in a gas. These assumptions reflect a rather simple view of a gas but they do represent the essential features of real gases that are far from the liquefaction point and at low pressures. Under these circumstances, gases obey the ideal gas law; and we shall refer to such a gas as an *ideal gas*. The assumptions, which are the basic postulates of the kinetic theory, are as follows. (1) There are a large number of molecules, N, each of mass m, moving in every which direction with a variety of speeds. This assumption is in accord with our observation that a gas fills its container and, in the case of air on the earth, is kept from escaping only by the force of gravity. (2) The molecules are, on the average, far apart from one another; that is, their average separation is much greater than the diameter of one molecule. (3) The molecules are presumed to interact with one another only when they collide.† (4) Collisions with another molecule or the wall of the vessel are assumed to be perfectly elastic, like the collisions of perfectly elastic billiard balls (Section 5-6).

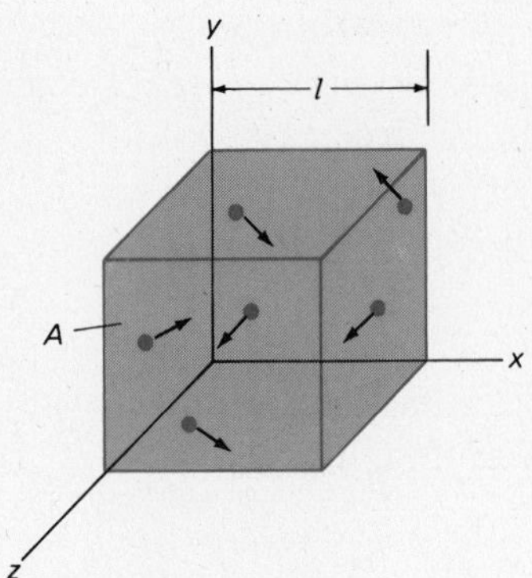

FIGURE 7-10
Molecules of a gas moving about in a cubical container.

We can see immediately how this kinetic view of a gas can explain Boyle's law: the pressure exerted on a wall of a container of gas is due to the constant bombardment of molecules; if the volume is reduced by (say) half, the molecules are closer together and twice as many will be striking a given area of the wall per second. Hence we expect the pressure to be twice as great, which is Boyle's law.

Now let us calculate quantitatively the pressure in a gas based on kinetic theory. For purposes of argument, we imagine that the molecules are contained in a cubical container whose ends have area A and whose length is l as shown in Figure 7-10. The pressure exerted by the gas on the walls of its container is, according to our model, due to the collisions of the molecules with the walls. Let us focus our attention on the wall, of area A, at the left end of the container and examine what happens when one molecule strikes this wall, as shown in Figure 7-11. This molecule exerts a force on the wall and the wall exerts an equal and opposite force back on the molecule. The magnitude of this force, according to Newton's second law, is equal to the molecule's rate of change of momentum (Section 5-1). Assuming the collision takes place without loss of kinetic energy, only the x component of the molecule's momentum changes, and it changes from $-mv_x$ (it is moving in the negative x direction) to $+mv_x$. Thus the change in momentum $\Delta(mv)$, which is the final momentum minus the initial momentum, is

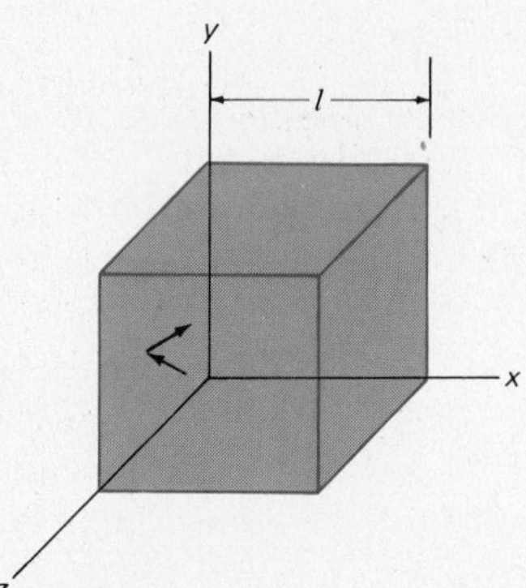

FIGURE 7-11
Arrows indicate the momentum of one molecule as it rebounds from the end wall.

$$\Delta(mv) = mv_x - (-mv_x) = 2mv_x$$

for one collision. This molecule will make many collisions with the wall,‡ each separated by a time t which is the time it takes the molecule to travel across the box and back again, a distance equal to $2l$. Thus $2l = v_x t$ or $t = 2l/v_x$. The time between collisions is very small so the number of collisions per second is very large. Thus the average force—averaged over many collisions—will be equal to the force exerted during one collision

† Although molecules exert weak attractive forces on each other between collisions, the potential energy associated with these forces is small compared to the kinetic energy, and can be ignored.

‡ During its passage back and forth across the container, the molecule may collide with the tops and sides of the container, but this does not alter its x component of momentum and thus does not alter our result. It may also collide with other molecules, which may change its v_x; however, any loss (or gain) of momentum is acquired by the molecules it collides with, and because we will sum over all the molecules in the end, this effect will be included.

divided by the time between collisions (Newton's second law):

$$F = \frac{\Delta(mv)}{t}$$

$$= \frac{2mv_x}{2l/v_x} = \frac{mv_x^2}{l} \qquad \begin{bmatrix}\text{due to one}\\ \text{molecule}\end{bmatrix}.$$

Of course the actual force due to one molecule is intermittent, but because a huge number of molecules are striking the wall per second, the force is pretty steady. To calculate the force due to all the molecules in the box, we have to add the contributions of each. Thus the net force on the wall must be:

$$F = \frac{m}{l}(v_{x1}^2 + v_{x2}^2 + \cdots + v_{xN}^2),$$

where v_{x1} means v_x for molecule number 1 (we arbitrarily assign each molecule a number) and the sum extends over the total of N molecules. Now the average value of the square of the x component of velocity is

$$\overline{v_x^2} = \frac{v_{x1}^2 + v_{x2}^2 + \cdots + v_{xN}^2}{N}. \qquad (7\text{-}7)$$

Thus we can write the force as

$$F = \frac{m}{l} N\overline{v_x^2}.$$

We know that the square of any vector is equal to the sum of the squares of its components (theorem of Pythagoras; see Section 1-6). Thus $v^2 = v_x^2 + v_y^2 + v_z^2$ for any velocity v. Taking averages we obtain

$$\overline{v^2} = \overline{v_x^2} + \overline{v_y^2} + \overline{v_z^2}.$$

Since the velocities of the molecules in our gas are assumed to be random, there is no preference to one direction or another; hence

$$\overline{v_x^2} = \overline{v_y^2} = \overline{v_z^2}$$

and therefore

$$\overline{v^2} = 3\,\overline{v_x^2}.$$

We substitute this into the equation for the force F:

$$F = \frac{m}{l} N \frac{\overline{v^2}}{3}.$$

The pressure on the wall is then

$$P = \frac{F}{A} = \frac{1}{3}\frac{Nm\overline{v^2}}{Al}$$

or

$$P = \frac{1}{3}\frac{Nm\overline{v^2}}{V}, \qquad (7\text{-}8)$$

where $V = lA$ is the volume of the container. This is the result we were seeking, the pressure in a gas expressed in terms of molecular properties.

Equation 7-8 can be rewritten in a clearer form by multiplying both sides by V and slightly rearranging the right side:

$$PV = \tfrac{2}{3}N(\tfrac{1}{2}m\overline{v^2}). \tag{7-9}$$

The quantity $\frac{1}{2}m\overline{v^2}$ is the average kinetic energy ($\overline{\text{KE}}$) of the molecules in the gas. If we compare Equation 7-9 with Equation 7-6, the ideal gas law, we see that the theoretical equation agrees with the experimental law if

$$\tfrac{2}{3}(\tfrac{1}{2}m\overline{v^2}) = kT,$$

or,

$$\tfrac{1}{2}m\overline{v^2} = \tfrac{3}{2}kT. \tag{7-10}$$

This equation tells us that the average kinetic energy of molecules in a gas is directly proportional to the absolute temperature. The higher the temperature, according to kinetic theory, the faster the molecules are moving on the average. This relation is one of the most important in physics and is one of the triumphs of the kinetic theory.

▌EXAMPLE 7-11 What is the average KE of molecules in a gas at 37°C?

SOLUTION We use Equation 7-10 and change 37°C to 310 K:

$$\overline{\text{KE}} = \tfrac{3}{2}kT$$

$$= \tfrac{3}{2}(1.38\times 10^{-23}\ \text{J/K})(310\ \text{K}) = 6.42\times 10^{-21}\ \text{J}.$$

▌▌

Equation 7-10 holds not only for gases, but also applies reasonably accurately to liquids and solids. Thus the result of Example 7-11 would apply to molecules within living cells at body temperature (37°C).

We can also use Equation 7-10 to calculate how fast molecules are moving on the average. Notice that the average in Equations 7-7 through 7-10 is over the *square* of the velocity. The square root of $\overline{v^2}$ is called the "root mean square" velocity, v_{rms} (since we are taking the square *root* of the *mean* of the *square* of the velocity):

$$v_{rms} = \sqrt{\overline{v^2}} = \sqrt{\frac{3kT}{m}}. \tag{7-11}$$

The *mean speed*, v_m, is the average of the magnitudes of the speeds themselves; v_m is generally not equal to v_{rms}. To see the difference between the mean speed and the rms speed, we take the following example.

▌EXAMPLE 7-12 Eight particles have the following speeds, given in m/s: 1.0, 6.0, 4.0, 2.0, 6.0, 3.0, 2.0, 5.0. Calculate (*a*) the mean speed, and (*b*) the rms speed.

SOLUTION (*a*) The mean speed is

$$v_m = \frac{1.0+6.0+4.0+2.0+6.0+3.0+2.0+5.0}{8}\ \text{m/s} = 3.6\ \text{m/s}.$$

(*b*) The rms speed is

$$v_{rms} = \sqrt{\frac{(1.0)^2+(6.0)^2+(4.0)^2+(2.0)^2+(6.0)^2+(3.0)^2+(2.0)^2+(5.0)^2}{8}}\ \text{m/s}$$

$$= 4.1\ \text{m/s}.$$

▌▌

We see in this example that v_m and v_{rms} are not necessarily equal. In fact, for an ideal gas they differ by about 8 percent. We usually won't be concerned with this small difference however, and when we say "average speed," we will generally mean v_{rms}.

▌EXAMPLE 7-13 What is the rms speed of air molecules at room temperature (20°C)?

SOLUTION We must apply Equation 7-11 to oxygen and nitrogen separately since they have different masses. Using the result of Example 7-10, the masses of one molecule of O_2 (molecular mass = 32) and N_2 (molecular mass = 28) are:

$$m(O_2) = (32)(1.67\times10^{-27}\ \text{kg}) = 5.3\times10^{-26}\ \text{kg}$$

$$m(N_2) = (28)(1.67\times10^{-27}\ \text{kg}) = 4.7\times10^{-26}\ \text{kg}.$$

Thus, for oxygen

$$v_{rms} = \sqrt{\frac{3kT}{m}} = \sqrt{\frac{(3)(1.38\times10^{-23}\ \text{J/K})(293\ \text{K})}{(5.3\times10^{-26}\ \text{kg})}}$$

$$= 480\ \text{m/s},$$

and for nitrogen the result is $v_{rms} = 510$ m/s. (This is over 1500 km/h or 1000 mi/h.) Air molecules are moving incredibly fast. ▌▌

7-8 ▌Distribution of molecular speeds

Maxwell distribution

The molecules in a gas are assumed to be in random motion, which means that many molecules have speeds less than the average speed and others have speeds greater than the average. In 1860, James Clerk Maxwell (1831–1879) derived, on the basis of kinetic theory, that the speeds of molecules in a gas are distributed according to the graph shown in Figure 7-12. This is known as the *Maxwell distribution of speeds.*† The speeds vary

† Mathematically, the distribution is given by $\Delta N = Cv^2 \exp(-\frac{1}{2}mv^2/kT)\,\Delta v$ where ΔN is the number of molecules with speed between v and $v + \Delta v$, C is a constant, and exp means the expression in parenthesis is an exponent on the natural number $e = 2.718\ldots$.

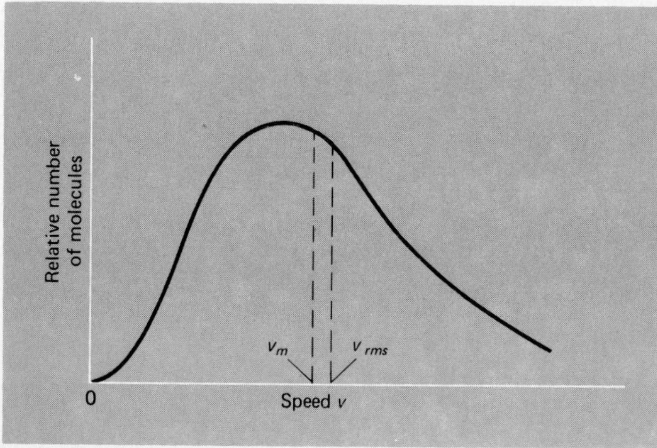

FIGURE 7-12
Distribution of speeds of molecules in an ideal gas. Note that v_m and v_{rms} are not at the peak of the curve (that speed is called the "most probable speed," v_p). This is because the curve is skewed to the right: It is not symmetrical.

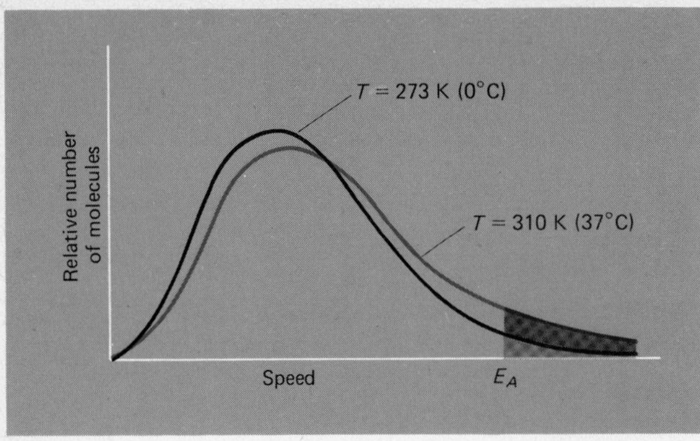

FIGURE 7-13
Distribution of molecular speeds for two different temperatures.

from zero up to many times the average speed, but as can be seen from the graph, most molecules have speeds that are not far from the average. Less than one percent of the molecules exceed four times v_{rms}. Experiments to determine the distribution were first performed in the 1920s. They confirmed with considerable accuracy the Maxwell distribution and the direct proportion between average kinetic energy and absolute temperature, Equation 7-10.

Figure 7-13 shows the distribution for two different temperatures; just as v_{rms} increases with temperature, so the whole distribution curve shifts to the right at higher temperatures.

Why chemical reactions and life processes depend on temperature

Figure 7-13 illustrates how kinetic theory can be used to explain the fact that many chemical reactions take place more rapidly as the temperature increases, a fact that is very important in life processes. Two molecules will chemically react only if their kinetic energy is great enough so that when they collide they penetrate into one another somewhat. This minimum energy is called the *activation energy*, E_A, and it has a specific value for each chemical reaction. The molecular speed corresponding to a kinetic energy of E_A for a particular reaction is indicated in Figure 7-13. The number of molecules with energy greater than this value is given by the area under the curve to the right of E_A. In Figure 7-13, the respective areas (which are proportional to the number of molecules) for two different temperatures are shown by gray shadings. It is clear that the number of molecules that have kinetic energies in excess of E_A increases greatly for only a small increase in temperature. The rate at which a chemical reaction occurs is proportional to the number of molecules with energy greater than E_A, and thus we see why reaction rates increase rapidly with increased temperature.

Biological activity varies significantly with temperature, and this is a result of the temperature dependence of chemical reaction rates. Indeed, most biological organisms can operate only over a limited range of temperature. For humans and other warm-blooded animals, this temperature range is very small indeed, within one or two degrees from 37°C internally. The exterior temperature can vary a great deal more, but

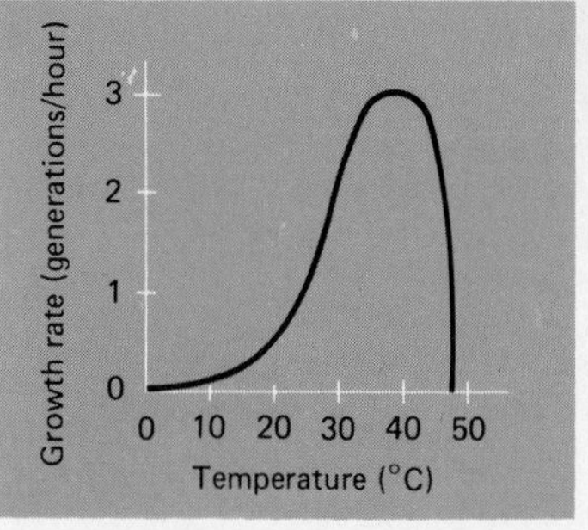

FIGURE 7-14
Growth rate as a function of temperature for the bacterium *Escherichia coli* (which lives in the human intestine and helps in the digestion of food).

only because the body has a variety of mechanisms (feedback processes) to maintain its internal temperature.

Even bacteria are sensitive to temperature and their rate of growth varies considerably with temperature. Between 20°C and 30°C bacterial growth rates typically increase by a factor of 2 or 3. This increased metabolism is a reflection of the increased chemical reaction rate. This is why the low temperatures of refrigeration retard food spoilage. However, a maximum growth rate is reached at some temperature (37°C for many bacteria) above which the growth rate declines drastically, as shown in Figure 7-14. This decline is due to the fact that at higher temperatures important molecules in the cell are rendered inactive through collisions with other molecules that are now moving faster. An important example is enzymes. Enzymes function in the cell to lower the activation energy of particular chemical reactions that would not otherwise take place at normal temperatures. Each enzyme is folded into a characteristic shape or "conformation," and it is this shape that allows it to hold two (or more) molecules close enough together so that they can react chemically.† At higher temperatures, however, the bombardment by high speed molecules in the cell can cause the enzyme to lose its shape; it can then no longer perform its proper function. Thus metabolism in the cell must slow down. The genetic material of the cell, DNA, is similarly deactivated by high temperatures. If the temperature is high enough, the high-speed collisions can cause molecules to be broken down to the point where they cannot recover when the temperature is later lowered; in this case the cell dies. Pastuerization of milk products is a heating process that kills pathogenic bacteria.

*Thermal-gradient precipitator

A thermal-gradient particle precipitator is a device used to remove particulate matter from polluted air. When a cold surface and a hot surface are close to one another, particles will collect on the cold surface. This is due to the fact that air molecules that have been in contact with the hot surface have more kinetic energy on the average than molecules that have recently been in contact with the cold surface. A particle between the two surfaces is thus subject to unequal bombardment by air molecules on its two sides. This results in a net force pushing it toward the cold surface

† We will discuss how this is done in Chapter 13.

as shown in Figure 7-15. This effect can be noticed on the wall behind a radiator, particularly if the wall is an outside wall and therefore cool.

Thermal-gradient precipitators are mainly used to monitor air quality. The cold surface is often a glass slide that can be examined under a microscope. The thermal gradient—the increase in temperature per unit distance—is normally about 1000 C°/cm. The flow of gas must be fairly slow, about 10 cm^3/s, for there to be significant precipitation of large molecules.

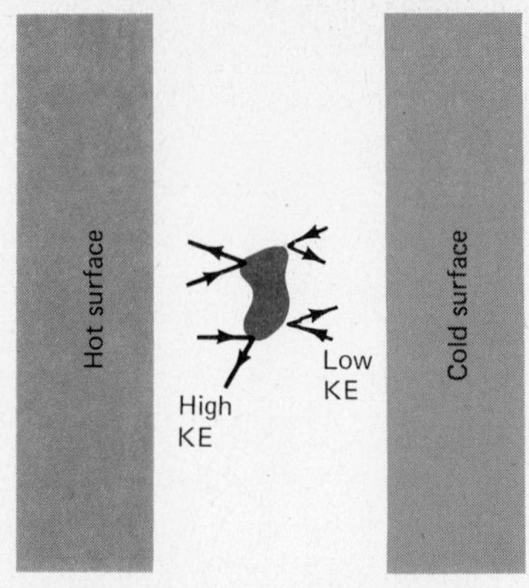

FIGURE 7-15
Thermal-gradient precipitator. Particle struck by high-speed molecules from the left and low-speed molecules from the right.

▮ Summary

The atomic theory of matter postulates that all matter is made up of tiny entities called *atoms* which are typically 10^{-10} m in diameter. Some substances are made up of only one type of atom, and these are called *elements.* Atoms can combine to form *molecules,* and substances made up of a single type of molecule are called *compounds.* A substance made up of more than one type of molecule is called a mixture. *Atomic* and *molecular masses* are specified on a scale where ordinary carbon (^{12}C) is arbitrarily given the value 12.000 u (atomic mass units). The distinction between solids, liquids, and gases can be attributed to the strength of the attractive forces between the atoms or molecules, and depends on their average speed.

Temperature is a measure of how hot or cold a body is. *Thermometers* are used to measure temperature on the Celsius (°C), Fahrenheit (°F), and Kelvin (K) scales. Two standard points on each scale are the freezing point of water (0°C, 32°F, 273.15 K) and the boiling point of water (100°C, 212°F, 373.15 K). A change in temperature of one kelvin equals a change of one Celsius degree or 9/5 Fahrenheit degrees.

The change in length, ΔL, of a solid when its temperature changes by an amount ΔT is directly proportional to the temperature change and to its original length, L_0. That is, $\Delta L = \alpha L_0 \, \Delta T$ where α is the coefficient of linear expansion. The change in volume of most solids, liquids, and gases is proportional to the temperature change and to the original volume V_0: $\Delta V = \beta V_0 \, \Delta T$; the coefficient of volume expansion, β, is approximately equal to 3α for solids. Water is unusual because, unlike most materials whose volume increases with temperature, its volume actually decreases as the temperature increases from 0°C to 4°C.

The *ideal gas law* relates the pressure P, volume V, and temperature T (in kelvins) of n moles of gas by the equation $PV = nRT$, where $R = 8.314$ J/mol · K for all gases. Real gases obey the ideal gas law quite accurately if they are not at too high a pressure or near their liquefaction point. One *mole* of a substance is defined as that many grams which is numerically equal to its molecular mass. *Avogadro's number,* $N_A = 6.02 \times 10^{23}$, is the number of atoms or molecules in one mole of any pure substance. The ideal gas law can be written in terms of the number of molecules, N, in the gas as $PV = NkT$ where $k = R/N_A = 1.38 \times 10^{-23}$ J/K is Boltzmann's constant. Dalton's law of partial pressures states that each gas in a mixture exerts a partial pressure proportional to its molecular concentration; the total pressure is the sum of the partial pressures.

According to the kinetic theory of gases, which is based on the idea that a gas is made up of molecules that are moving rapidly and at random, the average kinetic energy of the molecules is proportional to the kelvin temperature. At any moment, there exists a wide distribution of molecular speeds within a substance.

▌Problems

▌Questions

1 Name several properties of materials that could be exploited to make a thermometer.

*2 Explain how food supply acts as negative feedback on world population.

3 A circular ring is heated from 20°C to 80°C. Will the hole in the ring become larger or smaller?

4 In the relation $\Delta L = \alpha L_0 \,\Delta T$, should L_0 be the initial length, the final length, or does it matter?

5 Why is it sometimes easier to remove the lid from a tightly closed jar after warming it under hot running water?

6 Long steam pipes often have a section in the shape of a U. Why?

7 Why is it incorrect to say that air expands at a rate of 3.4 ml/l per Celsius degree? When is this true?

8 In view of the fact that fireplaces become hot, why should chimneys not be used for support of a house or building?

9 Why doesn't the size of different molecules enter into the gas laws?

10 When a gas is rapidly compressed, say, by pushing down a piston, its temperature increases. When a gas expands against a piston, it cools. Explain these changes in temperature using the kinetic theory, in particular noting what happens to the momentum of molecules when they strike the moving piston.

11 In Section 7-5 we assumed the gas molecules made perfectly elastic collisions with the walls of the container. This assumption is not necessary as long as the walls are at the same temperature as the gas. Why?

12 Explain in words how Charles' law follows from kinetic theory and the relation between average kinetic energy and the absolute temperature.

13 Explain in words how Gay-Lussac's law follows from kinetic theory and Equation 7-10.

14 As you go higher in the earth's atmosphere, the ratio of N_2 molecules to O_2 molecules increases. Why?

15 Escape velocity for the earth refers to the minimum speed an object must have to leave the earth and never return. The escape velocity for the moon is about one-tenth what it is for the earth since the moon is smaller. Explain, then, why the moon has practically no atmosphere.

16 The planet Mercury is smaller and hotter than the earth. Explain why it therefore has very little atmosphere.

▌Problems

SECTION 7-1

1 (I) What mass of gold has the same number of atoms as 1 kg of silver?

2 (I) Which has more atoms: 1 kg of aluminum or 1 kg of iron (see the periodic table inside the cover)?

SECTION 7-2

3 (I) "Room temperature" is often considered to be 68°F. What is this on the Celsius scale.

4 (I) The melting point of iron is 1530°C. What is this in °F?

5 (I) The original Celsius temperature scale [due to Anders Celsius (1701–1744)] defined the freezing point of water as 100° and the boiling point as 0°. What temperature on this scale corresponds to 25°C?

6 (II) At what temperature will the Fahrenheit and Centigrade scales yield the same numerical value?

SECTION 7-4

7 (I) A concrete highway is built of slabs 30 m long. How wide should the expansion cracks between the slabs be to prevent buckling if the range of temperature is −10°C to +50°C?

8 (I) How far apart should steel railroad rails 20 m long be placed so that they will not touch even if the temperature were to increase by 50 C°?

9 (II) At what temperature will the ultimate compressive strength of concrete be exceeded for the blocks discussed in Example 7-3?

10 (II) A horizontal steel I-beam of cross-sectional area 0.015 m^2 is rigidly connected to two vertical steel girders. If the beam was installed when the temperature was 28°C, how much force does it exert on each girder when the temperature drops to −24°C? Is the ultimate strength of the steel exceeded? Assume that the vertical supports are perfectly rigid.

11 (II) Repeat the above problem assuming the beam is concrete and has a cross section of 0.30 m^2. Will it fracture?

12 (II) A precise steel tape measure has been calibrated at 20°C. At 40°C, (*a*) will it read high or low, and (*b*) what will be the percentage error?

13 (II) An iron plug is to be placed in a ring made of aluminum. At room temperature the diameter of the plug is 10.12 cm and that of the inside of the ring is 10.09 cm. They must be brought to what common temperature in order to fit?

14 (II) If the density of mercury is $13.59 \times 10^3 \text{ kg/m}^3$ at 20°C, what will its density be at 80°C?

15 (II) To make a secure fit, rivets are often used that are larger than the rivet hole and the rivet is cooled (usually in dry ice) before it is placed in the hole. A steel rivet 2.400 cm in diameter is to be placed in a hole 2.398 cm in diameter (at 20°C). To what temperature must the rivet be cooled if it is to fit in the hole?

16 (II) Show that if a rod of original length L_0 has its temperature changed by ΔT, its new length will be $L = L_0(1 + \alpha\,\Delta T)$ where α is the coefficient of linear expansion.

17 (II) Show that if an object originally has a volume V_0, when its temperature is changed by ΔT its new volume is $V = V_0(1 + \beta\,\Delta T)$ where β is the volume coefficient of expansion.

18 (III) Show that for an isotropic solid, $\beta = 3\alpha$ as long as the amount of expansion is small. β and α are the coefficients of volume and linear expansion, respectively.

19 (III) A wagon wheel of diameter 112.860 cm at 20°C is to be enclosed by an iron rim. The circular rim has an inside diameter of 112.848 cm at 20°C. It is 8.0 cm wide and 0.50 cm thick. (*a*) To what temperature must the rim be heated so that it will fit over the wheels? (*b*) What will be the tension in the rim when it cools to 20°C?

20 (III) The tube of a mercury thermometer has an inside diameter of 0.110 mm. The bulb has a volume of 0.250 cm^3. How far will the thread of mercury move when the temperature changes from 10.0°C to 20.0°C? Take into account expansion of the glass (Pyrex).

SECTION 7-5

21 (I) What are the following temperatures on the Kelvin scale: (*a*) 37°C, (*b*) 90°F, (*c*) −196°C?

22 (I) Absolute zero is what temperature on the Fahrenheit scale?

23 (I) How many molecules of water are there in 1 kg?

24 (I) The pressure in a helium gas cylinder is initially 30 atm. After blowing up many balloons, the pressure has decreased to 6 atm. What fraction of the original gas remains in the cylinder?

25 (II) Calculate the number of molecules/m^3 in a gas at STP.

26 (II) How many moles of water are then in 1.0 liter? How many molecules?

27 (II) Calculate the density of oxygen at STP using the ideal gas law.

28 (II) If an atom is measured to have a mass of 6.7×10^{-27} kg, what atom do you think it is?

29 (II) If 3.0 moles of helium gas is at 20°C and a gauge pressure of 0.90 atm, calculate (*a*) the volume of the helium gas under these conditions and (*b*) the temperature if the gas is compressed to half the volume at a gauge pressure of 3.10 atm.

30 (II) A house has a volume of 600 m^3. (*a*) What is the total mass of air inside the house at 0°C? (*b*) If the temperature rises to 25°C, how much air leaves the house?

31 (II) What is the pressure inside a 10-liter container holding 14 kg of argon gas at 10°C?

32 (II) A storage tank contains 30.2 kg of nitrogen at an absolute pressure of 2.45 atm. What will the pressure be if the nitrogen is replaced by an equal mass of CO_2?

33 (II) A tank contains 18.0 kg of oxygen at a gauge pressure of 6.10 atm. If the oxygen is replaced by helium, how many kg of the latter will be needed to produce a gauge pressure of 7.40 atm?

34 (II) The correct inflation of a tire at 20°C is 2.0 kg/cm^2. After driving several hours, the driver checks the tires. If the tire temperature is now 48°C, what should the pressure read?

35 (II) At 10°C and an absolute pressure of 1.88 atm, 50.0 liters of oxygen are compressed to 36.6 liters at the same time that temperature is raised to 80°C. What will the pressure be now?

36 (II) If a skin diver filled his lungs to full capacity of 5.5 liters when 10 m below the surface, to what volume would his lungs expand if he quickly rose to the surface? Is this advisable?

37 (III) An air bubble at the bottom of a lake 20 m deep has a volume of 1.0 cm^3. If the temperature at the bottom is 5°C and at the top it is 18°C, what is the volume of the bubble just before it reaches the surface?

38 (III) Compare the value for the density of water vapor at 100°C and 1 atm with the value predicted from the ideal gas law. Why would you expect a difference?

SECTION 7-6

39 (II) The partial pressure of CO_2 in the lungs is about 35 mm Hg, somewhat higher than outside in the air. What is the percentage of CO_2 in air in the lungs?

SECTION 7-7

40 (I) Calculate the rms speed of a seed of mass 1.0×10^{-2} gram suspended in water at room temperature (20°C).

41 (I) What is the average KE of an oxygen molecule at STP?

42 (I) Calculate the rms speed of helium atoms near the surface of the sun at a temperature of about 6000°K.

43 (I) Twelve molecules have the following speeds, given in arbitrary units: 5, 2, 2, 6, 0, 4, 1, 3, 5, 1, 7, 3. Calculate (*a*) the mean speed and (*b*) the rms speed.

44 (II) (*a*) Calculate the approximate rms speed of an amino acid whose molecular mass is 89 u in a living cell at 37°C. (*b*) What would be the average speed of a protein of molecular mass 30,000 u at 37°C?

45 (II) Derive an equation relating the average speeds of two kinds of molecules whose masses are M_1 and M_2 which are kept at the same temperature.

46 (II) A space vehicle returning from the moon enters the atmosphere at a speed of about 40,000 km/h. Molecules (assume nitrogen) striking the nose of the vehicle with this speed correspond to what temperature? (Because of this high temperature, the nose of a space vehicle requires a very special design; indeed, part of it does vaporize, and this is seen as a bright blaze upon reentry.)

47 (II) The lowest pressure attainable using the best vacuum pumps is around 10^{-10} N/m^2. At such a pressure, how many molecules are there per cm^3?

48 (II) What is the rms speed of nitrogen molecules contained in a 3.0 m^3 volume at 2.4 atm if the total amount of nitrogen is 1300 moles?

49 (II) By what factor will the rms speed of particular gas molecules increase if the temperature is increased from 100°C to 200°C?

50 (II) In outer space the density of matter is about one atom per cm^3, mainly hydrogen atoms, and the temperature is about 3.4 K. Calculate the average speed of these hydrogen atoms, and the pressure (in atmospheres).

51 (II) What is the total kinetic energy of all the atoms in 2.0 kg of helium at 25°C?

52 (II) Show that the pressure P in a gas can be written $P = \frac{1}{3}\rho v^2$ where ρ is the density of the gas and v is the rms speed of the molecules.

53 (III) Calculate (*a*) the rms speed of an oxygen molecule at 25°C, and (*b*) determine how many times per second it would move back and forth across a 10-m-long room, on the average, assuming it made very few collisions with other molecules.

54 (III) What is the average distance between oxygen molecules at STP?

55 (III) The two isotopes of uranium, ^{235}U and ^{238}U (the superscripts refer to their atomic mass), can be separated by a gas diffusion process by combining them with fluorine to make the gaseous compound UF_6. Calculate the ratio of the rms speeds of these molecules for the two isotopes.

For additional reading

Conant, J. B., ed., *Robert Boyle's Experiments in Pneumatics*, Harvard University Press, Cambridge, Mass., 1950.

Heller, H. C., L. I. Crawshaw, and H. T. Hammel, "The Thermostat of Vertebrate Animals," *Sci. American*, **239**:2 (Aug., 1978), p. 102.

Magie, W. F., *Source Book in Physics*, Harvard University Press, Cambridge, Mass., 1963. Excerpts from Boyle and Gay-Lussac on behavior of gases; from Bernoulli, Maxwell, and Joule on the kinetic theory of gases; and from Kelvin on the absolute temperature scale.

Nash, L. K., *Atomic-Molecular Theory*, Harvard University Press, Cambridge, Mass., 1950.

8 ▮ Applications of kinetic theory

The material in this chapter is all considered optional. We will discuss a number of important phenomena that are nicely explained by the kinetic theory. First we will cover certain aspects of liquid–gas changes: namely, evaporation, vapor pressure, humidity, and boiling. Then we will discuss some properties of real gases and how they differ from ideal gases. Finally we will discuss the processes of diffusion and osmosis which are so important in life processes and other situations.

*8-1 ▮ Vapor pressure and humidity

*Evaporation

If a glass of water is left out overnight, the water level will have dropped by morning. We say the water has evaporated, meaning that some of the water has changed to the vapor or gas phase.

This process of evaporation can be explained on the basis of kinetic theory. The molecules in a liquid move past one another with a variety of speeds that follow, approximately, the Maxwell distribution (Figure 7-12). There are strong attractive forces between these molecules, which is why they stay together in the liquid phase. A molecule in the upper regions of the liquid may, because of its speed, leave the liquid momentarily. But just as a rock thrown into the air returns to the earth, so the attractive forces of the other molecules pull the vagabond molecule back to the liquid surface—at least if its velocity is not too large. If the molecule has a high enough velocity, it will escape from the liquid entirely and become part of the gas phase. Only those molecules that have kinetic energy above a particular value can escape to the gas phase. We have already seen (in Section 7-8 and Figure 7-13) that kinetic theory predicts that the number of molecules with kinetic energy above a particular value (such as E_A) increases with temperature; this is in accord with the well-known observation that the evaporation rate is greater at higher temperatures.

Since it is the fastest molecules that escape from the surface, the average speed of those remaining is less. When the average speed is less, the absolute temperature is less (Equation 7-10). Thus, kinetic theory predicts that evaporation is a cooling process. Experiment bears this out: you have no doubt noticed, for example, that when you step out of a warm shower and the water on your body begins to evaporate you feel cold; and after working up a sweat on a hot day, you will feel cool from even a slight breeze.

*Vapor pressure

Air normally contains water vapor (water in the gas phase) and it comes mainly from evaporation. To look at this process using kinetic theory, consider a closed container that is partially filled with water (it could just as well be any other liquid), and from which the air has been removed (Figure 8-1). The fastest-moving molecules quickly evaporate into the space above. As they move about some of these molecules will strike the liquid surface and return to the liquid phase (this is called *condensation*). The number of molecules in the vapor increases for a time, until a point is reached where the number returning to the liquid equals the number leaving per unit time interval. Equilibrium then exists and the space is said to be *saturated.* The pressure of the vapor when it is saturated is called the **vapor pressure**.

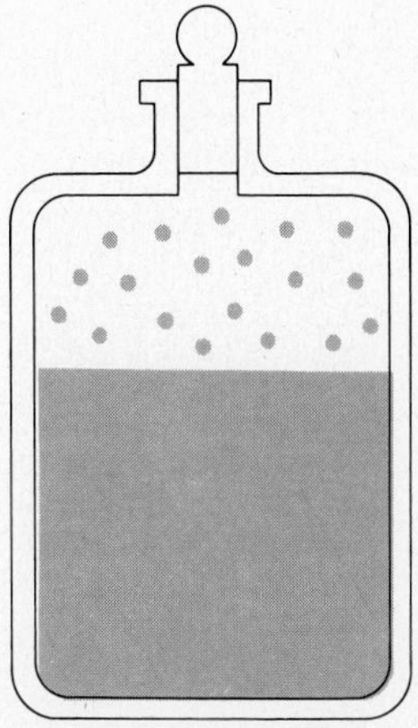

FIGURE 8-1
Vapor appears above a liquid in a closed container.

The vapor pressure does not depend on the volume of the container. For if the volume were reduced, the density of molecules in the vapor phase would be increased. More molecules would then strike the liquid surface per second. There would be a net flow of molecules back to the liquid phase until equilibrium was again reached and this would occur at the same value of vapor pressure.

The vapor pressure of any substance depends on the temperature. At higher temperatures, a greater number of molecules have sufficient kinetic energy to break from the liquid surface into the vapor phase. Hence equilibrium will be reached at a higher pressure. The vapor pressure of water at various temperatures is given in Table 8-1. Notice that even solids—in this case, ice—have a measurable vapor pressure.

TABLE 8-1
Vapor pressure of water

Temperature (°C)	Vapor pressure: mm Hg or torr	Vapor pressure: N/m²	Temperature (°C)	Vapor pressure: mm Hg or torr	Vapor pressure: N/m²
−50	0.030	4.0	40	55.3	7.37×10^3
−10	1.95	2.60×10^2	50	92.5	1.23×10^4
0	4.58	6.11×10^2	60	149	1.99×10^4
5	6.54	8.72×10^2	70	234	3.12×10^4
10	9.21	1.23×10^3	80	355	4.73×10^4
15	12.8	1.71×10^3	90	526	7.01×10^4
20	17.5	2.33×10^3	100	760	1.01×10^5
25	23.8	3.17×10^3	120	1489	1.99×10^5
30	31.8	4.24×10^3	150	3570	4.76×10^5

In everyday situations, evaporation from a liquid takes place into the air above it rather than into a vacuum. This does not materially alter the discussion above relating to Figure 8-1. Equilibrium will still be reached when there are sufficient molecules in the gas phase that the number reentering the liquid equals the number leaving. This number is not affected by the presence of air (or other gases) although collisions with air molecules may lengthen the time needed to reach equilibrium. Thus equilibrium occurs at the same value of the vapor pressure as if air weren't there.

Of course, if the container is large or is not closed, all the liquid may evaporate before saturation is reached. If the container is not sealed—as, for example a room in your house—it is not likely that the air will become saturated with water vapor, unless of course, it is raining outside.

*Humidity

When we refer to the weather as dry or humid, we are referring to the water-vapor content of the air. We use the concept of partial pressure (Section 7-6) in order to specify this quantitatively. The partial pressure of water in the air can be as low as zero, and can vary up to a maximum equal to the vapor pressure of water at the given temperature. Thus at 20°C the partial pressure cannot exceed 17.5 mm Hg (see Table 8-1). The **relative humidity** is defined as the ratio of the partial pressure to the vapor pressure at a given temperature. It is usually expressed as a percentage:

$$\text{Relative humidity} = \frac{\text{partial pressure of } H_2O}{\text{vapor pressure of } H_2O} \times 100\%.$$

Thus, when the humidity is close to 100 percent, the air holds nearly all the water vapor it can.

▌ EXAMPLE 8-1 On a particular hot day, the temperature is 30°C and the partial pressure of water vapor in the air is 21.0 mm Hg. What is the relative humidity?

SOLUTION From Table 8-1, the vapor pressure of water at 30°C is 31.8 mm Hg. Hence the relative humidity is

$$\frac{21.0 \text{ torr}}{31.8 \text{ torr}} \times 100\% = 66\%.$$ ▌▌

Humans are sensitive to humidity. A relative humidity of 40 to 50 percent is generally optimum for both health and comfort. High humidity, particularly on a hot day, reduces the evaporation of moisture from the skin which is one of the body's vital mechanisms for regulating body temperature. Very low humidity, on the other hand, can have a drying effect on the skin and mucous membranes.

Proper humidity must be maintained to prevent damage to paintings, tape recordings, and a wide variety of sensitive objects. Thus the design of heating and air-conditioning systems for buildings must take into account not only heating and cooling but the control of relative humidity as well.

Air is saturated with water vapor when the partial pressure of water in the air is equal to the vapor pressure at that temperature. If the partial pressure of water exceeds the vapor pressure, the air is said to be supersaturated. This situation can occur when a temperature decrease occurs. For example, suppose the temperature is 30°C and the partial pressure of water is 21 mm Hg, which is a humidity of 66 percent as we saw in the example above. Suppose now, the temperature falls to, say, 20°C, such as might happen at nightfall. From Table 8-1, we see that the vapor pressure of water at 20°C is 17.5 mm Hg. Hence the relative humidity would be greater than 100 percent and the supersaturated air cannot hold this much water. The excess water condenses and appears as dew; this process can also be responsible for the formation of fog, clouds, and rain.

When air containing a given amount of water is cooled, a temperature is reached where the partial pressure of water equals the vapor pressure. This temperature is the *dew point.* Measurement of dew point is the most accurate means of determining the relative humidity. Normally, a polished metal surface in contact with air is gradually cooled down. The temperature at which moisture begins to appear on the surface is the dew point, and the partial pressure of water can then be obtained from vapor-pressure tables. For example, suppose on a given day the temperature is 20°C and a meteorologist determines that the dew point is 5°C. This means that the partial pressure of water (Table 8-1) in the original air was 6.54 mm Hg whereas its vapor pressure was 17.5 mm Hg. Hence the relative humidity was $6.54/17.5 = 37$ percent.

A more convenient but less accurate method for measuring relative humidity is the so-called wet bulb–dry bulb technique which makes use of two thermometers. One thermometer bulb is fitted with a snug cloth jacket that is soaking wet. The apparatus is usually swung in the air, and the lower the humidity, the more evaporation takes place from the wet bulb; this causes its temperature reading to be less. A comparison of the temperature readings on the wet bulb thermometer and the dry (ordinary) thermometer can then be compared to special tables that have been compiled to obtain the relative humidity.

*Boiling

The vapor pressure of a liquid increases with temperature. When the temperature is raised to the point where the vapor pressure equals the external pressure, boiling occurs. Let us examine this process. As the boiling point is approached, tiny bubbles tend to form in the liquid, which indicates a change from the liquid to the gas phase. However, if the vapor pressure inside the bubbles is less than the external pressure, the bubbles immediately are crushed. As the temperature is increased, the vapor pressure inside a bubble eventually becomes equal to or exceeds the external air pressure. The bubble will not then collapse (Figure 8-2), but will increase in size and rise to the surface. Boiling has then begun. A liquid boils when its vapor pressure equals the external pressure. This occurs for water under 1 atm of pressure at 100°C as can be seen from Table 8-1.

The boiling point of a liquid clearly depends on the external pressure. At high elevations, the boiling point of water is somewhat less than at sea level since the air pressure is less. For example, on the summit of Mt.

FIGURE 8-2
Boiling of water begins when bubbles of steam form at the bottom of a heated pot and then rise to the surface.

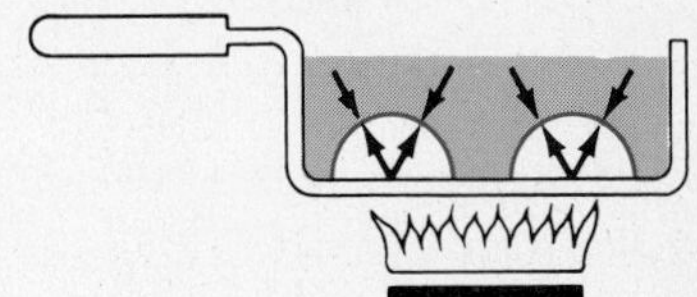

Everest (8,850 m) the air pressure is about one-third of what it is at sea level, and from Table 8-1 we can see that water will boil at about 70°C. Cooking food by boiling takes longer at high elevations, since the temperature is less. Pressure cookers, however, build up a pressure as high as 2 atm and thus reduce cooking time.

*8-2 ▌ Real gases and changes of phase

The ideal gas law, as mentioned before, is an accurate description of the behavior of a gas as long as the pressure is not too high, and as long as the temperature is far from the liquefaction point. But what happens to real gases when these two criteria are not satisfied?

To answer this question, let us look at a graph of pressure plotted against volume for a given amount of gas. On such a "*PV* diagram," Figure 8-3, each point on a curve represents an equilibrium state of the given substance. The various curves show how the pressure varies as the volume is changed at constant temperature for several different values of the temperature. The dashed curve A' represents the behavior of a gas as predicted by the ideal gas law; that is, $PV = \text{constant}$. The solid curve A represents the behavior of a real gas at the same temperature. Notice that, at high pressure, the volume in a real gas is less than that predicted by the ideal gas law. At lower temperatures, curves B and C in Figure 8-3, the behavior deviates even more (such as B') from the curves predicted by the ideal gas law, and the deviation is greater the closer the gas is to liquefying.

To explain this, we note that at higher pressures we expect the molecules to be closer together; and, particularly at lower temperatures, the potential energy associated with the attractive forces between the molecules that we ignored before is no longer negligible compared to the

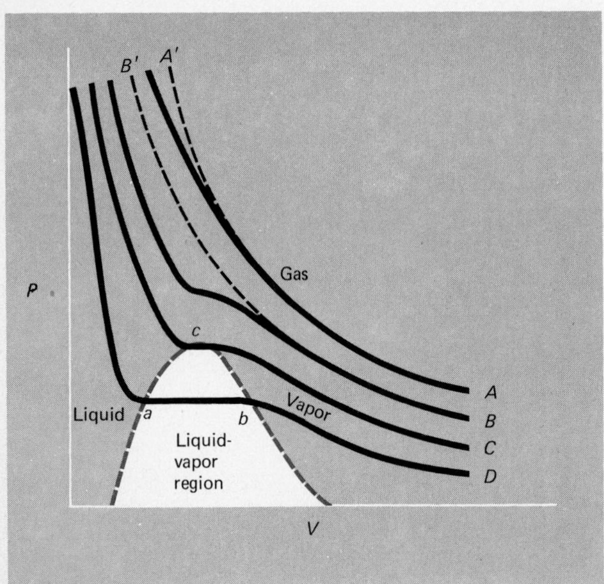

FIGURE 8-3
A *PV* diagram for a real substance.

TABLE 8-2

Critical temperatures

Substance	Critical temperature (°C)
Iodine	**512**
Water	**374**
Carbon dioxide	**31**
Methane	**−82**
Oxygen	**−119**
Nitrogen	**−147**
Hydrogen	**−240 (33 K)**
Helium	**−267 (5 K)**

now reduced kinetic energy of the molecules. These forces tend to pull the molecules closer together so that at a given pressure, the volume is less than expected from the ideal gas law. At still lower temperatures, these forces will cause liquefaction and the molecules become very close together.

Curve *D* represents the situation when liquefaction occurs. At low pressure on curve *D* (on the right in the graph), the substance is a gas and occupies a large volume. As the pressure is increased, the volume decreases until point *b* is reached. Beyond *b*, the volume decreases with no change in pressure; the substance is gradually changing from the gas to the liquid phase. At point *a*, all of the substance has changed to liquid. Further increase in pressure reduces the volume only slightly—liquids are nearly incompressible—so the curve is very steep as shown. The area within the tongue-shaped dashed line represents the region where the gas and liquid phases coexist together in equilibrium.

Curve *C* in Figure 8-3 represents the behavior of the substance at its **critical temperature**; and the point *c* (the one point where this curve is horizontal) is called the *critical point.* At temperatures less than the critical temperature (and this is the definition of the term), a gas will change to the liquid phase if sufficient pressure is applied. But at temperatures above the critical temperature, no amount of pressure can cause a gas to change phase and become a liquid. What happens instead is that the gas becomes denser and denser as the pressure is increased and gradually it acquires properties resembling a liquid, but no liquid surface forms. Critical temperature for various gases are given in Table 8-2. Scientists tried for many years to liquefy oxygen without success. It was not until the discovery of the behavior of substances associated with the critical point that it was realized that oxygen can only be liquefied if first cooled below its critical temperature of −119°C.

Often a distinction is made between the terms "gas" and "vapor": a substance below its critical temperature in the gaseous state is called a *vapor*; when above the critical temperature, it is called a *gas*; this is indicated in Figure 8-3.

The behavior of a substance can be diagrammed not only on a *PV* diagram but also on a *PT* diagram. A *PT* diagram, often called a *phase diagram*, is particularly convenient for comparing the different phases of a

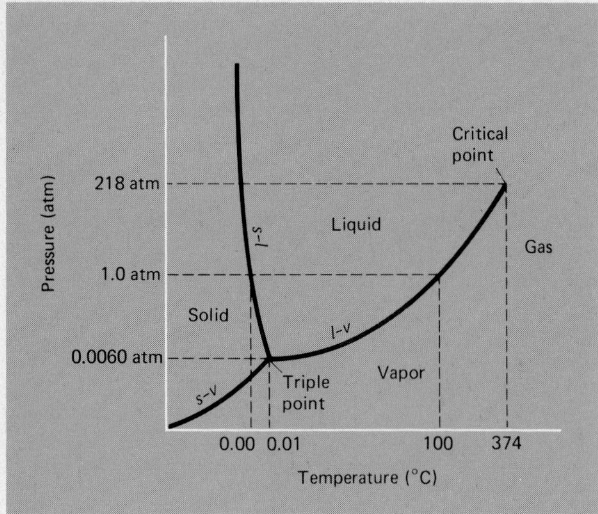

FIGURE 8-4
Phase diagram for water (note that scales are not linear).

substance. Figure 8-4 is the phase diagram for water. The curve labeled *l-v* represents those points where the liquid and vapor phases are in equilibrium— it is thus a graph of the boiling point versus pressure. Note that the curve correctly shows that at a pressure of 1 atm the boiling point is 100°C and that the boiling point is lowered for a decreased pressure. The curve *s-l* represents points where solid and liquid exist in equilibrium and thus is a graph of the freezing point versus pressure. At 1 atm, the freezing point of water is of course 0°C as shown. Notice also in Figure 8-4 that, at a pressure of 1 atm, the substance is in the liquid phase if the temperature is between 0°C and 100°C but is in the solid or vapor phase if the temperature is below 0°C or above 100°C. The curve labeled *s-v* represents points where solid and vapor are in equilibrium and is thus the *sublimation point* versus pressure curve. *Sublimation* refers to the process whereby at low pressures (in the case of water, at pressures less than 4.5 mm Hg) a solid changes directly into the vapor phase without passing through the liquid phase. A common example of sublimation is carbon dioxide—in the solid phase it is often called Dry Ice—which sublimates even at atmospheric pressure.

The intersection of the three curves is called the **triple point**. The triple point represents a unique temperature and pressure and it is only at

TABLE 8-3

Triple-point data

Substance	Temperature (°K)	Pressure	
		N/m^2	atm
Water	**273.16 (0.01°C)**	**6.10×10^2**	**6.03×10^{-3}**
Carbon dioxide	**216.6**	**5.16×10^5**	**5.10**
Ammonia	**195.40**	**6.06×10^3**	**6.00×10^{-2}**
Nitrogen	**63.2**	**1.25×10^4**	**1.24×10^{-1}**
Oxygen	**54.4**	**1.52×10^2**	**1.50×10^{-3}**
Hydrogen	**13.8**	**7.03×10^3**	**6.95×10^{-2}**

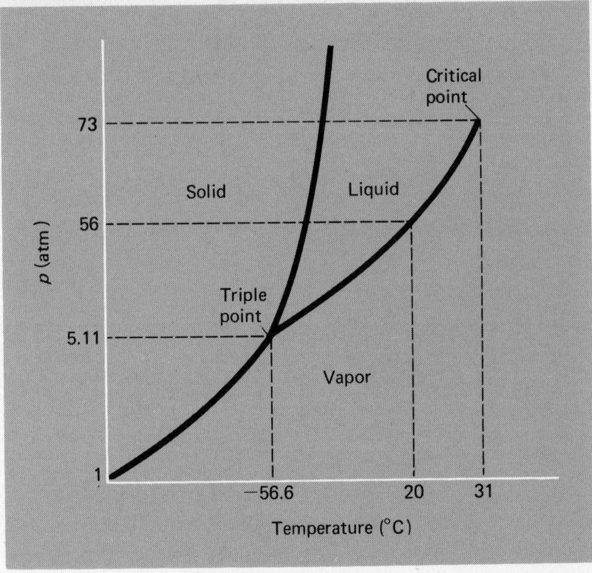

FIGURE 8-5
Phase diagram for carbon dioxide.

this point that the three phases can exist together in equilibrium (see Table 8-3). Because the triple point corresponds to a unique value of temperature and pressure, it is precisely reproducible and is often used as a point of reference. For example, the standard of temperature is often specified as exactly 273.16 K at the triple point of water, rather than 273.15 K at the freezing point of water at 1 atm.

Because the *l-v* curve in Figure 8-4 is the boiling point curve, it also represents the vapor pressure of the substance at a given temperature (remember boiling takes place when the external pressure equals the vapor pressure). Below the triple point, the *s-v* curve represents the vapor pressure as a function of temperature. Thus the vapor pressure at any temperature can be read off an accurate phase diagram.

Notice that the *s-l* curve for water slopes upward to the left. This is true only of substances that *expand* upon freezing; for at a higher pressure, a lower temperature is needed to cause the liquid to freeze. More commonly, substances contract upon freezing and the *s-l* curve slopes upward to the right, as shown for CO_2 in Figure 8-5.

FIGURE 8-6
Superfluid liquid helium at 1.5 K. A thin film has crept up and over the walls of the bowl and collected into a drop hanging from the bottom. (Courtesy of A. Leitner.)

The phase transitions we have been discussing are the common ones. Some substances, however, can exist in several forms in the solid phase. A transition from one of these phases to another occurs at a particular temperature and pressure, just like ordinary phase changes. For example, ice has been observed in at least eight different modifications at very high pressure. Ordinary helium is a unique substance in that it has two distinct liquid phases, called helium I and II. They exist only at temperatures within a few degrees of absolute zero. Helium II exhibits very unusual properties referred to as *superfluidity*. It has extremely low viscosity and exhibits strange properties, such as actually climbing up the sides of an open container and over the top (Figure 8-6).

Some materials, like glass, certain resins, and sulfur, are called *amorphous solids*. They do not have a crystal structure like most solids and they do not have a definite melting point. When heated, they gradually soften and there is no phase transition. Amorphous solids are thus often considered to be extremely viscous liquids rather than true solids.

*8-3 ▮ Diffusion

If you carefully place a drop of food coloring in a glass of water, you will find that the color spreads throughout the water. The process may take several hours (assuming you don't shake the glass), but eventually the color will become uniform. This mixing occurs because of the random movement of the molecules, and is called **diffusion**. Diffusion occurs in gases too, and much more quickly. When a perfume bottle is opened, the scent can be smelled across the room only a few moments later even though the room air is still. And if you burn something on the stove, the smell—as well as visible smoke—diffuses throughout the house. In each case the diffusing substance moves from a region where its concentration is high to one where its concentration is low.

Diffusion can be readily understood on the basis of kinetic theory and the random motion of molecules. Consider a tube of cross-sectional area A containing molecules in a higher concentration on the left than on the right, Figure 8-7. We assume the molecules are in random motion. Yet there will be a net flow of molecules to the right. To see why this is true, let us consider the small section of tube of length d as shown. Molecules from both regions 1 and 2 cross into this central section as a result of their random motion. The more molecules there are in a region, the more will strike a given area or cross a boundary. Since there is a greater concentration of molecules in region 1 than in region 2, more molecules cross into the central section from region 1 than from region 2. There is, then, a net flow of molecules from left to right, from high concentration toward low concentration. The flow stops only when the concentrations become equal.

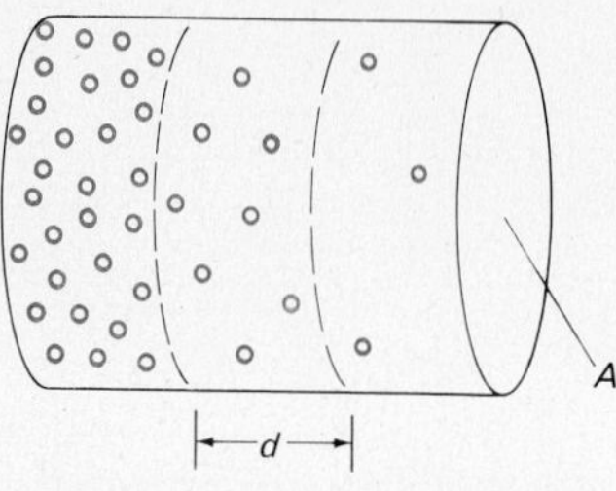

FIGURE 8-7
Diffusion occurs from a region of high concentration to one of lower concentration.

*Fick's law of diffusion

You might expect that the greater the difference in concentration, the greater the flow rate. This is indeed the case. In 1855, the physiologist Adolf Fick (1829–1901) determined experimentally that the rate of diffusion (J) is directly proportional to the change in concentration per unit distance $(C_1 - C_2)/d$ (which is called the *concentration gradient*), and to the cross-sectional area A (see Figure 8-7):

$$J = DA\frac{C_1 - C_2}{d}. \tag{8-1}$$

D is a constant of proportionality called the *diffusion constant*. Equation 8-1 is known as the *diffusion equation*, or *Fick's law*. If the concentrations are given in mol/m^3, then J is the number of moles passing a given point per second; if the concentrations are given in kg/m^3, then J is the mass movement per second (kg/s). The length d, of course, is given in meters.

Equation 8-1 applies not only to the simple situation of a gas diffusing as shown in Figure 8-7, which is called *self-diffusion*. The same considerations apply for a gas diffusing in a second gas (perfume vapor in air), or for a substance dissolved in a liquid; these are the more common situations. The rate of diffusion will be slower, particularly in liquids, because there will be collisions with the other molecules. Thus, the diffusion

TABLE 8-4

Diffusion coefficients

Diffusing molecules†	Diffusion coefficient (m^2/s)
Glycine (an amino acid)	95×10^{-11}
Cytochrome c	10.1×10^{-11}
Hemoglobin	6.9×10^{-11}
DNA (molecular mass 6×10^6)	0.13×10^{-11}
Tobacco mosaic virus	0.3×10^{-11}
Oxygen, through tissue	about 1×10^{-11}
Simple gases (O_2, N_2, CO_2) through gas at 1 atm	about 1×10^{-5}

† Through water at 20°C, for the first five entries.

constant D will depend on the properties of the substances involved, and also on the temperature and the external pressure. The values of D for a variety of substances diffusing through water at 20°C are given in Table 8-4.

For some purposes it is convenient to write Fick's law, Equation 8-1, in terms of partial pressures rather than concentrations. We saw earlier (Section 7-6) that the partial pressure of each component in a mixture of gases is directly related to its concentration. This followed from the ideal gas law. If we let P_i be the partial pressure of a particular component, and C_i its concentration in moles/m^3 ($C_i = n_i/V$), we can write

$$P_i = \frac{n_i RT}{V} = C_i RT.$$

Thus $C_i = P_i/RT$ and Fick's law can be written as

$$J_i = \frac{DA}{RT}\left(\frac{\Delta P_i}{d}\right), \tag{8-2}$$

where ΔP_i means the change in partial pressure of substance i over the distance d. Thus a gas diffuses from a region where its partial pressure is high to a region where its pressure is low, irrespective of the pressure due to the other components.

*Diffusion processes in living organisms

Diffusion is extremely important for living organisms. Within cells, for example, molecules produced in certain chemical reactions must diffuse through water to other areas where they take part in other reactions.

Gas diffusion is important too. Plants require carbon dioxide for photosynthesis. The CO_2 diffuses into leaves from the outside through tiny openings called stomata. As CO_2 is utilized by the cells, its concentration drops below that in the air outside, and more diffuses inward as described by Fick's law. Water vapor and oxygen produced by the cells diffuses outward into the air.

Animals also exchange oxygen and CO_2 with the environment. Oxygen is required for energy producing reactions and must diffuse into cells; CO_2 is produced as an end product of many metabolic reactions and must diffuse out of cells. To get an idea of the time t required for diffusion, we can use Equation 8-1, setting $\Delta C = C_1 - C_2$ and $J = n/t$ where n is the number of moles; then

$$\frac{n}{t} = DA \frac{\Delta C}{d}.$$

Solving for t we find

$$t = \frac{nd}{DA\,\Delta C} = \frac{nd^2}{DV\,\Delta C}$$

$$= \frac{1}{D}\frac{\bar{C}}{\Delta C} d^2, \qquad (8\text{-}3)$$

where the volume $V = Ad$, and n/V is the average concentration $\bar{C}$ of the gas as it diffuses; that is, $\bar{C} = (C_1 + C_2)/2$.

▮ EXAMPLE 8-2 Oxygen diffuses from the surface of insects to the interior through tiny tubes called tracheae. An average trachea is about 2 mm long and has cross-sectional area of 2×10^{-9} m^2. Assuming the concentration of oxygen inside is half what it is outside in the atmosphere, calculate (a) the diffusion rate J and (b) the average time for a molecule to diffuse in. Assume the diffusion constant is 1×10^{-5} m^2/s (Table 8-4).

SOLUTION (a) Since one mole of a gas has a volume of 22.4×10^{-3} m^3 at STP (Example 7-5) and oxygen makes up 21 percent of air, the concentration of oxygen outside, C_1, is

$$C_1 = \left(\frac{1\text{ mol}}{22.4 \times 10^{-3}\text{ m}^3}\right)(0.21) = 9.4\text{ mol/m}^3;$$

C_2 is half this value so

$$J = DA\frac{C_1 - C_2}{d}$$

$$= (1 \times 10^{-5}\text{ m}^2/\text{s})(2 \times 10^{-9}\text{ m}^2)\left(\frac{4.7\text{ mol/m}^3}{2 \times 10^{-3}\text{ m}}\right)$$

$$= 0.5 \times 10^{-10}\text{ mol/s}.$$

(b) We use Equation 8-3 and note that

$$\frac{\bar{C}}{\Delta C} = \frac{\frac{3}{4}C_1}{\frac{1}{2}C_1} = \frac{3}{2}.$$

Then

$$t = \left(\frac{1}{1 \times 10^{-5}\text{ m}^2/\text{s}}\right)\left(\frac{3}{2}\right)(2 \times 10^{-3}\text{ m})^2$$

$$= 0.6\text{ s}.$$

▮▮

Carbon dioxide produced within the insect diffuses outward via the tracheae, and similar considerations apply. From Equation 8-3 we can see that the diffusion time increases with the square of the distance, independent of the area involved. Thus for an insect with tracheae 10

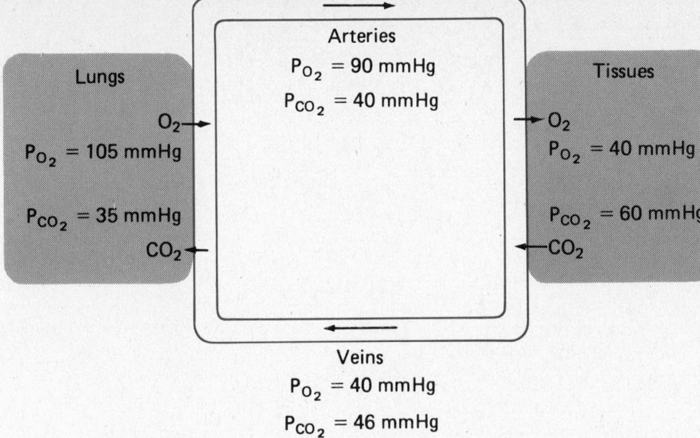

FIGURE 8-8
Schematic diagram of human circulation system. Partial pressures of O_2 and CO_2 in the blood, lungs, and tissue are indicated.

times longer, the diffusion time would be 100 times longer! This would seriously limit the rate at which oxygen could be used and is one reason why few large insects have evolved.

Because of the slowness of diffusion over longer distances, all but the smallest animals have developed complex respiratory and circulatory systems. Humans at rest, for example, receive only about 2 percent of their oxygen by diffusion through the skin. The rest is taken in by the lungs and distributed to the cells throughout the body by the blood. The blood also carries carbon dioxide produced by the cells back to the lungs where it diffuses outward. We can examine this in the light of Fick's law as it is written in terms of partial pressures, Equation 8-2. Figure 8-8 is a schematic representation of the human circulation system. Blood leaving the lungs and traveling in the arteries contains oxygen at a partial pressure of about 90 mm Hg and carbon dioxide at a partial pressure of 40 mm Hg. Since the partial pressures of O_2 and CO_2 in tissue are typically 40 mm Hg and 60 mm Hg respectively, when the blood passes through the capillaries that bathe the tissues of the body, oxygen diffuses from the blood into the tissue and CO_2 diffuses from the tissue into the blood. The blood flowing back to the heart via the veins contains oxygen at a reduced partial pressure and CO_2 at an increased partial pressure (Figure 8-8). When the blood returns to the lungs, it is recharged with oxygen since the partial pressure of O_2 in the lungs is quite high, typically 105 mm Hg.† At the same time, CO_2 diffuses out from the blood into the lungs where the partial pressure is only about 35 mm Hg.

Because O_2 and CO_2 diffuse into and out of the blood through a liquid and across membranes (see next section), the rate is much less than for diffusion through gases. However, the distance d over which diffusion takes place is very small, about 10^{-8} m. This is possible because of the close spacing of capillaries in tissue as well as in the lungs. The interior surface of the lungs is very large due to the convoluted structure of the lungs. The human windpipe, or trachea, branches many times into smaller and smaller tubes which end in tiny sacs called alveoli. Gas exchange with blood capillaries takes place in these alveoli. The total surface area

† You might expect that since air contains about 20 percent oxygen that the partial pressure of O_2 would be (0.20)(760 mm Hg) = 150 mm Hg in the lungs. It is less than this on the average because the air does not get completely replenished with each breath.

available for gas exchange in the lungs is 60 to 100 m^2, almost 50 times the surface area of the skin! This large area, and the smallness of d, means that the diffusion rate is quite large.

*Diffusion through membranes; permeability

Diffusion often takes place through a membrane or a porous wall rather than simply through a fluid as discussed above. This situation is of great importance in the life sciences since cells and cellular organelles are surrounded by membranes. There are applications in other areas as well.

We can think of a membrane as being porous, and the size of the pores obviously has a great effect on diffusion through the membrane (see Figure 8-9). Molecules which are larger than the pores will generally be prevented from passing through (although some molecules may pass through by dissolving in the membrane itself). If the molecules are smaller than the pores, diffusion can occur. The effect of the membrane is to restrict the area through which the molecules can diffuse and thus reduce the diffusion rate. Fick's law can be used with membranes where d is the thickness of the membrane and the concentrations (or partial pressures) are just those on the two sides of the membrane. The value of the diffusion constant, D, obviously depends strongly on the structure of the membrane and on the size of the diffusing molecules.

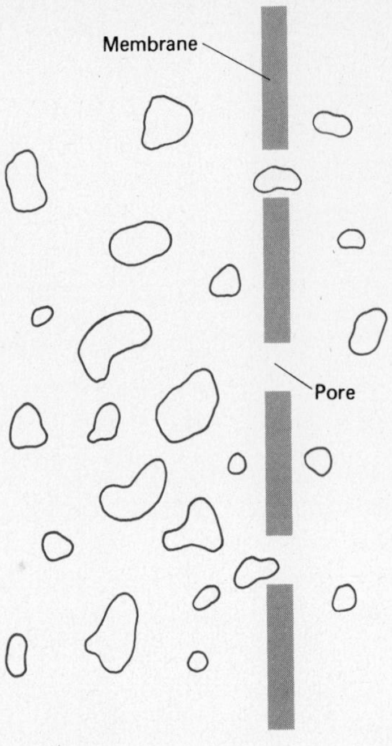

FIGURE 8-9
Schematic diagram of a porous membrane.

When discussing diffusion through a membrane (or the similar but more subtle case of diffusion across a boundary between two different fluids), the thickness d is not always known accurately. It is then convenient to use the *permeability*, P, which is defined in terms of the diffusion constant D as:

$$P = \frac{D}{d}.$$

Then the diffusion equation (Fick's law) becomes:

$$J = PA(C_1 - C_2).$$

The permeability P is often more readily measurable than are D and d separately, and thus is a more useful idea from an experimental point of view. It directly relates the rate of diffusion J to the concentration difference.

*Active transport

The concentration of certain substances is often far different inside a cell than outside. This cannot always be explained by assuming the membrane to be totally impermeable to those molecules. In these cases a mechanism other than diffusion is operating. Such mechanisms are referred to as *active transport,* in comparison to diffusion which is then referred to as "passive transport." Active transport refers to any process whereby molecules are made to move (as if "pumped") from a region where their concentration is low to where it is high. Electrical forces play a role in these processes and we will look at some aspects of this situation in a later chapter.

*8-4 ▮ Osmosis

If a membrane allows the passage of certain (small) molecules, but not other (larger) molecules, it is called a *semipermeable membrane.* For example, many animal membranes, and even some kinds of paper, are permeable to water but impermeable to, say, sugar. When such a membrane separates a solution of pure water from one containing a high concentration of sugar, an interesting thing happens as shown in Figure 8-10. Water flows into the sugar solution. This process is known as **osmosis**.

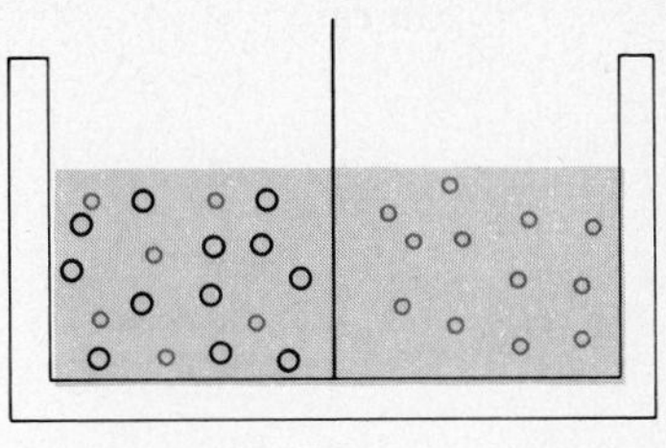

(a)

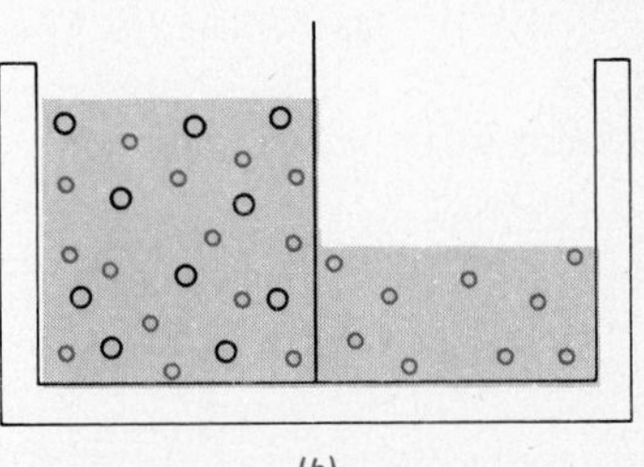

(b)

FIGURE 8-10
Sugar (black circles) solution separated from pure water by a semipermeable membrane (*a*) initially and (*b*) after some time has passed and water has diffused through the membrane into the sugar solution.

*Osmotic pressure

Osmosis can be understood on the basis of kinetic theory using the concept of partial pressure. In Figure 8-10*a*, the total pressure on either side of the membrane is the same. On the left side of the membrane, the sugar exerts a partial pressure but on the right it does not (and if it weren't for the impermeability of the membrane to the sugar, it would diffuse to the right). Because the sugar contributes to the total pressure on the left, the partial pressure of the water on the left must be less than it is on the right which is pure water. Because of this difference in partial pressure, water flows from the right compartment to the left, resulting in the situation shown in Figure 8-10*b*. The extra height of the solution on the left is said to be due to **osmotic pressure**. Specifically, the osmotic pressure of the solution is defined as the pressure that would have to be exerted on the solution to prevent osmosis from taking place. This is equal to ρgh where ρ is the density of the fluid on the left and h is its height above the fluid on the right.

Although osmosis differs from diffusion in that a semipermeable membrane is involved, nonetheless the two processes resemble one another. The sugar is unable to pass through the membrane, but the water can pass through easily. That is, the water moves from where its own concentration (or partial pressure) is high to where it is lower.

Osmotic pressure is generally denoted by Π (upper-case Greek letter pi). Careful experiments show that for reasonably dilute solutions, the osmotic pressure is given by

$$\Pi = \frac{nRT}{V}, \tag{8-4a}$$

where n is the number of moles of solute (the substance dissolved) in the volume V, R is the gas constant, and T is the absolute temperature. In other words, the solute behaves like an ideal gas. Since n/V is the concentration C, Equation 8-4*a* can be written

$$\Pi = CRT. \tag{8-4b}$$

If a solution contains more than one substance, the concentration c (or n in Equation 8-4*a*) must include all substances that do not permeate the membrane. Furthermore, if both sides of the membrane contain nonpermeating molecules, then C refers to *the difference in concentration* between the two sides and Π is the difference in osmotic pressure.

EXAMPLE 8-3 100 g of glucose ($C_6H_{12}O_6$) are dissolved in a liter of water at 20°C. What is the osmotic pressure of this solution relative to pure water?

SOLUTION The molecular mass of glucose is $(6)(12)+(12)(1)+(6)(16)=180$. Hence $n=(0.100\,\text{kg})/(0.180\,\text{kg})=0.55$ mol, and using Equation 8-4*a* we find:

$$\Pi=\frac{(0.55\text{ mol})(8.3\text{ J/mol}\cdot\text{K})(293\text{ K})}{(1.00\times10^{-3}\text{ m}^3)}$$

$$=1.3\times10^6\text{ N/m}^2$$

which is about 13 atm!

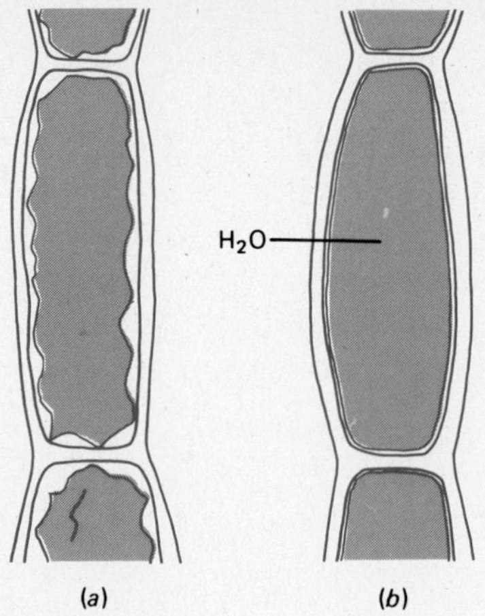

FIGURE 8-11
A plant cell (*a*) lacking water and (*b*) with sufficient water so that it becomes turgid.

*Osmosis in life processes

Individual biological cells can exchange water with the surrounding fluid, depending on the relative concentrations of solute in the fluid and in the cell. If the concentration of solutes is greater in the fluid than in the cell (in which case the fluid is said to be *hypertonic*), water flows out of the cell until the concentrations become equal. If, on the other hand, the surrounding fluid has a lower concentration of solutes than the interior of the cell (*hypotonic* fluid), water will flow into the cell. In this case pressure builds up inside the cell. For example, red blood cells placed in distilled water quickly swell and then burst. When cells are to be examined, they are generally placed in an *isotonic* solution—that is, one whose concentration of solutes is the same as the interior of the cell.

In plants, the pressure buildup due to inflow of water by osmosis is called *turgor pressure.* Because plant cells have fairly rigid cell walls, the cells do not swell excessively (see Figure 8-11). Instead they become firm, or turgid, and this helps support the plant. However, when water is lacking, the cells lose their firmness and the plant looks wilted. Similarly osmotic pressure is responsible for the opening and closing of blossoms. The openings of the stomata on the underside of a leaf are likewise controlled by osmotic pressure, and this is shown in Figure 8-12. To see this same process in action, observe a dried prune put in water.

Osmosis plays an important role in many processes in the human body. One example is the removal of small "waste" molecules from the blood by the kidneys. The membranes involved allow the passage of these small solute molecules, but do not allow the passage of the large molecules the cell must retain; hence it is only the latter that give rise to the osmotic pressure. This process whereby only smaller molecules can pass through a membrane whereas larger ones do not is called *dialysis* and is often used in the laboratory to separate large solute molecules from small ones.

It should be recognized that a true semipermeable membrane is an ideal. Real membranes do not completely prevent large molecules from passing through. However, the passage of large molecules is at a much smaller rate than the passage of small molecules such as water. Real membranes are thus often referred to as *differentially permeable membranes.*

FIGURE 8-12
The guard cells at stoma openings possess two walls of varying thickness. When water is available, it passes into the guard cells by osmosis, and the uneven expansion of the walls causes the opening to widen.

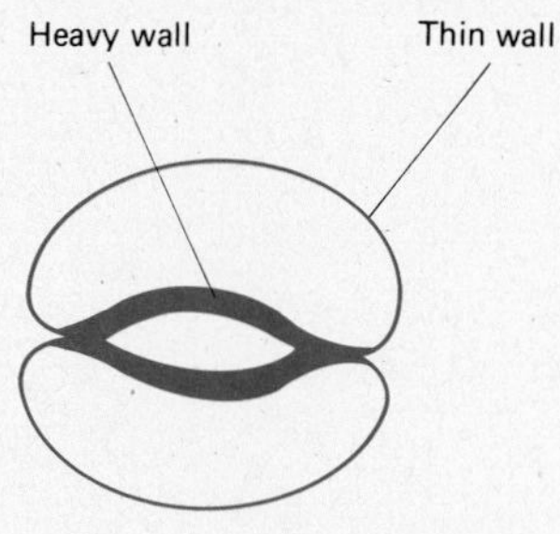

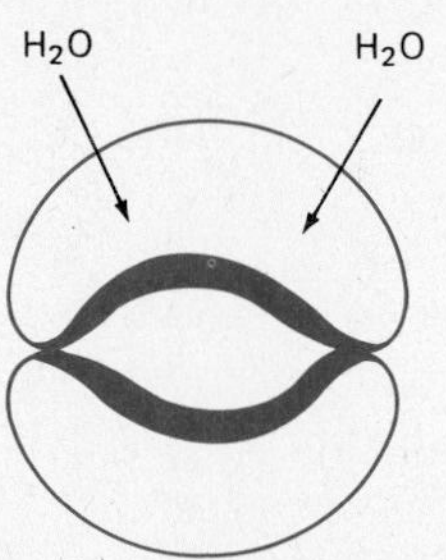

*Reverse osmosis

One method of desalinating seawater is by reverse osmosis. In this process, an external pressure is applied to the salt water, Figure 8-13. If

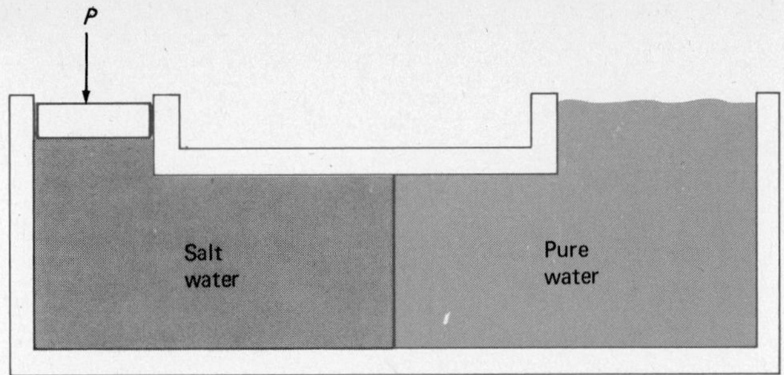

FIGURE 8-13
Reverse osmosis used to desalinate seawater.

the external pressure is greater than the osmotic pressure, water will be forced across the membrane out of the concentrated solution. The total concentration of solute in seawater is $1.08 \times 10^3\ mol/m^3$. This corresponds (see Equation 8-4) to an osmotic pressure of 26 atm, and a pressure greater than this would be required to obtain pure water. Unfortunately, a practical difficulty exists in that an efficient long-lasting membrane has not yet been found.

* Transport processes

It is important to distinguish the processes of diffusion and osmosis from the transport of fluid as discussed in the previous chapter (see Poiseuille's law). Diffusion and osmosis take place because of a difference in concentration (or partial pressure). It is essentially a random process. Ordinary flow of a fluid as described by Poiseuille's law is due to a difference in hydrostatic pressure and hence is often called *hydrostatic flow*; it is not a random process, for all the molecules have a velocity (added onto their random thermal velocity) in the direction of the flow. It is worth noting that many natural processes involve both kinds of flow.

Nonetheless, these different kinds of flow have one thing in common. They are each due to a change in some quantity with distance—that is, a gradient. Diffusion and osmosis depend on a concentration gradient; hydrostatic flow is due to a hydrostatic pressure gradient. In later chapters we will see other types of flow. The flow of heat by conduction will be seen to be due to a temperature gradient. And the flow of electric current is a result of a gradient in the electric potential.

Because of the similarities of these processes, they are often referred to collectively as *transport processes.*

▌Summary

Evaporation of a liquid is the result of the fastest moving molecules escaping from the surface. Because the average molecular velocity is less after the fastest molecules escape, the temperature decreases when evaporation takes place. *Vapor pressure* refers to the pressure of the vapor above a liquid when the two phases are in equilibrium. The vapor pressure of a substance (such as water) depends strongly on temperature and is equal to atmospheric pressure at the boiling point. *Relative humidity* is the ratio of the partial pressure of water vapor in the air in a given area to the vapor pressure at that temperature; it is usually expressed as a percentage.

The behavior of real gases at high pressure, and when near their liquefaction point, deviates from the ideal gas law. These deviations are due to the attractive forces between molecules which become more important in these two situations. Below the *critical temperature*, a gas can change to a liquid if sufficient pressure is applied; but if the temperature is higher than the critical temperature, no amount of pressure will cause a liquid surface to form. The *triple point* of a substance is that unique temperature and pressure at which all three phases—solid, liquid and gas—can coexist in equilibrium. Because of its precise reproducibility, the triple point of water is often taken as a standard reference point.

The movement of molecules from regions of high concentration to regions of low concentration because of their random motions is called *diffusion*. Fick's law describes the rate of diffusion (J) as being proportional to the concentration gradient ($[C_1-C_2]/d$) and to the cross-sectional area $A: J=DA(C_1-C_2)/d$, where D is the diffusion constant. Fick's law, also called the diffusion equation, can be written in terms of the difference in partial pressure of the diffusing substance between the two regions. The rate of diffusion through a permeable membrane depends on the permeability, P, of the membrane: $J=PA(C_1-C_2)$. Active transport across a permeable membrane allows the concentration of a given substance to be higher on one side than the other.

A semipermeable membrane permits passage of certain (usually small) molecules but not others. When such a membrane allows water molecules, but not solute molecules, to pass through, the water molecules will pass through in an attempt to equalize the solute concentrations. This process is called *osmosis* and proceeds until the difference in pressure on the two sides (called *osmotic pressure*) is equal to the difference in concentration times RT. Osmosis plays an important role in life processes.

▮ Questions

*1 We saw that the vapor pressure of a liquid (say, water) does not depend on the external pressure. Yet the temperature of boiling does depend on the external pressure. Is there a contradiction? Explain.

*2 Alcohol evaporates more quickly than water at room temperature. What can you infer about the molecular properties of one relative to the other?

*3 Explain why a hot humid day is far more uncomfortable than a hot dry day at the same temperature.

*4 Is it possible to boil water at room temperature (20°C) without heating it? Explain.

*5 What exactly does it mean when we say that oxygen boils at −183°C?

*6 A loop of thin wire is hung over a block of ice (or an ice cube) at 0°C and weights are hung from the ends of the wire. It is found that the wire cuts its way through the ice cube, but leaves a solid block of ice behind it. This process is called *regelation*. Explain how this happens using the fact that the freezing point of water is less at higher pressures.

*7 How do a gas and a vapor differ?

*8 Liquid CO_2 can exist under what conditions? Be specific. Can it exist as a liquid at normal room temperature?

*9 Why doesn't Dry Ice last long at room temperature?

*10 Why must we breathe? That is, why can't we simply rely on diffusion of oxygen into our lungs?

*11 The swelling around a sprained ankle is due to an inflow of water into the region. The swelling can be reduced by soaking in a strong solution of Epsom salts. Explain.

Problems

*SECTION 8-1

*1 (I) What is the dew point (approximately) if the humidity is 40 percent on a day when the temperature is 30°C.

*2 (I) If the air pressure at a particular place in the mountains is 0.7 atm, at what temperature does water boil?

*3 (I) What is the air pressure at a place where water boils at 80°C?

*4 (I) What is the partial pressure of water on a day when the temperature is 25°C and the relative humidity is 65 percent?

*5 (II) What is the approximate pressure inside a pressure cooker if the water is boiling at a temperature of 110°C?

*6 (II) If the humidity in a room of volume 600 m^3 at 25°C is 90 percent, what mass of water must be removed to reduce the humidity to 50 percent?

*7 (II) If the relative humidity in a room 3.0 m by 6.5 m by 8.8 m is 60 percent, what mass of water can still evaporate from an open pan if the temperature is 20°C?

*8 (II) An *autoclave* is a device used to sterilize laboratory instruments. It is essentially a high pressure steam boiler and operates on the same principle as a pressure cooker. However, because hot steam under pressure is more effective in killing microorganisms than moist air at the same temperature and pressure, the air is removed and replaced by steam. Typically, the gauge pressure inside the autoclave is 1 atm; what is the temperature of the steam? Assume the steam is in equilibrium with boiling water.

*9 (III) Air that is at its dew point of 5°C is drawn into a building where it is heated to 25°C. What will be the relative humidity at this temperature? Take into account the expansion of the air.

*10 (III) On a day when the temperature is 30°C, the wet bulb thermometer of a humidity tester falls to 10°C. What is the relative humidity?

*SECTION 8-2

*11 (I) (*a*) At atmospheric pressure, in what phases can CO_2 exist? (*b*) For what range of pressures and temperatures can CO_2 be a liquid. Refer to Figure 8-5.

*12 (I) CO_2 exists in what phase when the pressure is 30 atm and the temperature is 30°C? (Figure 8-5).

*13 (I) Water is in which phase when the temperature is 90°C and the pressure is 0.01 atm?

*SECTION 8-3

*14 (II) Calculate the approximate rate of diffusion of O_2 and CO_2 between the lungs and the blood using values given in the text.

*15 (II) What is the time needed for a glycine molecule (see Table 8-4) to diffuse a distance of 10^{-3} cm in water at 20°C if its concentration varies over that distance from 1.0 mol/m^3 to half this value. Compare this "speed" to its average thermal speed. The molecular mass of glycine is about 75 u.

*16 (II) A particular membrane allows urea to diffuse at a rate of 6.0×10^{-10} moles per second per square meter of surface when the concentration difference is 100 mol/m^3. What is the permeability of this membrane?

*17 (III) (*a*) Derive *Graham's law*, which states that "the rate of diffusion of gas molecules is inversely proportional to the square root of the molecular mass." (*b*) Which would diffuse faster, N_2 gas or O_2 gas, and by how much (percentage)?

*SECTION 8-4

*18 (II) Approximately what concentration of solute would be necessary if water at 20°C is to rise to the top of a 30-m tree by osmosis?

*19 (II) The solutions on either side of a semipermeable membrane contain 350 and 150 mol/m^3 of non-permeating substance respectively. What is the difference in osmotic pressure between the two sides? Assume $T = 25$°C.

*20 (II) The concentration of proteins in blood plasma is approximately 840 mol/m^3. Calculate the osmotic pressure in blood plasma due to these proteins.

For additional reading

Burstyn, H. P., and A. A. Bartlett, "Critical Point Drying: Applications of the Physics of the *PVT* Surface to Electron Microscopy," *Am. J. Phys.*, **43** (1975), p. 414.

Gaines, J. L., "Dunking Duck," *Am. J. Phys.*, **27** (1959), p. 189.

Peusner, L., *Concepts in Bioenergetics*, Prentice-Hall, Inc., Englewood Cliffs, N.J., 1974. Diffusion and osmosis in biology.

Stanford, A. L., Jr., *Foundations of Biophysics*, Academic Press, New York, 1975.

9 Heat

When a pot of cold water is placed on a hot burner of a stove, the temperature of the water increases. We say that heat flows from the hot burner to the cold water. Whenever two objects at different temperatures are put in contact, heat flows from the hotter one to the colder one. The flow of heat is in the direction that will tend to equalize the temperature. If the two objects are kept in contact long enough for their temperatures to become equal, the two bodies are said to be in equilibrium, and there is no further heat flow between them. For example, when the mercury in a fever thermometer is still rising, heat is flowing from the patient's mouth to the thermometer; when the mercury stops, the thermometer is in equilibrium with the person's mouth, and they are at the same temperature.

But just what is heat? Is it the same thing as temperature? Does heat have to do with energy? Let us try to answer these questions.

9-1 Heat as energy transfer

The caloric theory

It is common to speak of the flow of heat—heat flows from a stove burner to a pot of coffee, from the sun to the earth, from a person's mouth into a fever thermometer. It flows from an object at higher temperature to one of lower temperature. Indeed, early theories of heat, before the nineteenth century, pictured heat as a fluid substance called *caloric*. According to the caloric theory, any object contained a certain amount of caloric; if more caloric flowed into the object, its temperature increased; and if caloric flowed out, the object's temperature decreased. When matter was broken apart, such as during burning, a great deal of caloric was believed to be released.

However, no change in mass was ever detected as a result of the flow of heat, nor could the caloric be detected by any other means. Hence caloric was assumed to be massless, odorless, tasteless, and transparent. In spite of the mysterious nature of this fluid, the caloric theory did explain many observations, such as the "flow" of heat from a hot object to a cold one. However, other phenomena came to light that could not be satisfactorily explained, as we shall see shortly.

Units of heat

Although the caloric theory has long since been discarded, remnants of the theory still remain, such as in the expression "flow of heat," as if heat were a fluid. The common unit for heat that is still used is named after caloric. It is called the *calorie* (cal) and is defined as *the amount of heat necessary to raise the temperature of* 1 *gram of water by* 1 *Celsius degree, from* 14.5°C *to* 15.5°C. This particular temperature range is specified since the heat required is very slightly different at different temperatures. (The difference is less than 1 percent over the range 0 to 100°C and can be ignored for most purposes.) More often used than the calorie is the *kilocalorie* (kcal), which is 1000 calories. Thus 1 *kcal is the heat needed to raise* 1 *kg of water by* 1 C°, *from* 14.5°C *to* 15.5°C. Sometimes a kilocalorie is called a Calorie (with a capital C), and it is by this unit that the energy value of food is specified. (In the British system of units, heat is measured in British thermal units (Btu). One Btu is defined as the heat needed to raise the temperature of 1 lb of water by 1°F. It can be shown (Problem 5) that 1 Btu = 0.252 kcal.)

▌ EXAMPLE 9-1 How much heat is required to heat 2.0 liters of water from 20°C to 100°C?

SOLUTION A liter of water has a mass of 1.0 kg. Thus it takes 2.0 kcal to heat 2.0 liters of water by 1.0 C°. To heat the 2.0 liters by 80 C° requires (80) (2.0 kcal) = 160 kcal. ▌▌

In this example we used the obvious (and experimentally verified) fact that the heat required is directly proportional to the mass being heated and to the temperature rise.

Heat as energy

One of the main problems with the caloric theory was its inability to account for all the heat generated by friction. You can, for example, rub your hands or two pieces of metal together for a long time and generate heat indefinitely. The American-born Benjamin Thompson (1753–1814), who later became Count Rumford of Bavaria, was acutely aware of this problem when he supervised the boring out of cannon barrels. Water was placed in the bore of the cannon to keep it cool during the cutting process, and as the water boiled away, it was replenished. The "caloric" that caused the boiling of the water was assumed to be due to the breaking apart of the metal. But Rumford noticed that even when the cutting tools were so dull that they did not cut the metal, heat was still generated and the water boiled away. Thus caloric was being released even though subdivision of matter was not occurring. Furthermore, Rumford realized, this process could go on indefinitely and produce a limitless amount of heat. This was not consistent with the idea that heat is a substance and that only a finite amount of it could thus be contained within an object. Rumford therefore rejected the caloric theory and proposed instead that heat is a kind of motion. He claimed that in some circumstances, at least, heat is produced by doing mechanical work (for example, rubbing two objects together). This idea was pursued by others in the early 1800s, particularly by an English brewer, James Prescott Joule (1818–1889).

Joule performed a number of experiments that were crucial in establishing our present view that heat, like work, represents a transfer of energy. One of Joule's experiments is illustrated in Figure 9-1. The falling weight causes the paddle wheel to turn; the friction between the water and the paddle wheel causes the temperature of the water to rise. Of course, the same temperature rise could also be obtained by heating the water on a hot stove. In this and a great many other experiments (some involving electrical energy) Joule found that a given amount of work was always equivalent to a particular amount of heat. Quantitatively, 4.18 joules (J) of work was found to be equivalent to 1 calorie of heat. This is known as the *mechanical equivalent of heat*:

$$4.18\text{ J} = 1\text{ cal}$$

$$4.18 \times 10^3\text{ J} = 1\text{ kcal}.$$

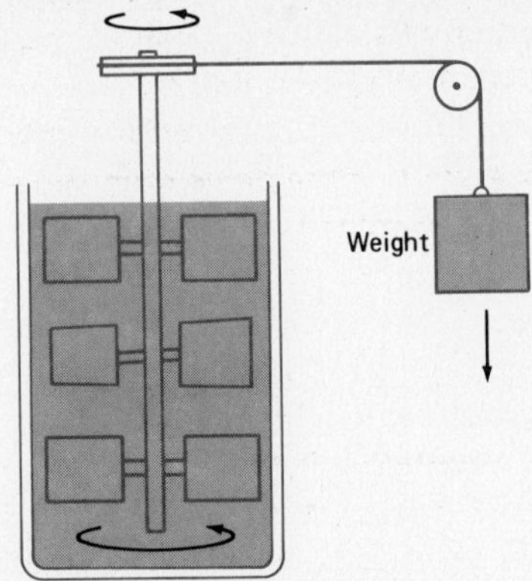

FIGURE 9-1
Joule's experiment on the mechanical equivalent of heat.

As a result of these and other experiments, scientists came to interpret **heat** not as the flow of a substance, but as a transfer of energy; when heat flows from a hot object to a cooler one, it is energy that is being transferred from the hot to the cold object. Thus in SI units the official unit for heat is the joule. Nonetheless, calories and kcal are still often used.

The development of kinetic theory fully supported, and indeed nicely explains, the idea of heat as a transfer of energy. Let us examine the process of heating a pot of water on a hot stove. According to kinetic theory, the average kinetic energy of molecules increases with temperature; thus the molecules of the stove burner have much more kinetic energy on the average than those of the cold water or the pot. When the high-KE stove molecules collide with the lower-KE molecules of the pot, some of their kinetic energy is transferred to the pot molecules, just as a fast-moving billiard ball transfers some of its kinetic energy to a ball it collides with. The molecules of the pot gain in kinetic energy. The now-higher-KE pot molecules, in turn, transfer some of their kinetic energy, by collision, to the lower-KE water molecules. The temperature of the water and the pot consequently increases. Thus we see how heat flow is a transfer of energy.

The sum total of all the energy of all the molecules in an object is called its **internal energy**. Internal energy is sometimes called **thermal energy**; we will use the two terms interchangeably. Occasionally, the term "heat content" of a body is used for this purpose. However, this is not a good term to use, for it can be confused with heat itself. Heat, as we have seen, is not the energy a body contains, but rather refers to the amount of energy transferred from a hot to a cold body.

Distinction between temperature, heat, and internal energy

Using the kinetic theory, we can now make a clear distinction between temperature, heat, and internal energy. Temperature is a measure of the *average* kinetic energy of individual molecules. Thermal or internal energy refers to the *total* energy of all the molecules in the object. Thus two equal-mass hot ingots of iron may have the same temperature, but two of them have twice as much thermal energy as one does. Heat, finally, refers to a *transfer* of energy (usually thermal energy) from one object to a second which is at a lower temperature. Heat, as Count Rumford saw, can be generated indefinitely, but the thermal energy of a body is strictly limited.

Notice that the direction of heat flow between two objects depends on their temperatures, not on how much internal energy they each have. Thus, if 50 g of water at 30°C is placed in contact (or mixed) with 200 g of water at 25°C, heat flows *from* the water at 30°C *to* the water at 25°C even though the internal energy of the 25°C water is much greater because there is so much more of it.

Internal energy of an ideal gas

As an example, let us calculate the internal energy of n moles of an ideal monatomic (one atom per molecule) gas. The internal energy, U, is the sum of the kinetic energies of all the atoms. This sum is just equal to the average kinetic energy per molecule times the total number of molecules, N:

$$U = N(\tfrac{1}{2}m\overline{v^2}).$$

From Equation 7-10 we then have

$$U = \tfrac{3}{2}NkT$$

or

$$U = \tfrac{3}{2}nRT, \tag{9-1}$$

where n is the number of moles. Thus the internal energy depends only on temperature and the number of moles of gas.

If the gas molecules contain more than one atom, the rotational and vibrational energy of the molecules (Figure 9-2) must also be taken into account. The internal energy will be greater at a given temperature than for a monatomic gas, but it will still be a function of temperature only.

The internal energy of real gases also depends mainly on temperature, but where they deviate from ideal gas behavior, it depends also on pressure and volume; we usually do not need to worry about this, however.

The internal energy of liquids and solids is quite complicated, for it includes electrical potential energy associated with the forces (or "chemical bonds") between atoms and molecules.

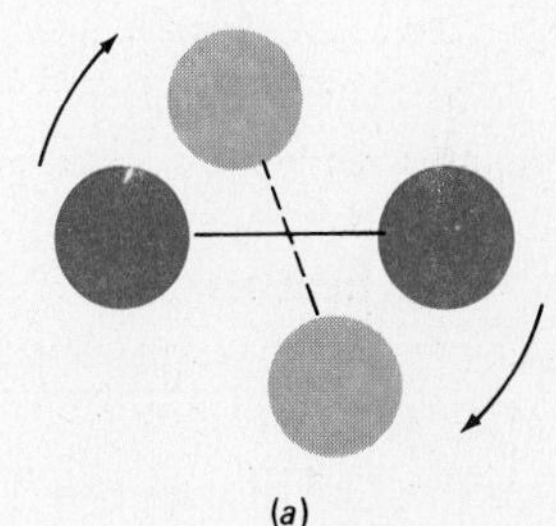

(*a*)

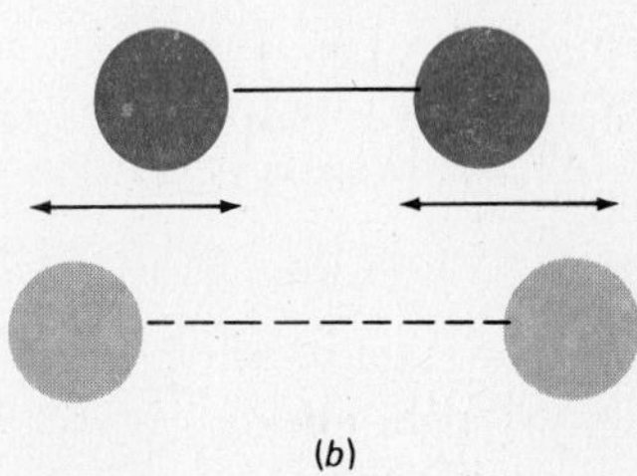

(*b*)

FIGURE 9-2
Molecules can have (*a*) rotational and (*b*) vibrational energy.

9-2 ▮ Heat capacity and calorimetry

Specific heat capacity

The amount of heat Q required to raise the temperature of a system is found to be proportional to the mass m of the system and to the temperature rise ΔT. This is expressed in the equation

$$Q = mc\,\Delta T, \tag{9-2}$$

where c is a quantity characteristic of the material called the **specific heat capacity**. For water at 15°C, $c = 1.00$ kcal/kg · C° or 4180 J/kg · C°; this corresponds to a fact we saw earlier—that it takes 1 kcal of heat to raise the

TABLE 9-1

Specific heat capacities† (20°C)

Substance	Specific heat capacity kcal/kg · C°	J/kg · C°
Aluminum	**0.22**	**900**
Copper	**0.093**	**390**
Glass	**0.20**	**840**
Ice (−5°C)	**0.50**	**2100**
Iron or steel	**0.11**	**450**
Lead	**0.031**	**130**
Marble	**0.21**	**860**
Silver	**0.056**	**230**
Wood	**0.4**	**1700**
Alcohol (ethyl)	**0.58**	**2400**
Mercury	**0.033**	**140**
Water	**1.00**	**4180**
Human body (average)	**0.83**	**3470**
Protein	**0.4**	**1700**

† For gases, see Table 10-2.

temperature of 1 kg of water by 1 C°. Table 9-1 gives the values of specific heat capacity for other substances at 20°C. The values of c depend to some extent on temperature, but for temperature changes that are not too great, c can usually be considered constant.

▌EXAMPLE 9-2 How much heat is required to raise the temperature of 25 kg of iron from 10°C to 90°C?

SOLUTION From Table 9-1 the specific heat of iron is 0.11 kcal/kg · C°. The change in temperature is (90°C − 10°C) = 80 C°. Thus

$$Q = mc\,\Delta T = (25\text{ kg})(0.11\text{ kcal/kg}\cdot\text{C}°)(80\text{ C}°) = 220\text{ kcal.}$$ ▌▌

If the iron had been cooled from 90°C to 10°C, 220 kcal of heat would have flowed out of the iron. In other words, Equation 9-2 is valid for heat flow either in or out, with a corresponding increase or decrease in temperature. If 25 kg of water had been heated from 10°C to 90°C in Example 9-2, the heat required would have been 2000 kcal. Water has one of the highest heat capacities of all substances, which makes it an ideal substance for hot-water radiator-heating systems and other uses that require a minimal drop in temperature.

Notice that the larger the value of c, the more heat a substance can absorb for a given temperature difference. This is perhaps the reason c is called specific heat "capacity." Another way of saying this is that if the heat capacity is high, the temperature rise is less for a given mass and input of heat.

Calorimetry

When different parts of an isolated system are at different temperatures, heat will flow from the part at higher temperature to the part at lower temperature. In this case the conservation of energy tells us that the heat lost by one part of the system is equal to the heat gained by the other part:

$$(\text{heat lost}) = (\text{heat gained}).$$

Let us take an example.

▮ EXAMPLE 9-3 If 200 cm^3 of tea at 95°C is poured into a 300-g glass cup initially at 25°C, what will be the final temperature of the mixture when equilibrium is reached, assuming that no heat flows to the surroundings?

SOLUTION Since tea is mainly water, its specific heat capacity is 1.00 kcal/kg · C° and its mass m is its density times its volume: $m = \rho V = (1.0 \times 10^3\ \text{kg/m}^3)(200 \times 10^{-6}\ \text{m}^3) = 0.20$ kg. First we set

$$(\text{heat lost by tea}) = (\text{heat gained by cup}),$$

and then using Equation 9-2, we have

$$m_{\text{tea}} \cdot c_{\text{tea}}(95°\text{C} - T) = m_{\text{cup}} \cdot c_{\text{cup}}(T - 25°\text{C}),$$

where T is the as-yet-unknown final temperature. We solve for T by putting in numbers and using Table 9-1, and we find that

$$(0.20\ \text{kg})(1.00\ \text{kcal/kg} \cdot \text{C°})(95°\text{C} - T)$$
$$= (0.30\ \text{kg})(0.20\ \text{kcal/kg} \cdot \text{C°})(T - 25°\text{C})$$
$$19 - 0.20T = 0.06T - 1.5$$
$$T = 79°\text{C}.$$ ▮▮

The exchange of heat as discussed in this example is the basis for a technique known as *calorimetry*, which is the quantitative measurement of heat exchange. To make such measurements, a *calorimeter* is used; a simple water calorimeter is shown in Figure 9-3. It is very important that the calorimeter be well insulated so that only a negligible amount of heat is exchanged with the outside.

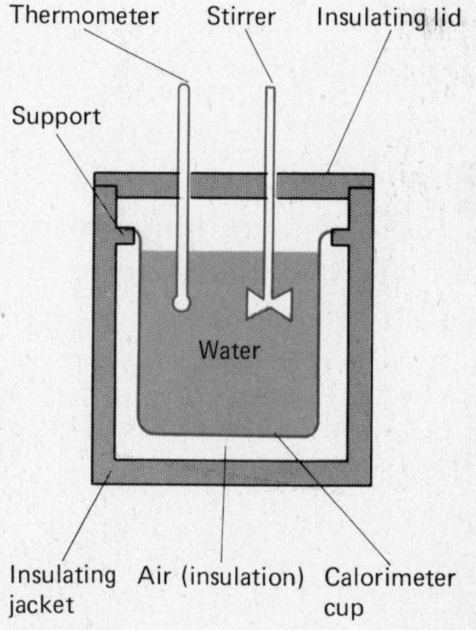

FIGURE 9-3
Simple water calorimeter.

One important use of the calorimeter is in the determination of specific heats of substances. In the technique known as the "method of mixtures," a sample of the substance is heated to a high temperature, which is accurately measured; then the sample is quickly placed in the cool water of the calorimeter. The heat lost by the sample will be gained by the water and the calorimeter; by measuring the final temperature of the mixture, the specific heat can be calculated as illustrated in the following example.

▮ EXAMPLE 9-4 We wish to determine the specific heat capacity of a new alloy. A 0.150-kg sample of the alloy is heated to 540°C. It is then quickly placed in 400 g of 10.0°C water, which is contained in a 205-g aluminum calorimeter cup. (We do not need to know the mass of the insulating jacket, since we assume that it does not change significantly in temperature.) The final temperature of the mixture is 30.5°C. Calculate the specific heat capacity of the alloy.

SOLUTION We use the fact that the heat lost equals the heat gained; that is

$$\begin{pmatrix}\text{heat lost}\\ \text{by sample}\end{pmatrix} = \begin{pmatrix}\text{heat gained}\\ \text{by water}\end{pmatrix} + \begin{pmatrix}\text{heat gained}\\ \text{by calorimeter cup}\end{pmatrix}$$

$$m_s c_s\,\Delta T_s = m_w c_w\,\Delta T_w + m_{\text{cal}} c_{\text{cal}}\,\Delta T_{\text{cal}}.$$

Putting in values and using Table 9-1, this equation becomes

$$\begin{aligned}(0.150\text{ kg})(c_s)(540°\text{C} - 30.5°\text{C}) &\\ = (0.400\text{ kg})(1.0\text{ kcal/kg}\cdot\text{C°})(30.5°\text{C} - 10.0°\text{C})&\\ + (0.205\text{ kg})(0.22\text{ kcal/kg}\cdot\text{C°})(30.5°\text{C} - 10.0°\text{C})&\\ 76.5 c_s = (8.2 + 0.9)\text{ kcal/kg}\cdot\text{C°}&\\ c_s = 0.12\text{ kcal/kg}\cdot\text{C°}.&\end{aligned}$$

▮▮

In making this calculation, we ignored any heat transferred to the thermometer and the stirrer (which is needed to quicken the heat transfer process and thus reduce heat loss to the outside). These can be taken into account by adding additional terms to the right side of the preceding equation and will result in a slight correction to the value of c_s (see the Problems). It should be noted that the quantity $m_{\text{cal}}c_{\text{cal}}$ is often called the *water equivalent* of the calorimeter—that is, $m_{\text{cal}}c_{\text{cal}}$ is numerically equal to the mass of water (in kilograms) that would absorb the same amount of heat.

A *bomb calorimeter* is used to measure the heat released when a substance burns. Important applications are the burning of foods to determine their Calorie content, and burning of seeds and other substances to determine their energy content, or heat of combustion. A carefully weighed sample of the substance, together with an excess amount of oxygen at high pressure, is placed in a sealed container (the "bomb"). The bomb is placed in the water of the calorimeter and a fine wire passing into the bomb is then heated briefly, which causes the mixture to ignite.

▮ EXAMPLE 9-5 Determine the Calorie content of 100 g of a certain kind of cake from the following measurements. A 10-g sample of the cake is allowed to dry before putting it in the bomb. The aluminum bomb has a mass of 0.615 kg and is placed in 2.00 kg of water contained in an aluminum calorimeter cup of mass 0.524 kg. The initial temperature of the mixture is 15.0°C and its temperature after ignition is 36.0°C.

SOLUTION In this case the heat Q released in the burning is absorbed by the entire mixture:

$$\begin{aligned}Q &= (m_w c_w + m_{\text{cal}}c_{\text{cal}} + m_{\text{bomb}}c_{\text{bomb}})\,\Delta T\\ &= [(2.00\text{ kg})(1.0\text{ kcal/kg}\cdot\text{C°}) + 0.524\text{ kg})(0.22\text{ kcal/kg}\cdot\text{C°})\\ &\quad + 0.615\text{ kg})(0.22\text{ kcal/kg}\cdot\text{C°})][36.0°\text{C} - 15.0°\text{C}]\\ &= 47\text{ kcal}.\end{aligned}$$

Since 47 kcal is released in the burning of 10 g of cake, a 100-g portion would contain 470 "Calories." ▮▮

9-3 ▌ Latent heat

We now investigate how a material changes phase from solid to liquid and from liquid to gas (see also Section 8-2); and we will see that energy is involved in such a *change of phase.*

To begin, let us trace what happens when 1.0 kg of water is heated at a steady rate from, say, −20°C, where it is ice, until it has all changed to water and then to steam above 100°C. As shown in Figure 9-4, as heat is added to the ice, its temperature rises at a rate of about 2 C°/kcal of heat added (since $c = 0.50$ kcal/kg · C°). However, when 0°C is reached, the temperature stops increasing even though heat is still being added. Instead, as heat is added, the ice is gradually observed to change to water in the liquid state. After 40 kcal has been added at 0°C, half the ice remains and half has changed to water. After 80 kcal has been added, all the ice has changed to water. Further addition of heat causes the temperature to increase again. When 100°C is reached, the temperature remains constant as the heat added changes the water to vapor (steam). It requires 539 kcal to change the 1.0 kg of water completely to steam, after which the graph rises again, indicating that the temperature of the steam now rises as heat is added.

The heat required to change 1.0 kg of a substance from the solid to the liquid state is called the **heat of fusion**; it is denoted by l_F. As we have just seen, the heat of fusion of water is 80 kcal/kg. The heat required to change a substance from the liquid to the vapor phase is called the **heat of vaporization**, l_V, and for water it is 539 kcal/kg. Other substances follow graphs similar to Figure 9-4, although the melting-point and boiling-point temperatures are different, as are the heats of fusion and vaporization. Values for the heats of fusion and vaporization, also called the *latent heats*, of a number of substances are given in Table 9-2.

The heats of vaporization and fusion also refer to the amount of heat released by a substance when it changes from a gas to a liquid, or from a liquid to a solid. Thus steam releases 540 kcal/kg when it changes to water, and water releases 80 kcal/kg when it becomes ice.

Of course, the heat involved in a change of phase depends not only

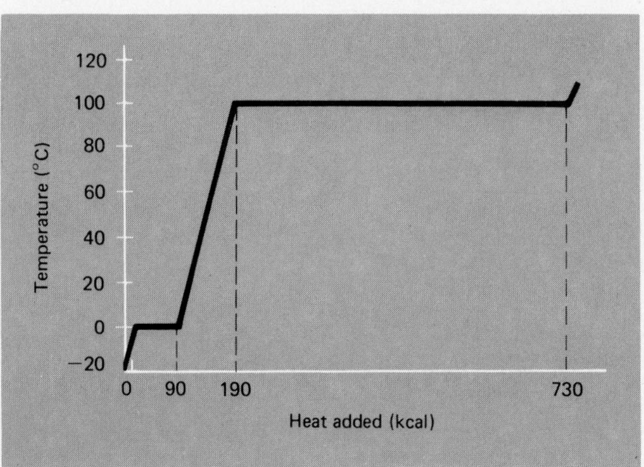

FIGURE 9-4
Heat added to bring 1.0 kg of ice at −20°C to steam.

TABLE 9-2

Latent heats

Substance	Melting point (°C)	Heat of fusion kcal/kg†	Heat of fusion J/kg	Boiling point (°C)	Heat of vaporization kcal/kg†	Heat of vaporization J/kg
Oxygen	**−218.8**	**3.3**	**0.14×10^5**	**−183**	**51**	**2.2×10^5**
Ethyl alcohol	**−114**	**25**	**1.05×10^5**	**78**	**204**	**8.7×10^5**
Water	**0**	**80**	**3.34×10^5**	**100**	**539**	**22.5×10^5**
Lead	**327**	**5.9**	**0.25×10^5**	**1750**	**208**	**9.0×10^5**
Silver	**961**	**21**	**0.88×10^5**	**2193**	**558**	**23×10^5**
Tungsten	**3410**	**44**	**1.84×10^5**	**5900**	**1150**	**48×10^5**

† Numerical values in kcal/kg are the same in cal/g.

on the latent heat but also on the total mass of the substance. That is,

$$Q = ml,$$

where l is the latent heat of the particular process, m the mass of substance, and Q the heat required or given off. For example, when 5.0 kg of water freezes at 0°C, (5.0 kg)(80 kcal/kg) = 400 kcal of energy is released.

Calorimetry sometimes involves a change of state, as the following examples show. Indeed, latent heats are often measured using calorimetry.

▮ EXAMPLE 9-6 How much energy does a refrigerator have to remove from 1.5 kg of water at 20°C to make ice at −12°C?

SOLUTION Heat must flow out to reduce the water from 20°C to 0°C, to change it to ice, and then to lower the ice from 0°C to −12°C:

$$\begin{aligned} Q &= mc_{\text{water}}(20°\text{C} - 0°\text{C}) + ml_F + mc_{\text{ice}}[0° - (-12°\text{C})] \\ &= (1.5\text{ kg})(1.0\text{ kcal/kg}\cdot\text{C}°)(20\text{ C}°) + (1.5\text{ kg})(80\text{ kcal/kg}) \\ &\qquad + (1.5\text{ kg})(0.50\text{ kcal/kg}\cdot\text{C}°)(12\text{ C}°) \\ &= 160\text{ kcal}, \end{aligned}$$

or 6.7×10^5 J. ▮▮

▮ EXAMPLE 9-7 A 0.50-kg chunk of ice at −10°C is placed in 3.0 kg of water at 20°C. At what temperature and in what phase will the final mixture be?

SOLUTION In this situation, before we can write down an equation, we must first check to see if the final mixture will be ice, a mixture of ice and water at 0°C, or all water. To bring the 3.0 kg of water at 20°C down to 0°C would require an energy release of $mc\,\Delta T = (3.0\text{ kg})(1.00\text{ kcal/kg}\cdot\text{C}°)(20\text{ C}°) = 60$ kcal. To raise the ice from −10°C to 0°C would require (0.50 kg)(0.50 kcal/kg · C°)(10 C°) = 2.5 kcal, and to change the ice to water at 0°C will require (0.50 kg)(80 kcal/kg) = 40 kcal, for a total of 42.5 kcal. This is not enough energy to bring the 3.00 kg of water at 20°C down to 0°C, so we know that the mixture must end up all water, somewhere between 0°C and 20°C. Now we can determine the final temperature by writing

$$\begin{pmatrix}\text{heat to raise}\\ \text{ice from}\\ -10°\text{C to }0°\text{C}\end{pmatrix} + \begin{pmatrix}\text{heat to}\\ \text{change ice}\\ \text{to water}\end{pmatrix} + \begin{pmatrix}\text{heat to raise the}\\ 0.50\text{ kg of water}\\ \text{from }0°\text{C to }T\end{pmatrix} = \begin{pmatrix}\text{heat lost by }3.0\text{ kg}\\ \text{of water cooling}\\ \text{from }20°\text{C to }T\end{pmatrix}$$

$$2.5\text{ kcal} + 40\text{ kcal} + (0.50\text{ kg})(1.0\text{ kcal/kg}\cdot\text{C°})(T) = (3.0\text{ kg})(1.0\text{ kcal/kg}\cdot\text{C°})(20°\text{C} - T)$$

$$42.5 + 0.50\,T = 60 - 3.0\,T$$

$$T = 5.0°\text{C}.$$

▮▮

▮ **EXAMPLE 9-8** The specific heat capacity of mercury is 0.033 kcal/kg · C°. When 1.0 kg of solid mercury at its melting point of −39°C is placed in a 0.50-kg aluminum calorimeter filled with 1.2 kg of water at 20.0°C, the final temperature of the mixture is found to be 16.5°C. What is the latent heat of fusion of mercury?

SOLUTION The heat gained by the mercury (Hg) equals the heat lost by the water and calorimeter:

$$m_{\text{Hg}}l_{\text{Hg}} + m_{\text{Hg}}c_{\text{Hg}}[16.5°\text{C} - (-39°\text{C})] = m_w c_w(20°\text{C} - 16.5°\text{C}) + m_{\text{Al}}c_{\text{Al}}(20.0°\text{C} - 16.5°\text{C})$$

$$(1.0\text{ kg})(l_{\text{Hg}}) + (1.0\text{ kg})(0.033\text{ kcal/kg}\cdot\text{C°})(55.5\text{ C°}) = (1.2\text{ kg})(1.0\text{ kcal/kg}\cdot\text{C°})(3.5\text{ C°}) + (0.50\text{ kg})(0.22\text{ kcal/kg}\cdot\text{C°})(3.5\text{ C°})$$

$$l_{\text{Hg}} = (4.2 + 0.4 - 1.8)\text{ kcal/kg} = 2.8\text{ kcal/kg}.$$

▮▮

The latent heat to change a liquid to a gas is needed not only at the boiling point. Water can change from the liquid to the gas phase even at room temperature. This process is called *evaporation*; and evaporation at room temperature still requires about 540 kcal/kg. As a result, when water evaporates, it cools, since the energy required (the latent heat of vaporization) comes from the water itself; so its internal energy, and therefore its temperature, must drop.†

Evaporation of water from the skin is one of the most important methods the body uses to control its temperature; when the temperature of the blood rises slightly above normal, the hypothalamus gland detects this temperature increase and sends a (feedback) signal to the sweat glands to increase their production. The energy required to vaporize this water comes from the body, and hence the body cools.

We can make use of kinetic theory to see why energy is needed to melt or vaporize a substance. At the melting point, the latent heat of fusion does not increase the kinetic energy (and the temperature) of the molecules in the solid but, instead, is used to overcome the potential energy associated with the forces between the molecules. That is, work must be done against these attractive forces to break the molecules loose from their relatively fixed positions in the solid so they can freely roll over one another in the liquid phase. Similarly, energy is required for

† According to kinetic theory, evaporation is a cooling process because it is the fastest-moving molecules that escape from the surface (Section 8-1); hence the average speed of the remaining molecules is less, so by Equation 7-10 the temperature is less.

molecules held close together in the liquid phase to escape into the gaseous phase. This process is a more violent reorganization of the molecules than is melting (the average distance between the molecules is greatly increased); hence the heat of vaporization is generally much greater than the heat of fusion for a given substance.

9-4 ▌ Heat transfer: conduction, convection, and radiation

Heat is transferred from one place or body to another in three different ways: by *conduction, convection,* and *radiation.* We shall now discuss each of these in turn; but in practical situations, any two or all three may be operating at the same time.

Conduction

When a metal poker is put in a hot fire, or a silver spoon is placed in a hot bowl of soup, the exposed end of the poker or spoon soon becomes hot as well, even though it is not directly in contact with the source of heat. We say that heat has been *conducted* from the hot end to the cold end.

Heat **conduction** can be visualized as the result of molecular collisions. As one end of the object is heated, the molecules there move faster and faster. As they collide with their slower-moving neighbors, they transfer some of their energy to these molecules, whose speeds thus increase. These, in turn, transfer some of their energy by collision with molecules still farther down the object. Thus the energy of thermal motion is transferred by molecular collision down the rod.

Heat conduction takes place only if there is a difference in temperature. Indeed, it is found experimentally that the rate of heat flow through a substance is proportional to the difference in temperature between its ends. It also depends on the size and shape of the object, and to investigate this quantitatively, let us consider the heat flow through a uniform object, as illustrated in Figure 9-5. It is found experimentally that the heat flow Q per time t is given by the relation

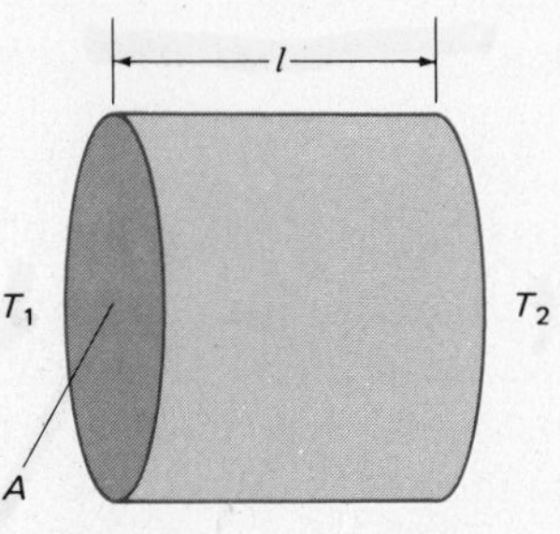

FIGURE 9-5
Heat conduction.

$$\frac{Q}{t} = kA\frac{T_1 - T_2}{l}, \qquad (9\text{-}3)$$

where A is the cross-sectional area of the object; l is the distance between the two ends, which are at temperatures T_1 and T_2; and k is a proportionality constant called the *thermal conductivity,* which is characteristic of the material. From Equation 9-3 we see that the rate of heat flow† is directly proportional to the cross-sectional area and to the temperature gradient $(T_1 - T_2)/l$.

The thermal conductivities, k, for a variety of substances are given in

† This is quite similar to the relations describing diffusion (Chapter 8) and the flow of fluids through a pipe (Chapter 6). In those cases the flow of matter was found to be proportional to the pressure gradient $(P_1 - P_2)/l$. This close similarity is one reason we speak of the flow of heat; yet we must keep in mind that no substance is flowing in this case—it is energy that is being transferred.

TABLE 9-3

Thermal conductivities

Substance	Thermal conductivity, k kcal/s · m · C°	J/s · m · C°
Silver	10×10^{-2}	420
Copper	9.2×10^{-2}	380
Aluminum	5.0×10^{-2}	200
Steel	1.1×10^{-2}	40
Glass (typical)	2.0×10^{-4}	0.84
Brick and concrete	2.0×10^{-4}	0.84
Water	1.4×10^{-4}	0.56
Human tissue (without blood)	0.5×10^{-4}	0.2
Asbestos	0.4×10^{-4}	0.16
Wood	$0.2\text{–}0.4 \times 10^{-4}$	0.08–0.16
Cork and glass wool	0.1×10^{-4}	0.042
Air	0.055×10^{-4}	0.023
Down	0.06×10^{-4}	0.025

Table 9-3. Substances for which k is large conduct heat rapidly and are said to be *good conductors*. Most metals fall in this category, although there is a wide range even among them, as you may observe by holding the ends of a silver spoon and a stainless steel spoon which are in the same hot cup of coffee or soup. Substances for which k is small, such as asbestos and down, are poor conductors of heat and are therefore good *insulators*. The relative magnitude of k can explain simple phenomena such as why a tile floor is much colder on the feet than a rug floor at the same temperature: tile is a better conductor of heat than the rug; heat transferred from your foot to the rug is not conducted away rapidly, so the rug quickly heats up on its surface to the temperature of your foot; but the tile conducts the heat away more rapidly and thus can take more heat from your foot, making it feel colder.

▌ EXAMPLE 9-9 An important source of heat loss from a house is through the windows. Calculate the rate of heat flow through a glass window 2.0 m×1.5 m in area and 3.2 mm thick if the temperatures at the inner and outer surfaces are 15.0°C and 14.0°C, respectively.

SOLUTION Since $A = (2.0\text{ m})(1.5\text{ m}) = 3.0\text{ m}^2$, $l = 3.2 \times 10^{-3}\text{m}$, and using Table 9-3 to get k, we have, from Equation 9-3:

$$\frac{Q}{t} = \frac{(0.84\text{ J/s}\cdot\text{m}\cdot\text{C}^\circ)(3.0\text{ m}^2)(15.0^\circ\text{C} - 14.0^\circ\text{C})}{3.2\times10^{-3}\text{ m}}$$

$$= 800\text{ W}.$$

This is equivalent to $(800\text{ J/s})/(4.18\times10^3\text{ J/kcal}) = 0.19\text{ kcal/s}$, or $(0.19\text{ kcal/s})(3600\text{ s/h}) = 700\text{ kcal/h}$. ▌▌

You might notice that 15°C is not very warm for the living room of a house. The room itself may indeed be much warmer, and the outside might be colder than 14°C. But the temperatures of 15°C and 14°C were specified as those at the window surfaces, and there is often a considerable drop in temperature of the air in the vicinity of the window both on the inside and the outside. That is, the layer of air on either side of the window acts as an insulator and part of the temperature drop between the inside and outside of the house takes place across the air. But if there is a heavy wind, the air next to the window on the exterior will constantly be replaced with cold air; the temperature gradient across the glass will be greater and there will be a much greater rate of heat loss.

The insulating properties of clothing comes from a similar effect. Without clothes, our bodies would heat the air in contact with the skin and soon become reasonably comfortable, since air is a very good insulator; but since air moves—there are breezes and drafts, and people themselves move about—the warm air would be replaced by cold air, thus increasing the heat loss from the body. The purpose of clothes is to hold the air so that it cannot move readily. Thus it is not the cloth that insulates us but the air that the cloth traps. Down is such a good insulator because even a small amount of it fluffs up and traps a great amount of air.

Convection

Although liquids and gases are generally not very good conductors of heat, they can transfer heat quite rapidly by convection. **Convection** is the process by which heat is transferred by the mass movement of molecules from one place to another. Whereas conduction involves molecules moving only over very small distances and colliding, convection involves the movement of molecules over large distances.

A forced-air furnace, in which air is heated and then blown by a fan into a room, is an example of *forced convection. Natural convection* occurs as well, and one familiar example is that hot air rises. For instance, the air above a radiatior (or other type of heater) expands as it is heated, and hence its density decreases; because its density is less, it rises. Warm or cold ocean currents, such as the Gulf Stream, represent natural convection on a large scale. Wind is another example of convection, and weather in general is a result of convective air currents.

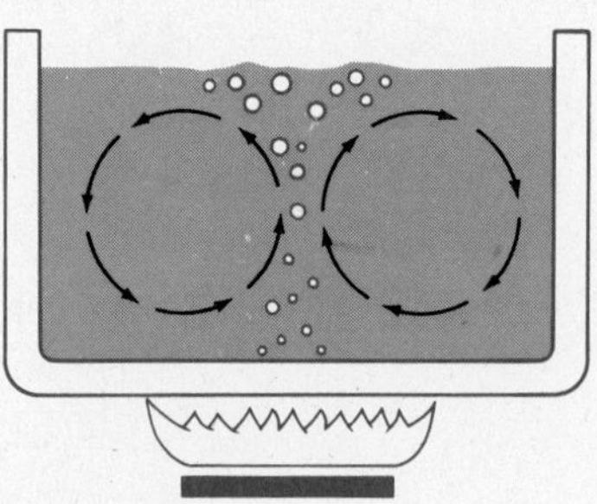

FIGURE 9-6
Convection currents in a pot of water being heated on a stove.

When a pot of water is heated (Figure 9-6), convection currents are set up as the heated water at the bottom of the pot rises because of its reduced density and is replaced by cooler water from above. This principle is used in many heating systems, such as the hot-water-radiator system shown in Figure 9-7. Cold water is heated in the furnace and as its temperature rises, it expands and rises as shown. This causes the water to circulate in the system; hot water then enters the radiators, heat is transferred by conduction to the air, and the cooled water returns to the furnace. Thus the water circulates because of convection; however, pumps are sometimes used to facilitate circulation. The air in the room also becomes uniformly heated as a result of convection. The air heated by the radiator rises and is replaced by cooler air, resulting in convective air currents, as shown.

Other types of furnace also depend on convection. Hot-air furnaces with registers (openings) near the floor often do not have fans but depend

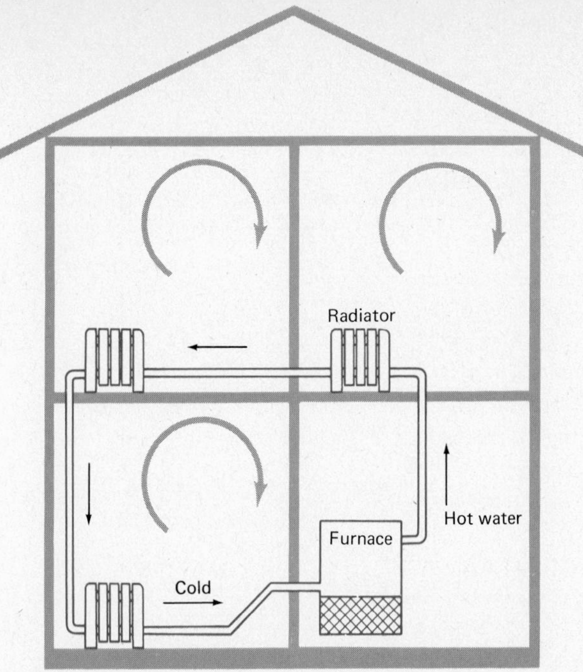

FIGURE 9-7
Convection plays a role in heating a house.

on natural convection, which can be appreciable. In other systems a fan is used. In either case it is important that cold air can return to the furnace so that convective currents circulate throughout the room if the room is to be uniformly heated.

Another example of convection and its effects is given in the following excerpt from Francois Matthes' "The Winds of Yosemite Valley"†:

> It happens to be so ordained in nature that the sun shall heat the ground more rapidly than the air. And so it comes that every slope or hillside basking in the morning sun soon becomes itself a source of heat. It gradually warms the air immediately over it, and the latter, becoming lighter, begins to rise. But not vertically upward, for above it is still the cool air pressing down. Up along the warm slope it ascends, much as shown by the arrows in the accompanying diagram [Figure 9-8*a*]. Few visitors to the valley but will remember toiling up some neverending zigzags on a hot and breathless day, with the sun on their backs and their own dust floating upward with them in an exasperating, choking cloud. Perhaps they thought it was simply their misfortune that the dust should happen to rise on that particular day. It always does on a sun-warmed slope.
>
> But again, memories may arise of another occasion when, on coming down a certain trail the dust ever descended with the travelers, wafting down upon them from zigzag to zigzag as if with malicious pleasure. That, however, undoubtedly happened on the shady side of the valley. For there the conditions are exactly reversed. When the sun leaves a slope the latter begins at once to lose its heat by radiation, and in a short time is colder than the air. The layer next to the ground then gradually chills by contact, and, becoming heavier as it condenses, begins to creep down along the slope [Figure 9-8*b*]. There is, thus, normally a warm

FIGURE 9-8
Convection on a hiking trail: (*a*) upward movement of air in the morning because it is heated; (*b*) downward movement in evening because it is cooled.

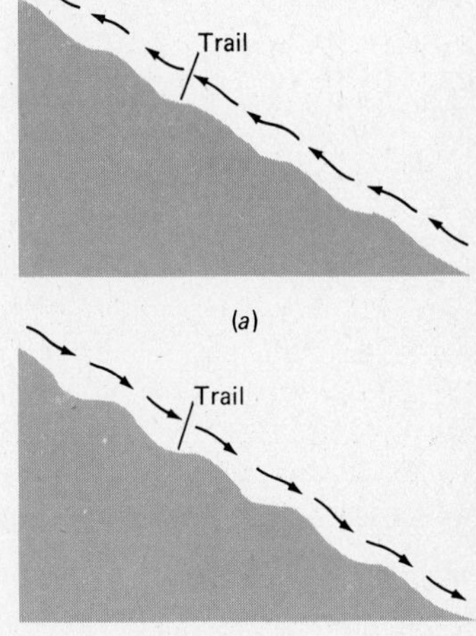

† Reprinted from the *Science Club Bulletin*, June, 1911, pp. 91–92.

updraft on a sunlit slope and a cold downdraft on a shaded slope—and that rule one may depend on almost any day in a windless region like the Yosemite. Indeed, one might readily take advantage of it and plan his trips so as to have a dust-free journey.

The human body generates a great deal of heat; of the food energy transformed within the body, at best 20 percent is used to do work, so over 80 percent appears as heat. During light activity, for example, if this heat were not dissipated, the body temperature would rise about 3 C° per hour. Clearly, the heat generated by the body must be transferred to the outside. Is the heat transferred by conduction? The temperature of the skin in a comfortable environment is 33 to 35°C, whereas the interior of the body is at 37°C. A simple calculation (see the Problems) shows that, because of this small temperature difference, plus the low thermal conductivity of tissue, conduction is responsible for very little of the heat that must be dissipated. Instead, the heat is carried to the surface by the blood. In addition to all its other important responsibilities, blood acts as a convective fluid to transfer heat to just beneath the surface of the skin. It is then conducted (over a very short distance) to the surface. Once at the surface, the heat is transferred to the environment by convection, evaporation, and radiation (see the next subsection).

It is difficult to deal with convection quantitatively since the heat transferred in this way depends on many complex factors. Yet we can make some estimates. The rate at which heat is convected away from an object is approximately proportional to the temperature difference ($T_1 - T_2$) between the object and the convective fluid (for example, air blowing by), and to the surface area A exposed to the fluid; this can be written

$$\frac{Q}{t} = hA(T_1 - T_2), \tag{9-4}$$

where h is a coefficient that depends on the properties of the fluid including its speed, and on the shape and orientation of the object. For the human body in still air, h is about 6 J/m^2 · s · C°. It rises to double this value if the air speed is 1 m/s and to about five times this if the air speed is 5 m/s.

▮ EXAMPLE 9-10 Calculate the rate of convective heat loss from an unclothed person standing in air at 23°C. Assume that the skin temperature is 34°C and the body surface area is 1.5 m^2.

SOLUTION From Equation 9-4, we have

$$\frac{Q}{t} = (6\,\text{J/m}^2 \cdot \text{s} \cdot \text{C°})(1.5\,\text{m}^2)(11\,\text{C°})$$

$$= 100\,\text{J/s} = 100\,\text{W},$$

or about 90 kcal/h. ▮▮

We shall see in Section 10–3 that a person doing moderate or heavy work puts out over 400 W; thus from this example, we see that convection alone cannot be responsible for cooling of the body when moderate or heavy work is being done. Evaporation of sweat from the skin is also important—and this factor could be included in the values of h above, although it is often considered separately. Radiation is also important.

Radiation

Convection and conduction require the presence of matter. Yet all life on earth depends on the transfer of energy from the sun, and this energy is transferred to the earth over empty (or nearly empty) space. This form of energy transfer is heat—since the sun's temperature is much higher (6000 K) than earth's—and is referred to as **radiation**. The warmth we receive from a fire is mainly radiant energy; most of the heated air rises by convection up the chimney and does not reach us.

As we shall see in later chapters, radiation consists essentially of electromagnetic waves. Suffice it to say for now that radiation from the sun consists of visible light plus many other wavelengths that the eye is not sensitive to; this includes infrared (IR) radiation, which is mainly responsible for heating the earth.

The rate at which an object radiates energy has been found to be proportional to the fourth power of the Kelvin temperature, T. That is, a body at 2000 K as compared to one at 1000 K radiates energy at a rate $2^4 = 16$ times greater. The rate of radiation is also proportional to the area A of the object, so the rate at which energy leaves the object, Q/t, is

$$\frac{Q}{t} = e\sigma A T^4, \tag{9-5}$$

where σ is a universal constant called the *Stefan–Boltzmann constant*, which has the value $5.67 \times 10^{-8}\ \mathrm{J/s \cdot m^2 \cdot K^4}$, and e, the *emissivity*, is a number between 0 and 1 that is characteristic of the material. Very black surfaces, such as charcoal, have e close to 1, whereas shiny surfaces have e close to zero and thus emit correspondingly less radiation. Skin has surprisingly high emissivity: about 0.6 for light skin and 0.8 for dark skin.

Not only do shiny surfaces emit less radiation, but they absorb little of the radiation that falls upon them (most is reflected). Black and very dark objects, on the other hand, absorb nearly all the radiation that falls on them—which is why light-colored clothing is preferable to dark clothing on a hot day. Thus *a good absorber is also a good emitter.*

Any object not only emits energy by radiation, but it also absorbs energy radiated by other bodies. If an object of emissivity e and area A is at a temperature T_1, it radiates energy at a rate $e\sigma A T_1^4$ as given by Equation 9-5. If the object is surrounded by an environment at temperature T_2, the rate the surroundings radiate energy is proportional to T_2^4 and thus the rate that energy is absorbed by the object is proportional to T_2^4. The *net* rate of radiant heat flow from the object is then given by the equation

$$\frac{Q}{t} = e\sigma A (T_1^4 - T_2^4), \tag{9–6}$$

where A is the area of the object, T_1 its temperature, e its emissivity, and T_2 the temperature of the surroundings. Notice in this equation that the rate of heat absorption by an object was taken to be $e\sigma A T_2^4$; that is, the proportionality constant is the same for both emission and absorption. This must be true to correspond with the experimental fact that equilibrium between object and surroundings is reached when they come to the same temperature; that is, Q/t must equal zero when $T_1 = T_2$, so the coefficient of each term must be the same.

Because both the object and its surroundings radiate energy, there is a net transfer of energy from one to the other unless everything is at the same temperature. From Equation 9-6 it is clear that if $T_1 > T_2$, the net flow of heat is from the body to the surroundings, so the body cools; but if $T_1 < T_2$, the net heat flow is from the surroundings into the body and its temperature rises. If different parts of the surroundings are at different temperatures, Equation 9-6 becomes more complicated.

▮ EXAMPLE 9-11 A person is sitting unclothed in a room whose walls are at a temperature of 15°C. Calculate the rate of heat loss by radiation assuming a skin temperature of 34°C and $e = 0.70$. Take the surface area of the body not in contact with the chair to be 1.5 m^2.

SOLUTION From Equation 9-6 we have

$$\frac{Q}{t} = (0.70)(5.67 \times 10^{-8}\ \text{J/s} \cdot \text{m}^2 \cdot \text{K}^4)(1.5\ \text{m}^2)[(307\ \text{K})^4 - (288\ \text{K})^4]$$

$$= 120\ \text{W},$$

or about 100 kcal/h. ▮▮

We shall see in Chapter 10 that a resting person produces heat internally at a lower rate than the heat loss by radiation as calculated in this example. Because of this, and because other forms of heat loss are also present, the person's temperature would drop, causing considerable discomfort. The body responds to excessive heat loss by increasing its metabolic rate (Section 10-3) and shivering is one method by which the body increases its metabolism. Naturally, clothes help a great deal. Example 9-11 illustrates the fact that a person may be uncomfortable even if the temperature of the air were, say 25°C, which is quite a warm room. If the walls or floor are cold, radiation to them occurs no matter how warm the air is. Indeed, it is estimated that radiation accounts for about 50 percent of the heat loss from a sedentary person in a normal room. Rooms are most comfortable when the walls and floor are warm and the air not so warm. Floors and walls can be heated by means of hot-water conduits or electric heating elements. Although such first-rate heating systems are not common today, it is interesting to note that 2000 years ago the Romans, even in houses in the remote province of Great Britain, made use of hot water and steam conduits in the floor to heat their houses.

Windows contribute greatly to heat loss since they not only conduct heat out of a room, but at night, when the outside temperature is low, considerable radiation occurs from your body to the outside. Closing the curtains at night helps to reduce both these losses.

Heating of an object by radiation from the sun cannot be calculated using Equation 9-6 since T_2 in this equation refers to the temperature of the environment surrounding the object, whereas the sun is essentially a point source; hence it must be treated as an additional source of energy. Heating by the sun is calculated using the fact that 1350 J of energy strikes the atmosphere of the earth from the sun per second per meter squared of area at right angles to the sun's rays. This number, 1350 J/s · m^2 (or 1350 W/m^2), is called the *solar constant.* The atmosphere may absorb as much as 70 percent of this energy before it reaches the ground, depending on the cloud cover. On a reasonably clear day, perhaps 1000 W/m^2

reaches the earth's surface; an object of emissivity e with area A facing the sun then absorbs heat equal to $1000eA \cos\theta$ watts, where θ is the angle between the sun's rays and a line perpendicular to area A (Figure 9-9). That is, $A \cos\theta$ is the "effective" area, at right angles to the sun's rays. The factors that explain the seasons, the polar ice caps, and why the sun heats the earth more at midday than at sunrise or sunset are also rooted in this $\cos\theta$ factor.

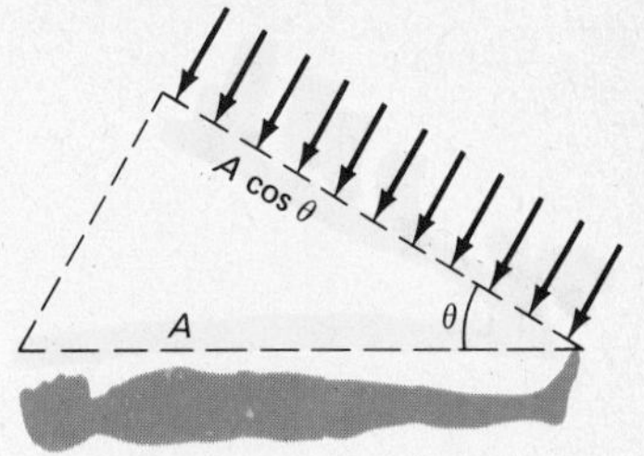

FIGURE 9-9
Radiant energy striking a body at an angle θ.

▮ EXAMPLE 9-12 What is the rate of energy absorption from the sun by a person lying flat on the beach on a clear day if the sun makes a 30° angle with the vertical? Assume that $e = 0.70$, the area of the body exposed to the sun is $0.80\ m^2$, and that $1000\ W/m^2$ reaches the earth's surface.

SOLUTION Since $\cos 30° = 0.866$, we have

$$\frac{Q}{t} = (1000\ J/s \cdot m^2)(0.70)(0.80\ m^2)(0.866)$$

$$= 490\ W,$$

or 420 kcal/h. ▮▮

Notice that if a person wears light-colored clothing, e is much smaller, so the energy absorbed is less.

An interesting application of thermal radiation to diagnostic medicine is *thermography*. A special instrument, the thermograph, scans the body, measuring the intensity of radiation from many points and forming a picture that resembles an x-ray (Figure 9-10). Areas where metabolic activity is high, such as in tumors, can often be detected on a thermogram as a result of their higher temperature and thereby increased radiation.

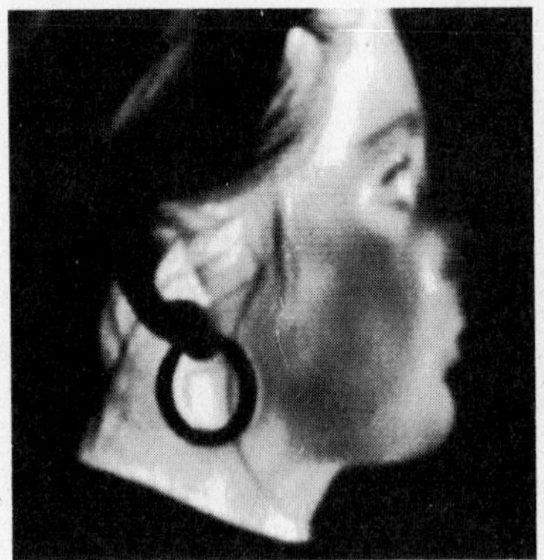

FIGURE 9-10
Thermogram of a girl's head. The shading represents variations of temperature from hot to cold, the white areas being the warmest and the black the coolest. (Courtesy of AGA Corporation.)

*9-5 ▮ Space heating and solar energy

A large fraction of the energy consumption in the United States is used for space heating—that is, keeping the inside of homes and buildings warm. Most heating systems burn gas, oil, or coal, which are of limited supply since the earth's reserves are diminishing and costs are rising (see Section 18-6). The extraction and burning of these fuels causes environmental problems: destruction of the landscape, rendering farmland less useful, and air and water pollution. A typical home furnace is about 60 percent efficient: that is, 60 percent of the energy released in the burning of the fuel actually heats the house; the other 40 percent goes out the vent or chimney, or is wasted as unburned or poorly burned fuel.

Electric heating is generally less efficient and more wasteful since most electric energy today is generated from the burning of fuel and the efficiency of generating electricity is somewhat lower—about 30 to 35 percent.

It is clearly desirable to find ways to reduce use of our fuel supplies. We can help by doing less driving so less fuel is consumed by automobiles. A great deal can also be done to reduce the fuel needed for space heating. First by setting the thermostat of the heating system lower, particularly at night; and second, by better insulation of buildings. Indeed, if all build-

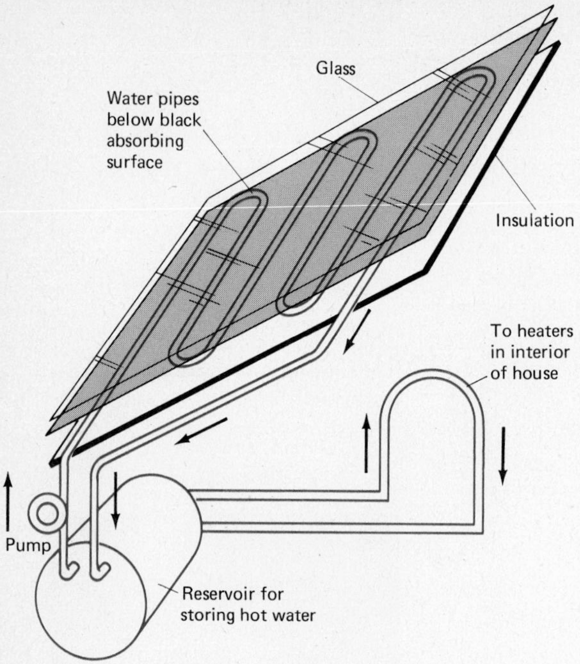

FIGURE 9-11
Solar heating system for a house. Either natural convection or forced convection (pumps) can move the water in the two parts of the system.

ings were well insulated, it has been estimated that this could save roughly one-third of the fuel used for space heating.

An important alternative to burning fuel for space heating (or cooling) is to use solar energy—that is, radiation from the sun. Such solar energy heating systems can be installed in individual houses, although some form of backup system is needed in many climates to serve when there are prolonged periods of heavy clouds. In a solar heating system for a house (Figure 9-11), water-carrying tubes are located on the roof in contact with a large black surface which absorbs the radiant energy and heats the water. The surface is covered with a piece of glass to prevent heat loss by convection, and the other sides of the tubes are well insulated to reduce conductive heat losses. The heated water is then circulated to a large well-insulated reservoir (kept in the basement perhaps), where it is stored and recirculated to heaters in the house. The reservoir can also serve as the source for the hot-water supply.

Greater efficiency and higher temperatures can be achieved using "concentrators," such as large reflectors, to focus the radiation from a large area onto a smaller absorbing surface. The possible use of solar energy for generating electricity will be discussed later (Chapter 18).

The cooling of buildings in summer requires energy if air conditioning (Section 10-5) is used. Unfortunately, much modern architecture has ignored the problem of wasting fuel in this regard. For example, many modern buildings do not have windows that open so that natural ventilation can be used for cooling; consequently, a great deal of energy is wasted on air conditioning. Also, modern office buildings often have large areas of exposed (unshaded) windows; the consequent overheating by the sun in summer results in energy waste through extra air conditioning. Shades and eaves that are high in winter (to let in radiant energy) and low in summer (to keep it out) are very practical devices for reducing energy use.

Summary

Thermal energy, or *internal energy*, refers to the total energy of all the molecules in a body. *Heat* refers to the transfer of energy from one body to another because of a difference of temperature. Heat is thus measured in energy units, such as joules; heat and thermal energy are also sometimes specified in calories or kilocalories, where 1 cal = 4.18 J is the amount of heat needed to raise the temperature of 1 g of water by 1 C°.

The *specific heat capacity*, *c*, of a substance is defined as the energy (or heat) required to change the temperature of unit mass of substance by 1 degree; as an equation, $Q = mc\,\Delta T$, where Q is the heat absorbed or given off, ΔT the temperature rise or decline, and m the mass of the substance. When heat flows within an isolated system, the heat gained by one part of the system is equal to the heat lost by the other part of the system; this is the basis of calorimetry, which is the quantitative measurement of heat exchange.

An exchange of energy occurs, without a change in temperature, whenever a substance changes phase; this happens because the potential energy of the molecules changes as a result of the changes in the relative positions of the molecules. The *heat of fusion* is the heat required to melt 1 kg of a solid into the liquid phase; it is also equal to the heat given off when the substance changes from liquid to solid. The *heat of vaporization* is the energy required to change 1 kg of a substance from the liquid to the vapor phase; it is also the energy given off when the substance changes from vapor to liquid.

Heat is transferred from one place (or body) to another in three different ways. In *conduction*, energy is transferred from higher-KE molecules to lower-KE neighboring molecules when they collide. *Convection* is the transfer of energy by the mass movement of molecules over considerable distances. *Radiation*, which does not require the presence of matter, is energy transfer by electromagnetic waves, such as from the sun. All bodies radiate energy in an amount that is proportional to their surface area and to the fourth power of their Kelvin temperature; the energy radiated (or absorbed) also depends on the nature of the surface (dark and absorbing versus brightly reflecting), which is characterized by the emissivity.

Questions

1 What happens to the work done when a jar of orange juice is vigorously shaken?

2 The specific heat capacity of water is quite large. Explain why this fact makes water particularly good for heating systems (that is, hot-water radiators).

3 Explain why burns caused by steam on the skin are often so severe.

4 Why does water in a canteen stay cooler if the cloth jacket surrounding the canteen is kept moist?

5 In warmer areas where tropical plants grow but the temperature may drop below freezing a few times in the winter, the destruction of sensitive plants due to freezing can be reduced by watering them in the evening. Explain.

6 Explain why air-temperature readings are always taken with the thermometer in the shade.

7 Sea breezes are often encountered on sunny days at the shore of a large body of water. Explain in light of the fact that the temperature of the land rises more rapidly than that of the nearby water.

8 Down sleeping bags and parkas are often specified as so many inches or centimeters of *loft*, the actual thickness of the garment when it is fluffed up. Explain.

9 List all the factors you can think of that might affect the convective coefficient h.

10 Why are light-colored clothes more comfortable in hot climates than dark clothes are?

11 Why do Bedouins wear several layers of clothing in their desert environment even when the temperature reaches 50°C (122°F) or higher?

12 A premature baby in an incubator can be dangerously cooled even when the air temperature in the incubator is warm. Explain.

13 Explain why cold drafts are so uncomfortable.

14 In the northern hemisphere, the amount of heat required to heat a room where the windows face north is much higher than that required where the windows face south. Explain.

15 Suppose that you are designing one of the following (choose one): house, concert hall, medical office building. List as many sources of heat as you can think of. Estimate the heat produced by each.

16 On hot sunny days, a house can be kept cooler if the windows are closed and the curtains pulled to cover the windows. Explain.

17 Heat loss occurs through windows by the following processes: (1) ventilation around edges; (2) through the frame, particularly if it is metal; (3) through the glass panes; (4) radiation. (*a*) For the first three, what is (are) the mechanism(s): conduction, convection, or radiation? (*b*) Heavy curtains reduce which of these heat losses? Explain in detail.

18 A piece of wood lying in the sun absorbs more heat than does a piece of shiny metal. Yet the wood feels less hot than the metal when you pick it up. Explain.

19 The floor of a house on a foundation under which the air can flow is often cooler than a floor that rests directly on the ground (slab). This is a result of what process?

20 The earth cools off at night much more quickly when it is clear than when it is cloudy. Why?

Problems

SECTION 9-1

1 (I) How much heat (kcal) is needed to raise 5.0 kg of water from 20°C to boiling?

2 (I) How much work (joules) is required to raise the temperature of 2.0 kg of water from 0°C to 20°C?

3 (I) To what temperature will 9500 J of work raise 2.0 kg of water initially at 10.0°C?

4 (I) How much work must a person do to offset eating a 400-Cal piece of cake?

5 (II) A British thermal unit (Btu) is a unit of heat in the British system of units. One Btu is defined as the heat needed to raise 1 pound of water by 1 F°. Show that 1 Btu = 0.252 kcal = 1055 J.

6 (II) A water heater can generate 9500 kcal/h. How much water can it heat from 10°C to 60°C per hour?

7 (II) The 0.50 kg head of a hammer has a speed of 5.0 m/s just before it strikes a nail and is brought to rest. How much heat (kcal) is generated by 10 such hammer blows?

8 (II) How many kilocalories of heat are generated when the brakes are used to bring a 1400-kg car to rest from a speed of 80 km/h?

9 (II) If coal gives off 7000 kcal/kg when it is burned, how much coal will be needed to heat a house that requires 4.2×10^7 kcal for the whole winter? Assume that an additional 30 percent of the heat is lost up the chimney.

10 (II) It has been estimated that for an average hiker the energy E required for a hike (in kcal) is given by the formula

$$E = 60(15 + R + 2C + 2H),$$

where R is the distance (in kilometers) traveled on roads or trails, C is the distance (in kilometers) traveled cross-country, and H is the gain in altitude (in hundreds of meters), counting only uphill. How much energy is expended on a 22-km round-trip hike, half of which is on trails and half cross-country, when a 3900-m peak is ascended starting from an elevation of 2700 m?

SECTION 9-2

11 (I) An automobile cooling system holds 12 liters of water. How much heat does it absorb if its temperature rises from 20°C to 70°C?

12 (I) What is the specific heat capacity of a metal substance if 36 kcal of heat is needed to raise 4.0 kg of the metal from 20°C to 32°C?

13 (I) What is the temperature rise of 100 kg of iron if 9.5×10^6 J of heat is transferred to it?

14 (I) What is the water equivalent of 0.280 kg of glass?

15 (I) What is the specific heat capacity of water in Btu/lb · F°? (See Problem 5.)

16 (I) How many kilograms of copper will experience the same temperature rise as 10 kg of water when the same amount of heat is absorbed?

17 (II) A 30-g lead bullet traveling 200 m/s is brought to rest when it penetrates a 1.0-kg block of wood. What is the temperature rise of the block and bullet (assumed to have reached thermal equilibrium) as a result of this collision?

18 (II) What will be the equilibrium temperature when a 200-g block of copper at 210°C is placed in an 80-g aluminum calorimeter cup containing 300 g of water at 10°C?

19 (II) When a 290-g piece of iron at 190°C is placed in a 100-g aluminum calorimeter cup containing 250 g of glycerin at 10°C, the final temperature is observed to be 38°C. What is the specific heat capacity of glycerin?

20 (II) A large 300-kg marble rock falls vertically a height of 200 m before striking the ground. What will be the temperature rise of the rock if 50 percent of the heat generated remains in the rock?

21 (II) During light activity, a 70-kg person may generate 250 kcal/h. Assuming that 20 percent of this goes into useful work and the other 80 percent is converted to heat, calculate the temperature rise of the body after 1.0 h if none of this heat were transferred to the environment.

22 (II) How long does it take a 500-W coffee pot to bring to a boil 0.50 liter of water initially at 10°C? Assume that the part of the pot which is heated with the water is made of 400 g of aluminum.

23 (II) When 120 g of a substance is heated to 330°C and then plunged into a 70-g aluminum calorimeter cup containing 110 g of water at 15°C, the final temperature, as registered by a 17-g glass thermometer, is 38°C. What is the specific heat of the substance?

24 (III) A 1.0-cm-diameter lead rod absorbs 62 kcal of heat. By how much does its length change?

SECTION 9-3

25 (I) How much heat must a refrigerator remove from 0.50 liters of water at 20°C to make ice at −10°C?

26 (I) A person gives off 200 kcal of heat by evaporation of water from the skin. How much water has been lost?

27 (I) How much heat is required to change 2.0 kg of ice at 0°C into steam at 100°C?

28 (I) How much heat is needed to melt 300 g of silver that is initially at 20°C?

29 (I) If 6.5 kcal of heat is released when 0.10 kg of mercury condenses at 357°C, what is the heat of vaporization for mercury?

30 (I) If 1.0×10^5 J of energy is supplied to a flask of oxygen at −183°C, how much oxygen will evaporate?

31 (II) An iron boiler of mass 28 kg contains 180 kg of water at 10°C. A heater supplies energy at the rate of 16,000 kcal/h. How long does it take for the water (*a*) to reach the boiling point, and (*b*) to all have changed to steam?

32 (II) A 10-kg ingot of iron at 580°C is placed in an insulated room filled with ice at 0°C. How much ice is melted to cool the iron to 0°C?

33 (II) How much water would have to evaporate from the skin per minute to take away all the heat generated by the basal metabolism (60 kcal/h) of a 65-kg person?

34 (II) The specific heat capacity of mercury is 0.033 kcal/kg · °C. Determine the latent heat of fusion of mercury using the following calorimeter data: 1.00 kg of solid Hg at its melting point of −39.0°C is placed in a 0.620-kg aluminum calorimeter with 0.400 kg of water at 12.80°C; the resulting equilibrium temperature is 5.06°C.

35 (II) What will be the final temperature when 0.30 kg of ice at 0°C is mixed with 0.10 kg of steam at 100°C?

36 (II) What will be the final result when equal amounts of ice at 0°C and steam at 100°C are mixed together?

37 (II) A 50-kg ice skater moving at 8.0 m/s glides to a stop. Assuming the ice is at 0°C and that 50 percent of the heat generated by friction is absorbed by the ice, how much ice melts?

38 (III) A 30-g lead bullet traveling at 500 m/s passes through a thin iron wall and emerges at a speed of 250 m/s. If the bullet absorbs 50 percent of the heat generated, (*a*) what will be the temperature rise of the bullet? (*b*) If the ambient temperature is 20°C, will any of the bullet melt, and if so, how much?

SECTION 9-4

39 (I) How much power is radiated by a tungsten sphere (emissivity $e = 0.35$) of radius 10 cm at a temperature of 20°C?

40 (I) If the sphere in Problem 39 is enclosed in a room whose walls are kept at −5°C, what is the *net* flow of energy out of the sphere?

41 (I) Calculate the rate of heat flow by conduction in Example 9-9 assuming there are strong gusty winds and that the external temperature is −5°C.

42 (I) What is the relative rate of heating of the earth by the sun at a latitude of 45°C as compared to the equator on March 21?

43 (I) Over what distance must the heat flow by conduction from the blood capillaries beneath the skin to the surface if the temperature difference is 0.50°C. Assume that 200 kcal/h must be transferred through the whole body's surface area of 1.5 m^2.

44 (II) The convective coefficient for wind blowing across the skin at 5 m/s is about five times that used in Example 9-10 for nearly still air. Calculate at what temperature in still air the heat loss by convection will be equal to the heat loss when the wind is blowing 5 m/s at 20°C.

45 (II) Calculate the rate that heat can be conducted from the interior of the body to the surface. Assume that the thickness of tissues is 4.0 cm, that the skin is at 34°C and the interior at 37°C, and that the surface areas is 1.5 m^2. Compare this to the measured value of 200 kcal/h that must be dissipated by a person working lightly (Section 10-3). This clearly shows the necessity of convective cooling by the blood.

46 (II) A 100-W light bulb generates 95 W of heat, which is dissipated through a glass bulb of radius 3.0 cm and is 1.0 mm thick. What is the difference in temperature between the inner and outer surfaces of the glass?

47 (II) The 1.0 kg of water initially in an aluminum pot boils away on a stove at a uniform rate in 1.0 h. The bottom of the pot is 0.60 cm thick and has an area of 270 cm^2. It is directly in contact with the electric burner of the stove. What is the surface temperature of the burner? Assume that all the heat is conducted through the pan bottom.

48 (II) A mountain climber wears down clothing 2.0 cm thick whose total surface area is 1.7 m^2. The temperature at the surface of the clothing is 0°C and at the skin is 34°C. Determine the rate of heat flow by conduction through the clothing (*a*) assuming it is dry and that k, the thermal conductivity, is that of down, and (*b*) assuming the clothing is wet, so that the k is that of water and the jacket has matted down to 0.50 cm thick.

49 (II) Approximately how much radiation does a person, of total area 1.5 m^2, absorb per hour from the sun when it is directly overhead on a clear day?

50 (II) Write an equation for the total rate of heat flow through the wall of a house if the wall consists of material with thermal conductivity k_1, of total area A_1 and thickness l_1, and with windows with thermal conductivity k_2, area A_2, and thickness l_2. The temperature difference is ΔT.

51 (III) A house has well-insulated walls 20 cm thick (assume that the conductivity rate is that of air) and area 300 m^2, a roof of wood 5.0 cm thick and area 220 m^2, and uncovered windows 0.80 cm thick and total area 32 m^2. (*a*) Assuming that the heat loss is only by conduction, calculate the rate at which heat must be supplied to this house to maintain its temperature at 20°C if the outside temperature is −5°C. (*b*) If the house is initially at 10°C, estimate how much heat must be supplied to raise the temperature to 20°C within 30 min. Assume that only the air needs to be heated and that its volume is 700 m^3. (*c*) If natural gas costs \$0.03 per kilogram and its heat of combustion is 5.4×10^7 J/kg, how much is the monthly cost to maintain the house as in part (a) for 16 hours each day, assuming that 50 percent of the heat produced is used to heat the house? Assume the specific heat capacity of air is 0.17 kcal/kg · C°.

52 (III) A house thermostat is normally set to 22°C, but at night it is turned down to 12°C for 8 h. Estimate how much more heat would be needed (state as a percentage of daily usage) if the thermostat were not turned down at night. Assume that the outside temperature averages 0°C for the 8 h at night and 8°C for the remainder of the day, and that the heat loss from the house is proportional to the difference in temperature inside and out. To obtain an estimate from the data, you will have to make other simplifying assumptions; state what these are.

53 (III) A double-glazed window is one with two panes of glass separated by an air space (Figure 9-12). (*a*) Show that the rate of heat flow, considering only conduction, is given by

$$\frac{Q}{t} = \frac{A(T_2 - T_1)}{l_1/k_1 + l_2/k_2 + l_3/k_3},$$

where k_1, k_2, and k_3 are the thermal conductivities for glass, air, and glass, respectively. (*b*) Generalize this expression for any number of materials placed next to one another.

FIGURE 9-12

54 (III) *Newton's law of cooling* states that for small temperature differences, if a body at a temperature T_1 is in surroundings at a temperature T_2, the body cools at a rate given by

$$\frac{Q}{t} = C(T_1 - T_2),$$

where C is a constant. It includes the effects of conduction, convection, and radiation. That this linear relationship should hold is obvious if only conduction and convection are considered (see Equations 9-3 and 9-4). Show that it is also approximately true for radiation by showing that Equation 9-6 reduces to

$$\frac{Q}{t} = \sigma e A T_2^3 (T_1 - T_2) = \text{constant} \times (T_1 - T_2)$$

if $(T_1 - T_2)$ is small.

55 (III) Approximately how long should it take 10 kg of ice at 0°C to melt when it is placed in a carefully sealed styrofoam "icebox" of dimensions $30 \times 40 \times 60$ cm whose walls are 1.5 cm thick. Assume that the conductivity of styrofoam is equal to that of air and that the outside temperature is 23°C.

56 (III) A *thermal transmission coefficient*, U, defined by the equation

$$\frac{Q}{t} = AU(T_2 - T_1),$$

is often used in practical work. Note that U is the rate of heat flow per unit area per degree of temperature. (*a*) How is U related to the thermal conductivity coefficient k? (*b*) Is U a characteristic property of a material? If not, what else must be known about the material before U can be determined? (*c*) Write an equation describing the total heat loss by conduction from a room in terms of the values of U for the different materials; assume that the windows are of total area A_1 and have thermal transmission coefficient U_1. The walls have a total area A_2 and consist of a layer of brick (U_2) separated by an air gap (U_3) from a layer of wood (U_4). (*d*) What is the thermal transmission coefficient U for a brick wall 12 cm thick and 8.0 m wide by 2.5 m high?

57 (III) A leaf of area 40 cm^2 and mass 4.0×10^{-4} kg directly faces the sun on a clear day. The leaf has an emissivity of 0.80 and a specific heat capacity of 0.80 kcal/kg · K. (*a*) Calculate the rate of rise of the leaf's temperature. (*b*) Calculate the temperature the leaf would reach if it lost all its heat by radiation (the surroundings are at 20°C). (*c*) What other ways can the heat be dissipated by the leaf?

58 (III) Use the result of part (*a*) in Problem 57 and take into account both convection ($h = 4 \times 10^{-3}$ kcal/m^2 · s · C° for a wind speed of, say, 1 m/s) and radiation from the leaf to calculate how much water must be transpired (evaporated) by the leaf per hour to maintain a temperature of 35°C.

*SECTION 9-5

59 (II) Suppose that a single-family house requires about 1.0×10^9 J/day to keep it warm. If only 35 percent of the 1350 W/m^2 of energy reaching the upper atmosphere from the sun (see Section 9-4) actually reaches the solar collector, how large must the solar collector be in area if it collects energy at this rate for 8 h?

60 (III) If the solar collector in Problem 59 is to raise the temperature of the water by 15°C, at what rate must the water pass through the collector (kg/min)?

61 (III) A solar heating system for use in the home makes use of a black absorbing material placed on the roof; bonded to this material are pipes that carry water heated by solar radiation back into the house. Assuming a flat roof area 20×15 m and a daily average rate of radiation of 200 W/m^2, with 50 percent of this being transferred to the water, calculate (*a*) how much water per day can be heated from 20°C to 80°C, and (*b*) estimate if this is sufficient heat on the average to maintain the interior of the house at 22°C when the outside temperature is 0°C. Assume that the heat loss from the house is entirely by conduction through 20 cm of brick walls (area = 200 m^2) and 0.80 cm of glass (area = 20 m^2).

For additional reading

Anderson, B., *The Solar Home Book: Heating, Cooling, and Designing with the Sun*, M. Riordan and L. Goodman, eds., Cheshire Books, Harrisville, N.H., 1976.

Brown, S. C., "Benjamin Thomson, Count Rumford," *The Physics Teacher*, **14** (1976), p. 270.

Marston, E. H., *The Dynamic Environment*, John Wiley & Sons, Inc., New York, 1975.

McKenzie, D. P., and F. Richter, "Convection Currents in the Earth's Mantle," *Sci. American*, **235**:5 (Nov., 1976), p. 72.

Roller, D., ed., *Early Development of the Concepts of Heat and Temperature*, Harvard University Press, Cambridge, Mass., 1950.

Young, M., "Solar Energy—The Greenhouse Effect," *The Physics Teacher*, **14** (1976), p. 226.

10 ▮ The first and second laws of thermodynamics

Thermodynamics is the name we give to the study of how natural processes are affected by changes in temperature. It deals with the laws describing energy transformations in natural processes and so involves the relations between heat, work, and energy.

In Chapter 5 we saw that work is done when energy is transferred from one body to another by mechanical means. In Chapter 9 we saw that heat is also a transfer of energy from one body to a second body which is at a lower temperature. Thus heat is much like work. To distinguish them, *heat* is defined as a *transfer of energy due to a difference in temperature*, whereas work results from a transfer of energy that is not because of a temperature difference.

In discussing thermodynamics we shall often refer to a particular system. A *system* is any object or set of objects that we wish to consider. Everything else in the universe is called the "environment." There are several kinds of systems. A *closed system* is one whose mass is constant. In an *open system*, mass may enter or leave. Most systems we study in physics are closed systems; but many systems, including plants and animals, are open systems since they exchange materials with the environment (food, oxygen, waste products). A closed system is said to be *isolated* if no energy in any form passes across its boundaries; otherwise it is not isolated.

10-1 ▮ The first law of thermodynamics

The internal energy of a closed system can be increased in two ways: heat can be added to the system, or work can be done on the system. The internal energy is decreased if heat flows out of the system, or if work is done by the system on something else.

The outcome of Joule's experiments (plus a great many others) is that experimentally it is always found that the change in internal energy of a closed system ΔU, is given by

$$\Delta U = Q - W, \tag{10-1}$$

where Q is the heat *added to* the system and W is the work done *by* the system. Note that if work is done *on* the system, W will be negative and in Equation 10-1 ΔU will increase. (Of course, we could have defined W as the work done *on* the system, in which case there would be a plus sign in Equation 10-1, but it is conventional to define W, and Q, as we have done.) Similarly, if heat leaves the system, Q is negative. Equation 10-1 is known as the **first law of thermodynamics**. Since Q and W represent energy transferred into or out of the system, the internal energy changes accordingly; thus the first law of thermodynamics is simply a statement of the *conservation of energy*! It is worth noting that the conservation of energy law was not formulated until the nineteenth century, for it depended on the interpretation of heat as a transfer of energy.

Equation 10-1 applies to a closed system. It also applies to an open system if we take into account the change in internal energy due simply to the increase or decrease in the amount of matter. For an isolated system, $W = Q = 0$ and $\Delta U = 0$.

Although the kilocalorie is still sometimes used as the unit of heat flow, it is becoming more common to express heat in the SI unit of energy, the joule: 1 kcal $= 4.18 \times 10^3$ J or 1 J $= 2.4 \times 10^{-4}$ kcal.

As an example of Equation 10-1, let us briefly analyze the boring out of a cannon barrel (which we discussed in Section 9-1). The work done in turning the cutting tool goes into increasing the internal energy of the barrel. At the same time, some of the increased internal energy of the cannon flows from the cannon to the cooling water, and because this flow of energy is due to a temperature difference, we say that it is a flow of heat. Thus the work done on the cannon goes partly into increasing the internal energy of the cannon and partly into heat flow out of the cannon.

10-2 ▮ Applications of the first law of thermodynamics

Some simple thermodynamic processes

Let us analyze some simple processes in the light of the first law. First we consider an idealized process that is carried out at constant temperature. Such a process is called an **isothermal** process (from the Greek meaning "same temperature"). If the system is an ideal gas, the process follows the curve AB on the PV diagram shown in Figure 10-1. This is the curve for $PV =$ constant, which is Boyle's law (see Chapter 7). We assume that the gas is enclosed in a container fitted with a movable piston. The gas is assumed to be in contact with a *heat reservoir* (an object whose mass is so large that its temperature does not change significantly when heat is exchanged with the system under discussion), and the process is done very slowly to make certain that all of the gas stays at the same temperature. If the gas is initially at point A in Figure 10-1, an amount of heat Q must be added to bring the system to point B. If the temperature is to remain constant, the gas must expand and do an amount of work W on the environment (it exerts a force through a distance on the piston). Since the temperature is kept constant, the internal energy does not change (see Equation 9-1). Hence, by Equation 10-1, $W = Q$: the work done by the gas equals the heat put into it. We can also consider the reverse process, taking the gas from point B to point A. In this case, we have to do work $-W$ on the gas to compress it ($W < 0$ since it is done *on* the gas). At the

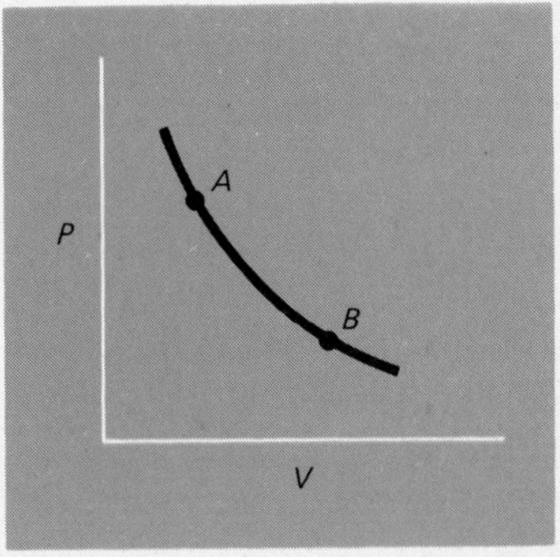

FIGURE 10-1
PV diagram for an ideal gas undergoing an isothermal process.

same time, an amount of heat $-Q$ flows out of the gas to the environment; since $\Delta U=0$, we again have $W=Q$.

An **adiabatic** process is one in which no heat is allowed to flow into or out of the system: $Q=0$. This situation can occur if the system is extremely well insulated, or the process happens so quickly that heat—which flows slowly—has no time to flow in or out. The expansion of gases in an internal combustion engine is one example of a process that is very nearly adiabatic. An adiabatic expansion of an ideal gas follows a curve like that labeled AC in Figure 10-2. Since $Q=0$, we have from Equation 10-1 that $\Delta U=-W$. That is, the internal energy decreases; hence the temperature decreases as well. This should be evident in Figure 10-2, where the product $PV(=nRT)$ is less at point C than at point B. In an adiabatic compression, work is done on the gas; hence the internal energy increases and the temperatures rises. In a diesel engine, air is compressed adiabatically by a factor of 15 or more; the temperature rise is so great that, when the fuel is injected, the mixture ignites.

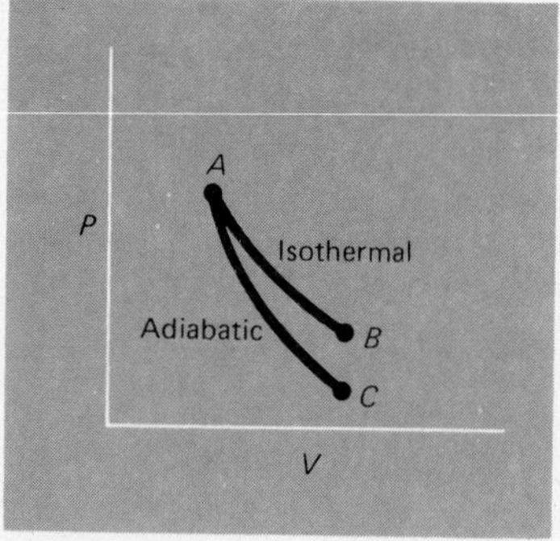

FIGURE 10-2
***PV* diagram for adiabatic and isothermal processes.**

Isothermal and adiabatic processes are just two possible processes that can occur. They are simple, yet important from a practical point of view. In these, and in all other processes (we shall meet a few of them in due time), the first law of thermodynamics holds.

Calculating work

It is often valuable to calculate the work done in a process. If the pressure is kept constant during a process (such a process is called *isobaric*, meaning "same pressure"), the work done is easily calculated. For example, if the gas in Figure 10-3 expands against the piston, the work done is the force F times the distance d. But the force is just the pressure of the gas times the area of the piston, $F=PA$. Thus

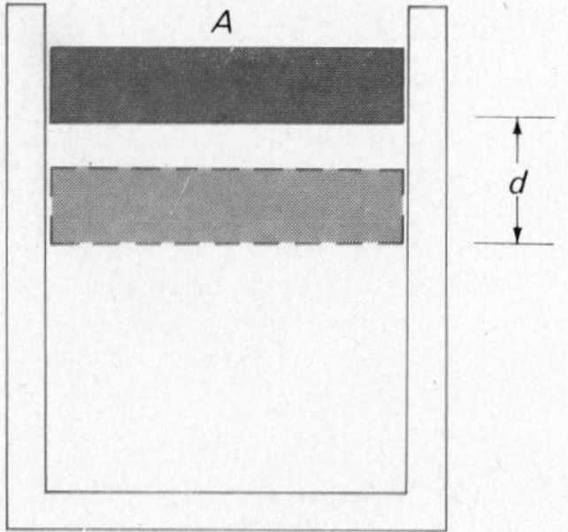

FIGURE 10-3
Work is done on the piston when the gas expands, moving the piston a distance *d*.

$$W=Fd$$

$$=PAd$$

or

$$W=P\,\Delta V, \tag{10-2}$$

where $\Delta V=Ad$ is the change in volume of the gas. This equation also holds if the gas is compressed, in which case ΔV is negative; W is then negative, which indicates that work is done *on* the gas, not by it. Equation 10-2 is also true for liquids and solids, as long as the pressure is constant during the process.

Suppose that a gas, instead of being taken isothermally from point A to point B, is taken by way of the path ADB as shown in Figure 10-4. In going from A to D, the gas does no work since the volume does not change; but in going from D to B, the gas does work equal to $P_B(V_B-V_A)$. This is the total work done in the process.

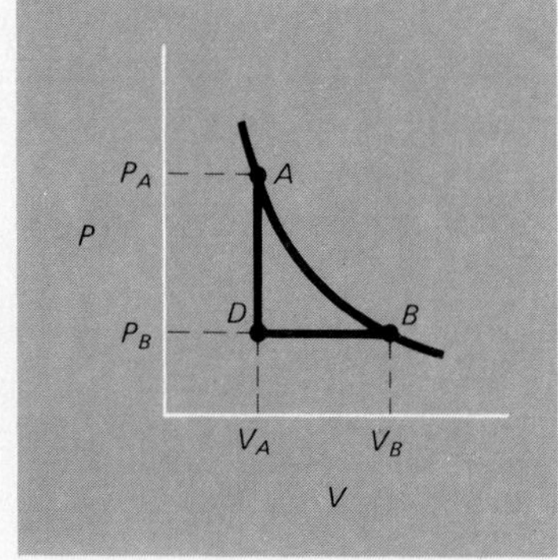

FIGURE 10-4
***PV* diagram for several processes (see the text).**

If the pressure varies during a process, such as for the isothermal process AB in Figure 10-1, Equation 10-2 cannot be used directly. A rough estimate can be obtained, however, by using an "average" value for P in Equation 10-2. More accurately, the work done is equal to the area under the PV curve. This is obvious when the pressure is constant, for as can be seen from Figure 10-5a, the crosshatched area is just $P_B(V_B-V_A)$ and this is the work done. Similarly, the work done during an isothermal process is equal to the crosshatched area shown in Figure 10-5*b*. The calculation of work done in this case can be carried out using calculus.

▮ EXAMPLE 10-1 An ideal gas is compressed at a constant pressure of 2.0 atm from 10.0 liters to 2.0 liters. (In this process some heat flows out and the temperature drops.) Heat is then added to the gas, holding the volume constant, and the pressure and temperature are allowed to rise until the temperature reaches its original value. This total process is shown as *BDA* in Figure 10-4. Calculate (*a*) the total work done in the process and (*b*) the total heat flow into the gas.

SOLUTION (*a*) Work is done only in the first part of the compression (*BD*):

$$W = P\,\Delta V = (2.0\times10^5\ \mathrm{N/m^2})(2.0\times10^{-3}\ \mathrm{m^3} - 10.0\times10^{-3}\ \mathrm{m^3})$$
$$= -1.6\times10^3\ \mathrm{J}.$$

The sign is negative because the work is done *on* the gas. (*b*) Since the temperatures at the beginning and at the end of the process are the same, there is no change in internal energy, $\Delta U = 0$. From the first law of thermodynamics we have

$$Q = W$$
$$= -1.60\times10^3\ \mathrm{J},$$

or, using the fact that 4.18 J equals 1 cal, $Q = -380$ cal. Since Q is negative, we know that 380 cal or 1600 J of heat flow out of the gas. Note that this is for the whole process, *BDA*. ▮▮

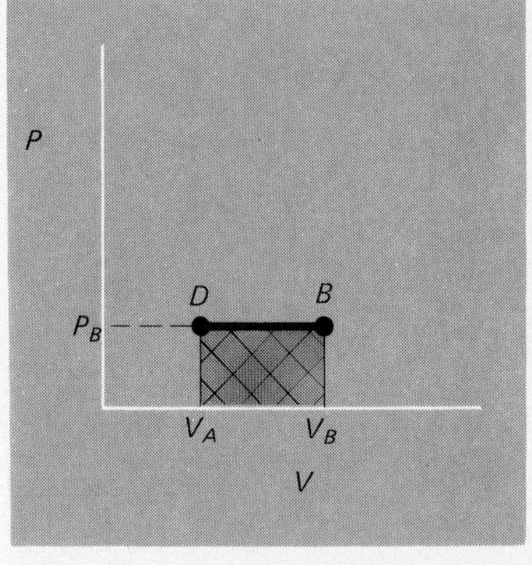

(*a*)

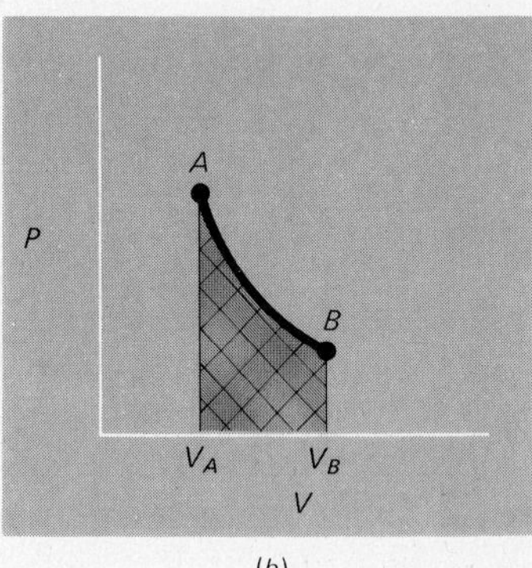

(*b*)

FIGURE 10-5
Work done by a gas is equal to the area under the *PV* curve.

*10-3 ▮ Human metabolism and the first law

Human beings and other animals do work. Work is done when a person walks or runs, or lifts a heavy object. Work requires energy. Energy is also needed for growth—to make new cells, and to replace old cells that have died. A great many energy-transforming processes occur within an organism, and they are referred to as *metabolism*.

We can apply the first law of thermodynamics

$$\Delta U = Q - W$$

to an organism, say the human body. Work W is done by the body, and this would result in a decrease in the body's internal energy (and temperature), which must be replenished. The body's internal energy is not maintained by a flow of heat Q into the body, however. Normally, the body is at a higher temperature than its surroundings, so heat usually flows *out* of the body. (See Examples 9-10 and 9-11.) Even on a very hot day when heat is absorbed, the body has no way of utilizing this heat to support its vital processes.

What then is the source of energy for animals? It is the internal energy (chemical potential energy) stored in foods. Now in a closed system the internal energy changes only as a result of heat flow or work done; in an open system, such as an animal, internal energy itself can flow into or out of the system. When we eat food, we are bringing internal energy into our bodies directly, which thus increases the total internal

TABLE 10-1

Metabolic rate for 65-kg human being

Activity	Approximate metabolic rate† kcal/h	W
Sleeping	**60**	**70**
Sitting upright	**100**	**115**
Light activity (eating, dressing, household chores)	**200**	**230**
Moderate work (bicycling slowly, walking)	**400**	**460**
Heavy work	**600**	**690**
Running, very heavy activity	**1000**	**1150**

† 1 kcal/h = 4180 J/3600 s = 1.15 W.

energy U in our bodies. This energy eventually goes into work and heat flow from the body according to the first law, Equation 10-1.

The internal energy contained in food molecules is transferred not as heat (temperature differences within the body are almost nonexistent), but by means of chemical reactions to build other needed molecules, or into work done by muscles. Determination of the actual processes involved are still active fields of research in biophysics and biochemistry. But the net result of all these metabolic processes is fully in accord with the first law of thermodynamics. The internal energy from food is transformed into useful work W and heat Q, with some internal energy leaving the body in waste products. On the average, the internal energy of the body remains practically constant as long as the person is not gaining or losing weight. Of course, the internal energy gradually increases when an organism is young and growing.

The metabolic rate is the rate at which internal energy is transformed within the body. It is usually specified in kcal/h or in watts. The metabolic rates for a variety of human activities is given in Table 10-1 for an "average" 65-kg adult.

Even when a person is sleeping, energy is being transformed; this is essentially the *basal metabolic rate*, which is defined as the metabolic rate when the organism is fasting and completely at rest. The basal metabolic rate depends on the mass m of the organism, and for mammals varies approximately as $m^{3/4}$ over a wide range from mouse to elephant.‡ That is, the basal metabolic rate for a mammal of mass m (in kilograms) is then approximately (see Table 10-1)

$$(70\ \text{W})\left(\frac{m}{65\ \text{kg}}\right)^{3/4}.$$

Thus, a mammal of mass 130 kg has a basal metabolic rate of about $(70\ \text{W})(2^{3/4}) = (70\ \text{W})\left(\sqrt{\sqrt{8}}\right) = (70\ \text{W})(\sqrt{2.828}) = 120\ \text{W}$. This same relation ($m^{3/4}$) can also be used to obtain the approximate metabolic rates of human beings of different masses performing various activities, starting from the values given in Table 10-1.

‡ $m^{3/4}$ equals the square root of the square root of m^3; that is, $\sqrt{\sqrt{m^3}}$. Logarithms are often useful in calculating the value of numbers to a power since if $A = B^c$, $\log A = c \log B$.

▮ EXAMPLE 10-2 How much energy is transformed in 24 h by a 65-kg person who spends 8.0 h sleeping, 4.0 h at moderate physical labor, 4.0 h in light activity, and 8.0 h working at a desk or relaxing.

SOLUTION From Table 10-1 the energy used is (60 kcal/h)(8.0 h) + (400 kcal/h)(4.0 h) + (200 kcal/h)(4.0 h) + (100 kcal/h)(8.0 h) = 3700 kcal. If this person eats food totaling 3700 Cal, there will be no weight change. ▮▮

▮ EXAMPLE 10-3 Calculate the average power production by the person of Example 10-2, in watts.

SOLUTION Power is defined as energy transformed per unit time. Thus

$$P = \frac{(3700\ \text{kcal})(4.18 \times 10^3\ \text{J/kcal})}{(24\ \text{h})(3600\ \text{s/h})} = 180\ \text{W}.$$ ▮▮

The metabolism of a person at rest or doing light work mainly appears as heat that flows out to the environment. Very little appears as work. If a person's intake of energy from food is excessive, the extra energy is stored as internal energy in tissue. If exercise accompanies the excessive food intake, the energy is used to increase muscle tissue. Otherwise, the extra energy is used to manufacture fatty tissue. In other words, fatty tissue is used to store excess energy, and it can later be used as a source of energy if food is lacking.

*10-4 ▮ Heat capacities of gases and the equipartition of energy

In Section 9-2 we discussed the concept of heat capacity and applied it to solids and liquids. Unlike solids and liquids, the values of the specific heat capacity for gases depends on how the process is carried out. Two important processes are those in which either the volume or the pressure is kept constant. Although for solids and liquids it matters little, Table 10-2 shows that the specific heats of gases at constant volume (c_V) and at constant pressure (c_P) are quite different.

TABLE 10-2

Heat capacities of gases at 15°C

	Specific heat capacities (kcal/kg · K)		Molar heat capacities (cal/mol · K)		
Gas	c_V	c_P	C_V	C_P	$C_P - C_V$ (cal/mol · K)
Monatomic					
He	**0.75**	**1.15**	**2.98**	**4.97**	**1.99**
Ne	**0.148**	**0.246**	**2.98**	**4.97**	**1.99**
Diatomic					
$\mathbf{N_2}$	**0.177**	**0.248**	**4.96**	**6.95**	**1.99**
$\mathbf{O_2}$	**0.155**	**0.218**	**5.03**	**7.03**	**2.00**
Triatomic					
$\mathbf{CO_2}$	**0.153**	**0.199**	**6.80**	**8.83**	**2.03**
$\mathbf{H_2O}$ **(100°C)**	**0.350**	**0.482**	**6.20**	**8.20**	**2.00**
Polyatomic					
$\mathbf{C_2H_6}$	**0.343**	**0.412**	**10.30**	**12.35**	**2.05**

This is easily explained in terms of the first law of thermodynamics and kinetic theory. Indeed, the values of the specific heats can be calculated using the kinetic theory, and the results are in close agreement with experiment. Before showing this, we introduce the *molar heat capacities*, C_V and C_P, which are defined as the heat required to raise 1 mole of the gas by 1 C° at constant volume and at constant pressure, respectively. That is, in analogy to Equation 9-2, the heat Q needed to raise n moles of gas by ΔT degrees is

$$Q = nC_V\,\Delta T \qquad \text{(volume constant)} \tag{10-3a}$$

$$Q = nC_P\,\Delta T \qquad \text{(pressure constant).} \tag{10-3b}$$

It is also clear from the definition of molar heat capacity (or by comparing Equations 9-2 and 10-3) that

$$C_V = Mc_V$$

$$C_P = Mc_P$$

where M is the molecular mass of the gas. The values for molar heat capacities are included in Table 10-2, and we see that the values are nearly the same for different gases that have the same number of atoms per molecule.

Let us now make use of the kinetic theory, and see first why the heat capacities of gases are higher for constant-pressure processes than for constant-volume processes. Let us imagine that a gas is heated by these two different processes—first at constant volume and then at constant pressure. In both processes, we let the temperature increase by the same amount, ΔT. In the process done at constant volume, no work is done, since (Equation 10-2) $W = P\,\Delta V$ and $\Delta V = 0$. Thus, according to the first law of thermodynamics, the heat added (which we now denote by Q_V) all goes into increasing the internal energy of the gas:

$$Q_V = \Delta U.$$

In the process carried out at constant pressure, work is done and hence the heat added, Q_P, must not only increase the internal energy but is also used to do the work $W = P\,\Delta V$. Thus more heat must be added in this process than in the first process at constant volume, and from the first law of thermodynamics

$$Q_P = \Delta U + P\,\Delta V.$$

Since ΔU is the same in the two processes (ΔT was chosen to be the same) we have

$$Q_P - Q_V = P\,\Delta V.$$

From the ideal gas law, $V = nRT/P$, so for a process at constant pressure, we have† $\Delta V = (nR/P)\,\Delta T$. Putting this into the equation above and using Equation 10-3, we find

$$nC_P\,\Delta T - nC_V\,\Delta T = P\left(\frac{nR\,\Delta T}{P}\right)$$

† If $\Delta V = V_2 - V_1$ and $\Delta T = T_2 - T_1$, then $\Delta V = nRT_2/P - nRT_1/P = nR\,\Delta T/P$.

or $$C_P - C_V = R. \tag{10-4}$$

Since the gas constant $R = 8.314\ \text{J/mol}\cdot\text{K} = 1.99\ \text{cal/mol}\cdot\text{K}$, our prediction is that the C_P will be larger than C_V by $1.99\ \text{cal/mol}\cdot\text{K}$. Indeed, this is very close to what is obtained experimentally, as can be seen from the final column in Table 10-2.

Let us now calculate the molar heat capacity of a monatomic gas using the kinetic theory model of gases. First we consider a process carried out at constant volume. Since no work is done in this process, the first law of thermodynamics tells us that if heat Q_V is added to the gas, the internal energy of the gas changes by

$$\Delta U = Q_V.$$

For a monatomic gas, the internal energy U is simply the total kinetic energy of all the molecules,

$$U = N(\tfrac{1}{2}m\overline{v^2}),$$

where N is the total number of molecules and $\frac{1}{2}m\overline{v^2}$ is the average kinetic energy of one molecule. As we saw earlier, we can use Equation 7-10 and write

$$U = \tfrac{3}{2}NkT = \tfrac{3}{2}nRT.$$

Then, using Equation 10-3a, we can write $\Delta U = Q_V$ in the form

$$\tfrac{3}{2}nR\ \Delta T = nC_V\ \Delta T$$

or

$$C_V = \tfrac{3}{2}R. \tag{10-5}$$

Since $R = 8.314\ \text{J/mol}\cdot\text{K} = 1.99\ \text{cal/mol}\cdot\text{K}$, kinetic theory predicts that $C_V = 2.98\ \text{cal/mol}\cdot\text{K}$ for a monatomic gas. This is very close to the experimental values for monatomic gases such as helium and neon (see Table 10-2). From Equation 10-4, c_P is predicted to be about $4.97\ \text{cal/mol}\cdot\text{K}$, also in agreement with experiment.

The measured molar specific heats for more complex gases (Table 10-2) such as diatomic (two atoms) and triatomic (three atoms) gases increase with increased number of atoms per molecule. This is apparently because of the fact that the internal energy includes not only translational kinetic energy, but also other forms of energy as well. Take, for example, a diatomic gas. As shown in Figure 10-6, the two atoms can rotate about two different axes (but rotation about a third axis passing through the two atoms would give rise to very little energy, since the moment of inertia is so small; see Equation 5-8). Thus the molecules can have rotational kinetic energy as well. It is useful to introduce the idea of "degrees of freedom," by which we mean the number of independent ways molecules can possess energy. For example, a monatomic gas is said to have three degrees of freedom, since an atom can have velocity along the x axis, the y axis, and the z axis; these are considered to be three independent motions, since change in any one of the components would not affect one of the others. A diatomic molecule has the same three degrees of freedom associated with translational kinetic energy plus two more degrees of freedom associated with rotational kinetic energy, for a

FIGURE 10-6
A diatomic molecule can rotate about two different axes.

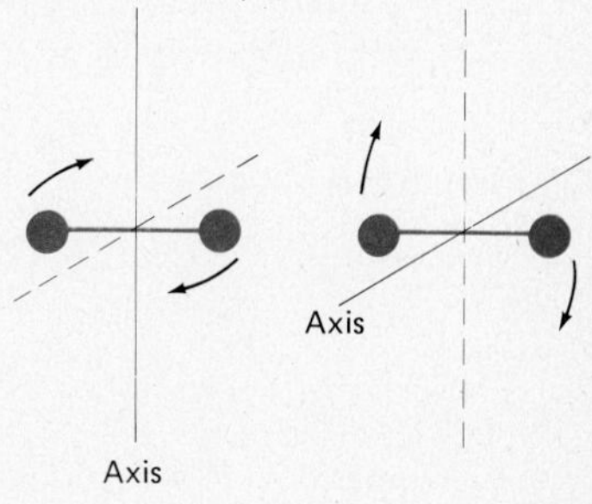

total of five degrees of freedom. A quick look at Table 10-2 indicates that the C_V for diatomic gases is about $\frac{5}{3}$ times as great as for a monatomic gas—that is, in the same ratio as their degrees of freedom. This led nineteenth-century physicists to an important idea, the *principle of equipartition of energy.* This principle states that energy is shared equally among the active degrees of freedom, and in particular each active degree of freedom of a molecule has on the average an energy equal to $\frac{1}{2}kT$. Thus, the average energy for a molecule of a monatomic gas would be $\frac{3}{2}kT$ (which we already knew) and of a diatomic gas $\frac{5}{2}kT$. Hence the internal energy of a diatomic gas would be $U = N(\frac{5}{2}kT) = \frac{5}{2}nRT$, where n is the number of moles; using the same argument we used for monatomic gases, we find that for diatomic gases the molar heat capacity would be $\frac{5}{2}R = 4.97$ cal/mol · K, in accordance with measured values. More complex molecules have even more degrees of freedom and thus greater molar heat capacities.

The situation was complicated, however, by measurements that showed that for diatomic gases at very low temperatures, C_V has a value of only $\frac{3}{2}R$ as if it had only three degrees of freedom. And at very high temperatures, C_V was about $\frac{7}{2}R$, as if there were seven degrees of freedom. The explanation for this is that at low temperatures, the molecules have only translational kinetic energy; that is, no energy goes into rotational energy, so only three degrees of freedom are "active." At very high temperatures, on the other hand, all five degrees of freedom are active, plus two additional ones. The two new degrees of freedom are associated with the two atoms vibrating as if they were connected by a spring, as shown in Figure 10-7; one degree of freedom comes from the kinetic energy of the vibrational motion, and the second from the potential energy. At room temperature, these two degrees of freedom are apparently not active. Just why fewer degrees of freedom are "active" at lower temperatures was eventually explained by Albert Einstein using the quantum theory. Thus calculations based on kinetic theory and the principle of equipartition of energy give numerical results in accord with experiment.

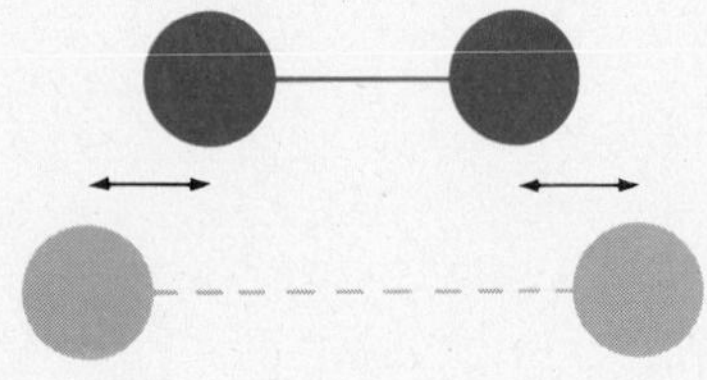

FIGURE 10-7
A diatomic molecule can vibrate.

The principle of equipartition of energy can be applied to solids as well, for the molar heat capacities of all solids is close to $3R$ (6.0 cal/mol · K) at high temperatures (Figure 10-8). (Note that Table 9-1

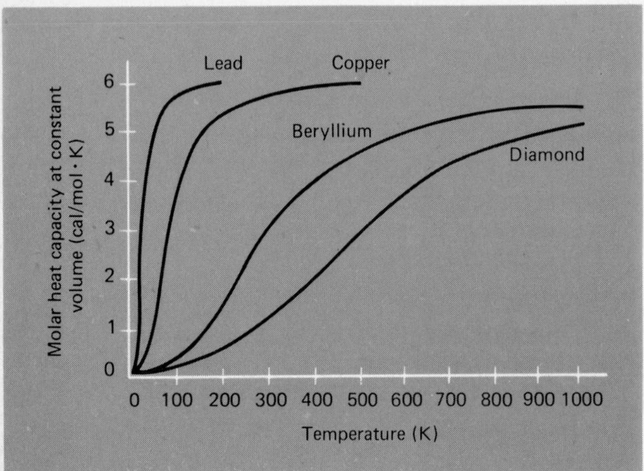

FIGURE 10-8
Molar heat capacities as a function of temperature.

gave the heat capacities per kilogram, not per mole.) At high temperatures, each atom apparently has six degrees of freedom, although some are not active at low temperatures. Each atom in a crystalline solid can vibrate about its equilibrium position as if it were connected by springs to each of its neighbors (Figure 10-9). Thus it can have three degrees of freedom for translational kinetic energy and three more associated with potential energy of vibration in each of the x, y, and z directions, which is in accord with the measured values.

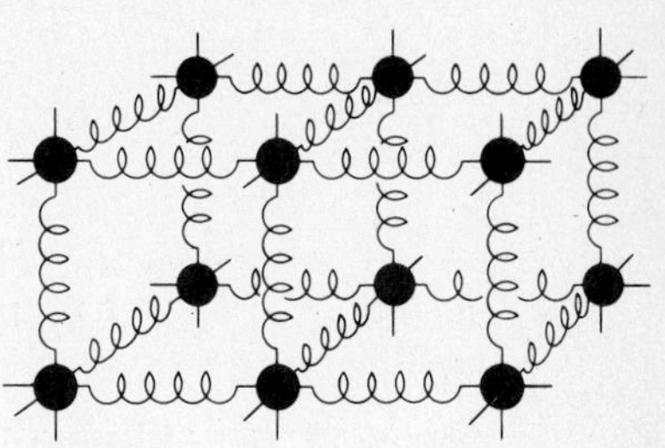

FIGURE 10-9
The atoms in a crystalline solid can vibrate about their equilibrium positions as if they were connected to their neighbors by springs. (The forces between atoms are actually electrical in nature.)

10-5 ▌ The second law of thermodynamics; heat engines

Need for a new law

The first law of thermodynamics states that energy is conserved. There are, however, many processes we can imagine that conserve energy but are not observed to occur in nature. For example, when a hot object is placed in contact with a cold object, heat flows from the hotter one to the colder one, never the reverse. If heat were to leave the colder and pass to the hotter one, energy would still be conserved. Yet it does not happen. As another example, consider what happens when you drop a rock and it hits the ground. The initial potential energy of the rock changes to kinetic as the rock falls, and when the rock hits the ground this energy is, in turn, transformed into thermal energy of the rock and the ground in the vicinity of the impact; the molecules move faster and the temperature rises very slightly. But have you seen the reverse happen—a rock at rest on the ground suddenly rises into the air because the thermal energy of molecules is transformed into kinetic energy of the rock as a whole? Energy would be conserved in this process, yet we never see it happen.

There are many other examples of processes that occur in nature whose reverse processes do not occur. Here are two more. If you put a layer of salt in a jar and cover it with a layer of pepper, when you shake it you get a thorough mixture; no matter how long you shake it, you will not get it to separate again into two layers. Coffee cups and glasses break spontaneously if you drop them; but they do not go back together spontaneously.

This lack of reversibility led scientists in the latter half of the nineteenth century to the formulation of a new principle known as the **second law of thermodynamics**. This law is a statement about which processes occur in nature and which do not. It can be stated in a variety of ways, all of which are equivalent. One statement, due to R. J. E. Clausius (1822–1888), is that *heat flows naturally from a hot object to a cold object*; *heat will not flow spontaneously from a cold object to a hot object.* Since this statement applies to a particular process, it is not obvious how it applies to other processes. A more general statement must be found that will include other possible processes in a more obvious way.

The development of a general statement of the second law was based partly on the study of heat engines. By a **heat engine** we mean any device that changes thermal energy into mechanical work. We now examine heat engines, both from a practical point of view and to show their importance in developing the second law of thermodynamics.

Heat engines

It is easy to produce thermal energy by doing work—for example, by simply rubbing your hands together briskly, or indeed in any frictional process. But to get work from thermal energy is more difficult, and the invention of a practical device to do this did not occur until about 1700, with the development of the steam engine. In the early nineteenth century, the French engineer N. L. Sadi Carnot (1796–1832) studied in detail the process of transforming heat into mechanical energy. Originally, his aim was to determine how to increase the efficiency of heat engines, but he soon found himself investigating the foundations of thermodynamics itself.

The basic idea behind any heat engine, as Carnot knew, is that mechanical energy can be obtained from heat only when heat is allowed to flow from a high temperature to a low temperature; in the process some of the heat can then be transformed to mechanical work, as diagrammed in Figure 10-10. The high and low temperatures are called the *operating temperatures* of the engine. We will be interested only in engines that run in a cycle, and thus run continuously.

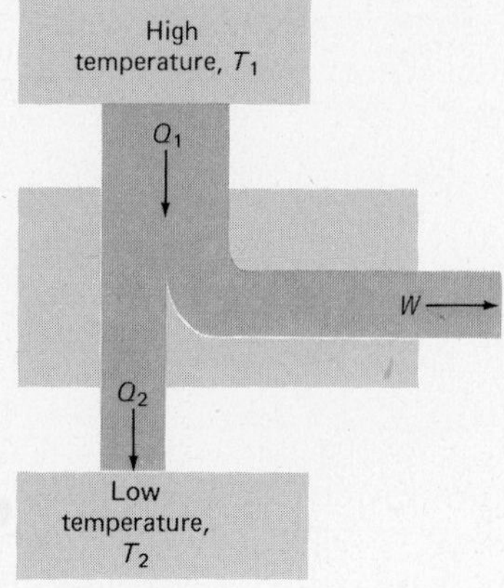

FIGURE 10-10
Schematic diagram of a heat engine.

The operation of two practical engines, the steam engine and the internal combustion engine (used in most automobiles), are illustrated in Figures 10-11 and 10-12. Steam engines are of two principal types. In the reciprocating type (Figure 10-11*a*), the heated steam passes through the intake valve and expands against a piston, forcing it to move; as the piston returns to its original position, it forces the gases out the exhaust valve. In a steam turbine (Figure 10-11*b*), everything is essentially the same except that the reciprocating piston is replaced by a rotating turbine; this resembles a paddlewheel but usually has several sets of blades. Most electricity today is generated using steam turbines (see Chapter 18).† In a steam engine, the high temperature is obtained by burning coal, oil, or

† Even nuclear power plants utilize steam turbines; the nuclear fuel—uranium—merely serves as fuel to heat the steam.

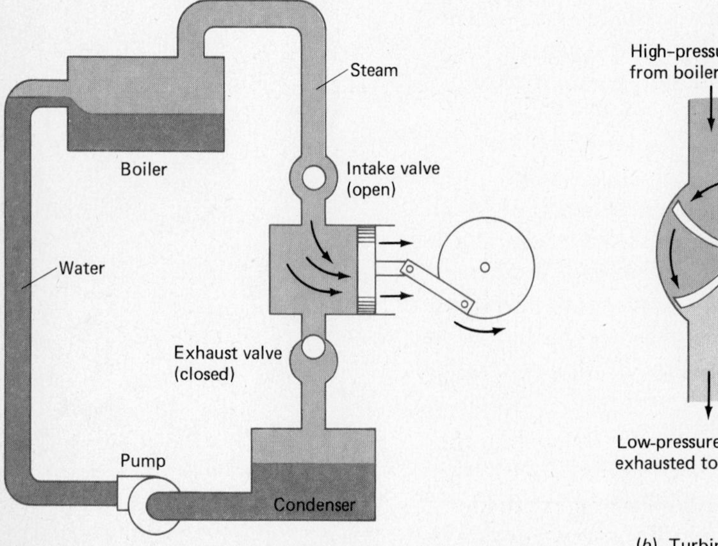

(*a*) Reciprocating type

(*b*) Turbine (boiler, condenser not shown)

FIGURE 10-11
Steam engines.

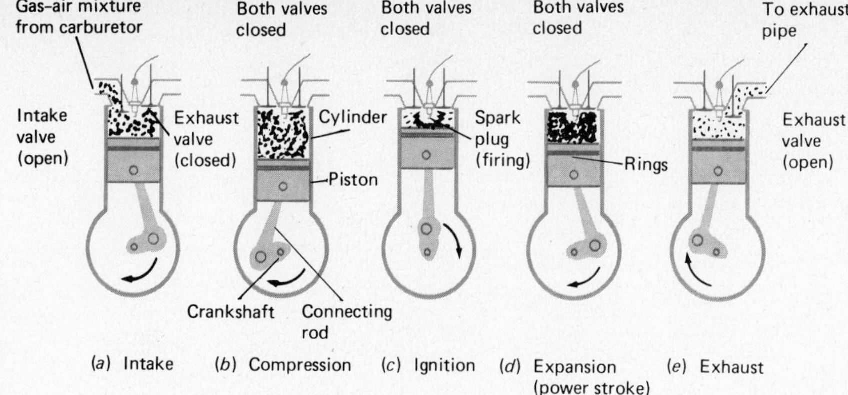

FIGURE 10-12
Four-cycle internal combustion engine: (*a*) the gasoline–air mixture flows into the cylinder as the cylinder moves down; (*b*) the piston moves upward and compresses the gas; (*c*) firing of the spark plug ignites the gasoline–air mixture, raising it to a high temperature; (*d*) the gases, now at high temperature and pressure, expand against the piston in this, the power stroke; (*e*) the burned gases are pushed out to the exhaust pipe; the intake valve then opens, and the whole cycle repeats.

other fuel to heat the steam. In an internal combustion engine, the high temperature is achieved by burning the gasoline–air mixture in the cylinder itself (ignited by the spark plug).

To see just why a temperature difference is required to run an engine, let us examine the steam engine. In the reciprocating engine, for example, suppose that there were no condenser or pump and that the steam was at the same temperature throughout the system. This would mean that the pressure of the gas being exhausted would be the same as on the intake. Thus, although work would be done by the gas *on* the piston when it expands, an equal amount of work would have to be done *by* the piston to force the steam out the exhaust; hence no net work would be done. In a real engine, the exhausted gas is cooled or condensed to a lower temperature, which means that the pressure on the exhaust is less than on intake. Thus, although the piston must do some work on the gas to expel it on the exhaust stroke, this work is much less than the work done by the gas on the piston during the intake. So a net amount of work can be obtained, but only if there is a difference of temperature. Similarly, in the gas turbine if the gas were not cooled, the pressure on each side of the blades would be the same; by cooling the gas on the exhaust side, the pressure on the front side of the blade is greater, so the turbine turns.

The efficiency of any heat engine is defined as the work it does, W, divided by the heat input Q_1 (see Figure 10-10). Since energy is conserved, the heat input Q_1 must equal the work done plus the heat output: $Q_1 = W + Q_2$. Thus $W = Q_1 - Q_2$, and the efficiency e of an engine is

$$e = \frac{W}{Q_1} = \frac{Q_1 - Q_2}{Q_1}$$

$$= 1 - \frac{Q_2}{Q_1}. \tag{10-6}$$

It is clear from Equation 10-6 that the efficiency will be greater if Q_2 can be made small.

To see how Q_2 might be reduced, Carnot theorized regarding the characteristics of an ideal engine (now called a Carnot engine). Each of the processes of heat addition and exhaust, of gas expansion or compression, were considered to be done *reversibly*. That is, each of the processes (say, during expansion of the gases against a piston) was done so slowly that the process could be considered a series of equilibrium states, and the whole process could be done in reverse with no change in the magnitude of work done or heat exchanged. A real process, on the other hand, would occur more quickly; there would be turbulence present in the gas, friction would be present, and so on; because of these factors, the process could not be done precisely in the reverse—the turbulence would be different and the heat lost to friction would not reverse itself. Thus real processes are called *irreversible.*

Carnot found that for his idealized reversible engine the efficiency could be written in terms of the operating temperatures of the engine, T_1 and T_2 (where $T_1 > T_2$), specified in degrees Kelvin:

$$e = \frac{T_1 - T_2}{T_1} = 1 - \frac{T_2}{T_1}. \tag{10-7}$$

And real engines, he found, could not have an efficiency even this high because of losses due to friction and the like. Real engines that are well designed reach 60 to 80 percent of the Carnot efficiency. (Note that to give the efficiency in percent, Equations 10-6 and 10-7 must be multiplied by 100 percent.)

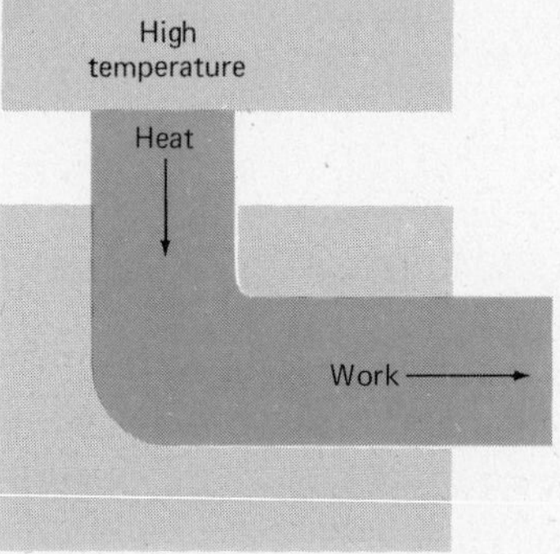

FIGURE 10-13
Schematic diagram of a hypothetical perfect heat engine in which all the heat input is used to do work. Such a perfect heat engine is not possible to construct.

▌EXAMPLE 10-4 A steam engine operates between 500°C and 270°C. What is the maximum possible efficiency of this engine?

SOLUTION We must first change the temperature to degrees Kelvin. Thus $T_1 = 773$ K and $T_2 = 543$ K, and

$$e = 1 - \frac{543}{773} = 0.30.$$

Thus the maximum (or Carnot) efficiency is 30 percent. Realistically, an engine might attain 0.70 of this value or 21 percent. Note in this example that the exhaust temperature is still rather high, 270°C. Steam engines are often arranged in series so that the exhaust of one engine is used as intake by a second or third engine. ▌▌

It is quite clear from Equation 10-7 that at normal temperatures, a 100 percent efficient engine is not possible. Only if the exhaust temperature, T_2, were at absolute zero could 100 percent efficiency be obtained; but this is a practical (as well as theoretical) impossibility.† Thus Carnot was able to state that *no device is possible whose sole effect is to transform a given amount of heat completely into work.* That is, there can be no perfect (100 percent efficient) heat engine such as the one diagrammed in Figure 10-13. This statement is another way of expressing the second law of thermodynamics. But it is not yet the most general statement.

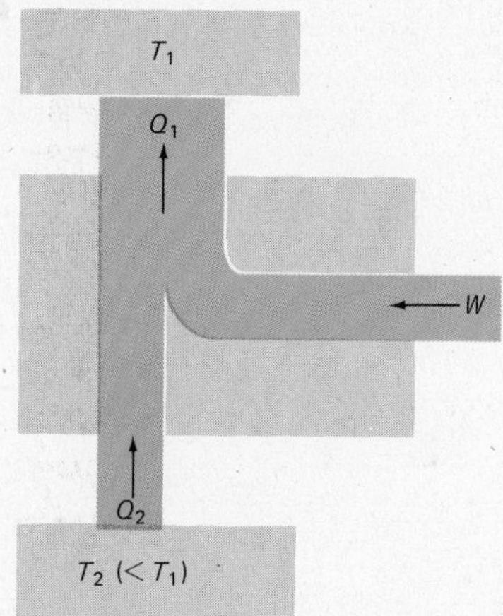

FIGURE 10-14
Schematic diagram of the operation of a refrigerator or air conditioner.

Refrigerators and air conditioners

The operating principles of refrigerators and air conditioners are just the opposite of a heat engine. As shown in Figure 10-14, by doing work W an amount of heat Q_2 is taken from a low-temperature area, T_2 (the inside of a refrigerator), and heat Q_1 is exhausted at a high temperature, T_1 (the room). You can often feel this heat blowing out beneath a refrigerator.

† It seems, from careful experimentation, that absolute zero is unattainable. This result is known as the *third law of thermodynamics.*

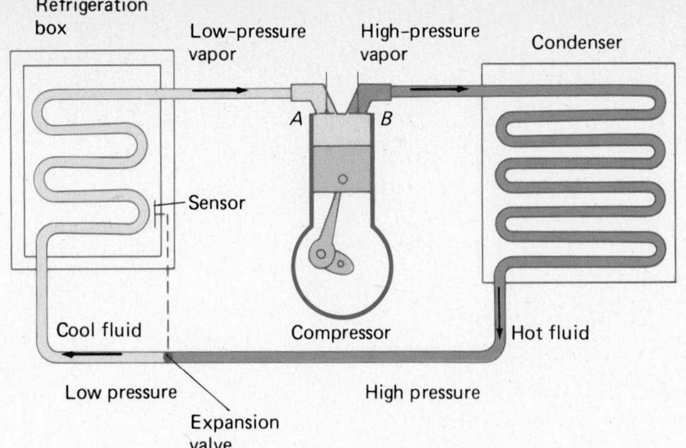

FIGURE 10-15
Typical refrigeration system.

The work W is usually done by a compressor motor which compresses a fluid, as illustrated in Figure 10-15. A perfect refrigerator—one in which no work is required to take heat from the low-temperature region to the high-temperature region—is not possible. This was implicit in our first statement of the second law of thermodynamics at the beginning of this section—that heat does not flow spontaneously from a cold object to a hot object. To accomplish such a task, work must be done.

Thermal pollution from heat engines

We are all aware that heat engines, particularly the internal combustion engine (ICE) of an automobile, are responsible for a great deal of air pollution. This pollution is a result of the burning of fuel. It is particularly bad in the ICE because the burning takes place so quickly that complete combustion does not take place and more noxious gases are thus produced. To help reduce air pollution, special devices are used (for example, to reburn the exhaust gases), and several new engines have been developed. One of these is the stratified charge engine. It is basically an ICE with an extra small chamber near the spark plug (Figure 10-16). A special rich mixture (higher gas-to-air ratio) is drawn into this extra chamber during intake. When the spark plug fires after compression, this rich mixture explodes quickly and readily ignites the leaner mixture in the main chamber. Fuller combustion takes place in this case and fewer pollutants are emitted. Electric cars and steam cars are other possibilities for the future.

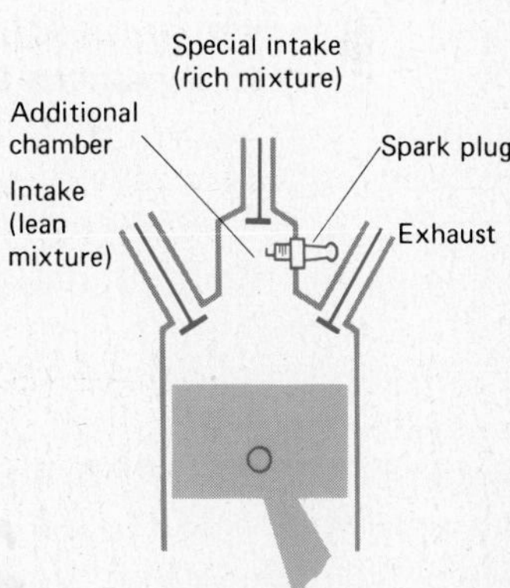

FIGURE 10-16
Stratified charge engine.

Another type of environmental pollution is thermal pollution. Every heat engine, from automobiles to power plants, exhausts heat to the environment (Q_2 in Figure 10-10). Most power plants today make use of a heat engine to transform thermal energy into electricity and the exhaust heat is generally absorbed by a coolant such as water. If the engine is run efficiently (at best, 30 to 40 percent today), the temperature T_2 must be kept as low as possible. Hence a great deal of water must flow as coolant through a power plant. The water is usually obtained from a nearby river or lake, or from the ocean. As a result of the transfer of heat to the water, its temperature rises. This can cause great damage to aquatic life in the vicinity, in large part because the warmed water holds less dissolved oxygen. The lack of oxygen can adversely affect fish and other organisms and at the same time may encourage excessive growth of other (perhaps alien) organisms, such as algae, thus greatly disrupting the ecology of an area. Another way of exhausting waste heat is to discharge it into the atmosphere by means of large cooling towers. Unfortunately this method may also have an environmental effect, for the heated air can alter the weather and climate of a region.

Although careful controls may eventually reduce air pollution to a reasonably low level, thermal pollution cannot be avoided. All we can do, in light of the second law of thermodynamics, is to use less energy and try to build more efficient engines. One hope is that eventually we can develop an engine that can transform nuclear energy, say, directly into electricity without using a heat engine. This would greatly improve efficiency and reduce pollution, but such an engine is not yet a real possibility.

10-6 ▌ Entropy and probability

Order to disorder: the second law of thermodynamics

We have seen several aspects of the second law of thermodynamics, but we have not yet arrived at a general statement of it. Indeed, it was not until the end of the nineteenth century that the **second law of thermodynamics** was finally stated in a general way, namely that

▌ **natural processes tend to move toward a state of greater disorder.**

Let us see how this general statement relates to the examples of nonreversible processes that we discussed at the beginning of Section 10-5. A jar containing separate layers of salt and pepper is more orderly than when the salt and pepper are all mixed up. Shaking a jar containing separate layers results in a mixture, and no amount of shaking restores the original layers. The natural process is from a state of relative order (layers) to one of relative disorder (a mixture), not the reverse; that is, disorder increases. Similarly, an unbroken tea cup is a more "orderly" object than the pieces of a broken cup. Cups break when they fall but do not spontaneously mend themselves. Again, the normal course of events is an increase of disorder.

When a hot object is put in contact with a cold object, heat flows from the high temperature to the low until the two objects reach the same intermediate temperature. At the beginning of the process we can distinguish two classes of molecules: those with a high average kinetic energy and those with a low average kinetic energy. After the process, all the molecules are in one class with the same average kinetic energy, and we no longer have the more orderly arrangement of molecules in two classes. Order has gone to disorder. To see this more clearly, note that the separate hot and cold objects could serve as the hot- and cold-temperature regions of a heat engine and thus could be used to obtain useful work. But once the two objects are put in contact and reach the same temperature, no work can be obtained. Disorder has increased.

The remaining example of those we discussed earlier was that of a stone falling to the ground, its kinetic energy being transformed to thermal energy; and we noted that the reverse never happens: a stone never rises into the air of its own accord. This is another example of order changing to disorder. For although thermal energy is associated with the disorderly random motion of molecules, the molecules in the falling stone all have the same velocity downward in addition to their own random velocities. Thus the more orderly kinetic energy of the stone is changed to disordered thermal energy when it strikes the ground. Disorder increases in this process, as it does in all processes that occur in nature.

This last example of order going to disorder illustrates another very important aspect of the second law of thermodynamics—that thermal energy can be considered as a lower and less useful form of energy than mechanical energy. Mechanical energy can be used to do work. But after it is changed to thermal energy, the possibility of doing work is lost—the energy is no longer available to do work. Thus we can consider mechanical and other forms of energy as more ordered than thermal energy is. In natural processes, energy is in a sense *degraded*, going from more orderly forms to less orderly forms, eventually ending up as thermal energy. Although no energy is lost in natural processes, the energy becomes less available to do work.

Entropy: a measure of disorder

In the 1860s, Clausius introduced a new quantity called **entropy**, which is a quantitative measure of the disorder of a system. How much entropy a system has depends on its state. However, when we deal with entropy—as with potential energy—it is the *change* in entropy during a process that is important, not the absolute amount. When heat is added to an object, its entropy increases—for increasing the thermal energy of an object represents an increase in the disordered motion of molecules—whereas removal of heat from a body represents a decrease in entropy. According to Clausius, the change in entropy S of an object when an amount of heat Q is *added* to it by a reversible process at constant temperature is given by

$$\Delta S = \frac{Q}{T}, \qquad (10\text{-}8)$$

where T is the Kelvin temperature. If the temperature varies during a process, a summation of the heat flow over the changing temperature must be done, using calculus. However, if the temperature change is not too great, a reasonable approximation can be made using the average value of the temperature, as indicated in the next example.

▌ EXAMPLE 10-5 A sample of 50.0 kg of water at 20.0°C is mixed with 50.0 kg of water at 24.0°C. Calculate the change in entropy.

SOLUTION The final temperature of the mixture will be 22.0°C, since we started with equal amounts of water. A quantity of heat $Q = mc\,\Delta T = (50.0\text{ kg})(1.0\text{ kcal/kg}\cdot\text{K})(2.0\text{ C}^\circ) = 100\text{ kcal}$ flows out of the hot water as it cools down from 24°C to 22°C, and into the cold water as it warms from 20°C to 22°C. The total change in entropy, ΔS, will be the sum of the changes in entropy of the hot water, ΔS_H, and that of the cold water, ΔS_C:

$$\Delta S = \Delta S_H + \Delta S_C.$$

We now make use of Equation 10-8; for the hot water we use an average temperature of 23.0°C (286 K) and for the cold water an average temperature of 21.0°C (284 K). Thus

$$\Delta S = -\frac{100\,\text{kcal}}{286\text{ K}} + \frac{100\text{ kcal}}{284\text{ K}}$$

$$= +0.002\text{ kcal/K}.$$

Notice that ΔS_H is negative since heat flowed out of the hot water. ▮▮

In this example we see that although the entropy (disorder) of one part of the system decreased, the entropy of the other part increased by a greater amount so that the net change in entropy of the whole system is positive. Indeed, the total entropy is found to increase in all natural processes, as we might expect since entropy is a measure of disorder. The second law of thermodynamics can be stated in terms of entropy as: *The entropy of an isolated system never decreases. It can only stay the same or increase.* That is, in any process the change in entropy ΔS is greater than or equal to zero:

$$\Delta S \geq 0. \tag{10-9}$$

The equal sign applies only when no change takes place. If the system is not isolated, then the change in entropy of the system ΔS_s plus the change in entropy of the environment, ΔS_{env} must be greater than or equal to zero:

$$\Delta S = \Delta S_s + \Delta S_{\text{env}} \geq 0. \tag{10-10}$$

In other words, the total entropy of the universe continually increases. It never decreases. It is possible that in a given process the entropy of a system decreases, but when this happens the entropy of the surroundings increases by a greater amount.

Evolution and growth

An interesting example of the increase in entropy relates to biological evolution and to growth of organisms. Clearly, a human being is a highly ordered organism. The process of evolution from the early macromolecules and simple forms of life to *Homo sapiens* is a process of increasing order. So, too, the development of an individual from an egg to a grown person is a process of increasing order. Do these processes violate the second law of thermodynamics? No, they do not. In the processes of evolution and growth, and even during the mature life of an individual, waste products are eliminated. These small molecules that remain as a result of metabolism are simple molecules without much order. Thus they represent relatively great disorder or entropy. Indeed, the total entropy of the molecules cast aside by organisms during the processes of evolution and growth is greater than the decrease in entropy associated with the order of the growing individual or evolving species.

Entropy as "time's arrow"

If you were to see a film being run backward, you would undoubtedly be able to tell that it *was* run backward. For you would see odd occurrences, such as a broken coffee cup rising from the floor and reassembling on a table, a bouncing ball flying higher with each bounce and arriving at a person's hand, or a torn balloon suddenly becoming whole again and filled with air. We know that these things don't happen in real life; they are processes in which order increases—or entropy decreases. They violate the second law of thermodynamics. Indeed, we can actually tell which way time is flowing by whether the processes we see are occurring in accordance with the second law of thermodynamics, or in violation of it. Of course, time only flows forward in our real world. But when watching a movie (or imagining that time could go backward), we are tipped off to a reversal of time by observing whether entropy is increasing or not. As a result, entropy has been called "time's arrow." For it tells in which direction time is going.

Heat death

A prediction of the second law of thermodynamics is that as time goes on, the universe will approach a state of maximum disorder. Matter will become a uniform mixture, heat will have flowed from high-temperature regions to low-temperature regions until the whole universe is at one temperature. No work can then be done. All the energy of the universe will have become degraded to thermal energy. All change will cease. This, the *heat death* of the universe, has been much discussed by philosophers. This final state seems an inevitable consequence of the second law of thermodynamics, although very far in the future. Yet it is based on the assumption that the universe is finite, which cosmologists are not really sure of. Furthermore, there is some question as to whether the second law of thermodynamics, as we know it, actually applies in the vast reaches of the universe. The answers are not yet in.

Statistical interpretation of entropy and the second law

The ideas of entropy and disorder are made clearer with the use of a statistical or probabilistic analysis of the molecular state of a system. This statistical approach, which was first applied toward the end of the nineteenth century by Ludwig Boltzmann (1844–1906), makes a distinction between the "macrostate" and the "microstate" of a system. The *microstate* of a system is specified when the position and velocity of every molecule is given. The *macrostate* of a system is specified by giving the macroscopic properties of the system—the temperature, pressure, number of moles, and so on. In reality, we can only know the macrostate of a system. There are generally far too many molecules in a system to be able to know the velocity and position of every one at a given moment. Nonetheless, it is important to recognize that a great many different microstates can correspond to the *same* macrostate.

Let us take a simple example. Suppose that you repeatedly shake four coins in your hand and drop them on the table. The number of heads and the number of tails that appear on a given throw is the macrostate of this system. Specifying each coin as being a head or a tail is the microstate of the system. In the following table we see the number of microstates that correspond to each macrostate:

Macrostate	Possible microstates (H = heads, T = tails)	Number of microstates
4 heads	**HHHH**	**1**
3 heads, 1 tail	**HHHT, HHTH, HTHH, THHH**	**4**
2 heads, 2 tails	**HHTT, HTHT, THHT, HTTH, THTH, TTHH**	**6**
1 head, 3 tails	**TTTH,TTHT, THTT, HTTT**	**4**
4 tails	**TTTT**	**1**

A basic principle behind the statistical approach is that each microstate is equally probable. Thus the number of microstates that give the same macrostate corresponds to the relative probability of that macrostate occurring. The macrostate of two heads and two tails is the most probable one in our case; out of the total of 16 possible microstates, 6 correspond to two heads and two tails. Thus the probability of throwing two heads and two tails is 6 out of 16 (or $6/16 = 38$ percent). The probability of throwing one head and three tails is 4 out of 16 (or 25 percent); the probability of four heads is only 1 in 16 (or 6 percent). Of course, if you threw the coins 16 times, you might not find that two heads and two tails appears exactly 6 times, or four tails exactly once. These are only probabilities or averages. But if you made 1600 throws, very nearly 25 percent of them would be one head and three tails. The greater the number of tries, the closer the percentages are to the calculated probabilities.

If we consider more coins, say 100, the relative probability of throwing all heads (or all tails) is greatly reduced. There is only one microstate corresponding to all heads. For 99 heads and 1 tail, there are 100 microstates, since each of the coins could be the one tail. The relative probabilities for other macrostates are given in Table 10-3. There are a total of 10^{30} microstates possible.† Thus the relative probability of finding all heads is 1 in 10^{30}, an incredibly unlikely event! The probability of obtaining 50 heads and 50 tails (see Table 10-3) is $1.0 \times 10^{29}/10^{30} = 0.10$, or 10 percent. The probability of obtaining between 45 and 55 heads is 90 percent.

Thus we see that as the number of coins increases, the probability of obtaining an orderly arrangement (all heads or all tails) becomes extremely unlikely. And the least orderly arrangement (half heads, half tails) is the most probable. The same ideas can be applied to the molecules of a system. For example, the most probable state of a gas (say, the air in a room) is one in which the molecules take up the whole space and move about randomly: this corresponds to the Maxwellian distribution (Figure 10-17*a*; see Chapter 7). On the other hand, the very orderly arrangement of all the molecules located in one corner of the room and all moving with nearly the same velocity (Figure 10-17*b*) is extremely unlikely, and indeed it is never observed.

FIGURE 10-17
(*a*) Most probable distribution of speeds in a gas (random); (*b*) unlikely, but orderly, distribution of speeds, in which all molecules have nearly the same speed.

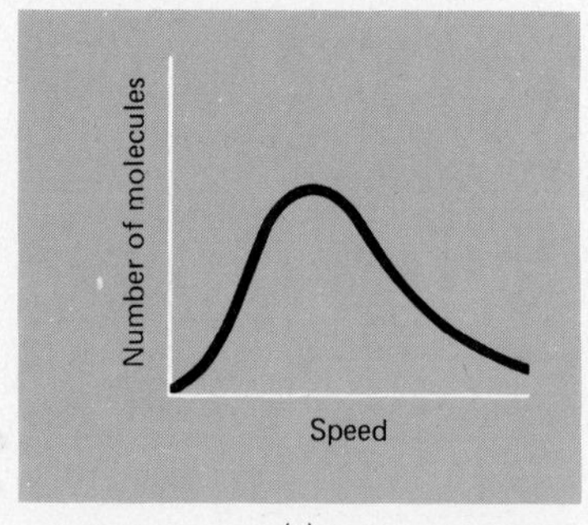

(*a*)

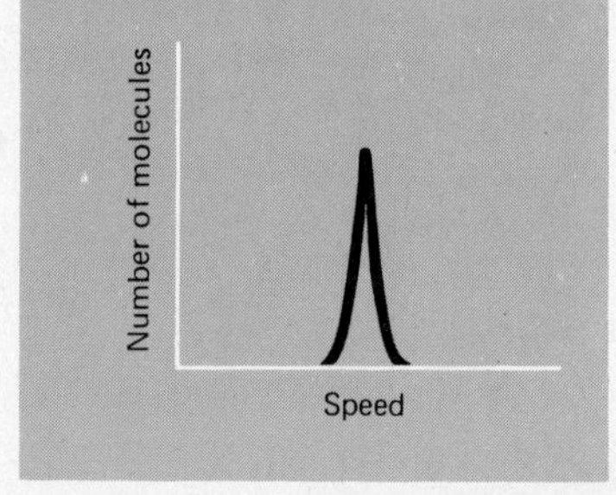

(*b*)

† Each coin has two possibilities, heads or tails. Then the possible number of microstates is $2 \times 2 \times 2 \times \cdots = 2^{100}$. Using logarithms, we see that $2 = 10^{\log 2} = 10^{0.30}$, so $2^{100} = (10^{0.30})^{100} = 10^{30}$.

TABLE 10-3

Probabilities of various macrostates for 100 coin flips

Macrostate: Heads	Macrostate: Tails	Number of microstates = W
100	0	1
99	1	1.0×10^2
90	10	1.7×10^{13}
80	20	5.4×10^{20}
60	40	1.4×10^{28}
55	45	6.1×10^{28}
50	50	1.0×10^{29}
45	55	6.1×10^{28}
40	60	1.4×10^{28}
20	80	5.4×10^{20}
10	90	1.7×10^{13}
1	99	1.0×10^2
0	100	1

From these examples it is clear that probability is directly related to disorder and hence to entropy. That is, the most probable state is the one with greatest entropy or disorder. Boltzmann showed, consistent with Clausius's definition ($\Delta S = Q/T$), that the entropy of a system in a given (macro) state can be written†

$$S = 2.3k \log W,$$

where k is Boltzmann's constant ($k = 1.38 \times 10^{-23}$ J/K) and W is the number of microstates corresponding to the given macrostate; that is, W is the probability of the occurrence of that state. In terms of probability, the second law of thermodynamics—which tells us that entropy increases in any process—reduces to the statement that those processes occur which are most probable. The second law thus becomes a trivial statement. However, there is an additional element now. The second law in terms of probability does not *forbid* a decrease in entropy. Rather, it says that the probability is extremely low. It is not forbidden that salt and pepper separate spontaneously into layers, or a broken tea cup mend itself. It is even possible that a lake could freeze over on a hot summer day. But the probability of such events occurring is extremely small. In our coin examples, we saw that increasing the number of coins from 4 to 100 reduced drastically the probability of large deviations from the average or most probable arrangement. In ordinary systems we are dealing with incredibly large numbers of molecules: in 1 mole alone there are 6×10^{23} molecules. Hence the probability of deviation from the average is incredibly tiny. For example, it has been calculated that the probability that a stone resting on the ground would transform 1 cal of thermal energy into

† The factor 2.3 comes from the fact that this equation is usually written $S = k \ln W$. Here the "natural logarithm," ln, to the base e, where $e = 2.718\ldots$, is used instead of the usual base 10. Since $\log e = \log 2.718 = 2.3$, $\ln W = 2.3 \log_{10} W$.

mechanical energy and rise into the air is much less likely than that a group of monkeys typing randomly by chance would produce the complete works of Shakespeare.

▮ Summary

The *first law of thermodynamics* states that the change in internal energy ΔU of a system is equal to the heat added to the system, Q, minus the work, W, done by the system: $\Delta U = Q - W$. This is simply a restatement of the conservation of energy, and is found to hold for all types of processes. Two simple thermodynamic processes are *isothermal*, which is a process carried out at constant temperature; and *adiabatic*, a process in which no heat is exchanged. The work W done by (or on) a gas at constant pressure, P, is given by $W = P\,\Delta V$, where ΔV is the change in volume of the gas.

A *heat engine* is a device for changing thermal energy, by means of heat flow, into useful work. The efficiency of a heat engine is defined as the ratio of the work W done by the engine to the heat input Q_1. Because of conservation of energy, the work input equals $Q_1 - Q_2$, where Q_2 is the heat exhausted to the environment; hence the efficiency $e = W/Q_1 = 1 - Q_2/Q_1$. The efficiency can also be written in terms of the upper and lower operating temperatures (in kelvins) of the engine, T_1 and T_2: $e = 1 - T_2/T_1$. All heat engines give rise to thermal pollution because they exhaust heat to the environment. The operation of refrigerators and air conditioners is the reverse of that of a heat engine: work is done to extract heat from a cold region and exhaust it to a region at a higher temperature.

The *second law of thermodynamics* can be stated in several equivalent ways: (1) heat flows spontaneously from a hot object to a cold one, but not the reverse; (2) there can be no 100 per cent efficient heat engine—that is, one that can change a given amount of heat completely into work; and (3) natural processes tend to move toward a state of greater disorder or greater entropy. *Entropy* is a quantitative measure of the disorder of a system. From a statistical point of view, the most probable state of a system is that with the most entropy or disorder.

▮ Questions

1 What happens to the internal energy of water vapor in the air that condenses on the outside of a cold glass of water? Is work done or heat exchanged? Explain.

2 Use the conservation of energy to explain why the temperature of a gas increases when it is compressed, say by pushing down on a cylinder; and vice versa, the temperature decreases when the gas expands.

3 In Figure 10-2, will more work be done in the isothermal process AB or in the adiabatic process AC? In which process will there be a greater change in internal energy? In which will there be a greater flow of heat?

4 For the processes illustrated in Figure 10-4, is more work or less work done in the isothermal process AB or in the process ADB?

5 Is it possible to cool down a room on a hot summer day by leaving the refrigerator door open?

6 The oceans contain an incredible amount of thermal energy. Why, in general, is it not possible to put this energy to useful work?

7 It has been suggested that a heat engine could be developed that made use of the fact that the temperature several hundred meters beneath the surface of the ocean is several degrees cooler than

the temperature at the surface. In the tropics, the temperatures may be 5°C and 25°C, respectively. What is the maximum efficiency such an engine could have? Why might such an engine be feasible in spite of the low efficiency? Can you imagine any adverse environmental effects that might occur?

8 What happens if you remove the lid of a bottle containing chlorine gas? Does the reverse process ever happen? Why or why not?

9 Give three examples, other than those mentioned in this chapter, of naturally occurring processes in which order goes to disorder.

10 Which do you think has the greater entropy, 1 kg of solid iron or 1 kg of liquid iron? Why?

11 Give three examples of naturally occurring processes that illustrate the degradation of usable energy into heat.

12 Think up several processes (other than those already mentioned) that would obey the first law of thermodynamics, but, if they actually occurred, would violate the second law.

13 A gas is allowed to expand (*a*) adiabatically, and (*b*) isothermally. In each process, does the entropy increase, decrease, or stay the same?

14 Entropy is often called "time's arrow" because it tells us in which direction natural processes occur. If a movie film were run backward, name some processes that you might see that would tell you that time was "running backward."

Problems

SECTIONS 10-1 AND 10-2

1 (I) In an isothermal process, 3700 J of work is done by an ideal gas. Is this enough information to tell how much heat has been added to the system? If so, how much?

2 (I) One liter of air is cooled at constant pressure until its volume is halved, and then it is allowed to expand isothermally back to its original volume. Draw the process on a *PV* diagram.

3 (I) In Example 10-1, if the heat lost from the gas in the process *BD* is 2.3×10^3 J, what is the change in internal energy of the gas?

4 (II) When 540 kcal of heat is added to 1.0 kg of water at 100°C, it is completely changed to steam at 100°C. Calculate (*a*) the work done in this process (the pressure is 1 atm), and (*b*) the change in internal energy of the water. One kilogram of steam occupies 1.67 m^3 at 100°C and 1 atm.

5 (II) An ideal gas was slowly compressed at constant temperature to one-half its original volume. In the process, 80 kcal of heat was given off. (*a*) How much work was done (in joules)? (*b*) What was the change in internal energy of the gas? (*Hint*: See Equation 9-1.)

6 (II) An ideal gas is allowed to expand adiabatically to twice its volume. In doing so, the gas does 850 J of work. (*a*) How much heat flowed into the gas? (*b*) What is the change in internal energy of the gas? (*c*) Did its temperature rise or fall?

7 (II) When 400 kcal of heat is added to a gas enclosed in a cylinder fitted with a light frictionless piston maintained at atmospheric pressure, the volume is observed to increase from 2.0 m^3 to 3.2 m^3. Calculate (*a*) the work done by the gas, and (*b*) the change in internal energy of the gas. (*c*) Graph this process on a *PV* diagram.

*8 (III) A sample of 800 mol of nitrogen gas is maintained at a constant pressure of 1 atm in a flexible container. The gas is heated from 0°C to 80°C. Calculate (*a*) the heat added to the gas, (*b*) the work done by the gas, and (*c*) the change in internal energy. (*Hint*: Refer to the discussion in Section 10-4 on the heat capacities of gases.)

* SECTION 10-3

*9 (I) How much energy would the person of Example 10-2 transform if 2.0 h of sleeping were used instead to read?

*10 (I) What is the average power output of a person who spends 10 h each day sleeping, 4 h at light activity, and 10 h watching television or loafing?

*11 (II) What is the basal metabolic rate for a 2.0-kg monkey?

*12 (II) A person decides to lose weight by sleeping $\frac{1}{2}$ h less per day, using the time for light activity. How much weight (or mass) can this person expect to lose in 1 year, assuming no change in food intake? Use the fact that 1 kg of fat stores about 9000 kcal of energy.

*13 (III) Calculate the average power output of the person in Example 10-2 assuming that the person's mass is 40 kg.

SECTION 10-4

*14 (I) By how much will the temperature rise if 80 kcal of heat is added to 300 mol of CO_2 held at constant pressure?

*15 (II) If a heater supplies 1.5×10^6 J/h to a room 6.5 by 5.0 by 3.0 m containing air, by how much will the temperature rise in 1.0 h, assuming no heat losses to the outside?

*16 (II) Show that if the molecules of a gas have f degrees of freedom, theory predicts that $C_V = (f/2)R$ and $C_P = [(f+2)/2]R$.

*17 (II) In Example 9-5, we ignored the heat absorbed by the gases evolved in the burning of the cake (CO_2 and H_2O). Calculate approximately how big a correction (in percent) this would make. Assume 2.0 g of each gas is produced.

*18 (II) The specific heat at constant volume of a particular gas is 0.182 kcal/kg·C° at room temperature, and its molecular mass is 34. (*a*) What is its specific heat at constant pressure? (*b*) What do you think is the molecular structure of this gas?

*19 (II) Estimate the molar and specific heat capacities at both constant pressure and constant volume for hydrogen gas (H_2) at room temperature.

*20 (III) How much heat must be added to 2.0 m^3 of nitrogen gas initially at 20°C to double its volume at a pressure of 1 atm?

*21 (III) The capacity of a concert hall of volume 30,000 m^3 is 2500 people. If there were no ventilation, by approximately how much would the temperature rise over a period of 2.0 h as a result of the metabolism of the people?

SECTION 10-5

22 (I) What is the maximum efficiency of a heat engine whose operating temperatures are 450°C and 310°C?

23 (I) A heat engine produces 9500 J of heat while performing 2500 J of useful work. What is the efficiency of this engine?

24 (I) The exhaust temperature of a heat engine is 270°C. What must be the high temperature if the Carnot efficiency is to be 35 percent?

25 (II) An engine that operates at half its theoretical (Carnot) efficiency operates between 545°C and 310°C while producing work at the rate of 1000 kW. How much heat is wasted per hour?

26 (II) A Carnot engine performs work at the rate of 500 kW while using 960 kcal of heat per second. If the temperature of the heat source is 620°C, at what temperature is the waste heat exhausted?

27 (II) A heat engine utilizes a heat source at 550°C and has an ideal (Carnot) efficiency of 30 percent. To increase the efficiency to 40 percent, what must be the temperature of the heat source?

28 (II) The burning of gasoline in a car releases about 3.0×10^4 kcal/gal. If a car averages 35 km/gal when driving 90 km/h, which requires 25 hp, what is the efficiency of the engine under those conditions?

29 (III) At a steam power plant, steam engines work in pairs, the output of heat from one being the heat input of the second. The operating temperatures of the first are 700°C and 440°C, and of the second 430°C and 310°C. If the heat of combustion of coal is 2.8×10^7 J/kg, at what rate must coal be burned if the plant is to put out 500 MW of power? Assume the efficiency of the engines is 70 percent of the ideal (Carnot) efficiency.

30 (III) Water is used to cool the power plant in Problem 29. If the water temperature is allowed to increase by no more than 7 C°, estimate how much water must pass through the plant per hour.

31 (III) A 40 percent efficient power plant puts out 700 MW of work (electrical energy). Cooling towers are used to take away the exhaust heat. If the air temperature is allowed to rise 8.0 C°, what volume of air (km^3) is heated per day? Will the local climate be heated significantly? (The heat capacity of air is about 7.0 cal/mol · C° at constant pressure.)

32 (III) A "perfect" refrigerator absorbs heat from the freezer compartment at a temperature of −20°C and exhausts it into the room at 25°C. How much work must be done by the refrigerator to change 0.50 kg of water at 25°C into ice at −20°C? (*Hint*: Think of the refrigerator as a heat engine, and then reverse the process.)

SECTION 10-6

33 (I) One kilogram of water is heated from 0°C to 100°C. Calculate approximately the change in entropy of the water.

34 (I) What is the change in entropy of 2.0 kg of water at 100°C when it is vaporized to steam at 100°C?

35 (I) What is the change in entropy of 10 kg of water at 0°C when it is frozen to ice at 0°C?

36 (II) If the water in Problem 35 were frozen by being in contact with a great deal of ice at −10°C, what would be the total change in entropy of the process?

37 (II) An aluminum rod conducts 180 cal/s from a heat source maintained at 250°C to a large body of water at 27°C. Calculate the rate entropy increases in this process.

38 (II) Calculate the probabilities, when you throw two dice, of obtaining (*a*) a 2, and (*b*) a 7.

39 (II) Suppose that you repeatedly shake six coins in your hand and drop them on the table. Construct a table showing the number of microstates that correspond to each macrostate. What is the probability of obtaining (*a*) three heads and three tails, and (*b*) six heads?

40 (II) One kilogram of water at 20°C is mixed with 1 kg of water at 80°C in a well-insulated container. Calculate (approximately) the net change in entropy of the system.

For additional reading

Blum, H. R., *Time's Arrow and Evolution,* rev. ed., Princeton University Press, Princeton, 1969.

Cole, D. E., "The Wankel Engine," *Sci. American,* **227**:2 (Aug., 1972), p. 14.

Davidovits, P., *Physics in Biology and Medicine,* Prentice-Hall, Inc., Englewood Cliffs, N.J., 1975.

Layzer, D., "Arrow of Time," *Sci. American,* **233**:6 (Dec., 1975), p. 56.

Lehninger, A. L., *Bioenergetics: The Molecular Basis of Biological Energy Transformations,* 2nd ed., W. A. Benjamin, Reading, Mass., 1971. Thermodynamics applied to the cell.

Magie, W. R., *Source Book in Physics,* Harvard University Press, Cambridge, Mass., 1963. Excerpts (1840s) on the conservation of energy—the first law—by Mayer, Joule, and Helmholtz; on the second law by Clausius, Kelvin, and Boltzmann.

Marston, E. H., *The Dynamic Environment,* John Wiley & Sons, Inc., New York, 1975.

Schrödinger, E., *What is Life? The Physical Aspect of the Living Cell,* Cambridge University Press, Cambridge, 1969.

Shonle, J. I., *Environmental Applications of General Physics,* Addison-Wesley Publishing Co., Inc., Reading, Mass., 1974.

Wilson, D. G., "Alternative Automobile Engines," *Sci. American,* **239**:1 (Jul., 1978), p. 39.

11 ▌ Vibrations and waves

Many objects vibrate or oscillate—an object on the end of a spring, a tuning fork, the balance wheel of a watch, a pendulum, a plastic ruler held firmly over the edge of a table and gently struck, the strings of a guitar or piano. Spiders detect prey by the vibrations of their webs, cars oscillate up and down when they hit a bump, buildings and bridges vibrate when heavy trucks pass or the wind is fierce. Indeed, because most solids are elastic (see Chapter 4), most material objects vibrate (at least briefly) when given an impulse. Electrical oscillations occur in radio and television sets. And at the atomic level, atoms vibrate within a molecule, and the atoms of a solid vibrate about their relatively fixed positions.

Vibrations and wave motion are intimately related subjects. Waves—whether ocean waves, waves on a string, earthquake waves, or sound waves in air—have as their source a vibration. In the case of sound, not only is the source a vibrating object, but so is the detector—the ear drum or the membrane of a microphone. Indeed, the medium through which a wave travels itself vibrates (such as air for sound waves). In the second half of this chapter, after we discuss vibrations, we will discuss simple waves such as those on water and on a string. In Chapter 12 we will study sound waves, and in later chapters we will encounter other forms of wave motion, including light.

Vibrations and wave motion are not really "new" phenomena, however, because they are fully explained on the basis of Newtonian mechanics.

Let us begin by examining vibrations.

FIGURE 11-1
Mass vibrating at the end of a spring.

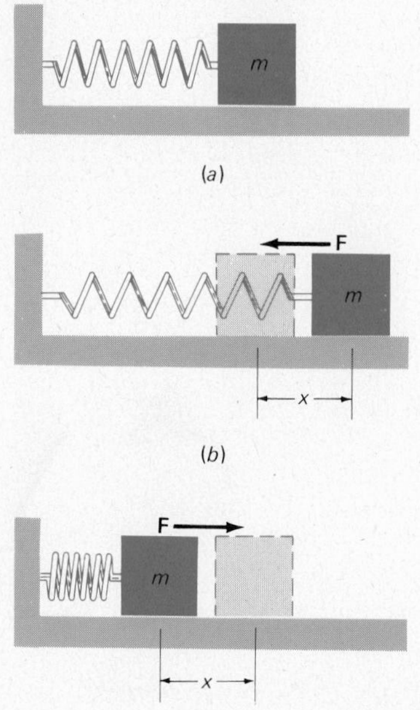

11-1 ▌ Simple harmonic motion

Oscillations of a spring

When we speak of a *vibration* or an *oscillation*, we mean the motion of an object that repeats itself over and over. That is, the motion is *periodic*. The simplest form of periodic motion is represented by an object oscillating on the end of a coil spring. Because many other types of vibrational motion closely resemble this system, we will look at it in detail. We assume that the spring is mounted horizontally, as shown in Figure 11-1*a*, and that the object of mass m slides without friction on the horizontal surface. Any spring has a natural length at which it exerts no force on the mass m, and this is called the *equilibrium position.* If the mass is moved either to the left, which compresses the spring, or to the right, which stretches it, the spring exerts a force on the mass which acts in the direction of returning it to the equilibrium position. This is called a "restoring force." The magnitude of the restoring force F is found to be directly proportional to the distance x the spring has been stretched or compressed (Figure 11-1*b* and *c*):

$$F = -kx. \tag{11-1}$$

The minus sign is present because the restoring force is always in the direction opposite the displacement x. Suppose we chose the positive direction to the right in Figure 11-1 (notice that we have taken $x = 0$ at the equilibrium position). Then x is positive when the spring is stretched; but the direction of the restoring force is to the left, or negative. And when the spring is compressed, x is negative but the force F acts toward the right and so is positive. The proportionality constant k in Equation 11-1 is called the "spring constant" or the "force constant." Notice that to stretch the spring a distance x, one has to exert an (external) force on the spring at least equal to $F = +kx$. The greater the value of k, the greater is the force needed to stretch a spring a given distance. That is, the stiffer the spring, the greater the spring constant k.

Let us examine what happens when the spring is initially stretched a distance $x = A$ as shown in Figure 11-2*a* and then released. The spring exerts a force on the mass that pulls it toward the equilibrium position; but because the mass has been accelerated by the force, it passes the equilibrium position with considerable speed. Notice that as the mass reaches the equilibrium position, the force on it decreases to zero, but its speed at this point is a maximum, Figure 11-2*b*. As it moves farther to the left, the force on it acts to slow it down, and it stops momentarily at $x = -A$, Figure 11-2*c*. It then begins moving back in the opposite direction, (*d*), until it reaches the original starting point, $x = A$ (*e*). It then repeats the motion.

To discuss vibrational motion, we need to define a few terms. The distance x of the mass from the equilibrium point at any moment is called the *displacement.* The maximum displacement—the greatest distance from the equilibrium point—is called the **amplitude**, A. A *cycle* or *vibration* refers to the complete to-and-fro motion from $x = A$ to $x = -A$ and back to $x = A$. The **period** T is defined as the time required for one complete cycle. Finally, the **frequency** f is the number of complete cycles per second. Frequency is usually specified in hertz (Hz), where 1 Hz = 1 cycle per second. It should be evident that

$$f = \frac{1}{T} \quad \text{and} \quad T = \frac{1}{f}. \tag{11-2}$$

For example, if the frequency is 5 cycles per second, each cycle takes $\frac{1}{5}$ s.

The oscillation of a spring from which a mass is hung vertically is basically the same as that of a horizontal spring. Because of the force of gravity, the length of the vertical spring at equilibrium will be longer than when it is horizontal. If x is measured from the new equilibrium position, Equation 11-1 can be used directly with the same value of k. (See the Problems.)

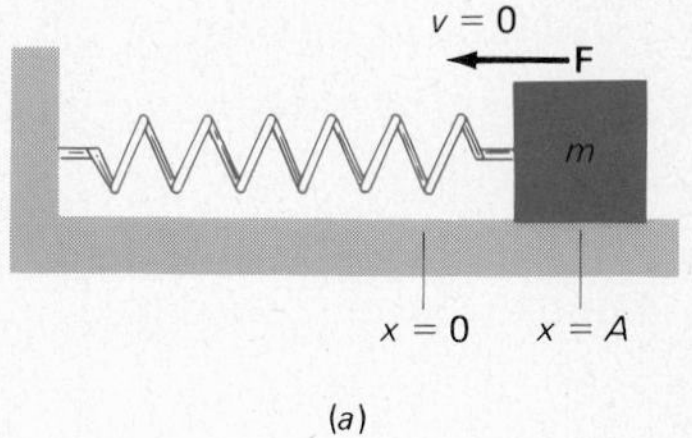

(*a*)

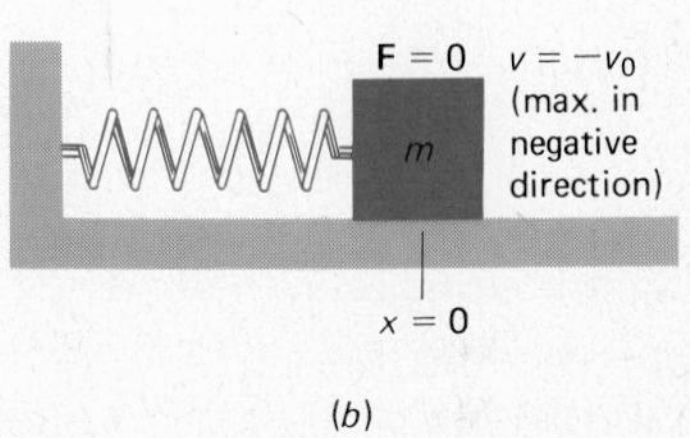

(*b*)

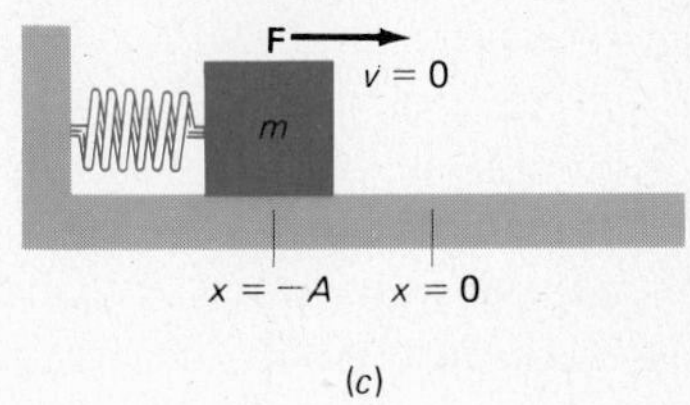

(*c*)

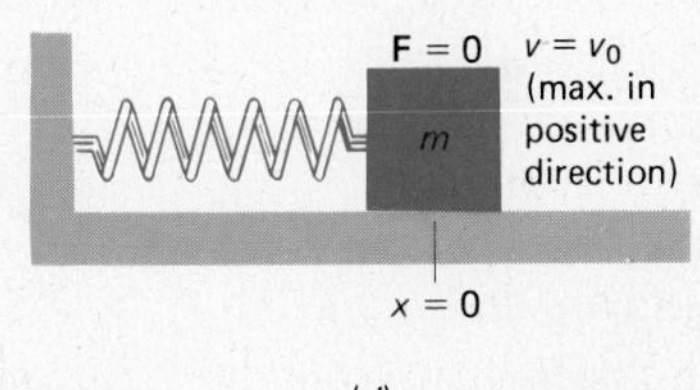

(*d*)

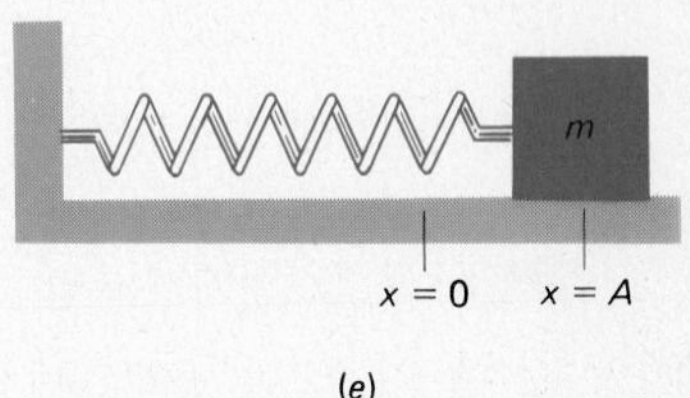

(*e*)

FIGURE 11-2
Force on, and velocity of, mass at different positions of its oscillation.

Any vibrating system for which the restoring force is given by Equation 11-1 is said to exhibit **simple harmonic motion** (SHM). Such a system is often called a *simple harmonic oscillator* (SHO). We saw in Section 4-5 that most solid materials stretch or compress according to Equation 11-1, which is known as Hooke's law, as long as the displacement is not too great. Because of this, most natural vibrations are essentially simple harmonic or close to it.

▌ EXAMPLE 11-1 When a family of four people with a total mass of 200 kg step into their 1200-kg car, the car's springs compress 3.0 cm. (*a*) What is the spring constant of the car's springs (taken together)? (*b*) How far will the car descend if loaded with 300 kg?

SOLUTION (*a*) The added force of $(200\text{ kg})(9.8\text{ m/s}^2) = 1960\text{ N}$ causes the springs to compress 3.0×10^{-2} m. Therefore, by Equation 11-1, we have

$$k = \frac{F}{x} = \frac{1960\text{ N}}{3.0 \times 10^{-2}\text{ m}} = 6.5 \times 10^4\text{ N/m}.$$

(*b*) If the car is loaded with 300 kg, $x = F/k = (300\text{ kg})(9.8\text{ m/s}^2)/(6.5 \times 10^4\text{ N/m}) = 4.5 \times 10^{-2}$ m, or 4.5 cm. We could have obtained this answer without solving for k: since x is proportional to F, if 200 kg compresses the spring 3.0 cm, then 1.5 times the force will compress the spring 1.5 times as much, or 4.5 cm. ▌▌

FIGURE 11-3
Energy changes from kinetic energy to potential energy and back again as the spring oscillates.

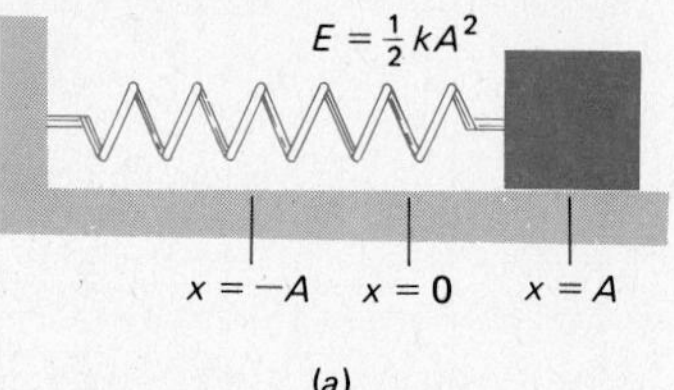

(*a*)

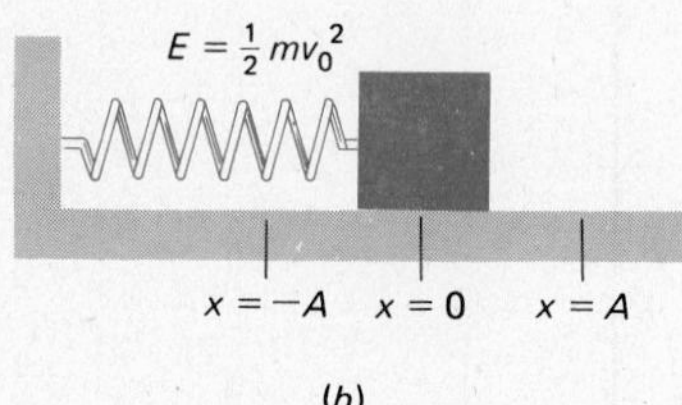

(*b*)

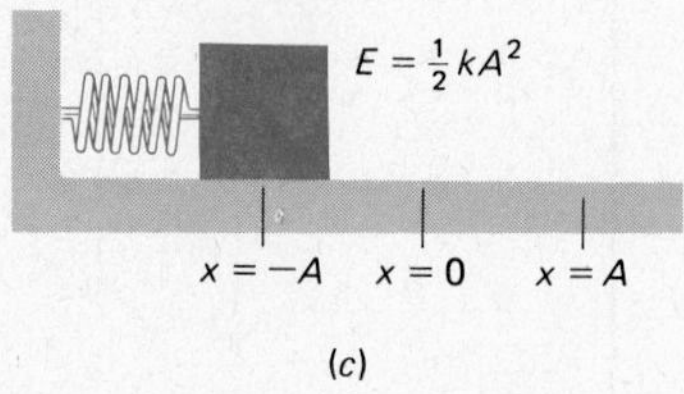

(*c*)

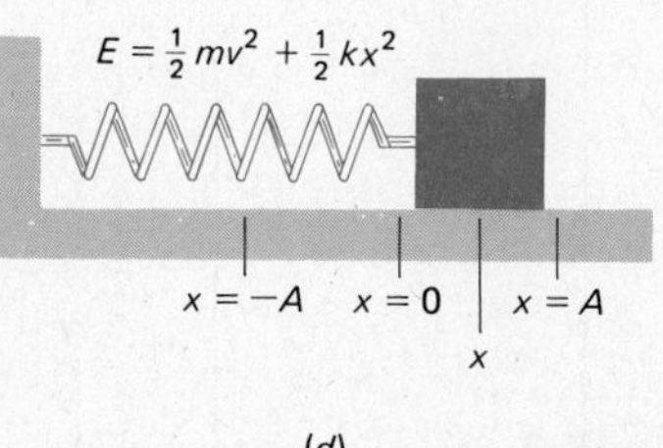

(*d*)

Energy in the simple harmonic oscillator

To stretch or compress a spring, work has to be done; hence potential energy must be stored in a stretched or compressed spring. Indeed, we have already seen in Section 5-5 that the potential energy is given by

$$\text{PE} = \tfrac{1}{2}kx^2.$$

Thus, since the total mechanical energy E is the sum of the kinetic and potential energies, we have

$$E = \tfrac{1}{2}mv^2 + \tfrac{1}{2}kx^2, \tag{11-3}$$

where v is the velocity of the mass m when it is a distance x from the equilibrium position. As long as there is no friction, the total mechanical energy E remains constant. As the mass oscillates back and forth, the energy continuously changes from potential energy to kinetic energy, and back again (Figure 11-3). At the extreme points, $x = A$ and $x = -A$, all the energy is potential energy (and is the same whether the spring is compressed or stretched to the full amplitude), so

$$E = \tfrac{1}{2}m(0)^2 + \tfrac{1}{2}kA^2 = \tfrac{1}{2}kA^2. \tag{11-4a}$$

At the equilibrium point, $x = 0$, so all the energy is kinetic:

$$E = \tfrac{1}{2}mv_0^2 + \tfrac{1}{2}k(0)^2 = \tfrac{1}{2}mv_0^2, \tag{11-4b}$$

where v_0 represents the maximum velocity during the motion (which occurs at $x = 0$). At intermediate points the energy is part kinetic and part potential. By combining Equation 11-4*a* with Equation 11-3, we can find a useful equation for the velocity as a function of the position x:

$$\tfrac{1}{2}mv^2+\tfrac{1}{2}kx^2=\tfrac{1}{2}kA^2.$$

Solving for v^2, we have

$$v^2=\frac{k}{m}(A^2-x^2)=\frac{k}{m}A^2\left(1-\frac{x^2}{A^2}\right).$$

Now, from Equations 11-4*a* and 11-4*b*, we have $\frac{1}{2}mv_0^2=\frac{1}{2}kA^2$, so $v_0^2=(k/m)A^2$. Inserting this in the equation above and taking the square root, we have

$$v=v_0\sqrt{1-x^2/A^2}. \tag{11-5}$$

This gives the velocity of the object at any position x.

▮ EXAMPLE 11-2 A spring stretches 0.150 m when a 0.30-kg mass is hung from it. The spring is then stretched an additional 0.100 m from this equilibrium point and released. Determine (*a*) the spring constant k, (*b*) the maximum velocity v_0, (*c*) the velocity v when the mass is 0.050 m from equilibrium, and (*d*) the maximum acceleration of the mass.

SOLUTION (*a*) Since the spring stretches 0.150 m when 0.30 kg is hung from it, we find k from Equation 11-1 to be

$$k=\frac{F}{x}=\frac{mg}{x}=\frac{(0.30\text{ kg})(9.8\text{ m/s}^2)}{0.150\text{ m}}=19.6\text{ N/m}.$$

(*b*) The maximum velocity v_0 is attained as the mass passes through the equilibrium point and all the energy is kinetic. By conservation of energy, $\frac{1}{2}mv_0^2=\frac{1}{2}kA^2$. Solving for v_0, we have

$$v_0=A\sqrt{\frac{k}{m}}=(0.100\text{ m})\sqrt{\frac{19.6\text{ N/m}^2}{0.30\text{ kg}}}=0.81\text{ m/s}.$$

(*c*) We use Equation 11-5 and find that

$$v=v_0\sqrt{1-x^2/A^2}$$

$$=(0.81\text{ m/s})\sqrt{1-\frac{(0.050\text{ m})^2}{(0.100\text{ m})^2}}=0.70\text{ m/s}.$$

(*d*) By Newton's second law, $F=ma$. So the maximum acceleration occurs where the force is greatest—that is, when $x=A=0.100$ m. Thus

$$a=\frac{kA}{m}=\frac{(19.6\text{ N/m})(0.100\text{ m})}{0.30\text{ kg}}=6.5\text{ m/s}^2.$$ ▮▮

The period of SHM; the reference circle

The period of a simple harmonic oscillator is found to depend on the stiffness of the spring and also on the mass m that is oscillating; but—strange as it may seem—the period does not depend on the amplitude; you can find this out for yourself by using a watch and counting 10 or 20 cycles of a spring for a small amplitude and then for a large amplitude.

We can derive a formula for the period of SHM, and this can be done by comparing SHM to an object rotating in a circle. From this same "reference circle" we can obtain a second useful result—a formula for the position of an oscillating mass at any instant of time.

Consider now a mass m rotating in a circle of radius A with speed v_0 on top of a table as shown in Figure 11-4. As viewed from above, the motion is a circle. But a person who looks at the motion from the edge of the table, sees an oscillatory motion back and forth, and this corresponds precisely to SHM, as we shall now see. What the person sees, and what we are interested in, is the projection of the circulatory motion onto the x axis (Figure 11-4*b*). To see that this motion is analogous to SHM, let us calculate the x component of the velocity v_0, which is labeled v in Figure 11-4. The two triangles shown are similar, so

$$\frac{v}{v_0}=\frac{\sqrt{A^2-x^2}}{A}$$

or

$$v=v_0\sqrt{1-\frac{x^2}{A^2}}.$$

This is exactly the equation for the speed of a mass oscillating with SHM (Equation 11-5). Thus the projection on the x axis of a ball rotating in a circle has the same motion as a mass at the end of a spring. That is, if the two motions are placed side by side, the two objects will move in unison. The period of the SHO is equal to that of the rotating object making one complete revolution; the time for one revolution is equal to the circumference of the circle (distance) divided by the speed

$$T=\frac{2\pi A}{v}.$$

From equations 11-4*a* and *b*, we have $\frac{1}{2}kA^2=\frac{1}{2}mv_0^2$, so $A/v_0=\sqrt{m/k}$. Thus

$$T=2\pi\sqrt{\frac{m}{k}}. \tag{11-6}$$

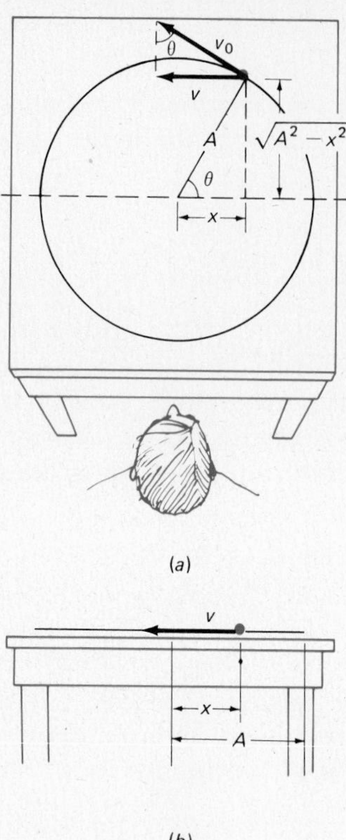

FIGURE 11-4
Analysis of simple harmonic motion as a side view (*b*) of circular motion (*a*).

This is the formula we were looking for. The period depends on the mass m and the spring constant k, but not on the amplitude. We see from Equation 11-6 that the greater the mass, the longer the period; and the stiffer the spring, the shorter the period. This makes sense since a greater mass means more inertia and therefore slower response (or acceleration); and larger k means greater force and therefore quicker response. Notice that Equation 11-6 is not a direct proportion: the period varies as the square root of m/k. For example, the mass must be quadrupled to double the period. Equation 11-6 is fully in accord with experiment and is valid not only for a spring, but for all kinds of SHM that obey Hooke's law, Equation 11-1.

▮ EXAMPLE 11-3 What are the period and frequency of the spring in Example 11-2?

SOLUTION From Equation 11-6,

$$T=2\pi\sqrt{\frac{m}{k}}=6.28\sqrt{\frac{0.30\text{ kg}}{19.6\text{ N/m}}}=0.78\text{ s}.$$

The frequency $f=1/T=1.28$ Hz.

▮ EXAMPLE 11-4 A small cockroach of mass 0.30 g is caught in a spider's web. The web vibrates predominately with a frequency of 15 Hz. (*a*) What is the value of the spring constant k for the web? (*b*) At what frequency would you expect the web to vibrate if an insect of mass 0.10 g were trapped?

SOLUTION (*a*) Using Equation 11-6 and $f = 1/T$, we find

$$f = \frac{1}{2\pi}\sqrt{\frac{k}{m}}.$$

We solve for k and find that

$$k = (2\pi f)^2 m$$

$$= (6.28 \times 15\ \mathrm{s}^{-1})^2(3.0 \times 10^{-4}\ \mathrm{kg}) = 2.7\ \mathrm{N/m}.$$

(*b*) We could substitute $m = 1.0 \times 10^{-4}$ kg in the above equation for f; but instead, we notice that the frequency decreases with the square root of the mass. Since the new mass is one-third the first mass, the frequency increases by a factor of $\sqrt{3}$. That is, $f = (15\ \mathrm{Hz})(\sqrt{3}) = 26$ Hz. ▮▮

SHM is sinusoidal

We now use the reference circle to find the position of a mass undergoing SHM as a function of time. From Figure 11-4 we see that $\cos\theta = x/A$, so the projection of the ball's position on the x axis is

$$x = A \cos\theta.$$

Since the ball is rotating with angular velocity ω, we can write $\theta = \omega t$ (see Chapter 3). Thus

$$x = A\cos\omega t. \tag{11-7a}$$

Furthermore, since the angular velocity ω (specified in radians per second) can be written as $\omega = 2\pi f$, where f is the frequency, we also can write

$$x = A\cos 2\pi ft, \tag{11-7b}$$

or in terms of the period T:

$$x = A\cos\frac{2\pi t}{T}. \tag{11-7c}$$

Notice in Equation (11-7*c*) that when $t = T$ (that is, after a time equal to the period), we have the cosine of 2π, which is the same as the cosine of zero. This is just a reflection of the fact that the motion repeats itself after a time T.

As we have seen, the projection on the x axis of the rotating object corresponds precisely to the motion of a simple harmonic oscillator. Thus, Equations 11-7 give the position of an object oscillating under simple harmonic motion. Since the cosine function varies between 1 and -1, x varies between A and $-A$, as it must. If a pen is attached to a vibrating mass as shown in Figure 11-5 and a sheet of paper is moved at a steady rate beneath it, a curve will be drawn that accurately follows Equations 11-7.

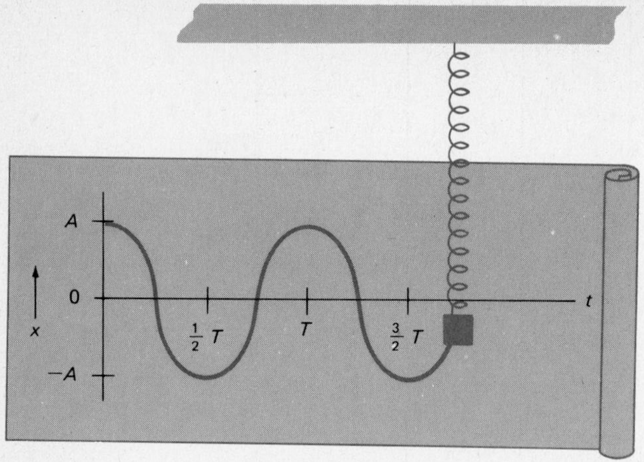

FIGURE 11-5
Sinusoidal nature of SHM as a function of time; in this case, $x = A\cos(2\pi t/T)$.

Other equations for SHM are also possible depending on the initial conditions. For example, if at $t = 0$ the oscillations are begun by giving the mass a push when it is at the equilibrium position, the equation would be

$$x = A\sin\frac{2\pi t}{T}.$$

This curve has precisely the same shape as the cosine curve shown in Figure 11-5, except it will be shifted by a quarter cycle so that it starts out at $x = 0$ instead of at $x = A$.

Either of the curves, sine or cosine, are referred to as being **sinusoidal**. Thus simple harmonic motion is said to be sinusoidal since the position varies as a sinusoidal function of time. It is easily shown that the velocity and acceleration of SHM are also sinusoidal.

It is worth noting that the equations of uniformly accelerating motion, Equations 1-9, do *not* apply to SHM since the acceleration of SHM is not constant.

▮ EXAMPLE 11-5 (*a*) What is the equation describing the motion of a spring that is stretched 20 cm from equilibrium and then released, and whose period is 1.5 s? (*b*) What will be its displacement after 1.8 s?

SOLUTION (*a*) The amplitude $A = 0.20\,\text{m}$, and $2\pi/T = 6.28/(1.5\,\text{s}) = 4.2\,\text{s}^{-1}$. Thus

$$x = 0.20\cos 4.2t\ \text{m}.$$

(*b*) At $t = 1.80$ s, $x = 0.20\cos 7.56$. The cosine repeats itself after $2\pi = 6.28$ rad so 7.56 rad gives the same result as $7.56 - 6.28 = 1.28$ rad; thus $x = 0.20\cos 1.28$. From the tables, cos 1.28 is 0.29, so $x = (0.20)(0.29) = 0.058\,\text{m}$. ▮▮

11-2 ▮ The pendulum

Simple pendulum

A simple pendulum consists of an object (the pendulum "bob") suspended from the end of a light string, Figure 11-6. We assume that the mass of the string is much less than that of the "bob," and we ignore it. A simple pendulum moving back and forth resembles simple harmonic motion: it oscillates along the arc of a circle with equal amplitude on either side of its equilibrium point (where it hangs vertically) and as it passes through the equilibrium point it has its maximum speed. But, is it really undergoing SHM? That is, is the restoring force proportional to its displacement? Let us find out.

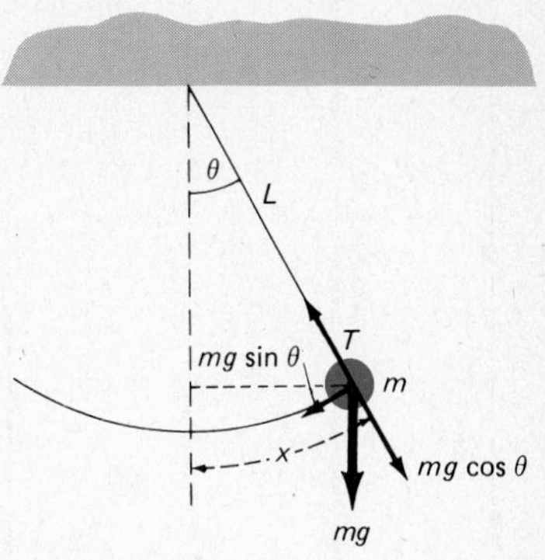

FIGURE 11-6
Simple pendulum.

The displacement of the pendulum along the arc, x, is given by $x = L\theta$, where θ is the angle the string makes with the vertical and L is the length of the string (Figure 11-6). Thus, if the restoring force is proportional to x or to θ, the motion will be simple harmonic. The restoring force is the component of the weight, mg, tangent to the arc:

$$F = -mg \sin \theta.$$

Since F is proportional to the sine of θ and not to θ itself, the motion is *not* SHM. However, if θ is small, then $\sin \theta$ is very nearly equal to θ if the latter is specified in radians. This can be seen by looking at the trigonometry table inside the back cover or by noting that in Figure 11-6 the arc length x $(= L\theta)$ is nearly the same length as the chord $(= L \sin \theta)$ indicated by the dashed line if θ is small. For angles less than 15°, the difference between θ and $\sin \theta$ is less than 1 percent. Thus, to a very good approximation for small angles,

$$F \approx -mg\theta;$$

using the fact that $x = L\theta$, we have

$$F \approx -\frac{mg}{L}x.$$

Thus the motion is essentially simple harmonic, with an effective force constant of $k = mg/L$ (see Equation 11-1). The period of a simple pendulum can be found using Equation 11-6, where for k we substitute mg/L:

$$T = 2\pi\sqrt{\frac{m}{mg/L}}$$

$$= 2\pi\sqrt{\frac{L}{g}} \qquad (\theta \text{ small}). \qquad (11\text{-}8)$$

A surprising result is that the period does not depend on the mass of the pendulum bob! You may have noticed this if you pushed a small child and a large one on the same swing.

We saw in the previous section that the period of any SHM, including a pendulum, does not depend on the amplitude. Galileo is said to have first noted this fact while watching a swinging lamp in the cathedral at Pisa; this discovery led to the pendulum clock, the first really precise timepiece, which became the standard for centuries.

Because a pendulum is not precisely SHM, the period does depend slightly on the amplitude, mainly for large amplitudes. The period of the pendulum in a clock would thus be affected slightly by the decrease in amplitude due to friction; but the mainspring (or falling weight) in a pendulum clock supplies energy to compensate for the friction and to maintain the amplitude constant, and therefore the timing remains accurate.

The pendulum finds use in geology, for geologists are interested in surface irregularities of the earth and frequently need to measure very accurately the acceleration of gravity at a given location. They often use a carefully designed pendulum to do this, as illustrated in the next example.

▮ EXAMPLE 11-6 A geologist's simple pendulum, whose length is 37.10 cm, has a frequency of 0.8190 Hz at a particular location on the earth. What is the acceleration of gravity at this location?

SOLUTION From Equation 11-8, we have

$$f=\frac{1}{T}=\frac{1}{2\pi}\sqrt{\frac{g}{L}}.$$

Solving for g, we obtain

$$\begin{aligned} g &= (2\pi f)^2 L \\ &= (6.283\times 0.8190\ \mathrm{s}^{-1})^2(0.3710\ \mathrm{m}) \\ &= 9.824\ \mathrm{m/s}^2. \end{aligned}$$

▮▮

The physical pendulum

The term "physical pendulum" refers to any real extended body which oscillates back and forth, in contrast to the rather idealized simple pendulum where all the mass is assumed concentrated in the tiny pendulum bob. A physical pendulum of arbitrary shape is shown in Figure 11-7 suspended from the point O. The weight of the object acts at the center of gravity which is a distance h from the pivot point. The physical pendulum is best analyzed using the equations of rotational motion.

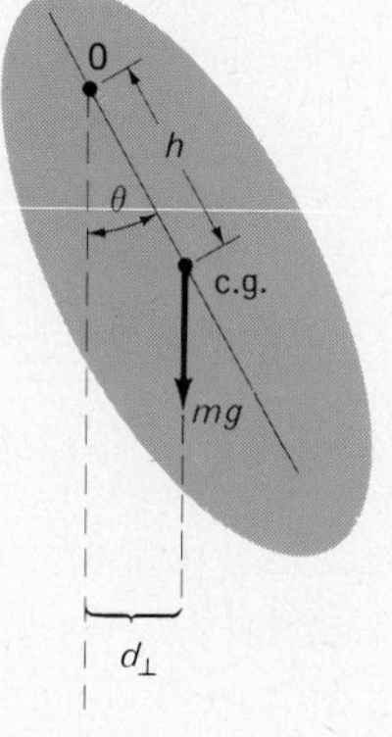

FIGURE 11-7
Physical pendulum suspended from point *O*.

For rotational SHM, the motion will be SHM if the restoring torque τ is proportional to the displacement angle θ:

$$\tau=-k'\theta. \tag{11-9}$$

For the physical pendulum, the torque is

$$\begin{aligned} \tau &= Fd_{\perp} \\ &= -(mg)(h\sin\theta), \end{aligned}$$

where $d_{\perp}=h\sin\theta$ is the lever arm. For small angles, $\sin\theta\approx\theta$, so

$$\tau\approx -mgh\theta.$$

Thus, to a first approximation, the motion is SHM and the effective value of k' is $k'=mgh$. The period can now be found by using the appropriate angular quantity for the mass in Equation 11-6, which is the moment of inertia I. Thus

$$T=2\pi\sqrt{\frac{I}{k'}}, \tag{11-10}$$

which for a physical pendulum becomes $T=2\pi\sqrt{I/mgh}$.

Equations 11-9 and 11-10 are valid not only for a physical pendulum but also for any angular harmonic motion such as the balance wheel of a watch (in this particular case they are valid even for large angles). For such other cases k' will not necessarily be equal to mgh.

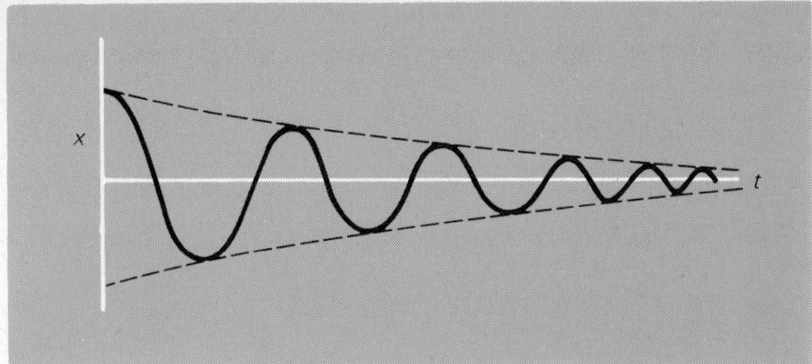

FIGURE 11-8
Damped harmonic motion.

It is interesting to note that the pace of ordinary walking is determined by the functioning of the leg as a physical pendulum. Each step is basically a half period of a swinging pendulum. In an effortless walk, the leg is allowed to swing at its natural frequency. To walk at a faster or slower rate, or to run, requires extra muscle exertion. However, you can cover more ground, often with less effort, by lengthening the stride, since the period is nearly independent of amplitude.

11-3 ▮ Nonharmonic motion

Damping

The amplitude of any real oscillating spring or swinging pendulum slowly decreases in time until the oscillations stop altogether. Figure 11-8 shows a typical graph of the displacement as a function of time. This is called **damped** harmonic motion. The damping† is generally due to the resistance of air and to internal friction within the oscillating system. The energy that is thus dissipated to thermal energy is reflected in a decreased amplitude of oscillation.

Since natural oscillating systems are damped in general, why do we even talk about (undamped) simple harmonic motion? The answer is that SHM is much easier to deal with mathematically. And if the damping is not large, the oscillations can be thought of as simple harmonic motion on which is superposed the damping—that is, the decrease in amplitude represented by the dashed curves in Figure 11-8. Although frictional damping does alter the frequency of vibration, the effect is usually small unless the damping is large; thus Equation 11-6 can still be used in most cases.

Heavily damped systems

Sometimes the damping is so large that the motion no longer resembles simple harmonic motion. Three common cases of heavily damped systems are shown in Figure 11-9. Curve C represents the situation when the damping is so large it takes a long time to reach equilibrium; the system is *overdamped.* Curve A represents an *underdamped* situation in which the system makes several swings before coming to rest. Curve B represents *critical damping*; in this case equilibrium is reached the

FIGURE 11-9
Underdamped (A), critically damped (B), and overdamped (C) motion.

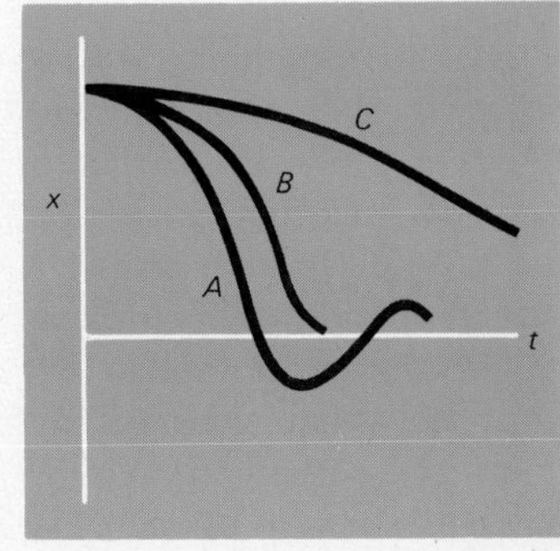

† To "damp" means to diminish, restrain, or extinguish, as to "dampen one's spirits."

quickest. These terms all derive from the use of practical damped systems such as door-closing mechanisms and shock absorbers in a car. These are usually designed to give critical damping; but as they wear out, underdamping occurs: a door slams and a car bounces up and down several times every time it hits a bump. Needles on electronic instruments (voltmeters, ammeters, level indicators on tape recorders) are usually critically damped or slightly underdamped. If they were very underdamped, they would swing back and forth excessively before arriving at the correct value; and if overdamped they would take too long to reach equilibrium, so rapid changes in the signal (say, recording level) would not be detected.

* Fourier's theorem

It is often the case that systems vibrate at more than one frequency at the same time. The vibration is then said to be a *superposition* of sinusoidal motions of various frequencies. The amplitude at any instant of time is then the sum of the individual motions.

Some kinds of oscillatory motion do not resemble simple harmonic motion at all. For example, a bouncing elastic ball is an oscillating object as it moves up and down repeatedly. The force on it is strongly upward each time it hits the ground and is equal to the constant force of gravity when it is in the air; clearly, it does not satisfy Hooke's law, Equation 11-1. Yet this, or any other periodic oscillatory motion, can be written as a superposition (or sum) of sinusoidal motions of various frequencies and amplitudes. This is known as *Fourier's theorem*, after the French mathematician who first proposed it in the nineteenth century.

As an example, Figure 11-10*a* is a graph of the voltage across a beating heart. The most obvious frequency is the usual heart beat at a frequency of about one per second (that is, 1 Hz). But superposed on this are other frequencies. An analysis of what frequencies are present is called a "Fourier analysis" and Figure 11-10*b* shows such an analysis for the heart vibration shown in part *a*.

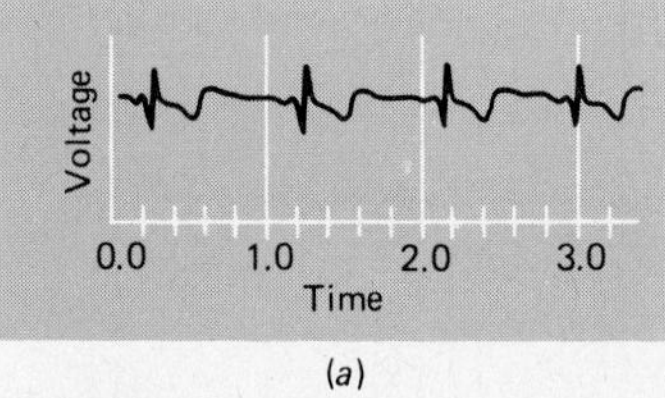

(*a*)

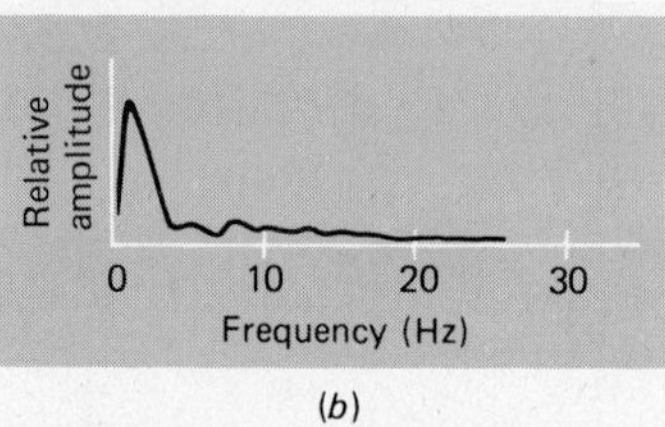

(*b*)

FIGURE 11-10
(*a*) Heart-beat trace (ECG); (*b*) frequencies present in (*a*) showing relative amplitudes.

11-4 ▌ Forced vibrations; resonance

When a vibrating system is set into motion, it vibrates at its natural frequency. Earlier we developed formulas (Equations 11-6, 11-8, and 11-10) that relate the natural frequency (or period) to the properties of the system for elastic objects (like springs) and pendulums.

However, a system is often not left to merely oscillate on its own, but may have an external force applied to it which itself oscillates at a particular frequency. For example, we might pull the mass on the spring of Figure 11-1 back and forth at a frequency f. The mass thus vibrates at the frequency f of the external force, even if this frequency is different from the natural frequency of the spring, which we will now denote by f_0 (where $f_0 = (1/2\pi)\sqrt{m/k}$; see Equation 11-6).

This is an example of *forced vibration*. The amplitude of vibration, and

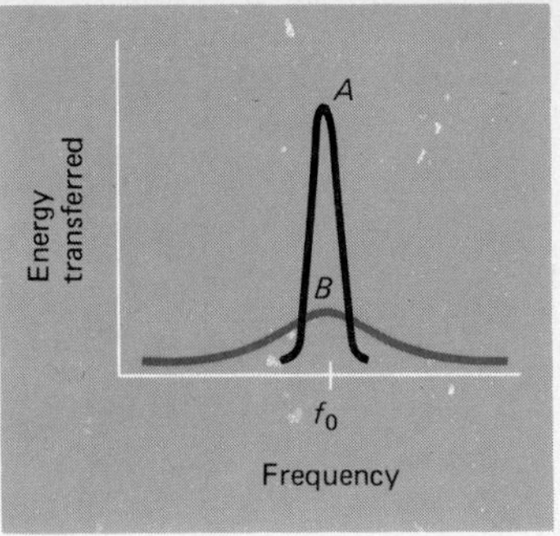

FIGURE 11-11
Resonance for lightly damped (*A*) and heavily damped (*B*) systems.

hence the energy transferred to the vibrating system, is found to depend on the difference between f and f_0. It is a maximum when the frequency of the external force equals the natural frequency of the system; that is, when $f = f_0$. The energy transferred to the system (proportional to the amplitude squared) is plotted in Figure 11-11 as a function of the external frequency f. Curve A represents light damping and curve B heavy damping. The amplitude becomes very large when $f = f_0$, particularly when the damping (friction) is small. This is known as **resonance**, and the natural vibrating frequency of a system is often called its *resonant frequency*.

A simple illustration of resonance is pushing a child on a swing. A swing, like any pendulum, has a natural frequency of oscillation. If you were to close your eyes and push on the swing at a random frequency, the swing would bounce around and reach no great amplitude. But if you push with a frequency equal to the natural frequency of the swing, the amplitude increases greatly. This clearly illustrates the fact that at resonance, relatively little effort is required to obtain a large amplitude.

The great tenor Enrico Caruso was said to be able to break a goblet by singing a note of just the right frequency at full voice. This is an example of resonance, for the sound waves emitted by the voice act as a forced vibration on the glass. At resonance, the resulting vibration of the goblet may be large enough in amplitude that the glass exceeds its elastic limit and breaks.

Since material objects are, in general, elastic, resonance is an important phenomenon in a variety of situations. It is particularly important in building, although the effects are not always to be foreseen. For example, it has been reported that a railway bridge collapsed because a nick in one of the wheels of a passing train set up a resonant vibration in the bridge. Indeed, marching soldiers break step when crossing a bridge to avoid the possibility of a similar catastrophe. And the famous collapse of the Tacoma Narrows bridge (Figure 11-12) in 1940 was due in part to the bridge resonating with the gusting of the wind.

We will meet important examples of resonance later in this chapter and in succeeding chapters. We will also see that vibrating objects often have not one, but many resonant frequencies.

FIGURE 11-12
Collapse of the Tacoma Narrows bridge due to heavy gusty winds (November 7, 1940). (Wide World Photo.)

11-5 ▮ Wave motion

Characteristics of wave motion

When you throw a stone into a lake or pool of water, circular waves form and move outward (Figure 11-13). Waves will also travel along a cord (or a "slinky") that is stretched out straight on a table if you vibrate one end back and forth as shown in Figure 11-14. Water waves, and waves on a cord, are two common examples of wave motion. We will meet other kinds of wave motion later, but for now we will concentrate on these simple "mechanical" waves.

If you have ever watched ocean waves moving toward shore, you may have wondered if the waves were carrying water into the beach. This is, in fact, not the case.† Water waves move with a recognizable velocity. But the water itself remains essentially in place, merely moving up and down as the waves pass. This is clearly demonstrated by observing leaves on a pond as waves move by. The leaves (or a cork) are not carried forward by the waves, but simply oscillate up and down, because the water itself simply moves up and down. Similarly, the wave on the rope of Figure 11-14 moves to the right, but each piece of the rope only vibrates back and forth. This is a general feature of waves: waves can move over large distances, but the medium (the water or the rope) itself has only a limited movement. Thus a wave is not matter, although it may move in matter. A wave consists of oscillations that move without carrying matter with them.

Waves carry energy from one place to another. Energy is given to a water wave, for example, by a rock thrown into the water, or by wind far out at sea. The energy is transported by waves to the shore. If you have been under an ocean wave when it breaks, you know the energy it carries. The oscillating hand in Figure 11-14 transfers energy to the rope, which is then transported down the rope and can be transferred to an object at the other end. All forms of wave motion transport energy.

Let us look a little more closely at how a wave is formed and how it comes to "travel." We first look at a single wave bump or *pulse*. A single pulse can be formed on a rope by a quick up-and-down motion of the hand (Figure 11-15). The hand pulls up on one end of the rope and because the end piece is attached to adjacent pieces, these also feel an upward force and they, too, begin to move upward. As each succeeding piece of rope moves upward, the wave crest moves outward along the rope. Meanwhile, the end piece of rope has been returned to its original position by the hand, and as each succeeding piece of rope reaches its peak position, it, too, is pulled back down again. Thus the source of a

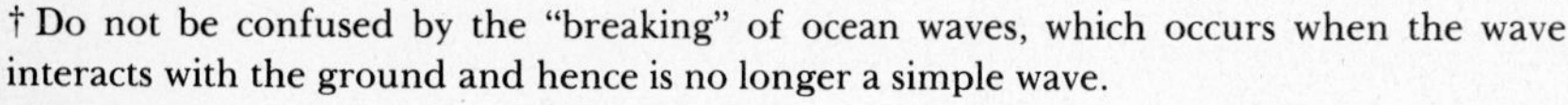
† Do not be confused by the "breaking" of ocean waves, which occurs when the wave interacts with the ground and hence is no longer a simple wave.

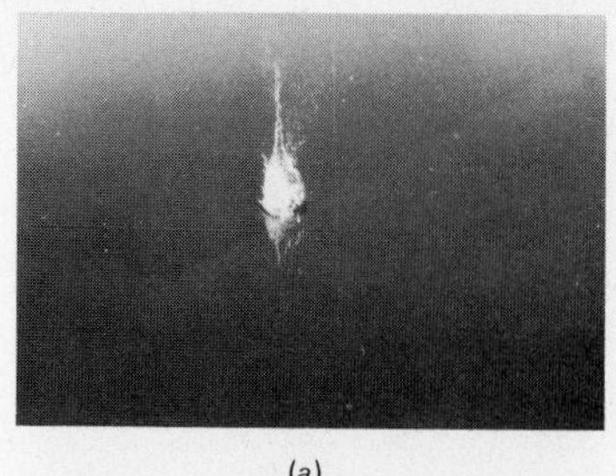
(*a*)

(*b*)

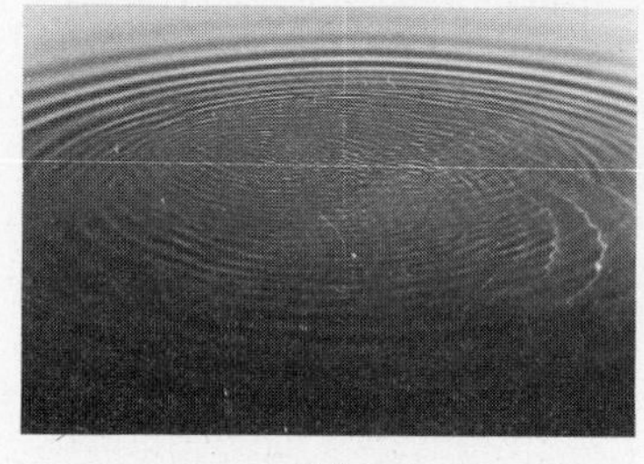
(*c*)

(*d*)

FIGURE 11-13
Water waves spreading outward from a source. (D. C. Giancoli.)

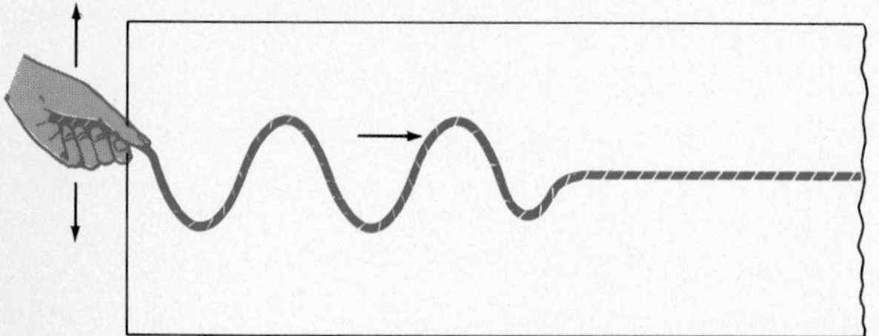

FIGURE 11-14
Wave traveling on a rope.

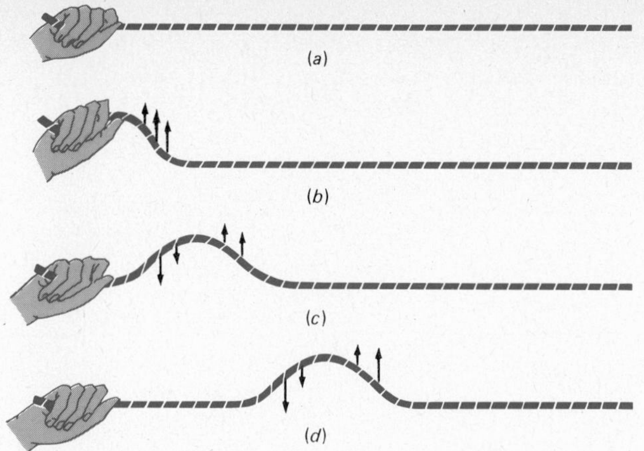

FIGURE 11-15
Motion of a wave pulse. Arrows indicate velocity of rope particles.

traveling wave pulse is a disturbance, and cohesive forces between adjacent pieces of rope cause the pulse to travel outward. Waves in other media are created and propagate outward in a similar fashion.

A *continuous* or *periodic wave*, such as that shown in Figure 11-14, has as its source a disturbance that is continuous and oscillating; that is, the source is a *vibration* or *oscillation*. In Figure 11-14 a hand oscillates one end of the rope. Water waves may be produced by any vibrating object placed at the surface such as your hand; or the water itself is made to vibrate when wind blows across it or a rock is thrown into it. A vibrating tuning fork or drum membrane gives rise to sound waves in air; and we will see later that oscillating electric charges give rise to light waves. Indeed, almost any vibrating object sends out waves.

The source of any wave, then, is a vibration. And it is the vibration that propagates outward and thus constitutes the wave. If the source vibrates sinusoidally in SHM, the wave itself will have a sinusoidal shape both in space and in time. That is, if you take a picture of the wave spread throughout space at a given instant of time, the wave will have the shape of a sine or cosine function. On the other hand, if you look at the motion of the medium at one place over a long period of time, for example if you look between two closely spaced posts of a pier or out of a ship's porthole as water waves pass by, the up-and-down motion of that small segment of water will be simple harmonic motion—the water moves up and down sinusoidally in time.

Some of the important quantities used to describe a simple periodic wave are shown in Figure 11-16. The high points on a wave are called

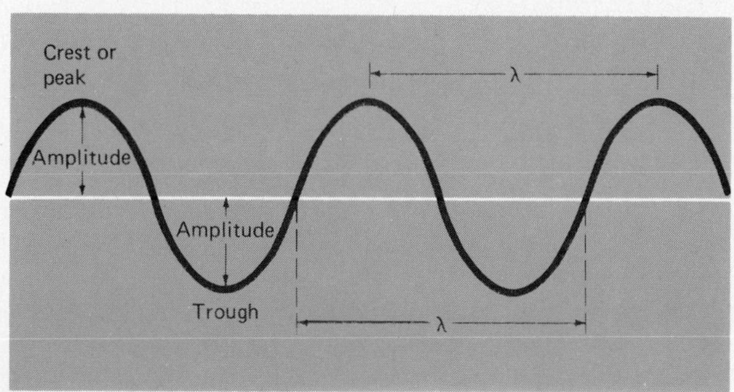

FIGURE 11-16
Characteristics of a continuous wave.

crests, the low points troughs. The **amplitude** is the distance from the midpoint of the wave to a crest or a trough; the total swing from a crest to a trough is twice the amplitude. The distance between two successive crests is called the **wavelength** λ. The wavelength is also equal to the distance between any two successive identical points on the wave. The **frequency** f (sometimes called ν, the Greek lowercase nu) is the number of crests, or complete cycles, that pass a given point per unit time. The period T, of course, is just $1/f$.

The **wave velocity**, v, is the velocity at which wave crests appear to move. (The wave velocity must be distinguished from the velocity of a particle of the medium itself.) Since a wave crest travels a distance of one wavelength, λ, in one period, T, the wave velocity $v = \lambda/T$, or

$$v = \lambda f. \tag{11-11}$$

For example, suppose that a wave has a wavelength of 5 m and a frequency of 3 Hz. Since three crests pass a given point per second, and the crests are 5 m apart, the first crest (or any other part of the wave) must travel a distance of 15 m during the 1 s, so its speed is 15 m/s.

The velocity of a wave depends on the properties of the medium in which it travels. The velocity of a wave on a string, for example, depends on the tension in the string, F_T, and on the mass per unit length, m/L; the exact relationship is

$$v = \sqrt{\frac{F_T}{m/L}}. \tag{11-12}$$

Before giving a derivation of this formula, it is worth noting that at least qualitatively it makes sense on the basis of Newtonian mechanics: that is, we do expect the tension to be in the numerator and the mass per unit length in the denominator because when the tension is greater, we expect the velocity to be greater since each segment of string is in better contact with its neighbor; and, the greater the mass per unit length, the more inertia the string has and the more slowly the wave would be expected to propagate.

▮ EXAMPLE 11-7 A wave whose wavelength is 0.50 m is traveling down a 300-m-long wire whose total mass is 30 kg. If the wire is under a tension of 4000 N, what is the velocity and frequency of this wave?

SOLUTION From Equation 11-12, the velocity is

$$v = \sqrt{\frac{4000\ \text{N}}{30\ \text{kg}/300\ \text{m}}} = 200\ \text{m/s}.$$

The frequency then is

$$f = \frac{v}{\lambda} = \frac{200\ \text{m/s}}{0.50\ \text{m}} = 400\ \text{Hz}.$$

▮▮

* Derivation of velocity equation

We can make a simple derivation of Equation 11-12 using a simple model of a string under a tension F_T as shown in Figure 11-17a. The string is

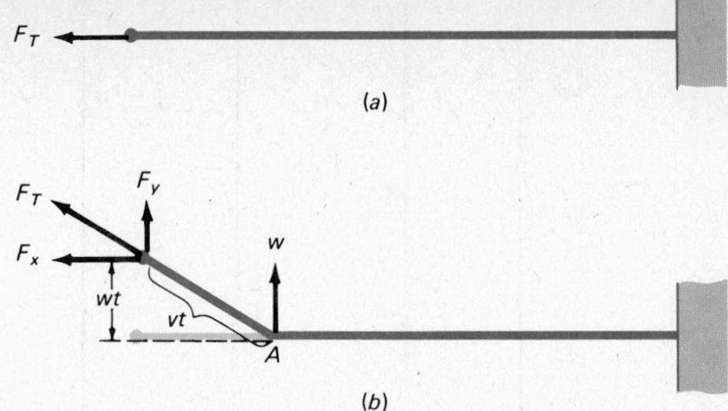

FIGURE 11-17
Diagram of simple wave pulse for derivation of Equation 11-12.

pulled upward at a speed w by the force F_y; as shown in Figure 11-17*b*, all points of the string to the left of point A move upward at the speed w, and those to the right are still at rest. The speed of propagation v of this wave pulse is the speed of point A, the leading edge of the pulse.

Point A moves to the right a distance vt in a time t, whereas the end of the rope moves upward a distance wt. By similar triangles we have the relation

$$\frac{F_T}{F_y} = \frac{v}{w}.$$

As we saw in Equation 5-3, the impulse given to an object is equal to its change in momentum. During time t the total upward impulse is $F_y t = (w/v)F_T t$; the change in momentum of the string Δp is the mass of rope moving upward times its velocity. Since the upward moving segment of rope has mass equal to the mass per unit length m/L times its length vt, we have

$$F_y t = \Delta p$$

$$\frac{w}{v}F_T t = \left(\frac{m}{L}vt\right)w.$$

Solving for v, we find that $v = \sqrt{F_T/(m/L)}$, which is Equation 11-12. Although it was derived for this special case, Equation 11-12 is valid for any wave shape. Experiment is fully in accord with this result derived from Newtonian mechanics.

11-6 ▮ Types of waves

We saw earlier that although waves may travel over long distances, the particles of the medium vibrate only over a limited region of space. When a wave travels down a rope, say from left to right, the particles of the rope vibrate up and down in a direction transverse (or perpendicular) to the motion of the wave itself. Such a wave is called a **transverse wave**. There exists another type of wave known as a **longitudinal wave.** In a longitudinal wave, the vibration of the particles of the medium is along the *same* direction as the motion of the wave. Longitudinal waves are readily formed on a stretched spring or "slinky" by alternately compressing and

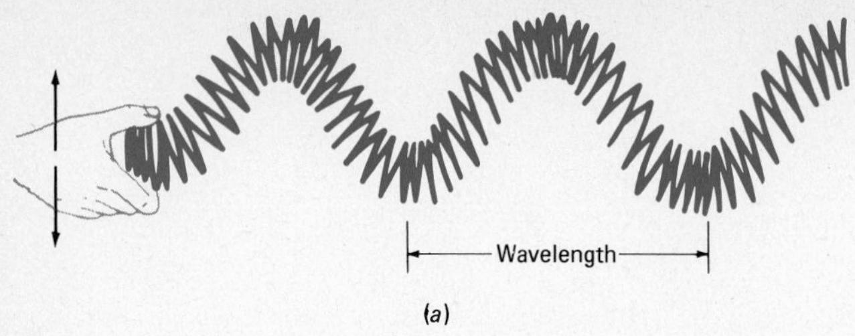

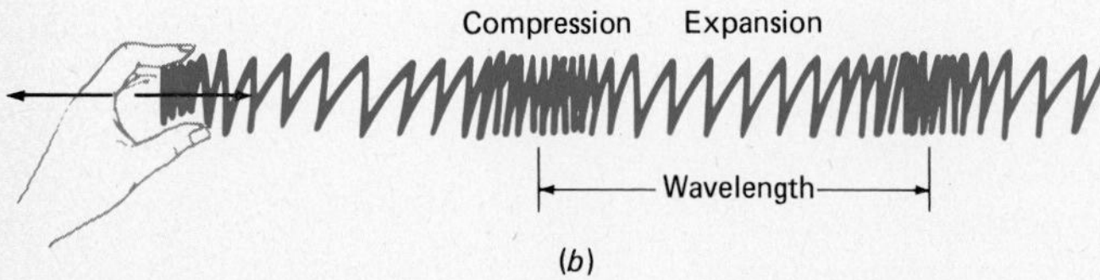

FIGURE 11-18
(*a*) Transverse wave; (*b*) longitudinal wave.

expanding one end. This is shown in Figure 11-18*b* and can be compared to the transverse wave in Figure 11-18*a*. A series of compressions and expansions propagate along the spring. The compressions are those areas where the coils are momentarily close together. Expansions (sometimes called "rarefactions") are regions where the coils are momentarily far apart. Compressions and expansions correspond to the crests and troughs of a transverse wave.

An important example of a longitudinal wave is a sound wave in air. A vibrating drum head, for example, alternately compresses and rarefies the air and produces a longitudinal wave that travels outward in the air, as shown in Figure 11-19.

As in the case of transverse waves, each section of the medium in which a longitudinal wave passes oscillates over a very small distance, whereas the wave itself can travel large distances. Wavelength, frequency, and wave velocity all have meaning for a longitudinal wave. The wavelength is the distance between successive compressions (or between successive expansions), and frequency is the number of compressions that pass a given point per second. The wave velocity is the velocity with which each compression appears to move and is equal to the product of wavelength and frequency.

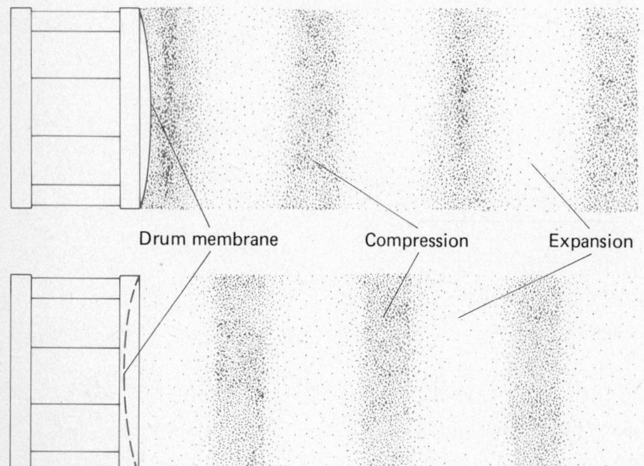

FIGURE 11-19
Production of a sound wave, which is longitudinal.

The velocity of a longitudinal wave has a form similar to that of Equation 11-12 for a transverse wave on a string; that is,

$$v = \sqrt{\frac{\text{elastic force factor}}{\text{inertia factor}}}.$$

In particular, for a longitudinal wave traveling down a long solid rod,

$$v = \sqrt{\frac{E}{\rho}}, \tag{11-13a}$$

where E is the elastic modulus (Chapter 4) of the material and ρ its density. For a longitudinal wave traveling in a liquid or gas,

$$v = \sqrt{\frac{B}{\rho}}, \tag{11-13b}$$

where B is the bulk modulus and ρ the density.

▮ EXAMPLE 11-8 You can often hear an approaching train by putting your ear to the track. How long does it take for the wave to travel down the steel track if the train is 1.0 km away?

SOLUTION Referring to Tables 4-1 and 6-1 for the elastic modulus and density of steel, respectively, we have

$$v = \sqrt{\frac{2.0 \times 10^{11}\ \text{N/m}^2}{7.8 \times 10^3\ \text{kg/m}^3}} = 5.1 \times 10^3\ \text{m/s}.$$

Then the time $t = \text{distance/velocity} = (1.0 \times 10^3\ \text{m})/(5.1 \times 10^3\ \text{m/s}) = 0.20\ \text{s}$. ▮▮

A longitudinal wave can be represented graphically by plotting the density of air molecules (or coils of a slinky) versus position as shown in Figure 11-20. We will often use such a graphical representation because it is much easier to illustrate what is happening. Note that the graph looks much like a transverse wave.

Both transverse and longitudinal waves are produced when a disturbance known as an earthquake occurs. The transverse waves that

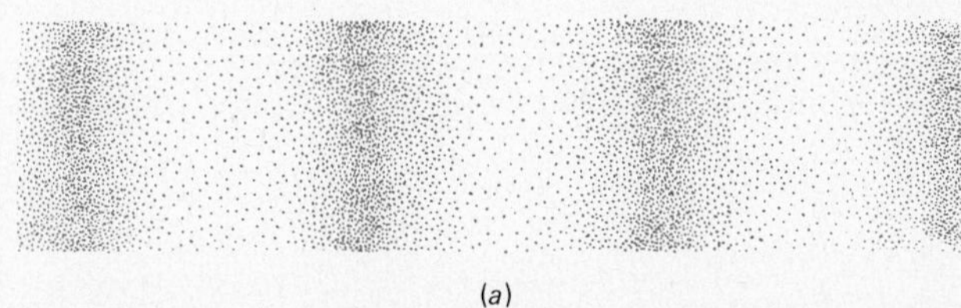

(a)

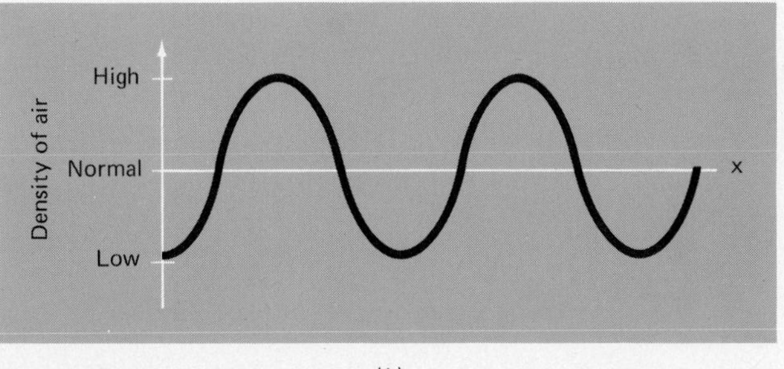

(b)

FIGURE 11-20
Graphical representation (b) of a longitudinal wave (a).

travel through the body of the earth are called S waves and the longitudinal waves are called P waves. Both longitudinal and transverse waves can travel through a solid since the atoms or molecules can vibrate about their relatively fixed positions in any direction. But in a fluid, only longitudinal waves can propagate; this is because any transverse motion would experience no restoring force since a fluid can flow. This fact was used by geophysicists in their discovery that the earth's core is molten. Longitudinal waves are detected diametrically across the earth, but never transverse waves; the only explanation for this was that the core of the earth must be liquid.

There is a third kind of wave, called a *surface wave*, which occurs at the boundary between two materials. A wave on water is actually a surface wave that moves on the boundary between water and air. The motion of each particle of water is roughly circular (Figure 11-21), so it is a combination of transverse and longitudinal motions. Surface waves are also set up on the earth when an earthquake occurs. The waves that travel along the surface (Figure 11-22) are mainly responsible for the damage caused by earthquakes.

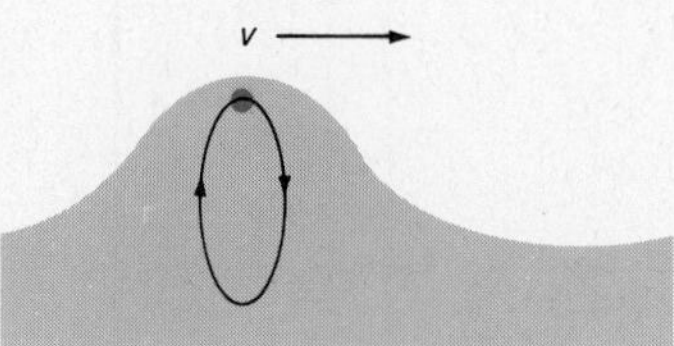

FIGURE 11-21
Motion of a single water particle as a wave passes by.

*11-7 ▎Energy transmitted by waves

FIGURE 11-22
Earthquake wave.

Waves transmit energy from one place to another. As waves travel through a medium, the energy is transmitted as vibrational energy from particle to particle of the medium. Since the particles move in SHM as a wave passes, each particle has an energy $E = \frac{1}{2}kx_0^2$, where x_0 is the amplitude of its motion, either transversely or longitudinally (see Equation 11-4*a*; we have replaced A by x_0.) Using Equation 11-6, we can write k in terms of the frequency, $k = 4\pi^2 m/T^2 = 4\pi^2 mf^2$, so that

$$E = 2\pi^2 mf^2 x_0^2.$$

The mass $m = \rho V$, where ρ is the density of the medium and V its volume. Also, the volume $V = Al$, where A is the cross-sectional area through which the wave travels, and we can write l as the distance the wave travels in a time t: $l = vt$; here v is the speed of the wave. Thus $m = \rho Al = \rho Avt$ and

$$E = 2\pi^2 \rho Avtf^2 x_0^2. \qquad (11\text{-}14)$$

From this equation we have the important result that the *energy transported by a wave is proportional to the square of the amplitude.* The rate of energy transferred is the power P:

$$P = \frac{E}{t} = 2\pi^2 \rho Avf^2 x_0^2. \qquad (11\text{-}15)$$

Finally, the *intensity* I of a wave is defined as the power transferred across unit area perpendicular to the direction of energy flow:

$$I = \frac{E}{tA} = 2\pi^2 v\rho f^2 x_0^2. \qquad (11\text{-}16)$$

If the waves flow out from the source in all directions, it is a three-dimensional wave. Examples are sound traveling in the open air, earthquake waves, and light waves. If the medium is isotropic (same in all directions), the wave is a *spherical wave* (Figure 11-23). As the wave moves outward, it is spread over a larger and larger area since the surface area of a sphere of radius r is $4\pi r^2$. Because energy is conserved, we can see from Equation 11-14 or 11-15 that as the area A increases, the amplitude x_0 must decrease; that is, at two different distances from the source, r_1 and r_2 (see Figure 11-23), $A_1x_{01}^2 = A_2x_{02}^2$, where x_{01} and x_{02} are the amplitudes of the wave at r_1 and r_2, respectively. Since $A_1 = 4\pi r_1^2$ and $A_2 = 4\pi r_2^2$, we have $(x_{01}^2 r_1^2) = (x_{02}^2 r_2^2)$, or

$$x_{01}r_1 = x_{02}r_2.$$

Thus the amplitude decreases inversely as the distance from the source. When the wave is twice as far from the source, the amplitude is half as large, and so on.

The intensity I also decreases with distance. Since I is proportional to x_0^2 (Equation 11-16), I must decrease as the *square* of the distance from the source. This inverse-square law applies to sound and light and other types of waves. Another way to view this is to consider two points r_1 and r_2 at the same time. If the power output remains constant, the intensity at r_1 is $I_1 = E/A_1t = P/4\pi r_1^2$ and $I_2 = P/4\pi r_2^2$. Thus

$$\frac{I_2}{I_1} = \frac{r_1^2}{r_2^2}. \tag{11-17}$$

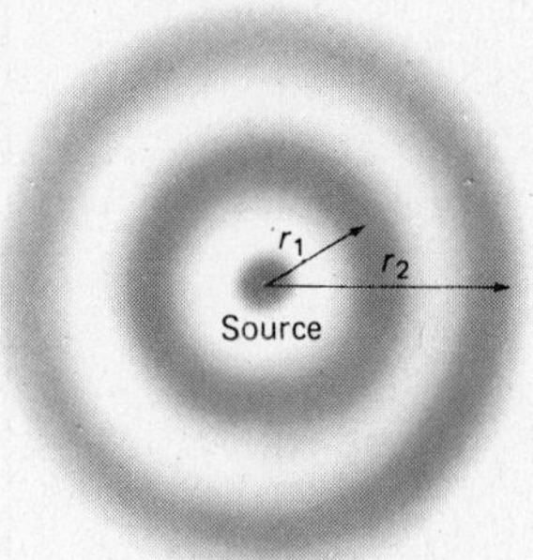

FIGURE 11-23
Wave traveling outward from source has spherical shape; two different crests (or compressions) are shown, of radius r_1 and r_2.

▌ EXAMPLE 11-9 If the intensity of an earthquake P wave 100 km from the source is 1.0×10^6 W/m^2, what is the intensity 400 km from the source?

SOLUTION The intensity decreases as the square of the distance from the source. Therefore, at 400 km, the intensity will be $(\frac{1}{4})^2 = \frac{1}{16}$ what it is at 100 km, or 6.2×10^4 W/m^2. Alternatively, Equation 11-17 could be used: $I_2 = I_1 r_1^2/r_2^2 = (1.0 \times 10^6\ \text{W/m}^2)(100\ \text{km})^2/(400\ \text{km})^2 = 6.2 \times 10^4\ \text{W/m}^2$. ▌▌

The situation is different for a one-dimensional wave, such as a transverse wave on a string or a longitudinal wave pulse traveling down a uniform metal rod. The area A remains constant so the amplitude x_0 also remains constant; thus the intensity as well as the amplitude do not decrease with distance.

In practice, frictional damping is generally present and some of the energy is transformed into thermal energy. Thus the amplitude and intensity of a one-dimensional wave decrease with distance from the source, and for a three-dimensional wave the decrease will be greater than that discussed above, although the effect is often small.

11-8 ▌ Behavior of waves: reflection, refraction, interference, and diffraction

Reflection

When a wave strikes an obstacle, or comes to the end of the medium it is traveling in, at least a part of the wave is reflected. You have probably seen water waves reflect off a rock or the side of a swimming pool. And you may

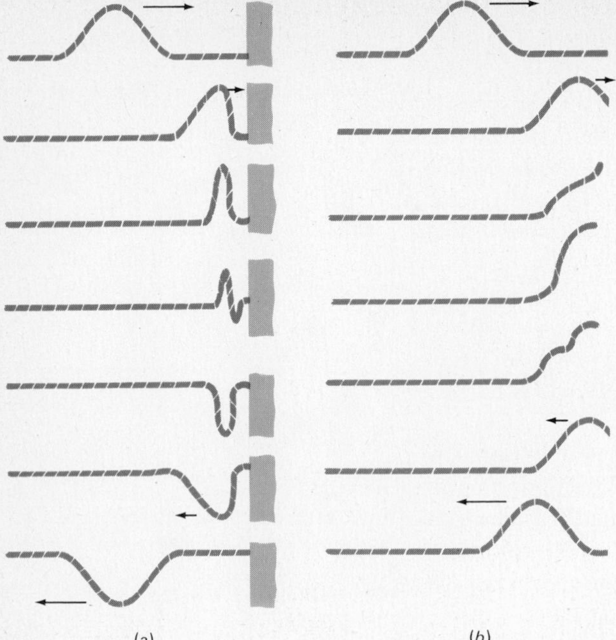

FIGURE 11-24
Reflection of a wave pulse on a rope when end of rope is (*a*) fixed and (*b*) free.

have heard a shout reflected from a distant cliff—which we call an "echo."

A wave pulse traveling down a rope (or slinky) is reflected as shown in Figure 11-24. You can observe this for yourself and see that the reflected pulse is inverted as in part *a* if the end of the rope is fixed, and returns right side up if the end is free as in part *b*. In the case of two-dimensional waves (for example, waves on the surface of water) or three-dimensional waves, we are concerned with **wave fronts**, by which we mean the whole width of a wave crest. A line drawn in the direction of motion, perpendicular to the wave front, is called a *ray*. As shown in Figure 11-25, the angle that the incoming or *incident wave* makes with the reflecting surface is equal to the angle made by the reflected wave. That is, *the angle of reflection equals the angle of incidence,* where the "angle of incidence" is defined as the angle the incident ray makes with the perpendicular to the reflecting

FIGURE 11-25
Law of reflection. (*a* courtesy of Battelle-Northwest Laboratories.)

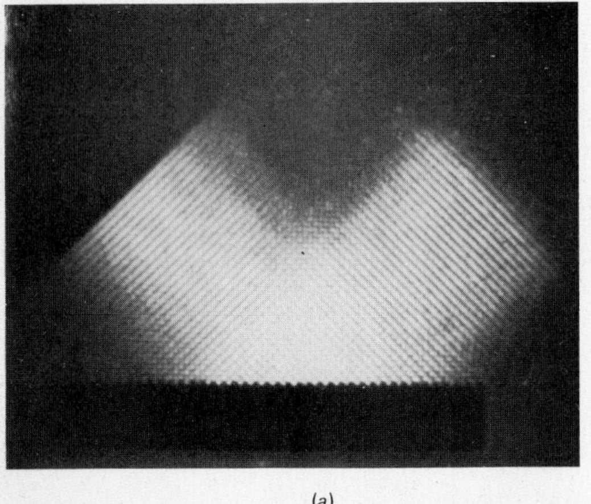

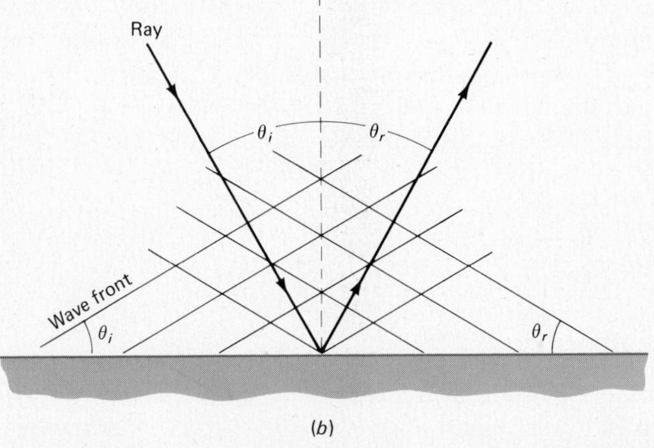

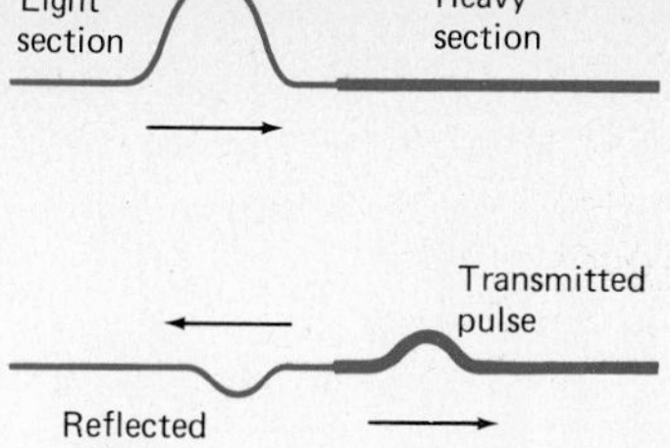

FIGURE 11-26
When a wave pulse reaches the discontinuity, part is reflected and part transmitted.

surface (or the wave front makes with a tangent to the surface) and the "angle of reflection" is the corresponding angle for the reflected wave (Figure 11-25*b*).

Refraction

When the wave pulse in Figure 11-24*a* reaches the wall, not all of the energy is reflected. Some of it is absorbed by the wall. Part of the absorbed energy is transformed into thermal energy, and part continues to propagate through the material of the wall. This is more clearly illustrated by considering a pulse that travels down a rope which consists of a light section and a heavy section as shown in Figure 11-26. When the wave reaches the boundary between the two sections, part of the pulse is reflected and part is transmitted as shown. The heavier the second section, the less is transmitted; and when the second section is a wall or rigid support, very little is transmitted.

When any wave strikes a boundary, some of the energy is reflected and some is transmitted or absorbed. When a two- or three-dimensional wave traveling in one medium crosses a boundary into a medium where its velocity is different, the transmitted wave may move in a different direction than the incident wave, as shown in Figure 11-27. This phenomenon is known as **refraction**. One example is a water wave; the velocity

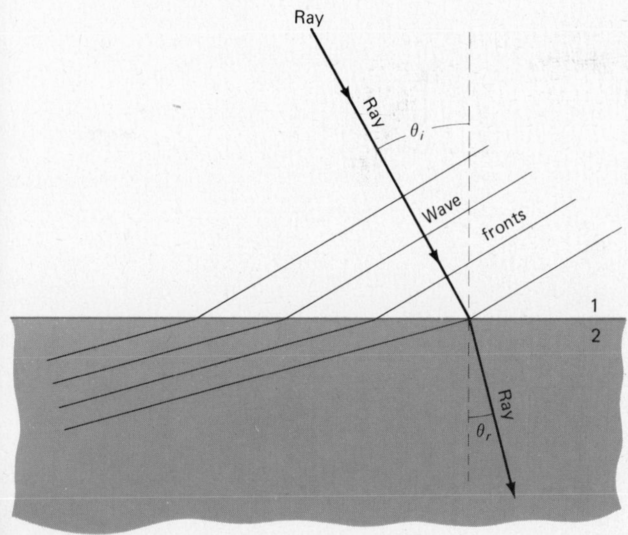

FIGURE 11-27
Refraction of waves passing a boundary.

decreases in shallow water and the waves refract. This is shown in Figure 11-28. But let us go back to the simpler case shown in Figure 11-27, where the change is abrupt and hence there is a distinct boundary. In this figure, the velocity of the wave in medium 2 is less than in medium 1. And in this case, the direction of the wave bends so it moves more nearly perpendicular to the boundary. That is, the angle of refraction, θ_r is less than the angle of incidence θ_i. To see why this is so, and to help us get a quantitative relation between θ_r and θ_i, let us think of each wave front as a row of soldiers; the soldiers are marching from firm ground (medium 1) into mud (medium 2) and hence are slowed down. The soldiers that reach the mud first are slowed down first and the row bends as shown in Figure 11–29*a*. Let us consider the wave front (or row of soldiers) labeled A in Figure 11-29*b*. In the same time t that A_1 moves a distance $l_1 = v_1 t$, we see that A_2 moves a distance $l_2 = v_2 t$. Now the two triangles shown have the side labeled a in common. Thus

$$\sin\theta_1 = \frac{l_1}{a} = \frac{v_1 t}{a}$$

and

$$\sin\theta_2 = \frac{l_2}{a} = \frac{v_2 t}{a}.$$

Dividing these two equations, we find that

$$\frac{\sin\theta_2}{\sin\theta_1} = \frac{v_2}{v_1}. \tag{11-18}$$

Since $\theta_1 = \theta_i$, the angle of incidence, and $\theta_2 = \theta_r$, the angle of refraction, Equation 11-18 gives the quantitative relation between the two. Of course, if the wave were going in the opposite direction, the argument would not be changed; only θ_1 and θ_2 would change roles: θ_1 would be the angle of refraction and θ_2 the angle of incidence. Clearly, then, if the wave travels into a medium where it can move faster, it will bend in the opposite way,

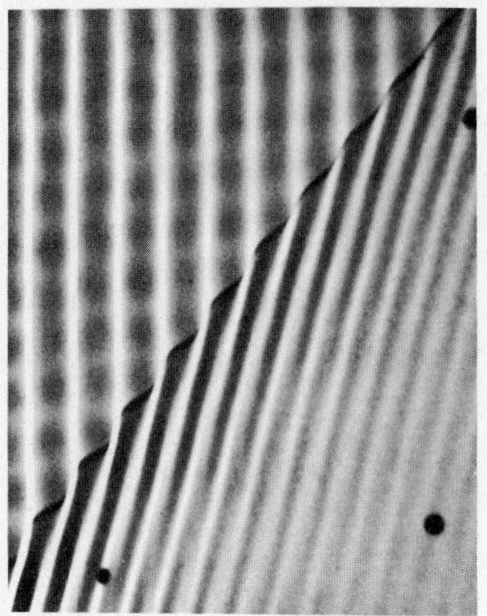

FIGURE 11-28
Refraction of water waves (proceeding from left) in a ripple tank at a boundary at which the wave speed changes because the water depth changes. Reflection also occurs at the boundary. (Courtesy of Education Development Center, Newton, Mass. 02160.)

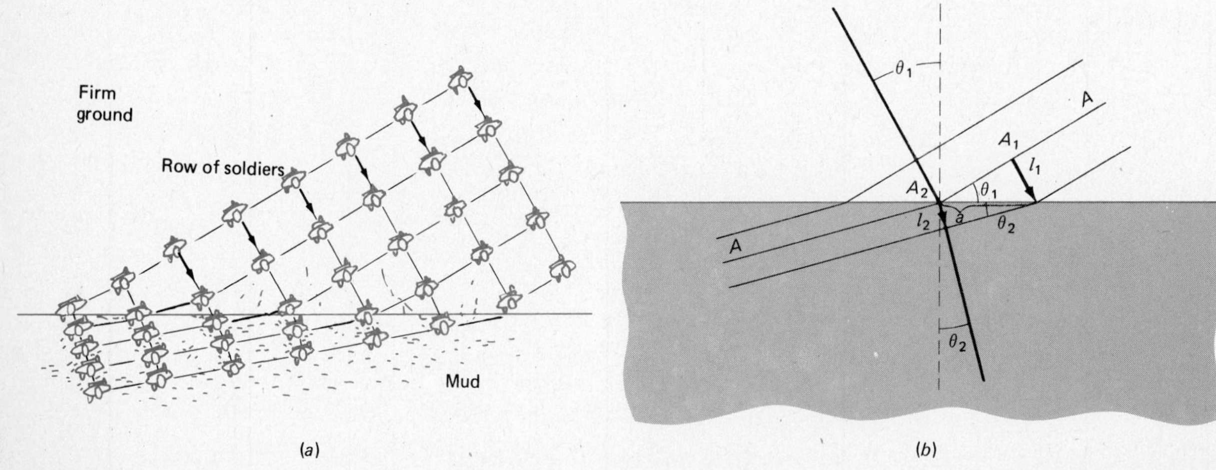

FIGURE 11-29
Soldier analogy (*a*) to derive law of refraction for waves (*b*).

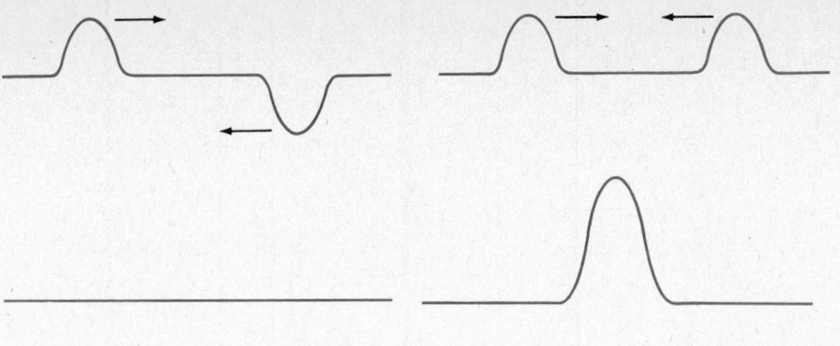

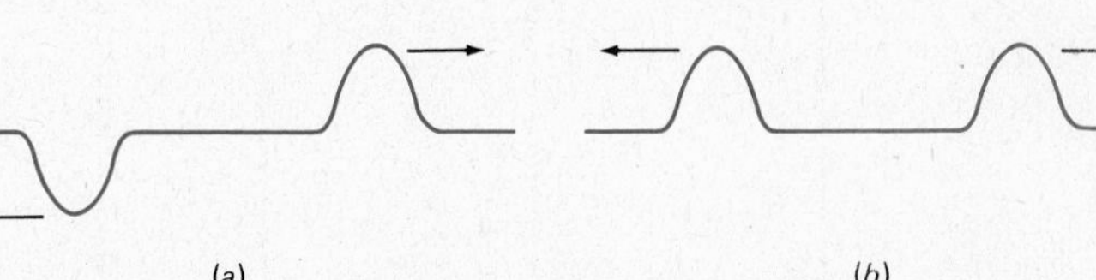

(a) (b)

FIGURE 11-30
Two wave pulses pass each other. Where they overlap, interference occurs: (*a*) destructive; (*b*) constructive.

$\theta_r > \theta_i$. We see from Equation 11-18 that if the velocity increases, the angle increases and vice versa.

Earthquake waves refract within the earth as they travel through rock of different densities (and therefore the velocity is different) just as water waves do. Light waves refract as well, and in Chapter 21 we shall find Equation 11-18 very useful.

▌ EXAMPLE 11-10 An earthquake *P* wave passes across a boundary in rock where its velocity increases from 6.5 km/s to 8.0 km/s. If it strikes this boundary at 30°, what is the angle of refraction?

SOLUTION Since sin 30° = 0.50, Equation 11-18 yields

$$\sin\theta_2 = \frac{8.0\ m/s}{6.5\ m/s}(0.50) = 0.62.$$

Using the trigonometry tables, we find that the sine of 38° is 0.62. Hence $\theta_2 = 38°$. ▌▌

Interference

Interference refers to what happens when two waves pass through the same region of space at the same time. Consider, for example, the two wave pulses on a string traveling toward each other, as shown in Figure 11-30. In Figure 11-30*a* the two pulses have the same amplitude, but one is a crest and the other a trough, and in Figure 11-30*b* they are both crests. In both cases, the waves meet and pass right on by each other. However, in the region where they overlap, the resultant displacement is the *algebraic sum of their separate displacements* (a crest is considered positive and a trough negative). This is called the *principle of superposition.* In Figure 11-30*a* the two waves oppose one another as they pass by and the result is called **destructive interference.** In Figure 11-30*b*, the resultant displacement is greater than that of either pulse and the result is called **constructive interference**.

(a)

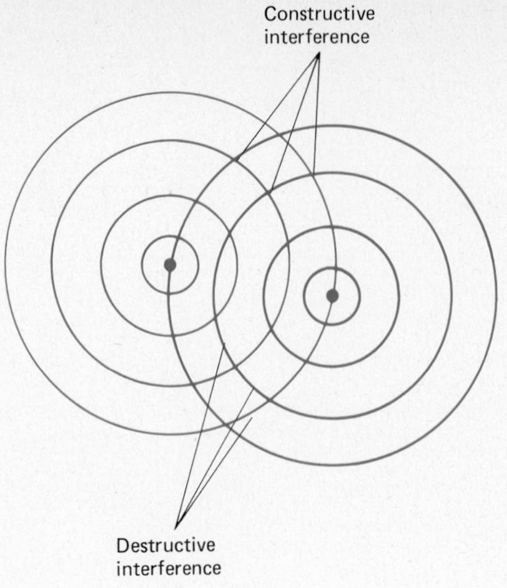

(b)

FIGURE 11-31
Interference of water waves. (D. C. Giancoli.)

When two rocks are thrown into a pond simultaneously, the two sets of circular waves interfere with one another as shown in Figure 11-31. In some areas of overlap, crests of one wave meet crests of the other (and troughs meet troughs); this is constructive interference and the water oscillates up and down with greater amplitude than either wave separately. In other areas, destructive interference occurs where the water actually does not move at all—this is where crests of one wave meet troughs of the other, and vice versa. In the first case, we say the two waves are *in phase* (by *phase* we mean the relative positions of the crests), whereas in destructive interference the two waves are *out of phase*. Of course, the relative phases of the two waves in most areas is intermediate between these two extremes, and this results in *partially destructive* interference. All three of these situations are shown in Figure 11-32, where the amplitudes are plotted versus time at a given point in space.

FIGURE 11-32
Two waves interfere: (*a*) constructively, (*b*) destructively, (*c*) partially destructively.

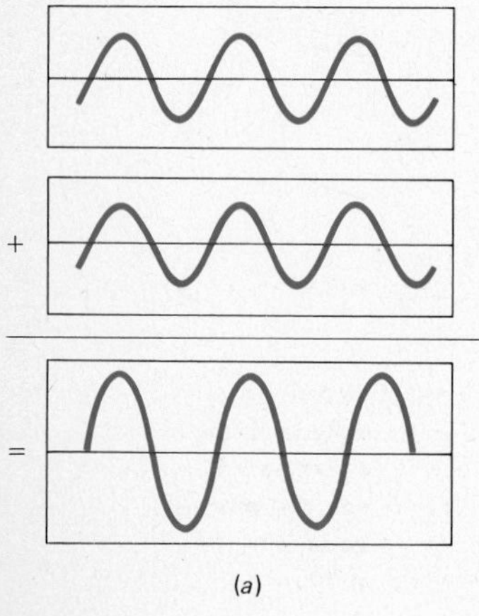

(a)

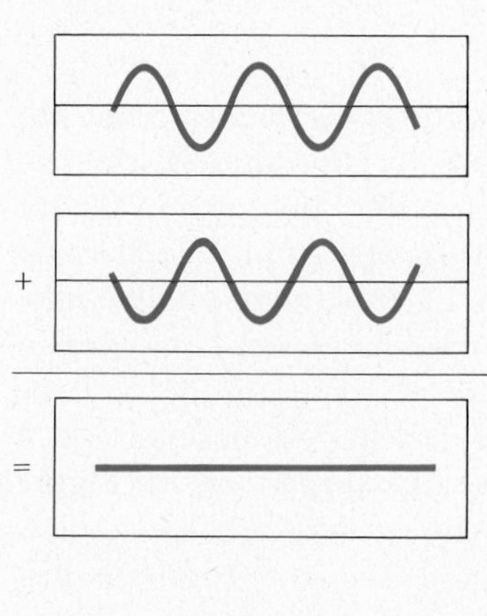

(b)

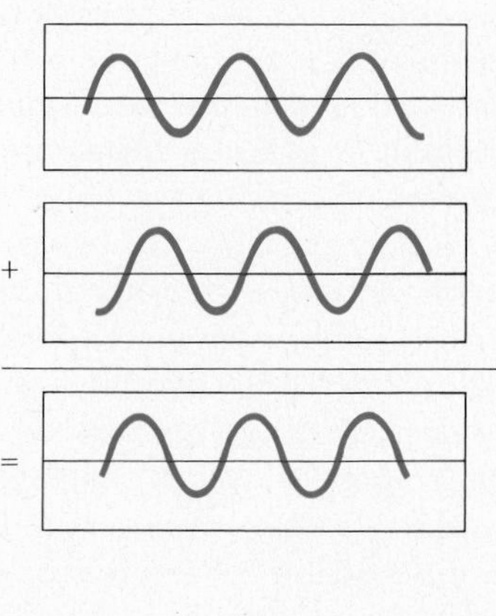

(c)

FIGURE 11-33
Wave diffraction. (From R. L. Wiegel, *Oceanographical Engineering*, Prentice-Hall, Inc., Englewood Cliffs, N.J., 1964.)

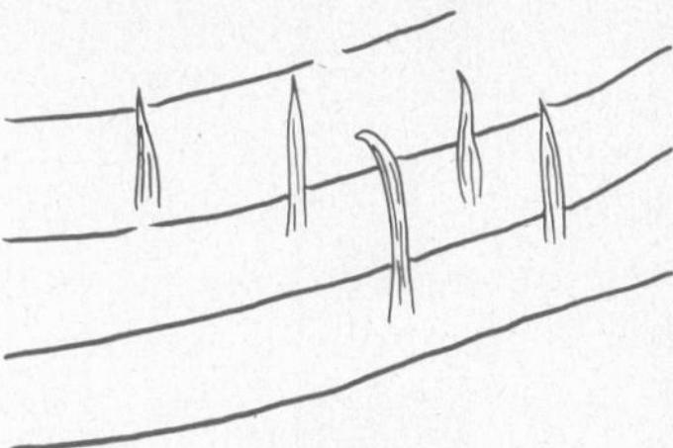

(*a*) Water waves passing blades of grass

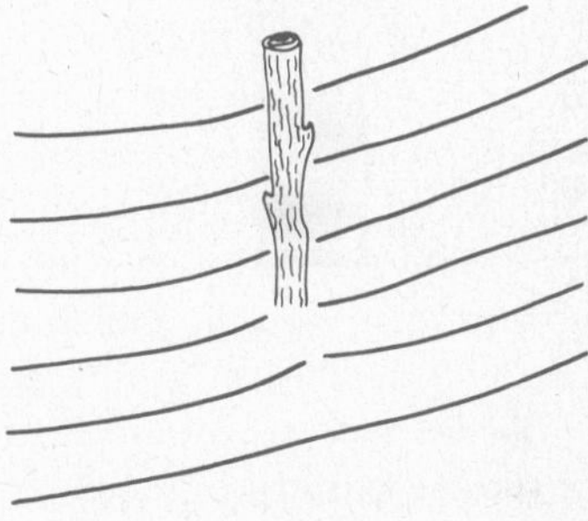

(*b*) Stick in water

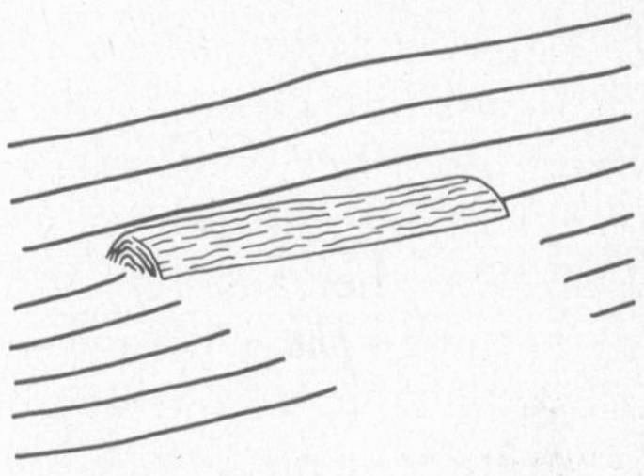

(*c*) Short wavelength waves passing log

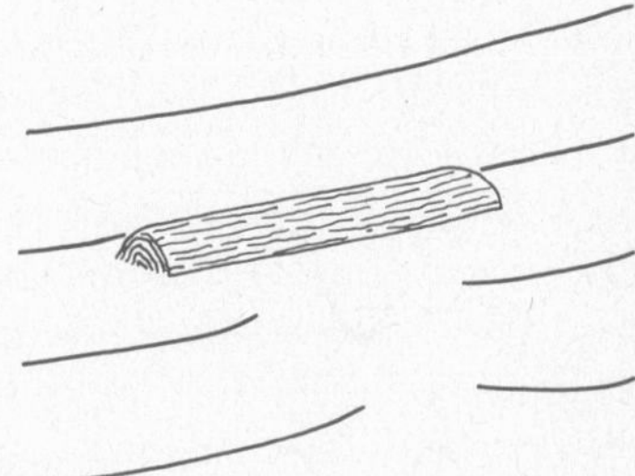

(*d*) Long wavelength waves passing log

FIGURE 11-34
Water waves passing objects of various sizes. Note that the larger the wavelength compared to the size of the object, the more diffraction there is into the "shadow region."

Diffraction

Waves exhibit another important characteristic known as **diffraction**. This refers to the fact that waves spread as they travel, and when they encounter an obstacle they bend around it somewhat and pass into the region behind as shown in Figure 11-33 for water waves.

The amount of diffraction depends on the wavelength of the wave and on the size of the obstacle. This is shown in Figure 11-34. If the wavelength is much larger than the object, such as the grass blades, Figure 11-34*a*, the wave bends around them almost as if they were not there. For larger objects (parts *b* and *c*) there is more of a "shadow" region behind the obstacle. But notice in part *d*, where the obstacle is the same as in part *c* but the wavelength is longer, that there is more diffraction into the shadow region. As a rule of thumb, only if the wavelength is less than the size of the object will there be a significant shadow region.

It is worth noting that this rule applies to *reflection* from the obstacle as well. Very little of the wave is reflected unless the wavelength is less than the size of the obstacle.

The fact that waves can carry energy to areas behind obstacles is in clear distinction to energy carried by material particles. A clear example is the following: if you are standing behind a wall, you cannot be hit by a baseball thrown from the other side, but you can hear a shout or other sound because the sound waves diffract around the edges.

Both interference and diffraction occur only for energy carried by waves and not for energy carried by material particles. This distinction was important for the determination of the nature of the light, and of matter itself, as we shall see in later chapters.

11-9 ▮ Standing waves

If you shake one end of a rope (or slinky) and the other end is kept fixed, a continuous wave will travel down to the fixed end and be reflected back. As you continue to vibrate the rope, there will be waves traveling in both directions, and the wave traveling down the rope will interfere with the reflected wave coming back. Usually, there will be quite a jumble. But if you vibrate the rope at just the right frequency, these two waves will interfere in such a way that a large-amplitude **standing wave** will be produced (Figure 11-35). It is called a "standing wave" because it does not appear to be moving. The points of destructive interference, called **nodes**, and of constructive interference, called **antinodes**, remain in fixed positions. Standing waves occur at more than one frequency. The lowest frequency of vibration that produces a standing wave gives rise to the pattern shown in Figure 11-35*a*. The standing waves shown in parts *b* and *c* are produced at precisely twice and three times the lowest frequency (assuming that the tension in the rope is the same). The rope can also vibrate with four loops at four times the lowest frequency, and so on.

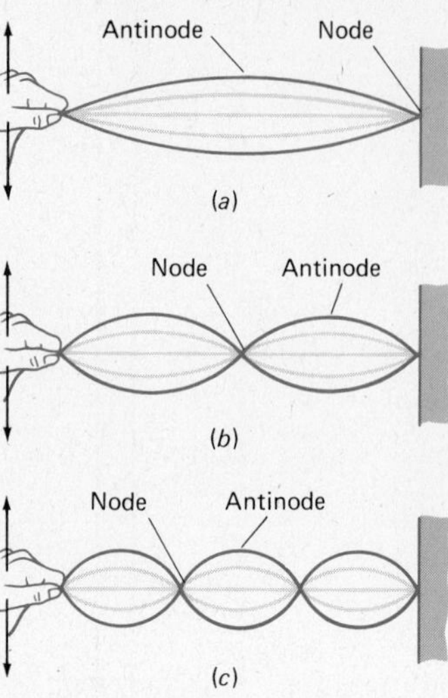

FIGURE 11-35
Standing waves corresponding to three resonant frequencies.

The frequencies at which standing waves are produced are the *natural frequencies* or **resonant frequencies** of the rope, and the different standing-wave patterns shown in Figure 11-35 are different "resonant modes of vibration." For although a standing wave is the result of the interference of two waves traveling in opposite directions, it is also an example of a vibrating object at resonance (Section 11-4). When a standing wave exists on a rope, the rope is vibrating in place; and at the frequencies at which resonance occurs, little effort is required to achieve a large amplitude. Standing waves then represent the same phenomenon as the resonance of a vibrating spring or pendulum which we discussed earlier. The only difference is that a spring or pendulum has only one resonant frequency, whereas the rope has an infinite number of resonant frequencies, each of which is a whole-number multiple of the lowest frequency.

Now let us consider a string stretched between two supports that is plucked like a guitar string (Figure 11-36*a*). Waves of a great variety of frequencies† will travel in both directions along the string, will be reflected at the ends, and will travel back in the opposite direction. Most of these waves interfere in a random way with each other and quickly die away. However, those waves that correspond to the resonant frequencies of the string will persist. The ends of the string, since they are fixed, will be nodes. There may be other nodes as well. Some of the possible resonant modes of vibration (standing waves) are shown in Figure 11-36*b*. Generally, the motion will be a combination of these different resonant modes; but only those frequencies that correspond to a resonant frequency will be present.

To determine the resonant frequencies, we first note that the wavelengths of the standing waves bear a simple relationship to the length L of the string. The lowest frequency, called the *fundamental* frequency, corresponds to one antinode (or loop); and as can be seen in Figure

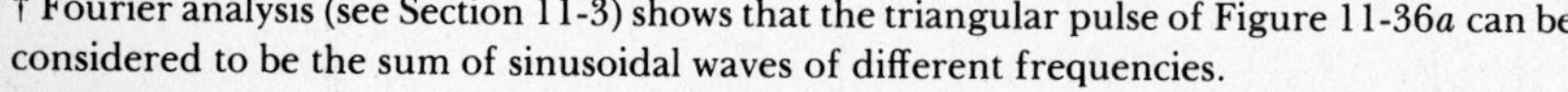
† Fourier analysis (see Section 11-3) shows that the triangular pulse of Figure 11-36*a* can be considered to be the sum of sinusoidal waves of different frequencies.

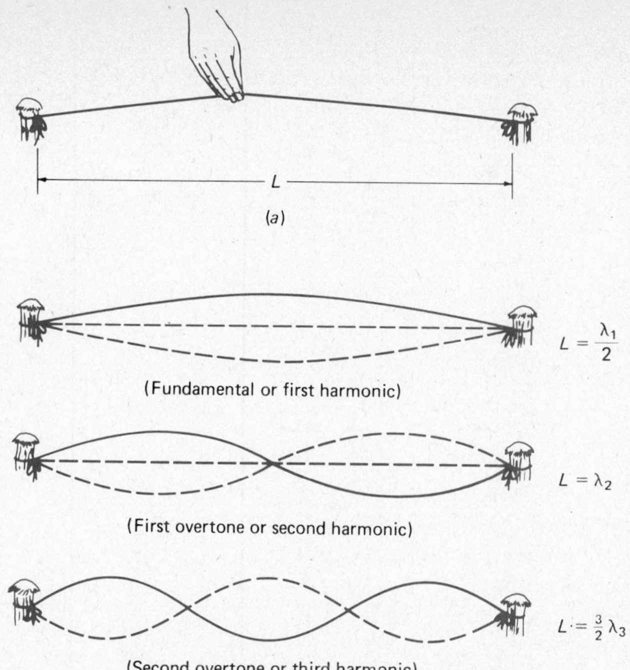

FIGURE 11-36
(*a*) A string is plucked. (*b*) Only standing waves corresponding to resonant frequencies persist for long.

11-36*b*, the whole length corresponds to one half of a wavelength. Thus $L = \frac{1}{2}\lambda_1$, where λ_1 stands for the wavelength of the fundamental. The next mode has two loops and is called the *first overtone* (a term derived from music); the length of the string L corresponds to one complete wavelength: $L = \lambda_2$. For the second and third overtones, $L = \frac{3}{2}\lambda_3$ and $L = 2\lambda_4$, respectively, and so on. In general, we can write

$$L = \frac{n\lambda_n}{2}, \qquad \text{where } n = 1, 2, 3, \ldots.$$

That is, n is an integer, and $n = 1$ corresponds to the fundamental, $n = 2$ corresponds to the first overtone, and so on. Solving for λ_n, we find that

$$\lambda_n = \frac{2L}{n}, \qquad n = 1, 2, 3, \ldots. \qquad (11\text{-}19)$$

In order to find the frequency f of each vibration, we use Equation 11-11, $f = v/\lambda$.

The various resonant frequencies of standing waves are often called *harmonics* (because in music they "harmonize"—see Chapter 12). The fundamental is called the "first harmonic," and the overtones are the second, third, etc., harmonics. The letter n in Equation 11-19 labels the number of the harmonic (see Figure 11-36*b*). Because a standing wave is equivalent to two traveling waves moving in opposite directions, the concept of velocity still makes sense and is given by Equation 11-12 in terms of the tension F_T and mass per unit length m/L: $v = \sqrt{F_T/(m/L)}$.

▎EXAMPLE 11-11 A piano string is 1.40 m long and has a mass of 110 g. (*a*) How much tension must the string be under if it is to vibrate at a fundamental frequency of 131 Hz? (*b*) What are the frequencies of the first three overtones?

SOLUTION (*a*) The wavelength of the fundamental is $\lambda_1 = 2L = 2.80$ m (Equation 11-19). The velocity is then $v = \lambda f = (2.80\text{ m})(131\text{ s}^{-1}) = 367$ m/s. Then, from Equation 11-12, we have

$$F_T = \frac{m}{L}v^2$$

$$= \frac{0.110\text{ kg}}{1.40\text{ m}}(367\text{ m/s})^2 = 1.1 \times 10^4\text{ N}.$$

(*b*) The frequencies of the first three overtones are two, three, and four times the fundamental frequency: 262, 393, and 524 Hz. ▮▮

Standing waves are produced not only on strings but on any object that is set into vibration. Even when a rock or a piece of wood is struck with a hammer, standing waves are set up that correspond to the natural resonant frequencies of that object. In general, the resonant frequencies depend on the dimensions of the object, just as for a string they depend on its length. For example, a small object does not produce as low frequencies as a large object. All musical instruments depend on standing waves to produce their musical sounds, from string instruments to wind instruments (in which a column of air vibrates as a standing wave) to drums and other percussion instruments. We shall look at this in more detail in Chapter 12.

▮ Summary

A vibrating object undergoes *simple harmonic motion* (SHM) if the restoring force is proportional to the displacement (it obeys Hooke's law). The force constant k is the ratio of restoring force to displacement, $k = F/x$. The maximum displacement is called the *amplitude*. The *period* T is the time required for one complete cycle (back and forth) and the *frequency* f is the number of cycles per second; they are related by $f = 1/T$. The period of vibration for a mass m on the end of a spring is given by $T = 2\pi\sqrt{m/k}$. SHM is *sinusoidal*, which means that the displacement as a function of time follows a sine or cosine curve. A simple pendulum of length L approximates SHM if the amplitude is not too great; its period is given by $T = 2\pi\sqrt{L/g}$, where g is the acceleration due to gravity.

During a vibration, the energy continually alternates between kinetic and potential. When friction is present, the motion is said to be *damped*: the displacement decreases in time and the energy is eventually all transformed to heat.

When an oscillating force is applied to a system capable of vibrating, the amplitude of vibration is very large if the frequency of the applied force equals (or nearly equals) the natural frequency of vibration for the object. This is called *resonance*.

A vibrating object can give rise to waves that travel outward. Waves on water or on a rope are simple examples. The wave may be a pulse (a single crest) or it may be continuous (many crests and troughs). The *wavelength* of a continuous wave is the distance between two adjacent crests; and its *frequency* is equal to the number of crests that pass a given point per second. The *velocity* of a wave (how fast a crest moves) is equal to the product of wavelength and frequency: $v = \lambda f$. The *amplitude* of a wave is defined as the height of a crest or depth of a trough, or as the amplitude of vibration within the wave. Waves carry energy from place to place without matter being carried.

In a *transverse wave*, the oscillations are perpendicular to the direction in which the wave travels. In a *longitudinal wave*, the oscillations are along the line of travel; sound is an example of a longitudinal wave.

Waves reflect off objects in their path. When a *wave front* strikes an object obliquely, the angle of reflection equals the angle of incidence. When a wave strikes a boundary between two materials in which it can travel, part of the wave is reflected and part is transmitted. A transmitted wave front may undergo *refraction* or bending. When two waves pass through the same region at the same time, they *interfere*. The resultant displacement or amplitude is the sum of their separate displacements; this can result in *constructive interference*, *destructive interference*, or something in between, depending on the amplitude and relative phases of the waves. Waves also undergo *diffraction*, which means that they tend to spread as they travel; for example, a wave passing through a small slit spreads out beyond the slit.

Waves traveling on a string (or other medium) of fixed length can interfere with waves that have reflected off the end and are traveling in the opposite direction. At certain frequencies *standing waves* can be produced in which the wave seems to be standing still instead of traveling. The string (or other medium) is vibrating as whole; this is essentially a resonance phenomenon and the frequencies at which standing waves occur are called *resonant frequencies*. The points of complete destructive interference (no vibration) are called *nodes*. Points of maximum vibrational amplitude are called *antinodes*.

Questions

1 Give some examples of everyday vibrating objects. Which follow SHM, at least approximately?

2 Is the motion of a piston in an automobile engine simple harmonic? Explain. (See Figure 10-12.)

3 Is the acceleration of a simple harmonic oscillator ever zero? If so, where?

4 If a pendulum clock is accurate at sea level, will it gain or lose time when taken to the mountains?

5 The fact that harmonic motion is damped is said to be another example of the second law of thermodynamics. Explain.

6 A tuning fork of natural frequency 264 Hz sits on a table at the front of a room. At the back of the room, two tuning forks, one of natural frequency 260 Hz and one of 420 Hz, are initially silent, but when the tuning fork at the front of the room is set into vibration, the 260-Hz fork spontaneously begins to vibrate, but the 420-Hz fork does not. Explain.

7 Give several everyday examples of resonance.

8 Is a rattle in a car ever a resonance phenomenon? Explain.

9 Why can you make water slosh back and forth in a pan only if you shake the pan at a certain frequency?

10 Is the frequency of a simple periodic wave equal to the frequency of its source? Why or why not?

11 Explain the difference between the speed of a transverse wave traveling down a rope and the speed of a tiny piece of the rope.

12 Over the years, buildings have been able to be built out of lighter and lighter materials. How has this

affected the natural vibration frequencies of buildings and the problems of resonance due to passing trucks, airplanes, or natural sources of vibration?

13 What kind of waves do you think will travel down a horizontal metal rod if you strike its end (*a*) vertically from above, and (*b*) horizontally parallel to its length?

14 Since the density of air decreases with an increase in temperature, but the bulk modulus B is nearly independent of temperature, how would you expect the speed of sound in air to vary with temperature?

15 Give two reasons why the waves which form when a stone is dropped into a lake decrease in amplitude as they travel away from the source.

16 AM radio signals can usually be heard behind a hill, but FM often cannot. That is, AM signals bend more than FM. Explain. Radio signals, as we shall see, are carried by electromagnetic waves whose wavelength for AM is typically 200 to 600 m and for FM about 3 m.

17 If a string is vibrating in three segments, are there any places one can touch it with a knife blade without disturbing the motion?

18 When a standing wave exists on a string, the vibrations of incident and reflected waves cancel at the nodes. Does this mean that energy was destroyed?

Problems

SECTION 11-1

1 (I) A piece of rubber is 45 cm long when a weight of 8.0 N hangs from it and is 58 cm long when a weight of 12.5 N hangs from it. What is the "spring" constant of this piece of rubber?

2 (I) An unloaded spring is 38 cm long and its spring constant is 62 N/m. How much will it stretch when 0.50 kg is hung from the end of it?

3 (I) When an 80-kg person climbs into an 1100-kg car, the car's springs compress vertically by 1.2 cm. What will be the frequency of vibration when the car hits a bump? (Ignore damping.)

4 (I) A spring vibrates at 3.0 Hz when a weight of 0.82 kg is hung from it. What is the spring constant?

5 (I) A fisherman's scale stretches 2.5 cm when a 2.1-kg fish hangs from it. What is the spring constant and what will be the frequency of vibration if the fish is pulled down and released so that it vibrates up and down?

6 (I) A watch spring oscillates with a frequency of 3.22 Hz. How long does it take to make 100 vibrations?

7 (I) Construct a table indicating the position of the mass in Figure 11-2 at the times $t = 0, \frac{1}{4}T, \frac{1}{2}T, \frac{3}{4}T, T, \frac{5}{4}T$, where T is the period of oscillation. On a graph of x versus t, plot these six points. Now connect these points with a smooth curve. Does your curve, based on these simple considerations, resemble that of a sine or cosine wave (Figure 11-5)?

8 (II) A 0.350-kg mass at the end of a spring vibrates 2.0 times per second with an amplitude of 0.12 m. Determine (*a*) the velocity when it passes the equilibrium point, (*b*) the velocity when it is 0.10 m from equilibrium, (*c*) the total energy of the system, and (*d*) the equation describing the motion of the mass.

9 (II) A spring vibrates with a frequency of 2.4 Hz when a weight of 0.60 kg is hung from it. What will its frequency be if only 0.30 kg hangs from it?

10 (II) A 300-kg wooden raft floats on a lake. When a 75-kg man stands on the raft, it sinks deeper into the water by 5.0 cm. When the man jumps off, the raft vibrates briefly. (*a*) What is the frequency of vibration? (*b*) What is the total energy of vibration (ignoring damping)?

11 (II) A mass of 2.0 kg stretches a spring 0.25 m. If the spring is stretched an additional 0.10 m and released, how long does it take to reach the equilibrium position again?

12 (II) It takes a force of 60 N to compress the spring of a popgun 0.10 m to "load" a 0.200-kg ball. With what speed will the ball leave the gun?

13 (II) A mass on the end of a spring is stretched 10.0 cm from equilibrium and released. At what distance from equilibrium will it have (*a*) velocity equal to half its maximum velocity and (*b*) acceleration equal to half its maximum acceleration?

14 (II) Show that if a simple harmonic oscillator at rest at $x = 0$ is given a push at $t = 0$, the position x is given by $x = x_0 \sin[(2\pi/T)t]$.

15 (II) A spring of force constant 250 N/m vibrates with an amplitude of 8.0 cm when 0.30 kg hangs from it. (*a*) What is the equation describing this motion as a function of time? Assume that the mass passes through the equilibrium point at $t = 0$. (*b*) At what times will the spring have its maximum and minimum extensions?

16 (II) A 1.0-kg mass vibrates according to the equation $x = 0.42 \cos 7.40t$, where x is in meters and t in seconds; determine (*a*) the amplitude, (*b*) the frequency, and (*c*) the total energy.

17 (II) A 0.012-kg bullet strikes a 0.300-kg block attached to a fixed horizontal spring whose spring constant is 5.2×10^3 N/m and sets it into vibration with an amplitude of 12.4 cm. What was the speed of the bullet if the two objects move together after impact?

18 (II) A 70-kg person jumps from a window to a fire net 15 m below, which stretches the net 1.2 m. Assume that the net behaves like a simple spring, and calculate how much it would stretch if the same person were lying in it. How much would it stretch if the person jumped from 30 m?

19 (II) If one vibration has 10 times the energy of a second, but their frequencies are the same, how do their amplitudes compare?

20 (II) The water in a U-shaped tube is displaced an amount Δx from equilibrium. (The level in one side is $2\Delta x$ above the level in the other side.) If friction is neglected, will the water oscillate harmonically? Determine a formula for the equivalent of the spring constant k. Does k depend on the density of the liquid, the cross section of the tube, or the length of the water column?

21 (III) If a mass m hangs from a vertical spring, show that the equilibrium length of the spring is a distance $l = mg/k$ longer than when the spring is horizontal, as in Figure 11-1. Show also that Hooke's law, $F = -kx$, holds for a vertical spring, where x is the displacement from the (vertical position) equilibrium point.

22 (III) Use the reference circle to show that the velocity and acceleration of an object undergoing SHM are given by:

$$v = -v_0 \sin \omega t, \qquad a = -a_0 \cos \omega t,$$

where $v_0 = x_0\sqrt{k/m}$ and $a_0 = kx_0/m$ refer to the maximum speed and acceleration.

23 (III) A mass m is gently placed on the end of a freely hanging spring. The mass then falls 30 cm before it stops and begins to rise. What is the frequency of the motion?

24 (III) A rectangular block of wood floats in a calm lake. Show that, if friction is ignored, when the block is pushed gently down into the water, it will then vibrate with SHM. Also, determine an equation for the force constant.

SECTION 11-2

25 (I) A pendulum makes 24 vibrations in 30 s. What is its (a) period, and (b) frequency?

26 (I) How long must a clock pendulum be if it is to make exactly one complete vibration per second?

27 (II) What is the period of a simple pendulum 50 cm long (a) on the earth, and (b) when it is in a freely falling elevator.

28 (II) A pendulum oscillates at a frequency of 2.0 Hz. At $t = 0$ it is released starting at an angle of 20°. Ignoring friction, what will be the position (angle) of the pendulum at (a) $t = 0.25$ s, (b) $t = 1.60$ s, and (c) $t = 5.00$ s?

29 (II) The length of a simple pendulum is 0.25 m and it is released at an angle of 20° to the vertical. (a) With what frequency does it vibrate? (b) What is the pendulum bob's speed when it passes through the lowest point of the swing? (Assume SHM.)

30 (II) The balance wheel of a watch is a thin ring of radius 1.0 cm and oscillates with a frequency of 3.0 Hz. If a torque of 1.0×10^{-5} N · m causes the wheel to rotate 60°, calculate the mass of the balance wheel.

31 (II) Treating the leg as a physical pendulum, determine the period (a) by measuring it for your own leg, and (b) by using Equation 11-10, assuming the leg is a long rod pivoted about one end (see Table 3-1). For part (b), assume that the leg's mass is 12.0 kg, that it is 0.80 m long, and that its cg is 0.50 m from the floor. (See also Table 3-2.)

32 (II) What is the period of a pendulum on Mars, where the acceleration of gravity is about 0.37 that on earth, if the pendulum has a period of 0.50 s on earth?

33 (III) A pendulum vibrates with an amplitude of 10°. What fraction of the time does it spend between +5° and −5°? Assume SHM.

34 (III) The pendulum of a clock consists of a small heavy bob at the end of a brass rod. The clock keeps accurate time at 15°C, at which point the pendulum has a period of 0.4220 s. When the temperature is 35°C, will the clock be fast or slow? By how much will it be in error after 1 day at 35°C?

SECTION 11-5

35 (I) A fisherman notices that wave crests pass the bow of his anchored boat every 6.0 s. He measures the distance between two crests to be 20 m. How fast are the waves traveling?

36 (I) A sound wave in air has a frequency of 262 Hz and travels with a speed of 330 m/s. How far apart are the wave crests (or compressions)?

37 (I) AM radio signals have frequencies between 550 kHz and 1600 kHz (kilohertz) and travel with a speed of 3.0×10^8 m/s. What are the wavelengths of these signals? On FM the frequencies range from 88 MHz to 108 MHz (megahertz) and travel at the same speed; what are their wavelengths?

38 (II) A rope of mass 0.75 kg is stretched between two supports 30 m apart. If the tension in the rope is 1200 N, how long will it take a pulse to travel from one support to the other?

39 (II) A 0.40-kg rope is stretched between two supports 8.0 m apart. When one support is struck by a hammer, a transverse wave travels down the rope and reaches the other support in 0.85 s. What is the tension in the rope?

40 (II) The ripples in a certain groove 12 cm from the center of a 33-rpm phonograph record have a wavelength of 2.4 mm. What will be the frequency of the sound emitted?

SECTION 11-6

41 (I) Calculate the speed of longitudinal waves in (*a*) water, (*b*) granite.

42 (I) Two solid rods have the same bulk modulus, but one is twice as dense as the other. In which rod will the speed of longitudinal waves be greater, and by what factor?

43 (II) Determine the wavelength of a 10,000-Hz sound wave traveling along an iron rod. (*Hint:* See Table 4-1.)

44 (II) A sailor strikes the side of his ship just below the surface of the sea. He hears the echo of the wave reflected from the ocean floor directly below 1.5 s later. How deep is the ocean at this point?

45 (II) S and P waves from an earthquake travel at different speeds, and this fact helps in the determination of the earthquake "focus" (where the disturbance took place). (*a*) Assuming typical speeds of 9.0 km/s and 5.0 km/s for P and S waves, respectively, how far away did the earthquake occur if a particular seismic station detects the arrival of these two types of waves 2.0 min apart? (*b*) Is one seismic station sufficient to determine the position of the focus? Explain.

* SECTION 11-7

*46 (I) Two earthquake waves have the same frequency as they travel through the same portion of the earth, but one is carrying twice the energy. What is the ratio of the amplitudes of the two waves?

*47 (I) Two waves have the same amplitude but the frequency of one is 1.8 times as large as the other. Does one transmit more energy than the other? If so, which one carries more energy and how much more does it carry?

*48 (I) Leaves on the surface of a pond are observed to move up and down a total vertical distance of 0.20 m as a wave passes. (*a*) What is the amplitude of the wave? (*b*) If the amplitude were changed to 0.15 m, by what factor would the energy in the wave change?

*49 (II) Compare (*a*) the intensities and (*b*) the amplitudes of an earthquake wave as it passes two points 10 km and 20 km from the source.

*50 (II) The intensity of a particular earthquake wave is measured to be $1.0 \times 10^6\ \text{J/m}^2 \cdot \text{s}$ at a distance of 100 km from the source. (*a*) What was the intensity when it passed a point only 1.0 km from the source? (*b*) What was the total energy passing through an area of 5.0 m^2 at 1.0 km?

*51 (III) Show that the amplitude x_0 of circular water waves decreases as the square root of the distance r from the source: $x_0 \propto 1/\sqrt{r}$.

SECTION 11-8

52 (I) The two pulses shown in Figure 11-37 are moving toward each other. Sketch the shape of the string at the moment they directly overlap.

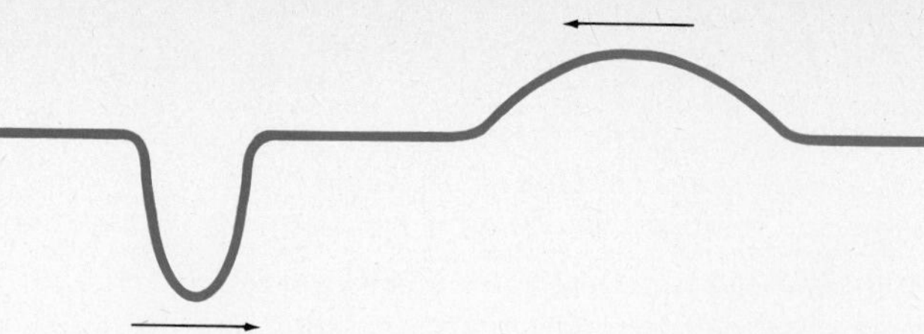

FIGURE 11-37

53 (I) An earthquake P wave traveling 13.9 km/s strikes a boundary within the earth between two kinds of material. If it approaches the boundary at an incident angle of 42° and the angle of refraction is 23°, what is the speed in the second medium?

54 (I) Water waves approach a shelf where the velocity changes from 2.8 m/s to 1.8 m/s. If the incident waves make a 25° angle with the shelf, what will be the angle of refraction?

55 (II) A longitudinal earthquake wave strikes a boundary between two types of rock at a 10° angle. As it crosses the boundary, the specific gravity of the rock changes from 3.8 to 4.7. Assuming that the elastic modulus is the same for both types of rock, determine the angle of refraction.

56 (III) It is found for any type of wave, say an earthquake wave, that if it reaches a boundary beyond which its speed is increased, there is a maximum incident angle if there is to be a transmitted refracted wave. This maximum incident angle θ_{iM} corresponds to an angle of refraction equal to 90°. If $\theta_i > \theta_{iM}$, all the wave is reflected at the boundary and none is refracted (because this would correspond to $\sin\theta_r > 1$, where θ_r is the angle of refraction, which is impossible); this is referred to as *total internal reflection*. (*a*) Find a formula for θ_{iM} using Equation 11-18. (*b*) At what angles of incidence will there be only reflection and no transmission for an earthquake *P* wave traveling 6.5 km/s when it reaches a different kind of rock where its speed is 8.2 km/s?

SECTION 11-9

57 (I) A violin string vibrates at 196 Hz when unfingered. At what frequency will it vibrate if it is fingered one-fourth of the way down from the end?

58 (I) If a violin string vibrates at 440 Hz as its fundamental frequency, what are the frequencies of the first four overtones?

59 (I) A particular string resonates in four loops at a frequency of 120 Hz. Name at least three other frequencies at which it will resonate.

60 (I) The velocity of waves on a string is 400 m/s. If the frequency of standing waves is 360 Hz, how far apart are the nodes?

61 (II) If two successive overtones of a vibrating string are 400 Hz and 450 Hz, what is the frequency of the fundamental?

62 (II) A 1.0-m-long string whose mass per unit length is 0.010 kg/m must be placed under what tension if it is to vibrate in four segments at a frequency of 120 Hz?

63 (II) A guitar string is 90 cm long and has a mass of 3.6 g. From the bridge to the support post ($= L$) is 60 cm and the string is under a tension of 550 N. What are the frequencies of the fundamental and first two overtones?

64 (II) A particular violin string plays at a frequency of 294 Hz. If the tension is increased 10 percent, what will the frequency be now?

65 (II) Derive an equation for the natural frequencies of standing waves on a string in terms of the tension and mass per unit length of the string.

66 (II) When you slosh the water back and forth in a tub, at just the right frequency, the water alternately rises and falls at each end. Suppose that the frequency required to produce such a standing wave in a 50-cm-long tub is 0.80 Hz. What is the speed of the water wave?

For additional reading

Alexander, R. M., *Animal Mechanics*, University of Washington Press, Seattle, 1968. Examples of vibration in organisms.

Boore, D. M., "Motion of the Ground in Earthquakes," *Sci. American*, **237**:6 (Dec., 1977), p. 68.

Magie, W. F., *Source Book in Physics*, Harvard University Press, Cambridge, Mass., 1963. Includes Galileo's work on the pendulum.

Schelleng, J. C., "Physics of the Bowed String," *Sci. American*, **230**:1 (Jan., 1974), p. 87.

12 ▮ Sound

Sound is of great importance in our lives. We communicate with other people by means of the sound of speech, and we obtain satisfaction and enjoyment from listening to music. Sound is important for most animals in a variety of ways. Many animals communicate through sound—in the case of whales, over distances of hundreds of kilometers. Bats, on the other hand, use sound waves to guide them in their flight. Human beings, too, have come to use sound in different ways, such as to detect the depth of the ocean using sonar, and in some kinds of medical diagnosis.

12-1 ▮ Characteristics of sound

Sound is associated with our sense of hearing, and therefore with the physiology of our ears and the psychology of our brain, which interprets the sensations that reach our ears. The term *sound* refers also to the physical entity that stimulates our ears: longitudinal waves.

We can distinguish three aspects of any sound. There must be a *source* for a sound; and as with any wave, the source of a sound wave is a vibrating object. Second, the energy is transferred from the source in the form of longitudinal sound *waves*. And third, the sound is *detected* by an ear or an instrument. We shall discuss sources and detectors of sound later in this chapter, but first we look at some aspects of sound waves themselves.

We have seen in Figure 11-19 how a vibrating drumhead produces a sound wave in air. Indeed, we usually think of sound waves traveling in the air, for normally it is air that is in contact with our ear drums and the vibrations of the air force our ear drums to vibrate. But sound waves can also travel in other materials such as water. Two stones struck together under water can be heard by a swimmer beneath the surface, for the vibrations are carried to the ear by the water. When you put the ear flat against the ground, you can hear an approaching train or truck. In this case the ground does not actually touch your ear drum but we call the longitudinal wave transmitted by the ground a sound wave just the same, for its vibrations cause the outer ear and the air within it to vibrate. Indeed, longitudinal waves traveling in any material medium are often referred to as sound waves. Clearly sound cannot travel in the absence of matter. This was first demonstrated by Robert Boyle, who showed that a bell ringing inside an evacuated jar could not be heard (Figure 12-1).

FIGURE 12-1

Sound does not travel in a vacuum. (Lester V. Bergman & Assoc., Inc.)

The speed of sound

The speed of sound is different in different materials. In air at 0°C and 1 atm, sound travels at a speed of 331 m/s. We saw in Equation 11-13 that the speed depends on the elastic modulus and the density of the material. Thus for helium, whose density is much less than that of air but whose elastic modulus is not greatly different, the speed is about three times as large as in air. In liquids and solids, which are much less compressible and therefore have much greater elastic moduli, the speed is larger still. The speed of sound in various materials is given in Table 12-1. The values depend somewhat on temperature, but this is significant mainly for gases. For example, in air, the speed increases approximately 0.60 m/s for each Celsius degree increase in temperature:

$$v = 331 + 0.60T \quad \text{m/s},$$

where T is the temperature in °C. At 20°C, for instance, $v = [331 + (0.60)(20)]$ m/s = 343 m/s.

TABLE 12-1

Speed of sound in various materials

Material	Speed (m/s)
Air (0°C)	**331**
Helium (20°C)	**1005**
Hydrogen (20°C)	**1300**
Water	**1440**
Seawater	**1560**
Iron and steel	**around 5000**
Glass	**around 4500**
Aluminum	**around 5100**
Hard wood	**around 4000**

Pitch

Two aspects of any sound are immediately evident to a human listener. These are "loudness" and "pitch," and each refers to a sensation in the consciousness of the listener. But to each of these subjective sensations there corresponds a physically measurable quantity. Loudness is related to the energy in the sound wave and we shall discuss it in the next section. The **pitch** of a sound refers to whether it is high, like the sound of a piccolo or violin, or low, like the sound of a bass drum or string bass. The physical quantity that determines pitch is the frequency, a fact that was first noted by Galileo. The lower the frequency, the lower the pitch, and the higher the frequency, the higher the pitch.†

The human ear responds to frequencies in the range from about 20 Hz to about 20,000 Hz. This is called the *audible range.* These limits vary somewhat from one individual to another. One general trend is that as people age, they are less able to hear the high frequencies, so that the high-frequency limit may be 10,000 Hz or less.

Sound waves whose frequencies are outside the audible range may reach the ear, but we are not generally aware of them. Frequencies above 20,000 Hz are called *ultrasonic* (do not confuse with "supersonic," which means a speed faster than the speed of sound). Many animals can hear ultrasonic frequencies; dogs, for example, can hear sounds as high as 50,000 Hz and bats can detect frequencies as high as 100,000 Hz. Ultrasonic waves have a number of applications in medicine and other fields, which we will discuss later in this chapter.

Sound waves whose frequencies are below the audible range, that is, less than 20 Hz, are called *infrasonic.* Sources of infrasonic waves are earthquakes, thunder, volcanoes, and waves produced by vibrating heavy machinery. This last source can be particularly troublesome to workers, for infrasonic waves—even though inaudible—can cause damage to the human body. These low-frequency waves act in a resonant fashion, causing considerable motion and irritation of internal organs of the body.

† Although pitch is determined mainly by frequency, it also depends to a slight extent on loudness; for example, a very loud sound may seem slightly lower in pitch than a quiet sound of the same frequency.

12-2 ▮ Sound intensity

Intensity and intensity level (decibels)

Like pitch, **loudness** is a sensation in the consciousness of a human being. It, too, is related to a physically measurable quantity, the **intensity** of the wave. Intensity is defined as the energy transported by a wave per unit time across unit area. Since energy per unit time is power, intensity has units of power per unit area, or watts/meter^2 (W/m^2). The intensity depends on the amplitude of the wave (actually it is proportional to the square of the amplitude, as was shown in Section 11-7).

The human ear can detect sounds with an intensity as low as 10^{-12} W/m^2 and as high as 1 W/m^2 (and even higher, although above this it is painful). This is an incredibly wide range of intensity, spanning a factor of 10^{12} from lowest to highest. Presumably because of this wide range, what we perceive as loudness is not directly proportional to the intensity. True, the greater the intensity, the louder the sound. But to produce a sound that sounds about twice as loud requires a sound wave that has about 10 times the intensity. This is true to a first approximation at any sound level. For example, a sound wave of intensity 10^{-9} W/m^2 sounds to an average human being as if it is about twice as loud as one whose intensity is 10^{-10} W/m^2; and an intensity of 10^{-2} W/m^2 sounds about twice as loud as 10^{-3} W/m^2.

Because of this relationship between the subjective sensation of loudness and the physically measurable quantity intensity, it is usual to specify sound intensity levels using a logarithmic scale. The unit on this scale is a *bel*, or much more commonly, the *decibel* (dB), which is $\frac{1}{10}$ bel (1 dB = 0.1 bel). The **intensity level**, β, of any sound is defined in terms of its intensity, I, as follows:

$$\beta \text{ (in dB)} = 10 \log \frac{I}{I_0} \tag{12-1}$$

where I_0 is the intensity of some reference level and the logarithm is to the base 10. I_0 is usually taken as the minimum intensity audible to an average person, the "threshold of hearing," which is $I_0 = 1.0 \times 10^{-12}$ W/m^2. The intensity level of a sound, for example, whose intensity is $I = 1 \times 10^{-10}$ W/m^2 will be (logarithms are discussed in Appendix A)

$$\begin{aligned}\beta &= 10 \log \frac{10^{-10}}{10^{-12}} \\ &= 10 \log 100 \\ &= 20 \text{ dB},\end{aligned}$$

since log 100 is exactly equal to 2; notice that the intensity level at the threshold of hearing is 0 dB; that is, $\beta = 10 \log 10^{-12}/10^{-12} = 10 \log 1 = 0$, since $\log 1 = 0$. Notice, too, that an increase in intensity by a factor of 10 corresponds to a level increase of 10 dB. An increase in intensity by a factor of 100 corresponds to a level increase of 20 dB (since log 100 = 2). Thus a 40-dB sound is 100 times more intense than a 20-dB sound.

The intensities and intensity levels for a number of common sounds are listed in Table 12-2.

TABLE 12-2

Intensity of various sounds

Source of the sound	Intensity level (dB)	Intensity (W/m^2)
Jet plane at 30 m	**140**	**100**
Threshold of pain	**120**	**1**
Loud indoor rock concert	**120**	**1**
Siren at 30 m	**100**	1×10^{-2}
Auto interior, moving at 90 km/h	**75**	3×10^{-5}
Busy street traffic	**70**	1×10^{-5}
Ordinary conversation	**65**	3×10^{-6}
Quiet radio	**40**	1×10^{-8}
Whisper	**20**	1×10^{-10}
Rustle of leaves	**10**	1×10^{-11}
Threshold of hearing	**0**	1×10^{-12}

▮ EXAMPLE 12-1 A high-quality loudspeaker is advertised to reproduce, at full volume, frequencies from 30 Hz to 18,000 Hz with uniform intensity ±3 dB. That is, over this frequency range, the intensity level does not vary by more than 3 dB from the average. By how much does the intensity change for the maximum intensity-level change of 3 dB?

SOLUTION Let us call the average intensity I_1 and the average level β_1. Then the maximum intensity, I_2 corresponds to a level $\beta_2=\beta_1+3$ dB. Thus

$$\beta_2-\beta_1=10\log\frac{I_2}{I_0}-10\log\frac{I_1}{I_0}$$

$$3\text{ dB}=10\left(\log\frac{I_2}{I_0}-\log\frac{I_1}{I_0}\right)$$

$$=10\log\frac{I_2}{I_1}$$

since $(\log a-\log b)=\log a/b$. Thus

$$\log\frac{I_2}{I_1}=0.30.$$

We now must look in the log table to find the logarithm of what number equals 0.30, and we find that log 2.0 = 0.30. Thus

$$\frac{I_2}{I_1}=2.0,$$

or I_2 is twice as intense as I_1. ▮▮

It is worth noting that a sound-level difference of 3 dB (which corresponds to a doubled intensity as we just saw) corresponds to only a very small change in the subjective sensation of apparent loudness. Indeed, the average human being can distinguish a difference in level of only about 1 dB.

* Intensity related to amplitude and pressure amplitude

The intensity I is proportional to the square of the wave amplitude x_0, as discussed in Section 11-7. Indeed, using Equation 11-16 we can relate the amplitude quantitatively to the intensity I or level β, as the following example shows.

▮ EXAMPLE 12-2 Calculate the maximum displacement of air molecules for a sound at the threshold of hearing, having a frequency of 1000 Hz.

SOLUTION We use Equation 11-16 and solve for x_0:

$$x_0 = \frac{1}{\pi f}\sqrt{\frac{I}{2\rho v}}$$

$$= \frac{1}{(3.14)(1.0\times 10^{3}\ \mathrm{s}^{-1})}\sqrt{\frac{10^{-12}\mathrm{W/m^2}}{(2)(1.29\ \mathrm{kg/m^3})(331\ \mathrm{m/s})}}$$

where we have taken the density of air to be $1.29\ \mathrm{kg/m^3}$ and the speed of sound in air (assumed 0°C) as 331 m/s. When we do the arithmetic, we find that $x_0 = 1.0\times 10^{-11}$ m. ▮▮

The result of this example illustrates just how sensitive the human ear is. For it can detect displacements of air molecules which are actually less than the diameter of atoms (about 10^{-10} m)!

We have so far described sound waves in terms of the vibration of the molecules of the medium. But they can also be viewed from the point of view of pressure. Indeed, longitudinal waves are often referred to as *pressure waves*. Referring to Figure 11-19, it should be clear that the pressure of the air in a compression is higher than normal, whereas in a rarefaction (or expansion) the pressure is less than normal. Figure 12-2 shows a graphical representation of a sound wave in air both in terms of displacement (part a) and pressure (part b). The pressure in a 1000-Hz sound wave at the threshold of hearing differs from atmospheric by at most $3\times 10^{-5}\ \mathrm{N/m^2}$ ($=3\times 10^{-10}$ atm). This is called the ***pressure amplitude***. Just as the intensity is proportional to the square of the displacement amplitude, x_0, it is also proportional to the square of the pressure amplitude. Indeed, it can be shown that the intensity is related to the pressure amplitude P_0 by the formula

$$I = \frac{P_0^2}{2\rho v},$$

where ρ is the density of the medium and v the velocity of sound in that medium.

Normally, the loudness or intensity of a sound decreases as you get farther from the source of the sound. In interior rooms, this effect is reduced because of reflections from the walls. However, if a source is in the open so that sound can radiate out freely in all directions, the intensity decreases as the inverse square of the distance,

$$I \propto \frac{1}{r^2},$$

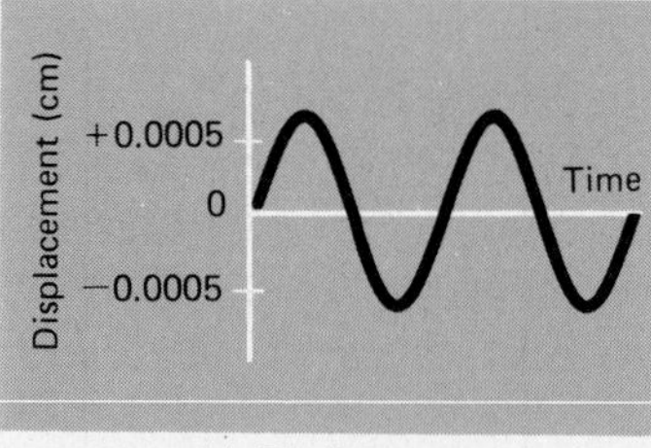

(a) Displacement

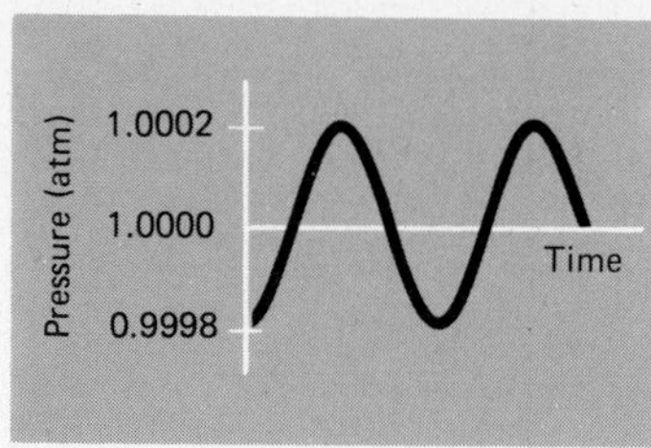

(b) Pressure

FIGURE 12-2
Representation of a sound wave in terms of (a) displacement; (b) pressure.

as we saw in Equation 11-17. Of course, if there is a significant reflection from structures or the ground, the situation will be more complicated.

▌ EXAMPLE 12-3 The intensity level of a jet plane at a distance of 30 m is 140 dB. What is the intensity level at 300 m? (Ignore reflections from the ground.)

SOLUTION The intensity I at 30 m is found from Equation 12-1: $140\text{ dB} = 10 \log I/(10^{-12}\text{ W/m}^2)$. Reversing the log equation to solve for I, we have $10^{14} = I/(10^{-12}\text{ W/m}^2)$, so $I = 10^2\text{ W/m}^2$. At 300 m, 10 times as far away, the intensity will be $(\frac{1}{10})^2 = \frac{1}{100}$ as much, or 1 W/m^2. Hence the intensity level $\beta = 10 \log (1/10^{-12}) = 120$ dB. Even at 300 m, the sound is at the threshold of pain. This is why workers at airports wear ear covers to protect their ears from damage. ▌▌

12-3 ▌ Sound detection; the ear

The human ear, as we have seen, is a remarkably sensitive detector of sound. Mechanical detectors of sound, namely microphones, can barely match the ear in detecting low-intensity sounds.

The function of the ear is to efficiently transform the vibrational energy of sound waves into electrical signals which are carried to the brain by way of nerves. A microphone performs essentially the same task. Sound waves striking the diaphragm of a microphone set it into vibration. A "transducer" transforms this vibration into an electrical signal of the same frequency, which can then be amplified and sent to a loudspeaker or tape recorder. We shall discuss the operation of transducers when we study electricity. Here we shall discuss the structure and response of the ear.

Structure and functioning of the ear

Figure 12-3 is a diagram of the human ear. The internal structure of other vertebrate ears is similar. The external part of the ear is called the pinna

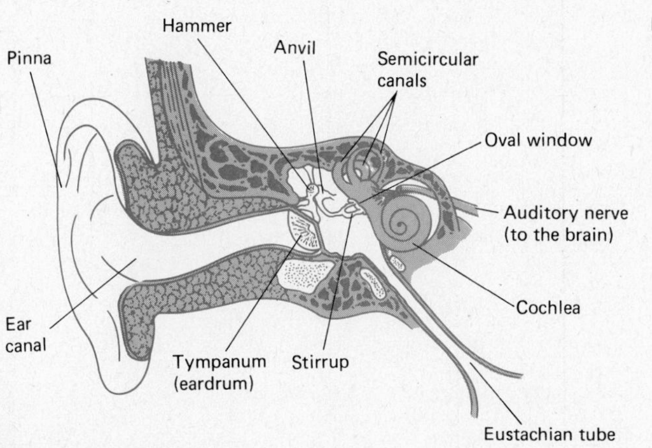

FIGURE 12-3
Diagram of the human ear.

and in human beings serves little purpose; in some animals, however, it can be rotated to help locate the source of a sound.

Internally, the ear has three main divisions, the outer ear, the middle ear, and the inner ear. The outer and middle ears conduct the sound to the inner ear, where the vibrational energy is transformed into electrical energy.

The outer ear consists of the ear canal, which is open to the atmosphere at one end and terminates at the other end in the eardrum or tympanum. Sound waves traveling down the ear canal set the tympanum into vibration. The middle ear consists of three small bones called ossicles (the hammer, the anvil, and the stirrup) in an air-filled cavity. The middle ear is connected to the throat by the eustachian tube, which opens when a person swallows or yawns; its purpose is to maintain the middle ear at the outside pressure. The three small bones transfer the vibrations of the eardrum to the inner ear at the oval window. This delicate system of levers offers a mechanical advantage of about two. The amplification of the pressure in the wave is much greater than this, however, since the area of the eardrum is about 20 times the area of the oval window. Thus the middle ear amplifies the pressure by a factor of about 40.

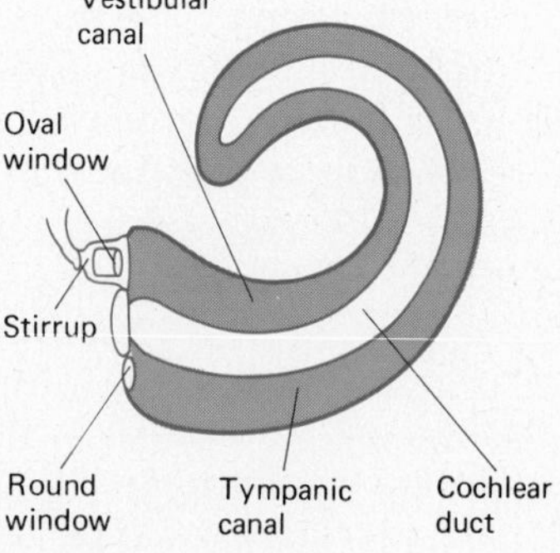

FIGURE 12-4
Diagram of cochlea.

The inner ear consists of the semicircular canals, which are important for controlling balance, and the liquid-filled cochlea, where the vibrational energy of sound waves is transformed into electrical energy and sent to the brain. Figure 12-4 is a diagrammatic representation of the cochlea. The sound vibration travels from the oval window down the vestibular canal and back up the tympanic canal. Because of the viscosity of the liquid, considerable damping occurs, but any remaining energy is dissipated at the round window at the end of the tympanic canal. Between these two canals is a third, known as the cochlear duct. The membrane separating the cochlear duct from the tympanic canal is called the basilar membrane; it is on this membrane that the organ of Corti is found, which contains some 30,000 nerve endings. As a pressure wave passes along the tympanic canal, it causes ripples in the basilar membrane and the attached organ of Corti; this is where the energy is transformed into electrical impulses and sent to the brain by way of the auditory nerve.

The basilar membrane is under tension, but is less taut and becomes thicker as it goes from the middle ear to the apex of the cochlea. From our earlier consideration we might expect that the thicker, less taut end would be more sensitive to low frequencies and the tighter, thinner end more sensitive to the higher frequencies. Careful experiments indicate that this is indeed true, and this fact is important for our sensing of pitch.

* Response of the ear; loudness

The ear is not equally sensitive to all frequencies. The lowest curve in Figure 12-5 represents the intensity level, as a function of frequency, in the softest sound that is just audible. As can be seen, the ear is most sensitive to sounds of frequency between 2000 and 3000 Hz. Whereas a 1000-Hz sound is audible at a level of 0 dB, a 100-Hz sound must be at least 30 dB to be heard. This lowest (solid) curve represents a very good ear; only about 1 percent of the population, mostly young people, have such a low "threshold of hearing." The middle (dashed) curve represents

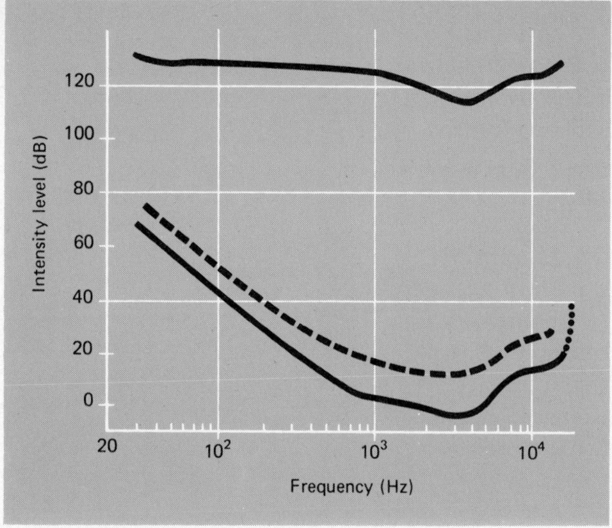

FIGURE 12-5
Sensitivity of the human ear as a function of frequency (see text). Note that the frequency scale is "logarithmic," in order to cover a wide range of frequencies.

a more typical curve; 50 percent of the population have a threshold of hearing equal to or better than this. The top curve represents the "threshold of feeling or pain." Sounds above this level can actually be felt and cause pain. As can be seen, it does not vary much with frequency.

Our subjective sensation of loudness obviously depends not only on the intensity but also on frequency. For example, as seen in Figure 12-5, an average person will detect a 30-dB sound at 1000 Hz as reasonably loud but a 30-dB sound at 50 Hz would not be heard at all.

To obtain the same loudness for sounds of different frequencies requires different intensities. Studies averaged over large numbers of people have produced the curves shown in Figure 12-6. On this graph,

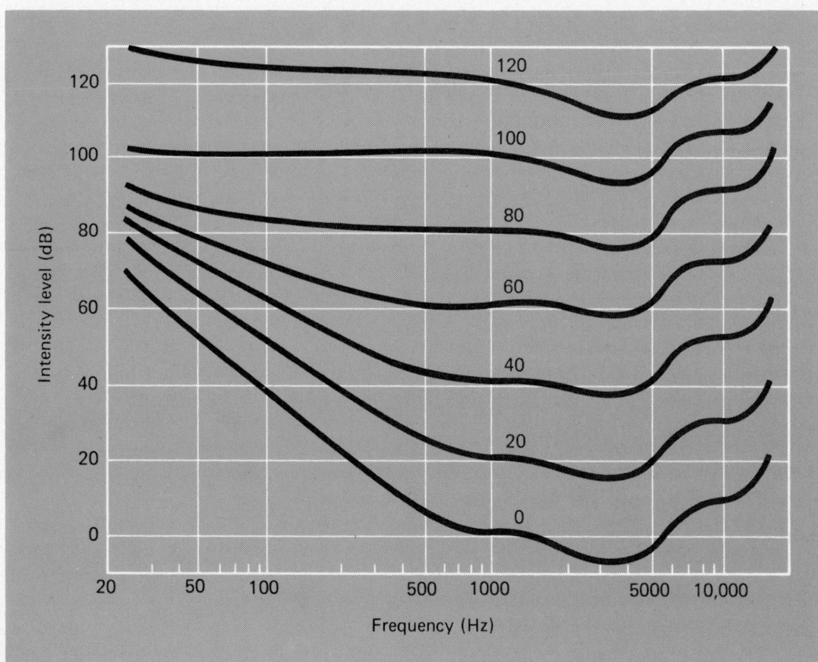

FIGURE 12-6
Loudness-level curves (see text).

each curve represents sounds that seemed to be equally loud. The number labeling each curve represents the *loudness level* (the units are called *phons*), which is numerically equal to the intensity level at 1000 Hz for that curve. For example, the curve labeled 40 represents sounds that are heard to have the same loudness as a 1000-Hz sound with an intensity level of 40 dB. From this curve we see that a 100-Hz tone must have an intensity of about 62 dB to sound as loud (to an average person) as a 1000-Hz tone of only 40 dB.

It is quite clear that at lower intensity levels, our ears are less sensitive to the high and low frequencies relative to middle frequencies. The "loudness" control on many stereo and hi-fi systems is intended to compensate for this. As the volume is turned down, the loudness control boosts the high and low frequencies relative to the middle frequencies so that the sound will be more uniform. Many listeners, however, find the sound more pleasing or natural without the loudness control.

12-4 ▌ Sources of sound; musical instruments

The source of any sound is a vibrating object. Almost any object can vibrate and hence be a source of sound. We shall now discuss some simple sources of sound, particularly as they apply to those most pleasant of sounds, music. We will also discuss briefly nonmusical sound—that is, noise.

Musical instruments make use of a variety of vibrating materials to produce sound. A drum has a stretched membrane that vibrates. Xylophones and marimbas have metal or wood bars that can be set into vibration. Bells, cymbals, and the gong also make use of a vibrating metal. The most widely used instruments make use of vibrating strings, such as the violin, guitar, and piano, or make use of vibrating columns of air, such as the flute, trumpet and pipe organ.

In each musical instrument, the source is set into vibration by striking, plucking, bowing, or blowing. Standing waves are produced and the object vibrates at its natural resonant frequencies.

Strings

We saw in Figure 11-36 how standing waves are established on a string. This is the basis for all stringed instruments. The pitch of a note is normally determined by the lowest resonant frequency, the **fundamental**, which corresponds to nodes occurring only at the ends. As we saw in Figure 11-36, the wavelength of the fundamental is equal to twice the length of the string; therefore, the fundamental frequency $f = v/\lambda = v/2L$, where v is the velocity of the wave on the string. When a finger is placed on the string of, say, a guitar or violin, the effective length of the string is shortened; when it is plucked or a bow pulled across it, the pitch is higher, since the wavelength of the fundamental is now shorter (Figure 12-7). The strings on a guitar or violin are all the same length when unfingered. They each sound at a different pitch because the strings have different mass per unit length, m/L, which affects the velocity as seen in Equation

FIGURE 12-7
The wavelength of a fingered string (*b*) is shorter than that of an unfingered string (*a*). Hence, the frequency of the fingered string is higher. Only the simplest standing wave, the fundamental, is shown.

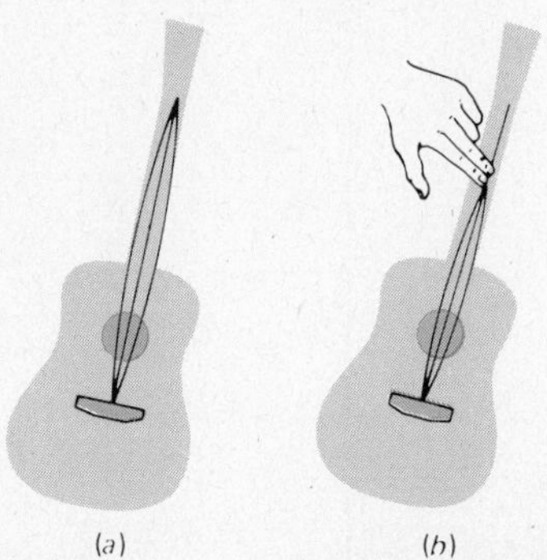

11-12, $v = \sqrt{F_T/(m/L)}$. Thus the velocity on a heavier string is less, so the frequency will be less for the same wavelength. Different notes or chords are obtained by fingering the appropriate string or strings.

The piano and harp, unlike other stringed instruments, have one string for each note and thus require no fingering. The strings are each of different length; for the lower notes they are not only longer but heavier as well, for reasons illustrated in the following example.

▌ EXAMPLE 12-4 The highest key on a piano corresponds to a frequency about 150 times that of the lowest key. If the string for the highest note is 10 cm long, how long would the string for the lowest note have to be if it had the same mass per unit length and was under the same tension?

SOLUTION The velocity would be the same on each string, so the frequency is inversely proportioned to the length L of the string ($f = v/\lambda = v/2L$). Thus

$$\frac{L_l}{L_h} = \frac{f_h}{f_l},$$

where the subscripts l and h refer to the lowest and highest notes, respectively. Thus $L_l = L_h(f_h/f_l) = (10\text{ cm})(150) = 1500\text{ cm}$, or 15 m. This would require a ridiculously long piano. The longer lower strings are made heavier partly to avoid this, so that even grand pianos are no more than about 3 m long. ▌▌

Stringed instruments would not be very loud if they relied on their vibrating strings to produce the sound waves. The strings on a stringed instrument are simply too thin to compress and expand very much air. Because of this, all stringed instruments make use of a kind of a mechanical amplifier known as a *sounding board* or *sounding box*, which is sometimes called a *resonator*. The strings on a piano are attached to a large wooden board—the sounding board. On a guitar, a violin, and a cello, they are attached through the "bridge" to the sounding box (Figure 12-8). When the strings are set into vibration, the sounding board or box is set into vibration as well. Since it has much greater area in contact with the air, it can produce a much stronger sound wave, and thus acts to amplify the sound. On an electric guitar the sounding box is not so important, since the vibrations of the strings are amplified electrically.

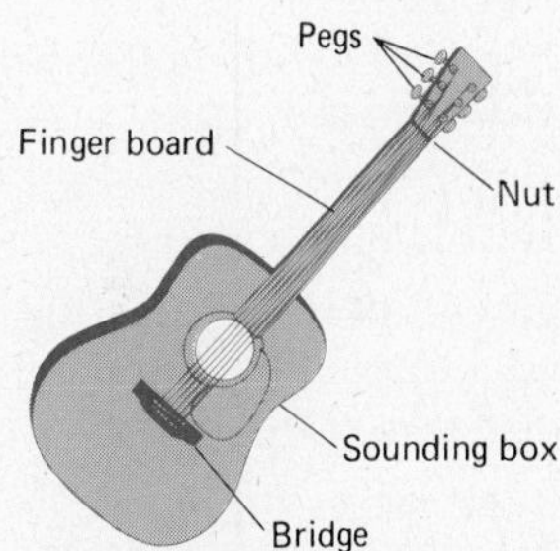

FIGURE 12-8
Stringed instruments have a sounding board or chamber to amplify the vibrations.

The sounds of human speech and song are produced by two "strings" in the throat, the vocal chords. The human voice is capable of a wide range of pitch, which is accomplished by muscles that affect the thickness and tension of these two vocal chords. The throat, mouth, and nose play a role similar to that of a sounding box on a violin.

Wind instruments: vibrating columns of air

Many instruments produce sound from the vibrations of standing waves in a column of air within the tube or pipes of the instrument. These include the woodwinds, the brasses, and the pipe organ. The amplification by the sounding box of a violin or guitar and that of the human voice also depends to some extent on vibrations of the air within the box or oral cavity.

Standing waves can occur in the air of any cavity, but the frequencies are difficult to calculate for any but very simple shapes, such as a long narrow tube. Fortunately, and for good reason, this is the situation for most wind instruments.

In some instruments, such as some woodwinds and brasses, a vibrating reed or the vibrating lip of the player helps to set up the vibrations of the air column. In other instruments, such as the flute and the organ, a stream of air is directed against one edge of the opening or mouthpiece; the deflection of the air leads to turbulence, which directly sets up vibrations in the column of air. Because of the disturbance, whatever its source, the air within the tube vibrates with a variety of frequencies, but only certain frequencies persist, which correspond to standing waves. For a string fixed at both ends, we saw in Section 11-9 that the standing waves have nodes (no movement) at the two ends, and one or more antinodes (large amplitude of vibration) in between; a node also separates each antinode. The lowest-frequency standing wave, the *fundamental*, corresponds to a single antinode. The higher-frequency standing waves are the *overtones* or **harmonics;** specifically, the first harmonic is the fundamental, the second harmonic has twice the frequency of the fundamental, and so on (see Figure 11-36).

The situation is similar for a column of air, but we must remember that it is now air itself that is vibrating. Thus the air at the closed end of a tube must be a (displacement) node, since the air is not free to move there, whereas at the open end of a tube there will be an antinode, since the air can move freely. The air within the tube vibrates in the form of longitudinal standing waves. The possible modes of vibration for a tube open at both ends (called an *open tube*) and for one that is open at one end but closed at the other (called a *closed tube*) are shown graphically in Figure 12-9. The graphs represent the amplitude of motion of the air molecules within the tube. Let us look first at the open tube (Figure 12-9*a*). An open tube has antinodes at both ends.† Notice that there must be at least one node within an open tube if there is to be a standing wave at all. This corresponds to the fundamental frequency of the tube; since the distance between two successive nodes, or between two successive antinodes, is $\frac{1}{2}\lambda$, there is one-half a wavelength within the length of the tube in this case: $L = \frac{1}{2}\lambda$, so the fundamental frequency is $f_1 = v/\lambda = v/2L$, where v is the velocity of sound in air. The standing wave with two nodes is the first overtone or second harmonic and has half the wavelength ($L = \lambda$), and therefore twice the frequency. Indeed, the frequency of each overtone is an integral multiple of the fundamental frequency. This is just what is found for a string.

For a closed tube (Figure 12-9*b*) there is always a node at the closed end and an antinode at the open end. Since the distance between a node and the nearest antinode is $\frac{1}{4}\lambda$, we see that the fundamental in this case corresponds to only one-fourth of a wavelength within the length of the

†Actually, the antinodes do not occur precisely at the ends of a tube; their position depends on the diameter of the tube but if this is small compared to the length, which is the usual case, the antinode occurs very close to the end as shown.

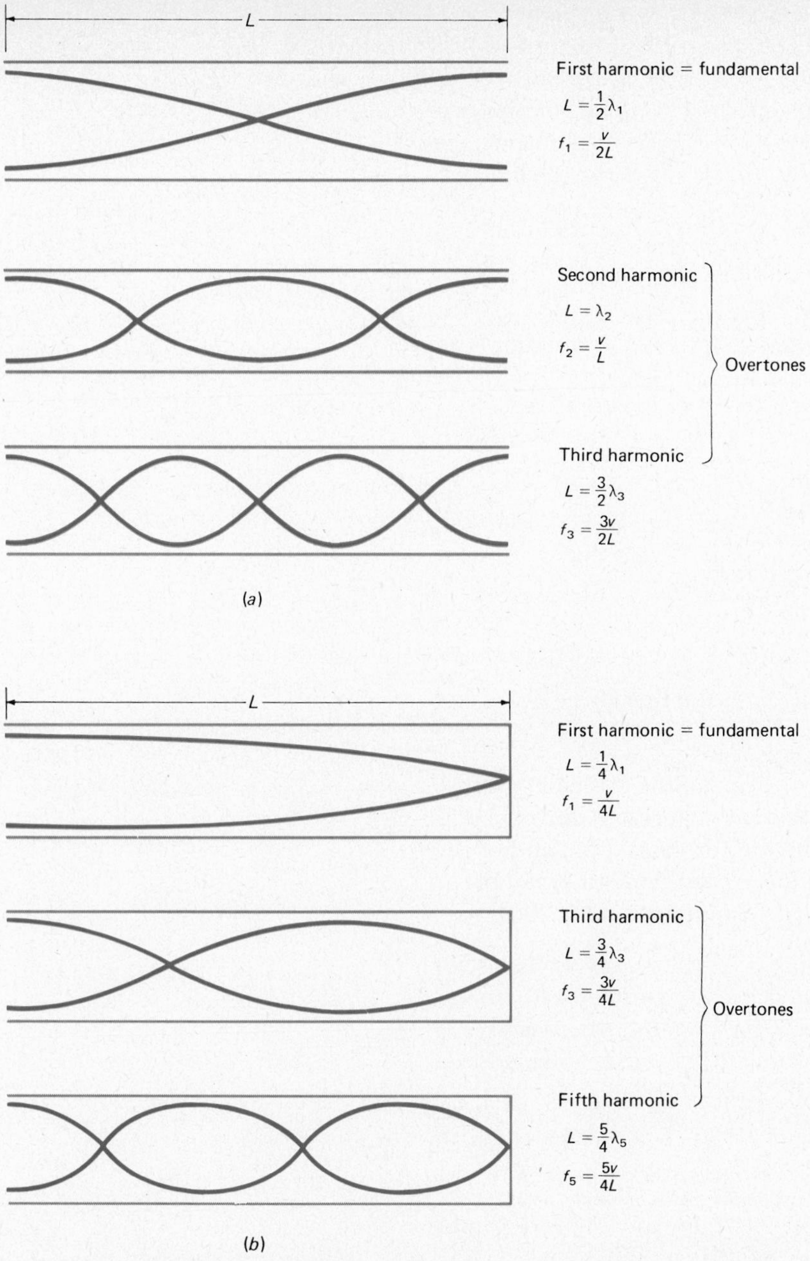

L

First harmonic = fundamental

$L = \frac{1}{2}\lambda_1$

$f_1 = \frac{v}{2L}$

Second harmonic

$L = \lambda_2$

$f_2 = \frac{v}{L}$

Overtones

Third harmonic

$L = \frac{3}{2}\lambda_3$

$f_3 = \frac{3v}{2L}$

(a)

L

First harmonic = fundamental

$L = \frac{1}{4}\lambda_1$

$f_1 = \frac{v}{4L}$

Third harmonic

$L = \frac{3}{4}\lambda_3$

$f_3 = \frac{3v}{4L}$

Overtones

Fifth harmonic

$L = \frac{5}{4}\lambda_5$

$f_5 = \frac{5v}{4L}$

(b)

FIGURE 12-9
Modes of vibration (standing waves) for (*a*) an open tube; (*b*) a closed tube.

tube: $L = \lambda/4$. The fundamental frequency is thus $f_1 = v/4L$, or half what it is for an open tube of the same length. There is another difference, for as we can see from Figure 12-9*b* only the odd harmonics are present in a closed tube. That is, the overtones have frequencies equal to 3, 5, 7, . . . times the fundamental frequency. There is no way for a wave with 2, 4, . . . times the fundamental to have a node at one end and an antinode at the other; thus they cannot exist as standing waves in a closed tube.

Organs make use of both open and closed pipes. Notes of different pitch are sounded using different pipes which vary in length from a few centimeters to 5 m or more. Other musical instruments can act like a closed tube or like an open tube. A flute, for example, is an open tube, for it is open not only where you blow into it, but also at the opposite end as well. The different notes on a flute and many other instruments are obtained by shortening the length of the tube—that is, by uncovering holes along its length. In a trumpet, on the other hand, the pushing down of the valves opens additional lengths of tube. In all these instruments, the longer the length of the vibrating air column, the lower the pitch.

▮ EXAMPLE 12-5 What will be the fundamental frequencies and first three overtones for a 25-cm-long organ pipe if it is (*a*) open and (*b*) closed?

SOLUTION (*a*) For the open pipe, the fundamental frequency is (at 0°C)

$$f_1 = \frac{v}{2L} = \frac{331 \text{ m/s}}{2(0.25 \text{ m})} = 660 \text{ Hz}.$$

The overtones, which include all harmonics, are 1320 Hz, 1980 Hz, 2640 Hz, and so on. (*b*) Referring to Figure 12-9, we see that

$$f_1 = \frac{v}{4L} = \frac{331 \text{ m/s}}{4(0.25 \text{ m})} = 330 \text{ Hz}.$$

But only the odd harmonics will be present, so the first three overtones will be 990, 1650, and 2310 Hz. ▮▮

▮ EXAMPLE 12-6 A flute is designed to play middle C (264 Hz) as the fundamental frequency when all the holes are covered. Approximately how long should the distance be from the mouthpiece to the end of the flute? (*Note*: This is only approximate since the antinode does not occur precisely at the mouthpiece.) Assume that the temperature is 20°C.

SOLUTION Since the speed of sound in air increases 0.60 m/s for each C° above 0°C, at 20°C the speed is $v = [331 + (0.60)(20)]$ m/s $=$ 343 m/s. Then, from Figure 12-9, the fundamental frequency f_1 is related to the length of the vibrating air column by $f = v/2L$. Solving for L, we find that

$$L = \frac{v}{2f} = \frac{343 \text{ m/s}}{(2)(264 \text{ s})} = 0.650 \text{ m},$$

or 65 cm. ▮▮

▮ EXAMPLE 12-7 If the temperature is only 10°C, what will be the frequency of the note played when all the openings are covered in the flute of Example 12-6?

SOLUTION The length L is still 65 cm. But now the velocity of sound is less, since it changes by 0.60 m/s per C°. For a drop of 10 C°, the velocity decreases by 6 m/s, to 337 m/s. The frequency will be

$$f = \frac{v}{2L} = \frac{337 \text{ m/s}}{2(0.650 \text{ m})} = 259 \text{ Hz}.$$

This example illustrates why players of wind instruments take time to "warm up" their instruments so they will be in tune. The effect of temperature on stringed instruments is much smaller. ▮▮

12-5 ▮ Quality; music and noise

Quality or timbre

Whenever we hear a sound, particularly a musical sound, we are aware of its loudness, its pitch, and also of a third aspect called "quality." For example, when a piano and then an oboe play a note of the same loudness and the same pitch (say middle C), there is a clear difference in the overall sound. We would never mistake a piano for an oboe. This is what is meant by the **quality**† of a sound; for musical instruments the terms *timbre* or *tone color* are also used.

Just as loudness and pitch can be related to physically measurable quantities, so can quality. The quality of a sound depends on the presence of overtones—their number and their relative amplitudes. Generally, when a note is played on a musical instrument, the fundamental as well as the overtones are present simultaneously. Figure 12-10 illustrates how the

† Note that quality in this sense does not refer to the goodness or badness of a sound or to the craftsmanship that went into building the instrument.

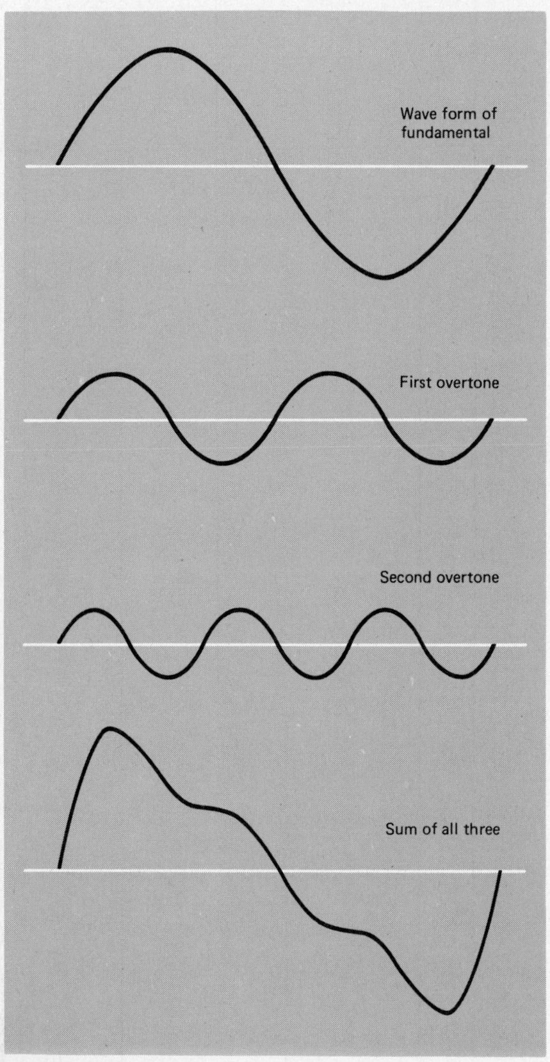

FIGURE 12-10
The amplitudes of the fundamental and first two overtones are added at each point to get the "sum," or composite waveform.

superposition of the fundamental and first two overtones would combine to give a composite *wave form*. Of course, more than two overtones are usually present.

The relative amplitudes of the various overtones are different for different instruments, and this is what gives each instrument its characteristic quality or timbre. A graph showing the relative amplitudes of the harmonics emitted by an instrument is called a sound spectrum; several typical examples for different instruments are shown in Figure 12-11. Normally, the fundamental has the greatest amplitude and its frequency is what is heard as the pitch.

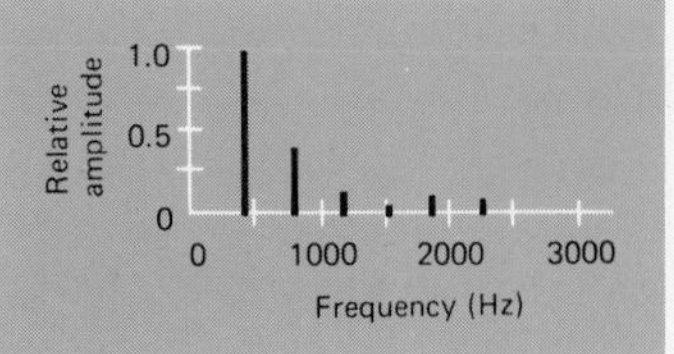

Violin

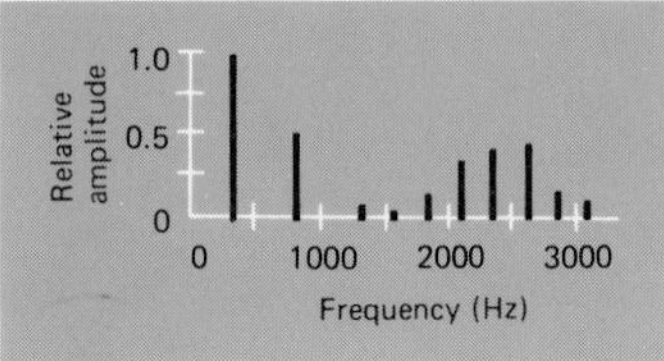

Clarinet

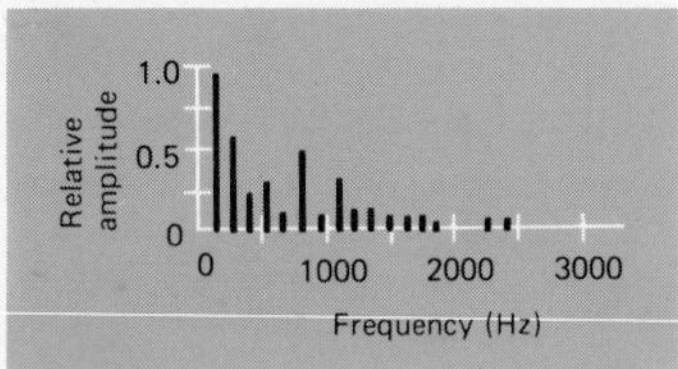

Piano

FIGURE 12-11
Sound spectra for several instruments.

The manner in which an instrument is played strongly influences the sound quality. Plucking a violin string, for example, makes a very different sound than pulling a bow across it. The pitch of a note can also be changed in some instruments by exciting the vibrations in a different way. For example, a flutist can direct the stream of air more directly into the mouthpiece rather than across it. When this is done, the fundamental is hardly excited at all and the first overtone, at twice the frequency, predominates. The pitch, then, in musical terms, is an octave higher.

The sound spectrum of an instrument is determined largely by the shape of the instrument and the material from which it is made, although of course the way it is played has an influence as well. Often small factors in the construction of an instrument can affect the quality of the sound significantly. This is especially true of stringed instruments with a sounding box; the sounding box not only amplifies the sound, but, because it has resonant frequencies of its own, helps to determine which overtones will be emphasized.

Music versus noise

An ordinary sound, such as that made by striking two stones together, is a noise that has a certain quality, but a clear pitch is not discernible. A noise such as this is a mixture of many frequencies which bear little relation to one another. If a sound spectrum were made of this noise, it would not show discrete lines such as those of Figure 12-11; instead it would show a continuous, or nearly continuous, spectrum of frequencies.

Our ears and minds interpret sounds which include frequencies that are simple multiples of one another as harmonious or mellow, whereas a random mixture of frequencies is less pleasant and we call it noise. Musical instruments sound harmonious because the frequencies they produce are whole-number multiples of the fundamental. What makes musical instruments different is that the vibrations are set up essentially in one dimension only—along a string or a narrow column of air. And this is why the frequencies can be integral multiples of the fundamental.

But complex objects, such as stones struck together, can vibrate in all three dimensions, and the dimensions (length, width, height) can be quite variable. The frequencies of standing waves, since they depend on the dimensions of the object, will be of great variety and will not bear a simple relationship to one another. A drum is an interesting intermediate example. A drum does not sound as harmonious as other instruments because the drum head vibrates in two dimensions. The frequencies of standing waves, while bearing some relation to one another, are not integral multiples of the fundamental. Some drums (for example, tympani) sound quite mellow because the enclosed air resonates with only certain of the frequencies and accentuates only those.

The distinction between music and noise is, of course, not sharp. Sounds vary continuously from harmonious musical tones when the frequencies are simply related to one another, to sounds of a wide mixture of random frequencies known as "white noise." To draw a line between music and noise requires a subjective judgment. Indeed, the judgment of the public on this matter has varied throughout history. The complex harmonies of Beethoven were once regarded as noise compared to the simpler mixtures of frequencies found in the music of Mozart or Haydn; in a later period, the music of Stravinsky sounded harsh compared to that of his predecessors. Today there is little question about Beethoven or Stravinsky. Instead, many people wonder whether the sounds introduced by contemporary avant-garde composers and by rock musicians is music or noise. Of course, the decision is a personal one, not scientific. Science can tell us what distinguishes a harmonious sound from a nonharmonious or noisy sound, but it does not tell us where to draw the line. We each make that decision ourselves.

Noise pollution

Today we are constantly surrounded by noise: the neighbor running a power mower on a relaxing Sunday afternoon, or starting a power saw at 6:00 A.M. when you're fast asleep; the noise of garbage disposals and electric dishwashers, of loud parties and jet planes taking off, of freeways and factories. Noise affects us in various ways, particularly psychologically. Sometimes it is a mere annoyance, but in some circumstances noise can make one irritable or anxious.

Loud noise can also cause loss of hearing, and this is particularly a problem in factories and other industrial works, where the sound level may be high for long periods. Pop musicians, too, can suffer hearing loss, for levels as high as 120 dB are commonly produced. Hearing loss due to excessive noise levels was recorded by the ancient Romans. Regardless of source, hearing loss due to noise is particularly large in the frequency range from about 2000 to 5000 Hz, an important region for speech and music.

Although research is beginning in this area, not a great deal is known. Standards for industrial noise have been set, and the limits depend on how long one is subjected to a given noise level. For example, 85 dB experienced over a period of six hours per day is regarded as a maximum by the U.S. government. But even at this limit, people subjected to this sound level for 20 years will experience a hearing loss perhaps 20 percent greater than the population as a whole.

The problem of noise control is a difficult one. Isolation of a source of sound with barriers is sometimes attempted, although it is expensive and not always convenient. Attacking the source of the sound is often the best remedy. Reducing the area of vibration of machinery is important, for the greater the area, the more air that can be "pushed" and the louder the sound. Making the surface out of a stiffer material, thus reducing the amplitude, or coating the surface with energy-absorbing material, helps a great deal. The placement of machinery is also important, for the floor, a wall, or other object may resonate with the machine's vibration, causing increased amplitude. Careful maintenance is important, too, for lack of lubrication, loose bolts, and worn parts can cause vibrations. The noise produced by jet planes is also a serious problem and can affect a whole community; much work remains to be done in this area.

On the other side of the coin, the adaptability of the human organism is nothing short of amazing. Having become used to the urban roar, many people are uncomfortable in a truly quiet environment. In the design and construction of new office buildings architects and engineers often introduce some background noise—perhaps by making the heating and ventilation system noisy—so that workers will feel more comfortable. It is even said that although a number of quiet home appliances have been invented, people will not buy them. Apparently, a vacuum cleaner is supposed to make noise if it is to get things clean!

*12-6 ▮ Musical scales and harmony

Nearly all music is based on a basic set of notes or tones called a *scale*, in which the notes bear a definite relationship to one another. Many scales have been used throughout the ages. The simplest one in Western music is the *diatonic scale*, which consists of the eight familiar notes "do-re-mi-fa-sol-la-ti-do." Each note is named by a letter from A to G, and each corresponds to a particular pitch or frequency. The higher "do" has twice the frequency of the lower "do." The difference in pitch between each tone on the diatonic scale is called a *whole interval*, except for those between *mi* and *fa* and between *ti* and *do*, which are *half-intervals*. The ratio of the successive frequencies for whole intervals is 9/8 or 10/9 and for half-intervals it is 16/15. The pitch of the first *do* can be chosen arbitrarily; but the pitch of the remaining notes then conform to the regular sequence indicated in Table 12-3; in this table the initial *do* is taken as middle C at 264 Hz, and this is the *C major scale*.

The interval from middle C to the next C, which is called "C above middle C," or C′, is called an octave from the Latin word for "eight," because there are eight notes counting both ends (but only seven intervals). C′ has twice the frequency of C. This is always the case as you go up

TABLE 12-3

Diatonic C major scale†

Note	Letter name	Frequency (Hz)	Frequency ratio	Interval
do	**C**	**264**		
			9/8	**Whole**
re	**D**	**297**		
			10/9	**Whole**
mi	**E**	**330**		
			16/15	**Half**
fa	**F**	**352**		
			9/8	**Whole**
sol	**G**	**396**		
			10/9	**Whole**
la	**A**	**440**		
			9/8	**Whole**
ti	**B**	**495**		
			16/15	**Half**
do′	**C′**	**528**		

† Only one octave is included.

TABLE 12-4

Diatonic D major scale†

Note	Letter name	Frequency (Hz)	Frequency ratio	Interval
do	D	297		
			9/8	Whole
re	E	334		
			10/9	Whole
mi	F$^\sharp$	371		
			16/15	Half
fa	G	396		
			9/8	Whole
sol	A	445		
			10/9	Whole
la	B	495		
			9/8	Whole
ti	C$^\sharp$	557		
			16/15	Half
do	D′	594		

† Only one octave is included.

the scale. Each C has twice the frequency of the preceding C, each G has twice the frequency of the preceding G, and so on.†

A diatonic scale can start on any note, and the one beginning on D, called D major, is given in Table 12-4. As can be seen, the spacing of half and whole intervals is the same. Some of the notes must thus be raised or lowered a half-interval; when raised a half-interval it is called a *sharp* (♯) and when lowered, a *flat* (♭). Sharps and flats are also used in the "minor scales," in which the spacing arrangement of intervals is slightly different than for a major scale.

A scale that includes all the sharps and flats is a *chromatic scale.* The *equally tempered chromatic scale* is indicated in Table 12-5. On this scale, F$^\sharp$

† At high frequencies, a doubling of frequency sounds like slightly more than one octave. Because instruments are intended to sound harmonious, the precise relationships given above are not followed exactly at high frequencies.

TABLE 12-5

Equally tempered chromatic scale†

Note	Frequency (Hz)	Frequency ratio	Interval
C	262		
		1.06	Half
C$^\sharp$ or D$^\flat$	277		
		1.06	Half
D	294		
		1.06	Half
D$^\sharp$ or E$^\flat$	311		
		1.06	Half
E	330		
		1.06	Half
F	349		
		1.06	Half
F$^\sharp$ or G$^\flat$	370		
		1.06	Half
G	392		
		1.06	Half
G$^\sharp$ or A$^\flat$	415		
		1.06	Half
A	440		
		1.06	Half
A$^\sharp$ or B$^\flat$	466		
		1.06	Half
B	494		
		1.06	Half
C′	524		

† Only one octave is included.

and G^b, for example, are taken to have the same frequency. Originally, the frequencies of these two notes (and other like pairs) were slightly different, and string soloists often play them that way today. However, taking the sharp and corresponding flat to be the same frequency greatly reduces the number of keys needed on a piano and other fixed-note instruments. Note also that the note A does not have precisely the same frequency on the C major and D major scales. This is true of many notes when all the different diatonic scales are considered. The equally tempered chromatic scale, in the interests of reducing the number of keys on fixed-note instruments, ignores these slight differences and thus does not reproduce precisely a diatonic scale. The equally tempered scale is made up of 12 equally spaced intervals, so each note has a frequency that is 1.059 times that of the previous note. This particular number arises because $(1.059)^{12} = 2$, as required for an octave.

Many other scales have been used in the West and in other cultures. One that is common in Africa and the Orient is the pentatonic scale, based on just five tones.

Harmony, which we have already mentioned briefly, is an important aspect of music. Two notes are said to be harmonious if they sound pleasant when sounded together and are not harsh. In general, people find notes harmonious if the ratio of their frequencies equals the ratio of two small whole numbers. Thus a ratio of $\frac{2}{1}$, which is an octave, sounds very harmonious. A ratio of $\frac{3}{2}$ (such as G = 396 Hz and C = 264 Hz) is also very harmonious; it is called a "fifth," because G is five notes above C (counting both ends). Ratios of $\frac{4}{3}$ and $\frac{5}{4}$ are also harmonious, but a little less so, whereas the ratios of larger numbers such as $\frac{9}{8}$ or $\frac{10}{9}$ are quite dissonant; you can check the latter by playing two successive notes (say C and D) on the piano.

The "progress" of music over the past several hundred years has been in part the acceptance of less harmonious tones played together. At one time, music consisted only of single notes, or of octaves and fifths played at the same time. Later, "thirds" became acceptable (such as C and E) and today almost any dissonance will be found in musical scores.

It is interesting to note that the ancients first found the relation between harmonious notes and the ratio of small whole numbers. They did not, however, relate it to frequency but rather to the lengths of strings. If the ratio of the lengths of two equally heavy and equally taut strings was $\frac{2}{1}$ or $\frac{3}{2}$, the sound was harmonious. Today we can see how this discovery relates to frequency because of our understanding of standing waves.

FIGURE 12-12
Sound waves from two loudspeakers interfere.

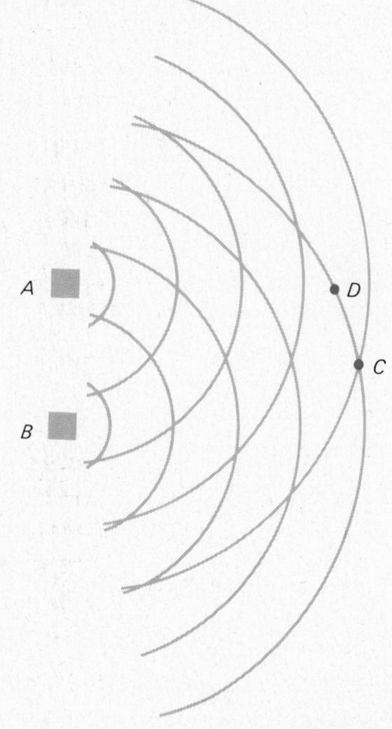

12-7 ▮ Interference of sound waves; beats

We saw in Section 11-8 that when two waves simultaneously pass through the same region of space, they interfere with one another. Since this is true of any kind of wave, we should expect that interference will occur with sound waves, and indeed it does.

As a simple example, consider two large loudspeakers, A and B, a distance d apart on the stage of an auditorium, as shown in Figure 12-12.

Let us assume that the two speakers are emitting sound waves of the same single frequency and that they are in phase: that is, when one speaker is forming a compression, so is the other. The curved lines in the diagram represent the crests of sound waves from each speaker. Of course, we must remember that for a sound wave, a crest is a compression in the air, whereas a trough—which falls between two crests—is a rarefaction. A person or detector at a point such as C, which is the same distance from each speaker, will experience a loud sound because the interference will be constructive. On the other hand, at a point such as D in the diagram, little if any sound will be heard because destructive interference occurs—compressions of one wave meet rarefactions of the other and vice versa (see the similar diagram and related discussion on water waves in Section 11-8). An analysis of this situation is clearer if we graphically represent the wave forms as in Figure 12-13. There it can be seen that at point C constructive interference occurs since both waves simultaneously have crests or simultaneously have troughs. In part b we see the situation for point D. The wave from speaker B must travel a greater distance than the wave from A. Thus the wave from B lags behind that from A. In Figure 12-13b, point E is chosen so that the distance ED is equal to AD. Thus we see that if the distance BE is equal to precisely one-half the wavelength of the sound, the two waves will be exactly out of phase when they reach D, and destructive interference occurs. This then is the criterion for determining at what point(s) destructive interference occurs: destructive interference occurs at any point whose distance from one speaker is greater than its distance from the other speaker by exactly one-half wavelength. Notice that if this extra distance (BE in the diagram) is equal to a whole wavelength (or 2, 3, . . . wavelengths), the two waves will be in phase and *constructive interference* occurs. If the distance BE equals $\frac{1}{2}$, $1\frac{1}{2}$, $2\frac{1}{2}$, . . . wavelengths, *destructive interference* occurs.

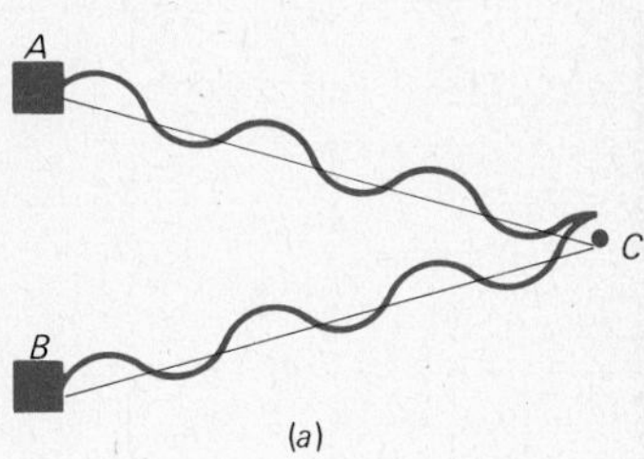

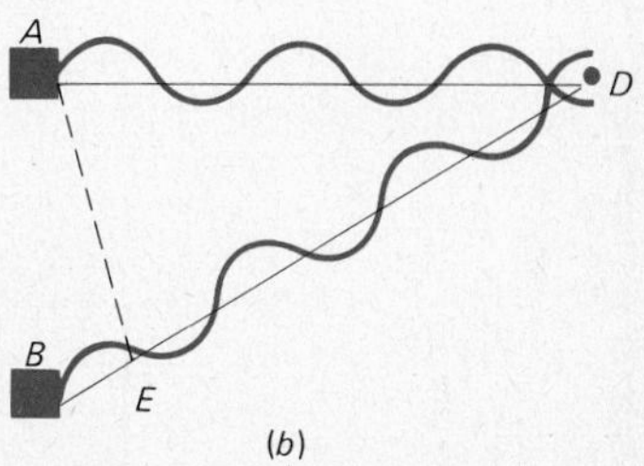

FIGURE 12-13
Sound waves from loudspeakers *A* and *B* (see Figure 12-12) constructively interfere at *C* and destructively interfere at *D*.

If a speaker emits a whole range of frequencies, not all wavelengths will destructively interfere at any one point such as D; rather only specific wavelengths will destructively interfere completely, according to the criteria above.

▮ EXAMPLE 12-8 The two speakers in Figure 12-12 are 1.00 m apart. A person stands 4.00 m from one speaker. How far must this person be from the second speaker in order to detect destructive interference when the speakers emit a 1100-Hz sound?

SOLUTION The wavelength of this sound is

$$\lambda = \frac{v}{f} = \frac{331\ \text{m/s}}{1100\ \text{s}^{-1}} = 0.30\ \text{m}.$$

For destructive interference to occur, the person must be one-half a wavelength farther from one speaker than the other, or 0.15 m. Thus the person must be 4.15 m (or 3.85 m) from the second speaker. Notice in this example that if the speakers were less than 0.15 m apart, there would be no point that was 0.15 m farther from one speaker than the other, and there would be no point where destructive interference would occur. ▮▮

An interesting and important example of interference occurs in the phenomenon known as **beats**. This is the phenomenon that occurs if two

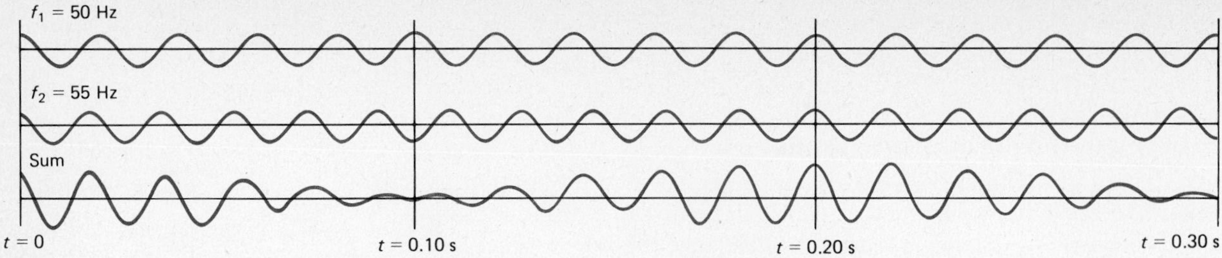

FIGURE 12-14
Beats occur as a result of the superposition of two sound waves of slightly different frequency.

sources of sound, say two tuning forks, are close in frequency but not exactly the same. Sound waves from the two sources interfere with each other and the sound level alternately rises and falls; the regularly spaced intensity changes are called beats and often sound eerie.

To see how beats arise, consider two equal-intensity sources of sound of frequency $f_1 = 50$ Hz and $f_2 = 55$ Hz, respectively. In 1.00 s, the first source makes 50 vibrations, whereas the second makes 55. We now examine the waves at one point in space equidistant from the two sources. The wave forms for each wave as a function of time are shown on the first two lines of Figure 12-14; the third line shows the sum of the two waves. At the time $t = 0$ the two waves are shown to be in phase and interfere constructively. Because the two waves vibrate at different rates, at the time $t = 0.10$ s they are completely out of phase and destructive interference occurs. At $t = 0.20$ s, they are again in phase and the resultant amplitude again large. Thus the resultant amplitude is large every 0.20 s and in between it drops drastically. This rising and falling of the intensity is what is heard as beats. In this case the beats are 0.20 s apart. That is, the *beat frequency* is five per second, or 5 Hz. From our analysis it is clear that in general the beat frequency equals the difference in frequency of the two waves.

The phenomenon of beats can occur with any kind of wave and is a very sensitive method for comparing frequencies. For example, to tune a piano, a piano tuner listens for beats produced between his standard tuning fork and that of a particular string on the piano and knows it is in tune when the beats disappear; he also listens for beats between the different piano strings. The members of an orchestra tune up by listening for beats between their instruments and that of a standard tone (usually A above middle C at 440 Hz) produced by a piano or an oboe.

The formation of beats is what determines harmony or dissonance. We can hear beat frequencies up to almost 10 per second. Beyond that they give rise to a sense of dissonance and in the audible range (>20 Hz) can actually be heard. The fact discussed earlier that two tones are harmonious if their frequencies are in the ratio of small whole numbers, but sound dissonant if this is not the case, is due essentially to beats. For example, two tones whose frequency ratio is $\frac{2}{1}$ will give rise to a beat frequency equal to the lower tone itself. It is the lack of additional beats that is so pleasing. But two tones in the ratio $\frac{10}{9}$ (say, 1000 and 900) give rise to a much smaller beat frequency (in this case 100) and such additional frequencies make a sound dissonant.

*12-8 ▌ Acoustic and medical applications of sound and ultrasound

*Reflection and reverberation; acoustics

We saw in Chapter 11 that when a wave traveling in one medium comes to a barrier or the boundary of another medium, some of the energy in the wave is reflected and the rest is absorbed or transmitted.

In the case of sound waves we call a reflection an echo. The fraction of the incident energy that is reflected from a surface is large if the surface is rigid and smooth; if the surface is soft and irregular, less of the sound is reflected and more is absorbed. In an enclosed room or hall, sound reflects from the walls, ceilings, and floor. Therefore, in the design of the interior of buildings, an understanding of the reflective properties of sound is important, from office buildings to factories to concert halls and auditoriums. This is the field of *acoustics*.

In an auditorium it is often advantageous to place behind the stage a surface that has high reflective properties to direct the sound out to the audience. However, in the rest of the auditorium, and indeed for many other building interiors, the surface should absorb a good part of the energy. Otherwise, the many reflections from the walls, ceiling, and floor would cause the sound to be garbled. Earlier sound waves, after reflection, would interfere with ones emitted slightly later. Speakers could not be understood and music would sound muddy and unclear. On the other hand, there must be some reflection or the sound level would be low and the hall would sound dull and lifeless. Indeed, reflection of sound in a room makes it sound lively and full. Singing in the shower is a good example of this. In the design of an auditorium or concert hall, a balance must be found between the reflection of the sound, which is called *reverberation*, and its absorption, so that the sound will be full and yet not unclear and garbled.

This balance is characterized by the *reverberation time*, which is defined as the time required for a sound to decrease in intensity level by 60 dB. It is approximately the time required for a sound at normal levels to become essentially inaudible, and is typically in the range $\frac{1}{2}$ s to perhaps 2 s in most rooms and halls.

The reverberation time depends on the reflective properties of the walls and other surfaces. In this regard, the *absorption coefficient* β of a surface is defined as the fraction of the incident sound energy that is absorbed (that is, not reflected) by the surface; if E_i is the incident sound energy and E_r is the energy reflected,

$$\beta = \frac{E_i - E_r}{E_i}.$$

The absorption coefficient, which is dimensionless, depends on the material and also to some extent on the frequency of the sound. Its values for a number of materials are given in Table 12-6 for a frequency of 500 Hz. Note that for an open window, none of the sound is reflected, so $\beta = 1.0$. Marble, with an absorption coefficient of 0.01, absorbs only 1 percent of the sound striking it; the other 99 percent is reflected.

TABLE 12-6

Absorption coefficients β at 500 Hz

Material	β
Open window	1.00
Felt	0.20
Heavy drapes	0.50
Heavy carpets	0.40
Acoustic plaster	0.25
Acoustic tile	0.55
Wood, unpainted	0.08
Glass	0.04
Brick	0.03
Concrete	0.02
Plaster, linoleum	0.02
Marble	0.01
Absorbing power, a_p (m^2)	
Upholstered seat	0.2
Wood seat	0.03
Each person (in addition to seat)	about 0.4

The absorbing power a_p of a surface is defined as the product of the absorption coefficient β and the total area A of the surface:

$$a_p = \beta A.$$

It has been experimentally determined that the reverberation time τ, in seconds, is given approximately by the formula

$$\tau = \frac{V}{6.25 \sum a_p},$$

where V is the volume of the room and $\sum a_p$ is the sum of the absorbing powers of all surfaces in the room; all measurements are in meters.

▮ EXAMPLE 12-9 Calculate the reverberation time for a small hall 40 m deep, 30 m wide, and 10 m high, whose ceiling is plaster; the walls have 700 m^2 of wood paneling and 800 m^2 of heavy drapes, the floor has 1200 m^2 of heavy carpet, and there are 1000 upholstered seats, of which 800 are occupied.

SOLUTION It is easiest if we make a table to determine the total absorbing power:

	β	A (m^2)	a_p
Plaster ceiling	**0.02**	**1200**	**24**
Wood	**0.08**	**700**	**56**
Heavy drapes	**0.05**	**800**	**400**
Heavy carpet	**0.40**	**1200**	**480**
Seats (1000 × 0.2)			**200**
People (800 × 0.4)			**320**
			1480

Thus

$$\tau = \frac{V}{6.25 \sum a_p} = \frac{12{,}000}{(6.25)(1480)} = 1.3 \text{ s.}$$ ▮▮

Similar considerations are important in offices and factories because it is important for them to keep the intensity level from becoming too high, although completely absorbing walls in an office would probably make the sound level too low for psychological comfort. For special testing purposes, nearly perfectly absorbing rooms called *anechoic chambers* (meaning "without echo") are sometimes built (Figure 12-15). Speaking in such a room gives one an uncanny feeling, for it is almost impossible to hear yourself; it shows how much sound actually is reflected in ordinary situations, even outdoors.

FIGURE 12-15
Anechoic chamber. (Courtesy of Bell Labs.)

*Sonar and soundings

The reflection of sound is used in many applications to determine distance. The *sonar* or pulse-echo technique is used to locate underwater objects. A transmitter sends out a sound pulse through the water and a detector receives its reflection, or echo, a short time later. This time is carefully measured, and from it the distance to the reflecting object can be determined since the speed of sound in water is known. The depths of the sea and the location of reefs, sunken ships, submarines, or schools of fish can be determined in this way.

Our knowledge of the interior structure of the earth depends in part on the reflection of longitudinal waves traveling in the earth—earthquake waves, or waves produced by a deliberate explosion which are called "soundings." The techniques are similar to sonar. An analysis of waves reflected from various structures and boundaries within the earth reveals characteristic patterns that are also useful in the exploration for oil and minerals.

*Ultrasound and its applications

Sound waves whose frequencies are beyond the audible, that is, above 20,000 Hz, are called *ultrasonic*. Ultrasonic waves are used in a variety of medical and other applications. For example, the frequencies used in sonar are often in the ultrasonic range 20 to 100 kHz (1 kilohertz = 10^3 Hz). This has the advantage that the sounds are inaudible; another advantage is that for shorter wavelengths, there is less diffraction, so the beam spreads less and smaller objects can be detected. For as we saw in Chapter 11, particularly in Figure 11-34, an obstacle intercepts and reflects a portion of a wave significantly only if the wavelength is less than the size of the object. With the higher frequencies of ultrasound, the wavelength is smaller, so smaller objects can be detected.

In medicine, ultrasonic waves are used both in diagnosis and in treatment. For diagnosis, a pulse-echo technique much like sonar is used. A high-frequency sound pulse is directed into the body and its reflection from the boundaries or interfaces between different regions of the body and its organs are then detected. In this way, tumors and other abnormal growths, or pockets of fluid, can be distinguished. With this technique, the action of heart valves and the development of a fetus can be examined. Information about various organs of the body, such as the brain, the liver, and the kidneys, can also be obtained in this way. Although ultrasound does not replace x-rays, for certain kinds of diagnosis it is more helpful; some kinds of tissue or fluid are not detected in x-ray photographs, but ultrasound waves are reflected from the boundaries.

The frequencies used in ultrasonic diagnosis are in the range 1 to 10 MHz (1 megahertz = 10^6 Hz). The wavelength of a 1-MHz wave is $\lambda = v/f = (331\ \text{m/s})/(10^6\ \text{s}^{-1}) = 3 \times 10^{-4}\ \text{m} = 0.3\ \text{mm}$, and thus objects as small as this (which is very small indeed) can be detected. The generation of a pulse of such high frequency is done electronically. The electric energy is transformed by a transducer into a sound wave of the same frequency. The action of a transducer is like that of a loudspeaker but operates at higher frequencies (we will discuss their operation in a later chapter). The detector of the reflected ultrasonic waves is also a transducer that acts like a microphone; it transforms the sound wave into an electrical wave that can be displayed on an oscilloscope.

Ultrasonic waves are also used as a means of treating unwanted tissue. Waves of very high intensity (as high as $10^7\ \text{W/m}^2$) are focused on the unwanted tissue and the heating effect of the absorbed energy destroys the tissue.

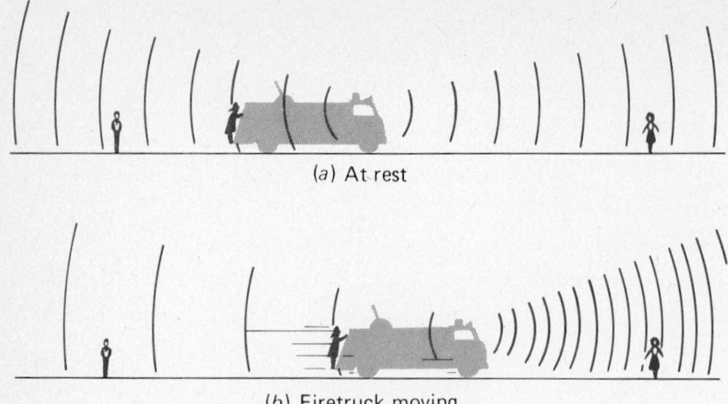

FIGURE 12-16
(a) Both observers on the sidewalk hear the same frequency from the firetruck at rest. (b) Doppler effect: observer toward whom the firetruck moves hears a higher-frequency sound, and observer behind the firetruck hears a lower frequency.

12-9 ▌ Doppler effect

You may have noticed that the pitch of the siren on a speeding firetruck drops abruptly as it passes you. Or you may have noticed the change in pitch of a blaring horn on a fast-moving car as it passes by. The pitch of the sound from the engine of a race car also changes as it passes an observer—aaaaahhrroooooomm. When a source of sound is moving toward an observer, the pitch is higher than when the source is at rest; and when the source is traveling away from the observer, the pitch is less. This phenomenon is known as the **Doppler effect** and occurs for all types of waves. Let us now see why it occurs, and calculate the change in frequency for sound waves.

The siren of a firetruck at rest emits sound of a particular frequency in all directions, as shown in Figure 12-16*a*. If the firetruck is moving, the siren emits sound at the same frequency. But the sound waves it emits forward are closer together than normal, as shown on the right in Figure 12-16*b*. This is because the firetruck, as it moves, is "catching up" with the previously emitted waves. Thus an observer on the sidewalk will detect more wave crests passing per second, so the frequency is higher. The waves emitted behind the truck are, on the other hand, farther apart than normal because the truck is speeding away from them. Since fewer wave crests per second pass by an observer behind the truck, the pitch is lower.

To calculate the change in frequency, we make use of Figure 12-17. In part *a*, the source of the sound (say a siren) is at rest; two successive wave crests are shown, the second of which is just in the process of being emitted. The distance between these crests is λ, the wavelength. If the frequency of the source is f, then the time between emissions of wave crests is

$$T = \frac{1}{f}.$$

In part *b* the source is moving with a velocity v_s. In a time T (as just defined) the first wave crest has moved a distance $d = vT$, where v is the velocity of the sound wave in air (which is, of course, the same whether the source is moving or not). In this same time, the source has moved a

FIGURE 12-17
Determination of frequency change in the Doppler effect (see text).

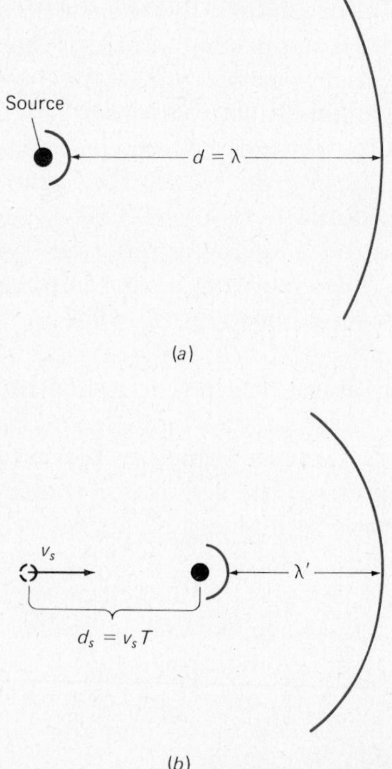

distance $d_s = v_s T$. Then the distance between successive wave crests, which is the new wavelength λ', is

$$\begin{aligned}\lambda' &= d - d_s \\ &= (v - v_s)T \\ &= (v - v_s)\frac{1}{f},\end{aligned}$$

since $T = 1/f$. The frequency f' of the wave is

$$\begin{aligned}f' &= \frac{v}{\lambda'} \\ &= \frac{v}{v - v_s} f\end{aligned}$$

or

$$f' = \frac{1}{1 - v_s/v} f \quad \text{(source moving toward stationary observer).} \quad (12\text{-}2a)$$

Because the denominator is less than 1, $f' > f$. For example, if a siren emits a sound of frequency 400 Hz when at rest, then when the firetruck moves toward an observer with a speed of 30 m/s, the observer hears a frequency

$$f' = \frac{400 \text{ Hz}}{1 - \dfrac{30 \text{ m/s}}{330 \text{ m/s}}} = 440 \text{ Hz}$$

For a source that is moving *away* from the observer at a speed v_s, the new wavelength will be

$$\lambda' = d + d_s,$$

and the frequency f' will then be

$$f' = \frac{1}{1 + v_s/v} f \quad \text{(source moving away from stationary observer).} \quad (12\text{-}2b)$$

In this case, an observer behind the fire truck traveling at 30 m/s and whose siren vibrates at 400 Hz would hear a frequency of about 366 Hz.

The Doppler effect also occurs when the source is at rest and the observer is in motion. If the observer is traveling toward the source, the pitch is higher; and if the observer is traveling away from the source, the pitch is lower. Quantitatively, the change in frequency is slightly different than for the case of a moving source. In this case the distance between wave crests, the wavelength λ, is not changed. But the velocity of the crests with respect to the observer is changed. If the observer is moving toward the source, the speed of the waves relative to the observer is $v' = v + v_o$, where v is the velocity of sound in the air (we assume the air is still) and v_o is the velocity of the observer. Hence the new frequency is

$$f' = \frac{v'}{\lambda} = \frac{v + v_o}{\lambda},$$

or, since $\lambda = v/f$,

$$f' = \left(1 + \frac{v_o}{v}\right) f \qquad \text{(observer moving toward stationary source).} \tag{12-3a}$$

If the observer is moving away from the source, the relative velocity is $v' = v - v_o$ and

$$f' = \left(1 - \frac{v_o}{v}\right) f \qquad \text{(observer moving away from stationary source).} \tag{12-3b}$$

When a sound wave is reflected from a moving obstacle, the frequency of the reflected wave will, because of the Doppler effect, be different from that of the incident wave. This is illustrated in the following example.

▮ EXAMPLE 12-10 A 5000-Hz sound wave is directed toward an object moving 3.3 m/s toward the source. What is the frequency of the reflected wave?

SOLUTION There are actually two Doppler shifts in this situation. First, the object acts like a moving observer and "detects" a sound wave of frequency

$$f' = \left(1 + \frac{v_o}{v}\right) f = \left(1 + \frac{3.3 \text{ m/s}}{330 \text{ m/s}}\right)(5000 \text{ Hz}) = 5050 \text{ Hz}.$$

Second, the object acts like a moving source in reemitting (reflecting) the sounds, so the reflected frequency is

$$f'' = \frac{1}{1 - v_s/v} f' = \frac{5050 \text{ Hz}}{1 - \dfrac{3.3 \text{ m/s}}{330 \text{ m/s}}} = 5100 \text{ Hz}.$$

Thus the frequency shifts by 100 Hz. ▮▮

The incident wave and the reflected wave, when mixed together (say, electronically), interfere with one another and beats are produced. The beat frequency is equal to the difference in the two frequencies, and in Example 12-10 would be 100 Hz.

This Doppler technique is used in a variety of medical applications, usually with ultrasonic waves in the megahertz frequency range. For example, ultrasonic waves reflected from red blood cells can be used to determine the velocity of blood flow. Similarly, the technique can be used to detect the movement of the chest of a young fetus and also to monitor its heartbeat. The problems at the end of the chapter deal with some of these applications quantitatively.

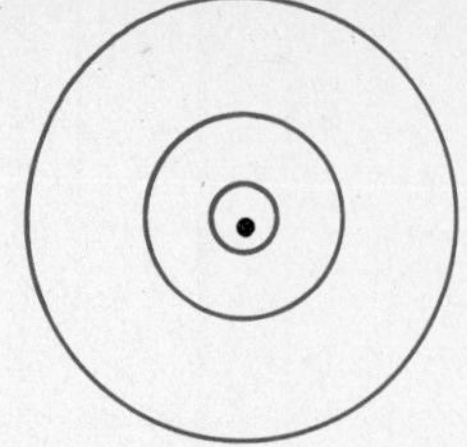

(a) $v_{\text{obj}} = 0$

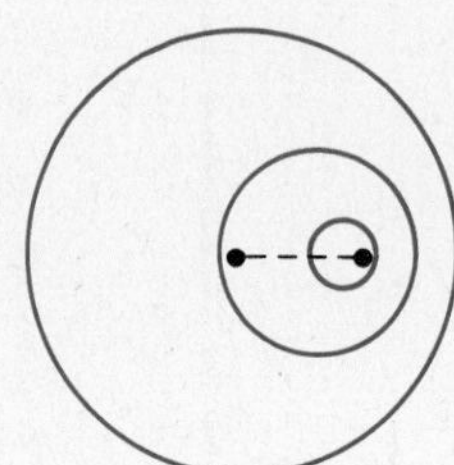

(b) $v_{\text{obj}} < v_{\text{snd}}$

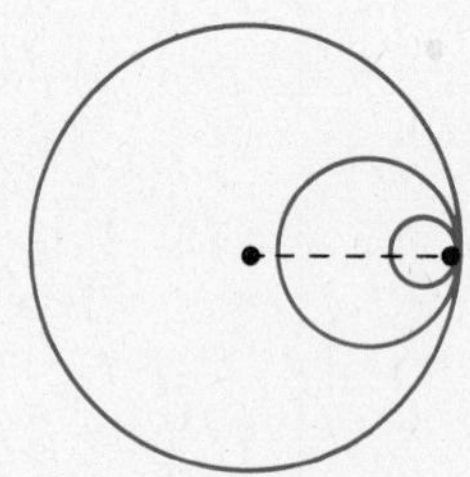

(c) $v_{\text{obj}} = v_{\text{snd}}$

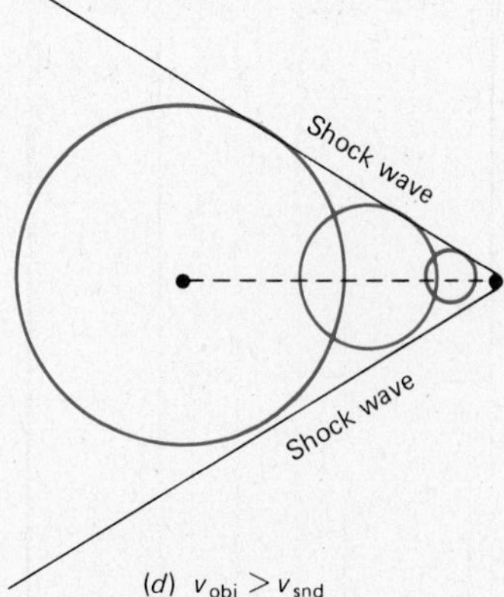

(d) $v_{\text{obj}} > v_{\text{snd}}$

FIGURE 12-18
Sound waves emitted by an object at rest (*a*) or moving (*b, c, d*). If the object's velocity is less than the velocity of sound, the Doppler effect occurs (*b*); if its velocity is greater than the velocity of sound, a shock wave is produced (*d*).

*12-10 ▮ Shock waves and the sonic boom

An object such as an airplane traveling faster than the speed of sound is said to have *supersonic speed*. Such speed is often given as a *Mach number*, which is the ratio of the object's speed to that of sound in the medium at that location. For example, a plane traveling 900 m/s high in the atmosphere, where the speed of sound is only 300 m/s, has a speed of Mach 3.

When a source of sound moves at subsonic speeds, the pitch of the sound is altered, as we have seen (the Doppler effect). But if a source of sound moves faster than the speed of sound, a more dramatic effect occurs known as a *shock wave.* In this case the source is actually "outrunning" the waves it produces. As shown in Figure 12-18*c*, when the source is traveling at the speed of sound, the waves it emits in the forward direction "pile up" directly in front of it. When the object moves at a supersonic speed, the waves pile up on one another along the sides, as shown in Figure 12-18*d*. The different wave crests overlap one another and form a single very large crest, which is the shock wave. Behind this very large crest there is usually a very large trough. A shock wave is essentially due to the constructive interference of a large number of waves. A shock wave in air is analogous to the bow wave of a boat traveling faster than the speed of the water waves it produces (Figure 12-19).

FIGURE 12-19
Bow wave produced by a boat. (Courtesy of U.S. Navy.)

When an airplane travels at supersonic speeds, the noise it makes and its disturbance of the air form into a shock wave containing a tremendous amount of sound energy. When the shock wave passes a listener, it is heard as a loud "sonic boom." A sonic boom lasts only a fraction of a second, but the energy it contains is often sufficient to break windows and cause other damage. It can be psychologically unnerving as well. Actually the sonic boom from a supersonic airplane is a double boom, since a shock wave forms at both the front and rear of the plane (Figure 12-20).

When an aircraft approaches the speed of sound, it encounters a barrier of sound waves in front of it (see Figure 12-18*c*). To exceed the speed of sound, extra thrust is needed to pass through this "sound barrier." This is called "breaking the sound barrier." Once a supersonic speed is attained, this barrier no longer impedes the motion.

It is sometimes erroneously thought that a sonic boom is produced only at the moment an aircraft is breaking through the sound barrier. Actually, a (double) shock wave follows the aircraft whenever it is traveling at supersonic speeds. A series of observers on the ground will each hear a loud double "boom" as the shock wave passes (Figure 12-20). The two shock waves consist of cones whose apex are at the aircraft. The angle of this cone, θ (see Figure 12-18*d*) is given by

$$\sin\theta = \frac{v_{\text{snd}}}{v_{\text{obj}}} \tag{12-4}$$

where v_{obj} is the velocity of the object (the aircraft) and v_{snd} is the velocity of sound in the medium. (The proof is left as a problem.)

There are many environmental problems that accompany supersonic flight, particularly for large aircraft. We have already mentioned

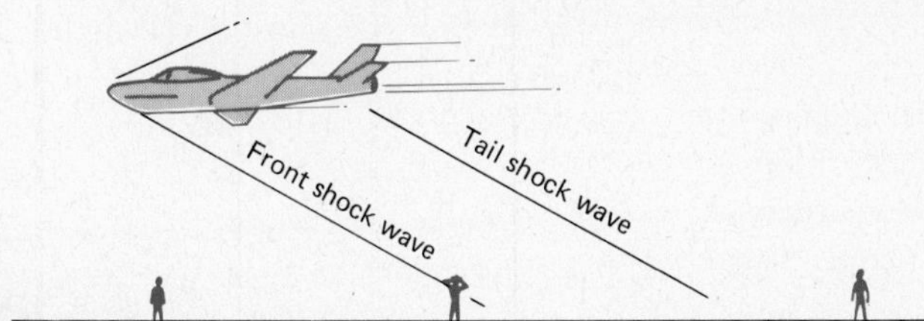

FIGURE 12-20
The (double) sonic boom has already been heard by the person on the right; it is just being heard by the person in the center; and it will shortly be heard by the person on the left.

that a sonic boom can cause damage to structures on the earth. It is also possible that supersonic transport planes may cause enough disruption of the upper atmosphere to influence climate and to affect the layer of ozone there that protects us from cosmic radiation. Also, at subsonic speeds during takeoff and landing, the noise level of a supersonic plane is considerably higher than for ordinary airplanes. A higher speed and greater thrust (which means more noise) is required of the engines at takeoff, since the wings are designed more smoothly for the high speeds of supersonic flight and consequently provide less lift. Finally, the amount of energy used per passenger mile is somewhat higher (and therefore wasteful) than for conventional aircraft or train travel.

Summary

Sound travels as a longitudinal wave in air and other materials. In air, the speed of sound increases with temperature; at 0°C it is about 330 m/s.

The *pitch* of a sound is determined by the frequency; the higher the frequency, the higher the pitch. The audible range of frequencies is roughly 20 to 20,000 Hz (1 Hz = 1 cycle per second). The *loudness* or *intensity* of a sound is related to the amplitude of the wave. Because the human ear can detect sound intensities from 10^{-12} W/m^2 to over 1 W/m^2, intensity levels are specified on a logarithmic scale. The *intensity level* β, specified in decibels, is defined in terms of intensity I as $\beta = 10 \log I/I_0$, where the reference intensity I_0 is usually taken to be 10^{-12} W/m^2. An increase in intensity by a factor of 100, for example, corresponds to a level increase of 20 dB. The *quality* or timbre of a sound depends on the mixture of frequencies present. The simpler the ratio of frequencies present, the more harmonious the sound.

Musical instruments are simple sources of sound in which standing waves are produced. The strings of a stringed instrument may vibrate as a whole with nodes only at the ends; the frequency at which this occurs is called the *fundamental*. The string can also vibrate at higher frequencies, called overtones or *harmonics*, in which there are one or more additional nodes. The frequency of each harmonic is a whole-number multiple of the fundamental. In wind instruments, standing waves are set up in the column of air within the tube. The vibrating air in an open tube (open at both ends) has antinodes at both ends; the fundamental frequency corresponds to a wavelength equal to twice the tube length. The harmonics have frequencies that are 2, 3, 4, . . . times the fundamental frequency. For a closed tube (closed at one end) the fundamental corresponds to a wavelength four times the length of the tube; only the odd harmonics are present, equal to 1, 3, 5, 7, . . . times the fundamental frequency.

Sound waves from different sources can interfere with each other. If two sounds are at slightly different frequencies, *beats* can be heard at a frequency equal to the difference in frequency of the two sources.

The *Doppler effect* refers to the change in pitch of a sound due to either the motion of the source or of the listener. If they are approaching each other, the pitch is higher; if they are moving apart, the pitch is lower.

Questions

1 What is the evidence that sound travels as a wave?

2 What is the evidence that sound is a form of energy?

3 Country folk have a rule of thumb that the time delay between seeing lightning and hearing the thunder is an indication of how far away the lightning hit, and in particular, that each 5 s corresponds to 1 mile. Explain.

4 When boating on a lake or river at night, one can often hear clearly the voices or radios of people quite a distance away. Yet this rarely happens in the daytime. This phenomenon can be explained by considering the refraction of sound due to different layers of air having different densities (because of differences in temperature). Draw a diagram and explain this phenomenon and determine if the layer of air nearest the water surface is denser than the one above it at nighttime or in the daytime.

5 When a sound wave passes from air into water, do you expect the frequency or wavelength to change?

6 Explain why the larger area of the eardrum, as compared to that of the oval window, leads to an amplification of the pressure.

7 The voice of a person who has inhaled helium sounds very high pitched. Why?

8 Whistle through your lips and describe how the pitch of the whistle is controlled.

9 What is the reason that catgut strings on some musical instruments are wrapped with fine wire?

10 How does high C played on a violin differ from high C played on a flute?

11 Why are the frets on a guitar farther apart as you move down the neck?

*12 Which do you think should sound more harmonious: A and D played together or F# and B? Why?

*13 Will the reverberation time in a room be affected by whether the windows are open or not? Explain.

*14 It was noted, in a famous concert hall, that the acoustics in recent months had been much less lively than in the past. Acoustic specialists traced the problem to the fact that the coat checkroom had been removed. Explain.

*15 If you see heavy drapes hanging in a large hall, what would you guess the use of the room to be?

16 If a wind is blowing, will this alter the frequency of the sound heard by a person at rest with respect to the source? Is the wavelength or velocity changed?

*17 A sonic boom sounds much like an explosion. Explain the similarity between the two.

Problems

SECTION 12-1

1 (I) Ultrasonic waves with frequencies as high as 250,000 Hz are emitted by dolphins. What would be the wavelength of such a wave (*a*) in water, and (*b*) in air? Use Table 12-1.

2 (I) A hiker determines the length of a lake by listening for the echo of her shout reflected by a cliff at the far end of the lake. She hears the echo about 1 s after shouting. How long is the lake?

3 (II) What is the wavelength of a sound wave in air whose frequency is 1000 Hz if the temperature is 30°C?

4 (II) A flute is "in tune" at 20°C. By what percentage will its notes be off at 0°C?

SECTION 12-2

5 (I) What is the intensity level of a sound whose intensity is 10^{-8} W/m^2?

6 (I) What is the intensity of a sound whose intensity level is 50 dB?

7 (I) A stereo tape recorder is said to have a signal-to-noise ratio of 55 dB. What is the ratio of intensities of the signal and the background noise?

8 (I) Human beings can detect a difference in sound intensity level of 1.0 dB. What is the ratio of the intensities of two sounds whose levels differ by this amount?

*9 (I) If the amplitude of a sound wave is tripled, (*a*) by what factor will the intensity increase? (*b*) By how many decibels will the intensity level increase?

*10 (I) Calculate the maximum displacement of air molecules when a loud sound wave passes whose intensity is 1.0 W/m^2 and frequency is 80 Hz.

*11 (I) What would be the intensity level (in dB) of a sound wave in air that corresponds to a displacement amplitude of vibrating air molecules of 1.0 mm at 50 Hz?

12 (II) If two firecrackers produce an intensity level of 80 dB at a certain place, what will be the intensity if only one is exploded?

13 (II) A single mosquito 10 m from a person makes a sound close to the threshold of human hearing (0 dB). What will be the intensity level of 10,000 such mosquitoes?

14 (II) A 90-dB sound wave strikes an eardrum whose area is 5.0×10^{-5} m^2. How much energy is absorbed by the eardrum per second?

15 (II) A stereo amplifier is rated at 30 W output at 1000 Hz. The output drops by 2 dB at 20 Hz. What is the power output at 20 Hz?

16 (II) What is the resultant sound level (in dB) when an 80-dB and a 75-dB sound are heard simultaneously?

*17 (II) In audio and communications sytems, the *gain* β in decibels is defined as

$$\beta = 10 \log \frac{P_{out}}{P_{in}},$$

where P_{in} is the power input to the system and P_{out} is the power output. A particular stereo amplifier puts out 30 W of power for an input of 1 mW. What is its gain in decibels?

*18 (II) Use the fact that the intensity of a sound wave is proportional to the square of the pressure amplitude, and the fact that at the threshold of sound ($I = 10^{-12}$ W/m^2), the pressure amplitude is 3.0×10^{-5} N/m^2, to calculate the pressure amplitude for the loudest sounds the ear can tolerate without pain ($I = 1.0$ W/m^2).

*19 (III) The intensity level 50 m from a loudspeaker is 100 dB. What is the acoustic power output (W) of the speaker?

*20 (III) (*a*) Use the fact that the intensity of a sound wave is proportional to the square of the pressure amplitude P to show that the intensity level β can be written

$$\beta \text{ (dB)} = 20 \log \frac{P}{P_0},$$

where P_0 is the pressure amplitude at some reference level. (*b*) The reference pressure P_0 is often taken to be 3.0×10^{-5} N/m^2, corresponding to an intensity of 1.0×10^{-12} W/m^2. What would the intensity level be if P were 1 atm?

*21 (III) A jet plane emits 10^5 J of sound energy per second. Show that the intensity level 90 m away is 120 dB.

*22 (III) Air absorbs sound at a rate of about 7 dB/km. For the jet plane in Problem 21, calculate what the intensity level will be (*a*) 1.0 km and (*b*) 5.0 km away, taking into account air absorption.

SECTION 12-3

*23 (I) What is the lowest intensity level that can be heard by an average ear (middle curve of Figure 12-5)?

*24 (I) An 8000-Hz tone must have what intensity level to seem as loud as an 80-Hz tone that has a 60-dB intensity level? (See Figure 12-6.)

*25 (I) What are the lowest and highest frequencies that the average ear (middle curve in Figure 12-5) can hear when the intensity level is 25 dB?

SECTION 12-4

26 (I) The G string on a violin has a fundamental frequency of 196 Hz. The length of the vibrating portion is 32 cm and has a mass of 0.50 g. Under what tension must the string be placed?

27 (I) An unfingered guitar string is 0.70 m long and is tuned to play E above middle C (330 Hz). How far from the end of this string must the finger be placed to play A above middle C (440 Hz)?

28 (I) Determine the length of a closed organ pipe that emits middle C (264 Hz) when the speed of sound in air is 345 m/s.

29 (I) How far from the end of the flute in Example 12-6 should the hole be that must be uncovered to play D above middle C at 294 Hz?

30 (I) An organ pipe is 80 cm long. What are the fundamental and first three audible overtones (*a*) if the pipe is closed at one end and (*b*) if it is open at both ends?

31 (II) How many overtones are present within the audible range for a 100-cm-long organ pipe (*a*) if it is open, and (*b*) if it is closed?

32 (II) (*a*) How long must an open organ pipe be if it is to have a fundamental frequency of 264 Hz? (*b*) If this pipe was filled with helium (see Table 12-1), what would its fundamental frequency be?

33 (II) Calculate the resonant frequency of the column of air in the outer ear of a human being, which is about 2.5 cm long. Does this correspond to a region of high sensitivity of the ear? Explain.

34 (II) Each string on a violin is tuned to a frequency $1\frac{1}{2}$ times that of its neighbor. If all the strings are to be placed under the same tension, what must be the mass per unit length of each string relative to that of the lowest string?

35 (II) A tuning fork is set into vibration above a vertical open tube filled with water. The water level is allowed to drop slowly; as it does so, the air in the tube above the water level is heard to resonate with the tuning fork when the distance from the tube opening to the water level is 0.12 m and again at 0.38 m. What is the frequency of the tuning fork?

36 (III) An 80-cm-long guitar string of mass 1.50 g is placed near a tube open at one end, also 80 cm long. How much tension should be in the string so that its fourth harmonic has the same frequency as the fifth harmonic of the tube?

*SECTION 12-6

*37 (I) What is the frequency of the third F above middle C?

*38 (I) What note has a frequency of 1320 Hz?

*39 (II) An organ is in tune at 25°C. By what fraction of a tone will the notes be off at 0°C?

SECTION 12-7

40 (I) Two horns emitting sounds of frequency 665 and 671 Hz, respectively, will produce beats of what frequency?

41 (I) Two flutes are slightly out of tune when they play middle C so that beats are heard every 2 s. By how much do the two flutes differ in frequency?

*42 (I) What will be the "beat frequency" if middle C and $C^{\#}$ are played together? (See Table 12-5).)

43 (II) Two piano strings are supposed to be vibrating at 132 Hz, but a piano tuner hears 4 beats per second when they are played together. (*a*) If one is vibrating at 132 Hz, what must be the frequency of the other? (Is there only one answer?) (*b*) By how much (in percent) must the tension be increased or decreased to bring them in tune?

44 (II) How many beats will be heard if two identical flutes each try to play middle C (264 Hz), but one is at 5°C and the other at 20°C?

45 (II) Two violin strings are tuned to the same frequency, 440 Hz. The tension in one string is then decreased by 2 percent. What will be the frequency of beats heard when the two strings are played together?

46 (II) A person stands 3.0 m from one loudspeaker and 3.5 m from a second loudspeaker, as shown in Figure 12-12. (*a*) What is the lowest frequency at which destructive interference will occur at this point? (*b*) Calculate two other frequencies that also result in destructive interference at this point (give the next two highest).

47 (II) A person hears a pure tone coming from two sources that seems to be in the range 500 to 1000 Hz. The sound is loudest at points equidistant from the two sources. In order to determine exactly what the frequency is, the person moves about and finds that the sound level is minimal at a point 0.22 m farther from one source than the other. What is the frequency of the sound?

48 (III) Show that the two speakers in Figure 12-12 must be separated by at least a distance d equal to one-half the wavelength λ of sound if there is to be any place where destructive interference occurs.

49 (III) A source emits sounds of wavelength 2.30 and 2.50 m, respectively. (*a*) How many beats per second will be heard? (Assume that $T = 20°C$.) (*b*) How far apart in space are the regions of maximum intensity?

*SECTION 12-8

*50 (I) What is the absorption coefficient of a wall that reflects 20 percent of the incident sound energy?

*51 (II) Estimate the reverberation time of a hall 60 m by 80 m by 18 m if the wall surfaces are half wood and half covered with drapes, the ceiling is plaster, the floor is carpeted, and there are 1000 upholstered seats and 100 wood seats. Assume that (*a*) all the seats are occupied, and (*b*) the hall is empty.

*52 (II) The smallest object a bat can detect with its ultrasonic "radar" is limited by diffraction. What, approximately, is the smallest object you would expect a bat could detect? The highest-frequency waves emitted by a bat are about 10^5 Hz.

SECTION 12-9

53 (I) A factory whistle emits a sound at 900 Hz. What frequency will be heard by an observer in an automobile traveling at 70 km/h (*a*) away from the source, and (*b*) toward the source?

54 (I) The predominant frequency of a certain police car's siren is 1800 Hz when at rest. What is the frequency if the car is (*a*) moving toward an observer at 30 m/s, and (*b*) if moving away at the same speed?

*55 (II) How fast must an observer be moving in order that musical notes will sound one-half interval lower than normal on the equally tempered chromatic scale? (See Section 12-6.)

56 (II) Sound waves of frequency 1.0×10^6 Hz are directed at the chest of a fetus and travel with a speed in the body of 1.5×10^3 m/s. What will be the expected shift in frequency if the chest of a normal fetus moves at a maximum speed of 0.10 m/s?

57 (II) Derive a general formula for the changed frequency f' due to the Doppler effect when both the source and the observer are moving.

58 (II) Two trains emit whistles of the same frequency, 380 Hz. If one train is at rest and the other is traveling at 90 km/h away from an observer at rest, what will the observer detect as the beat frequency?

59 (III) If the velocity of blood flow in the aorta is normally about 0.30 m/s, what beat frequency would you expect if 4.0 MHz ultrasound waves were directed along the flow and reflected from the red blood cells? Assume that the waves travel with a speed of 1.5×10^3 m/s (which is about the speed of sound in water).

60 (III) The Doppler effect using ultrasonic waves of frequency 2.0×10^6 Hz is used to monitor the heartbeat of a fetus. A (maximum) beat frequency of 600 Hz is observed. Assuming that the speed of sound in tissue is 1.5×10^3 m/s, calculate the maximum velocity of the surface of the beating heart.

61 (III) In Problem 60, the beat frequency is found to appear and then disappear 180 times per minute which reflects the fact that the heart is beating and its surface changes speed. What is the heartbeat rate?

*SECTION 12-10

*62 (I) How fast is an object moving on the earth if its speed is specified as Mach 0.5?

*63 (II) Show that the angle θ a sonic boom makes with the path of a supersonic object is given by Equation 12-4.

*64 (II) An airplane travels 1000 m/s where the speed of sound is 310 m/s. (*a*) What is its Mach number? (*b*) What is the angle the shock wave makes with the direction of the airplane's motion?

*65 (II) A boat traveling 10 m/s makes a bow wave at an angle of 16° to its direction of motion. What is the speed of the water waves?

*66 (III) The pressure P in a sonic boom decreases with the altitude of an airplane approximately as

$$P \propto \frac{1}{h^{3/4}}.$$

By what factor does the pressure decrease if a plane flies twice as high? Compare to an inverse-square law.

For additional reading

Ackerman, E., *Biophysical Science,* Prentice-Hall, Inc., Englewood Cliffs, N.J., 1962. Ear and hearing, ultrasound.

Backus, J., *Acoustical Foundations of Music,* W. W. Norton & Co., Inc., New York, 1969.

Benade, A. H., *Fundamentals of Musical Acoustics,* Oxford University Press, New York, 1976.

Benade, A. H., "Physics of Woodwinds," *Sci. American,* **203**:4 (Oct. 1960), p. 144.

Blacham, E. D., "The Physics of the Piano," *Sci. American,* **213**:6 (Dec., 1965), p. 88.

Bliven, B., "Annals of Architecture (Avery Fisher Hall)," *The New Yorker,* Nov. 8, 1976, p. 51.

Coltman, J. W., "Acoustics of the Flute," *Physics Today,* **21** (Nov., 1968), pp. 25–32.

Devey, G. B., and P. N. T. Wells, "Ultrasound in Medical Diagnosis," *Sci. American,* **238**:5 (May, 1978) p. 98.

Doelle, L., *Environmental Acoustics,* McGraw-Hill Book Company, New York, 1972. Detailed and readable text in acoustics and acoustical design of buildings.

Few, A. A., "Thunder," *Sci. American,* **233**:1 (Jul., 1975), p. 80.

Rossing, T. D., "Resource Letter: Musical Acoustics," *Am. J. Phys.,* **43** (1975), p. 944. Contains many references.

Shankland, R. S., "Acoustics of Greek Theatres," *Physics Today,* **26** (Oct. 1973), p. 30.

Shaw, E. A. G., "Noise Pollution: What Can Be Done?" *Physics Today,* **28** (Jan., 1975), p. 46.

Sundberg, J., "Acoustics of the Singing Voice," *Sci. American,* **236**:3 (Mar., 1977), p. 82.

Taylor, C. A., *Sounds of Music,* BBC, London, 1976.

13 ▌ Electric charge and electric field

The word "electricity" may evoke an image of complex modern technology. Yet the electric force plays a far more important role in our lives than at first may seem true. The forces that act between atoms and molecules to hold them together to form liquids and solids are now understood to be electrical forces. The electric force is responsible for the metabolic processes that occur within our bodies. Even ordinary pushes and pulls are the result of the electric force between the molecules of your hand and those of the object being pushed or pulled. Indeed, most of the forces we have dealt with so far, such as elastic forces and the normal force acting on a body are electric forces acting at the atomic level. This does not include gravity, however, which is a separate force. Indeed, physicists today recognize only four different forces in nature: (1) gravitational force, (2) electromagnetic force (we will see later that electric and magnetic forces are intimately related), (3) strong nuclear force, and (4) weak nuclear force. The last two forces operate at the level of the nucleus of an atom, and although they manifest themselves in such phenomena as radioactivity and nuclear energy, they are much less obvious in our daily lives. We shall discuss them in later chapters on nuclear physics.

It has been said that nature keeps her most important secrets well hidden, and this does seem to apply to electricity; for in spite of its great importance, it took a great deal of investigation and research to realize the role of electricity in so many phenomena—which culminated at the beginning of this century with an understanding of the structure of atoms and molecules.

FIGURE 13-1
Rub a plastic ruler (*a*) and bring it close to some tiny pieces of paper. (*b*).

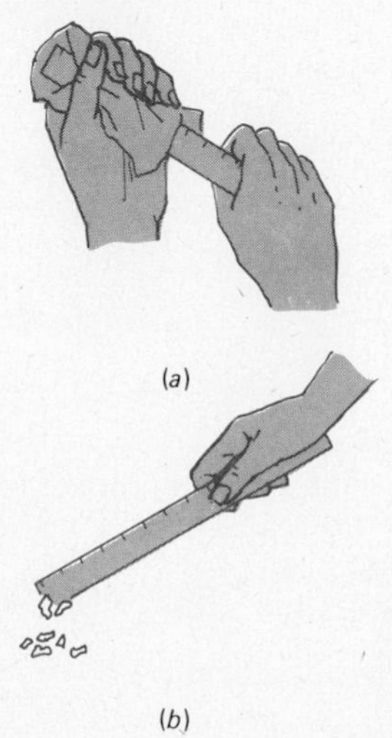

13-1 ▌ Electric charges and forces

Static electricity and electric charge

The word *electricity* comes from the Greek word "elektron," which means "amber." Amber is petrified tree resin and the ancients knew that if you rub an amber rod with a piece of cloth, the amber attracts small pieces of leaves or dust. A piece of hard rubber, a glass rod, or a plastic ruler rubbed with a cloth will also display this "amber effect"—or *static electricity*, as we call it today. You can readily pick up small pieces of paper with a plastic ruler that you have just vigorously rubbed with even a paper towel, Figure 13-1. You are probably familiar with static electricity when combing your hair, or upon taking a synthetic blouse or shirt from a clothes dryer. And you may have felt a shock when you touched a metal door knob after sliding across a car seat or walking across a nylon carpet. In each case we say the object becomes "charged" due to a rubbing process, and it is said to possess an **electric charge**.

Is all electric charge the same, or is it possible that there is more than one type? In fact, there are two types of electric charge, as the following experiments show. A plastic ruler is suspended by a thread and rubbed vigorously to charge it; when a second ruler, which has also been charged by rubbing, is brought close to the first, it is found that the one ruler *repels* the other. This is shown in Figure 13-2*a*. Similarly, if a rubbed glass rod is brought close to a second charged glass rod, again a repulsive force is seen to act, Figure 13-2*b*. Hower, if the charged glass rod is brought close to the charged plastic ruler, it is found that they *attract* each other, Figure 13-2*c*. The charge on the glass must therefore be different than that on the plastic. Indeed, it is found experimentally that all charged objects fall into one of two categories. Either they are attracted to the plastic and repelled by the glass, just as glass is; or they are repelled by the plastic and attracted to the glass, just as the plastic ruler is. Thus there seem to be two, and only two, types of electric charge. Each type of charge repels the same type but attracts the opposite type. That is: *unlike charges attract*; *like charges repel.*

These two types of charge were given the names *positive* and *negative* by the great American statesman, philosopher, and scientist, Benjamin Franklin (1706–1790). The choice of which name went with what type of charge was of course arbitrary. Franklin chose the charge on the rubbed glass rod to be positive charge, so the charge on a rubbed plastic ruler (or amber) is called negative charge; we still follow this convention today.

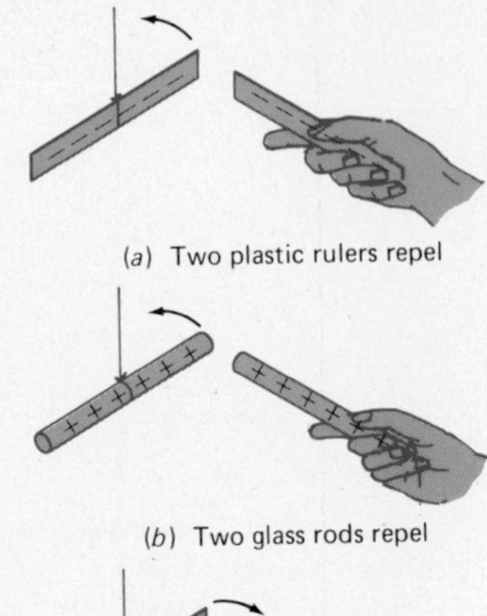

(*a*) Two plastic rulers repel

(*b*) Two glass rods repel

(*c*) Glass rod attracts plastic ruler

FIGURE 13-2
Unlike charges attract, whereas like charges repel one another.

Conservation of electric charge

Franklin contributed to our understanding of electricity in a number of ways. His demonstration of the electrical nature of lightning using a kite is well known. He also was the first to realize that whenever a certain amount of charge is produced in a process, an equal amount of the opposite type of charge is produced at the same time. The names "positive" and "negative" were intended to be taken *algebraically*, so that the net amount of charge produced in any process is zero. For example, when a plastic ruler is rubbed with a paper towel, the plastic acquires a negative charge and the towel an equal amount of positive charge; the sum of the two is zero, and so the *net* amount of charge produced is zero. This is an example of a law that is now well established: the **law of conservation of electric charge**, which states that

▮ **the net amount of electric charge produced in any process is zero.**

No violations have ever been found, and this conservation law is as firmly established as those for energy and momentum.

Origin of electric charge in the atom

Only within the past century has it become clear that electric charge has its origin within the atom itself. In later chapters we will discuss atomic structure and the ideas that led to our present view of the atom in detail, but it will help our understanding of electricity if we discuss it briefly now.

Today's view, slightly simplified, shows the atom as having a heavy, positively charged nucleus surrounded by one or more negatively charged electrons (Figure 13-3). In its normal state, the positive and negative charges within the atom are equal, and the atom is electrically neutral. Sometimes, however, an atom may lose one or more of its electrons, or may gain extra electrons. In this case the atom will have a net positive or negative charge and is called an *ion*.

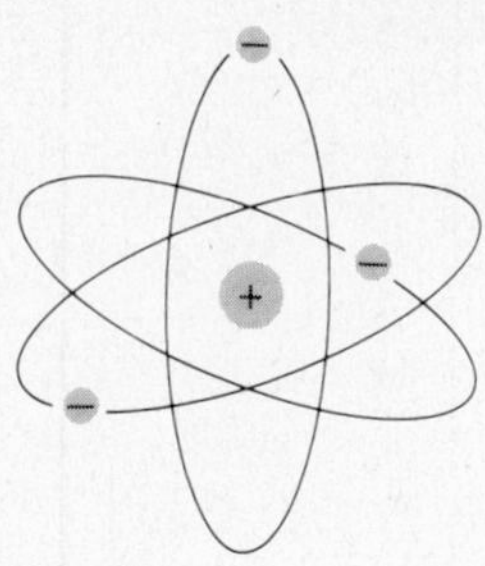

FIGURE 13-3
Simple view of the atom.

The nuclei in a solid material are fairly rigidly fixed in position, whereas the electrons move more freely. The charging of an object by rubbing is explained by considering how strongly the electrons are held by the nuclei of the material. When a plastic ruler becomes negatively charged by rubbing with a towel, the electrons in the towel are evidently held less tightly than those in the plastic; thus some of the electrons are pulled off the towel onto the plastic. This leaves the towel with a positive charge equal in magnitude to the negative charge given the plastic.

Normally, when objects are charged by rubbing, they hold their charge only for a limited time and eventually return to the neutral state. Where does the charge go? It "leaks off" onto water molecules in the air. This is because water molecules are *polar*—that is, even though they are neutral, their charge is not distributed uniformly (Figure 13-4). Thus the extra electrons on, say, a charged plastic ruler leak off because they are attracted to the positive end of water molecules. A positively charged object, on the other hand, is neutralized by transfer of loosely held electrons from water molecules in the air. On dry days, static electricity is much more noticeable since there are fewer water molecules to allow leakage. On humid or rainy days, it is difficult to make any object hold its charge for any appreciable amount of time.

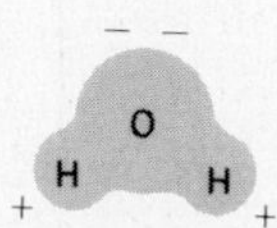

FIGURE 13-4
Diagram of a water molecule. Because it has opposite charges on different ends, it is called a "polar" molecule.

Insulators and conductors

Suppose that we have a highly charged glass rod and an electrically neutral object, say a small metal ball. If we now place an iron nail so that it touches both the charged glass rod and the neutral ball, it is found that the ball becomes charged. If, however, we had connected the two objects together with a wooden rod or a piece of rubber, the ball would *not* have become charged. Materials like the iron nail are said to be *conductors* of electricity, whereas wood and rubber are called *nonconductors* or *insulators*.

Metals are generally good conductors, whereas most other materials are insulators. Although even insulators conduct electricity very slightly, there is a great difference in the amount of conductivity between insulators and conductors. (We will discuss this quantitatively in Chapter 15.) It is interesting that nearly all natural materials fall into one or the other of these two quite distinct categories. Yet, there are a few materials—notably silicon, germanium and carbon—that fall into an intermediate (although still fairly distinct) category known as *semiconductors*.

From the atomic point of view, the electrons in an insulating material are held very tightly, whereas in a conductor they are held loosely and many of them move about freely within the material. When a positively charged object is brought close to or touches a conductor, the free electrons move quickly toward this positive charge. On the other hand, the free electrons move swiftly away from a negative charge that is brought close. In a semiconductor, there are a very few free electrons, and in an insulator, almost none.

Induced charge

When a positively charged glass rod touches a metal object, the free electrons in the metal are attracted to the positive glass rod and some will pass over to the glass. Since the metal is now missing some of its negative electrons, it will have a net positive charge. This process is called "charging by conduction."

But suppose that the charged glass rod is brought close to a metal rod, but does not touch it. Although the electrons of the metal do not jump over to the glass rod, they still move toward it within the metal; this leaves a positive charge at the opposite end (Figure 13-5). A charge is said to have been *induced* at the two ends of the metal rod. Of course, no charge has been created; it has merely been *separated*; the net charge on the metal rod is still zero. However, if the metal were now cut in half, we would have two charged objects, one positive and one negative.

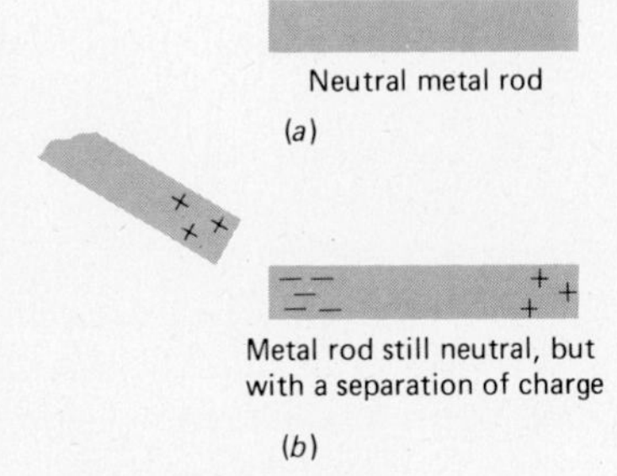

FIGURE 13-5
Charging by induction.

Another way to induce a net charge on a metal object is to connect it with a conducting wire to the ground (or a pipe leading into the ground) as shown in Figure 13-6*a* ($\frac{\perp}{=}$ means "ground"). The object is then said to be "grounded" or "earthed." Now the earth, since it is so large, can easily accept or give up electrons; hence it acts like a reservoir for charge. If a charged object—let us say negative this time—is brought up close to the metal, free electrons in the metal are repelled and many of them move down the wire into the earth (Figure 13-6*b*). This leaves the metal positively charged. If the wire is now cut, the metal will have a positive induced charge on it. If the wire were cut after the negative object is moved away, the electrons would all have moved back into the metal and it would be neutral.

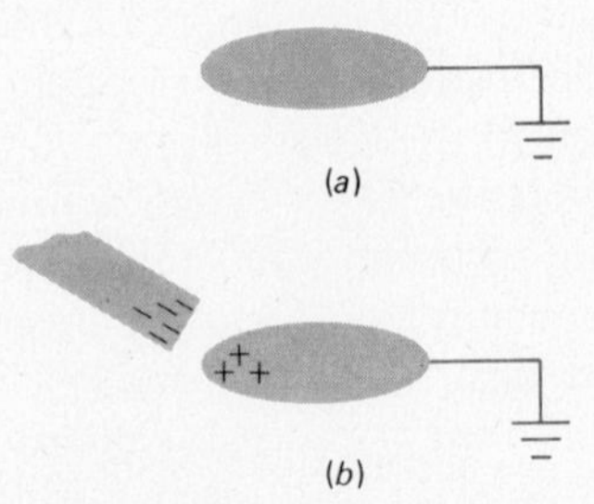

FIGURE 13-6
Inducing a charge on an object connected to ground.

The electroscope

An *electroscope* is a simple device used for measuring amounts of charge. As shown in Figure 13-7, it consists of a case inside of which are two movable leaves often made of gold. (Sometimes only one leaf is movable.) The leaves are connected by a conductor to a metal ball on the outside of the case, but are insulated from the case itself. If a charged object is brought close to the knob, a charge is induced on it (Figure 13-8*a*); the two leaves become charged and repel each other as shown. If, instead, the knob is charged by conduction, the whole apparatus becomes charged, as shown in Figure 13-8*b*. In either case, the greater the amount of charge, the greater the separation of the leaves. Thus the electroscope can be used to measure the amount of charge, at least roughly.

An electroscope can also be used to determine the sign of the charge if it is first charged by conduction, say negatively, as in Figure 13-9*a*. Now if a negative object is brought close, as in part *b*, electrons are induced to move farther down into the leaves and they separate further. On the

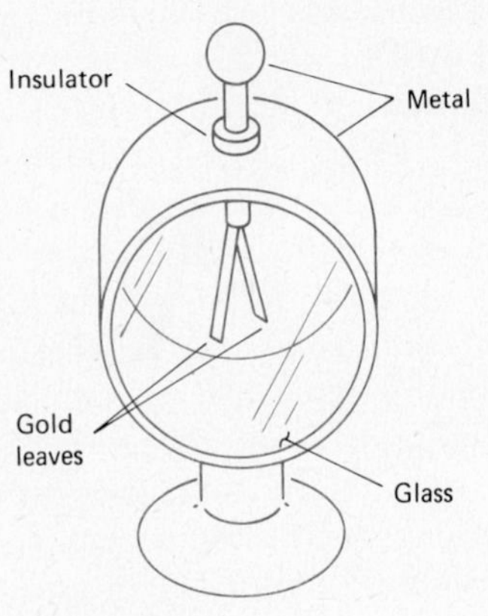

FIGURE 13-7
Electroscope.

other hand, if a positive object is brought close, the electrons are induced to flow upward, leaving the leaves less negative and their separation is reduced (Figure 13-9*c*).

The electroscope was much used in the early days of electricity. The same principle, aided by some electronics, is used in many modern *electrometers*, which can measure charge very precisely.

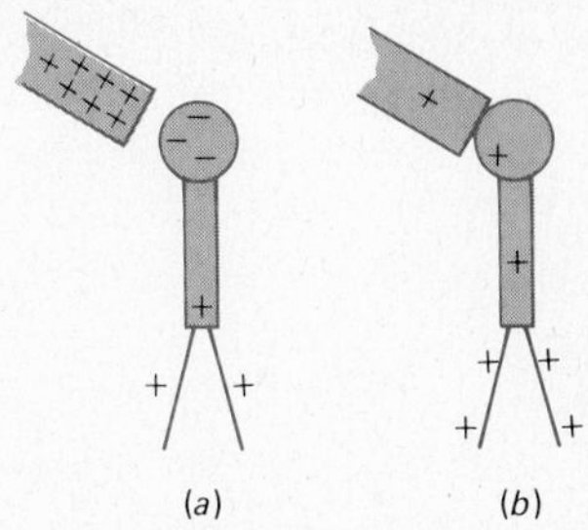

FIGURE 13-8
Electroscope charged (*a*) by induction, (*b*) by conduction.

13-2 ▮ Coulomb's law

The force between two charged objects—Coulomb's law

We have seen that an electric charge exerts a force on other electric charges. But how does the magnitude of the charges and other factors affect the magnitude of this force? To answer this, the French physicist Charles Coulomb (1736–1806) investigated electric forces using a torsion balance (Figure 13-10), much like that used by Cavendish for his studies of the gravitational force.

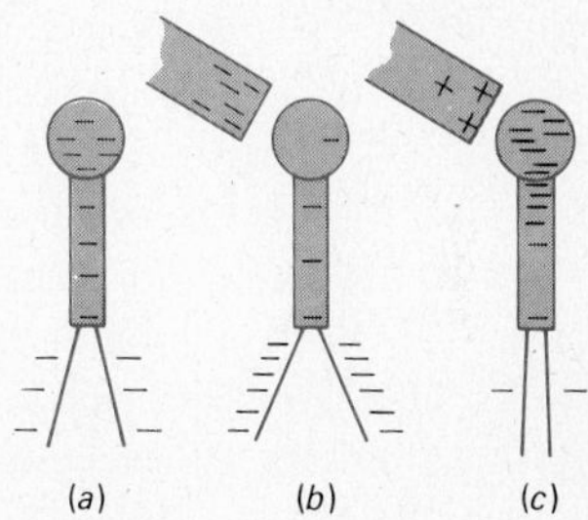

FIGURE 13-9
A previously charged electroscope can be used to determine the sign of a given charge.

Although precise instruments for the measurement of electric charge were not available in Coulomb's time, he was able to prepare spheres with different amounts of charge in which the *ratio* of the charges was known. He reasoned, if a charged conducting sphere were put in contact with an identical uncharged sphere, the charge on the first would then be shared equally by the two of them because of symmetry. He thus had a way to produce charges equal to $\frac{1}{2}$, $\frac{1}{4}$, and so on, of the original charge. Although he had some difficulty with induced charges, Coulomb was able to conclude that the force one charged object exerted on a second charged object is directly proportional to the charge on each of them. That is, if the charge on either of the objects was doubled, the force was doubled; and if the charge on both of the spheres was doubled, the force increased to four times the original value. This was the case when the distance between the two spheres remained the same. If the distance was allowed to change, he found that the force decreased with the square of the distance between them. That is, if the distance was doubled, the force fell to only one-fourth of its original value. Thus, Coulomb concluded, the force one charged objects exerts on a second is proportional to the product of the amount of charge on one, Q_1, times the amount of charge on the other, Q_2, and inversely proportional to the square of the distance r between them; that is,

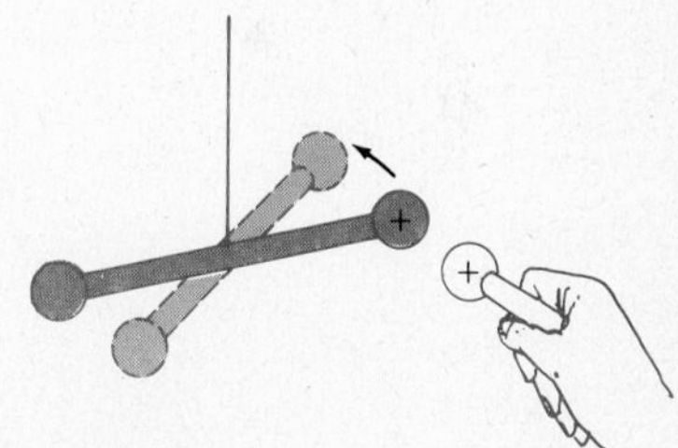

FIGURE 13-10
Schematic diagram of Coulomb's apparatus. It is similar to Cavendish's, which was used to measure the gravitational force. When a charged sphere is placed close to the one on the suspended bar, the bar rotates slightly. The suspending fiber resists the twisting motion and the angle of twist is proportional to the force applied. By use of this apparatus, Coulomb was able to investigate how the electric force varied as a function of the magntude of the charges and of the distance between them.

$$F = k\frac{Q_1 Q_2}{r^2} \tag{13-1}$$

where k is a proportionality constant. This is known as **Coulomb's law**. The unit of charge could be chosen so that the proportionality constant k would be 1. Indeed, such a system of units was once common.† However, the most widely used unit now is the *coulomb* (C), which is the SI unit. The precise definition of the coulomb today is given in terms of electric

† It is a cgs system of units, and its unit of electric charge is called the *electrostatic unit* (esu). One esu is defined as that charge, on each of two tiny objects 1 cm apart, that gives rise to a force of 1 dyne.

current and magnetic field and will be discussed in Section 17-7. In SI units, k has the value

$$k = 9.0 \times 10^9\ \mathrm{N \cdot m^2/C^2}.$$

Thus, 1 coulomb is that amount of charge which, if it exists on each of two objects placed 1 m apart, will result in each object exerting a force of $(9.0 \times 10^9\ \mathrm{N \cdot m^2/C^2})(1\ \mathrm{C})(1\ \mathrm{C})/(1\ \mathrm{m})^2 = 9.0 \times 10^9\ \mathrm{N}$ on the other.

Charges produced by rubbing ordinary objects (such as a comb or plastic ruler) are typically a microcoulomb ($1\ \mu\mathrm{C} = 10^{-6}\ \mathrm{C}$) or less. The charge on one electron, on the other hand, is measured to be $1.602 \times 10^{-19}\ \mathrm{C}$. This is the smallest known charge. Since an object cannot gain or lose a fraction of an electron, the net charge on any object must be an integral multiple of this charge. Because of its fundamental nature, the charge on the electron is often called the *elementary charge* and is given the symbol $-e$, where $e = 1.602 \times 10^{-19}\ \mathrm{C}$.

Equation 13-1 gives the magnitude of the force that either object exerts on the other. The direction of this force is along the line joining the two objects. If the two charges have the same sign, the force on each object is directed away from the other; if the two objects have opposite charges, the force on one is directed toward the other (Figure 13-11). Notice that the force one charge exerts on the second is equal but opposite to that exerted by the second on the first; this is in accord with Newton's third law.

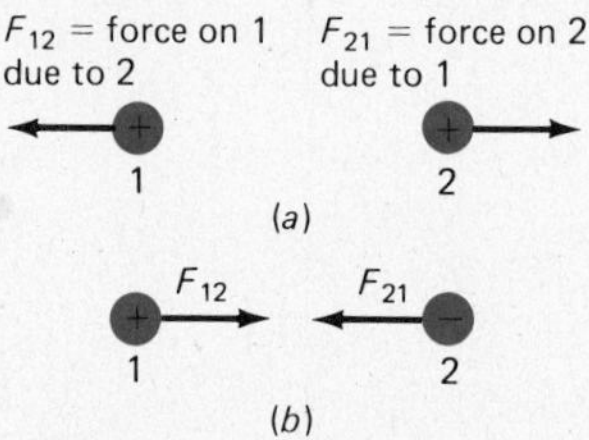

FIGURE 13-11
Direction of force depends on whether the charges have (*a*) same sign or (*b*) opposite sign.

It should be recognized that Equation 13-1 applies to objects whose size is much smaller than the distance between them. Ideally, they act like "point" charges. If the objects are large, it is not always clear what value to use for r, particularly since the charge may not be distributed uniformly on the objects. If the two objects are spheres and the charge is known to be distributed uniformly on each, then r would be the distance between centers. It is important to keep in mind that Equation 13-1 gives the force on a charge due to only *one* other charge. If several (or many) charges are present, the net force on any one of them will be the vector sum of the forces due to each of the others.

▌ EXAMPLE 13-1 What is the magnitude of the electric force on an electron exerted by a second electron 1.0 mm away?

SOLUTION We use Equation 13-1 with $Q_1 = Q_2 = -1.6 \times 10^{-19}\ \mathrm{C}$ and $r = 1.0 \times 10^{-3}\ \mathrm{m}$:

$$F = \frac{(9.0 \times 10^9\ \mathrm{N \cdot m^2/C^2})(-1.6 \times 10^{-19}\ \mathrm{C})(-1.6 \times 10^{-19}\ \mathrm{C})}{(1.0 \times 10^{-3}\ \mathrm{m})^2}$$
$$= +2.3 \times 10^{-22}\ \mathrm{N}.$$

The plus sign indicates the force is repulsive, in the direction to increase the separation r. If the force had been attractive, F would have been negative. ▌▌

▌ EXAMPLE 13-2 Calculate the net electric force on particle 2 in Figure 13-12*a* due to the other two charges.

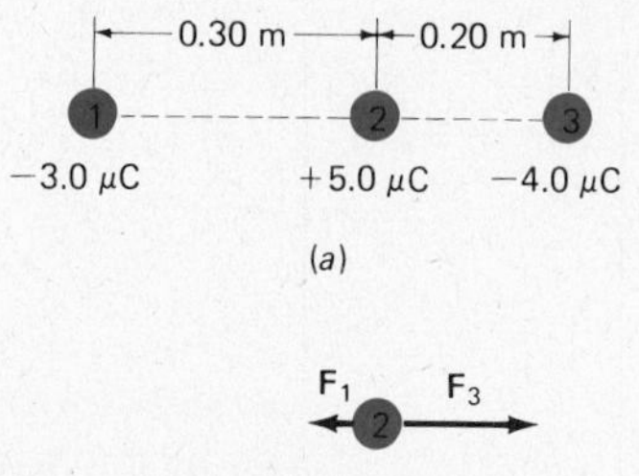

FIGURE 13-12
Diagram for Example 13-2.

SOLUTION The net force on particle 2 will be the sum of the force $\mathbf{F}_1$ exerted by particle 1 and the force $\mathbf{F}_3$ exerted by particle 3. These are both attractive forces, so they will be directed as shown in Figure 13-12*b*. The magnitudes of these two forces are

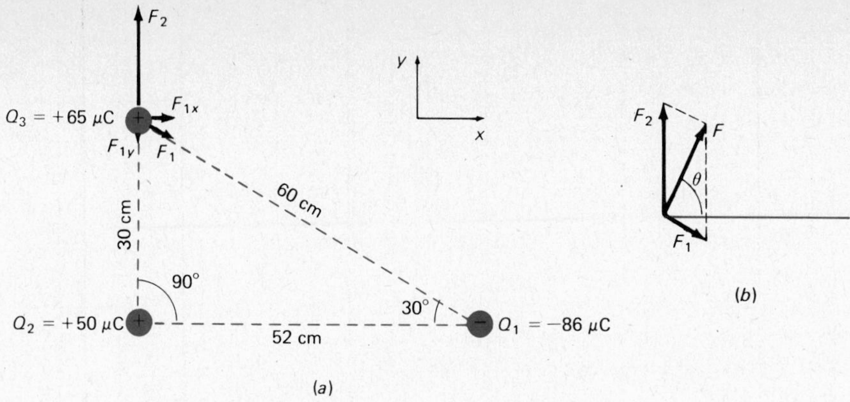

FIGURE 13-13
Determining the forces for Example 13-3.

$$F_1 = \frac{(9.0\times10^9\ \mathrm{N\cdot m^2/C^2})(5.0\times10^{-6}\ \mathrm{C})(3.0\times10^{-6}\ \mathrm{C})}{(0.30\ \mathrm{m})^2} = 1.5\ \mathrm{N}$$

$$F_3 = \frac{(9.0\times10^9\ \mathrm{N\cdot m^2/C^2})(5.0\times10^{-6}\mathrm{C})(4.0\times10^{-6}\ \mathrm{C})}{(0.20\ \mathrm{m})^2} = 4.5\ \mathrm{N}.$$

The net force on particle 2 is then

$$F = F_3 - F_1 = 4.5\ \mathrm{N} - 1.5\ \mathrm{N} = 3.0\ \mathrm{N},$$

and it acts to the right since F_3 is greater than F_1. Note that we omitted the signs of the charges when we used Equation 13-1 since they only indicate directions; and we determined the directions of each force beforehand when we drew Figure 13-12*b*. ▌▌

▌ EXAMPLE 13-3 Calculate the force on charge Q_3 shown in Figure 13-13*a* due to the charges Q_1 and Q_2.

SOLUTION The forces $\mathbf{F}_1$ and $\mathbf{F}_2$ have the directions shown in the diagram since Q_1 exerts an attractive force and Q_2 a repulsive force. The magnitudes of $\mathbf{F}_1$ and $\mathbf{F}_2$ are

$$F_1 = \frac{(9.0\times10^9\ \mathrm{N\cdot m^2/C^2})(6.5\times10^{-5}\ \mathrm{C})(8.6\times10^{-5}\ \mathrm{C})}{(0.60\ \mathrm{m})^2} = 140\ \mathrm{N}$$

$$F_2 = \frac{(9.0\times10^9\ \mathrm{N\cdot m^2})(6.5\times10^{-5}\ \mathrm{C})(5.0\times10^{-5}\ \mathrm{C})}{(0.30\ \mathrm{m})^2} = 330\ \mathrm{N}.$$

We resolve $\mathbf{F}_1$ into its components along the x and y axes as shown:

$$F_{1x} = F_1 \cos 30° = 120\ \mathrm{N}$$

$$F_{1y} = -F_1 \sin 30° = -70\ \mathrm{N}.$$

The force $\mathbf{F}_2$ has only a y component. So the net force $\mathbf{F}$ on Q_3 has components $F_x = F_{1x} = 120$ N, and $F_y = F_2 + F_{1y} = 330\ \mathrm{N} - 70\ \mathrm{N} = 260$ N. Thus we find the magnitude of the net force to be $F = \sqrt{F_x^2 + F_y^2} = \sqrt{(120\ \mathrm{N})^2 + (260\ \mathrm{N})^2} = 290$ N; and it acts at an angle θ (see Figure 13-13*b*) given by $\tan\theta = F_y/F_x = 260\ \mathrm{N}/120\ \mathrm{N} = 2.2$, so $\theta = 65°$. ▌▌

The constant k in Equation 13-1 is often written in terms of another constant, ε_0, the *permittivity of free space*; it is related to k by $k = 1/4\pi\varepsilon_0$. Coulomb's law can then be written

$$F = \frac{1}{4\pi\varepsilon_0}\frac{Q_1 Q_2}{r^2},$$

where $\varepsilon_0 = 1/4\pi k = 8.85 \times 10^{-12}\ \mathrm{C^2/N \cdot m^2}$.

Coulomb's law describes the force between two charges when they are at rest. Additional forces come into play when charges are in motion, and these will be discussed in later chapters. In this chapter we discuss only charges at rest, and this is called *electrostatics.*

Dielectric constant; the dissolving of salt

The value of the constant k given above, and that just given for ε_0, corresponds to the situation when the two charges are in vacuum or in air. In other materials, the force between two charges is often weaker. To express this fact, a second constant, K, called the *dielectric constant* of the material, is inserted into Coulomb's law:

$$F = \frac{k}{K}\frac{Q_1 Q_2}{r^2}$$

$$= \frac{1}{4\pi\varepsilon_0 K}\frac{Q_1 Q_2}{r^2}.$$

We can also define $\varepsilon = K\varepsilon_0$, where ε is the *permittivity of the material.* For vacuum, $K = 1$ and $\varepsilon = \varepsilon_0$. For air and other gases, K is very close to 1 and is generally taken to be so. However, for liquids and solids it can be quite large. For water, $K = 80$. This large value has a profound effect on materials, in that it allows many to be readily dissolved. Consider, for example, ordinary table salt, NaCl (sodium chloride), which arranges itself in a crystal structure as shown in Figure 13-14*a*; the attractive forces between Na^+ ions and Cl^- ions hold the crystal together. But when salt is placed in water, the water molecules that come between a Na^+ ion and a Cl^- ion cause the force between them to be reduced by a factor of 80. Because the force is now so weak, the ions move about the liquid freely (Figure 13-14*b*). The high dielectric constant of water thus accounts for its ability to dissolve many substances.† The reason behind the large value of K for water is the fact that it is a polar molecule and can surround an ion as shown in the figure.

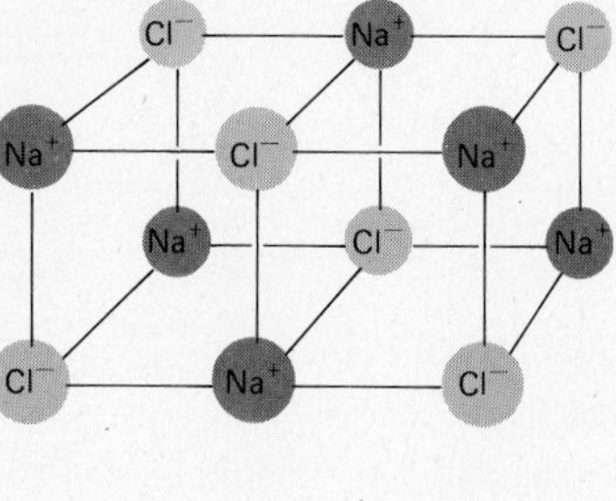

(*a*)

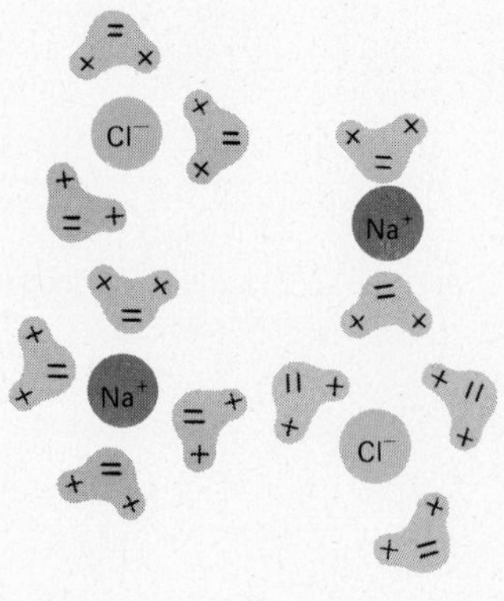

(*b*)

FIGURE 13-14
(*a*) Sodium chloride crystal; (*b*) sodium chloride dissolving in water.

13-3 ▮ Electric field

The concept of electric field

Many common forces might be referred to as "contact forces." That is, you exert a force on an object by coming into contact with it; for example,

† A situation such as this can be analyzed in a different but equivalent way by considering all the molecules involved as being in a vacuum, so that $K = 1$. Then the net force on, say, a sodium ion is the sum of the forces due to chlorine ions plus those due to neighboring water molecules. This would clearly be a more complicated analysis. The analysis is far simpler when we treat the water as a medium with dielectric constant $K = 80$, so that contributions of large numbers of individual water molecules are already taken account of. However, this analysis would not be accurate (or K would take on a value less than 80) over very short distances, where only one or a few water molecules would be involved.

you push or pull on a wagon, a television cart, or a stalled car. Similarly, a tennis racket exerts a force on a tennis ball when it makes contact; or a ball exerts a force on a window when it makes contact.

On the other hand, both the gravitational force and the electrical force act over a distance; that is, there is a force even when the two objects are not in contact. The idea of a force *acting at a distance* was a difficult one for early thinkers. Newton himself felt uneasy with this idea when he published his law of universal gravitation. The conceptual difficulties can be overcome with the idea of the *field*, introduced by the great British scientist Michael Faraday (1791–1867). In the electrical case, according to Faraday, an **electric field** extends outward from every charge and permeates all of space (Figure 13-15). When a second charge is placed near the first charge, we say it feels a force because of the electric field that is there (say, at point P in Figure 13-15). Because the electric field at the location of the second charge is considered to interact directly with this charge to produce the force, we are then a little closer to the idea of a "contact force." However, it must be emphasized that a field is *not* a kind of matter. It is a concept, invented by people, to help us analyze the world.

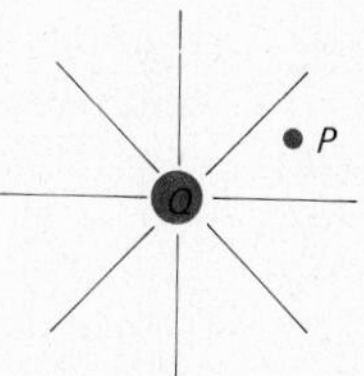

FIGURE 13-15
An electric field surrounds every charge. *P* is an arbitrary point.

We can investigate the electric field surrounding a charge or group of charges by measuring the force on a small positive *test* charge. By a test charge we mean a charge so small that the force it exerts does not significantly alter the distribution of the other charges, the ones that cause the field being measured. The force on a tiny positive test charge q placed at various locations in the vicinity of a single positive charge Q would be as shown in Figure 13-16. The force at b is less than at a because the distance is greater (Equation 13-1); and the force at c is smaller still. In each case the force is directed radially outward from Q. The electric field is defined in terms of the force on such a positive test charge. In particular, the **electric field intensity, E**, at any point in space is defined as the force exerted on a positive test charge at that point divided by the magnitude of the test charge q:

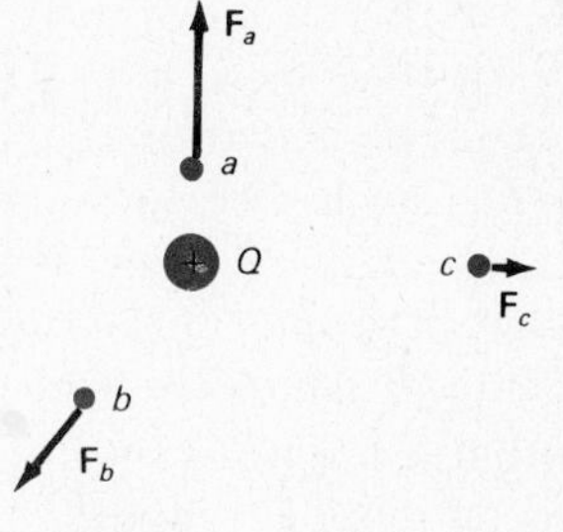

FIGURE 13-16
Force exerted by charge *Q* on a small test charge *q* placed at points *a*, *b*, and *c*.

$$\mathbf{E} = \frac{\mathbf{F}}{q}. \tag{13-2}$$

From this we see that the direction of the electric field intensity at any point in space is defined as the direction of the force on a positive test charge at that point. And the magnitude of the electric field intensity is the *force per unit charge*. Thus E is measured in units of newtons per coulomb (N/C).

The reason for defining $\mathbf{E} = \mathbf{F}/q$ is so that $\mathbf{E}$ does not depend on the test charge q. This means $\mathbf{E}$ describes only the effect of the charges creating it (the electric field).

The electric field at any point in space can be measured, based on the definition Equation 13-2. For simple situations involving one or several point charges, we can calculate what $\mathbf{E}$ will be. For example the magnitude of the electric field at a distance r from a single point charge Q would have magnitude

$$E = k\frac{qQ}{r^2}\frac{1}{q}$$

$$= k\frac{Q}{r^2} \quad \text{(single point charge).} \tag{13-3}$$

▮ EXAMPLE 13-4 Calculate the magnitude and direction of the electric field intensity at a point P 30 cm to the right of a charge $Q = -3.0 \times 10^{-4}$ C.

SOLUTION The magnitude of the electric field intensity is

$$E = \frac{kQ}{r^2} = \frac{(9.0 \times 10^9\ \mathrm{N \cdot m^2/C^2})(3.0 \times 10^{-4}\ \mathrm{C})}{(0.30\ \mathrm{m})^2} = 3.0 \times 10^7\ \mathrm{N/C}.$$

The direction of the electric field is *toward* the charge Q, as shown in Figure 13-17*a*, since we defined the direction as that of the force on a positive test charge. If Q had been positive, the electric field would have pointed away, as in Figure 13-17*b*. ▮▮

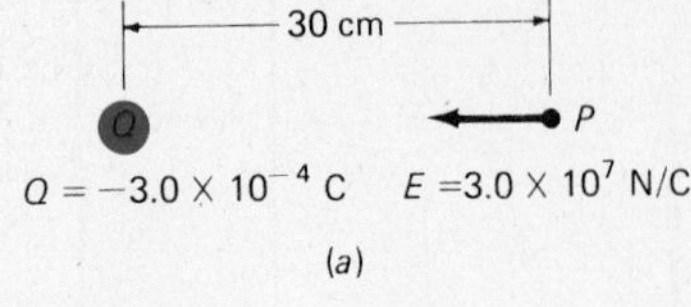

(*a*)

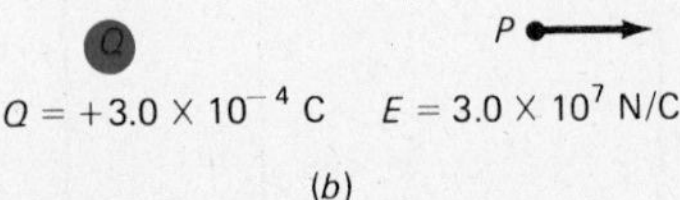

(*b*)

FIGURE 13-17

Electric field at point *P* due to (*a*) a negative charge Q and (*b*) a positive charge *Q* (Example 13-4).

If the field is due to more than one charge, the contributions from each must be calculated as a vector sum.

▮ EXAMPLE 13-5 Calculate the electric field intensity at points A and B in Figure 13-18.

SOLUTION (*a*) The calculation is much like that of Example 13-3, but now we are dealing with electric fields. The electric field at A is the vector sum of the fields $\mathbf{E}_{A1}$ due to Q_1 and $\mathbf{E}_{A2}$ due to Q_2: $E_{A1} = (9.0 \times 10^9\ \mathrm{N \cdot m^2/C^2})(50 \times 10^{-6}\ \mathrm{C})/(0.60\ \mathrm{m})^2 = 1.25 \times 10^6$ N/C, and similarly $E_{A2} = 5.0 \times 10^6$ N/C. The directions are as shown, so the total electric field at A, $\mathbf{E}_A$, has components $E_{Ax} = E_{A1} \cos 30° = 1.1 \times 10^6$ N/C and $E_{Ay} = E_{A2} - E_{A1} \sin 30° = 4.4 \times 10^6$ N/C. Thus the magnitude of $\mathbf{E}_A$ is $E_A = \sqrt{(1.1)^2 + (4.4)^2} \times 10^6\ \mathrm{N/C} = 4.5 \times 10^6$ N/C. It will act at an angle ϕ (see the diagram), where $\tan \phi = E_{Ay}/E_{Ax} = 4.4/1.1 = 4.0$, so $\phi = 76°$. (*b*) Since B is equidistant (40 cm by the Pythagorean theorem) from the two equal charges, the magnitudes of E_{B1} and E_{B2} are the same; that is to say, $E_{B1} = E_{B2} = (9.0 \times 10^9\ \mathrm{N \cdot m^2/C^2})(50 \times 10^{-6}\ \mathrm{N})/(0.40\ \mathrm{m})^2 = 2.8 \times 10^6$ N/C. Also, because of the symmetry, the y components of E_{B1} and E_{B2} are equal and opposite. Hence the total field E_B is horizontal and equals $E_{B1} \cos \theta + E_{B2} \cos \theta = 2E_{B1} \cos \theta$; from the diagram, $\cos \theta = 26\ \mathrm{cm}/40\ \mathrm{cm} = 0.65$. Then $E_B = 2(2.8 \times 10^6\ \mathrm{N/C})(0.65) = 3.6 \times 10^6$ N/C, and is along the x direction. ▮▮

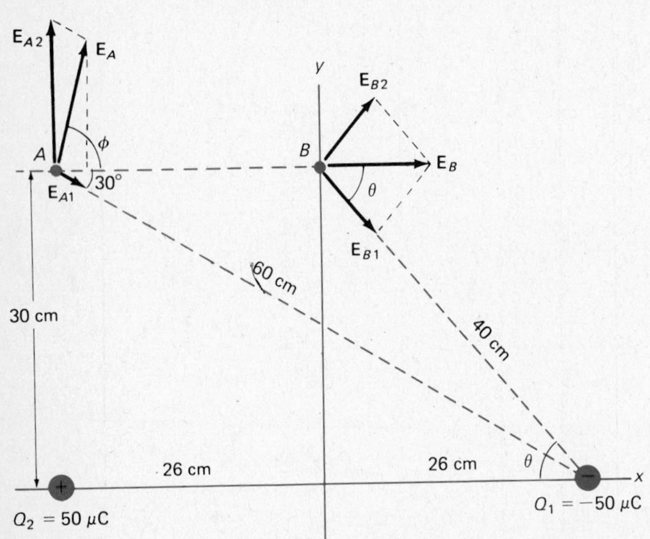

FIGURE 13-18

Calculation of the electric field at points *A* and *B* for Example 13-5.

Since the electric field is a vector, it is sometimes referred to as a *vector field.* We could indicate the electric field with arrows at various points in a given situation, such as at *a*, *b*, and *c* in Figure 13-16; the directions of $\mathbf{E}_a$, $\mathbf{E}_b$, and $\mathbf{E}_c$ would be the same as that of the forces already shown, but the lengths (magnitudes) would be different (since we divide by q); however, the relative length of $\mathbf{E}_a$, $\mathbf{E}_b$, and $\mathbf{E}_c$ would be the same as for the forces, since we divide by the same q each time. However, to indicate the electric field in such a way at *many* points would result in many arrows, which would appear confusing. To avoid this, we use another technique, that of lines of force.

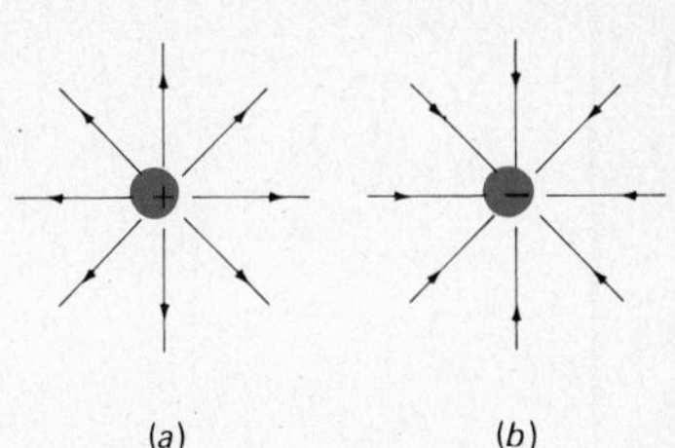

FIGURE 13-19
Electric field lines near (*a*) a positive point charge and (*b*) a negative point charge.

Lines of force

In order to visualize the electric field, we can draw a series of lines to indicate the direction of the electric field at each point in space. These *electric field lines* or **lines of force** are drawn so that they indicate the direction of the force due to the given field on a positive test charge. The lines of force due to a single positive charge are shown in Figure 13-19*a*, and those for a single negative charge in Figure 13-19*b*. In part *a*, the lines point radially outward from the charge, and in part *b*, they point radially inward toward the charge; these are the directions of the force that would be exerted on a positive test charge in each case. Only a few representative lines have been shown. We could just as well draw lines in between those shown since the electric field exists there as well. Notice that near the charge, where the force is greatest, the lines are closer together. This is a general property of electric field lines; *the closer the lines are together, the stronger the electric field in that region.*†

Figure 13-20*a* shows the electric field surrounding two charges of opposite sign. The electric field lines are curved in this case and are directed from the positive charge to the negative charge. The direction of the field at any point is directed tangentially as shown by the dashed arrow at point *P*. To satisfy yourself that this is the correct pattern for the electric field lines, you can make a few calculations such as those done in Example 13-5 for just this case (see Figure 13-18). Figures 13-20*b* and *c* show the electric field lines surrounding two positive charges and between two oppositely charge parallel plates. Notice that the electric lines between the two plates in part *c* are parallel except near the edges. Thus, in the central region, the electric field has the same magnitude at all points and we can write

$$E = \text{constant} \qquad \text{(between two parallel plates).} \qquad (13\text{-}4)$$

This should be compared to the field of a single point charge where the field decreases as the square of the distance, Equation 13-3.

We summarize the properties of lines of force as follows:

1 The lines of force indicate the direction of the electric field; the field points in the direction tangent to the field lines at any point.
2 In regions where the lines are close together, the electric field is strongest.
3 Electric field lines start on positive charges and end on negative charges.

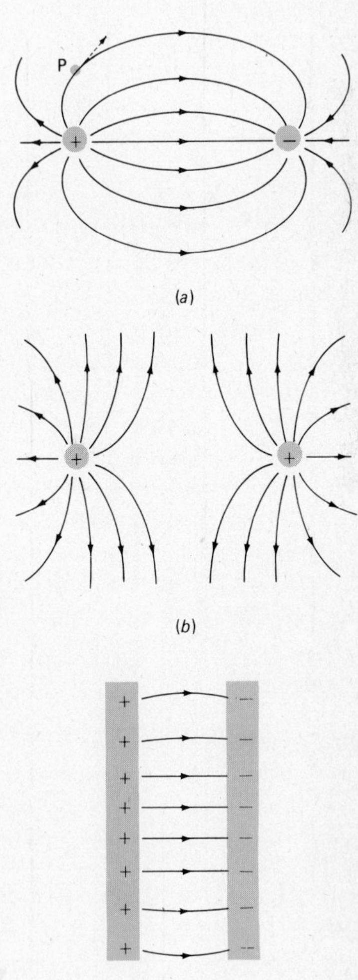

FIGURE 13-20
Electric field lines for three arrangements of charge.

† It can be shown that the lines can always be drawn so that the number crossing unit area perpendicular to **E** is proportional to the electric field intensity.

Another way of visualizing the electric field lines is that each represents the path that would be followed by a tiny test charge placed on it. (This is strictly true only if the test charge has no inertia, or moves slowly because of, say, a frictional force.)

Gravitational field

The field concept can be applied to the gravitational force as well. Thus we can say that a *gravitational field* exists for every object that has mass. One object attracts another by means of the gravitational field. The earth, for example, can be said to possess a gravitational field (Figure 13-21), which is responsible for the force on objects that we drop. The *gravitational field intensity* is defined as the *force per unit mass*. The magnitude of the earth's gravitational field intensity at any point is then (GM_e/r^2), where M_e is the mass of the earth, r the distance of the point from the earth's center, and G the gravitational constant (Chapter 3); at the earth's surface, r is simply the radius of the earth and the gravitational field intensity is simply equal to g, the acceleration due to gravity (since $F/m = mg/m = g$). In outer space the gravitational field intensity can be calculated at any point as a sum of terms due to earth, sun, moon, or other bodies that contribute significantly.

FIGURE 13-21
The earth's gravitational field.

Electric fields and conductors

The electric field inside a good conductor is zero in the static situation—that is, when the charges are at rest. If there were an electric field within a conductor, there would be a force on its free electrons since $\mathbf{F} = q\mathbf{E}$; the electrons would move until they reached positions where the electric field, and therefore the force on them, did become zero.

This reasoning has some interesting consequences. For one, the charge on a conductor distributes itself on the outer surface. For a negatively charged conductor you can imagine that the negative charges repel one another and race to the surface to get as far away from one another as possible; similar considerations apply to a positively charged conductor. Another consequence is the following. Suppose that a positive charge is placed inside the cavity of an uncharged conductor whose shape is a spherical shell (Figure 13-22). Since the field within the metal is zero, lines leaving the positive charge must end on negative charges on the inner surface of the metal. Thus an equal amount of negative charge is induced on the inner surface of the sphere; and a positive charge of the same magnitude must then exist on the outer surface (since the sphere is neutral). Thus, although no field exists in the metal itself, an electric field exists outside the sphere as shown in Figure 13-22 as if the metal were not even there.

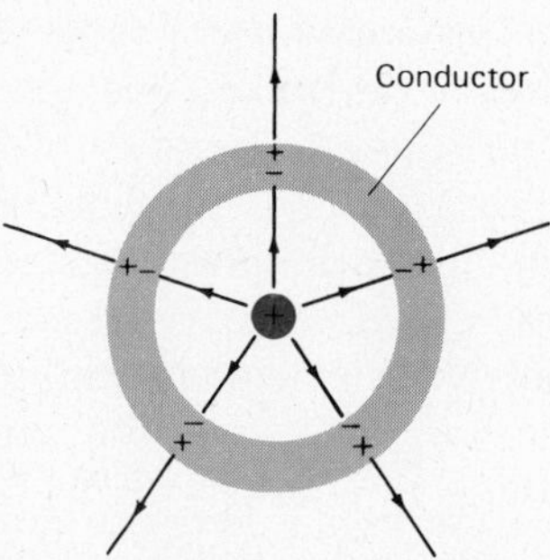

FIGURE 13-22
A charge placed inside a spherical shell. Charges are induced on the conductor surfaces, the electric field exists even beyond the shell but not within the conductor itself.

A related property of electric fields and conductors is that the electric field is always perpendicular to the surface of the conductor. If there were a component of $\mathbf{E}$ parallel to the surface, electrons at the surface would move along the surface in response to this force, until they reached positions where no force was exerted on them, that is, until the electric field was perpendicular to the surface.

These properties pertain only to conductors. In a nonconductor, which does not have free electrons, an electric field can exist; and the electric field does not necessarily make an angle of 90° to its surface.

*13-4 ▌ Electric forces in molecular biology

The study of the structure and functioning of a living cell at the molecular level is known as molecular biology. It is an important area for application of physics.

Since the interior of a cell is fluid (mostly water), we can imagine it as a vast sea of molecules continually in motion, colliding into one another with various amounts of kinetic energy. These molecules interact with one another in various ways—chemical reactions (making and breaking of covalent bonds) and more brief interactions or unions that occur because of electrostatic attraction between molecules.

The many activities that occur within the cell, and which help to differentiate living matter from nonliving, are now considered to be the result of random ("thermal") molecular motion plus the ordering effect of the electrostatic force. We will now use these ideas to analyze some cellular processes involving macromolecules (large molecules) such as DNA and proteins. The picture we present here has not been seen "in action." Rather, it is a model of what happens based on accepted physical theories and a great variety of experimental results.

*DNA structure and replication

The genetic information that is passed on from generation to generation in all living objects is contained in the chromosomes, which are made up of genes. Each gene contains the information needed to produce a particular type of protein molecule. A protein consists of one or more chains of small molecules called amino acids; it is used either as part of a structure (cell wall or muscle fiber, for example) or as an enzyme to catalyze a chemical reaction needed for the growth or survival of the organism. The genetic information contained in a gene is built into the principal molecule of a chromosome, the DNA (deoxyribonucleic acid). A DNA molecule consists of a long chain of many small molecules known as nucleotide bases. There are only four types of bases: adenine (A), cytosine (C), guanine (G), and thymine (T). These are arranged along the molecule according to a code, the "genetic code," which is "translated" into the amino acids that form the protein molecule.

A chromosome actually consists of two long DNA molecules wrapped about one another in the shape of a "double helix." As shown in Figure 13-23, the two strands are held together by electrostatic forces— that is, by the attraction of positive charges to negative charges. We see in part *a*, that an A (adenine) on one strand is always opposite a T on the other strand; similarly, a G is always opposite a C. This happens because the shape of the four molecules A, T, C, and G is such that a T fits closely only into an A, and a G into a C, and only in the case of this close proximity is the

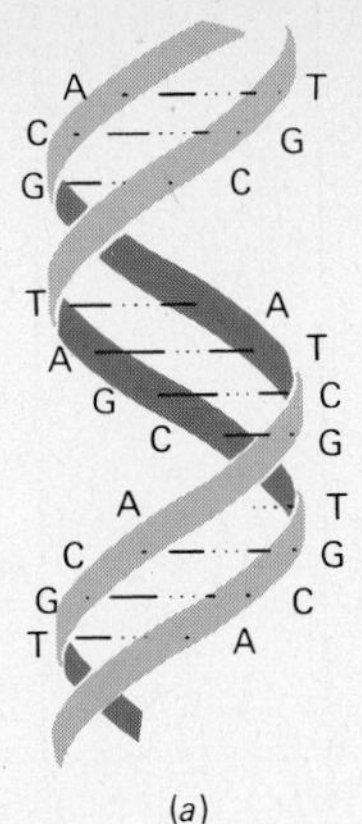

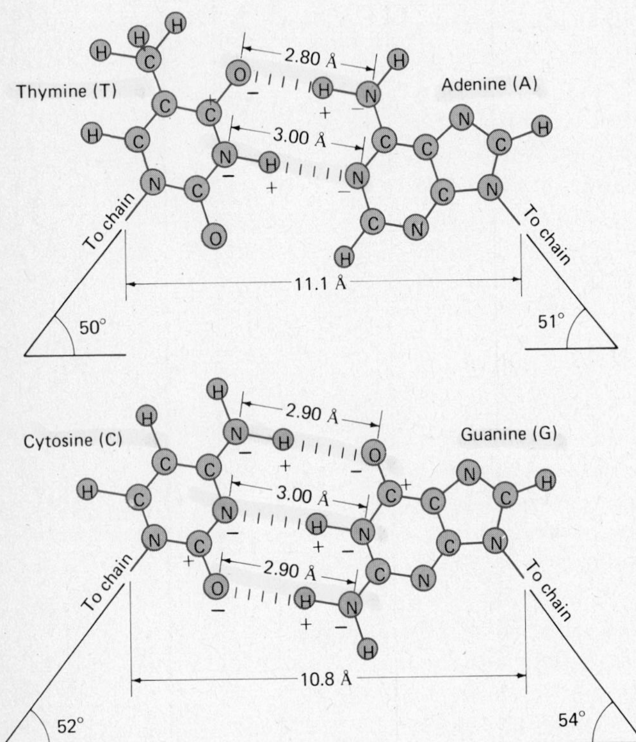

FIGURE 13-23
(*a*) Section of a DNA double helix. (*b*) "Close-up" view of the helix, showing how A and T attract each other and how G and C attract each other through electrostatic forces; this is how the double helix is held together. The vertical bars (|||||) are used to indicate the electrostatic attraction (or "weak bond"). Note that there are two weak bonds between A and T, and three between C and G. The distance unit is the angstrom (1 Å = 10^{-10} m). (After J. D. Watson, *Molecular Biology of the Gene*, 3rd ed., W. A. Benjamin, Inc., Menlo Park, Calif., 1976.)

electrostatic force great enough to hold them together even for a short time (Figure 13-23*b*). The electrostatic force between A and T, and between C and G, arises because some electrons in each of these molecules spend more time orbiting one atom than another; for example, the electron normally on the H atom of adenine spends some of its time orbiting the adjacent N atom,† so the N has a net negative charge and the H a positive charge (upper part of Figure 13-23*b*). This H^+ atom is then attracted to the O^- atom of the thymine.

How does the arrangement shown in Figure 13-23 come about? It occurs when the chromosome replicates (duplicates) itself just before cell

† Why this happens will be discussed in Chapter 27.

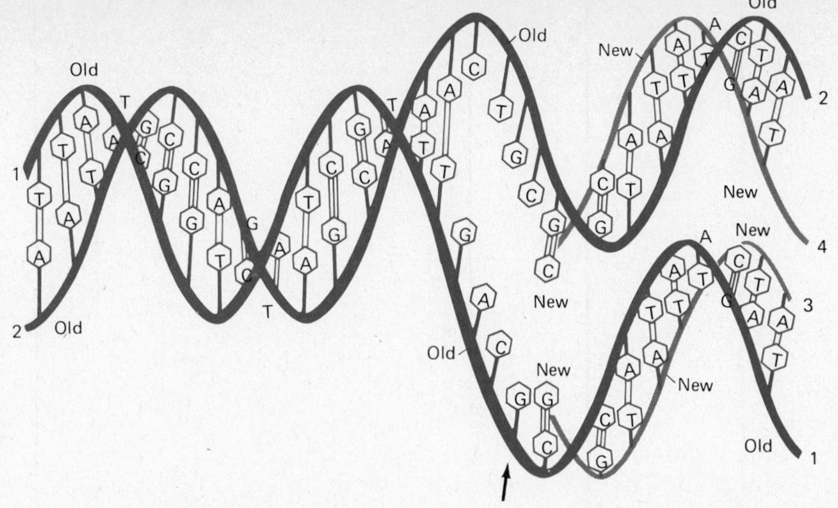

FIGURE 13-24
Replication of DNA. (After J. D. Watson, *Molecular Biology of the Gene*, 3rd ed., W. A. Benjamin, Inc., Menlo Park, Calif., 1976.)

division; indeed, the arrangement of A opposite T and G opposite C ensures that the genetic information is passed on accurately to the next generation. The process of replication is shown in a simplified form in Figure 13-24. The two strands of the DNA chain become unwound at one end, leaving the charged parts of the bases exposed. To see how the correct order of bases occurs, let us focus our attention on the G molecule indicated by the arrow on the bottom strand in the figure. Now there are many unattached nucleotide bases of all four kinds bouncing around in the cellular fluid. The only one of the four bases that will experience attraction to our G, if it bounces close to it, will be a C. The charges on the other three bases are not arranged so they can get close to those on the G; they will thus have no significant force exerted on them—remember that the force decreases rapidly with distance. The electrostatic forces will often hold the C opposite our G long enough so that an enzyme can attach the C to the growing end of the new chain.

Thus we see that electrostatic forces not only hold the two chains together; they also operate to select the bases in the proper order during replication so the genetic information is passed on accurately to the next generation. Note in Figure 13-24 that the new number 4 strand has the same order of bases as the old number 1 strand; and the new number 3 strand is the same as the old number 2; so the two new helixes, 1–3 and 2–4, are identical to the original 1–2 helix. The error rate—say a T being incorporated in a new chain opposite a G—is less than 1 in 10^8. Such an error would correspond to a spontaneous mutation and a possible change in some characteristic of the organism.

*Protein synthesis

Protein molecules play two very important roles in the cell: as structural units of the cell (cell walls, membranes, organelles); and as enzymes which catalyze the many chemical reactions that occur within the cell.

Protein molecules are long chains of single small molecules known as amino acids. There are 20 different amino acids of importance. These chains are generally linear, without branches, although several chains may be linked together. The order of the different amino acids determines the structure and role of each protein molecule.

The mechanism by which protein molecules are synthesized is illustrated in Figure 13-25. It begins at the DNA double helix, for each gene on a chromosome contains the information for one protein. The ordering of the four bases, A, C, G, and T, provides the "code" for the order of amino acids in the protein. First the DNA double helix unwinds and a new molecule called messenger-RNA (m-RNA) is synthesized; m-RNA is a chain molecule containing four different bases, just like those of DNA except that thymine (T) is replaced by a very similar molecule called uracil (U). In the diagram a C has just been added to the growing m-RNA chain in much the same way that DNA replicates; and an A, attracted to the T on the DNA chain, will soon be attached to the C by an enzyme. The order of the bases, and thus the genetic information, is preserved in the m-RNA. The m-RNA is buffeted about by all the moving molecules in the cell (remember kinetic theory) until it gets close to a tiny organelle known as a ribosome, to which it becomes attached by electrostatic attraction, Figure 13-25. Also held by the electrostatic force to the ribosome are one or two "transfer-RNA" (t-RNA) molecules. These t-RNA molecules "translate" the genetic code of nucleotide bases into amino acids. There are different t-RNA molecules for each combination of three bases. At one end of the t-RNA molecule is the "anticodon," which is a set of three nucleotide bases. If all three bases match the three (the "codon") on that part of the m-RNA attached to the ribosome, the anticodon is attracted to the ribosome and becomes attached briefly. Attached to the other end of a t-RNA molecule is the appropriate amino acid; for example, the CAU codon codes for the amino acid histidine. There are actually two codon sites on the ribosome; these hold two t-RNA molecules while enzymes link the two amino acids together to lengthen the amino acid chain. As each amino acid becomes added, the old t-RNA molecule—having lost its amino acid—is removed, probably by a random collision with some molecule in the cellular fluid. A new one then becomes attached as the ribosome moves along the m-RNA molecule.

FIGURE 13-25

Protein synthesis.

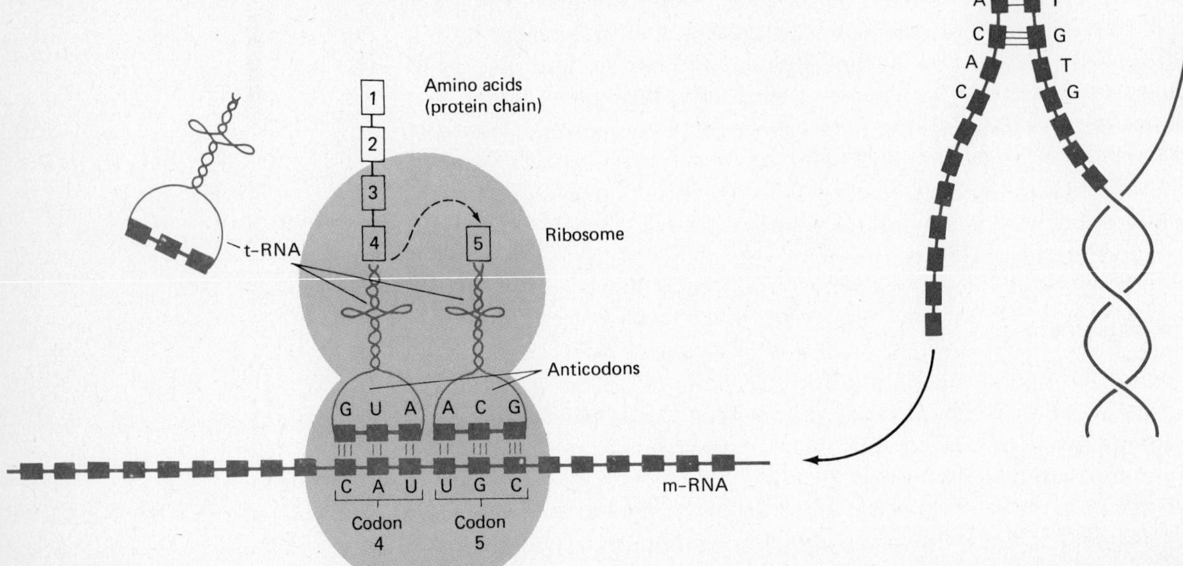

This process of protein synthesis is often presented as if it occurred in clockwork fashion—as if each molecule knew its role and went to its assigned place, like bees in a hive. But this is not the case. The forces of attraction between the electric charges of the molecules are rather weak; hence they become significant only when the molecules can come close together and several weak "bonds" can be made. Indeed, if the shapes are not just right, there is almost no attraction, which is why there are few mistakes. The point is that there are many molecules in the cell, all jostling about. Many different bases approach the end of the growing m-RNA chain, but only the A (see Figure 13-25) will be held long enough to the T on the DNA to be enzymatically attached to the m-RNA chain. And many different t-RNA molecules will approach an open spot on the ribosome but only that one which has the proper anticodon will remain opposite the codon long enough for its amino acid to become attached to the growing protein chain. Thus, out of the random motion of the molecules, the electrostatic force between opposite electric charges acts to bring order out of chaos.

The fact that the attraction is weak is very important. If it were strong, collisions with other molecules would not allow a t-RNA molecule to be released from the ribosome, or the m-RNA to be released from the DNA. If they were not temporary encounters, metabolism would grind to a halt.

*Protein structure

As amino acid is added to amino acid, the protein molecule grows in length until the last one is added and the protein is complete. Even as it is being made, this chain is being buffeted about in the cellular sea—we might think of a wiggling worm. But a protein molecule has charged polar groups along its length; and as it takes on various shapes, the electric forces of attraction between different parts of the molecule will eventually lead to a particular configuration which is quite stable. Each type of protein has its own peculiar shape depending on the location of charged atoms; in the last analysis, this depends on the order of the amino acids.

Amino acids all have the same basic structure except for a side group, which we designate R (Figure 13-26*a*); R can be H, CH_3, or many other combinations of atoms. In many proteins, particularly in structural proteins, electrical attraction occurs between the regular parts of the amino acids and produces a repeating structure. A common example is the "α-helix structure," shown schematically in Figure 13-26*b*. This structure is held together by the forces of attraction between the charged C—O and N—H groups.

Not all proteins have an α-helix structure. Often attraction between the R groups dominates. The usual situation is that a protein molecule has a twisted irregular shape, although it may have short sections of α-helix as well.

The *action* of proteins as enzymes, or as structureal units, also involves electrical forces of attraction between unlike electric charges, and so do many other cellular processes, such as the action of antibodies on antigens. The "lock-and-key" mechanism used to explain some of these processes involves molecules whose shapes are complementary, so they fit closely enough that electrostatic forces can hold them together.

FIGURE 13-26
(*a*) Amino acid. (*b*) α-Helix structure.

H N H C R O C

(*a*)

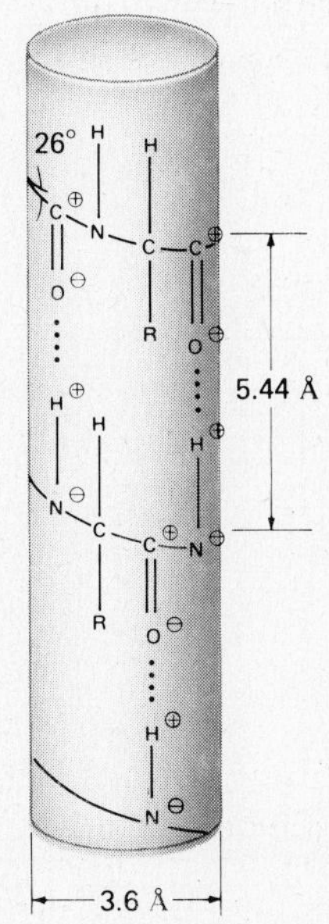

(*b*)

*13-5 Muscle action

In Chapter 4 we dealt with the forces that muscles exert on our limbs. Now we investigate the physics of what makes a muscle contract. Modern research indicates that the electrostatic force plays a fundamental role. There are three main types of muscle: skeletal muscle (that moves limbs), smooth or visceral muscle (that surrounds internal organs, blood vessels, etc.) and cardiac (heart) muscle. We will mainly be concerned with skeletal muscle.

Muscles can be made to contract in the laboratory by the application of an electric stimulus using metal electrodes. In the body, the stimulus is an electrical signal that has traveled down a nerve whose tip is attached to the muscle. (We discuss nerve conduction in Chapter 15.) Experiment shows that a short electric stimulus causes a single tiny contraction known as a "twitch": the tension in the muscle increases to a maximum and then relaxes in the space of less than 0.2 s. If repeated stimuli are applied in quick succession before the muscle relaxes from the previous one, the tension will increase to a new maximum and remain in tension as long as the signals continue. This is the usual situation when we use a muscle. In the body the nervous system sends the repeating signals to the muscle.

Skeletal muscle is made of many long thin *fibers*. High-magnification electron microscope photographs ("electron micrographs") show that within a fiber there are many tiny strands or filaments (Figure 13-27). The arrangement of these filaments, which are believed to be directly involved in the contraction mechanism, is diagrammed in Figure 13-28*a*. The letters A, I, and H refer to different regions (or bands), as shown. The region between the structures known as Z lines is called a sarcomere, and it is the basic unit of a muscle fiber. A sarcomere is typically 25 μm long in the resting state. There are two sizes of filament. The thick filaments contain a protein called myosin, and the thin filaments contain the protein actin.

The most widely accepted theory of muscle contraction is the "sliding filament theory." According to this theory, contraction occurs when the

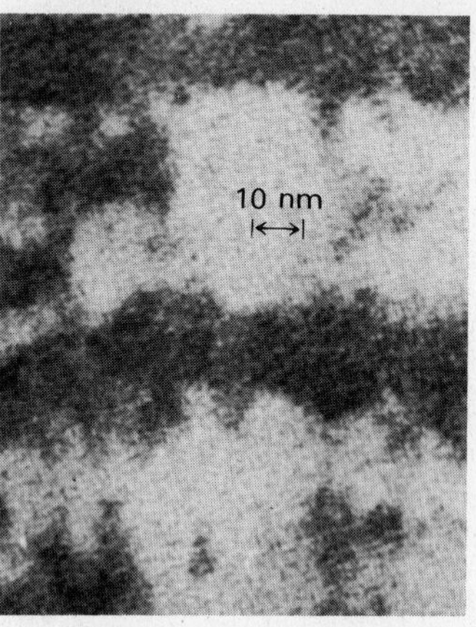

FIGURE 13-27
Electron micrographs of muscle at two different magnifications, showing filaments in the muscle fiber. (Courtesy of H. E. Huxley.)

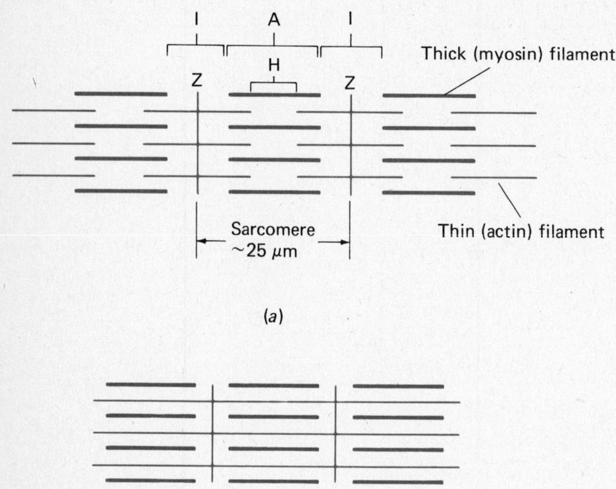

FIGURE 13-28
Arrangement of filaments in (*a*) uncontracted muscle; (*b*) contracted muscle.

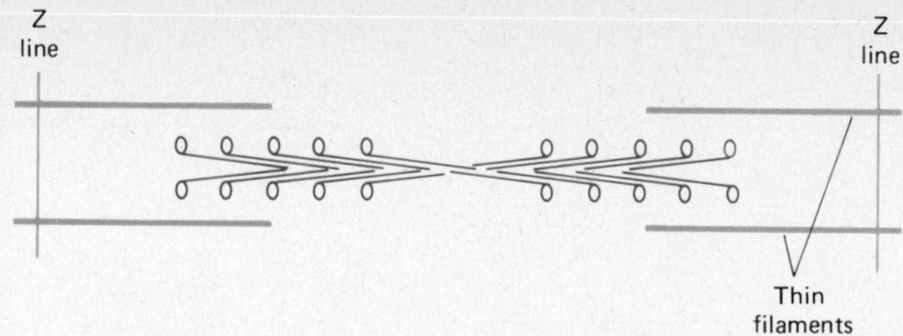

FIGURE 13-29
Arrangement of tadpole-shaped myosin molecules in a thick filament.

filaments slide past each other so that the Z lines are pulled closer together and the sarcomere (and thus the whole muscle) shortens, Figure 13-28*b*. Electron microscopy, x-ray diffraction, and other studies have shown that myosin molecules have a tadpole-like shape with a head and long tail and that they are arranged in the thick filament as shown in Figure 13-29. Notice that the heads point outward from the filament, and that the molecules are oriented in opposite directions on either side of the center. This is so that the myosin can pull the actin filaments and Z lines toward the center of the sarcomere.

It is difficult to obtain conclusive evidence on the actual molecular nature of the interaction, although there is little doubt that tiny "cross-bridges" are formed between the filaments (these can be seen in Figure 13-27) and that the electrostatic force plays a fundamental role. Models have been proposed and the most likely at present is illustrated in Figure 13-30. It is believed that on the head of the myosin molecule there is firmly

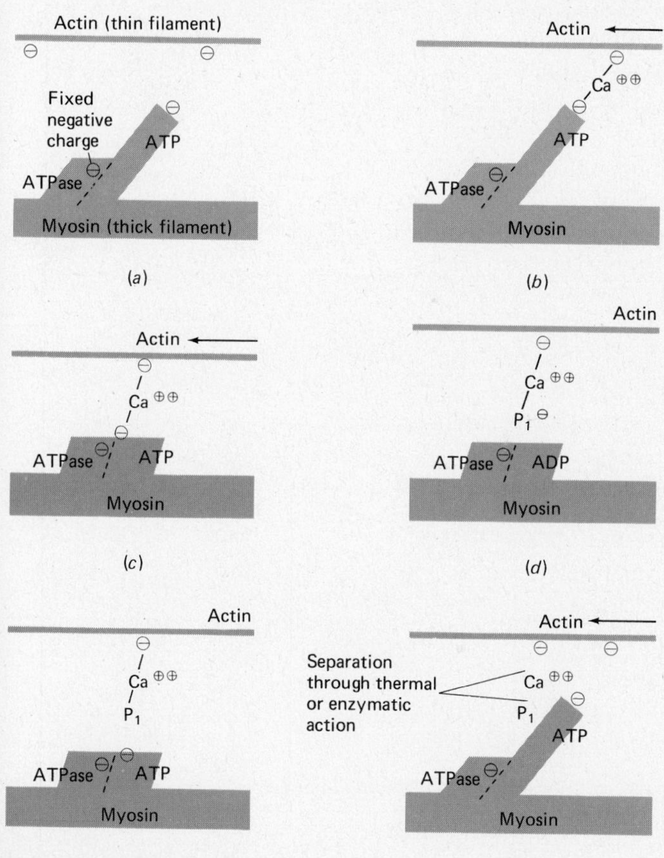

FIGURE 13-30
Action of muscle contraction at the molecular level. (See text for details.)

attached an ATP molecule. (ATP, adenosine triphosphate, contains three phosphate groups with a negative charge; we shall discuss ATP more later.) Below this the myosin molecules can be in an α-helix structure, but in this model the α-helix structure is extended because of repulsion between the negative charge on the ATP and a fixed negative charge assumed to reside at the base, as shown in Figure 13-30*a*. The stimulation of a muscle results in the release of calcium ions (Ca^{2+}) into the fluid of the sarcomere. By electrostatic attraction, a Ca^{2+} ion attaches to the ATP and simultaneously is attracted to a negatively charged site on the actin filament; thus the Ca^{2+} ion attracts both filaments and forms a link between them (part *b*). The negative charge on the ATP is neutralized by the Ca^{2+} ion so that the protein chain can contract (part *c*) into its α-helix configuration by forming hydrogen bonds (as in Figure 13-26*b*). The shortening of the protein chain pulls the actin filament towards the center of the sarcomere. It also brings the ATP close to the site where an ATPase molecule is presumed to be attached. ATPase is an enzyme that acts to shear off the end phosphate from ATP; when it does this, the Ca^{2+} ion goes, too, and the link is broken (part *d*). The last step in the cycle is reattachment of a phosphate to the remaining ADP (adenosine diphosphate) by an enzyme to remake an ATP, and the tip of the myosin molecule reextends because of the repulsion of the negative charges. It is now in position to form a link with the next actin site.

According to this model, a series of miniature contractions and reextensions of myosin molecules, mediated by Ca^{2+} ion linkages ("cross-bridges"), pulls the actin filaments toward the center of the sarcomere. This process repeats itself as long as calcium is present. Relaxation of the muscle occurs by removal of Ca^{2+} ions (by active transport) so that bridges are no longer formed and the filaments can slide back to the uncontracted positions, as when you let your arm fall.

Under "isometric" tension—that is, when tension is maintained but there is no shortening of the muscle and hence no work done, as when you hold a heavy bag of groceries—the repeated formation and breakage of ATP–Ca linkages continues. This is similar to your hands slipping when pulling on a rope that is attached to a heavy load. Thus energy is consumed without work being done.

When a muscle becomes fatigued, it is the relaxation mechanism (removal of Ca^{2+} ions) that fails. When completely fatigued, a muscle is in a fixed contracted state called "rigor," in which tight bridges are formed between the actin and myosin filaments; this is what we popularly call a cramp. Upon death, muscles can no longer relax, and they remain in a rigidly contracted state.

▮ Summary

There are two kinds of *electric charge*, positive and negative. These designations are to be taken algebraically—that is, any charge is plus or minus so many coulombs (C). Electric charge is *conserved*: if a certain amount of one type of charge is produced in a process, an equal amount of the opposite type is also produced on the same body or on a different body; thus the *net* charge produced is zero. According to the atomic theory, electric charge originates in the atom, which consists of a positively charged nucleus surrounded by negatively charged electrons. Each electron has a charge $-e=-1.6\times10^{-19}$ C. Conductors are those materials in which many electrons are relatively free to move, whereas electric insulators are those in which very few electrons are free to move. An object is negatively charged when it has an excess of electrons, and positively charged when it has less than its normal amount of electrons. An object can become charged in three ways: by rubbing, in which electrons are transferred from one material to another; by conduction, which is transfer of charge from one charged object to another by touching; or by induction, the separation of charge within an object because of the close approach of another charged object but without touching.

Electric charges exert a force on each other. If two charges are of opposite types, one positive and one negative, they each exert an attractive force on the other. If the two charges are the same type, each repels the other. The magnitude of the force one point charge exerts on another is proportional to the product of their charges and inversely proportional to the square of the distance between them; this is called *Coulomb's law*.

An *electric field* is imagined to exist in space due to any charge or group of charges. The force on another charged object is then conceived to be due to the electric field present at its location. The *electric field intensity*, **E**, at any point in space due to one or more charges is defined as the force per unit charge that would act on a test charge q placed at that point: $\mathbf{E}=\mathbf{F}/q$. Electric fields are represented by *lines of force* which start on positive charges and end on negative charges. Their direction indicates the direction the force would be on a tiny positive test charge placed at a point, and the closeness of the lines indicates the strength of the field. The electric field inside a good conductor is zero, and the electric field lines just outside a charged conductor are perpendicular to its surface.

Questions

1 Why does a plastic ruler that has been rubbed with a cloth have the ability to pick up small pieces of paper? Why is this difficult to do on a humid day?

2 When an electroscope is charged, the two leaves repel each other and remain at an angle. What balances the electric force of repulsion so that the leaves do not separate further?

3 If you charge a pocket comb by rubbing with a silk scarf, how can you determine if the comb is positively or negatively charged?

4 Why does a shirt or blouse taken from a clothes dryer sometimes cling to your body?

5 Why does a phonograph record attract dust just after it has been wiped clean?

6 Explain why fog or rain droplets tend to form around ions or electrons in the air. (*Hint*: See Figure 13-4.)

7 A positively charged rod is brought close to a neutral piece of paper, which it attracts. Draw a diagram showing the separation of charge and explain why attraction occurs.

8 Can you guess why trucks carrying flammable fluids drag a chain along the ground? (*Hint*: Have you ever experienced a slight shock when getting out of your car?)

9 The form of Coulomb's law is very similar to that for Newton's law of universal gravitation. What are the differences between these two laws?

10 We are not normally aware of the gravitational or electrical force between two ordinary objects. What is the reason in each case? Give an example where we are aware of each one and state why.

*11 Is the electric force a conservative force? (See Chapter 5 for a definition of conservative force.) Why or why not?

12 Draw the electric field lines surrounding two negative electric charges a distance l apart.

13 Explain why we use *small* test charges when measuring electric fields.

14 Assume that the two opposite charges in Figure 13-20*a* are 20 cm apart. Consider the magnitude of the electric field 5 cm above the positive charge. On which side of this charge—top, bottom, left, or right—is the electric field the greatest? The smallest?

15 Why can lines of force never cross?

*16 Explain why the maximum force a muscle can exert is proportional to its cross-sectional area.

Problems

SECTION 13-2

1 (I) How many electrons make up a charge of 1 μC?

2 (I) What is the total charge on 1.0 kg of electrons?

3 (I) What is the magnitude of the force a 10-μC charge exerts on a 3.0-mC charge 2.0 m away? ($1\ \mu C = 10^{-6}$ C, $1\ mC = 10^{-3}$ C.)

4 (I) Two charged bodies exert a force of 420 mN on each other. What will be the force if they are moved so they are only one-third as far apart?

5 (I) How far apart must two electrons be if the force between them is to be 1×10^{-12} N?

6 (I) What is the magnitude of the electric force of attraction between an iron nucleus ($q = +26e$) and its innermost electron if the distance between them is 1.0×10^{-12} m?

7 (II) A hydrogen atom consists of an electron which orbits a proton (charge $+e$) at an average distance of 0.53×10^{-10} m. Calculate the electrical and gravitational forces of attraction between them. Comment on the relative significance of electrical and gravitational forces at the atomic level. (The proton's mass is 1.67×10^{-27} kg and the electron's is 9.1×10^{-31} kg.)

8 (II) How close must two electrons be if the electric force between them is equal to the weight of either at the earth's surface?

9 (II) Three positive particles of charges 3.0 μC are located at the corners of an equilateral triangle of side 10 cm. Calculate the magnitude and direction of the net force on each particle.

10 (II) Particles of charge +36, +48, and −90 μC are placed in a line. The center one is 1.5 m from each of the others. Calculate the net force on each due to the other two.

11 (II) A charge of 0.050 C is placed at each corner of a square 15 m on a side. Determine the magnitude and direction of the force on each charge.

12 (II) Repeat Problem 11 for the case when two of the charges, on opposite corners, are replaced by negative charges of the same magnitude.

13 (III) In one model of the hydrogen atom the electron revolves in a circular orbit around the proton with a speed of 1.1×10^6 m/s. What is the radius of the electron's orbit?

14 (III) A large electroscope is made with "leaves" that are wires 50 cm long, with 30-g balls at the ends. When charged, nearly all the charge resides on the balls. If the wires each make a 30° angle with the vertical, what total charge must have been applied to the electroscope?

15 (III) A 3.0- and a −1.3-μC charge are placed 18 cm apart. Where can a third charge be placed so that it experiences no net force?

16 (III) Two nonconducting spheres have a total charge of 850 μC. When placed 1.0 m apart, the force each exerts on the other is 28 N and is repulsive. What is the charge on each? What if the force were attractive?

SECTION 13-3

17 (I) A force of 2.4 N is exerted on a −1.8-μC charge in a downward direction. What is the magnitude and direction of the electric field at this point?

18 (I) An electron in an electric field experiences a force of 8.0×10^{-16} N. What is the magnitude and direction of the electric field at this point?

19 (I) What is the magnitude of the force on an electron in an electric field of 600 N/C?

20 (I) What is the magnitude and direction of the electric field 2.0 m directly above a 3.0×10^{-3} C charge?

21 (II) Use Coulomb's law to determine the electric field at A and B in Figure 13-31 due to the two positive charges ($Q = 10\ \mu$C) shown. Calculate magnitude and direction. Is your result consistent with Figure 13-20*b*?

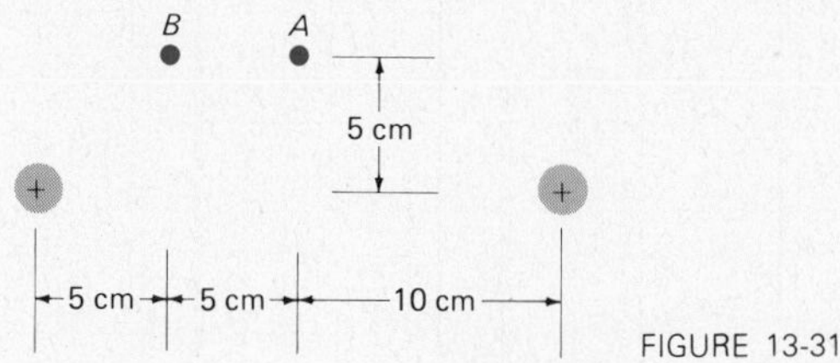

FIGURE 13-31

22 (II) What is the magnitude and direction of the electric field at a point midway between a 10-μC and a 50-μC charge 40 cm apart?

23 (II) Measurements indicate that there is an electric field surrounding the earth. Its magnitude is about 100 N/C at the earth's surface and points inward toward the earth's center. What is the magnitude of the electric charge on the earth? Is it positive or negative? (*Hint*: The electric field due to a uniformly charged sphere is the same as if all the charge were concentrated at its center.)

24 (II) Calculate the electric field at the center of a square 20 cm on a side if one corner is occupied by a 3.0-μC charge and the other three are occupied by -2.0-μC charges.

25 (II) Calculate the electric field at one corner of a square 50 cm on a side if the other three corners are occupied by 8.2×10^{-4} C charges.

26 (II) What is the acceleration of an electron in a 2200-N/C electric field?

27 (II) A spaceship makes a trip from earth to the moon, 380,000 km away. At what point in the trip will the gravitational field be zero? The mass of the moon is about $\frac{1}{81}$ that of earth.

28 (III) A water droplet of radius 0.020 mm remains stationary in the air. If the electric field of the earth is 100 N/C, how many excess electron charges must the water droplet have?

*SECTION 13-4

***29** (II) Estimate the net force between a thymine and an adenine base assuming (Figure 13-23*b*) that the distance between atoms on each molecule is about 1.0 Å, and that the charge on the O atom is 8.0×10^{-20} C and on the N and H atoms is 3.0×10^{-20} C.

***30** (II) Repeat Problem 29 for a G-C pair.

***31** (II) Hair consists of protein molecules that have an α-helix structure. If hair grows 15 cm/yr, what is the rate of synthesis of the protein molecules? Specify in terms of number of turns of the α-helix per second (see Figure 13-26*b*).

*SECTION 13-5

***32** (II) Calculate approximately (*a*) the number of myosin filaments per cross-sectional area and (*b*) the total number of thick filaments there are in, say, your biceps muscle. Assume that a sarcomere is 25 μm long and that the thick filaments are on the average 40 nm apart in a regular two-dimensional array.

***33** (II) X-ray diffraction data indicate that along each half-length of myosin filament there are about 100 cross-bridge linking sites. From the fact that the maximum tension a muscle can sustain is about 2×10^5 N/m^2 of cross-sectional area, show that the average force developed per cross bridge is about 3×10^{-12} N. (*Hint*: Use the result of Problem 32.)

***34** (III) Each molecule of ATP releases about 0.72×10^{-19} J of energy when one of its phosphates is sheared off. If 40 percent of this energy is used to do work, what is the approximate distance moved by the actin molecule per link? Use whatever data you need from the previous two problems.

For additional reading

Aidley, D. J., *Physiology of Excitable Cells*, Cambridge University Press, Cambridge, 1971. Muscle action.

Dickerson, R. E., and I. Geis, *Structure and Action of Proteins*, Harper & Row, Publishers, Inc., New York, 1970.

Einstein, A., and L. Infeld, *Evolution of Physics*, Simon & Schuster, New York, 1938.

Heibron, J. L., "Franklin's Physics," *Physics Today*, **29**:7 (Jul., 1976), p. 32.

Huddart, H., *Comparative Structure and Function of Muscle*, Pergamon Press, Inc., Elmsford, N.Y., 1975.

Katz, B., *Nerve, Muscle and Synapse*, McGraw-Hill Book Company, New York, 1966.

Magie, W. F., *Source Book in Physics*, Harvard University Press, Cambridge, Mass., 1963. Original writings of Coulomb and others.

Meyer, H. W., *History of Electricity and Magnetism*, The M.I.T. Press, Cambridge, Mass., 1971.

Roller, D. E., and D. H. D. Roller, *Development of the Concept of Electric Charge: Electricity from the Greeks to Coulomb*, Harvard University Press, Cambridge, Mass., 1954.

Stanford, A. L., Jr., *Foundations of Biophysics*, Academic Press, Inc., New York., 1975.

Watson, J. D., *Molecular Biology of the Gene*, 3rd ed., W. A. Benjamin, Inc., Menlo Park, Calif., 1976. DNA proteins.

14 ▮ Electric potential and electric energy

We saw in Chapter 5 that the concept of energy was extremely valuable in dealing with mechanical problems. For one thing, energy is a conserved quantity and thus an important aspect of nature. Furthermore, we saw that many problems could be solved using the energy concept even though a detailed knowledge of the forces involved was not possible, or when a calculation involving Newton's laws would have been too difficult. The energy point of view can be used in electricity, and it is especially useful.

14-1 ▮ Electric potential and potential difference

Energy will be conserved as long as we define electric potential energy just as for any type of potential energy. That is, the difference in electric potential energy of a charge Q at two different points in an electric field is defined as the work needed to move the charge from one point to the other. For example, consider the electric field between two equally but oppositely charged parallel plates which are large compared to their separation so the field will be uniform over most of the region (Figure 14-1). Now consider a positive charge Q placed at point B very near the negative plate, as shown. It will require work to move this charge to point A near the positive plate. In the process the potential energy of Q will be increased by an amount equal to the work required to move it from B to A. Now suppose that the charge has reached A and is then released; it will be attracted to the negative plate. The electric field does work on the charge to accelerate it, and as it moves to the right, its potential energy will decrease and its kinetic energy will increase. This is in accord with the conservation of energy; the electric potential energy is transformed into kinetic energy and the total energy will be conserved. Note that the positive charge Q has its greatest potential energy (PE) when near the positive plate. The reverse is true for a negative charge: its potential energy is greatest near the negative plate.

FIGURE 14-1
Work is required to move a positive charge from position *B* to position *A*.

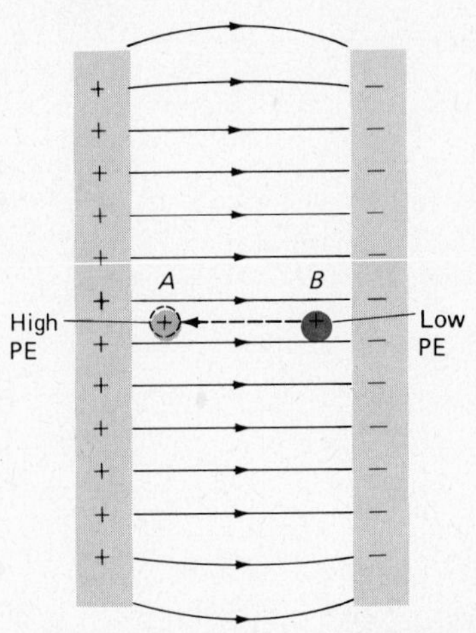

We defined the electric field (Chapter 13) as the force per unit charge; similarly, it is useful to define the **electric potential** (or simply the *potential* when "electric" is understood clearly) as the *potential energy per unit charge.* Electric potential is given the symbol V, so we define the electric potential, V_A, at some point A as

$$V_A = \frac{PE_A}{Q},$$

where PE_A is the potential energy of a charge Q placed at point A.

Generally, we will be interested in the **potential difference** between two points A and B (such as between A and B in Figure 14-1). Since the difference in potential energy between A and B is equal to the work done to get from one point to the other, W_{AB}, we have that the potential difference V_{AB} is

$$V_{AB} = V_A - V_B = \frac{W_{AB}}{Q}.$$

The unit of electric potential, and of potential difference, is joules per coulomb and is given a special name, the *volt,* in honor of Alessandro Volta (1745–1827), who invented the electric battery; it is abbreviated V, so 1 V = 1 J/C. Note from our definition that the positive plate in Figure 14-1 is at a higher potential than the negative plate. Thus a positively charged object moves naturally from a high potential to a low potential; a negative charge does the reverse.

The absolute potential, V_A, at a point A depends on where the potential is chosen to be zero; and the zero point for electric potential in a given situation, just as for potential energy (Section 5-5), can be chosen arbitrarily. This is because only differences in potential energy can be measured. As a result, we almost always speak of the difference of potential or *potential difference* between two points rather than an absolute potential. However, the ground (or a conductor connected directly to the ground) is often taken as zero potential. Potential difference, since it is measured in volts, is often referred to as **voltage**. And, in the rest of this book, we will use the symbol V for potential difference; if we want to speak of the absolute potential, we will so specify.

Since the electric potential is defined as the potential energy per unit charge, the change in potential energy of a charge Q when moved between two points A and B is

$$\Delta PE = QV_{AB}. \tag{14-1}$$

That is, if an object with charge Q moves through a potential difference V_{AB}, its potential energy changes by an amount QV_{AB}. For example, if the potential difference between the two plates in Figure 14-1 is 6 V, then a 1-C charge moved from B to A will gain (1 C) (6 V) = 6 J of potential energy (and it will lose 6 J of PE if it moves from A to B); similarly, a 2-C charge will gain 12 J, and so on. Thus, electric potential is a measure of how much energy an electric charge can acquire in a given situation, or how much work it can do; the exact amount depends both on the potential and on the charge.

To better understand electric potential, let us make a comparison to the gravitational case when a rock falls from the top of a cliff. The greater the height of a cliff, the more potential energy the rock has and the more kinetic energy it will have when it reaches the bottom. The actual amount of kinetic energy it will acquire, and the amount of work it can do, depends both on the height of the cliff and the mass of the stone. Similarly in the electrical case: the potential energy change, or the work that can be done, depends both on the potential difference (corresponding to height of the cliff) and on the charge (corresponding to mass), Equation 14-1.

Sources of electrical energy such as batteries and electric generators maintain a particular potential difference; the actual amount of energy used or transformed depends on how much charge flows. For example, consider a flashlight containing two 1.5-V batteries, so the total voltage applied to the bulb is 3.0 V. The amount of energy transformed (into light, and of course heat) is proportional to how much charge flows, which in turn depends on how long the light is on. If over a given period 0.50 C of charge flows through the bulb, the total energy transformed is (0.50 C) (3.0 V) = 1.5 J. If the flashlight is left on twice as long, 1.00 C of charge will flow and the energy transformed is (1.00 C) (3.0 V) = 3.0 J.

▮ EXAMPLE 14-1 An electron in a vacuum tube of a television set is accelerated from rest through a potential difference $V = 200$ V.

(*a*) What is the change in potential energy of the electron?

(*b*) What is the speed of the electron as a result of this acceleration?

SOLUTION (*a*) The charge on an electron is $e = -1.6 \times 10^{-19}$ C. Therefore, the potential energy is

$$\begin{aligned} \Delta PE = QV &= (-1.6 \times 10^{-19}\ \text{C})(200\text{V}) \\ &= -3.2 \times 10^{-17}\ \text{J}. \end{aligned}$$

The minus sign indicates that the potential energy decreases.

(*b*) The potential energy lost by the electron becomes kinetic energy. (Alternatively, we could say the work done by the field on the electron, QV, equals the increase in kinetic energy.) Thus

$$\Delta KE = \Delta PE$$

$$\tfrac{1}{2}mv^2 - 0 = QV,$$

where the initial $KE = 0$ since we assume that the electron started from rest. We solve for v and put in the mass of the electron, $m = 9.1 \times 10^{-31}$ kg:

$$\begin{aligned} v &= \sqrt{\frac{2QV}{m}} \\ &= \sqrt{\frac{2(1.6 \times 10^{-19}\ \text{C})(200\ \text{V})}{9.1 \times 10^{-31}\ \text{kg}}} \\ &= 8.4 \times 10^{6}\ \text{m/s}. \end{aligned}$$

▮▮

Relation between electric potential and uniform electric field

You may be wondering if there is a relation between potential differences and electric field. The answer is yes, and it is particularly simple in the case of a uniform electric field. Suppose that the potential difference between the two plates in Figure 14-1 is V_{AB}. Then the work needed to move a positive charge Q from B to A is

$$W = QV_{AB}.$$

We can also write the work as the force times distance and use the fact that the force on Q (Equation 13-2) is $F = QE$, where E is the uniform electric field between the plates. Thus

$$W = Fd = QEd,$$

where d is the distance between the plates. We now set these two expressions for W equal and find that $QV_{AB} = QEd$, or

$$V_{AB} = Ed \qquad (E \text{ uniform}). \tag{14-2}$$

If we solve for E, we find that $E = V_{AB}/d$. From this equation we can see that the units for electric field intensity can be written as volts per meter (V/m) as well as newtons per coulomb. These are equivalent in general, since 1 N/C = 1 N · m/C · m = 1 J/C · m = 1 V/m.

▮ EXAMPLE 14-2 Two parallel plates are charged to a voltage of 50 V. If the separation between the plates is 0.050 m, calculate the electric field between them.

SOLUTION We have from Equation 14-2,

$$E = \frac{V}{d} = \frac{50\ \text{V}}{0.050\ \text{m}} = 1000\ \text{V/m}.$$ ▮▮

* Relation between *E* and *V* for nonuniform *E*

In a region where E is not uniform, the connection between E and V takes on a slightly different form than Equation 14-2. In general, it is possible to show that the electric field in a given direction at any point in space is equal to the *rate at which the electric potential changes over distance in that direction.* Actually, if we take into account direction, this gives the negative of the electric field. For example, the x component of the electric field is given by $E_x = -\Delta V/\Delta x$, where ΔV is the change in potential over the very short distance Δx. Note that this relation resembles Equation 14-2† except that the distance Δx must be very small—so small that V does not change appreciably over this distance. Similar relations apply for the y and z components of **E**. Another way of stating the relation is this: if we plot V on a graph versus x, the slope of the graph at any point equals the magnitude of the x component of the electric field at that point. (And we must insert a minus sign if we want the direction to come out right.)

Equipotential lines

The electric potential can be represented diagrammatically by drawing the **equipotential lines** or, in three dimensions, the *equipotential surfaces.* An equipotential surface is one on which all points are at the same potential. That is, the potential differences between any two points on the surface is zero, and no work is required to move a charge from one point to the other. An *equipotential surface must be perpendicular to the electric field* at any point. If this were not so—that is, if there were a component of **E** parallel to the surface—it would require work to move the charge along

† Strictly speaking, Equation 14-2 should have a minus sign; but there we dealt only with magnitudes and did not worry about signs, which only have to do with the direction of **E**.

the surface against this component of **E**; and this would contradict the idea that it is an *equi*potential surface.

The fact that the electric field lines and equipotential surfaces are mutually perpendicular helps us locate the equipotentials when the electric field lines are known. In Figure 14-2, a few of the equipotential lines are drawn (dashed lines) for the field between two parallel plates at a potential difference of 20 V (the lines are only a part of the equipotential surface that extends into and out of the paper). The negative plate is arbitrarily chosen to be zero volts and the potential of each equipotential line is indicated. The equipotential lines for the case of two equal but oppositely charged particles are shown in Figure 14-3 as dashed lines.

We saw in Section 13-3 that there can be no electric fields within a conductor in the static case, for otherwise the free electrons would feel a force and would move. Thus a conductor must all be at the same potential. The surface of a conductor is then an equipotential surface. (If it weren't, the free electrons at the surface would move; for whenever there is a potential difference between two points, there is an electric field acting between the two points.) This is fully consistent with the fact, discussed earlier, that the electric field leading out from a conductor must be perpendicular to the surface.

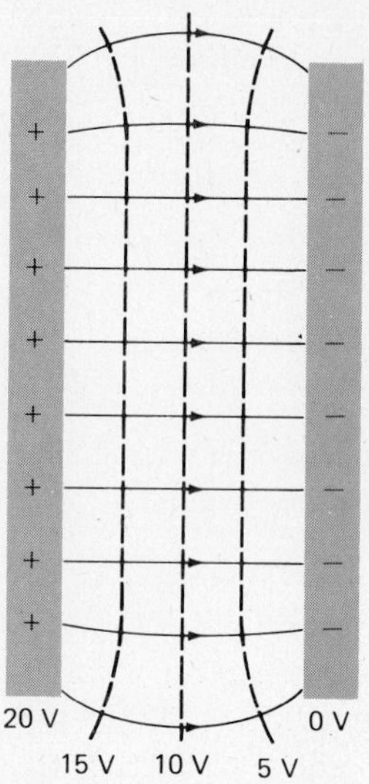

FIGURE 14-2
Equipotential lines (dashed) between two charged parallel plates; note they are perpendicular to the electric field lines (solid lines).

The electron volt, a unit of energy

The joule is a very large unit for dealing with energies of electrons, atoms, or molecules, whether in molecular biology or atomic and nuclear physics (see Example 14-1). For this purpose, the *electron volt* is used (abbreviated eV). One electron volt is defined as the energy acquired by an electron when moving through a potential difference of 1 V. Since the charge on an electron is 1.6×10^{-19} C, and since $\Delta PE = QV$, 1 eV is equal to $(1.6 \times 10^{-19}\text{ C})(1.0\text{ V}) = 1.6 \times 10^{-19}$ J, we thus have

$$1\text{ eV} = 1.6 \times 10^{-19}\text{ J}.$$

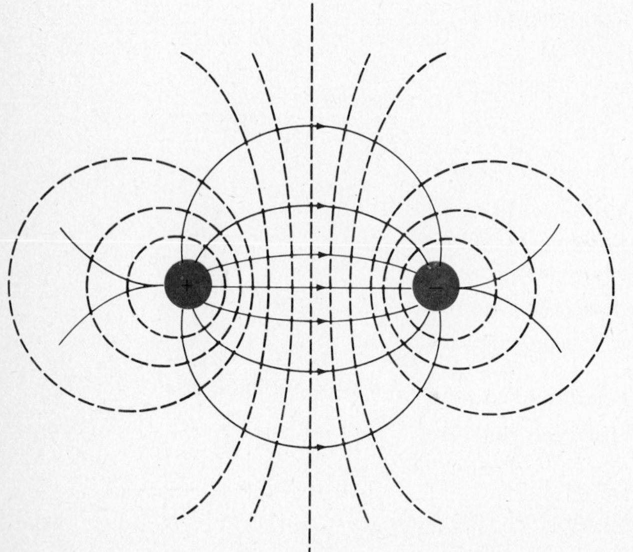

FIGURE 14-3
Equipotential (dashed) lines and electric field (solid) lines for two oppositely charged particles.

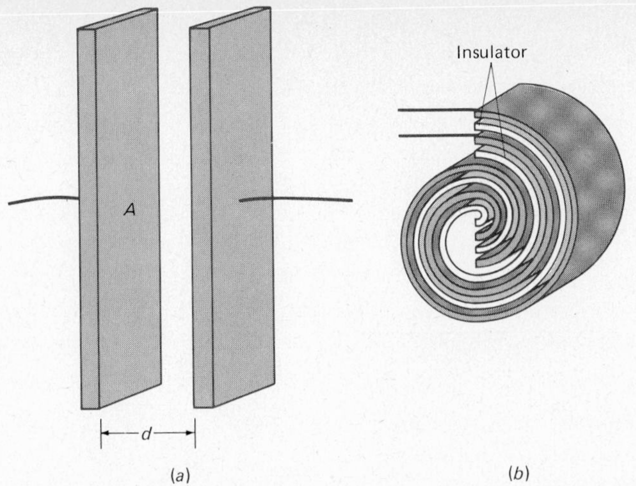

FIGURE 14-4
Capacitors.

An electron that falls through a potential difference of 1000 V will lose 1000 eV of potential energy and will thus gain 1000 eV or 1 keV of kinetic energy. On the other hand, if a particle has a charge equal to twice the charge on the electron ($=2e = 3.2 \times 10^{-19}$ C), when it moves through a potential difference of 1000 V its energy will change by 2000 eV.

Although the electron volt is handy for *stating* the energies of molecules and elementary particles, it is not a proper SI unit; for calculations it should be converted to joules using the conversion factor given above.

14-2 ▍ Capacitors

A **capacitor**, sometimes called a *condenser*, is a device for storing electric charge, and consists of two conducting objects placed near one another but not touching. A typical capacitor consists of a pair of parallel plates of area A separated by a small distance d (Figure 14-4*a*). Often the two plates are rolled into the form of a cylinder with paper or other insulator between the plates (Figure 14-4*b*).

Capacitance

If a voltage is applied to a capacitor, say by connecting the capacitor to a battery as in Figure 14-5, it quickly becomes charged. One plate acquires a negative charge, the other an equal amount of positive charge. For a given capacitor, the amount of charge Q acquired by each plate is found to be proportional to the potential difference:

$$Q = CV. \tag{14-3}$$

The constant of proportionality, C, in this relation is called the **capacitance** of the capacitor. The units of capacitance are coulombs per volt, and this unit is called a *farad* (F). Most capacitors have capacitance in the range 1 pF (picofarad $= 10^{-12}$ F) to 1 μF (microfarad $= 10^{-6}$ F).

The capacitance C is a constant for a given capacitor. Its value depends on the structure of the capacitor itself. For a parallel-plate

FIGURE 14-5
Parallel-plate capacitor connected to a battery.

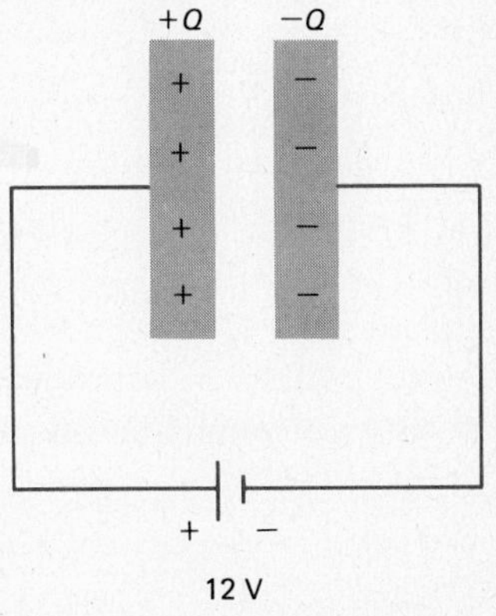

capacitor whose plates have area A and are separated by a distance d of air (Figure 14-4a), the capacitance is given by

$$C = \varepsilon_0 \frac{A}{d}. \qquad (14\text{-}4a)$$

This makes sense since a larger area A means that charges can be further apart, resulting in less repulsion, so that more charge can be held; and a greater separation d means the charge on each plate exerts less attractive force on the other plate, so less charge is drawn from the battery, and the capacitance is less. The constant ε_0 is the *permittivity of free space* and is the same one that appears in Coulomb's law; it has the value $8.85 \times 10^{-12}\ \text{C}^2/\text{N} \cdot \text{m}^2$.

▍ EXAMPLE 14-3 (a) Calculate the capacitance of a capacitor whose plates are 20 cm × 3.0 cm and are separated by a 1.0-mm air gap. (b) What is the charge on each plate if the capacitor is connected to a 12-V battery?

SOLUTION (a) The area A equals $(20 \times 10^{-2}\ \text{m})\ (3.0 \times 10^{-2}\ \text{m}) = 6.0 \times 10^{-3}\ \text{m}^2$. The capacitance C is then

$$C = (8.85 \times 10^{-12}\ \text{C}^2/\text{N} \cdot \text{m}^2) \frac{6.0 \times 10^{-3}\ \text{m}^2}{1.0 \times 10^{-3}\ \text{m}} = 53\ \text{pF}.$$

(b) The charge on each plate is

$$Q = CV = (53 \times 10^{-12}\ \text{F})\ (12\ \text{V}) = 6.4 \times 10^{-10}\ \text{C}.$$

▍▍

Dielectrics

In most capacitors there is an insulating sheet of paper or plastic called a *dielectric* between the plates. This serves several purposes. First, dielectrics break down less readily than air, so higher voltages can be applied without charge passing across the gap. Furthermore, a dielectric allows the plates to be placed closer together to increase the capacitance without fear of them touching.† Finally, the dielectric itself increases the capacitance by a factor K known as the *dielectric constant*. Thus, for a parallel-plate capacitor

$$C = K\varepsilon_0 \frac{A}{d}. \qquad (14\text{-}4b)$$

This can also be written

$$C = \varepsilon \frac{A}{d},$$

where $\varepsilon = K\varepsilon_0$ is the *permittivity of the material*. The values of the dielectric constant for various materials are given in Table 14-1. This is the same dielectric constant (and permittivity) discussed in Section 13-2; there we saw that the force between two charges is reduced by a factor equal to the dielectric constant. This is basically what happens when a dielectric is placed between the two plates of a capacitor. Let us see why, from the molecular point of view, the capacitance should increase.

† As we saw in Example 14-3, a simple parallel-plate capacitor of ordinary dimensions has a capacitance of only a few picofarads. A capacitance of 1 μF or more would require huge plates if dielectrics were not used.

TABLE 14-1

Dielectric constants (at 20°C)

Material	Dielectric constant K
Vacuum	**1.0000**
Air	**1.0006**
Paraffin	**2.2**
Rubber, hard	**2.8**
Vinyl (plastic)	**2.8–4.5**
Paper	**3–7**
Quartz	**4.3**
Glass	**4–7**
Porcelain	**6–8**
Mica	**7**
Ethyl alcohol	**24**
Water	**80**

Consider a capacitor whose plates are separated by an air gap. This capacitor has a charge $+Q$ on one plate and $-Q$ on the other (Figure 14-6*a*). The capacitor is isolated (not connected to a battery). The potential difference between the plates, V_0, is given by Equation 14-3: $Q = C_0 V_0$; the subscripts ($_0$) refer to the situation when only air is between the plates. Now we insert a dielectric between the plates (Figure 14-6*b*). The molecules of the dielectric may be *polar*; that is, although the molecules are neutral, the electrons may not be evenly distributed so that one part of the molecule is positive and another negative.† Because of the electric field between the plates, the molecules will tend to become oriented as shown. Even if the molecules are not polar, the electric field between the plates will induce a separation of charge in the molecules. Although the electrons do not leave the molecules, they will move slightly within the molecules toward the positive plate. So the situation is still as illustrated in Figure 14-6*b*. The net effect in either case is as if there were a net negative charge on the outer edge of the dielectric facing the positive plate, and a net positive charge on the opposite side as shown.

Now imagine a positive test charge within the dielectric. The force that it feels is reduced by a factor K, the dielectric constant (Section 13-2). This is reflected by the fact that some of the electric field lines actually do not pass through the dielectric but end (and restart) on the charges induced on the surface of the dielectric (Figure 14-6*c*). Because the force on our test charge is reduced by a factor K, the work needed to move it from one plate to the other is reduced by a factor K. (We assume that the dielectric fills all the space between the plates.) The voltage, which is the work done per unit charge, must therefore also have decreased by the factor K. That is, the voltage between the plates is now

$$V = \frac{V_0}{K}.$$

Now the charge Q on the plates has not changed, because they are isolated. So we have

$$Q = CV,$$

where C is the capacitance when the dielectric is present. When we combine this with the relation, $V = V_0/K$, we obtain

$$C = \frac{Q}{V} = \frac{Q}{V_0/K} = \frac{QK}{V_0} = KC_0,$$

since $C_0 = Q/V_0$. Thus we see, from an atomic point of view, why the capacitance is increased by the factor K.

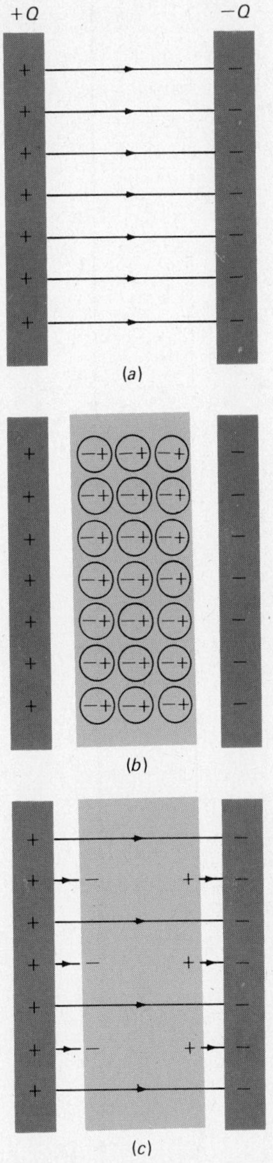

+Q −Q

(a)

(b)

(c)

FIGURE 14-6
Effect of dielectric in a capacitor.

14-3 ▌ Electric energy storage

Energy stored in a capacitor

A charged capacitor stores electric energy. The energy stored in a capacitor will be equal to the work done to charge it up. The net effect of charging a capacitor is to remove charge from one plate and add it to the

† They are said to have a *dipole moment*; see Section 14-5.

other plate. This is what a battery does when it is connected to a capacitor. When some charge is on each plate, it requires work to add more charge of the same sign. The more charge already on a plate, the more work is required to add more. The work needed to add a small amount of charge Δq, when a potential difference V is across the plates, is $\Delta W = \Delta q V$. Initially, when the capacitor is uncharged, it requires no work to move the first bit of charge over. By the end of the charging process, however, the work needed to add a charge Δq will be much greater because the voltage across the capacitor, which is proportional to the charge on the plates ($V = Q/C$), is now large. The voltage increases during this process from zero to its final value, V_f, at the end; then the total work done, W, will be equivalent to moving all the charge Q at once across a voltage equal to the average during the whole process; the average voltage is $V_f/2$, so

$$W = Q\frac{V_f}{2}.$$

Thus we can say that the energy stored in a capacitor is

$$\text{energy} = \tfrac{1}{2}QV,$$

where V is the potential difference between the plates (we have dropped the subscript) and Q is the charge on each plate. Since $Q = CV$, we can also write

$$\text{energy} = \frac{1}{2}QV = \frac{1}{2}CV^2 = \frac{1}{2}\frac{Q^2}{C}. \qquad (14\text{-}5)$$

▮ EXAMPLE 14-4 A 12-V battery is connected to a 20-μF capacitor which uses a paper dielectric. How much electric energy can be stored in the capacitor?

SOLUTION We do not need the dielectric constant since the value $C = 20 \times 10^{-6}$ F includes its effect. From Equation 14-5, we have

$$\text{energy} = \tfrac{1}{2}CV^2 = \tfrac{1}{2}(20 \times 10^{-6}\ \text{F})(12\ \text{V})^2 = 1.4 \times 10^{-3}\ \text{J}.$$ ▮▮

Energy stored in an electric field

As we have seen, energy is stored in a capacitor, and it is useful to think of this energy as being stored in the electric field between the plates. Let us then calculate the energy stored in a capacitor in terms of the electric field.

We saw in Equation 14-2 that the electric field E between two parallel plates is constant (approximately) and is related to the potential difference by $V = Ed$, where d is the separation. Also, Equation 14-4a tells us that $C = \varepsilon_0 A/d$. Thus

$$\text{energy} = \frac{1}{2}CV^2 = \frac{1}{2}\frac{\varepsilon_0 A}{d}E^2 d^2$$

$$= \frac{1}{2}\varepsilon_0 E^2 Ad.$$

The quantity Ad is simply the volume between the plates in which the electric field E exists. If we divide both sides by the volume, we obtain an expression for the energy per unit volume or *energy density*:

$$\text{energy density} = \tfrac{1}{2}\varepsilon_0 E^2. \qquad (14\text{-}6)$$

The energy stored per unit volume is proportional to the square of the electric field in that region. If a dielectric is present, ε_0 is replaced by ε. We derived Equation 14-6 for the special case of a capacitor. But it can be shown to be true for any region of space where there is an electric field.

14-4 ▌ Electric potential of single point charges

The electric potential at a distance r from a single point charge Q can be derived from the expression for its electric field (Equation 13-3) using calculus. The potential in this case is usually taken to be zero at infinity (∞), where the electric field is also zero. The result is

$$V = k\frac{Q}{r} \quad \text{(single point charge)}. \qquad (14\text{-}7)$$

Notice that the potential† V decreases with the first power of the distance, whereas the electric field (Equation 13-3) decreases as the *square* of the distance. The potential near a positive charge is large and it decreases toward zero at very large distances. For a negative charge, the potential is *less* than zero and increases toward zero at large distances.

▌ EXAMPLE 14-5 How much work is required to bring a charge $q = 3.0\ \mu\text{C}$ from a great distance away (take $r = \infty$) to a point 0.50 m from a charge $Q = 20.0\ \mu\text{C}$?

SOLUTION The work required is

$$W = qV_{AB} = q\left(\frac{kQ}{r_A} - \frac{kQ}{r_B}\right),$$

where $r_A = 0.50$ m and $r_B = \infty$. Therefore,

$$W = (3.0\times10^{-6})\frac{(9.0\times10^9\ \text{N}\cdot\text{m}^2/\text{C}^2)(2.0\times10^{-5}\ \text{C})}{0.50\ \text{m}}$$

$$= 1.08\ \text{J}.$$

▌▌

To determine the electric field surrounding a collection of two or more point charges requires adding up the electric fields due to each charge. Since the electric field is a vector, this can often be a chore. To find the electric potential due a collection of point charges is far easier, since the electric potential is a scalar and hence you only need to add numbers together without concern for direction.

▌ EXAMPLE 14-6 Calculate the electric potential at points A and B in Figure 13-18 due to the two charges shown. (See also Example 13-5, where we calculated the electric field at these points.)

† We can think of V here as representing the absolute potential, where $V = 0$ at $r = \infty$, or we can think of V as the potential differences between r and infinity.

SOLUTION The potential at point A is the sum of the potentials due to the + and − charges, and we use Equation 14-7 for each:

$$\begin{aligned} V_A &= V_{A2} + V_{A1} \\ &= \frac{(9.0\times10^9\ \text{N}\cdot\text{m}^2/\text{C}^2)(5.0\times10^{-5}\ \text{C})}{0.30\ \text{m}} \\ &\quad + \frac{(9.0\times10^9\ \text{N}\cdot\text{m}^2/\text{C}^2)(-5.0\times10^{-5}\ \text{C})}{0.60\ \text{m}} \\ &= 7.5\times10^5\ \text{V}. \end{aligned}$$

At point B:

$$\begin{aligned} V_B &= V_{B2} + V_{B1} \\ &= \frac{(9.0\times10^9\ \text{N}\cdot\text{m}^2/\text{C}^2)(5.0\times10^{-5}\ \text{C})}{0.40\ \text{m}} \\ &\quad + \frac{(9.0\times10^9\ \text{N}\cdot\text{m}^2/\text{C}^2)(-5.0\times10^{-5}\ \text{C})}{0.40\ \text{m}} \\ &= 0\ \text{V}. \end{aligned}$$

It should be clear that the potential will be zero everywhere on the plane between the two charges that is equidistant from each. ▮▮

A simple summation like these can easily be performed for any number of point charges.

*14-5 ▮ Electric dipoles

Two equal point charges Q, of opposite sign, separated by a distance l, are called an *electric dipole*. The two charges in Figure 13-18 constitute an electric dipole. The electric field lines and equipotential surfaces for a dipole were shown in Figure 14-3. Because electric dipoles occur often in physics, as well as in other fields such as molecular biology, it is useful to examine them more closely.

Let us calculate the electric potential at an arbitrary point P for the dipole shown in Figure 14-7. Since V is the sum of the potentials due to each of the two charges, we have

$$V = \frac{kQ}{r} + \frac{k(-Q)}{r+\Delta r} = kQ\left(\frac{1}{r} - \frac{1}{r+\Delta r}\right) = kQ\frac{\Delta r}{r(r+\Delta r)},$$

where r is the distance from P to the positive charge and $r+\Delta r$ is the distance to the negative charge. This equation becomes simpler if we consider points P whose distance from the dipole is much larger than the separation of the two charges, that is, for $r \gg l$. From the diagram we can see that in this case, $\Delta r \approx l\cos\theta$; and since $r \gg \Delta r = l\cos\theta$, we can neglect Δr in the denominator as compared to r. Therefore, we obtain

$$V = \frac{kQl\cos\theta}{r^2} \qquad \text{(dipole; } r \gg l\text{).} \qquad (14\text{-}8)$$

When θ is between 0° and 90°, V is positive; if θ is between 90° and 180°, V

FIGURE 14-7

Electric dipole. Calculation of potential V at point P.

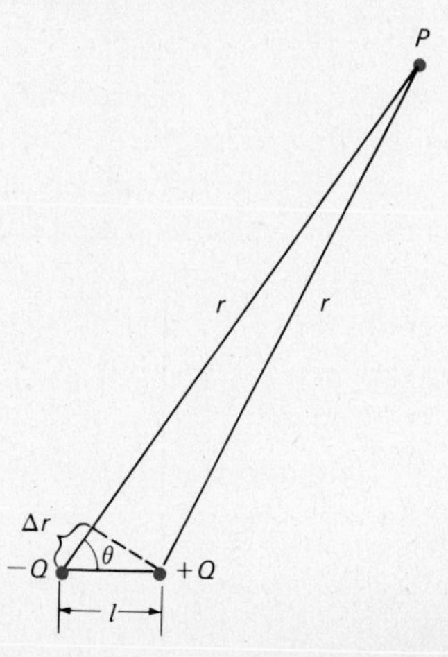

is negative (since cos θ is then negative). This makes sense since in the first case P is closer to the positive charge and in the second case it is closer to the negative charge. At $\theta = 90°$, the potential is zero ($\cos 90° = 0$), in accordance with the result of Example 14-6. From Equation 14-8 we see that the potential decreases as the *square* of the distance from the dipole, whereas for a single point charge the potential decreases with the first power of the distance (Equation 14-7). It is not surprising that the potential should fall off faster for a dipole; for when you are far from a dipole, the two equal but opposite charges appear so close together as to tend to neutralize each other.

The product Ql which occurs in Equation 14-8 is referred to as the *dipole moment, p*, of the dipole. Equation 14-8 can be written in terms of the dipole moment as $V = kp \cos \theta / r^2$. A dipole moment has units of coulomb-meters (C · m), although for molecules a smaller unit called a debye is sometimes used: 1 debye $= 3.33 \times 10^{-30}$ C · m.

In many molecules, even though they are electrically neutral, the electrons spend more time in the vicinity of one atom than another, which results in a separation of charge. Such molecules have a dipole moment and are called *polar molecules*. We have already seen that water (Figure 13-4) is a polar molecule and we met others in our discussion of molecular biology (Section 13-4). Table 14-2 gives the dipole moments for several molecules. The + and − signs indicate on which atoms these charges lie. The last two entries are a part of many organic molecules and play an important role in molecular biology.

TABLE 14-2

Dipole moments of selected molecules

Molecule	Dipole moment (C · m)
$H_2^{(+)}O^{(-)}$	6.1×10^{-30}
$H^{(+)}Cl^{(-)}$	3.4×10^{-30}
$N^{(-)}H_3^{(+)}$	5.0×10^{-30}
$>N^{(-)}{-}H^{(+)}$†	$\approx 3.0 \times 10^{-30}$
$>C^{(+)}{=}O^{(-)}$†	$\approx 8.0 \times 10^{-30}$

† These groups often appear on larger molecules; hence the value for the dipole moment will vary somewhat, depending on the rest of the molecule.

EXAMPLE 14-7 The distance between the carbon (+) and oxygen (−) atoms in the group C=O is about 1.2×10^{-10} m. Calculate (*a*) the net charge Q on the C (carbon) and O (oxygen) atoms, and (*b*) the potential 9.0×10^{-10} m from the dipole along its axis with the oxygen being the nearer atom (that is, to the left in Figure 14-7, so $\theta = 180°$). (*c*) What would the potential be at this point if only the oxygen (O) were charged?

SOLUTION (*a*) The dipole moment $p = Ql$. Therefore, from Table 14-2:

$$Q = \frac{p}{l} = \frac{8.0 \times 10^{-30}\ \text{C} \cdot \text{m}}{1.2 \times 10^{-10}\ \text{m}} = 6.6 \times 10^{-20}\ \text{C}.$$

(*b*) Since $\theta = 180°$, we have, using Equation 14-8:

$$V = \frac{kp \cos \theta}{r^2} = \frac{(9.0 \times 10^9\ \text{N} \cdot \text{m}^2/\text{C}^2)(8.0 \times 10^{-30}\ \text{C} \cdot \text{m})(-1.00)}{(9.0 \times 10^{-10}\ \text{m})^2}$$

$$= -0.088\ \text{V}.$$

(*c*) If we assume that the oxygen has charge $Q = -6.6 \times 10^{-20}$ C (as in part *a*) and that the carbon is not charged, we must use the formula for a single charge, Equation 14-7:

$$V = \frac{kQ}{r} = \frac{(9.0 \times 10^9\ \text{N} \cdot \text{m}^2/\text{C}^2)(-6.6 \times 10^{-20}\ \text{C})}{9.0 \times 10^{-10}\ \text{m}}$$

$$= -0.66\ \text{V}.$$

Of course, we expect the potential of a single charge to have greater magnitude than that of a dipole of equal charge at the same distance. ▮▮

*14-6 ▮ Interaction energy and weak bonds in molecular biology

In Section 13-4 we saw how processes in a cell can be described in terms of electric forces between molecules. It is often more convenient to consider the interaction energy between attracting molecules. The *interaction energy* is defined as the energy needed to pull two molecules apart once they are held together, such as for an adenine (A) and a thymine (T) in a DNA helix (Figure 13-23). It is thus equal to the potential energy of one in the presence of the other, relative to zero potential energy when they are far apart ($r=\infty$). Let us focus our attention on the attraction between the C=O dipole of the thymine to the N—H dipole of the adenine, which is at the top of Figure 13-23*b*. Now it should be pointed out that this calculation, done exactly, would be quite complicated. For one thing, there is not only the electrostatic attraction between these dipoles, but there is also a repulsive force when the two dipoles come close enough together that their outer electrons begin to overlap. When the molecules are in the position shown, this repulsion is just balancing the electrostatic attraction between the two dipoles. We will simply calculate the electrostatic energy in the following example.

▮ EXAMPLE 14-8 Calculate the interaction energy between the C=O of thymine and the H—N of adenine, assuming that the two dipoles are lined up as shown in Figure 14-8. The distance between the O and the N (2.80 Å) is taken from Figure 13-23*b*; all the distances are from experimental measurements ($1\ \text{Å} = 1\times 10^{-10}$ m).

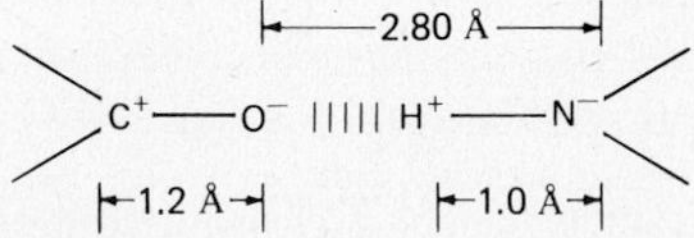

FIGURE 14-8
Diagram for calculation of interaction energy between C=O dipole of thymine and H—N dipole of adenine.

SOLUTION The interaction energy, for which we use the symbol U, will be equal to the potential energy of one dipole in the presence of the other, since this will be equal to the work needed to pull them infinitely far apart. However, we cannot use the dipole approximation (Equation 14-8) since the distances involved are too small. Hence we must use Equation 14-7 for the potential between two point charges and we will have four terms:

$$U = U_{CH} + U_{CN} + U_{OH} + U_{ON}.$$

Here U_{CH} means the potential energy of C in the presence of H, and similarly for the other terms. We do not have terms corresponding to C and O or N and H because the two dipoles are assumed to be stable entities. Since the potential energy $U = QV$, where V is the electric potential, then for two point charges $U_{12} = kQ_1Q_2/r$, where r is the distance between them. Then

$$U = k\frac{Q_C Q_H}{r_{CH}} + k\frac{Q_C Q_N}{r_{CN}} + k\frac{Q_O Q_H}{r_{OH}} + k\frac{Q_O Q_N}{r_{ON}}.$$

Now $Q_C = -Q_O = 6.6\times 10^{-20}$ C from Example 14-7, and the charges on the H and N can be calculated similarly:

$$Q_H = -Q_N = \frac{p}{l} = \frac{3.0\times 10^{-30}\ \text{C}\cdot\text{m}}{1.0\times 10^{-10}\ \text{m}} = 3.0\times 10^{-20}\ \text{C},$$

where the value of the dipole moment p was obtained from Table 14-2. The distances are obtained from Figure 14-8: $r_{CH} = 3.0\times 10^{-10}$ m, $r_{CN} =$

4.0×10^{-10} m, $r_{OH} = 1.8 \times 10^{-10}$ m, and $r_{ON} = 2.8 \times 10^{-10}$ m. Putting in these numbers and factoring out the powers of 10, we find that

$$U = (9 \times 10^9 \text{ N} \cdot \text{m}^2/\text{C}^2)\left(\frac{(6.6)(3.0)}{3.0} + \frac{(6.6)(-3.0)}{4.0} + \frac{(-6.6)(3.0)}{1.8} + \frac{(-6.6)(-3.0)}{2.8}\right)\left(\frac{10^{-40} \text{ C}^2}{10^{-10} \text{ m}}\right)$$

$$= (5.9 - 4.5 - 9.9 + 6.4) \times 10^{-20} \text{ J} = -2.1 \times 10^{-20} \text{ J}$$

$$= -0.13 \text{ eV}.$$

The potential energy is negative, which means that work is required to separate the molecules. ▮▮

This and other biologically important dipole interactions† are found to be in the range 0.1 to 0.3 eV. Thus, once two such dipoles are in the position shown in Figure 14-8 (or 13-23*b*), it requires about 0.1 to 0.3 eV to pull them apart. How are they pulled apart? By collision with a fast-moving molecule with at least this much kinetic energy. Now the average KE of molecules in a cell at 37°C (310 K) is, according to Equation 7-10,

$$\text{KE} = \tfrac{3}{2}kT = \tfrac{3}{2}(1.38 \times 10^{-23} \text{ J/K})(310 \text{ K}) = 6.4 \times 10^{-21} \text{ J}$$

$$= 0.04 \text{ eV}.$$

The molecules in a cell will have kinetic energies distributed around this average (see Figure 7-13), and only a small fraction of them would have enough kinetic energy to knock our two molecules apart; thus the two dipoles stay close together for a reasonably long time, but they will separate after a time when struck by some fast-moving molecule. Hence this interaction between dipoles is called a *dipole–dipole bond.* Biophysicists also refer to it as one type of *weak bond,*‡ since it is weak in comparison to the usual covalent and ionic bonds that hold atoms together to form molecules. These "strong bonds" have much larger interaction energies, often 2 to 4 eV, and thus are extremely stable, since almost no molecules in the cell will have kinetic energy great enough to break them.

The importance of weak bonds is twofold. First, since the energies involved are not too much larger than the average KE of molecules in the cell, *they are made and broken fairly easily.* Second, *they are very short range*; that is, the forces are strong enough to hold the two molecules together, even for a short time, only if they are very close together.

These two aspects are important, for example, for the nearly flawless replication of DNA. First, the two strands of a DNA chain are held firmly

† Interaction energies are sometimes specified in kilocalories per mole; that is, the number of kilocalories of energy for a mole (6.02×10^{23}) of molecules. It can be shown that 1 eV = 23 kcal/mol. Thus the interaction energies for two biological dipoles range from about 2 to 7 kcal/mol.

‡ The dipole–dipole bond is perhaps the most important weak bond in the cell. Another kind of weak bond is the *dipole–induced dipole* bond, in which a dipole can "induce" a dipole moment in an otherwise electrically balanced molecule just as a single charge can induce a separation of charge in a nearby object (Figure 13-5). When one of the atoms in a dipole–dipole bond is hydrogen, it is often called a "hydrogen bond." A hydrogen bond is generally the strongest of the weak bonds; this is because the hydrogen atom is the smallest atom and thus can be approached more closely. Hydrogen bonds also have a partial "covalent" character; that is, electrons between the two dipoles may be shared to a small extent. Chapter 27 has a discussion of covalent bonds.

together because there are very many weak bonds between dipoles on the two strands. However, a few chance collisions with molecules whose KE is somewhat above average can cause the double helix to unravel slightly. Normally, the two chains quickly wind back up again. At the time DNA replication begins, however, enzymes appear to hold the chains apart; and once replication has begun, the two new chains keep the original helix separated as in Figure 13-24. Let us again focus on the guanine base (G) indicated by the arrow in Figure 13-24. If a T, an A, or a G comes close to it, the interaction energy will be quite small since, owing to their shapes, the oppositely charged atoms cannot get very close together; and the next molecule that collides with any of these three bases will knock it away. However, when a cytosine (C) approaches our G, the oppositely charged atoms can come close together, and weak bonds occur that may last just long enough for the proper enzyme to link the new C to the end of the growing chain.

*14-7 ▮ The electrocardiogram

Each time the heart beats, changes in electrical potential occur on its surface that can be detected using metal contacts, called "electrodes," which are attached to the skin. The changes in potential are small, on the order of millivolts (mV), and must be amplified; they are displayed either with a chart recorder on paper, or on a cathode-ray-tube oscilloscope (discussed in Chapter 19). The record of the potential changes for a given person's heart is called an electrocardiogram (EKG or ECG); an example is shown in Figure 14-9. The instrument itself is called an electrocardiograph. We are not so interested now in the electronics, but in the source of these potential changes and their relation to heart activity.

Muscle cells and nerve cells are similar in that both have an electric dipole layer across the cell wall; that is, in the normal situation there is a net positive charge on the exterior surface and a net negative charge on the interior surface as shown in Figure 14-10*a*. The amount of charge depends on the size of the cell, but is approximately 10^{-3} C/m^2 of surface. For a cell whose surface area is 10^{-5} m^2, the total charge on either surface is thus $\pm 10^{-8}$ C. Just before the contraction of heart muscles, changes occur in the cell wall (Section 15-7) so that positive ions on the exterior of the cell are able to pass through the wall and neutralize those on the inside, or even make the inside surface slightly positive compared to the

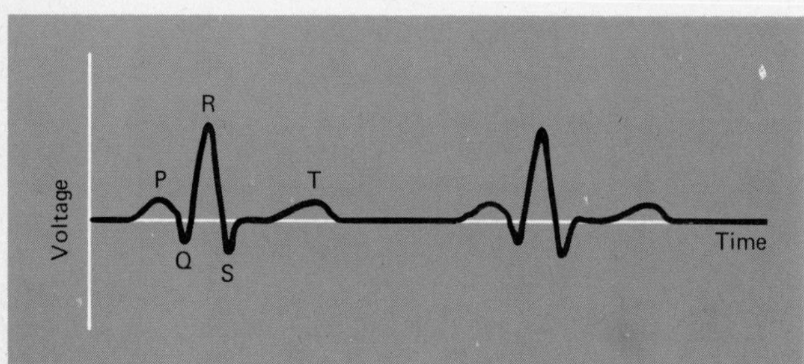

FIGURE 14-9
Typical ECG. Two heart beats are shown.

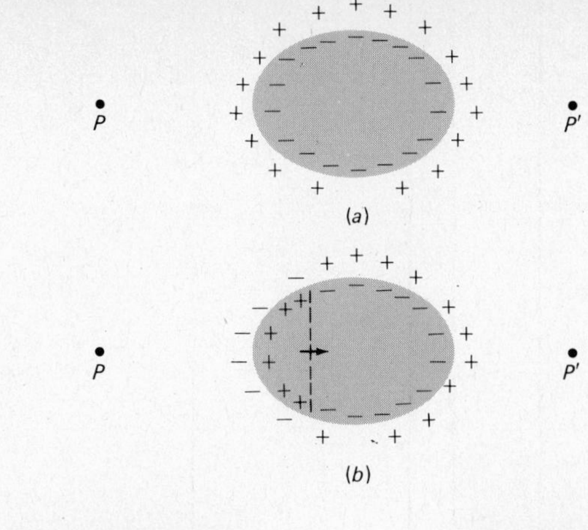

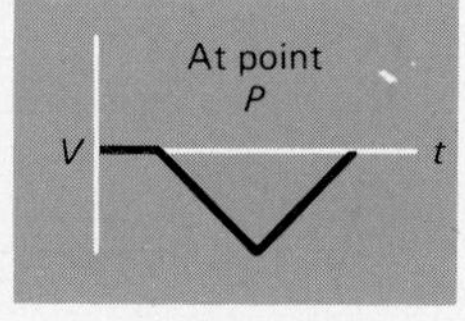

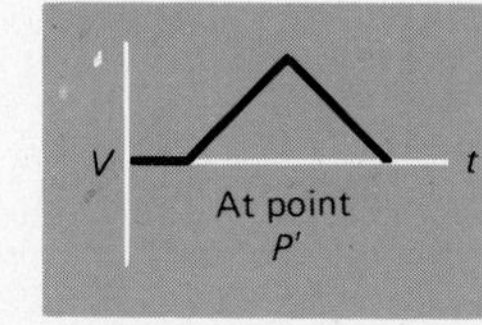

(c)

FIGURE 14-10
Heart muscle cell showing (*a*) charge dipole layer in resting state; (*b*) depolarization of cell progressing as muscle begins to contract; (*c*) potential *V* at points *P* and *P*′ as a function of time.

exterior, as shown in Figure 14-10*b*. This depolarization, as it is called, starts at one end of the cell and progresses toward the opposite end, as indicated by the arrow in part *b*, until the whole muscle is depolarized. (A similar process occurs in nerve cells and will be discussed in Chapter 15.) The muscle then slowly repolarizes to its original state (Figure 14-10*a*). This whole process requires less than a second.

The equal but opposite charges on the two sides of a cell wall can be considered as a set of electric dipoles. When the cell is in the resting state (Figure 14-10*a*) the potential at any point *P* outside the cell (see the figure) is zero; this is due to the fact that on opposite sides of the cell the individual dipoles are oriented in opposite directions, so the net effect of all of them is to give zero potential (the $\cos\theta$ factor in Equation 14-8 is important since it takes on values of -1 to $+1$). The potential at P' will also be zero, so the potential difference between electrodes at P and P' will be zero. When the cell begins to depolarize, however, the potential at P will decrease while that at P' will increase. This is because there is no longer symmetry, for there are more dipoles oriented with the positive charge on the right than vice versa. Point P is nearer the negative ends of the dipoles; so the potential there becomes negative (V is due to a sum of terms in which the θ of Figure 14-7 is $\approx 180°$, and therefore the $\cos\theta$ term in Equation 14-8 is -1). Point P', however, is closer to the positive ends of the dipoles, so V will be positive (these terms have $\theta \approx 0°$, so $\cos\theta = +1$ predominantly). As the depolarization is completed, the potentials at P and P' return to zero. This is shown in Figure 14-10*c*.

In the heart, the path of depolarization is complicated. Furthermore, after depolarization the muscles repolarize to the resting state (*a*). Thus the potential difference as a function of time is quite complicated (Figure 14-9).

It is standard procedure to divide a typical electrocardiogram into regions corresponding to the various deflections (or "waves" as they are called), as shown in Figure 14-9. Each of the deflections corresponds to the activity of a particular part of the heart beat (Figure 6-11). The P wave corresponds to contraction of the atria. The QRS group corresponds to contraction of the ventricles; this group has three main phases because the depolarization follows a complicated path from left to right, and toward the front, then downward to the left and toward the rear. The T wave corresponds to recovery (repolarization) of the heart in preparation for the next cycle.

Electrocardiograms make use of three basic electrodes, one placed on either side of the heart on the hands and one on the left foot. Sometimes six additional electrodes are placed at other locations. The measurement of so many potential differences provides additional information (some of it redundant), since the heart is a three-dimensional object and depolarization takes place in all three dimensions. A complete electrocardiogram may include as many as 12 graphs.

The ECG is a powerful tool in identifying heart defects. For example, the right side of the heart enlarges if the right ventricle must push against an abnormally large load (as when blood vessels become hardened or clogged). This problem is readily observed on an ECG, since the S wave becomes very large (negatively). *Infarcts,* which are dead regions of the heart muscle that result from a heart attack, are also detected on an ECG because they reflect the depolarization wave.

The interpretation of an ECG depends to a great extent on experience obtained with many patients rather than on theoretical understanding. A good deal of scientific research remains to be done.

▌ Summary

The *electric potential* at any point in space is defined as the electric potential energy per unit charge. The *electric potential difference* between any two points is defined as the work required to move a 1-C electric charge between the two points. Potential difference is measured in volts (1 V = 1 J/C) and is sometimes referred to as *voltage*. The potential difference V between two points where a uniform electric field E exists is given by $V = Ed$, where d is the distance between the two points. An *equipotential line* or surface is all at the same potential, and is perpendicular to the electric field at all points. The electric potential due to a single point charge Q, relative to zero potential at infinity, is given by $V = kQ/r$.

A *capacitor* is a device used to store charge and consists of two nontouching conductors. The two conductors generally hold equal and opposite charges, Q, and the ratio of this charge to the potential difference V between the conductors is called the *capacitance*, C; so $Q = CV$. The capacitance of a parallel-plate capacitor is proportional to the area of each plate and inversely proportional to their separation. The space between the conductors contains a nonconducting material such as air, paper, or plastic; the latter materials are referred to as dielectrics, and the capacitance is proportional to a property of dielectrics called the dielectric constant.

A charged capacitor stores an amount of electric energy given by $\frac{1}{2}QV = \frac{1}{2}CV^2 = \frac{1}{2}Q^2/C$. This energy can be thought of as stored in the electric field between the plates. The energy stored in any electric field E has a density (energy per unit volume) of $\frac{1}{2}\varepsilon_0 E^2$.

Problems

Questions

1 If two points are at the same potential, does this mean that no work is done in moving a test charge from one point to the other? Does this imply that no force must be exerted?

2 Is it possible for a plastic comb to acquire a charge of 0.5 V?

3 Draw in a few equipotential lines in Figure 13-20*b*.

4 Can two equipotential lines cross? Explain.

5 Is there a point between two equal positive charges where the electric field is zero? Where the electric potential is zero? Explain.

6 If the potential at a point is zero, must the electric field also be zero? Give an example.

7 What can you say about the electric field in a region of space that has the same potential throughout?

8 How does the earth's gravitational field change with distance? What about its gravitational potential?

9 A satellite orbits the earth along an equipotential line. What shape must the orbit be?

10 If a dielectric such as paper is placed between the plates of a small air-gap capacitor, how will the charge on each plate change if the same potential difference is maintained?

Problems

SECTION 14-1

1 (I) The electric field between two parallel plates connected to a 45-V battery is 500 V/m. How far apart are the plates?

2 (I) How strong is the electric field between two parallel plates 5 mm apart if the potential difference between them is 110 V?

3 (I) An electric field of 300 V/m is desired between two parallel plates 6.0 cm apart. How large a voltage should be applied?

4 (I) How much work is needed to move a -3.0-μC charge from ground to a point whose potential is $+60$ V?

5 (I) How much energy will an electron gain (joules) if it falls through a potential difference of 300 V?

6 (I) What potential difference is needed to give a helium nucleus ($Q = 3.2 \times 10^{-19}$ C) 8.0 keV of KE?

7 (II) An electron is accelerated by a potential difference of, say, 100 V. How much greater would its final speed be if it were accelerated with four times as much voltage?

8 (II) What is the speed of a 350-eV electron?

9 (II) What is the KE in electron volts of an oxygen molecule at STP (0 °C, 1 atm)?

10 (II) A lightning flash transfers 20 C of charge to earth from a potential difference of 5.0×10^7 V. (*a*) How much energy is dissipated? (*b*) How much water at 0°C could be brought to boiling?

11 (II) The work done to move a 1.0-μC charge from point P to point Q is 3.0×10^{-4} J. If the charge was started from rest and had 1.0×10^{-4} J of KE when it reached point Q, what must be the potential difference between P and Q?

12 (II) In a television picture tube, electrons are accelerated by thousands of volts through a vacuum. If a television set were laid on its back, would electrons be able to move upward against the force of gravity? What potential difference, acting over a distance of 20 cm, would be needed to balance the downward force of gravity so that an electron would remain stationary? Assume that the electric field is uniform.

13 (III) An electron is accelerated horizontally from rest in a television picture tube by a potential difference of 10,000 V. It then passes between two horizontal plates 5.0 cm long and 1.0 cm apart which have a potential difference of 200 V (Figure 14-11). At what angle θ will the electron be traveling after it passes between the plates?

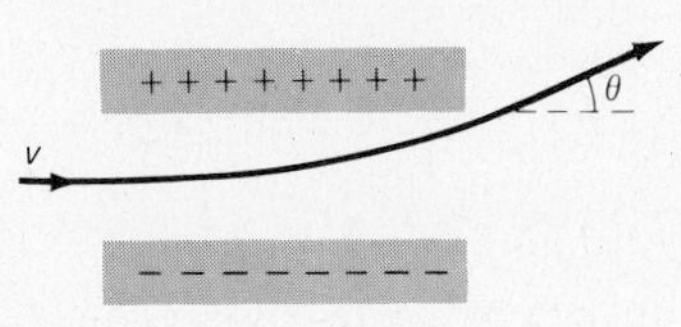

FIGURE 14-11

SECTION 14-2

14 (I) How much charge flows from a 12-V battery when it is connected to a 2.0-μF capacitor?

15 (I) The two plates of a capacitor hold $+1500\ \mu$C and $-1500\ \mu$C of charge, respectively, when the potential difference is 300 V. What is the capacitance?

16 (I) A 1.0-F capacitor is desired. What area must the plates have if they are to be separated by a 5.0-mm air gap?

17 (I) What is the capacitance of two square parallel plates 6.0 cm on a side that are separated by 2.0 mm of paraffin?

18 (II) An electric field of 6.0×10^6 V/m is desired between two parallel plates each of area 110 cm^2 and separated by 2.0 cm of air. What charge must be on each plate?

19 (II) How strong is the electric field between the plates of a 20-μF air-gap capacitor if they are 2.0 mm apart and each has a charge of 300 μC?

20 (II) It takes 6.0 J of energy to move a 2.0-mC charge from one plate of a 50-μF capacitor to the other. How much charge is on each plate?

21 (II) A 3000-pF air-gap capacitor is connected to a 12-V battery. If a piece of mica is placed between the plates, how much charge will then flow from the battery?

22 (III) Dry air will break down if the electric field exceeds 3.0×10^6 V/m. What amount of charge can be placed on a capacitor if the area of each plate is 10 cm^2?

23 (III) A 4.0-μF capacitor is charged to 100 V and a 10-μF capacitor is charged to 50 V. The positive plates are now connected to each other and the negative plates are connected to each other. What will be the potential difference across each and the charge on each?

SECTION 14-3

24 (I) 100 V is applied to a 600-pF capacitor. How much energy is stored?

25 (I) There is an electric field near the earth's surface of intensity 100 V/m. How much energy is stored per cubic meter in this field?

26 (II) How much energy is stored by the electric field between two square plates, 10 cm on a side, separated by a 3.0-mm air gap? The charges on the plates are equal and opposite and of magnitude 300 μC. What if the gap were filled with mica?

27 (II) A parallel-plate capacitor has a fixed charge Q. The separation of the plates is then doubled. By what factor does the energy stored in the electric field change?

SECTION 14-4

28 (I) What is the electric potential 10 cm from a 4.2-μC point change?

29 (I) What is the electric potential 1.0×10^{-10} m from a proton (charge $+e$)?

30 (I) What is the potential energy of an electron that is 1.0×10^{-10} m from a proton?

31 (II) A $+20$-μC charge is placed 50 cm from an identical $+20$-μC charge. How much work would be required to move a $+0.20$-μC test charge from a point midway between them to a point 10 cm closer to either of the charges?

32 (II) How much voltage must be used to accelerate a proton so that it has sufficient energy to penetrate an iron nucleus? An iron nucleus has a charge of $+26e$ and its radius is about 4×10^{-15} m.

33 (III) A 30-μC and a -2.0-μC charge are placed 2.0 cm apart. At what point along the line joining them is (*a*) the electric field zero, and (*b*) the potential zero?

34 (III) How much work must be done to bring three electrons from a great distance apart to within 1.0×10^{-10} m from one another?

*SECTION 14-5

*35 (I) An electron and a proton are 0.50×10^{-10} m apart. What is their dipole moment if they are at rest?

*36 (II) Calculate the electric potential due to a dipole whose dipole moment is 4.8×10^{-30} C · m at a point 1.0×10^{-9} m away if this point is (*a*) along the axis of the dipole nearer the positive charge, (*b*) 45° above the axis but nearer the positive charge, and (*c*) 45° above the axis but nearer the negative charge.

*SECTION 14-6

*37 (III) (*a*) Show that if two dipoles with dipole moments p_1 and p_2 are in line with one another (Figure 14-12) the potential energy of one in the presence of the other (their "interaction energy") is given by

$$\text{PE} = -\frac{2kp_1p_2}{r^3},$$

where r is the distance between the two dipoles. Assume that r is much greater than the length of either dipole. (*b*) Use this expression to calculate the interaction energy of the two dipoles of Example 14-8.

FIGURE 14-12

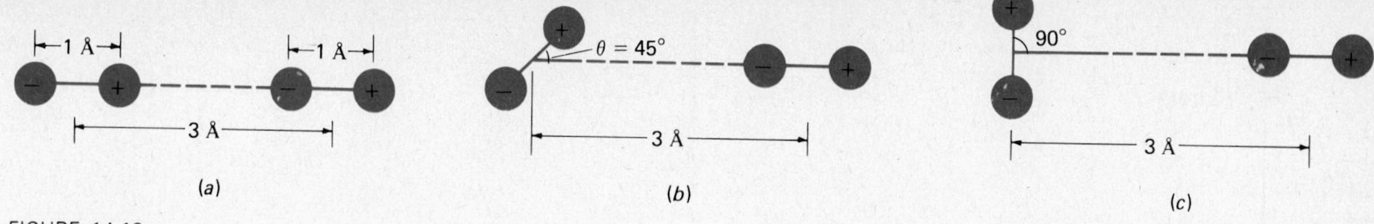

FIGURE 14-13

*38 (III) Two dipoles interact most strongly when they are directly lined up. To see that this is so, calculate the interaction energy for the two dipoles shown in Figure 14-13 (a) when they are aligned, (b) when the angle between their axes is 45°, and (c) when the angle is 90°. Assume that the distance from the center of one dipole to the other is 3.0 Å in each case and that the charge on each is $0.40e$.

*39 (III) Estimate roughly, using the result of Problem 33 in Chapter 13, the electrostatic interaction energy for the two "bonds" formed by the calcium ion between the ATP in the myosin and the negative charge on the actin. Assume that the negative charges are each $-1e$ and that the charge on the calcium is $+2e$. (See Figure 13-30.)

For additional reading

Ackerman, E., *Biophysical Science*, Prentice-Hall, Inc., Englewood Cliffs, N.J., 1962. Electrocardiogram.

Einstein, A., and L. Infeld, *Evolution of Physics*, Simon & Schuster, Inc., New York, 1938.

Setlow, R. B., and E. C. Pollard, *Molecular Biophysics*, Addison-Wesley Publishing Co., Inc., Reading, Mass., 1962.

Stanford, A. L., Jr., *Foundations of Biophysics*, Academic Press, Inc., New York, 1975. Dipoles, electrocardiogram.

15 ▌ Electric currents

Until the year 1800, the technical development of electricity consisted mainly of producing a static charge by friction. Beginning in the late 1600s a number of machines were built that could produce rather large amounts of charge by frictional means; one example is shown in Figure 15-1. Large sparks could be produced by these machines, but they had little practical value.

In nature itself there were grander displays of electricity, such as lightning and "St. Elmo's fire"—a glow that appeared around the yardarms of ships during storms. That these phenomena were electrical in origin was not recognized, however, until the eighteenth century. For example, it was only in 1752 that Franklin, in his famous kite experiment, showed that lightning was an electrical discharge—a giant electric spark.

FIGURE 15-1
An electrostatic machine. A sulfur ball rotated at high speed acquired a charge when another object was held against it. (Burndy Library, Norwalk, Conn.)

Finally, in 1800, an event of great practical importance occurred. In this year Alessandro Volta (1745–1827) invented the electric battery and with it produced the first steady flow of electric charge—that is, a steady electric current. This discovery opened a new era; for today's electrical technology is based on electric current.

15-1 ▌ The electric battery

The events that led to the discovery of the battery are interesting; for not only was this an important discovery, but it gave rise to a famous scientific debate. The importance of the battery is that it can produce a steady electric current that has many practical uses; and the discovery of electric current is generally attributed to the two men who were involved in the famous controversy, Volta and Luigi Galvani (1737–1798).

The story begins at the University of Bologna, thought to be the world's oldest university still in existence. As long ago as the thirteenth century it had 10,000 students and many women professors.† At this

† A thirteenth-century historian reported that one of the women professors, Novella d'Andrea, was so beautiful that she had to deliver her lectures from behind a curtain to keep from distracting the students.

university, in the year 1780, Luigi Galvani was a professor of anatomy. One day, while he was dissecting a frog with his wife's help, they made an astonishing observation. When a knife was in contact with the frog's nerve center and at the same time a spark was produced by a nearby electrical machine, the muscles of the frog's legs contracted violently as if in a convulsion.

Galvani proceeded to investigate this effect in detail. He was able to produce contraction of the muscles by other means as well. One of the most curious methods was discovered quite by accident (as, it seems, many important discoveries are): when a copper or brass hook was pressed into the frog's flesh and then hung from an iron trellis which also touched the frog, again the leg muscles would twitch. With further investigation Galvani found that it worked for other pairs of metals as well.

Galvani believed that the contractions were somehow due to electricity. Two explanations seemed possible: first, that the source of the electric charge was in the frog muscle or nerve itself and that the wire merely transmitted the charge to the proper points; or second, that the source of the electricity was instead in the contact of the two different metals and the muscles merely responded to the stimulus. Galvani chose the first—that the source was within the animal itself, and in 1791 he published his work which he termed "animal electricity." Many wondered, including Galvani himself, if he had discovered the long-sought "life force." Was electricity perhaps the secret of life?

Volta, at the University of Pavia 200 km away, was one of many scientists who hurried to confirm Galvani's work and was at first a believer in Galvani's theory. But soon Volta began to doubt Galvani's idea of animal electricity and he came to believe that the source of the electricity was not within the animal but rather in the contact between the two metals. One of the experiments Volta tried is illustrated in Figure 15-2*a*. When two dissimilar metals were placed in contact and the opposite ends were touched to the tongue as shown, the person felt a tingling sensation on his tongue. In another experiment in which two dissimilar metal strips were placed in contact, Volta put one of the free ends on his tongue and the other on the skin near his eye (Figure 15-2*b*); when he did this, he sensed a flash of light in his eye. These and other experiments led him to believe that it was the contact of dissimilar metals that caused Galvani's frog muscle to twitch, and not only could this stimulate muscular contraction, but the electricity could also stimulate the senses.

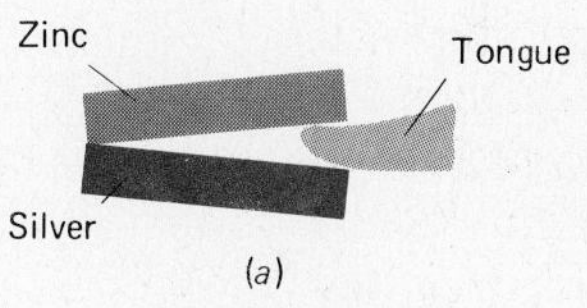

(*a*)

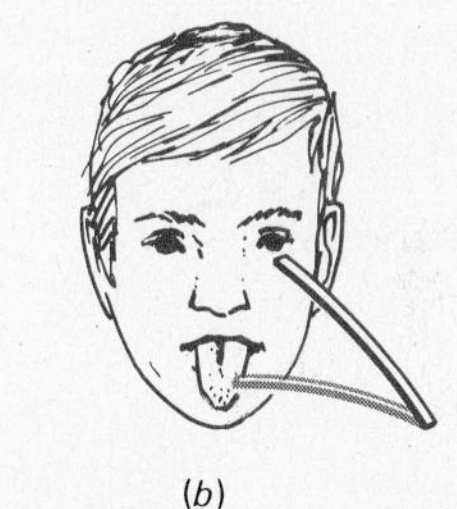
(*b*)

FIGURE 15-2
How Volta tingled his tongue and flashed his eye.

Volta made public his views and soon he had many followers. But many others sided with Galvani and claimed that the source was within the animal. Indeed, scientists from all over Europe took sides in this controversy. Time, we must point out, has sided with Volta.

Volta pursued his experiments and found that moisture was necessary at the contact point of the two dissimilar metals. He also found that certain combinations of metals produced a greater effect than others, and he listed them in order of effectiveness. (This "electrochemical series" is still used by chemists today.) And he found that carbon could be used in place of one of the metals.

Volta then conceived what has come to be his greatest contribution to science. Between a disk of zinc and one of silver, he placed a piece of cloth or paper soaked in salt solution or dilute acid. He then piled a "battery" of

(a)

(b)

FIGURE 15-3
Two types of voltaic battery: (*a*) a pile; (*b*) "crown of cups." Z stands for zinc and A for silver (*argentum* in Latin). Taken from Volta's original publication. (Burndy Library, Norwalk, Conn.)

such couplings one on top of another as shown in Figure 15-3, with the intention of producing a greater electrical effect. Indeed, when strips of metal connected to the two ends of the pile were brought close, a spark was produced. Volta had designed and built the first electric battery! A second design, known as the "crown of cups," is also shown in Figure 15-3.

Volta made public his great discovery in 1800. He had shown that dissimilar metals produced electricity without a frog. He considered this proof of his own theory over Galvani's.†

The charge produced by Volta's battery was weak compared to that produced by the best friction machines of the time. But it had one great advantage. It was "self-renewing"—it could produce electricity continuously. It was not long before even more powerful batteries were constructed.

After Volta's discovery of the electric battery, it was eventually recognized that a battery produces electricity by chemical action. Today a great variety of electric cells and batteries are available, from flashlight batteries (sometimes called "dry cells") to the storage battery of a car, which can be recharged. The simplest batteries contain two plates or rods made of dissimilar metals (one can be carbon), called *electrodes*, and the electrodes are immersed in a solution, such as a dilute acid, called the *electrolyte*. Such a device is properly called an *electric cell*, and several cells connected together is called a battery. The chemical reactions involved in most electric cells are quite complicated. A detailed description of different electric cells can be found in chemistry textbooks. Here we describe how one very simple cell works.

+ terminal
− terminal
Carbon electrode (anode)
Zinc electrode (cathode)
Acid

FIGURE 15-4
Simple electric cell.

The simple electric cell shown in Figure 15-4 uses dilute sulfuric acid as the electrolyte. One of the electrodes is made of carbon, the other of zinc. (In a typical flashlight battery, one electrode is carbon and the outer casing is the zinc electrode; the electrolyte is absorbed in a powdery material such as sawdust.) Those parts of the electrodes that stick out of the solution are called the *terminals*. The acid attacks the zinc electrode

† Galvani was not, however, completely wrong. For a century later it was shown that a beating heart produces tiny electric impulses. And we will see later in this chapter how nerves and other cells do produce a potential difference across their membranes. Galvani's extensive work on muscle contraction and electricity is considered the beginning of neurophysiology. Volta's explanation dealt with muscle contraction due to contact of dissimilar metals. But the explanation of Galvani's very first (1780) experiment, in which the frog's legs twitched when a spark was produced by an electrical machine was not found for nearly a century; contractions in that case were apparently due to transmission of an electromagnetic wave. Although he was not aware of it, Galvani was the first to detect a "man-made" electromagnetic wave.

and tends to dissolve it. But each zinc atom leaves two electrons behind, so it enters the solution as a positive ion. The zinc electrode thus acquires a negative charge. As more zinc ions enter solution, the electrolyte becomes increasingly positively charged. Because of this, and through other chemical reactions, electrons are pulled off the carbon electrode. Thus the carbon electrode becomes positively charged. The positive electrode is called the *anode*, and the negative electrode the *cathode*. Because there is an opposite charge on the two electrodes, there is a potential difference between the two terminals. In a cell whose terminals are not connected, only a small amount of the zinc is dissolved; for as the zinc electrode becomes increasingly negative, any new positive zinc ions produced are attracted back to the electrode. Thus a particular potential difference or voltage is produced between the two terminals. If charge is allowed to flow between the terminals, say through a wire or a light bulb, then more zinc can be dissolved. The carbon, too, suffers disintegration. After a time, one or the other electrode is used up and the cell becomes "dead."

The voltage that exists between the terminals of a battery depends on what the electrodes are made of and their relative ability to be dissolved or give up electrons. This potential difference is called the **electromotive force**, or **emf**, of the battery. The symbol $\mathscr{E}$ is often used for emf (do not confuse it with E for electric field), although sometimes V is used, as for any voltage (but see "Emf and terminal voltage" in Section 16-1). The emf of typical cells is 1.0 to 2.0 V. When two or more cells are connected so that the positive terminal of one is connected to the negative terminal of the next, they are said to be connected in *series* and their voltages add up. Thus the voltage between the ends of the two flashlight batteries in Figure 15-5*a* is 3 V while the six 2-V cells of the storage battery in part *b* give 12 V.

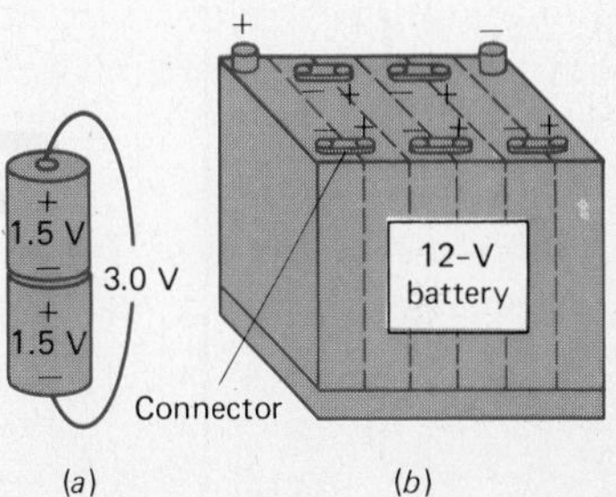

FIGURE 15-5
(*a*) Two 1.5-V flashlight batteries connected in series produces 3.0 V. (*b*) The six 2-V cells of an automobile storage battery are connected in series to produce 12 V.

15-2 ▌ Electric current and Ohm's law

Electric current

When a continuous conducting path, such as a wire, is connected to the terminals of a battery, we have an electric *circuit* (Figure 15-6*a*). When drawing a circuit (Figure 15-6*b*), we will represent a battery by the symbol "⊣⊢" rather than draw in the shape of a particular battery; the longer line on this symbol represents the positive terminal and the shorter line the negative terminal (you can remember this by noting that to make a + sign you can break the long line in half and cross the two pieces). When such a circuit is formed, charge can flow through the circuit from one terminal of the battery to the other. A flow of charge such as this is called an **electric current**.

FIGURE 15-6
(*a*) Very simple electric circuit. (*b*) Schematic drawing of the circuit in part *a*.

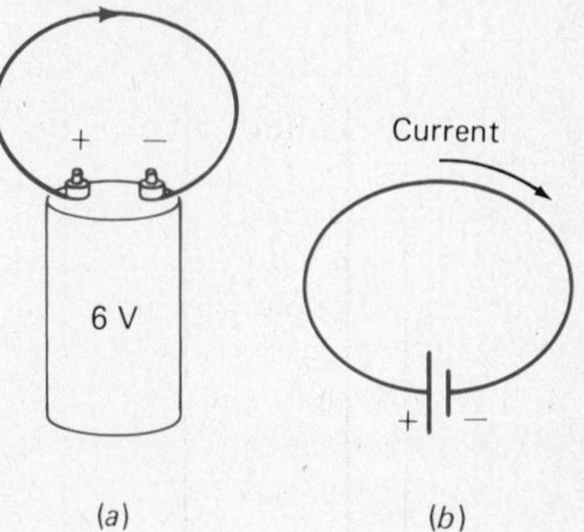

More precisely, the electric current is defined as the amount of charge that passes a given point in a circuit per unit time. Thus the average current I is defined as

$$I = \frac{Q}{t}, \tag{15-1}$$

where Q is the amount of charge that passes a given point during the time t. Electric current is measured in coulombs per second; this is given a special name, the *ampere* (A), after the French physicist André Ampère (1775–1836). Thus 1 A = 1 C/s.

▮ EXAMPLE 15-1 A steady current of 2.5 A flows in a wire connected to a battery. But after 4.0 min, the current suddenly ceases because the wire is disconnected. How much charge passed through the circuit?

SOLUTION Since the current was 2.5 A, or 2.5 C/s, then in 4.0 min (240 s) the total charge that flowed was, from Equation 15-1,

$$Q = It$$
$$= (2.5\ \text{C/s})(240\ \text{s}) = 600\ \text{C}.$$ ▮▮

We saw earlier that electrical conductors contain many free electrons; thus, when a conducting wire is connected to the terminals of a battery as in Figure 15-6, it is actually the negatively charged electrons that flow in the wire. When the wire is first connected, free electrons at one end of the wire are attracted into the positive terminal. At the same time, electrons leave the negative terminal of the battery and enter the wire at the other end. Thus there is a continuous flow of electrons through the wire that begins as soon as the wire is connected to *both* terminals. However, when the conventions of positive and negative charge were invented two centuries ago, it was assumed that positive charge flowed in a wire. Actually, for nearly all purposes, positive charge flowing in one direction is exactly equivalent to negative charge flowing in the opposite direction (Figure 15-7). Today, we still use the historical convention of positive current flow when discussing the direction of a current. (It also fits in well with our definition of electrical potential in terms of a positive test charge.) Thus when we speak of the current flowing in a circuit, we normally mean the direction positive charge would flow. This is sometimes referred to as *conventional current.* When we want to speak of the direction of electron flow (in a wire, say) we will specifically state that it is the electron current. In liquids and gases, both positive and negative charges (ions) move.

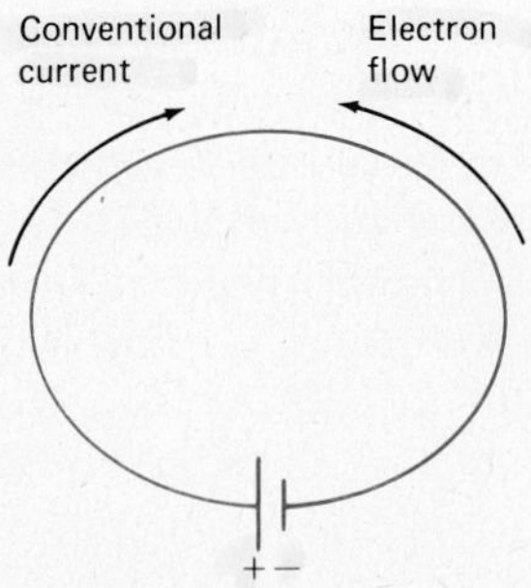

Conventional current

Electron flow

\+ −

FIGURE 15-7
Conventional current from + to − is equivalent to a negative (electron) current flowing from − to +.

Sparks and lightning

Although the discovery of electric current is often attributed to Galvani or Volta, the sparks produced by the early friction machines—and which you sometimes feel on a dry day—were actually produced by electric currents passing through the air. A modern example of a spark is that which occurs across the electrodes of the spark plug to ignite the gas–air mixture in a car; and lightning is a giant "spark" between two clouds of opposite charge, or between a cloud and the earth. In each case, a large potential difference is needed; if it is great enough, the electric field actually pulls the electrons off atoms of the air. Normally, air is a good insulator; but when the atoms become ionized, the air becomes conducting. Electrons and ions flow rapidly across the gap, colliding with atoms and causing more ionization, but the difference in potential is quickly reduced and the "discharge" stops. We see a spark, or bright flash because electric energy is transformed to light energy as the electrons recombine with their atoms.

Ohm's law

To produce an electric current in a circuit, a difference in potential is required. One way of producing a potential difference is by a battery. It was Georg Simon Ohm (1787–1854) who established the experimental fact that the current in a wire is proportional to the potential difference V applied to its ends:

$$I \propto V.$$

For example, if we connect a wire to a 6-V battery, twice the current will flow in it than if we connect it to a 3-V battery.

It is helpful to compare an electric current to the flow of water in a river or a pipe. If the pipe (or river) is nearly level, the flow rate is small. But if one end is somewhat higher than the other, the flow rate—or current—is much greater. The greater the difference in height, the greater the current. We saw in Chapter 14 that electric potential is analogous, in the gravitational case, to the height of a cliff; and this applies in the present case to the height through which the fluid flows. Just as an increase in height causes a greater flow of water, so a greater electric potential difference causes a greater current flow.

Exactly how much current flows in a wire depends not only on the potential difference but also on the resistance the wire offers to the flow of electrons. The walls of a pipe, or the banks of a river and rocks in the middle, offer resistance to the flow of current. Similarly, electrons are slowed down because of interactions with the atoms of the wire. The higher this resistance, the less the current for a given voltage V. So we define resistance so that the current is inversely proportional to it. When we combine this with the proportion above, we have

$$I = \frac{V}{R}; \qquad (15\text{-}2)$$

R is the *resistance* of some object or device, V the potential difference across the object, and I the current that flows through it. This is known as **Ohm's law** and is often written

$$V = IR.$$

It is one of the basic laws of electricity. The unit for resistance is called the *ohm* and is abbreviated Ω (Greek capital omega). Because $R = V/I$, we see that $1\ \Omega$ is equivalent to 1 V/A.

▮ EXAMPLE 15-2 A plate on the bottom of an electric toaster specifies that it should be connected to 120 V and will draw 4 A. What is the resistance of the toaster?

SOLUTION From Ohm's law we have

$$R = \frac{V}{I} = \frac{120\ \text{V}}{4\ \text{A}} = 30\ \Omega. \qquad ▮▮$$

Resistors

All electric devices, from heaters to light bulbs to stereo amplifiers, offer resistance to the flow of current. Generally, the connecting wires have

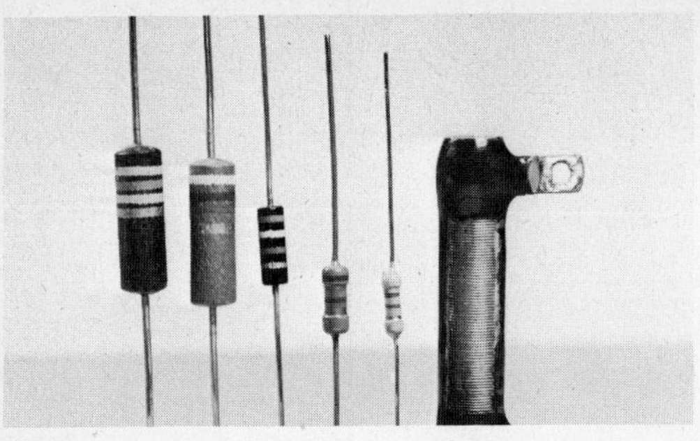

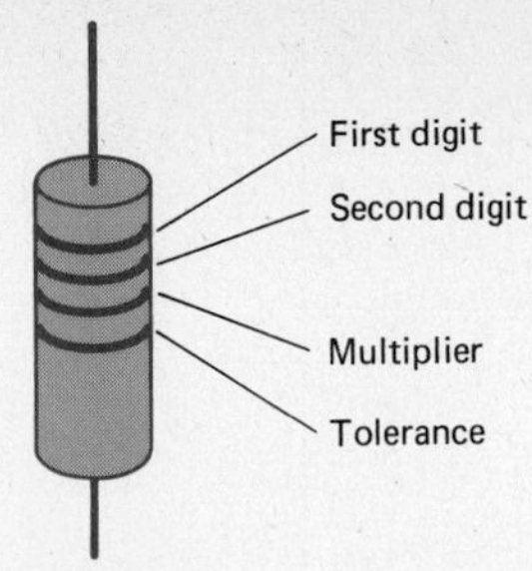

Resistor color code

Color	Number	Multiplier	Tolerance (%)
Black	0	1	
Brown	1	10^1	
Red	2	10^2	
Orange	3	10^3	
Yellow	4	10^4	
Green	5	10^5	
Blue	6	10^6	
Violet	7	10^7	
Gray	8	10^8	
White	9	10^9	
Gold		10^{-1}	5%
Silver		10^{-2}	10%
No color			20%

FIGURE 15-8
Some resistors. (Photo by Lester V. Bergman & Assoc., Inc.)

very low resistance. In many circuits, particularly in electric devices, *resistors* are used to control the amount of current. Resistors have resistances from less than an ohm to millions of ohms. Several kinds are shown in Figure 15-8. The two main types are "wire-wound" resistors, which consist of a coil of fine wire, and "composition" resistors, which are usually made of the semiconductor carbon. The value of a given resistor is written on the exterior or may be given a color code as shown in the figure. The first two colors represent the first two digits in the value of the resistance, the third represents the power of 10 that it must be multiplied by, and the fourth is the manufactured tolerance. For example, a resistor whose four colors are red, green, orange, and silver has a resistance of 25,000 Ω (25 kΩ), give or take 10 percent.

When we draw a diagram of a circuit, we will indicate a resistance with the symbol —⁄\/\/—. Wires whose resistance is negligible, however, will be shown simply as straight lines.

15-3 ▌ Resistivity and superconductivity

Resistivity and its temperature dependence

It is found experimentally that the resistance R of a wire is directly proportional to its length L and inversely proportional to the cross-sectional area A. That is,

$$R = \rho \frac{L}{A}, \tag{15-3}$$

where ρ, the constant of proportionality, is called the *resistivity* and depends on the material used. This relation makes sense since you would expect that the resistance of a thick wire would be less than that of a thin one, since a thicker wire has more area for the electrons to pass through; and you might expect the resistance to be greater if the length is greater, since there would be more obstacles to obstruct electron flow.

Typical values of ρ for various materials are given in the first column of Table 15-1. (The values depend somewhat on purity, heat treatment, temperature, etc.) Notice that silver has the lowest resistivity and is thus the best conductor; but it is expensive. Copper is not far behind, so it is clear why most wires are made of copper. Aluminum, although it has a higher resistivity, is much less dense than copper (see Table 6-1); it is preferable to copper in some situations, such as transmission lines, because its resistance for the same weight is less than for copper.

TABLE 15-1

Resistivity and temperature coefficients (at 0°C)

Material	Resistivity, ρ $(\Omega \cdot m)$	Temperature coefficient, α $(C°)^{-1}$
Conductors		
Silver	1.6×10^{-8}	0.0041
Copper	1.7×10^{-8}	0.0068
Aluminum	2.7×10^{-8}	0.0043
Tungsten	5.6×10^{-8}	0.0045
Iron	10×10^{-8}	0.0065
Platinum	11×10^{-8}	0.003927
Mercury	98×10^{-8}	0.0009
Nichrome (alloy of Ni, Fe, Cr)	100×10^{-8}	0.0004
Semiconductors		
Carbon	3.5×10^{-5}	−0.0005
Germanium	0.46	−0.05
Silicon	2300†	−0.07
Insulators		
Glass	10^{10}–10^{14}	
Hard rubber	10^{13}–10^{16}	

† Theoretical value; practical values of ρ are much less because of impurities, usually less than about 20 $\Omega \cdot m$, as essentially pure Si has proved difficult to obtain.

▌ EXAMPLE 15-3 Suppose that you want to connect your stereo set to a remote speaker. If each wire must be 20 m long, what diameter of copper wire should you use to keep the resistance less than 0.10 Ω per wire?

SOLUTION We solve Equation 15-3 for A and use Table 15-1:

$$A = \rho \frac{L}{R} = \frac{(1.7 \times 10^{-8}\ \Omega \cdot \text{m})(20\ \text{m})}{0.10\ \Omega} = 3.4 \times 10^{-6}\ \text{m}^2.$$

The cross-sectional area A of a circular wire is related to its diameter d by $A = \pi d^2/4$. The diameter must then be at least $d = \sqrt{4A/\pi} = 2.1 \times 10^{-3}\ \text{m} = 2.1\ \text{mm}$. ▌▌

The resistivity of a material depends somewhat on temperature. In general, the resistance of metals increases with temperature. This is not surprising, for at high temperatures the atoms are moving more rapidly and are arranged in a less orderly fashion. So they might be expected to interfere more with the flow of electrons. If the temperature change is not too great, the resistivity of metals increases nearly linearly with temperature. That is,

$$\rho_T = \rho_0(1 + \alpha T), \tag{15-4}$$

where ρ_0 is the resistivity at 0°C, ρ_T the resistivity at a temperature T (in °C), and α the *temperature coefficient of resistance*. Values for α are given in Table 15-1. Note that the temperature coefficients of semiconductors are negative; the resistance of a semiconductor *decreases* with an increase in temperature. This is apparently because at higher temperatures some of the electrons that are not normally free become free and can contribute to the current.

The resistance thermometer

The variation in electrical resistance with temperature can be used to make precise temperature measurements. Platinum is usually used since it is relatively free from corrosive effects and has a high melting point.

▮ EXAMPLE 15-4 At 0°C the resistance of a platinum resistance thermometer is 164.2 Ω. When placed in the tissue of a particular animal, the resistance is 187.4 Ω. What is the temperature of this tissue?

SOLUTION Since the resistance R is directly proportional to the resistivity ρ, we can combine Equation 15-3 with Equation 15-4 and write

$$R = R_0(1+\alpha T).$$

Here $R_0 = \rho_0 L/A$ is the resistance of the wire at 0°C. We solve this equation for T and find

$$T = \frac{R - R_0}{\alpha R_0} = \frac{187.4\ \Omega - 164.2\ \Omega}{(3.927\times 10^{-3}\ (\mathrm{C}°)^{-1})(164.2\ \Omega)} = 35.9°\mathrm{C}.$$ ▮▮

At very high or low temperatures, Equation 15-4 is not adequate and terms proportional to the square and cube of the temperature are needed: $\rho_T = \rho_0(1+\alpha T+\beta T^2+\gamma T^3)$; the coefficients β and γ are generally very small, but when T is large, their terms become significant.

More convenient than a resistance thermometer for some applications is a *thermistor*, which consists of a metal oxide or semiconductor whose resistance also varies in a repeatable way with temperature. Thermistors can be made quite small and respond very quickly to temperature changes.

Superconductors

The resistivity of certain metals and their compounds or alloys becomes zero (unmeasurably small) at very low temperatures. Materials in such a state are said to be *superconducting*. This phenomenon was first observed by H. K. Onnes (1853–1926) in 1911 when he cooled mercury below 4.2 K. In general, superconductors become superconducting only below a certain *transition temperature*, which is usually within a few degrees of absolute zero (as high as about 20 K). Currents in ring-shaped superconducting materials have been observed to flow for years in the absence of a potential difference, with no apparent decrease.

Much research has been done on superconductivity in recent years to try to understand why it occurs and to find materials that superconduct at higher temperatures to reduce the cost and inconvenience of refrigeration. The first successful superconductivity theory was published in 1957 by Bardeen, Cooper, and Schrieffer (the BCS theory).

Superconductors are already being used in large magnets (we shall see in Chapter 17 that electric currents produce magnetic fields), which sharply reduces the electric power needs. Energy is needed, of course, to refrigerate the superconducting material at a low temperature. Superconductors are being planned for use in large generating plants and for large motors where the negligible resistance will allow very large currents to be used. Also being developed are superconducting transmission cables that will reduce energy loss tremendously and will allow electric generating plants to be placed close to fuel sources (coal, gas, etc.) rather than close to population centers, and thus reduce fuel transportation costs. Superconductors may also be used in high-speed ground transportation: the magnetic fields produced by superconducting magnets would be used to "levitate" vehicles over the tracks so that there is essentially no friction; the levitation arises from the repulsive force between the magnet (on the train) and the eddy currents produced in the track below.

15-4 ▌ Electric power

Electric energy is useful to us because it can easily be transformed into other forms of energy. Motors, whose operation we will examine in Chapter 17, transform electric energy into mechanical work.

In other devices, such as electric heaters, stoves, toasters, and hair dryers (Figure 15-9), electric energy is transformed into heat in a wire resistance known as a "heating element." In an ordinary light bulb, the tiny wire filament (Figure 15-10) becomes so hot it glows; only a few percent of the energy is transformed into light and the rest, over 90 percent, into heat. Light-bulb filaments and heating elements in household appliances have a resistance typically of a few ohms to a few hundred ohms.

Electric energy is transformed into heat (or light) in such devices because the current is usually rather large, and there are many collisions between the moving electrons and the atoms of the wire. In each collision, part of the electron's kinetic energy is transferred to the atom with which it collides. As a result, the kinetic energy of the atoms increases and hence the temperature of the wire element increases. The increased thermal energy can be transferred as heat by conduction and convection to the air in a heater or food in a pan, by radiation to toast in a toaster, or radiated as visible light.

To find the power transformed by an electric device we use the fact that the energy transformed when a charge Q moves through a potential difference V is QV (Equation 14-1). Then the power P, which is the rate energy is transformed, is

$$P = \text{power} = \frac{\text{energy transformed}}{\text{time}} = \frac{QV}{t}.$$

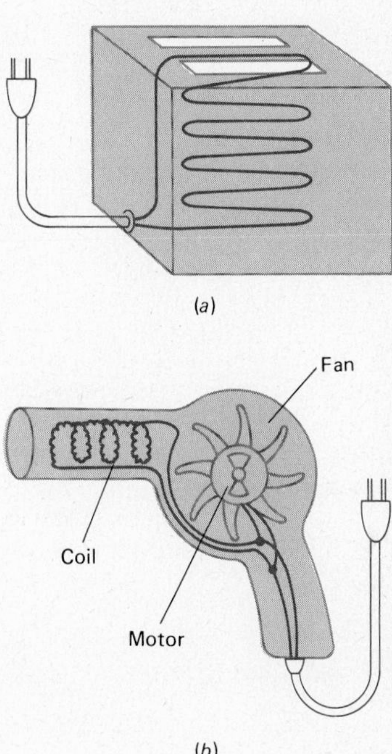

FIGURE 15-9
Electric toaster (*a*) and hair dryer (*b*), showing heating elements (resistors) in each.

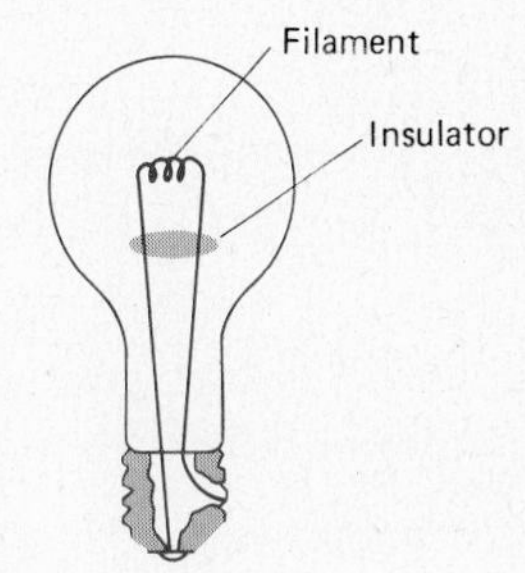

FIGURE 15-10
Incandescent light bulb.

The charge that flows per second, Q/t, is simply the electric current I. Thus we have

$$P = IV. \tag{15-5}$$

This general relation gives us the power transformed by any device, where I is the current passing through it and V is the potential difference across it. It also gives the power delivered by a source such as a battery. The SI unit of electric power, the *watt*, is the same as for any kind of power.

The rate of energy transformation in a resistance R can be written, using Ohm's law ($V = IR$), in two other ways:

$$P = IV \tag{15-5a}$$

$$= I(IR) = I^2R \tag{15-5b}$$

$$= \left(\frac{V}{R}\right)V = \frac{V^2}{R}. \tag{15-5c}$$

▌ EXAMPLE 15-5 Calculate the resistance of a 60-W light bulb designed for 120 V (average).

SOLUTION Since $P = 60$ W and $V = 120$ V, we use Equation 15-5c and solve for R:

$$R = \frac{V^2}{P} = \frac{(120\ \text{V})^2}{60\ \text{W}} = 240\ \Omega.$$

This is the resistance when the bulb is burning brightly at 60 W; when the bulb is cold, the resistance is much lower. ▌▌

It is energy, not power, you must pay for on your electric bill. Since power is the *rate* energy is transformed, the total energy used by any device is simply its power consumption multiplied by the time it is on, given in seconds. This gives the energy in joules since $1\ \text{W} = 1\ \text{J/s}$. Electric companies usually specify the energy with a much larger unit, the kilowatt-hour (kWh). One kWh $= (1000\ \text{W})\ (3600\ \text{s}) = 3.6 \times 10^6\ \text{J}$.

▌ EXAMPLE 15-6 An electric heater draws 15 A on a 120-V line. How much power does it use and how much does it cost per month (30 days) if it operates 3.0 h per day and the electric company charges \$0.060 per k Wh?

SOLUTION The power is $P = IV = (15\ \text{A})(120\ \text{V}) = 1800\ \text{W}$ or 1.8 kW. To operate it for 90 h would cost (1.8 kW) (90 h)(\$0.060) = \$9.72. ▌▌

The electric wires that carry electricity to lights and other electric appliances have some resistance, although usually it is quite small. Nonetheless, if the current is large enough, the wires will heat up and produce heat equal to I^2R, where R is the wire's resistance. One possible hazard is that the current-carrying wires in the wall of a building may become so hot that they will start a fire. Thicker wires have less resistance (see Equation 15-3) and thus can carry more current without becoming too hot. When a wire carries more current than is safe, it is said to be "overloaded." To prevent overloading, fuses or circuit breakers are installed in circuits. They are basically switches (Figure 15-11) that open

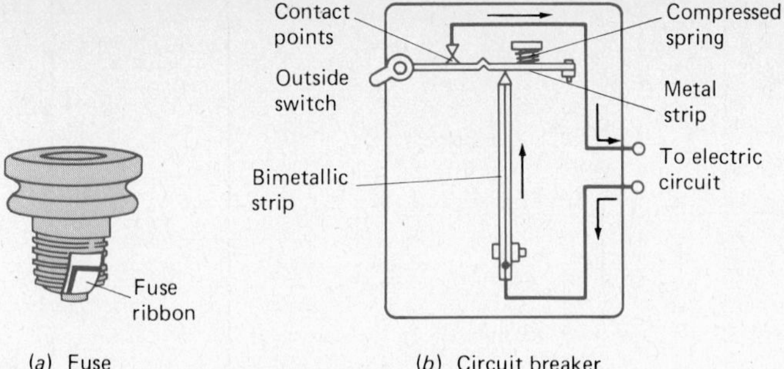

FIGURE 15-11
(a) A fuse. When the current exceeds a certain value, the ribbon melts and the circuit opens. Then the fuse must be replaced. (b) A circuit breaker. Electric current passes through a bimetallic strip. When the current is great enough, the bimetallic strip heats sufficiently to bend so far to the left that the notch in the spring-loaded metal strip drops down over the end of the bimetallic strip. The circuit then opens at the contact points (one is attached to the metal strip) and flips the outside switch. As the device cools down, it can be reset using the outside switch.

the circuit when the current exceeds a particular value. A 20-A fuse or circuit breaker, for example, opens when the current passing through it exceeds 20 A.

If a circuit in your house repeatedly burns out a fuse or opens a circuit breaker, there are two possibilities: (*a*) you have too many appliances drawing current in that circuit, or (*b*) there is a fault somewhere, such as a "short." A short, or "short circuit," means that two wires have crossed, perhaps because the insulation has worn down; so the path of the current is shortened. The resistance of the circuit is then very small, so the current will be very large. Short circuits should, of course, be remedied immediately.

Household circuits are designed so that the various devices are connected as shown in Figure 15-12; each receives the 120 V or so from the electric company. When a fuse blows or circuit breaker opens, the total current being drawn should be checked first. For example, the circuit in Figure 15-12 draws the following currents: the light bulb draws $I = P/V =$ 100 W/120 V = 0.8 A, the heater 1800 W/120 V = 15 A, and the frying pan 1300 W/120 V = 10.8 A, for a total of 26.6 A. If the circuit has a 20-A fuse, no wonder it blows! If it has a 30-A fuse, it should not blow, so we would look for a short. The most likely place is in the cord of one of the appliances. These types of considerations must also be used when designing a building. For example, the diameter of all electrical wires must be great enough to carry the electric currents that will be needed.

In electric circuits, heat dissipation by resistors must be considered. The physical size of a resistor is a rough indicator of the maximum permissible power it can dissipate ($=I^2R$) without an appreciable rise in temperature. Common values are $\frac{1}{4}$ W, $\frac{1}{2}$ W, and 1 W; the higher the wattage, the larger the size.

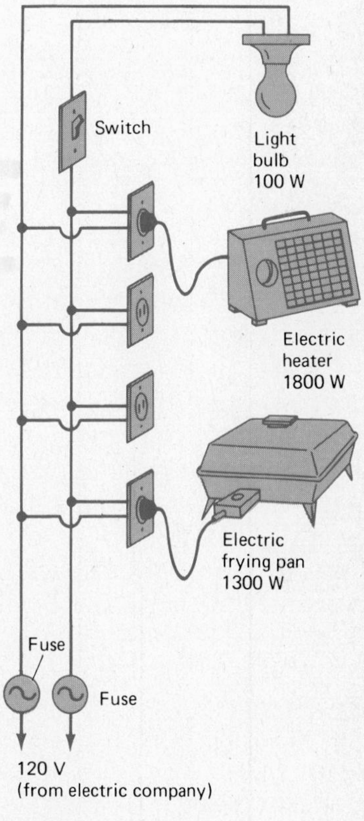

Switch
Light bulb 100 W
Electric heater 1800 W
Electric frying pan 1300 W
Fuse
Fuse
120 V (from electric company)

FIGURE 15-12
Connection of household appliances.

15-5 ▌ Alternating current

When a battery is connected to a circuit, the current flows steadily in one direction. This is called a *direct current* (dc). Electric generators at electric power plants, however, produce *alternating current* (ac).† An alternating current reverses direction many times per second. The electrons in a wire

† Although it is redundant, we sometimes say "ac current" or "ac voltage," which really means "alternating current" or "alternating voltage."

first move in one direction and then in the other. See Figure 15-13. The current supplied to homes and businesses by electric companies is ac in nearly all areas of the United States and in many other countries.

The emf produced by an ac electric generator is sinusoidal, as we shall see in Chapter 18. The current it produces is thus sinusoidal (Figure 15-13*b*). We can write the voltage as a function of time:

$$V = V_0 \sin 2\pi ft.$$

This is just like the equation for simple harmonic motion and for waves which we discussed in Chapter 11. The potential V oscillates between $+V_0$ and $-V_0$; V_0 is referred to as the *peak voltage.* The frequency f refers to how many complete oscillations are made per second. In most areas of the United States and Canada, f is 60 Hz (cycles per second). In some countries 50 Hz is used.

From Ohm's law, if an alternating voltage $V = V_0 \sin 2\pi ft$ exists across a resistance R, the current I is

$$I = \frac{V}{R} = \frac{V_0}{R} \sin 2\pi ft = I_0 \sin 2\pi ft. \tag{15-6}$$

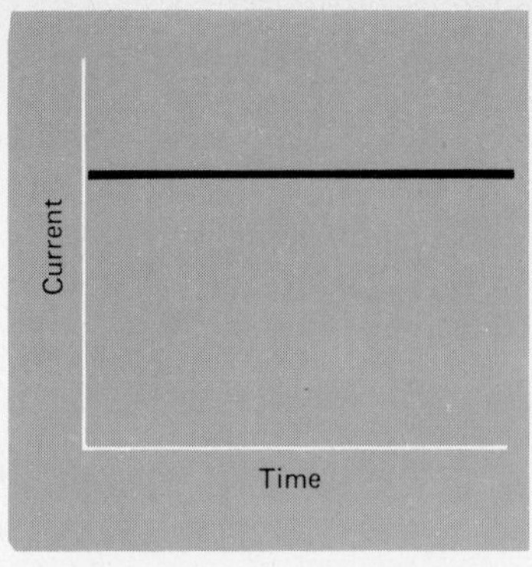

(*a*) dc

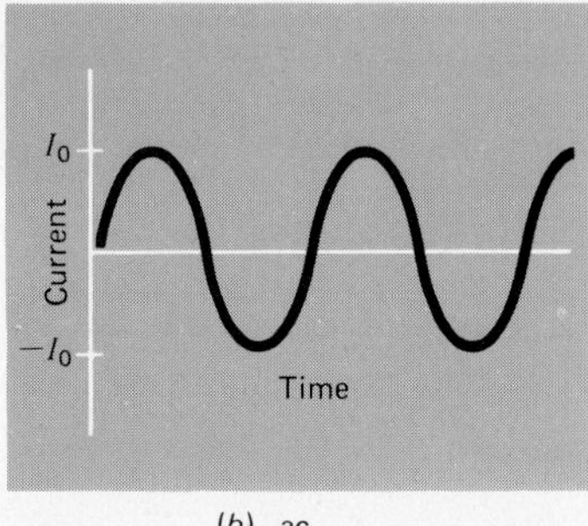

(*b*) ac

FIGURE 15-13
(*a*) Direct current. (*b*) Alternating current.

The quantity $I_0 = V_0/R$ is the *peak* current. The current is considered positive when the electrons flow in one direction and negative when they flow in the opposite direction. It is clear from Figure 15-13*b* that an alternating current is positive as often as it is negative. Thus the average current is zero. This does not mean, however, that no power is needed or that no heat is produced in a resistor. Electrons do move back and forth in a resistor and do produce heat. Indeed, the power delivered to a resistance R at any instant is

$$P = I^2R = I_0^2 R \sin^2 2\pi ft.$$

Because the current is squared, we see that the power is always positive. This is plotted in Figure 15-14. The quantity $\sin^2 2\pi ft$ varies between 0 and 1; and it is not too difficult to show that its average value is $\frac{1}{2}$. Thus the *average power* developed, $\bar{P}$, is

$$\bar{P} = \tfrac{1}{2} I_0^2 R.$$

Since power can also be written $P = V^2/R = (V_0^2/R)\sin^2 2\pi ft$, we can similarly show that the average power is

$$\bar{P} = \frac{1}{2}\frac{V_0^2}{R}.$$

The average value of the *square* of the current or voltage is thus what is important for calculating average power: $\overline{I^2} = \frac{1}{2}I_0^2$ and $\overline{V^2} = \frac{1}{2}V_0^2$. The square root of each of these is defined as the rms (root mean square) value of the current or voltage:

$$I_{\rm rms} = \sqrt{\overline{I^2}} = \frac{1}{\sqrt{2}} I_0 = 0.707 I_0 \tag{15-7a}$$

$$V_{\rm rms} = \sqrt{\overline{V^2}} = \frac{1}{\sqrt{2}} V_0 = 0.707\, V_0. \tag{15-7b}$$

FIGURE 15-14
Power delivered to a resistor in an ac circuit.

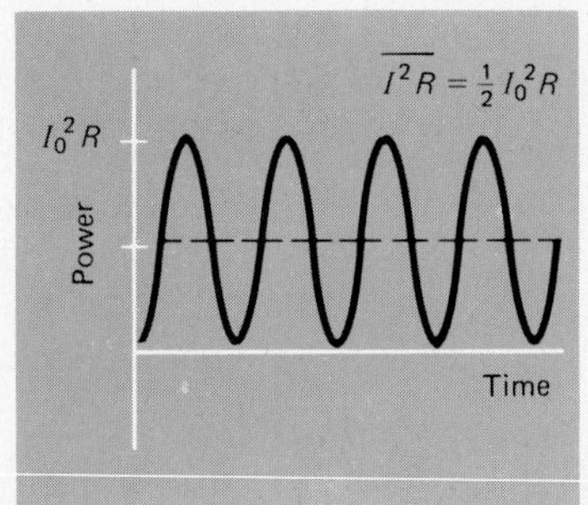

The rms values of V and I are sometimes called the "effective values." They are useful because they can be substituted directly into the power formulas (Equation 15-5) to get the average power. For example, $\bar{P} = I_{\rm rms}^2 R = \frac{1}{2} I_0^2 R$, which agrees with our calculation above. Thus a direct current whose values of I and V equal the rms values of I and V for an alternating current will produce the same power. Hence it is usually the rms value of current that is specified or measured. For example, in the U.S. and Canada, standard line voltage is 120 V ac. The 120 V is $V_{\rm rms}$. Thus the peak voltage V_0 is

$$V_0 = \sqrt{2}\, V_{\rm rms} = 170 \text{ V}.$$

In most of Europe the rms voltage is 240 V, so the peak voltage is 340 V.

▌ EXAMPLE 15-7 Calculate the resistance and the peak current in a 1000-W hair dryer connected to a 120-V line. What happens if it is connected to a 240-V line in Great Britain?

SOLUTION The rms current is

$$I_{\rm rms} = \frac{\bar{P}}{V_{\rm rms}} = \frac{1000 \text{ W}}{120 \text{ V}} = 8.3 \text{ A}.$$

Thus $I_0 = \sqrt{2} I_{\rm rms} = 11.8$ A. The resistance is

$$R = \frac{V_{\rm rms}}{I_{\rm rms}} = \frac{120 \text{ V}}{8.3 \text{ A}} = 14.5\ \Omega.$$

The resistance could equally well be calculated using peak values: $R = V_0/I_0 = 170 \text{ V}/11.8 \text{ A} = 14.5\ \Omega$. When connected to a 240 V line, the average power delivered would be

$$\bar{P} = \frac{V_{\rm rms}^2}{R} = \frac{(240 \text{ V})^2}{14.5\ \Omega} = 4000 \text{ W}.$$

This would undoubtedly melt the heating element or the wire coils of the motor. ▌▌

We will discuss alternating current and ac circuits in more detail in Chapter 19.

* 15-6 ▌ Electrolysis and electrophoresis

*Electrolysis

If two metal electrodes are placed in an electrolyte and connected to a battery as shown in Figure 15-15, several interesting things can happen. If the electrolyte is a "salt" such as silver nitrate ($AgNO_3$), it is found that silver is deposited on the negative electrode. If the electrolyte is a dilute acid in water, it is found that the water is broken down with hydrogen gas given off at the negative electrode and oxygen at the positive electrode. The process in each case is called *electrolysis* (meaning "to break apart with electricity").

Let us look at how electrolysis works from the atomic point of view.

FIGURE 15-15
Electrolysis cell.

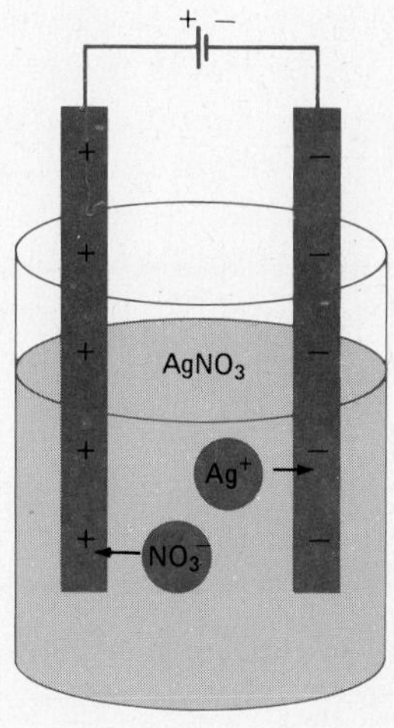

Suppose that the electrolyte is silver nitrate dissolved in water, as shown in Figure 15-15. The $AgNO_3$ becomes ionized, so there are Ag^+ and NO_3^- ions in the solution. The Ag^+ ions are attracted to the electrode that is kept negative by the battery. When a silver ion reaches the negative electrode, it acquires an electron to become neutral and remains attached to the electrode. Eventually, this electrode becomes coated with silver. This is the basis of *electroplating*, which was discovered in 1802 by one of Volta's students. What happens at the negative electrode is more complicated, but the net effect is that electrons are given by negative ions to the positive electrode.

If the electrolyte is a dilute acid such as HCl or H_2SO_4, positive hydrogen ions attracted to the negative electrode become hydrogen gas when they acquire electrons; at the positive electrode, oxygen gas is produced. The source for the hydrogen and oxygen is largely from the small amount of naturally ionized water and not simply from the acid. Electrolysis is used in this way to decompose water.

*Electrophoresis

When two electrodes are placed in a solution, as in electrolysis, ions are accelerated toward the electrode of opposite sign. The ions, however, collide with other ions and soon reach an average terminal velocity. The value of the terminal velocity depends on the strength of the electric field, the properties of the fluid, and the charge and shape of the ions themselves. Highly charged small ions clearly move more quickly than large extended ones or those with low charge. The former are said to have greater *mobility* than the latter.

This difference in mobility is made use of in the technique known as *electrophoresis*. It is used a great deal in biological and medical research to determine whether certain molecules are present in a given sample. Or it can be used to separate different constituents in a particular sample. It works because nearly all biological molecules of interest can carry an electric charge.

In paper electrophoresis, a long strip of filter paper is kept moist and a potential difference is applied to the ends. A few drops of the sample are placed on the paper and after a time the different molecules are spread out on the paper according to their mobilities. Often a dye is applied to the paper to better reveal the presence of particular types of molecules. Sometimes a second potential difference is applied at right angles to the first to produce better separation. In acrylamide gel electrophoresis, the sample is placed on top of a glass tube filled with a thick (acrylamide) gel. When a voltage is applied with electrodes to the ends of the tube, different molecules again move at different rates, which provides separation.

* 15-7 ▮ The nervous system and nerve conduction

The nervous system of humans and higher animals is remarkable and complex. It provides us with the means for being aware of the world, for communication within the body, and to control the body's muscles.

Although the detailed functioning of the nervous system is still not well understood, we do have a reasonably good understanding of how messages are transmitted within the nervous system: they are electrical signals passing along the basic element of the nervous system, the neuron.

*Neurons

Neurons are living cells of unusual shape (Figure 15-16). Attached to the main cell body are several small appendages known as dendrites and a long tail called the axon. Signals are received by the dendrites and are propagated along the axon; when a signal reaches the nerve endings, it is transmitted to the next nerve or to a muscle at a connection called a synapse. Some neurons have separate cells (called Schwann cells) wrapped around their axons; they form a layered sheath called a myelin sheath and help to insulate neurons from one another. In the short spaces between the Schwann cells (the nodes of Ranvier) the axon is in contact with the intercellular fluid.

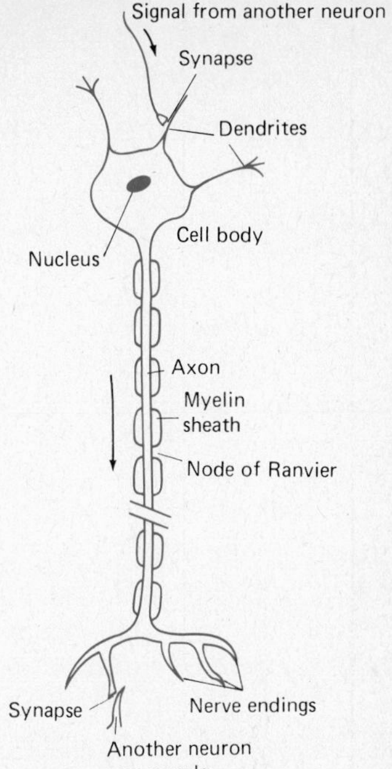

FIGURE 15-16
Neuron.

Neurons serve in three capacities. "Sensory neurons" carry messages from the eye, ear, skin, and other organs to the central nervous system, which consists of the brain and spinal cord. "Motor neurons" do the opposite; they carry signals from the central nervous system to particular muscles and can signal them to contract. These two types of neurons make up the "peripheral system" as distinguished from the central nervous system. The third type of neuron is the interneuron. These transmit signals between neurons. They are the neurons in the brain and spinal column, and often are connected in an incredibly complex array.

Most sensory neurons lead to the brain; there a complex analysis is carried out, and then a signal is sent to particular muscles to react. However, there are a few very simple connections in the body which are known as "reflex arcs." One example is the reflex action when you touch a hot stove; your hand is already withdrawing before the sensation of pain reaches the brain. In such reflex arcs, the sensing signal does not pass to the brain but only to the spinal column and then directly back to muscles, which are stimulated to contract. This is shown in Figure 15-17.

Axons vary in length. Those that reach the toes and fingers can be a meter long. In humans, axons are typically 10 to 20 μm in diameter. However, in certain squid they are 500 μm ($\frac{1}{2}$ mm) in diameter. This makes squid axions relatively easy to work with, and much of our

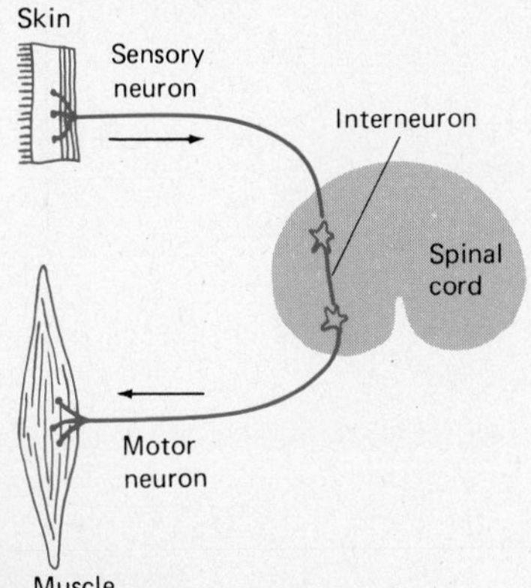

FIGURE 15-17
A reflex arc. In some reflex arcs, such as the familiar "knee jerk," only two neurons are involved (there is no interneuron).

knowledge of nerve conduction has come from research on these "giant" axons of the squid.

Before we examine how signals are transmitted along a neuron, we first examine the electrical properties of a neuron in the resting state.

TABLE 15-2

Concentrations of ions inside and outside a typical axon

	Concentration inside axon, C_I (mol/m^3)	Concentration outside axon, C_O (mol/m^3)
K^+	140	5
Na^+	15	140
Cl^-	9	125

*Resting potential and the Nernst equation

In nearly all living cells, including neurons, there is a net positive charge on the outer surface of the cell membrane and a negative charge on the inner surface. This was already mentioned in Section 14-7 with regard to heart muscles and the ECG. This difference in charge, or "dipole layer" as it is called, means that a potential difference exists across the cell membrane. When a neuron is not transmitting a signal, this potential difference is called the "resting potential." It is typically 60 to 90 mV, depending on the type of organism. The magnitude of the resting potential is related to the concentration of ions, as we shall see next.

The most common ions in a cell are K^+, Na^+, and Cl^-. There are large differences in the concentrations of these ions inside and outside a cell, as indicated by the typical values given in Table 15-2. Other ions are also present, so the fluids both inside and outside the axon are electrically neutral. Because of the differences in concentration, there is a tendency for ions to diffuse across the membrane. However, in the resting state the cell membrane prevents any net flow of Na^+ and most other ions, including the many large organic ions present. The membrane is, however, permeable to Cl^- ions and less so to K^+ ions, and it is these two ions that produce the dipole charge layer on the membrane. Because there is a greater concentration of K^+ inside the cell than outside, more K^+ ions tend to diffuse outward across the membrane than diffuse inward. A K^+ ion that passes through the membrane becomes attached to the outer surface of the membrane; it leaves behind an equal negative charge which lies on the inner surface of the membrane (Figure 15-18). The fluids themselves remain neutral. Indeed, what keeps the ions on the membrane is their attraction for each other across the membrane. Independent of this process, Cl^- ions tend to diffuse *into* the cell since their concentration outside is higher. Both K^+ and Cl^- diffusion tends to charge the interior surface of the membrane negatively and the outside positively. As charge accumulates on the membrane surface, it becomes increasingly difficult for more ions to diffuse: K^+ ions trying to move outward, for example, are repelled by the positive charge already there.

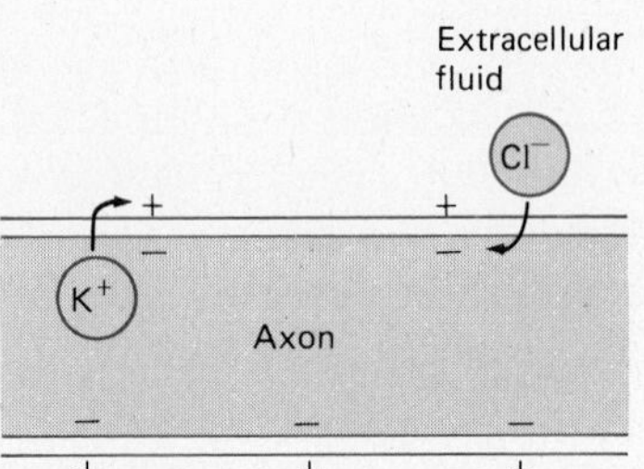

FIGURE 15-18
How a dipole layer of charge forms on a cell membrane.

Equilibrium is reached when the tendency to diffuse because of the concentration difference is just balanced by the electrical potential difference across the membrane. The greater the concentration difference, the greater the potential difference across the membrane. The quantitative relation between these is given by the *Nernst equation*:

$$V = V_{\text{inside}} - V_{\text{outside}} = 2.30\frac{kT}{ze}\log\frac{C_O}{C_I}. \qquad (15\text{-}8)$$

Here V is the potential difference (inside minus outside the cell), C_I and C_O the concentrations of a particular ion inside and outside the cell, k the Boltzmann constant, T the absolute temperature, e the charge on the electron, and z the valence (number of electron charges) on the ion.

The Nernst equation is not a cause-and-effect relationship. Rather it relates the resting potential to the concentrations of a particular ion species; it is valid *only* for an ion species that can freely diffuse through the membrane. Let us now check this for the ions given in Table 15-2. The measured value of V for this particular organism is -70 mV. The constant in the Nernst equation, with $z=-1$ and $T=310$ K, is $(2.30)\times(1.38\times10^{-23}\ \mathrm{J/K})(310\ \mathrm{K})/(-1)(1.6\times10^{-19}\ \mathrm{C}) = -61.4$ mV. So, for Cl^- ions, the Nernst equation gives

$$V = (-61.4\ \mathrm{mV})\log\tfrac{125}{9} = -(61.4\ \mathrm{mV})(\log 13.9)$$
$$= -(61.4)(1+0.143)\ \mathrm{mV} = -70\ \mathrm{mV},$$

where we used the log table of Appendix A-7. Because this agrees with the measured value, we conclude that the membrane is fully permeable to Cl^- ions. From this calculation we see that for any monovalent ion, the ratio of concentration outside to that inside must be 13.9 for negative ions and 1/13.9 for positive ions if they are to be in equilibrium at a resting potential of -70 mV. This is not the case for K^+ or Na^+. Indeed, for K^+ the predicted potential is $V = (61.4\ \mathrm{mV})\log 0.0357 = 61.4\ \mathrm{mV}\ (-2+0.553) = -88$ mV; and for Na^+, $V = (61.4\ \mathrm{mV})\log 9.33 = +59$ mV. Thus K^+ is not far from equilibrium and must diffuse to some extent through the membrane, but the membrane is highly impermeable to Na^+.

An important question is how the membrane is able to prevent Na^+ from entering the cell, and also keep K^+ from leaving the cell. Tracer studies show that K^+ does diffuse out and Na^+ does diffuse in. Since equilibrium is not reached, we conclude that there must be an active transport process occurring. K^+ is actively "pumped" back into the cell, and Na^+ is pumped out of the cell. How these potassium and sodium "pumps" work is not yet fully understood.

*The action potential and propagation of nerve impulses

The most important aspect of a neuron is not that it has a resting potential (most cells do), but rather that it can respond to a stimulus and conduct an electrical signal along its length. A nerve can be stimulated in various ways. The stimulus could be thermal (when you touch a hot stove) or chemical (as in taste buds); it could be pressure (as on the skin or at the eardrum), or light (as in the eye); or it could be the electric stimulus of a signal coming from the brain or another neuron. In the laboratory, the stimulus is usually electrical and is applied by a tiny probe at some point on the neuron. If the stimulus exceeds some threshold, a voltage pulse will travel down the axon. This voltage pulse can be detected at a point on the axon using a voltmeter or an oscilloscope (Chapter 19) connected as in Figure 15-19. This voltage pulse has the shape shown in Figure 15-20 and is called an "action potential." As can be seen, the potential increases from the resting potential of -70 mV so that it becomes positive 30 or 40 mV. The action potential lasts for about 1 ms and travels down an axon with a speed of 30 m/s to 150 m/s.

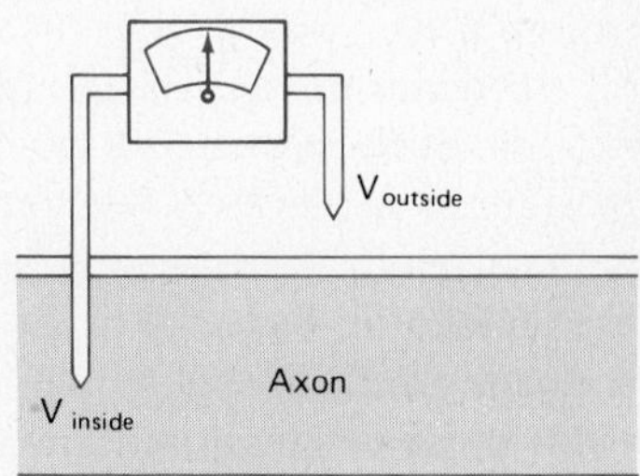

FIGURE 15-19
Measuring the potential difference between the inside and outside of a nerve.

The size of the action potential is independent of the strength of the stimulus. Either the neuron "fires" or it does not. A stronger stimulus does

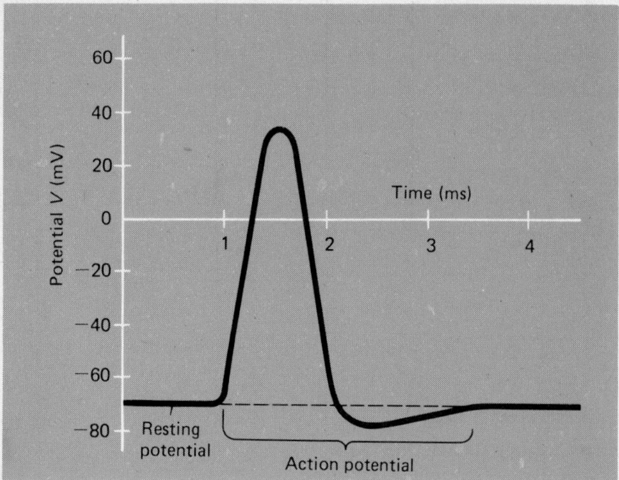

FIGURE 15-20
Action potential.

not produce a greater action potential, but instead produces more pulses, one after the other. The peaks of subsequent pulses never overlap—they are no closer than 2 ms and typical rates are 10 to 100 per second. This fits in with how muscle contraction occurs; for, as we saw in Section 13-5, each pulse received from a nerve causes a muscle to "twitch," and repeated pulses cause the muscle to contract more strongly, up to a maximum limit. Also, the depolarization of the heart muscle we discussed in Section 14-7 is basically an action potential passing across the heart.

But what causes the action potential? Apparently, the cell membrane has the ability to alter its permeability properties. At the point where the stimulation occurs, the membrane suddenly becomes much more permeable to Na^+ than to K^+ and Cl^- ions. Thus Na^+ ions rush into the cell and the inner surface of the wall becomes positively charged. The Na^+ would be in equilibrium at about $+59$ mV, as we calculated a few paragraphs back. Indeed, the potential reaches almost this point (Figure 15-20). But then the membrane suddenly returns to its original characteristics: it becomes impermeable to Na^+ and in fact pumps out Na^+ ions. The diffusion of Cl^- and K^+ ions again predominates and the original resting potential is restored.

What, then, causes the action potential to travel along the axon? The action potential occurs at the point of stimulation, as shown in Figure 15-21*a*. The membrane momentarily is positive on the inside and negative on the outside at this point. Nearby charges are attracted toward this region as shown in part *b*. The potential in these adjacent regions then drops, causing an action potential there. Thus, as the membrane returns to normal at the original point, nearby it experiences an action potential, so the action potential moves down the axon (Figure 15-21*c* and *d*).

You may wonder if the number of ions that pass through the membrane would significantly alter the concentrations. The answer is no; and we can show this by treating the axon as a capacitor in the following example.

FIGURE 15-21
Propagation of an action potential along an axon membrane.

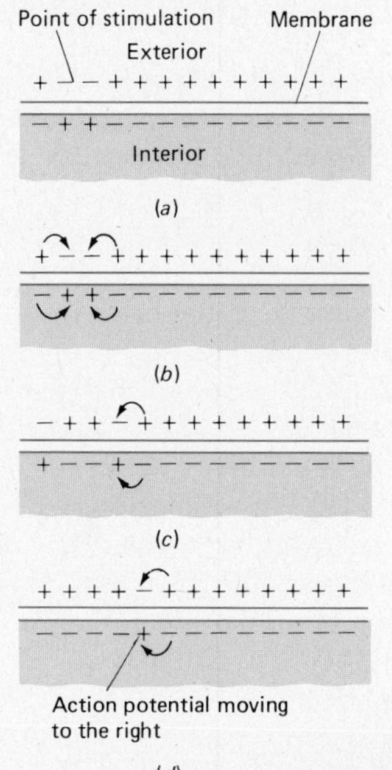

▮ EXAMPLE 15-8 (*a*) Determine the capacitance of an axon 10 cm long of radius 10 μm. The thickness of the membrane is about 10^{-8} m and the dielectric constant is about 3. (*b*) By what factor does the concentration of Na^+ ions in the cell change as a result of one action potential?

SOLUTION This will be an order-of-magnitude calculation.

(*a*) The membrane of an axon acts like a cylindrically shaped parallel-plate capacitor, with opposite charges on each side. The separation of the "plates" is the thickness of the membrane, $d = 10^{-8}$ m. The area A is the area of a cylinder of radius r and length l: $A = 2\pi rl = (6.28)(10^{-5}\text{ m})(0.1\text{ m}) \approx 6\times 10^{-6}\text{ m}^2$. From Equation 14-4*b* we have

$$C = K\varepsilon_0 \frac{A}{d} = (3)(8.85\times 10^{-12}\text{ C}^2/\text{N}\cdot\text{m}^2)\frac{6\times 10^{-6}\text{ m}^2}{10^{-8}\text{ m}} \approx 10^{-8}\text{ F}.$$

(*b*) Since the voltage changes from -70 mV to about $+30$ mV, the total change is about 100 mV. The amount of charge that moves is then $Q = CV = (10^{-8}\text{ F})(0.1\text{ V}) = 10^{-9}\text{ C}$. This corresponds to $Q/e = 10^{-9}\text{ C}/1.6\times 10^{-19}\text{ C} \approx 10^{10}$ ions that must flow per action potential. The volume of our cylindrical axon is $V = \pi r^2 l \approx (3)(10^{-5}\text{ m})^2(0.1\text{ m}) = 3\times 10^{-11}\text{ m}^3$; the concentration of Na^+ ions inside the cell (Table 15-2) is $15\text{ mol/m}^3 = 15\times 6.02\times 10^{23}\text{ ions/m}^3 \approx 10^{25}\text{ ions/m}^3$. Thus the cell contains $10^{25}\times(3\times 10^{-11}) \approx 3\times 10^{14}$ Na^+ ions. One action potential, then, will change the concentration of Na^+ ions by at most 1 part in 30,000; this tiny change would not be measurable. ▮▮

Thus even 1000 action potentials will not alter the concentration significantly. The sodium pump does not, therefore, have to remove Na^+ ions quickly after an action potential, but can operate slowly over time to maintain a relatively constant concentration.

The propagation of a nerve pulse as described here applies to an unmyelinated axon. Myelinated axons, on the other hand, are insulated from the extracellular fluid by the myelin sheath except at the nodes of Ranvier. Thus an action potential cannot be generated where there is a myelin sheath. Once such a neuron is stimulated, the pulse will still travel along the membrane, but there is resistance and the pulse becomes smaller as it moves down the axon. Nonetheless, the weakened signal can still stimulate a full-fledged action potential when it reaches a node of Ranvier. Thus the signal is repeatedly amplified at these points. Compare this to an unmyelinated neuron, in which the signal is continually amplified by repeated action potentials all along its length. This naturally requires much more energy. Development of myelinated neurons was a significant evolutionary step. For it meant reliable transmission of nerve pulses with less energy required; and the pulses travel more quickly, since ordinary conduction is faster than the repeated production of action potentials, whose speed depends on the flow of ions across the membrane.

Summary

By transforming chemical energy into electric energy, an electric battery serves as a source of potential difference or *electromotive force* (emf). A simple battery consists of two electrodes, made of different metals, immersed in a solution or paste known as an electrolyte.

Electric current, I, refers to a flow of electric charge, and is measured in amperes (A): 1 A equals a flow of 1 coulomb per second past a given point. The direction of current flow is generally taken as being that of positive charge; in a wire, it is actually electrons (which are negatively charged) that move, so they flow in a direction opposite to the direction of the "conventional current." Positive conventional current always flows from a high potential to a low potential. *Ohm's law* states that the current in a conductor is proportional to the potential difference applied to its two ends; the proportionality constant is called the resistance R of the material, so $V = IR$. The unit of resistance is the ohm (Ω), where $1\ \Omega = 1\ \text{V/A}$.

The resistance of a wire is inversely proportional to its cross-sectional area A and directly proportional to its length L and a property of the material called its resistivity, $\rho : R = \rho(L/A)$. The resistivity increases with temperature for metals, but for some semiconductors it decreases. A superconductor is a material for which the electrical resistance is essentially zero.

The rate at which energy is transformed in a resistance R from electric to other forms of energy (such as to heat and light) is equal to the product of current and voltage; that is, the power transformed, measured in watts, is given by $P = IV$ and can be written with the help of Ohm's law as $P = I^2R = V^2/R$. The total electric energy transformed in any device equals the product of power and the time during which the device is operated: $E = Pt$. In SI units, energy is given in joules ($1\ \text{J} = 1\ \text{W} \cdot \text{s}$), but electric companies use a larger unit, the kilowatt-hour ($1\ \text{kWh} = 3.6 \times 10^6\ \text{J}$).

Electric current can be direct (dc), in which the charges flow in only one direction; or it can be alternating (ac), in which the charges reverse direction at a particular frequency, typically 60 Hz. Alternating currents are often sinusoidal in time, $I = I_0 \sin 2\pi ft$, and are produced by an alternating voltage. The rms values of an alternating current and voltage are given by $I_{\text{rms}} = I_0/\sqrt{2}$ and $V_{\text{rms}} = V_0/\sqrt{2}$, where I_0 and V_0 are the peak values. The power relationship, $P = IV = I^2R = V^2/R$, is valid for alternating currents when the rms values are used.

▮ Questions

1 When you turn on a water faucet, the water usually flows immediately. You do not have to wait for water to flow from the faucet valve to the spout. Why not? Is the same thing true when you connect a wire to the terminals of a battery?

2 Car batteries are often rated in ampere-hours (A · h). What does this rating mean?

3 When an electric cell is connected to a circuit, electrons flow away from the negative terminal in the circuit. But within the cell, electrons flow *to* the negative terminal. Explain.

4 Is current used up in a resistor?

5 Develop an analogy between blood circulation and an electrical circuit. Discuss what plays the role of the heart for the electric case, and so on.

6 Design a circuit in which two different switches of the type shown in Figure 15-22 can be used to operate the same light bulb from opposite sides of a room.

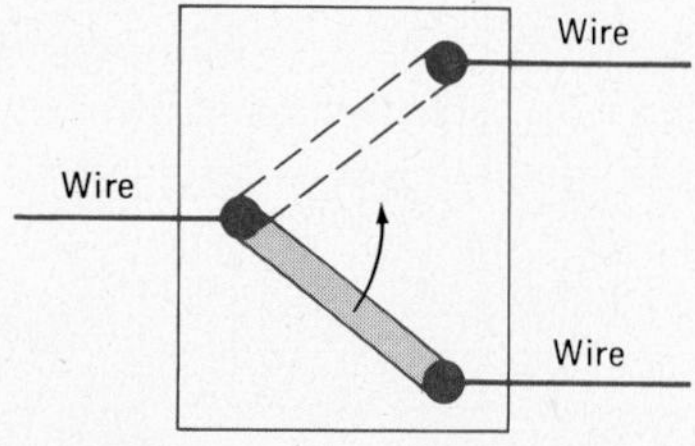

FIGURE 15-22

7 Can a copper wire and an aluminum wire of the same length have the same resistance? Explain.

8 The equation $P = V^2/R$ indicates that the power dissipated in a resistor decreases if the resistance is increased, whereas the equation $P = I^2R$ implies the opposite. Is there a contradiction here? Explain.

9 What happens when a light bulb burns out?

10 Which draws more current, a 100-W light bulb or a 75-W bulb?

11 Electric power is transferred over large distances at very high voltages. Explain how this reduces power losses in the transmission lines.

12 Why is it dangerous to replace a 15-A fuse that blows repeatedly with a 25-A fuse?

13 Electric lights operated on low-frequency ac (say, 10 Hz) flicker noticeably. Why?

14 In a car, one terminal of the battery is said to be connected to "ground." Since it is not really connected to the ground, what is meant by this expression?

Problems

SECTION 15-2

1 (I) A service station charges a battery using a current of 5 A for 6 h. How much charge passes through the battery?

2 (I) A current of 1.0 A flows in a wire. How many electrons are flowing past any point in the wire per second? The charge on one electron is 1.6×10^{-19} C.

3 (I) How many coulombs are there in an ampere-hour?

4 (I) What is the current in amperes if 1000 Na^+ ions were to flow across a cell membrane in 1.0 μs? The charge on the sodium is the same as on an electron, but positive.

5 (I) What is the current in a 2.0-Ω resistor connected to a 9.0-V battery?

6 (I) What voltage will produce 2.0 A of current through a 15 Ω resistor?

7 (I) What is the resistance of a toaster if 110 V produces a current of 4.0 A?

8 (II) Copper has approximately 10^{29} free electrons per cubic meter. What is the approximate average speed of electrons in a 1-mm-radius wire carrying 1.0 A?

9 (II) Two 1.5-V batteries are connected together as in Figure 15-5*a* to a bulb whose resistance is 10 Ω. How many electrons leave each battery per minute?

10 (II) A bird stands on an electric transmission line carrying 1200 A. The line has 1.0×10^{-5} Ω resistance per meter and the bird's feet are 3.0 cm apart. What voltage does the bird feel?

11 (II) The *conductance* G of an object is defined as the reciprocal of the resistance $R: G = 1/R$. The unit of conductance is a mho ($=\text{ohm}^{-1}$), which is also called the siemens (S). What is the conductance (in siemens) of an object that draws 200 mA of current at 2.0 V?

SECTION 15-3

12 (I) What is the resistance of a 2.0-m length of copper wire 1.5 mm in diameter?

13 (I) What is the diameter of a 20-cm length of tungsten wire whose resistance is 0.10 Ω?

14 (II) A 30-m length of wire 1.0 mm in diameter has a resistance of 2.0 Ω. What is the resistance of a 50-m length of wire 2.0 mm in diameter made of the same material?

15 (II) Can a 1.0-mm-diameter copper wire have the same resistance as a tungsten wire of the same length? Give numerical details.

16 (II) A length of wire is cut in half, and the two lengths are wrapped together side by side to make a thicker wire. How does the resistance of this new combination compare to the resistance of the original wire?

17 (II) A 100-W light bulb has a resistance of about 12 Ω when cold and 140 Ω when "on" (hot). Estimate the temperature of the filament when "on," assuming an average temperature coefficient of resistivity $\alpha = 0.0060/\text{C}°$.

18 (III) A 0.50-Ω wire is drawn out ("stretched") to four times its original length. What is its resistance now?

19 (III) For some applications it is important that the value of a resistance not change with temperature. For example, suppose that you made a 1.0-kΩ resistor from a carbon resistor and a nichrome wire-wound resistor connected together. The total resistance will be the sum of their separate resistances. What value should each of these resistors have (at 0°C) so that the combination is temperature independent?

20 (III) A 33-Ω resistor is made from a coil of copper wire whose total mass is 12 g. What is the diameter of the wire and how long is it?

SECTION 15-4

21 (I) The element of an electric oven is designed to produce 3.0 kW of heat when connected to a 240-V source. What must be the resistance of the element?

22 (I) What is the current through a 1300-W electric frying pan if its resistance is 11 Ω?

23 (I) What is the maximum power consumption of a 9-V transistor radio that draws a maximum of 400 mA of current?

24 (I) What is the maximum voltage that can be applied to a 200-Ω resistor rated at $\frac{1}{4}$ W?

25 (I) An automobile starter motor draws 150 A from the 12-V battery. How much power is this?

26 (I) How many kWh does a 1300-W electric frying pan use in 15 min of operation?

27 (I) At $0.08 per kWh, what does it cost you if you forget to turn off a 200-W (input) stereo receiver before going to sleep for 8 h?

28 (II) The heating element of a 120-V 1800-W heater is 8.2 m long. If it is made of iron, what must its diameter be?

29 (II) A person accidentally leaves a car with the lights on. If each of the two front lights uses 40 W and each of the two rear lights 6 W, how long will a fresh 12-V battery last if it is rated at 90 A · h? Assume the full 12 V appears across each bulb.

30 (II) How many 100-W light bulbs, operated at 120 V can be used without blowing a 10-A fuse?

31 (II) What is the total amount of energy stored in a 12-V 60-A · h car battery when it is fully charged?

32 (II) A particular household uses a 5.0-kW heater 2.0 h/day ("on" time), six 100-W light bulbs 5.0 h/day, a 3.0-kW electric stove element for a total of 1.5 hours, and miscellaneous power amounting to 1.5 kWh/day. If electricity costs $0.10 per kWh, what will be their monthly bill (30 days)?

33 (II) How much coal (which produces 7000 kcal/kg) must be burned by a 30-percent-efficient power plant to provide the yearly needs of the household in Problem 32?

34 (II) At $0.10 per kWh, what does it cost to leave a 40-W porch light on all day for a year?

35 (II) A power station delivers 250 kW of power to a factory through 4.0-Ω lines. How much less power is wasted if the electricity is delivered at 40,000 V rather than 10,000 V?

36 (II) The wiring in a house must be thick enough so that it does not become hot enough to start a fire. Of what diameter must a copper wire be if it is to carry a maximum current of 60 A and produce no more than 2.0 W of heat per meter of length?

37 (III) An electric car makes use of a storage battery as its source of energy. Suppose that a small, 800-kg postman's delivery car is to be powered by ten 100-A · h 12-V batteries. Assume that the car is driven on the level at an average speed of 30 km/h, and the average friction force is 200 N. Assume 100 percent efficiency and neglect energy used for acceleration; note that no energy is consumed when the vehicle is stopped since the engine does not need to idle. (*a*) Determine the horsepower required. (*b*) After approximately how many kilometers must the batteries be recharged? (*c*) Can such a car go uphill?

38 (III) The current in an electromagnet connected to a 240-V line is 60 A. At what rate must cooling water pass over the coils if the water temperature is to rise by no more than 10 C°?

39 (III) An electric heater is used to heat a room of volume 40 m^3. Air is brought into the room at 5°C and is changed completely twice per hour. Heat loss through the walls amounts to approximately 500 kcal/h. If the air is to be maintained at 22°C, what minimum wattage must the heater have? (The specific heat capacity of air is approximately 0.17 kcal/kg · C°.)

40 (III) A small immersion heater can be used in a car to heat a cup of water for coffee. If the heater can heat 200 ml of water from 5°C to 95°C in 5 min, approximately how much current does it draw from the 12-V battery?

SECTION 15-5

41 (I) An ac electric appliance draws an average current of 4.0 A. What is the peak value of the current?

42 (I) An ac voltage, whose peak value is 90 V, is across a 35-Ω resistor. What is the value of the rms and peak currents in the resistor?

43 (I) Calculate the peak current in a 2.0-kΩ resistor connected to a 600-V ac source.

44 (II) What is the maximum instantaneous value of the power dissipated in a 100-W light bulb?

45 (II) A 10-Ω heater is connected to a 240 V ac line. What is the average power used? What are the maximum and minimum values of the instantaneous power?

46 (II) Calculate the peak current passing through a 60-W light bulb connected to a 120-V ac line.

47 (II) Suppose that a current is given by the equation $I = 2.5 \sin 120t$. (*a*) What is the frequency? (*b*) What is the rms value of the current? (*c*) If this is the current in amps through a 50-Ω resistor, what is the equation that describes the voltage as a function of time (I is in amperes and t is in seconds)?

*SECTION 15-7

*48 (I) What is the magnitude of the electric field across an axon membrane 1.0×10^{-8} m thick if the resting potential is -70 mV?

*49 (II) If the concentration of K^+ outside the axon of Table 15-2 were 5 mol/m³ as stated, what would the K^+ concentration be inside the axon if K^+ were in equilibrium and the resting potential were -70 mV? Assume $T = 310$ K.

*50 (II) In the axon of the giant squid the concentration of Cl^- is about 55 mol/m³, whereas outside it is 540 mol/m³. What would you expect the resting potential to be? Assume $T = 310$ K.

*51 (II) The measured resting potential of the giant squid axon is -60 mV. The concentration of K^+ is

10 mol/m³ outside and 400 mol/m³ inside. Is K^+ in equilibrium? Is there a K^+ pump? If so, what would the inside concentration have to be to produce the measured resting potential?

*52 (II) A nerve is stimulated with an electric pulse. The action potential is detected at a point 3.40 cm down the axon 0.0052 s later. When the action potential is detected 7.20 cm from the point of stimulation, the time required is 0.0063 s. What is the speed of the electric pulse along the axon? (Why are two measurements needed instead of only one?).

*53 (III) Estimate how much energy is required to transmit one action potential along the axon of Example 15-8. (*Hint*: One pulse is equivalent to charging and discharging the axon capacitance; see Section 14-3.) What minimum average power is required for 10^4 neurons transmitting 100 pulses per second?

*54 (III) During the action potential, Na^+ moves into the cell at a rate of about 3×10^{-7} mol/m$^2 \cdot$ s. How much power must be produced by the active transport system to produce this flow against a +30-mV potential difference? Assume that the axon is 10 cm long and 20 μm in diameter.

For additional reading

Aidley, D. J., *The Physiology of Excitable Cells*, Cambridge University Press, Cambridge, 1971. Nerve action.

Brophy, J. J., *Basic Electronics for Scientists*, McGraw-Hill Book Company, New York, 1977.

Davidovits, P., *Physics in Biology and Medicine*, Prentice-Hall, Inc., Englewood Cliffs, N.J., 1975.

Katz, B., *Nerve, Muscle and Synapse*, McGraw-Hill Book Company, New York, 1966.

Magie, W. F., *Source Book in Physics*, Harvard University Press, Cambridge, Mass., 1963. Excerpts from original work of Galvani, Volta, Ohm, and others.

Meyer, H. W., *History of Electricity and Magnetism*, The M.I.T. Press, Cambridge, Mass., 1972.

Orville, R. E., "The Lightning Discharge," *Physics Teacher*, **14**, p. 7, 1976.

Shepherd, G. M., "Microcircuits in the Nervous System," *Sci. American* **238**:2 (Feb., 1978), p. 92.

Stanford, A. L., Jr., *Foundations of Biophysics*, Academic Press, Inc., New York, 1975. Nervous system.

16 ▌ Dc circuits and instruments

In Chapter 15 we discussed the basic principles of current electricity. Now we shall apply these principles to analyze dc circuits and to understand the operation of a number of useful instruments.†

When we draw a diagram of a circuit, we represent a battery by the symbol "⊣⊢," a capacitor by the symbol "⊣⊢," and a resistor by the symbol "—ʌʌʌ—." Wires whose resistance is negligible are drawn simply as straight lines.

16-1 ▌ Circuits involving resistors; internal resistance

Resistors in series and in parallel

When two or more resistors are connected end to end so that the same current passes through each in turn, as in Figure 16-1, they are said to be connected in *series*. On the other hand, if the resistors are connected so that the current from the source splits into separate branches as shown in Figure 16-2, the resistors are said to be in *parallel*; in this case the same potential difference exists across each. In either case we shall want to calculate the net resistance of the set of resistors. In other words, we want to find what single resistance R could replace the combination of given resistors without altering the rest of the circuit. The resistors could be like those shown in Figure 15-8, or they could be light bulbs or other electrical devices.

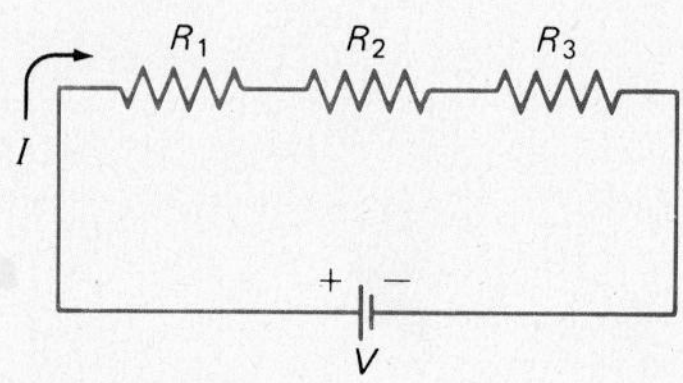

FIGURE 16-1
Resistors connected in series: $R = R_1 + R_2 + R_3$.

First we consider the series case, Figure 16-1. The same current I passes through each resistor. If it did not, this would imply that charge was accumulating at some point in the circuit, which does not happen in the steady situation. We let V represent the voltage across all three resistors; we assume that all other resistance in the circuit can be ignored, so V equals the emf of the battery. We let V_1, V_2, and V_3 be the potential differences across each of the resistors R_1, R_2, and R_3. By Ohm's law, $V_1 = IR_1$, $V_2 = IR_2$, and $V_3 = IR_3$. Because of conservation of energy, the

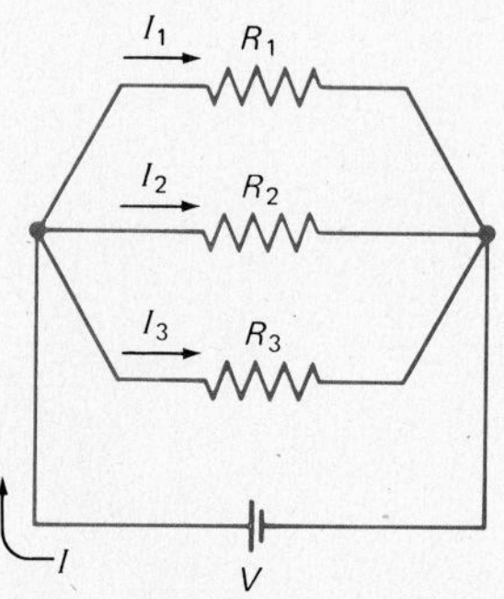

FIGURE 16-2
Resistors connected in parallel: $1/R = 1/R_1 + 1/R_2 + 1/R_3$.

† Ac circuits that contain only a source of emf and resistors can be analyzed like the dc circuits in this chapter. However, ac circuits in general are more complex—for example if they contain capacitors—and we shall discuss them in Chapter 19.

total voltage V is equal to the sum of the voltages across each resistor, so we have

$$V = V_1 + V_2 + V_3 = IR_1 + IR_2 + IR_3.$$

For the equivalent single resistor R that draws the same current, we have

$$V = IR.$$

We equate these two expressions for V and find that

$$R = R_1 + R_2 + R_3 \qquad \text{(series).} \tag{16-1}$$

This is, in fact, what we expect. When we put several resistances in series, the net resistance is the sum of the separate resistances. This applies for any number of resistances, not simply for three. Clearly, when you add more resistance to the circuit, the current will decrease. If a 12-V battery is connected to a 4-Ω resistor, the current will be 3 A. But if the 12-V battery is connected to three 4-Ω resistors in series, the total resistance is 12 Ω and the current only 1 A.

The situation is quite different for the parallel case, Figure 16-2. Again we want to find what single resistance R is equivalent to the three resistors, this time in parallel. In this situation the total current I that leaves the battery breaks into three branches, and because charge is conserved, the current flowing into a junction must equal the current flowing out; so

$$I = I_1 + I_2 + I_3,$$

where I_1, I_2, and I_3 are the currents through each of the resistors R_1, R_2, and R_3. The full voltage of the battery is applied to each resistor, so

$$I_1 = \frac{V}{R_1}, \qquad I_2 = \frac{V}{R_2}, \qquad I_3 = \frac{V}{R_3}.$$

Furthermore, for the single resistor R that will draw the same current I as these three in parallel, we must have

$$I = \frac{V}{R}.$$

We now combine these equations:

$$I = I_1 + I_2 + I_3$$

$$\frac{V}{R} = \frac{V}{R_1} + \frac{V}{R_2} + \frac{V}{R_3}.$$

When we cancel out the V from each term, we have

$$\frac{1}{R} = \frac{1}{R_1} + \frac{1}{R_2} + \frac{1}{R_3} \qquad \text{(parallel).} \tag{16-2}$$

For example, if three 30-Ω resistors are put in parallel, the net resistance R offered by this network is

$$\frac{1}{R} = \frac{1}{30\ \Omega} + \frac{1}{30\ \Omega} + \frac{1}{30\ \Omega} = \frac{3}{30\ \Omega} = \frac{1}{10\ \Omega},$$

and so $R = 10\ \Omega$. Thus the net resistance is *less* than that of each single resistance. This may at first seem surprising. But remember that when you put resistors in parallel, you are giving the current additional paths to follow; hence the net resistance will be less.

▌ EXAMPLE 16-1 How much current flows from the battery shown in Figure 16-3*a*?

SOLUTION First we find the equivalent resistance, R_p, of the 500-Ω and 700-Ω resistors that are in parallel:

$$\frac{1}{R_p}=\frac{1}{500\ \Omega}+\frac{1}{700\ \Omega}=0.0020\ \Omega^{-1}+0.0014\ \Omega^{-1}$$

$$=0.0034\ \Omega^{-1}.$$

This is $1/R$, so we must take the reciprocal to find R. (It is a common mistake to forget to do this. Notice that the units of reciprocal ohms, Ω^{-1}, help to remind us of this.) Thus

$$R_p=\frac{1}{0.0034\ \Omega^{-1}}=290\ \Omega.$$

This 290 Ω is the equivalent resistance of the two parallel resistors and is in series with the 400-Ω resistor; the equivalent circuit is shown in Figure 16-3*b*. To find the total resistance R_T, we add the 400-Ω and 290-Ω resistance together, since they are in series, and find that $R_T=400\ \Omega+290\ \Omega=690\ \Omega$. The total current flowing from the battery is then

$$I=\frac{V}{R}=\frac{12\ \text{V}}{690\ \Omega}=17\text{mA}.$$

▌▌

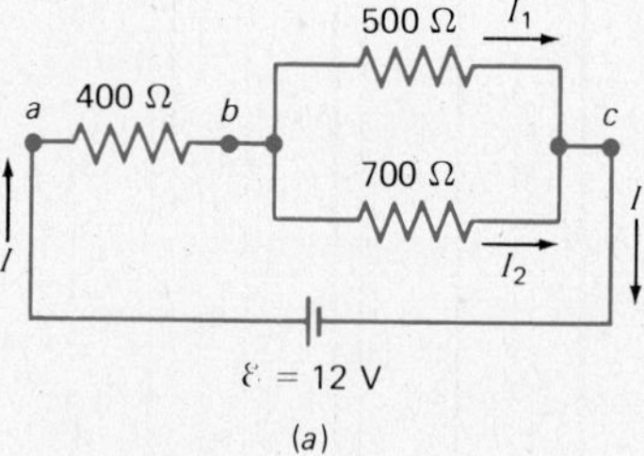

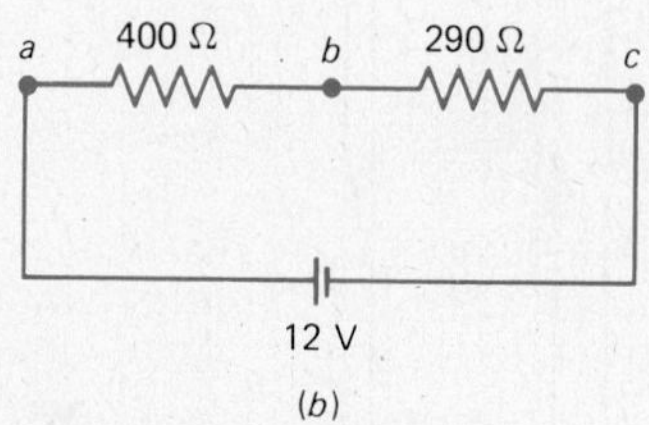

FIGURE 16-3
(*a*) Circuit for Examples 16-1 and 16-2. (*b*) Equivalent circuit, showing the equivalent resistance of 290Ω for the two parallel resistors in (*a*).

▌ EXAMPLE 16-2 What is the current flowing through the 500-Ω resistor in Figure 16-3*a*?

SOLUTION To solve this problem we must find the voltage across the 500-Ω resistor; this is the voltage between points *b* and *c* in the diagram and we call it V_{bc}; once V_{bc} is known, we can apply Ohm's law to get the current. First, we find the voltage across the 400-Ω resistor, V_{ab}. Since 17 mA passes through this resistor, the voltage across it can be found using Ohm's law, $V=IR$:

$$V_{ab}=(0.017\ \text{A})(400\ \Omega)=6.8\ \text{V}.$$

Since the total voltage across the network of resistors is $V_{ac}=12$ V, then V_{bc} must be 12 V − 6.8 V = 5.2 V. Then Ohm's law tells us that the current I through the 500-Ω resistor is

$$I_1=\frac{5.2\ \text{V}}{500\ \Omega}=10\ \text{mA}.$$

This is the answer we wanted. However, we can also calculate the current I_2 through the 700-Ω resistor, since the voltage across it is also 5.2 V:

$$I_2=\frac{5.2\ \text{V}}{700\ \Omega}=7\ \text{mA}.$$

Notice that when I_1 combines with I_2 to form the total current I (at point *c* in Figure 16-3*a*), their sum is 10 mA + 7 mA = 17 mA. This is, of course, the total current as calculated in Example 16-1.

▌▌

Emf and terminal voltage

You may have noticed in your own experience that when a current is drawn from a battery, the voltage across its terminals drops below its rated emf. For example, when you have started your car with the headlights on, you may have noticed that the headlights dimmed. This is because the starter draws a great deal of current and the battery voltage drops as a result. The voltage drop occurs because the chemical reactions in a battery cannot supply charge fast enough to maintain the full emf. For one thing, charge must flow within the electrolyte between the electrodes of the battery, and there is always some resistance to this flow. Thus a battery itself has some resistance, which is called its *internal resistance*; it is usually designated r. The internal resistance acts as if it were in series with the emf, as shown in Figure 16-4. Since this resistance r is inside the battery, we can never separate it from the battery. The two points a and b in the diagram represent the two terminals. What we measure is the *terminal voltage* V_{ab}. When no current is drawn from the battery, the terminal voltage equals the emf, which is determined by the chemical reactions in the battery: $V_{ab} = \mathscr{E}$. However, when a current I flows from the battery, there is a drop in the terminal voltage equal to Ir. Thus the terminal voltage is

$$V_{ab} = \mathscr{E} - Ir.$$

For example, if a 12-V battery has an internal resistance of 0.1 Ω, then when 10 A flows from the battery, the terminal voltage is 12 V − (10 A)(0.1 Ω) = 11 V.

The internal resistance of a battery is usually small. For example, an ordinary flashlight battery when fresh may have an internal resistance of perhaps 0.05 Ω. However, as it ages and the electrolyte dries out, the internal resistance increases to many ohms. Car batteries have even lower internal resistance.

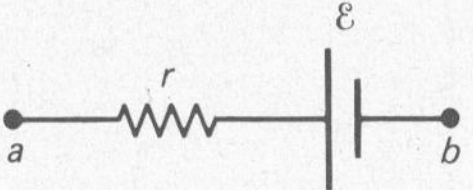

FIGURE 16-4
Diagram for an electric cell or battery.

❙ EXAMPLE 16-3 A 9.0-V battery whose internal resistance r is 0.50 Ω is connected in the circuit shown in Figure 16-5a. (a) How much current is drawn from the battery? (b) What is the terminal voltage of the battery?

SOLUTION (a) The 4.0-Ω and 8.0-Ω resistors in parallel have an equivalent resistance R_{I} given by

$$\frac{1}{R_{\mathrm{I}}} = \frac{1}{8.0\ \Omega} + \frac{1}{4.0\ \Omega} = \frac{3}{8.0\ \Omega},$$

so $R_{\mathrm{I}} = 2.7\ \Omega$. This is in series with the 6.0-Ω resistor, so the net resistance of the lower arm of the circuit is 6.0 Ω + 2.7 Ω = 8.7 Ω, as shown in Figure 16-5b. The equivalent resistance R_{II} of the 10.0-Ω and the 8.7-Ω resistors in parallel is given by

$$\frac{1}{R_{\mathrm{II}}} = \frac{1}{10.0\ \Omega} + \frac{1}{8.7\ \Omega} = 0.21\ \Omega^{-1},$$

so $R_{\mathrm{II}} = 4.8\ \Omega$. This 4.8 Ω is in series with the 5.0-Ω resistor and the 0.50-Ω internal resistance of the battery, so the total resistance R of the circuit is $R = 4.8\ \Omega + 5.0\ \Omega + 0.50\ \Omega = 10.3\ \Omega$. Hence the current drawn is

$$I = \frac{\mathscr{E}}{R} = \frac{9.0\ \mathrm{V}}{10.3\ \Omega} = 0.87\ \mathrm{A}.$$

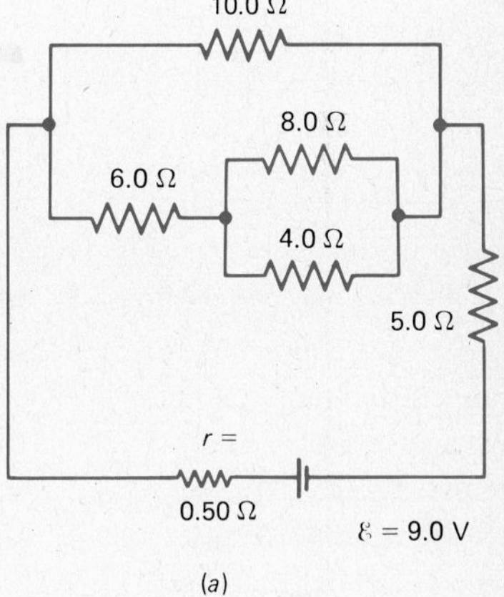

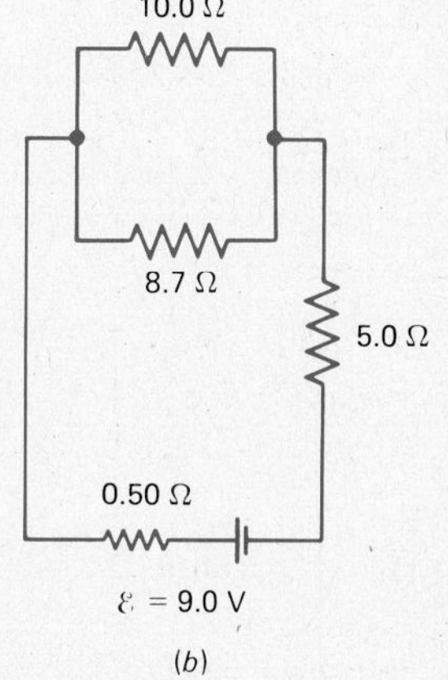

FIGURE 16-5
Circuit for Example 16-3; *r* is the internal resistance of the battery.

(*b*) The terminal voltage of the battery is $V_{ab} = \mathscr{E} - Ir = 9.0\ \text{V} - (0.87\ \text{A})(0.50\ \Omega) = 8.6\ \text{V}$. ▮▮

* 16-2 ▮ Kirchhoff's rules

*Kirchhoff's rules and their application to circuits

In the last few examples we have been able to find the currents flowing in circuits by combining resistances in series and parallel, and using Ohm's law. This technique can be used for many circuits. However, you sometimes encounter a circuit that is too complicated for this analysis. For example, we cannot find the currents flowing in each part of the circuit shown in Figure 16-6 simply by combining resistances as we did before.

To deal with such complicated circuits, we use Kirchhoff's rules; there are two of them. *Kirchhoff's first* or *junction rule* is based on the conservation of charge and we already used it in devising the rule for parallel resistors; it states that *at any junction point, the sum of all currents entering the junction must equal the sum of all currents leaving the junction.* For example, at point *a* in Figure 16-6, I_3 is entering whereas I_1 and I_2 are leaving. Thus Kirchhoff's junction rule states that $I_3 = I_1 + I_2$. We already saw an instance of this at the end of Example 16-2, where the currents passing through the 400-Ω and 700-Ω resistors were 10 mA and 7 mA, respectively, and they added at point *c* in Figure 16-3*a* to give the outgoing current of 17 mA.

Kirchhoff's junction rule is based on the conservation of charge. Charges that enter a junction must also leave—none is lost or gained. *Kirchhoff's second* or *loop rule* is based on the conservation of energy. It states that *the algebraic sum of the changes in potential around any closed path must be zero.* To see why this should hold, consider the analogy of a roller coaster on its track. When it starts from the station, it has a particular potential energy. As it climbs the first hill, its potential energy increases and reaches a peak at the top. As it descends the other side, its potential energy decreases and reaches a low value at the bottom of the hill. As the roller coaster continues on its path, its potential energy goes through more changes. But when it arrives back at the starting point, it has exactly as much potential energy as it had when it started at this point. Another way of saying this is that there was as much uphill as there was downhill.

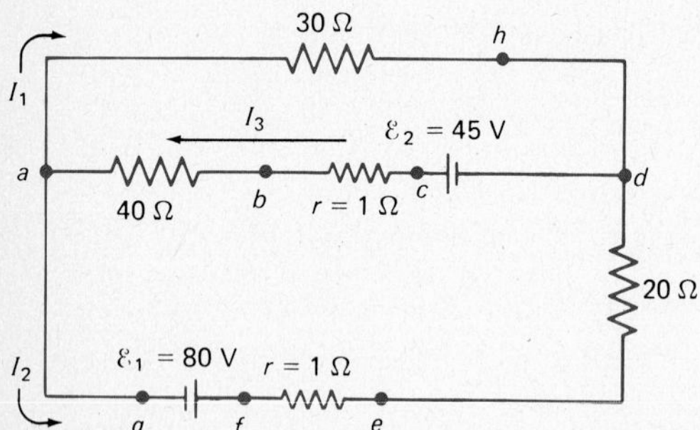

FIGURE 16-6
Currents in this circuit can be calculated using Kirchhoff's rules.

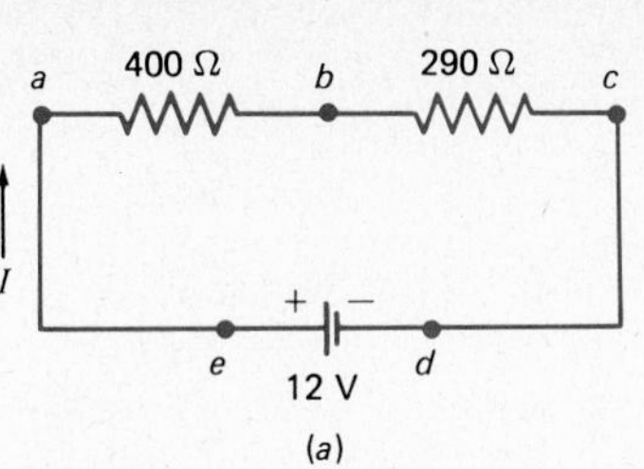

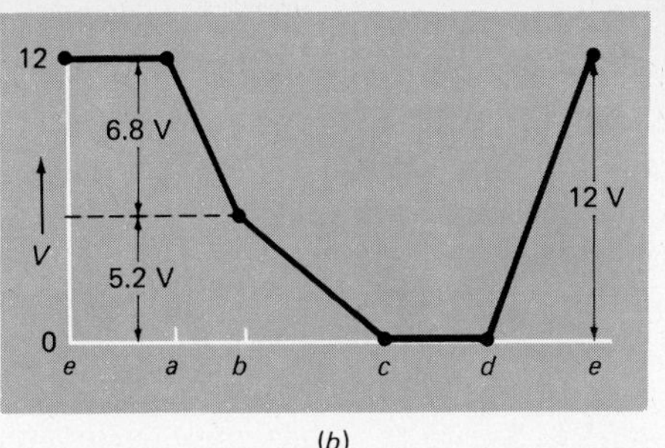

FIGURE 16-7
Changes in potential around the circuit in (*a*) are plotted in (*b*).

The same reasoning can be applied to an electric circuit. As an example, consider the simple circuit in Figure 16-7. It happens to be the same as the equivalent circuit of Figure 16-3*b* we have already dealt with. The current flowing in this circuit is $I = (12\text{ V})/(690\ \Omega) = 0.017$ A, as we calculated in Example 16-1. (Note that we are ignoring the internal resistance of the battery.) The positive side of the battery, point *e* in the figure, is at a high potential compared to point *d* at the negative side of the battery. That is, point *e* is like the top of a hill for a roller coaster. We can now follow the current around the circuit, starting at any point we choose. Let us start at point *e* and follow a positive test charge completely around this circuit. As we go, we will note all changes in potential; when the charge returns to point *e*, the potential there will be the same as when we started, so the total change in potential will be zero. It may be helpful to plot the changes in voltage around the circuit as in Fig. 16.7*b*. (Point *d* is arbitrarily taken as zero.) As our positive test charge goes from point *e* to point *a*, there is no change in potential, since there is no source of potential nor any resistance. However, as the charge passes through the 400-Ω resistor to get to point *b*, there is a decrease in potential of $V = IR = (0.017\text{ A})(400\ \Omega) = 6.8$ V. In effect, the charge is flowing "downhill," since it is heading toward the negative terminal of the battery. This is indicated in the graph of Figure 16-7*b*. The decrease in voltage between the two ends of a resistor ($= IR$) is called a *voltage drop*. Because this is a *decrease* in voltage, we use a negative sign when applying Kirchhoff's loop rule; that is,

$$V_{ba} = -6.8\text{ V}.$$

As the charge proceeds from *b* to *c*, there is a further voltage drop of $(0.017\text{ A})(290\ \Omega) = 5.2$ V, and since this is a decrease, we write

$$V_{cb} = -5.2\text{ V}.$$

There is no change in potential as our test charge moves from *c* to *d*. But when it moves from *d*, which is the negative or low potential side of the battery, to point *e*, which is the positive terminal, the voltage *increases* by 12 V. That is,

$$V_{ed} = +12\text{ V}.$$

The sum of all the changes in potential in going around the circuit of Figure 16-7 is then

$$-6.8\text{ V} - 5.2\text{ V} + 12\text{ V} = 0.$$

And this is exactly what Kirchhoff's loop rule said would be true.

We already knew the details of this simple circuit, and we merely gave this example to show how the loop rule is used. Before we use Kirchhoff's rules to determine the currents in a more difficult circuit, we must point out that when using Kirchhoff's rules, the current in each separate branch of the circuit is designated by a different subscript, such as I_1, I_2, and I_3 in Figure 16-6. You do not have to know in advance in which direction these currents are actually moving. You make a guess and calculate the potentials around the circuit as if you were right; if you made the wrong guess for the direction of a current, your answer will merely have a negative sign. This and other details of using Kirchhoff's laws will become clearer in the following example.

▌ EXAMPLE 16-4 Calculate the currents I_1, I_2, and I_3 in each of the branches of the circuit in Figure 16-6.

SOLUTION We choose the directions of the currents as shown in the figure. Since (positive) current moves away from the positive terminal of a battery, we expect I_2 and I_3 to have the directions shown. It is hard to tell the direction of I_1 in advance, so we choose the direction shown arbitrarily. We have three unknowns and therefore we need three equations. We first apply Kirchhoff's junction rule to the currents at point a: I_3 enters, whereas I_2 and I_1 leave. Therefore,

$$I_3 = I_1 + I_2. \qquad (a)$$

This same equation holds at point d, so we get no new information there. We now apply Kirchhoff's loop rule to two different closed loops. First, we apply it to the loop a–h–d–c–b–a. From a to h we have a voltage change $V_{ha} = -(I_1)(30\ \Omega)$. From h to d there is no change, but from d to c the potential increases by 45 V: $V_{dc} = +45$ V. From c to a the voltage drops through the two resistors by an amount $V_{ac} = -(I_3)(40\ \Omega + 1\ \Omega)$. Thus we have $V_{ha} + V_{cd} + V_{ac} = 0$, or

$$-30I_1 - 41I_3 + 45 = 0, \qquad (b)$$

where we have omitted the units. For our second loop we take the complete circuit $ahdefga$. (We could have just as well taken $abcdefg$ instead.) Again we have $V_{ha} = -(I_1)(30\ \Omega)$, and $V_{dh} = 0$. But when we take our positive test charge from d to e, it is actually going uphill, against the natural flow of the current—or at least against the *assumed* direction of the current, which is what counts in this calculation. Thus $V_{ed} = I_2(20\ \Omega)$ has a *positive* sign. Similarly, $V_{fe} = I_2(1\ \Omega)$. From f to g there is a decrease in potential of 80 V, since we go from the high potential terminal to the low. Thus $V_{fg} = -80$ V. Finally, $V_{ag} = 0$, and the sum of the potentials around this loop is then

$$-30I_1 + 21I_2 - 80 = 0. \qquad (c)$$

We now have three equations—labeled (a), (b), and (c)—in three unknowns. From (c) we have

$$I_2 = \frac{80 + 30I_1}{21} = 3.80 + 1.41I_1. \qquad (d)$$

From Equation (b), above, we have

$$I_3 = \frac{45 - 30I_1}{41} = 1.10 - 0.73I_1. \qquad (e)$$

We substitute these into Equation (a) and solve for I_1:

$$\begin{aligned} I_1 &= I_3 - I_2 \\ &= -2.70 - 2.14I_1, \end{aligned}$$

so

$$\begin{aligned} 3.14I_1 &= -2.70 \\ I_1 &= -0.86\ \text{A} \end{aligned}$$

I_1 has magnitude −0.86 A; the negative sign indicates that its direction is actually opposite to that assumed and shown in the figure. Note that the answer automatically comes out in amperes since all numbers were in volts and ohms. From Equation (d) we have

$$I_2 = 3.80 + 1.41I_1 = 2.59\ \text{A},$$

and from Equation (e)

$$I_3 = 1.10 - 0.73I_1 = 1.73\ \text{A}.$$

*Emfs in series and parallel

When two or more sources of emf, such as batteries, are arranged in series, the total voltage is the algebraic sum of their respective voltages. For example, if two 1.5 V flashlight batteries are connected as shown in Figure 16-8a, the voltage V_{ca} across a light bulb, represented by the resistor R, is 3.0 V. (To be absolutely correct, we should also take into account the internal resistance of the batteries, but we assume it to be small.) On the other hand, when a 20-V and a 12-V battery are connected oppositely, as shown in part b, the net voltage V_{ca} is 8 V; that is, a positive test charge moved from a to b gains in potential by 20 V, but when it goes from b to c it drops by 12 V. So the net change is 20 V − 12 V = 8 V. You might think that connecting batteries in reverse like this would be wasteful. And for most purposes that would be true. But such a reverse arrangement is precisely how a battery charger works: in Figure 16-8b, the 20-V source is charging up the 12-V battery. Because of its greater voltage, the 20-V source is forcing charge back into the 12-V battery: electrons are being forced into its negative terminal and removed from its positive terminal. Not all batteries can be recharged, however, since the chemical reaction in many cannot be reversed; in this case, the arrangement of Figure 16-8b would simply waste energy.

Sources of emf can also be arranged in parallel (Figure 16-8c); this is not done to increase voltage but to provide more energy when large currents are needed. Each of the cells in parallel only has to produce a fraction of the current, so the loss due to internal resistance is less than for a single cell.

FIGURE 16-8
Batteries in series, (a) and (b), and in parallel, (c).

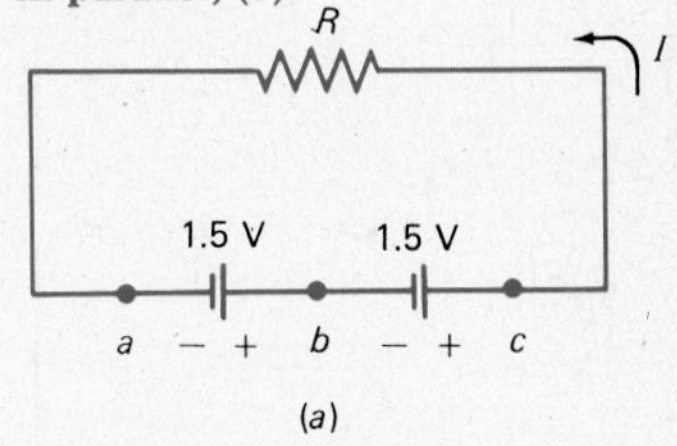

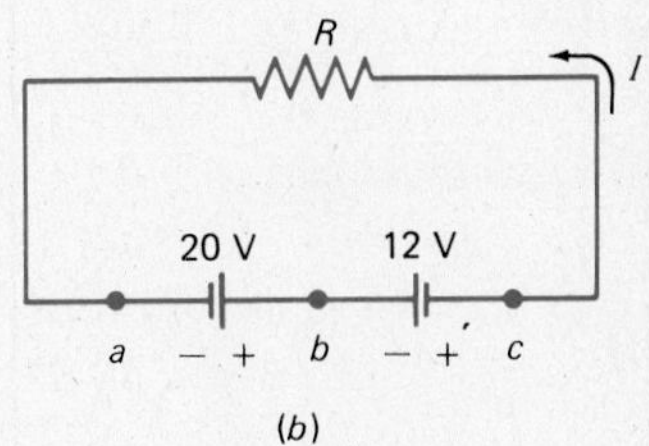

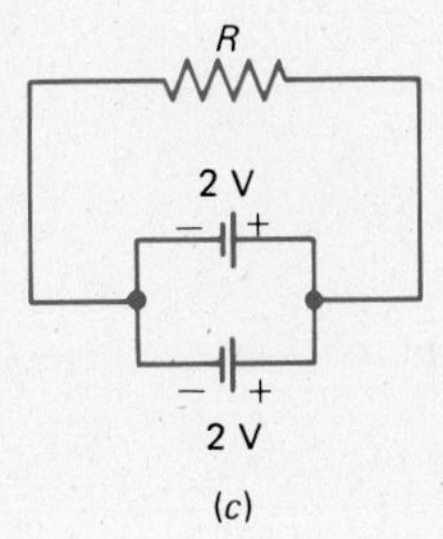

16-3 ▮ Ammeters and voltmeters

An *ammeter* is used to measure current, and a voltmeter measures potential difference or voltage. We will discuss how these meters are constructed, not only because it is interesting in itself, but also because a meter has resistance and thus has some effect, often significant, on the quantities that it is measuring.

The crucial part of each of these meters is a *galvanometer*. A galvanometer works on the principle of the force between a current and a magnetic field; we will discuss this in Chapter 17. For now, we merely need to know that the deflection of the needle attached to the galvanometer is proportional to the current flowing through it. The *current sensitivity* of the galvanometer is the current needed to make the needle deflect full scale (Figure 16-9). For example, suppose that the sensitivity is 50 μA (1 μA = 1 microampere = 10^{-6} A). This means that a current of 50 μA will cause the needle to go to the end of the scale. A current of 25 μA will make it go only halfway, and so on. Of course, if there is no current, the needle should be on zero; usually, there is an adjustment screw that can be used to make it so. Since the galvanometer is basically a coil of wire, it is important also to know its resistance.

Many meters we meet with in everyday life are galvanometers connected as ammeters or voltmeters. These include the VU meter on a tape recorder and some of the meters on automobile dashboards.

The ammeter

A galvanometer can be used directly to measure small currents. For example, a galvanometer whose sensitivity is 50 μA can measure currents from about 1 μA (currents smaller than this would be hard to read on the scale) up to 50 μA. Currents above this would drive the needle beyond full scale and could damage the meter. Often, we want to measure larger currents. To do so, a resistor is placed in parallel with the galvanometer. Thus an ammeter (represented by the symbol •—Ⓐ—•) consists of a galvanometer (•—Ⓖ—•) in parallel with a resistor called the *shunt* resistor. ("Shunt" is a synonym for "in parallel.") This is shown in Figure 16-10*a*. The shunt resistance is R and the resistance of the galvanometer

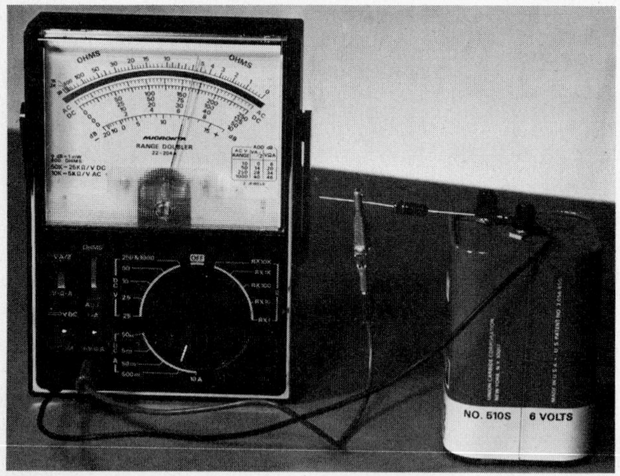

FIGURE 16-9
Multimeter used as an ammeter. Needle is about two-thirds of full-scale deflection. (Lester V. Bergman & Assoc., Inc.)

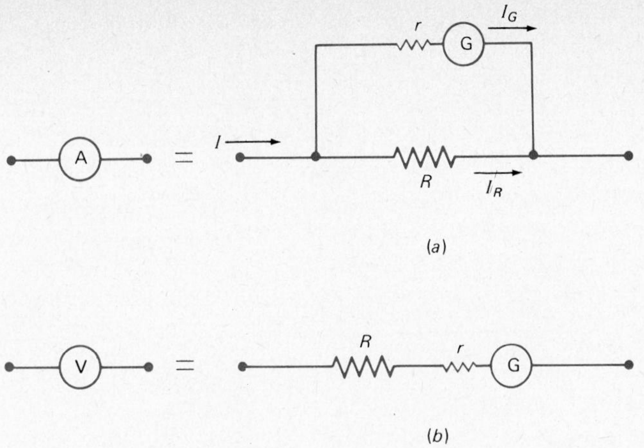

FIGURE 16-10
(a) An ammeter is a galvanometer in parallel with a small (shunt) resistor, R. (b) A voltmeter is a galvanometer in series with a large resistor, R.

coil itself is r. The value of R is chosen according to what full-scale deflection you want.

Suppose you want the meter to read 1.0 A at full scale. And suppose the sensitivity of the galvanometer is 50 μA and its resistance is $r = 30\ \Omega$. This means that when the total current entering the ammeter is 1.0 A, we want precisely 50 μA (call this I_G) to flow through the galvanometer in order to give full-scale deflection. Thus, when 1.0 A flows into the meter, we want 0.99995 A ($= I_R$) to pass through the shunt resistor R. Now the potential difference across the shunt is the same as across the galvanometer. Thus

$$I_R R = I_G r$$

and

$$R = \frac{I_G r}{I_R} = \frac{(5.0 \times 10^{-5}\ \text{A})(30\ \Omega)}{0.99995\ \text{A}} = 1.5 \times 10^{-3}\ \Omega$$

or 0.0015 Ω. The shunt resistor must thus have a very low resistance so that most of the current passes through it.

The face of the dial is marked off in this case from 0 to 1.0 A. It is easy to see that if the current I into the meter is 0.50 A, this will produce a 25-μA current in the galvanometer and give a deflection half of full scale, which is what we want. The shunt resistor is usually located inside the case of the ammeter, out of sight.

The voltmeter

A voltmeter (•—Ⓥ—•) also consists of a galvanometer and a resistor. But the resistor R is connected in series, Figure 16-10b; and it is usually large. Let us see how to calculate the value of R. Suppose that we have the same galvanometer with internal resistance $r = 30\ \Omega$ and full-scale sensitivity of 50 μA. And suppose we want a voltmeter that reads from 0 to 15 V. Thus, when a potential difference of 15 V exists across the terminals of our voltmeter, we want 50 μA to be passing through it so as to give a full-scale deflection. From Ohm's law we have

$$15\ \text{V} = (50\ \mu\text{A})(r + R).$$

Electronic meters are now available which use transistors. Their sensitivity is often 10^6 to $10^7\ \Omega$. Hence they have very little effect on most circuits, and their readings are reliable for nearly any circuit,

* 16-4 ▮ The potentiometer

The *potentiometer* is a device to measure precisely potential differences, particularly the emf of a battery or other source of emf. They are considerably more precise than ordinary voltmeters. And they have the further advantage of being able to measure a voltage *without drawing current.* Hence they can be used to directly measure an emf. And they will not alter the currents in an operating circuit as a standard voltmeter will. Potentiometers are often used to measure potential differences produced in the human body or in animals.

A diagram of a potentiometer is shown in Figure 16-12. A fresh battery, termed the "working battery," is connected in series to a variable resistor R (variable-resistor symbol), which can be varied to give a convenient voltage between points A and B. The potential difference between A and B, which we call V, produces a steady current through the wire AB. The wire AB is a precisely made uniform wire which is called the "slide wire" because of the sliding contact C that moves along it. The emf to be measured, $\mathscr{E}_x$, is placed in the circuit as shown; the slide contact is moved until the galvanometer indicates zero current when the switch S_2 is closed momentarily. In this position no current flows from the unknown source, so the voltage between points A and C will be equal to $\mathscr{E}_x$, the emf of the unknown:

$$V_{AC} = \mathscr{E}_x.$$

This relationship holds only when no current flows from the unknown emf through the galvanometer, since otherwise there would be potential drops across the internal resistances of the galvanometer and the unknown source. Next, a standard cell replaces the unknown source, often by moving a switch S_1 as shown in the diagram. A standard cell is one whose emf $\mathscr{E}_s$ is accurately known in advance; it is never used to draw

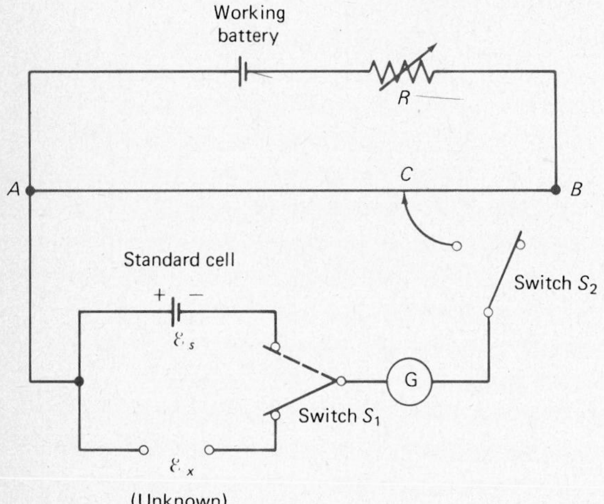

FIGURE 16-12
Potentiometer circuit.

current, for this will affect its emf over time. The slide wire is again adjusted to a new position C' so that again no current passes through the galvanometer. In this case we have

$$V_{AC'} = \mathscr{E}_s.$$

The voltage between A and B is the same for both operations and the wire AB is uniform. Therefore, the voltage V_{AC} due to the current in the slide wire AB is proportional to the length AC; similarly $V_{AC'}$ is proportional AC'. Thus

$$\frac{\mathscr{E}_x}{\mathscr{E}_s} = \frac{V_{AC}}{V_{AC'}} = \frac{AC}{AC'},$$

or

$$\mathscr{E}_x = \mathscr{E}_s\left(\frac{AC}{AC'}\right). \tag{16-3}$$

In practice, a potentiometer is enclosed in a box and the two lengths AC and AC' can be read off a dial that is connected to the sliding contact. Sometimes the standard cell, working battery, and galvanometer are enclosed within; in such instruments the emf of the unknown source can often be read directly from a dial. Other forms of potentiometer have external connections, particularly for the standard cell, since precise work requires a freshly calibrated standard cell. Of course, there are always external terminals to connect to the source whose emf is to be measured. In many potentiometers the slide wire is replaced by, or supplemented with, a number of precisely made fixed resistors. Many of the recording devices that are used to plot an electrocardiogram or other kinds of electrical signals make use of the potentiometer principle.

Finally, it should be noted that the word "potentiometer" is sometimes also used to refer to a variable resistance; and colloquially it is sometimes called a "pot."

* 16-5 ▮ The Wheatstone bridge

The Wheatstone bridge is a device used to make precise measurements of resistance. Figure 16-13 is a diagram of a Wheatstone bridge circuit. The resistances of R_1, R_2, and R_3 are accurately known. At least one of these—R_3 in our diagram—is variable; its resistance is usually read from a dial. The resistance R_X which is to be measured (the "unknown" resistance) is connected so as to form the fourth arm of the bridge. When the battery is connected, current flows through each of the resistors. In operation, the variable resistor R_3 is adjusted until the points B and D are at the same potential. This is checked by momentarily closing the switch to the galvanometer, which will not deflect when B and D are at the same potential. In this situation the bridge is said to be "balanced" and we have $V_{AB} = V_{AD}$, or (see Figure 16-13)

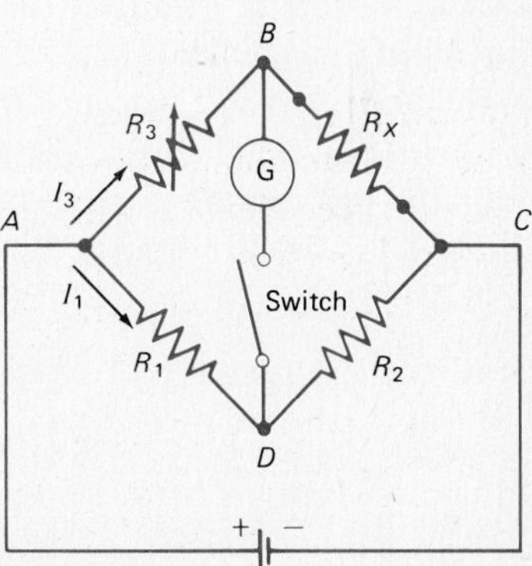

FIGURE 16-13
Wheatstone bridge.

$$I_3R_3 = I_1R_1.$$

I_1 is the current that passes through R_1 and also through R_2 when the bridge is balanced; I_3 is the current passing through R_3 and R_X. When the bridge is balanced, the voltage across R_X equals that across R_2, so

$$I_3R_X = I_1R_2.$$

We now divide these two equations and find that

$$\frac{I_3R_X}{I_3R_3} = \frac{I_1R_2}{I_1R_1}$$

or

$$R_X = \frac{R_2}{R_1}R_3 \quad \text{(when balanced).} \tag{16-4}$$

This equation allows a precise calculation of R_X when the other three resistances are precisely known. In practice, the galvanometer is very sensitive, so when R_3 is being adjusted, the switch is closed only momentarily to check if the current is zero or not.

A Wheatstone bridge is used in many situations where resistance must be accurately measured. One example is to measure the resistance of a resistance thermometer, which we discussed in Chapter 15.

* 16-6 ▌ Transducers and the thermocouple

A *transducer* is a device that converts one type of energy into another. A high-fidelity loudspeaker is one kind of transducer—it transforms electric energy into sound energy (see Chapter 17); so is a microphone, which changes sound into an electrical signal.

Transducers are often used for measuring particular quantities. In this case they can be thought of as changing one kind of signal into another kind, usually electrical. The *resistance thermometer* and the *thermistor* we discussed in Chapter 15 are examples; basically they convert temperature into an electrical property, resistance.

We shall discuss next some other transducers that are useful in a variety of fields. We shall meet other examples later.

*The strain gauge and pressure transducers

We saw in Chapter 4 that a wire will stretch an amount proportional to the stress (or force) applied to it. When stretched, its resistance must increase since it is longer and its cross-sectional area is reduced (see Equation 15-3). This fact is made use of in the wire strain gauge illustrated in Figure 16-14. The fine wire is usually bonded to a flexible backing material. When fixed tightly to a structure, the resistance of the strain gauge changes in direct proportion to any change in stress on the structure. Since the wire must not be strained beyond its elastic limit, the change in length is generally quite small. Hence the change in resistance is quite small (less than 1 part per thousand) and a very sensitive Wheatstone bridge is used to measure it. Of course, strain gauges must be carefully calibrated before use.

FIGURE 16-14
Wire strain gauge.

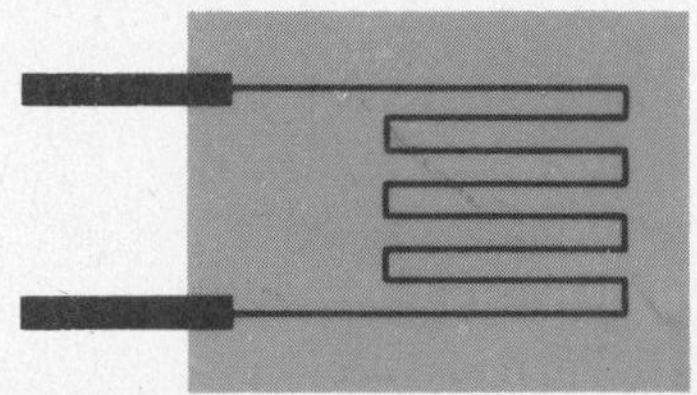

Strain gauges are applied by architects and engineers to models of a proposed structure. In this way they can determine the stress at critical points. Strain gauges are used in many other applications as well. For example, strain gauges have been used to determine the stress in the bones of the human skull when biting, by attaching them at various points on dry skulls.

A wire strain gauge can be attached to a membrane or diaphragm. A change in pressure against the membrane causes strain in the wire. Thus a strain gauge can be used to measure pressure. In this case it is called a *pressure transducer*. Strain gauges can be made very small (2 to 3 mm wide). When attached to a diaphragm, they can be swallowed to measure internal pressure, or inserted into a vein or artery to measure blood pressure. In these and other situations there will also be variation in resistance due to temperature changes. To correct for this, an identical gauge must also be used which is subjected to the same temperature but no stress.

Another kind of pressure transducer makes use of the *piezoelectric* effect. This effect occurs in certain crystals, such as quartz, that produce an emf proportional to the force applied to them. Piezoelectric crystals are used in some phonograph cartridges; they transform the movement of the needle in a record groove into an electric signal (an emf).

*The thermocouple

The *thermocouple* is another device that produces an electrical signal when subjected to differing temperatures. Unlike the resistance thermometer (in which it is the resistance that changes with temperature), a thermocouple produces an emf. Indeed, a thermocouple is one of the most useful of thermometers.

The thermocouple is based on the "thermoelectric effect." When two dissimilar metals, say iron and copper, are joined at the ends as shown in Figure 16-15*a*, it is found that an emf is produced if the two junctions are at different temperatures. Apparently this is due to the fact that the electrons in one metal are held more tightly than in the other—so they

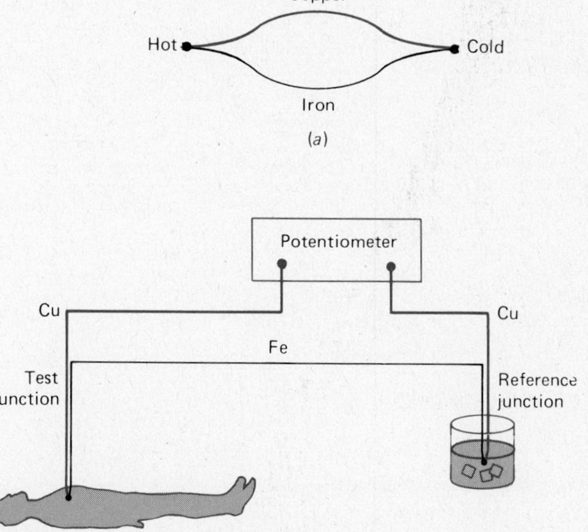

FIGURE 16-15
Thermocouple.

have a lower energy in one metal than in the other; hence some of them flow across the junction. This leaves one metal slightly more positive than the other, so a *contact potential* exists between them. If the two junctions are at the same temperature, the same contact potential exists at each; they balance each other and no current will flow. However, when one of the junctions is at a higher temperature, the energy of the electrons in the metals is altered and the contact potential will be different. In this case there will be a net emf and a current will flow. The magnitude of this emf depends on the temperature difference.

In operation, one junction of a thermocouple is kept at a known temperature. This "reference temperature" is often 0°C. The other junction, called the "test junction," is placed where the desired temperature is to be measured. The emf is measured most accurately with a potentiometer. It is important that the potentiometer terminals be of the same metal, and kept at the same temperature, so that no additional emfs are produced. Connections to lead wires, if used, must also be kept at the same temperature.

Because thermocouple junctions can be made very small, they can be used to measure temperature in small objects such as particular parts of the body and even in insects (Figure 16-15*b*).

A number of different combinations of dissimilar metals are used in thermocouples. The emf as a function of temperature can be found in tables for each set of metals and for a particular reference temperature. Over small temperature ranges, the emf is approximately proportional to the temperature difference. However, this linear relationship cannot be assumed to be true in general, and tables should be consulted.

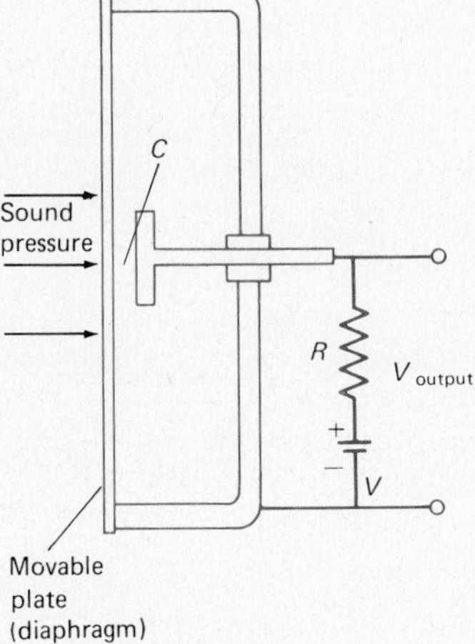

FIGURE 16-16
Diagram of a capacitor microphone.

*Capacitor microphone

A microphone is a transducer that changes a sound wave into an electrical signal. One type of microphone transducer is the capacitor (sometimes called condenser) microphone (Figure 16-16). The changing air pressure in a sound wave causes one plate of the capacitor C to move back and forth. We saw in Equation 14-4 that the capacitance is inversely proportional to the separation of the plates. Thus a sound wave causes the capacitance to change. This, in turn, causes the charge Q on the plates to change ($Q = CV$) so that an electric current is generated at the same frequencies as the incoming sound wave.

16-7 ▮ Circuits containing capacitors

Capacitors in series and parallel

Just as resistors can be placed in series or parallel in a circuit, so can capacitors. We first consider a parallel connection as shown in Figure 16-17*a*. If a battery of voltage V is connected to points a and b, this voltage exists across each of the capacitors. Each acquires a charge given by $Q_1 = C_1V$, $Q_2 = C_2V$, and $Q_3 = C_3V$. The total charge Q that must leave the battery is then

$$Q = Q_1 + Q_2 + Q_3 = C_1V + C_2V + C_3V.$$

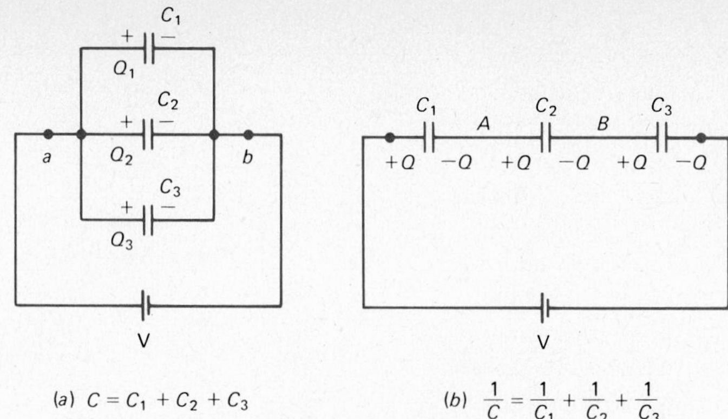

(a) $C = C_1 + C_2 + C_3$ (b) $\frac{1}{C} = \frac{1}{C_1} + \frac{1}{C_2} + \frac{1}{C_3}$

FIGURE 16-17
Capacitors (*a*) in parallel, (*b*) in series.

A single equivalent capacitor that will hold the same charge Q at the same voltage V will have a capacitance C given by

$$Q = CV.$$

Thus we have

$$CV = C_1V + C_2V + C_3V,$$

or

$$C = C_1 + C_2 + C_3 \qquad \text{(parallel)}. \tag{16-5}$$

The net effect of connecting capacitors in parallel is thus to increase the capacitance. This is just what we should expect, since we are essentially increasing the area of the plates for charge to accumulate on (see Equation 14-4).

If the capacitors are connected in series, Figure 16-17*b*, a charge $+Q$ flows from the battery to one plate of C_1, and $-Q$ flows to one plate of C_3. The regions A and B in between the capacitors were originally neutral, so the net charge there must still be zero. The $+Q$ on the left plate of C_1 attracts a charge of $-Q$ on the other plate of C_1; because region A must have a zero net charge, there is thus $+Q$ on the left plate of C_2. The same considerations apply to the other capacitors, so we see the charge on each capacitor is the same, namely Q. A single capacitor that could replace these three in series without affecting the circuit would have a capacitance C, given by

$$Q = CV.$$

Now the total voltage V across the three capacitors in series must equal the sum of the voltages across each capacitor:

$$V = V_1 + V_2 + V_3.$$

We also have $Q = C_1V_1$, $Q = C_2V_2$, and $Q = C_3V_3$, so from the preceding equation we get

$$\frac{Q}{C} = \frac{Q}{C_1} + \frac{Q}{C_2} + \frac{Q}{C_3},$$

or

$$\frac{1}{C} = \frac{1}{C_1} + \frac{1}{C_2} + \frac{1}{C_3} \qquad \text{(series)}. \tag{16-6}$$

Notice that the form of the equations for capacitors in series or in parallel are just the reverse of their counterparts for resistance. That is, the formula for capacitors in series resembles the formula for resistors in parallel, and vice versa.

Circuits containing resistor and capacitor

Capacitors and resistors are often found together in a circuit. A simple example is shown in Figure 16-18*a*. Let us analyze this circuit. To begin with, we leave the switch S_2 open so we can ignore the right-hand part of the circuit. When the switch S_1 is closed, current will immediately begin to flow through the circuit. Electrons will flow out of the negative terminal of the battery, through the resistor R, and accumulate on the upper plate of the capacitor. And electrons will flow into the positive terminal of the battery, leaving a positive charge on the lower plate of the capacitor. As the charge accumulates on the capacitor, the current is reduced until eventually the voltage across the capacitor equals the emf of the battery

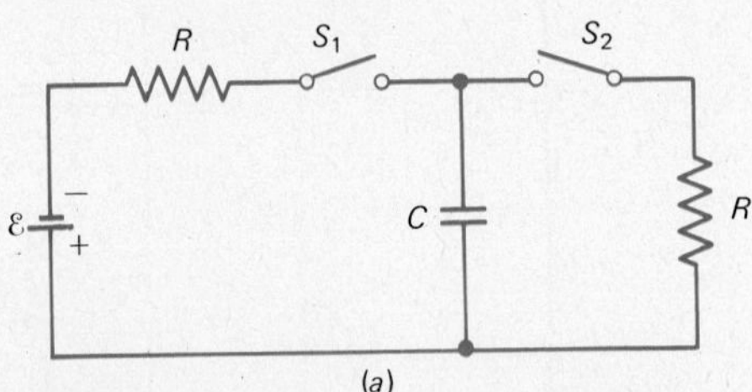

(*a*)

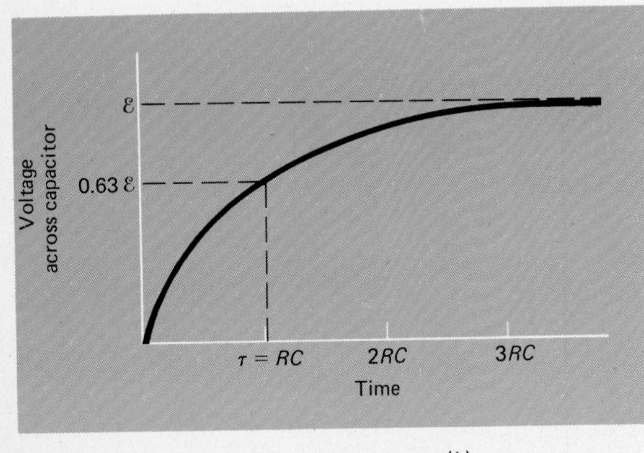

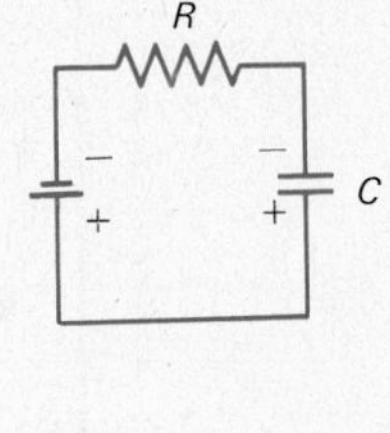

(*b*)

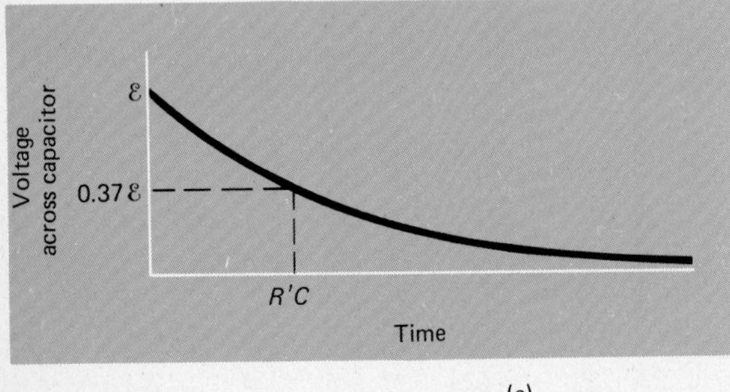

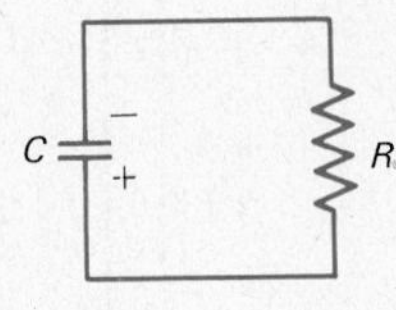

(*c*)

FIGURE 16–18
***RC* circuits.**

and no further current flows. The potential difference across the capacitor thus increases gradually, as shown in Figure 16-18*b*. The product of the value of the resistance times that of the capacitor is called the *time constant* τ of the circuit: $\tau = RC$. It is a measure of how quickly the capacitor becomes charged. Specifically, it can be shown (using calculus) that the product RC gives the time† required for the capacitor to reach 63 percent of full voltage. For example, if $R = 200\ \text{k}\Omega$ and $C = 3.0\ \mu\text{F}$, the time constant is $(2.0 \times 10^5\ \Omega)(3.0 \times 10^{-6}\ \text{F}) = 0.60$ s. If the resistance is low, the time constant is much smaller and the capacitor becomes charged almost instantly. This makes sense, since a lower resistance will retard the flow of charge less.

Now let us refer back to Figure 16-18*a*. Suppose now that the capacitor is fully charged and we open switch S_1 and close switch S_2. As soon as S_2 is closed, charge begins to flow through resistor R' from one side of the capacitor toward the other side, until it is fully discharged. The voltage across the capacitor decreases, as shown in Figure 16-18*c*. The voltage falls 63 percent of the way to zero (to $0.37\,\mathscr{E}$) in a time $\tau' = R'C$.

One kind of circuit that finds wide use is illustrated in Figure 16-19*a*. The gas-filled tube on the right acts like a switch. For low voltages it has essentially infinite resistance; but above a certain voltage the gas breaks down and becomes conducting, so the tube's resistance becomes nearly zero. The breakdown voltage, V_0, is designed to be somewhat less than the emf of the battery. Thus the voltage across the capacitor rises nearly linearly (lower part of the graph of Figure 16-18*b*) until it reaches the breakdown voltage of the gas-filled tube. The tube's resistance suddenly drops to near zero and the capacitor discharges. The value of R' is usually much smaller than R, so the discharge is much quicker than the charging process. Once the capacitor has discharged, the gas-filled tube returns to its normal state of nearly infinite resistance. The capacitor is thus repeatedly charged and discharged. This produces a "sawtoothed" voltage as a function of time (Figure 16-19*b*).

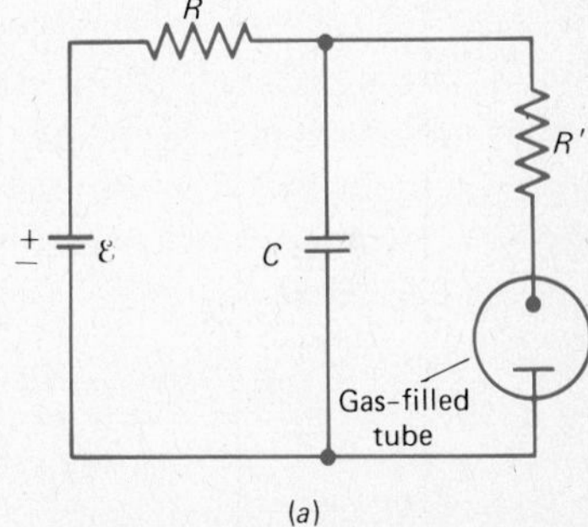

(*a*)

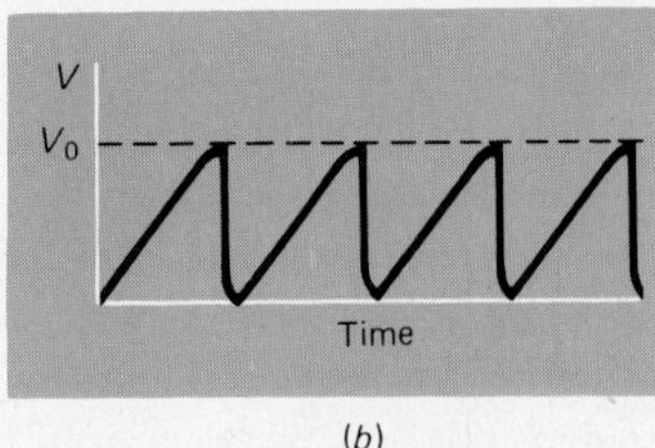

(*b*)

FIGURE 16-19
RC circuit, coupled with a gas-filled tube as a switch, can produce a sawtoothed voltage.

This kind of voltage is useful in many kinds of electronic gear. It is one way to produce the pulsed voltage in artificial heart pacemakers, as will be mentioned in the next section. It is also used in many high-voltage situations, such as in a television set, in industry, and in research. A gas-filled tube designed to break down at high voltage is called a *thyratron*.

Capacitors are used a great deal in ac circuits, and this will be discussed in Chapter 19.

16-8 ▌ Effects of current on the human body

Hazards

An electric shock can cause damage to the body and can even be fatal. The severity of a shock depends on the magnitude of the current, how long it acts, and through what part of the body it passes. A current that flows from the thumb to a finger on the same hand is much less damaging than one that flows from one hand to the other and passes through the heart. Indeed, a current that passes through vital organs such as the heart and

† In terms of units, $RC = (\Omega)(\text{F}) = (\text{V/A})(\text{C/V}) = \text{C/(C/s)} = \text{s}$.

brain is especially serious, for it can interfere with their operation. A current heats tissue and can cause burns, particularly on the skin, where the resistance is high. Also a current stimulates the nerves and muscles of the body. We feel a "shock" because our muscles contract.

Most people can "feel" a current of about 1 mA. Currents of a few milliamperes cause pain but rarely much damage in a healthy person. However, currents above 10 mA cause severe contraction of the muscles; in this case a person may not be able to release the source of the current (say, a faulty appliance or wire). Death from paralysis of the respiratory system could then occur; artificial respiration, however, can often revive a victim. If a current above about 70 mA passes in the region of the heart for a second or more, the heart muscles will begin to contract irregularly and blood is not properly pumped. This condition is called *ventricular fibrillation*. If it lasts for long, death results. Strangely enough, if the current is much larger, on the order of 1 A, the damage may be less and death less likely. Apparently larger currents bring the entire heart to a standstill. Upon release of the current, the heart returns to its normal rhythm. This may not happen when fibrillation occurs, for it is often hard to stop once it starts.†

Since it is current not voltage that causes damage, shocks produced by static electricity are rarely very harmful. Although the potential may be high (even kilovolts), the charge is usually small and flows quickly. (Lightning is an exception, however.) The main danger from static electricity is that in an explosive atmosphere (such as a volatile substance mixing with air) a spark could cause an explosion. Similarly, a voltage source that cannot produce much current will not be able to cause much damage. Such a source would have a high internal resistance, so the terminal voltage would drop greatly if much current were drawn. Ordinary house current is another matter, since there is little internal resistance and the voltage is maintained steadily.

The seriousness of a shock depends on the resistance of the body. Living tissue has quite a low resistance, since the fluid of cells contains ions that can conduct quite well. However, the outer layers of skin, when dry, offer considerable resistance. The resistance of a human body when the skin is dry is in the range 10^4 to $10^6\ \Omega$. However, when the skin is wet, the resistance may be $1000\ \Omega$ or less. A person in good contact with the ground who touches a 120-V line with wet hands can suffer a current

$$I = \frac{120\ \text{V}}{1000\ \Omega} = 120\ \text{mA}.$$

As we saw above, this could be lethal.

Figure 16-20 shows how the circuit is completed when a person touches an electric wire. One side of a 120-V source is connected to ground by a wire connected to a buried conductor (say, a water pipe). Thus the current passes from the high-voltage wire, through the person, to the ground; it passes through the ground back to the other terminal of the source to complete the circuit. If the person in Figure 16-20 stands on a

FIGURE 16-20
A person receives an electric shock when the circuit is completed.

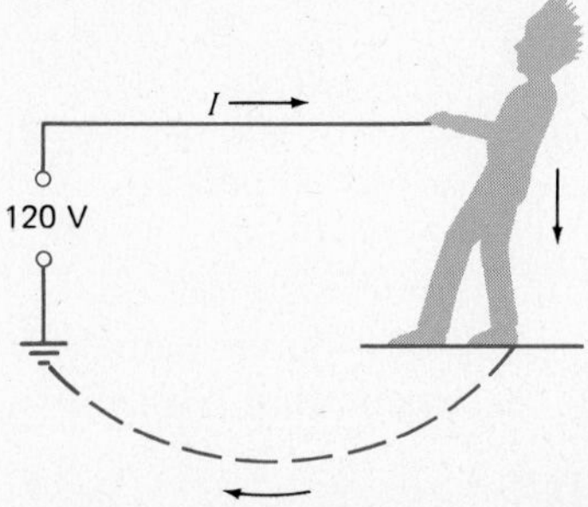

† Fibrillation may also occur as a result of a heart attack or during heart surgery. A device known as a *defibrillator* can apply a brief high current to the heart; this causes complete heart stoppage and is often followed by resumption of normal beating.

good insulator—thick-soled shoes or a dry wood floor—there will be much more resistance in the circuit and consequently much less current will flow. However, if the person stands with bare feet on the ground, or is sitting in a bathtub, there is considerable danger. In a bathtub, not only are you wet, but the water is in contact with the drain pipe that leads to the ground. That is why it is strongly recommended not to touch anything electrical in such a situation.

A principal danger comes from touching a bare wire whose insulation has worn off, or from a bare wire inside an appliance when you're tinkering with it. (Always unplug an appliance before investigating its insides!) Sometimes a wire inside an appliance breaks or loses its insulation and comes in contact with the case. If the case is metal, it will conduct electricity. A person could then suffer a severe shock merely by touching the case. To prevent an accident, most appliance cases are connected by a wire to the ground. Then if a "hot" wire touches the case, a short circuit to ground immediately occurs and the fuse or circuit breaker opens the circuit.

An appliance can be connected to ground in one of several ways: (1) a separate wire can be connected to a metal water pipe that leads to the ground; (2) the cord can contain three wires and the plug three prongs (the third hole in an outlet is connected to ground); (3) if the outlet accepts only two prongs, it may have an extra screw connection that leads to ground and an adapter for a three-prong plug can be connected to this screw; and (4) some appliances come with plugs whose two prongs are of different sizes; if the holes in the outlet are of two sizes, the plug can (hopefully) be put in only one way; if the proper outlet lead in the building has been grounded nearby, the case of the appliance is connected to this lead and thus is grounded.

The currents discussed above relate to potential differences applied to the exterior of the body. In hospitals where electrodes may be placed inside the body, damaging effects may occur at only 1/1000 of the currents mentioned above. This is because the currents may be concentrated over a small area. Thus great care must be taken under these circumstances.

* Pacemakers

If the heart stops, it can be made to start beating again by applying an electric stimulus through electrodes attached to the chest. The stimulus can be repeated at the normal heartbeat rate if necessary.

The heart itself contains a *pacemaker*, which sends out tiny electric pulses at a rate of 60 to 80 per second. These are the signals that induce the start of each heartbeat. In some forms of heart disease, the pacemaker cells fail to function properly and the heart loses its beat. *Electronic pacemakers* are now commonplace for people suffering this ailment. They can be external or can be inserted beneath the skin. They produce a regular voltage pulse that starts and controls the frequency of the heartbeat. The "fixed-rate" type produces signals continuously. The "demand" type operates only when the natural pacemaker fails. The electrodes are implanted in or near the heart and the circuit usually contains a capacitor and a resistor. The charge on the capacitor increases to a certain point and then discharges; then it starts charging again. The pulsing rate depends on the values of R and C.

Generally, the power source is a battery which must be replaced every year or two; this requires a surgical operation for an internal pacemaker. Some pacemakers obtain their energy from the heat produced by a radioactive element; the thermal energy is transformed to electricity by a thermocouple. A third type has a rechargeable battery. It must be recharged about once a week and can be done externally. This type of battery may last 20 years. A fourth type uses the heart's own contractions as its source; the critical element is a piezoelectric crystal (see Section 16-6), which produces an emf in response to the pressure of the contracting heart. One problem with this type is that the body tends to encapsulate it and impair its contact with the heart's contractions.

▮ Summary

When resistances are connected in series, the net resistance is the sum of the individual resistances. When connected in parallel, the reciprocal of the total resistance equals the sum of the reciprocals of the individual resistances; in a parallel connection, the net resistance is less than any of the individual resistances.

A battery behaves like a source of emf in series with an internal resistance. The emf is the potential difference determined by the chemical reactions in the battery and equals the terminal voltage only when no current is drawn. When a current is drawn, the voltage at the battery's terminals is less than its emf by an amount equal to the Ir drop across the internal resistance.

An ammeter, which measures current, consists of a low resistance in parallel with a galvanometer whose deflection is proportional to the current passing through it. An ammeter is placed into a circuit, in series with the other elements. A voltmeter measures potential difference and consists of a high resistance in series with a galvanometer. A voltmeter is not connected directly into a circuit but is placed in parallel with the element across which the voltage is to be measured.

When capacitors are connected in parallel, the net capacitance is the sum of the individual capacitances. When connected in series, the reciprocal of the net capacitance equals the sum of the reciprocals of the individual capacitances. When a circuit containing a resistor R in series with a capacitance C is connected to a dc source of emf, the voltage across the capacitor rises gradually in time characterized by the time constant $\tau = RC$; this is the time it takes for the voltage to reach 63 percent of its maximum value. A capacitor discharging through a resistor is characterized by the same time constant: in a time $\tau = RC$, the voltage across the capacitor drops to 37 percent of its initial value.

Electric shocks are caused by current passing through the body. To avoid shocks, the body must not become part of a circuit by allowing different parts of the body to touch objects at different potentials; commonly, one part of the body may be touching ground and another part a high or low potential.

Questions

1 Discuss the advantages and disadvantages of Christmas tree lights connected in parallel versus those connected in series.

2 If all you have is a 120-V line, would it be possible to light several 6-V lamps without burning them out? How?

3 Why are batteries connected in series? Why in parallel? Does it matter if the batteries are nearly identical or not in either case?

4 The internal resistance of an electric cell is not actually constant. Why not?

5 Describe a situation in which the terminal voltage of a battery is greater than its emf.

6 The 18-V source in Figure 16-21 is "charging" the 12-V battery. Explain how it does this.

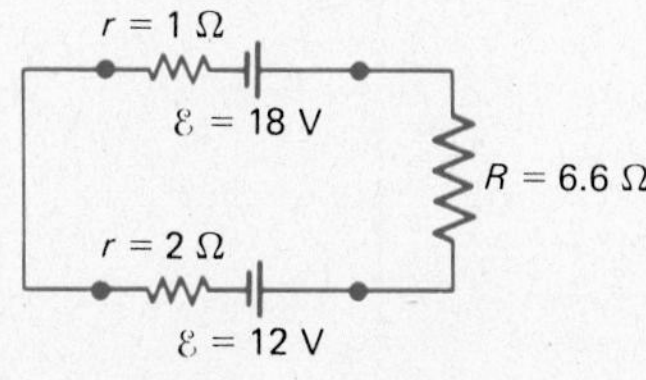

FIGURE 16-21

7 What is the main difference between a voltmeter and an ammeter?

8 Explain why an ideal ammeter would have zero resistance and an ideal voltmeter infinite resistance.

9 What are the advantages of a potentiometer over a voltmeter? What are the advantages of a voltmeter over a potentiometer?

10 If the battery of a Wheatstone bridge wears down so that its terminal voltage drops a little, will this affect the determination of the unknown resistance?

11 Suppose that three identical capacitors are connected to a battery. Will they store more energy if connected in series or in parallel?

12 What might be the consequences of cutting off the third prong of a three-prong plug? If the outlet only accepts a two-prong plug, what should you do?

13 Why is it more dangerous to turn on an electric appliance when you are standing outside in bare feet than when you are inside wearing shoes with thick soles?

Problems

SECTION 16-1

In these problems neglect the internal resistance of a battery unless it is given.

1 (I) Six 200-Ω light bulbs are connected in series. What is the total resistance of the circuit? What is their resistance if they are connected in parallel?

2 (I) Suppose that you have a 500-Ω, an 800-Ω, and a 1.2 kΩ resistor. What is (*a*) the maximum, and (*b*) the minimum resistance you can obtain by combining these?

3 (I) A battery whose emf is 6.0 V and internal resistance is 1.0 Ω is connected to a circuit whose net resistance is 16 Ω. What is the terminal voltage of the battery?

4 (I) Four 1.5-V cells are joined in series to a 12-Ω device. The internal resistance of each cell is 0.3 Ω. What current flows to the device?

5 (I) The internal resistance of a 1.35-V mercury cell is 0.04 Ω, whereas that of a 1.5-V dry cell is 0.50 Ω. Explain why three mercury cells can more effectively power a 2-W hearing aid that requires 4 V than can three dry cells.

6 (I) A 1.5-V dry cell can be tested by connecting it to a low-resistance ammeter; it should be able to supply at least 30 A. What is the internal resistance of the cell in this case?

7 (II) Suppose that you have a 6-V battery and you wish to apply a voltage of only 1 V. How could you connect resistors so that you could make a "voltage divider" that produced a 1-V output for a 6-V input?

8 (II) Three 100-Ω resistors can be connected together in four different ways, making combinations of series and/or parallel circuits. What are these four ways and what is the net resistance in each case?

9 (II) What is the net resistance of the circuit connected to the battery in Figure 16-22? Each resistance has $R = 1.0$ kΩ.

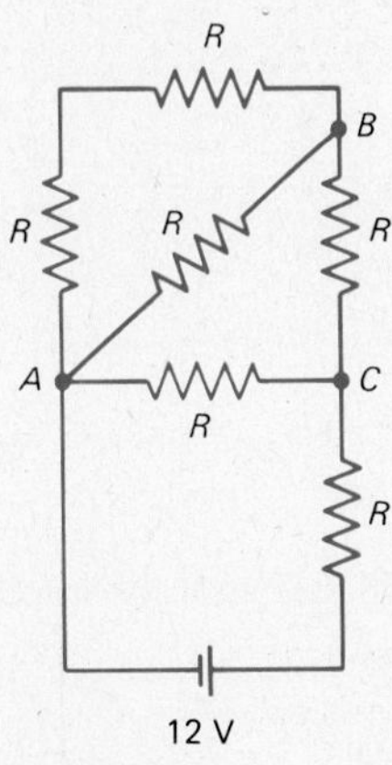

FIGURE 16-22

10 (II) Eight lights are connected in series across a 120-V line. (*a*) What is the voltage across each bulb? (*b*) If the current is 0.50 A, what is the resistance of each bulb and the power dissipated in each?

11 (II) Eight lights are connected in parallel to a 120-V source by two leads of total resistance 2.0 Ω. If 100 mA flows through each bulb, what is the resistance of each, and what fraction of the total power is wasted in the leads?

12 (II) Seven 10-W Christmas tree lights are connected in series to a 120-V source. What is the resistance of each bulb?

13 (II) Suppose that a person's body resistance is 1000 Ω. (*a*) What current passes through the body when the person accidentally is connected to 100 V? (*b*) If there is an alternative path to ground whose resistance is 20 Ω, what current passes through the person? (*c*) If the voltage source can produce at most 2.0 A, how much current passes through the person in case (*b*)?

14 (II) What is the internal resistance of a 12-V battery whose terminal voltage drops to 7.5 V when the starter draws 80 A?

15 (II) Two resistors when connected in series to a 120-V line use one-fourth the power that is used when they are connected in parallel. If one resistor is 2.0 kΩ, what is the resistance of the other?

16 (II) Suppose that you wish to apply a 0.25-V potential difference between two points on the body. The resistance is about 2000 Ω, and you only have a 9.0-V battery. How can you connect up one or more resistors so that you can produce the desired voltage?

17 (II) A three-way light bulb can produce 50 W, 100 W, or 150 W at 120 V. Such a bulb contains two filaments that can be connected to the 120 V individually or in series. Describe how the connections to the two filaments are made to give each of the three wattages; what must be the resistance of each filament?

18 (II) A 75-W 120-V bulb is connected in parallel with a 40-W 120-V bulb. What is the net resistance?

19 (II) Suppose you want to run some apparatus that is 200 m from an outlet. Each of the wires connecting your apparatus to the 120-V source has a resistance per unit length of 0.0050 Ω/m. If your apparatus draws 2.0 A, what will be the voltage drop across the lead cord and what voltage will be applied to your apparatus?

20 (III) What is the current in the 8.0 Ω resistor in Figure 16-5?

21 (III) A 2.2-kΩ and a 1.3-kΩ resistor are connected in parallel; this combination is connected in series with a 1.7-kΩ resistor. If each resistor is rated at $\frac{1}{2}$ W, what is the maximum voltage that can be applied across the whole network?

22 (III) How many $\frac{1}{2}$-W resistors, each of the same resistance, must be used to produce an equivalent 2.0-kΩ 5-W resistor? What is the resistance of each, and how must they be connected?

23 (III) A battery produces 50.0 V when 5.0 A is drawn from it and 48.5 V when 20.0 A is drawn. What is the emf and internal resistance of the battery?

*SECTION 16-2

***24** (II) Determine the terminal voltage of each battery in Figure 16-21.

***25** (II) What is the terminal voltage of each battery in Figure 16-6?

***26** (II) What is the potential difference between points *a* and *d* in Figure 16-6?

***27** (III) Determine the current through each of the resistors in Figure 16-23.

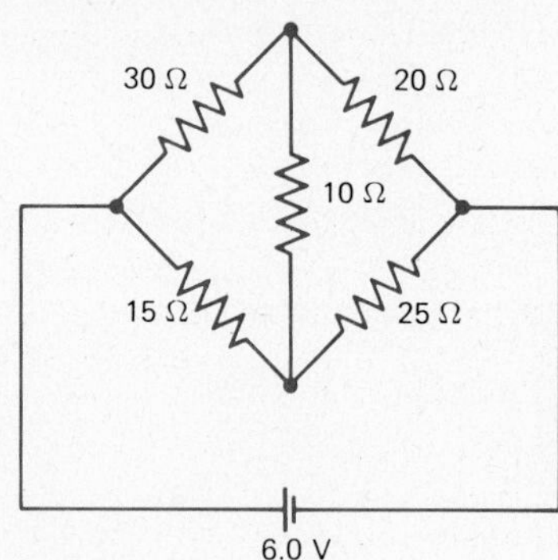

FIGURE 16-23

***28** (III) If the 30-Ω resistor in Figure 16-23 were shorted out (resistance = 0), what would be the current through the 10-Ω resistor?

***29** (III) Determine the currents I_1, I_2, and I_3 in Figure 16-24. Assume that the internal resistance of each battery is $r = 1.0$ Ω. What is the terminal voltage of the 6.0-V battery?

***30** (III) What would the current I_1 be in Figure 16-24 if the 12-Ω resistor were shorted out ($r = 1.0$ Ω)?

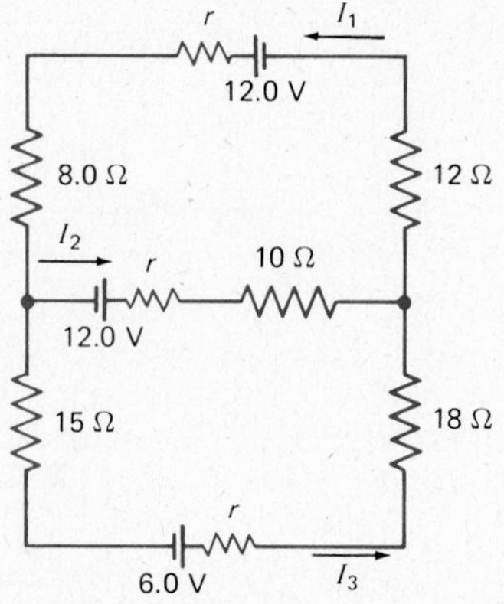

FIGURE 16-24

SECTION 16-3

31 (I) What is the resistance of a voltmeter on the 50-V scale if the meter sensitivity is 20,000 Ω/V?

32 (I) An ammeter has a sensitivity of 5000 Ω/V. What current passing through the galvanometer produces full-scale deflection?

33 (II) A galvanometer has an internal resistance of 30 Ω and deflects full-scale for a 70-μA current. Describe how to use this galvanometer to make (*a*) an ammeter to read currents up to 15 A; and (*b*) a voltmeter to give a full-scale deflection of 5000 V.

34 (II) A galvanometer has a sensitivity of 40,000 Ω/V and internal resistance 20 Ω. How could you make this into (*a*) an ammeter reading 10 mA full scale; (*b*) a voltmeter reading 100 mV full scale?

35 (II) A milliammeter reads 10 mA full scale. It consists of a 0.20-Ω resistor in parallel with a 30-Ω galvanometer. How can you change this ammeter to a voltmeter giving a full-scale reading of 10 V without taking the ammeter apart? What will be the sensitivity (Ω/V) of your voltmeter?

36 (II) A 45-V battery of negligible internal resistance is connected to a 30-kΩ and a 20-kΩ resistor in series. What reading will a voltmeter give when used to measure the voltage across each resistor if its internal resistance is 100 kΩ?

37 (II) A battery with $\mathscr{E} = 9.0$ V and internal resistance $r = 1.2$ Ω is connected to two 5.0-kΩ resistors in series. An ammeter of internal resistance 0.50-Ω measures the current and at the same time a voltmeter with internal resistance 15 kΩ measures the voltage across one of the 5.0-kΩ resistors in the circuit. What do the ammeter and voltmeter read?

38 (II) An ammeter whose internal resistance is 100 Ω reads 1.60 A when connected in a circuit containing a battery and two resistors in series whose values are 300 Ω and 500 Ω. What is the actual current when the ammeter is absent?

39 (III) What internal resistance should the voltmeter have to be in error by less than 3 percent for the situation of Example 16-5?

40 (III) A 9.0-V battery is connected to two resistors in series. A voltmeter whose internal resistance is 10.0 kΩ measures 3.0 V and 4.0 V, respectively, when connected across each of the resistors. What is the resistance of each resistor?

41 (III) Two 5.0-kΩ resistors are placed in series and connected to a battery. A voltmeter of sensitivity 1000 Ω/V is on the 5.0-V scale and reads 3.0 V when placed across either of the resistors. What is the emf of the battery? (Ignore the battery's internal resistance.)

42 (III) The voltage across a 100-kΩ resistor in a circuit containing additional resistance (R_2) in series with a battery (V) is measured, by a 20,000-Ω/V meter on the 100-V scale, to be 25 V. On the 30-V scale, the reading is 23 V. What is the actual voltage in the absence of the voltmeter? What is the value of R_2?

43 (III) An ohmmeter consists of an ammeter connected to a series resistor and a battery (Figure 16-25). The scale is different from ammeters and voltmeters. Zero resistance corresponds to full-scale deflection (because maximum current flows from the battery), whereas infinite resistance corresponds to no deflection (no current flow). Suppose that a 1.5-V battery is used, and the galvanometer has internal resistance of 30 Ω and deflects full scale for a current of 30 μA. What values of shunt resistance, R_{sh}, and series resistance, R_{ser}, are needed to make an ohmmeter that registers a midscale deflection (that is, half of maximum) for a resistance of 30 kΩ? (*Note*: An additional series resistor, which is variable, is also needed so the meter can be zeroed; this should be checked frequently be touching the leads, since the battery voltage can vary. Because of battery voltage variation, such ohmmeters are not considered precision instruments but are useful to obtain approximate values.)

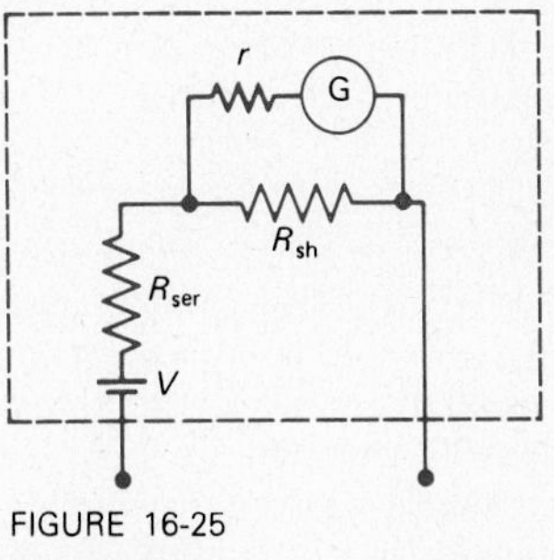

FIGURE 16-25
Ohmmeter.

44 (III) (*a*) A voltmeter and an ammeter can be connected as shown in Figure 16-26*a* to measure a resistance *R*. The value of *R* will not quite be V/I, where V is the voltmeter reading and I is the ammeter reading since some of the current actually goes

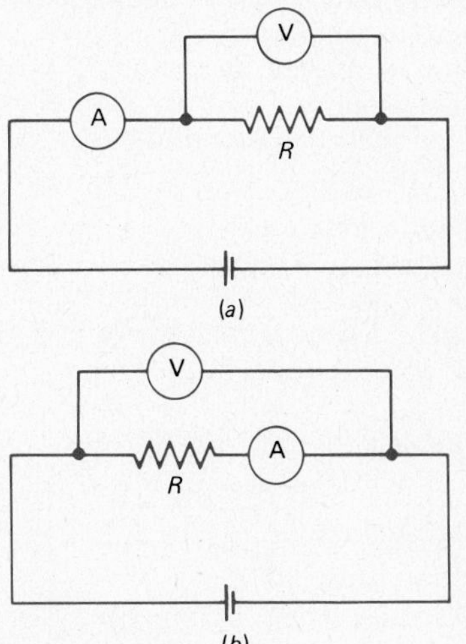

FIGURE 16-26

through the voltmeter. Show that the actual value of R is given by

$$\frac{1}{R}=\frac{I}{V}-\frac{1}{R_V},$$

where R_V is voltmeter resistance. Note that $R \approx V/I$ if $R_V \gg R$. (*b*) A voltmeter and an ammeter can also be connected as shown in Figure 16-26*b* to measure a resistance R. Show that in this case

$$R=\frac{V}{I}-R_A,$$

where V and I are the volt- and ammeter readings and R_A is the resistance of the ammeter. Note that $R \approx V/I$ if $R_A \ll R$.

*SECTION 16-4

***45** (I) A potentiometer is balanced against a 1.0182 V standard cell when the slide wire is set at 40.2 cm out of a total length of 100.0 cm. For an unknown source, the setting is 11.9 cm. What is the emf of the unknown?

***46** (I) The galvanometer of a potentiometer has an internal resistance of 40 Ω and can detect a current as small as 0.015 mA. What is the minimum uncertainty possible in measuring an unknown voltage?

***47** (II) The uniform 30.00-cm-long slide wire of a potentiometer has a total resistance of 3.224 Ω. When a 1.0183-V standard cell is connected, a 22.40-cm length of wire is needed to achieve balance. When an unknown source is connected, balance is achieved only when a 9.050-Ω resistor (previously connected between the working battery and point A in Figure 16-12) is now included along with the slide wire resistance, and the slide wire is set at 14.16 cm. What is the value of the unknown emf?

***48** (III) A potentiometer slide wire must be set at a 48.0-cm length to balance the emf of a cell. When a 5.0-Ω resistor is placed in parallel across the terminals of the cell, a 44.0-cm length is needed. What is the internal resistance of the cell?

*SECTION 16-5

***49** (I) Figure 16-27 shows a slide-wire form of Wheatstone bridge. If $R_3 = 34.5$ Ω and the lengths $AC = 28.2$ cm and $CB = 71.8$ cm when the galvanometer reads zero current, what is the value of R_X? Assume that the slide wire is uniform.

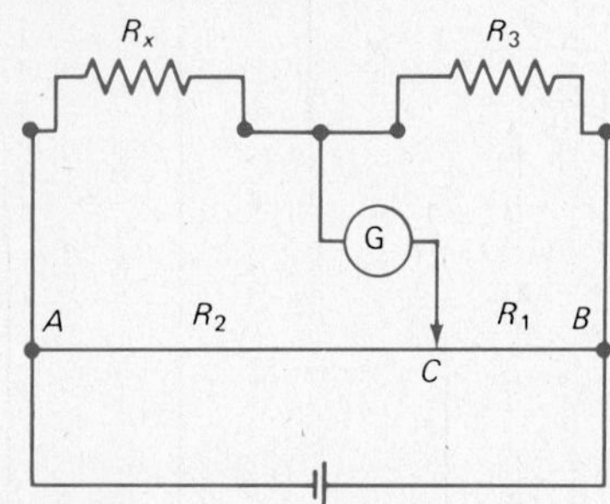

FIGURE 16-27

Slide-wire form of Wheatstone bridge.

***50** (I) A Wheatstone bridge is balanced when $R_1 = 710$ Ω, $R_2 = 317$ Ω, and $R_3 = 2.24$ kΩ. What is the value of the unknown resistance in the fourth arm?

***51** (II) An unknown length of platinum wire 1.2 mm in diameter is placed as the unknown resistance in a Wheatstone bridge. Arms 1 and 2 have resistance of 36.0 Ω and 84.0 Ω, respectively. Balance is achieved when R_3 is 4.16 Ω. How long is the platinum wire?

*SECTION 16-6

***52** (I) A copper-constantan thermocouple produces emfs of about 40 μV/C°. If the reference temperature is 25°C, what must be the temperature of the test junction if the emf produced is 2.12 mV?

***53** (I) For iron-copper junctions near room temperature, the emf produced by a thermocouple is about 14 μV/C°. If a potentiometer can detect emfs as low as 0.50 μV, to what accuracy can the temperature be read?

***54** (II) The emf of a thermocouple can be written $\mathscr{E} = aT + bT^2$, where T is the temperature (°C) to be measured and the reference temperature is 0 °C; a and b are constants, and for a purely linear response b would be zero (usually it is small). (*a*) Determine the values for a and b if the emf is 3.62 mV at $T = 80$ °C and is 6.80 mV at $T = 170$ °C. (*b*) What is the temperature if the emf is 4.90 mV?

***55** (II) The *strain factor K* of a strain gauge is defined as the fractional change in resistance ($\Delta R/R$) divided by the fractional change in length,

$$K=\frac{\Delta R/R}{\Delta L/L},$$

and is relatively constant at a given temperature. In other words, the change in resistance ΔR is proportional to the change in length of the wire ΔL. A strain gauge with a strain factor of 2.0 is connected crosswise to a small muscle about 4.5 mm wide. The gauge is connected as the unknown arm in a Wheatstone bridge. When the muscle is relaxed, a Wheatstone bridge is balanced when $R_2/R_1 = 1.480$ and $R_3 = 40.70$ Ω. When the muscle contracts, the bridge balances for $R_3 = 40.73$ Ω. How much has the muscle widened?

SECTION 16-7

56 (I) Six 1.0-μF capacitors are connected in parallel. What is the net capacitance? What if they are connected in series?

57 (I) A circuit contains a 3.0-μF capacitor. However, a technician decides 4.0 μF would be better. What size capacitor should be added to the circuit and how should it be connected?

58 (I) The capacitance of a portion of a circuit is to be reduced from 3600 pF to 1000 pF. What capacitance can be added to the circuit to produce this effect without removing anything from the circuit? How should the extra capacitor be connected?

59 (I) Electrocardiographs are often connected as shown in Figure 16-28. The leads are said to be capacitatively coupled. A time constant of 3.0 s is typical and allows rapid changes in potential to be accurately recorded. If $C = 3.0\ \mu F$, what value must R have?

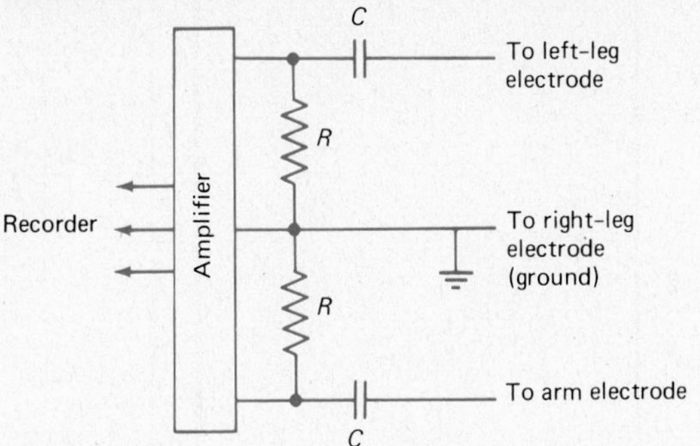

FIGURE 16-28

60 (II) You have three capacitors, of capacitance 2000 pF, 5000 pF, and 0.010 μF. What is the maximum and minimum capacitance that you can form from these? How do you make the connection in each case?

61 (II) A 0.20-μF and a 0.10-μF capacitor are connected in series to a 9.0-V battery. Calculate (*a*) the potential difference across each capacitor, and (*b*) the charge on each. (*c*) Repeat, assuming the capacitors are in parallel.

62 (II) Suppose a 4.0 μF capacitor were placed across the 6.0 Ω resistor in Figure 16-5*a*. Calculate the charge on the capacitor in the steady state—that is, after the capacitor reaches its maximum charge.

63 (II) A stage designer wishes a light to slowly dim when it is switched off. A possible circuit is that shown in Figure 16-29. When the switch is opened, the capacitor discharges through the lamp which is rated at 350 W at 120 V. Approximately what value of capacitance C should be used so that the power going to the bulb drops to 50 W after 5.0 s? Assume that the resistance of the bulb remains constant.

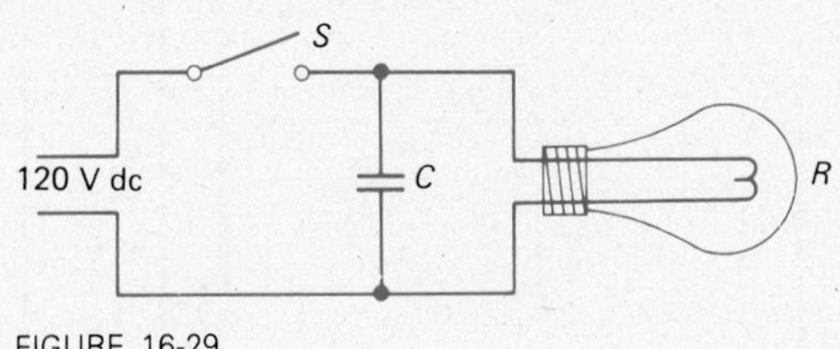

FIGURE 16-29

64 (III) A 3.0-μF and a 4.0-μF capacitor are connected in series, and this combination is connected in parallel with a 2.0-μF capacitor. (*a*) What is the net capacitance? (*b*) If 50 V is applied across the whole network, calculate the voltage across each capacitor.

SECTION 16-8

*65 (II) A pacemaker is designed to operate at 70 beats/min using a 7.0-μF capacitor. What value of resistance should be used if the pacemaker is to fire when the voltage reaches 63 percent of maximum?

66 (II) Electricity can be a hazard in hospitals, particularly to patients who are connected to electrodes, such as an ECG. For example, suppose that the motor of a motorized bed shorts out to the bed frame, and the bed frame's connection to a ground has broken (or was not there in the first place). If a nurse touches the bed and the patient at the same time, she becomes a conductor and a complete circuit can be made through the patient to ground through the ECG apparatus. This is shown schematically in Figure 16-30. Calculate the current through the patient.

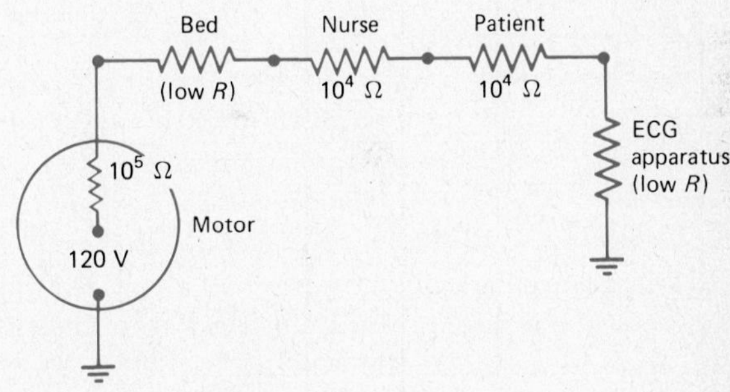

FIGURE 16-30

For additional reading

Brophy, J. J., *Basic Electronics for Scientists*, McGraw-Hill Book Company, New York, 1977.

17 ▮ Magnetism

Today it is clear that magnetism and electricity are closely related. This relationship was not discovered, however, until the nineteenth century. The history of magnetism begins much earlier with the ancient civilizations in Asia Minor. It was in a region of Asia Minor known as Magnesia that rocks were found that would attract each other. These rocks were called "lodestones" or "magnets" after their place of discovery.

17-1 ▮ Magnets and magnetism

Magnets

Nearly every child discovers that a magnet will attract paper clips, nails, and other objects made of iron. Each magnet, whether it is in the shape of a bar or a horseshoe, has two ends or faces, called poles; this is where the magnetic effect is strongest. If a magnet is suspended from a fine thread, it is found that one pole of the magnet will always point toward the north. It is not known for sure when this fact was discovered, but it is known that the Chinese were making use of it as an aid to navigation by the eleventh century and perhaps much earlier. This is, of course, the principle of a compass. A compass needle is simply a magnet that is supported at its center of gravity so it can rotate freely. That pole of a freely suspended magnet which points toward the north is called the *north pole* of the magnet. The other pole points toward the south and is called the *south pole*.

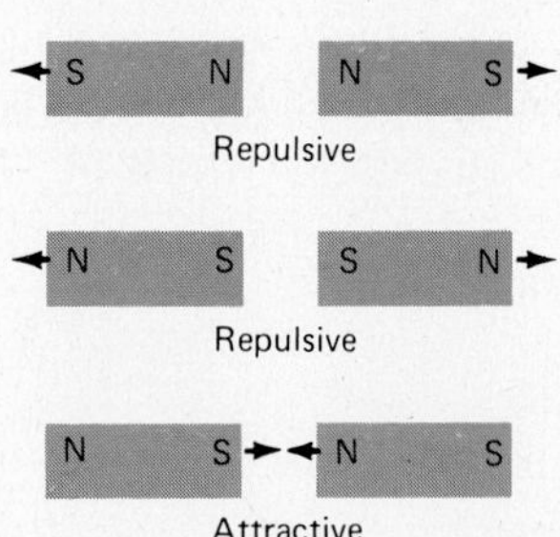

FIGURE 17-1
Like poles of a magnet repel; unlike poles attract.

It is a familiar fact that when two magnets are brought near one another, each exerts a force on the other. The force can be either attractive or repulsive and can be felt even when they do not touch. If the north pole of one magnet is brought near the north pole of a second magnet, the force is repulsive. Similarly, if two south poles are brought close, the force is repulsive. But when a north pole is brought near a south pole, the force is attractive. See Figure 17-1. This is reminiscent of the force between electric charges; for like poles repel and unlike poles attract. But do not confuse magnetic poles with electric charge. They are not the same thing.

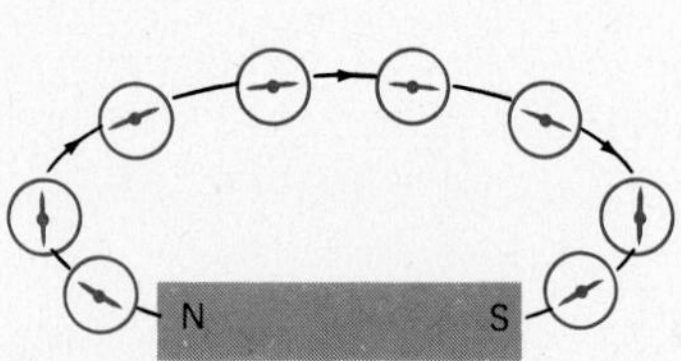

FIGURE 17-2
Plotting a magnetic field line of a bar magnet.

Only iron and a few other materials such as cobalt, nickel, and gadolinium show strong magnetic effects. They are said to be *ferro-*

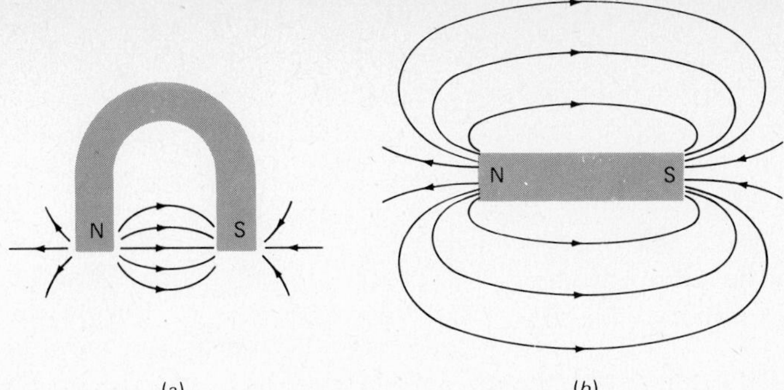

FIGURE 17-3
Magnetic field lines of (*a*) a horseshoe magnet and (*b*) a bar magnet.

magnetic (from the Latin word *ferrum* for iron). All other materials show a slight, but extremely small, magnetic effect.

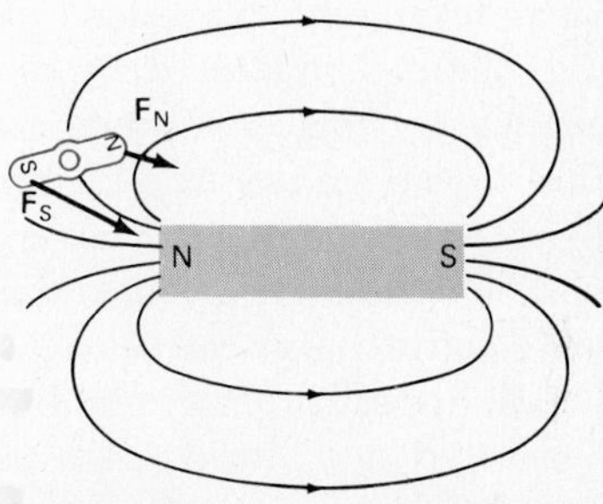

FIGURE 17-4
Forces on a compass needle that produce a torque to orient it parallel to the magnetic field lines. The torque will be zero when the needle is parallel to the magnetic field line at that point. (Only the "attractive" forces are shown; can you draw in the repulsive forces and show that they produce a similar torque?)

Magnetic fields

We found it useful to speak of an electric field surrounding an electric charge. In the same way we can imagine a magnetic field surrounding a magnet. The force one magnet exerts on another can then be described as the interaction between one magnet and the magnetic field of the other. Just as we drew electric field lines, we can also draw magnetic field lines.

The direction of the magnetic field at a given point is defined as the direction that the north pole of a compass needle would point when placed at that point. Figure 17-2 shows how one magnetic field line around a bar magnet is found using compass needles. The magnetic fields determined in this way for a horseshoe magnet and a bar magnet are shown in Figure 17-3. Notice that because of our definition, the lines always point from the north toward the south pole of a magnet (the north pole of a magnetic compass needle is attracted to the south pole of another magnet).

As in the case of the electric field, the magnetic field is strongest where the lines are closest together. We define the **magnetic field strength** as a vector, represented by the symbol **B**, whose direction is as defined above; the magnitude of **B** can be defined in terms of the torque exerted on a compass needle when it is not aligned with the magnetic field, as in Figure 17-4. That is, the greater the torque, the greater the magnetic field strength. We could use such a definition for magnetic field strength now, but a more precise definition will be given in Section 17-3. The terms "magnetic flux density" or "magnetic induction" are sometimes used for **B** rather than our term "magnetic field strength."

The earth's magnetic field

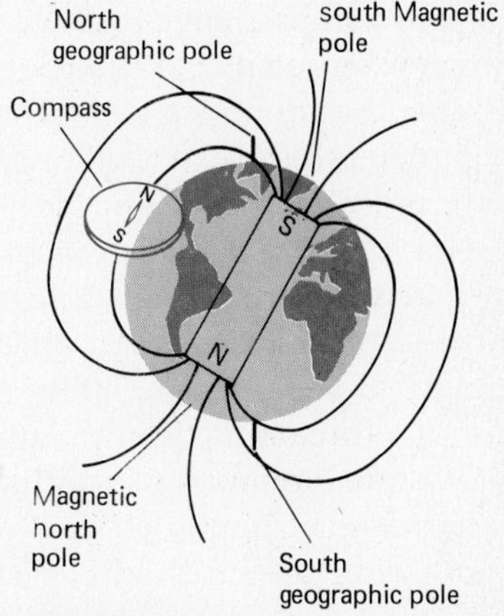

FIGURE 17-5
The earth acts like a huge magnet with its magnetic south pole near the geographic north pole.

The earth's magnetic field is shown in Figure 17-5. Since the north pole of a compass needle points north, it is the earth's *south* magnetic pole which is in the geographic north (remember, the north pole of one magnet is attracted to the south pole of a second); and the earth's north magnetic pole is near the geographic south pole. The earth's magnetic poles do not, however, coincide with the geographic poles (which are on the earth's axis of rotation). The south magnetic pole, for example, is in northern Canada, about 1500 km from the geographic north pole. This must be taken into account when using a compass. The angular difference between magnetic north and true (geographical) north is called the "magnetic declination." In the U.S. it varies from 0° to about 25°, depending on location. On the east coast, magnetic north is west of true north, whereas on the west coast it is east of true north. The declination also varies over the years; the magnetic poles apparently move slowly in time. For example, the declination at Baltimore was 6°W in 1900 but 8°W in 1960. Notice in Figure 17-5 that the earth's magnetic field is not tangent to the earth's surface at all points. The angle **B** makes to the horizontal at any point is referred to as the angle of dip.

Domains

A bar magnet resembles an electric dipole (equal-magnitude positive and negative charges separated by a distance). Indeed, a bar magnet is sometimes referred to as a "magnetic dipole." There are opposite "poles" separated by a distance. And the magnetic field of a bar magnet is much like the electric field of an electric dipole: compare Figure 13-20*a* with Figure 17-3*b*. One important difference, however, is that a positive or negative electric charge can be isolated. But it has not been found possible to isolate a single magnetic pole. If a bar magnet is cut in half, you do not obtain isolated north and south poles; instead, two new magnets are produced (Figure 17-6). If the cutting operation is repeated, more magnets are produced, each with a north and a south pole. Physicists have tried various ways to isolate a single magnetic pole, but there is no evidence today that single poles exist.

FIGURE 17-6
If you break a magnet in half, you do not obtain isolated north and south poles; instead, two new magnets are produced, each with a north and south pole.

Microscopic examination reveals that a magnet is actually made up of tiny regions known as **domains**, which are at most 1 mm in length or width. Each domain behaves as a tiny magnet with a north and a south pole. In an unmagnetized piece of iron, these domains are arranged randomly as shown in Figure 17-7*a*. The magnetic effects of the domains cancel each other out, so this piece of iron is not a magnet. In a magnet, the domains are preferentially aligned in one direction as shown in Figure 17-7*b* (downward in this case).

A magnet can be made from an unmagnetized piece of iron by placing it in a strong magnetic field. (You can make a needle magnetic by stroking it with one pole of a strong magnet.) Careful observations show in this case that domains may actually rotate slightly so they are more nearly parallel to the external field; or, more commonly, the borders of domains move so that those domains whose magnetic orientation is parallel to the external field grow in size at the expense of other domains. This can be seen by comparing Figure 17-7*a* and *b*.

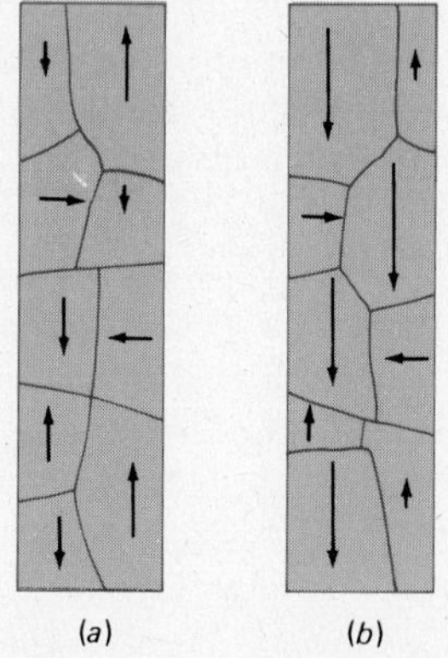

FIGURE 17-7
(*a*) Unmagnetized pieces of iron are made up of domains that are randomly arranged. (*b*) In a magnet, the domains are preferentially aligned in one direction. (The tip of each arrow represents the north pole of the domain.)

Now we can explain why a magnet can pick up unmagnetized pieces of iron such as paper clips or bobby pins. The magnet's field causes a slight alignment of the domains in the unmagnetized object so that the object becomes a temporary magnet with its north pole facing the south pole of the permanent magnet, and vice versa. Thus attraction results. In the same way, elongated iron filings will arrange themselves in a magnetic field just as a compass needle does and will reveal the shape of the magnetic field (Figure 17-8).

An iron magnet can remain magnetized for a long time, in which case it is referred to as a "permanent magnet." However, if you drop a magnet

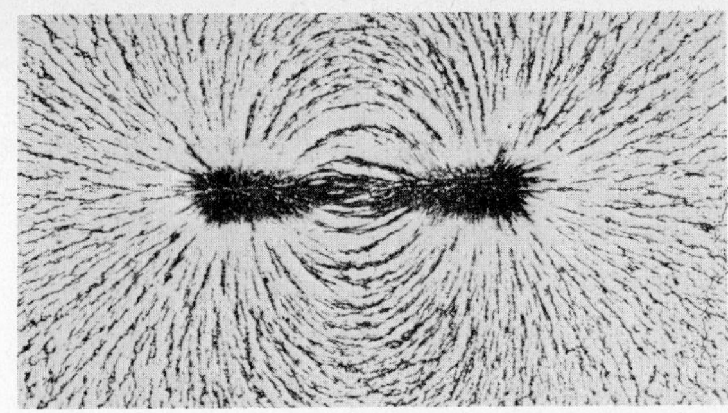

FIGURE 17-8
Iron filings line up along magnetic field lines.

on the floor or strike it with a hammer, you may jar the domains into randomness, and the magnet can thus lose some or all of its magnetism.

Heating a magnet can also cause a loss of magnetism; for raising the temperature increases the random thermal motion of the atoms, which tends to randomize the domains. Indeed, above a certain temperature known as the *Curie temperature* (1043 K for iron), a magnet cannot be made at all.†

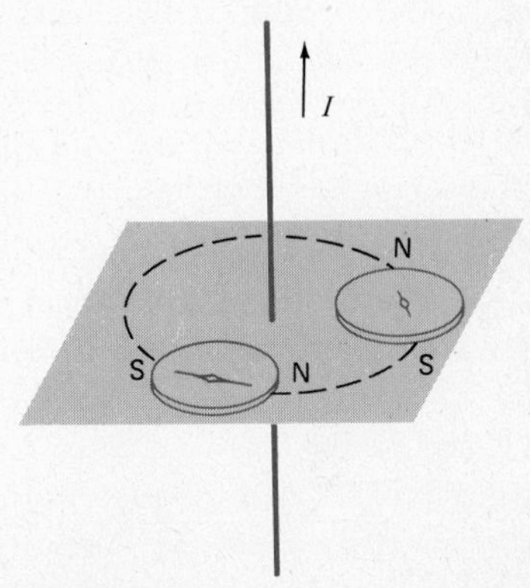

FIGURE 17-9
Deflection of a compass needle near a current-carrying wire, showing the presence and direction of the magnetic field.

17-2 ▌ Electric currents produce magnetism

The history of magnetism developed slowly until the nineteenth century. The real turning point was an accidental discovery made in 1820 by a Dane, Hans Christian Oersted (1777–1851). Oersted had believed for a long time in the unity of nature. Philosophically, he felt there ought to be a connection between magnetism and electricity. However, a stationary electric charge and a magnet had been shown not to have any influence on each other. But Oersted found that when a compass needle is placed near an electric wire, the needle deflects as soon as the wire is connected to a battery and a current flows.

As we have seen, a compass needle can be deflected by a magnetic field. What Oersted discovered was that **an electric current produces a magnetic field.** He had found the long-sought-after connection between electricity and magnetism.

Magnetic field of a straight wire

A compass needle placed near a straight section of wire aligns itself so that it is tangent to a circle drawn around the wire (Figure 17-9). Thus the magnetic field lines of the wire are in the form of circles, with the wire at their center (Figure 17-10*a*). The direction of these lines is indicated by the north pole of the compass in Figure 17-9. There is a simple way to remember the direction of the magnetic field lines in this case. It is called a *right-hand rule*. You grasp the wire with your right hand so that your thumb points in the direction of the conventional (positive) current; then your fingers will encircle the wire in the direction of the magnetic field, Figure 17-10*b*.

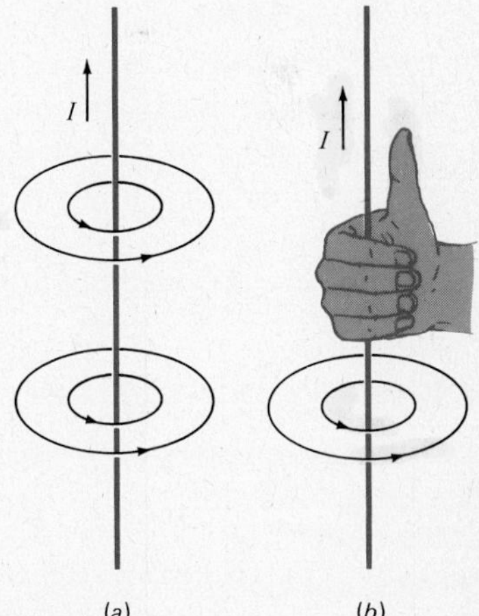

FIGURE 17-10
(*a*) Magnetic field lines around a straight wire. (*b*) Right-hand rule for remembering the direction of the magnetic field: When the thumb points in the direction of the conventional current, the fingers wrapped around the wire point in the direction of the magnetic field.

† Iron, nickel, cobalt, gadolinium, and certain alloys are ferromagnetic at room temperature; several other elements and alloys have low Curie temperatures and thus are ferromagnetic only at low temperatures.

Field of a circular loop

The magnetic field lines due to a circular loop of current-carrying wire can be determined in a similar way using a compass. The result is shown in Figure 17-11. Again the right-hand rule can be used, as shown in Figure 17-12.

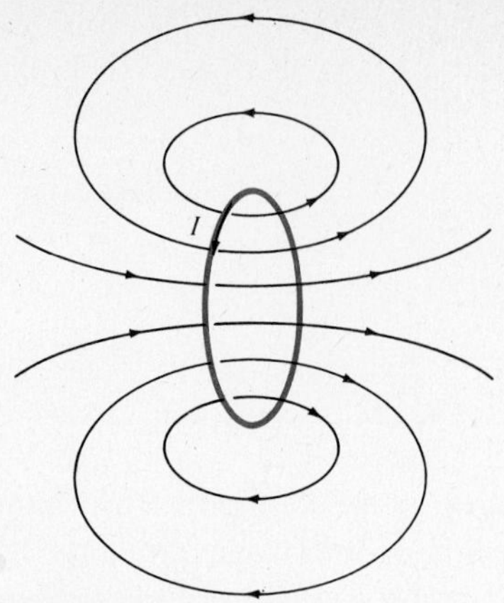

FIGURE 17-11
Magnetic field due to a circular loop of wire.

Ferromagnetism

Notice that the magnetic field of a current loop (Figure 17-11), is much like that produced by a bar magnet (Figure 17-3*b*). This suggests that the magnetic field produced by a current may have something to do with ferromagnets. The atoms that make up any material can be visualized as containing electrons that orbit around a central nucleus. Since the electrons are charged, they constitute an electric current and therefore produce a magnetic field. But the electron orbits in different atoms are arranged randomly, so the magnetic effects due to the many orbits of all the atoms in a material will cancel out. However, electrons produce an additional magnetic field as if they were spinning about their own axes; their charge would thus be rotating and producing a magnetic field.† It is the magnetic field due to electron spin that is now believed to produce ferromagnetism. In most materials the magnetic fields due to electron spin cancel out because they are oriented at random. However, in iron and other ferromagnetic materials, a complicated cooperative mechanism operates; the result is that all the electrons contributing to the ferromagnetism in a domain "spin" in the same direction. Thus the tiny magnetic fields due to each of the electrons add up to give the magnetic field of a domain; and when the domains are aligned, as we have seen, a strong magnet results.

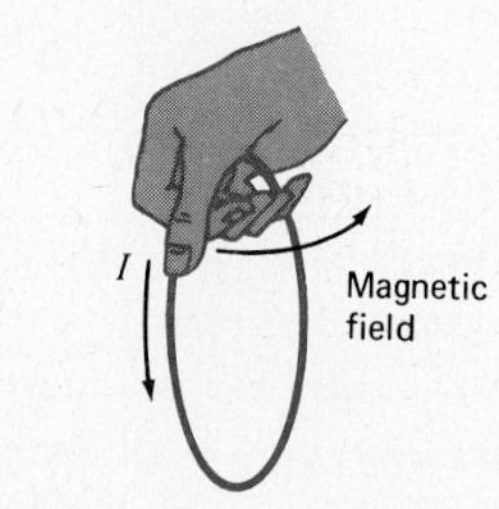

FIGURE 17-12
Right-hand rule for determining direction of magnetic field relative to current.

It is believed today that *all* magnetic fields are caused by electric currents. If this is true, it explains why it has not been possible to find a single magnetic pole. Each "spinning" electron is itself like a tiny current loop or magnet, with a north and a south pole. There is no way to divide up a current and obtain a single magnetic pole. Of course, if an isolated pole ever were to be found, we would have to change our theory that all magnetic fields are produced by currents.

Solenoids and electromagnets

A long coil of wire consisting of many loops of wire is called a *solenoid.* The magnetic field within a solenoid can be fairly large since it will be the sum of the fields due to the current in each loop (see Figure 17-13). The solenoid acts like a magnet; one end can be considered the north pole and the other the south pole, depending on the direction of the current in the loops (use the right-hand rule). Since the magnetic field lines leave the north pole of a magnet, the north pole of the solenoid in Figure 17-13 is on the right.

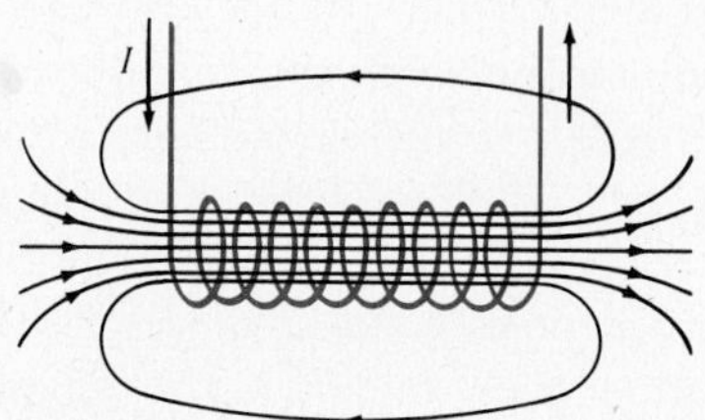

FIGURE 17-13
Magnetic field of a solenoid. The north pole of this solenoid, thought of as a magnet, is on the right, and the south pole is on the left.

If a piece of iron is placed inside a solenoid, the magnetic field is increased greatly because the domains of the iron are aligned by the magnetic field of the current. The resulting magnetic field is the sum of that due to the current and that due to the iron; it can be hundreds or

† This visual picture of electron spin is not completely valid, however; see Chapter 27.

thousands of times that due to the current alone. This arrangement is called an *electromagnet.*

The iron used in electromagnets acquires and loses its magnetism quite readily when the current is turned on or off; this is called "soft iron." (It is only "soft" in a magnetic sense.) Iron that holds its magnetism even when there is no externally applied field is called "hard iron." Hard iron is used in permanent magnets. Soft iron is usually used in electromagnets so that the field can be turned on and off readily. Whether iron is hard or soft depends on heat treatment and other factors.

Electromagnets find use in many practical applications, from lifting cars in junkyards, to use in motors and generators, to producing large magnetic fields for research. Because the current flows continually, a great deal of waste heat (I^2R power) is often produced. Cooling coils, which are tubes carrying water, must be used to absorb the heat in bigger installations. For some applications, superconducting magnets are used. The current-carrying wires are made of superconducting material (Section 15-3) kept below its transition temperature. Essentially no electric power is needed to maintain large current, which means large savings of electricity. Of course, some energy is needed to keep the superconducting coils at the necessary low temperature.

Another useful device consists of a solenoid into which a rod of iron is partially inserted. This combination is also referred to as a solenoid. One use is as a doorbell (Figure 17-14). When the circuit is closed by pushing the button, the coil effectively becomes a magnet and exerts a force on the iron rod. The rod is pulled into the coil and strikes the bell. A larger solenoid is used in the starters of cars; when you push the starter button, you are closing a circuit that not only turns the starter motor but activates a solenoid that first moves the starter into contact with the engine. Solenoids are used as switches in many other devices, such as tape recorders. They have the advantage of moving mechanical parts quickly and accurately.

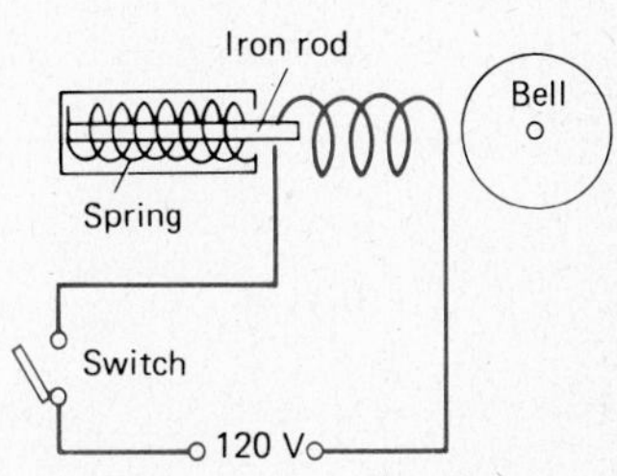

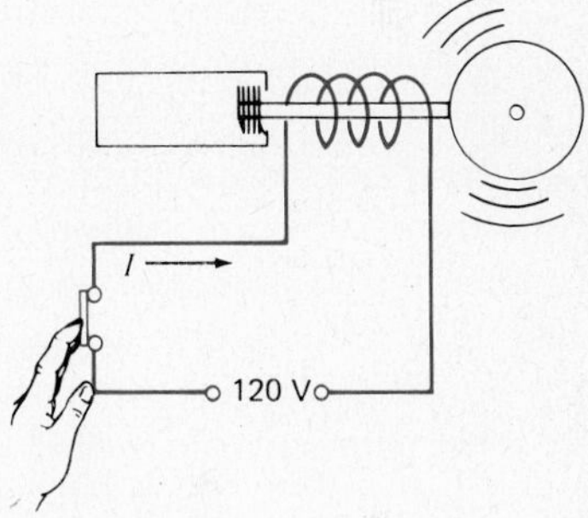

FIGURE 17-14
Solenoid used as a doorbell.

17-3 ▮ Magnetic force on currents and moving charges

Force on a current; definition of magnetic field strength

In the previous section we saw that an electric current exerts a force on a magnet, such as a compass needle. By Newton's third law we might expect the reverse to be true as well, and this is indeed the case: **a magnet exerts a force on a current-carrying wire.** This effect was also discovered by Oersted.

Let us look at the force exerted on a wire in detail. Suppose that a straight wire is placed between the pole pieces of a magnet as shown in Figure 17-15. When a current flows in the wire, a force is exerted on the wire, but this force is *not* toward one or the other poles of the magnet. Instead, the force is directed *at right angles to the magnetic field direction.* If the current is reversed in direction, the force is in the opposite direction. It is found that the direction of the force is always perpendicular to the direction of the current I and also perpendicular to the direction of the magnetic field, **B**. Depending on the relative orientation of I and **B**, the

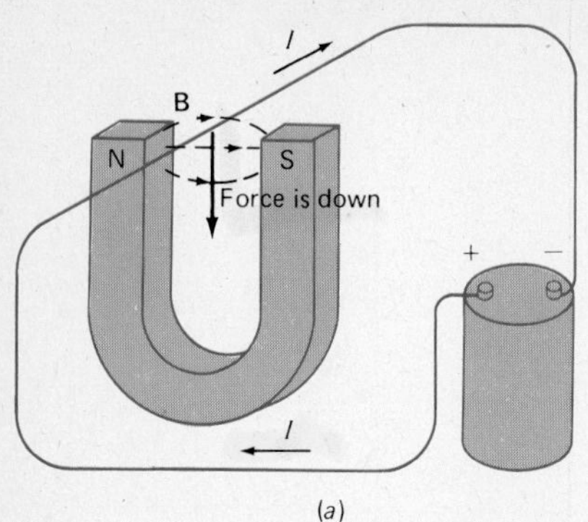

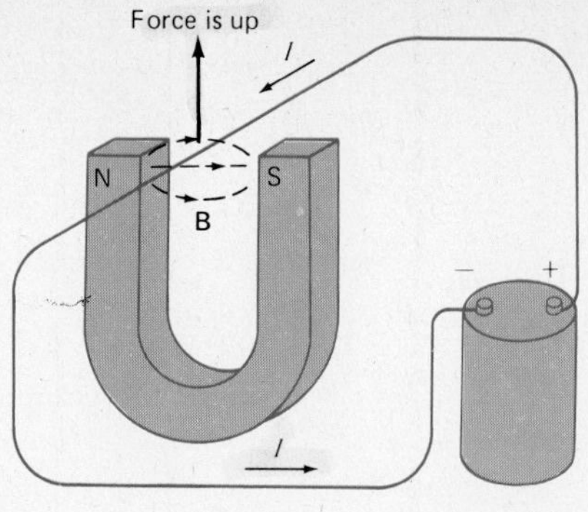

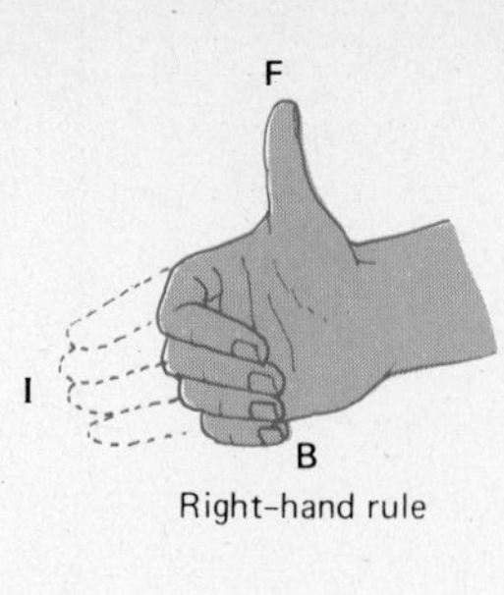

FIGURE 17-15
Force on a current-carrying wire placed in a magnetic field B.

force can be either up or down in Figure 17-15, as shown. The actual direction of the force is given by another right-hand rule, as illustrated in Figure 17-15*b*. First you orient your right hand so that the outstretched fingers point in the direction of the (conventional) current; from this position, when you bend your fingers, they should point in the direction of the magnetic field lines; if they do not, rotate your hand and arm about the wrist until they do, remembering that straightened fingers must point along I. When your hand is oriented in this way so outstretched fingers point along I and bent fingers along B, then the extended thumb points in the direction of the force on the wire.

This describes the direction of the force. Now what about its magnitude? It is found experimentally that the magnitude of the force is directly proportional to the current I in the wire, to the length l of wire in the magnetic field, and to the magnetic field strength B. The force also depends on the angle θ between the wire and the magnetic field. When the wire is perpendicular to the field lines, the force is strongest; when the wire is parallel to the magnetic field lines, there is no force at all. And at other angles it is proportional to sin θ (see Figure 17-16). Thus we have

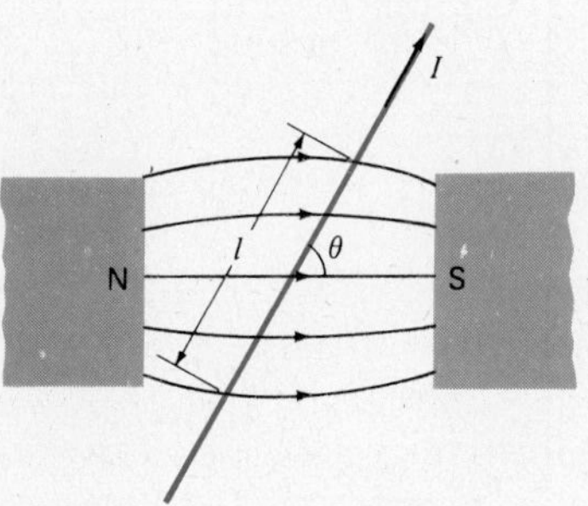

FIGURE 17-16
Current-carrying wire in a magnetic field.

$$F \propto IlB \sin\theta.$$

The precise definition of magnetic field strength is actually made using this relation, by setting the constant of proportionality equal to precisely 1. Thus we have

$$F = IlB \sin\theta. \tag{17-1}$$

If the wire is perpendicular to the field ($\theta = 90°$), the force is

$$F = IlB \qquad (I \perp B). \tag{17-2}$$

If the wire is parallel to the field ($\theta = 0°$), the force is zero.†

The SI unit for magnetic field strength B is the *tesla* (T). From Equation 17-1 it is clear that $1\ \text{T} = 1\text{N/A}\cdot\text{m}$. An older name for the tesla is the "weber per meter squared" ($1\ \text{Wb/m}^2 = 1\ \text{T}$). Another unit commonly used to specific magnetic field strength is the gauss (G): $1\ \text{G} = 10^{-4}\ \text{T}$. (It is the cgs unit.) A field given in gauss should always be changed to teslas

† In this discussion we have (implicitly) assumed that the magnetic field is uniform. If it is not, then B in Equations 17-1 and 17-2 is the average field over the length l of the wire. In practical cases one can consider a wire as made up of many short segments Δl and the force on each segment is proportional to the length Δl of that segment and the magnetic field intensity B at that segment; the total force is the sum of the individual forces.

before using with other SI units, as in Equation 17-1 or 17-2. To get a "feel" for these units, we note that the magnetic field of the earth at its surface is about $\frac{1}{2}$ G or 0.5×10^{-4} T. Electromagnets can produce fields on the order of several teslas.

▌ EXAMPLE 17-1 A wire carrying a 30-A current has a length $l = 12$ cm between the pole faces of a magnet at an angle $\theta = 60°$, as shown in Figure 17-16. The magnetic field is approximately uniform at 0.90 T. We ignore the field beyond the pole pieces. What is the force on the wire?

SOLUTION We use Equation 17-1 and find that

$$F = IlB \sin \theta$$
$$= (30 \text{ A})(0.12 \text{ m})(0.90 \text{ T})(0.866) = 2.8 \text{ N}.$$ ▌▌

Force on a moving charge

Since a current-carrying wire experiences a force when placed in a magnetic field, and since a current in a wire consists of moving electric charges, we might expect that freely moving charged particles (not in a wire) would also experience a force when passing through a magnetic field. Indeed, this is the case.

From what we already know, let us calculate the force on a single moving electric charge. If n such particles of charge q pass by a given point in time t, they constitute a current $I = nq/t$. We let t be the time for a charge q to travel a distance l in a magnetic field B; then $l = vt$, where v is the velocity of the particle. Thus the force on these n particles is, by Equation 17-1, $F = IlB \sin \theta = (nq/t)(vt)B \sin \theta = nqvB \sin \theta$. The force on *one* of the n particles is then

$$F = qvB \sin \theta. \tag{17-3}$$

This equation gives the magnitude of the force on a particle of charge q moving with velocity v in a magnetic field of strength B; θ is the angle between **v** and **B**. The force is greatest when the particle moves perpendicular to **B** ($\theta = 90°$):

$$F = qvB \qquad (\mathbf{v} \perp \mathbf{B}). \tag{17-4}$$

The force is *zero* if the particle moves *parallel* to the field lines ($\theta = 0°$). The *direction* of the force is always perpendicular to the magnetic field **B** and to the velocity **v** of the particle. It is again given by a right-hand rule: you orient your right hand so that your outstretched fingers point along the direction of motion of the particle (**v**) and when you bend your fingers they must point along the direction of **B**; then your thumb will point in the direction of the force. This is true only for *positively* charged particles, and will be "down" for the situation of Figure 17-17. For negatively charged particles the force is in exactly the opposite direction ("up" in Figure 17–17).†

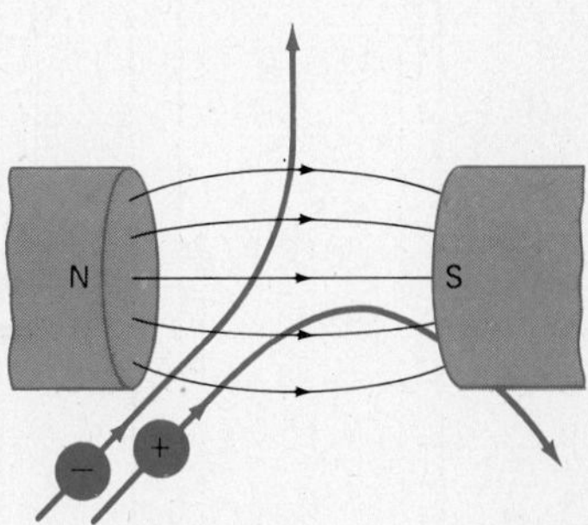

FIGURE 17-17
Force on charged particles due to a magnetic field.

▌ EXAMPLE 17-2 A proton having a speed of 5.0×10^6 m/s in a magnetic field feels a force of 8.0×10^{-14} N toward the west when it moves vertically upward. When moving horizontally in a northerly direction, it feels zero force. What is the magnitude and direction of the magnetic field in this region?

† Alternatively, one can use the *left hand* to determine the force on a *negative* particle; the left thumb then will point in the direction of the force on a negatively charged particle if the fingers are oriented as discussed for the right-hand rule.

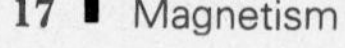

SOLUTION Since the proton feels no force when moving north, the field must be in a north–south direction. The right-hand rule tells us that B must point toward the north in order to produce a force to the west when the proton moves upward. (Your thumb points west when your right hand is oriented so that your outstretched fingers point upward and your bent fingers point north.) The magnitude of **B**, from Equation 17-4, is

$$B=\frac{F}{qv}=\frac{8.0\times10^{-14}\text{ N}}{(1.6\times10^{-19}\text{ C})(5.0\times10^{6}\text{ m/s})}=0.10\text{ T}.$$

▮▮

The path of a charged particle moving in a plane perpendicular to a uniform magnetic field is a circle (or the arc of a circle if the particle leaves the magnetic field region). This can be seen from Figure 17-18. In this drawing the magnetic field is directed *into* the paper and this is represented by ×'s. (The × is meant to resemble the end of an arrow going away; if the magnetic field were pointing out of the paper, we would represent this using ⊙'s to resemble the tip of an arrow coming at us.) An electron at point P is moving to the right, and the force on it is therefore downward as shown. The electron is thus deflected downward. When it reaches point Q, the force is still perpendicular to the velocity and is in the direction shown. Since the force is of constant magnitude and always perpendicular to **v**, the particle moves in a circular path. The force is directed toward the center of this circle at all points. Note that the electron moves clockwise in Figure 17-18. A positive particle would feel a force in the opposite direction and would thus move in a counterclockwise circle.

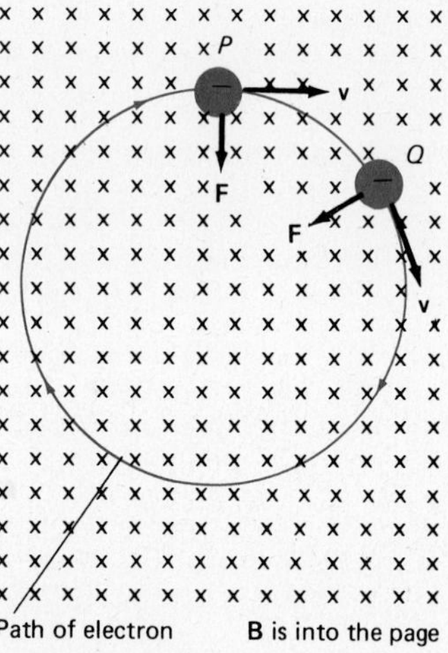

FIGURE 17-18
Force exerted by a uniform magnetic field on a moving charged particle (in this case an electron) produces a curved path.

▮ EXAMPLE 17-3 An electron travels with a speed of 2.0×10^{7} m/s in a plane perpendicular to a 0.10-T magnetic field. Describe its path.

SOLUTION The radius of the curved path is found using Newton's second law, $F = ma$, and the fact that the acceleration of a particle moving in a circle is $a = v^2/r$. Thus

$$F = ma$$

$$qvB = \frac{mv^2}{r}.$$

We solve for r and find that

$$r=\frac{mv}{qB}.$$

Since **F** is perpendicular to **v**, the magnitude of **v** does not change. From this equation we thus see that if **B** is constant, then r is constant and the curve must be a circle. To get r we put in the figures:

$$r=\frac{(9.1\times10^{-31}\text{ kg})(2.0\times10^{7}\text{ m/s})}{(1.6\times10^{-19}\text{ C})(0.10\text{ T})}=1.1\times10^{-3}\text{ m},$$

or 1.1 mm.

▮▮

The Hall effect

When a current-carrying conductor is held firmly in a magnetic field, the field exerts a sideways force on the charges moving in the conductor. For example, if electrons move to the right in the rectangular conductor shown in Figure 17–19*a*, the inward magnetic field will exert a downward force on the electrons; so they will tend to move nearer face *S* than face *R*. There will thus be a potential difference between faces *R* and *S* of the conductor. This potential difference builds up until the electric field it produces exerts a force on the moving charges that is equal and opposite to the magnetic force. This effect is called the *Hall effect* after E. H. Hall, who discovered it in 1879. The difference of potential produced is called the *Hall emf*.

Now a current of negative charges moving to the right is equivalent to positive charges moving to the left, at least for most purposes. But the Hall effect can distinguish these two. As can be seen in Figure 17-19*b*, positive particles moving to the left would be deflected downward, so the bottom surface would be positive relative to the top surface. This is the reverse of part *a*. Indeed, the direction of the emf in the Hall effect first revealed that it is negative particles that move in most conductors. In some semiconductors, however, the Hall effect reveals that the carriers of current are positive. (More on this in Chapter 19.)

The magnitude of the Hall emf is proportional to the current and to the strength of the magnetic field. The Hall effect can thus be used to measure magnetic field strengths. First the conductor, called a *Hall probe*, is calibrated with known magnetic fields. Then, for the same current, its emf output will be a measure of *B*. Hall probes can be made very small and are convenient and accurate to use.

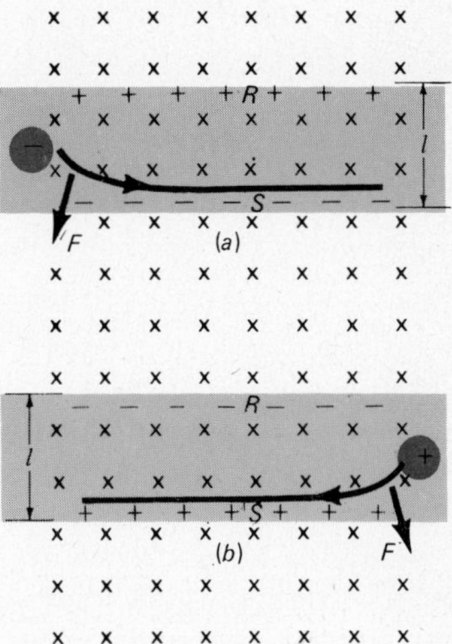

FIGURE 17-19
Hall effect.

*Electromagnetic pumping

The force on a particle in a magnetic field is the idea behind "electromagnetic pumping." It is used to pump metallic fluids (such as sodium) and more recently to pump blood in artificial heart machines. The basic design is shown in Figure 17-20. An electric field is applied perpendicular to a blood vessel and to a magnetic field. The ions in the blood move vertically in the drawing as a result of the electric field. The magnetic field then exerts a sideways force on the ions that causes the fluid to flow along the blood vessel. Do positive and negative ions feel a force in the same direction?

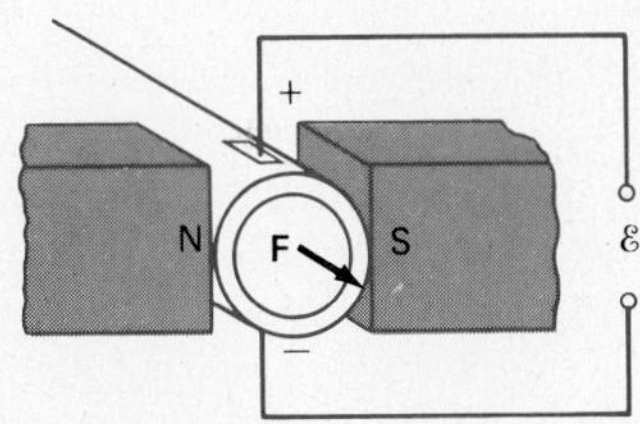

FIGURE 17-20
Electromagnetic pumping in a blood vessel.

17-4 ▮ Discovery and properties of the electron; the mass spectrograph

The electron plays a basic role in our present understanding of electricity and magnetism. But its existence was not known until the 1890s. The use of magnetic fields was crucial for measuring its properties.

Measurement of *e*/*m* for the electron

This is how the discovery of the electron came about. Toward the end of the nineteenth century, many studies were done on the discharge of electricity through rarefied gases. One piece of apparatus used is shown in Figure 17-21. A glass tube fitted with electrodes was evacuated so that only

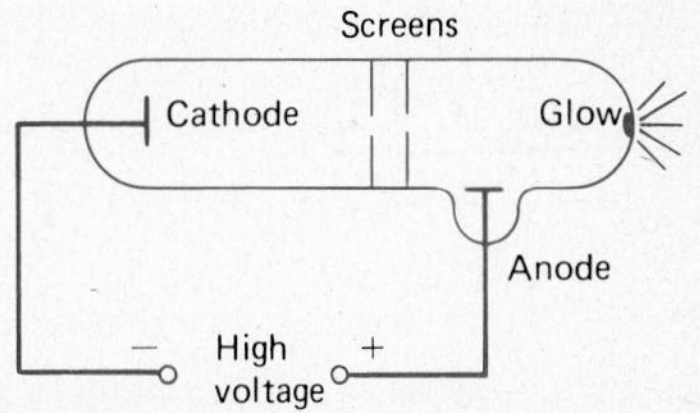

FIGURE 17-21
Discharge tube. In some models one of the screens is the anode (positive plate).

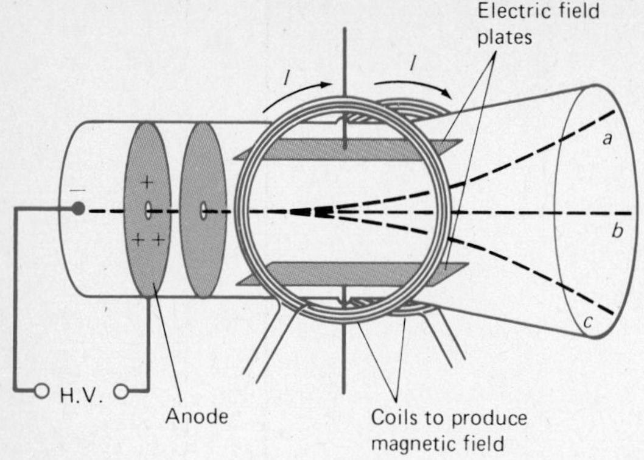

FIGURE 17-22
Cathode rays deflected by electric and magnetic fields.

a small amount of gas remained inside. When a very high voltage was applied to the electrodes, a dark area was observed to extend outward from the cathode toward the opposite end of the tube, and that end of the tube would glow. If one or more screens containing a small hole were inserted as shown, the glow was restricted to a tiny spot on the end of the tube. It seemed as though something being emitted by the cathode traveled to the opposite end of the tube; these were named *cathode rays*.

There was much discussion at the time about what these rays might be. Some scientists thought they might resemble light. In 1897, J. J. Thomson (1856–1940) reported that the bright spot at the end of the tube was deflected to one side if an electric or magnetic field was applied across the tube. This indicated to him that cathode rays must be charged particles, and the direction of the deflection was consistent with a negative charge. Furthermore, if the tube contained certain types of rarefied gas, the path of these objects was made visible by a slight glow.

Thomson then proceeded to measure the ratio of the charge to the mass, e/m, of these objects. This can be done as shown in Figure 17-22. A pair of parallel plates is built into the tube; when a voltage is applied to them, an electric field is produced. The pair of coils produces a magnetic field. When only an electric field is present, say with the upper plate positive, the cathode rays are deflected upward, as in path a in the figure. If only a magnetic field exists, say inward in the figure, the rays are deflected downward along path c. These observations are just what is expected for a negatively charged particle.

When only the magnetic field B is applied, the force on the rays is

$$F = evB,$$

where e is the charge and v the velocity of the cathode rays. They are bent into a curved path, so we have, from $F = ma$,

$$evB = \frac{mv^2}{r}.$$

Thus

$$\frac{e}{m} = \frac{v}{Br}.$$

The radius of curvature r can be measured and so can B. The velocity v is found by applying an electric field in addition to the magnetic field. The electric field E is adjusted so that the cathode rays are undeflected and follow path b in Figure 17-22. In this situation, the force due to the electric field, $F = eE$, is just balanced by the force due to the magnetic field, $F = evB$. Thus we have $eE = evB$ and

$$v = \frac{E}{B}.$$

Combining this with the equation above, we have

$$\frac{e}{m} = \frac{E}{B^2 r}. \tag{17-5}$$

The quantities on the right side can all be measured so that, although e and m could not be determined separately, the ratio could be determined.

Thomson obtained a value very close to the presently accepted value of $e/m = 1.76 \times 10^{11}$ C/kg. These cathode rays, whose properties Thomson measured, soon came to be called *electrons*, a name very familiar to us today.

"Discovery" of the electron

It is worth noting that the "discovery" of the electron, like many other discoveries in science, is not quite so obvious as discovering gold or oil. Should the discovery of the electron be credited to the person who first saw a glow in the tube—or to the person who first called them cathode rays? Perhaps neither one, for they had no conception of the electron as we know it today. In fact, the credit for the discovery is generally given to Thomson—but not because he was the first to see the glow in the tube. Rather, he believed this phenomenon to be due to tiny negatively charged particles, and actually measured the value of e/m as if they were particles. This view is close to what we accept today, and it is for this reason that Thomson is credited with the discovery. Note, too, that neither he nor anyone else ever actually saw an electron.

We discussed this (briefly) for it illustrates the fact that discovery in science is not always a clear-cut matter. In fact, some philosophers of science think the word "discovery" is not always appropriate, such as in this case.

Millikan oil-drop experiment

Thomson also believed that these electrons were not atoms but rather a constituent or part of an atom. Convincing evidence for this came about 10 years later when Robert A. Millikan (1868–1953) measured the charge on the electron in his famous *oil-drop experiment*. In this experiment tiny droplets of mineral oil were allowed to fall under gravity between two parallel plates (Figure 17-23). The electric field E between the plates was adjusted until the drop was suspended in midair; the downward pull of gravity, mg, was then just balanced by the upward force due to the electric field. Thus $qE = mg$, so the charge $q = mg/E$. The mass of the droplet was determined by measuring its terminal velocity in the absence of the electric field and using Stoke's law (the drop is too small to measure its

FIGURE 17-23
Millikan's oil-drop experiment.

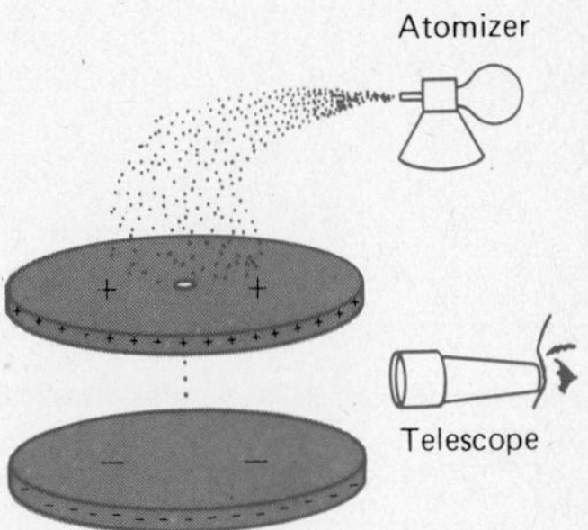

radius directly; see Section 6-9). Millikan found that the charge on each drop was a small whole-number multiple of 1.60×10^{-19} C. A charge of less than 1.60×10^{-19} C was never observed. Sometimes the charge was negative and sometimes positive, indicating that the drop had acquired or lost electrons (presumably through friction when ejected by the atomizer). Millikan interpreted his experiment to mean that the smallest charge to be found in nature was 1.60×10^{-19} C, and this was ascribed to the electron and given the symbol e.

This result, combined with the measurement of e/m, gave the mass of the electron to be $(1.6 \times 10^{-19}\ \text{C})/(1.76 \times 10^{11}\ \text{C/kg}) = 9.1 \times 10^{-31}$ kg. This is less than a thousandth the mass of the smallest atom, and thus confirmed the idea that the electron is only a part of an atom.

Mass spectrograph

A number of methods were developed in the early part of this century to measure the masses of atoms. One of the most accurate was the *mass spectrograph*† of Figure 17-24. Ions are produced by heating, or by an electric current, in the source S. Those that pass through slit S_1 enter a region where there are crossed electric and magnetic fields, as in Thomson's device. Only those ions whose speed is $v = E/B$ will pass through undeflected and emerge through slit S_2 (see equation just before Equation 17-5). In this region there is only a magnetic field B' and the ions follow a circular path. The radius of their path can be measured because the ions expose the photographic plate where they strike. Since $qvB' = mv^2/r$ and $v = E/B$, we have

$$m = \frac{qB'r}{v} = \frac{qBB'r}{E}.$$

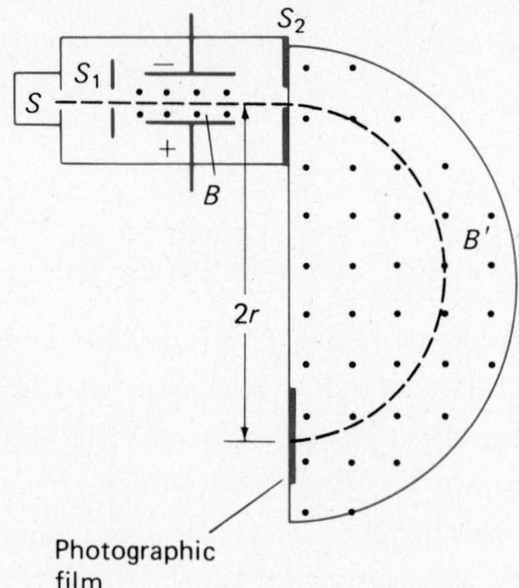

FIGURE 17-24
Bainbridge mass spectrograph. The magnetic fields B and B' point out of the paper (indicated by the dots).

All the quantities on the right can be measured, and thus m can be determined. Note that for ions of the same charge, the mass of each is proportional to the radius of its path.

The masses of many atoms were measured in this way. Furthermore, when a pure substance was used, it was sometimes found that two or more closely spaced marks would appear on the film. For example, neon produced two marks, whose radii corresponded to atoms of mass 20 and 22 atomic mass units. Impurities were ruled out and it was concluded that there must be two types of neon with different mass. These different forms are called *isotopes*. It was soon found that most elements are mixtures of isotopes. (We shall see in Chapter 28 that the difference in mass is due to different numbers of neutrons.)

Mass spectrographs can be used not only to separate different elements and isotopes, but molecules as well. They are used in physics and chemistry, and in biological and biomedical laboratories.

▌ EXAMPLE 17-4 Carbon atoms of atomic mass 12.0 are found to be mixed with another, unknown, element. In a mass spectrograph, the carbon traverses a path of radius 22.4 cm and the unknown's path has a 26.2-cm radius. What is the unknown element? Assume they have the same charge.

† The term *mass spectrometer* is also used.

SOLUTION Since mass is proportional to the radius, we have

$$\frac{m_x}{m_C} = \frac{26.2\text{ cm}}{22.4\text{ cm}} = 1.17.$$

Thus $m_x = 1.17 \times 12.0 = 14.0$. The other element is probably nitrogen (see the periodic table, inside front cover). However, it could also be an isotope of carbon or oxygen. Further physical or chemical analysis would be needed. ▮▮

17-5 ▌ Applications—meters, motors, and loudspeakers

A number of important practical devices make use of the force that exists between a current and a magnetic field. In most of these devices, the current flows in a coil of wire.

The galvanometer and the torque on a current loop

The basic component of most meters, including ammeters, voltmeters, and ohmmeters, is a *galvanometer*. We have already seen how these meters are designed (Section 16-3), and now we can examine how the crucial element, a galvanometer, itself works.

As shown in Figure 17-25, a galvanometer consists of a loop or coil of wire suspended in the magnetic field of a permanent magnet. When current flows through the loop, which is usually rectangular, the magnetic field exerts a force on the vertical sections of wire as shown. Notice that, by the right-hand rule (Figure 17-15), the force on the upward current on the left is inward, whereas that on the descending current on the right is outward. These forces give rise to a net torque that tends to rotate the coil about its vertical axis. The greater the current I, the more the coil and its attached pointer will turn against the resistance of the small spring.

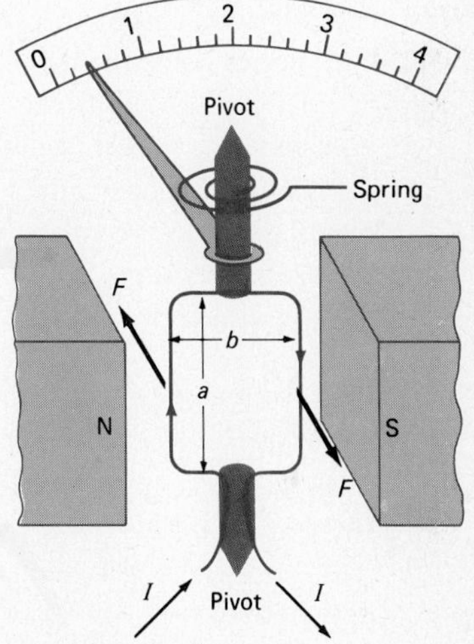

FIGURE 17-25
Galvanometer.

Let us calculate the magnitude of this torque. From Equation 17-2, the force $F = IaB$, where a is the length of the vertical arm of the coil; the lever arm for each force is $b/2$, where b is the width of the coil and the "axis" is at the midpoint. The total torque is the sum of the torques due to each of the forces, so

$$\tau = IaB\frac{b}{2} + IaB\frac{b}{2} = IabB = IAB,$$

where $A = ab$ is the area of the coil. If the coil consists of N loops of wire, the current is then NI, so the torque becomes

$$\tau = NIAB. \tag{17-6}$$

This formula, derived here for a rectangular coil, is valid for any shape of flat coil. The quantity NIA is referred to as the *magnetic moment* of the coil.

This torque is opposed by the spring. By Hooke's law, the torque τ_s exerted by the spring is proportional to the angle ϕ through which it is turned. That is,

$$\tau_s = k\phi,$$

where k is the stiffness constant of the spring. Thus the coil and the attached pointer will rotate only to the point where the spring torque balances that due to the magnetic field. From Equation 17-6 we then have $k\phi = NIAB$, or

$$\phi = \frac{NIAB}{k}. \tag{17-7}$$

Thus the deflection of the pointer, ϕ, is directly proportional to the current I flowing in the coil; this is what we want for a galvanometer.

However, there is a problem. When the loop of Figure 17-25 is not parallel to the magnetic field, the torque will be less than that given by Equation 17-6 since the lever arm is less. Indeed, when the loop is rotated 90° from the position shown in Figure 17-25, the two forces act along the same line and the torque is zero. Thus the torque would depend on the angle, and ϕ would not be proportional to I as in Equation 17-7. To solve this problem, curved pole pieces are used and the galvanometer coil is wrapped around a cylindrical iron core as shown in Figure 17-26. The iron tends to concentrate the magnetic field lines so that B always points across the face of the coil at the wire. The force is then always perpendicular to the face of the coil and the torque will not vary with angle. Thus ϕ will be proportional to I, as required.

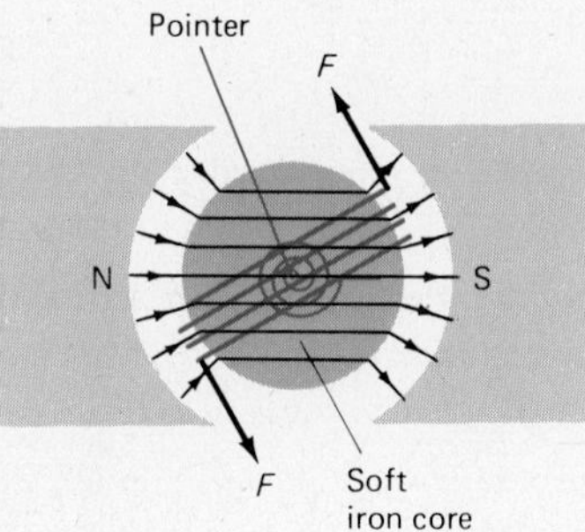

FIGURE 17-26
Galvanometer coil wrapped on an iron core.

A *chart recorder*, in which a pen graphs a signal such as an ECG on a moving roll of paper, is basically a galvanometer. The pen is attached to an arm, which is connected to the galvanometer coil. The instrument could record either voltage or current, just as any galvanometer can be connected as a voltmeter or ammeter.

Electric motor

An electric motor changes electric energy into (rotational) mechanical energy. Motors are used to turn fans, grinding wheels, and washing-machine tubs, and they are found in a great many other devices, such as refrigerators, hair dryers, mixers, pumps, and electric cars.

A motor works on the same principle as a galvanometer, except that the coil is larger and is mounted on a large cylinder called the *rotor* or *armature* (Figure 17-27). Actually, there are several coils, although only one is indicated in the figure. The armature is mounted on a shaft or axle.

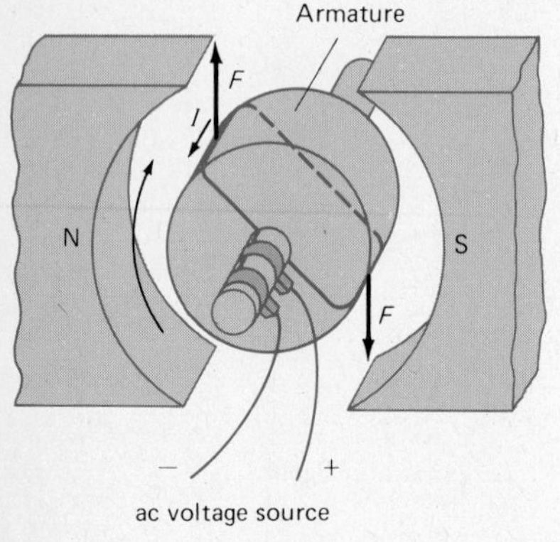

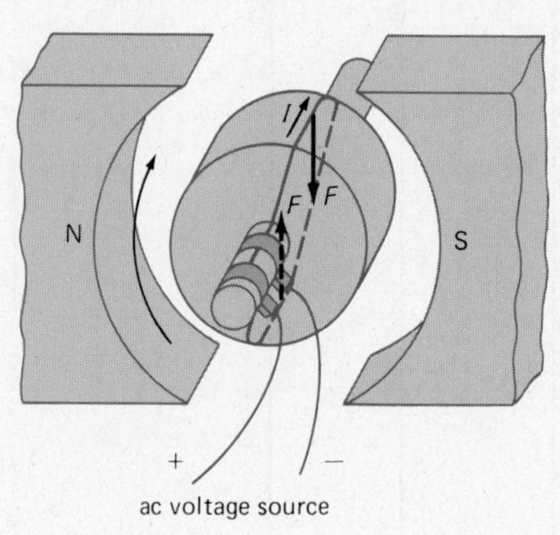

FIGURE 17-27
Simplified diagram of an ac motor.

The permanent magnet is replaced in some motors by an electromagnet. Unlike a galvanometer, a motor must turn continously in one direction. This presents a problem. When the coil, which is rotating clockwise in Figure 17-27*a*, passes beyond the vertical position, the forces would then act to return the coil back to vertical. However, if an ac source is used, the current is timed so that it starts flowing in the opposite direction just as the loop reaches the vertical position. The force is then reversed so that the motor continues to rotate in the same direction (Figure 17-27*b*). This is the principle of an *ac motor.* Since the axle rotates, something must be done to keep the wires from tangling. The armature wires are connected to separate "slip rings," which rotate on the axle. The source is applied to stationary contacts called "brushes," which rub against the rotating sliprings.

Alternation of the current is necessary if a motor is to turn continuously in one direction. This can be achieved in a dc motor with the use of *commutators* and *brushes,* as shown in Figure 17-28. The commutators are mounted on the motor shaft, and every half-revolution each commutator changes its connection to the other brush. Thus the current in the coil reverses every half-revolution as required for continuous rotation. Most motors contain several coils, each located in a different place on the armature (Figure 17-29). These are referred to as the "windings" of the motor. Current flows through each coil only during a small part of a revolution, at the time when its orientation results in the maximum torque. In this way a motor produces a much steadier torque than can be obtained from a single coil.

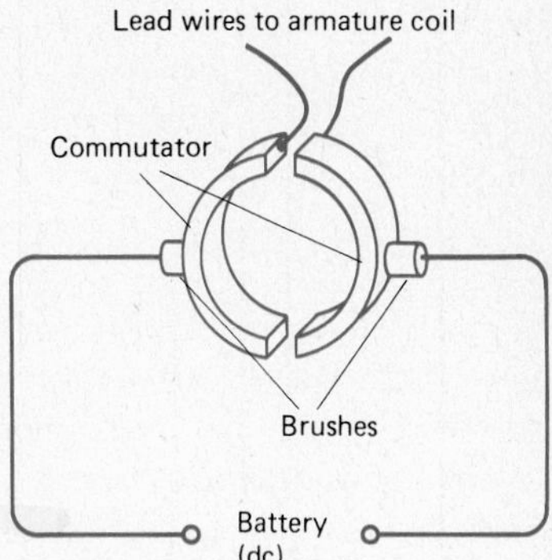

FIGURE 17-28
This commutator-brush arrangement in a dc motor changes dc to ac. The commutators are attached to the motor shaft and turn with it while the brushes remain stationary.

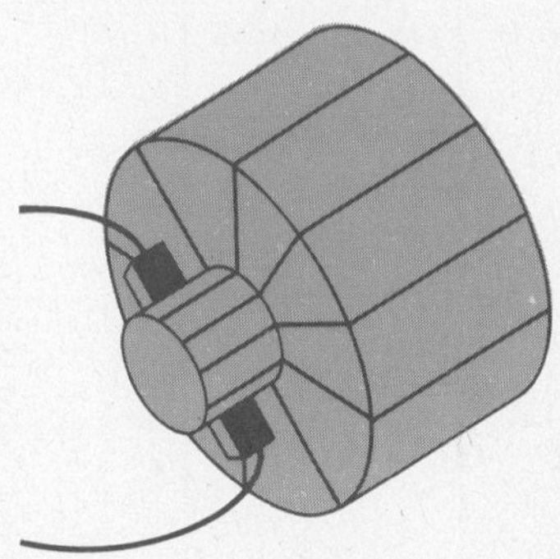

FIGURE 17-29
Motor with many windings.

Loudspeakers

A loudspeaker also works on the principle that a magnet exerts a force on a current-carrying wire. The electrical output of a radio or television set is connected to the wire leads of the speaker. These, in turn, are joined internally to a coil of wire attached to the speaker cone (Figure 17-30). The speaker cone, usually made of stiffened cardboard, is mounted so that it can move freely backward and forward. In line with the coil of wire is a permanent magnet. When the audio signal (an alternating current) flows in the wire coil, the coil and speaker cone move back and forth at the frequency of the signal because of the force on the coil due to the magnetic field. The cone's motion causes alternate compressions and rarefactions of the adjacent air, and sound waves are produced. A speaker thus changes electrical energy into sound energy, and the frequencies of the emitted sound waves are an accurate reproduction of the electrical input.

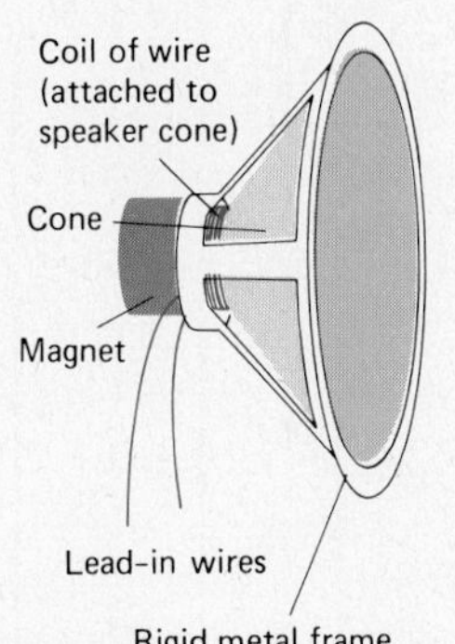

FIGURE 17-30
Loudspeaker.

17-6 ▮ Determination of magnetic field strengths; Ampère's law

The simplest magnetic field is one that is uniform—that is, it does not change from one point to another. A perfectly uniform field over a large area is not easy to produce. But the field between two flat pole pieces of a magnet is nearly uniform if the area of the pole faces is large compared to their separation (Figure 17-31). At the edges, the field "fringes" out somewhat and is not uniform. The parallel, evenly spaced field lines in the drawing indicate that the field is uniform at points not too near the edge.

Nonuniform magnetic fields are more common. It is often important to know how a magnetic field varies from point to point. We shall see next how magnetic field strengths are determined for some simple situations.

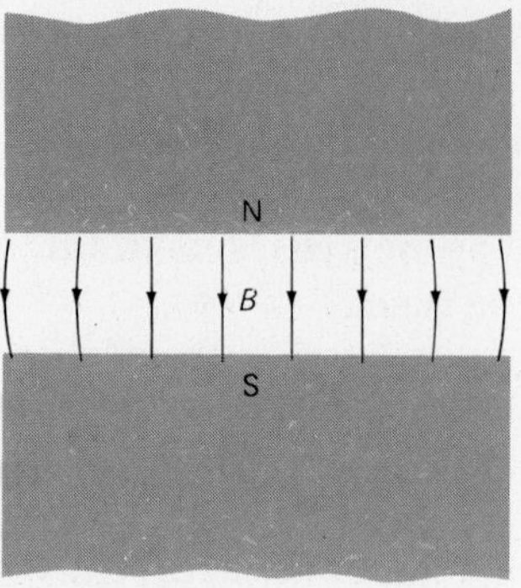

FIGURE 17-31
Magnetic field between two large poles of a magnet is nearly uniform except at the edges.

Magnetic field of a straight wire

We have already seen (Figure 17-10) that the magnetic field due to the electric current in a long straight wire is such that the field lines are circles with the wire at the center. You might expect that the field strength at a given point would be greater if the current flowing in the wire were greater; and that the field would be less at points farther from the wire. This is indeed the case. Careful experiments show that the magnetic field strength B at a point near the wire is directly proportional to the current I in the wire and inversely proportional to the distance r from the wire:

$$B \propto \frac{I}{r}.$$

This relation is valid as long as r, the distance to the closest point on the wire, is much less than the distance to the ends of the wire.

The proportionality constant is written, for historical reasons, as $\mu_0/2\pi$; thus

$$B = \frac{\mu_0}{2\pi}\frac{I}{r}. \tag{17-8}$$

The value of the constant μ_0, which is called the "permeability of free space," is $\mu_0 = 4\pi \times 10^{-7}\ \mathrm{T{\cdot}m/A}$.

▌ EXAMPLE 17-5 A vertical electric wire in the wall of a building carries a dc current of 25 A upward. What is the magnetic field at a point 10 cm due north of this wire?

SOLUTION According to Equation 17-8:

$$B = \frac{(4\pi \times 10^{-7}\ \mathrm{T \cdot m/A})(25\ \mathrm{A})}{(2\pi)(0.10\ \mathrm{m})} = 5.0 \times 10^{-5}\ \mathrm{T},$$

or 0.50 G. By the right-hand rule (Figure 17-10*b*), the field points to the west at this point. Since this field has about the same magnitude as the earth's, a compass would not point north but in a northwesterly direction. ▌▌

*Ampère's law

Equation 17-8 gives the relation between the current in a long, straight wire and the magnetic field it produces. This equation is valid only for a long, straight wire. The following question arises: Is there a general relation between a current in a wire of any shape and the magnetic field due to it? The answer is yes.

The French scientist André Marie Ampère (1775–1836) proposed such a relation shortly after Oersted's discovery. Consider any closed path around a current as shown in Figure 17-32, and imagine this path as being made up of short segments each of length Δl. First we take the product of the length of each segment times the component of B parallel to that segment. If we now sum all these terms, according to Ampère the result will be equal to μ_0 times the net current I that passes through the surface enclosed by the path. This is known as **Ampère's law** and can be written mathematically as

$$\sum B_{\parallel} \Delta l = \mu_0 I. \tag{17-9}$$

The symbol $\sum$ means "the sum of" and $B_{\parallel}$ means the component of B parallel to that particular Δl. The lengths Δl are chosen so that $B_{\parallel}$ is essentially constant on each length. The sum must be made over a closed path; and I is the net current passing through the surface bounded by this closed path.

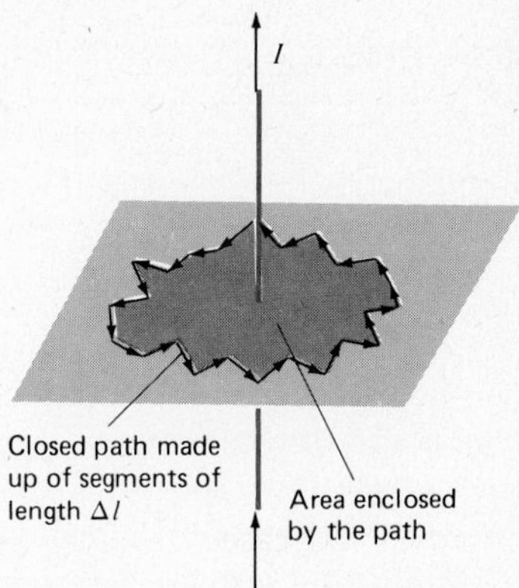

FIGURE 17-32
Arbitrary path enclosing a current, for Ampère's law. The path is broken down into segments of equal length Δl.

We can understand Ampère's law more easily by applying it to a simple case. We take the case of a long, straight wire carrying a current I which we have already examined and which served as an inspiration for Ampère himself. Suppose that we want to find the magnitude of B at point A, a distance r from the wire in Figure 17-33. We know that the magnetic field lines are circles with the wire at their center. For the path to be used in Equation 17-9 we choose a circle of radius r. (The choice of path is ours—so we choose one that will be convenient.) At any point on this path, B will be tangent to this circle. Thus for any short segment of the circle (Figure 17-33), B will be parallel to that segment, so $B_{\parallel} = B$. Suppose that we break the circular path down into 100 segments.† Then Ampère's law states that

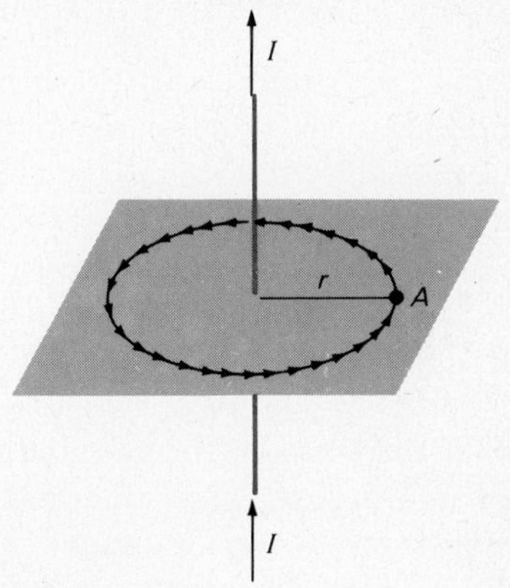

FIGURE 17-33
Circular path of radius r.

$$(B\,\Delta l)_1 + (B\,\Delta l)_2 + (B\,\Delta l)_3 + \cdots + (B\,\Delta l)_{100} = \mu_0 I.$$

The dots represent all the terms we did not write down. Since all the segments are the same distance from the wire, we expect B to be the same at each segment. We can then factor out B from the sum:

$$B(\Delta l_1 + \Delta l_2 + \Delta l_3 + \cdots + \Delta l_{100}) = \mu_0 I.$$

The sum of the segment lengths is just the circumference of the circle, $2\pi r$. Thus we have

$$B(2\pi r) = \mu_0 I,$$

or

$$B = \frac{\mu_0 I}{2\pi r}.$$

This is just Equation 17-8 for the field near a long, straight wire, as discussed earlier.

Ampère's law thus works for this simple case. A great many experiments indicate that Ampère's law is true in general. However, it can be used to calculate the magnetic field mainly for simple cases (see the next

† Actually, Ampère's law is precisely accurate when there are an infinite number of infinitesimally short segments, but this leads into calculus.

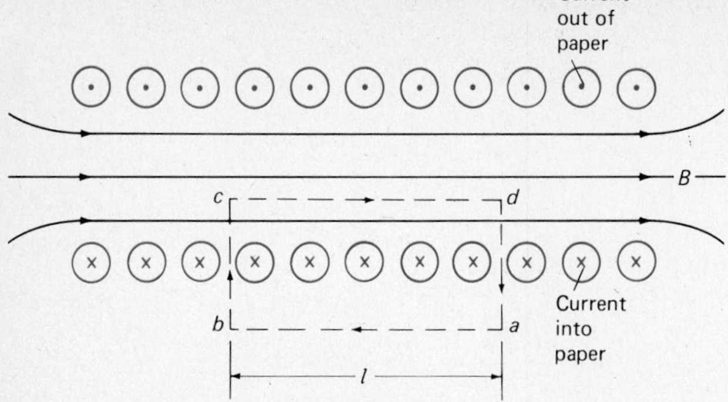

FIGURE 17-34
Magnetic field inside a solenoid is straight except at the ends. Dashed lines indicate the path chosen for use in Ampère's law.

subsection and the problems). Its importance lies in the fact that it relates the magnetic field to the current in a direct and mathematically elegant way. Ampère's law is thus considered one of the basic laws of electricity and magnetism. It is valid for any situation where the currents and fields are steady and not changing in time.

We now can see why the constant in Equation 17-8 is written $\mu_0/2\pi$; this is done so that only μ_0 appears in Equation 17-9 (rather than, say, $2\pi k$ if we had used k in Equation 17-8). In this way, the more general equation, Ampère's law, has the simplest form.

*Ampère's law applied to a solenoid

We now use Ampère's law to calculate the magnetic field inside a long solenoid. If the coils of the solenoid are very closely spaced, the field inside will be parallel to the axis except at the ends (Figure 17-34). We choose the path *abcd* shown in the figure, far from either end, for applying Ampère's law. We will consider this path as made up of four segments, the sides of the rectangle: *ab*, *bc*, *cd*, *da*. Then the left side of Equation 17-9 becomes

$$(B_{\parallel}\,\Delta l)_{ab} + (B_{\parallel}\,\Delta l)_{bc} + (B_{\parallel}\,\Delta l)_{cd} + (B_{\parallel}\,\Delta l)_{da}.$$

Now the field outside the solenoid is so small as to be negligible compared to the field inside. Thus the first term will be zero. Furthermore, B is perpendicular to the segments bc and da inside the solenoid, and is essentially zero outside. Thus these terms are zero. Therefore, the left side of Equation 17-9 is simply $(B_{\parallel}\,\Delta l)_{cd} = Bl$, where B is the field inside the solenoid and l is the length cd. Now for the right side of Equation 17-9. If a current I flows in the wires of the solenoid, the total current enclosed by our path *abcd* is NI, where N is the number of loops our path encircles (five in Figure 17-34). Thus Ampère's law gives us

$$Bl = \mu_0 NI.$$

If we let $n = N/l$ be the number of loops per unit length, then

$$B = \mu_0 nI. \tag{17-10}$$

This is the field within a solenoid. Note that B depends only on the number of loops per unit length, n, and the current I. The field does not depend on the position within the solenoid. Thus B is uniform. This is strictly true, of course, only for points not close to the ends.

▌ EXAMPLE 17-6 A 10-cm-long solenoid has a total of 400 turns of wire and carries a current of 2.0 A. Calculate the field inside near the center.

SOLUTION The turns per unit length are $n = 400/0.10\ \text{m} = 4.0 \times 10^3\ \text{m}^{-1}$. Thus

$$B = \mu_0 n I = (12.57 \times 10^{-7}\ \text{T} \cdot \text{m/A})(4.0 \times 10^3\ \text{m}^{-1})(2.0\ \text{A}) = 1.0 \times 10^{-2}\ \text{T}.$$

▌▌

*17-7 ▌ Operational definition of the ampere and coulomb; force between two parallel conductors

You may have wondered how the constant μ_0 can be exactly $4\pi \times 10^{-7}$ T·m/A. With an older definition of the ampere, μ_0 was measured experimentally to be very close to this value. Today, however, μ_0 is *defined* to be exactly $4\pi \times 10^{-7}$ T·m/A. This, of course, could not be done if the ampere were defined independently. The ampere, the unit of current, is now defined in terms of the magnetic field B it produces using the defined value of μ_0.

Let us be more precise. Consider two long parallel conductors a distance L apart (Figure 17-35*a*). They carry currents I_1 and I_2, respectively. Each current produces a magnetic field that is "felt" by the other so that a force is exerted on each. For example, the magnetic field B_1 produced by I_1 is given by Equation 17-8; at the location of the second conductor, the magnitude of this field is

$$B_1 = \frac{\mu_0}{2\pi}\frac{I_1}{L}.$$

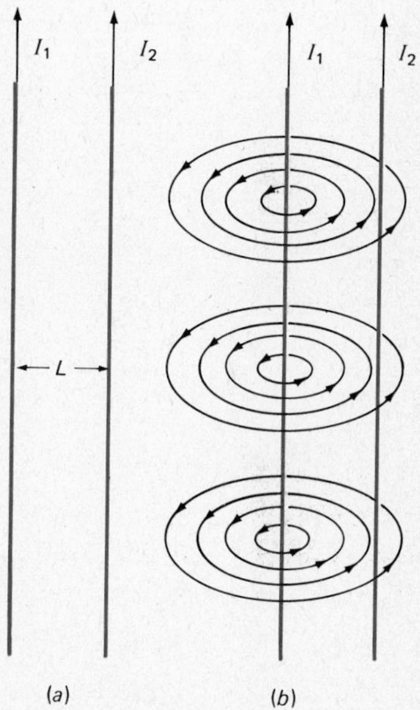

FIGURE 17-35
(*a*) Two parallel conductors carrying currents I_1 and I_2. (*b*) Magnetic field produced by I_1.

See Figure 17-35*b*, where only the field due to I_1 is shown. According to Equation 17-2, the force F per unit length l on the conductor carrying current I_2 is

$$\frac{F}{l} = I_2 B_1.$$

We substitute for B_1 in the formula above and find that

$$\frac{F}{l} = \frac{\mu_0}{2\pi}\frac{I_1 I_2}{L}. \qquad (17\text{-}11)$$

If we use the right-hand rule of Figure 17-10*b*, we see that the lines of B are as shown in Figure 17-35*b*. Then using the right-hand rule of Figure 17-15*b*, we see that the force exerted on I_2 will be to the left in the figure. That is, I_1 exerts an attractive force on I_2. This is true as long as the currents are in the same direction. If I_2 is in the opposite direction, the right-hand rule indicates that the force is in the opposite direction; that is, I_1 exerts a repulsive force on I_2.

Similar reasoning to that above shows that the magnetic field produced by I_2 exerts an equal but opposite force on I_1. This must be true of course, by Newton's third law.

Equation 17-11 can be used to define the ampere precisely. If $I_1 = I_2 = 1\text{A}$, and the two wires are 1 m apart, then

$$\frac{F}{l} = \frac{(4\pi \times 10^{-7}\ \text{T} \cdot \text{m/A})}{2\pi} \frac{(1\ \text{A})(1\ \text{A})}{1\ \text{m}} = 2 \times 10^{-7}\ \text{N/m}.$$

Thus, *1 ampere is defined as that current flowing in each of two long parallel conductors 1 meter apart, which results in a force of exactly* 2×10^{-7} *N/m of length of each conductor.*

This, then, is the precise definition of the ampere. The *coulomb* is then defined as being *exactly* 1 ampere-second: 1 C = 1 A·s. The value of k or ε_0 in Coulomb's law (Section 13-2) is obtained from experiment.

You may think that this is a rather roundabout way of defining quantities. Indeed, it does seem so. The reason behind it is that physicists seek *operational definitions* of quantities. That is, definitions of quantities that can actually be measured given a definite set of operations to carry out. For example, the unit of charge, the coulomb, could be defined in terms of the force between two equal charges after defining a value for ε_0 or k in Equation 13-1. However, to carry out an actual experiment to measure the force between two charges is very difficult. For one thing, any desired amount of charge is not easily obtained precisely; and charge tends to leak from objects into the air. On the other hand, the amount of current can be varied continuously by putting a variable resistor in a circuit. Thus the force between two current-carrying conductors is far easier to measure precisely. And this is why the ampere is defined first and then the coulomb in terms of the ampere. In the National Bureau of Standards, Washington, D.C., precise measurement of current is made using circular coils of wire rather than straight lengths because it is more convenient and accurate. The principle, of course, is not altered.

▌ EXAMPLE 17-7 The two wires of a 2.0-m-long appliance cord are 3.0 mm apart and carry 8.0 A dc. Calculate the force between these wires.

SOLUTION Equation 17-11 gives us

$$F = \frac{(2.0 \times 10^{-7}\ \text{T} \cdot \text{m/A})(8.0\ \text{A})(8.0\ \text{A})(2.0\ \text{m})}{3.0 \times 10^{-3}\ \text{m}} = 8.5 \times 10^{-3}\ \text{N}.$$

Since the currents are in opposite directions, the force would tend to spread them apart. ▌▌

The electric and magnetic field strengths are also defined operationally: the electric field in terms of the measureable force on a charge, via Equation 13-2; and the magnetic field in terms of the force per unit length on a current-carrying wire, via Equation 17-1.

*17-8 ▌ Hysteresis

The field of a long solenoid is directly proportional to the current. Indeed, Equation 17-10 tells us that the field B_0 inside a solenoid is given by

$$B_0 = \mu_0 n I.$$

This is valid if there is only air inside the coil. If we put a piece of iron or other ferromagnetic material inside the solenoid, the field will be greatly increased, often by hundreds or thousands of times. We saw in our discussion of electromagnets (Section 17-2) that this happens because the domains in the iron become preferentially aligned by the external field. It is sometimes convenient to write the total field in this case as a sum of two terms:

$$B = B_0 + B_M. \tag{17-12}$$

Here B_0 refers to the field due to the current in the wire (the "external field"); it is equal to the field that would be present in the absence of a ferromagnetic material. Then B_M represents the additional field due to the ferromagnetic material; usually $B_M \gg B_0$.

The total field inside a solenoid in such a case can also be written by replacing the constant μ_0 in Equation 17-10 by another constant, μ, characteristic of the material inside the coil:

$$B = \mu n I; \tag{17-13}$$

μ is called the *magnetic permeability* of the material. For ferromagnetic materials μ is much greater than μ_0. For all other materials, its value is very close to μ_0.† The value of μ, however, is not constant for ferromagnetic materials; it depends on the value of the external field B_0, as the following experiment shows.

Suppose that an iron core is initially unmagnetized and there is no current in the solenoid. Then the current is slowly increased and B_0 increases according to Equation 17-10. The total field B also increases, but follows the curved line shown in the graph of Figure 17-36. (Note the different scales: $B \gg B_0$.) Initially (point *a*), no domains are aligned. As B_0 increases, the domains become more and more aligned until at point *b*, nearly all are aligned. The iron is said to be approaching *saturation*.‡ Now suppose that the external field B_0 is reduced by decreasing the current in the solenoid. As the current is reduced to zero, point *c* in Figure 17-37, the domains do not become completely unaligned. Some permanent magnetism remains. If the current is then reversed in direction, enough domains can be turned around so that $B = 0$ (point *d*). As the reverse current is increased further, the iron approaches saturation in the opposite direction (point *e*). Finally, if the current is again reduced to zero and then increased in the original direction, the total field follows the path *efgb*, again approaching saturation at point *b*.

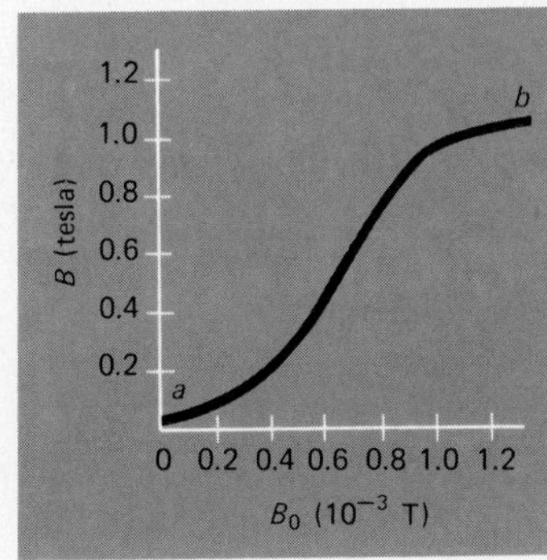

FIGURE 17-36
Total magnetic field B of an iron core solenoid as a function of the external field B_0.

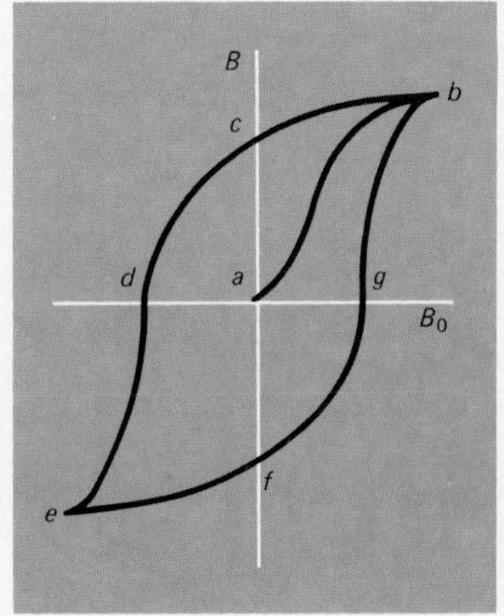

FIGURE 17-37
Hysteresis curve.

† All materials are slightly magnetic. Nonferromagnetic materials fall into two classes: *paramagnetic*, in which μ is very slightly larger than μ_0; and *diamagnetic*, in which μ is very slightly less than μ_0. Paramagnetic materials apparently contain atoms that have a net magnetic dipole moment due to orbiting electrons, and these become slightly aligned with an external field just as the galvonometer coil in Figure 17-25 experiences a torque that tends to align it. Atoms of diamagnetic materials have no net dipole moment. The spins and orbital motions of their electrons exactly cancel. However, in the presence of an external field, electrons revolving in one direction are caused to increase in speed slightly, whereas those revolving in the opposite direction are reduced in speed; the result is a slight net magnetic effect which actually opposes the external field.

‡ Point *b* is typically 70 percent of full saturation. The curve continues to rise very slowly, and reaches 98 percent saturation only when B_0 is increased by about a thousandfold above that at point (*b*). The last few domains are very difficult to align.

Notice that the field did not pass through the origin (point a) in this cycle. The fact that the curves do not retrace themselves on the same path is called *hysteresis*. The curve *bcdefgb* is called a *hysteresis loop*. In such a cycle, much energy is lost to heat (friction) as a result of realignment of the domains; it can be shown that the energy lost in this way is proportional to the area of the hysteresis loop.

At points c and f, the iron core is magnetized even though there is no current in the solenoid. This point corresponds to a permanent magnet. For a permanent magnet, it is desired that ac and af be as large as possible. Materials for which this is true are said to have high *retentivity* and may be referred to as "hard." A hysteresis curve such as that in Figure 17-38 occurs for "soft iron." This is preferred for electromagnets since the field can be more readily switched off, and the field can be reversed with less loss of energy. Whether iron is "soft" or "hard" depends on how it is alloyed, heat treatment, and other factors.

You may wonder how a material can be demagnetized—that is, made unmagnetized. This can be done by reversing the magnetizing current repeatedly while decreasing its magnitude. This results in the curve of Figure 17-39. If your watch becomes magnetized, it can be demagnetized by placing it in a coil of wire carrying an ac current, and then slowly pulling the watch away.

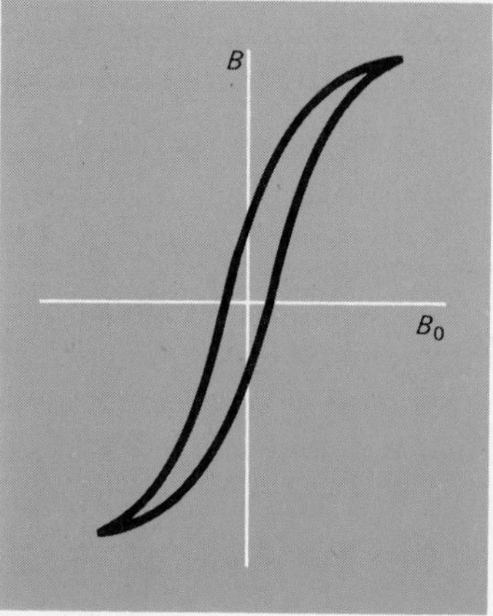

FIGURE 17-38
Hysteresis curve for soft iron.

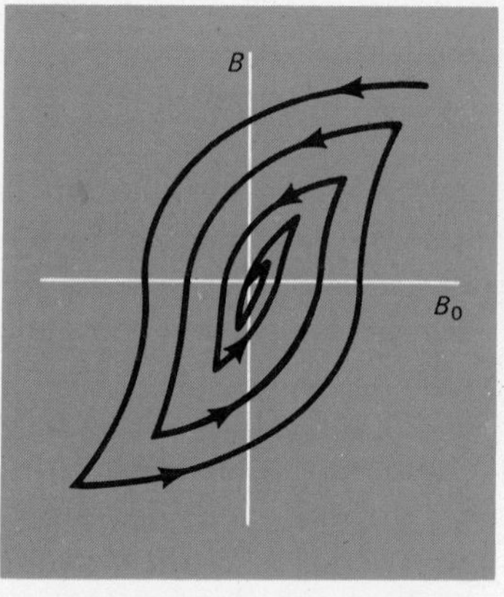

FIGURE 17-39
Successive hysteresis loops during demagnetization.

▮ Summary

A magnet has two poles, north and south. The north pole is that end which points toward the north when the magnet is freely suspended. Unlike poles of two magnets attract each other, whereas like poles repel. Permanent magnets are made up of tiny *domains*—each a tiny magnet—which are aligned. In an unmagnetized piece of iron, the domains are randomly oriented.

We can apply the field concept to magnetism and imagine that a magnetic field surrounds every magnet. The SI unit for *magnetic field strength* is the tesla (T). The force one magnet exerts on another is said to be an interaction between one magnet and the magnetic field produced by the other.

Electric currents produce magnetic fields. For example, the lines of magnetic field due to a current in a straight wire form circles around the wire and the field exerts a force on magnets placed near it.

A magnetic field exerts a force on an electric current. For a straight wire of length l carrying a current I the force has magnitude $F = IlB \sin\theta$; θ is the angle between the magnetic field of strength B and the wire; the direction of the force is perpendicular to the wire and to the magnetic field, and is given by the right-hand rule. Similarly, a magnetic field exerts a force on a charge q moving with velocity v of magnitude $F = qvB \sin\theta$, where θ is the angle between $\mathbf{v}$ and $\mathbf{B}$; the direction of $\mathbf{F}$ is perpendicular to $\mathbf{v}$ and to $\mathbf{B}$. The path of a charged particle moving perpendicular to a uniform magnetic field is a circle.

The measurement of the charge to mass ratio (e/m) of the electron was done using magnetic and electric fields. The charge e on the electron was measured in the Millikan oil-drop experiment, and then its mass was obtained from the measured value of the e/m ratio. Electric and magnetic fields are employed in the mass spectrograph, which is used to determine the masses of atoms.

The force exerted on a current-carrying wire by a magnetic field is the basis for operation of many devices, such as meters, motors, and loudspeakers.

Questions

1 A compass needle is not always balanced parallel to the earth's surface but one end may dip downward. Explain.

2 How do you suppose the first magnets found in Magnesia were formed?

3 Why will either pole of a magnet attract an unmagnetized piece of iron?

4 Suppose that you have three iron rods, two of which are magnetized but the third is not. How would you determine which two are the magnets without using any additional objects?

5 Will a magnet attract any metallic object or only those made of iron? (Try it and see.) Why is this so?

6 An unmagnetized nail will not attract an unmagnetized paper clip. However, if one end of the nail is in contact with a magnet, the other end *will* attract a paper clip. Explain.

*7 Note that the pattern of magnetic field lines surrounding a bar magnet is similar to that of the electric field around an electric dipole. From this fact predict how the magnetic field will change with distance (*a*) when near one pole of a very long bar magnet, and (*b*) when far from a magnet as a whole.

8 Draw the magnetic field lines around a straight section of wire carrying a current horizontally to the left.

9 In what direction are the magnetic field lines surrounding a straight wire carrying a current that is moving directly toward you?

10 Another type of magnetic switch similar to a solenoid is a *relay*. A relay is an electromagnet (the iron rod inside the coil does not move) which, when activated, attracts a piece of soft iron on a pivot. Design a relay (*a*) to make a doorbell, and (*b*) to close an electrical switch. A relay is used in the latter case when you need to switch on a circuit carrying a very large current but you do not want that large current flowing through the main switch. For example, the starter switch of a car is connected to a relay so that the large currents needed for the starter do not pass to the dashboard switch.

11 The magnetic field due to current in wires in your home can affect a compass. Discuss the problem in terms of currents, depending on whether they are ac or dc.

12 What kind of field or fields surround a moving electric charge?

13 A horseshoe magnet is held vertically with the north pole on the left and south pole on the right. A wire passing perpendicularly between the poles carries a current away from you. In what direction is the force on the wire?

14 Can you set a resting electron into motion with a magnetic field? With an electric field?

15 A charged particle is moving in a circle under the influence of a uniform magnetic field. If an electric field that points in the same direction as the magnetic field is turned on, describe the path the charged particle will take.

16 Each of the right-hand rules you learned in this chapter can be changed to *left-hand rules* if you are specifying the direction of movement of *negative* particles, such as electrons in a wire. Show, for each right-hand rule, that the same operations using the left hand give the same results if the direction of charge flow is for negative charges.

*17 In the electromagnetic pumping device shown in Figure 17-20 does the direction of the force, and therefore of the blood flow, depend on the sign of the ions?

18 Two ions have the same mass, but one is singly ionized and the other is doubly ionized. How will their positions on the film of the mass spectrograph of Figure 17-24 differ?

19 What factors determine the sensitivity of a galvanometer?

20 How can you make a compass without using iron or other ferromagnetic material?

21 A beam of electrons is directed toward a horizontal wire carrying a current from left to right. In what direction is the beam deflected?

Problems

SECTION 17-3

1 (I) (*a*) What is the force per meter on a wire carrying a 0.50-A current when perpendicular to a 1.0-T magnetic field? (*b*) What if the angle between the wire and field is 45°?

2 (I) The force on a wire carrying 20 A is a maximum of 3.6 N when placed between the pole faces of a magnet. If the pole faces are 15 cm in diameter, what is the approximate strength of the magnetic field?

3 (I) How much current is flowing in a wire 2.0 m long if the force on it is 0.70 N when placed in a uniform 0.030-T field?

4 (I) Calculate the magnetic force on a 140-m length of wire stretched between two towers carrying a 200-A current. The earth's magnetic field of 5.0×10^{-5} T makes an angle of 60° with the wire.

5 (I) Determine the magnitude and direction of the force on an electron traveling 3.4×10^5 m/s horizontally to the east in a vertically upward magnetic field of strength 1.5 T.

6 (I) Describe the path of a proton ($q = e$, $m = 1.67 \times 10^{-27}$ kg) that moves perpendicular to a 0.020-T magnetic field with a speed of 9.2×10^6 m/s. The field points directly toward the observer.

7 (I) Calculate the force on an airplane that has acquired a net charge of 130 C and moves with a speed of 250 m/s perpendicular to the earth's magnetic field of 5.0×10^{-5} T.

8 (II) An electron is accelerated by a voltage of 5.0 kV. What will be the radius of its path in a 3.0×10^{-3} T magnetic field?

9 (II) An electron experiences the greatest force as it travels 2.1×10^5 m/s in a magnetic field when it is moving southward. The force is upward and of magnitude 5.6×10^{-13} N. What is the magnitude and direction of the magnetic field?

10 (II) A proton moves in a circular path perpendicular to a 1.10-T magnetic field. The radius of its path is 4.5 cm. Calculate the energy of the proton in eV.

11 (II) A doubly charged helium atom whose mass is 6.7×10^{-27} kg is accelerated by a voltage of 800 V. What will be its radius of curvature in a uniform 0.070-T field?

12 (II) A particle of charge q moves in a circular path of radius r in a uniform magnetic field B. Show that its momentum is $p = qBr$.

13 (II) An electron is accelerated through a potential difference of 5000 V. What is the strength of the magnetic field if the radius of its path is 3.4 mm?

14 (II) Show that the emf produced by the Hall effect is given by $\mathscr{E} = vBl$, where v is the speed of the charged particles in the conductor of width l (Figure 17-19).

15 (III) A 10-g bullet moves with a speed of 300 m/s perpendicular to the earth's magnetic field 5.0×10^{-5} T. If the bullet possesses a net charge of 5.0×10^{-9} C, by what distance will it be deflected from its path due to the magnetic field after it has traveled 500 m?

16 (III) An electron enters a large solenoid at a 5° angle to the axis. If the field is a uniform 0.23 T, determine the radius and pitch (distance between loops) of the electron's helical path if its speed is 2.8×10^5 m/s.

SECTION 17-4

17 (I) In a mass spectrograph, germanium atoms have radii of curvature equal to 21.0, 21.6, 21.9, 22.2, and 22.8 cm. The largest radius corresponds to an atomic mass of 76 u. What are the atomic masses of the other isotopes?

18 (II) What is the value of e/m for a particle that moves in a circle of radius 5.0 mm in a 0.60-T magnetic field if a crossed 200-V/m electric field will make the path straight?

19 (II) A mass spectrometer is being used to monitor air pollutants. It is difficult, however, to separate molecules with nearly equal mass such as CO (28.0106 u) and N_2 (28.0134). How large (radius of curvature) must the spectrograph be if these two molecules are to be separated on the film by 0.40 mm?

20 (II) Protons move in a circle of radius 6.0 cm in a 0.80-T magnetic field. What value of electric field could make their path straight? In what direction must it point?

21 (II) What is the velocity of a beam of electrons that go undeflected when passing through crossed electric and magnetic fields of magnitude 3.8×10^3 V/m and 2.5×10^{-3} T, respectively? What is the radius of the electron orbit if the electric field is turned off?

22 (II) Suppose that the electric field between the electric plates in the mass spectrograph of Figure 17-24 is 2.5×10^4 V/m and the magnetic fields $B = B' = 0.42$ T. The source contains boron isotopes of mass number 10 and 11 (to get their masses, multiply by 1.66×10^{-27} kg). How far apart are the lines formed by the singly charged ions of each type on the photographic film?

23 (II) An oil drop whose mass is determined to be 3.3×10^{-15} kg is held at rest between two large plates separated by 1.0 cm when the potential difference between them is 340 V. How many excess electrons does this drop have?

24 (III) The cyclotron (Figure 17-40) is a device used to accelerate elementary particles such as protons to high speeds. Particles starting at point A with some initial velocity travel in circular orbits in the magnetic

field B. The particles are accelerated to higher speeds each time they pass in the gap between the metal "dees," where there is an electric field E. (There is no electric field within the cavity of the metal dees.) The electric field changes direction each half-cycle, owing to an ac voltage $V = V_0 \sin 2\pi ft$ so that the particles are increased in speed at each passage through the gap. (*a*) Show that the frequency f of the voltage must be $f = Bq/2\pi m$, where q is the charge on the particles and m their mass. (*b*) Show that the kinetic energy of the particles increases by $2qV_0$ each revolution, assuming that the gap is small. (*c*) If the radius of the cyclotron is 2.0 m and the magnetic field strength is 0.50 T, what will be the maximum kinetic energy of accelerated protons in MeV? (*d*) How is a cyclotron like a swing?

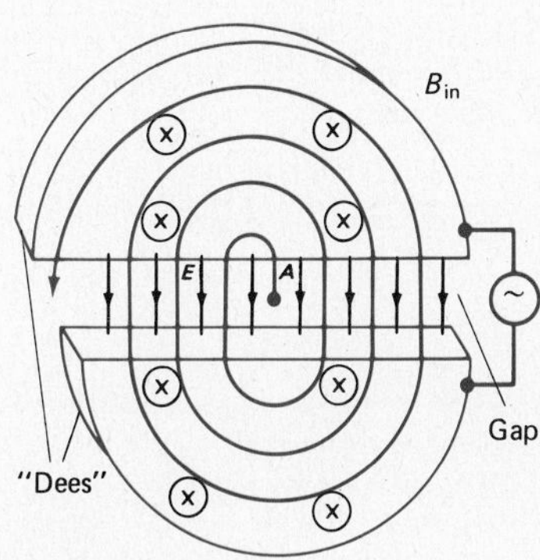

FIGURE 17-40
Cyclotron.

25 (III) One form of mass spectrometer accelerates ions by a voltage V before they enter a magnetic field B. The ions are assumed to start from rest. Show that the mass of an ion is $m = qB^2R^2/2V$, where R is the radius of the ions' path in the magnetic field and q is their charge.

*26 (III) In Millikan's oil-drop experiment, the mass of the oil drop is obtained by observing the terminal speed v_T of the freely falling drop in the absence of an electric field. Under these circumstances, the "effective" weight equals the viscous force given by Stokes's law (Section 6-9), $F = 6\pi\eta r v_T$, where η is the viscosity of air and r the radius of the drop. Also, the actual weight $mg = \frac{4}{3}\pi r^3 \rho g$ must be corrected for the buoyant force of the air; this is done by replacing ρ with $\rho - \rho_A$, where ρ is the density of the oil and ρ_A the density of the air. With these preliminaries, show that the charge on the drop is given by

$$q = 18\pi \frac{d}{V} \sqrt{\frac{\eta^3 v_T^3}{2(\rho - \rho_A)g}},$$

where d is the separation of the plates (Figure 17-23) and V is the voltage across them that just keeps the drop stationary. All the quantities on the right side of this equation are known or can be measured. The terminal velocity v_T is determined by measuring the time it takes the drop to fall a measured distance, which is observed through a small telescope.

SECTION 17-5

27 (I) A galvanometer needle deflects full scale for a 40-μA current. What current will give full-scale deflection if the magnetic field weakens to 0.90 of its original value?

28 (I) If the restoring spring of a galvanometer weakens by 20 percent over the years, what current will give full-scale deflection if it originally required 50 μA?

29 (I) If the current to a motor drops by 10 percent, by what factor does the output torque change?

30 (I) A circular loop of wire is placed with its face parallel to the magnetic field between the pole pieces of a large magnet. When 6.0 A flows in the coil, the torque on it is 3.6 N·m. What is the magnetic field strength? The loop is 14 cm by 14 cm.

31 (II) Show that the torque acting on the current loop of Figure 17-25 is given by $\tau = NIAB \cos\theta$, where θ is the angle between the face of the coil and direction of the (uniform) magnetic field.

32 (II) A circular coil 20 cm in diameter and containing five loops lies flat on the ground. The earth's magnetic field at this location is 6.0×10^{-5} T; it points into the earth at an angle of 60° below a line pointing due north. If a 5.0-A clockwise current passes through the coil (*a*) determine the torque on the coil; (*b*) which edge of the coil rises up—north, east, south, or west?

33 (III) A square loop of aluminum wire is 25 cm on a side. It is to carry 20 A and rotate in a 2.0-T magnetic field. (*a*) Determine the minimum diameter of the wire so that it will not fracture from tension or shear. Assume a safety factor of 10. (See Table 4-2.) (*b*) What is the resistance of a single loop of this wire?

SECTION 17-6

34 (I) How strong is the magnetic field 10 cm from a long, straight wire carrying 8.0 A?

35 (I) How much current does a wire carry if $B = 0.030$ T at a point 12 cm from it?

36 (I) If a magnetic field of no more than 10^{-3} T is to be allowed 25 cm from an electrical wire, what is the maximum current the wire can carry?

37 (I) How far from a long straight wire carrying 10 A will the magnetic field be 1.0×10^{-2} T?

38 (II) What is the maximum acceleration (in g's) of a 230-g model airplane charged to 3.0 C and traveling 1.5 m/s if it passes within 15 cm of a wire, perpendicular to its path, carrying a 20-A current?

39 (II) An electron traveling 4.0×10^6 m/s is directed at a 45° angle to a wire carrying a 5.0-A current. If the electron passes within 10 cm of the wire, what maximum force does it feel?

40 (II) A compass is placed 20 cm north of a straight vertical wire carrying 30 A current downward. In what direction does a compass needle point at this location? Assume that the horizontal component of the earth's field at this point is 0.50×10^{-4} T and the magnetic declination is 0°.

41 (II) Lighting strikes a solid 10-m-tall, 8.0-cm-diameter iron flagpole such that electrons flow up the pole. If the potential difference between the top and bottom of the pole is 40 kV, what will be the magnitude and direction of the magnetic field 1.0 m north of the pole?

42 (II) Determine the magnetic field midway between two long straight wires 10 cm apart if one carries 10.0 A and the other 8.0 A and these currents are (*a*) in the same direction, and (*b*) in opposite directions.

43 (II) A compass needle in a particular location points 15°E of N outdoors. However, when it is placed 10 cm to the east of a vertical wire inside a building, it points 60°E of N. What is the magnitude and direction of the current in the wire? The earths' field there is 0.50×10^{-4} T and is horizontal.

*44 (II) A toroid is a solenoid in the shape of a circle (Figure 17-41). Use Ampère's law along the circular path shown (dashed) to determine that the magnetic field strength inside the toroid is $B = \mu_0 NI/2\pi R$, where N is the total number of turns. Is the field inside a toroid uniform like a solenoid's? If not, how does it vary?

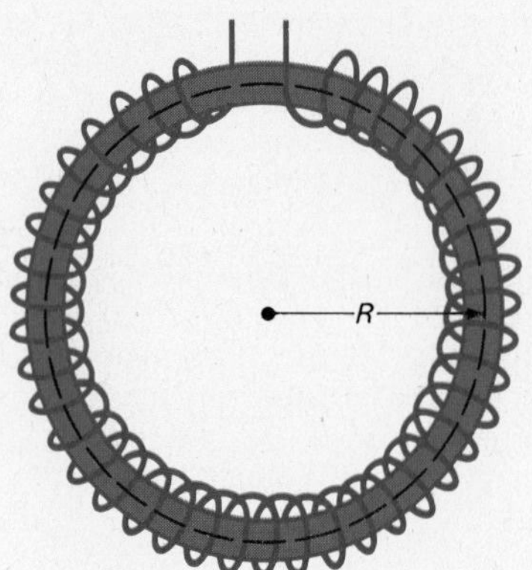

FIGURE 17-41
Toroid.

*45 (III) Suppose that a current I flows uniformly through a long cylindrical conductor of radius r_0. Use Ampère's law to show that the magnetic field inside the conductor at a distance r from the center of the conductor is

$$B = \frac{\mu_0 I r}{2\pi r_0^2} \qquad (r < r_0).$$

Assume that the field lines are circles, just as they are outside the conductor.

46 (III) Two long parallel wires 12 cm apart carry 15-A currents in the same direction. Determine the magnetic field strength at a point 10 cm from one wire and 20 cm from the other. (*Hint*: Make a drawing in a plane containing the field lines and recall the rules for vector addition.)

47 (III) The magnetic field strength B at the center of a circular coil of wire carrying a current I is

$$B = \frac{\mu_0 NI}{2R},$$

where N is the number of loops in the coil and R is its radius. Suppose that an electromagnet uses a coil 2.0 m in diameter made from square copper wire 2.0 mm on a side. The power supply produces 50 V at a maximum power output of 1.0 kW. (*a*) How many turns are needed to run the power supply at maximum power? (*b*) What is the magnetic field strength at the center of the coil? (*c*) If you use a greater number of turns and this same power supply, will a greater magnetic field strength result? Explain.

*SECTION 17-7

*48 (I) What is the magnitude and direction of the force between two parallel wires 80 m long and 30 cm apart, each carrying 65 A in the same direction?

*49 (I) A vertical, straight wire carrying a 5.0-A current exerts an attractive force per unit length 6.0×10^{-4} N/m on a second parallel wire 8.0 cm away. What current (magnitude and direction) flows in the second wire?

For additional reading

Cox, A., et al., "Reversals of the Earth's Magnetic Field," *Sci. American,* **216**:2 (Feb., 1967), p. 44.

Einstein, A., and L. Infeld, *Evolution of Physics*, Simon & Schuster, Inc., New York, 1938.

Magie, W. F., *Source Book in Physics*, Harvard University Press, Cambridge, Mass., 1964. Writings of Oersted, Ampère, and others; J. J. Thomson's discovery of the electron.

Meyer, H. W., *History of Electricity and Magnetism*, The M.I.T. Press, Cambridge, Mass., 1972.

18 ▌ Electromagnetic induction and power generation

In Chapter 17 we discussed two ways in which electricity and magnetism are related: (1) an electric current produces a magnetic field, and (2) a magnetic field exerts a force on an electric current or moving electric charge. These discoveries were made in 1820–1821. Scientists then began to wonder if the reverse of (1) might be true. That is, if electric currents produce a magnetic field, is it possible that a magnetic field can produce an electric current? Ten years later the American Joseph Henry (1797–1878) and the Englishman Michael Faraday (1791–1867) independently found that it was possible. Henry actually made the discovery first, but Faraday published his results earlier and investigated the subject in more detail. We now discuss this phenomenon and some of its world-changing applications.

18-1 ▌ Induced emf and Faraday's law

Induced emf

In his attempt to produce an electric current from a magnetic field, Faraday used the apparatus shown in Figure 18-1. A coil of wire, X, was connected to a battery. The current that flowed through X produced a magnetic field that was intensified by the iron core. Faraday hoped that by using a strong-enough battery, a steady current in X would produce a great-enough magnetic field to produce a current in a second coil Y. This second circuit, Y, contained a galvanometer to detect any current but contained no battery. At first he met no success. Then quite by accident he noticed that the galvanometer in circuit Y did deflect strongly at the

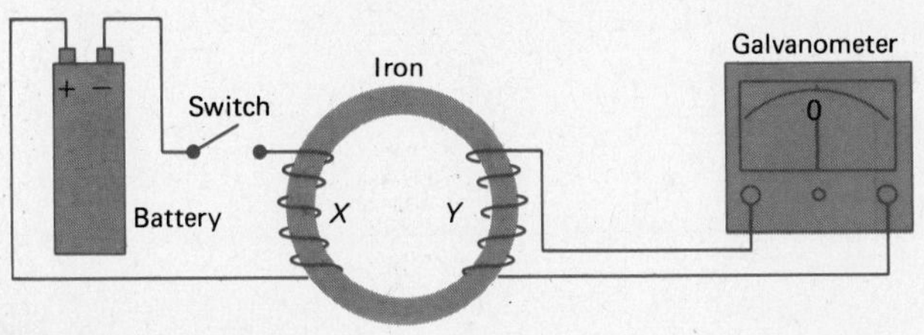

FIGURE 18-1
Faraday's experiment to induce an emf.

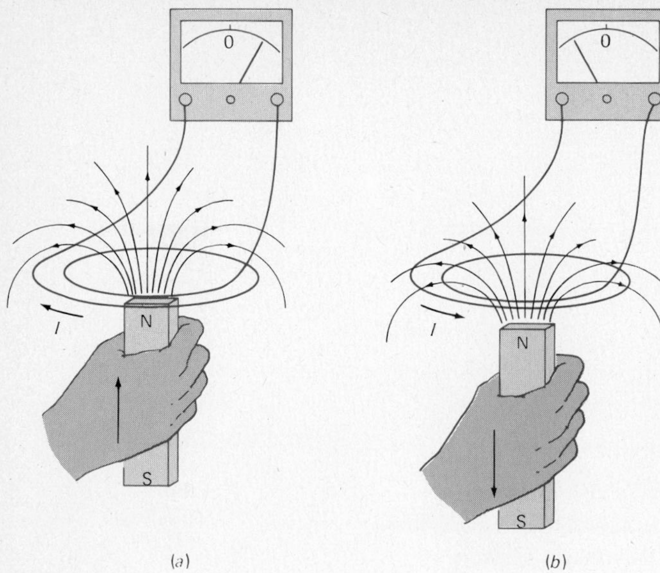

FIGURE 18-2
A current is induced when a magnet is moved toward a coil (*a*). The induced current is opposite when the magnet is removed (*b*). Note that the galvanometer zero is in the center of the scale and deflects left or right, depending on direction of the current.

moment he closed the switch in circuit X. And the galvanometer deflected strongly in the opposite direction when he opened the switch. A *steady* current in X had produced *no* current in Y. Only when the current in X was starting or stopping was a current produced in Y.

Faraday concluded that although a steady magnetic field produced no current, a *changing* magnetic field can produce an electric current! When the magnetic field through coil Y changes, a current flows as if there were a source of emf in the circuit. We therefore say that an **induced emf is produced by a changing magnetic field**.

Faraday did further experiments on *electromagnetic induction*, as this phenomenon is called. For example, Figure 18-2 shows that if a magnet is moved quickly into a coil of wire, a current is induced in the wire. If the magnet is quickly removed, a current is induced in the opposite direction. Furthermore, if the magnet is held steady and the coil of wire is moved toward or away from the magnet, again an emf is induced and a current flows. Motion or change is required to induce an emf. It does not matter whether the magnet moves or the coil. What matters is that the magnetic field that passes through the coil must be changing if an emf is to be induced.

Magnetic flux

Faraday investigated quantitatively what factors influence the magnitude of the emf induced. He found, first of all, that it depends on time: the more rapidly the field changes, the greater the induced emf. But, he found, it is not simply a change in the field but rather a change in *flux* that is important. The **magnetic flux**, Φ, passing through a loop of wire of area A is defined as

$$\Phi = B_{\perp}A = BA\cos\theta. \qquad (18\text{-}1)$$

Here $B_{\perp}$ is the component of the magnetic field strength $\mathbf{B}$ perpendicular to the face of the coil; θ is the angle between $\mathbf{B}$ and a line drawn perpendicular to the face of the coil. These quantities are shown in Figure 18-3 for a square coil of side l whose area $A = l^2$. When the face of the coil is parallel to B, $\theta = 90°$ and $\Phi = 0$; and when B is perpendicular to the coil, $\theta = 0°$ and

$$\Phi = BA \qquad (B_\perp \text{ coil face}).$$

It will be helpful if we establish the following convention: the lines of B are drawn such that the number of lines per unit area is proportional to the field strength. (This is just what we did for the electric field lines in Chapter 13.) If we do this, the flux Φ can be thought of as the *total number of lines passing through the coil.* This is illustrated in Figure 18-4. For $\theta = 90°$, no lines pass through the coil and $\Phi = 0$; Φ is a maximum when $\theta = 0°$. The unit of flux is the tesla-meter^2; this is called† a *weber* (Wb): $1\ \text{Wb} = 1\ \text{T}\cdot\text{m}^2$.

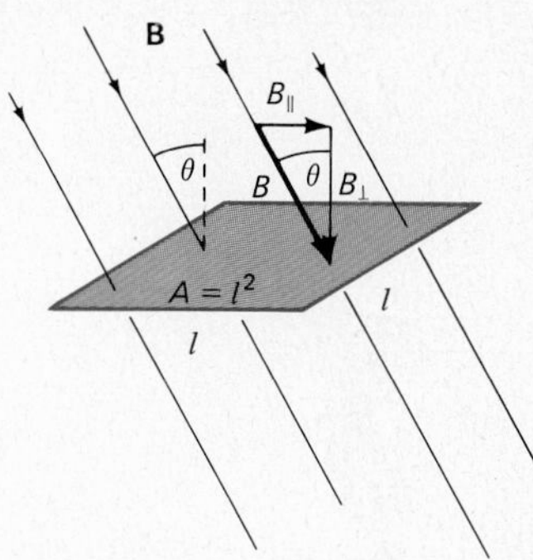

FIGURE 18-3
Determining the flux through a flat loop of wire. This loop is square, of side l and area $A = l^2$.

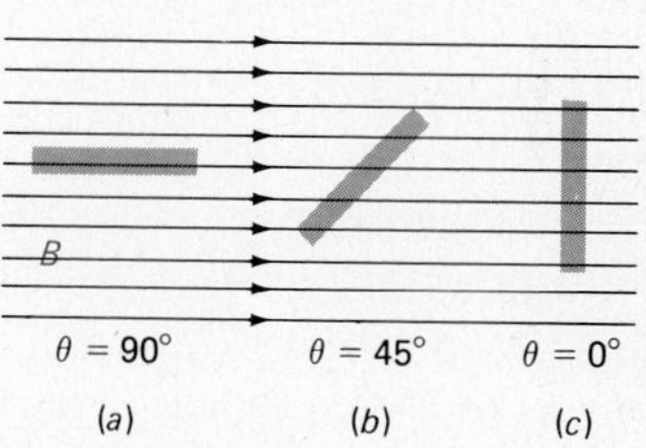

FIGURE 18-4
Magnetic flux Φ is proportional to the number of lines of B that pass through the loop.

Faraday's law

With this definition of the flux, we can now return to Faraday's investigations. He found that the magnitude of the emf produced depends on the rate at which the magnetic flux changes. Faraday found that if the flux through N loops of wire changes by an amount $\Delta\Phi = \Phi_2 - \Phi_1$, during a time Δt, the average induced emf during this time is

$$\mathscr{E} = -N\frac{\Delta\Phi}{\Delta t}. \qquad (18\text{-}2)$$

This fundamental result is known as **Faraday's law of induction**.

The minus sign in Equation 18-2 is placed there to remind us in which direction the induced emf acts. Experiment shows that *an induced emf always gives rise to a current whose magnetic field opposes the original change in flux.* This is known as **Lenz's law**. Let us apply it to the case of relative motion between a magnet and a coil, Figure 18-2. The changing flux induces an emf, which produces a current in the coil; and this induced current produces its own magnetic field. In Figure 18-2*a* the distance between the coil and the magnet decreases; so the magnetic field, and therefore the flux, through the coil increases. The magnetic field of the magnet points upward. To oppose this upward increase, the field produced by the induced current must point *downward.* Thus Lenz's law tells us that the current must move as shown (use the right-hand rule). In Figure 18-2*b*, the flux *decreases,* so the induced current produces an *upward* magnetic field that is "trying" to maintain the status quo. Thus the current must be as shown.

Let us consider what would happen if Lenz's law were just the reverse. The induced current would produce a flux in the same direction as the original change; this greater change in flux would produce an even larger current, followed by a still greater change in flux, and so on. The current would continue to grow indefinitely, producing power $(= I^2R)$ even after the original stimulus ended. This would violate the conservation of energy. Such "perpetual-motion" devices do not exist.

It is important to note that Faraday's investigations, as summarized in Faraday's law (Equation 18-2), says that an emf is induced whenever there

† Note also that $1\ \text{T} = 1\ \text{Wb/m}^2$; this is where the old SI unit for magnetic field strength, the Wb/m^2, comes from. The magnetic field strength B, since it equals Φ/A, is often called the *flux density.*

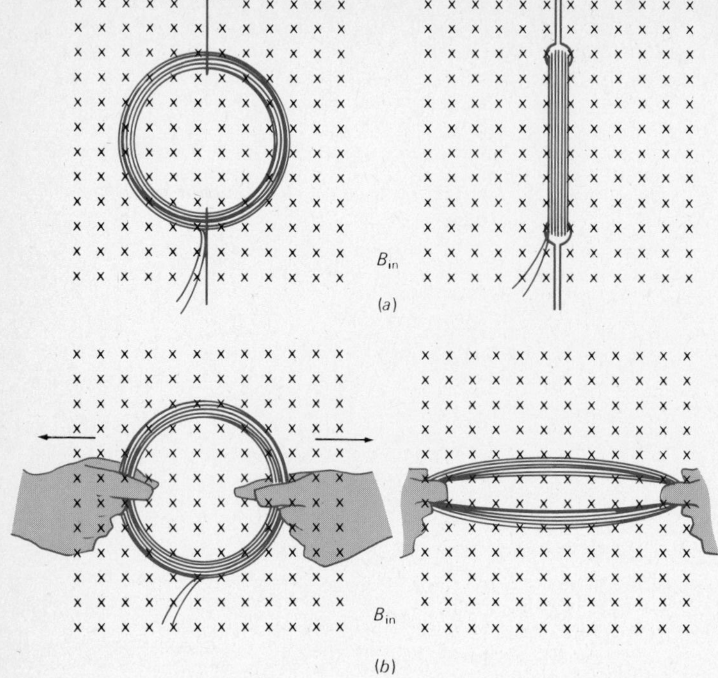

FIGURE 18-5
A current can be induced (a) by rotating the coil in the magnetic field or (b) by changing the area of the coil. In both cases the flux through the coil is reduced.

is a change in *flux*. Thus (see Equation 18-1) an emf can be induced in two ways: (1) by changing the magnetic field B; or (2) by changing the area A of the loop or its orientation θ with respect to the field. Figures 18-1 and 18-2 illustrated case (1). Examples of case (2) are illustrated in Figure 18-5 and the following example.

▌ EXAMPLE 18-1 A circular coil of radius 5.0 cm contains 100 loops and is positioned perpendicular to a 0.60-T magnetic field. It is quickly rotated, in 0.10 s, so its face is parallel to the field as in Figure 18-5*a*. How much energy is dissipated in the coil if its resistance is 100 Ω?

SOLUTION The area of the coil is $A = \pi r^2 = (3.14)(0.050\text{ m})^2 = 7.9\times10^{-3}\text{ m}^2$. The flux is initially $\Phi = BA = (0.60\text{ T})(7.9\times10^{-3}\text{ m}^3) = 4.7\times10^{-3}\text{ Wb}$. After 0.10 s the flux is zero. Therefore, the average induced emf is (Equation 18-2)

$$\mathscr{E} = -(100)\frac{0-4.7\times10^{-3}\text{ Wb}}{0.10\text{ s}} = 4.7\text{ V}.$$

The current $I = \mathscr{E}/R = 4.7\text{ V}/100\ \Omega = 47\text{ mA}$. The total energy dissipated is $I^2Rt = (4.7\times10^{-2}\text{ A})^2\,(100\ \Omega)\,(0.10\text{ s}) = 2.2\times10^{-2}\text{ J}$. From the conservation-of-energy principle, this is just equal to the work needed to rotate the coil. ▌▌

Changing magnetic field produces an electric field

Another way to induce an emf is shown in Figure 18-6, and this situation helps illuminate the nature of the induced emf. Assume that a uniform magnetic field B is perpendicular to the area bounded by the U-shaped

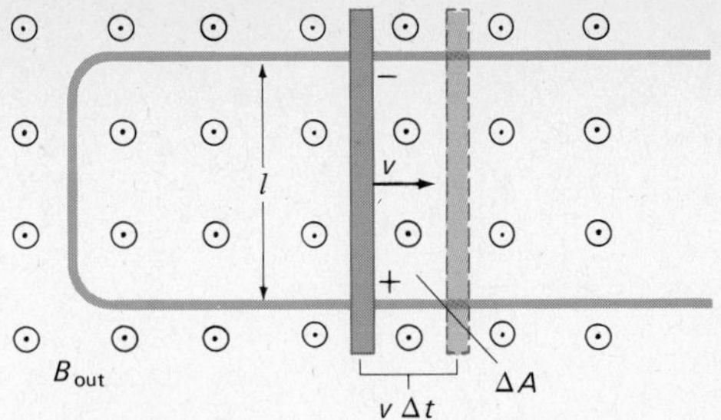

FIGURE 18-6
A conducting rod is moved to the right on a U-shaped conductor in a uniform magnetic field B that points out of the paper.

conductor and the movable rod resting on it. If the rod moves at a speed v, it travels a distance $\Delta x = v\,\Delta t$ in a time Δt. Therefore, the area of the loop increases by an amount $\Delta A = lv\,\Delta t$ in a time Δt. By Faraday's law, there is an induced emf $\mathcal{E}$ given by

$$\mathcal{E} = \frac{\Delta\Phi}{\Delta t} = \frac{B\,\Delta A}{\Delta t} = \frac{Blv\,\Delta t}{\Delta t} = Blv, \tag{18-3}$$

where we have omitted the minus sign. This equation is valid as long as B, l, and v are mutually perpendicular. (If they are not, we use only the components of each that are mutually perpendicular.)

We can also obtain Equation 18-3 without the use of Faraday's law. We saw in Chapter 17 that a charged particle moving perpendicular to a magnetic field B with speed v experiences a force $F = qvB$. When the rod of Figure 18-6 moves to the right with speed v, the electrons in the rod move with this velocity. Therefore, each feels a force $F = qvB$, which acts upward in the figure. If the rod was not in contact with the U-shaped conductor, electrons would collect at the upper end of the rod, leaving the lower end positive. There must thus be an induced emf. If the rod does slide on the U-shaped conductor, the electrons will flow into it; there will thus be a clockwise (conventional) current flowing in the loop. To calculate the emf, we determine the work W needed to move a charge q from one end of the rod to the other against this potential difference: $W =$ force × distance $= (qvB)(l)$. The emf equals the work done per unit charge, so $\mathcal{E} = qvBl/q = Blv$, just as above.†

▮ EXAMPLE 18-2 An airplane travels 1000 km/h in a region where the earth's field is 5.0×10^{-5} T and is nearly vertical. What is the potential difference induced between the wing tips that are 70 m apart?

SOLUTION Since $v = 1000\text{ km/h} = 280\text{ m/s}$, we have $\mathcal{E} = Blv = (5.0 \times 10^{-5}\text{ T})(70\text{ m})(280\text{ m/s}) = 1.0\text{ V}$. Not much to worry about. ▮▮

The fact that the electrons in the moving rod of Figure 18-6 feel a force implies that there must be an electric field present. Since the electric field E is the electric force per unit charge, $E = F/q$, the electric field in the rod must be $E = qvB/q$, or

$$E = vB. \tag{18-4}$$

In our earlier examples, in which a changing magnetic field induced an emf, there must also have been an electric field present in the wire.

† This argument, which is basically the same as for the Hall effect, explains this one way of inducing an emf. It does not explain the general case of electromagnetic induction, however.

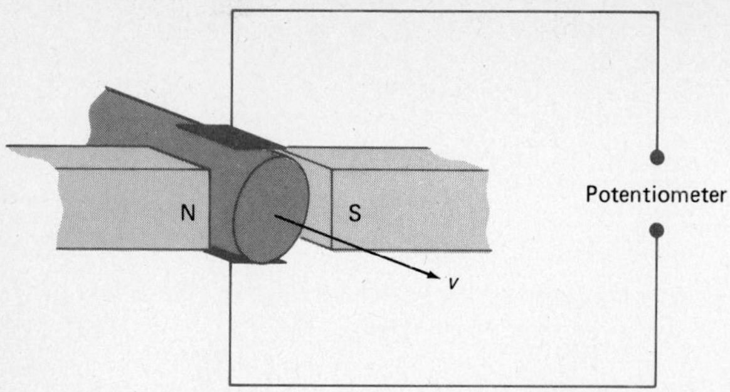

FIGURE 18-7
Measurement of blood velocity from the induced emf.

Thus we can say that **a changing magnetic field produces an electric field**. This applies to a conductor. But it is a general result and can also apply to empty space: an electric field will be produced at any point in space where there is a changing magnetic field. We shall see later that this has important ramifications.

▌ EXAMPLE 18-3 *Electromagnetic blood flow measurement.* The rate of blood flow can be measured using the apparatus shown in Figure 18-7 since blood contains ions. Suppose that the blood vessel is 2.0 mm in diameter, the magnetic field is 0.080 T, and the measured emf is 0.10 mV. What is the flow velocity of the blood?

SOLUTION We solve for v in Equation 18-3 and find that $v = \mathscr{E}/Bl = (1.0\times10^{-4}\ \text{V})/(0.080\ \text{T})(2.0\times10^{-3}\ \text{m}) = 0.63\ \text{m/s}$. (In actual practice, an alternating current is used to produce an alternating magnetic field; the induced emf is then alternating.) ▌▌

*Magnetic microphone and phonograph cartridges

Many microphones work on the principle of induction. In one form, a microphone is just the inverse of a loudspeaker (Figure 17-30). A small coil connected to a membrane is suspended close to a small permanent magnet. When sound waves strike the membrane, the coil moves in the magnetic field; the frequency of the induced emf will be just that of the impinging sound waves.

In a "ribbon" microphone, a thin metal ribbon is suspended between the poles of a permanent magnet. The ribbon vibrates in response to sound waves and the emf induced in the ribbon is proportional to its velocity.

When playing a phonograph record, a needle moves back and forth in the grooves. In one type of phonograph cartridge, the needle is connected to a tiny magnet inside the cartridge in the arm of the record player. The magnet is suspended inside a tiny coil, and an emf is induced in the coil according to the motion of the magnet. This tiny emf, at frequencies corresponding to the recorded material, must be amplified electronically before sending it to a loudspeaker. Most good-quality cartridges are the magnetic type described. Others, called "crystal pickups," make use of the piezoelectric effect (Section 16-6).

Microphones and cartridges are two more types of transducer. Another is the head of a tape recorder. Recorded tape has been magnetized at different intensities along its length corresponding to the program material. As tape passes the head, a tiny emf is induced in a coil within the head due to the changing magnetic field passing the head. This signal, too, is amplified before going to the speaker.

18-2 ▮ The electric generator

Probably the most important practical result of Faraday's great discovery was the development of the electric **generator** or *dynamo*. A generator generates electricity from mechanical work. This is just the opposite of what a motor does. Indeed, a generator is basically the inverse of a motor.† An ac generator is shown in Figure 18-8. Like a motor, a generator consists of a coil of wire wound on an armature that can rotate in a magnetic field. In a motor, as we saw, an input of electric current causes the axle to turn. In a generator, the axle is turned by some mechanical means and an emf is induced in the rotating coil. An electric current is thus the *output* of a generator. In Figure 18-8, the right-hand rule tells us that the (conventional) current in the wire M on the armature is inward; therefore, it is inward at brush M. After one half-revolution, wire M will be where wire N is now in the drawing and the current then coming from M will be outward. Thus the current output is alternating. Let us look at this in more detail.

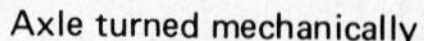

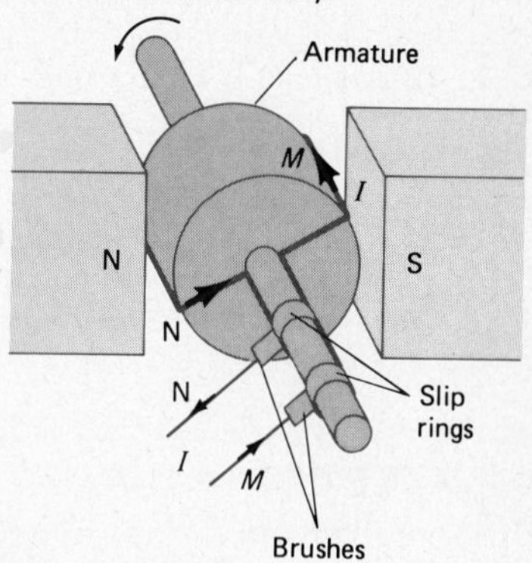

FIGURE 18-8
Ac generator.

In Figure 18-9 the loop is being made to rotate clockwise in a uniform magnetic field B. The velocity of the two lengths ab and cd at this instant are shown. Although the sections of wire bc and da are moving, the force on electrons in these sections is toward the side of the wire, not along its length. The emf generated is then due only to the force on charges in the sections ab and cd. From the right-hand rule we see that the direction of the induced current in ab is from a toward b. And in the lower section it is from c to d; so the flow is continuous in the loop. The magnitude of the emf generated in ab is given by Equation 18-3, except that we must take the component of the velocity perpendicular to B:

$$\mathscr{E} = Blv_{\perp},$$

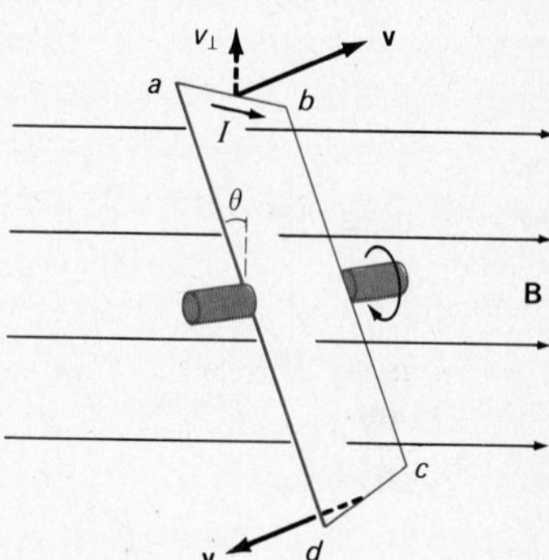

FIGURE 18-9
The emf is induced in the segments ab and cd whose velocity components perpendicular to field B equal $v \sin \theta$.

where l is the length of ab. From the diagram we can see that $v_{\perp} = v \sin \theta$, where θ is the angle the face of the loop makes with the vertical. The emf induced in cd has the same magnitude and is in the same direction. Therefore, they add and the total emf is

$$\mathscr{E} = 2NBlv \sin \theta,$$

where we have multiplied by N, the number of loops in the coil (if there is more than one). If the coil is rotating with constant angular velocity ω, then the angle $\theta = \omega t$; we also have from the angular equations (Chapter 3) that $v = \omega r = \omega(h/2)$, where h is the length of bc or ad. Thus $\mathscr{E} = 2NlB\omega(h/2) \sin \omega t$, or

$$\mathscr{E} = NAB\omega \sin \omega t, \tag{18-5}$$

where $A = lh$ is the area of the loop. (This equation holds for any shape

† You can, for example, actually run a car generator backward as a motor by connecting its output terminals to a battery.

power output can be no greater than the power input. A well-designed transformer can be greater than 99 percent efficient, so little energy is lost to heat. The power input thus essentially equals the power output; since power $P = VI$, we have

$$V_p I_p = V_s I_s,$$

or

$$\frac{I_s}{I_p} = \frac{N_p}{N_s}. \tag{18-7}$$

▌ EXAMPLE 18-6 A transformer for a transistor radio reduces 120 V ac to 9 V ac. (Such a device also contains diodes to change the 9 V ac to dc; see Chapter 19.) The secondary contains 30 turns and the radio draws 400 mA. Calculate (*a*) the number of turns in the primary, (*b*) the current in the primary, and (*c*) the power transformed.

SOLUTION (*a*) This is a step-down transformer, and from Equation 18-6 we have

$$N_p = N_s \frac{V_p}{V_s} = (30)(120\text{ V})/(9.0\text{ V}) = 400\text{ turns}.$$

(*b*) From Equation 18-7,

$$I_p = I_s \frac{N_s}{N_p} = (0.40\text{ A})\left(\frac{30}{400}\right) = 0.030\text{ A}.$$

(*c*) The power transformed is

$$P = I_s V_s = (9.0\text{ V})(0.40\text{ A}) = 3.6\text{ W},$$

which is, assuming 100 percent efficiency, the same as the power in the primary, $P = (120\text{ V})(0.030\text{ A}) = 3.6\text{ W}$. ▌▌

It is important to recognize that a transformer operates only on ac. A dc current in the primary does not produce a changing flux and therefore induces no emf in the secondary. However, if a dc voltage is applied to the primary through a switch, at the instant the switch is opened or closed there will be an induced current in the secondary. For example, if the dc is turned on and off as shown in Figure 18-16*a*, the voltage induced in the secondary is as shown in part *b*. Notice that the secondary voltage drops to zero when the dc voltage is steady.

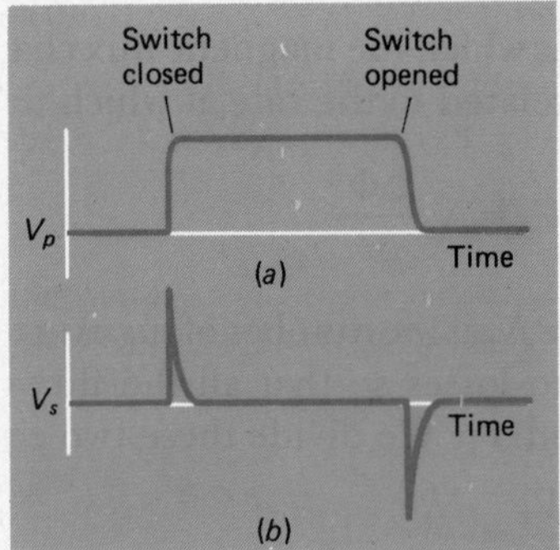

FIGURE 18-16
Dc voltage turned on and off as shown in (*a*) produces voltage pulses in the secondary (*b*).

This is basically how the ignition system of a car works (Figure 18-17). The "ignition points" serve as the switch in the primary. When they open a high voltage is induced in the secondary of the ignition coil (a step-up transformer). The secondary voltage, often as great as 25 kV, is sent to a particular spark plug to ignite the gas–air mixture at the moment when the gases are compressed. The distributor not only houses the points but also the revolving rotor, which sends the high voltage to the right plug at the right time.

Transmission of power; ac versus dc

Power plants are often situated some distance from metropolitan areas. Hydroelectric plants are located at a dam site and nuclear plants need

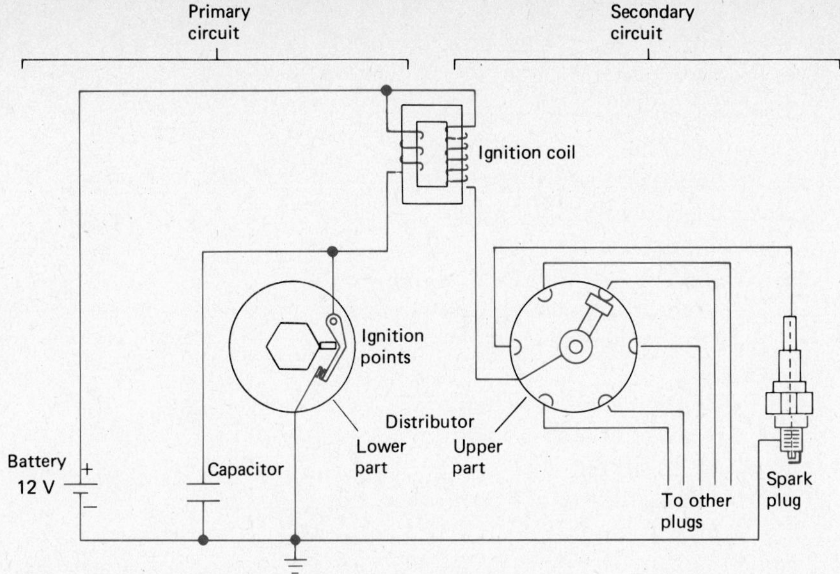

FIGURE 18-17
Ignition system for a six-cylinder car.

much cooling water. Fossil-fuel plants, too, are often situated far from a city because of lack of availability of land or to avoid contributing to air pollution.

In any case electricity must often be transmitted over long distances, and there is always some power loss in the transmission lines. This loss can be minimized if the power is transmitted at high voltage, as the following example shows.

❚ EXAMPLE 18-7 An average of 120 kW of electric power is supplied to a small town from a power plant 10 km away. The transmission lines have a total resistance of 0.40 Ω. Calculate the power loss if the power is transmitted at (*a*) 240 V, and (*b*) 24,000 V.

SOLUTION (*a*) If 120 kW is supplied at 240 V, the total current will be

$$I = \frac{P}{V} = \frac{1.2 \times 10^5 \text{ W}}{2.4 \times 10^2 \text{ V}} = 500 \text{ A}.$$

The power loss in the lines, P_L, is then

$$P_L = I^2R = (500 \text{ A})^2(0.40\ \Omega) = 100 \text{ kW}.$$

Thus over 80 percent of all the power is wasted as heat in the power lines! (*b*) When $V = 24{,}000$ V,

$$I = \frac{P}{V} = \frac{1.2 \times 10^5 \text{ W}}{2.4 \times 10^4 \text{ V}} = 5.0 \text{ A}.$$

The power loss is then

$$P_L = I^2R = (5.0 \text{ A})^2(0.40\ \Omega) = 10 \text{ W},$$

which is insignificant. ❚❚

It should be clear that the greater the voltage, the less the current and thus the less power is wasted in the transmission lines. It is for this reason that power is transmitted at very high voltages, as high as 750 kV.

Power is generated at somewhat lower voltages than this, and the voltages in homes and factories is also much lower. The great advantage of ac, and the principal reason it is in nearly universal use, is that the voltage can easily be stepped up and down by a transformer. The output voltage of an electric generating plant is stepped up prior to transmission. Upon arrival in a city, it is stepped down in stages at electric substations prior to distribution. The voltage in lines along city streets is typically 2400 V. It is stepped down to 240 V or 120 V for home use by transformers, often located in boxes on telephone poles.

Dc transmission has gained in popularity recently. Although changing voltage with dc is more difficult and expensive, it offers some advantages over ac. A few of these are as follows. Ac produces alternating magnetic fields which induce current in nearby wires and reduce transmitted power; this is absent in dc. Dc can be transmitted at a higher average voltage than ac since for dc the rms value equals the peak; and breakdown of insulation or of air must be avoided for the peak voltage. Alternating currents tend to flow only near the surface of a conductor, a phenomenon called the "skin effect"; thus large conductors offer less resistive losses to dc than to ac.

*18-5 ▌ Inductance

*Mutual inductance

If two coils of wire are placed near one another, as in Figure 18-18, a changing current in one will induce an emf in the other. According to Faraday's law, the emf $\mathscr{E}_2$ induced in coil 2 is proportional to the rate of change of flux passing through it. Since the flux is proportional to the current flowing in coil 1, $\mathscr{E}_2$ must be proportional to the rate of change of the current in coil 1, $\Delta I_1/\Delta t$. Thus we can write

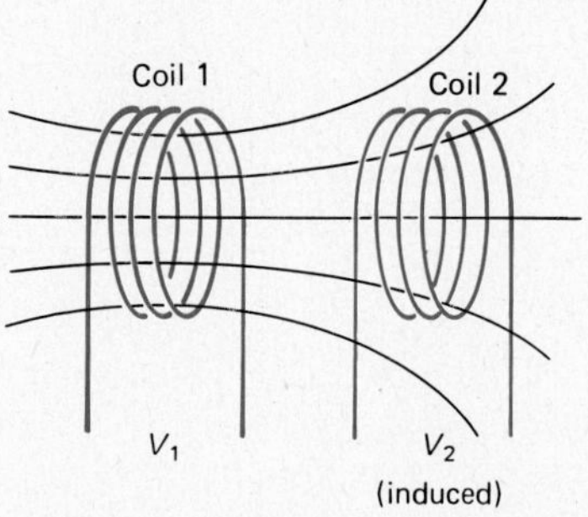

FIGURE 18-18
A changing current in one coil will induce a current in the second coil.

$$\mathscr{E}_2 = M\frac{\Delta I_1}{\Delta t},$$

where the constant of proportionality, M, is called the **mutual inductance**. It has units of $\mathrm{V \cdot s/A = \Omega \cdot s}$, which is called the *henry* (H): $1\ \mathrm{H} = 1\ \Omega \cdot \mathrm{s}$. The value of M depends on whether iron is present or not, on the size of the coils, on the number of turns, and on their separation; for example, the closer the coils, the more lines of flux will pass through coil 2, so M will be greater. In some arrangements M can be calculated; more usually it is measured experimentally.

A transformer is an example of mutual inductance in which the coupling is maximized so that nearly all flux lines pass through both coils. However, mutual inductance has other uses as well. For example, some pacemakers, which are used to maintain blood flow in heart patients (Section 16-8), are powered externally. Power in an external coil is transmitted via mutual inductance to a second coil in the pacemaker at the heart. This has the advantage over battery-powered pacemakers that surgery is not needed to replace a battery when it wears out.

Mutual inductance can sometimes be a problem, however. Any alternating current in a circuit can induce an emf in another part of the same circuit or in a different circuit even though the conductors are not in the shape of a coil. The mutual inductance M is usually small unless multiturn coils and/or iron cores are involved. However, in situations where small signals are present, problems due to mutual inductance often arise; shielded cable, in which an inner conductor is surrounded by a cylindrical grounded conductor, is often used to reduce the problem.

*Self-inductance

Similar considerations apply to a single coil. When a changing current passes through a coil or solenoid, a changing magnetic flux is produced inside the coil, and this in turn induces an emf. This induced emf opposes the change in flux (Lenz's law); it is much like the back emf generated in a motor. For example, if the current through the coil is increasing, the increasing magnetic flux induces an emf that opposes the original current and tends to retard its increase. If the current is decreasing in the coil, the decreasing flux induces an emf in the same direction as the current, thus tending to maintain the original current. In either case, the average induced emf $\mathscr{E}$ is proportional to the rate of change in current:

$$\mathscr{E} = -L\frac{\Delta I}{\Delta t}. \tag{18-8}$$

The constant of proportionality L is called the **self-inductance**, or simply the *inductance* of the coil. It, too, is measured in henries. The magnitude of L depends on the geometry and on the presence of an iron core.

An ac circuit always contains some inductance, but often it is quite small unless the circuit contains a coil of many turns. A coil that has a high self-inductance L is called an *inductor* or a *choke coil.* It can serve a useful purpose in certain circuits. Often, inductance is to be avoided in a circuit. Precision resistors are normally wire-wound and thus would have inductance as well as resistance; the inductance can be minimized by winding the wire back on itself so that the current going in the two directions cancels and little magnetic flux is produced; this is called a "noninductive winding."

If an inductor has negligible resistance, it is the inductance, or the back emf, that controls the current. If a source of alternating voltage is applied to the coil, this applied voltage will just be balanced by the induced emf of the coil given by Equation 18-8. Thus we can see from Equation 18-8 that if the inductance L is large, the change in the current—and therefore the current itself—will be small. The greater the inductance, the less the current. An inductance thus acts something like a resistance to impede the flow of alternating current; we use the term **impedance** for this quality of an inductor. We shall discuss impedance more fully in Chapter 19 and we shall see that it depends not only on L but also on the frequency. Here we mention one example of its importance. The resistance of the primary in a transformer is usually quite small, perhaps less than 1 Ω. If resistance alone limited the current, tremendous currents would flow when a high voltage is applied. Indeed, a dc voltage applied to a transformer can burn it out. It is the impedance of the coil to an alternating current (or its "back" emf) that limits the current to a reasonable value.

▌ EXAMPLE 18-8 (*a*) Determine a formula for the self-inductance L of a solenoid (a long coil) containing N turns of wire in its length l and whose cross-sectional area is A. (*b*) Calculate the value of L if $N=100$, $l=5.0$ cm, $A=3.0\ \text{cm}^2$, and the solenoid is air filled. (*c*) Calculate L if the solenoid has an iron core with $\mu=4000\ \mu_0$.

SOLUTION (*a*) According to Equation 17-10, the magnetic field inside a solenoid is $B=\mu_0 nI$, where $n=N/l$. From Equations 18-2 and 18-8 we have $\mathscr{E}=-N(\Delta\Phi/\Delta t)=-L(\Delta I/\Delta t)$. Thus $L=N(\Delta\Phi/\Delta I)$. Since $\Phi=BA=\mu_0 NIA/l$, then any change in I causes a change in flux $\Delta\Phi=\mu_0 NA\ \Delta I/l$. Thus

$$L=N\frac{\Delta\Phi}{\Delta I}=\mu_0\frac{N^2A}{l}.$$

(*b*) Since $\mu_0=4\pi\times10^{-7}\ \text{T}\cdot\text{m/A}$

$$L=\frac{(4\pi\times10^{-7}\ \text{T}\cdot\text{m/A})(100)^2(3.0\times10^{-4}\ \text{m}^2)}{(5.0\times10^{-2}\ \text{m})}=75\ \mu\text{H}.$$

(c) Here we replace μ_0 by $\mu=4000\mu_0$, so L will be 4000 times larger: $L=0.30\ \text{H}=300\ \text{mH}$. ▌▌

*Energy stored in a magnetic field

In Section 14-3 we saw that the energy stored in a capacitor is equal to $\frac{1}{2}CV^2$. By using a similar argument, we can show that the energy stored in an inductance L, carrying a current I, is

$$\text{energy}=\tfrac{1}{2}LI^2.$$

Just as the energy stored in a capacitor can be considered to reside in the electric field between its plates, so the energy in an inductor can be considered to be stored in its magnetic field.

To write the energy in terms of the magnetic field, let us use the result of the example above, that the inductance of a solenoid is $L=\mu_0 N^2A/l$. Now the magnetic field B in a solenoid is related to the current I by $B=\mu_0 NI/l$. Thus

$$\text{energy}=\frac{1}{2}LI^2=\frac{1}{2}\frac{\mu_0 N^2A}{l}\left(\frac{Bl}{\mu_0 N}\right)^2$$
$$=\frac{1}{2}\frac{B^2}{\mu_0}Al.$$

We can think of this energy as residing in the volume enclosed by the windings, which is Al. Then the energy per unit volume, or *energy density*, is

$$\text{energy density}=\frac{1}{2}\frac{B^2}{\mu_0}. \qquad (18\text{-}9)$$

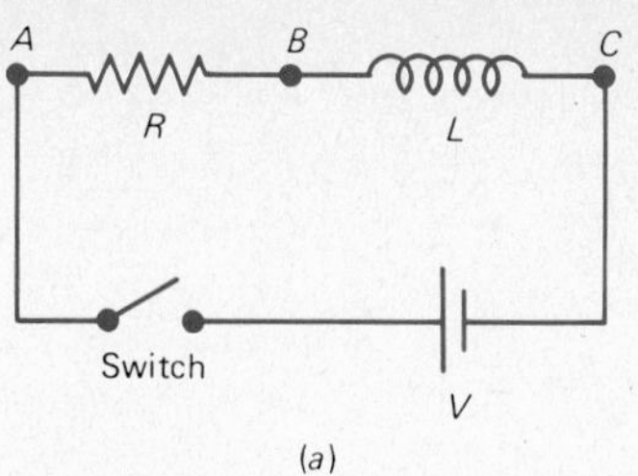

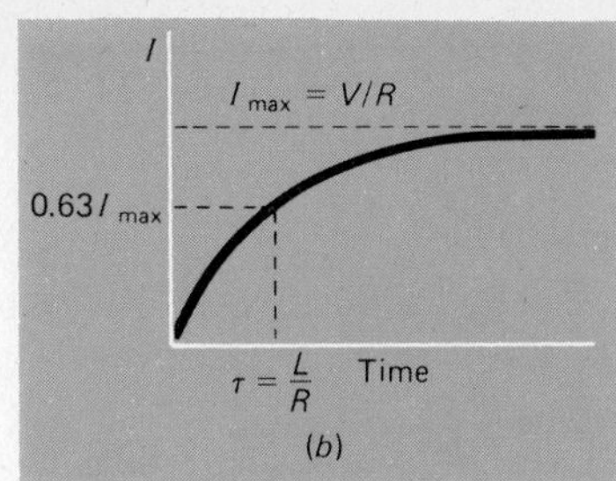

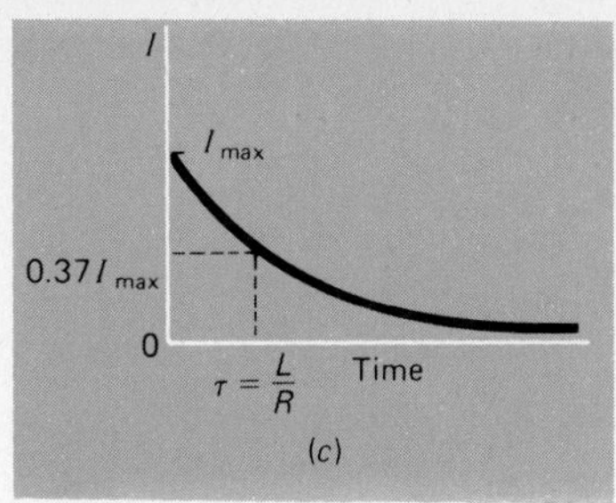

FIGURE 18-19
(*a*) *LR* circuit; (*b*) growth of current when connected to battery; (*c*) decay of current when battery is shorted out.

This formula, which was derived for the special case of a solenoid, can be shown to be valid for any region of space where a magnetic field exists. If a ferromagnetic material is present, μ_0 is replaced by μ. This equation is analogous to that for an electric field, $\frac{1}{2}\varepsilon_0 E^2$, Equation 14-6.

*An *LR* circuit

Any inductor will have some resistance. We represent this situation by drawing the inductance L and the resistance R separately, as in Figure 18-19*a*. The resistance R could also include a separate resistor connected in series. Now we ask what happens when a dc source is connected in series. At the instant the switch connecting the battery is closed, the current starts to flow. It is, of course, opposed by the induced emf in the inductor. However, as soon as current starts to flow, there is a voltage drop across the resistance. Hence the voltage drop across the inductance is reduced and there is then less impedance to the current flow from the inductance. The current thus rises gradually as shown in Figure 18-19*b*; it approaches the steady value $I_{max} = V/R$ when all the voltage drop is across the resistance. It can be shown that the time required to reach $0.63I_{max}$ is $\tau = L/R$. This is called the *time constant* of the circuit.

If the battery is suddenly shorted out, the current drops off as shown in Figure 18-19*c*. The time constant is the time for the current to drop to 37 percent of the original value.

These graphs show that there is always some "reaction time" when an electromagnet, for example, is turned on or off. We also see that an *LR* circuit has properties similar to an *RC* circuit (Section 16-7). Unlike the capacitor case, however, the time constant here is *inversely* proportional to R.

*18-6 ▌ Power plants and resources

Earlier in this chapter we saw how electric generators work. In this section we discuss the practical aspects of electric power generation and various means of doing so. This is of great practical importance, and combines a number of aspects of physics and other disciplines.

At this time most of the energy used in the U.S. comes from the burning of fossil fuels—that is, oil, coal, and natural gas. The direct

TABLE 18-1

Electric energy production

Process	Approximate percent of total U.S. electricity production† 1970s
Generator, turbine	
Fossil-fuel steam plants	**80**
Nuclear power	
Fission	**5**
Fusion	**—**
Geothermal power	**<1**
Tropical seas	**—**
Hydroelectric power	**15**
Tidal power	**—**
Wind power	**—**
Solar power	**—**
Other generators	
Solar cells	**—**
MHD	**—**

† Fossil fuel and nuclear plants are 30 to 40 percent efficient, whereas hydroelectric plants are nearly 100 percent efficient. So the percent of energy used (including waste heat produced) is about 90%, 5%, and 5% for fossil fuel, nuclear, and hydroelectric plants, respectively.

burning of fuels is used to heat our houses, offices, and factories; to run our cars and other vehicles; and in industrial production of manufactured goods from raw material, such as blast furnaces for steel, production of chemicals, and refining of petroleum. This accounts for about 75 percent of the energy used. The remaining 25 percent produces electrical energy, which is used for many of the same purposes.

We now briefly discuss many of the ways currently used to produce electricity as well as some possible future methods; see Table 18-1.

*Electricity produced by generators

At this time an electric generator represents the most practical method of producing electric power (Section 18-2). A generator is usually connected to a large turbine (Figure 18-20) containing many large blades. The various means to turn the turbine will now be discussed, along with some of the advantages and disadvantages of each.

Fossil-fuel steam plants At such a plant, coal, oil, or natural gas is burned to boil water and produce high-pressure steam that turns the turbine. The basic principles of this sort of steam engine were covered in Chapter 10. The advantages of such plants are that we know how to build

FIGURE 18-20

Mechanical or heat energy is transformed to electric energy with a turbine and generator.

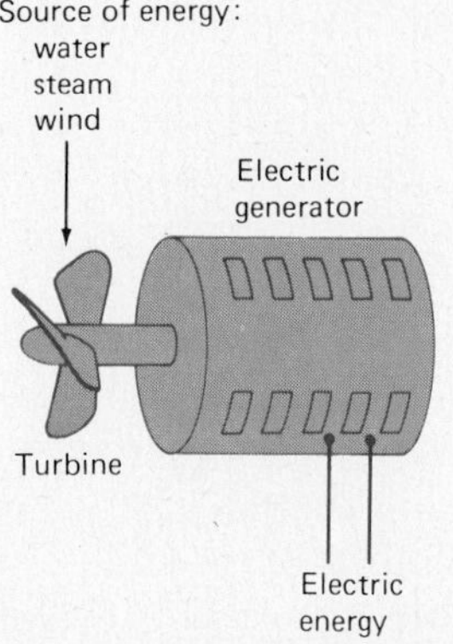

them and they are not too expensive to run (yet). The disadvantages are that the products of combustion create air pollution or "smog"; like all heat engines, their efficiency is limited (30 to 40 percent is typical); the waste heat produces thermal pollution; the extraction of the raw materials can be devastating to the land, especially in coal strip mining and oil-shale recovery; accidents are possible, such as oil spills in the sea; and there is not a great deal of fossil fuel left—estimates range from a supply lasting a few centuries to only a few decades.

Nuclear power Two nuclear processes release energy: fission and fusion. In fission, the nuclei of uranium or plutonium atoms are made to split ("fission") with the release of energy. In fusion, energy is released when small nuclei, such as those of hydrogen, combine ("fuse"); these processes are described in greater detail in Chapter 29.

The fission process is used in all present nuclear power plants, since fusion has not yet been controlled. Nuclear energy is used to heat steam just as fossil fuels are. A nuclear power plant is thus essentially a steam engine using uranium as its fuel; it suffers from the low efficiency characteristic of all heat engines and the accompanying thermal pollution. Although nuclear power plants produce practically no air pollution (unless an accident occurs), they do present problems: radioactive substances are produced that are difficult to dispose of; a serious accident could result in the release of radioactive material into the air; there is the possibility of diversion of nuclear material by terrorists; and the fuel supply is limited. However, the energy produced per kilogram of fuel is very large, and extraction is less damaging to the land than for fossil fuels. The fusion process, on the other hand, has fewer disadvantages and has the advantage of a vast supply of fuel—the hydrogen in the water molecules (H_2O) in the oceans. Unfortunately, this process cannot yet be controlled sufficiently, but holds promise for the future. (See Chapter 29.)

Geothermal power Both fossil-fuel and nuclear plants heat water to steam for the steam turbine. Natural steam can be obtained from the earth itself. In many places, water beneath the ground is in contact with the hot interior of the earth and is raised to high temperature and pressure. It comes to the surface as hot springs, geysers, or steam vents. We can not only use natural vents but can drill down to trapped steam beds. Already a large geothermal plant operates at the Geysers in northern California. A similar plant has been successfully operating for over 60 years in Italy, and a number of others are functioning in various areas around the world. There are a limited number of sites where high temperature and pressure exist naturally. Another possibility, however, is to drill two parallel wells down to hot dry rock in contact with the earth's interior and to pass cold water under pressure down one well, the heated water (or steam) returning up the other well. Geothermal energy appears to be clean in that it produces little air pollution, although there is some (nonsteam) gas emission; there is also the spent hot water to be disposed of, leading to thermal pollution; the mineral content of the water (often high) may not only be environmentally polluting (for example, water pollution) but can be corrosive to the parts of the apparatus itself. Nonetheless, it is a reasonably inexpensive means of production, and with better technology (such as better drilling techniques) it may become a more common source.

Tropical seas There is a difference in temperature of perhaps 20 C° between the surface and the depths of the sea in the tropics; although not yet tried, a number of designs have been made in which this temperature difference could be used to drive a heat engine. (You will recall from Chapter 10 that a heat engine—such as a steam engine—must exhaust heat at a lower temperature than that of the heat input.) The working fluid, of course, could not be steam; rather, it would have to be a substance with a lower boiling point that could drive the turbines. Before this method is tried, many difficulties need to be overcome. The efficiency (Equation 10-7) would be low because of the small temperature difference (at best, 7 percent); losses due to heat transfer must be kept very low, so the efficiency does not drop to zero. The ocean environment causes problems, such as corrosion and fouling by biological organisms caught in the intake water. Furthermore, on a large scale the ocean currents of the world could be affected. Yet there would be little problem with air pollution or radioactive disposal, so the advantages, if done on a limited scale, might outweigh disadvantages.

Hydroelectric power plants Hydroelectric plants use falling water instead of steam to turn the turbines. They are usually located at the base of a dam. Hydroelectric power plants produce practically no air or water pollution; furthermore, they are nearly 100 percent efficient, since very little waste heat is produced. However, there are not many good locations remaining in which to build dams, and the resulting reservoirs inundate land that may be fertile or of great scenic beauty. It is not expected that many new hydroelectric dams will be built in the future in the U.S. However, another form of water power is the following.

Tidal power The earth's tides have been put to work at a plant in France. As shown in Figure 18-21, a reservoir behind a dam is filled at high tide and the water is released at low tide to drive turbines. At the next high tide the reservoir is filled again and the inrushing water also turns turbines. Good sites for tidal power (where there is a large difference between high and low tides) are not plentiful and would require large dams across natural or artificial bays. The rather abrupt changes of water level could have an effect on wildlife, but otherwise tidal power would seem to have a minimal environmental effect. Unfortunately, however, reasonable estimates of available sites indicate that tidal power could at best produce only a small fraction of the world's energy needs. Nonetheless, since our future energy needs may have to come from a variety of small sources, rather than mainly one, as is the case now (fossil fuels), tidal power could become more important.

Wind power Windmills were once a very practical device. Their resurrection—on a much grander scale—is a possibility as a means of turning a generator to produce electricity. One specific proposal projects some 300,000 of them, each over 800 ft high with blades 50 ft in diameter, dispersed throughout the midwestern U.S. where winds are strong and steady. Such an array might produce a good portion of U.S. energy needs. Of course, such a vast project might be something of an eyesore; it also might affect the weather. But wind power, on a modest scale, offers possibilities.

Solar power Many kinds of solar energy are already in use: fossil fuels

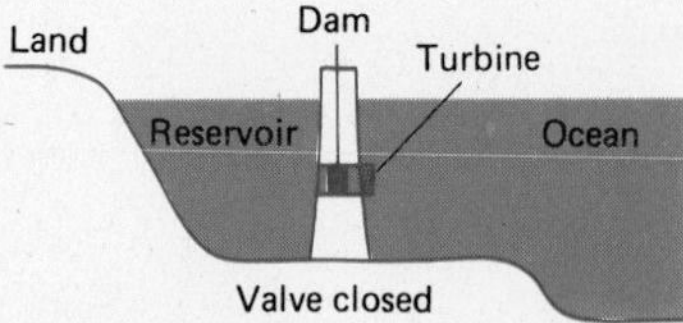

High tide

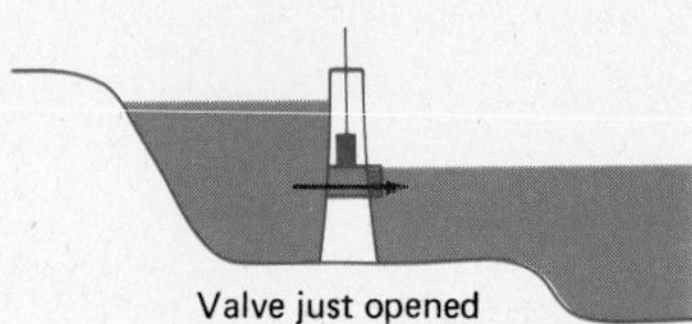

Low tide: Water is beginning to flow out of reservoir to ocean, driving turbines.

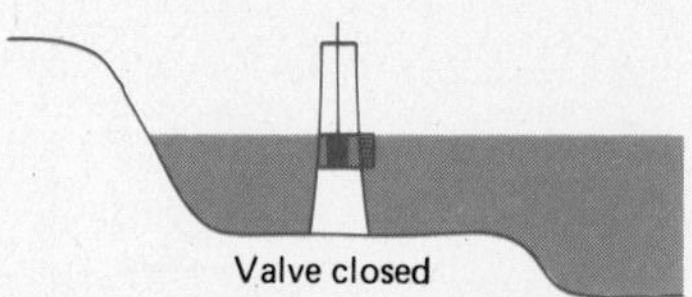

Water level equalized.

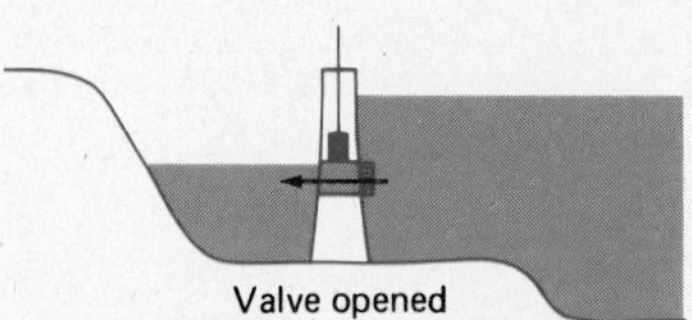

High tide: Water is allowed to flow back into reservoir, driving turbines.

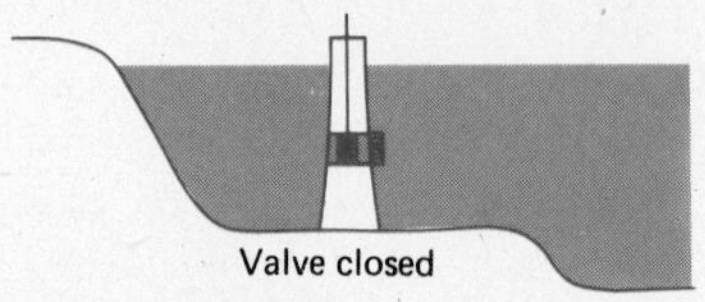

Water level equalized.

FIGURE 18-21
Tidal power plant. Turbines are located inside dam.

are the remains of plant life that grew by photosynthesis of light from the sun; hydroelectric power depends on the sun to evaporate water that later comes down as rain; and wind power depends on convection currents produced by the sun heating the atmosphere. But now let us see what we can do with the sun's rays directly. We saw in Chapter 9 how the absorption of the sun's energy by a black surface can be used to raise the temperature of water to heat a house. The same principle can run electric generators. The sun's rays must be concentrated by large mirrors or lenses onto a small surface in order to produce temperatures high enough to produce steam to drive a turbine. Such a system could be used for a home generating system, although a backup system would probably be needed for cloudy days. On a larger scale, large areas of land would be used to collect sufficient sunlight; indeed, it has been proposed to use large areas of Arizona and New Mexico—where the skies are clear nearly all year. The principal disadvantages are that the sun does not shine brightly every day, and not at all at night; thus a safe storage method would be needed. Also, large areas of land are needed for the collectors and concentrators. Although the ubiquitous thermal pollution would exist and the climate might be affected, there would be essentially no air or water pollution and no radioactivity, and the technology would not be too difficult. Thus, although problems still exist, solar energy is one of the most promising future sources of energy.

*Direct conversion: solar cells and MHD

Two other processes that may in the future be used to produce electricity are solar cells and magnetohydrodynamics (MHD). Both of these techniques produce electrical energy directly without the use of a turbine or electric generator. *Solar cells*—or more correctly *photovoltaic cells*—were developed by the U.S. space program to directly convert sunlight into electricity. The best cells today can do this with about 15 percent efficiency. Unfortunately, they are very expensive, costing about 10 to 100 times per unit of power as ordinary means. However, if the cost should decrease, they would be very desirable, since thermal (and other) pollution would be very low. (Remember, no heat engine is involved.) They might be placed on roofs for home use. On a large scale they would again require a large land area since the sun's energy is not very concentrated. Another possibility is to place solar cells and concentrators in orbit around the earth. The sun's radiation would be greater there, before the atmosphere absorbs some of it, and the satellites would spend a minimal time in the earth's shadow. The electricity produced might be fed to a microwave generator and transmitted to receivers on earth.

In an *MHD generator*, a gas is heated to such high temperatures (above 2000°C) that electrons are pulled off the molecules and the gas becomes partly ionized. This collection of positive and negative ions and electrons, called a *plasma*, is directed between two parallel plates in the presence of a magnetic field. The magnetic field exerts a force on the moving charged particles; positive particles are directed toward one plate, negative particles toward the other. Thus the two plates become oppositely charged and a potential difference results which can drive a current (Figure 18-22). Although conventional means would be used to

FIGURE 18-22
MHD generator.

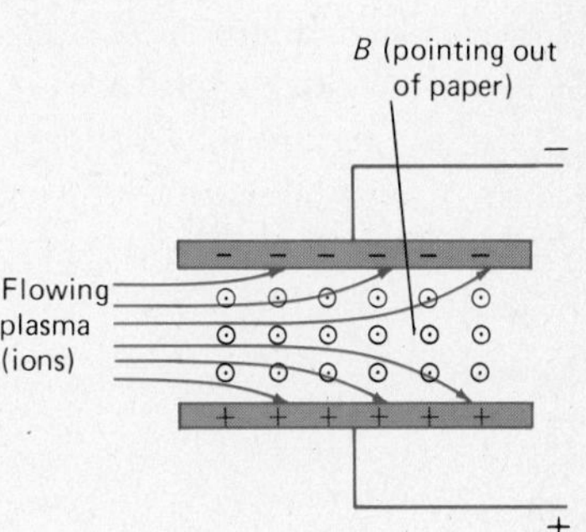

heat the gas—say, fossil fuels—the efficiency would be somewhat higher than conventional heat engines, perhaps 50 to 60 percent; so more electricity is produced, and less pollution generated, per unit of fuel burned. Also, some of the waste heat could perhaps be used to turn an ordinary turbine. Although the idea of MHD generators dates back to the early part of this century, it was only in 1958 that an operational device was successfully built. No large-scale MHD generators have yet been built, but they may be in the future. MHD may be the ideal method for obtaining energy from nuclear fusion when and if it is controlled (Chapter 29).

*Problems and the future

It should be clear that all forms of energy production have undesirable side effects. Some are worse than others, of course, and not all problems can be anticipated. The need for energy must be balanced against the "cost" of it. When electricity demand was small, the problems were also small. Now the demand has grown so large that the problems are great as well. Society as a whole must decide whether the environmental and health problems posed by energy production are worth it. It is unlikely that energy production will decrease significantly in the near future, but we can do something to prevent its rapid increase—by stopping wasteful use of energy.

New forms of energy production will be needed in the future as old reserves of fuel are used up. It is likely that many different methods—including those we have just discussed—will have to be used in the future to satisfy the needs of our energy-consuming society. The problems posed by any one form of energy production might not loom as large since there would be diverse fuel sources available to us.

▌ Summary

The *magnetic flux* passing through a loop is equal to the product of the area of the loop times the perpendicular component of the magnetic field strength. If the magnetic flux through a loop of wire changes in time, an emf is induced in the loop; the magnitude of the induced emf in a loop equals the time rate of change of the magnetic flux through the loop. This result is called *Faraday's law of induction.* The induced emf produces a current whose magnetic field opposes the original change in flux (*Lenz's law*). Faraday's law also tells us that a changing magnetic field produces an electric field; and that a straight wire of length l moving with speed v perpendicular to a magnetic field of strength B has an emf induced between its ends equal to Blv.

An electric *generator* changes mechanical energy into electrical energy. Its operation is based on Faraday's law: a coil of wire is made to rotate by mechanical means in a magnetic field, and the changing flux through the coil induces a sinusoidal current, which is the output of the generator.

A motor, which operates in the reverse of a generator, acts like a generator, in that a counter emf is induced in its rotating coil; since it opposes the input voltage, this counter emf acts to limit the current in the motor coil. Similarly, a generator acts somewhat like a motor in that a counter torque acts on its rotating coil.

A transformer, which is a device to change the magnitude of an ac voltage, consists of a primary and a secondary coil. The changing flux due to an ac voltage in the primary induces a voltage in the secondary. In a 100 percent efficient transformer, the ratio of output to input voltages (V_s/V_p) equals the ratio of the number of turns N_s in the secondary to the number N_p in the primary: $V_s/V_p = N_s/N_p$. The ratio of secondary to primary current is in the inverse ratio of turns: $I_s/I_p = N_p/N_s$.

A changing current in a coil of wire will induce an emf in a second coil placed nearby. The *mutual inductance* M is defined as the proportionality constant between the induced emf $\mathscr{E}_2$ in the second coil and the time rate of change of current in the first: $\mathscr{E}_2 = M(\Delta I_1/\Delta t)$. Within a single coil, a changing current induces a back emf, $\mathscr{E}$, so a coil has a *self-inductance* L defined by $\mathscr{E} = -L(\Delta I/\Delta t)$. This back emf acts as an *impedance* to the flow of an alternating current.

A magnetic field stores energy, like an electric field, with an energy density (J/m^3) equal to $B^2/2\mu_0$.

Questions

1 What is the difference between magnetic flux and magnetic field strength?

2 Suppose you are holding a circular piece of wire and suddenly thrust a magnet, south pole first, toward the center of the circle; is a current induced in the wire? Is a current induced when the magnet is held steady within the loop? Is a current induced when you withdraw the magnet? In each case, if your answer is yes, specify the direction.

3 Show, using Lenz's law, that the emf induced in the moving rod in Figure 18-6 is positive at the bottom and negative at the top, so that the current flows clockwise in the circuit.

4 What direction will the current flow in Figure 18-6 if the rod moves to the left, which decreases the area?

5 In situations where a small signal must travel over a distance, a "shielded cable" is used in which the signal wire is surrounded by an insulator and then enclosed by a cylindrical conductor. Why is a "shield" necessary?

6 What is the advantage of placing the two electric wires carrying ac close together?

7 In some early automobiles, the starter motor doubled as a generator to keep the battery charged once the car was started. Explain how this might work.

8 Explain why, exactly, the lights may dim briefly when a refrigerator motor starts up. (*Hint*: Consider "terminal voltage.")

9 When a refrigerator starts up, the lights in older houses may dim briefly. When an electric heater is turned on, the lights may stay dimmed as long as it is on. Explain the difference.

10 Explain what is meant by the statement "A motor acts as a motor and generator at the same time." Can the same be said for a generator?

11 Use Figure 18-8 and the right-hand rules to show why the counter torque in a generator *opposes* the motion.

12 Will an eddy current brake (Figure 18-14) work on a copper or aluminum wheel, or must it be ferromagnetic?

13 It has been proposed that eddy currents be used to help sort solid waste for recycling. The waste is first ground into tiny pieces and iron removed with a dc magnet. The waste then is allowed to slide down an incline over permanent magnets. How will this aid in the separation of nonferrous metals (Al, Cu, Pb, brass) from nonmetallic materials?

14 The primary of a transformer on a telephone pole has a resistance of 0.1 Ω and the input voltage is 2400 V ac. Can you estimate the current that will flow? Will it be 24,000 A? Explain.

15 A transformer designed for a 120-V ac input will often "burn out" if connected to a 120-V dc source. Explain. (*Hint*: The resistance of the primary coil is usually very low.)

16 The use of larger voltage lines in homes, say 600 V or 1200 V, would reduce energy waste. Why are they not used?

*17 In what ways are nuclear and fossil fuel plants similar? In what ways are they different?

Problems

SECTION 18-1

1 (I) A 10-cm-diameter circular loop of wire is in a 0.50-T magnetic field. It is removed from the field in 0.10 s. What is the average induced emf?

2 (I) A square wire loop, 6.0 cm on a side, lies between the pole faces of an electromagnet. Calculate the average emf induced in the wire when the electromagnet is turned on if its field increases from zero to its maximum of 1.5 T in $\frac{3}{4}$ s.

3 (I) The magnetic field perpendicular to a circular loop of wire 12 cm in diameter is changed from +0.35 T to −0.15 T in 90 ms, where + means the field points away from an observer and − toward the observer. (*a*) Calculate the induced emf. (*b*) In what direction does the induced current flow?

4 (I) A rectangular loop of wire 10 cm on a side is rotated 90° in a 2.0×10^{-2} T magnetic field in 10 ms. What is the average emf induced?

5 (I) The moving rod in Figure 18-6 is 20 cm long and moves with a speed of 50 cm/s. If the magnetic field is 0.80 T, calculate (*a*) the emf developed, and (*b*) the electric field strength in the rod.

6 (I) The magnetic flux through a coil of wire containing two loops changes from −10 Wb to +5 Wb in 1.4 s. What is the emf induced in the coil?

7 (II) A 10-cm-diameter circular loop of wire has a resistance of 3.5 Ω. It is initially in a 0.40-T magnetic field but is removed from the field in 100 ms. Calculate the electric energy dissipated in this process.

8 (II) In Figure 18-6, the rod moves with a speed of 2.3 m/s, is 40 cm long, and has negligible resistance. The magnetic field is 0.21 T and the resistance of the U-shaped conductor is 25 Ω at a given instant. Calculate (*a*) the emf induced, and (*b*) the current flowing in the U-shaped conductor.

9 (II) The moving rod in Figure 18-6 has negligible resistance but that of the U-shaped conductor is R. Show that mechanical power needed to move the rod is $P = B^2 l^2 v^2 / R$ and that this equals the electric power dissipated in the circuit.

10 (III) If the U-shaped conductor in Figure 18-6 has resistivity ρ, whereas that of the moving rod is negligible, show that if the rod moves outward at a uniform speed, the current decreases in time. In particular, derive a formula for the current I as a function of time, assuming that the rod starts at the bottom of the U at $t = 0$.

11 (III) The magnetic field perpendicular to a single 20-cm-diameter circular loop of copper wire decreases uniformly from 0.25 T to zero. If the wire is 2.0 mm in diameter, how much charge moves through the coil during this operation?

SECTION 18-2

12 (I) A car generator produces 12 V when the armature turns at 600 rev/min. What will be its output at 1500 rev/min, assuming that nothing else changes?

13 (I) Calculate the peak output voltage of a simple generator whose square armature windings are 8.0 cm on a side if the armature contains 75 loops and rotates in a field of 0.30 T at a rate of 120 rev/s.

14 (I) The armature of a 60-Hz ac generator rotates in a 0.12-T magnetic field. If the area of the coil is $2.0 \times 10^{-2}\ \mathrm{m}^2$, and it contains 100 loops, what will be the peak output voltage?

15 (I) Show that the rms output of an ac generator is $V_{rms} = NAB\omega/\sqrt{2}$.

16 (II) A 100-loop square armature coil 20 cm on a side rotates at 60 rev/s in a uniform magnetic field. If the rms output is 120 V, what is the strength of the magnetic field?

17 (II) A simple generator has a 100-loop square coil 8.0 cm on a side. How fast must it turn in a 0.50-T field to produce a 20-V peak output?

SECTION 18-3

18 (I) A motor has an armature resistance of 2.5 Ω. If it draws 4.0 A when running at full speed and connected to a 120-V line, how large is the counter emf?

19 (I) The counter emf in a motor is 90 V when operating at 1500 rpm. What would the counter emf be at 1800 rpm if the magnetic field is unchanged?

20 (II) What will be the current in the motor of Example 18-5 if the load causes it to run at half-speed?

21 (II) The magnetic field of a "shunt-wound" dc motor is produced by field coils placed in parallel to the armature coils. Suppose that the field coils have a resistance of 60 Ω and the armature coils 4.0 Ω. The back emf at full speed is 105 V when the motor is connected to a 115-V line. (*a*) Draw the equivalent circuit for the situations when the motor is just starting and when it is running full speed. (*b*) What is the total current drawn by the motor at start up? (*c*) What is the total current drawn when the motor runs at full speed?

22 (III) A small electric car overcomes a 300-N friction force when traveling 30 km/h. The electric motor is powered by ten 12-V batteries connected in series and is coupled directly to the wheels whose

diameters are 40 cm. The 200 armature coils are rectangular, 20 cm by 15 cm, and rotate in a 0.40-T magnetic field. (*a*) How much current does the motor draw to produce the required torque? (*b*) What is the back emf? (*c*) How much power is dissipated in the coils? (*d*) What percent of the input power is used to drive the car?

SECTION 18-4

23 (I) A transformer changes 12 V to 18,000 V, and there are 6000 turns in the secondary. How many turns are there in the primary? Assume 100 percent efficiency.

24 (I) A transformer has 145 turns in the primary and 55 in the secondary. What kind of transformer is this and, assuming 100 percent efficiency, by what factor does it change the voltage?

25 (I) A step-up transformer increases 30 V to 120 V. What is the current in the secondary as compared to the primary? Assume 100 percent efficiency.

26 (I) Neon signs require 12 kV for their operation. To operate from a 120-V line, what must be the ratio of secondary to primary turns of the transformer? What would the voltage be if the transformer were connected backward?

27 (I) Describe a transformer that could be used to light a 6-V light bulb from a 120-V 60-Hz source.

28 (I) A transformer has 1500 primary turns and 120 secondary turns. The input voltage is 120 V and the output current is 8.0 A. What is the secondary voltage and primary current?

29 (II) The output voltage of a 180-W transformer is 6.0 V and the input current is 10 A. (*a*) Is this a step-up or a step-down transformer? (*b*) By what factor is the voltage multiplied?

30 (II) A pair of power transmission lines each have a 0.60-Ω resistance and carry 800 A over 10 km. If the input voltage is 12 kV, calculate (*a*) the voltage at the other end, (*b*) the power input, (*c*) power loss in the lines, and (*d*) the power output.

31 (II) If 20 MW of power at 60 kV arrives at a town from a generator via 4.0-Ω transmission lines, calculate (*a*) the emf at the generator end of the lines, and (*b*) the fraction of the power generated that is lost in the lines.

32 (II) Show that the power loss in transmission lines, P_L, is given by $P_L = (P_T)^2 R_L / V^2$, when P_T is the power transmitted to the user, V is the delivered voltage, and R_L is the resistance of the power lines.

33 (II) If 30 kW is to be transmitted over lines whose total resistance is 0.10 Ω, estimate how much power is saved if the voltage is stepped up from 120 V to 1200 V and then down again rather than simply transmitting at 120 V. Assume that the transformers are each 99 percent efficient.

34 (III) Design a dc transmission line that can transmit 400 MW of electricity 500 km with only a 2 percent loss. The wires are to be made of aluminum and the voltage is 500 kV.

*SECTION 18-5

***35** (I) If the current in a 30-mH coil changes steadily from 20 mA to 80 mA in 40 ms, what is the induced emf?

***36** (I) What is the inductance L of a 2.2-m-long air-filled coil 15 cm in diameter containing 20,000 loops?

***37** (I) What is the inductance of a coil if it produces an emf of 5.0 V when the current in it changes from −20 mA to +30 mA in 10 ms?

***38** (I) The magnetic field inside an air-filled solenoid 10 cm long and 2.0 cm in diameter is 0.50 T. Approximately how much energy is stored in this field?

***39** (I) How much energy is stored in 400-mH inductor at an instant when the current is 2.0 A?

***40** (II) How many turns does an air-filled coil have if it is 1.2 cm diameter, 4.0 cm long, and its inductance is 0.25 mH. How many turns are needed if it has an iron core and $\mu = 10^3 \mu_0$?

***41** (II) A 25-V emf is induced in a 0.20-H coil by a current that rises uniformly from zero to I_0 in 1.0 ms. What is the value of I_0?

***42** (II) A coil has 2.0-Ω resistance and 0.80-H inductance. If the current is 3.0 A and is increasing at a rate of 4.0 A/s, what is the potential difference across the coil?

***43** (III) Calculate the equivalent inductance L of the motor windings in Example 18-5 if full speed is 2400 rpm.

***44** (II) A toroid is a long coil of wire bent into a circular shape (Figure 17-41). Show that the self-inductance L of a toroid of radius R containing N loops each of radius r is

$$L = \frac{\mu_0 N^2 r^2}{2R}$$

if $R \gg r$. Assume that the field is uniform inside the toroid; is this actually true?

***45** (II) Calculate the inductance L of a large toroid (Problem 44) if the diameter of the coils is 2.0 cm and the diameter of the whole ring is 50 cm. Assume that the field inside the toroid is uniform. There are a total of 300 loops of wire.

***46** (III) A 32-cm-long coil with 500 loops is wound on an iron core ($\mu = 2000\ \mu_0$) along with a second coil of 200 loops. The loops of each coil have a radius of 2.0 cm. If the current in the first coil drops uniformly from 4.0 A to zero in 80 ms, determine (*a*) the emf induced in the second coil, and (*b*) the mutual inductance M.

*SECTION 18-6

***47** (I) Solar cells can produce about 40 W of electricity per square meter of surface area if directly facing the sun. How large an area is needed to supply the needs of a house that requires 100 kWh/day? Would this fit on the roof of an average house? (Assume that the sun shines 12 h/day.)

***48** (II) Water falls 80 m over a dam at a rate of 15,000 kg/s. How many megawatts of electric power could be produced by a power plant using this energy?

***49** (II) The bay behind a dam used for tidal power has an area of 2.2×10^8 m^2. The water level between high and low tide average 3.2 m. About how much electric energy can be produced per day?

***50** (II) One way of storing energy for use during peak demand periods is to pump water to a high reservoir when the demand is low, and then release it to drive turbines when needed. Suppose that water is pumped to a lake 100 m above the turbines at a rate of 1.0×10^6 kg/s for 8.0 h during the night. (*a*) How much energy (kWh) is needed to do this each night? (*b*) If all this energy is released during a 16-h day, what is the average power output? Assume that the process is 80 percent efficient.

For additional reading

Barnea, J., "Geothermal Power," *Sci. American*, **226**:1 (Jan., 1972), p. 70.

Chalmers, B., "Photovoltaic Generation of Electricity," *Sci. American*, **235**:4 (Oct., 1976), p. 34.

Einstein, A., and L. Infeld, *Evolution of Physics*, Simon & Schuster, Inc., New York, 1938.

Glaser, P. E., "Solar Power from Satellites," *Physics Today*, **30**:2 (Feb., 1977), p. 30.

Magie, W. F., *Source Book in Physics*, Harvard University Press, Cambridge, Mass., 1963. Excerpts from Faraday's and Henry's work.

Marston, E. H., *The Dynamic Environment*, John Wiley & Sons, Inc., New York, 1975.

Meyer, H. W., *History of Electricity and Magnetism*, The M.I.T. Press, Cambridge, Mass., 1972.

Schwartz, B. B., and S. Foner, "Large-Scale Applications of Superconductivity," *Physics Today*, **30**:7 (Jul., 1977), p. 34. Superconductors used in the generation, transformation, and transmission of power as well as other uses.

Shonle, J. I., *Environmental Applications of General Physics*, Addison-Wesley Publishing Co., Inc., Reading, Mass., 1974.

19 ▌ Ac circuits and electronics

We deal, in this chapter, with some practical physics. First we examine circuits containing resistors, capacitors, and/or inductors and see how these behave when connected individually and in series to a source of alternating voltage. We will see that each offers a kind of resistance, or impedance, to the flow of current.

Then, in the second part of the chapter, we deal with the subject of electronics: from vacuum tubes to television picture tubes to transistors, and explain how they are used in many practical devices.

*19-1 ▌ Ac circuits and impedance

We now discuss, one at a time, how a resistor, a capacitor, and an inductor behave when connected to a source of alternating emf, represented by the symbol —⊝—, which produces a sinusoidal voltage of frequency f; and we assume in each case that it gives rise to a current

$$I = I_0 \cos 2\pi ft$$

where t is time and I_0 is the peak current. We must rember (Section 15-5) that $V_{rms} = V_0/\sqrt{2}$ and $I_{rms} = I_0/\sqrt{2}$ (Equation 15-7).

FIGURE 19-1
Resistor connected to ac source. Current is in phase with the voltage.

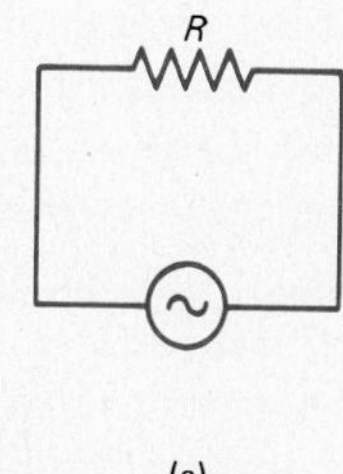

(a)

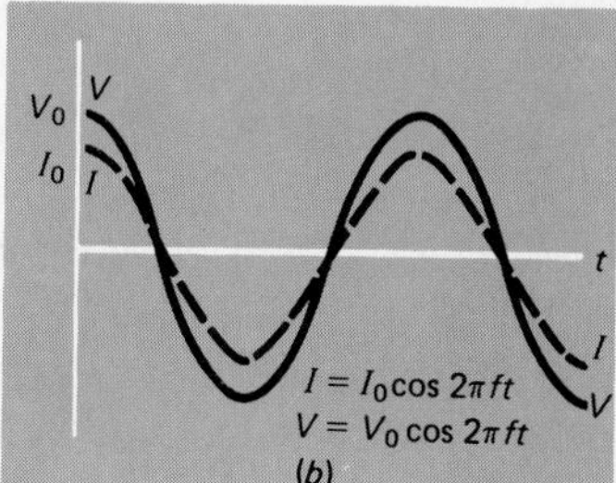

(b)

*Resistor

When an ac source is connected to a resistor as in Figure 19-1*a*, the current increases and decreases with the alternating emf according to Ohm's law: $I = V/R$. This is graphed in Figure 19-1*b*, where the voltage is the solid curve and the current is the dashed curve. Because the current is zero when the voltage is zero and the current reaches a peak when the voltage does, we say that the current and voltage are *in phase*.

Energy is transformed into heat, as discussed in Section 15-5, at an average rate $P = \overline{IV} = I_{rms}^2 R = V_{rms}^2/R$.

*Inductor

In Figure 19-2*a* an inductor of inductance L, represented by the symbol -ꝏꝏ-, is connected to the ac source. We ignore any resistance it might have (it is usually small). The voltage applied to the inductor will be equal to the "back" emf generated in the inductor by the changing current as given by Equation 18-8. This is because the sum of the emfs around any closed circuit must be zero, as Kirchhoff's rule tells us. Thus

$$V - L\frac{\Delta I}{\Delta t} = 0$$

or

$$V = L\frac{\Delta I}{\Delta t},$$

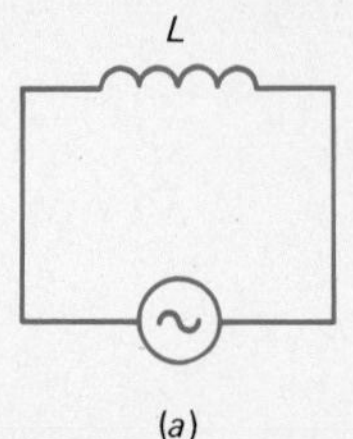

(a)

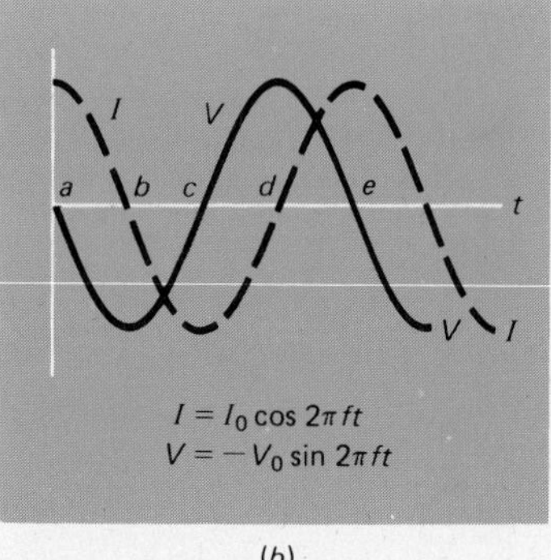

(b)

FIGURE 19-2
Inductor connected to ac source. Current lags voltage by a quarter-cycle, or 90°.

where V is the sinusoidally varying voltage of the source and $L\,\Delta I/\Delta t$ is the voltage induced in the inductor. According to this equation, I is increasing most rapidly when V has its maximum value, $V = V_0$. And I will be decreasing most rapidly when $V = -V_0$. These two instants correspond to points d and b on the graph of voltage versus time in Figure 19-2b. At points a and c, $V = 0$; the equation above tells us that $\Delta I/\Delta t = 0$ at these instants, so I is not changing; thus the curve of I versus t at these points is neither increasing nor decreasing and therefore corresponds to the maximum and minimum values of the current I. These correspond to points a and c on the graph. Going point by point in this manner, the curve of I versus t as compared to that for V versus t can be constructed, and they are shown by the dashed and solid lines respectively in Figure 19-2b.

Notice that the current reaches its peaks (and troughs) one-quarter of a cycle after the voltage does. We say that *in an inductor, the current lags the voltage by* 90°. (Remember that 360° corresponds to a full cycle, so 90° is a quarter cycle.) Alternatively, we can say that the voltage leads the current by 90°. (Since we originally chose $I = I_0 \cos 2\pi ft$, we see from the graph that V must vary as $V = -V_0 \sin 2\pi ft$.)

Because the current and voltage are out of phase by 90°, no power is dissipated in an inductor on the average; that is, energy is not transformed to other forms. This can be seen as follows. From point c to d, the voltage is increasing from zero to its maximum. The current, however, is in the opposite direction to the voltage and is approaching zero. The average power over this interval, VI, is negative. From d to e, however, both V and I are positive so VI is positive; this contribution just balances the negative contribution of the previous quarter-cycle. Similar considerations apply to the rest of the cycle. Thus the average power dissipated over one or many cycles is zero. We can see that energy from the source passes into the magnetic field of the inductor, where it is stored momentarily; then the field decreases and the energy is transferred back to the source. None is lost in this process. Compare this to a resistor where the current is always in the same direction as the voltage and energy is transferred out of the source and never back into it. (The product VI is never negative.) This energy is not stored, but is transformed to heat in the resistor.

As mentioned in Section 18-5, the back emf of an inductor impedes the flow of an ac current. Indeed, it is found that the magnitude of the current in an inductor is directly proportional to the applied ac voltage at a given frequency. We can therefore write an equation for a pure inductor (no resistance) that is the equivalent of Ohm's law:

$$V = IX_L, \tag{19-1a}$$

where X_L is called the *inductive reactance* or *impedance* of the inductor. Normally, we use the term "reactance" to refer solely to the inductive properties. We then reserve the term "impedance" to include the total "impeding" qualities of the coil—its inductance plus any resistance it may have (more on this in the next section). In the absence of resistance, the impedance is the same as the reactance.

The quantities V and I in Equation 19-1a can refer either to rms or to peak values. Note, however, that although this equation may relate the peak values, the peak current and voltage are not reached at the same time, so this equation is *not* valid at a particular instant, as is the case for Ohm's law; it is valid only on the average.

From the fact that $V = L\,\Delta I/\Delta t$, we see that the larger L is, the less will be the change in current ΔI during the time Δt; hence I itself will be smaller at any instant for a given frequency. Thus $X_L \propto L$. The reactance X_2 also depends on the frequency; the greater the frequency, the more rapidly the magnetic flux changes in the inductor. If the emf induced by this field is to remain equal to the source emf, as it must, the magnitude of the current must then be less. Hence the greater the frequency, the greater is X_L, so $X_L \propto fL$. This is also consistent with the fact that if the frequency f is zero (so the current is dc), there is no back emf and no impedance to the flow of charge. Careful calculation (using calculus), as well as experiment, shows that the constant of proportionality is 2π. Thus

$$X_L = 2\pi fL. \tag{19-1b}$$

▌ EXAMPLE 19-1 A coil has a resistance $R = 1.0\ \Omega$ and an inductance of 0.300 H. Determine the current in the coil if (a) 120 V dc is applied to it; (b) 120 V ac (rms) at 60 Hz is applied.

SOLUTION (a) There is no inductive impedance ($X_L = 0$ since $f = 0$), so we apply Ohm's law for the resistance:

$$I = \frac{V}{R} = \frac{120\ \mathrm{V}}{1.0\ \Omega} = 120\ \mathrm{A}.$$

(b) The inductive reactance in this case is

$$X_L = 2\pi fL = (6.28)(60\ \mathrm{s}^{-1})(0.300\ \mathrm{H}) = 113\ \Omega.$$

In comparison to this, the resistance can be ignored. Thus

$$I_{\mathrm{rms}} = \frac{V_{\mathrm{rms}}}{X_L} = \frac{120\ \mathrm{V}}{113\ \Omega} = 1.1\ \mathrm{A}.$$

(It might be tempting to say that the total impedance is $113\ \Omega + 1\ \Omega = 114\ \Omega$; this might imply that about 1 percent of the voltage drop is across the resistor, or about 1 V; and that across the impedance is 119 V. Although the 1 V_{rms} across the resistor is correct, the other statements are *not* true because of the alteration in phase in an inductor. This will be discussed in the next section.) ▌▌

*Capacitor

When a capacitor is connected to a battery, the capacitor plates quickly acquire equal and opposite charges; but no steady current flows in the circuit. A capacitor prevents the flow of a dc current (unless the capacitor is "leaky," in which case a leakage current would exist across the gap). However, if a capacitor is connected to an alternating source of voltage, as in Figure 19-3*a*, an alternating current will flow continuously. This can happen because when the ac voltage is first turned on, charge begins to flow so that one plate acquires a negative charge and the other a positive charge. But before the capacitor becomes fully charged, the voltage reverses itself, and the charges flow in the opposite direction. Thus for an alternating applied voltage, an ac current is present in the circuit continuously.

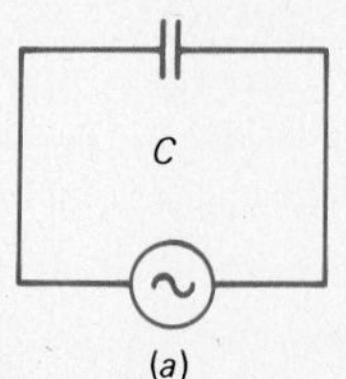

(*a*)

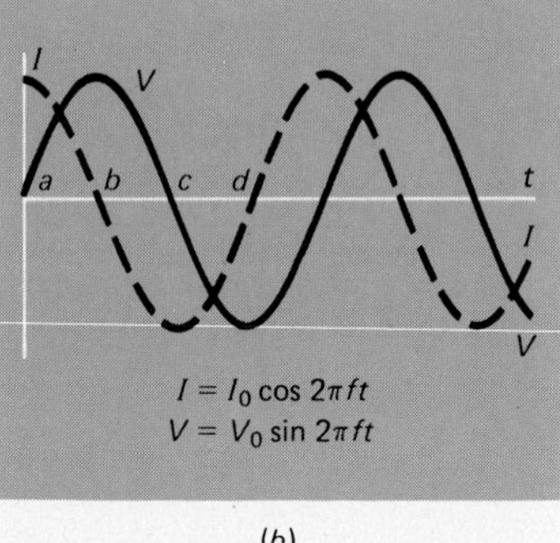

(*b*)

FIGURE 19-3
Capacitor connected to ac source. Current leads voltage by a quarter-cycle, or 90°.

Let us look at this process in more detail. First, we recall that the applied voltage must equal the voltage across the capacitor: $V = Q/C$, where C is the capacitance and Q the charge on the plates. Thus the charge Q on the plates follows the voltage; when the voltage is zero, the charge is zero; when the voltage is a maximum, the charge is a maximum. But what about the current I?

At point a in Figure 19-3*b*, when the voltage starts increasing, the charge on the plates is zero; thus charge flows readily toward the plates and the current is large. As the voltage approaches its maximum of V_0 (point b), the charge that has accumulated on the plates tends to prevent more charge from flowing, so the current drops to zero at point b. The charge that has accumulated now starts to flow off the plates and the current again increases (dashed curve), but in the opposite direction; it reaches a maximum (negatively) when the voltage is at point c. Thus the current follows the graph shown dashed in Figure 19-3*b*.

Like an inductor, the voltage and current are out of phase by 90°. But for a capacitor, the *current leads the voltage by* 90°. (Or the voltage lags the current by 90°.) This is the opposite of what happens for an inductor. (Again we have chosen $I = I_0 \cos 2\pi ft$, and we see from the graph that $V = V_0 \sin 2\pi ft$.)

Because the current and voltage are out of phase, the average power dissipated is zero just as for an inductor. Energy from the source is fed to the capacitor, where it is stored in the electric field between its plates. As the field decreases the energy returns to the source. Thus, on the average in an ac circuit, *only a resistance will dissipate energy* to heat or other forms of energy.

A relationship between the applied voltage and the current can be written just as for an inductance:

$$V = IX_C, \qquad (19\text{-}2a)$$

where X_C is the *capacitive reactance* (or *impedance*) of the capacitor. This equation relates the rms or peak values for the voltage and current but is not valid at a particular instant because I and V are out of phase.

X_C depends on both the capacitance C and the frequency f. The larger the capacitance, the more charge it can handle, so the less it will retard the flow of an alternating current; hence X_C will be inversely proportional to C. It is also inversely proportional to the frequency f, since, when the frequency is higher, there is less time per cycle for the charge to build up on the plates and impede the flow. Again there is a factor of 2π, and

$$X_C = \frac{1}{2\pi fC}. \tag{19-2b}$$

Notice that for dc conditions, $f = 0$ and X_C becomes infinite. This is as it should be, since a capacitor does not pass dc current.

It is interesting to note that the reactance of an inductor increases with frequency, but that of a capacitor decreases with frequency.

▌ EXAMPLE 19-2 What are the peak and rms currents in the circuit of Figure 19-3*a* if $C = 1.0\ \mu\text{F}$ and $V_{\text{rms}} = 120$ V? Calculate for (*a*) $f = 60$ Hz, and then for (*b*) $f = 6.0 \times 10^5$ Hz.

SOLUTION (*a*) $V_0 = \sqrt{2}\ V_{\text{rms}} = 170$ V. Then

$$X_C = \frac{1}{2\pi fC} = \frac{1}{(6.28)(60\ \text{s}^{-1})(1.0 \times 10^{-6}\ \text{F})} = 2.7\ \text{k}\Omega.$$

Thus

$$I_0 = \frac{V_0}{X_C} = \frac{170\ \text{V}}{2.7 \times 10^3\ \Omega} = 63\ \text{mA},$$

$$I_{\text{rms}} = \frac{V_{\text{rms}}}{X_C} = \frac{120\ \text{V}}{2.7 \times 10^3\ \Omega} = 44\ \text{mA}.$$

(*b*) For $f = 6.0 \times 10^5$ Hz, X_C will be $0.27\ \Omega$, $I_0 = 630$ A, and $I_{\text{rms}} =$ 440 A. ▌▌

Capacitors are used for a variety of purposes, some of which we have already described. Two other applications are illustrated in Figure 19-4. In part *a*, circuit *A* is said to be capacitively coupled to circuit *B*. The purpose of the capacitor is to prevent a dc voltage from passing from *A* to *B* but allowing an ac signal to pass relatively unimpeded. If *C* is sufficiently large, the ac signal will not be significantly attenuated. The capacitor in part *b* also passes ac but not dc. In this case, a dc voltage can be maintained between circuits *A* and *B*; and if *C* is large enough, the capacitor offers little impedance to an ac signal leaving *A*. Such a signal then passes to ground instead of into *B*. Thus the capacitor in part *b* acts like a filter when a constant dc voltage is required; any sharp variation in voltage will pass to ground instead of into circuit *B*. Capacitors used in these two ways are very common in circuits.

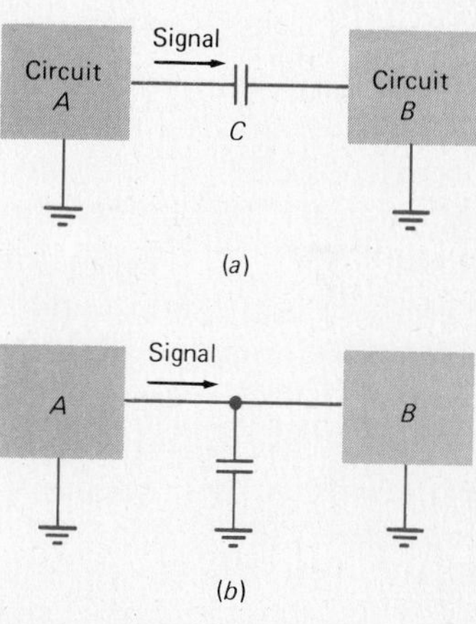

FIGURE 19-4
Two common uses for a capacitor.

*Ac hazards; leakage currents

All the hazards we discussed in Section 16-8 for dc currents are present for ac currents, plus some additional ones we will discuss now.

The human body acts as if it had capacitance in parallel with its resistance. A dc current can only pass through the resistance, and if this is large the current will be small. An ac current can pass through both the resistance and the capacitance; since they are in parallel, the net impedance is less and the current will be greater. At 60 Hz, the impedance may be less than half what it is for dc. Thus an ac voltage is more dangerous than an equal dc voltage.

Another danger is *leakage current* in electrical devices. By this we mean a current along an unintended path, most commonly capacitively or inductively coupled. For example, a wire in a lamp forms a capacitor with the metal case; charges moving in one conductor attract or repel charge in the other. Typical electrical codes limit leakage currents to 1 mA for any device. A 1-mA leakage current is usually harmless. It can be very dangerous, however, to a hospital patient with implanted electrodes connected to ground through the apparatus. This is because the current can pass directly through the heart as compared to the usual situation, where it spreads out through the body. Although 70 mA may be needed to cause heart fibrillation when passing through the hands (very little of it actually passes through the heart), as little as 0.02 mA has been known to cause fibrillation when passing directly to the heart. Thus a "wired" patient is in considerable danger from leakage current even from as simple an act as touching a lamp.

*19-2 ▮ *LRC* series circuit

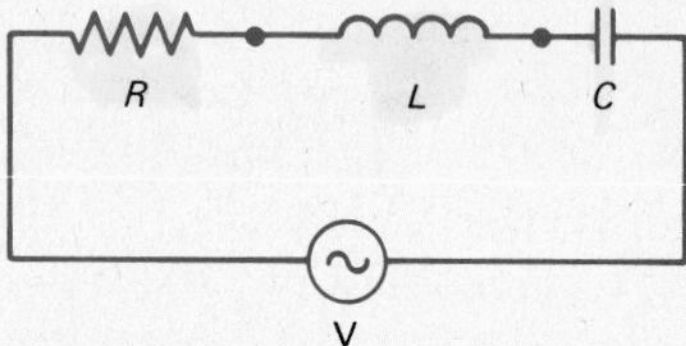

FIGURE 19-5
An *LRC* circuit.

We now examine a circuit containing all three elements in series, a resistor R, an inductor L, and a capacitor C (Figure 19-5). If a given circuit contains only two of these elements, we can still use the results of this section by setting $R=0$, $L=0$, or $C=\infty$ (infinity) as needed. We let V_R, V_L, and V_C represent the voltage across each element at a *given instant* in time; and V_{R0}, V_{L0}, and V_{C0} represent the *maximum* values of these voltages. The voltage across each of the elements will follow the phase relations we just discussed in the last section. That is, V_R will be in phase with the current, V_L will lead the current by 90°, and V_C will lag behind the current by 90°. Also, at any instant the total voltage V supplied by the source will be $V = V_R + V_L + V_C$. However, because the various voltages are not in phase, the rms voltages (which is what ac voltmeters usually measure) will not simply add up to give the rms voltage of the source. And V_0 will *not* equal $V_{R0} + V_{L0} + V_{C0}$. Let us now examine the circuit in detail. What we would like to find in particular is the impedance of the circuit as a whole, the rms current that flows, and the phase difference between the source voltage and the current.

First we note that the current at any instant must be the same at all points in the circuit. Thus the currents in each element are in phase with each other, although the voltages are not. For convenience in our analysis, we therefore choose our origin in time, $t=0$, so that the current I at any time t is $I = I_0 \cos 2\pi ft$. That is, the current is a maximum, $I = I_0$, at $t=0$.

It is convenient to analyze an *LRC* circuit using a *phasor* diagram. Arrows (like vectors) are drawn in an *xy* coordinate system. The lengths of these arrows represent the magnitudes of the peak voltages across each element: $V_{R0} = I_0R$, $V_{L0} = I_0X_L$, and $V_{C0} = I_0X_C$. V_{R0} is in phase with the current and is initially drawn along the positive x axis. Since V_{L0} leads the current by 90°, it leads V_{R0} by 90°, so is initially drawn along the positive y axis. V_{C0} lags the current by 90°, so lags V_{R0} by 90°; hence V_{C0} is drawn initially along the negative y axis. Such a diagram is shown in Figure 19-6*a*. If we let the whole diagram rotate counterclockwise at frequency f, we get

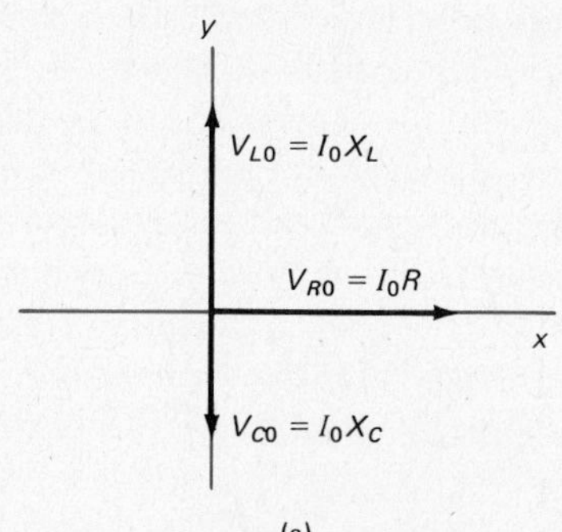

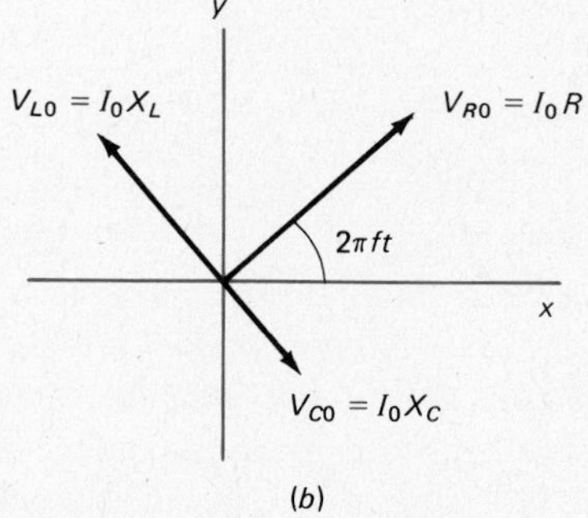

FIGURE 19-6
Phasor diagram for an *LRC* circuit.

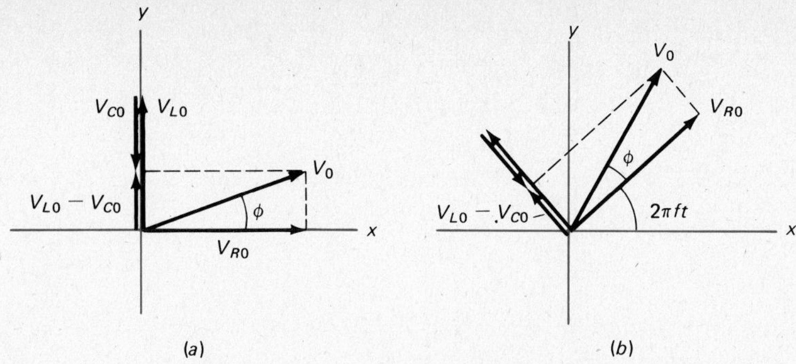

FIGURE 19-7
Phasor diagram for an *LRC* circuit showing the sum vector, V_0.

the diagram shown in Figure 19-6*b*; after a time, t, each arrow has rotated through an angle $2\pi ft$. Then the *projections of each arrow on the x axis represent the voltages across each element at the instant t.* For example, the projection of V_{R0} on the x axis is $V_{R0}\cos 2\pi ft$ (as in Figure 19-1); and the projections of V_{L0} and V_{C0} on the x axis are $-V_{L0}\sin 2\pi ft$ and $V_{C0}\sin 2\pi ft$ as in Figures 19-2*b* and 19-3*b*. Maintaining the 90° angle between each vector ensures the correct phase relations. Although these facts show the validity of a phasor diagram, what we are really interested in is how to add the voltages.

The sum of the projections of the three vectors on the x axis is equal to the projection of their sum. But the sum of the projections represents the instantaneous voltage across the whole circuit, V (equal to the source voltage). Therefore, the vector sum of these vectors will be the vector that represents the peak source voltage, V_0. This is shown in Figure 19-7, where it is seen that V_0 makes an angle ϕ with V_{R0}. As time passes, V_0 rotates with the other vectors, so the instantaneous voltage V (projection of V_0 on the x axis) is

$$V = V_0 \cos(2\pi ft + \phi).$$

Thus we see that the voltage from the source is out of phase† with the current by an angle ϕ.

From this analysis we can now draw some useful conclusions. First, we determine the total impedance Z of the circuit, which is defined by the relation

$$V_{\mathrm{rms}} = I_{\mathrm{rms}}Z, \qquad \text{or} \qquad V_0 = I_0 Z. \tag{19-3}$$

From Figure 19-7 we see that

$$\begin{aligned} V_0 &= \sqrt{V_{R0}^2 + (V_{L0} - V_{C0})^2} \\ &= \sqrt{I_0^2R^2 + (I_0X_L - I_0X_C)^2} \\ &= I_0\sqrt{R^2 + (X_L - X_C)^2}. \end{aligned}$$

Thus, from Equation 19-3,

$$Z = \sqrt{R^2 + (X_L - X_C)^2} \tag{19-4a}$$

$$= \sqrt{R^2 + (2\pi fL - 1/2\pi fC)^2}. \tag{19-4b}$$

† If $R = 0$ and $C = \infty$, then $\phi = 90°$, and V_0 would lead the current by 90°, as it must for an inductor alone. Similarly, if $R = L = 0$, $\phi = -90°$ and V_0 would lag the current by 90°, as it must for a capacitor alone.

This gives the total impedance of the circuit. Also from Figure 19-7 we can find the phase angle ϕ:

$$\cos\phi = \frac{V_{R0}}{V_0} = \frac{I_0 R}{I_0 Z},$$

or

$$\cos\phi = \frac{R}{Z}. \tag{19-5}$$

Finally, we can determine the power dissipated in the circuit. We saw earlier that power is only dissipated by a resistance; none is dissipated by inductance or capacitance. Therefore, the power $P = I_{\rm rms}^2 R$. But from Equation 19-5, $R = Z\cos\phi$. Therefore,

$$\begin{aligned} P &= I_{\rm rms}^2 Z\cos\phi \\ &= I_{\rm rms} V_{\rm rms}\cos\phi. \end{aligned} \tag{19-6}$$

The factor $\cos\phi$ is referred to as the *power factor* of the circuit. For a pure resistor, $\cos\phi = 1$ and $P = I_{\rm rms} V_{\rm rms}$. For a capacitor or inductor, $\cos\phi = 0$ and no power is dissipated.

The test of this analysis is, of course, in experiment; and experiment is in full agreement with these results.

▮ EXAMPLE 19-3 Suppose that $R = 25\ \Omega$, $L = 30$ mH, and $C = 12\ \mu$F in Figure 19-5, and that they are connected to a 90-V ac (rms) 500-Hz source. Calculate (*a*) the current in the circuit, (*b*) the voltmeter readings (rms) across each element, (*c*) the phase angle ϕ, and (*d*) the power dissipated in the circuit.

SOLUTION (*a*) First, we find the individual impedances at $f = 500\ \text{s}^{-1}$:

$$X_L = 2\pi f L = 94\ \Omega$$

$$X_C = \frac{1}{2\pi f C} = 27\ \Omega.$$

Then

$$\begin{aligned} Z &= \sqrt{R^2 + (X_L - X_C)^2} = \sqrt{(25\ \Omega)^2 + (94\ \Omega - 27\ \Omega)^2} \\ &= 72\ \Omega. \end{aligned}$$

From Equation 19-3,

$$I_{\rm rms} = \frac{V_{\rm rms}}{Z} = \frac{90\ \text{V}}{72\ \Omega} = 1.25\ \text{A}.$$

(*b*) The rms voltage across each element is

$$(V_R)_{\rm rms} = I_{\rm rms} R = (1.25\ \text{A})(25\ \Omega) = 31\ \text{V}$$

$$(V_L)_{\rm rms} = I_{\rm rms} X_L = 117\ \text{V}$$

$$(V_C)_{\rm rms} = I_{\rm rms} X_C = 34\ \text{V}.$$

Notice that these do *not* add up to give the source voltage, 90 $\text{V}_{\rm rms}$. Indeed, the rms voltage across the inductance *exceeds* the source voltage. This is due to the fact that the different voltages are out of phase with each other. The instantaneous voltages at any time do add up, of course; this can happen because at any instant one voltage can be negative, to compensate for a large positive voltage of another. The rms voltages, however, are always positive by definition.

(*c*) Since $\cos\phi = R/Z = 25\ \Omega/72\ \Omega = 0.35$, then $\phi = 70°$.

(*d*) $P = I_{rms}V_{rms}\cos\phi = (1.25\ \text{A})(90\ \text{V})(0.35) = 39\ \text{W}$. ▌▌

*19-3 ▌ Resonance in ac circuits

The rms current in an *LRC* circuit is given by (see Equations 19-3 and 19-4):

$$I_{rms} = \frac{V_{rms}}{Z} = \frac{V_{rms}}{\sqrt{R^2 + (2\pi fL - 1/2\pi fC)^2}}. \qquad (19\text{-}7)$$

Because the impedance of inductors and capacitors depends on the frequency *f* of the source, the current in an *LRC* circuit will depend on the source frequency. From Equation 19-7 we can see that the current will be maximum at a frequency such that

$$2\pi fL - \frac{1}{2\pi fC} = 0.$$

We solve this for *f*, and call the solution f_0:

$$f_0 = \frac{1}{2\pi}\sqrt{\frac{1}{LC}}. \qquad (19\text{-}8)$$

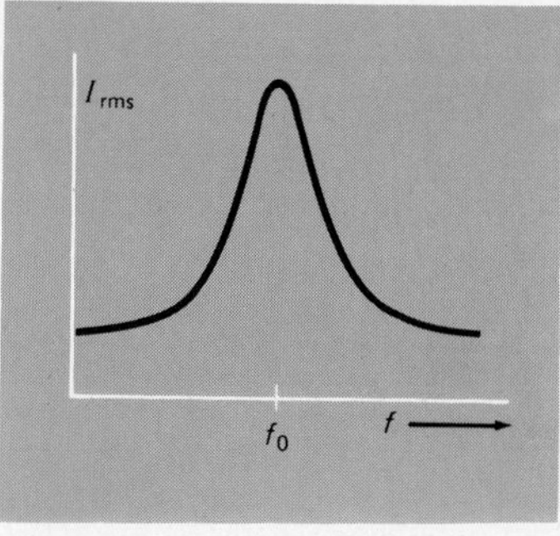

FIGURE 19-8
Current in an *LRC* circuit as a function of frequency, showing resonance peak at $f = f_0 = (1/2\pi)\sqrt{1/LC}$.

This is called the *resonant frequency* of the circuit. At this frequency, $X_C = X_L$, so the impedance is purely resistive and $\cos\phi = 1$. A graph of I_{rms} versus *f* is shown in Figure 19-8 for particular values of *R*, *L*, and *C*. For smaller *R* compared to X_L and X_C, the resonance peak will be higher and sharper. When *R* is small, we speak of an *LC* circuit; the energy in such a circuit oscillates, at frequency f_0, between the inductor and the capacitor, with some being dissipated in *R* (some resistance is unavoidable).

This electrical resonance is analogous to mechanical resonance, which we discussed in Chapter 11 (see Figure 11-11). The energy transformed to the system is a maximum at resonance whether it is electrical resonance, the oscillation of a spring or pushing a child on a swing (Section 11-4). That this is true in the electrical case can be seen from Equation 19-6. At resonance, $\cos\phi = 1$, and I_{rms} is a maximum. A graph of power versus frequency looks much like that for the current (Figure 19-8).

Electric resonance is used in many circuits. Radio and television sets, for example, use resonant circuits for tuning in a station. Many frequencies reach the circuit, but a significant current flows only for frequencies at or near the resonant frequency. Either *L* or *C* is variable so that different stations can be tuned in (more on this in Chapter 20). *LC* circuits are also used in oscillators (Section 19-9).

An *LRC* circuit can have the elements arranged in parallel instead of in series. Resonance will occur in this case, too, but the analysis of such circuits is more involved.

*19-4 ▮ Thermionic emission; vacuum tubes

*Thermionic emission

In the course of his experiments on the electric light bulb, Thomas Edison (1847–1931) made an interesting discovery in 1883. He inserted into an evacuated glass bulb an electrode or plate in addition to the bulb filament. When a positive potential was applied by a battery to the plate, he found that a current would flow in the external circuit as long as the filament was hot and glowing (due to a separate current passing through it). However, if a negative potential was applied to the plate, no current was observed to flow. When the filament was cold, no current flowed in either case. This phenomenon is known as *thermionic emission*.

Although Edison was unaware of the significance of his discovery, it eventually led to the invention of the electronic vacuum tube. Indeed, an explanation for this effect had to await the discovery of the electron more than a decade later. Apparently what happens is that electrons are being "boiled off" the filament when it is hot. When the plate is positive, the electrons are attracted to it, so a current flows. If the plate is negative, the electrons are repelled and no current flows.

We can understand how electrons might be "boiled off" a hot metal filament if we treat the electrons like molecules in a gas. This makes sense if the electrons are relatively free to move about, which is consistent with the fact that metals are good conductors. However, electrons do not readily escape from the metal. There are forces that keep them in. For example, if an electron were to escape outside the metal surface, a net positive charge would remain behind, and this would attract the electron back. To escape, an electron would have to have a certain minimum kinetic energy, just as a molecule in a liquid must have a minimum kinetic energy if it is to evaporate into the gas phase. We saw in Chapter 7 that the average kinetic energy (KE) of molecules in a gas is proportional to the absolute temperature T: $\overline{\mathrm{KE}} = \frac{3}{2}kT$. Of course, some have more KE and others less. We can apply this roughly to electrons, and at room temperature very few electrons would have sufficient energy to escape. At higher temperatures $\overline{\mathrm{KE}}$ is larger and many can escape; this is just like evaporation from liquids, which occurs more readily at high temperatures. Thus significant thermionic emission occurs only at elevated temperatures—and Edison's results are readily understood.

*Vacuum-tube diodes

Edison's tube was basically the first "diode." However, credit for its invention goes to Sir John Fleming (1849–1945), who, about 1904, developed the tube and understood its possibilities.

The term *diode* refers to the fact that two ("di") electrodes are sealed inside an evacuated glass or metal container or tube (Figure 19-9). The current used to heat the cathode is usually carried in a separate circuit, as shown. An important use of a diode is for changing ac voltage into dc. This is called *rectification*, and a diode tube is often called a *rectifier*. We will see how this is done when we discuss semiconductor diodes in Section 19-6.

FIGURE 19-9
Diagram of a vacuum-tube diode.

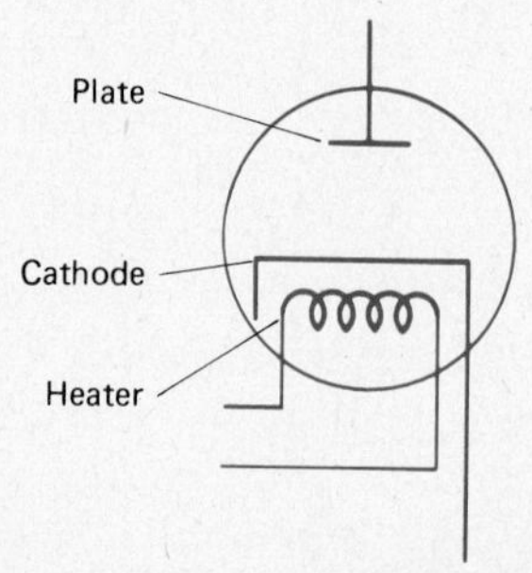

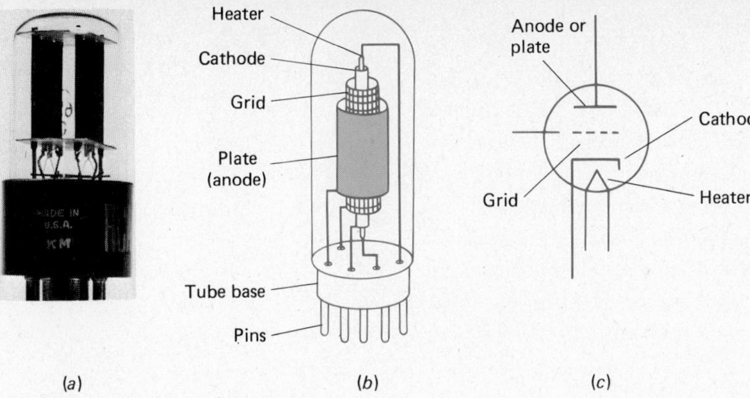

FIGURE 19-10
(*a*) Photograph of a tube that contains two triodes. (*b*) Parts of a triode. (*c*) Schematic drawing of a triode. (Photo by Lester V. Bergman & Assoc., Inc.)

*Triodes

A *triode*, as its name implies, contains three electrodes (Figure 19-10). The *cathode* is usually heated (Figure 19-10*c*) by a separate "heater" placed close to it. The *plate* or *anode* is at room temperature. When the plate is at a positive voltage with respect to the cathode, electrons emitted by the heated cathode are attracted to the plate and a large current can pass through the tube. The third electrode, placed between the cathode and plate, is the *control grid.* It is usually a fine wire grid, hence its name. The inventor of the triode, Lee de Forest (1873–1961), showed that voltage applied to the grid could be used to control the current flow across the tube. To see how this is done and why it is important, we study the circuit of Figure 19-11.

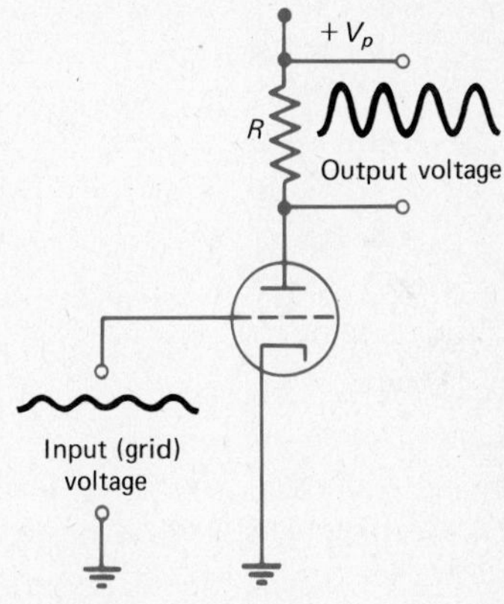

FIGURE 19-11
Simple triode circuit.

The cathode is heated by a heater so that thermionic emission occurs. The plate is kept at some fixed positive dc voltage, usually somewhere between 50 and 300 V. If the voltage on the grid is zero relative to the cathode, most of the electrons will pass through the grid and reach the plate. If the grid voltage is made negative with respect to the cathode, the electrons emitted by the cathode will be repelled. The number of electrons that have sufficient speed to pass through the grid and reach the cathode depends on the magnitude of the negative voltage on the grid. If the grid voltage is sufficiently negative, no electrons will get through and the plate current will be zero; this is called the *cutoff point.* A graph of the plate current, I_p, versus grid voltage, V_g, is shown in Figure 19-12 for two

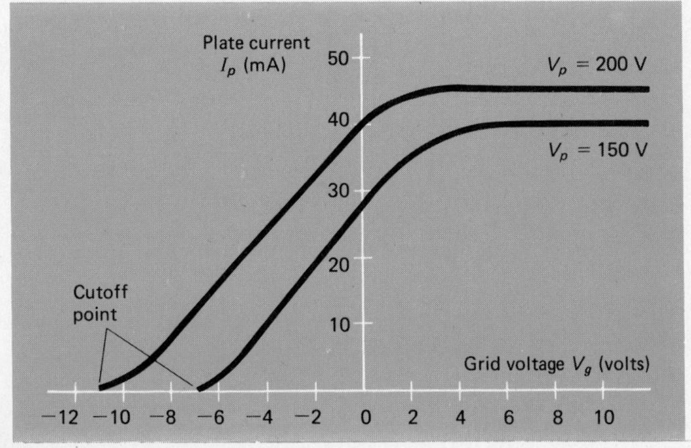

FIGURE 19-12
Plate current, I_p, of a typical triode as a function of the voltage, V_g, applied to the grid for two cases: when the plate voltage V_p is 200 V and when it is 150 V.

different plate voltages, V_p. Because the flow of current can be controlled, vacuum tubes are sometimes referred to as *valves*, particularly in Britain.

Probably the most important use of a triode is to amplify a small signal. Normally, a dc voltage is applied to the grid, which is called the *bias voltage.* It is usually set between zero and the cutoff voltage, where the response of the tube (Figure 19-12) is linear. A slight change in the grid voltage due to a small ac signal will result in a large change in the current to the plate, I_p. This results in a large change in the voltage across the output resistor R ($V_{\text{output}} = I_pR$). Thus a small input signal voltage to the grid results in a large output voltage of the same frequency. This amplification is indicated by the sinusoidal curves in Figure 19-11.

*19-5 ▌ Cathode ray tube; television and oscilloscope

Although vacuum-tube diodes and triodes have largely been supplanted by semiconductor devices, another type of vacuum tube remains very common. This is the *cathode ray tube* (CRT), which is used in oscilloscopes and as the picture tube in television sets. It derives its name from the fact that it directs a beam of cathode rays, or electrons as we call them now, to various parts of a screen to produce a "picture."

A simple CRT is diagrammed in Figure 19-13. Electrons emitted by the heated cathode are accelerated by a high voltage applied to the anode (5000 to 50,000 V). The electrons pass out of this "electron gun" through a small hole in the anode. The inside face of the tube is coated with a fluorescent material that glows when struck by electrons. A tiny bright spot is thus visible where the electron beam strikes the screen. Two horizontal and two vertical plates deflect the beam of electrons when a voltage is applied to them. The electrons are deflected toward whichever plate is positive. By varying the voltage on the deflection plates, the bright spot can be placed at any point on the screen.

In the picture tube of a television set, magnetic deflection coils are usually used instead of electric plates. But the effect is the same. The electron beam is made to sweep over the screen in the manner shown in Figure 19-14. The beam is swept horizontally by the horizontal deflection plates or coils. When the horizontal deflecting field is maximum in one

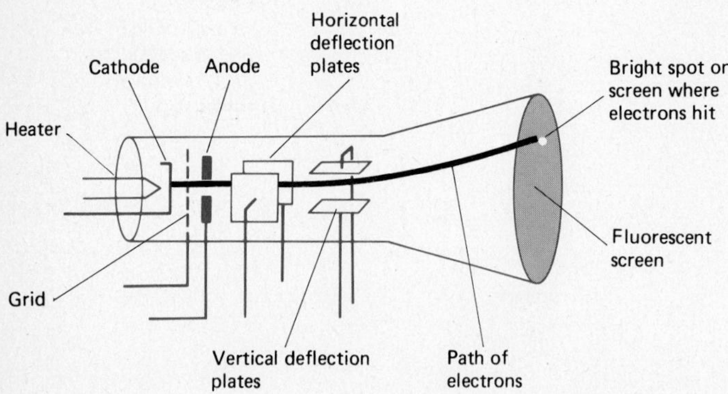

FIGURE 19-13
Cathode ray tube.

direction, the beam is at one edge of the screen. As the field decreases to zero, the beam moves to the center; and as it increases to a maximum in the opposite direction, the beam approaches the opposite edge. When it reaches this edge, the voltage or current abruptly changes to return the beam to the opposite side of the screen. Simultaneously, the beam is deflected downward slightly by the vertical deflection plates or coils, and then another horizontal sweep is made. In the United States, 525 lines constitutes a complete sweep over the entire screen. (Some European systems use almost twice that number.) The complete picture of 525 lines is swept out in $\frac{1}{30}$ s. Actually, a single vertical sweep takes $\frac{1}{60}$ s and involves every other line; the lines in between are then swept out over the next $\frac{1}{60}$ s. We see a picture because the image is retained by the fluorescent screen and by our eyes for about $\frac{1}{20}$ s. The picture we see consists of the varied brightnesses of the spots on the screen. The brightness at any point is controlled by the grid (Figure 19-13), which can limit the flow of electrons. The voltage on the grid is determined by the video signal sent out by the station and received by the set. Accompanying this signal are signals that synchronize the grid voltage to the horizontal and vertical sweeps.

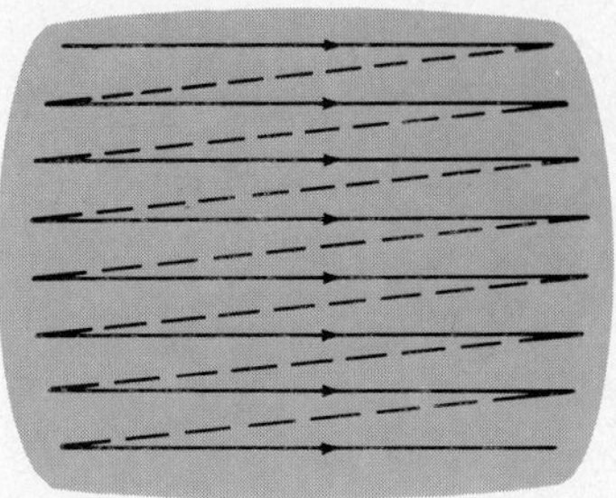

FIGURE 19-14
Electrons sweep across a television screen in a succession of horizontal lines.

A color television screen contains an orderly array of tiny phosphor lines (or spots), alternating among the three primary colors: red, green, and blue. The phosphor lines are excited by a single gun whose intensity depends on the particular phosphor it is aimed at (or, in some cases, by three separate guns).

An *oscilloscope* is a device for amplifying, measuring, and visually observing an electrical signal. Indeed, it provides the principal means for measuring rapidly changing voltages or currents, since galvanometers (even on a chart recorder) have too much inertia to follow such a signal well. The signal is displayed on the screen of a CRT. In normal operation, the electron beam is swept horizontally at a uniform rate in time by the horizontal deflection plates. A "sawtoothed" voltage is needed for this, and the means of producing it, using an *RC* circuit, was discussed in Section 16-7. The signal to be displayed is applied, after amplification, to the vertical deflection plates. The line made on the screen by the moving electron beam is called the "trace." The oscilloscope trace is thus a plot of the signal voltage (vertically) versus time (horizontally). The horizontal sweep rate can be varied to give different time scales. The amplification of the input signal can be varied as well. A typical oscilloscope face with the various controls is shown in Figure 19-15. The screen normally has a grid of 1-cm squares so measurements can be made on the trace displayed.

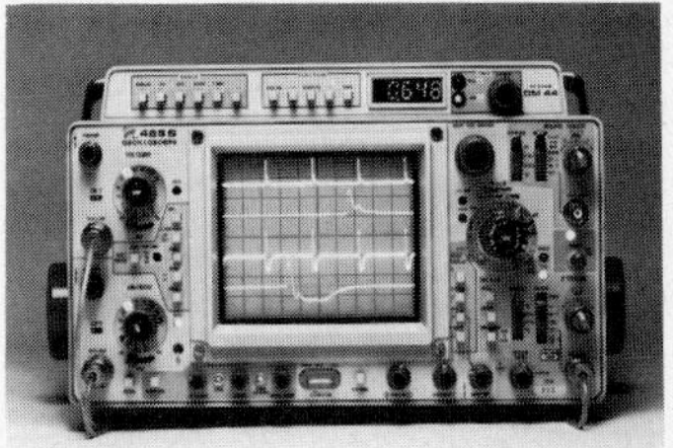

FIGURE 19-15
Photograph of the face of an oscilloscope, showing the controls. Four different traces (signals) are displayed on the CRT screen. (Courtesy of Tektronix, Inc.)

*19-6 ▮ Semiconductors; diodes and transistors

Diode and triode tubes have been largely replaced by devices made of semiconductor materials. A semiconductor diode can perform the same functions as a diode tube, and a transistor can perform the same functions as a triode tube. Because semiconductors are solid at normal temperatures, devices made from them are often called "solid-state" devices. Semiconductor devices offer several advantages over vacuum tubes: they

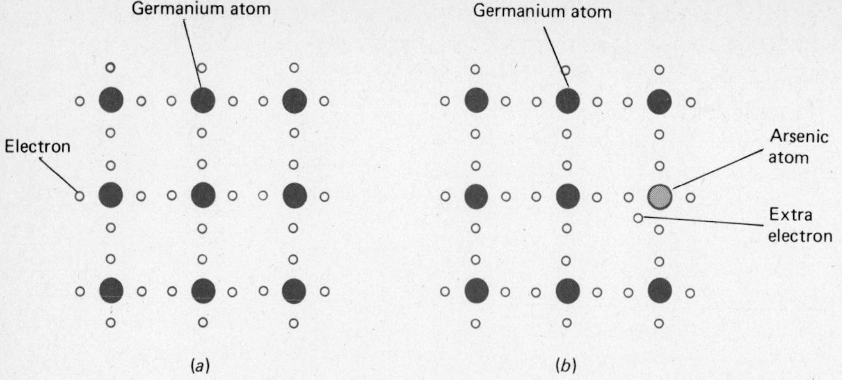

FIGURE 19-16
A germanium crystal. (*a*) Four (outer) electrons surround each germanium atom. (*b*) Germanium crystal doped with a few arsenic atoms; the extra electron doesn't fit into the crystal lattice and so is free to move about. This is an *n*-type semiconductor.

are smaller, need less dc voltage to operate, have a longer life, and use less power because there are no heating elements. Before treating these devices, we first look at some properties of semiconductors.

**n*- and *p*-type semiconductors

The two main semiconductors used in diodes and transistors are silicon and germanium. An atom of silicon or germanium has four outer electrons which act to hold the atoms in the regular lattice structure of the crystal (Figure 19-16*a*). Germanium and silicon become useful in diodes and transistors only when a minute amount of impurity is introduced into the crystal structure (about 1 part in 10^6). This is called "doping" the semiconductor. Two kinds of semiconductors can be made, depending on the type of impurity introduced. If the impurity consists of atoms which have five outer electrons, such as arsenic, we have the situation illustrated in Figure 19-16*b*. Since only four of arsenic's electrons fit into the crystal structure, the fifth is able to move about relatively freely, much like the electrons in a conductor. Because of this small number of extra electrons, a doped semiconductor becomes highly conducting. An arsenic-doped germanium crystal is called an *n-type semiconductor* because electrons (*n*egative charge) carry the electric current.

In a *p-type semiconductor*, a small amount of impurity with three outer electrons—such as gallium—is added to the semiconductor. As shown in Figure 19-17*a*, there is a "hole" in the lattice structure next to a

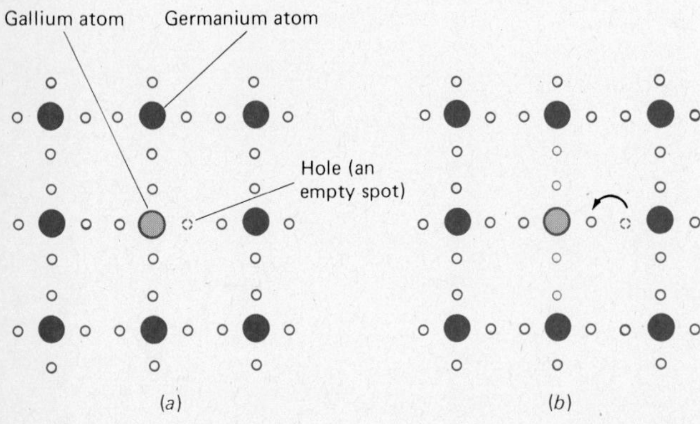

FIGURE 19-17
A *p*-type semiconductor, gallium–doped germanium. (*a*) Gallium has only three outer electrons, so there is an empty spot or *hole* in the structure. (*b*) Electrons from germanium atoms can jump into the hole and fill it. As a result, the hole moves to a new location (to the right in this figure).

gallium atom since it has only three outer electrons. Electrons from nearby germanium atoms can jump into this hole and fill it. But this leaves a "hole" where that electron had previously been (Figure 19-17*b*). The vast majority of atoms are germanium, so the hole is almost always next to a germanium atom. Since germanium atoms require four outer electrons to be neutral, this means there is a net positive charge at a hole. Whenever an electron moves to fill a hole, the positive hole is then at the previous position of that electron. Another electron can then fill this hole, and the hole thus moves to a new location; and so on. This type of semiconductor is called *p-type* because it is the *p*ositive holes that seem to carry the electric current. Note, however, that *p*-type and *n*-type semiconductors have *no net charge* on them.

*Energy bands

Semiconductor behavior can be explained at the atomic level using the "electron band theory" of solids which has been developed only over the past few decades. We shall see in a later chapter that the electrons in a single atom can occupy only certain energy levels. The lowest possible energy level is called the ground state and higher ones are called excited states. In a solid, these energy levels spread out into wide "bands" because of the interaction between the atoms. The outer electrons can be in either of two bands: the lower *valence band,* which corresponds to the ground state; or the upper *conduction band.* Normally, the electrons reside in the valence band, where they are held rather tightly to individual atoms. Given sufficient energy (say, by collision with another electron) an electron can jump up to the higher conduction band. Once there, an electron is no longer bound to a particular atom but can move about freely in the lattice. No electron can have an energy in the "forbidden" energy gap between the two bands.

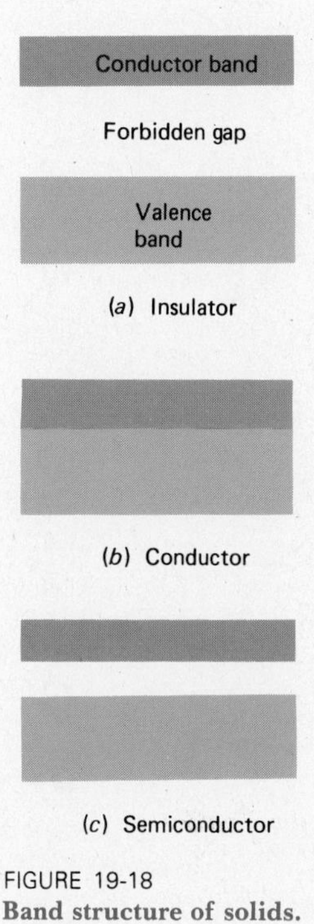

FIGURE 19-18
Band structure of solids.

In an insulator, the gap between the bands is large and very few electrons have sufficient energy to reach the conduction band (see Figure 19-18*a*). In a good conductor, on the other hand, there is no gap (Figure 19-18*b*) and the bands may overlap. Electrons can readily be in the conduction band; they can thus move about freely and carry an electric current.

In a pure (or *intrinsic*) semiconductor such as carbon, germanium, or silicon, the forbidden energy gap is small (Figure 19-18*c*). A small number of electrons will have enough energy to jump the gap, so there will be a very slight amount of conduction. If the temperature is raised, more electrons will have enough energy to jump the gap; thus the resistivity of semiconductors *decreases* with temperature (see Table 15-1).

In a doped semiconductor, the impurity provides additional energy states between the bands as shown in Figure 19-19. In an *n*-type semiconductor, the impurity energy level lies just below the conduction band. Electrons in this energy level can easily acquire sufficient energy to reach the conduction band. Since this level supplies electrons to the conduction band, it is called a *donor* level. In *p*-type semiconductors, the impurity energy level is just above the valence band. It is called an *acceptor* level, because electrons from the valence band can easily jump into it. A positive hole is left behind; and as other electrons move into this hole, the hole moves about as discussed earlier.

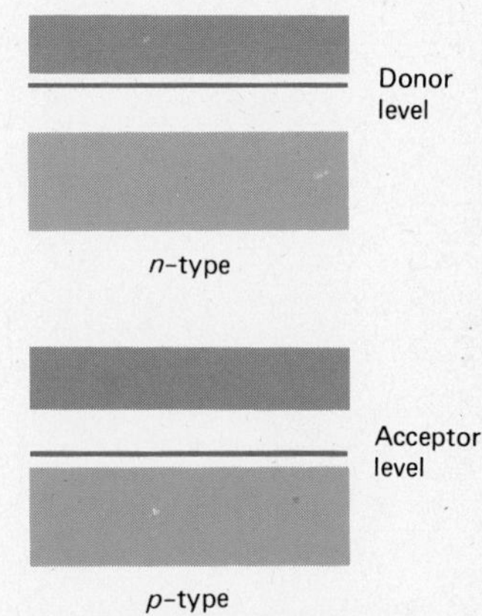

FIGURE 19-19
Impurity levels in doped semiconductors.

*19-7 ▌ Semiconductor diodes and rectification

*Diodes

When an n-type semiconductor is joined to a p-type, a *pn junction diode* is formed. Separately, the two semiconductors are electrically neutral. When joined, a few electrons in the region near the junction diffuse from the n-type semiconductor into the p-type, where they fill a few of the holes. The n-type is thus left with a slight positive charge, and the p-type acquires a net negative charge. Thus a potential difference is established, with the n side positive relative to the p side, and this prevents further diffusion of electrons.

If a battery is connected to a diode with the positive terminal to the p side and the negative terminal to the n side as in Figure 19-20a, such a voltage opposes the internal potential difference and the diode is said to be *forward-biased.* If the voltage is great enough (about 0.3 V for Ge, 0.6 V for Si at room temperature) a current will flow. The positive terminal of the battery repels the positive holes in the p-type semiconductor; and the negative terminal of the battery repels the electrons in the n-type. The holes and electrons meet at the junction in the middle, and the electrons go across and fill the holes. Meanwhile, the positive terminal of the battery continues to pull electrons off the p end, which creates new holes; and electrons continue to be supplied by the negative terminal at the n end. As a result, a large current flows through the diode.

When the diode is *reverse-biased* (Figure 19-20b) the battery's positive terminal attracts the holes in the p end and the negative terminal attracts the electrons in the n end. The current carriers are thus pulled away from the junction so they cannot meet and, ideally, no current flows.

A graph of current versus voltage for a typical diode is shown in Figure 19-21. As can be seen, a real diode does allow a small amount of reverse current to flow. For most practical purposes, this is negligible.† The symbol for a semiconductor diode is shown in Figure 19-22. The arrow points in the direction that a diode conducts conventional (positive) current.

▌ EXAMPLE 19-4 The diode whose current–voltage characteristics are shown in Figure 19-21 is connected in series with a 4.0-V battery and a

† Reverse current in Ge at room temperature is typically a few μA; for Si, a few pA. The reverse current increases rapidly with temperature, however, and may render a diode ineffective above 200°C.

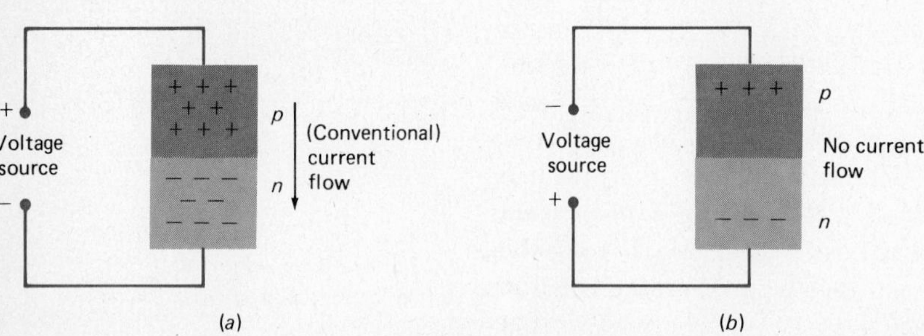

FIGURE 19-20
Schematic diagram showing how a semiconductor diode operates. Current flows when the voltage is connected in forward bias, as in (*a*), but not when in reverse bias, as in (*b*).

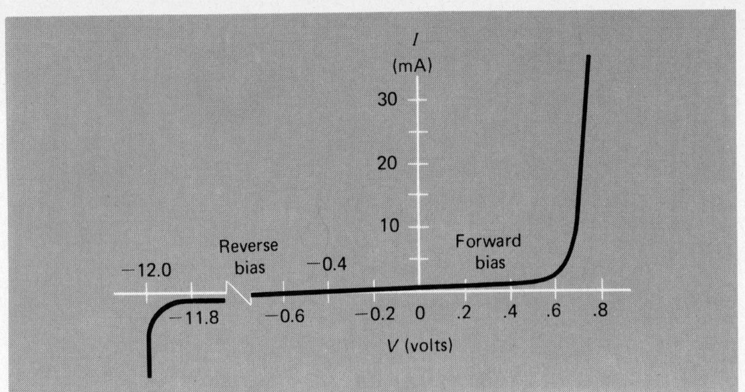

FIGURE 19-21
Current through a diode as a function of applied voltage.

resistor. If a current of 10 mA is to pass through the diode, what resistance must the resistor have?

SOLUTION From Figure 19-21 we see that the voltage drop across the diode is about 0.7 V when the current is 10 mA. Therefore, the voltage drop across the resistor is $4.0\text{ V} - 0.7\text{ V} = 3.3\text{ V}$, so $R = V/I = (3.3\text{ V})/(1.0 \times 10^{-2}\text{ A}) = 330\ \Omega$. ▮▮

FIGURE 19-22
Symbol for a semiconductor diode. The diode readily passes conventional current in the direction of the arrow but not in the reverse direction.

If the voltage across a diode connected in reverse bias is increased greatly, a point is reached where breakdown occurs. The electric field across the junction becomes so large that ionization of atoms results. The electrons thus pulled off of their atoms contribute to a larger and larger current as breakdown continues. The voltage remains constant over a wide range of currents. This is shown on the far left in Figure 19-21. This property of diodes can be used to accurately regulate a voltage supply. A diode designed for this purpose is called a *zener diode.* When placed across the output of an unregulated power supply, a zener diode can maintain the voltage at its own breakdown voltage as long as the supply voltage is always above this point. Zener diodes can be obtained corresponding to voltages of a few volts to hundreds of volts.

A diode is called a *nonlinear device* since the current is not proportional to the voltage; that is, a graph of current versus voltage (Figure 19-21) is not a straight line as it is for a resistor (which *is* linear). Triodes and transistors are also *nonlinear* devices.

*Rectification

Since a *pn* junction diode allows current to flow only in one direction, it can serve as a rectifier—to change ac into dc. Several rectifier circuits are common, and can be designed for either semiconductor or vacuum-tube diodes.

A simple rectifier circuit is shown in Figure 19-23, using a semiconductor diode, but a vacuum-tube diode could also be used. The ac source applies a voltage across the diode that is alternately positive and negative. Only during the half-cycle when the voltage source can produce an upward current through the diode in the diagram will a current in fact flow, so only then is there a current through the resistor R. During the half-cycle when the voltage is reversed, no current flows through the diode. Hence the voltage V_{ab} across R as a function of time looks like that

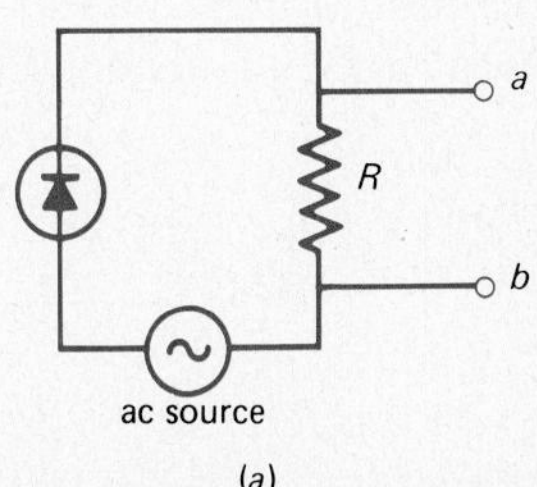

(*a*)

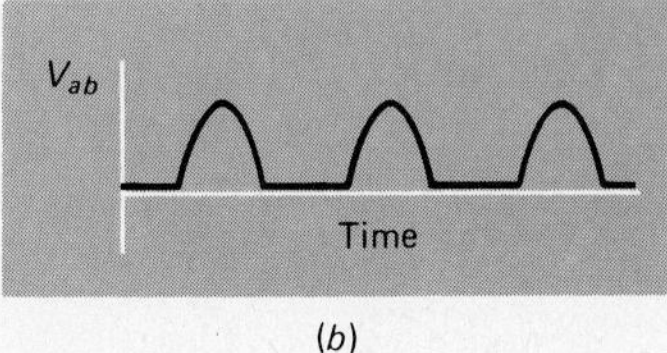

(*b*)

FIGURE 19-23
(*a*) Simple rectifier circuit, using a semiconductor diode. (*b*) Output voltage across *R* as a function of time.

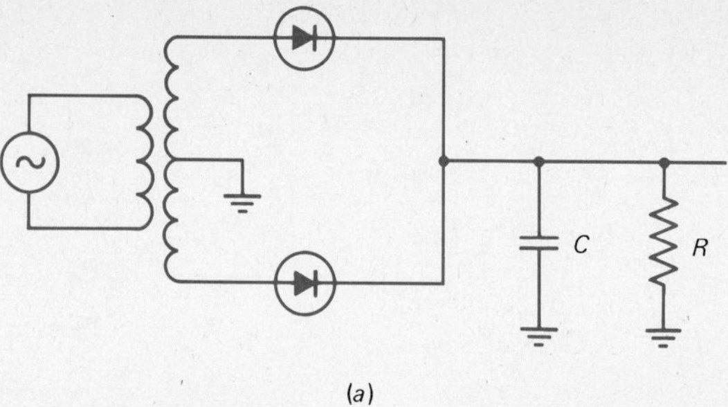

(a)

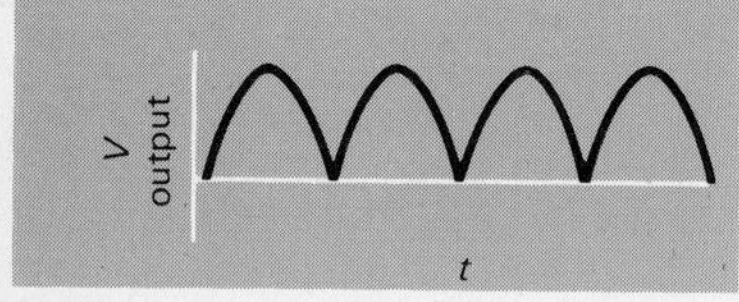

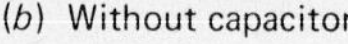

(b) Without capacitor

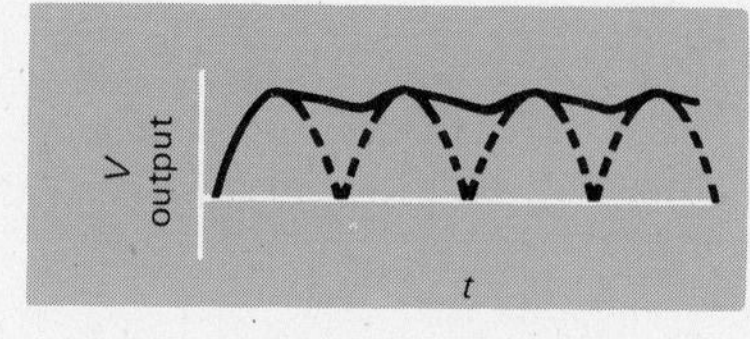

(c) With capacitor

FIGURE 19-24
(a) Full-wave rectifier circuit. (b) Output voltage in the absence of capacitor C. (c) Output voltage with capacitor in the circuit.

shown in Figure 19-23*b*. This is not exactly dc, but it is unidirectional or "pulsating dc." This is called *half-wave rectification.*

A *full-wave rectifier* circuit uses two diodes (or sometimes four), as shown in Figure 19-24*a*. The ac source is shown coupled to the rectifier through a transformer so that the magnitude of the voltage can also be changed. At any given instant, either one diode or the other will conduct a current to the right. Therefore, the output across the load resistor R will be as shown in Figure 19-24*b*; actually, this is the voltage if the capacitor C were not in the circuit. The capacitor tends to store charge, and thus helps to smooth out the current. This is shown in part *c*.

Rectifier circuits are important because most line voltage is ac, and other tubes and transistors require dc voltage for their operation. Hence diodes are found in nearly all electronic devices including radio and television sets, and today they are mainly semiconductor diodes.

*19-8 ▮ Transistors; integrated circuits

The *junction transistor*, developed at the Bell Telephone laboratories in 1948, consists of a crystal of one type of semiconductor sandwiched between two crystals of the opposite type. Transistors can be either *pnp* or *npn* (Figure 19-25*b*). The three semiconductors are given the names emitter, base, and collector. They play the same roles as the cathode, grid, and anode in a triode tube. The symbols for *npn* and *pnp* transistors are shown in Figure 19-25*c*. The arrow is always placed on the emitter and indicates the direction of (conventional) current flow in normal operation.

A transistor can operate like a triode. Consider an *npn* transistor connected as shown in Figure 19-26. A voltage V_{CE} is maintained between the collector and emitter by the battery ε_C. (This is equivalent to the anode voltage in a triode.) The voltage applied to the base is called the *base bias voltage,* V_{BE}: if it is positive, electrons in the emitter are attracted into the base. Since the base region is very thin (about 1 μm), most of the electrons

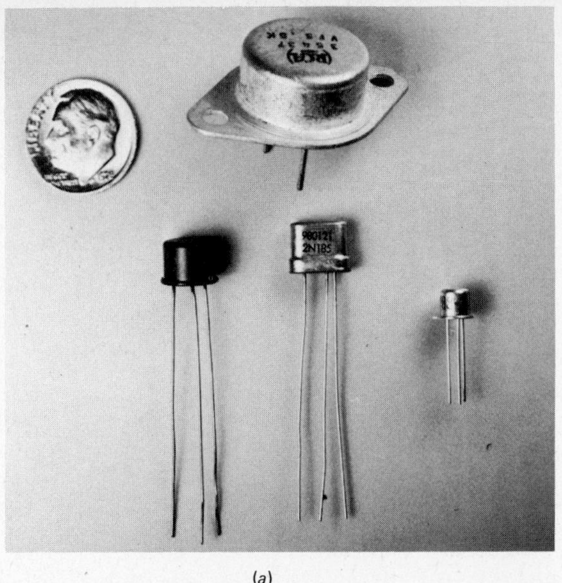

(a)

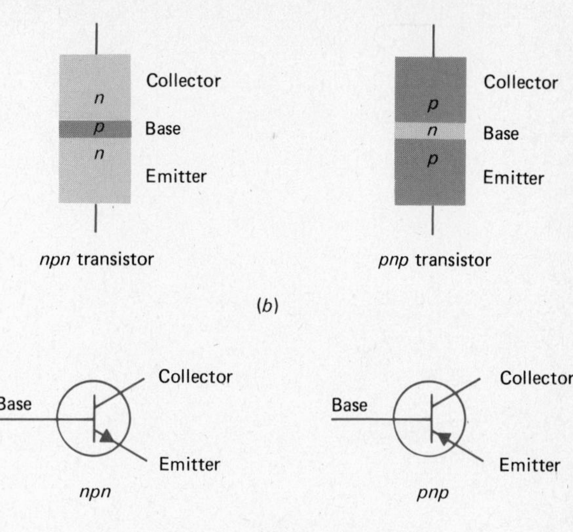

(c)

FIGURE 19-25
(*a*) Some typical transistors. (Courtesy of M. Lubin.) (*b*) Schematic diagram of *npn* and *pnp* transistors. (*c*) Symbols for *npn* and *pnp* transistors.

flow right across into the collector which is maintained at a positive voltage. A large current, I_C, flows to the collector, and a much smaller current, I_B, through the base. A small variation in the base voltage due to an input signal (an ac voltage or current) causes a large change in the collector current and therefore a large change in the voltage drop across the output resistor R_C. Hence a transistor can amplify a small signal into a large one.

A *pnp* transistor operates in the same fashion except that holes move instead of electrons. The collector voltage is negative and so is the base voltage in normal operation.

The operation of a transistor is commonly examined using graphs called *characteristic curves.* For example, it is common to plot the collector current I_C versus collector voltage V_{CE} for different values of the base current I_B. Such a graph is shown in Figure 19-27 for a typical *npn* transistor. As can be seen, a few microamps in the base corresponds to milliamps in the collector. If the base current is zero, there is no collector current. The greater the base current (caused by a greater base voltage), the greater the collector current will be. Normally, a small ac signal is to be amplified. Such a signal, when added to the base bias voltage, causes the voltage and current at the base to vary. Thus, what is important for amplification is the *change* in collector current for a given *change* in base

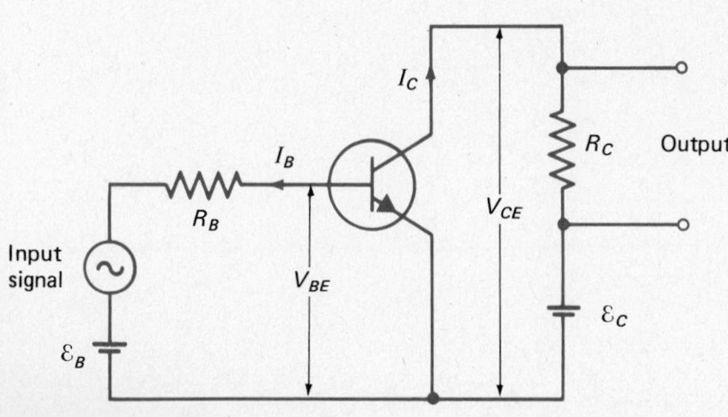

Figure 19-26
An *npn* transistor used as an amplifier.

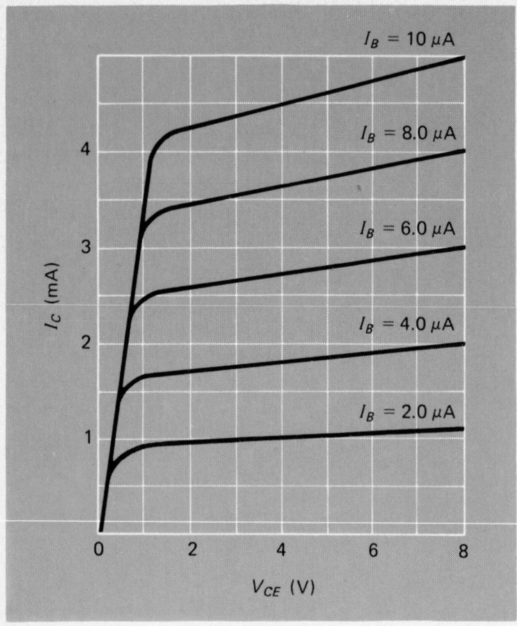

FIGURE 19-27
Typical transitor characteristics. Collector current I_C is plotted against collector voltage V_{CE} for various values of the base current I_B.

current. *Gain* is defined as the ratio output/input, and to see how to calculate it from a graph such as Figure 19-27, we take an example.

▮ EXAMPLE 19-5 Refer to Figure 19-27. Suppose that the base bias voltage is such that the base current is 6.0 μA when there is no input signal. Suppose also that the collector voltage, V_{CE}, is kept at 5.0 V. Then from the graph we see that the collector current is 2.8 mA. Now let an ac input signal, whose voltage and current we represent by v and i, be applied to the base. Suppose that i alternates from 2.0 μA to −2.0 μA. Calculate the current gain and voltage gain assuming that $R_B = 5.0\ \text{k}\Omega$ and $R_C = 1.0\ \text{k}\Omega$ (in Figure 19-26).

SOLUTION The total base current oscillates between 4.0 μA and 8.0 μA (6.0 ± 2.0 μA). From Figure 19-27 we see that if V_{CE} remains at 5.0 V, the collector current varies from 1.8 mA to 3.8 mA. That is, I_B varies ± 1.0 mA from its original value of 2.8 mA. Thus a 2.0-μA alternating current in the base causes a 1.0-mA alternating current in the collector. *The current gain β* is given by

$$\beta = \frac{\text{output ac current}}{\text{input ac current}} = \frac{1.0\times10^{-3}\ \text{A}}{2.0\times10^{-6}\ \text{A}} = 500.$$

The voltage, too, will be amplified. The input voltage v of the signal is related to the current i by $v = iR_B$. (The transistor offers little resistance, since the base–emitter junction is like a forward-biased diode.) $R_B = 5.0\ \text{k}\Omega$ and $R_C = 1.0\ \text{k}\Omega$; then the peak ac input voltage is $v = (2.0\times10^{-6}\ \text{A})(5.0\times10^{3}\ \Omega) = 10\ \text{mV}$; the peak ac output voltage is $(1.0\times10^{-3}\ \text{A})(1.0\times10^{3}\ \Omega) = 1.0\ \text{V}$. Thus the voltage gain is $(1.0\ \text{V})/(10\times10^{-3}\ \text{V}) = 100$. ▮▮

We assumed in this example that V_{CE} does not change. In more typical situations, a battery or other dc source ($\mathscr{E}_C$ in Figure 19-26) maintains a constant voltage across a resistor R_C and the transistor in series; so the voltage V_{CE} varies, depending on the current I_C that passes through R_C. This is dealt with in the Problems at the end of this chapter.

In Figure 19-26 two batteries are shown: $\mathscr{E}_C$ supplies the collector voltage and $\mathscr{E}_B$ the base bias voltage. In practice, only one source is usually used, and the base bias voltage is obtained using a resistance voltage divider as in Figure 19-28. Transistors can be connected in many other ways as well, and many new and innovative uses have been found for them.

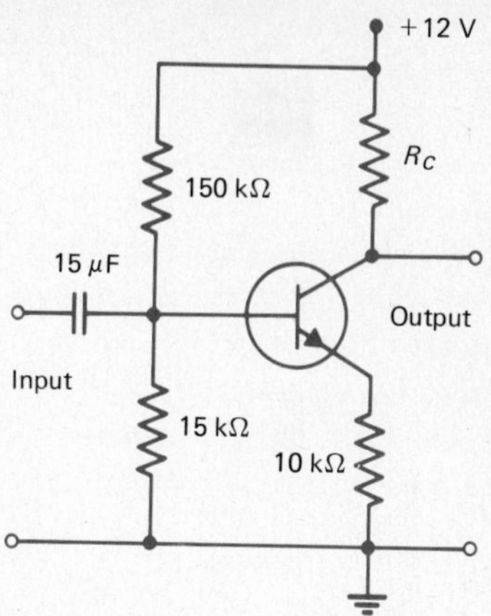

FIGURE 19-28
Typical transistor circuit involving an *npn* transistor. $\mathscr{E}_C = +12$ V and $\mathscr{E}_B$ is determined by the resistors.

Transistors were a great advance in miniaturization of electronic circuits. Although individual transistors are very small compared to vacuum tubes, they are huge compared to the latest development, *integrated circuits.* Tiny amounts of impurities can be placed at particular locations within a single silicon crystal. These can be arranged to form diodes, transistors, and resistors (which are simply undoped semiconductors). Capacitors and inductors can also be formed, although they are usually connected separately. A tiny "chip," as it is sometimes called, only 1 cm on a side, may contain over 1000 transistors (Figure 19-29). Integrated circuits are used today in many electronic devices.

*19-9 Instrumentation; amplifiers

FIGURE 19-29
Integrated circuit with fingers to show scale. (Courtesy of IBM.)

*Basic instrumentation

Except for the very simplest cases, the measurement of a physical quantity involves an instrument with several parts, as illustrated schematically in Figure 19-30. A very simple example is when you weigh yourself. In this case the springs inside the scale serve as a transducer that changes the force (weight) into a change in length. The *amplification* stage corresponds to the linkage connected to the rotating or moving dial. The *display* is the dial itself, which indicates your weight.

The same basic components are present in laboratory instruments that involve electronics rather than mechanical parts. The input transducer of Figure 19-30 typically converts the quantity to be measured, often referred to as the "signal," into an electrical signal. We have dealt with several types of transducers in earlier chapters. Pressure transducers, for example (Section 16-6), change a pressure into an electrical signal, and microphones do the same for a sound wave. A strain gauge puts out an electrical signal for very small changes in length. Temperatures are converted into electrical signals by resistance thermometers and thermistors. We will see in a later chapter how light levels are changed into an electrical signal using a phototube. These are just a few of the transducers that can be used.

The second stage, amplification, is usually necessary since the signals are often very small. We shall discuss amplifiers later in this section. The display unit is also a transducer, which changes the electrical signal into a

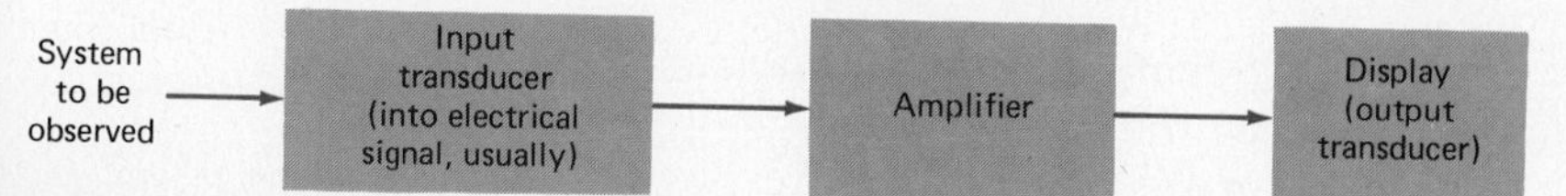

FIGURE 19-30
Block diagram of the observation and measurement of a quantity electronically.

usable form. It may be a simple meter such as a volt or ammeter. Or it may be a chart recorder, in which a pen connected to a galvanometer writes on a moving roll of paper; in this way a permanent record can be made. The display can also be on a cathode ray tube of an oscilloscope (Section 19-5); a permanent record can be obtained by photographing the trace on the tube face. Besides the amplifier, there may be additional electronics between the input and output stages for "processing" the signal (not indicated in Figure 19-30). This may involve adding or subtracting two or more signals. Or a signal may be amplified only if its voltage is above (or below) a certain level. Sometimes "logic" circuits are used; for example, a signal may be displayed only if a second signal from some other source is present.

*Basic amplifier

A *basic amplifier* is a device that takes a small input signal and makes it larger without significantly altering its shape or the frequencies present. We have already seen in Figures 19-11 and 19-26 how triodes and transistors can be used to amplify a signal. These devices, particularly transistors, can be designed and connected in a number of other ways so that proper impedance matching (Section 19-11) can be accomplished, or in some situations to reduce distortion or increase sensitivity. Most amplifiers contain several stages (several transistors or tubes), each of which multiply the output of the previous stage. Total gains of 10^3 to 10^6 are common. Signals of less than a millivolt can, for instance, be amplified to many volts.

Amplifiers are designed for particular purposes. There are *dc amplifiers* to amplify dc voltage; *voltage amplifiers*, which produce a large voltage gain but do not necessarily produce much power (output current is limited); *power amplifiers*, which maximize the power amplification, as in a radio receiver; *pulse amplifiers*, which are designed to respond accurately to a short pulse; *audio amplifiers*, which operate linearly in the frequency range of human hearing; and other amplifiers, which amplify signals with minimal distortion over particular frequency ranges.

*Differential amplifier

A *differential amplifier* is designed to amplify the *difference* between two signals. It is particularly useful in situations where background noise may be larger than the signals to be detected.

One example is in taking a simple electrocardiogram (ECG). Background noise radiated from lights and other electric devices in a room is usually larger than the ECG signals themselves. To overcome this problem, three ECG leads are used (and sometimes more). One is connected to each arm and the third to one leg; the last serves as a common ground. An ECG measures the potential difference between the two arms. If electrodes were connected only to the two arms, however, the potential difference would include all the noise, which would swamp the desired signal. This is why the third lead is needed. Let V_L represent the ECG signal of the left arm with respect to the leg (ground). Then the potential difference between the left arm and the leg is $V_L + \text{noise}$. Similarly, the potential difference between the right arm and ground is V_R + noise. The noise picked up by the two circuits is nearly the same (since they are close), so the difference in these two signals is $(V_L + \text{noise}) - (V_R + \text{noise}) = (V_L - V_R)$, which is the desired signal. The noise is thus subtracted out. Such an arrangement is used for many laboratory measurements of small signals.

*Other devices

An *oscillator* is a device that puts out a sine wave (or square wave) of particular frequency. One way of doing this is to use a resonant *LC* circuit (Section 19-3). Usually, the circuit contains a tube or transistor that supplies power to keep the oscillations from dying out.

Although batteries are used to supply power to operate a few devices, electronic *power supplies* are more common. A power supply is a basic part of radio and television receivers, for example. The main ingredient of a power supply is a transformer that changes the line voltage (from the wall outlet) to the desired voltage. If it is a dc power supply, the transformer is followed by a rectifier circuit using tube or semiconductor diodes. Line voltage varies, and if the power supply voltage must be accurately maintained at a particular value, it must be "regulated." One way of doing this is to use a zener diode, as mentioned in Section 19-7. Another way is to use a feedback circuit (see the next section).

*19-10 ▍ Feedback and control

In Chapter 7 we discussed the ideas of feedback and control. Much of the terminology used in those more general systems derives from the analysis of feedback in electronic circuits.

In a simple amplifier without feedback, the output voltage V_o is related to the input voltage V_i by the amplification or gain A:

$$V_o = AV_i.$$

Feedback is introduced by "feeding back" a portion of the output to the input, as shown in Figure 19-31. If a fraction β of the output (β is usually much less than 1) is fed back to the input, the voltage at the input terminals is then

$$V_i' = V_i + \beta V_o.$$

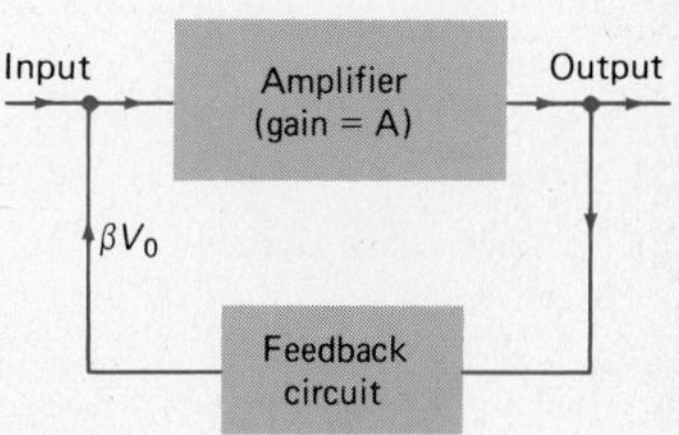

FIGURE 19-31
Feedback amplifier.

Thus we have $V_o = AV_i' = A(V_i + \beta V_o)$. We rearrange this equation to find V_o:

$$V_o - A\beta V_o = AV_i$$

$$V_o = \frac{A}{1 - \beta A} V_i.$$

Thus the net gain, A_f, of a feedback amplifier is

$$A_f = \frac{A}{1 - \beta A}. \tag{19-9}$$

If β is positive, the feedback increases the gain; this is referred to as *positive* feedback. Positive feedback, because it causes an increasing signal to increase even further, tends to lead to instability and runaway conditions (this was discussed in Chapter 7). However, it does find a few uses, such as in oscillators.

More widely used is *negative* feedback, in which β is negative. This implies that the feedback is out of phase with the input (negative when the input is positive, or vice versa). As can be seen from Equation 19-9, negative feedback reduces the gain of an amplifier. This reduction in gain is compensated by an important advantage. A problem with any amplifier is that the characteristics of the tubes or transistors change if the supply voltage or the temperature changes. This alters the gain of the amplifier and also increases distortion.† Negative feedback acts to stabilize an amplifier against these outside influences. For example, suppose that the gain of an amplifier increases due to temperature or supply-voltage variation. The output signal then increases; but part of this is fed back to the input, and since it is out of phase, the net input is thereby reduced. This, in turn, reduces the output toward its normal level.

▮ EXAMPLE 19-6 To see how negative feedback brings stability, consider an amplifier for which $A = 2.0 \times 10^4$ and $\beta = -1.0 \times 10^{-3}$. Suppose conditions change so that A increases by 10 percent, to 2.2×10^4. Calculate the change in A_f.

SOLUTION Before the alteration, Equation 19-9 gives us

$$A_f = \frac{2.0 \times 10^4}{1+20} = 952.$$

When A changes to 2200, A_f becomes

$$A_f = \frac{2.2 \times 10^4}{1+22} = 957,$$

which is less than a 1 percent effect on the system (in fact, $5/952 = 0.5$ percent). ▮▮

If $-\beta A \gg 1$, then from Equation 19-9 we see that

$$A_f \approx -\frac{1}{\beta}. \tag{19-10}$$

This has the advantage that the gain is controlled by the feedback circuit. Feedback circuits usually consist only of resistors and/or capacitors, which are much less sensitive to external influence than are transistors or tubes. Thus the net gain remains very stable.

Negative feedback is used to *control* an amplifier and keep its operation stable. Indeed, negative feedback is used to control a system in many other applications and is referred to as "automatic control." It was first applied to mechanical systems, such as the centrifugal governor of a steam engine (Chapter 7). However, the theory of automatic control systems developed rapidly only after the invention of the feedback amplifier in the 1920s, and particularly since World War II. The general

† Distortion means the output signal differs in shape (that is, the relative amplitude of the frequencies present) from the input signal and thus does not accurately reflect the input.

application of control theory is called *cybernetics*. This term was coined by Norbert Wiener (1894–1964) in 1947, who applied the ideas to a wide range of fields. He was particularly inspired by the many examples of feedback and control found in the human organism. We have already discussed (Chapter 7) some examples of how feedback is used to control body functions. For instance, the light intensity reaching the retina of the eye is controlled by feeding back a portion of it to the muscles that control the pupil size.

Control is achieved in each case by feeding back a part of the output to the input. The techniques of automatic control are widely used in industry. A thermostat (Section 7-3) is a simple example. In many situations a *servomechanism* operates. Some kind of sensor detects the output of a system and sends a signal to run a motor, called a "slave" or "servomotor," that controls the input. For example, at a steel mill, a device senses the thickness of steel sheets formed by passing through a set of rollers; if the thickness varies slightly, a signal is sent to a servomotor that changes the separation of the rollers.

*19-11 ▮ Impedance matching

It is common to connect one electric circuit to a second circuit. For example, a television antenna is connected to a television set; an FM tuner is connected to an amplifier; the output of an amplifier is connected to a speaker; electrodes for an ECG or EEG are connected to an amplifier or a recorder. In many cases it is important that the maximum power be transferred from one to the other, with a minimum of loss. This can be achieved when the output impedance of the one device matches the input impedance of the second.

To show why this is true, we consider (for simplicity) circuits that contain only resistance. In Figure 19-32 the source in circuit A could represent a power supply, the output of an amplifier, or the signal from an antenna, a laboratory probe, or a set of electrodes. R represents the resistance of this device and includes the internal resistance of the source. R is called the output impedance (or resistance) of circuit A. The output of circuit A is across the terminals x and y, which are connected to the input of circuit B. Circuit B may be very complicated. By combining the various resistors, we can find an equivalent resistance. This is represented by R_{in}, the "input resistance" of circuit B.

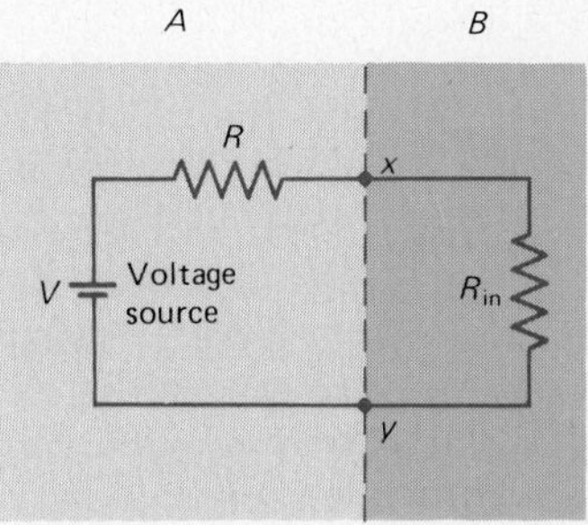

FIGURE 19-32
Output of the circuit on the left is input to the circuit on the right.

The power delivered to circuit B is $P = I^2 R_{in}$, where $I = V/(R + R_{in})$. Thus

$$P = I^2 R_{in} = \frac{V^2 R_{in}}{(R + R_{in})^2}.$$

We divide the top and bottom of the right side by R and find that

$$P = \frac{V^2}{R} \frac{R_{in}/R}{(1 + R_{in}/R)^2}. \qquad (19\text{-}11)$$

The question is: If the resistance of the source is R, what value should R_{in} have so that the maximum power is transferred to circuit B? To determine this we plot a graph of P versus (R_{in}/R). This is shown in Figure 19-33, where representative values are given in the table. For example, for $R_{in}/R = 1$, Equation 19-11 gives $P = V^2/4R$; for $R_{in}/R = 3$, $P = 3V^2/16 = 0.19\ V^2/R$; and so on.

As can be seen from the graph, P is a maximum when $R_{in} = R$. That is, when the *output impedance* of one device *equals the input impedance* of the second. This is called *matching the impedances.*

In an ac circuit that contains capacitors and inductors, the different phases are important and the analysis is more complicated. However, the same result holds: to maximize power transfer it is important to match impedances. In addition, one must be aware that it is possible to seriously distort a signal. For example, when a second circuit is connected, it may put the first circuit into resonance, or take it out of resonance.

Without proper consideration of the impedances involved, one can make measurements that are completely meaningless. These considerations are normally examined by engineers when producing an integrated set of apparatus. However, there have been cases when researchers connected several components together without regard for impedance matching, and made a "new discovery." Then, after their announcement of the "discovery," it was found that what they had observed was due to impedance mismatch rather than the natural phenomenon they had thought!

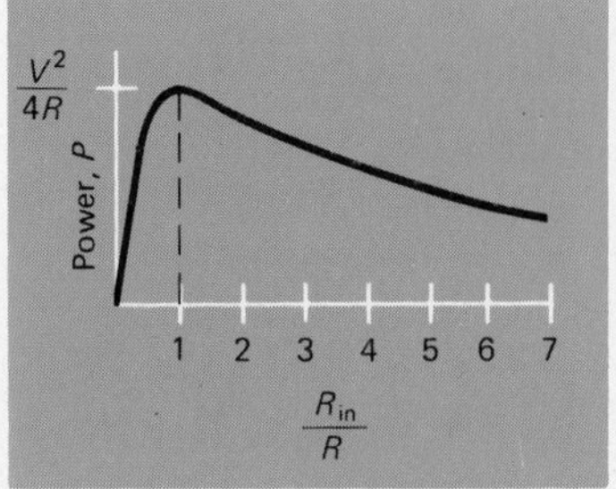

$\frac{R_{in}}{R}$	P
0	0
0.5	$0.22\ V^2/R$
1.0	$0.25\ V^2/R$
2.0	$0.22\ V^2/R$
5.0	$0.14\ V^2/R$
100	$0.01\ V^2/R$

FIGURE 19-33
Power transferred is at its maximum when $R_{in} = R$.

In some cases a transformer is used to alter an impedance. If Z_s is the secondary impedance and Z_p the primary impedance, then $V_s = I_sZ_s$ and $V_p = I_pZ_p$. Hence

$$\frac{Z_p}{Z_s} = \frac{V_pI_s}{V_sI_p} = \left(\frac{N_p}{N_s}\right)^2,$$

where we have used the transformer equations 18-6 and 18-7. Thus the impedance can be changed with a transformer. A transformer is used for this purpose in some stereo amplifiers; often it has several taps, corresponding to 4, 8, and 16 Ω, so that the output can be matched to the impedance of any loudspeaker.

Some instruments, such as oscilloscopes, require only a signal voltage, but no power. Maximum power transfer is not then important and such instruments can have a high input impedance. This has the advantage that the instrument draws very little current and disturbs the original circuit as little as possible. This is often the situation in a laboratory experiment.

▮ Summary

Capacitance and inductance offer impedance to the flow of alternating current just as resistance does. This impedance is referred to as reactance, X. For capacitance and inductance the reactance is defined as the equivalent of R in Ohm's law: as the proportionality constant between voltage and current (either the rms or peak values); across a capacitor, $V = IX_C$, and across an inductor, $V = IX_L$. The reactance of an inductor increases with frequency: $X_L = 2\pi fL$; the reactance of a capacitor decreases with frequency: $X_C = 1/(2\pi fC)$. Whereas the current through a resistor is always in phase with the voltage across it, this is not true for inductors and capacitors: in an inductor, the current lags the voltage by 90°, and in a capacitor the current leads the voltage by 90°.

In an LRC series circuit, the total impedance Z is defined by the equivalent of Ohm's law, $V = IZ$; it is related to R, C, and L by $Z = \sqrt{R^2 + (2\pi fL - 1/2\pi fC)^2}$. The current in the circuit lags the source voltage by an angle ϕ given by $\cos\phi = R/Z$. Only the resistor in an LRC circuit dissipates energy, and at a rate $P = I^2_{\text{rms}}R = I^2_{\text{rms}}Z\cos\phi$.

An LRC series circuit resonates at a frequency $f_0 = (1/2\pi)\sqrt{1/LC}$. The rms current in the circuit is largest when the applied voltage has a frequency equal to f_0. The lower the resistance R, the higher and sharper the resonance peak.

When a metal is heated, electrons are "boiled off" its surface; this is called thermionic emission and is the basis of operation of vacuum-tube diodes and triodes. Diodes, which contain two electrodes, allow current to flow in one direction only and can be used to change ac to dc. Triodes have three electrodes; one of these, the grid, controls the flow of current across the tube, so the tube can be used to amplify electrical signals.

A cathode ray tube (CRT) is used as the picture tube in television sets and in oscilloscopes. Thermionically emitted electrons are accelerated by a high voltage within the tube; the resulting electron beam strikes the fluorescent screen on the inside face of the tube, which glows at that point. The electron beam is made to move about the screen by electric deflecting plates or magnetic deflecting coils.

Semiconductor devices make use of two types of doped semiconductors: n-type, which contain impurities with extra electrons that can move rather freely; and p-type which contain impurities with fewer than the normal number of electrons so that positively charged "holes" can move about. A semiconductor diode consists of a pn-junction and allows current to flow in one direction only; it can be used as a rectifier to change ac to dc.

Common transistors consist of three semiconductor sections, either as pnp or npn. Transistors can amplify electrical signals and find many other uses. An integrated circuit consists of a tiny semiconductor crystal or "chip" on which many transistors, diodes, resistors, and other circuit elements have been constructed using careful placement of impurities.

Amplifiers use several transistors or vacuum tubes to amplify a signal many times. Negative feedback is used in amplifiers to stabilize them against changes in amplification and distortion due to supply voltage or temperature changes. When one circuit is connected to a second circuit, it is important that their impedances match so that losses and distortion are kept to a minimum.

Questions

1 Under what conditions is the impedance in an LRC circuit a minimum?

2 An LC resonance circuit is often called an *oscillator* circuit. What is it that oscillates?

3 Compare the oscillations of an LC circuit to the vibration of a mass m on a spring. What do L and C correspond to in the mechanical system?

4 Describe how a vacuum tube diode acts like a valve for the flow of electric current. Do the same for a triode and discuss differences.

5 Explain how a triode can be used to rectify as well as amplify a signal by setting the grid bias voltage near the cutoff point.

6 Bringing a magnet close to a television screen will distort the picture. Why? (Do not do this to a color set; permanent damage may result.)

7 The horizontal sweep of an oscilloscope can be produced using the sawtooth voltage of an RC circuit as discussed in Section 16-7. How would you alter such a circuit so that the sweep rate (the time axis) is faster? Slower?

8 A silicon semiconductor is doped wih phosphorus. Will these atoms be donors or acceptors? What type of semiconductor will this be?

9 Explain how a transistor could be used as a switch.

10 Describe how a *pnp* transistor can operate as an amplifier.

11 Discuss how the base of a transistor is like the grid of a triode.

12 Can a diode be used to amplify a signal?

13 Figure 19-34 shows a "bridge-type" full-wave rectifier. Explain how the current is rectified and how current flows during each half cycle.

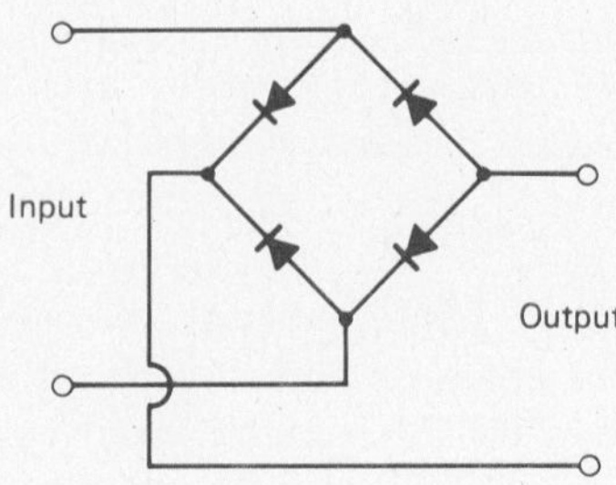

FIGURE 19-34

14 If $\mathscr{E}_C$ were reversed in Figure 19-26, how would the amplification be altered?

15 Compare the resistance of a *pn* junction diode when connected in forward bias to when connected in reverse bias.

16 In a transistor, the base–emitter junction and the base–collector junction are basically diodes. Are these junctions reverse-biased or forward-biased in the application shown in Figure 19-26?

17 What purpose does the capacitor in Figure 19-28 serve?

Problems

*SECTION 19-1

*1 (I) At what frequency will a 200-mH inductor have a reactance of 470 Ω?

*2 (I) At what frequency will a 4.0-μF capacitor have a reactance of 2.2 kΩ?

*3 (I) Plot a graph of the impedance of a 1.0 μF capacitor as a function of frequency from 10 to 1000 Hz.

*4 (I) Plot a graph of the imedance of a 1.0-mH inductor as a function of frequency from 100 to 10,000 Hz.

*5 (II) Calculate the impedance of, and rms current in, a 60-mH radio coil connected to a 400-V (rms) 3.3-kHz ac line. Ignore resistance.

*6 (II) What is the inductance of a coil whose input is 110 V at 60 Hz and the current drawn is 1.2 A?

*7 (II) (*a*) What is the impedance of a well-insulated 0.020-μF capacitor connected to a 1.0-kV (rms) 200-Hz line? (*b*) What will be the peak value of the current?

*8 (II) A current $I = 4.4 \cos 377t$ flows in an LR circuit in which $L = 0.50$ mH and $R = 60\ \Omega$. What is the average power dissipation?

*9 (II) A capacitor is placed in parallel across a load as in Figure 19-4*b* to filter out stray high-frequency signals, but to allow ordinary 60-Hz ac to pass through with little loss. Suppose that circuit B in the figure is a resistance $R = 100\ \Omega$ connected to ground, and that $C = 0.50\ \mu$F. What percent of the incoming current will pass through C rather than R if (*a*) it is 60 Hz? (*b*) 60,000 Hz?

*10 (II) Suppose that circuit B in Figure 19-4*a* is a resistance $R = 100\ \Omega$, connected to ground, and the capacitance $C = 1.0\ \mu$F. Will this capacitor act to eliminate 60 Hz ac but pass a high-frequency signal of frequency 60,000 Hz? To check this, determine the voltage drop across R for a 30-mV signal of frequency (*a*) 60 Hz, and (*b*) 60,000 Hz.

*11 (II) A current of 0.02 mA has been known to cause heart fibrillation (irregular pumping) in a patient with implanted electrodes. (*a*) What value of leakage capacitance would this correspond to for a 60-Hz 120-V line? (*b*) If the conductors were, say, 2 mm apart, how large an area would each have to have? Is this unreasonable?

* SECTION 19-2

*12 (I) A 1.0-kΩ resistor and a 2.0-μF capacitor are connected in series to an ac source. Calculate the impedance of the circuit if the source frequency is (*a*) 60 Hz, and (*b*) 1.0×10^4 Hz.

*13 (I) A 10-kΩ resistor is in series with a 0.10-mH inductor and an ac source. Calculate the impedance of the circuit if the source frequency is (*a*) 60 Hz, and (*b*) 1.0×10^4 Hz.

*14 (I) For a 120-V rms 60-Hz voltage, a current of 70 mA passing through the body for 1 s could be lethal. What must be the impedance of the body for this to occur?

*15 (II) (*a*) What is the rms current in an *RC* circuit if $R = 4.7$ kΩ, $C = 0.10$ μF, and the rms applied voltage is 120 V at 60 Hz? (*b*) What is the phase angle between voltage and current? (*c*) What is the power dissipated by the circuit? (*d*) What are the voltmeter readings across *R* and *C*?

*16 (II) (*a*) What is the rms current in an *RL* circuit when a 60-Hz 120-V rms ac voltage is applied, where $R = 250$ Ω and $L = 500$ mH? (*b*) What is the phase angle between voltage and current? (*c*) How much power is dissipated? (*d*) What are the rms voltage readings across *R* and *L*?

*17 (II) A voltage $V = 2.8 \sin 754t$ is applied to an *LRC* circuit. If $L = 2.0$ mH, $R = 1.0$ kΩ, and $C = 3.0$ μF, how much power is dissipated in the circuit?

*18 (II) A circuit contains two elements, but it is not known if they are *L*, *R*, or *C*. The current in this circuit when connected to a 120-V 60-Hz source is 6.4 A and lags the voltage by 30°. What are the two elements and what are their values?

*19 (II) A 50-mH inductor with a 2.0-Ω resistance is connected in series to a 200-μF capacitor and a 60-Hz 16-V source. Calculate for this circuit the (*a*) rms current, (*b*) phase angle, and (*c*) power dissipated.

*20 (II) A 30-mH coil whose resistance is 6.0 Ω is connected to a capacitor *C* and a 360-Hz source voltage. If the current and voltage are to be in phase, what value must *C* have?

*21 (II) What is the total impedance, phase angle, and rms current in an *LRC* circuit connected to a 2.0-kHz, 300-V (rms) source if $L = 3.0$ mH, $R = 1.1$ kΩ, and $C = 2000$ pF?

*22 (II) What is the resistance of a coil if its impedance is 25 Ω and its reactance is 20 Ω?

*23 (II) Show that for the *LRC* circuit of Figure 19-5, if we have $I = I_0 \cos \omega t$, then $V_R = I_0 R \cos \omega t$, $V_L = I_0 \omega L \cos (\omega t + \pi/2)$, and $V_C = (I_0/\omega C) \cos (\omega t - \pi/2)$, where $\omega = 2\pi f$.

*24 (III) An inductance coil draws 1.5 A dc when connected to a 36-V battery. When connected to a 60-Hz 120-V source, the current drawn is 2.8 A (rms). Determine the inductance and resistance of the coil.

* SECTION 19-3

*25 (I) A 200-pF capacitor is connected to a 60-μH coil of resistance 2.0 Ω. What is the resonant frequency of this circuit?

*26 (I) What is the resonant frequency of the *LRC* circuit of Example 19-3?

*27 (I) What value must *C* have if $L = 30$ mH and $R = 30$ Ω in a circuit that is to resonate at 830 kHz?

*28 (II) An *LRC* circuit has $L = 2.5$ mH and $R = 20$ Ω. (*a*) What value must *C* have to produce resonance at 40 kHz? (*b*) What will be the maximum current at resonance if the peak external voltage is 36 V?

*29 (II) What will be the peak current in Problem 28 if the capacitor is chosen so that the resonant frequency is twice the applied frequency of 40 kHz?

*30 (III) The *Q factor* of a resonant circuit is defined as the ratio of the voltage across the capacitor (or inductor) to the voltage across the resistor, at resonance. The larger the *Q* factor, the sharper the resonance curve will be and the sharper the tuning. (*a*) Show that the *Q* factor is given by the equation $Q = (1/R)\sqrt{L/C}$. (*b*) At a resonant frequency $f_0 = 1.0$ MHz, what must be the value of *L* and *R* to produce a *Q* factor of 1000; assume that $C = 0.010$ μF. (*c*) What is the *Q* value of the circuit in Example 19-3?

*31 (III) A resonant circuit using a 12-pF capacitor is to resonate at 8.0 MHz. The air-core inductor is to be a solenoid with closely packed coils made from 20 m of insulated wire 1.0 mm in diameter. How many loops will the inductor contain?

*SECTION 19-4

*32 (I) Determine the average speed of a free electron in a metal (*a*) at 300 K, and (*b*) at 2500 K (the typical temperature of the cathode in a tube). Assume the electrons behave like an ideal gas.

*33 (I) The *transconductance* of a triode is defined as the change in plate current per unit change in grid voltage at constant plate voltage. Determine the transconductance for the triode described by Figure 19-12 at $V_p = 200$ V in the linear region of the graph.

*34 (II) Suppose that the plate voltage for the triode of Figure 19-12 is 200 V and the grid bias is −5.0 V. Estimate from the graph what the ac rms output voltage will be for a 100-mV (rms) signal. Assume the output resistance is 82 kΩ.

* SECTION 19-5

*35 (II) Electrons in a CRT are accelerated by 20 kV and then pass through a 0.0250-T magnetic field. Calculate the magnetic force on the electrons.

*36 (II) What would be the approximate maximum deflections of the electron beam near the center of a television screen due to the earth's 5.0×10^{-5} T field? Assume that the screen is 20 cm from the electron gun, where the electrons are accelerated (*a*) by 1.0 kV, and (*b*) by 30 kV. Note that in color television sets, the beam must be directed accurately to less than 1 mm in order to strike the correct phosphor. Because the earth's field is significant here, mu-metal shields are used to reduce the earth's field in the CRT.

*37 (III) Electrons are accelerated by 5.0 kV in a CRT. The screen is 10 cm wide and is 25 cm from the 3.0-cm-long deflection plates. Over what range must the horizontally deflecting electric field vary to sweep the beam fully across the screen?

*38 (III) In a given CRT, electrons are accelerated horizontally by 30 kV. They then pass through a uniform magnetic field B for a distance of 4.0 cm, which deflects them upward so that they reach the top of the screen 25 cm away, 13 cm above the center. Estimate the value of B.

* SECTION 19-7

*39 (I) A silicon diode whose current–voltage characteristics are given in Figure 19-21 is connected in series with a battery and a 200-Ω resistor. What battery voltage is needed to produce a 10-mA current?

*40 (II) An ac voltage of 240 V rms is to be rectified. Determine the average current in the output resistor R (15 kΩ) for (*a*) a half-wave rectifier (Figure 19-23), and (*b*) a full-wave rectifier (Figure 19-24) without capacitor.

*41 (III) A 240 V rms 60-Hz voltage is to be rectified with a full-wave rectifier as in Figure 19-24, where $R = 15$ kΩ and $C = 30$ μF. (*a*) what will be the average current (approximately)? (*b*) What if $C = 0.10$ μF? (*Hint*: See Section 16-7.)

*42 (III) A full-wave rectifier (Figure 19-24) uses two diodes to rectify a 74-V (rms) ac voltage. If $R = 15$ kΩ and $C = 30$ μF, what will be the approximate percent variation in the output voltage? This is called *ripple voltage* (see Figure 19-24*c*). (*Hint*: See Section 16-7 and assume that the discharge of the capacitor is approximately linear.)

*43 (III) Suppose that the diode of Figure 19-21 is connected in series to a 100-Ω resistor and a 2.0-V battery. What current flows in the circuit? (*Hint*: Draw a line on Figure 19-21 representing the current in the resistor as a function of the voltage across the diode; the intersection of this line with the characteristic curve will give the answer.)

*44 (III) A zener diode voltage regulator is shown in Figure 19-35. Suppose that $R = 2.0$ kΩ and that the diode breakdown voltage is 130 V; the diode is rated at a maximum current of 100 mA. (*a*) If $R_{load} = 15$ kΩ, over what range of supply voltages will the circuit maintain the voltage at 130 V? (*b*) If the supply voltage is 200 V, over what range of load resistance will the voltage be regulated?

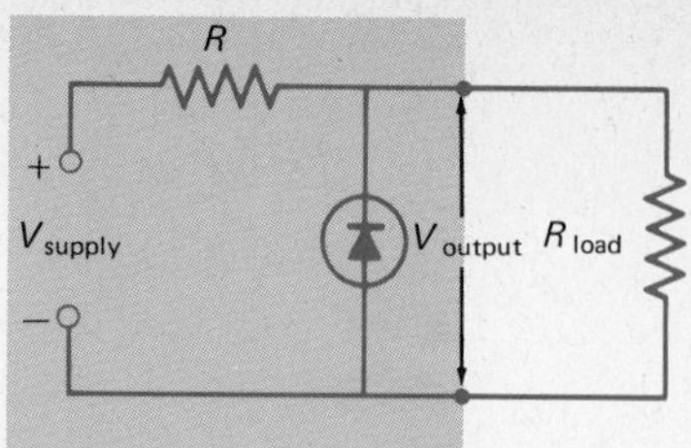

FIGURE 19-35

*45 (III) A silicon diode passes significant current only if the forward bias voltage exceeds about 0.6 V. Determine the average current in the output resistor R of (*a*) a half-wave rectifier (Figure 19-23), and (*b*) a full-wave rectifier (Figure 19-24) without a capacitor. Assume that $R = 200$ Ω in each case and that the ac voltage is 10.0 V (rms) in each case.

* SECTION 19-8

*46 (I) What collector current results if the collector voltage on the transistor of Figure 19-27 is 5.0 V and the base current is 5.0 μA?

*47 (I) What base current will produce a collector current of 3.5 mA for a collector voltage of 6.0 V for the transistor of Figure 19-27?

*48 (II) Suppose that the current gain of the transistor in Figure 19-26 is $\beta = 100$. If $R_C = 3.0$ kΩ, calculate the output voltage for a time-varying input current of 2.0 μA.

*49 (II) If the current gain of the transistor amplifier in Figure 19-26 is $\beta = 60$, what value must R_C have if a 1-μA base current is to produce an output voltage of 0.4 V?

*50 (II) A transistor, whose current gain $\beta = 80$, is connected as in Figure 19-26 with $R_B = 2.0$ kΩ and $R_C = 6.5$ kΩ. Calculate (*a*) the voltage gain, and (*b*) the power amplification.

*51 (II) An amplifier has a voltage gain of 90 and a 10-kΩ load (output) resistance. What is the peak output current through the load resistor if the input voltage is an ac signal with a peak of 0.080 V?

*52 (III) Suppose that the transistor represented by the curves in Figure 19-27 is connected as shown in Figure 19-26 with $R_C = 1.0$ kΩ, $R_B = 4.5$ kΩ, and $\mathscr{E}_C = 5.0$ V. (*a*) Show that $V_{CE} = 5.0\ \text{V} - I_C\ (10^3\ \Omega)$. (*b*) Draw a straight line representing this equation on the graph of Figure 19-27. (*Hint*: Note that $V_{CE} = 5.0$ V when $I_C = 0$ and that $I_C = 5$ mA when $V_{CE} = 0$.) This is called the *load line*; its intersection with the curve corresponding to the bias base current I_B is called the *operating point* (see Figure 19-36). (*c*) I_C and V_{CE} must satisfy this equation and therefore must lie on the line just drawn as well as on the proper characteristic curves. Use this fact to calculate the variation in I_C when I_B varies from 4.0 μA to

where $\Phi_E = EA$ is the *electric flux*, defined in analogy to magnetic flux. Then, Ampère's law becomes

$$\sum B_\parallel \Delta l = \mu_0 I_C + \mu_0 \varepsilon_0 \frac{\Delta \Phi_E}{\Delta t}. \tag{20-1}$$

This equation embodies Maxwell's idea that a magnetic field can be caused not only by a normal electric current, but also by a changing electric field or changing electric flux. Equation 20-1 is essentially Maxwell's fourth equation. (Actually, there is a third term on the right for the case when a magnetic field is produced by magnetized objects; but we assume in what follows that no magnets are present.)

20-2 ▌ Production of electromagnetic waves

According to Maxwell, we expect that a magnetic field will be produced in empty space if there is a changing electric field. From this, Maxwell derived another startling conclusion. If a changing magnetic field produces an electric field, the electric field will itself be changing. This changing electric field will, in turn, produce a magnetic field. The latter will be changing and so will produce a changing electric field; and so on. When Maxwell manipulated his equations, he found that the net result of these interacting changing fields was to produce a wave of electric and magnetic fields that can actually propagate (travel) through space. We now examine, in a simplified way, how such **electromagnetic waves** are produced.

Consider two conducting rods that will serve as an "antenna" (Figure 20-3*a*). Suppose that these two rods are connected by a switch to the opposite terminals of a battery. As soon as the switch is closed, the upper rod quickly becomes positively charged and the lower one negatively charged. Electric field lines are formed as indicated by the lines in Figure 20-3*b*. While the charges are flowing a current exists; the direction of conventional current is indicated by the arrows. A magnetic field is therefore produced; the magnetic field lines encircle the wires and therefore, in the plane of the paper, **B** points into the paper (⊗) on the right and out of the paper (⊙) on the left. Now we ask, how far out do these fields extend? In the static case, the fields extend outward indefinitely far. However, when the switch in Figure 20-3 is closed, the

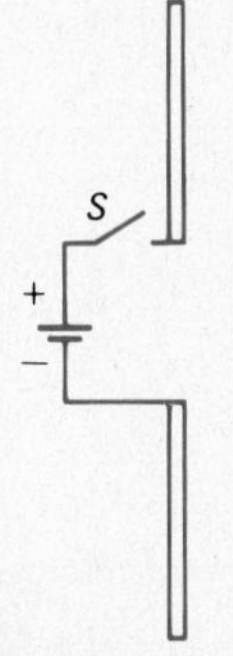

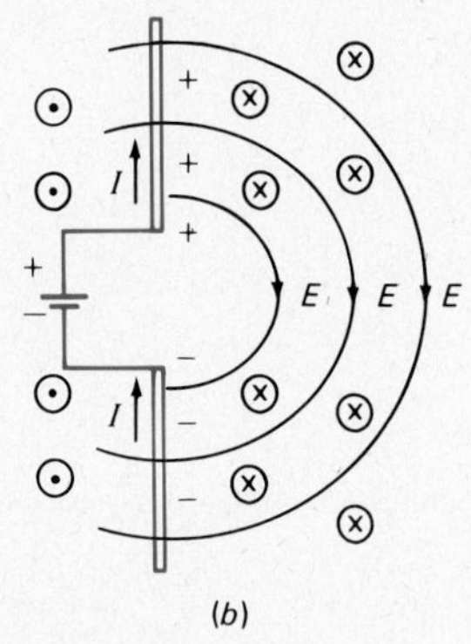

FIGURE 20-3
Fields produced by charge flowing into conductors. It takes time for the *E* and *B* fields to travel outward to distant points.

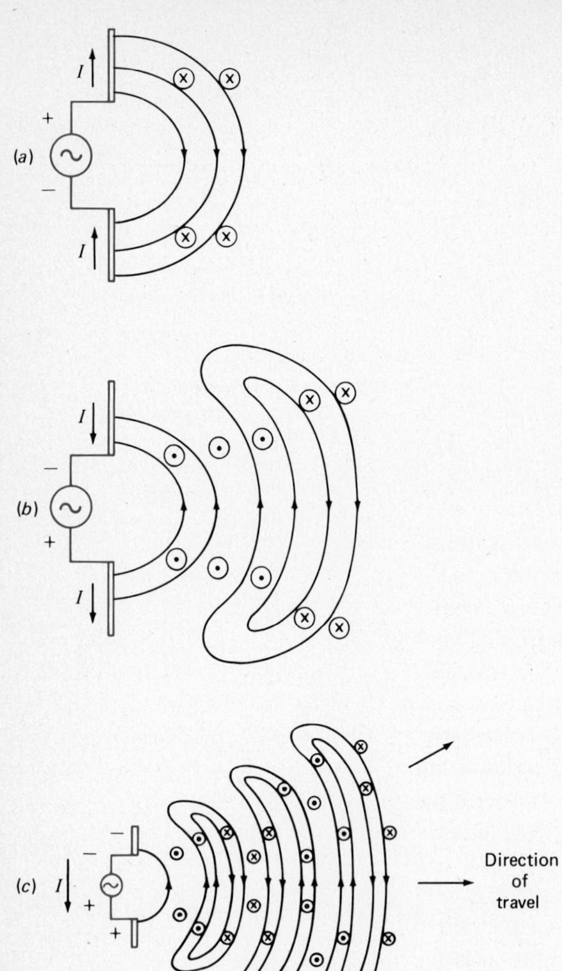

FIGURE 20-4
Sequence showing electric and magnetic fields that spread outward from oscillating charges on two conductors connected to an ac source (see the text).

fields quickly appear nearby, but it takes time for them to reach distant points. Both electric and magnetic fields store energy, and this energy cannot be transferred to distant points at infinite speed.

This example illustrates that it takes time for electric and magnetic fields, once they are formed, to reach distant points. Now we look at a more interesting situation. Suppose that the two vertical rods are connected to an ac generator (Figure 20-4). In part *a*, the connection has just been completed. Charge starts building up and fields form just as in Figure 20-3. The + and − signs indicate the net charge on each rod. The arrows indicate the direction of the current (upward). The electric field is represented by lines in the plane of the paper; and the magnetic field, according to the right-hand rule, is into (⊗) or out of (⊙) the paper, as shown. In part *b*, the alternating emf has changed sign; the current is downward, and charge builds up with the opposite sign. New fields are produced, which point in the opposite direction. The old fields, however, do not suddenly disappear; they are on their way to distant points. But because the new fields have changed direction, the old lines fold back to

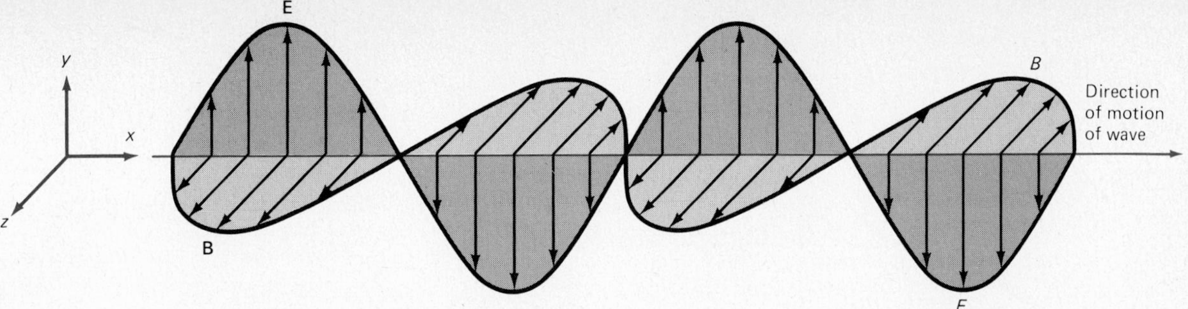

FIGURE 20-5
Electric and magnetic field strengths in an electromagnetic wave. E and B are at right angles to each other. The entire pattern moves in a direction perpendicular to both E and B.

connect up to some of the new lines and form closed loops as shown. In part *c* the situation is shown a little later when the field lines have formed several loops (note the change of scale). These field lines continue moving outward. The magnetic field lines also form closed loops; they are not shown since they are perpendicular to the paper.

The fields continue to move outward. Although the lines are shown only on the right of the source, fields also travel outward in other directions. (The field strengths are greatest in directions perpendicular to the oscillating changes; and they drop to zero along the direction of oscillation—above and below the antenna in Figure 20-4.)

Several things can be noted from Figure 20-4. First, *the electric and magnetic fields at any point are perpendicular to each other and to the direction of motion.* Second, we can see that the fields alternate in direction (**B** is into the paper at some points and out of the paper at others; similarly for **E**). Thus the field strengths vary from a maximum in one direction to zero to a maximum in the other direction.

If the source emf varies sinusoidally, the electric and magnetic field strengths far from the source also vary sinusoidally. This variation cannot be seen in Figure 20-4, which shows only the field *lines.* Figure 20-5 shows the field *strengths* plotted as a function of distance. Notice that **B** and **E** are perpendicular to each other and to the direction of travel, and that the electric and magnetic fields are "in phase"; that is, they are each zero at the same points and reach their maximum at the same points.

Electromagnetic (EM) waves are transverse waves and resemble other types of waves (Chapter 11). However, EM waves are waves of *fields,* not of matter as are waves on water or a rope.

We have seen in the analysis above that EM waves are produced by oscillating electric charges of objects that are oscillating and are undergoing acceleration. In fact, we can generalize and say that it is *accelerating electric charges that give rise to electromagnetic waves.*

20-3 Speed of EM waves and the electromagnetic spectrum

Maxwell's prediction that EM waves should exist was startling. Equally startling was the speed at which they were predicted to travel. We now calculate this speed.

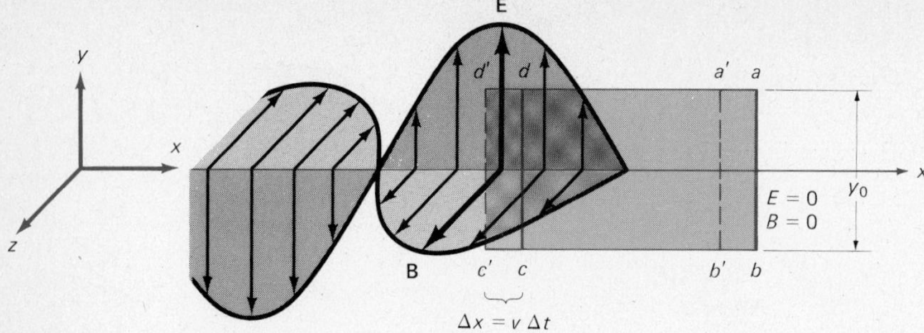

FIGURE 20-6
In a time Δt, the rectangle in the xy plane (actually stationary) moves a distance $\Delta x = v\,\Delta t$ relative to the wave (the wave moves to the right with speed v).

*Calculation of the speed of EM waves

We shall consider a region far from the source, so that the wave fronts (the field lines in Figure 20-4c) are essentially flat over a reasonable area instead of curved. They are then called *plane waves*, meaning that **E** and **B** are uniform over a plane at any given time. Let us assume that, in a particular coordinate system, the wave is traveling in the x direction with speed v; **E** is parallel to the y axis and **B** is parallel to the z axis as in Figure 20-5.

Now let us apply Faraday's law to an imaginary rectangle $abcd$ placed in the xy plane as shown in Figure 20-6. We assume that the side ab is in a region where the wave has not yet reached, so $E = B = 0$ there. As the wave moves, its magnetic flux through our rectangular loop changes. Instead of showing the motion of the wave in Figure 20-6, we show instead the relative position of the rectangle (it's easier to visualize). After a short time Δt, the wave moves to the right a distance $\Delta x = v\,\Delta t$, where v is the velocity of the wave; or, equivalently, the rectangular loop moves $\Delta x = v\,\Delta t$ to the left. Rectangle $a'b'c'd'$ represents the new position of the rectangle relative to the wave. The change in magnetic flux† $\Delta\Phi_B$ through the loop during the time Δt is just the flux that passes through the small rectangle $dd'c'c$. Thus $\Delta\Phi_B = B\,\Delta A = By_0\,\Delta x = By_0 v\,\Delta t$, where y_0 is the width ab or cd; $\Delta A = y_0\,\Delta x = y_0 v\,\Delta t$ is the area of $dd'c'c$; and B represents the magnetic field passing through this area (represented by the heavy arrow in Figure 20-6). We assume Δx is small so that B is essentially constant over this small area. According to Faraday's law, then, the emf induced around the loop $abcd$ equals the rate of change of magnetic flux:

$$\mathscr{E} = \frac{\Delta\Phi_B}{\Delta t} = \frac{B\,\Delta A}{\Delta t} = \frac{By_0 v\,\Delta t}{\Delta t} = By_0 v.$$

(We omitted the minus sign for convenience.) The emf around the loop $abcd$ is the sum of the emfs in each straight section:

$$\mathscr{E} = \mathscr{E}_{ab} + \mathscr{E}_{bc} + \mathscr{E}_{cd} + \mathscr{E}_{da}.$$

Now the emf is the work done per unit charge: $\mathscr{E} = W/q = Fd/q = Ed$, where F is the force exerted over the distance d and the electric field E is the force per unit charge, $E = F/q$. (The result, $\mathscr{E} = Ed$, is the same as Equation 14-2.) The term $\mathscr{E}_{ab}$ is zero since $E = 0$ in this region; and

† The subscript on Φ_B reminds us that it is *magnetic* flux, not electric.

$\mathscr{E}_{bc} = \mathscr{E}_{da} = 0$, since **E** is perpendicular to the path (Figure 20-6). Thus

$$\mathscr{E} = \mathscr{E}_{cd} = Ey_0,$$

where E is the magnitude of the electric field along cd (or $c'd'$). From Faraday's law as we wrote it above ($\mathscr{E} = By_0v$) we thus have

$$Ey_0 = By_0v$$

$$E = vB. \tag{20-2}$$

Thus we see that at any point in space, E and B are in the ratio $E/B = v$, the velocity of the wave.

Now we consider a rectangle in the xz plane, as shown in Figure 20-7. Again we show the rectangle moving to the left a distance $\Delta x = v\,\Delta t$ relative to the wave, although actually the rectangle is stationary and the wave is moving to the right; $abcd$ is the position initially and $a'b'c'd'$ after a time Δt. There is a changing electric flux through this loop equal to the electric field E (heavy arrow) times the increasing area $\Delta A = z_0\,\Delta x = z_0v\,\Delta t$ (where z_0 is the width $ab = cd$ of the rectangle). According to Ampère's law (Equation 20-1 with $I_C = 0$), since there are no conduction currents, we have

$$\sum B_{\parallel}\,\Delta l = \mu_0\varepsilon_0\frac{\Delta\Phi_E}{\Delta t}$$

$$= \mu_0\varepsilon_0\frac{(E)(z_0v\,\Delta t)}{\Delta t} = \mu_0\varepsilon_0 E z_0 v.$$

The sum of $B_{\parallel}\,\Delta l$ on the sides ab, bc, and da are all zero, because either $B = 0$ or B is perpendicular to these sides. But along side cd, the contribution is Bz_0, where B is the magnetic field (heavy arrow) parallel to cd. Thus

$$Bz_0 = \mu_0\varepsilon_0 E z_0 v$$

$$B = \mu_0\varepsilon_0 vE.$$

We combine this with Equation 20-2 and find that

$$B = \mu_0\varepsilon_0 v(vB) = \mu_0\varepsilon_0 v^2 B.$$

We cancel B on both sides and solve for v:

$$v = \frac{1}{\sqrt{\varepsilon_0\mu_0}}. \tag{20-3}$$

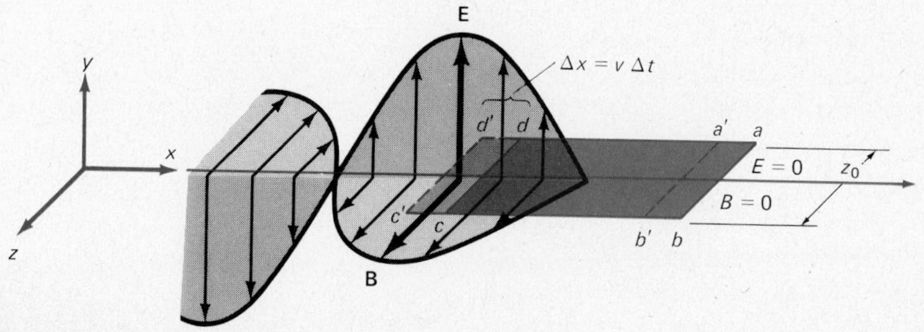

FIGURE 20-7
Rectangle in the xz plane (actually at rest) moves a distance $\Delta x = v\,\Delta t$ relative to a wave traveling to the right.

When we put in the values for ε_0 and μ_0, we find that

$$v = \frac{1}{\sqrt{\varepsilon_0\mu_0}} = \frac{1}{\sqrt{(8.85\times10^{-12}\ \mathrm{C^2/N\cdot m^2})(4\pi\times10^{-7}\ \mathrm{N\cdot s^2/C^2})}}$$

$$= 3.00\times10^8\ \mathrm{m/s}.$$

This is a remarkable result. For this is precisely equal to the measured speed of light!

Light as an electromagnetic wave

The calculation above gives Maxwell's result: that the speed of EM waves is 3.0×10^8 m/s, the same as the measured speed of light.

Light had been shown some 60 years previously to behave like a wave (we shall discuss this in Chapter 22). But nobody knew what kind of wave it was—that is, what is it that is oscillating in a light wave? Maxwell, on the basis of the calculated speed of EM waves, argued that light must be an electromagnetic wave. This soon came to be generally accepted by scientists, but not fully until after EM waves were experimentally detected.

EM waves were first generated and detected experimentally by Heinrich Hertz (1857–1894) in 1887, eight years after Maxwell's death. Hertz used a spark-gap apparatus in which charge was made to rush back and forth for a short time; this generated waves whose frequency was about 10^9 Hz. He detected them some distance away using a loop of wire in which an emf was produced when a changing magnetic field passed through. These waves were later shown to travel at the speed of light, 3.00×10^8 m/s, and to exhibit all the characteristics of light, such as reflection, refraction, and interference. The only difference was that they were not visible. Hertz's experiment was a strong confirmation of Maxwell's theory.

The electromagnetic spectrum

The wavelengths of visible light were measured in the first decade of the nineteenth century, long before anyone imagined that light was an electromagnetic wave. The wavelengths were found to lie between 4.0×10^{-7} and 7.5×10^{-7} m; or from 400 to 750 nm (1 nm $= 10^{-9}$ m). The frequencies of visible light waves can be found using Equation 11-11, which we rewrite here as

$$f\lambda = c. \qquad (20\text{-}4)$$

Here c is the velocity of light, 3.00×10^8 m/s; it gets the special symbol c because of its universality for all EM waves; f and λ are the frequency and wavelength of the wave. This equation tells us that the frequencies of visible light vary between 4.0×10^{14} and 7.5×10^{14} Hz.

But light is only one kind of EM wave. As we have seen, Hertz produced EM waves of much lower frequency, about 10^9 Hz. These are called *radio waves,* since frequencies in this range are used today to transmit radio and television signals.

Electromagnetic waves, or EM radiation as we sometimes say, have been produced or detected over a wide range of frequencies. They are

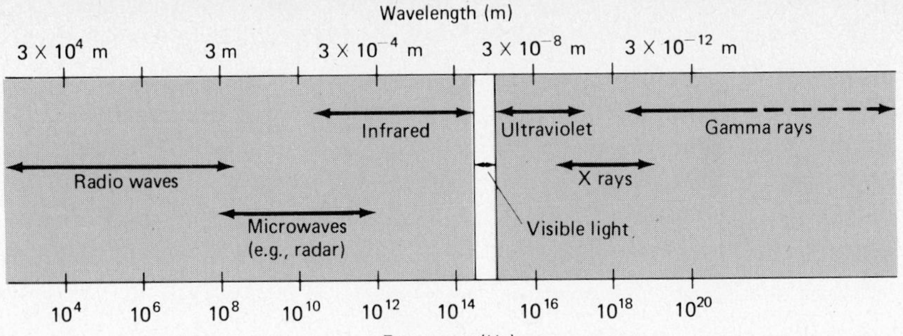

FIGURE 20-8
Electromagnetic spectrum.

usually categorized as shown in Figure 20-8. This is known as the **electromagnetic spectrum.**

Radio waves and microwaves can be produced experimentally using electronic equipment, as discussed earlier (Figure 20-4). Higher-frequency waves are very difficult to produce electronically. These and other types of EM waves are produced in natural processes, as emission from atoms, molecules, and nuclei (more on this later).

Generally, EM waves are produced by the acceleration of electrons or other charged particles, such as the electrons accelerating in the antenna of Figure 20-4. X rays are produced (see Chapter 25) when fast-moving electrons are rapidly decelerated when striking a metal target. Even the visible light emitted by an ordinary incandescent light is due to electrons undergoing acceleration within the hot filament.

We will meet various types of EM waves later. However, it is worth mentioning here that infrared radiation (EM waves whose frequency is just less than that of visible light) is mainly responsible for the heating effect of the sun. The sun emits not only visible light but substantial amounts of infrared and ultraviolet was well. However, the molecules of our skin tend to "resonate" at infrared frequencies, so it is these that are preferentially absorbed and thus warm us up.

▮ EXAMPLE 20-1 Calculate the wavelength (*a*) of a 60-Hz EM wave; (*b*) of a 1240-kHz AM radio wave.

SOLUTION We use Equation 20-4:

$$(a) \qquad \lambda = \frac{c}{f} = \frac{3.0 \times 10^{8}\ \mathrm{m/s}}{60\ \mathrm{s^{-1}}} = 5.0 \times 10^{6}\ \mathrm{m},$$

or 5000 km. One wavelength stretches all the way across the U.S.!

$$(b) \qquad \lambda = \frac{3.0 \times 10^{8}\ \mathrm{m/s}}{1.24 \times 10^{6}\ \mathrm{s^{-1}}} = 240\ \mathrm{m}.$$

▮▮

Transmission lines

When a source of emf is connected up to a transmission line—be it two parallel wires or a coaxial cable—the electric field within the wire is not set up immediately at all points along the wires. This is based on the same argument we used in Section 20-2 with reference to Figure 20-4. Indeed, it can be shown that if the wires are separated by air, the electrical signal travels along the wires at the speed $c = 3.0 \times 10^8$ m/s. For example, when you flip a light switch, the light actually goes on a tiny fraction of a second later. If the wires are in a medium whose electric permittivity is ε and magnetic permeability is μ, the speed is not given by Equation 20-3 but by $v = 1/\sqrt{\varepsilon\mu}$.

▮ EXAMPLE 20-2 When you speak on the telephone from Los Angeles to a friend in New York 4000 km away, how long does it take your voice to travel?

SOLUTION Since speed = distance/time, then time = distance/speed = $(4.0 \times 10^6\ \text{m})/(3.0 \times 10^8\ \text{m/s}) = 1.3 \times 10^{-2}$ s, or about $\frac{1}{100}$ s. ▮▮

* 20-4 ▮ Energy in EM waves

Electromagnetic waves carry energy from one region of space to another. This energy is associated with the moving electric and magnetic fields. In Section 14-3 we saw that the energy density (J/m^3) stored in an electric field E is $u = \frac{1}{2}\varepsilon_0 E^2$, where u is the energy per unit volume. The energy stored in a magnetic field B, as we discussed in Section 18-5 (Equation 18-9) is given by $u = \frac{1}{2}B^2/\mu_0$. Thus the total energy stored per unit volume in a region of space where there is an electromagnetic wave is

$$u = \frac{1}{2}\varepsilon_0 E^2 + \frac{1}{2}\frac{B^2}{\mu_0}. \tag{20-5}$$

In this equation E and B represent the electric and magnetic field strengths of the wave at any instant in a small region of space. We can write Equation 20-5 in terms of the E field only, since from Equation 20-3 we have $\sqrt{\varepsilon_0\mu_0} = 1/c$, and from Equation 20-2, $B = E/c$. We insert these into Equation 20-5 to obtain

$$u = \frac{1}{2}\varepsilon_0 E^2 + \frac{1}{2}\frac{\varepsilon_0\mu_0 E^2}{\mu_0}$$

$$= \varepsilon_0 E^2. \tag{20-6a}$$

Notice that the energy density associated with the B field is equal to that associated with the E field, so each contributes half to the total energy. We can also write the energy density in terms of the B field only, or in one term containing both: $u = \varepsilon_0 E^2 = \varepsilon_0 c^2 B^2 = \varepsilon_0 B^2/\varepsilon_0\mu_0$, or

$$u = \frac{B^2}{\mu_0}; \tag{20-6b}$$

and $u = \varepsilon_0 E^2 = \varepsilon_0 EcB = \varepsilon_0 EB/\sqrt{\varepsilon_0\mu_0}$, or

$$u = \sqrt{\frac{\varepsilon_0}{\mu_0}}\,EB. \tag{20-6c}$$

Equations 20-6 give the energy density in any region of space at any instant.

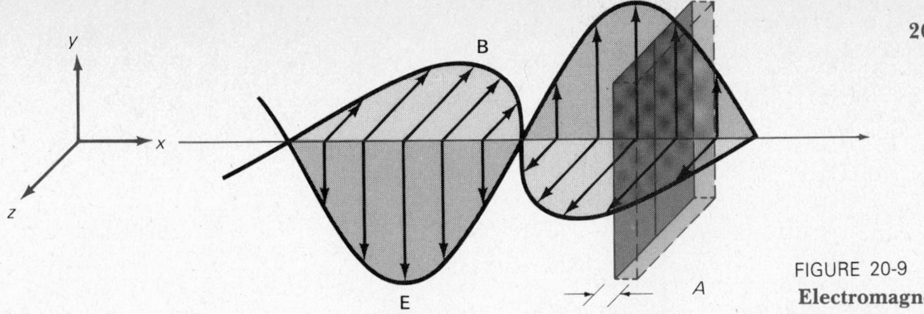

FIGURE 20-9
Electromagnetic wave carrying energy through area A.

Now let us determine the energy that is transported per unit time per unit area perpendicular to the wave direction. This is given by a vector **S**, which is called the *Poynting vector.* The units of **S** are J/s · m² (joules per second per square meter). The direction of **S** is the direction in which the energy is transported, which is the direction in which the wave is moving. Let us imagine that the wave is passing through an area A perpendicular to the x axis, as shown in Figure 20-9. In a short time Δt, the wave moves to the right a distance $\Delta x = c\,\Delta t$. The energy that has passed through A in the time Δt is the energy that now occupies the volume $\Delta V = A\,\Delta x = Ac\,\Delta t$. The energy density u is $u = \varepsilon_0 E^2$, where E is the electric field in this volume at the given instant. So the energy ΔU contained in this volume is the energy density u times the volume: $\Delta U = u\,\Delta V = (\varepsilon_0 E^2)(Ac\,\Delta t)$. Therefore, the energy crossing the area A per time Δt is

$$S = \frac{\Delta U}{A\,\Delta t} = \varepsilon_0 c E^2.$$

Since $E = cB$ and $c = 1/\sqrt{\varepsilon_0 \mu_0}$, this can also be written

$$S = \varepsilon_0 c E^2 = \frac{cB^2}{\mu_0} = \frac{EB}{\mu_0}. \tag{20-7}$$

Equation 20-7 gives the energy transported per unit area per unit time at any *instant.* We often want to know the *average* over an extended period of time. If E and B are sinusoidal, then $\overline{E^2} = E_0^2/2$ just as for electric currents and voltages (Section 15-5), where E_0 is the *maximum* value of E. Thus we can write

$$\bar{S} = \frac{1}{2}\varepsilon_0 c E_0^2 = \frac{1}{2}\frac{c}{\mu_0}B_0^2 = \frac{E_0 B_0}{2\mu_0}, \tag{20-8}$$

where B_0 is the maximum value of B. Equations 20-7 also hold on the average (as well as instantaneously) if for E and B we use the rms values.

▌ EXAMPLE 20-3 Radiation from the sun reaches the earth (above the atmosphere) at a rate of about 1350 J/s · m². Assume that this is a single EM wave and calculate the maximum values of E and B.

SOLUTION Since $\bar{S} = 1350\ \text{J/s}\cdot\text{m}^2 = \varepsilon_0 c E_0^2/2$, then

$$E_0 = \sqrt{\frac{2\bar{S}}{\varepsilon_0 c}} = \sqrt{\frac{2(1350\ \text{J/s}\cdot\text{m}^2)}{(8.85\times 10^{-12}\ \text{C}^2/\text{N}\cdot\text{m}^2)(3.0\times 10^8\ \text{m/s})}}$$

$$= 1.01\times 10^3\ \text{V/m}.$$

From Equation 20-2, $B = E/c$, so

$$B_0 = \frac{E_0}{c} = \frac{1.01 \times 10^3 \text{ V/m}}{3.0 \times 10^8 \text{ m/s}} = 3.4 \times 10^{-6} \text{ T}.$$

(B has a small numerical value compared to E because of the different units for E and B and the way these units are defined; as we saw earlier in the section, B contributes as much energy to the wave as does E.) ▮▮

* 20-5 ▮ Radio and television

Electromagnetic waves offer the possibility of transmitting information over long distances. The first to realize this and put it in practice was Guglielmo Marconi (1874–1937). In the 1890s Marconi invented and developed the wireless telegraph. With it, he was able to send messages hundreds of kilometers at the speed of light without the use of wires. The first signals were merely long and short pulses that could be translated into words by a code, such as the "dots" and "dashes" of the Morse code. The next decade saw the development of vacuum tubes (Section 19-4). Out of this early work radio and television were born, which we now discuss.

*Radio and television transmission

The process by which a radio station transmits information (words and music) is outlined in Figure 20-10. The audio (sound) information is changed into an electrical signal of the same frequencies by a transducer. The transducer may be a microphone, a phonograph cartridge, or a tape recorder head. This electrical signal is called an audio frequency (AF) signal, since the frequencies are in the audio range (20 to 20,000 Hz). The signal is small and must be amplified (Chapter 19). The AF signal is then mixed with a radio-frequency (RF) signal. Its frequency is determined by the values of L and C in a resonant LCR circuit (Section 19-3). L and C are chosen to produce a particular frequency for each station, called its *carrier frequency.* AM radio stations have carrier frequencies between 550 and 1500 kHz. For example, "710 on your dial" means a station whose carrier frequency is 710,000 Hz. FM radio stations have much higher frequencies, between 88 and 108 MHz. The carrier frequencies for TV stations in the U.S. lie between 54 and 88 MHz for channels 2 through 6, and between 174 and 216 MHz for channels 7 through 13; UHF (ultra-high-frequency) stations have even higher carrier frequencies, between 470 and 890 MHz.

FIGURE 20-10
Block diagram of a radio transmitter.

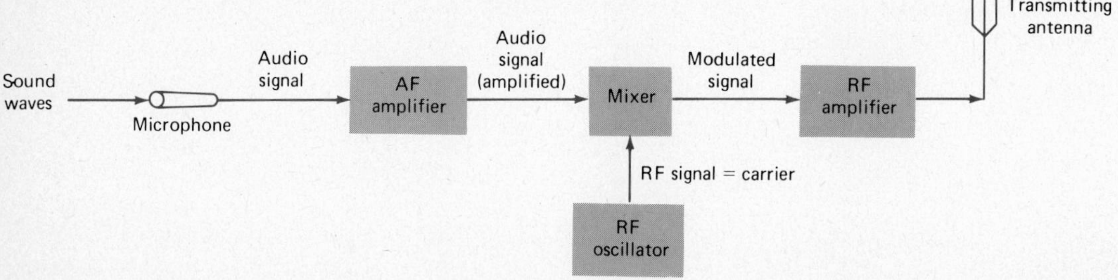

The mixing of the audio and carrier frequencies can be done in two ways. In amplitude modulation (AM) the amplitude of the audio signal is added to that of the much higher carrier frequency, as shown in Figure 20-11. It is called "amplitude modulation" because the amplitude of the carrier is altered ("modulate" means to change or alter). In frequency modulation (FM), the *frequencies* of the audio and carrier signals are added. Thus the *frequency* of the carrier wave is altered by the audio signal as shown in Figure 20-12.

The mixed signal is amplified further; since the signal contains radio frequencies, this amplifier is called an RF amplifier. The output of the RF amplifier is then sent to the antenna, where the complex mixture of frequencies is sent out in the form of EM waves.

A television transmitter works in a similar way, using frequency modulation, except that the carrier frequency is mixed with both an audio and a video (visual) signal.

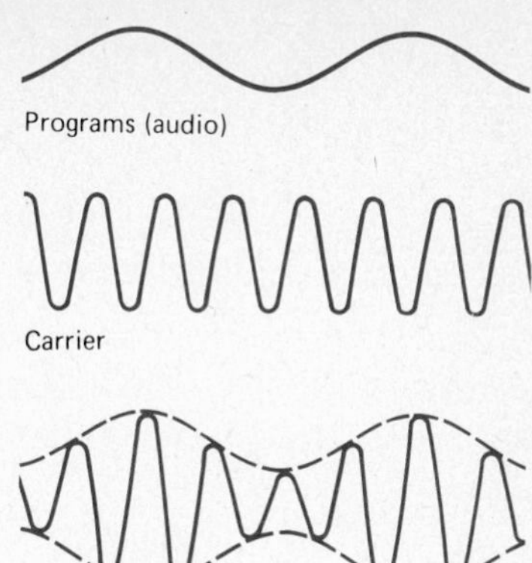

FIGURE 20-11
In amplitude modulation (AM), the amplitudes of the audio and carrier signals are added together.

*Receivers

A simple radio receiver is diagrammed in Figure 20-13. The EM waves sent out by all stations are received by the antenna. One kind of antenna consists of one or more conducting rods; the electric field in the EM waves exert a force on the electrons in the conductor, causing them to move back and forth at the frequencies of the waves. A second type of antenna, often found in AM radios, consists of a tubular coil of wire. This type of antenna detects the magnetic field of the wave, for the changing B field induces an emf in the coil.

The signal from the antenna is very small and contains frequencies from many different stations. The first stage of a receiver selects out a particular RF frequency (actually a narrow range of frequencies) corresponding to a particular station. This is done using a resonant LC circuit (Section 19-3) with a variable capacitor or inductor. A simple example is shown in Figure 20-14. A particular station is "tuned-in" by adjusting L and C so that the resonant frequency of the circuit equals that of the station's carrier frequency. The RF signal may be amplified both before and after the tuning is done.

The signal, containing both audio and carrier frequencies, next goes to the *detector* (Figure 20-13). This is where "demodulation" takes place; that is, the RF carrier frequency is separated from the audio signal. In a very simple receiver, this is done by a capacitor connected to ground. The capacitor offers very little impedance to the high carrier frequency, so it passes to ground, whereas little of the lower frequency audio signal passes to ground. Most modern receivers use diodes or transistors at this stage. The audio signal is then amplified considerably and sent to a loudspeaker or headphones.

Modern receivers have more stages than those shown. Various means

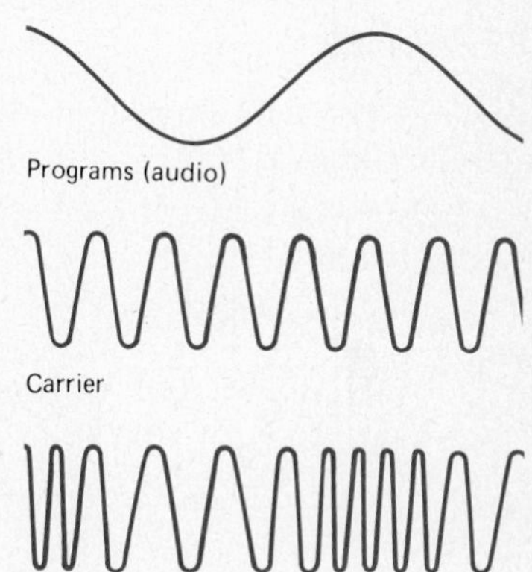

FIGURE 20-12
In FM, the frequencies of the audio and carrier signals are added together. This method is used by FM radio and television.

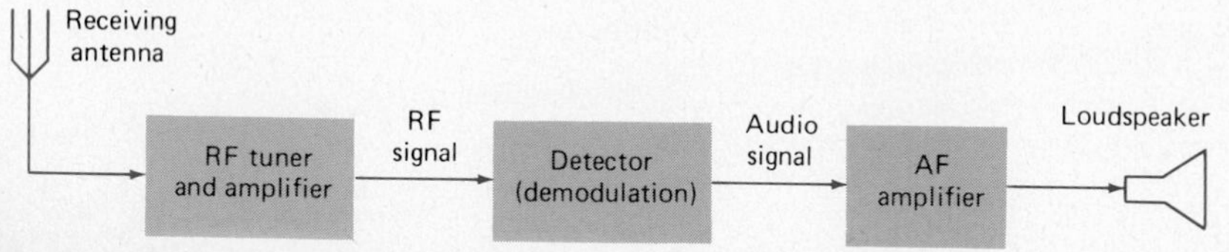

FIGURE 20-13
Block diagram of a simple radio receiver.

are used to increase the sensitivity and selectivity (ability to detect weak signals and distinguish them from other stations), and to minimize distortion of the original signal.†

A television receiver does similar things to both the audio and the video signals. The audio signal goes finally to the loudspeaker and the video signal to the picture tube or *cathode ray tube* (CRT) as it is called. The operation of a CRT is discussed in Section 19-5.

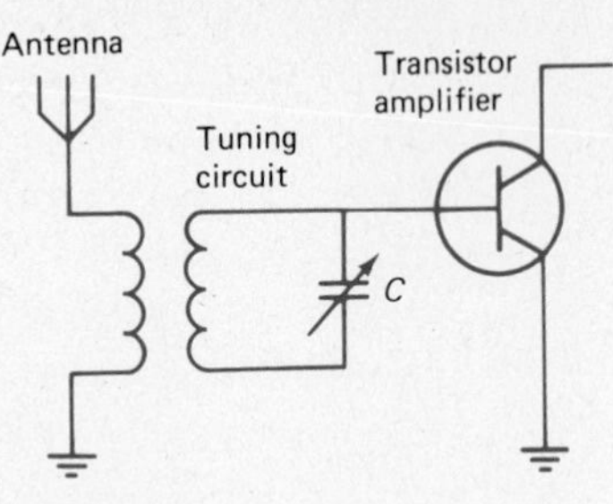

FIGURE 20-14
Simple tuning stage of a radio.

▮ EXAMPLE 20-4 An FM radio station transmits at 100 MHz. Calculate (*a*) its wavelength and (*b*) the value of the capacitance in the tuning circuit if $L = 0.40$ mH.

SOLUTION (*a*) From equation 20-4, $\lambda = c/f = (3.0\times 10^8\ \text{m/s})/(1.0\times 10^8\ \text{s}^{-1}) = 3.0$ m. The wavelengths of other FM signals (88 to 108 MHz) are close to this. FM antennas are typically 1.5 m long, or about a half wavelength. This length is chosen so that the antenna reacts in a resonant fashion and thus is more sensitive.

(*b*) According to Equation 19-8, the resonant frequency is $f_0 = 1/(2\pi\sqrt{LC})$. Therefore

$$C = \frac{1}{4\pi^2 f_0^2 L} = \frac{1}{4(3.14)^2(1.0\times 10^8\ \text{s}^{-1})^2(4.0\times 10^{-4}\ \text{H})}$$

$$= 0.63\ \mu\text{F}.$$

Of course, the capacitor is variable, so other stations can be selected. ▮▮

Citizens band (CB) receivers are similar to AM radios, except that the carrier frequencies are in the vicinity of 27 MHz. The various regions of the radio-wave spectrum are assigned by governmental agencies to various purposes. Besides those mentioned above, there are regions (or "bands") assigned for use by ships, airplanes, police, military, amateurs, satellites and space, and radar.

▮ Summary

James Clerk Maxwell synthesized an elegant theory in which all electric and magnetic phenomena could be described using four equations, now called Maxwell's equations. They are based on earlier ideas, but Maxwell added one more—that a changing electric field produces a magnetic field.

Maxwell's theory predicted that transverse *electromagnetic* (EM) *waves* would be produced by accelerating electric charges, and these waves would propagate through space at the speed of light. The oscillating

† For *FM stereo broadcasting*, two signals are carried by the carrier wave. One of these contains frequencies up to about 17,000 Hz, which includes most audio frequencies; the other signal includes the same range of frequencies, but 21,000 Hz is added to it. A stereo receiver subtracts this 21,000-Hz signal and distributes the two signals to the left and right channels. The first signal actually consists of the sum of left and right channels ($L+R$), so mono radios detect all the sound; the second signal is the difference between left and right ($L-R$). Hence the receiver must add and subtract the two signals to get pure left and right signals for each channel.

electric and magnetic fields in an EM wave are perpendicular to each other and to the direction of propagation.

Problems

After EM waves were experimentally detected in the late 1800s, the idea that light is an EM wave (although of much higher frequency than those detected directly) became generally accepted. The *electromagnetic spectrum* includes EM waves of a wide variety of wavelengths, from microwaves and radio waves to visible light to X rays and γ rays, all of which travel through space at a speed $c = 3.0 \times 10^8$ m/s.

Questions

1 The electric field in an EM wave traveling north oscillates in an east–west plane. Describe the direction of the magnetic field vector in this wave.

2 Is sound an electromagnetic wave? If not, what kind of wave is it?

3 Can EM waves travel through a perfect vacuum? Can sound waves?

4 How are light and sound alike? How are they different?

5 Are the wavelengths of radio and television signals longer or shorter than those detectable by the human eye?

6 What does the result of Example 20-1 tell you about the phase of an ac current that starts at a power plant as compared to its phase at a house 200 km away?

7 When you connect up two loudspeakers to a stereo amplifier, should you be sure the lead wires are equal in length so that there will not be a time lag between speakers? Explain.

*8 A lost person may signal by flashing a flashlight on and off using Morse code. This is actually a modulated EM wave. Is it AM or FM? What is the frequency of the carrier, approximately?

*9 Can two radio or TV stations broadcast on the same carrier frequency? Explain.

*10 If a radio transmitter has a vertical antenna, should a receiver's antenna be vertical or horizontal to obtain best reception?

*11 The carrier frequencies of FM broadcasts are much higher than for AM broadcasts. On the basis of what you learned about diffraction in earlier chapters, explain why AM signals can be detected more readily than FM signals behind low hills or buildings.

Problems

SECTION 20-1

*1 (I) Calculate the displacement current I_D between the square plates, 1.0 cm on a side, of a capacitor if the electric field is changing at a rate of 10^6 V/m · s.

*2 (II) At a given instant, a 2.4-A current flows in the wires connected to a parallel-plate capacitor. What is the rate at which the electric field is changing between the plates if the square plates are 2.1 cm on a side?

*3 (II) A 0.20-μF capacitor with parallel circular plates accumulates charge at a rate of 20 mC/s. What will be the magnetic field strength 10 cm radially outward from the center of the plates whose radii are 0.85 cm?

*4 (III) The electric field between two parallel circular capacitor plates (capacitance C) changes at a rate $\Delta E/\Delta t$. If the radius of the plates is R, show that the magnetic field B a distance r from the center of the plates if $r \leq R$ is $B = \frac{1}{2}\mu_0\varepsilon_0 r(\Delta E/\Delta t)$; and outside the plates, $r \geq R$, show that $B = (\mu_0\varepsilon_0 R^2/2r)(\Delta E/\Delta t)$.

SECTION 20-3

5 (I) What is the wavelength of a 10.85×10^9 Hz radar signal?

6 (I) What is the frequency of a microwave whose wavelength is 1.0 cm?

*7 (I) If the electric field in an EM wave has a peak of 1.5×10^{-4} V/m, what is the peak value of the magnetic field strength?

*8 (I) In an EM wave traveling west, the B field oscillates vertically and has a frequency of 800 kHz and an rms strength of 6.6×10^{-9} T. What is the frequency and rms strength of the electric field and what is its direction?

9 (II) Who will hear the voice of a singer first—a person in the balcony 50 m away from the stage, or a person 3000 km away at home whose ear is next to the radio? How much sooner? Assume that the microphone is a few centimeters from the singer and the temperature is 20°C.

*SECTION 20-4

*10 (I) The **E** field in an EM wave has a peak of 25 mV/m. What is the average rate at which this wave carries energy across unit area per unit time?

*11 (I) The magnetic field in an EM wave has an rms strength of 2.0×10^{-8} T. How much energy does this wave transport per meter squared per second?

*12 (II) How much energy is transported across a 1.0-cm^2 area per hour by an EM wave whose E field has an rms strength of 25 V/m?

*13 (II) What is the energy contained in a 1.0-cm^3 volume near the earth's surface due to radiant energy from the sun? See Example 20-3.

*SECTION 20-5

*14 (I) An FM station broadcasts at 91.5 MHz. What is the wavelength of this wave?

*15 (I) What is the wavelength of an AM station at 1500 on the dial?

*16 (I) Compare 940 on the AM dial to 94 on the FM dial. Which has the longer wavelength, and by what factor is it larger?

*17 (I) The variable capacitor in the tuner of an AM radio has a capacitance of 1500 pF when the radio is tuned to a station at 550 kHz. What must the capacitance be for a station at the other end of the dial, 1600 kHz?

*18 (II) The oscillator of a 106 MHz FM station has an inductance of 4.0 mH. What value must the capacitance be?

*19 (II) A certain FM radio tuning circuit has a fixed capacitor $C = 900$ pF. Tuning is done by a variable inductance. What range of values must the inductance have to tune stations from 88 to 108 MHz?

*20 (II) A 100-kW radio station emits EM waves uniformly in all directions. (*a*) How much energy per second crosses a 1.0-m^2 area 100 m from the transmitting antenna? (*Hint*: The area of a sphere is $4\pi r^2$.) (*b*) What is the rms magnitude of the E field at this point, assuming the station is operating at full power?

*21 (II) Repeat Problem 20 for a distance of 20 km from the station.

*22 (II) An amateur radio operator wishes to build a receiver that can tune a range from 14.0 MHz to 15.0 MHz. A variable capacitor has a minimum capacitance of 64 pF. (*a*) What is the required value of the inductance? (*b*) What is the maximum capacitance used on the variable capacitor?

*23 (II) A 1.40-m-long FM antenna is oriented parallel to the electric field of an EM wave. How large must the **E** field strength be to produce a 1.00-mV (rms) voltage between the ends of the antenna? What is the rate of energy transport per square meter?

*24 (II) How large an emf (rms) will be generated in an antenna that consists of a 500-loop circular coil of wire 2.0 cm in diameter if the EM wave has a frequency of 910 kHz and is transporting energy at an average rate of 1.0×10^{-4} W/m^2 at the antenna. (*Hint:* You can use Equation 18-5, since it could be applied to an observer moving with the coil so that the magnetic field is oscillating with the frequency $f = \omega/2\pi$.)

*25 (III) The variable capacitance of a radio tuner consists of six plates connected together placed alternately between six other plates, also connected together. Each plate is separated from its neighbor by 1.0 mm of air. One set of plates can move so that the area of overlap varies from 1.0 cm^2 to 9.0 cm^2. (*a*) Are these capacitors connected in series or in parallel? (*b*) Determine the range of capacitance values. (*c*) What value of inductor is needed if the radio is to tune AM stations from 550 to 1600 kHz?

For additional reading

Brodeur, P.. "Microwaves," *The New Yorker*, Dec. 13, 1976, p. 50, and Dec. 20, 1976, p. 43. Effects and dangers of microwaves.

Magie, W. F., *Source Book in Physics*, Harvard University Press, Cambridge, Mass., 1963. Original writings of Maxwell and Hertz.

Meyer, H. W., *History of Electricity and Magnetism*, The M.I.T. Press, Cambridge, Mass., 1972.

21 ▮ Light: geometrical optics

The sense of sight is extremely important to us. Our eyes are not only a source of pleasure, but the sense of sight provides us with a large part of our information about the world and the people around us. How do we see? What is this something called *light* that enters our eyes and causes the sensation of sight? And how does light behave so that we can see the great range of phenomena that we do? This subject of light will occupy us for the next three chapters, and we will also return to it in later chapters.

21-1 ▮ Characteristics of light; the ray model

We see an object in one of two ways. First we can detect light emitted directly from a source such as when we look at a light bulb, a flame, or the stars. The second, more common way in which we see an object is by light reflected off it. The light may have originated from the sun or from some other source. An understanding of how bodies emit light was not achieved until the 1920s, and this will be discussed in Chapters 25 and 26. How light is reflected from objects was understood much earlier, and we shall discuss this in the next section.

The ray model

A great deal of evidence suggests that light travels in straight lines. A source of light like the sun casts distinct shadows. We can hear sounds from around the side of a wall or building but we cannot see around it. The beam of a flashlight appears to be a straight line. Indeed, we infer the positions of objects in our environment by assuming that light moves from the object to our eyes in straight-line paths. Our whole orientation to the physical world is based on this assumption.

This reasonable assumption has led to a model for light known as the *ray model.* The straight-line paths that light follows are called light **rays**. When we see an object, we are aware that it occupies some space and that light reaches our eyes from each *point* on the object. Although light rays leave each point in many different directions, normally only a small bundle of these rays can enter an observer's eye as shown in Figure 21-1.

FIGURE 21-1
Light rays radiate from each single point on an object. A small bundle of rays leaving one point is shown entering a person's eye.

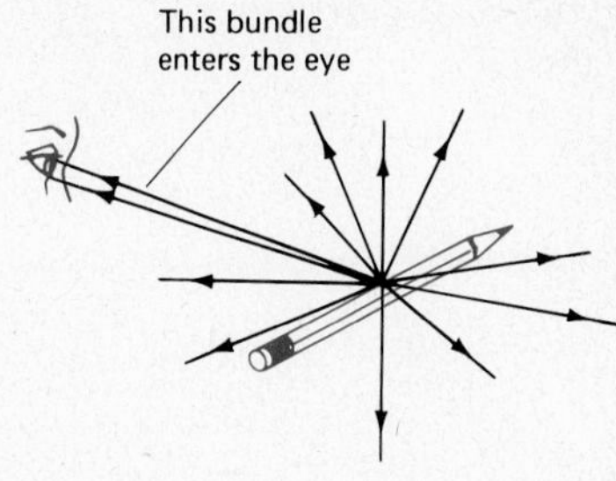

If the person's head moves to one side, a different bundle of rays will enter his eye from each point.

Although the ray model of light does not deal with the actual nature of light (we discuss this in Chapter 22), it has been very successful in dealing with many aspects of light, such as reflection, refraction, and the formation of images by mirrors and lenses. Because these explanations involve straight-line rays and analysis using geometry, this subject is referred to as *geometrical optics.*

The speed of light and index of refraction

Light seems to travel very rapidly. The first recorded attempt to measure the speed of light was made by Galileo. Galileo tried to measure the time required for light to travel a known distance between two hilltops. He stationed an assistant on one hilltop and himself on another and instructed the assistant to lift the cover from a lamp the instant he saw a flash from Galileo's lamp. Galileo then measured the time between the flash of his lamp and when he received the light from his assistant's lamp. The time was so short that Galileo concluded it merely represented human reaction time.

The first successful measurement of the speed of light was done accidentally by a Danish astronomer, Olaf Roemer (1644–1710). Roemer had carefully observed the moons of Jupiter and had determined the periods of their orbits quite accurately. Several years later, when Jupiter was farther from the earth (Jupiter takes about 12 years to make a complete orbit around the sun), Roemer noticed that the moons' orbits were behind schedule. He concluded that this delay must be due to the extra time needed for light to travel the extra distance. From his data he made a rough estimate of the speed of light to be about 2×10^8 m/s.

Since then a number of techniques have been used to measure the speed of light. Among the most important experiments were those carried out by an American, Albert A. Michelson (1852–1931). In a series of experiments carried out from 1880 to the 1920s, Michelson used the rotating-mirror apparatus shown in Figure 21-2. Light from a source was directed at one face of a rotating eight-sided mirror. The reflected light traveled to a stationary mirror a large distance away and back again as shown. If the rotating mirror was turning at just the right speed—which

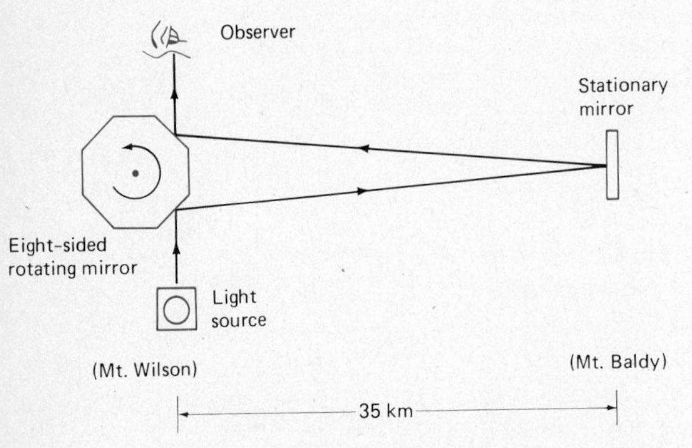

FIGURE 21-2
Michelson's speed-of-light apparatus.

could be varied—the returning beam of light would reflect from one face of the mirror into a small telescope through which the observer looked. At a different speed of rotation, the beam would be deflected to one side and would not be seen by the observer. From the required speed of the rotating mirror and the known distance to the stationary mirror, the speed of light could be calculated. In the 1920s, Michelson set up the rotating mirror on the top of Mt. Wilson in southern California and the stationary mirror on Mt. Baldy (Mt. San Antonio) 35 km away. He later measured the speed of light in vacuum using a long, evacuated tube.

The accepted value today for the speed of light, c, in vacuum is

$$c = 2.997924562 \pm 0.000000011 \times 10^8\ \text{m/s}.$$

We usually round this off to 3.00×10^8 m/s when extremely precise results are not required. In air, the speed is only slightly smaller. In other transparent materials, such as glass and water, the speed is always less than that in vacuum. For example, in water it travels at about $0.75c$. The ratio of the speed of light in vacuum to the speed v in a given material is called the **index of refraction**, n, of that material:

$$n = \frac{c}{v}. \tag{21-1}$$

The index of refraction for various materials is given in Table 21-1.†

TABLE 21-1

Indices of refraction†

Material	$n = c/v$
Air (0°C, 1 atm)	**1.0003**
Water	**1.33**
Ethyl alcohol	**1.36**
Glass	
Fused quartz	**1.46**
Crown glass	**1.52**
Light flint	**1.58**
Lucite or Plexiglas	**1.51**
Sodium chloride	**1.53**
Diamond	**2.42**

† $\lambda = 589$ nm.

▮ EXAMPLE 21-1 Calculate the speed of light in diamond.

SOLUTION Since $n = c/v$, then $v = c/n = (3.00 \times 10^8\ \text{m/s})/(2.42) = 1.24 \times 10^8\ \text{m/s}$. ▮▮

21-2 ▮ Reflection; the plane mirror

The law of reflection

When light strikes the surface of an object, some of the light is reflected. The rest is either absorbed by the object (and transformed to heat) or, if the object is transparent, like glass or water, part of it is transmitted through. For a very shiny object such as a silvered mirror, over 95 percent of the light may be reflected.

When light strikes a flat surface we define the *angle of incidence*, θ_i, to be the angle the incident ray makes with the normal to the surface ("normal" means perpendicular) and the *angle of reflection* to be the angle the reflected ray makes with the normal. For flat surfaces it is found that the incident and reflected rays lie in the same plane with the normal to the surface. Furthermore, it is found that:

▮ **The angle of incidence equals the angle of reflection.**

This is the **law of reflection** and is indicated in Figure 21-3. It was known to the ancient Greeks, and you can confirm it yourself by shining a narrow flashlight beam at a mirror in a darkened room.

FIGURE 21-3
Law of reflection.

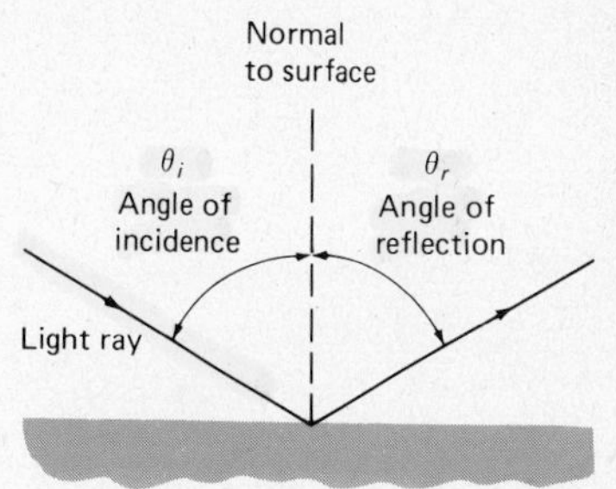

† As we shall see later, n varies somewhat with the wavelength of the light—except in vacuum—so a particular wavelength is specified, that of yellow light with wavelength $\lambda = 589$ nm.

When light is incident upon a rough surface, it is reflected in many directions (Figure 21-4). This is called *diffuse reflection.* The law of reflection still holds at each tiny area of the surface, as indicated in the figure. Because of diffuse reflection in all directions, an ordinary object can be seen from many different angles. When you move your head to the side, a different bundle of reflected rays reach your eye from each point on the object (Figure 21-5a). But when a beam of light (say from a flashlight) is shined on a mirror, the light will not reach your eye unless it is at just the right place, so the law of reflection is satisfied (Figure 21-5b). This is what gives rise to the unusual properties of mirrors.

FIGURE 21-4
Diffuse reflection from a rough surface.

Image formation by a plane mirror

When you look straight in a mirror, you see what appears to be yourself as well as various objects around and behind you. Your face and the other objects look as if they are in front of you, beyond the mirror; but, of course, they are not. What you see in the mirror is an **image** of the objects.

Figure 21-6 shows how an image is formed by a plane (that is, flat) mirror. Rays from two different points on an object are shown. Rays leave each point on the object going in many directions, but only those that enclose the bundle of rays that reach the eye from the two points are shown. The diverging bundle of rays that enters the eye appears to come from behind the mirror, as shown by the dashed lines. The point from which each bundle of rays seems to come is one point on the image. For each point on the object there is a corresponding image point. Let us concentrate on the two rays that leave the point A on the object and strike the mirror at points B and B'. The angle ADB and CDB are right angles. And angles ABD and CBD are equal because of the law of reflection. Therefore, the two triangles ABD and CDB are congruent and the length $AD = CD$. That is, the image is as far behind the mirror as the object is in front: the *image distance,* d_i (distance from mirror to image; Figure 21-6), equal the *object* distance, d_o. From the geometry we also see that the height of the image is the same as that of the object.

The light rays do not actually pass through the image itself. It merely *seems* like the light is coming from the image, because our brains interpret any light entering our eyes as coming from in front of us. Since the rays do

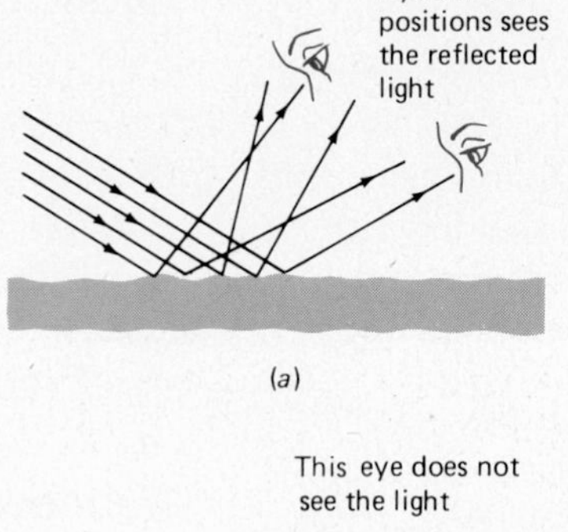

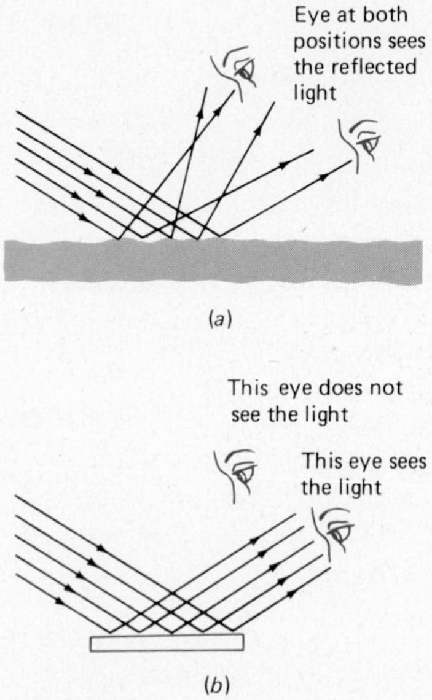

FIGURE 21-5
A beam of light from a flashlight is shined on (a) white paper; (b) a small mirror. In part a you can see the white light reflected at various points because of diffuse reflection. But in part b you see the reflected light only when your eye is placed correctly ($\theta_r = \theta_i$).

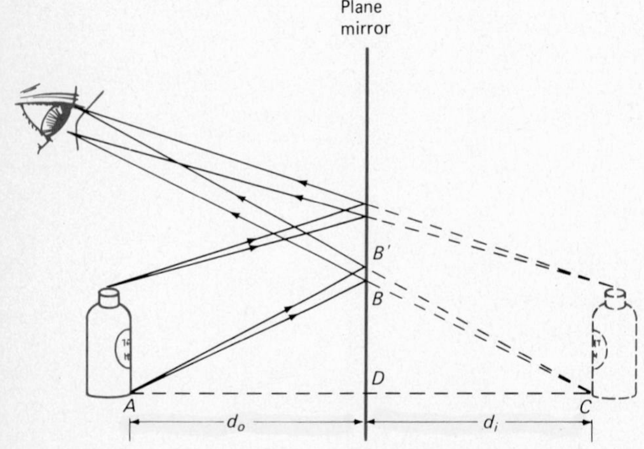

FIGURE 21-6
Formation of a virtual image by a plane mirror.

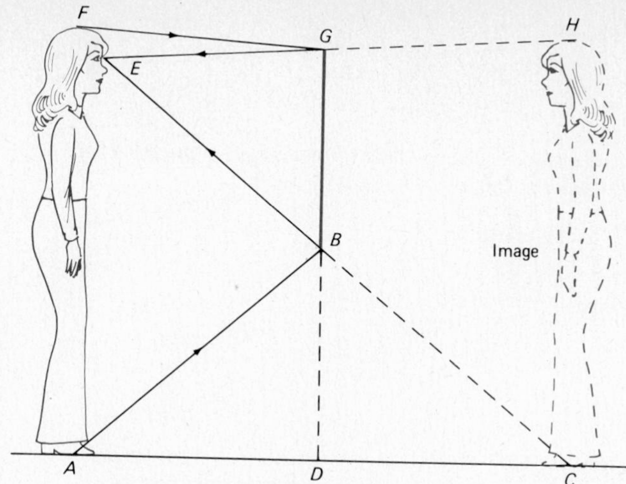

FIGURE 21-7
Seeing oneself in the mirror; Example 21-2.

not actually pass through the image, a piece of white paper or film placed at the image would not detect the image. Therefore, it is called a *virtual image*. This is to distinguish it from a *real image*, in which the light does pass through the image and which therefore can be detected physically. We will see that curved mirrors and lenses can form real images.

▌ EXAMPLE 21-2 A woman 1.60 m tall stands in front of a vertical plane mirror. What is the minimum height of the mirror and how high must its lower edge be above the floor if she is to be able to see her whole body? (Assume that her eyes are 10 cm below the top of her head.)

SOLUTION The situation is diagrammed in Figure 21-7. First consider the ray from the toe, AB, which upon reflection becomes BE and enters the eye E. Since light enters the eye from point A (the toes) after reflecting at B, the mirror needs to extend no lower than B. Because the angle of reflection equals the angle of incidence, the height BD is half of the height AE. Since $AE = 1.60\text{ m} - 0.10\text{ m} = 1.50\text{ m}$, $BD = 0.75\text{ m}$. Similarly, to see the top of her head, the top edge of the mirror only needs to reach point G, which is 5 cm below the top of her head (half of $FE = 10$ cm). Thus $DG = 1.55$ m and the mirror must be only $1.55\text{ m} - 0.75\text{ m} = 0.80\text{ m}$ high which is half the person's height; and its bottom edge must be 0.75 m above the floor. In general, a mirror need be only half as tall as a person for that person to see all of him- or herself. Does this result depend on the person's distance from the mirror? ▌▌

*21-3 ▌ Formation of images by spherical mirrors

Reflecting surfaces do not have to be flat. The most common curved mirrors are *spherical*, which means they form a section of a sphere. A spherical mirror is called *convex* if the reflecting material is on the outer surface of the spherical shape so that the center of the mirror bulges out toward the viewer; and it is called *concave* if the reflecting surface is on the inner surface of the sphere so that the center of the mirror is farther from

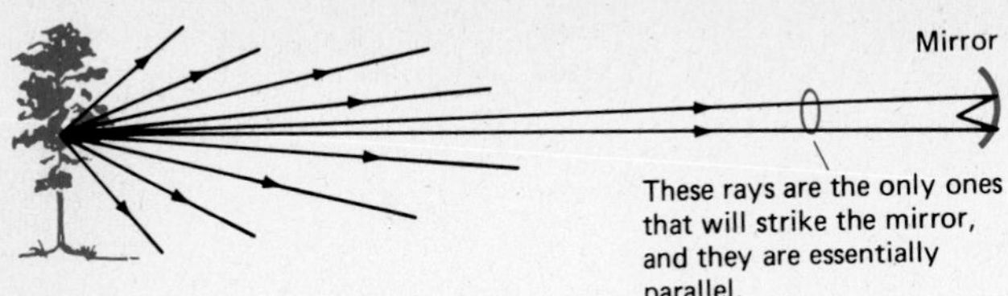

FIGURE 21-8
If the object's distance is large compared to the size of the mirror, the rays are nearly parallel.

the viewer than the edges.† Concave mirrors find use as shaving or makeup mirrors, and convex mirrors are sometimes used on cars and trucks and in shops (to watch for thieves), since they take in a wide field of view.

*Focal point of a concave mirror

Spherical mirrors, too, form images. Let us first consider an object that is very far from a concave mirror; in this case the rays from each point on the object that reach the mirror will be nearly parallel, as shown in Figure 21-8. For an object infinitely far away (the sun and stars approach this), the rays are precisely parallel. Now consider such parallel rays falling on a concave mirror as in Figure 21-9. The law of reflection holds for each of these rays at the point each strikes the mirror. As can be seen, they are not all brought to a single point. To form a sharp image, the rays must come to a point. Thus a spherical mirror will not make as sharp an image as a plane mirror will. However, if the mirror width is small compared to its radius of curvature, so that the reflected rays make only a small angle upon reflection, the rays will come to nearly a single point, or *focus*, as shown in Figure 21-10. In the case shown, the rays are parallel to the *principal axis*, which is defined as the straight line perpendicular to the curved surface at its center (line CA in the diagram). The point F where rays parallel to the principal axis come to a focus is called the *focal point* of the mirror; the distance between F and the center of the mirror, length FA, is called the *focal length* of the mirror. Another way of defining the focal point is to say that it is the image point for an object infinitely far away along the principal axis. For example, the image of the sun would be at F.

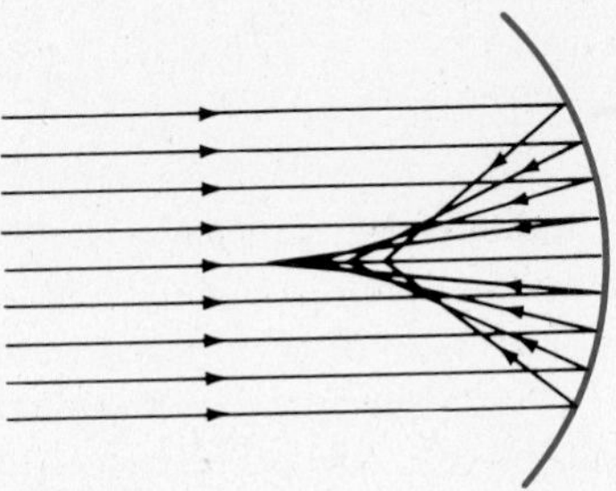

FIGURE 21-9
Parallel rays striking a concave spherical mirror do not focus at a single point.

Now we will show that the rays do indeed meet at a common point, F, and we calculate the focal length f. We consider a ray that strikes the mirror at B in Figure 21-10. The point C is the center of curvature of the mirror (the center of the sphere of which the mirror is a part). So the

† To help remember the names *convex* and *concave*, remember that a *concave* mirror is like a cave, or is "caved in" in the center.

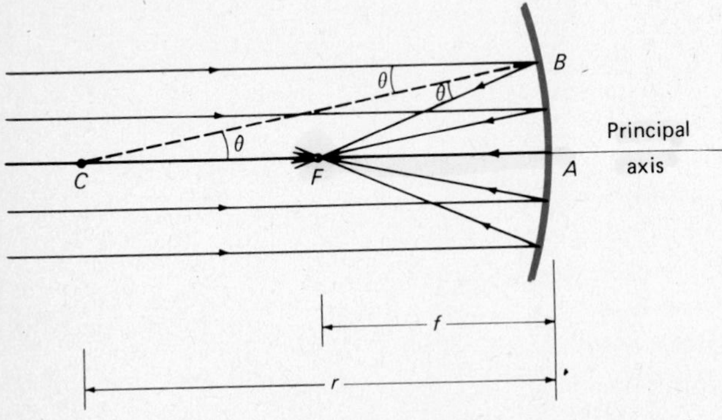

FIGURE 21-10
Rays parallel to the principal axis of a spherical mirror come to a focus at *F*, called the focal point, as long as the mirror is small in extent as compared to its radius of curvature *r*.

dashed line CB is equal to r, the radius of curvature, and CB is normal to the surface at B. Thus the angles θ are equal, as indicated. The triangle CBF is isosceles because two of its angles are equal. Thus length $FC = FB$. We assume that the mirror is small compared to its radius of curvature, so the angles are small. Hence the length FB is nearly equal to length FA. In this approximation, then, $FA = FC$. But $FA = f$, the focal length, and $CA = 2FA = r$. Thus the focal length is half the radius of curvature:

$$f = \frac{r}{2}. \tag{21-2}$$

This argument only assumed that the angle θ was small, so the same result applies for all the other rays and they all come to the same focus in this approximation of a mirror small compared to its radius of curvature.

Since it is only approximately true that the rays come to a perfect focus at F, the larger the mirror, the worse the approximation (Figure 21-9) and the blurrier the image. This "defect" of spherical mirrors is called *spherical aberration*; we shall discuss it more with regard to lenses in Chapter 23. A *parabolic* reflector, on the other hand, will reflect the rays to a perfect focus. However, parabolic shapes are much harder to make and thus much more expensive. Consequently, spherical mirrors are used for most purposes. We shall consider only spherical mirrors and shall assume that they are small compared to their radius of curvature, so the image is sharp and Equation 21-2 holds.

*Images and the mirror equation

We saw that for an object at infinity the image is located at the focal point of a concave spherical mirror, where $f = r/2$. Where does the image lie for an object not at infinity?

First consider the object shown in Figure 21-11 which is placed at point O between F and C. Let us determine where the image will be for a given point O' on the object. To do this we can draw several rays and make sure these reflect from the mirror such that the reflection angle equals the incidence angle. This can involve much work and our task is simplified if we deal with three particularly simple rays. These are rays 1, 2, and 3 in the diagram. Ray 1 is drawn parallel to the axis; therefore, it must pass through F after reflection. Ray 2 is drawn through F; therefore, it must

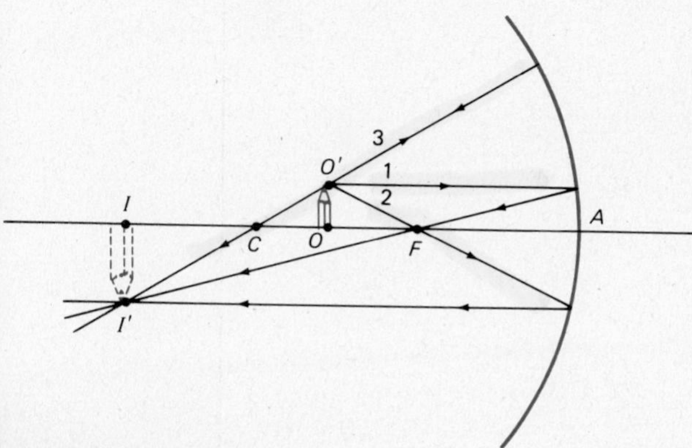

FIGURE 21-11
Rays from object at O form an image at I.

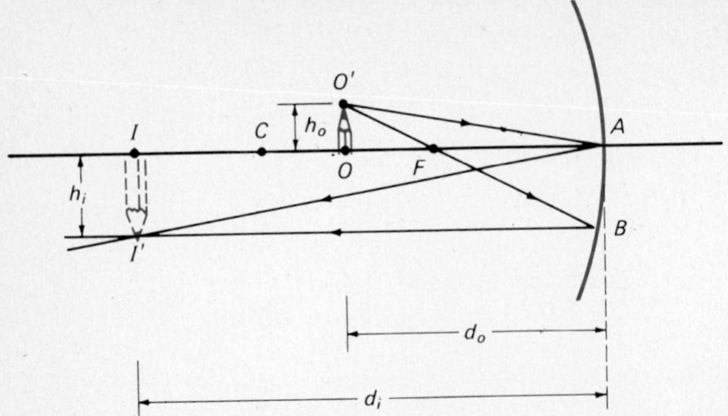

FIGURE 21-12
Diagram for deriving the mirror equation.

reflect so it is parallel to the axis. (The angles for rays 1 and 2 correspond to those for the rays in Figure 21-10.) Ray 3 is drawn so that it passes through *C*, the center of curvature; since ray 3 is along a radius of the spherical surface, it must be perpendicular to the mirror and thus will be reflected back on itself. The point at which all three rays cross is the image point I'. All other rays from the same object point will also pass through this image point. To find the image point for any object point, only these three types of rays need be used. Actually, only two of these rays are needed, but the third serves as a check.

We have only shown the image point in Figure 21-11 for a single point on the object. Other points on the object are imaged nearby so a complete image of the object is formed, as shown by the dashed outline. Because the light actually passes through the image itself, this is a *real image.* This can be compared to the virtual image formed by a plane mirror.

The image can always be formed by drawing the three rays as described above. However, it is possible to derive an equation that gives the image distance if the object distance and radius of curvature are known. To do this we refer to Figure 21-12. The distance of the object from the center of the mirror, called the *object distance,* is labeled d_o; and the *image distance* is labeled d_i. The height of the object OO' is called h_o and the height of the image, $I'I$, is h_i. Two rays are shown, $O'FBI'$ (same as ray 2 in the previous figure) and $O'AI'$. The ray $O'AI'$ obeys the law of reflection, of course, so the triangles $O'AO$ and $I'AI$ are similar. Therefore, we have

$$\frac{h_o}{h_i}=\frac{d_o}{d_i}.$$

For the other ray, $O'FBI'$, the triangles $O'FO$ and AFB are also similar. Therefore,

$$\frac{h_o}{h_i}=\frac{OF}{FA}=\frac{d_o-f}{f},$$

since the length $AB=h_i$ (in our approximation of a mirror small compared to its radius) and $FA=f$, the focal length of the mirror. Since the left side of the two preceding expressions are the same, we equate the right sides:

$$\frac{d_o}{d_i}=\frac{d_o-f}{f}.$$

We now divide both sides by d_o and rearrange to obtain

$$\frac{1}{d_o}+\frac{1}{d_i}=\frac{1}{f}. \tag{21-3}$$

This is the equation we were seeking. It is called the *mirror equation* and relates the object and image distances to the focal length f (where $f=r/2$).

The *magnification* m of a mirror is defined as the height of the image divided by the height of the object. From our first set of similar triangles above, we have:

$$\text{magnification} = m = -\frac{h_i}{h_o} = -\frac{d_i}{d_o}. \tag{21-4}$$

The minus sign is used, as a convention, to indicate that the magnification for an inverted image (upside down) as in Figure 21-12 is negative; that is, m is positive for an upright image and negative for an inverted one.

▌ EXAMPLE 21-3 A 1.5-cm-high object is placed 20 cm from a concave mirror whose radius of curvature is 30 cm. Determine (*a*) the position of the image and (*b*) its size.

SOLUTION (*a*) The focal length $f=r/2=15$ cm. Then, since $d_o=$ 20 cm, we have from Equation 21-3 that

$$\frac{1}{d_i}=\frac{1}{f}-\frac{1}{d_o}=\frac{1}{15\text{ cm}}-\frac{1}{20\text{ cm}}=0.0167\text{ cm}^{-1}.$$

So $d_i = 1/0.0167\text{ cm}^{-1} = 60$ cm; the image is 60 cm from the mirror on the same side as the object. (*b*) From Equation 21-4, the magnification $m=-60\text{ cm}/20\text{ cm}=-3.0$. Therefore, the image is $(3.0)(1.5\text{ cm}) = 4.5$ cm high and is inverted. ▌▌

▌ EXAMPLE 21-4 A 1.0-cm-high object is placed 10 cm from a mirror whose radius of curvature is 30 cm. (*a*) Draw a ray diagram to locate the position of the image. (*b*) Determine the position of the image and the magnification analytically.

SOLUTION (*a*) Since $f=15$ cm, the object is between the mirror and the focal point. We draw the three rays as described earlier (for Figure 21-11) and this is shown in Figure 21-13. The rays reflected from the mirror diverge and so never meet at a point. They appear, however, to be coming from a point behind the mirror. The image is thus behind the mirror and is *virtual*.

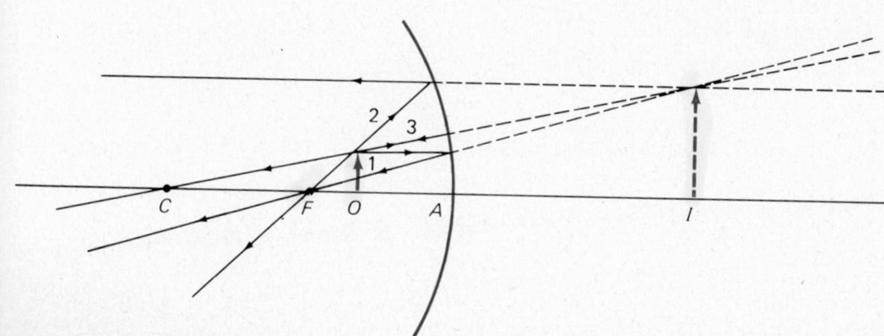

FIGURE 21-13
Object placed within the focal point *F*, Example 21-4. The image is *behind* the mirror and is *virtual*.

(*b*) We use Equation 21-3 to find d_i where $d_o = 10$ cm:

$$\frac{1}{d_i} = \frac{1}{15\text{ cm}} - \frac{1}{10\text{ cm}} = \frac{2-3}{30\text{ cm}} = -\frac{1}{30\text{ cm}}.$$

Therefore, $d_i = -30$ cm. The minus sign means that the image is *behind* the mirror. The magnification is $m = -(-30\text{ cm})/(10\text{ cm}) = 3.0$. So the image is 3 times larger than the object; the plus sign indicates that the image is upright. ▮▮

The preceding example illustrates a useful general rule: if the reflecting surface of the mirror faces to the left (as it has in all figures so far), distances are considered positive for points (object, image, and focal points) that are to the left; and they will be negative if they are to the right behind the mirror, as the image distance was in Example 21-4.

It is useful to compare Figures 21-11 and 21-13. We can see that if the object is within the focal point, as in Figure 21-13, the image is virtual, upright, and magnified. This is how a shaving or makeup mirror is used—you must place your head within the focal point if you are to see yourself right side up. If the object is *beyond* the focal point as in Figure 21-11, the image is real and inverted. Whether the magnification is greater or less than one in this case depends on whether the object is beyond point C or not.

*Convex mirrors

The analysis used for concave mirrors can also be applied to convex mirrors. Even the mirror equation (Equation 21-3) holds for a convex mirror, although the quantities must be carefully defined.

Figure 21-14*a* shows parallel rays falling on a convex mirror. Again spherical aberration will be present, but we assume that the mirror's size is small compared to its radius of curvature. The reflected rays diverge, but seem to come from point F behind the mirror. This is called the focal point, and its distance from the center of the mirror is the focal length f. Again $f = r/2$. We see that an object at infinity produces a virtual image in a convex mirror. Indeed, no matter where the object is placed on the reflecting side of the mirror, the image will be virtual and erect, as indicated in Figure 21-14*b*. To find the image, we draw rays 1 and 3 according to the rules used before on the concave mirror.

The mirror equation (21-3) holds for convex mirrors, but the focal length f must be considered negative, as must the radius of curvature. The proof is left as a problem. It is also left as a problem to show that Equation 21-4 for the magnification is valid.

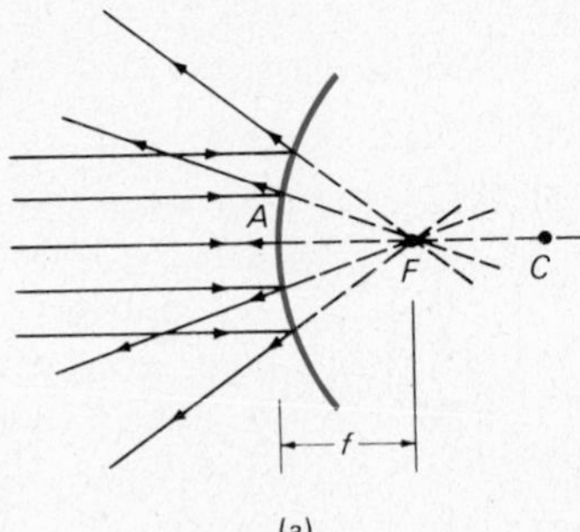

(*a*)

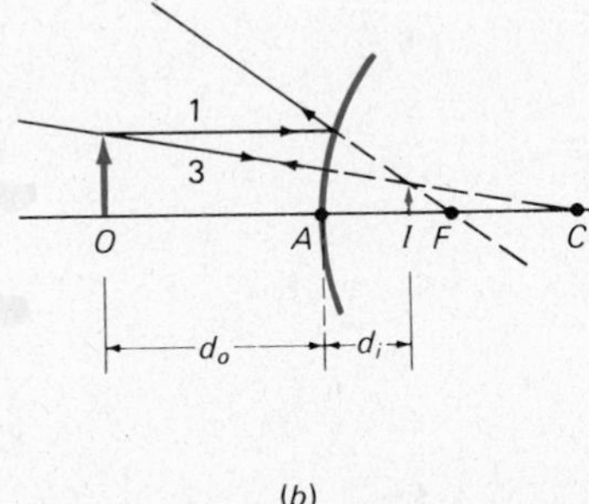

(*b*)

FIGURE 21-14
Convex mirror: (*a*) the focal point is at *F*, behind the mirror; (*b*) image *I* for the object at *O* is vertical, upright, and smaller than **the object.**

▮ EXAMPLE 21-5 A convex rearview car mirror has a radius of curvature of 40.0 cm. Determine the location of the image and its magnification for an object 10.0 m from the mirror.

SOLUTION Since $r = -40.0$ cm, then $f = -20.0$ cm, and the mirror equation gives

$$\frac{1}{d_i} = \frac{1}{f} - \frac{1}{d_o} = -\frac{1}{0.200\text{ m}} - \frac{1}{10.0\text{ m}} = -\frac{51.0}{10.0\text{ m}}.$$

So $d_i = -0.196$ m or 19.6 cm behind the mirror. The magnification is

$m = -d_i/d_o = -(-0.196\text{ m})/(10.0\text{ cm}) = 0.0196$ or $\frac{1}{51}$, so the upright image is reduced by a factor of 51. ▮▮

We summarize the rules for applying Equations 21-3 and 21-4 to concave and convex mirrors: *when the object, image, or focal point is on the reflecting side of the mirror* (on the left in all our drawings), *the corresponding distance is considered positive*; *if any of these points is behind the mirror* (on the right), *the corresponding distance must be considered negative.*†

21-4 ▮ Refraction

When light passes from one transparent medium into another, part of the incident light is reflected at the boundary. The remainder passes into the new medium. If a ray of light is incident at an angle to the surface (other than perpendicular), the ray is bent as it enters the new medium. This bending is called **refraction**. Figure 21-15*a* shows a ray passing from air into water. The angle θ_1 is the *angle of incidence* and θ_2 is the *angle of refraction.* Notice that the ray bends toward the normal when entering the water. This is always the case when the ray enters a medium where the light's speed is reduced. If light travels from one medium into a second where its speed is greater, the ray bends away from the normal; this is shown in Figure 21-15*b* for a ray traveling from water to air.

Refraction is responsible for a number of common optical illusions. For example, a person standing in waist-deep water appears to have shortened legs, Figure 21-16*a*. As shown in Figure 21-16*b*, the rays leaving the person's foot are bent at the surface; the observer's eye (and brain) assumes the rays to have traveled a straight-line path, so the feet

FIGURE 21-15
Refraction.

† We have seen examples where d_i and f are negative. The object distance for any material object is, of course, always positive. However, if the mirror is used in conjunction with a lens or another mirror, the image formed by the first mirror or lens becomes the object for our second mirror; it is then possible for such an "object" to be behind our mirror, in which case d_o would be negative. These rules are also consistent with considering the focal length of a concave mirror positive and that of a convex mirror negative.

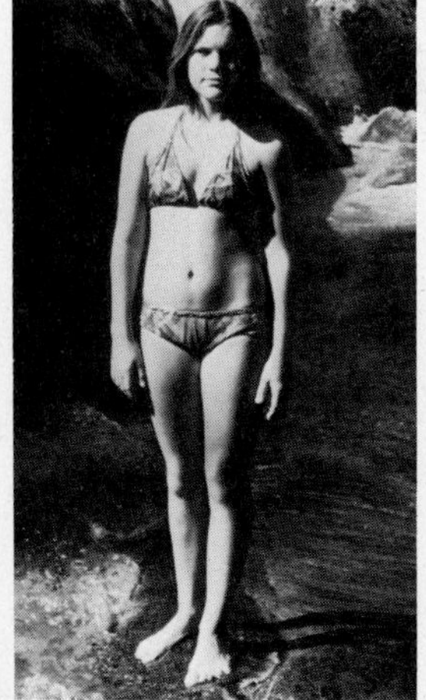

(*a*)

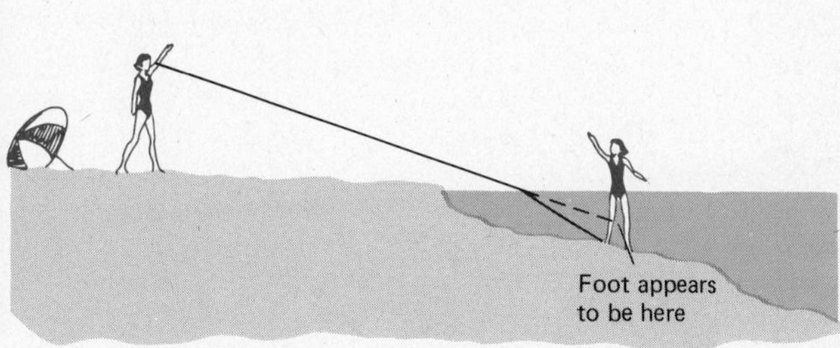

(*b*)

FIGURE 21-16
Because of refraction, when a person is standing in water, his or her legs look shorter. (Photos by D. C. Giancoli.)

appear to be higher than they really are. Similarly, when you put a stick in water, it appears to be bent (try it!).

The angle of refraction depends on the speed of light in the two media and on the incident angle. An analytical relation between θ_1 and θ_2 was discovered experimentally in 1621 by Willebrord Snell (1591–1626). This is known as **Snell's law** and is written

$$n_1 \sin\theta_1 = n_2 \sin\theta_2; \tag{21-5}$$

θ_1 is the angle of incidence and θ_2 is the angle of refraction; n_1 and n_2 are the respective indices of refraction in the materials (see Figure 21-15). Snell's law is the basic law of refraction.†

It is clear from Snell's law that if $n_2 > n_1$, then $\theta_2 < \theta_1$; that is, if light enters a medium where n is greater (and its speed less), the ray is bent toward the normal. And if $n_2 < n_1$, then $\theta_2 > \theta_1$, so the ray bends away from the normal. This is what we saw in Figure 21-15.

▌ EXAMPLE 21-6 Light strikes a flat piece of glass at an incident angle of 60° as in Figure 21-17. If the index of refraction of the glass is 1.50, (*a*) what is the angle of refraction θ_a in the glass, and (*b*) what is the angle θ_b at which the ray emerges from the glass?

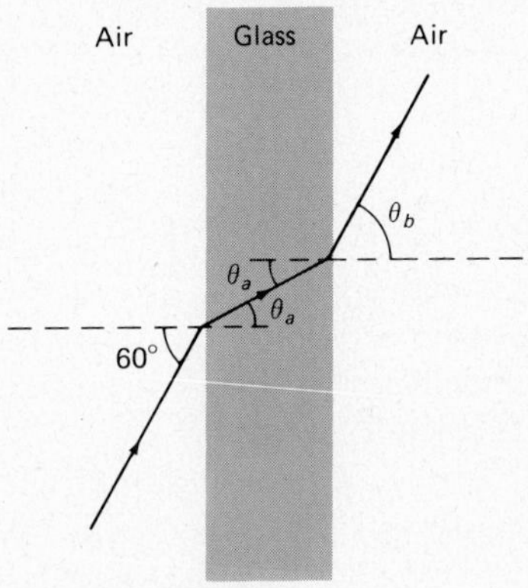

SOLUTION (*a*) We assume that the incident ray is in air, so $n_1 = 1.00$ and $n_2 = 1.50$. Then, from Equation 21-5 we have

$$\sin\theta_a = \frac{1.00}{1.50}\sin 60° = 0.577,$$

so $\theta_a = 35°$. (*b*) Since the faces of the glass are parallel, the incident angle at the second surface is just θ_a, so $\sin\theta_a = 0.577$. This time $n_1 = 1.50$ and $n_2 = 1.00$. Thus θ_b $(=\theta_2)$ is

$$\sin\theta_b = \frac{1.50}{1.00}\sin\theta_a = 0.866,$$

and $\theta_b = 60°$. The direction of the beam is thus unchanged by passing through a plane piece of glass. It should be clear that this would be true for any angle of incidence. The ray is displaced slightly to one side, however. You can observe this by looking (at an angle) through a piece of glass at an object and then moving your head to the side so you see the object directly. ▌▌

Air
Glass
Air
θ_b
θ_a
θ_a
60°

FIGURE 21-17
Light passing through a piece of glass, Example 21-6.

21-5 ▌ Total internal reflection; fiber optics

Total internal reflection

When light passes from one material into a second material where the index of refraction is less (say from water into air), the light bends away from the normal, as for ray A in Figure 21-18. At a particular incident

† Snell actually was not aware that the index of refraction is related to the speed of light in the particular medium. Rather, he found that Equation 21-5 worked for all angles if he assigned the appropriate value of n to each medium. It was only later (in the 1800s) that it was found that the index of refraction could be written as the ratio of the speed of light in vacuum to that in the given material, Equation 21-1. We shall see in the next chapter that light has a wave nature. Snell's law can easily be derived from the wave theory, and we already did so (for water waves) in Chapter 11 (Equation 11-18).

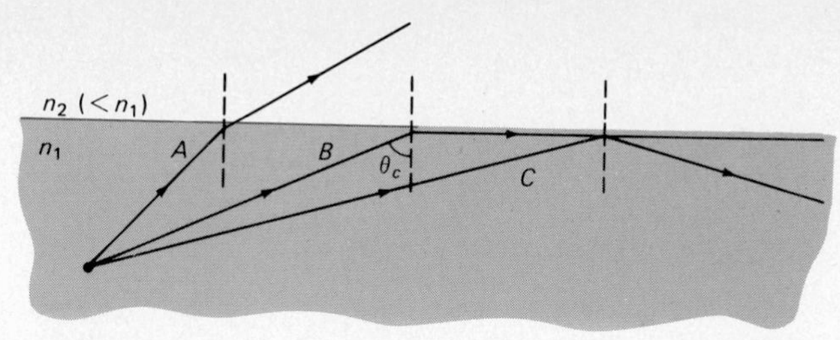

FIGURE 21-18
Since $n_2 < n_1$, light rays are totally internally reflected if $\theta > \theta_c$, as for ray C.

angle, the refracted angle will be 90°, and the refracted ray will skim the surface (ray B). The incident angle at which this occurs is called the *critical angle* θ_c; and from Snell's law it is given by

$$\sin \theta_c = \frac{n_2}{n_1} \sin 90° = \frac{n_2}{n_1}. \tag{21-6}$$

For any incident angle less than θ_c there will be a refracted ray. But for incident angles greater than θ_c, Snell's law would tell us that $\sin \theta_2$ is greater than 1.00. But the sine of an angle can never be greater than 1.00; there can be no refracted ray. Instead, *all the light is reflected*, as for ray C in the diagram. This is called **total internal reflection**. But note that total internal reflection can occur only when light strikes a boundary where the medium beyond is optically less dense—that is, has a lower index of refraction.

▮ EXAMPLE 21-7 Determine the critical angle for water at an interface with air.

SOLUTION For an air–water interface,

$$\sin \theta_c = \frac{1.00}{1.33} = 0.750.$$

Therefore, $\theta_c = 49°$. ▮▮

Total internal reflection produces some interesting phenomena. For example, when you look up from beneath the (smooth) surface of the water in a pool or lake, you will see the outside world compressed into a circle whose edge makes a 49° angle with the vertical. Beyond this angle, you will see reflections from the sides and bottom of the pool or lake.

Many optical instruments such as binoculars use total internal reflection within a prism to reflect light. The advantage is that very nearly 100 percent of the light is reflected, whereas even the best mirrors reflect somewhat less than 100 percent. Thus the image is brighter. For glass with $n = 1.50$, $\theta_c = 42°$. Therefore, 45° prisms will reflect all the light internally as shown in the binoculars of Figure 21-19.

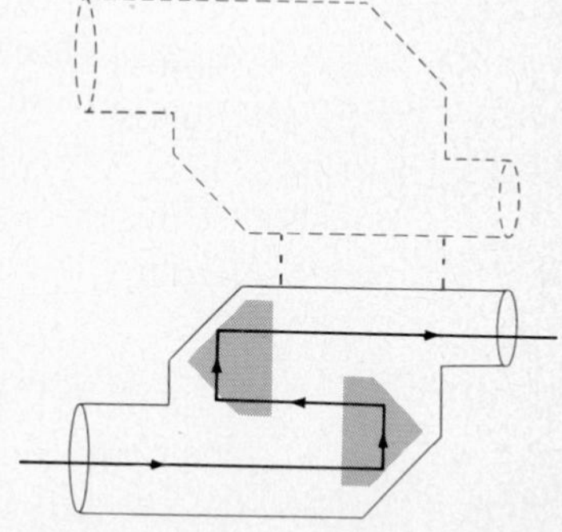

FIGURE 21-19
Prisms reflect light in binoculars.

Fiber optics

Total internal reflection is the principle behind *fiber optics.* Very thin glass and plastic fibers can now be made as small as a few micrometers in diameter. A bundle of such tiny fibers is called a *light pipe*, since light can be transmitted along it without loss. Figure 21-20 shows how light traveling down a thin fiber makes only glancing collisions with the walls, so that total internal reflection occurs. Even if the light pipe is bent into a

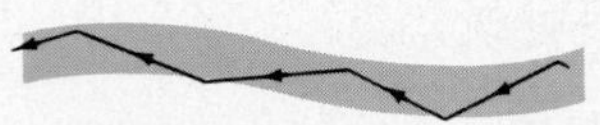

FIGURE 21-20
Light reflected totally at the interior surface of a glass or transparent plastic fiber.

complicated shape, the critical angle will not (usually) be exceeded, so light is transmitted practically undiminished to the other end.

This effect is used in decorative lamps and to illuminate water streams in fountains. It can be used to illuminate difficult places to reach, such as inside the human body. Light pipes are used to transmit telephone calls and other communication signals; the signal is a modulated light beam (the signal is the variable intensity of the light beam) and is transmitted with less loss than an electrical signal in a copper wire.

A more sophisticated use of fiber optics, particularly in medicine, is to transmit a picture, Figure 21-21. For example, a patient's stomach can be examined by inserting a light pipe through the mouth. Light is sent down one set of fibers to illuminate the stomach. The reflected light returns up another set of fibers. Light directly in front of each fiber travels up that fiber. At the opposite end, a viewer sees a series of bright and dark spots, much like a TV screen—that is, a picture of what lies at the opposite end. The fibers must be optically insulated from one another, usually by a thin coating of material whose refractive index is less than that of the fiber. The fibers must be arranged precisely parallel to one another if the picture is to be clear. The more fibers there are, and the smaller they are, the more detailed the picture. Examining the stomach for lesions can thus be done without surgery. Other internal tracts can be observed similarly.

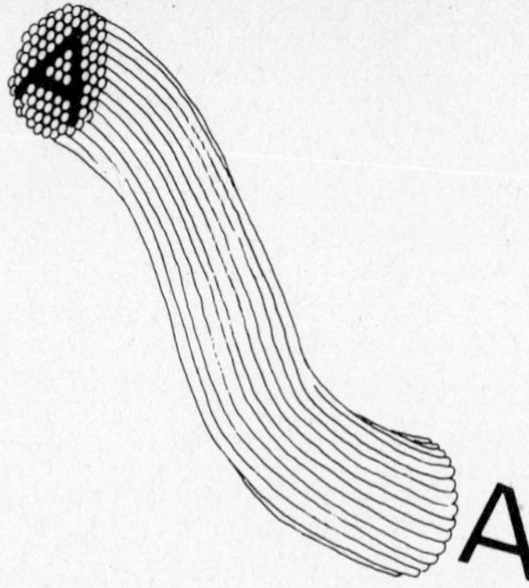

FIGURE 21-21
Fiber optic image.

21-6 ▌ Thin lenses; the lens equation

The most important simple optical device is no doubt the thin lens. The development of optical devices using lenses dates to the sixteenth and seventeenth centuries, although primitive eyeglasses were known as early as the fourteenth century. Today we find lenses in eyeglasses, cameras, magnifying glasses, telescopes, binoculars, microscopes, and medical instruments. A thin lens is usually circular, and its two faces are portions of a sphere. (Although cylindrical surfaces are also possible, we will concentrate on spherical.) The two faces can be concave, convex, or plane (Figure 21-22). The importance of lenses is that they form images of objects.

Focal point; converging and diverging lenses

Consider the rays parallel to the axis of the double convex lens shown in cross section in Figure 21-23a. We assume that the lens is made of glass or transparent plastic so its index of refraction is greater than that of the air outside. According to Snell's law, each ray is bent toward the axis at both lens surfaces (note the dashed lines, indicating the normals to each surface for the top ray). If the lens is thin, *rays parallel to the principal axis incident on the lens are focused to a point called the* **focal point**. This will not be precisely true for a lens with spherical surfaces. But if the lens is sufficiently thin compared to its diameter, the rays will be focused to a tiny region that is nearly a point. In what follows we assume the rays are focused to a point.

Since the rays from a distant object are essentially parallel (see Figure 21-8), we can also say that *the focal point is the image point for an object on the*

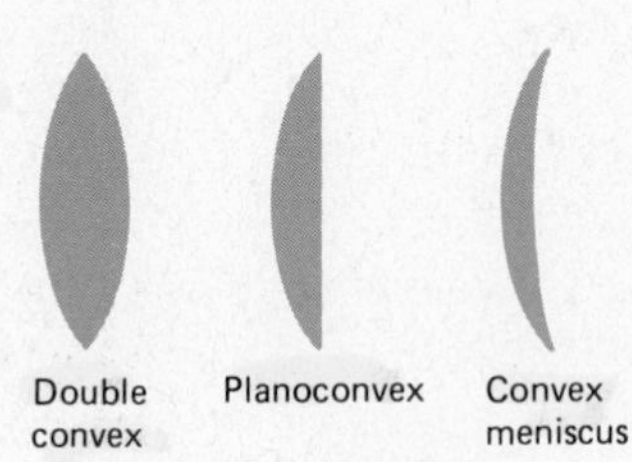

(*a*) Converging lenses

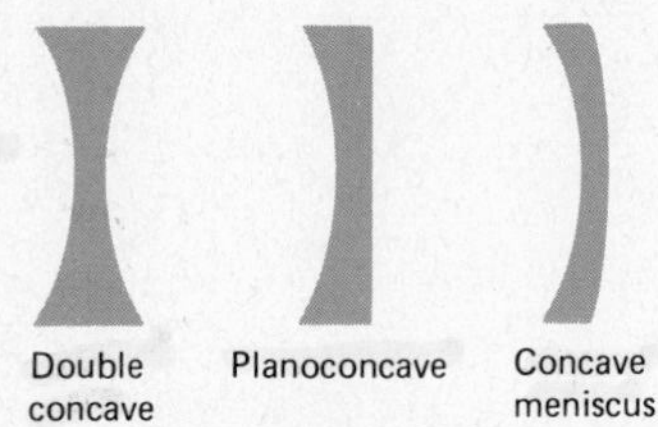

(*b*) Diverging lenses

FIGURE 21-22
Types of lenses.

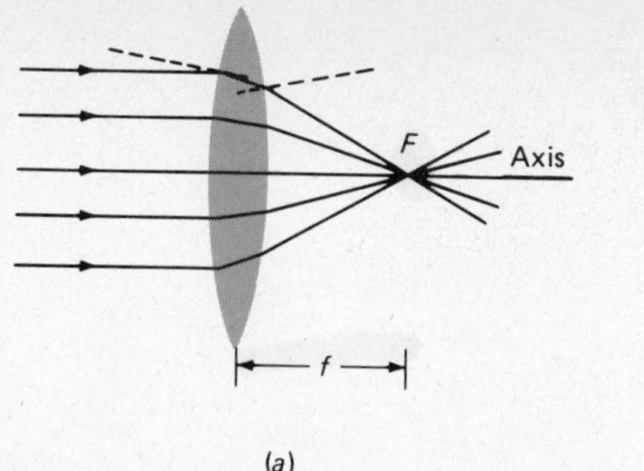

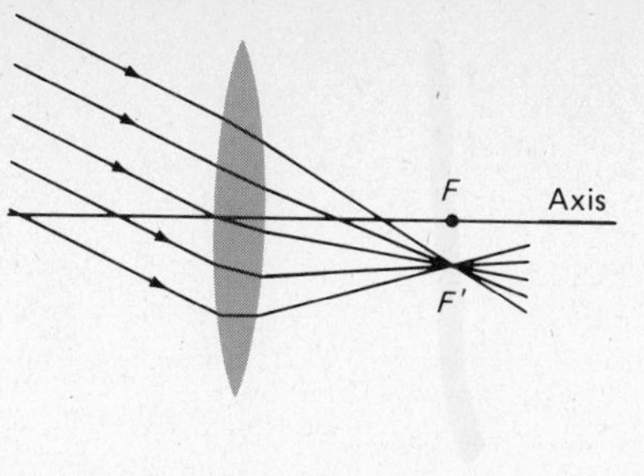

FIGURE 21-23
Parallel rays are brought to a focus by a converging lens.

principal axis at infinity. (An image is formed where the rays meet at a single point just as for a mirror.) Thus the focal point of a lens can be found by locating the point where the sun's rays (or those of some other distant object) are brought to a sharp image. The distance of the focal point from the center of the lens is called the **focal length**, f. A lens can be turned around so that light can pass through it from the opposite side; the focal length is the same on both sides (this is proved in Section 21-7). If parallel rays fall on a lens at an angle as in Figure 21-23*b*, they focus at point F'. The plane FF' is called the *focal plane* of the lens.

Any lens that is thicker in the center than at the edges will make parallel rays converge to a point. Such lenses are called *converging* and several types are shown in Figure 21-22*a*. Lenses that are thinner in the center than at the edges, such as those in Figure 21-22*b*, are called *diverging* lenses because they make parallel light diverge. This is shown in Figure 21-24. The *focal point F* of a diverging lens is defined as that point from which the refracted rays seem to emerge as shown in the figure. And the distance from F to the lens is called the focal length, just as for a converging lens.

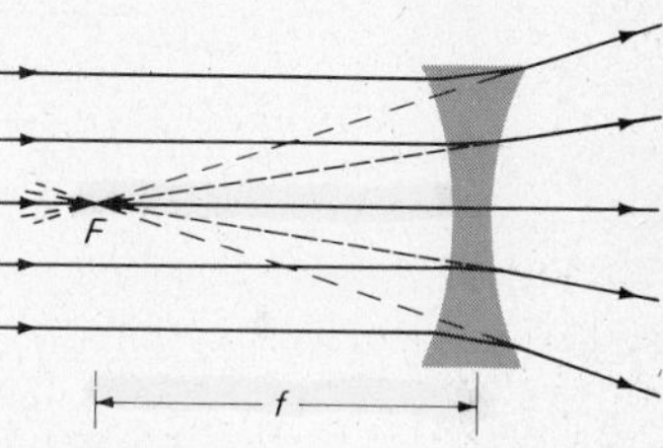

FIGURE 21-24
Diverging lens.

Diopters; power of a lens

Optometrists and ophthalmologists, instead of using the focal length, use the reciprocal of the focal length to specify the strength of eyeglass (or contact) lenses. This is called the *power, P,* of a lens:

$$P = \frac{1}{f}.$$

The unit for lens power is the diopter (D), which is an inverse meter: $1\ \mathrm{D} = 1\ \mathrm{m}^{-1}$. For example, a 20-cm-focal length lens has a power $P = 1/0.20\ \mathrm{m} = 5.0\ \mathrm{D}$.

Ray tracing

The most important parameter of a lens is its focal length f. For a converging lens, f is easily measured by finding the image point for the sun or other distant objects. Once f is known, the image position can be found for any object.

To find the image point by drawing rays would be difficult if we had to determine all the refractive angles. Instead, we can do it very simply by making use of certain facts we already know, such as that a ray parallel to the axis of the lens passes (after refraction) through the focal point. In fact, we need consider only the three rays indicated in Figure 21-25. All

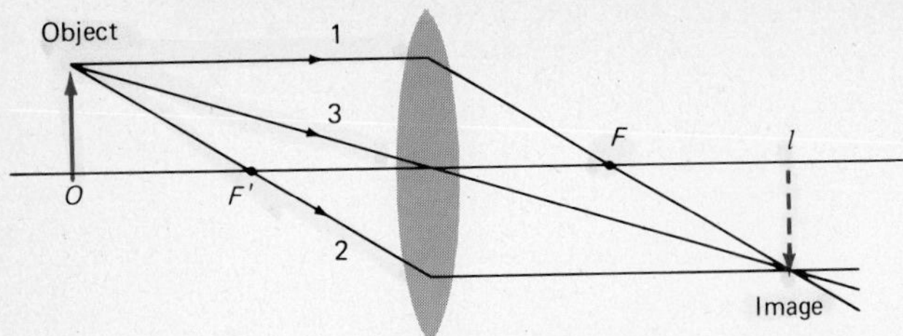

FIGURE 21-25
Finding the image by ray tracing—converging lens.

three rays emanate from a single point on the object. Ray 1 is drawn parallel to the axis; therefore it is refracted by the lens so it passes through the focal point F behind the lens. Ray 2 is drawn through the focal point F' on the same side of the lens as the object; it therefore emerges from the lens parallel to the axis. Ray 3 is directed toward the very center of the lens where the two surfaces are essentially parallel to each other. Therefore this ray emerges from the lens at the same angle as it entered; as we saw in Example 21-6, the ray would be displaced slightly to one side, but since we assume the lens is thin, we draw ray 3 straight through as shown. Actually, any two of these rays will suffice to locate the image point, which is the point where they intersect. However, drawing the third serves as a check.

In this way we have found the image point for the tip of the object. The image points for all other points on the object can be found similarly and will form a complete image of the object as shown. Because the rays actually pass through it, this is a *real image.*

By drawing the same three rays, we can determine the image position for a diverging lens, as shown in Figure 21-26. Note that ray 1 is drawn parallel to the axis, but does not pass through the focal point F behind the lens; instead, it seems to come from the focal point F' in front of the lens (dashed line; see Figure 21-24). Ray 2 is directed toward F and is refracted parallel by the lens. The three refracted rays seem to emerge from a point on the left of the lens. This is the image, I. Since the rays do not pass through it, this is a *virtual image.*

The lens equation

We now derive an equation that relates the image distance to the object distance and the focal length of the lens. This will make the determination of image position quicker and more accurate than doing ray tracing. Consider the two rays shown in Figure 21-27*a* for a converging lens; h_o and h_i refer to the heights of the object and image, and d_o and d_i are their distances from the lens. The triangles $FI'I$ and FBA are similar, so

$$\frac{h_i}{h_o}=\frac{d_i-f}{f},$$

since length $AB = h_o$. Triangles OAO' and IAI' are similar. Therefore,

$$\frac{h_i}{h_o}=\frac{d_i}{d_o}.$$

We equate these two, divide by d_i, and rearrange to obtain

$$\frac{1}{d_o}+\frac{1}{d_i}=\frac{1}{f}. \qquad (21\text{-}7)$$

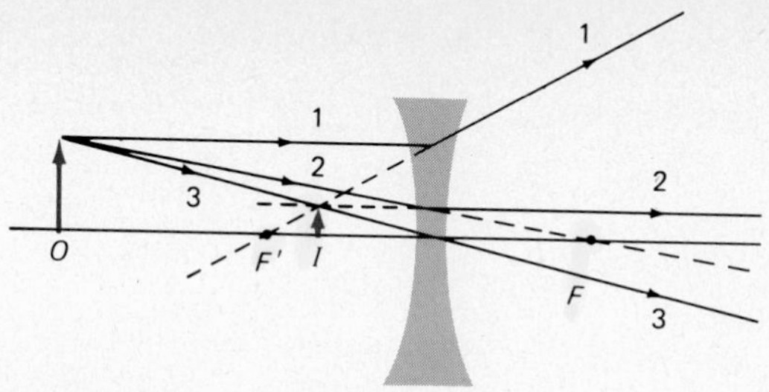

FIGURE 21-26
Finding image *I* by ray tracing—diverging lens.

This is called the **lens equation**. It relates the image distance d_i to the object distance d_o and the focal length f. It is the most useful equation in geometric optics. (Interestingly, it is exactly the same as the mirror equation, Equation 21-3). Note that if the object is at infinity, then $1/d_o = 0$, so $d_i = f$. Thus the focal length is the image distance for an object at infinity, as mentioned earlier.

From the rays in Figure 21-27*b*, we can derive the lens equation for a diverging lens. Triangles IAI' and OAO' are similar; and triangles IFI' and AFB are similar. Thus

$$\frac{h_i}{h_o} = \frac{d_i}{d_o} \quad \text{and} \quad \frac{h_i}{h_o} = \frac{f - d_i}{f}.$$

When these are equated and simplified, we obtain

$$\frac{1}{d_o} - \frac{1}{d_i} = -\frac{1}{f}.$$

This equation becomes the same as Equation 21-7 if we make f and d_i

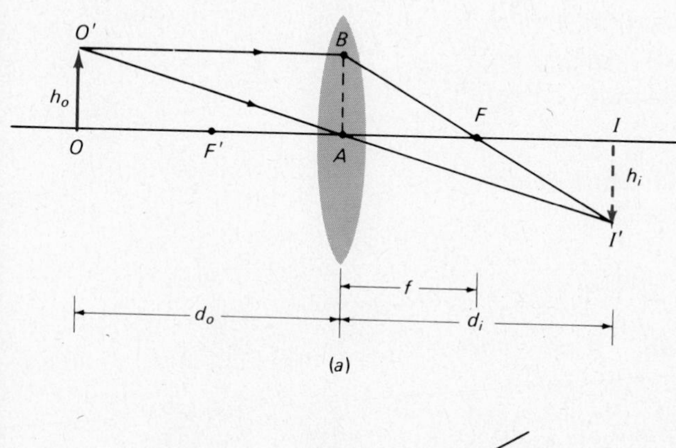

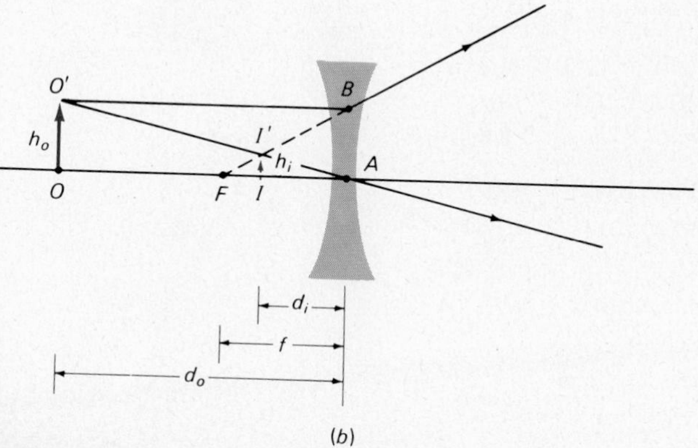

FIGURE 21-27
Deriving the lens equation for (*a*) converging and (*b*) diverging lenses.

negative; that is, we take f to be negative for a diverging lens, and d_i negative when the image is on the same side of the lens as the light comes from. Thus Equation 21-7 will be valid for both converging and diverging lenses, and for *all* situations, if we use the following conventions†:

1 The focal length is positive for converging lenses and negative for diverging lenses.

2 The object distance is positive if it is on the side of the lens from which the light is coming (this is normally the case, although when lenses are used in combination, it might not be true); otherwise, it is negative.

3 The image distance is positive if it is on the opposite side of the lens from where the light is coming; if it is on the same side, d_i is negative. Equivalently, the image distance is positive for a real image and negative for a virtual image.

From convention 1, it follows that the power of a converging lens, in diopters, is positive, whereas the power of a diverging lens is negative. A converging lens is sometimes referred to as a *positive lens*, and a diverging lens as a *negative lens.*

The *magnification* m of a lens is defined as the ratio of the image height to object height. From Figure 21-27 and the relations leading to the lens equation, we have

$$m = -\frac{h_i}{h_o} = -\frac{d_i}{d_o}. \tag{21-8}$$

The minus sign is used, by convention, so that for an upright image the magnification is positive, and for an inverted image m is negative. Note that a real image is always inverted, whereas a virtual image is always upright.

▮ EXAMPLE 21-8 What is the position and size of the image of a 24-cm-high book placed 1.50 m from a 50.0-mm-focal-length camera lens?

SOLUTION A camera lens is converging, so $f = 5.00$ cm; Equation 21-7 gives

$$\frac{1}{d_i} = \frac{1}{f} - \frac{1}{d_o} = \frac{1}{5.00\text{ cm}} - \frac{1}{150\text{ cm}} = \frac{30-1}{150\text{ cm}},$$

so $d_i = 150\text{ cm}/29 = 5.17$ cm or 51.7 mm behind the lens. The magnification $m = -d_i/d_o = -(5.17\text{ cm})/(150\text{ cm}) = -0.034$, so $h_i = -(0.034)(24\text{ cm}) = -0.83$ cm. The image is 8.3 mm high and is inverted as in Figure 21-27*a*. Notice that the image is 1.7 mm farther from the lens than the focal length. This is an example of the fact that when focusing a camera lens, the closer the object is to the camera, the farther the lens must be from the film. ▮▮

▮ EXAMPLE 21-9 An object is placed 10 cm from a 15-cm focal-length converging lens. Determine the image position and size (*a*) analytically, and (*b*) using a ray diagram.

SOLUTION: (*a*) Since $f = 20$ cm and $d_o = 10$ cm,

$$\frac{1}{d_i} = \frac{1}{15\text{ cm}} - \frac{1}{10\text{ cm}} = \frac{2}{30\text{ cm}} - \frac{3}{30\text{ cm}} = -\frac{1}{30\text{ cm}},$$

† These conventions are the same for mirrors, except for 3. See Section 21-3.

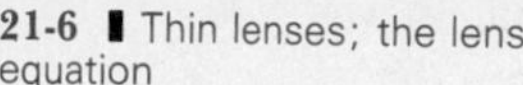

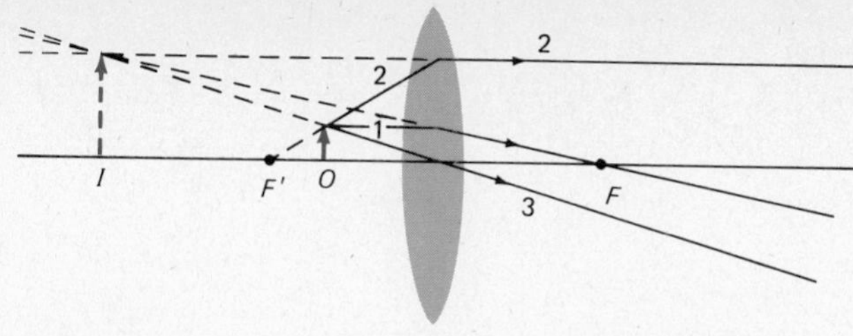

FIGURE 21-28
An object placed within the focal point of a converging lens produces a virtual image; Example 21-9.

so $d_i = -30$ cm. Since d_i is negative, the image must be virtual and on the same side of the lens as the object. The magnification $m = -(-30 \text{ cm})/(10 \text{ cm}) = 3.0$, so the image is three times as large as the object and is upright. (*b*) The ray diagram is shown in Figure 21-28 and confirms the result in (*a*). ▌▌

It is a general rule that when an object is placed between a converging lens and its focal point, the image is virtual.

▌EXAMPLE 21-10 Where must a small insect be placed if a 25-cm-focal-length diverging lens is to form a virtual image 20 cm in front of the lens?

SOLUTION Since $f = -25$ cm and $d_i = -20$ cm, then

$$\frac{1}{d_o} = -\frac{1}{25 \text{ cm}} + \frac{1}{20 \text{ cm}} = \frac{1}{100 \text{ cm}}.$$

So the object must be 100 cm in front of the lens. The ray diagram is basically that of Figure 21-27*b*. ▌▌

▌EXAMPLE 21-11 To measure the focal length of a diverging lens, a converging lens is placed in contact with it as in Figure 21-29. The sun's rays are focused by this combination at a point 28.5 cm behind them as shown. If the converging lens has a focal length f_C of 16.0 cm, what is the focal length f_D of the diverging lens?

SOLUTION Let $f_T = 28.5$ cm refer to the focal length of the total combination. If the diverging lens were absent, the converging lens would form the image at its focal point—that is, at a distance $f_C = 16.0$ cm behind it (dashed lines in Figure 21-29). When the diverging lens is placed next to the converging lens (we assume both lenses are thin and the space between them is small) we treat the image formed by the first lens as the *object* for the second (diverging) lens. Since this object lies to the right of the diverging lens, this is one of the unusual situations where d_o is

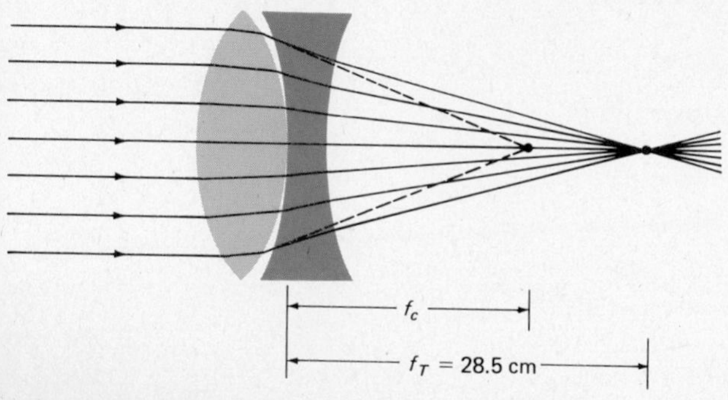

FIGURE 21-29
Determing the focal length of a diverging lens; Example 21-11.

negative. Thus, for the diverging lens, the object is virtual and $d_o = -16.0$ cm; and it forms the image a distance $d_i = 28.5$ cm away (this was given). Thus

$$\frac{1}{f_D} = \frac{1}{d_o} + \frac{1}{d_i} = \frac{1}{-16.0\text{ cm}} + \frac{1}{28.5\text{ cm}} = -0.0274\text{ cm}^{-1}.$$

So $f_D = -1/(0.0274\text{ cm}^{-1}) = -36.5$ cm. Note that the converging lens must have a focal length whose magnitude is less than that of the diverging lens if this technique is to work. ▮▮

This example is our first illustration of how to deal with lenses used in combination. In general, when light passes through several lenses, the image formed by one lens becomes the object for the next lens. The total magnification will be the product of the separate magnification of each lens. We will see more examples in Chapter 23, where we discuss telescopes and microscopes.

*21-7 ▮ The lens-maker's equation

In this optional section we will prove that parallel rays are brought to a focus at a *single* point for a thin lens. We will also derive an equation that relates the focal length of a lens to the radii of curvature of its two surfaces, which is known as the lens-maker's equation.

In Figure 21-30 a ray parallel to the axis of a lens is refracted at the front surface of the lens at point A_1 and is refracted at the back surface at point A_2; it then passes through point F, which we call the focal point for this ray. Point A_1 is a height h_1 above the axis, and point A_2 is height h_2 above the axis. C_1 and C_2 are the centers of curvature of the two lens surfaces, so the length $C_1A_1 = R_1$, the radius of curvature of the front surface; and $C_2A_2 = R_2$ is the radius of the second surface. The thickness of the lens has been grossly exaggerated so the various angles would be clear. But we will assume that the lens is actually very thin. In this approximation $h_1 \approx h_2$, and all the angles will be small. Thus the sines and

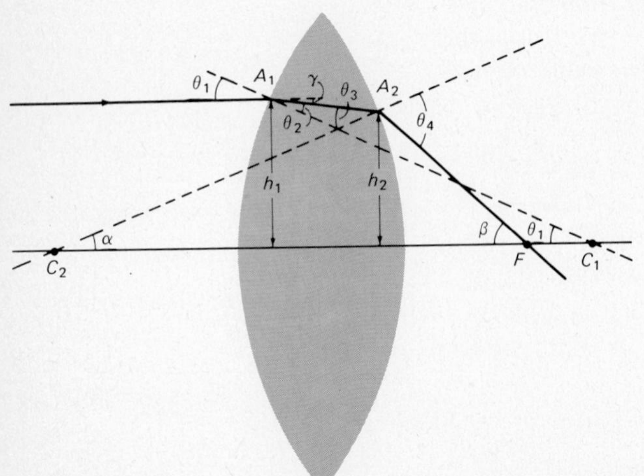

FIGURE 21-30
Diagram of ray passing through a lens for derivation of the lens-maker's equation.

tangents of all the angles will be equal to the angles themselves in radians; for example, $\sin\theta_1 \approx \tan\theta_1 \approx \theta_1$.

To this approximation, then, Snell's law tells us that

$$\theta_1 = n\theta_2$$

$$\theta_4 = n\theta_3$$

where n is the index of refraction of the glass, and we assume that the lens is surrounded by air ($n = 1$). Notice also that

$$\theta_1 = \sin\theta_1 = \frac{h_1}{R_1}$$

$$\alpha = \frac{h_2}{R_2}$$

$$\beta = \frac{h_2}{f}.$$

The last follows because the distance from F to the lens (assumed very thin) is f. From the diagram the angle γ is defined as

$$\gamma = \theta_1 - \theta_2.$$

A careful examination of the diagram shows also that

$$\alpha = \theta_3 - \gamma.$$

This can be seen by drawing a horizontal line to the left from point A_2, which divides the angle θ_3 into two parts. The upper part equals γ and the lower part equals α. (The opposite angles between an oblique line and two parallel lines are equal.) Thus $\theta_3 = \gamma + \alpha$. Finally, by drawing a horizontal line to the right from point A_2, we divide θ_4 into two parts. The upper part is α and the lower is β. Thus

$$\theta_4 = \alpha + \beta.$$

We now combine all these equations:

$$\begin{aligned}\alpha &= \theta_3 - \gamma\\ &= \frac{\theta_4}{n} - (\theta_1 - \theta_2)\\ &= \frac{\alpha}{n} + \frac{\beta}{n} - \theta_1 + \theta_2,\end{aligned}$$

or

$$\frac{h_2}{R_2} = \frac{h_2}{nR_2} + \frac{h_2}{nf} - \frac{h_1}{R_1} + \frac{h_1}{nR_1}.$$

Since the lens is thin, $h_1 \approx h_2$ and we can cancel this from all the numerators. We then multiply through by n and rearrange to find that

$$\frac{1}{f} = (n-1)\left(\frac{1}{R_1} + \frac{1}{R_2}\right). \tag{21-9}$$

This is the *lens-maker's equation.* It relates the focal length of a lens to the curvatures of the lens surfaces and the index of refraction. Notice that f does not depend on h_1 or h_2. Thus the position of the point F does not depend on where the ray strikes the lens. Hence all rays parallel to the axis of a thin lens will pass through the same point F, which we wished to prove.

In our derivation both surfaces are convex and R_1 and R_2 are considered positive.† Equation 21-9 also works for lenses with one or both surfaces concave; but for a concave surface, the radius must be considered *negative*.

Notice in Equation 21-9 that the equation is symmetrical in R_1 and R_2. Thus, if a lens is turned around so that light impinges on the other surface, the focal length is the same.

▮ EXAMPLE 21-12 A concave meniscus lens is one flat surface made from glass with $n = 1.50$. The radius of curvature of the convex surface is 22.4 cm and that of the concave surface is 46.2 cm. What is the focal length?

SOLUTION $R_1 = 22.4$ cm and $R_2 = -46.2$ cm; the latter is negative since it refers to the concave surface. Then

$$\frac{1}{f} = (1.50 - 1.00)\left(\frac{1}{22.4\text{ cm}} - \frac{1}{46.2\text{ cm}}\right) = 0.0114\text{ cm}^{-1}.$$

So $f = 89$ cm and is converging. Notice that if we turn the lens around so that $R_1 = -46.2$ cm and $R_2 = 22.4$ cm, we get the same result. ▮▮

▮ EXAMPLE 21-13 A Lucite planoconcave lens has one flat surface and the other has $R = -18.4$ cm. What is the focal length?

SOLUTION From Table 21-1, n for Lucite is 1.51. A plane surface has infinite radius of curvature; if we call this R_1, then $1/R_1 = 0$. Therefore,

$$\frac{1}{f} = (1.51 - 1.00)\left(-\frac{1}{18.4\text{ cm}}\right).$$

So $f = (-18.4\text{ cm})/0.51 = -36.0$ cm, and the lens is diverging. ▮▮

▮ Summary

Light appears to travel in straight-line paths, called *rays*, at a speed v that depends on the *index of refraction*, n, of the material: $v = c/n$, where c is the speed of light in vacuum.

When light reflects from a flat surface, the angle of reflection equals the angle of incidence. This *law of reflection* explains why mirrors can form *images*. In a plane mirror, the image is virtual, upright, the same size as the object, and as far behind the mirror as the object is in front.

When light passes from one transparent medium into another, the rays bend or refract. The law of *refraction* (*Snell's law*) states that

† Some books use a different convention—for example, R_1 and R_2 are considered positive if their centers of curvature are to the right of the lens, in which case minus signs appear in their equivalent of Equation 21-9.

$n_1 \sin \theta_1 = n_2 \sin \theta_2$, where n_1 and θ_1 are the index of refraction and angle with the normal to the surface for the incident ray, and n_2 and θ_2 are for the refracted ray.

When light rays reach the boundary of a material where the index of refraction decreases, the rays will be *totally internally reflected* if the incident angle, θ, is such that Snell's law would predict $\sin \theta_2 > 1$; this occurs if θ_1 exceeds the critical angle θ_c given by $\sin \theta_c = n_2/n_1$.

A lens uses refraction to produce a real or virtual image. Parallel rays of light are focused to a point, called the *focal point*, by a converging lens. The distance of the focal point from the lens is called the *focal length* f of the lens. After parallel rays pass through a diverging lens, they appear to diverge from a point, its focal point; and the corresponding focal length is considered negative. The power of a lens, equal to $1/f$, is given in diopters, which are units of inverse meters (m^{-1}). The position and size of the image formed by a lens of a given object can be found by ray tracing. Alternatively, the relation between image and object distances, d_i and d_o, and the focal length is given by the *lens equation*: $1/d_o + 1/d_i = 1/f$. The ratio of image height to object height, which equals the magnification, is given by $-d_i/d_o$.

Questions

1 When you look at yourself in a tall plane mirror, you see the same amount of your body whether you are close to the mirror or far away. (Try it and see.) Use a ray diagram for two different positions of the object (you) to show why this should be true.

*2 Show with diagrams that the magnification of a concave mirror is less than 1 if the object is beyond the center of curvature C and is greater than 1 if it is within this point.

*3 If a concave mirror produces a real image, is the image necessarily inverted?

*4 When you use a concave mirror, you cannot see an inverted image of yourself unless you place your head beyond the center of curvature C. Yet you can see an inverted image of another object placed between C and F as in Figure 21-11. Explain. (*Hint*: You can see a real image only if your eye is behind the image, so it can be formed.)

*5 Using the rules for the three rays discussed with references to Figure 21-11, draw ray 2 for Figure 21-14*b*.

*6 What is the focal length of a plane mirror?

*7 Does the mirror equation, Equation 21-3, hold for a plane mirror? Explain.

8 What is the magnification of a plane mirror?

9 What is the angle of refraction when a light ray meets the boundary between two materials perpendicularly?

10 When you look down into a swimming pool or a lake, are you likely to under- or overestimate its depth? Explain.

11 Draw a ray diagram to show why a stick looks bent when part of it is under water.

12 How are you able to "see" a round drop of water on a table even though the water is transparent and colorless?

13 Can a light ray traveling in air be totally reflected when it strikes a smooth water surface if the incident angle is right?

14 When you look up at an object in air from beneath the water in a pool, does the object appear to be the same size as when you see it directly in air? Explain.

15 Where must the film be placed if a camera lens is to make a sharp image of an object very far away?

16 Can a diverging lens form a real image under any circumstances? Explain.

17 Why must the converging lens have a shorter focal length than the diverging lens if the latter's focal length is to be determined by combining them as in Example 21-11?

18 A lens is made of a material with an index of refraction $n = 1.30$. In air it is a converging lens. Will it still be a converging lens if placed in water? Explain, using a ray diagram.

19 The thicker a double convex lens is in the center as compared to its edges, the shorter its focal length for a given lens diameter. Explain.

Problems

SECTION 21-1

1 (I) What is the speed of light in ethyl alcohol?

2 (I) The speed of light in ice is 2.29×10^8 m/s. What is the index of refraction of ice?

3 (I) What is the speed of light (in air) specified in miles per hour? In kilometers per hour?

4 (I) Light is emitted from an ordinary light-bulb filament in a wave-train burst about 10^{-8} s in duration. What is the length in space of such wave trains?

5 (I) How long does it take light to reach us from the sun, 1.5×10^8 km away?

6 (II) What is the minimum speed at which Michelson's eight-sided mirror would have had to rotate in order that light would be reflected into an observer's eye by succeeding mirror faces (Figure 21-2)?

SECTION 21-2

7 (I) Suppose that you want to take a photograph of yourself as you look at your image in a mirror 2.0 m away. For what distance should the camera lens be focused?

8 (I) What is the minimum-height plane mirror for which a person 1.84 m tall can see his or her whole self?

9 (II) Stand up two plane mirrors so they form a right angle as in Figure 21-31. When you look into this double mirror, you see yourself as others see you, instead of reversed as in a single mirror. Make a ray diagram to show why.

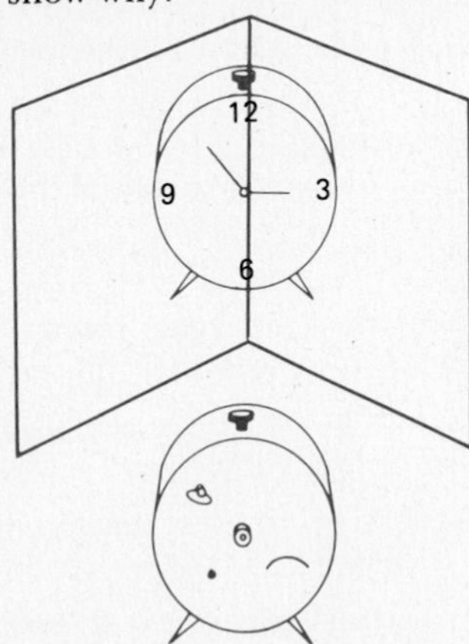

FIGURE 21-31

10 (II) A person whose eyes are 1.40 m above the floor stands in front of a plane mirror whose bottom edge is 40 cm above the floor. What is the horizontal distance from the wall supporting the mirror to the nearest point on the floor that can be seen reflected in the mirror? The person is 2.0 m from the mirror.

11 (III) Show that if two plane mirrors meet at an angle ϕ, a single ray reflected (successively) from both mirrors is deflected through an angle of 2ϕ, independent of the incident angle.

*SECTION 21-3

*12 (I) What is the radius of a concave reflecting surface that brings parallel light to a focus 13.7 cm in front of it?

*13 (I) What is the focal length of a 30-cm-diameter crystal ball whose surface reflects light?

*14 (I) How far from a concave mirror (radius 40 cm) must an object be placed if its image is to be at infinity?

*15 (II) You try to look at yourself in a silvered ball of diameter 60 cm when you are 2.0 m away. Where is your image? Is it real or virtual? Can you see yourself clearly?

*16 (II) A dentist wants a small mirror that, when 2.0 cm from a tooth, will produce a 6.0× upright image. What kind of mirror must be used, and what must its radius of curvature be?

*17 (II) Show, using a ray diagram, that the magnification m of a convex mirror is $m = -d_i/d_o$ just as for a concave mirror. (*Hint*: Consider a ray from the top of the object that reflects at the center of the mirror.)

*18 (II) Use ray diagrams to show that the mirror equation, Equation 21-3, is valid for a convex mirror as long as f is considered negative.

*19 (II) Use the mirror equation to show that the magnitude of the magnification of a concave mirror is less than one if the object is beyond the center of curvature C ($d_o > r$) and is greater than one if the object is within C ($d_o < r$).

*20 (II) A 2.0-cm-tall object is placed 20 cm from a spherical mirror. It produces a virtual image 3.0 cm high. (*a*) What type of mirror is being used? (*b*) Where is the image located? (*c*) What is the radius of curvature of the mirror?

*21 (II) The magnification of a convex mirror is $\frac{1}{4}\times$ for objects 5.0 m away. What is the focal length of this mirror?

*22 (III) What is the radius of a concave mirror that gives a 1.5× magnification of a face 25 cm from it?

*23 (III) A 1.50-m-tall person stands 6.0 m from a convex mirror and notices that he looks only half as tall as he does in a plane mirror placed at the same distance. What is the radius of curvature of the convex mirror? (Assume that $\sin\theta \approx \theta$.)

SECTION 21-4

24 (I) A flashlight beam strikes the surface of a pane of glass ($n = 1.50$) at a 45° angle. What is the angle of refraction?

25 (I) A diver shines a flashlight upward from beneath the water at a 30° angle to the vertical. At what angle does the light leave the water?

26 (I) The rays of the sun are seen to make an angle of 20° with the vertical beneath the water. At what angle did they strike the water surface?

27 (II) A bright light is 2.0 m below the surface of a

swimming pool and 1.5 m from one edge of the pool. At what angle does the light leave the water at the edge of the pool? Assume that the water reaches the top edge of the pool.

28 (II) Prove in general that for a light beam incident on a uniform layer of transparent material, as in Figure 21-17, the direction of the emerging beam is parallel to the incident beam, independent of the incident angle θ and the thickness of the material.

29 (II) An aquarium filled with water has flat glass sides whose index of refraction is 1.50. A beam of light from outside the aquarium strikes the glass at a 37° angle to the perpendicular. What is the angle of this light ray when it enters the glass and then the water? What would it be if the ray entered the water directly?

30 (II) Light is incident on an equilateral crown glass prism at a 45° angle to one face (Figure 21-32). Calculate the angle at which light emerges from the opposite face. Assume that $n = 1.52$.

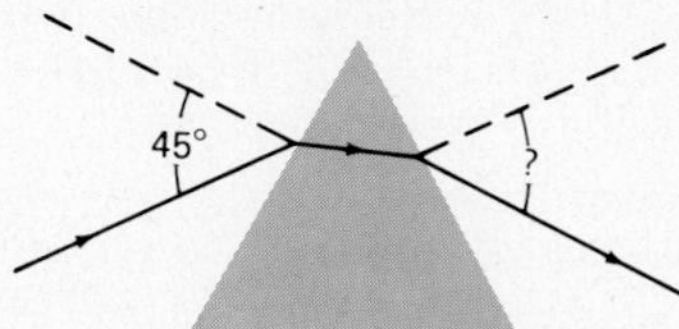

FIGURE 21-32

31 (III) When light passes through a prism, the angles that the refracted ray makes relative to the incident ray is called the deviation angle δ (Figure 21-33). This angle is a minimum when the ray passes through the prism symmetrically—that is, for a prism shaped like an isosceles triangle, when the ray inside the prism is parallel to the base. Show that this minimum deviation angle δ_m is related to the prism's index of refraction n by

$$n = \frac{\sin \frac{1}{2}(\phi + \delta_m)}{\sin \phi/2},$$

where ϕ is the apex angle.

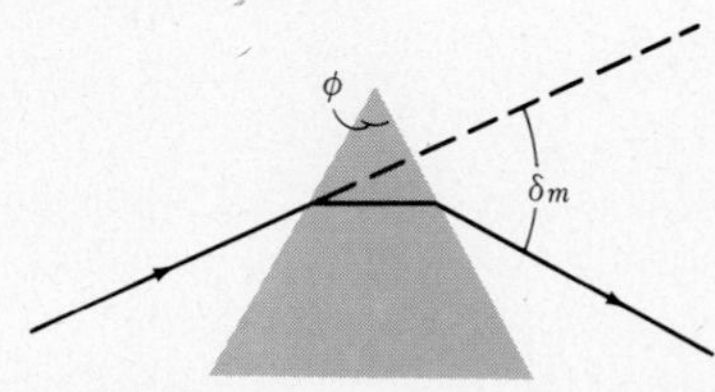

FIGURE 21-33

32 (III) A light ray is incident on a flat piece of glass as in Figure 21-17. Show that if the incident angle θ is small, the ray is displaced a distance $d = t\theta(n-1)/n$, where t is the thickness of the glass and θ is in radians.

SECTION 21-5

33 (I) What is the critical angle for a diamond–water interface?

34 (I) Would a 45° prism totally internally reflect light if made of a material for which (*a*) $n = 1.36$? (*b*) $n = 1.62$?

35 (I) The critical angle for a certain liquid–air surface is 57°. What is the index of refraction of the liquid?

36 (II) A beam of light is emitted 6.0 cm beneath the surface of a liquid and strikes the surface 4.0 cm from the point directly above the source. If total internal reflection occurs, what can you say about the index of refraction of the liquid?

37 (II) The end faces of a cylindrical glass rod ($n = 1.54$) are perpendicular to the sides. Show that a light ray entering an end face at any angle will be totally internally reflected inside the rod when it strikes the sides.

SECTION 21-6

38 (I) A sharp image is located 60 mm behind a 50-mm-focal-length converging lens. Calculate the object distance.

39 (I) A leaf is placed 70 cm in front of a −80-mm-focal-length lens. Where is the image? Is it real or virtual?

40 (I) A certain lens focuses an object 20 cm away as an image 30 cm on the other side of the lens. What type of lens is it and what is its focal length? Is the image real or virtual?

41 (I) (*a*) An object 40 cm in front of a certain lens is imaged 10 cm in front of that lens (on the same side as the object). What type of lens is this and what is its focal length? Is the image real or virtual? (*b*) What if the image were located, instead, 70 cm in front of the lens?

42 (I) What is the power of a 30.0-cm-focal-length lens?

43 (I) What is the focal length of a −8.5-diopter lens? Is this lens converging or diverging?

44 (II) A −3.0-diopter lens is held 20 cm from an ant 1.0 mm high. What is the position, type, and height of the image?

45 (II) (*a*) How far from a 50-mm-focal-length lens must an object be placed if its image is to be magnified 3.0× and be real? (*b*) What if the image is to be virtual and magnified 3.0×?

46 (II) Repeat Problem 45 for a −50-mm-focal-length lens.

47 (II) How large is the image of the sun on the film used in a camera with a 50-mm-focal-length lens? The sun's diameter is 1.4×10^6 km and it is 1.5×10^8 km away.

48 (II) A 65-cm-focal length converging lens is 50 cm behind a diverging lens. Parallel light strikes the diverging lens; after passing through the converging lens, the light is again parallel. What is the focal length of the diverging lens?

49 (II) (*a*) A 2.0-cm-high insect is 2.2 m from a 135-mm-focal-length lens. Where is the image, how high is it, and what type is it? (*b*) What if $f = -135$ mm?

50 (II) In a slide or movie projector, the film acts as the object whose image is projected on a screen. If a 100-mm-focal-length lens is to project an image on a screen 4.0 m away, how far from the lens should the slide be? If the slide is 35 mm wide, how wide will the picture be on the screen?

51 (II) A 35-mm slide (picture size is actually 24 by 36 mm) is to be projected on a screen 1.00 by 1.50 m placed 6.0 m from the projector. What focal-length lens should be used if the image is to cover the screen?

52 (II) Show analytically that the image formed by a converging lens is real and inverted if the object is beyond the focal point ($d_o > f$), and is virtual and upright if the object is within the focal point ($d_o < f$).

53 (III) A diverging lens is placed next to a converging lens of focal length f_c as in Figure 21-29. If f_T represents the focal length of the combination, show that the focal length of the diverging lens, f_D, is given by

$$\frac{1}{f_D} = \frac{1}{f_T} - \frac{1}{f_C}.$$

54 (III) (*a*) Show that if two lenses of focal lengths f_1 and f_2, respectively, are placed next to each other, the focal length of the combination f_T is given by $f_T = f_1 f_2/(f_1 + f_2)$. (*b*) Show that the power P of the combinations of two lenses is the sum of their separate powers, $P = P_1 + P_2$.

55 (III) How far apart are an object and an image formed by a 65-cm-focal-length converging lens if the image is 3.0× larger than the object and is real?

56 (III) Two 20-cm-focal-length converging lenses are placed 15 cm apart. An object is placed 50 cm in front of one. Where will the final image formed by the second lens be located? What is the total magnification?

57 (III) A diverging lens with $f = -25$ cm is placed 10 cm behind a converging lens with $f = 15$ cm. Where will an object at infinity be focused?

58 (III) A bright object is placed on one side of a converging lens of focal length f and a white screen for viewing the image is on the other side. This distance $d_T = d_i + d_o$ between the object and the screen is kept fixed, but the lens can be moved. (*a*) Show that if $d_T > 4f$, there will be *two* positions where the lens can be placed and a sharp image will be produced on the screen. (*b*) Show that if $d_T < 4f$, there will be no lens position where a sharp image is formed.

*SECTION 21-7

* **59** (I) A double concave lens has surface radii of 20.6 cm and 10.2 cm. What is the focal length if $n = 1.58$?

* **60** (I) Both surfaces of a double convex lens have radii of 28.0 cm. If the focal length is 26.2 cm, what is the index of refraction of the lens material?

* **61** (I) A planoconvex lens is to have a focal length of 18.0 cm. If made from fused quartz, what must be the radius of curvature of the convex surface?

* **62** (I) A glass ($n = 1.50$) planoconcave lens has a focal length of -26.6 cm. What is the radius of the concave surface?

* **63** (II) An object is placed 120 cm from a glass lens ($n = 1.56$) with one concave surface of radius 18.0 cm and one convex surface of radius 13.5 cm. Where is the final image? What is the magnification?

* **64** (III) A glass lens ($n = 1.55$) in air has a power of +3.0 diopters. What would its power be if submerged in water?

For additional reading

Boyle, W. S., "Light-wave Communications," *Sci. American*, **237**:2 (Aug. 1977), p. 40.

Fraser, A. B., and W. H. Mach, "Mirages," *Sci. American*, **234**:1 (Jan., 1976), p. 102.

Jaffe, B., *Michelson and the Speed of Light*, Doubleday & Co., New York, 1960.

Jenkins, F. A., *Fundamentals of Optics*, 4th ed., McGraw-Hill Book Company, New York, 1975.

Kopany, N. S., *Fiber Optics, Principles and Applications*, Academic Press, Inc., New York, 1967.

Lynch, D. K., "Atmospheric Halos," *Sci. American*, **238**:4 (Apr., 1978), p. 144.

Meyer-Arendt, J. R., *Introduction to Classical and Modern Optics*, Prentice-Hall, Inc., Englewood Cliffs, N.J., 1972.

Sears, F. W., *Optics*, Addison-Wesley Publishing Co., Inc., Reading, Mass., 1949.

22 ▌The wave nature of light

That light carries energy is obvious to anyone who has focused the sun's rays with a magnifying glass on a piece of paper and burned a hole in it. But how does light travel and in what form is this energy carried?

In our discussion of waves in Chapter 11, we noted that energy can be carried from place to place in basically two ways: by particles or by waves. In the first case, material bodies or particles can carry energy, such as a thrown baseball or rushing water. In the second case, water waves and sound waves, for example, can carry energy over long distances even though mass itself does not travel these distances. So now we come to the question of the nature of light: Does light travel as a stream of particles away from its source; or does it travel in the form of waves that spread outward from the source? Historically, this question has turned out to be a difficult one, for light does not reveal itself in any obvious way as being made up of tiny particles nor do we see tiny light waves passing by. Firm evidence for either view was not found until the nineteenth century, when the evidence pointed strongly to the wave theory. And by the end of the nineteenth century, light was seen to be an *electromagnetic* wave (Chapter 20). Although modifications had to be made in the twentieth century (Chapter 25), the wave theory of light has proved very successful. We now investigate the evidence for the wave theory and show how it has explained a wide range of phenomena.

22-1 ▌Waves versus particles; Huygens' principle

Huygens' principle

A Dutch scientist, Christian Huygens (1629–1695), a contemporary of Newton, proposed a wave theory of light that had much merit. Still useful today is a technique he developed for predicting the future position of a wavefront when an earlier position is known. This is known as *Huygens' principle* and can be stated as follows: every point on a wavefront can be considered as a source of tiny wavelets that spread out in the forward direction at the speed of the wave itself; the new wavefront is the envelope of all the wavelets (that is, the tangent to all of them).

As a simple example of the use of Huygens' principle, consider the wavefront AB in Figure 22-1, which is traveling away from a source S. We assume that the medium is *isotropic*—that is, the speed v of the waves is the same in all directions. To find the wavefront a short time t after it is at AB, tiny circles are drawn whose radius $r = vt$. The centers of these tiny circles are on the original wavefront AB and the circles represent Huygens' wavelets. The tangent to all these wavelets, the line CD, is the new position of the wavefront.

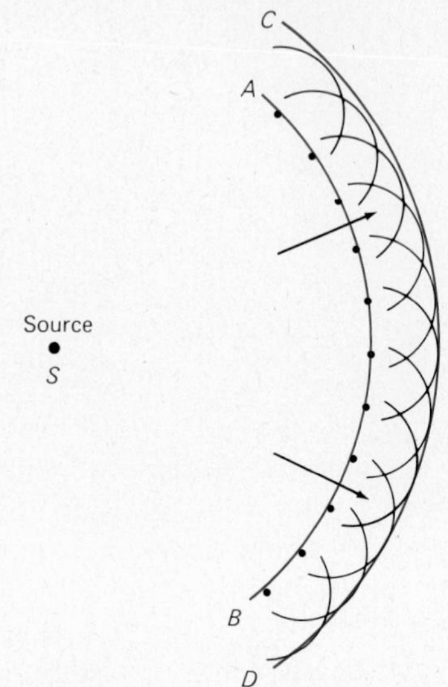

FIGURE 22-1
Huygens' principle used to determine wavefront *CD* when *AB* is given.

Huygens' principle is particularly useful when the waves impinge on an obstacle and the wavefronts are partially interrupted. Huygens' principle shows that the waves bend in behind the obstacles, as shown in Figure 22-2. This is just what water waves do, as we saw in Chapter 11 (Figures 11-33 and 11-34). The bending of waves behind obstacles into the "shadow region" is known as **diffraction**. Since diffraction occurs for waves, but not for particles, it can serve as one means for distinguishing the nature of light.

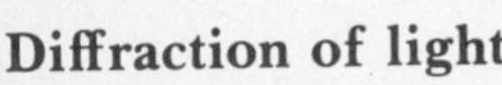

Diffraction of light

Does light exhibit diffraction? In the midseventeenth century, a Jesuit monk, Francesco Grimaldi (1618–1663), had observed that, when light is allowed to enter a darkened room through a tiny hole in a curtain, the spot on the opposite wall was larger than the hole in the curtain; and it was surrounded by colored fringes. Grimaldi attributed this to the diffraction of light.

Newton, who favored a particle theory, was aware of Grimaldi's result. Building on an earlier model by Descartes, Newton imagined that light traveled as tiny corpuscles ("little bodies"). He felt that Grimaldi's result was due to the interaction of corpuscles with the edges of the hole. If light were a wave, he said, the light waves should bend more than that observed. Newton's argument seems reasonable. Yet, as we saw in Chapter 11, diffraction is large only when the size of the obstacle or the hole is on the order of the wavelength of the wave (Figure 11-34). Newton had no way of knowing that the wavelengths of visible light were incredibly tiny, and thus that diffraction effects would be very small. (Indeed, this is why ray optics is so successful—normal openings and obstacles are much larger than the wavelength of the light, so little diffraction or bending occurs.)

Newton recognized, however, that the experimental evidence was not strong either for waves or for particles. So he was not dogmatic about the particle theory but did express his preference for it. But Newton, as a result of his great work in mechanics, was held in very high regard. And for the next century, the particle theory was dominant.

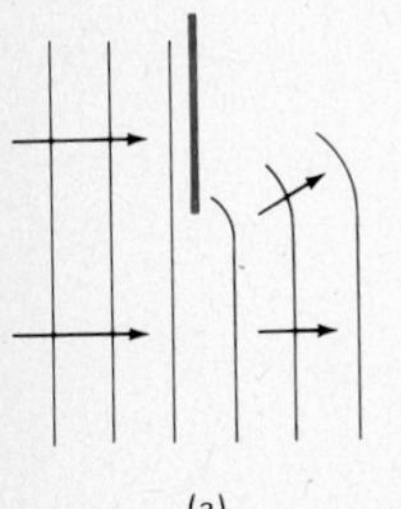

(a)

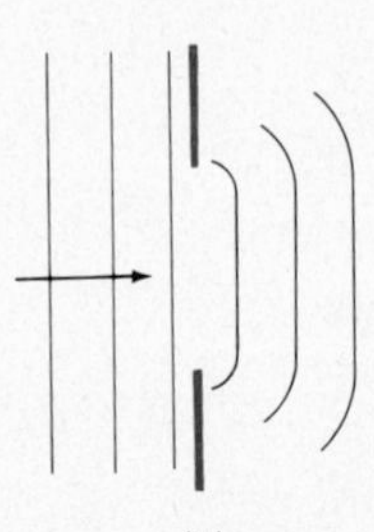

(b)

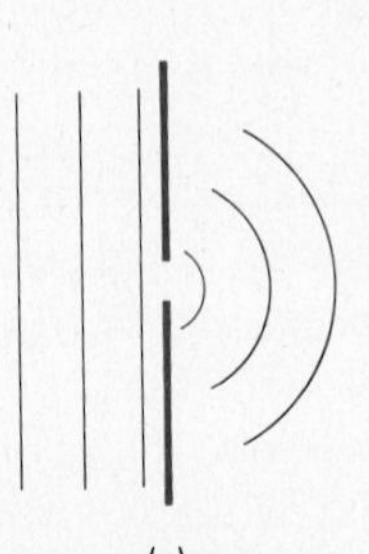

(c)

FIGURE 22-2
Huygens' principle is consistent with diffraction (*a*) around the edge of an obstacle, (*b*) through a large hole, (*c*) through a small hole whose size is on the order of the wavelength of the wave.

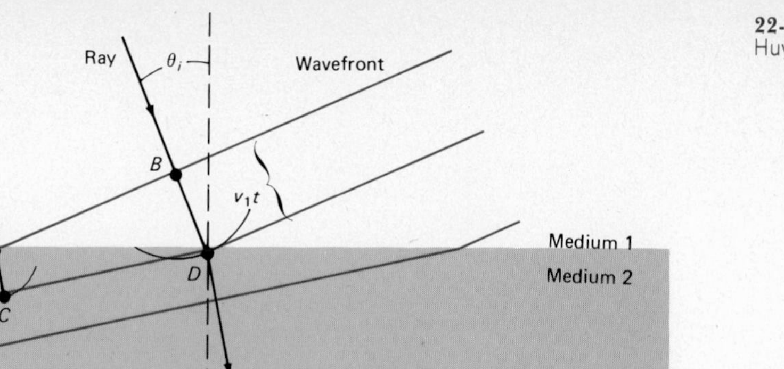

FIGURE 22-3
Refraction explained, using Huygens' principle.

Refraction

The laws of reflection and refraction were well known in Newton's time. The law of reflection could not distinguish the two theories. For when waves reflect from an obstacle, the angle of incidence equals the angle of reflection (Figure 11-25); the same is true of particles—think of a tennis ball without spin striking a flat surface.

The law of refraction is another matter. Consider light entering a medium where it is bent toward the normal as when it travels from air into water. As shown in Figure 22-3, this effect can be constructed using Huygens' principle if we assume that the speed of light is less in the second medium ($v_2 < v_1$). That is, in time t, the point B on wavefront AB goes a distance $v_1 t$ to reach point D; point A, on the other hand, travels a distance $v_2 t$ to reach point C. Huygens' principle is applied to points A and B to obtain the curved wavelets shown at C and D. The wavefront is tangent to these two wavelets, so the new wavefront is the line CD. Hence the rays (which are perpendicular to the wavefronts) bend toward the normal if $v_2 < v_1$ as drawn.† Newton's corpuscle theory predicted the opposite result. Newton argued that, if the path of light corpuscles entering a new medium changes direction, it must be because the medium exerts a force on the corpuscles at the boundary. This force was assumed to act perpendicular to the boundary and thus would affect only the perpendicular component of the corpuscles' velocity. When light enters a medium such as water where it is bent toward the normal, the force must accelerate the corpuscles so that the perpendicular component of velocity is increased; only in this way will the refracted angle be less than the incident. Thus, in Newton's theory, the speed of light is greater in the second medium ($v_2 > v_1$). So the wave theory predicts that the speed of light in water (say) is less than in air, and Newton's corpuscle theory predicts the reverse. An experiment to actually measure the speed of light in water and confirm the wave-theory prediction was not done until 1850 (by Foucault) and by then the wave theory was fully accepted, as we shall now see.

† This is basically the same as the discussion around Figure 11-29.

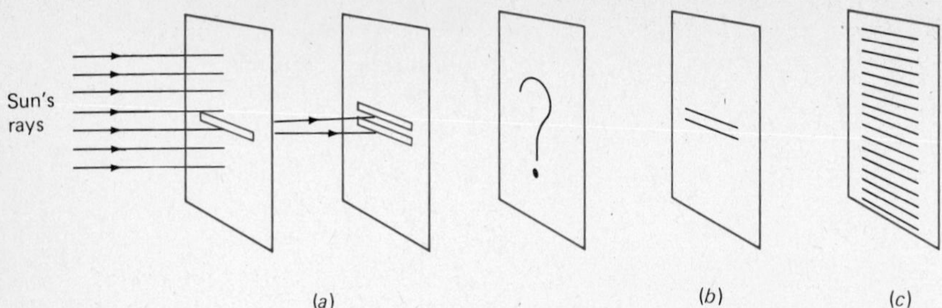

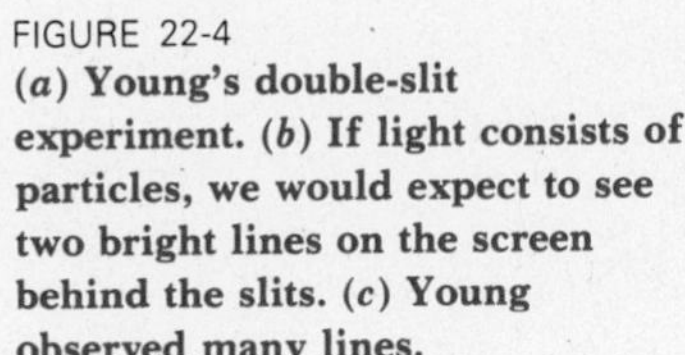

FIGURE 22-4
(*a*) Young's double-slit experiment. (*b*) If light consists of particles, we would expect to see two bright lines on the screen behind the slits. (*c*) Young observed many lines.

22-2 ▌ Interference—Young's double-slit experiment

In 1801, the Englishman Thomas Young (1773–1829) obtained convincing evidence for the wave nature of light and was even able to measure the wavelength. Figure 22-4*a* shows a diagram of Young's famous double-slit experiment. Light from a source (Young used the sun) passes through a slit S and then falls on a second screen containing two closely spaced slits S_1 and S_2. If light consists of particles, you would expect to see two bright lines on a screen placed behind the slits (part *b*). But Young observed instead a series of bright lines as shown in part *c*. Young was able to explain this result as a **wave-interference** phenomenon. To see this, imagine plane waves of light of a single wavelength ("monochromatic" light) falling on the two slits as shown in Figure 22-5. Because of diffraction, the waves leaving the two small slits spread out as shown. This is equivalent to the interference pattern produced when two rocks are thrown into a lake (Figure 11-31), or when sound from two loudspeakers interferes (Figure 12-12).

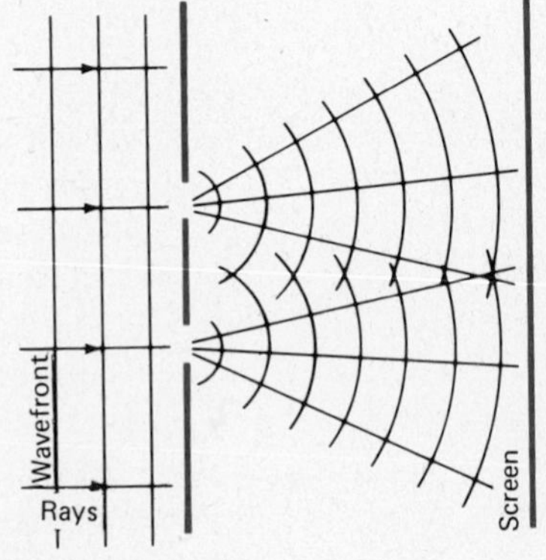

FIGURE 22-5
If light is a wave, light passing through one of two slits should interfere with light passing through the second slit.

To see how an interference pattern is produced on the screen, we make use of Figure 22-6. Waves of wavelength λ are shown entering slits S_1 and S_2, which are a distance d apart. The waves spread out in all directions after passing through the slits, but they are shown only for three different angles θ. In part *a* the waves reaching the center of the screen are shown ($\theta = 0$); the waves from the two slits travel the same distance, so they are in phase and constructive interference occurs; so there is a bright spot at the center of the screen. There will also be constructive interference when the paths of the two rays differ by one wavelength (or any whole number of wavelengths), as shown in part *b*. But if one ray travels an extra distance of one half a wavelength (or $\frac{3}{2}\lambda$, $\frac{5}{2}\lambda$, and so on) the two waves are exactly out of phase when they reach the screen, so destructive interference occurs and the screen is dark, Figure 22-6*c*. Thus there will be a series of bright and dark lines (or *fringes*).

FIGURE 22-6
How the wave theory explains the pattern of lines seen in the double-slit experiment.

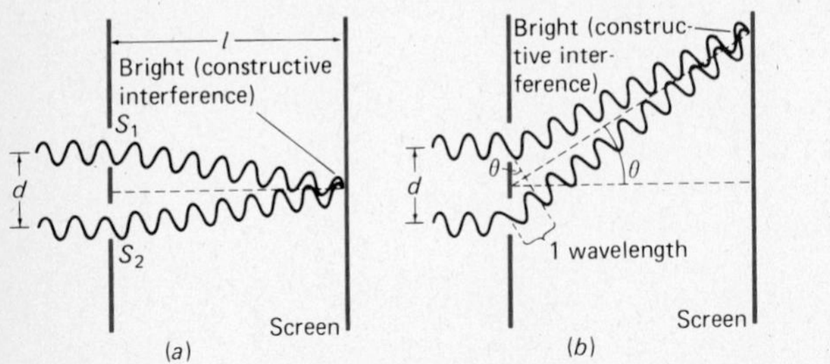

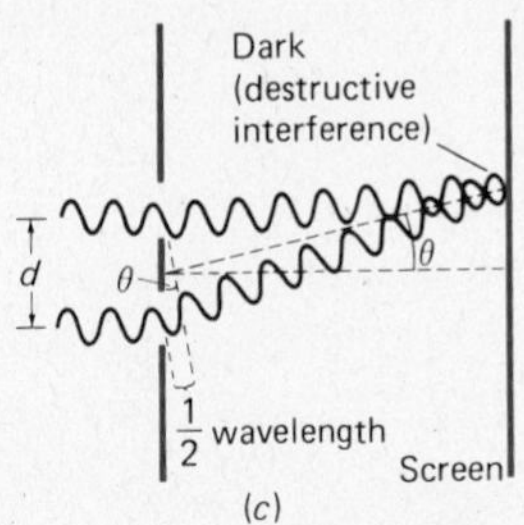

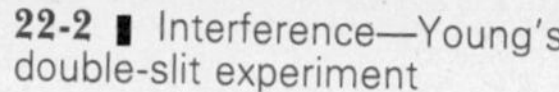

To determine exactly where the bright lines fall, first note that Figure 22-6 is somewhat exaggerated; in real situations, the distance d between the slits is very small compared to the distance l to the screen. The two rays shown for each case will therefore be very nearly parallel and θ is the angle they make with the horizontal. From the tiny right triangle drawn in Figure 22-6b, whose hypotenuse is d, we can see that the extra distance traveled by the lower ray is $d \sin \theta$. *Constructive interference* will occur on the screen when $d \sin \theta$ equals a whole number of wavelengths.

$$d \sin \theta = m\lambda, \quad m = 0, 1, 2, \ldots \quad \text{(constructive interference).} \quad (22\text{-}1)$$

The value of m is called the *order* of the interference fringe. Destructive interference occurs when the extra distance $d \sin \theta$ is $\frac{1}{2}$, $\frac{3}{2}$, etc., wavelengths:

$$d \sin \theta = (m + \tfrac{1}{2})\lambda, \quad m = 0, 1, 2, \ldots \quad \text{(destructive interference).}$$

The intensity of the bright fringes is greatest for the center one ($m = 0$) and decreases for higher orders, as shown in Figure 22-7.

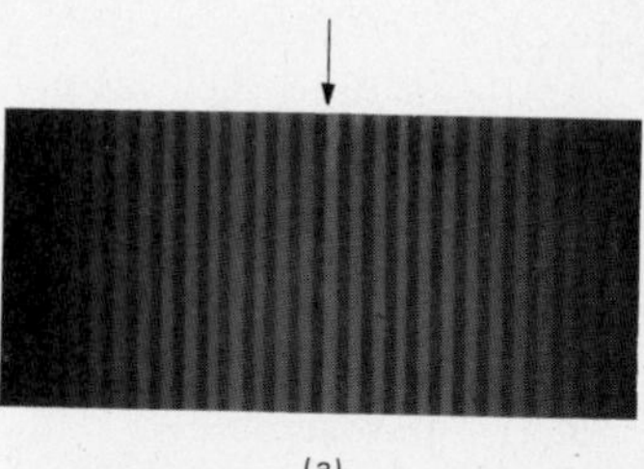

(a)

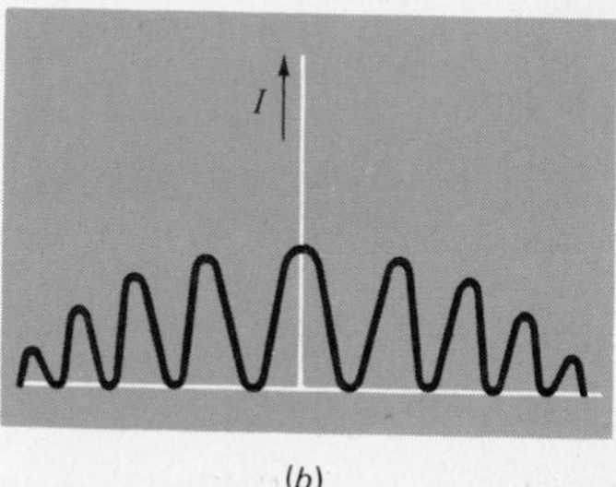

(b)

FIGURE 22-7
(a) Interference fringes produced by double-slit experiment detected by photographic film placed on the viewing screen. The arrow marks the zeroth fringe. (From F. W. Sears, M. W. Zemansky, and Hugh D. Young, *College Physics*, 4th ed., © 1974, Addison-Wesley Publishing Co., Inc., Reading, Mass., Figure 42-3. Reprinted with permission.) (b) Intensity of light in the interference pattern.

▮ EXAMPLE 22-1 A screen containing two slits 0.10 mm apart is 1.20 m from the viewing screen. Light of wavelength $\lambda = 500$ nm falls on the slits from a distant source. Approximately how far apart will the bright interference fringes be on the screen?

SOLUTION Since $d = 0.10 \text{ mm} = 1.0 \times 10^{-4}$ m and $l = 1.20$ m, first-order fringe ($m = 1$) occurs at an angle θ given by

$$\sin \theta_1 = \frac{m\lambda}{d} = \frac{(1)(500 \times 10^{-9} \text{ m})}{1.0 \times 10^{-4} \text{ m}} = 5.0 \times 10^{-3}.$$

This is a very small angle, so we can take $\sin \theta = \theta$. The first-order fringe will occur a distance x_1 above the center of the screen (see Figure 22-8) given by $x_1/l = \theta$, so

$$x_1 = l\theta_1 = (1.20 \text{ m})(5.0 \times 10^{-3}) = 6.0 \text{ mm}.$$

The second fringe ($m = 2$) will occur at

$$x_2 = l\theta_2 = l\frac{2\lambda}{d} = 12.0 \text{ mm}$$

above the center, and so on. Thus the fringes are 6.0 mm apart. ▮▮

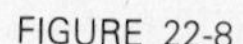

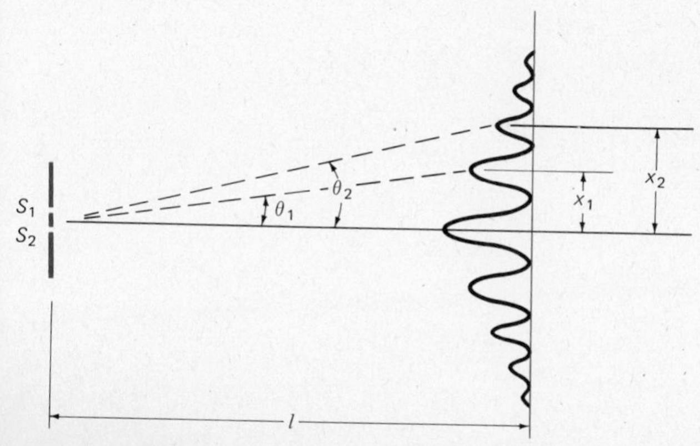

FIGURE 22-8
For small angles, the interference fringes occur at distances $x = \theta l$ above the center ($m = 0$) fringe.

From Equation 22-1 we can see that, except for the zeroth-order fringe at the center, the position of the fringes depends on wavelength. Thus, as Young found in his experiments, when white light falls on the two slits, the central fringe is white but the first- (and higher-) order fringes are like tiny rainbows; θ was found to be smallest for violet light and largest for red. By measuring the position of these fringes, Young became the first to determine the wavelength of visible light (using Equation 22-1). In doing so, he showed that what distinguishes different colors physically is their wavelength.

▮ EXAMPLE 22-2 White light passes through two slits 0.50 mm apart and an interference pattern is observed on a screen 2.5 m away. The first-order fringe resembles a rainbow with violet and red light at either end. The violet light falls about 2.0 mm and the red 3.5 mm from the center of the central white fringe. Estimate the wavelengths of the violet and red lights.

SOLUTION We use Equation 22-1 with $m = 1$ and $\sin\theta = \theta$. Then for violet light, $x = 2.0$ mm, so

$$\lambda = \frac{d\theta}{m} = \frac{d}{m}\frac{x}{l} = \left(\frac{5.0\times10^{-4}\ \text{m}}{1}\right)\left(\frac{2.0\times10^{-3}\ \text{m}}{2.5\ \text{m}}\right) = 4.0\times10^{-7}\ \text{m},$$

or 400 nm. For red light, $x = 3.5$ mm, so λ is about 700 nm. ▮▮

The two slits in Figure 22-6 act as if they were sources of light. They are called *coherent sources* because the waves leaving them bear the same phase relationship to each other at all times—that is, they are in phase. An interference pattern is observed only when the sources are coherent. If two tiny light bulbs replaced the two slits (or separate light bulbs illuminated each slit), an interference pattern would not be seen; the light emitted by one light bulb would have a random phase with respect to a second bulb. They would therefore be *incoherent sources.*

22-3 ▮ Visible spectrum and dispersion

The visible spectrum

The two most obvious properties of light are readily describable in terms of the wave theory of light: intensity (or brightness) and color. The *intensity* of light is related to the square of the amplitude of the wave, just as for any wave (see Section 11-7 and Equations 20-7 and 20-8). The *color* of the light, as Young showed, is related to the wavelength or frequency of the light. Visible light—that to which our eyes are sensitive—falls in the wavelength range of about 400 to 750 nm.† This is known as the **visible spectrum**, and within it lie the different colors of the rainbow from violet to red, as shown in Figure 22-9.

Light with wavelengths shorter than 400 nm is called ultraviolet (UV) and that with wavelength greater than 750 nm is called infrared (IR).‡

† Sometimes the angstrom (Å) unit is used when referring to light: $1\ \text{Å} = 10^{-10}$ m; then visible light falls in the wavelength range 4000 to 7500 Å.

‡ As discussed in Chapter 20, light is considered to be an electromagnetic wave. The complete electromagnetic spectrum is illustrated in Figure 20-8.

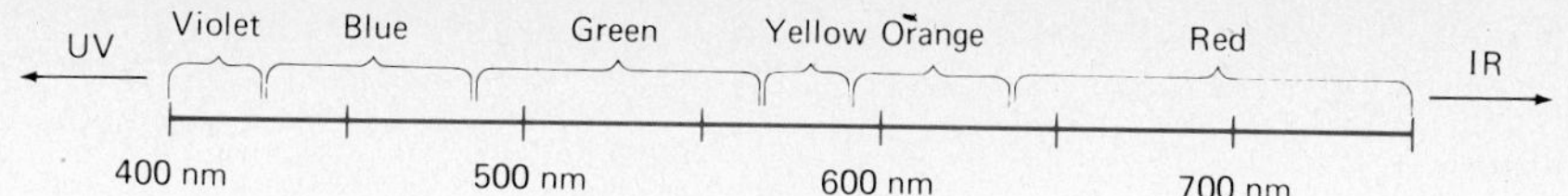

FIGURE 22-9
The spectrum of visible light, showing the range of wavelengths for the various colors. See also the lowest photograph of Figure 22-22 (color plate).

Although human eyes are not sensitive to UV or IR, these wavelengths can be detected by various instruments and some types of photographic film.

Dispersion; prisms and rainbows

It is a familiar fact that a prism separates white light into a rainbow of colors. This is due to the fact that the index of refraction of a material depends on the wavelength. This is shown for several materials in Figure 22-10. White light is a mixture of all visible wavelengths; and when incident on a prism as in Figure 22-11, the different wavelengths are bent to varying degrees. Since the index of refraction is greater for the shorter wavelengths, violet light is bent the most and red the least as indicated. This spreading of white light into the full spectrum is called **dispersion**.

Rainbows are a spectacular example of dispersion. Water's index of refraction depends on wavelength too. You can see rainbows when you look at falling water with the sun at your back. Figure 22-12 shows how red and violet rays are bent by spherical water droplets and are reflected off the back surface. Red is bent the least and so reaches the observer's eyes from droplets higher in the sky. Thus the top of the rainbow is red.

Diamonds achieve their brilliance from a combination of dispersion and total internal reflection. Since diamonds have an incredibly high index of refraction of 2.42, the critical angle for total internal reflection is only 24°. Incident light therefore strikes many of the internal surfaces before it strikes one at less than 24° and emerges. After many such reflections the light is well dispersed into colors that appear brilliant to the eye.

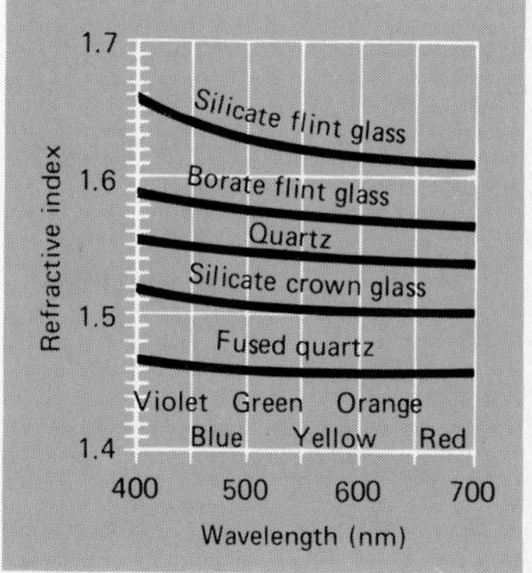

FIGURE 22-10
Index of refraction as a function of wavelength.

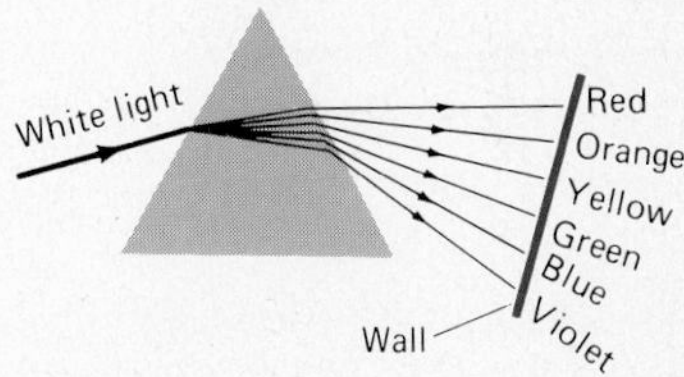

FIGURE 22-11
White light dispersed by a prism into the visible spectrum.

22-4 ▌ Diffraction by a single slit or disk

Young's double-slit experiment put the wave theory of light on a firm footing. But full acceptance did not come until studies of diffraction more than a decade later.

In 1816, Augustin Fresnel (1788–1827) presented to the French Academy a wave theory of light that predicted interference and

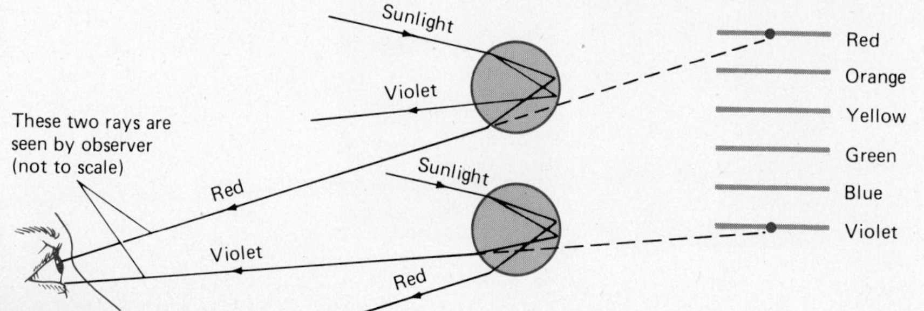

FIGURE 22-12
Formation of a rainbow.

diffraction effects. He was disappointed to learn that much of his work had been done earlier by Young. Worse, his theory was ridiculed by many of his colleagues, including the great Siméon Poisson (1781–1840). Poisson pointed out that according to Fresnel's wave theory, if light from a point source were to fall on a solid disk, then light diffracted around the edges should constructively interfere at the center of the shadow (Figure 22-13)—and that, of course, seemed ridiculous. But when the experiment was actually carried out in 1818 by François Arago, the bright spot was seen at the very center of the shadow! Little opposition to the wave theory remained.

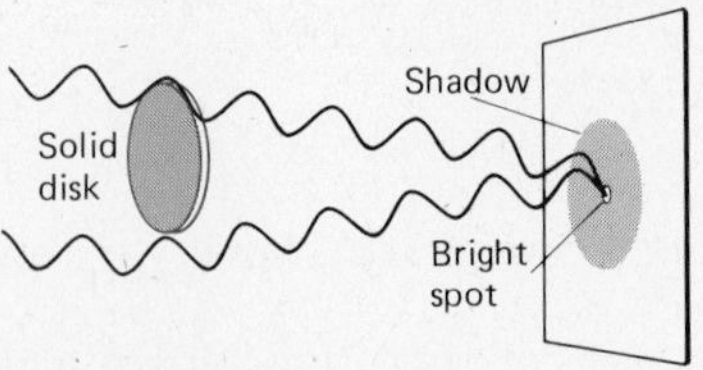

FIGURE 22-13
If light is a wave, a bright spot will appear at the center of the shadow of a solid disk illuminated by a point source of monochromatic light.

Figure 22-14*a* is a photograph of the shadow cast by a penny using a (nearly) point source of light (a laser in this case). The bright spot is clearly present at the center. Note that there also are bright and dark fringes beyond the shadow. These resemble the interference fringes of a double slit. Indeed, they are due to interference of waves diffracted around different parts of the disk, and the whole is referred to as a *diffraction pattern.* A diffraction pattern exists around any sharp object illuminated by a point source (Figure 22-14*b* and *c*). We are not always aware of them because most sources of light in everyday life are not points; so light from different parts of the source washes out the pattern.

To see how a diffraction pattern arises, we will analyze the important case of monochromatic light passing through a narrow slit. We will assume that parallel rays of light fall on the slit of width D, as shown (greatly enlarged) in Figure 22-15. If the viewing screen is infinitely far away, or a lens is placed behind the slit to focus parallel rays on the screen, the diffraction pattern is called *Fraunhofer diffraction.* If the screen is close and no lenses used, it is called *Fresnel diffraction.* For simplicity, we will assume that the screen is very far away, so that we have the former case. As we know from studying water waves and from Huygens' principle, the waves passing through the slit spread out in all directions. We will now examine how the waves passing through different parts of the slit interfere with each other.

First, we consider rays that pass straight through, as in Figure 22-15*a*.

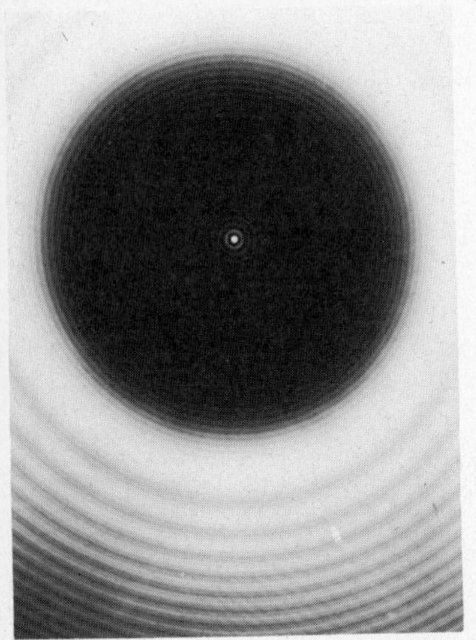

(*a*)

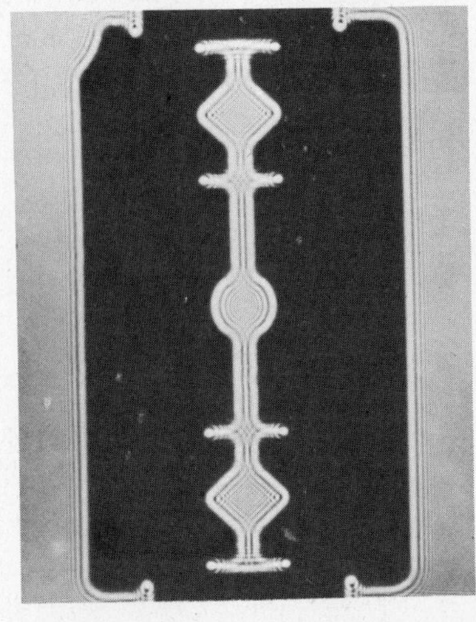

(*b*)

(*c*)

FIGURE 22-14
Diffraction pattern of (*a*) a penny (courtesy of P. M. Rinard, from *Am. J. Phys.*, 44, 1970, p. 70), (*b*) a razor blade (from F. W. Sears, *Optics*, © 1949, Addison-Wesley Publishing Co., Inc., Reading, Mass., Figure 9-8; reprinted with permission), (*c*) a single slit, illuminated by a nearly point source of monochromatic light (from M. Cagnet, M. Françon, and J. C. Thrierr, *Atlas of Optical Phenomena*, Springer-Verlag, Berlin, 1962).

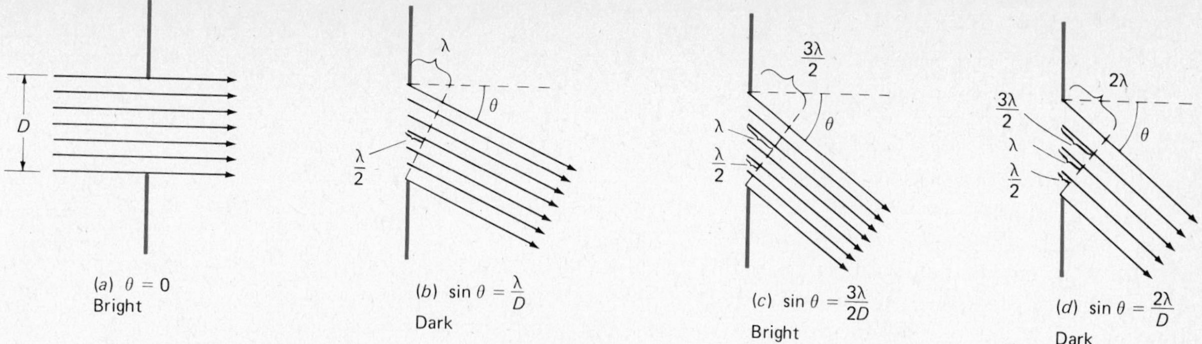

FIGURE 22-15
Analysis of diffraction patterns formed by light passing through a narrow slit.

They are all in phase, so there will be a central bright spot on the screen. In part *b*, we consider rays diffracted at an angle θ such that the ray from the top of the slit travels exactly one wavelength farther than the ray from the bottom edge. The ray passing through the very center of the slit will travel one half-wavelength farther than the ray at the bottom of the slit; these two rays will be exactly out of phase with one another, so will destructively interfere. Similarly, a ray slightly above the bottom one will cancel a ray that is the same distance above the central one. Indeed, each ray passing through the lower half of the slit will cancel with a corresponding ray passing through the upper half. Thus all the rays destructively interfere in pairs, so no light will reach the viewing screen at this angle. The angle θ at which this occurs can be seen from the diagram to occur when $\lambda = D\sin\theta$, or

$$\sin\theta = \frac{\lambda}{D} \qquad \text{(first minimum).} \tag{22-2}$$

The light intensity is a maximum at $\theta = 0°$ and decreases to a minimum (intensity = zero) at the angle θ given by Equation 22-2.

Now consider a larger angle θ such that the top ray travels $\frac{3}{2}\lambda$ farther than the bottom ray, as in Figure 22-15*c*. In this case the rays from the bottom third of the slit will cancel in pairs with those in the middle third, since they will be $\lambda/2$ out of phase. However, light from the top third of the slit will still reach the screen, so there will be a bright spot, but not as bright as the central spot at $\theta = 0°$. For an even larger angle θ such that the top ray travels 2λ farther than the bottommost ray (Figure 22-15*d*), rays from the bottom quarter of the slit will cancel with those in the quarter just above it since the path lengths differ by $\lambda/2$. And the rays through that quarter of the slit just above center will cancel with those through the top quarter. At this angle there will again be a minimum of zero intensity in the diffraction pattern. A plot of the intensity as a function of angle is shown in Figure 22-16. This corresponds well with Figure 22-14*c*. Notice that minima (zero intensity) occur at

$$D\sin\theta = m\lambda, \qquad m = 1, 2, 3, \ldots$$

but not at $m = 0$, where there is a strong maximum. Between the minima, smaller brightness maxima occur.

FIGURE 22-16
Intensity in diffraction pattern of a single slit as a function of sin θ.

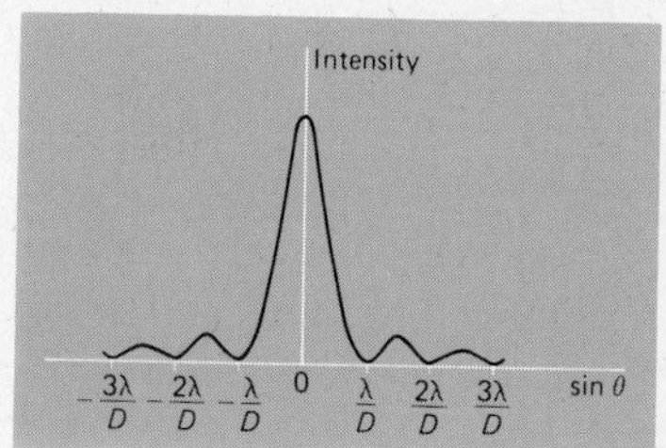

EXAMPLE 22-3 Light of wavelength 750 nm passes through a slit 1.0×10^{-3} mm wide. On a screen 20 cm away, how wide is the central maximum (*a*) in degrees, and (*b*) in centimeters?

SOLUTION (*a*) The first minimum occurs at

$$\sin\theta = \frac{\lambda}{D} = \frac{7.5\times10^{-7}\ \text{m}}{1.0\times10^{-6}\ \text{m}} = 0.75,$$

so $\theta = 49°$. This is the angle between the center and the first minimum (Figure 22-17). The angle subtended by the whole central maximum, between the minima above and below the center, is twice this, or 98°. (*b*) The width of the central maximum is $2x$, where $\tan 49° = x/20$ cm. So $2x = 2(20\ \text{cm})(\tan 49°) = 46$ cm, so a large width of the screen will be illuminated. It will not be terribly bright, normally, since the amount of light that passes through such a small slit will be small and it is spread over a large area. ▮▮

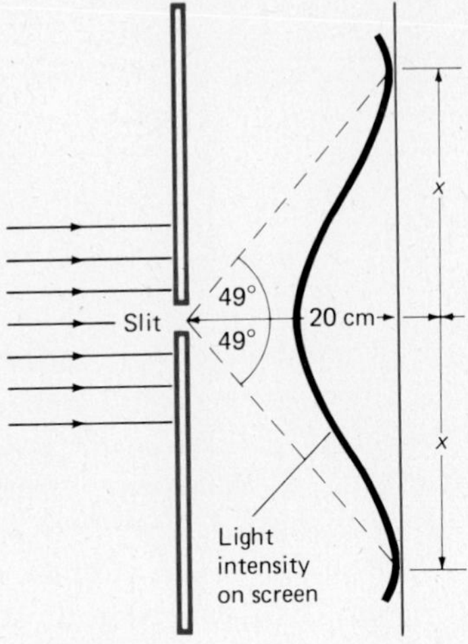

FIGURE 22-17
Example 22-3.

From Equation 22-2 we can see that the smaller the aperture D, the larger the central diffraction maximum. This is consistent with our earlier study of waves in Chapter 11. If the slit width is much larger than the wavelength, however, the diffraction pattern will not be clear and will not appear as in Figure 22-16. In this case the central diffraction maximum is smaller than the slit width, and it is overwhelmed by light passing straight through ($\theta = 0°$).

22-5 ▮ Diffraction grating

A large number of equally spaced parallel slits is called a **diffraction grating**, although the term "interference grating" might be as appropriate. Gratings are often made by ruling very fine lines on glass with a diamond tip. The untouched spaces between the lines serve as the slits. Photographic transparencies of an original grating serve as inexpensive gratings. Photographic reduction can be used to make very fine gratings. Gratings containing 10,000 lines per centimeter are common today, and are very useful for precise measurements of wavelengths.

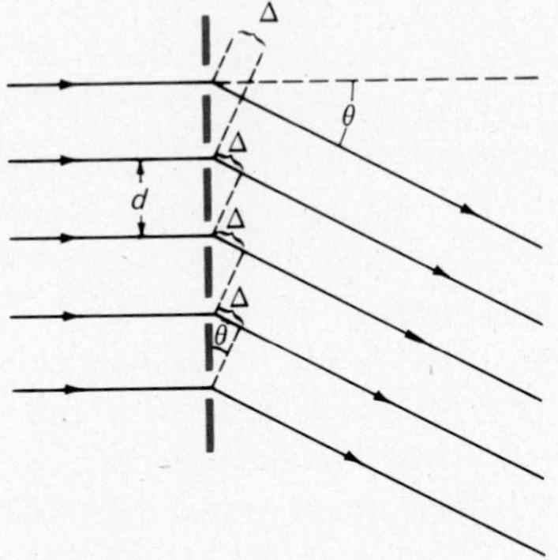

FIGURE 22-18
Diffraction grating.

The analysis of a diffraction grating is much like that of Young's double-slit experiment. We assume that parallel rays of light are incident on the grating, as shown in Figure 22-18. We also assume that the slits are narrow enough so that diffraction by each of them spreads light over a very wide angle on a distant screen behind the grating, and interference can occur with light from all the other slits. Light rays that pass through each slit with essentially no deviation ($\theta = 0°$) interfere constructively to produce a bright spot at the center of the distant screen. At an angle θ such that rays from adjacent slits must travel an extra distance (to reach the screen) of $\Delta = m\lambda$, where m is an integer, constructive interference again occurs. Thus, if d is the distance between slits,

$$\sin\theta = \frac{m\lambda}{d}, \qquad m = 0, 1, 2, \ldots \ \text{[maxima]} \tag{22-3}$$

in order to have a brightness maximum. This is the same equation as for the double-slit situation, and again m is called the *order* of the pattern.

There is an important difference between a double-slit and a multislit pattern, however. The bright maxima are much *sharper* and *narrower* for a grating. Why this happens can be seen as follows. Suppose that the angle θ is increased just slightly beyond that required for a maximum. In the case of only two slits, the two waves will be only slightly out of phase, so nearly full constructive interference occurs. This means the maxima are wide (Figure 22-7). For a grating, the waves from two adjacent slits will also not be significantly out of phase. But waves from one slit and those from a second a few hundred slits away may be exactly out of phase; all or nearly all the light will cancel in pairs in this way. For example, suppose that the angle θ is different from its first-order maximum so that the extra path length is not exactly λ but rather 1.001λ. The wave through one slit and another one 500 slits below will be out of phase by $1.500\lambda = 1\frac{1}{2}\lambda$ and the two will cancel. A pair of slits, one below each of these, will also cancel. That is, the light from slit 1 cancels with that from slit 501; light from slit 2 cancels with that from slit 502; and so on. Thus, the more lines there are in a grating, the sharper the peaks will be (Figure 22-19).† Because a grating produces much sharper (and brighter) lines than do two slits alone, it is a far more precise device for measuring wavelengths.

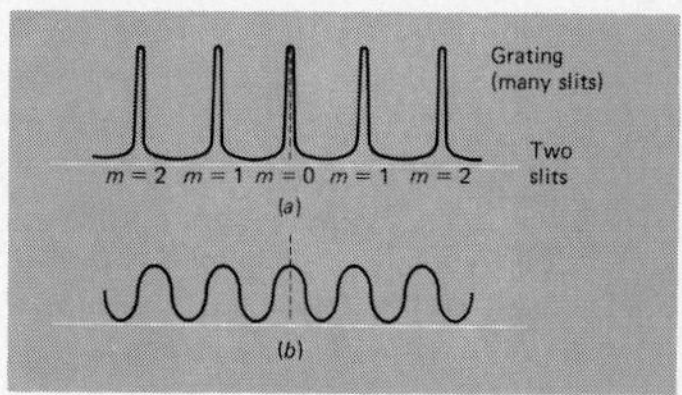

FIGURE 22-19
Intensity as a function of angle, or position, on a viewing screen, for (*a*) a grating and (*b*) two slits. Part (*b*) illustrates the same situation as Figure 22-7.

Suppose that the light striking a diffraction grating is not monochromatic, but consists of two or more distinct frequencies. Then for all orders other than $m = 0$, each wavelength will produce a maximum at a different angle (Figure 22-20*a*). If white light strikes a grating, the central ($m = 0$) maximum will be a sharp white peak; but for all higher orders, there will be a distinct rainbow spread out over a certain angular width (Figure 22-20*b*). Because a diffraction grating spreads out light into its component wavelengths, the resulting pattern is called a *spectrum*.

▌ **EXAMPLE 22-4** Calculate the angles for first- and second-order maxima for light of wavelength 400 and 700 nm if the grating contains 10,000 lines/cm.

† Depending on the number of slits, there may or may not be complete cancellation for such an angle, so there will be very tiny peaks between the main maxima, but they are usually much too small to be seen.

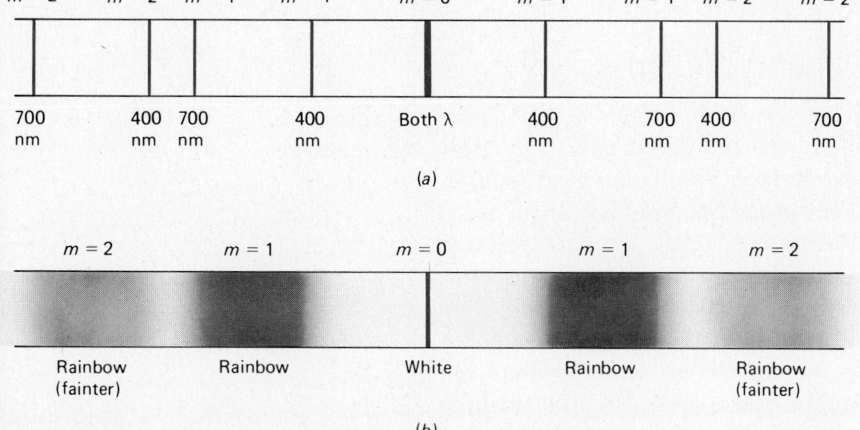

FIGURE 22-20
Spectra produced by a grating: (*a*) two wavelengths, 400 nm and 700 nm; (*b*) white light. The second order will be dimmer than the first order, normally; if grating spacing is small enough, second order may be missing. (Higher orders not shown.)

SOLUTION Since the grating contains 10^4 lines/cm $= 10^6$ lines/m, the separation between slits is $d = 1/10^6\ \text{m}^{-1} = 1.00 \times 10^{-6}$ m. In first order ($m = 1$), the angles are

$$\sin\theta_{400} = \frac{m\lambda}{d} = \frac{1(4.0\times 10^{-7}\ \text{m})}{1.00\times 10^{-6}\ \text{m}} = 0.400$$

$$\sin\theta_{700} = 0.700,$$

so $\theta_{400} = 23.6°$ and $\theta_{700} = 44.0°$. In second order,

$$\sin\theta_{400} = \frac{2(4.0\times 10^{-7}\ \text{m})}{1.00\times 10^{-6}\ \text{m}} = 0.800$$

$$\sin\theta_{700} = 1.40,$$

so $\theta_{400} = 53.0°$, but the second order does not exist for $\lambda = 700$ nm since $\sin\theta$ cannot exceed 1. No higher orders will appear. ▮▮

▮ EXAMPLE 22-5 White light containing wavelengths from 400 to 750 nm strikes a grating containing 4000 lines/cm. Show that the violet of the third-order spectrum overlaps the red of the second order.

SOLUTION The grating spacing is $d = 1/(4000\ \text{cm}^{-1}) = 2.50 \times 10^{-6}$ m. The violet of the third order occurs at an angle θ given by

$$\sin\theta = \frac{3(4.00\times 10^{-7}\ \text{m})}{2.50\times 10^{-6}\ \text{m}} = 0.480.$$

Red in second order occurs at

$$\sin\theta = \frac{2(7.50\times 10^{-7}\ \text{m})}{2.50\times 10^{-6}\ \text{m}} = 0.600,$$

which is clearly a greater angle. ▮▮

A diffraction grating containing slits, as we have discussed here, is called a *transmission grating*. *Reflection gratings* are also possible; they can be made by ruling fine lines on a metallic or glass surface from which light is reflected and examined. The analysis is basically the same as for a transmission grating.

22-6 ▮ The spectroscope and spectroscopy

A **spectroscope** (Figure 22-21) is a device to measure wavelengths accurately using a diffraction grating. Light from a source passes through a narrow slit S in the collimator. The slit is at the focal point of the lens L, so parallel light falls on the grating. The movable telescope can bring the rays to a focus. Nothing will be seen in the telescope unless it is positioned at an angle θ that corresponds to a diffraction peak (first order is usually used) of a wavelength emitted by the source. The angle θ can be measured† to very high accuracy, so the wavelength of a line can be determined to high accuracy using Equation 22-3. (The line you see in a

FIGURE 22-21
Spectroscope.

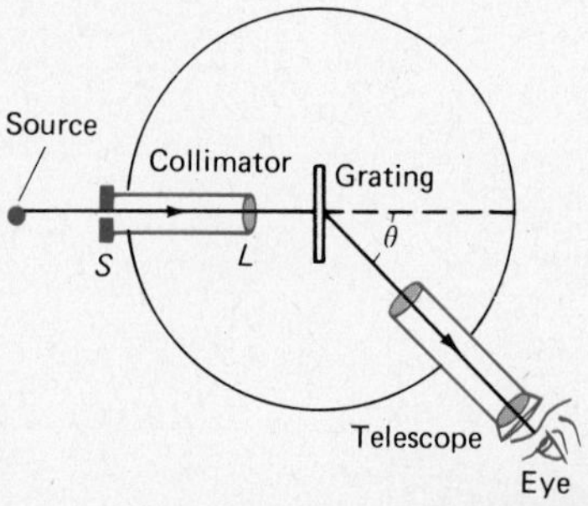

† The angle θ for a given wavelength is usually measured on both sides of center because the grating cannot generally be aligned precisely; the average value of θ is then used.

spectroscope corresponding to each wavelength is actually an image of the slit S; so the narrower the slit, the narrower—but dimmer—is the line, and the more precise its measurement. If the light contains a continuous range of wavelengths, you will then see a continuous spectrum in the spectroscope.)

In some spectroscopes, a reflection grating or a prism is used. A prism, because of dispersion (Section 22-3), bends light of different wavelengths into different angles. A prism has the advantage over a grating of deflecting more light, so is more useful for dim sources. However, it produces lines that are less sharp and it is less able to separate closely spaced lines.

With a normal diffraction grating, most of the light passes straight through to the central peak. However, reflection gratings can now be made that have shaped grooves, so that a large portion of the light is reflected at an appropriate angle to give a strong first-order peak. If the spectrum of a source is recorded on film rather than viewed by the eye, the device is called a *spectrometer*, as compared to a *spectroscope*, although these terms are sometimes used interchangeably. Devices that can also measure the intensity of light of a given wavelength are called *spectrophotometers*.

An important use of any of these related devices is for the identification of atoms or molecules. When a gas is heated or a large electric current is passed through it, the gas emits a characteristic **line spectrum**. That is, only certain wavelengths of light are emitted, and these are different for different elements and compounds.† Figure 22-22 (color plate) shows the line spectra for a number of elements in the gas state. Line spectra occur only for gases at high temperatures and low pressure. The light from heated solids, such as a light-bulb filament, and even from a dense gaseous object such as the sun, produces a *continuous spectrum* with a wide range of wavelengths.

As can be seen in Figure 22-22, the sun's spectrum contains a number of *dark* lines (only the most prominent are shown). These are called *absorption lines*. Atoms and molecules absorb light at the same wavelengths at which they emit light. The sun's absorption lines are due to absorption by atoms and molecules in the cooler outer atmosphere of the sun, as well as by atoms and molecules in the earth's atmosphere. A careful analysis of all these lines (there are thousands) reveals that at least two-thirds of all elements are present in the sun's atmosphere.

The presence of elements in the atmosphere of other planets, in interstellar space, and in stars is also determined by spectroscopy. Spectroscopy is useful in biology to determine the presence of certain types of molecules in specimens where chemical analysis would be difficult. For example, DNA and different types of protein absorb light in particular regions of the spectrum. The material to be examined, which is often in solution, is placed in a monochromatic light beam whose wavelength is chosen by placement of a prism or diffraction grating. The amount of absorption, as compared to a standard solution without the specimen, can reveal not only the presence of a particular type of molecule but also its concentration.

Light emission and absorption occur not only in the visible part of the

† Why atoms and molecules emit line spectra was a great mystery for many years and played a central role in the development of modern quantum theory, as we shall see in Chapter 26.

spectrum; they are often useful to investigate spectra in the UV or IR regions. Since glass absorbs light in these regions, reflection gratings and mirrors are used rather than lenses. Certain types of film, or photocell detectors, are used for detection.

22-7 ▮ Interference by thin films

Interference of light gives rise to many everyday phenomena, such as the bright colors reflected from soap bubbles and from thin oil films on water. In these and other cases, the colors are a result of constructive interference between light reflected from the two surfaces of the thin film. To see how this happens, consider a thin oil film lying on top of water (Figure 22-23). Part of the incident ray (say from the sun or street lights) is reflected at A on the top surface; and part of that transmitted is reflected at B on the lower surface. The part reflected at the lower surface must travel the extra distance ABC. If the distance ABC is equal to one or a whole number of wavelengths, the two waves will interfere constructively and the light will be bright. But if ABC equals $\frac{1}{2}\lambda$, $\frac{3}{2}\lambda$, and so on, the two waves will be out of phase and destructive inteference will occur. If white light falls on such a film, the path ABC will equal λ (or $m\lambda$ with $m =$ an integer) for only one wavelength at a given angle. This color will appear vivid since it is pure. For light reflected at a slightly different angle, the path ABC will be longer or shorter and a different color will undergo constructive interference. Thus a series of bright colors will be seen next to one another. Variations in thickness of the film will also alter the length ABC and therefore affect the color of light that is most strongly reflected.

A C Air
Oil
B Water

FIGURE 22-23
Light reflected from upper and lower surfaces of a thin film of oil lying on water.

When a curved glass surface is placed in contact with a flat glass surface (Figure 22-24), a series of concentric rings is seen when illuminated from above by *monochromatic light.* These are called *Newton's rings*† and they are due to interference between rays reflected by the top and bottom surfaces of the *air gap* between the two pieces of glass. Because this gap (which is equivalent to a thin film) increases in width from the central contact point out to the edges, the extra path length for the lower ray (equal to BCD) varies; where it equals 0, $\frac{1}{2}\lambda$, λ, $\frac{3}{2}\lambda$, 2λ, and so on, it corresponds to constructive and destructive interference; and this gives rise to the series of bright and dark lines seen in Figure 22-24*b*.

† Although Newton was the first to observe them, he did not realize their significance in support of a wave theory of light.

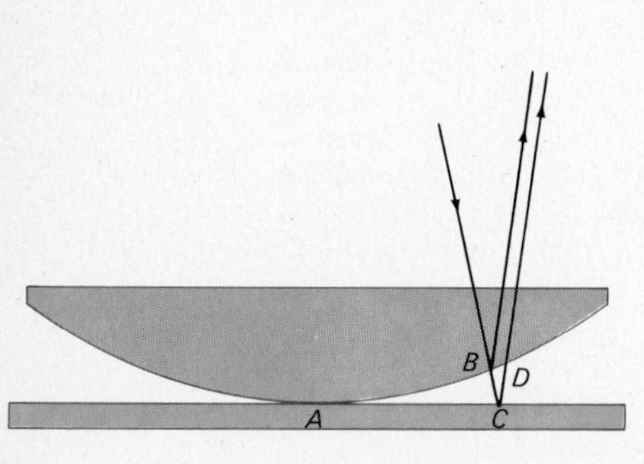

(*a*)

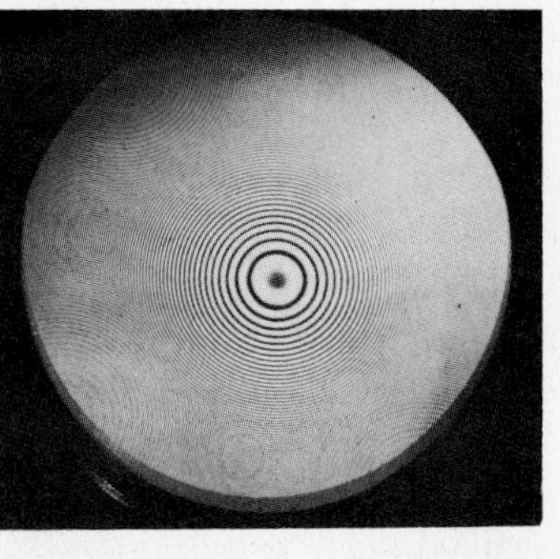

(*b*)

FIGURE 22-24
Newton's rings. (The circular patches at various places on the photograph of the rings are an artifact (moiré pattern) of the printing process.) (Part *b* courtesy of Bausch & Lomb Incorporated.)

Note that the point of contact of the two glass surfaces (A in Figure 22-24a) is dark (Figure 22-24b). Since the path difference is zero here, we might expect the rays reflected from each surface to be in phase and this point to be bright. But it is dark, which tells us that the two rays are not in phase; this happens because one of the waves undergoes a change in phase of 180° upon reflection. More specifically, it is found that *a beam of light reflected by a material whose index of refraction is greater than that in which it is traveling changes phase by* 180°. If the index is less than that of the material in which it is traveling, no phase change occurs. (This corresponds to the reflection of a wave traveling along a rope when it reaches the end; as we saw in Figure 11-24, if the end is tied down, the rope changes phase and the pulse flips over; but if the end is free, no phase change occurs.)

Thus the ray reflected by the curved surface above the air gap in Figure 22-24a undergoes no change in phase. That reflected at the lower surface, where the beam in air strikes the glass, undergoes a 180° change. Thus the two rays reflected at the point of contact A of the two glass surfaces will be 180°, or $\frac{1}{2}\lambda$, out of phase, and a dark spot occurs. Other dark bands will occur when the path difference BCD in Figure 22-24a is equal to an integral number of wavelengths. Bright bands will occur when the path difference is $\frac{1}{2}\lambda$, $\frac{3}{2}\lambda$, and so on, since the phase change at one surface effectively adds another $\frac{1}{2}\lambda$.

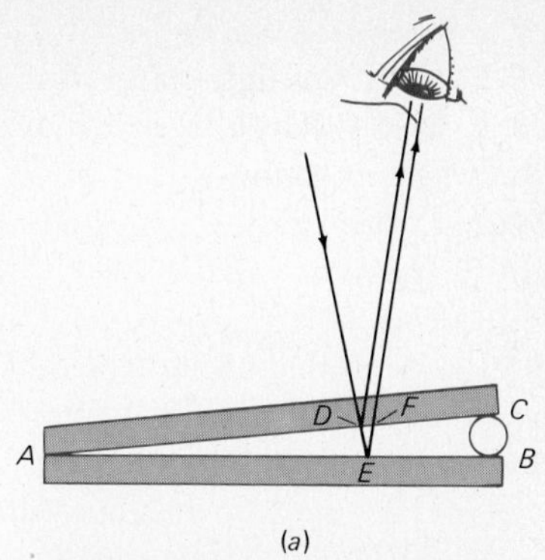

(a)

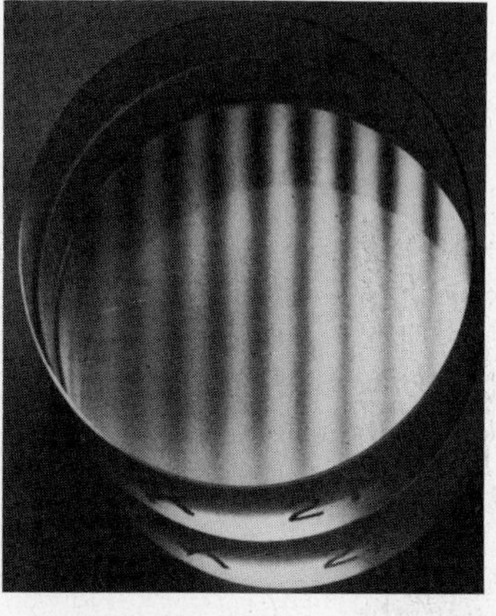
(b)

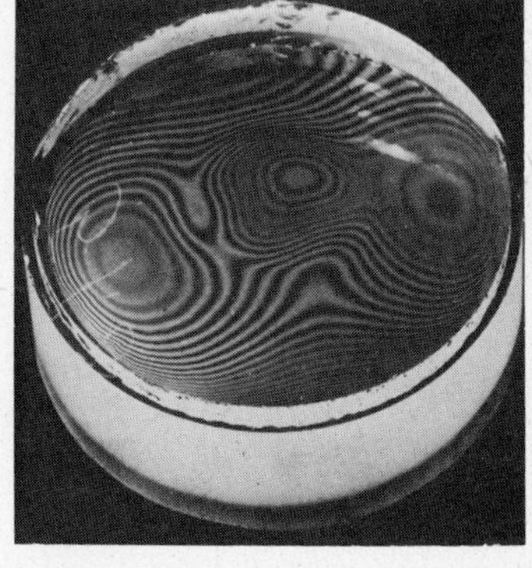
(c)

FIGURE 22-25
(a) Light rays reflected from upper and lower surfaces of a thin wedge of air interfere to produce bright and dark bands. (b) Pattern observed when glass plates are optically flat (courtesy of The Van Keuren Co.); (c) pattern when plates are not so flat (courtesy of Bausch & Lomb Incorporated.)

EXAMPLE 22-6 A very fine wire 7.35×10^{-3} mm in diameter is placed between two flat glass plates as in Figure 22-25a. Light whose wavelength is 600 nm falls perpendicularly on the plates, and a series of bright and dark bands is seen (Figure 22-25b). How many light and dark bands will there be in this case? Will the area next to the wire be bright or dark?

SOLUTION The thin film is the wedge of air between the two glass plates. Because of the phase change at the lower surface, there will be a dark band when the path difference is 0, λ, 2λ, 3λ, and so on. Since the light falls perpendicularly, the extra path length equals $2t$, where t is the thickness of the air gap at any point:

$$2t = m\lambda, \qquad m = 0, 1, 2, \ldots .$$

Bright bands occur when $2t = (m + \frac{1}{2})\lambda$, where m is an integer. At the position of the wire, $t = 7.35 \times 10^{-6}$ m. At this point there will be $2t/\lambda = (2)(7.35 \times 10^{-6}\ \text{m})/(6.00 \times 10^{-7}\ \text{m}) = 24.5$ wavelengths. Since this is a "half-integer," the area next to the wire will be bright. There will be a total of 25 dark lines along the plates, corresponding to path lengths of 0λ, 1λ, 2λ, 3λ, . . . , 24λ, including the one at the point of contact ($m = 0$). Between them there will be 24 bright lines plus the one at the end, or 25.

The bright and dark bands will be smooth only if the glass plates are extremely smooth. If they are not, the pattern is not smooth (Figure 22-25c)—a very precise way of testing a glass surface for smoothness.

If the wedge between the two glass plates of Example 22-6 is filled with some transparent substance other than air—say water—the pattern shifts. This is because the wavelength of the light is different in water than in air. (When light enters a medium where its speed is changed, the frequency is not altered; therefore, the wavelength must change.) In a material where the light speed is v, the wavelength $\lambda = v/f = c/nf = \lambda_0/n$, where λ_0 is the wavelength in vacuum. For instance, if the thin wedge of Example 22-6 were filled with water, $\lambda = 600\ \text{nm}/1.33 = 450\ \text{nm}$. Instead of 25 dark lines, there would be 33.

When white light (rather than monochromatic light as discussed above) is incident on the thin wedge of Figures 22-24*a* or 22-25*a*, a colorful series of fringes is seen. This is because constructive interference occurs in the reflected light at different locations along the wedge for different wavelengths. This difference in thickness is part of the reason bright colors appear when light is reflected from a soap bubble or a thin layer of oil on a puddle or lake. Which wavelengths appear brightest also depends on the viewing angle, as we saw earlier.

▮ EXAMPLE 22-7 A soap bubble appears green ($\lambda = 540$ nm) at its point nearest the viewer. What is its minimum thickness? Assume that $n = 1.35$.

SOLUTION The light is reflected perpendicularly from that point on a spherical surface nearest the viewer. Therefore, the path difference is $2t$, where t is the thickness of the soap film. Light reflected from the outer surface undergoes a 180° phase change, whereas that on the inner surface does not. Therefore, green light is bright for a (minimum) path difference of $\frac{1}{2}\lambda$. Thus $2t = \lambda/2n$, so $t = 540\ \text{nm}/(4)(1.35) = 100\ \text{nm}$. ▮▮

An important application of thin-film interference is in the coating of lenses. Glass reflects about 4 percent of the light incident upon it. Good-quality cameras, microscopes, and other optical devices may contain 6 to 10 thin lenses. Reflection from all these surfaces can reduce the light level considerably, and multiple reflections produce a background haze that reduces the quality of the image. A very thin coating on the lens surfaces can reduce these problems considerably. The amount of reflection at a boundary depends on the difference of index of refraction between the two materials. Ideally, the coating material should therefore have an index of refraction which is midway between air and glass. In this way, the amount of reflection at each surface is about the same; then destructive interference can occur nearly completely for some reflected wavelength, depending on the thickness of the coating. Nearby wavelengths will at least partially destructively interfere, but it is clear that a single coating cannot eliminate reflections for all wavelengths. Nonetheless, a coating can reduce reflection from 4 percent to 1 percent of the incident light. Often the coating is designed to eliminate the center of the spectrum, around 550 nm; the extremes of the spectrum—red and violet—will not be reduced as much; since a mixture of red and violet produces purple, the light seen reflected from such coated lenses is purple. Multiply-coated lenses, containing two or three separate coatings, can more effectively reduce a wider range of reflecting wavelengths.

▮ EXAMPLE 22-8 What is the thickness of an optical coating whose index of refraction is $n = 1.25$ and which is designed to eliminate wavelengths centered at 550 nm when incident perpendicularly?

SOLUTION Light reflected at both the front and back surfaces of the coating undergo a 180° phase shift. (Glass has $n \approx 1.50$, which is greater than that of the coating material.) Thus, for normal incidence, the thickness of the coating to produce destructive interference must be a half-integral number of wavelengths. That is, $2t = (m + \frac{1}{2})\lambda$, where m is an integer. The minimum thickness ($m = 0$) is usually chosen because destructive interference will then occur over the widest angle. Then $t = \lambda/4 = \lambda_0/4n = (550\text{ nm})/(4)(1.25) = 110\text{ nm}$.

*22-8 ▮ Michelson interferometer

Interference by thin films is the basis of an ingenious instrument invented by Albert A. Michelson. This is the *Michelson interferometer*,† shown in Figure 22-26. Monochromatic light strikes a half-silvered mirror M_s (also called the *beam splitter*). Half of the beam passes through to a fixed mirror M_2, where it is reflected back. The other half is reflected by M_s up to a movable mirror M_1, where it is also reflected back. Upon its return, part of beam 1 passes through M_s and reaches the eye; and part of beam 2, on its return, is reflected by M_s into the eye. (A compensator plate C of transparent glass is placed in the path of beam 2, so both beams pass through the same thickness of glass.) If the two path lengths are identical, the two beams entering the eye constructively interfere and brightness will be seen. If the movable mirror is moved a distance $\lambda/4$, one beam will travel an extra distance equal to $\lambda/2$ (because it travels back and forth over the distance $\lambda/4$). Hence the two beams will destructively interfere and darkness will be seen. As M_1 is moved farther, brightness will recur (when the path difference is λ), then darkness, and so on.

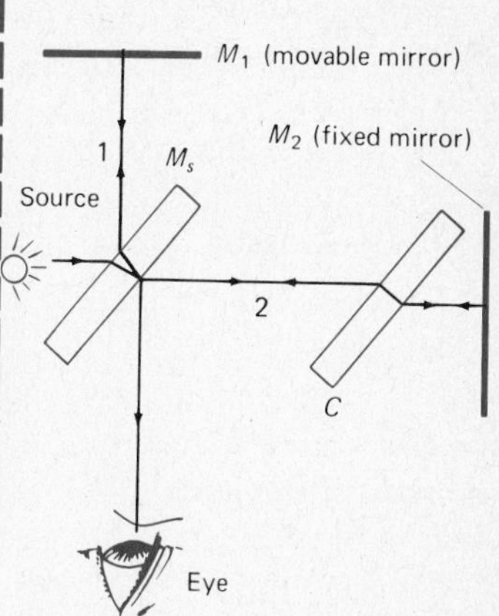

FIGURE 22-26
Michelson interferometer.

The eye sees the image of M_2 very close to the position of M_1; so, in effect, there is a thin layer of air (a "thin film") between this image and M_1. If M_1 is tilted slightly, the air gap is wedge-shaped, as in Figure 22-25, and a series of bright and dark lines are seen; when mirror M_1 is then moved, the bright and dark lines move to the left or right.‡

Very precise length measurements can be made with an interferometer. The motion of mirror M_1 by only $\frac{1}{4}\lambda$ produces a clear difference between brightness and darkness. For $\lambda = 400\text{ nm}$, this means a precision of $100\text{ nm} = 1 \times 10^{-4}\text{ mm}$! By observing the sideways motion of the fringes when the mirrors are not precisely aligned, even greater precision can be obtained. By counting the number of fringes, or fractions thereof, extremely precise length measurements can be made.

Michelson saw that the interferometer could be used to determine the length of the standard meter in terms of the wavelength of a particular light. Today, that standard is a particular orange-red line in the spectrum of krypton 86 (krypton atoms with atomic mass 86). Careful repeated measurements of the old standard meter (the distance between two marks on a platinum–iridium bar kept in Sèvres, near Paris) indicated that it was 1,650,763.73 wavelengths long. This number of wavelengths is now defined as the meter. Thus the primary standard of length can now be available in all laboratories of the world. And the accuracy of length measurement can be much higher.

† There are other types of interferometers, but Michelson's is the best known.

‡ When the path difference is many wavelengths, the observer no longer sees a nearly uniform brightness or darkness when the mirrors M_1 and M_2 are precisely aligned. Because the path difference is different for different angles of view, one sees a series of rings, much like the Newton ring pattern of Figure 22-24*b*. When mirror M_1 is moved, dark and bright spots seem to emerge from (or disappear into) the center of the pattern.

22-9 ▮ Polarization

An important and useful property of light is that it can be *polarized.* To see what this means, let us examine waves traveling on a rope. A rope may be set into vibration in a vertical plane, as in Figure 22-27*a*, or in a horizontal plane, as in part *b*. In either case the wave is said to be **plane-polarized**—that is, the oscillations are in a plane.

If we now place an obstacle containing a vertical slit in the path of the wave (Figure 22-28), a vertically polarized wave passes through, but a horizontally polarized wave will not. If a horizontal slit were used, the vertically polarized wave would be stopped. If both types of slit are used, neither wave could pass through. Note that polarization exists only for transverse waves, and not for longitudinal waves. The latter vibrate only along the direction of motion; and no kind of slit would affect them.

That light can be polarized was discovered in Newton's time. It was found that certain types of crystal, such as iceland spar, refract light into two rays; such crystals are called *doubly refracting*, and we will discuss them more in Section 22-11. Certain other crystals, such as tourmaline, would not transmit one or the other of these two rays, depending on the orientation of the crystal. Today we recognize that the two rays in a doubly refracting crystal are plane-polarized in mutually perpendicular directions, and that the tourmaline acts as a slit to eliminate one or the other if properly oriented. However, it was not until after the work of Young and Fresnel in the early 1800s that this phenomenon was recognized as evidence that light is a transverse wave. A half-century later, Maxwell's theory of light as electromagnetic (EM) waves was fully consistent with the facts of polarization since an EM wave is a transverse wave. The plane of polarization in a plane-polarized EM wave is taken as the direction of the electric field vector (Figure 20-5).

Light is not necessarily polarized. It can be **unpolarized**, which means that the source has vibrations in many planes at once (Figure 22-29). An ordinary incandescent light bulb emits unpolarized light, as does the sun.

Plane-polarized light can be obtained from unpolarized light using certain crystals such as tourmaline. Or, more commonly today, we can use a Polaroid sheet. Polaroid materials were invented in 1929 by Edwin Land and have been developed extensively since then. A Polaroid sheet consists of complicated long molecules arranged with their axes parallel. Such a Polaroid acts as a slit to allow one orientation of polarization to pass

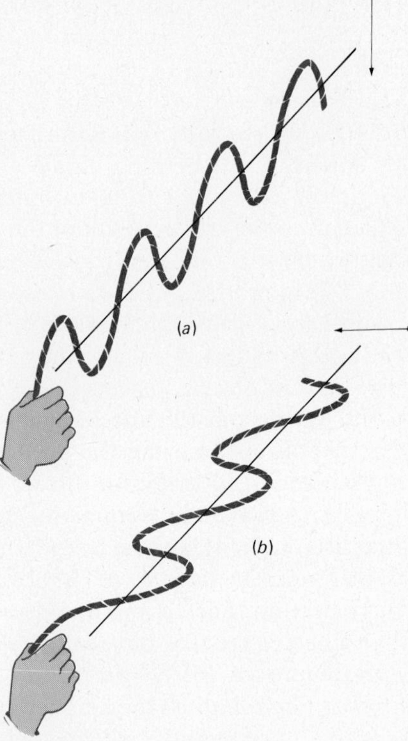

FIGURE 22-27
Transverse waves on a rope polarized (*a*) in a vertical plane and (*b*) in a horizontal plane.

FIGURE 22-28
Vertically polarized wave passes through a vertical slit, but a horizontally polarized wave will not.

through nearly undiminished (this direction is called the *axis* of the Polaroid), whereas a perpendicular polarization is absorbed almost completely. If a beam of plane-polarized light strikes a Polaroid whose axis is at an angle θ to the incident polarization direction, the beam will emerge plane-polarized parallel to the polaroid axis and its amplitude will be reduced by $\cos\theta$ (Figure 22-30). Thus a Polaroid passes only that component of polarization that is parallel to its axis. Since the intensity of a light beam is proportional to the square of the amplitude, we see that the intensity of a plane-polarized beam transmitted by a polarizer is

$$I = I_0 \cos^2 \theta, \tag{22-4}$$

where θ is the angle between the polarizer axis and the plane of the incoming wave and I_0 is the incoming intensity.

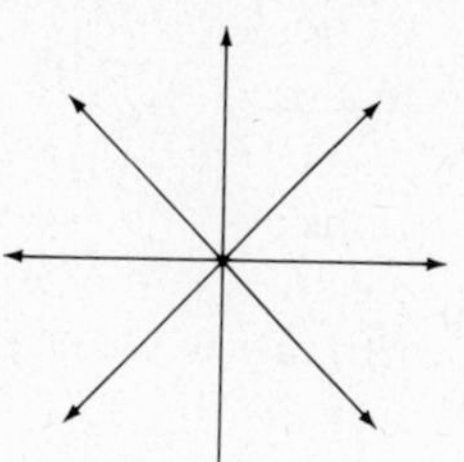

FIGURE 22-29

Vibrations of the electric field in unpolarized light.

A Polaroid can be used as a *polarizer* to produce plane-polarized light from unpolarized light, since only the component of light parallel to the axis is transmitted. A Polaroid can also be used as an *analyzer* to determine (1) if light is polarized, and (2) what is the plane of polarization. A Polaroid acting as an analyzer will pass the same amount of light independent of the orientation of its axis if the light is unpolarized; try rotating one lens of a pair of Polaroid sunglasses while looking through it at a light bulb. If the light is polarized, however, when you rotate the Polaroid the transmitted light will vary in intensity; it will be a maximum when the plane of polarization is parallel to the Polaroid's axis, and a minimum when perpendicular to it. If you rotate a Polaroid while looking through it at the sky, preferably at right angles to the sun's direction, you will see that skylight is polarized. (Direct sunlight is unpolarized, but do not look directly at the sun, even through a polarizer, for damage to the eye can occur.) If the transmitted light in this case falls to zero at one orientation of the analyzer Polaroid, then the light is 100 percent plane-polarized. If it merely reaches a minimum, the light is *partially polarized.*

Unpolarized light consists of light with random directions of polarizations. Each of these polarization directions can be resolved into components along two mutually perpendicular directions. Thus an unpolarized beam can be thought of as two plane-polarized beams of equal magnitude perpendicular to one another. When two Polaroids are *crossed*—that is, their axes are perpendicular to one another—unpolarized light can be entirely stopped (or nearly so—Polaroids are not

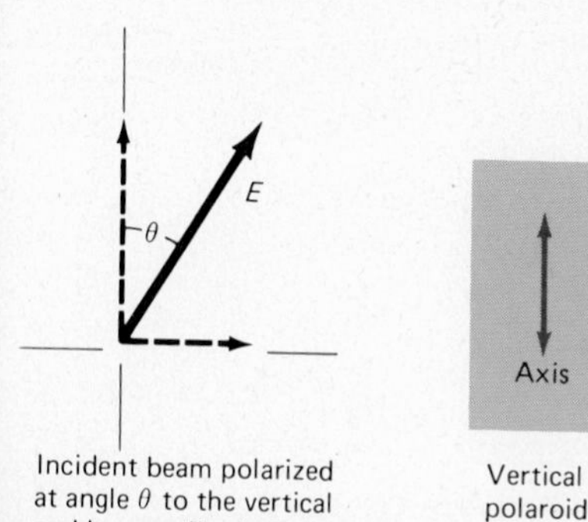

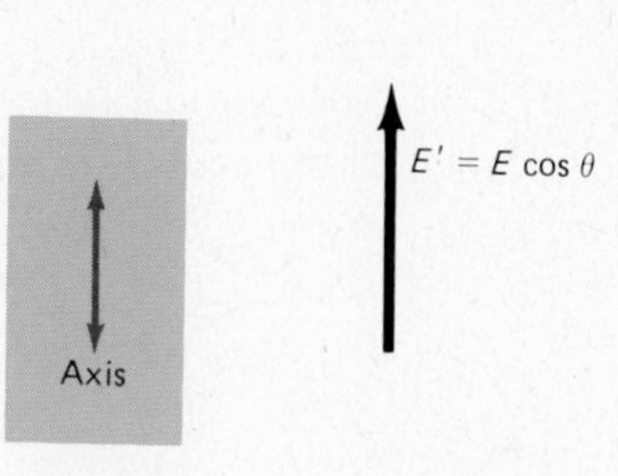

FIGURE 22-30

Vertical Polaroid transmits only the vertical component of a wave incident upon it.

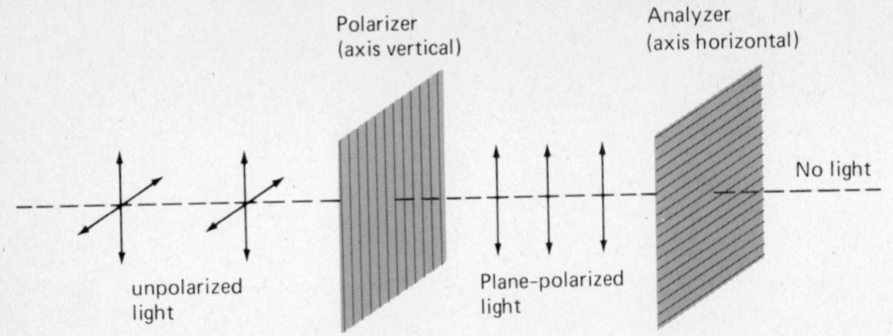

FIGURE 22-31
Crossed Polaroids completely eliminate light.

quite perfect). As shown in Figure 22-31, unpolarized light is made plane-polarized by the first Polaroid (the polarizer). The second polarizer (the analyzer) then eliminates this component, since its axis is perpendicular to the first. You can try this with Polaroid sunglasses. It should be clear that Polaroid sunglasses eliminate 50 percent of unpolarized light because of their polarizing property; they absorb even more because they are colored.

▌EXAMPLE 22-9 Unpolarized light passes through two Polaroids; the axis of one is vertical and that of the other is at 60° to the vertical. What is the orientation and intensity of the transmitted light?

SOLUTION The first Polaroid eliminates half the light, so the intensity is reduced by half: $I_1 = \frac{1}{2}I_0$. The light reaching the second polarizer is vertically polarized and so is reduced in intensity (see Equation 22-4) by

$$I_2 = I_1(\cos 60°)^2 = \tfrac{1}{4}I_1.$$

Thus $I_2 = \frac{1}{8}I_0$. The transmitted light has an intensity $\frac{1}{8}$ that of the original and is plane-polarized at a 60° angle to the vertical. ▌▌

Another means of producing polarized light from unpolarized light is by reflection. When light strikes a nonmetallic surface at any angle other than perpendicular, the reflected beam is polarized preferentially in the plane parallel to the surface (Figure 22-32). In other words, the component with polarization in the plane perpendicular to the surface is preferentially transmitted or absorbed. You can check this fact by rotating Polaroid sunglasses while looking through them at a flat surface of a lake or road. Since most outdoor surfaces are horizontal, Polaroid sunglasses are made with their axes vertical to eliminate the stronger component, and thus reduce glare. This is well known by fishermen who wear Polaroids to eliminate reflected glare from the surface of a lake or stream and thus see beneath the water more clearly. The amount of polarization by reflection depends on the angle. It varies from no polarization at normal incidence to 100 percent polarization at an angle† known as the *polarizing angle* θ_p. This angle is related to the index of refraction of the two materials on either side of the boundary by the equation

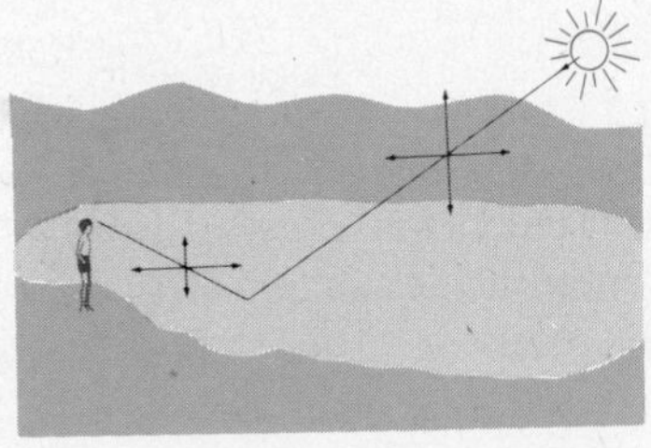

FIGURE 22-32
Light reflected from a nonmetallic surface is partially polarized parallel to the surface.

$$\tan \theta_p = \frac{n_2}{n_1}, \qquad (22\text{-}5a)$$

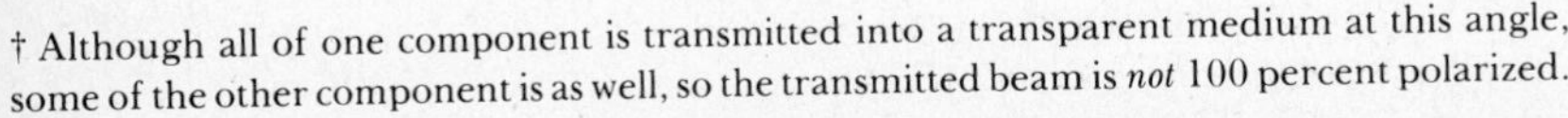

† Although all of one component is transmitted into a transparent medium at this angle, some of the other component is as well, so the transmitted beam is *not* 100 percent polarized.

Atomic hydrogen

Sodium

Helium

Neon

Mercury

Molecular hydrogen

Solar absorption spectrum

FIGURE 22 - 22
Line spectra for the elements indicated and spectrum of the sun, showing absorption lines. (Courtesy of Bausch & Lomb Incorporated.)

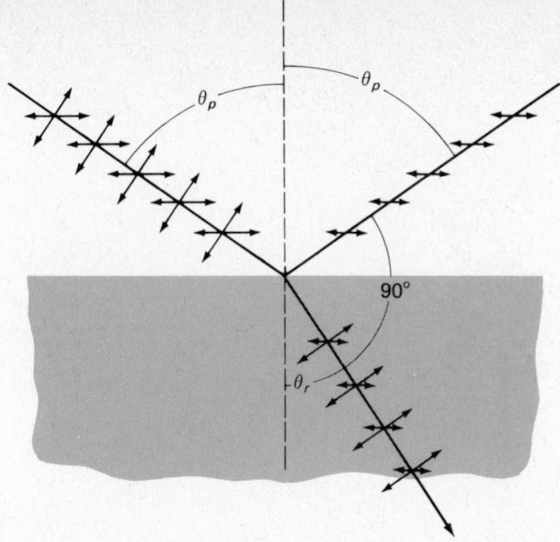

FIGURE 22-33
At θ_p the reflected light is plane-polarized parallel to the surface, and $\theta_p + \theta_r = 90°$, where θ_r is the refraction angle.

where n_1 is the index of refraction of the material in which the beam is traveling, and n_2 is that of the medium beyond the reflecting boundary. If the beam is traveling in air, $n_1 = 1$ and

$$\tan \theta_p = n. \tag{22-5b}$$

The polarizing angle θ_p is also called *Brewster's angle*, after the person who discovered Equation 22-5 experimentally in 1812. Equation 22-5 can be derived from the electromagnetic wave theory of light. It is interesting that at Brewster's angle, the reflected and transmitted rays make a 90° angle to each other; that is, $\theta_p + \theta_r = 90°$ (Figure 22-33). This can be seen by substituting Equation 22-5a, $n_2 = n_1 \tan \theta_p = n_1 \sin \theta_p / \cos \theta_p$, into Snell's law, $n_1 \sin \theta_p = n_2 \sin \theta_r$, and get $\cos \theta_p = \sin \theta_r$, which can only hold if $\theta_p = 90° - \theta_r$.

▮ EXAMPLE 22-10 At what incident angle is sunlight reflected from a lake plane-polarized?

SOLUTION We use Equation 22-5b with $n = 1.33$, so $\tan \theta_p = 1.33$ and $\theta_p = 53°$. ▮▮

*22-10 ▮ Optical activity

When a beam of polarized light is passed through a crystal or a solution, it is found that the plane of polarization is rotated through an angle. For example, Figure 22-34 shows light passing through a polarizer and then through a sugar solution. The crossed polaroid acting as analyzer does

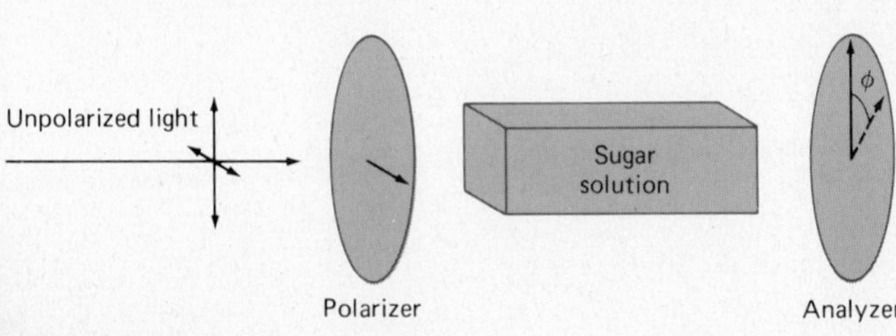

FIGURE 22-34
Sugar solution rotates the plane of polarization of incident light so that it is no longer horizontal but at an angle ϕ. The analyzer thus extinguishes the light when rotated ϕ from the vertical (crossed) position.

TABLE 22-1

Specific optical rotatory power, α†

Substance	(rad · m²/kg)‡
Alanine	**$+3.14\times10^{-4}$**
Leucine	**-19.2×10^{-4}**
Cysteine	**-28.8×10^{-4}**
Insulin	**-59×10^{-4}**
Collagen	**-500 to -700×10^{-4}**

† Substances in water solution at 20°C for $\lambda = 589$ nm
‡ + is dextrorotatory; − is levorotatory.

not cut out all the light. If the analyzer is rotated through an angle ϕ, however, there is no transmitted light; this indicates that the plane of polarization has been rotated through an angle ϕ by the intervening substance. Such substances are said to be *optically active*. Optical activity is due to asymmetry of the molecules, which may even have a spiral shape, such as some proteins (Section 13-4). Substances that rotate the polarization plane to the right as viewed along the direction of the beam as in Figure 22-34 are called *dextrorotatory* (also called *right-handed* or *positive*). Those that rotate it to the left are called *levorotatory* (*left-handed*, or *negative*). The most common sugar, dextrose or D-glucose, is dextrorotatory. Most naturally occurring amino acids and proteins are levorotatory.

The angle of rotation ϕ depends on the path length l (meters) through the substance and on the concentration C (kg/m^3) if it is in solution. For dilute solutions, this is a linear relationship and ϕ (in radians) is given by

$$\phi = \alpha l C.$$

The constant α is a property of the substance and is called the specific rotation or the *specific optical rotatory power* (α depends on the temperature and the wavelength of light used). Some typical values (at 20°C and for $\lambda = 589$, the wavelength of the sodium D lines) are given in Table 22-1.

Since ϕ is proportional to the concentration, optical activity is a standard method to measure concentrations of solutions such as for sugars. It is also helpful in determining the three-dimensional shape of protein molecules, or their change in shape when conditions are changed. For example, a protein in an α-helix shape will have a large (negative) value for α, but may alter its shape when the pH or temperature is changed, and this will change the value of α. Much information on the shape and properties of molecules has been obtained in this way.

Glass and plastics become optically active when put under stress; the plane of polarization is rotated most where the stress is greatest. Models of bones or machine parts made of plastic can be observed between crossed Polaroids to determine where the points of greatest stress lie. This is called "optical stress analysis."

*22-11 ▮ Double refraction: birefringence and dichroism

*Double refraction

In many transparent materials, the speed of light is the same in all directions. Such materials are called *isotropic*. In some crystals and solutions, however, the speed of light is different in different directions, and these are referred to as being *anisotropic*. Such substances are also said to be *doubly refracting* or *birefringent* because they give rise to an unusual phenomenon.

There is one particular direction in a birefringent crystal, such as calcite, known as its *optic axis* (this is not a single line, but a direction in the crystal). If ordinary light enters a crystal along its optic axis, nothing abnormal is noted. But if unpolarized light falls at an angle to the optic axis as in Figure 22-35, a very unusual thing happens: there are two refracted rays. In the case shown, the incident ray is normal to the surface. One refracted ray, called the *ordinary ray* (*o*) passes straight through in a normal way. But the other ray, called the *extraordinary ray* (*e*), is refracted at an angle.

Snell's law does not hold in this case for the *e* ray. It does hold, however, for the *o* ray.

The *e* ray and *o* ray are found to be plane-polarized in mutually perpendicular directions. This is indicated in Figure 22-35 by the dots on the *o* ray, indicating that the vibrations are perpendicular to the paper; the polarization of the *e* ray is indicated by the short arrows.

The phenomenon of double refraction can be explained if we make the following assumption: that the speed of light is different depending on the orientation of the polarization vector with respect to the crystal's optic axis. As can be seen in Figure 22-35, the polarization of the *o* ray is perpendicular to the optic axis, so its speed will be the same in all directions (as long as its polarization remains perpendicular to the optic axis). The *e* ray, on the other hand, has components of polarization both parallel and perpendicular to the optic axis, and it therefore travels with a different speed in different directions; it travels with the same speed as the *o* ray if its polarization vector is perpendicular to the optic axis; at other angles it has a higher speed (in certain crystals it is lower), which reaches a maximum (or minimum) when its polarization is parallel to the optic axis. (In what follows, we assume a crystal in which this speed is higher.) Thus the index of refraction n_o for the ordinary ray is the same for all directions, whereas that for the extraordinary ray, n_e, depends on angle; usually n_e is specified for the *e* ray traveling perpendicular to the optic

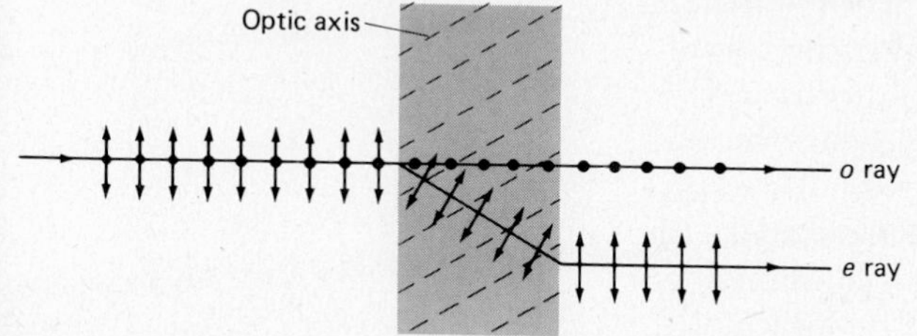

FIGURE 22-35
Unpolarized light, incident normally on a birefringent crystal such as calcite, is separated into two refracted beams. The dashed lines indicate the direction of the optic axis.

axis, where its polarization direction is parallel to the optic axis. These principal values for n_o and n_e are given in Table 22-2 for several crystals.

The preceding assumptions can be applied, using Huygens' principle, to explain double refraction. In Figure 22-36 we first examine the ordinary wave, in part *a*. Huygens' wavelets are shown for two points on the wavefront just as it has entered the crystal. Since these wavelets travel at the same speed in all directions, they are spheres just as before; and the wave continues moving straight. For the *e* ray, Figure 22-36*b*, the Huygens' wavelets are elliptically shaped since the speed depends on direction. Ray 1 on this wavelet has polarization perpendicular to the optic axis, so is traveling at the same speed as the *o* ray; but rays 2, 3, and 4 are traveling at higher speeds, with ray 4 having the maximum. (This is the case for calcite; for a crystal like ice or quartz, Table 22-2, rays 2, 3, and 4 would be slower.) The advancing wavefront, which is tangent to the Huygens' wavelets, will thus be displaced to one side, as shown. And this is the path taken by the *e* ray.

TABLE 22-2

Principal indices of refraction†

Crystal	n_o	n_e
Ice	**1.309**	**1.313**
Quartz	**1.544**	**1.553**
Calcite	**1.658**	**1.486**
Dolomite	**1.681**	**1.500**

† $\lambda = 589$ nm.

*Dichroism

Some birefringent crystals, such as tourmaline, absorb one of the polarized components more strongly than the other (Figure 22-37). Such a crystal is said to show *dichroism.* If the crystal is sufficiently thick, one component of unpolarized light will be completely eliminated, so the emerging light will be plane-polarized. Dichroism is the basic principle behind Polaroid sheets.

Tiny crystals and large molecules in solution are often birefringent and dichroic. The absorption of light polarized in two different directions can be measured. The ratio of the absorption for each of the two components is called the *dichroic ratio.* The dichroic ratio is a function of the amount of alignment of the crystals or molecules. Detailed studies using this phenomenon help in the determination of the structure of biological (and other) molecules.

FIGURE 22-36
Huygens' principle used to explain double refraction.

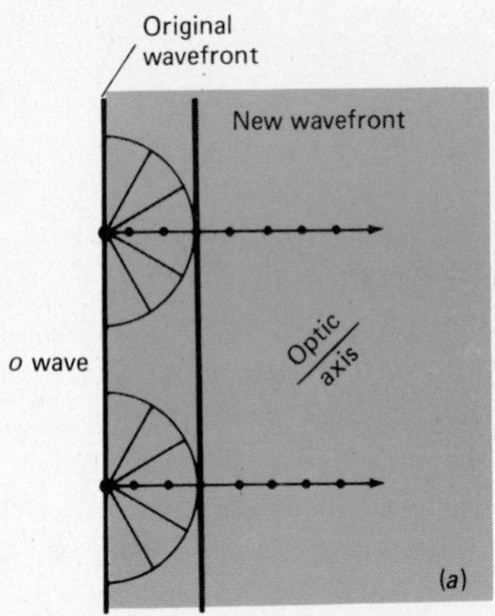

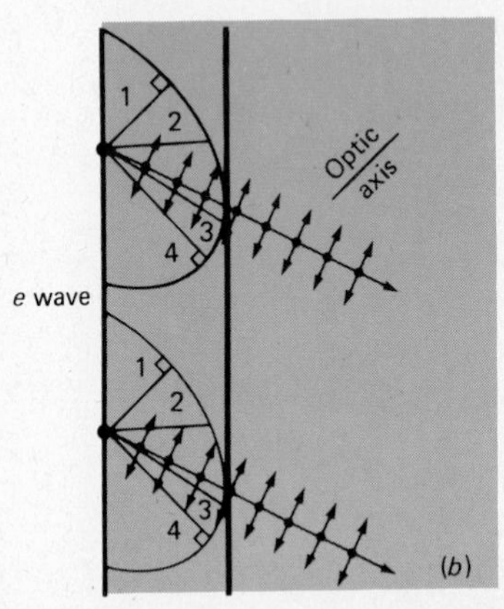

*22-12 ▌ Scattering of light by the atmosphere

Sunsets are red, the sky is blue, and skylight is polarized (at least partially). These phenomena can be explained on the basis of the *scattering* of light by the molecules of the atmosphere. In Figure 22-38 we see unpolarized light from the sun impinging on a molecule of the earth's atmosphere. The electric field of the EM wave sets the electric charges within the molecule into motion. Thus the molecule absorbs some of the incident radiation. But it quickly reemits this light since the charges are oscillating. As discussed in Section 20-2, oscillating electric charges produce EM waves. The electric field of these waves is in the plane that includes the line of oscillation. The intensity is strongest along a line perpendicular to the oscillation, and drops to zero along the line of oscillation (Figure 20-4). In Figure 22-38 the motion of the charges is resolved into two components. An observer at right angles to the direction of the sunlight, as shown, will see plane-polarized light since no light is emitted along the line of the

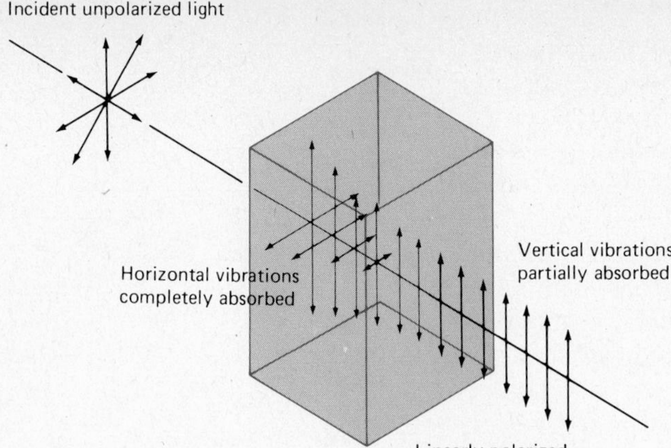

FIGURE 22-37
Dichroic crystal absorbs one polarized component more than the other. If the crystal is thick enough, the transmitted light is plane-polarized.

other component of the oscillation. At other viewing angles, both components will be present; one will be stronger, however, so the light appears partially polarized. Thus, the process of scattering explains the polarization of skylight. (It can also explain complete polarization by reflection, Figure 22-33. At Brewster's angle, as we saw, the angle between the reflected and refracted ray is 90°; thus the component of the oscillations of molecules in the medium, when struck by the incoming light, emits light only with the polarization vector as shown in Figure 22-33.)

Scattering of light by the earth's atmosphere is proportional to the fourth power of the frequency. Blue light and violet light are thus scattered much more than red and orange, so the sky looks blue. At sunset, the sun's rays pass through a maximum length of atmosphere. Much of the blue has been taken out by scattering. Thus light that reaches the surface of the earth and reflects off clouds and haze is lacking in blue which is why sunsets appear reddish.

The dependence of scattering on f^4 is valid only if the scattering objects are much smaller than the wavelength of the light. This is valid for oxygen and nitrogen molecules whose diameter are about 0.2 nm. Clouds, however, contain water droplets or crystals that are much larger than λ; they scatter all frequencies of light uniformly. Hence clouds appear white (or gray, if shadowed).

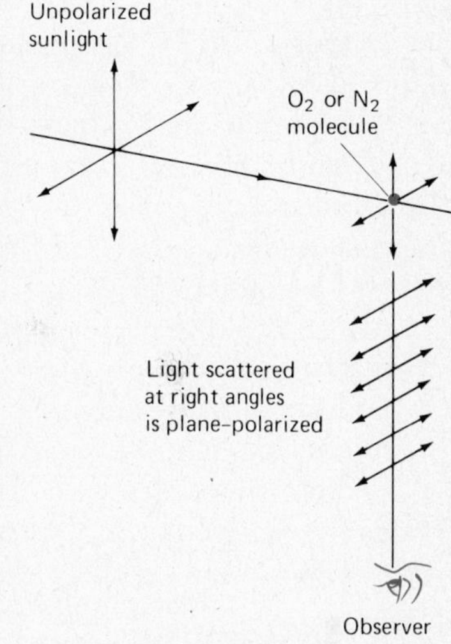

FIGURE 22-38
Unpolarized sunlight scattered by molecules of the air. An observer at right angles sees plane-polarized light, since the component of vibration along the line of sight emits no light along that line.

Summary

The wave theory of light is strongly supported by the fact that light exhibits interference and diffraction; this theory also explains the refraction of light and the fact that light travels more slowly in transparent solids and liquids than it does in air.

The wavelength of light determines its color; the *visible spectrum* extends from 400 nm (violet) to about 750 nm (red). Glass prisms break white light down into its constituent colors because the index of refraction varies with wavelength, a phenomenon known as *dispersion*.

Young's double-slit experiment clearly demonstrated the *interference* of light; the observed bright spots of the interference pattern were explained as constructive interference between the beams coming through the two slits, where they differ in path length by an integral number of wavelengths. The dark areas in between are due to destructive interference when the path lengths differ by $\frac{1}{2}\lambda$, $\frac{3}{2}\lambda$, and so on. The angles θ at which constructive interference occurs are given by $\sin\theta = m\lambda/d$, where λ is the wavelength of the light, d the separation of the slits, and m an integer (0, 1, 2, . . .).

The same formula holds for a *diffraction grating*, which consists of many parallel slits or lines, each separated from its neighbors by a distance d. The peaks of constructive interference are much brighter and sharper for a diffraction grating than for the simple two-slit apparatus. A diffraction grating (or a prism) is used in a *spectroscope* to separate different colors or to observe *line spectra*, since for a given order m, θ depends on λ; precise determination of wavelength can thus be done with a spectroscope by careful measurement of θ.

Diffraction refers to the fact that light, like other waves, bends around objects it passes and spreads out after passing through narrow slits. This bending gives rise to a diffraction pattern due to interference between rays of light that travel different distances. Light passing through a very narrow slit of width D will produce a pattern with a bright central maximum of half-width θ given by $\sin\theta = \lambda/D$, flanked by fainter lines to either side.

Light reflected from the front and rear surfaces of a thin film of transparent material can interfere. Such thin-film interference has many practical applications, such as lens coatings and Newton's rings.

In *unpolarized light* the electric field vectors vibrate at all angles. If the electric vector vibrates only in one plane the light is said to be *plane-polarized*. Light can also be partially polarized. When an unpolarized light beam passes through a Polaroid sheet or a doubly refracting crystal, the emerging beams are plane-polarized. That a light beam is polarized is determined experimentally by the fact that when it passes through a Polaroid, the intensity varies as the Polaroid is rotated. Thus a Polaroid can act as polarizer or as an analyzer. Light can also be partially or fully polarized by reflection; if light traveling in air is reflected from a medium of index of refraction n, it will be completely plane-polarized if the incident angle θ_p is given by $\tan\theta_p = n$. That light can be polarized shows that it must be a transverse wave.

▌Questions

1 Does Huygens' principle apply to sound waves? To water waves?

2 What is the evidence that light is energy?

3 Why is light sometimes described as rays and sometimes as waves?

4 Two rays of light from the same source destructively interfere if their path lengths differ by how much?

5 If Young's double-slit experiment were submerged in water, how would the fringe pattern be changed?

6 Why doesn't the light from the two headlights of a distant car produce an interference pattern?

7 When white light passes through a flat piece of window glass, it is not broken down into colors as it is by a prism. Explain.

8 We can hear sounds around corners, but we cannot see around corners; yet both sound and light are waves. Explain the difference.

9 Radio waves and light are both electromagnetic

waves. Why can we hear a radio behind a hill when we cannot see the transmitting antenna?

10 What is the difference in the interference patterns formed by two slits 10^{-4} cm apart and by a diffraction grating containing 10^4 lines/cm?

11 Explain why there are tiny peaks between the main peaks produced by a diffraction grating illuminated with monochromatic light. Why are the peaks so tiny?

12 Why is a diffraction grating preferable to a prism for use in a spectroscope?

13 Why are interference fringes noticeable only for a *thin* film like a soap bubble and not for a thick piece of glass, say?

14 Why are Newton's rings (Figure 22-24) closer together farther from the center?

15 Some coated lenses appear greenish yellow when seen by reflected light. What wavelengths do you suppose they are designed to eliminate completely?

16 A drop of oil on a pond appears bright at its edges, where its thickness is much less than the wavelengths of visible light. What can you say about the index of refraction of the oil?

*17 How could a Michelson interferometer be used to measure the index of refraction of air?

18 What does polarization tell us about the nature of light?

19 What is the difference between a polarizer and an analyzer?

20 How can you tell if a pair of sunglasses is polarizing or not?

*21 For which of the materials in Table 22-2 is the speed of the extraordinary ray less when its polarization vector is parallel to the optic axis than when it is perpendicular?

*22 What would be the color of the sky if the earth had no atmosphere?

*23 If the earth's atmosphere were 50 times denser than it is, would sunlight still be white, or would it be some other color?

Problems

SECTION 22-2

1 (I) The second-order fringe when 700 nm light falls on two slits is observed at a 15° angle to the initial beam direction. How far apart are the slits?

2 (I) Monochromatic light falling on two slits 0.030 mm apart produces the fourth-order fringe at a 22° angle. What is the wavelength of the light used?

3 (II) 480 nm light falls on two slits and produces an interference pattern in which the fourth-order fringe is 18 mm from the central fringe on a screen 1.0 m away. What is the separation of the two slits?

4 (II) Monochromatic light falls on two slits 0.020 mm apart. The fringes on a screen 3.00 m away are 9.2 cm apart. What is the wavelength of the light?

5 (II) A parallel beam of 600 nm light falls on two small slits 5.0×10^{-2} mm apart. How far apart are the fringes on a screen 5.0 m away?

6 (II) TV and radio waves can reflect from nearby mountains or from airplanes, and the reflections can interfere with the direct signal from the station. (*a*) Determine what kind of interference will occur when 75-MHz television signals arrive at a receiver directly from a station and reflected from an airplane 118 m directly above the receiver. (Assume no change in phase of the signal upon reflection.) (*b*) What kind of interference will occur when the plane has moved 22 m farther toward the station?

7 (II) The shortest-wavelength visible light falls on two slits 2.0×10^{-2} mm apart. The slits are immersed in water, as is a viewing screen 20 cm away. How far apart are the fringes on the screen?

8 (III) A screen contains two slits a distance d apart. If light of wavelength λ strikes it at an angle θ_i to the normal, determine the angle θ_m at which the mth-order maximum occurs.

SECTION 22-3

9 (II) A light beam strikes a piece of glass at a 50.00° incident angle. The beam contains two wavelengths, 500 nm and 700 nm, for which the index of refraction of the glass is 1.4810 and 1.4742, respectively. What is the angle between the two refracted beams?

10 (II) A parallel beam of light containing two wavelengths, $\lambda_1 = 400$ nm and $\lambda_2 = 650$ nm, strikes a piece of silicate flint glass at an incident angle of 60°. Calculate the angle between the two color beams inside the glass (see Figure 22-10).

11 (III) A double convex lens whose radii of curvature are both 12.0 cm is made of crown glass. Find the distance between the focal points for 400-nm and 700-nm light. (*Hint*: Use the lens-makers equation, Equation 21-9, and Figure 22-10.)

SECTION 22-4

12 (I) If 680-nm light falls on a slit 3.0×10^{-2} mm wide, what is the angular width of the central diffraction peak?

13 (I) Monochromatic light falls on a slit 2.2×10^{-3} mm wide. If the angle between the first dark fringes on either side of the central maximum is 30°, what is the wavelength of the light used?

14 (II) How wide is the central diffraction peak on a screen 3.50 m behind a 0.010 mm-wide slit illuminated by 400-nm light?

15 (II) When 450 nm light falls on a slit, the central diffraction peak on a screen 2.0 m away is 6.0 cm wide. Calculate the slit width.

16 (II) For a given wavelength λ, what is the maximum slit width for which there will be no diffraction minima?

17 (II) How wide is a slit if it diffracts 590-nm light so that its central diffraction peak is 2.0 cm wide on a screen 1.80 m away?

SECTIONS 22-5 AND 22-6

18 (I) The first-order line of 550-nm light falling on a diffraction grating is observed at a 12° angle. How far apart are the slits?

19 (I) At what angle will 710-nm light produce a third-order maximum when falling on a grating whose slits are 1.0×10^{-3} cm apart?

20 (II) How many lines per centimeter does a grating have if the third-order occurs at a 20° angle for 650-nm light?

21 (II) A grating has 5000 lines/cm. How many spectral orders can be seen when it is illuminated by white light?

22 (II) White light containing wavelengths from 400 to 700 nm falls on a grating with 8000 lines/cm. How wide is the first-order spectrum on a screen 2.0 m away?

23 (II) How many lines per centimeter must a grating have if there is to be no second-order spectrum for any visible wavelength?

24 (II) Show that the second- and third-order spectra of white light produced by a diffraction grating always overlap.

25 (II) The wings of a certain beetle have a series of parallel lines across it. When 600-nm light is reflected from the wing, the wing appears bright when viewed at an angle of 49°. How far apart are the lines? The light is incident normally.

26 (II) When yellow sodium light, $\lambda = 589$ nm, falls on a diffraction grating, its first-order peak on a screen 20 cm away falls 2.12 cm from the central peak. Another source produces a line 3.61 cm from the central peak. What is its wavelength?

27 (II) Light falling on a 10,000-line/cm spectrometer grating is revealed to contain three lines in the first-order spectrum at angles of 31.2°, 36.4°, and 47.5°. What wavelengths are these?

28 (III) Assume in Problem 27 that the spectrometer is submerged in the atmosphere of a planet whose index of refraction is 1.10. If the same angles were obtained, what would the wavelength be in vacuum?

SECTION 22-7

29 (I) How far apart are the dark fringes in Example 22-6 if the glass plates are each 10 cm long?

30 (II) A lens appears greenish yellow ($\lambda = 570$ nm is strongest) when white light reflects from it. What minimum thickness of coating ($n = 1.25$) do you think is used on such a (glass) lens, and why?

31 (II) A soap bubble appears blue ($\lambda = 480$ nm) at the center of its surface. What is the minimum thickness of the bubble there? Assume the index of refraction is 1.34.

32 (II) A total of 33 bright and 33 dark Newton's rings are observed when 450-nm light falls normally on a planoconvex lens resting on a flat glass surface (Figure 22-24). How much thicker is the center than the edges?

33 (II) A fine metal foil separates one end of two pieces of optically flat glass as in Figure 22-25. When light of wavelength 550 nm is incident normally, 37 dark lines are observed. How thick is the foil?

34 (II) If a soap bubble is 120 nm thick, what color will appear at the center when illuminated normally by white light? Assume that $n = 1.34$.

35 (II) How thick (minimum) should the air layer be between two flat glass surfaces if the glass is to appear bright when 540-nm light is incident normally? What if the glass is to appear dark?

36 (III) Show that the radius r of the nth dark Newton's ring, as viewed from directly above (Figure 22-24b) is given by $r = \sqrt{n\lambda R}$, where R is the radius of curvature of the curved glass surface and λ is the wavelength of light used. Assume $r \ll R$.

37 (III) A planoconvex Lucite lens 4.0 cm in diameter is placed on a flat piece of glass as in Figure 22-24. When 520-nm light is incident normally, 68 bright rings are observed, the last one right at the edge. What is the radius of curvature of the lens surface, and what is the focal length of the lens?

*SECTION 22-8

38 (I) What is the wavelength of the orange-red line of krypton 86 used to define the standard meter?

39 (II) What is the wavelength of the light entering an interferometer if 950 fringes are counted when the movable mirror moves 0.276 mm?

40 (II) A micrometer is connected to the movable mirror of an interferometer. When the micrometer bears on a thin metal foil, the net number of fringes that move, compared to the empty micrometer, is 218. What is the thickness of the foil? The wavelength of light used is 606 nm.

41 (II) How far must the mirror M_1 in a Michelson interferometer be moved if 1000 fringes of 510-nm light are to pass by a reference line?

42 (III) One of the beams of an interferometer passes

through a small glass container containing a cavity 1.20 cm wide. When a gas is allowed to slowly fill the container, a total of 188 dark fringes are counted to move past a reference line. The light used has a wavelength of 546 nm. Calculate the index of refraction of the gas, assuming that the interferometer is in vacuum.

*43 (III) The yellow sodium D lines have wavelengths of 589.0 and 589.6 nm. When they are used to illuminate a Michelson interferometer, it is noted that the interference fringes disappear and reappear periodically as the mirror M_1 is moved. Why does this happen? How far must the mirror move between one disappearance and the next?

SECTION 22-9

44 (I) What is Brewster's angle for an air–glass ($n = 1.50$) surface?

45 (I) What is Brewster's angle for a piece of glass ($n = 1.56$) submerged in water?

46 (I) Two polarizers are at 45° to one another. Unpolarized light falls on them. What fraction of the light intensity is transmitted?

47 (I) A plane-polarized beam of light strikes a polarizer whose axis makes a 60° angle with the polarization direction. What fraction of the beam is transmitted?

48 (II) Two Polaroids are aligned so that the light passing through them is a maximum. At what angle should one of them be placed so that the intensity is reduced by half?

49 (II) At what angle above the horizon is the sun when light reflecting off a smooth lake is polarized most strongly?

50 (II) Two polarizers are oriented at 40° to one another. Light polarized at a 20° angle to each polarizer passes through both. What reduction in intensity takes place?

51 (II) Unpolarized light falls on two polarizer sheets whose axes are at right angles. (*a*) What fraction of the incident light intensity is transmitted? (*b*) What fraction is transmitted if a third polarizer is placed between the first two so that its axis makes a 45° angle with each of their axes? (*c*) What if the third polarizer is in front of the other two?

52 (II) What would Brewster's angle be for reflections off the surface of water beneath the surface? Compare to the angle for total internal reflection, and to Brewster's angle from above the surface.

*SECTION 22-10

*53 (II) Calculate the concentration of an insulin solution if the plane of polarization is rotated 18.2° after passing through 11.0 cm of the solution.

*54 (II) A 1.1×10^3 kg/m^3 solution of the sugar fructose 8.0 cm thick produces a rotation in the plane of polarization of incident light by $-117°$. What is the optical rotatory power of fructose?

*SECTION 22-11

*55 (III) A region of a birefringent biological specimen is 1.5×10^{-3} cm thick with indices of refraction 1.322 and 1.331 for light whose wavelength in air is 590 nm. What phase difference arises between the two rays after passing through this region? That is, by what fraction of a wavelength are the two beams out of step when they exit?

For additional reading

Jenkins, F. A., *Fundamentals of Optics*, 4th ed., McGraw-Hill Book Company, New York, 1975.

Magie, W. F., *Source Book in Physics*, Harvard University Press, Cambridge, Mass., 1963. Excerpts from the writings of Huygens, Grimaldi, Young, Fresnel, and Arago.

Meyer-Arendt, J. R., *Introduction to Classical and Modern Optics*, Prentice-Hall, Inc., Englewood Cliffs, N.J., 1972.

Rinard, P. M., "Large-scale Diffraction Patterns from Circular Objects," *Am. J. Phys.*, **44** (1976), p. 70.

Sears, F. W., *Optics*, Addison-Wesley Publishing Co., Inc., Reading, Mass., 1949.

Shankland, R. S., "Michelson and His Interferometer," *Physics Today*, **27**:4 (Apr., 1974), p. 36.

Nussenzveig, H. M., "Theory of the Rainbow," *Sci. American*, **236**:4 (Apr., 1977), p. 116.

Wehner, R., "Polarized-light Navigation by Insects," *Sci. American*, **235**:1 (Jul., 1976), p. 106.

23 ▮ Optical instruments

In our discussion of the behavior of light in the two previous chapters, we also described a few instruments such as the spectroscope and the Michelson interferometer. In this chapter we will discuss some additional instruments. Most of these use lenses, such as the camera, telescope, microscope, and the human eye. To describe their operation, we will use ray diagrams. However, we will see that some aspects of their operation depend on the wave nature of light.

23-1 ▮ The camera

The basic elements of a camera are a lens, a lighttight box, a shutter to let light pass through the lens, and a sensitized plate or piece of film (Figure 23-1). When the shutter is opened briefly, light from external objects in the field of view are focused by the lens as an image, on the film. The film contains light-sensitive chemicals which undergo change when light strikes them. In the development process chemical reactions cause the changed areas to turn black so that the image is recorded on the film.† You can see an image yourself by removing the camera back and viewing through a piece of tissue or wax paper at the position of the film with the shutter open.

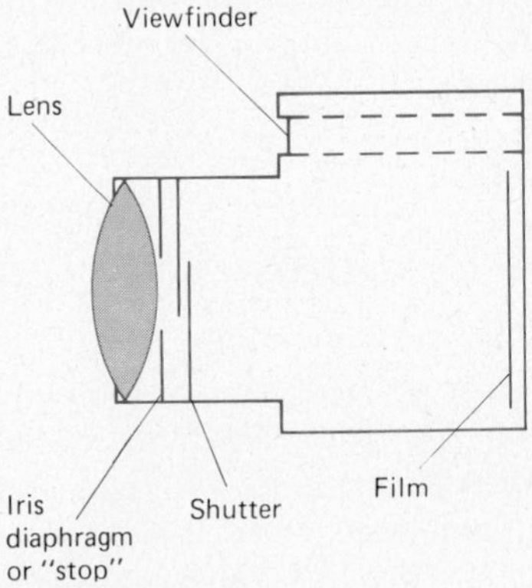

FIGURE 23-1
A simple camera.

There are three main adjustments on most good-quality cameras: shutter speed, f-stop, and focusing, and we now discuss them.

Shutter speed. This refers to how long the shutter is open and the film exposed. It may vary from a second or more ("time exposures") to 1/1000 s or less. To avoid blurring from camera movement, speeds faster than $\frac{1}{50}$ s are normally used; if the object is moving, faster shutter speeds are needed to "stop" the action.

f-stop. The amount of light reaching the film must be carefully controlled to avoid *underexposure* (too little light for any but the brightest objects to show up) or *overexposure* (too much light, so that all bright

† This is called a *negative*, since the black areas correspond to bright objects and vice versa. The reverse process occurs during printing to produce a black-and-white ("positive") picture from the negative. Color film makes use of three dyes corresponding to the primary colors.

objects look the same, with a consequent lack of contrast and a "washed-out" appearance). To control the exposure, a "stop" or iris diaphragm of variable diameter is placed behind the lens (Figure 23-1). The size of the opening is varied to compensate for bright or dark days, and for different shutter speeds. The opening is specified by the *f-stop*, defined as

$$f\text{-stop} = \frac{f}{D},$$

where f is the focal length of the lens and D the diameter of the opening. For example, a 50-mm-focal-length lens set at $f/2$ has an opening $D =$ 25 mm; when set at $f/8$, the opening is only $6\frac{1}{4}$ mm. The faster the shutter speed (or the darker the day), the greater the opening must be to get a proper exposure. This corresponds to a smaller f-stop number. The smallest f-number of a lens is referred to as the *speed* of the lens. It is common to find $f/2.0$ lenses today, and even some as fast as $f/1.0$. Fast lenses are expensive to make and require many elements in order to reduce the defects present in simple thin lenses (Section 23-6). The advantage of a fast lens is that it allows pictures to be taken under poorly lighted conditions.

Lenses normally stop down to $f/16$, $f/22$, or $f/32$. Although the lens opening can usually be varied continuously, there are nearly always markings for specific lens openings: the standard f-stop markings are 1.0, 1.4, 2.0, 2.8, 4.0, 5.6, 8, 11, 16, 22, and 32. Notice that each of these corresponds to a diameter of about $\sqrt{2} = 1.4$ times smaller. Since the amount of light reaching the film is proportional to the area of the opening (and therefore proportional to the diameter squared), we see that each standard f-stop corresponds to a factor of 2 in light intensity reaching the film.

Focusing is the operation of placing the film at the correct position for the sharpest image. The image distance is a minimum for objects at infinity (the symbol ∞ is used for infinity) and is equal to the focal length. For closer objects the image distance is greater than the focal length, as can be seen from the lens equation, $1/f = 1/d_o + 1/d_i$. To focus on nearby objects, the lens must therefore be moved away from the film, and this is usually done by turning a ring on the lens.

If the lens is focused on a nearby object, a sharp image of it will be formed. But the rays from an object far away will be out of focus—they will form a circle on the film as shown (exaggerated) in Figure 23-2; the distant object will thus produce an image consisting of overlapping circles and will be blurred. These circles are called *circles of confusion*. If you want to have near and distant objects sharp at the same time, you can set the lens focus at an intermediate position. Neither near nor distant objects will then be perfectly sharp, but the circles of confusion may be small enough that the blurriness is not noticeable.

For a given distance setting, there is a range of distances over which

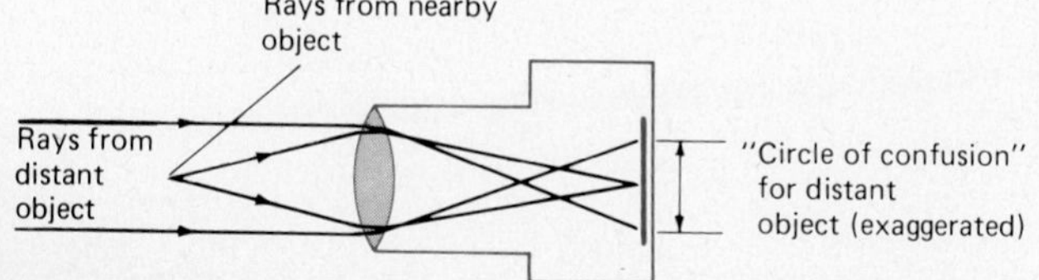

FIGURE 23-2

When the lens is positioned to focus on a nearby object, points on a distant object produce circles and are therefore blurred. (The effect is greatly exaggerated.)

the circles of confusion will be small enough that the images will be reasonably sharp. This is called the *depth of field.* For a particular circle of confusion diameter (typically taken to be 0.03 mm for 35-mm cameras), the depth of field depends on the lens openings. If the lens opening is smaller, the circles of confusion will be smaller, since only rays through the central part of the lens are accepted, and these form a smaller circle of confusion (Figure 23-2). Hence at smaller lens openings, the depth of field is greater.

Other factors also affect the sharpness of the image, such as the graininess of the film, diffraction, and lens aberrations relating to the quality of the lens itself. (Lens quality and diffraction effects will be discussed in Sections 23-6 and 23-7).

Camera lenses are categorized into normal, telephoto, and wide angle, according to focal length and film size. A *normal lens* is one that covers the film with a field of view that corresponds approximately to that of normal vision. A normal lens for 35-mm film has a focal length in the vicinity of 50 mm. A *telephoto* lens, as its name implies, acts like a telescope to magnify images. They have longer focal lengths than a normal lens. As we saw in Chapter 21 (Equation 21-8), the height of the image for a given object distance is proportional to the image distance; and the image distance will be greater for a lens with longer focal length. For distant objects, the image height is very nearly proportional to the focal length (can you prove this?). Thus a 200-mm telephoto lens for use with a 35-mm camera gives a 4× magnification over the normal 50-mm lens. A *wide-angle* lens has a shorter focal length than normal. A wider field of view is included and objects appear smaller. A *zoom lens* is one whose focal length can be changed so that you can zoom up to, or away from, the subject.

Two types of viewing systems are common in cameras today. In a *range-finder* camera, you view through a slot just above the lens as in Figure 23-1. The name derives from the fact that a dual-mirror system is usually used to determine the object distance for focusing. In a *single-lens reflex* camera (SLR), you view directly through the lens with the use of prisms and mirrors (Figure 23-3). A mirror hangs at a 45° angle behind the lens and flips up out of the way just before you take a picture. SLRs have the great advantage that you can see exactly what you will get on film.

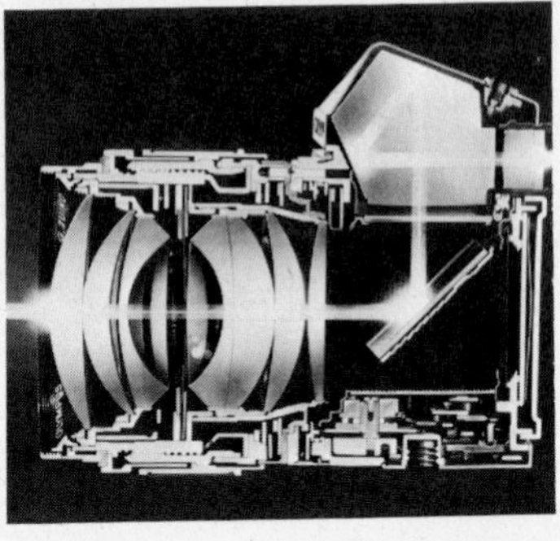

FIGURE 23-3
SLR, showing how light is viewed through the lens with the help of a movable mirror and a prism. (Courtesy of Olympus Camera Corporation.)

▮ EXAMPLE 23-1 How far must a 50-mm-focal-length camera lens be moved from its infinity setting in order to sharply focus an object 3.0 m away?

SOLUTION When focused at infinity, the lens is 50 mm from the film. When focused at $d_o = 3.0$ m, the image distance is given by the lens equation,

$$\frac{1}{d_i} = \frac{1}{f} - \frac{1}{d_o} = \frac{1}{50\text{ mm}} - \frac{1}{3000\text{ mm}}.$$

We solve for d_i and find $d_i = 50.8$ mm, so the lens moves 0.8 mm. ▮▮

▮ EXAMPLE 23-2 A light meter reads that a lens setting of $f/8$ is correct for a shutter speed of $\frac{1}{250}$ s under certain conditions. What would be the correct lens opening for a shutter speed of $\frac{1}{500}$ s?

SOLUTION The amount of light entering the lens is proportional to the area of the lens opening. Since the exposure is reduced by half (from

$\frac{1}{250}$ s to $\frac{1}{500}$ s), the area must be doubled. Since the area is proportional to D^2, where D is the diameter of the opening, then D must be increased by $\sqrt{2} = 1.41$. Thus the f stop must be $8.0/1.41 = f/5.6$. ▮▮

23-2 ▮ The human eye

Structure and functioning of the eye

The human eye resembles a camera in its basic structure (Figure 23-4). The eye is an enclosed volume into which light passes through a lens. There is a diaphragm called the iris (the colored part of your eye), which adjusts automatically† to control the amount of light entering the eye. The hole in the iris through which light passes is called the pupil; it is black because no light is reflected from it (it's a hole), and very little light is reflected back out from the interior of the eye. The retina, which plays the role of the film of a camera, is on the curved rear surface. It is a complex array of receptors known as rods and cones, and of nerves connected to them. The rods and cones act as transducers that change light energy into electrical signals that travel along the nerves. The construction of the image from all these tiny receptors is done mainly in the brain, although some analysis is apparently done in the complex interconnected nerve network at the retina itself. The nerves from all the receptors meet and form the optic nerve that leads to the brain. There are no receptors where the optic nerve leaves the eye, and this point is called the blind spot. We are not aware of it normally, since the brain "fills in" this location.

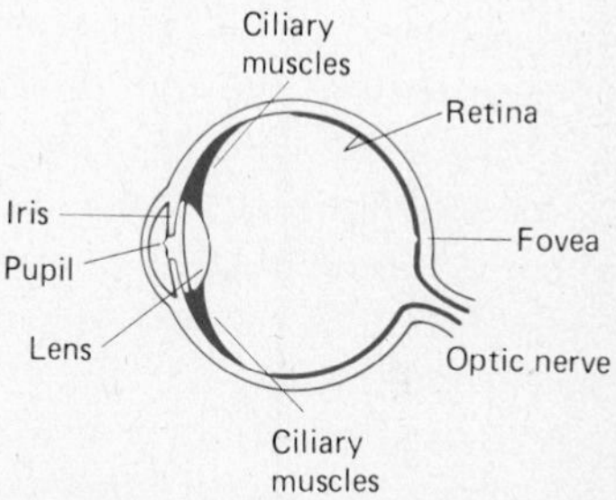

FIGURE 23-4
Diagram of a human eye.

The rods are very sensitive to light—they can detect very low levels of light—but they do not distinguish colors. The cones, on the other hand, are the receptors that distinguish colors, but they cannot detect as low a level of light as the rods. At the center of the retina is a small area called the fovea, where the cones are very closely packed. The sharpest image and best color discrimination is found here; we normally move our eyes so that the area of most interest is focused on the fovea. There are no rods in the fovea, but they are present in the area beyond it and predominate in the peripheral regions of the retina. Because the rods serve us best in dim light, we can detect objects in very dim light out of the corner of our eyes better than at the fovea. However, color distinction is poor on the periphery, which you can prove to yourself by having someone bring an unknown colored object from behind you just into your field of view. You will not be able to tell its color. When you see things directly, however, your brain usually remembers the colors even when they are viewed from the side of the eye.

Unlike a camera, the eye contains no shutter. The equivalent operation is carried out by the nervous system, which analyzes the signals to form images at the rate of about 30 per second. This can be compared to motion picture or television cameras, which operate by taking a series of still pictures at a rate of 24 (movies) or 30 (television) per second. The rapid projection of these on a screen or television set gives the appearance of motion.

† This is another example of *feedback*. If the light intensity reaching the retina is too large or small, feedback signals adjust the iris opening accordingly.

The eye, like a camera, adjusts to focus on objects near or far. But it does not do this by changing the distance from lens to retina as in a camera. Instead, the ciliary muscles (Figure 23-4) change the curvature of the lens so that its focal length is changed. To focus on a distant object, the muscles are relaxed and the lens is thin (Figure 23-5*a*). To focus on a nearby object, the muscles contract causing the center of the lens to be thicker (Figure 23-5*b*). Thus the focal length is shortened so that nearby objects can be focused on the retina. This adjustment by the eye is called *accommodation*.

Actually, the lens does little of the bending of the light rays. Most of the refraction is done at the front surface of the cornea, which also acts as a protective covering. The lens itself, which is a cellular structure, acts as a fine adjustment for different distances.

The lens cannot adjust to focus on objects very close to the eye. The closest distance at which the eye can focus clearly is called the *near point* of the eye, and for young adults is typically 25 cm. Children, however, can focus on objects as close as 10 cm, whereas as people grow older, the ability to accommodate is reduced and the near point increases. The *far point* is the farthest distance at which an object can be seen clearly. For some purposes it is useful to speak of a *normal eye* (a sort of average over the population), which is defined as one having a near point of 25 cm and a far point of infinity.

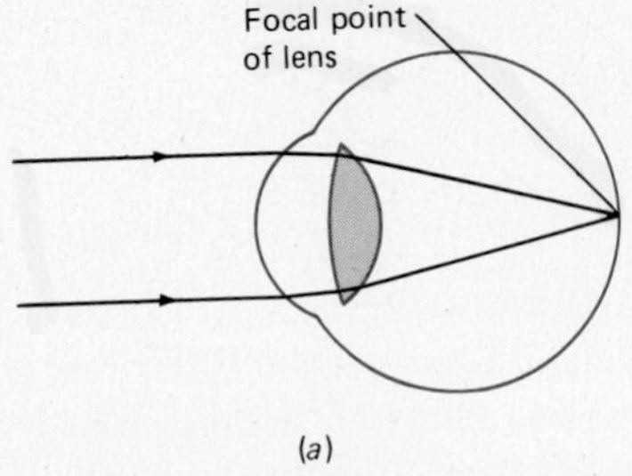

(*a*)

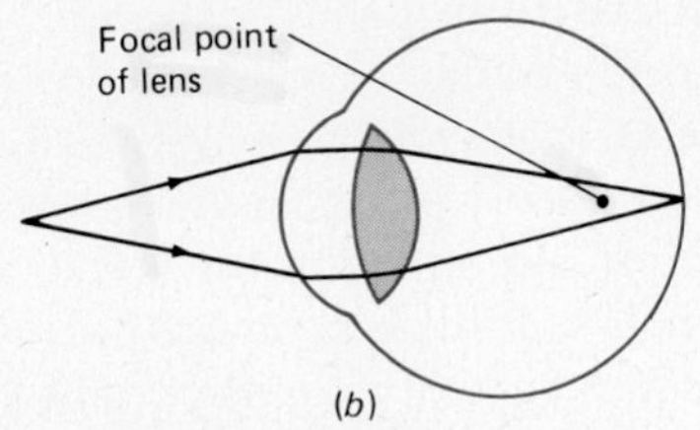

(*b*)

FIGURE 23-5
Accommodation by eye: (*a*) lens relaxed, focused on infinity; (*b*) lens thickened, focused on nearby object.

Defects of the eye

The "normal" eye is more of an ideal than a commonplace. A large part of the population have eyes that do not accommodate within the normal range of 25 cm to infinity, or have some other defect. Two common defects are nearsightedness and farsightedness. Both can be corrected to a large extent with lenses—either eyeglasses or contact lenses.

A *nearsighted* or *myopic* eye is one that can focus only on nearby objects; that is, the far point is not infinity but some shorter distance. It is usually caused by an eyeball that is too long, although sometimes it is the curvature of the cornea that is too great. In either case, images of distant objects are focused in front of the retina. A diverging lens, because it causes parallel rays to diverge, allows the rays to be focused at the retina (Figure 23-6*a*) and thus corrects this defect.

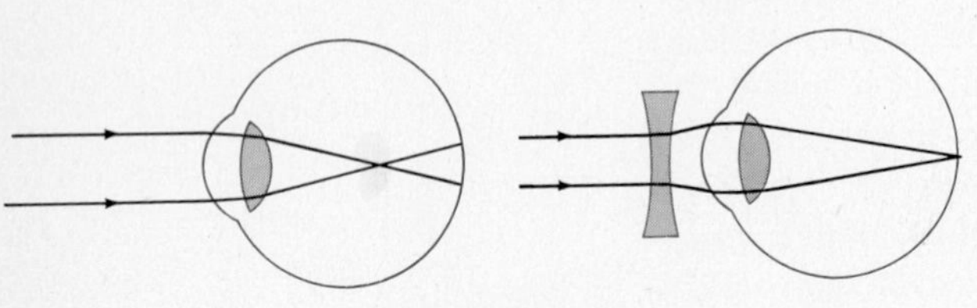

(*a*) Nearsighted eye

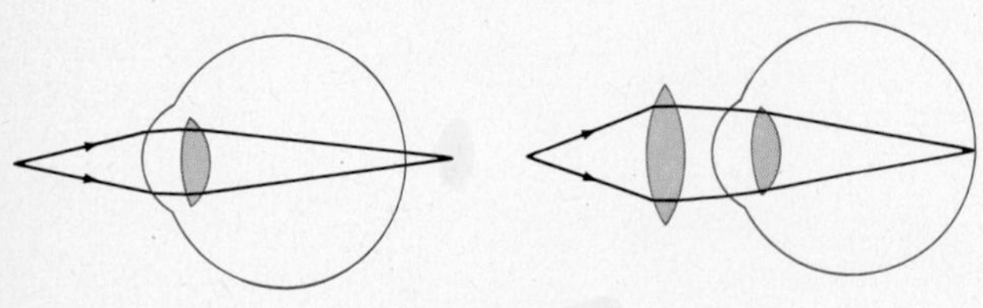

(*b*) Farsighted eye

FIGURE 23-6
(*a*) A nearsighted eye, which cannot focus clearly on distant objects, can be corrected by use of a diverging lens. (*b*) A farsighted eye, which cannot focus clearly on nearby objects, can be corrected by use of a converging lens.

Farsightedness or *hyperopia* refers to an eye that is not able to focus on nearby objects. Although distant objects are usually seen clearly, the near point is somewhat greater than the "normal" 25 cm which makes reading difficult. This defect is caused by an eyeball that is too short or (less often) by a cornea that is not sufficiently curved. It is corrected by a converging lens (Figure 23-6*b*).

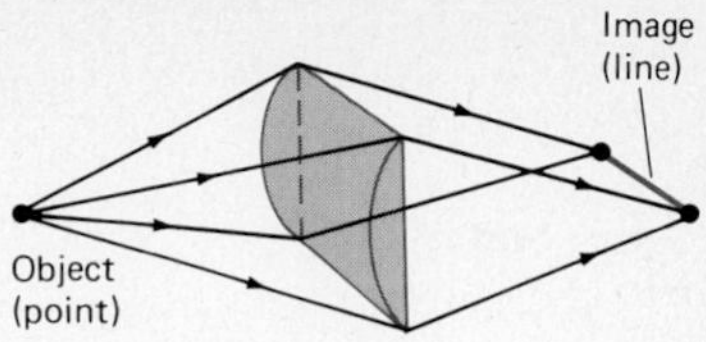

Very similar in effect to hyperopia is *presbyopia*. This refers to the fact that, as one ages, the eye is less able to accommodate and the near point moves out. Converging lenses compensate for this. Because lack of accommodation occurs even for people with nearsightedness (or other defect), bifocal (or even trifocal) lenses are common in which the upper and lower parts of the lens are of different focal length; the two parts are used for far and near vision, respectively.

Another common defect is *astigmatism*. This is usually caused by an out-of-round cornea or lens, and point objects are focused as short lines. It is as if the cornea were spherical, with a cylindrical section superimposed. As shown in Figure 23-7, a cylindrical lens focuses a point into a line parallel to its axis. Astigmatism is corrected with the use of a compensating cylindrical lens. Lenses for eyes that are near- or farsighted as well as astigmatic are ground with superimposed spherical and cylindrical surfaces. Thus the radius of curvature of the correcting lens is different in different planes. Astigmatism is tested for by looking with one eye at a pattern like that in Figure 23-8. An astigmatic eye focuses rays in a vertical plane, say, at a shorter distance than it does in a horizontal plane. Sharply focused lines appear dark, whereas those that are spread out slightly appear dimmer or gray.

Image (line)
Object (point)

FIGURE 23-7
A cylindrical lens forms a line image of a point object because it is converging in one plane only.

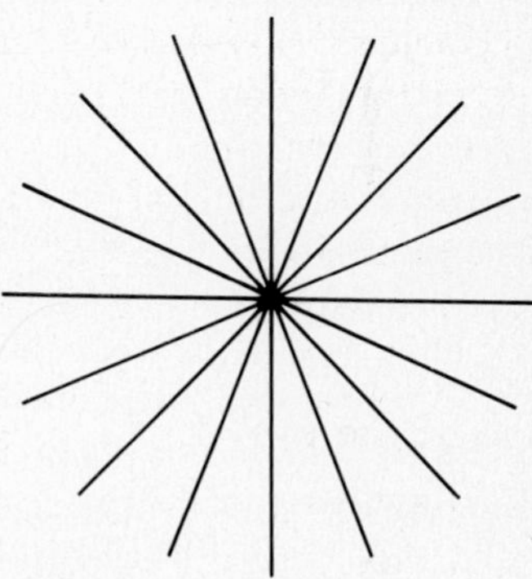

FIGURE 23-8
Test for astigmatism.

EXAMPLE 23-3 A farsighted person has a near point of 100 cm. Reading glasses must have what lens power so that this person can read a newspaper at a distance of 25 cm? Assume that the lens is very close to the eye.

SOLUTION When the object is placed 25 cm from the lens, we want the image to be 100 cm away on the *same* side of the lens, so it will be virtual (Figure 23-9). Thus $d_o = 25$ cm, $d_i = -100$ cm, and

$$\frac{1}{f} = \frac{1}{25\text{ cm}} + \frac{1}{-100\text{ cm}} = \frac{1}{33\text{ cm}},$$

so $f = 33$ cm $= 0.33$ m. The power P of the lens is $P = 1/f = +3.0$ diopters.

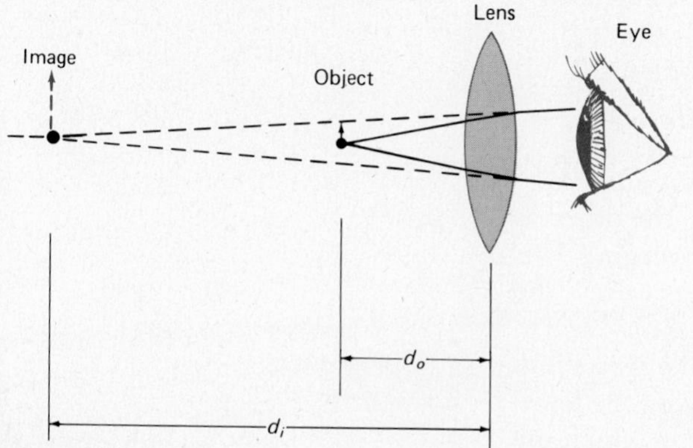

FIGURE 23-9
Example 23-3; lens of a reading glass.

The fact that it is positive indicates that it must be a converging lens. ❚❚

❚ EXAMPLE 23-4 A nearsighted eye has near and far points of 12 and 17 cm, respectively. What lens power is needed for this person to see distant objects clearly, and what then will be the near point? Assume that the lens is 2.0 cm from the eye.

SOLUTION The lens must image distant objects ($d_o = \infty$), so they are 17 cm from the eye, or 15 cm in front of the lens ($d_i = -15$ cm):

$$\frac{1}{f} = -\frac{1}{15\text{ cm}} + \frac{1}{\infty} = -\frac{1}{15\text{ cm}}.$$

So $f = -15\text{ cm} = -0.15\text{ m}$ or $P = 1/f = -6.7$ diopters; the minus sign indicates that it must be a diverging lens. For the near point, the image must be 12 cm from the eye or 10 cm from the lens, so $d_i = -0.10$ m and

$$\frac{1}{d_o} = \frac{1}{f} - \frac{1}{d_i} = -\frac{1}{0.15\text{ m}} + \frac{1}{0.10\text{ m}} = \frac{1}{0.30\text{ m}}.$$

The near point when the person is wearing glasses is thus 30 cm in front of the lens. ❚❚

Contact lenses could be used to correct the eye in Example 23-4. Since contacts are placed directly on the cornea, we would not subtract out the 2.0 cm. That is, for distant objects $d_i = -17$ cm, so $P = 1/f = -5.9$ diopters. Thus we see that a contact lens and an eyeglass lens will require slightly different focal lengths for the same eye because of their different placement relative to the eye.

23-3 ❚ The magnifying glass

Much of the remainder of this chapter will deal with optical devices that are used to produce magnified images of objects. We first discuss the *simple magnifier* or *magnifying glass* which is simply a converging lens.

How large an object appears, and how much detail we can see on it, depends on the size of the image it makes on the retina. This, in turn, depends on the angle subtended by the object at the eye. For example, a penny held 30 cm from the eye looks twice as high as one held 60 cm away, because the angle it subtends is twice as great (Figure 23-10). When we want to examine detail on an object, we bring it up close to our eyes so that it subtends a greater angle. However, our eyes can accommodate only up to a point (the near point), and we will assume a standard distance of 25 cm as the near point in what follows.

A magnifying glass allows us to place the object closer to our eye so that it subtends a greater angle. As shown in Figure 23-11*a*, the object is placed at the focal point or just within it. Then the converging lens produces a virtual image, which must be at least 25 cm from the eye if the eye is to focus on it. If the eye is relaxed, the image will be at infinity, and in this case the object is exactly at the focal point. (You make this slight adjustment yourself when you "focus" on the object by moving the lens.)

In a comparison of part *a* of Figure 23-11 with part *b*, in which the same object is viewed at the near point with the unaided eye, it is seen that

FIGURE 23-10
When the same object is viewed at a shorter distance, the image on the retina is greater; so the object appears larger and more detail can be seen. The angle θ the object subtends in (*a*) is greater than in (*b*).

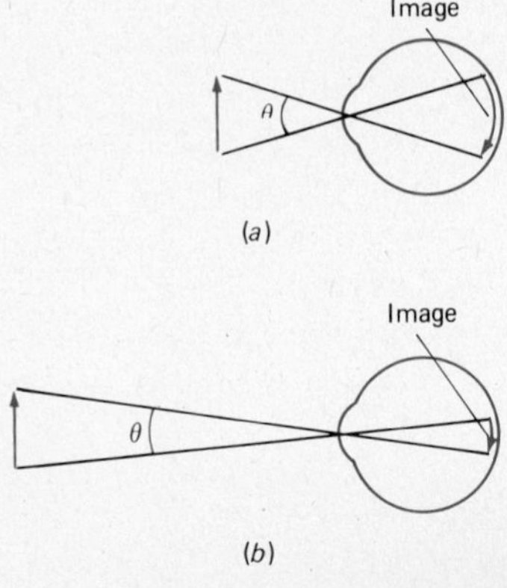

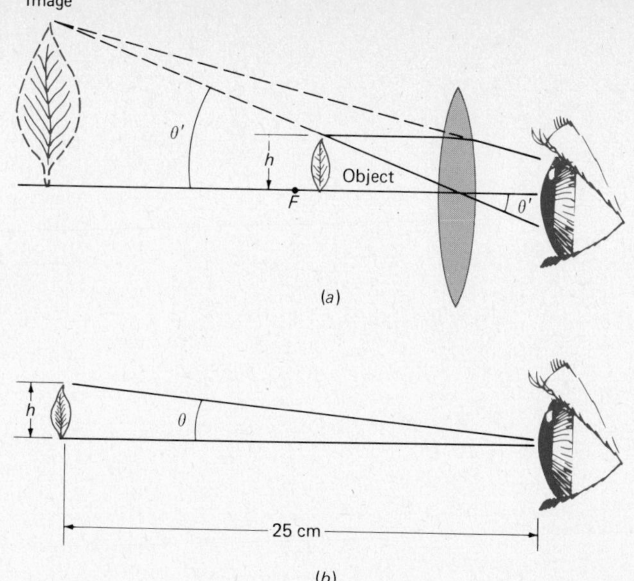

FIGURE 23-11
Leaf viewed (*a*) through a magnifying glass and (*b*) with the unaided eye focused at its near point.

the angle the object subtends at the eye is much larger when the magnifier is used. The **angular magnification** or *magnifying power, M,* of the lens is defined as the ratio of the angle subtended by an object with the lens to that subtended by the unaided eye without the lens at a distance of 25 cm:

$$M=\frac{\theta'}{\theta},$$

where θ and θ' are shown in Figure 23-11. This can be written in terms of the focal length f of the lens as follows. Suppose first that the image in Figure 23-11a is at the near point N of the eye, $d_i=-N$, where $N=25$ cm for the normal eye. Then the object distance d_o is given by

$$\frac{1}{d_o}=\frac{1}{f}-\frac{1}{d_i}=\frac{1}{f}+\frac{1}{N},$$

or $d_o=Nf/(f+N)$. (We see from this equation that $d_o<f$, as shown in Figure 23-11a, since $N/(f+N)$ must be less than 1.) Let h be the height of an object, and we assume that h is small, so the angles θ and θ' are approximately equal to their sines and tangents; then $\theta'=h/d_o=(f+N)h/Nf$ and $\theta=h/N$. Thus

$$M=\frac{\theta'}{\theta}=\frac{(f+N)h}{Nf}\left(\frac{N}{h}\right)$$

$$=1+\frac{N}{f} \qquad \text{(eye focused at near point which is } N=25 \text{ cm for the normal eye).} \qquad (23\text{-}1a)$$

If the eye is relaxed when using the magnifying glass, the image is then at infinity, and the object is then precisely at the focal point. In this case $\theta'=h/f$, so

$$M=\frac{\theta'}{\theta}=\left(\frac{h}{f}\right)\left(\frac{N}{h}\right)=\frac{N}{f} \qquad \text{(eye focused at } \infty\text{; } N=25 \text{ cm for normal eye).} \qquad (23.1b)$$

It is clear that the magnification is slightly greater when the eye is focused at its near point than when relaxed; and the shorter the focal length of the lens, the greater the magnification.

▮ EXAMPLE 23-5 An 8-cm-focal-length converging lens is used as a magnifying glass by a person with normal eyes ($N = 25$ cm). Calculate (*a*) the maximum magnification, and (*b*) the magnification when the eye is relaxed.

SOLUTION (*a*) The maximum magnification is obtained when the eye is focused at its near point:

$$M = 1 + \frac{N}{f} = 1 + \frac{25\text{ cm}}{8\text{ cm}} \approx 4\times.$$

(*b*) With the eye focused at infinity, $M = N/f = 25\text{ cm}/8\text{ cm} \approx 3\times$. ▮▮

23-4 ▮ Telescopes

A telescope is used to bring closer, or magnify, objects that are very far away; indeed, the object can usually be considered to be at infinity.

Several types of *astronomical telescope* exist. The "refracting" type contains two lenses located at opposite ends of a long tube (Figure 23-12). The lens closest to the object is called the *objective lens* and forms a real image I_1 of the object at its focal point F_o (or near it if the object is not at infinity). Although this image, I_1, is smaller than the original object, it is very close to the second lens, called the *eyepiece*, which acts as a magnifier. That is, the eyepiece magnifies the image I_1 produced by the objective to produce a second, greatly magnified image, I_2, which is virtual. If the viewing eye is relaxed, the eyepiece is adjusted so that the image I_2 is at infinity; this means that the real image I_1 is at the focal point F_e of the eyepiece, and that the distance between the lenses is $f_o + f_e$ for an object at infinity.

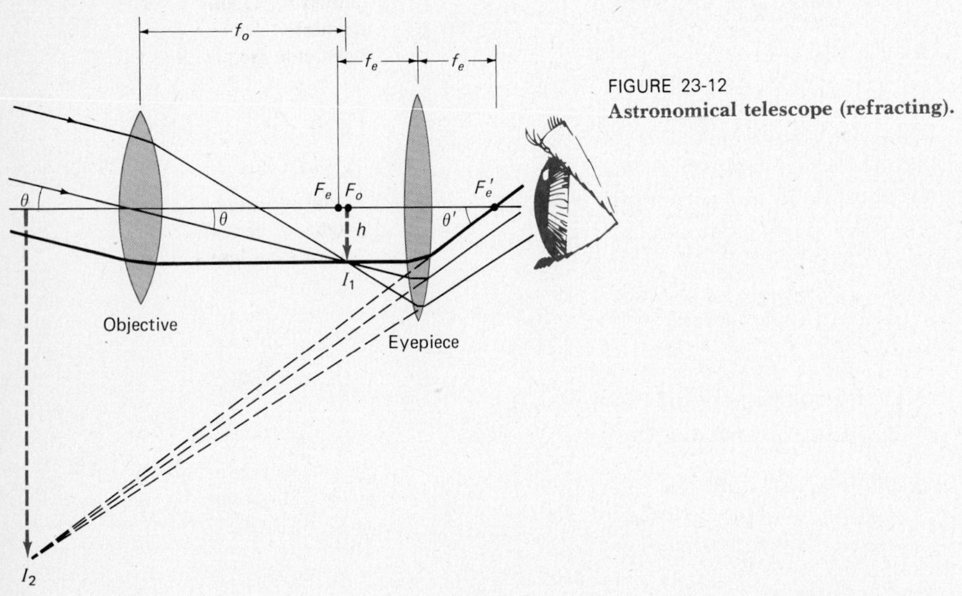

FIGURE 23-12
Astronomical telescope (refracting).

To find the total magnification of this telescope, we note that the angle an object subtends as viewed by the unaided eye is just the angle θ subtended at the telescope objective. From Figure 23-12 we can see that $\theta \approx h/f_o$, where h is the height of the image I_1 and we assume that θ is small, so $\tan\theta \approx \theta$. Note, too, that the darkest of the three rays drawn in the figure is parallel to the axis before it strikes the eyepiece and therefore passes through the focal point F'_e. Thus $\theta' \approx h/f_e$ and the total magnifying power of this telescope is

$$M = \frac{\theta'}{\theta} = -\frac{f_o}{f_e}, \tag{23-2}$$

where we have inserted a minus sign to indicate that the image is inverted. To achieve a large magnification, the objective lens should have a long focal length and the eyepiece a short one.

An important problem with astronomical telescopes is the fact that distant stars and planets are dim. To obtain a bright image, the objective lens must be large to allow in as much light as possible. The construction and grinding of large lenses is very difficult. Therefore, the largest telescopes use a curved mirror as the objective (Figure 23-13) and are referred to as "reflecting" telescopes. A mirror has only one surface to be ground and can be supported along its entire surface. (A large lens, supported at its edges, would sag under its own weight.†)

The large light-collecting ability of the world's great mirror telescopes has allowed very distant heavenly bodies to be seen. Normally, the eyepiece lens or mirror (Figure 23-13) is adjusted so that a real image is formed which can be recorded on film. Long exposures are often needed, which means that the telescope must move, or "track," its object, to compensate for the earth's rotation.

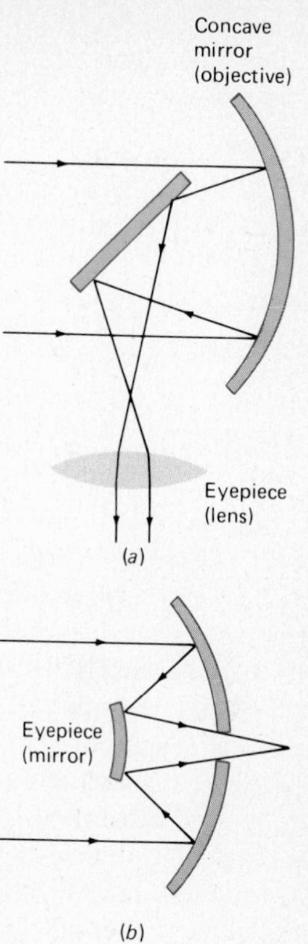

FIGURE 23-13
A concave mirror can be used as the objective of an astronomical telescope. Either a lens or a mirror can be used as the eyepiece. Arrangement (*a*) is called the Newtonian focus and (*b*) the Cassegrainian focus. Other arrangements are also possible.

Astronomical telescopes are not convenient for viewing objects on earth, since the final image is inverted. A *terrestrial telescope* (for use on earth) must provide an upright image. Two designs are shown in Figure 23-14. Galileo used the type shown in part *a*, and it is called a Galilean telescope. The eyepiece is a diverging lens which intercepts the converging rays from the objective before they reach a focus, and acts to form a virtual upright image. This design is used in some opera glasses. The tube is reasonably short, but the field of view is small. The second type, shown in part *b*, is often called a *spyglass* and makes use of a third lens (the "field lens"), which acts to make the image upright as shown. A spyglass must be quite long, however. The most practical design today is the *prism*

† Another advantage of mirrors is that they exhibit no chromatic aberration since the light does not pass through them; and they can be ground in a parabolic shape to correct for spherical aberration (see Section 23-6).

FIGURE 23-14
Terrestrial telescopes that produce an upright image: (*a*) Galilean; (*b*) spyglass or field-lens type.

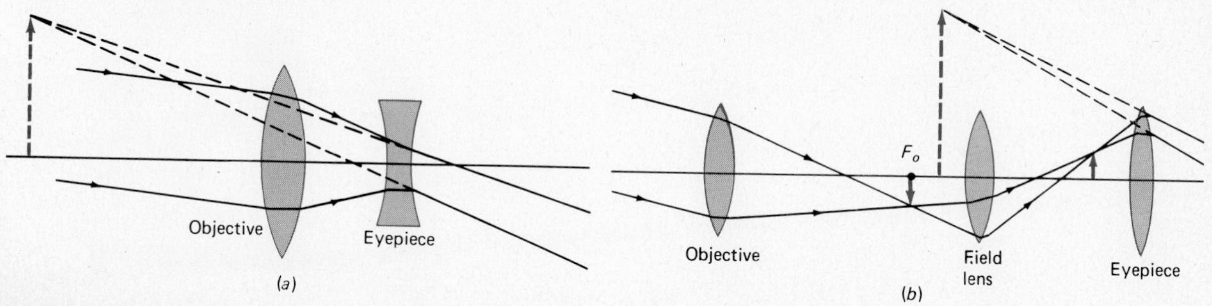

binocular, which was shown in Figure 21-19. The objective and eyepiece are converging lenses. The prisms reflect the rays by total internal reflection and shorten the physical size of the device. Furthermore, they act to produce an upright image. One prism reinverts the image in the vertical plane, the other in the horizontal plane.

▌ EXAMPLE 23-6 A Galilean telescope has an objective lens whose focal length is 28 cm and an eyepiece with focal length −8.0 cm. What is the magnification?

SOLUTION $M=-f_o/f_e=-(28\text{ cm})/(-8.0\text{ cm})=3.5\times$ ▌▌

23-5 ▌ Compound microscope

The compound microscope, like the telescope, uses two lenses, an objective and an eyepiece (Figure 23-15). The eyepiece is sometimes called the *ocular*. The design is different than a telescope, however, because a microscope is used to view objects that are very close. The use of two lenses in a microscope allows much greater magnification than is possible with a simple magnifier. The object is placed just beyond the objective's focal point, as shown in Figure 23-15*a*. The image I_1 formed by the objective lens is real, quite far from the lens, and much enlarged. This image is then magnified further by the eyepiece into a very large virtual image I_2 seen by the eye.

The total magnification of a microscope is the product of the magnifications produced by the two lenses. The image I_1 formed by the objective is a factor M_o greater than the object itself, where, from Equation 21-8 for the magnification of a simple lens,

$$M_o=\frac{d_i}{d_o}=\frac{l-f_e}{d_o},$$

where l is the distance between the lenses (equal to the length of the

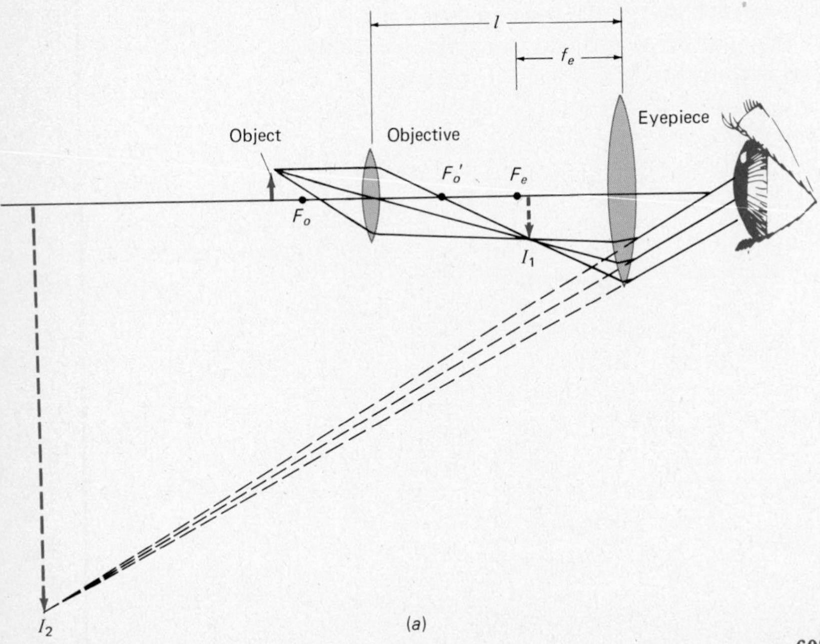

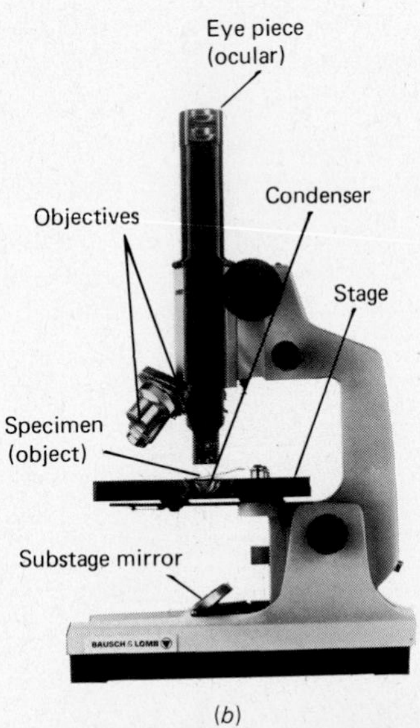

FIGURE 23-15
Compound microscope: (*a*) ray diagram; (*b*) actual microscope.

barrel), and we ignored the minus sign in Equation 21-8, which only tells us that the image is inverted. The eyepiece acts like a simple magnifier; if we assume that the eye is relaxed, its magnification M_e is (from Equation 23-1*b*)

$$M_e = \frac{N}{f_e}.$$

where the near point $N = 25$ cm for the normal eye. Since the eyepiece enlarges the image formed by the objective, the total magnification M is then

$$M = M_e M_o = \left(\frac{N}{f_e}\right)\left(\frac{l - f_e}{d_o}\right) \tag{23-3a}$$

$$\approx \frac{Nl}{f_e f_o}. \tag{23-3b}$$

The approximation, Equation 23-3*b*, is accurate when f_e and f_o are small compared to l, so $l - f_e \approx l$ and $d_o \approx f_o$ (Figure 23-15). This is a good approximation for large magnifications, since then f_o and f_e must be very small (since they are in the denominator of Equation 23-3). In order to make lenses of very short focal length, which can be done best for the objective, complex lenses involving several elements must be used to avoid serious aberrations, as discussed in the next section.

▌EXAMPLE 23-7 A compound microscope consists of a $10\times$ eyepiece and a $50\times$ objective 18.0 cm apart. Determine (*a*) the total magnification, (*b*) the focal length of each lens, and (*c*) the position of the object when the final image is in focus with the eye relaxed. Assume a normal eye so $N = 25$ cm.

SOLUTION (*a*) The total magnification is $10 \times 50 = 500\times$. (*b*) The eyepiece focal length is $f_e = N/M_e = 25\text{ cm}/10 = 2.5$ cm. It is easier to next find d_o (part *c*) before we find f_o since we can use the equation for M_o above: solving for d_o, we find $d_o = (l - f_e)/M_o = 15.5\text{ cm}/50 = 0.31$ cm. Then, from the lens equation,

$$\frac{1}{f_o} = \frac{1}{d_o} + \frac{1}{d_i} = \frac{1}{15.5\text{cm}} + \frac{1}{0.31\text{ cm}} = \frac{51}{15.5\text{ cm}};$$

so $f_o = 0.30$ cm. (*c*) We just calculated $d_o = 0.31$ cm, which is very close to f_o. ▌▌

Opaque objects are generally illuminated by a source placed above them. More commonly, the objects to be viewed are transparent, such as cells or tissue. In this case, light is normally passed through the object from a source beneath the microscope stage (see Figure 23-15*b*). The illumination system must be carefully designed if maximum sharpness and contrast are to be achieved. Usually, a *condenser* is employed, which is a set of two or three lenses, although inexpensive condensers may be a single lens. The purpose of the condenser is to gather light over a wide angle from the source, and to "condense" it down to a narrow beam that will illuminate the object strongly and uniformly. A number of different designs are employed. The source is often placed in the focal plane of the condenser so that light from each point on the source is parallel when it passes through the object.

Some specialized microscope types are described in Section 23-8.

23-6 ▮ Lens aberrations and compound lenses

Lens aberrations

In Chapter 21 we developed a theory of image formation by a thin lens. We found, for example, that all rays from each point on an object are brought to a single point as the image. This, and other results, were based on certain approximations such as that all rays make small angles with one another and that we can use $\sin\theta \approx \theta$. Because the approximations are not precisely true, we expect deviations from the simple theory, and these are referred to as **lens aberrations**. There are several types of aberration; we shall treat them separately, but all may be present at one time. The size of aberrations for a given lens can be determined experimentally, or analytically using Snell's law at each lens surface.

First, consider a point object on the axis of a lens. Rays from this point that pass through the outer regions of the lens are brought to a focus at a different point than those that pass through the center of the lens; this is shown exaggerated in Figure 23-16. Consequently, the image seen on a piece of film will not be a point but a tiny circular patch of light. If the film is placed at the point C as indicated, the circle will have its smallest diameter, which is referred to as the *circle of least confusion*. This aberration is called *spherical aberration* and is present whenever spherical surfaces are used. Spherical aberration can be corrected by using nonspherical lens surfaces, but to grind such lenses is very expensive. Spherical aberration can be minimized by choosing the curvatures of the lens surfaces so that equal amounts of bending occur at each; a lens can only be designed like this for one particular object distance. Spherical aberration can also be corrected (by which we mean reduced greatly) by the use of two or more lenses in combination.

FIGURE 23-16
Spherical aberration (exaggerated). Circle of least confusion obtained at *C*.

For object points off the lens axis, additional aberrations occur. Rays passing through different parts of the lens cause spreading of the image that is noncircular. We will not go into the details but merely point out that there are two effects: *coma* (because the image is comet-shaped rather than a circle) and *off-axis astigmation*.† Furthermore, the image points for objects off the axis but at the same distance from the lens do not fall on a flat plane, but on a curved surface. This aberration is known as *curvature of field* and is obviously a problem in cameras and other devices where the film is placed in a flat plane. The retina of the eye, however, is curved to compensate for this effect. Another aberration, known as *distortion*, is a result of variation of magnification at different distances from the lens axis. Thus a straight-line object not passing through the axis may form a curved image; a square grid of lines can be distorted to produce "pincushion distortion" or "barrel distortion" (Figure 23-17); the latter is common in extreme wide-angle lenses.

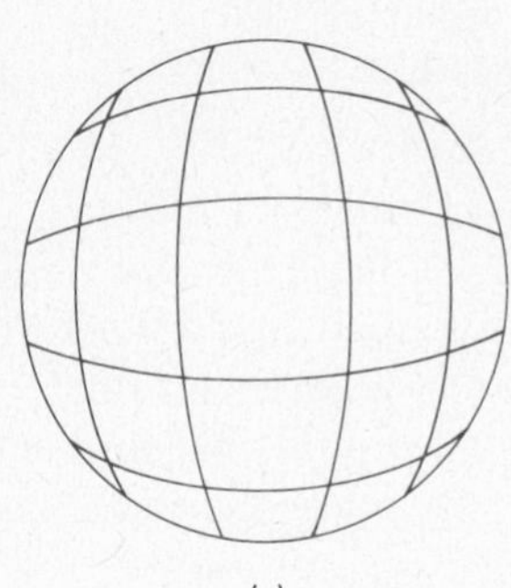

(a)

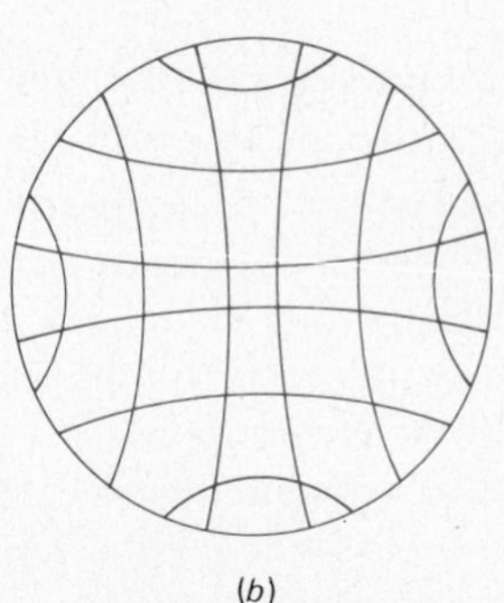

(b)

FIGURE 23-17
Distortion. Lenses may image a square grid of perpendicular lines to produce (*a*) barrel distortion or (*b*) pincushion distortion.

All the aberrations discussed above occur for monochromatic light, so are referred to as *monochromatic aberrations*. If the light is not monochromatic, there will also be *chromatic aberration*. This aberration arises

† Although the effect is the same as for astigmatism in the eye (see Section 23-2), the cause is different. Off-axis astigmation is no problem in the eye because objects are clearly seen only at the fovea, which is on the axis of the lens.

because of dispersion—the variation of index of refraction of transparent materials with wavelength (see Section 22-3). For example, blue light is bent more than red light by glass, so that if white light is incident on a lens, the different colors are focused at different points (Figure 23-18). Thus there will be colored fringes in the image. Chromatic aberration can be eliminated for any two colors (and reduced greatly for all others) by the use of two lenses. One lens is normally converging and the other diverging, and they are often cemented together (Figure 23-19). This is called an *achromatic doublet* (or "color-corrected" lens), and the two lenses are made of different materials with different indices of refraction and dispersion. In this way the combination can largely correct chromatic aberration, and the combination will be converging or diverging as needed.

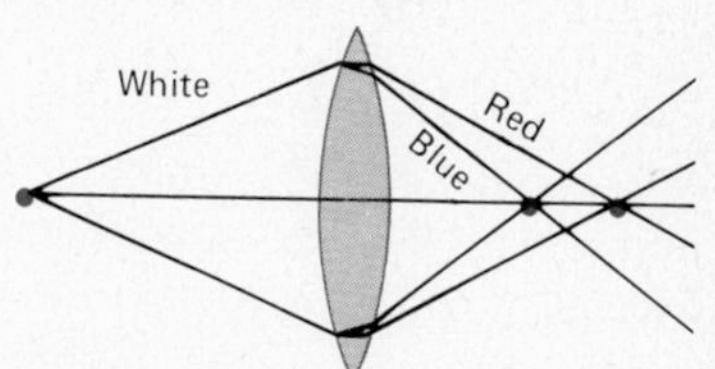

FIGURE 23-18
Chromatic aberration. Different colors are focused at different points.

Compound lenses

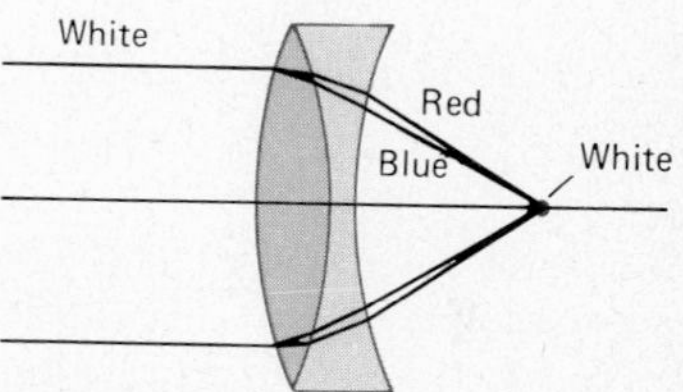

FIGURE 23-19
Achromatic doublet.

It is not possible to fully correct all aberrations. Combining two or more lenses can reduce some or all of them. High-quality lenses used in cameras, microscopes, and other devices contain many simple lenses, called elements. The combined lens is called a *compound lens.* A typical high-quality camera lens (Figure 23-20) may contain six to eight elements. Lenses are designed differently for different purposes, depending on which aberrations are most troublesome; for example, a normal camera lens which focuses objects at various distances will be designed differently from a microscope objective for which the object is always the same distance away. Today, lenses are designed with the help of computers, which trace literally millions of rays through different arrangements of elements.

The focal length of a compound lens is not measured from the center but from two *principal planes.* These are indicated in Figure 23-20. The lens equation, Equation 21-7, will hold for such a lens if the object distance is measured from the first principal plane and the image distance from the second principal plane.

The focal length, f, of a combination of two or more lenses in combination can be found by applying the lens formula, Equation 21-7, to each thin lens element. For example, if two lenses of focal length f_1 and f_2 are in contact (as in Figure 23-19), the net focal length f of the combination is found as follows. If the object distance for the first lens is $d_{o1} = \infty$,

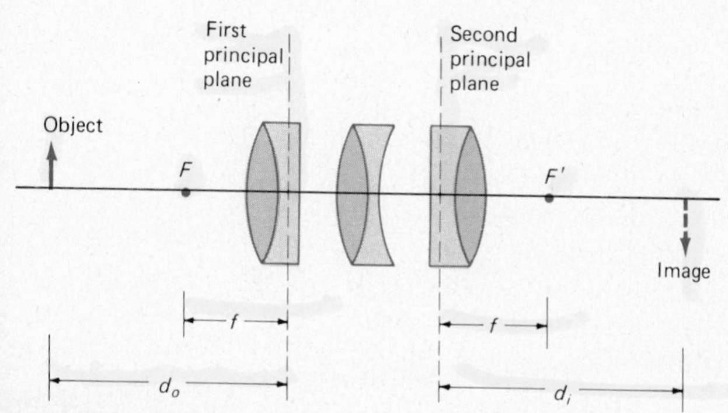

FIGURE 23-20
Compound lens, showing principal planes. Focal points *F* and *F′* are a focal length distance *f* from principal planes as shown, and object and image distances are measured from them.

the image distance d_{i2} of the second lens will be at the focal point of the combination, $d_{i2}=f$. For the first lens

$$\frac{1}{\infty}+\frac{1}{d_{i1}}=\frac{1}{f_1}.$$

Therefore, $d_{i1}=f_1$. This image becomes the object for the second lens, but if d_{i1} is positive, this image is behind the lens and is thus a virtual object for the second lens: $d_{o2}=-d_{i1}$. Thus the lens equation for the second lens is

$$\frac{1}{-d_{i1}}+\frac{1}{d_{i2}}=\frac{1}{f_2}.$$

Since $d_{i2}=f$ and $d_{i1}=f_1$, we have

$$\frac{1}{f}=\frac{1}{f_1}+\frac{1}{f_2}, \qquad (23\text{-}4)$$

where f is the focal length of the combination. If there are three or more lenses separated by distances much less than any of their focal lengths, Equation 23-4 can be extended by adding additional terms. However, if the lenses are separated by a significant distance, Equation 23-4 will not be valid (see the Problems). In terms of the power P of lenses, Equation 23-4 becomes

$$P=P_1+P_2.$$

❚ EXAMPLE 23-8 The two elements of the achromatic lens of Figure 23-19 have focal lengths $f_1=14.0$ cm and $f_2=-32.6$ cm. What is the focal length f of the combination?

SOLUTION From Equation 23-4, we have

$$f=\frac{f_1f_2}{f_1+f_2}=\frac{(14.0\text{ cm})(-32.6\text{ cm})}{14.0\text{ cm}-32.6\text{ cm}}=24.5\text{ cm}.$$

The combination is converging with a 24.5-cm focal length. ❚❚

For the most part, in this book we show lenses as if they were simple thin lenses. This is done for simplicity; it must be remembered that actual lenses, at least in good-quality instruments, are compound lenses.

Eye aberrations

The human eye is also subject to aberrations. In the course of evolution, however, these have been minimized. Spherical aberration, for example, has been largely corrected since (1) the cornea is less curved at the edges than at the center, and (2) the lens is less dense at the edges than at the center. Both effects cause rays at the outer edges to be bent less strongly, and thus help to reduce spherical aberration. Chromatic aberration is partially compensated for by the fact that the lens absorbs the shorter wavelengths appreciably and the retina is less sensitive to the blue and violet wavelengths. This is just the region of the spectrum where dispersion—and thus chromatic aberration—is greatest (Figure 22-10).

23-7 ▮ Limits of resolution

The ability of a lens to produce distinct images of two point objects very close together is called the **resolution** of the lens. The closer the two images can be and still be seen as distinct (rather than overlapping blobs), the higher the resolution. The resolution of a camera lens, for example, is often specified as so many lines per millimeter.† This is determined by using the lens to photograph a standard set of parallel lines using fine-grain film. The minimum spacing of lines that are distinguishable on film using the lens gives the resolution.

Two principal factors limit the resolution of a lens. The first is lens aberrations. As we saw, because of spherical and other aberrations, a point object is not a point on the image but a tiny blob. Careful design of compound lenses can reduce aberrations significantly, but they cannot be eliminated entirely. The second factor that limits resolution is diffraction, which cannot be corrected for because it is a natural result of the wave nature of light. We discuss it now.

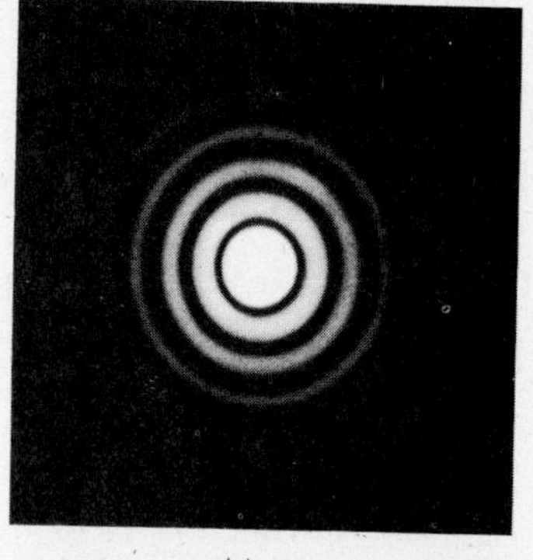

(a)

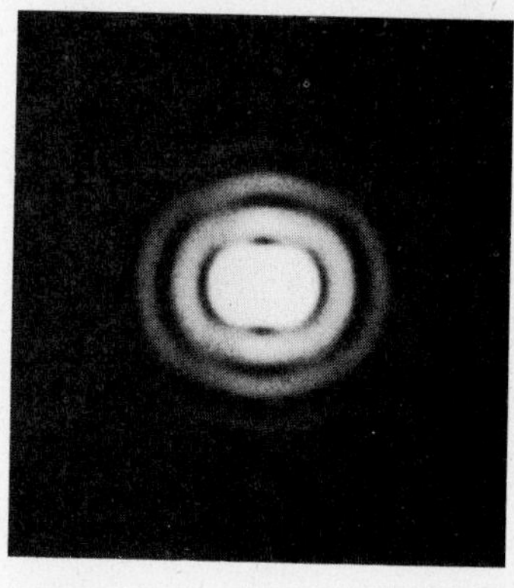

(b)

FIGURE 23-21
Photographs of images (greatly magnified) formed by a lens: (*a*) a single point object; (*b*) two point objects barely resolved. (From M. Cagnet, M. Françon, and J. C. Thrierr, *Atlas of Optical Phenomena*, Springer-Verlag, Berlin, 1962.)

Diffraction limit: the Rayleigh criterion

In Section 22-4 we saw that because light travels as a wave, light from a point source passing through a slit is spread out into a diffraction pattern (Figures 22-14 and 22-16). A lens, because it has edges, acts like a slit. When a lens forms the image of a point object, the image is actually a tiny diffraction pattern. Thus an image will be blurred even if aberrations were absent.

In the analysis that follows, we assume that the lens is perfect (has no aberrations), so we can focus our attention on diffraction effects and how much they limit the resolution of a lens. In Figure 22-16 we saw that the diffraction pattern produced by light passing through a rectangular slit has a central maximum in which most of the light falls. This central peak falls to a minimum on either side of its center at an angle $\theta \approx \sin\theta = \lambda/D$, where D is the diameter of the slit, λ the wavelength of light used, and we assume that θ is small (this is Equation 22-2). There are also low-intensity fringes beyond. For a circular hole, such as a lens, the image of a point object will consist of a *circular* central peak (called the *diffraction spot* or *Airy disk*) surrounded by faint circular fringes, as in Figure 23-21*a*. The central maximum has an angular half-width given by

$$\theta = \frac{1.22\lambda}{D}.$$

This differs from that for a slit (Equation 22-2) by the factor 1.22. This factor comes from the fact that the width of a circular hole—unlike a rectangular slit—is not uniform but varies from its diameter D to zero. A careful analysis shows that the "average" width is $D/1.22$; hence we get the equation above rather than Equation 22-2. The intensity of light in the diffraction pattern of a circular opening is shown in Figure 23-22.

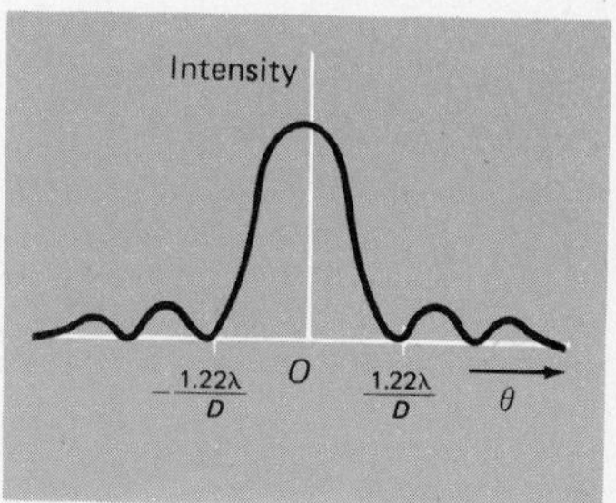

FIGURE 23-22
Intensity of light across the diffraction pattern of a circular hole.

If two point objects are very close, the diffraction patterns of their images will overlap as shown in Figure 23-21*b*. As the objects are moved

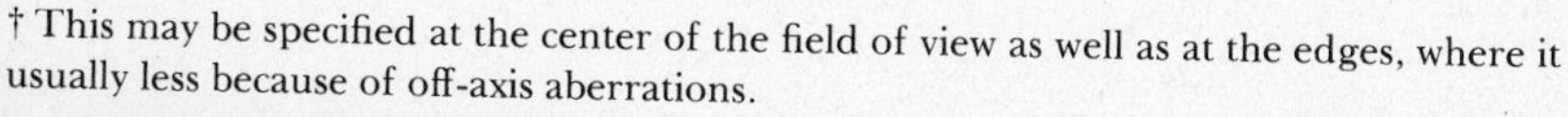
† This may be specified at the center of the field of view as well as at the edges, where it is usually less because of off-axis aberrations.

closer, a point is reached where you cannot tell if there are two overlapping images or a single image. Where this happens may be different for different observers. However, a generally accepted criterion is one proposed by Lord Rayleigh (1842–1919). The *Rayleigh criterion* states that *two images are just resolvable when the center of the diffraction disk of one is directly over the first minimum in the diffraction pattern of the other.* This is shown in Figure 23-23. Since the first minimum is at an angle $\theta = 1.22\lambda/D$ from the central maximum, Figure 23-23 shows us that two objects can be considered just resolvable if they are separated by this angle θ:

$$\theta = \frac{1.22\lambda}{D}. \tag{23-5}$$

This is the limit on resolution set by the wave nature of light due to diffraction.

The sharpness or resolution of a high-quality camera lens is usually limited by diffraction for small lens openings, say, $f/16$ or $f/22$. For large lens openings, D is large, so diffraction is less and the resolution is usually limited by spherical and other lens aberrations.

The resolution of a high-quality microscope is normally limited by diffraction. This is because microscope objectives are complex compound lenses containing 10 or more elements to reduce aberrations, and their diameter D is small (since f is small). Telescopes with large-diameter objectives are usually limited not by diffraction but by other effects, such as turbulence in the atmosphere.

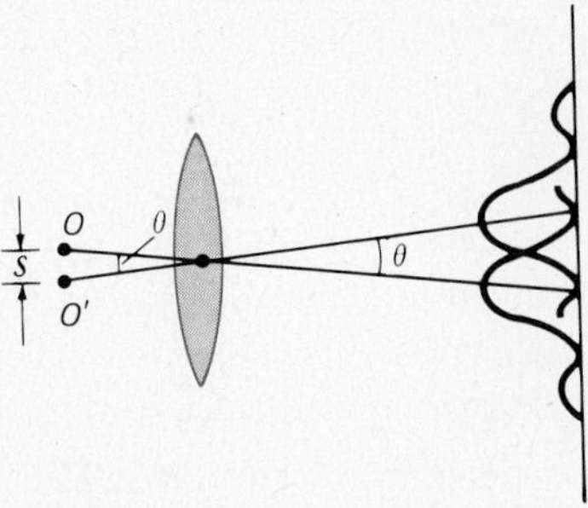

FIGURE 23-23
The *Rayleigh criterion.* Two images are just resolvable when the center of the diffraction peak of either is directly over the first minimum in the diffraction pattern of the other. The two point objects *O* and *O′* subtend an angle *θ* at the lens; one ray only is drawn for each point to indicate the center of the diffraction pattern of its image.

Resolution of telescopes and microscopes

It might be thought that a microscope or telescope could be designed to produce any desired magnification, depending on the choice of focal lengths. But because of diffraction, this is not possible. An increase in magnification above a certain point will merely result in magnification of the diffraction pattern. This would be very misleading, since we might think that we are seeing details of an object when we are really seeing details of the diffraction pattern. To examine this in detail, remember that two objects (or two nearby points on one object) are just resolvable if they are separated by an angle θ (Figure 23-23) given by Equation 23-5,

$$\theta = \frac{1.22\lambda}{D}.$$

This is valid for either a microscope or a telescope, where D is the diameter of the objective lens. For a telescope, the resolution is specified by stating θ as given by this equation. For a microscope it is more convenient to specify the actual distance s between two points that are just barely resolvable (Figure 23-23). Since objects are normally placed near the focal point of the objective, $\theta = s/f$ or $s = f\theta$. If we combine this with Equation 23-5, we obtain for the *resolving power* (RP)

$$\text{RP} = s = f\theta = \frac{1.22\lambda f}{D}.$$

This distance s is called the resolving power of the lens because it is the minimum separation of two object points that can just be resolved. The equation above is often written in terms of the angle of acceptance, α, of the objective lens, as defined in Figure 23-24. The derivation† is long and we only quote the result:

$$\mathrm{RP} = s = \frac{1.22\lambda}{2 \sin \alpha} = \frac{0.61\lambda}{\sin \alpha}.$$

The resolving power can be increased by placing a drop of oil that encloses the object and the front surface of the objective. This is called an *oil-immersion objective.* In the oil the wavelength of the light is reduced to λ/n, where n is the oil's index of refraction. Thus the resolving power becomes

$$\mathrm{RP} = \frac{0.61\lambda}{n \sin \alpha}. \tag{23-6}$$

The oil typically has $n \approx 1.5$, although n may be as great as 1.8. Thus oil immersion increases the resolution by 50 percent or more.

The quantity $n \sin \alpha$ is called the *numerical aperture* (NA) of the lens:

$$\mathrm{NA} = n \sin \alpha. \tag{23-7}$$

It is usually specified on the side of the objective lens housing together with the magnification. The larger the value of the NA, the finer the resolving power.

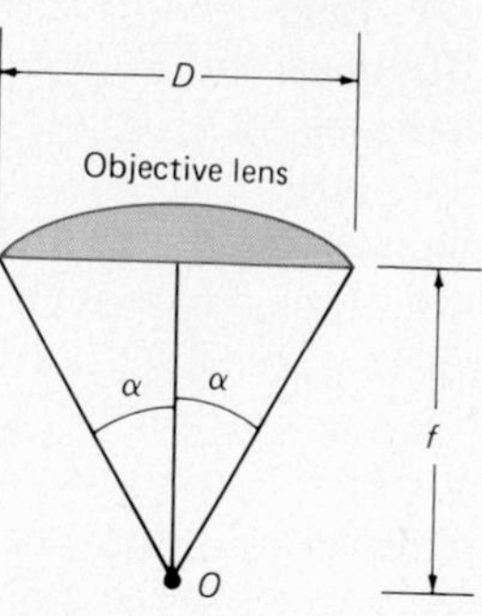

FIGURE 23-24
Objective lens of a microscope, showing the angle of acceptance, α.

▌ EXAMPLE 23-9 What is the theoretical minimum angular separation of two stars that can just be resolved by the 200-in telescope on Mt. Palomar? Assume that the stars emit white light of average wavelength 550 nm.

SOLUTION Since $D = 200$ in $= 5.1$ m, we have from Equation 23-5 that $\theta = 1.22\lambda/D = (1.22)(5.50 \times 10^{-7}\ \mathrm{m})/(5.1\ \mathrm{m}) = 1.3 \times 10^{-7}$ rad or 0.75×10^{-5} deg. This is the limit set by diffraction. In actual fact, the resolution is not this good because of aberrations and, more important, turbulence in the atmosphere. In fact, large-diameter objectives are not justified by increased resolution but by their greater light-gathering ability—they allow more light in so dimmer objects can be seen. ▌▌

▌ EXAMPLE 23-10 Determine the NA and RP of the best oil-immersion microscopes where the index of refraction of the oil is $n = 1.8$ and $\sin \alpha \approx 0.90$. Assume that $\lambda = 550$ nm.

SOLUTION The $\mathrm{NA} = n \sin \alpha = 1.6$. The resolving power is $\mathrm{RP} = 0.61\lambda/\mathrm{NA} = (0.61)(5.50 \times 10^{-7}\ \mathrm{m})/(1.6) \approx 2 \times 10^{-7}\ \mathrm{m} = 200$ nm. This is the best resolution that a visible light microscope can attain. ▌▌

Diffraction sets an ultimate limit on how clearly an object can be seen. Let us return to Equation 23-6 and assume that we are trying to magnify an object in air ($n = 1.0$). Since $\sin \alpha$ can never be greater than 1.0 (values for real microscopes are at best 0.6 to 0.9), we see that, within a factor of 2 or so,

$$\blacksquare\ \mathrm{RP} \approx \lambda. \tag{23-8}$$

† When the angle α is small, the derivation is easy, because in this case $\sin \alpha = \tan \alpha = \alpha = D/2f$ (see Figure 23-24) or $f/D = 1/(2 \sin \alpha)$. Thus $s = 1.22\lambda f/D = 1.22\lambda/2 \sin \alpha$. A more general derivation can be found in F. A. Jenkins and H. E. White, *Fundamentals of Optics*, 4th ed. (New York: McGraw-Hill Book Company, 1976), p. 332, and in F. W. Sears, *Optics*, 3rd ed. (Reading, Mass.: Addison-Wesley Publishing Company, Inc., 1949), pp. 257–260.

Thus we can say, approximately, that *it is not possible to resolve details smaller than the wavelength of the light being used.* This is an important and useful rule of thumb.

Compound lenses are now designed so well that the actual limit on resolution is often set by diffraction—that is, by the wavelength of the light used. To obtain greater detail, one must use radiation of shorter wavelength. The use of UV radiation can increase the resolution by a factor of perhaps 2. Far more important, however, was the discovery in the early twentieth century that electrons have wave properties (Chapter 25) and that their wavelengths can be very small; this fact is used in the electron microscope (Section 25-7), which can magnify 100 to 1000 times more than a visible light microscope because of the much shorter wavelengths. X rays, too, have very short wavelengths and are often used to study objects in great detail (Section 25-4).

Resolution of the human eye and useful magnification

The resolution of the human eye is limited by several factors, all of roughly the same order of magnitude. The resolution is best at the fovea, where the cone spacing is smallest, about 3 μm (=3000 nm). The diameter of the pupil varies from about 0.1 cm to about 0.8 cm; so for $\lambda = 550$ nm (where the eye's sensitivity is greatest), the diffraction limit is about $\theta \approx 1.22\lambda/D \approx 8 \times 10^{-5}$ rad to 6×10^{-4} rad. Since the eye is about 2 cm long, this corresponds to a resolving power of $s = (8 \times 10^{-5}\ \text{rad})(2 \times 10^{-2}\ \text{m}) \approx 2\ \mu\text{m}$ at best, to about 15 μm at worst (pupil small). Spherical and chromatic aberration also limit the resolution to about 10 μm. The net result is that the eye can resolve objects whose angular separation is about 5×10^{-4} rad at best. This corresponds to objects separated by about 1 cm at a distance of 20 m.

There is no point in constructing a microscope that produces an image greater than that resolvable by the eye. For then we would merely be seeing a magnified image of the diffraction pattern produced by the microscope objective. The typical near point of a human eye is about 25 cm. At this distance, the eye can just resolve objects that are $(25\ \text{cm})(5 \times 10^{-4}\ \text{rad}) \approx 10^{-4}$ m apart. Since the best light microscopes can resolve objects no smaller than about 200 nm (see Example 23-10), the useful magnification is limited to about

$$\frac{10^{-4}\ \text{m}}{200 \times 10^{-9}\ \text{m}} = 500\times.$$

In practice, magnifications of 1000× are often used to minimize eye-strain.

*23-8 ▮ Specialty microscopes and contrast

All the resolving power a microscope can attain will be useless if the object to be seen cannot be distinguished from the background. The difference in brightness between the image of an object and the image of the surroundings is called *contrast.* Achieving high contrast is an important problem in microscopy. The problem arises in biology, for example, because cells consist largely of water and are almost uniformly transparent to light.

One way of increasing contrast is by staining. Certain dyes or stains may be absorbed in varying amounts by different parts of objects, such as tissue or cells. The different absorption of light thus allows different parts of the object to be seen.

Staining usually involves killing the cell. Even then this technique is not always able to distinguish certain structures and may add artifacts (nonreal effects) itself. As an alternative, a number of special microscopes have been developed to increase contrast in a variety of situations. We now discuss some of these.

*UV microscope

A microscope using ultraviolet light will be able to resolve greater detail because of its shorter wavelengths. It can also increase contrast significantly in some cases because certain molecules, such as proteins and nucleic acids, absorb strongly in the UV region. Thus particular cells or cellular components may be more easily seen.

Unfortunately, glass does not transmit UV light, so quartz or fluorite lenses must be used. The added expense plus other difficulties, such as the need to record images photographically (the eye does not respond to UV and in fact is damaged by it), means that UV microscopes are used only for very special purposes.

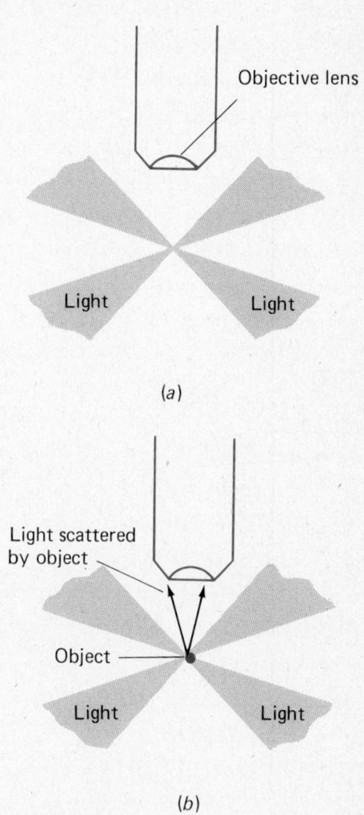

FIGURE 23-25
Dark-field microscope illumination.

*Dark-field microscopy

In an ordinary, or *bright-field*, microscope, light passes through the object into the lens. Where the light is absorbed or scattered sideways by the object, we see dark areas against a bright background.

In a *dark-field* microscope, the illumination system is designed so that light comes from the sides only, and no light enters the objective when an object is not present (Figure 23-25*a*). When an object is placed on the microscope stage, light is scattered by it into the objective lens (Figure 23-25*b*). The scattering is significant mainly in regions where there is a sharp change in the index of refraction, such as at the boundary of a cell or organelle. These regions will appear bright against a dark background. Although the contrast is greatly enhanced by this system, great care must be taken to avoid artifacts.

*Interference microscope

The interference microscope makes use of the wave properties of light in a direct way. It is one of the most effective means to increase contrast in a transparent object. To see how it works, let us consider a transparent object—perhaps a bacterium—in a water solution (Figure 23-26). Light enters uniformly from the left and is coherent (meaning in phase) at all points such as *a* and *b*. If the object is as transparent as the water solution, the beam leaving at *d* will be as bright as that at *c*. There will be no contrast and the object will not be seen. However, if the object's refractive index is slightly different from that of the surrounding medium, the wavelength

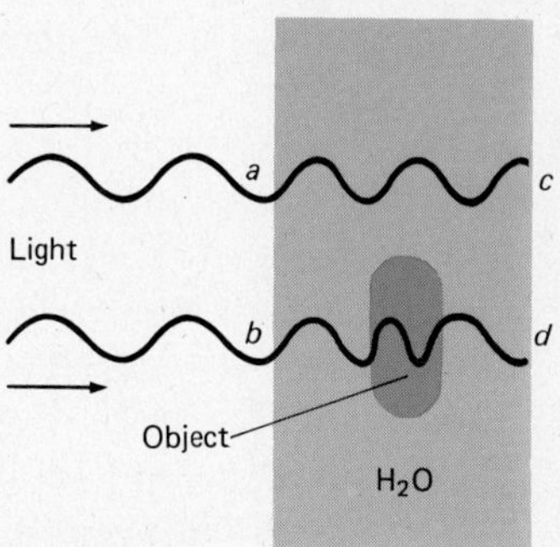

FIGURE 23-26
Object—perhaps a bacterium—in a water solution.

within the object will be altered as shown. Hence the waves at points *c* and *d* will differ in phase, if not in amplitude. This appears at first to be of no help, since the eye responds only to differences in amplitude or brightness, and does not detect this difference in phase. What the interference microscope does is to change this difference in phase into a difference of amplitude. It does this by superimposing the light that passes through the sample onto a reference beam that does not pass through the object, so that they interfere.

One way of doing this is shown in Figure 23-27. Light from a source is split into two equal beams by a half-silvered mirror, MS_1. One beam passes through the object and the second (comparison beam) passes through an identical system without the object. The two meet again and are superposed by the half-silvered mirror MS_2 before entering the eyepiece and the eye. The path length (and amplitude) of the comparison beam is adjustable. It can be adjusted, for example, so that the background is dark; that is, full destructive interference occurs. Light passing through the object (beam *bd* in Figure 23-26) will also interfere with the comparison beam; but because of its different phase, the interference will not be completely destructive. Thus it will appear brighter than the background. Where the object varies in thickness, the phase difference between beams *ac* and *bd* in Figure 23-26 will be different; and this will affect the amount of interference. Hence variation in the thickness of the object will appear as variations in brightness in the image. As an example, suppose that the object is a bacterium 1.0 μm thick and has a refractive index of 1.35; then if yellow light ($\lambda = 550$ nm in air) is used, there will be $(1.0\ \mu\text{m})/(550\ \text{nm}/1.35) = 2.46$ wavelengths in the bacterium and $(1.0\ \mu\text{m})/(550\ \text{nm}/1.33) = 2.42$ wavelengths in the water. Thus the two waves will be out of phase by 0.04 wavelengths, or 14° ($= 0.04 \times 360°$).

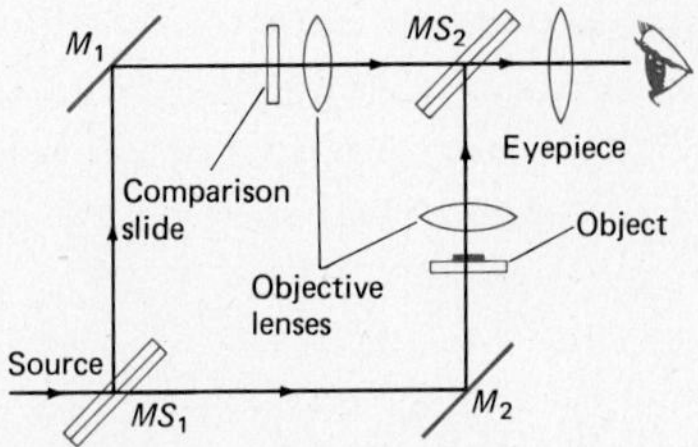

FIGURE 23-27
Diagram of an interference microscope.

*Phase-contrast microscope

The phase-contrast microscope also makes use of interference and differences in phase to produce a high-contrast image. Although it has certain limitations, it is far simpler to construct and operate than an interference microscope.

To describe the operation of a phase-contrast microscope in detail, we would have to discuss the diffraction theory of image formation—how the diffraction pattern produced by each point on the object contributes to the final image. This is quite complicated, so we give only a simplified description.

Figure 23-28 is a simplified diagram of a phase-contrast microscope. The object to be viewed is illuminated from below as usual; to be specific we assume that rays from each point on the source are made parallel by a set of condensing lenses (not shown). However, a plate with a ring-shaped hole is placed above the source, so that light can only pass through this annular ring. Light that is not deviated by the object (beam is shown shaded) is brought into focus by the objective lens in the *source image plane*; if the source is effectively at infinity (because of the condensing lenses), as is assumed here, the source image plane is at the focal point of the lens. Light that strikes the object, on the other hand, is diffracted or scattered by the object. Each point on the object then serves as a source for

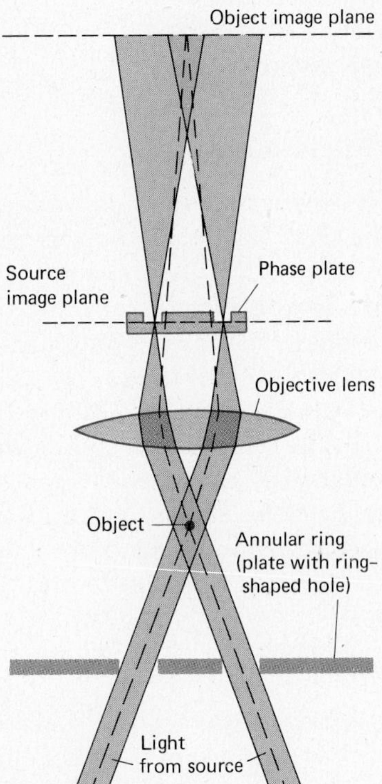

FIGURE 23-28
Phase-contrast microscope. Light beam from source that is undeviated is shown shaded for clarity. Rays deviated by object, and which form the image of the object, are shown dashed.

rays diverging from that point (dashed lines in the figure). These rays are brought to a focus in the *object image plane*, which is behind the source image plane (because the object is so close to the lens). The undeviated light from the source diverges, meanwhile, from its image plane and provides a broad bright background at the object image plane. The object is transparent, however, and the image will not be seen clearly since there will be little contrast. Contrast is achieved by inserting a circular glass *phase plate* at the source image plane. The phase plate has a groove as shown (or a raised portion) in the shape of a ring. This ring is positioned so that all the undeviated rays pass through it. Most of the rays deviated by the object, on the other hand, do not pass through this ring (see the diagram). Because the rays deviated by the object travel through a different thickness of glass than the undeviated source rays, the two can be out of phase and can interfere destructively at the object image plane. Thus the image of the object will contrast sharply with the background.

Actually, because only a small fraction of the light is deviated by the object, the background light will be much stronger and so the contrast will not be great. To compensate for this, the grooved ring on the phase plate is darkened to absorb a good part of the undeviated light so that its intensity is more nearly equal to that of the deviated light. Then nearly complete destructive interference can occur at particular points and the contrast will be very high.

The phase-contrast microscope has become a common laboratory tool for viewing transparent living objects such as microorganisms. As mentioned earlier, it is simpler to use than the interference microscope. Its chief limitation is that images tend to have "halos" around them as a result of diffraction from the phase plate opening. Because of this artifact, care must be taken in the interpretation of images.

*Polarizing microscope

Another way of obtaining a sharp image of a transparent specimen is by using crossed polarizers. One polarizer is placed beneath the microscope stage and the other above, often inside the microscope tube itself. When the polarizers are crossed, the field of view will be dark in the absence of an object, or if the specimen is isotropic. But what if the object is birefringent—that is, the index of refraction is different for different orientations of the polarization vector in the light wave? In this case, if the incident plane-polarized light is not parallel to the optic axis, it will have components both parallel and perpendicular to it; since these components travel at different speeds, the electric vector of the emerging light will be rotated. Thus it will not be extinguished by the second polarizer and will appear bright on a dark background.

Some molecules or structures within a cell are birefringent in themselves; these are said to be *intrinsically birefringent*, and nucleic acids and proteins fall in this category. However, a solution of isotropic (nonbirefringent) molecules may still exhibit birefringence if the refractive indices of the solution and the molecules is different, and the molecules are asymmetric and preferentially aligned in one direction. This is called textural birefringence. Both types will produce bright images.

▮ Summary

A camera lens forms an image on film by allowing light in through a shutter. The lens is focused by moving it relative to the film, and its f-stop (or lens opening) must be adjusted for the brightness of the scene and the chosen shutter speed. The f-stop is defined as the ratio of the focal length to the diameter of the lens opening.

The human eye also adjusts for the available light—by opening and closing the iris. It focuses not by moving the lens but by adjusting the shape of the lens to vary its focal length. The image is formed on the retina, which contains an array of receptors known as rods and cones. Diverging eyeglass or contact lenses are used to correct the defect of a nearsighted eye, which cannot focus well on distant objects; converging lenses are used to correct for defects in which the eye cannot focus on close objects.

A simple magnifier is a converging lens that forms a virtual image of an object placed at (or within) the focal point. The *angular magnification*, when viewed by a relaxed normal eye, is $M = N/f$, where f is the focal length of the lens and N is the near point of the eye (25 cm for a "normal" eye).

An astronomical telescope consists of an objective lens or mirror and an eyepiece, which magnifies the real image formed by the objective. The magnification is equal to the ratio of the objective and eyepiece focal lengths, but the image is inverted. A terrestrial telescope uses extra lenses, prisms, or a diverging lens as eyepiece, so that the final image is right side up.

A compound microscope also uses objective and eyepiece lenses, and the final image is also inverted. The total magnification is the product of the magnifications of the two lenses and is approximately $(N/f_e)(l/f_o)$ where l is the distance between the lenses, N is the nearpoint of the eye (usually taken as 25 cm), and f_o and f_e are the focal lengths of objective and eyepiece.

Microscopes, telescopes, and other optical instruments are limited in the formation of sharp images by *lens aberrations*. These include spherical aberration, in which rays passing through the edge of a lens are not focused at the same point as those that pass near the center; and chromatic aberration, in which different colors are focused at different points. Compound lenses, consisting of several elements, can largely correct for aberrations.

The wave nature of light also limits the sharpness or *resolution* of images. Because of diffraction, it is not possible to discern details much smaller than the wavelength of the radiation being used. This limits the useful magnification of a light microscope to about 1000×.

Questions

1 Why is the depth field greater, and the image sharper, when a camera lens is "stopped down" to a larger f number?

2 Why must a camera lens be moved farther from the film to focus on a closer object?

3 A "pinhole" camera uses a tiny pinhole instead of a lens. Show, using ray diagrams, how reasonably sharp images can be formed using such a pinhole camera. In particular, consider two point objects 2.0 cm apart that are 1.0 m from a 1-mm-diameter pinhole. Show that on a piece of film 5 cm behind the pinhole each object produces a tiny, easily resolvable spot.

4 Why are bifocals needed mainly by older persons and not generally by younger people?

5 Explain why swimmers with good eyes see distant objects as blurry when they are under water. Use a diagram and also show why goggles correct this problem.

6 Will a nearsighted person who wears corrective lenses be able to see clearly underwater when wearing glasses? Use a diagram to show why or why not.

7 Complete the ray diagram of Figure 23-14*a* by showing the intermediate image and the focal points.

8 Inexpensive microscopes sold for children's use usually produce images that are colored at the edges. Why?

9 Spherical aberration in a thin lens is minimized if rays are bent equally by the two surfaces. If a plano-convex lens is used to form a real image of an object at infinity, which surface should face the object? Use ray diagrams to show why.

10 Which aberrations present in a simple lens are not present (or are greatly reduced) in the human eye?

11 Atoms have diameters of about 10^{-8} cm. Can visible light be used to "see" an atom? Why or why not?

12 Which color of visible light would give the best resolution in a microscope?

13 Why can't a light microscope be used to observe molecules in a cell?

14 If monochromatic light were used in a microscope, would the color affect the resolution? Explain.

Problems

SECTION 23-1

1 (I) What is the maximum diameter of a 50-mm, $f/1.4$ camera lens?

2 (I) A 135-mm-focal-length lens has f stops ranging from $f/3.5$ to $f/32$. What is the corresponding range of lens diameters?

3 (I) A television camera lens has a 14-cm focal length and a lens diameter of 6.0 cm. What is its f number?

4 (I) A properly exposed photograph is taken at $f/11$ and $\frac{1}{125}$ s. What lens opening would be required if the shutter speed were $\frac{1}{1000}$ s?

5 (I) A light meter reports that a camera setting of $\frac{1}{100}$ s at $f/11$ will give a correct exposure. But the photographer wishes to use $f/16$ to increase the depth of field. What should the shutter speed be?

6 (II) Suppose that a correct exposure is $\frac{1}{250}$ s at $f/11$. Under the same conditions, what exposure time would be needed for a pinhole camera if the pinhole diameter is 1.0 mm and the film is 5.0 cm from the hole?

7 (II) A 135-mm-focal-length lens can be adjusted so that it is 135 to 140 mm from the film. For what range of object distances can it be adjusted?

8 (II) A nature photographer wishes to photograph a 4.5-m-high elephant from a distance of 100 m. What focal-length lens should be used if the image is to fill the 24-mm height of the film?

9 (II) Show that for objects at infinity, the magnification of a camera lens is proportional to its focal length.

SECTION 23-2

10 (I) A human eyeball is about 2.0 cm long and the pupil has a maximum diameter of about 5 mm. What is the "speed" of this lens?

11 (I) A person's left eye is corrected by a −8.5-diopter lens, 2.0 cm from the eye. (*a*) Is this person near- or farsighted? (*b*) What is this person's far point without glasses?

12 (I) Reading glasses of what power are needed for a person whose near point is 150 cm so that he can read at 25 cm? Assume a lens-eye distance of 2.0 cm.

13 (I) A man's left eye can see objects clearly only if they are between 15 and 35 cm away. (*a*) What power of contact lens is required so that objects far away are sharp? (*b*) What, then, will be his near point?

14 (II) About how much longer is the near-sighted eye in Example 23-4 than 2.0 cm of a normal eye?

15 (II) A woman can see clearly with her right eye only when objects are between 40 and 250 cm away. Prescription bifocals should have what powers so that she can see distant objects clearly (upper part) and be able to read a book 25 cm away (lower part)? Assume that the glasses will be 2.0 cm from the eye.

16 (II) One lens of a nearsighted person's eyeglasses has a focal length of −20.0 cm and the lens is 1.4 cm from the eye. If the person switches to contact lenses that are placed directly on the eye, what should be the focal length of the corresponding contact lens? (*Hint*: The image of an object at infinity should be the same distance from the eye.)

17 (II) What is the focal length of the eye lens system when viewing an object (*a*) at infinity? (*b*) 20 cm from the eye? Assume that the lens–retina distance is 2.0 cm.

18 (III) A 50-year-old man uses +2.5-diopter lenses to be able to read a newspaper 25 cm away. Ten years later, he finds that he must hold the paper 45 cm away to see clearly with the same lenses. What power lenses does he need now?

SECTION 23-3

19 (I) What is the focal length of a magnifying glass of 2.5× magnification for a relaxed normal eye?

20 (I) A magnifier is rated at 3× for a normal eye focusing on an image at its near point. What is its focal length?

21 (I) What is the magnification of a 6.0-cm-focal-length converging lens for a normal eye (*a*) focused at its near point and (*b*) relaxed.

22 (II) A 2.0-mm-wide stone is viewed with an 8.0-cm-focal-length lens. A normal eye views the image at its near point. Calculate (*a*) the angular magnification, (*b*) the width of the image, and (*c*) the object distance from the lens.

23 (II) A child has a near point of 10 cm. What is the maximum magnification the child can obtain using an 8.0-cm-focal-length magnifier? Compare to that for a normal eye.

24 (II) A small insect is placed 3.60 cm from a 4.00-cm-focal-length lens. Calculate (*a*) the position of the image, and (*b*) the angular magnification.

25 (II) A 5.0-cm-focal length lens is used as a simple magnifier. To obtain maximum magnification, where must the object be placed for a normal eye?

26 (II) What is the magnifying power of a +22-diopter lens? Assume a relaxed normal eye.

27 (III) A magnifying glass is rated at 3.0× for a normal eye that is relaxed. What would be the magnification for a relaxed eye whose near point is (*a*) 40 cm, and (*b*) 15 cm?

SECTION 23-4

28 (I) A 10× binocular has 3.0-cm-focal-length eyepieces. What is the focal length of the objective lenses?

29 (I) What is the magnification of an astronomical telescope whose objective lens has a focal length of 60 cm and whose eyepiece has a focal length of 2.4 cm? What is the overall length of the telescope when adjusted for a relaxed eye?

30 (II) An astronomical telescope has an objective with focal length 70 cm and a +40-diopter eyepiece. What is the total magnification?

*31 (II) What is the magnifying power of an astronomical telescope using a reflecting mirror whose radius of curvature is 4.00 m and an eyepiece whose focal length is 2.4 cm?

32 (II) A professor lost in the mountains tries to make a telescope using the lenses from his reading glasses. They have powers of +1.2 and +6.0 diopters, respectively. (*a*) What maximum magnification telescope is possible? (*b*) Which lens should be used as the eyepiece?

33 (II) A Galilean telescope adjusted for a relaxed eye is 28 cm long. If the objective lens has a focal length of 30 cm, what is the magnification?

34 (III) A 7.0× pair of binoculars has an objective focal length of 24 cm. If the binoculars are focused on an object 4.5 m away, what is the magnification? (The 7.0× refers to objects at infinity; Equation 23-2 holds only for objects at infinity, not for nearby ones.)

35 (III) A 50× astronomical telescope is adjusted for a relaxed eye when the two lenses are 84 cm apart. What is the focal length of each lens?

SECTION 23-5

36 (I) A 7.0× eyepiece is used in conjunction with a 25× objective lens in a compound microscope. What is the total magnification?

37 (I) A 900× microscope uses a 0.40-cm-focal-length objective lens. If the tube length is 16.8 cm, what is the focal length of the eyepiece? Assume a normal eye and that the final image is at infinity.

38 (II) A microscope has a 15× eyepiece and a 40× objective 17.5 cm apart. Calculate (*a*) the total magnification, (*b*) the focal length of each lens, and (*c*) where the object must be for a normal relaxed eye to see it in focus.

39 (II) A microscope has a 2.0-cm-focal-length eyepiece and 1.0-cm objective. Calculate (*a*) the position of the object if the distance between the lenses is 18.0 cm, and (*b*) the total magnification assuming a relaxed normal eye.

40 (II) Repeat Problem 39 assuming that the final image is located 25 cm from the eyepiece (near point of a normal eye).

41 (II) The eyepiece of a compound microscope has focal length 2.5 cm and the objective has $f = 0.80$ cm. If the object is placed 0.85 cm from the objective lens, calculate (*a*) the distance between the lenses when the microscope is adjusted for a relaxed eye, and (*b*) the total magnification.

SECTION 23-6

42 (I) An achromatic lens is made of two lenses with focal lengths $f_1 = -24$ cm and $f_2 = +15$ cm. (*a*) Is the

combination converging or diverging? (*b*) What is the net focal length?

43 (I) Two 10-cm-focal-length lenses are placed in contact. What is the focal length of the combination?

44 (II) A complex lens consists of three simple lenses in contact of focal lengths 20, 80, and −40 cm. What is the focal length of the combination? Is it converging or diverging?

45 (II) An optometrist finds that a patient sees clearly when lenses of focal length 60, −20, and 100 cm are placed in contact. A lens of what power (diopters) should be prescribed?

46 (II) Three lenses of focal length 20, 60, and −10 cm are each spaced 4.0 cm apart. How far from the last (−10 cm) lens will the final image be for an object at infinity?

SECTION 23-7

47 (I) What is the angle of acceptance α of a microscope oil-immersion objective and its resolving power if $n = 1.60$ and the NA = 1.35? Use $\lambda = 500$ nm.

48 (I) What is the angular resolution limit set by diffraction for the 100-in Mt. Wilson telescope ($\lambda = 500$ nm)?

49 (II) A microscope objective is immersed in oil ($n = 1.60$) and accepts light scattered from the object up to 60° on either side of vertical. (*a*) What is the numerical aperture? (*b*) What is the approximate resolution of the microscope if it uses 550-nm light?

50 (II) Two stars 10 light-years away are barely resolved by a 100-in (mirror diameter) telescope. How far apart are the stars? Assume $\lambda = 500$ nm.

51 (II) A certain sea organism has a pattern of dots on its surface with an average spacing of 0.40 μm. If the specimen is viewed using 550-nm light, what minimum value must the numerical aperture be in order that the dots be resolved?

52 (III) What minimum magnification would be required to see the dots on the organism of Problem 51?

53 (III) Suppose that you wish to construct a telescope that can resolve features 10 km across on the moon, 384,000 km away. You have a 2.0-m-focal-length objective lens whose diameter is 10 cm. What focal-length eyepiece is needed if your eye can resolve objects 0.10 mm apart at a distance of 25 cm? What is the resolution limit set by the size of objective lens (that is, by diffraction)? Use $\lambda = 500$ nm.

54 (III) Can a human eye distinguish the two headlights, 1.8 m apart, on a truck 10 km away? Consider only diffraction and assume an eye diameter of 5.0 mm and a wavelength of 500 nm.

* SECTION 23-8

***55** (II) Show that the phase difference (in radians) between the two waves *ac* and *bd* in Figure 23-26 is $\delta = (2\pi/\lambda)(n_2 - n_1)t$, where n_2 and n_1 are the refractive indices of the object and the medium, t the thickness of the object, and λ the wavelength of light used.

***56** (III) Birefringent spindles in a biological specimen are 4.50 μm thick. Light ($\lambda = 500$ nm in air) has indices of refraction in the material of 1.5110 and 1.5147 for the two orientations of the polarization vector. What is the phase difference (radians) between the two emerging beams?

For additional reading

Ackerman, E., *Biophysical Science*, Prentice-Hall, Inc., Englewood Cliffs, N.J., 1962. Resolving power, various microscopes, the eye, and vision.

Aidley, D. J., *The Physiology of Excitable Cells*, Cambridge University Press, Cambridge, 1971. Vision.

Cornsweet, T. N., *Visual Perception*, Academic Press, Inc., New York, 1970.

Horridge, G. A., "Compound Eye of Insects," *Sci. American*, **237**:1 (Jul., 1977), p. 108.

Jenkins, F. A., *Fundamentals of Optics*, 4th ed., McGraw-Hill Book Company, New York, 1975.

Land, E. H., "The Retinex Theory of Color Vision," *Sci. American*, **237**:6 (Dec., 1977), p. 108.

Meyer-Arendt, J. R., *Introduction to Classical and Modern Optics*, Prentice-Hall, inc., Englewood Cliffs, N.J., 1972.

Price, W. H., "Photographic Lens," *Sci. American*, **235**:2 (Aug., 1976), p. 72. Design of complex lenses.

Sears, F. W., *Optics*, 3rd ed., Addison-Wesley Publishing Co., Reading, Mass., 1949.

Setlow, R. B., and E. C. Pollard, *Molecular Biophysics*, Addison-Wesley Publishing Co., Reading, Mass., 1962. Specialty microscopes and other related topics.

24 ▮ Relativity

By the end of the nineteenth century, the physics community was rather self-satisfied. For the theories developed over the preceding 300 years had been very successful in explaining a wide range of natural phenomena. Newtonian mechanics beautifully explained the motion of objects on earth and in the heavens; furthermore, it formed the basis for successful treatments of fluids, wave motion, and sound. Kinetic theory, on the other hand, explained the behavior of gases and other materials. And Maxwell's great theory of electromagnetism not only brought together and explained electric and magnetic phenomena, but it predicted the existence of electromagnetic waves that would behave in every way just like light—so light came to be thought of as an electromagnetic wave. Indeed, it seemed that the natural world, as seen through the eyes of physicists, was very well explained; only a few small puzzles remained, and it was assumed these would soon be explained using already known principles.

Alas, it did not turn out to be so simple; the puzzles were only to be solved by the introduction of two revolutionary theories that changed our whole conception of the universe: the *theory of relativity* and *quantum theory*.

Physics as it was known at the end of the nineteenth century, which we have mainly been studying up to now, is referred to as *classical physics*. The great revolution that shook the world of physics at the turn of the twentieth century is referred to as *modern physics*. The remainder of this book deals with the fascinating subject of modern physics. In this chapter we discuss the theory of relativity. The following three chapters deal with the development of quantum theory, and how it explains atomic and molecular structure. The final three chapters deal with nuclear physics and elementary particles—those entities believed to be the basic building blocks of the universe.

24-1 ▮ Galilean–Newtonian relativity

Einstein's theory of relativity deals with how we observe events (anything that happens) in the physical world. In particular, it examines how objects and events are observed from different frames of reference. This had, of course, already been discussed by Galileo and Newton. We now briefly discuss these earlier ideas before tackling Einstein's revolutionary viewpoint.

We shall deal primarily with **inertial reference frames**. An inertial reference frame (as mentioned in Section 3-2) is one in which Newton's first law, the law of inertia, is valid. That is, if no net force acts on an object at rest, it will remain at rest; and if it is in motion, it will continue to move with constant speed in a straight line. Rotating or otherwise accelerating frames of reference are noninertial frames,† and we will not be concerned with them here. The earth is not quite an inertial frame (it rotates). But it is close enough that for most purposes we can consider it to be an inertial frame.

A reference frame that moves with constant velocity with respect to an inertial frame is itself an inertial frame. An airplane, a car, or a ship that moves with constant speed in one direction is an inertial frame just like the earth.

Both Galileo and Newton were aware of what we now call the **relativity principle**: *the laws of physics are the same in all inertial reference frames.* You may have recognized its validity in everyday life. For example, you may have noticed that objects move in just the same way in a smoothly moving (constant velocity) car or airplane as they do on earth. (Assume that there are no vibrations or rocking—for they would make the reference frame noninertial.) When you walk, drink a cup of soup, play Ping Pong, or drop a pencil on the floor, these bodies move just as they do on earth. Let us examine one case in more detail. Suppose that you are in a car traveling rapidly along a superhighway at constant velocity. If you hold a coin above your head inside the car and gently release it, how will it fall? If you try it, you will see that the coin falls straight downward and hits the floor directly below the point of release (Figure 24-1*a*). (If you drop the object out the car's window, this will not happen because the moving air drags the object backward.) This is just how objects fall on the earth—straight down—and thus our experiment in the moving car is in accord with the relativity principle.

Note in our example, however, that to an observer on the earth the object follows a curved path (Figure 24-1*b*). The actual path followed by the object is different from different frames of reference. This does not

† On a rotating platform, for example (say, a merry-go-round), an object at rest starts moving outward even though no force is exerted on it. This is therefore not an inertial frame.

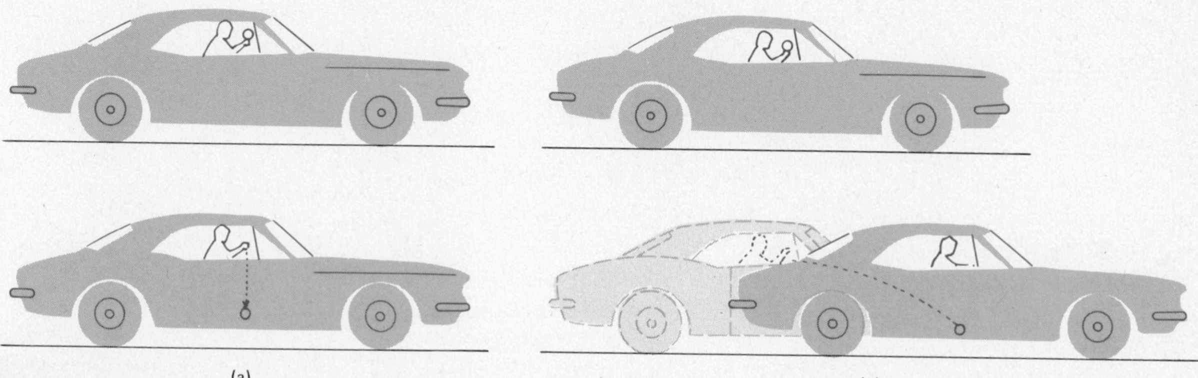

(*a*)
Reference frame = car

(*b*)
Reference frame = earth

FIGURE 24-1
A coin is dropped by a person in a moving car. (*a*) In the reference frame of the car, the coin falls straight down. (*b*) In a reference frame fixed on the earth, the coin follows a curved (parabolic) path. The upper views show the moment of the coin's release, and the lower views a short time later.

violate the relativity principle, because this principle states that the *laws* of physics are the same in all inertial frames. The same law of gravity, and the same laws of motion, apply in both reference frames. The difference between Figures 24-1*a* and *b* is that in the earth's frame of reference, the coin has an initial velocity (equal to that of the car); the laws of physics therefore predict that it will follow a parabolic path like any projectile. In the car's reference frame, there is no initial velocity and the laws of physics predict that it will fall straight down. Thus the laws are the same, although the specific paths are different.

Galileo, in his great book, *Dialogues Concerning Two New Sciences*, described a similar experiment and predicted the same results. Galileo's example involved a sailor dropping a knife from the top of the mast of a sailing vessel. If the vessel moves at constant speed, where will the knife hit the deck?

Galilean–Newtonian relativity involves certain unprovable assumptions that make sense from everyday experience. It is assumed that the lengths of objects are the same in one reference frame as in another. That is, a car 5.0 m long when at rest will be measured to be 5.0 m long when traveling at high speed. Time, too, is assumed to pass at the same rate in different reference frames. Thus in classical mechanics, space and time are considered to be *absolute*. The mass of an object is also assumed to be the same in different inertial reference frames; and forces are assumed to be unchanged by a change in reference frame.

The position of an object is, of course, different when specified in different reference frames, and so is velocity. For example, a person may walk toward the front of a bus with a speed of 5 km/h; but if the bus moves 40 km/h with respect to the earth, the person's speed is then 45 km/h with respect to the earth. The acceleration of a body, however, is the same in any inertial reference frame. This is because the change in velocity, and the time, will be the same. For example, the person in the bus may accelerate from 0 to 5 km/h in 1.0 s, so $a = 5$ km/h/s in the reference frame of the bus. With respect to the earth, the acceleration is (45 km/h − 40 km/h)/(1.0 s) = 5 km/h/s, which is the same.

Since neither F, m, nor a changes from one inertial frame to another, Newton's second law, $F = ma$, does not change. Thus it satisfies the relativity principle. It is easily shown that the other laws of mechanics—the laws of motion, the laws of conservation of energy, momentum, and angular momentum, and so on—also satisfy the relativity principle.

The fact that the laws of mechanics are the same in all inertial reference frames leads to an important conclusion: that *all inertial reference frames are equivalent.* No one inertial reference frame is any better than another. A reference frame fixed to a car or an aircraft traveling at constant velocity is as good as one fixed on the earth. When you travel smoothly at constant velocity in a car or airplane, it is just as valid to say that you are at rest and the earth is moving as it is to say the reverse. There is no experiment one can do to tell which frame is "really" at rest and which is moving. Thus there is no way to single out one particular reference frame as being at absolute rest. Indeed, as Galileo said, "everything moves."

The situation changed somewhat in the last half of the nineteenth century. When Maxwell presented his comprehensive and very successful theory of electromagnetism, he showed that light can be considered as an electromagnetic wave. Indeed, Maxwell's equations (Chapter 20) predicted that the velocity of light c would be 3.00×10^8 m/s; and this is just what is measured within experimental error. The question then arose: In what reference frame does light have precisely the value that is predicted by Maxwell's theory? For it was assumed that light, like other objects, would have a different speed in different frames of reference. For example, if an observer were traveling on a rocket ship at a speed of 1.0×10^8 m/s toward a source of light, we might expect that he would measure the speed of the light reaching him to be $3.0 \times 10^8 \text{ m/s} + 1.0 \times 10^8 \text{ m/s} = 4.0 \times 10^8$ m/s. But Maxwell's equations have no provision for relative velocity. They merely predicted the speed of light to be $c = 3.0 \times 10^8$ m/s. This seemed to imply that there must be a special reference frame where c could have this value.

We have seen that waves travel on water and along ropes or strings (Chapter 11). And sound waves travel in air and other materials (Chapter 12). Since nineteenth-century physicists viewed the material world in terms of the laws of mechanics, it was natural for them to assume that light, too, must travel in some kind of *medium*; they called this transparent medium the *ether* and assumed that it permeated all of space.† It was therefore presumed that the velocity of light given by Maxwell's equations must be with respect to this ether.

However, it appeared that Maxwell's equations did *not* satisfy the relativity principle. They were not the same in all inertial reference frames. They were simplest in the frame where $c = 3.0 \times 10^8$ m/s—that is, in a reference frame at rest in the ether. In any other reference frame, extra terms would have to be added to take into account the relative velocity. Thus, although most of the laws of physics obeyed the relativity principle, the laws of electricity and magnetism (or "electrodynamics") apparently did not. Instead, they seemed to single out one reference frame that was better than any other—a reference frame that could be considered to be absolutely at rest.

Scientists soon set out to determine the speed of the earth relative to this absolute frame, whatever it might be. A number of clever experiments were designed to do this. The most direct were performed by A. A. Michelson and E. W. Morley. The details of the Michelson–Morley experiments will be discussed in the next section. Briefly, what they did was measure the difference in the speed of light in different directions. They expected to find a difference depending on the orientation of their apparatus with respect to the ether. For just as a boat has a different speed going upstream, downstream, or across the stream as measured by an observer on the ground, so, too, light would be expected to have a different speed depending on its direction relative to the ether, which because the earth is in motion, was assumed to move relative to the earth.

Strange as it may seem, they detected no difference. This was a great puzzle. A number of explanations were put forth over a period of years, but they led to contradictions or were otherwise unsatisfactory (see the next section).

† The medium for light waves could not be air, since light travels from the sun to earth through nearly empty space. Therefore, another medium was postulated, the ether. The ether was not only transparent, but because of difficulty in detecting it, was assumed to have zero density. The properties of the ether were thus similar to caloric (the mysterious substance earlier assumed to be heat), whose existence had been rejected.

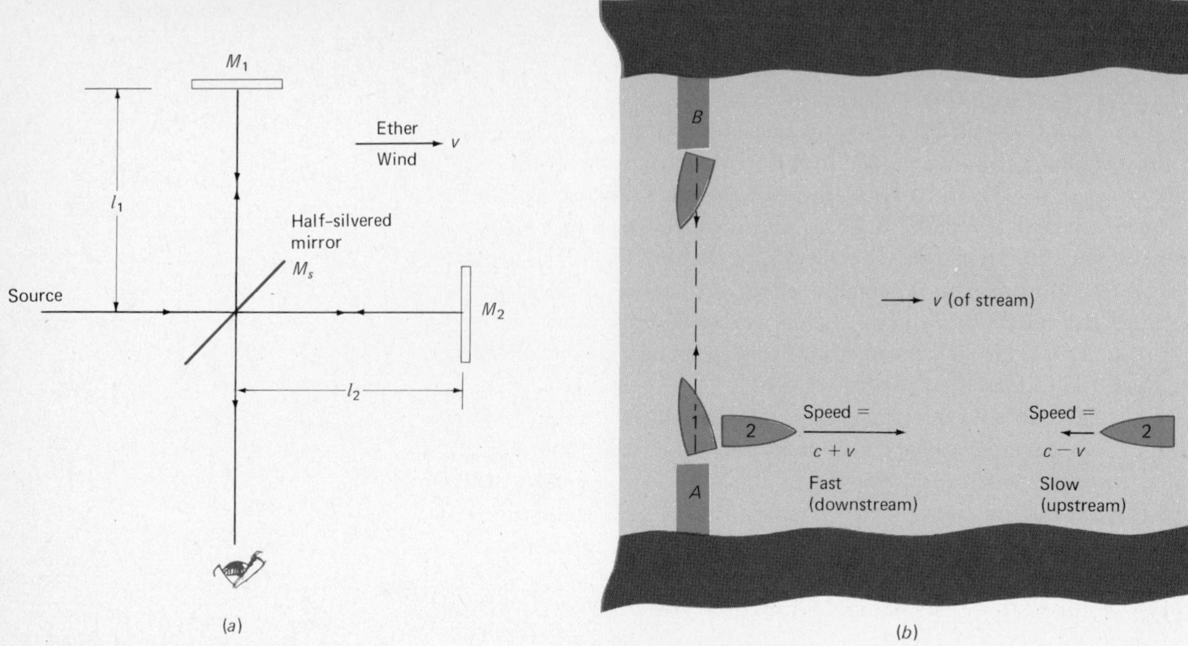

Then in 1905, Albert Einstein proposed a radical new theory that reconciled these many problems in a simple way. But at the same time, as we shall soon see, it completely changed our ideas of space and time.

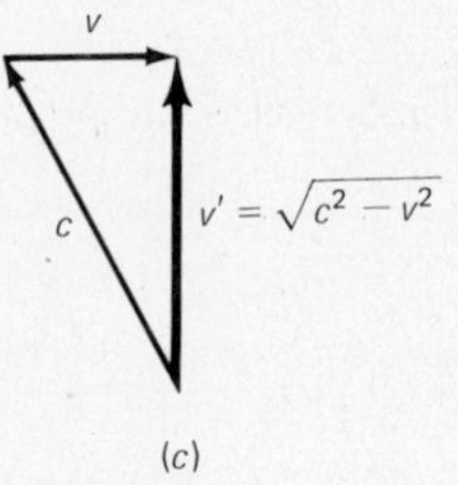

FIGURE 24-2

The Michelson–Morley experiment. (*a*) Michelson interferometer. (*b*) Boat analogy: boat 1 goes across the stream and back; boat 2 goes downstream and back upstream. (*c*) Calculation of velocity of boat (or light beam) traveling perpendicular to the current (or ether wind).

*24-2 ▌ The Michelson–Morley experiment

The Michelson–Morley experiment was designed to measure the speed of the ether—the medium in which light was assumed to travel—with respect to the earth. The experimenters thus hoped to find the absolute reference frame at rest.

Now the speed of sound (or other type of wave) depends on the observer's reference frame, as we saw when we discussed the Doppler effect (Section 12-9). For example, sound travels with a speed of about 330 m/s in air at 0°C. If an observer moves through the air at a speed of 30 m/s (driving in a car, say), that observer will measure the speed of a sound wave to be 330 m/s + 30 m/s = 360 m/s if he is moving toward the source, and 330 m/s − 30 m/s = 300 m/s if moving away from the source. Similarly, if the earth moves with a speed v with respect to the ether, an observer on earth should find that light traveling in the ether would have a speed of $c \pm v$ relative to earth.

One of the possibilities nineteenth-century scientists considered was that the ether is fixed relative to the sun, for even Newton had taken the sun as the center of the universe. If this were the case (there was no guarantee, of course), the earth's speed of 3.0×10^4 m/s in its orbit around the sun would produce a change of 1 part in 10^4 in the speed of light (3.0×10^8 m/s). Direct measurement of the speed of light to this accuracy was not possible. However, A. A. Michelson invented a device known as an interferometer (discussed in Chapter 22) which could measure the difference in the speed of light in different directions. With this device, Michelson, later with the help of E. W. Morley, performed their famous experiment based on the principle shown in Figure 24-2. Part a is a simplified diagram of the Michelson interferometer, and for simplicity it is assumed that the ether "wind" is moving with speed v to the right. (Alternatively, the earth is assumed to move to the left with respect to the ether at speed v.) The light from the source is split into two beams by the half-silvered mirror M_s; one beam travels to mirror M_1 and the other to mirror M_2. The beams are reflected by M_1 and M_2 and are joined again after passing through M_s. The now superposed beams interfere with one another and the resultant is viewed by the observer's eye as an interference pattern (discussed in Section 22-8).

Whether constructive or destructive interference occurs at the center of the interference pattern depends on the relative phases of the two beams after they have traveled their separate paths. This we examine next. To aid our analysis, we will use the analogy of a boat traveling up, down, and across a river whose current moves with speed v, as shown in Figure 24-2*b*. In still water the boat can travel with speed c (not necessarily the speed of light in this case.)

First, we consider beam 2 in Figure 24-2*a*, which travels parallel to the ether wind. In its journey from M_s to M_2, the light apparently travels with speed $c+v$, just as a boat traveling downstream (see part *b*) acquires the speed of the river current. Since the beam travels a distance l_2, the time it takes to go from M_s to M_2 is $t = l_2/(c+v)$. To make the return trip from M_2 to M_s, the light is moving against the ether wind (like the boat going upstream), so its relative speed would be $c - v$; the time for the return trip is thus $l_2/(c-v)$. The total time required for beam 2 to travel from M_s to M_2 and back to M_s is then

$$
\begin{aligned}
t_2 &= \frac{l_2}{c+v} + \frac{l_2}{c-v} \\
&= \frac{2l_2}{c(1-v^2/c^2)}.
\end{aligned}
$$

The second line was obtained from the first by finding the common denominator and factoring out c^2 in the denominator.

Now let us consider beam 1, which travels crosswise to the ether wind. Here the boat analogy (part *b*) is especially helpful. The boat wants to get from wharf A to wharf B directly across the stream. If it heads directly across, the stream's current will drag it downstream. To reach wharf B, the boat must head at an angle upstream. The precise angle depends on the magnitudes of c and v, but is of no interest to us in itself. However, part *c* does show us how to calculate the velocity v' of the boat relative to earth as it crosses the stream. Since c, v, and v' form a right triangle, we have that $v' = \sqrt{c^2 - v^2}$. The boat has the same velocity when it returns. If we now apply these principles to light beam 1 in Figure 24-2*a*, we see that the beam travels with a speed $\sqrt{c^2-v^2}$ in going from M_s to M_1 and back again. The total distance traveled is $2l_1$, so the time required for beam 1 to make the round trip is $2l_1/\sqrt{c^2-v^2}$ or,

$$
t_1 = \frac{2l_1}{c\sqrt{1-v^2/c^2}}.
$$

Notice that the denominator in this equation for t_1 involves a square root, whereas that for t_2 (above) does not.

If $l_1 = l_2 = l$, we see that beam 1 will lag behind beam 2 by an amount

$$\Delta t = t_2 - t_1 = \frac{2l}{c}\left(\frac{1}{1-v^2/c^2} - \frac{1}{\sqrt{1-v^2/c^2}}\right).$$

If $v = 0$, then $\Delta t = 0$ and the two beams will return in phase since they were initially in phase. But if $v \neq 0$, then $\Delta t \neq 0$, and the two beams will return out of phase. If this change of phase from the condition $v = 0$ to that for $v = v$ could be measured, then v could be determined. But the earth cannot be stopped. Furthermore, it is not possible to independently assume $l_1 = l_2$.

Michelson and Morley realized that they could detect the difference in phase (assuming that $v \neq 0$) if they rotated their apparatus by 90°, for then the interference pattern between the two beams should change. In the rotated position, beam 1 would now move parallel to the ether and beam 2 perpendicular to it. Thus the roles would be reversed and in the rotated positions, the times (designated by primes) would be

$$t_1' = \frac{2l_1}{c(1-v^2/c^2)} \quad \text{and} \quad t_2' = \frac{2l_2}{c\sqrt{1-v^2/c^2}}.$$

The time lag between the two beams in the nonrotated position (unprimed) would be

$$\Delta t = t_2 - t_1 = \frac{2l_2}{c(1-v^2/c^2)} - \frac{2l_1}{c\sqrt{1-v^2/c^2}}.$$

In the rotated position the time difference would be

$$\Delta t' = t_2' - t_1' = \frac{2l_2}{c\sqrt{1-v^2/c^2}} - \frac{2l_1}{c(1-v^2/c^2)}.$$

When the rotation is made, the fringes of the interference pattern (Section 22-8) will shift an amount determined by the difference in these:

$$\Delta t - \Delta t' = \frac{2}{c}(l_1 + l_2)\left(\frac{1}{1-v^2/c^2} - \frac{1}{\sqrt{1-v^2/c^2}}\right).$$

This can be considerably simplified if we assume that $v/c \ll 1$. For in this case we can use the binomial expansion† so that

$$\frac{1}{1-v^2/c^2} \approx 1 + \frac{v^2}{c^2} \quad \text{and} \quad \frac{1}{\sqrt{1-v^2/c^2}} \approx 1 + \frac{1}{2}\frac{v^2}{c^2}.$$

Then

$$\Delta t - \Delta t' \approx \frac{2}{c}(l_1 + l_2)\left(1 + \frac{v^2}{c^2} - 1 - \frac{1}{2}\frac{v^2}{c^2}\right)$$

$$\approx (l_1 + l_2)\frac{v^2}{c^3}. \qquad (24\text{-}1)$$

Now we take $v = 3.0 \times 10^4$ m/s, the speed of the earth in its orbit around the sun. In Michelson and Morley's experiments, the arms l_1 and l_2 were

† The binomial expansion states that $(1 \pm x)^n = 1 \pm nx + [n(n-1)/2]x^2 + \cdots$. In our case we have, therefore, $(1-x)^{-1} \approx 1 + x$, and $(1-x)^{-1/2} \approx 1 + \frac{1}{2}x$, where only the first term is kept, since $x = v/c$ is assumed to be small.

about 11 m long. The time difference would then be about $(22\text{ m})\times(3.0\times10^4\text{ m/s})^2/(3.0\times10^8\text{ m/s})^3\approx7.0\times10^{-16}$ s. For visible light of wavelength $\lambda=5.5\times10^{-7}$ m, say, the frequency would be $f=c/\lambda=3.0\times10^8\text{ m/s})/(5.5\times10^{-7}\text{ m})=5.5\times10^{14}$ Hz, which means that wave crests pass by a point every $1/(5.5\times10^{14}\text{ Hz})=1.8\times10^{-15}$ s. Thus, with a time difference of 7.0×10^{-16} s, Michelson and Morley should have noted a movement in the interference pattern of $(7.0\times10^{-16}\text{ s})/(1.8\times10^{-15}\text{ s})=0.4$ fringe. They could easily have detected this, since their apparatus was capable of observing a fringe shift as small as 0.01 fringe.

But they found *no significant fringe shift whatever*! They set their apparatus at various orientations. They made observations day and night, so that they would be at various orientations with respect to the sun (due to the earth's rotation). They tried at different seasons of the year (the earth at different locations due to its orbit around the sun). Never did they observe a significant fringe shift.

This "null" result was one of the puzzles of physics at the end of the nineteenth century. To explain the null result presented a difficult challenge to physicists. One possibility was that the ether is not at rest with respect to the sun or other stars, but instead is at rest with respect to the earth; in this case v would be zero and no fringe shift would be expected. But this implies that the earth is somehow a preferred object; only with respect to the earth would the speed of light be c as predicted by Maxwell's equations. This is tantamount to assuming that the earth is the central body of the universe, an ancient idea that had been rejected centuries earlier by Copernicus and Galileo. Another possibility was that the ether was dragged along by the earth and other bodies, so that the ether speed at the earth's surface would be zero. However, experiments in high-flying balloons, where at least some ether movement might be detected, also gave a null result.

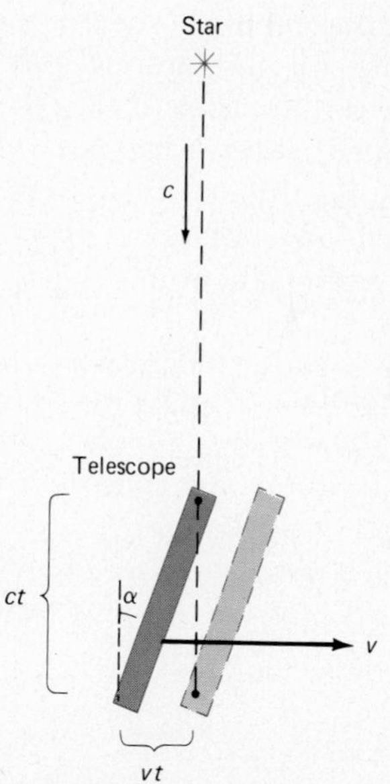

FIGURE 24-3
Aberration of starlight. If a telescope moves with the earth at speed v with respect to the ether, the telescope must be aimed slightly away from the star so that light entering the telescope will travel down the tube (since the tube moves to the right with speed v), rather than hitting the sides. Thus an overhead star will appear to be at an angle $\alpha=v/c$ to the vertical and, as the earth revolves, will seem to trace a circular path.

Both possibilities were negated by the observation of "aberration of starlight." This refers to the fact that over an observation period of a year, the stars appear to move in tiny ellipses; this is explained, if the earth is moving relative to the ether, by the fact that a telescope must be pointed not exactly at the star to be observed but rather at an angle (see Figure 24-3). If the ether were at rest compared to the earth, you could point directly at the star and no "aberration angle" would be present. But the aberration angle is there, so the ether could not be at rest with respect to the earth.

Another theory to explain the null result was put forth independently by G. F. Fitzgerald and H. A. Lorentz. They showed that it could be explained by assuming that any length (in particular, the arm of an interferometer) contracts by a factor $\sqrt{1-v^2/c^2}$ in the direction of motion through the ether. According to Lorentz, this could be due to the ether affecting the forces between the molecules of a substance, which were assumed to be electrical in nature.

Lorentz proposed this theory in 1904, and it might have been developed further. But a year later, a far more comprehensive theory was proposed by Albert Einstein—the special theory of relativity. It was a radical proposal. Although Einstein's thought was not actually based on the Michelson–Morley experiment, it explained the null result in a simple and straightforward way.

24-3 ▌ Einstein and the special theory of relativity

The problems that existed at the turn of the century with regard to electromagnetic theory and Newtonian mechanics were beautifully resolved by Einstein's introduction of the theory of relativity in 1905. Einstein, however, was apparently not influenced directly by the null result of the Michelson–Morley experiment (he may not even have been fully aware of it). What did motivate Einstein were certain questions regarding electromagnetic theory and light waves. For example, he asked himself: "What would I see if I rode a light beam?" The answer was that instead of a traveling electromagnetic wave, he would see electric and magnetic fields at rest whose magnitude changed in space but did not change in time. Such fields, he realized, had never been detected and indeed were not allowed by Maxwell's electromagnetic theory. He argued, therefore, that it was unreasonable to think that the speed of light relative to any observer could be reduced to zero. From there it was a small step to conclude that the relative speed of light could not be reduced at all and this became the second postulate of his theory of relativity (see below).

Einstein concluded that this and other inconsistencies that he found in electromagnetic theory were due to the assumption that an absolute space exists. In his famous 1905 paper, he proposed doing away completely with the idea of the ether and the accompanying assumption of an absolute reference frame at rest. This proposal was embodied in two postulates. The first postulate was an extension of the Newtonian relativity principle to include not only the laws of mechanics but also those of electricity and magnetism:

> ▌ **First postulate (the relativity principle): All the laws of physics have the same form in all inertial reference frames.**

The second postulate is consistent with the first:

> ▌ **Second postulate (constancy of the speed of light): Light propagates through empty space with a definite speed c independent of the speed of the source or observer.**

These two postulates form the foundation of Einstein's *special theory of relativity*. It is called "special" to distinguish it from his later "general theory of relativity," which deals with noninertial (accelerating) reference frames. The special theory, which is what we shall discuss here, deals only with inertial (nonaccelerating) reference frames.

The second postulate seems to be the hardest to accept, for it violates commonsense notions. First, we have to think of light traveling through empty space. Giving up the ether is not too hard, however; for after all, it could never be detected. But the second postulate also tells us that the speed of light in vacuum is always the same, 3.00×10^8 m/s, no matter what the speed of the observer or the source. Thus a person traveling toward a source of light will measure the same speed for that light as someone at rest with respect to the source (Figure 24-4). This conflicts with our everyday notions, for we would expect to have to add in the

FIGURE 24-4
Both observers, one at rest and one moving at high speed toward a source of light, measure the same velocity for that light.

velocity of the observer. But, according to Einstein, we do not do this for light. Part of the problem is that in our everyday experience we do not measure velocities anywhere near the speed of light; thus we cannot expect our everyday experience to be helpful when dealing with such a high velocity. On the other hand, the Michelson–Morley experiment is fully consistent with the second postulate. Indeed, the simplest and most straightforward interpretation of that experiment is that the speed of light is the same regardless of the velocity of the observer or source.†

The beauty of Einstein's proposal is evident. For by doing away with the idea of an absolute reference frame, Maxwell's electromagnetic theory could be reconciled with mechanics. The speed of light predicted by Maxwell's equation *is* the speed of light in vacuum in *any* reference frame.

Starting from the two postulates, Einstein derived a number of interesting but strange results. In the following sections we shall examine these consequences of Einstein's theory. Our arguments for the most part will be simple ones. (We will deal more deeply with the mathematics of relativity in Section 24-12.) Indeed, we will use a technique that Einstein himself used: we will imagine very simple experimental situations in which little mathematics is needed. In this way we can see what the consequences of relativity theory are without getting involved in detailed calculations. Einstein called these "gedanken" experiments, which is German for "thought" experiments.

The predictions of the special theory of relativity have been thoroughly tested in the years since 1905 and found to be valid. As a consequence, special relativity has become as much a working part of physics as the atomic theory or the conservation of energy.

24-4 ▮ Simultaneity

One of the important consequences of the theory of relativity is that we can no longer regard time as an absolute quantity. No one doubts that time flows onward and never turns back. But, as we shall see in this section and the next, the time between two events, and even whether two events are simultaneous, depends on the observer's reference frame. We now examine simultaneity, and will see that it is a relative concept.

Two events are said to occur simultaneously if they occur at exactly the same time. But how do we tell if two events occur precisely at the same time? If they occur at the same point in space—such as two apples falling on your head at the same time—it is easy. But if the two events occur at widely separated places, it is more difficult to tell, since we have to take into account the time it takes for the light to reach us. Because light travels at finite speed, a person who sees two events must calculate backward to find out when they actually occurred. For example, if two events were *observed* to occur at the same time, but one actually took place farther from the observer than the other, the former must have occurred earlier in time.

To avoid making calculations, we shall now make use of a simple thought experiment. We assume an observer, called O, is located exactly

† The Michelson–Morley experiment can also be considered as evidence for the first postulate, for it was intended to measure the motion of the earth relative to an absolute reference frame; its failure to do so implies the absence of any such preferred frame.

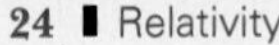

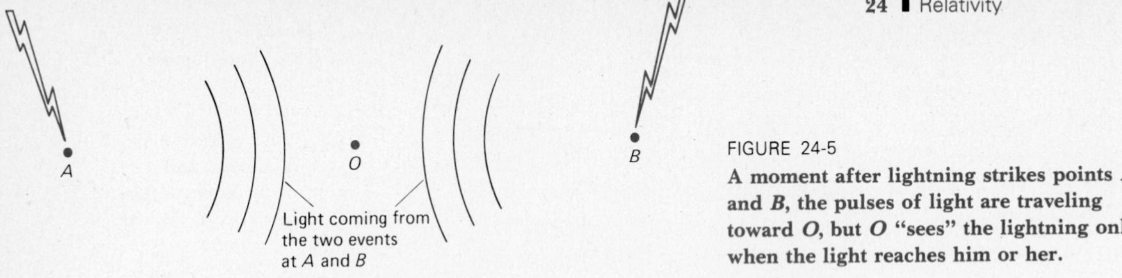

FIGURE 24-5
A moment after lightning strikes points A and B, the pulses of light are traveling toward O, but O "sees" the lightning only when the light reaches him or her.

halfway between points A and B where two events occur (Figure 24-5). The two events may be lightning that strikes the points A and B, as shown, or any other type of event. For brief events like lightning, only short pulses of light will travel outward from A and B and reach O. O "sees" the events when the pulses of light reach him. If the two pulses reach O at the same time, the two events had to be simultaneous. This is because the two light pulses travel at the same speed; and since the distance OA equals OB, the time for the light to travel from A to O and from B to O must be the same. Observer O can then definitely state that the two events occurred simultaneously. On the other hand, in this simple situation, if O sees the light from one event before that from the other, it is sure that the former event occurred first.

The question we really want to examine is this: If two events are simultaneous to one observer, are they also simultaneous to an observer moving with respect to the first? Let us call the observers O_1 and O_2, and they are fixed in reference frames 1 and 2 which move with speed v relative to one another. These two reference frames can be thought of as railroad cars (Figure 24-6). O_2 says that O_1 is moving to the right with speed v, as in part a; and O_1 says that O_2 is moving to the left, as in part b. Both viewpoints are legitimate according to the relativity principle. (There is, of course, no third point of view that will tell us which one is "really" moving.)

Now suppose that two events occur which are seen by both observers. Let us assume again that the two events are the striking of lightning and that the lightning marks both trains where it struck: at A_1 and B_1 on O_1's train, and at A_2 and B_2 on O_2's train. For simplicity, we assume that O_1 happens to be exactly halfway between A_1 and B_1 and that O_2 is halfway between A_2 and B_2. We must now put ourselves in one reference frame or the other. Let us put ourselves in O_2's reference frame, so we see O_1 moving to the right with speed v. Let us also assume that the two events occur *simultaneously* in O_2's frame, and just at the time when O_1 and O_2 are opposite each other (Figure 24-7a). A short time later (Figure 24-7b) the light reaches O_2 from A_2 and B_2 at the same time. Since O_2 knows (or

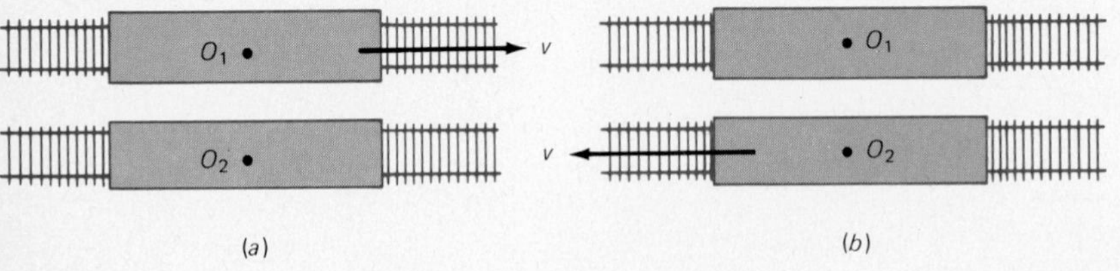

FIGURE 24-6
Observers O_1 and O_2, on two different trains (two different reference frames), are moving with relative velocity v. O_2 says that O_1 is moving to the right (a); O_1 says that O_2 is moving to the left (b). Both viewpoints are legitimate—it all depends on your reference frame.

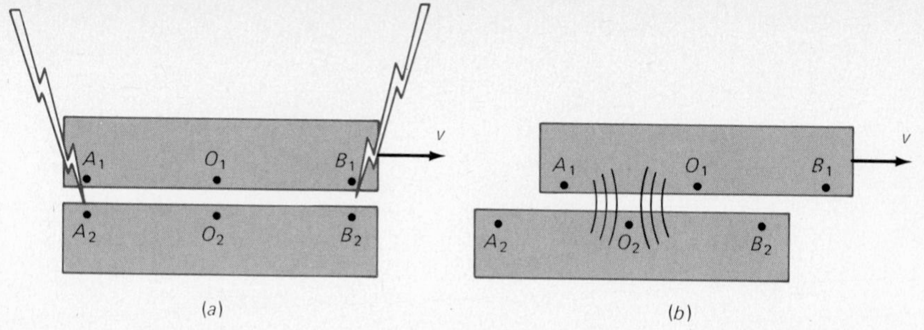

FIGURE 24-7
Thought experiment on simultaneity. To observer O_2, the reference frame of O_1 is moving to the right. In (*a*), one lightning bolt strikes the two reference frames at A_1 and A_2, and a second lightning bolt strikes at B_1 and B_2. According to observer O_2, the two bolts of lightning strike simultaneously. (*b*) A moment later the light from the two events reaches O_2 at the same time (simultaneously). But in O_1's reference frame, the light from B_1 has already reached O_1, whereas the light from A_1 has not yet reached O_1. So in O_1's reference frame the event at B_1 must have preceded the event at A_1. Time is not absolute!

measures) the distances O_2A_2 and O_2B_2 are equal, he knows that the two events are simultaneous.

But what does observer O_1 see? From our (O_2) reference frame, we can see that O_1 moves to the right during the time the light is traveling to O_1 from A_1 and B_1. As can be seen in Figure 24-7*b*, the light from B_1 has already passed O_1, whereas that from A_1 has not yet reached O_1. Therefore, it is clear that O_1 sees the light from B_1 before that from A_1. Now O_1's frame is as good as O_2's. Light travels at the same speed for O_1 as for O_2, and is the same from A_1 to O_1 as from B_1 to O_1. Since O_1 sees the light coming from B_1 first and since the distance O_1A_1 equals O_1B_1, observer O_1 must conclude that event B_1 occurred before event A_1.

We thus find that *two events which are simultaneous to one observer are not necessarily simultaneous to a second observer.*†

It may be tempting to ask: "Which observer is right, O_1 or O_2?" The answer is they are *both* right. There is no "best" reference frame we can choose to determine which observer is right. Both frames are equally good. We can only conclude that simultaneity is not an absolute concept, but is relative. It may be hard to accept the relative nature of simultaneity; at least in part this is because we are not aware of it in everyday life. The effect will be noticeable only when the relative speed of the two reference frames is very large (near c) or the distances involved are very large.

Finally, because of the principle of relativity, the argument we gave for the thought experiment of Figure 24-7 can be done from O_1's reference frame as well. In this case O_1 will be at rest and will see event B_1 occur before A_1. But O_1 will recognize (by drawing a diagram equivalent to Figure 24-7) that O_2, who is moving with speed v to the left, will see the two events as simultaneous. The analysis from the reference frame of O_1 is left as an exercise.

24-5 ▌ Time dilation and the twin paradox

Time dilation

The fact that two events simultaneous to one observer may not be simultaneous to a second observer suggests that time itself may pass differently in one reference frame than another. Indeed, this is just what Einstein's theory of relativity predicts, as the following thought experiment shows.

† Two events will be simultaneous to two observers who move with respect to each other only if the two events occur at the *same point* in each reference frame. Can you see why?

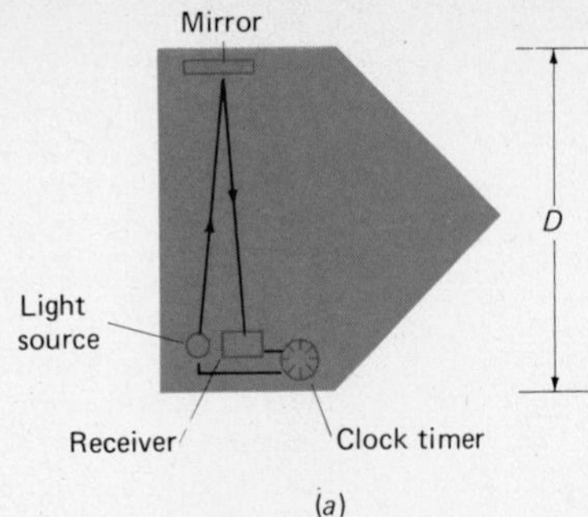

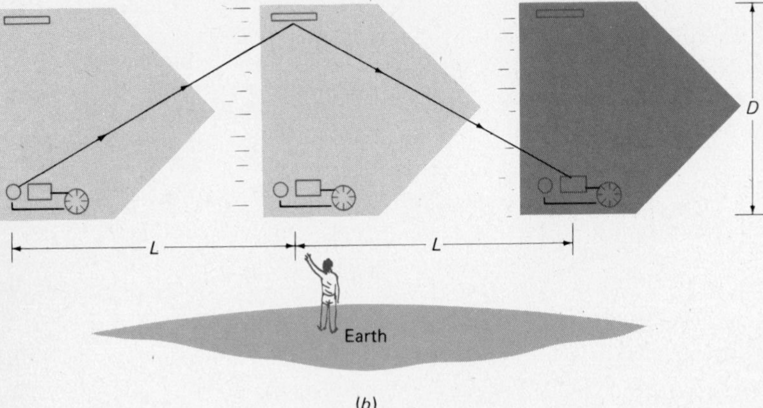

FIGURE 24-8
Time dilation can be shown by a thought experiment: the time it takes for the light to travel over and back on a spaceship is longer for the earth observer (*b*) than for the observer on the spaceship (*a*).

Figure 24-8 shows a spaceship traveling past earth at high speed. The point of view of an observer on the spaceship is shown in part *a*, and that of an observer on earth in part *b*. Both observers have accurate clocks. The person on the spaceship (*a*) flashes a light and measures the time it takes the light to travel across the spaceship and return after reflecting from a mirror. The light travels a distance $2D$ at speed c so the time required, which we call t_0, is

$$t_0 = \frac{2D}{c}.$$

The observer on earth (Figure 24-8*b*) observes the same process. But to this observer, the spaceship is moving; so the light travels the diagonal path shown in going across the spaceship, reflecting off the mirror, and returning to the sender. Although the light travels at the same speed to this observer (the second postulate), it travels a greater distance. Hence the time required, as measured by the earth observer, will be greater than that measured by the observer on the spaceship. The time interval, t, as observed by the earth observer, can be calculated as follows. In the time t, the spaceship travels a distance $2L = vt$, where v is the speed of the spaceship (Figure 24-8). Thus the light travels a total distance on its diagonal path of $2\sqrt{D^2 + L^2}$, and therefore

$$c = \frac{2\sqrt{D^2 + L^2}}{t} = \frac{2\sqrt{D^2 + v^2t^2/4}}{t}.$$

We square both sides and solve for t to find

$$c^2 = \frac{4D^2}{t^2} + v^2$$

$$t = \frac{2D}{c\sqrt{1 - v^2/c^2}}.$$

We combine this with the formula above for t_0 and find that

$$t = \frac{t_0}{\sqrt{1 - v^2/c^2}}. \tag{24-2}$$

Since $\sqrt{1 - v^2/c^2}$ is always less than 1, we see that $t > t_0$. That is, the time between the two events (the sending of the light and its reception at the receiver) is *greater* for the earth observer than for the traveling observer. This is a general result of the theory of relativity and is known as **time dilation**. Stated simply, the time-dilation effect says that *moving clocks are measured to run slowly*. However, we should not think that the clocks are somehow at fault. To the contrary, we assume that the clocks are good ones. Time is actually measured to pass more slowly in any moving reference frame as compared to your own. This remarkable result is an inevitable outcome of the two postulates of the theory of relativity.

The concept of time dilation may be hard to accept, for it violates our commonsense understanding. But, as Einstein pointed out, common sense is that which we learned prior to age 18. We can see from Equation 24-2 that the time-dilation effect is negligible unless v is reasonably close to c. The speeds we experience in everyday life are very much smaller than c, so it is little wonder we have not noticed time dilation as a real effect. Experiments have been done to test the time-dilation effect, and they have consistently confirmed Einstein's predictions (Equation 24-2). In 1971, for example, extremely precise atomic clocks were flown around the world in 600-mi/h jet planes. Since this speed is much less than c, the clocks had to be accurate to nanoseconds (10^{-9} s) to detect the time-dilation effect. They were this accurate and confirmed Equation 24-2 to within experimental error. Time dilation had been confirmed decades earlier, however, by observation on tiny "elementary particles" (see Chapter 30), which have very small masses (typically 10^{-30} to 10^{-27} kg) and so require little energy to be accelerated to speeds close to c. Many of these elementary particles are not stable and decay after a time into simpler particles. One example is the mu meson (or muon), whose mean lifetime is 2.0 μs when at rest. Careful experiments showed that, when traveling at high speeds, the lifetime of a muon increases just as predicted by the time-dilation formula.

▌ EXAMPLE 24-1 What will be the mean lifetime of a muon as measured in the laboratory if it is traveling $0.60c = 1.8 \times 10^8$ m/s with respect to the laboratory? Its mean life at rest is 2.0×10^{-6} s.

SOLUTION If an observer were to move along with the muon (the muon would be at rest to this observer), the muon would have a mean life of 2.0×10^{-6} s. To an observer in the laboratory the muon lives longer because of time dilation. From Equation 24-2 with $v=0.60c$, we have

$$t=\frac{t_0}{\sqrt{1-v^2/c^2}}=\frac{2.0\times10^{-6}\ \text{s}}{\sqrt{1-0.36c^2/c^2}}=\frac{2.0\times10^{-6}\ \text{s}}{\sqrt{0.64}}=2.5\times10^{-6}\ \text{s}.$$

▌▌

We need to make a comment about the use of Equation 24-2 and the meaning of t and t_0. The equation is true only when t_0 represents the time interval between the two events in a reference frame where the two events occur *at the same point in space* (as in Figure 24-8*a*). This time interval, t_0, is often called the *proper time*. Then t in Equation 24-2 represents the time interval between the two events as measured in a reference frame moving with speed v with respect to the first. In Example 24-1, t_0 (and not t) was set equal to 2.0×10^{-6} s because it is only in the rest frame of the muon that the two events ("birth" and "decay") occur at the same point in space.

Space travel

Time dilation has aroused interesting speculation about space travel. Under the old time regime, to reach a star 100 light-years away would not be possible by ordinary mortals (1 light-year is the distance light can travel in 1 year). Even if a rocket ship could travel at close to the speed of light, it would take over 100 years to reach such a star. But time dilation tells us that the time involved would be less for an astronaut. In a spaceship travelling at $v=0.999c$. the time for such a trip would be only about $t_0=t\sqrt{1-v^2/c^2}=(100\ \text{yr})(\sqrt{1-(0.999)^2})=4.5\ \text{yr}$. Thus a person could make such a trip. Time dilation allows such a trip, but the enormous practical problems of achieving such speeds will not be overcome in the foreseeable future.

Notice, in this example, that whereas 100 years pass on earth only 4.5 years pass for the astronaut on the trip. Is it just the clocks that slow down for the astronaut? The answer is no. All processes, including life processes, are seen to run more slowly for the astronaut according to the earth observer. But to the astronaut, time passes in the normal way. The astronaut experiences 4.5 years of normal sleeping, eating, reading, and so on; and people on earth experience 100 years of ordinary activity.

The twin paradox

Not long after Einstein proposed the special theory of relativity, an apparent paradox was pointed out. According to this "twin paradox," suppose that one of a pair of 20-year-old twins takes off in a spaceship traveling at very high speed to a distant star and back again, while the other twin remains on earth. According to the earth twin, the traveling twin will age less than himself. Whereas 20 years might pass for the earth twin, perhaps only 1 year (depending on the spacecraft's speed) passes for the traveler. Thus, when the traveler returns, the earthbound twin would expect to be 40 years old, whereas his twin would be only 21.

This is the viewpoint of the twin on earth. But what about the traveling twin? Since everything is relative, all inertial reference frames are equally good. Won't the traveling twin make all the claims the earth twin does, only in reverse? To the astronaut twin, the earth is moving away at high speed; to him, time passes more slowly on earth, so his twin on earth will age less than he will. This is the opposite of what the earth twin predicts. They cannot both be right, for after all the spacecraft returns to earth and a direct comparison of ages and clocks can be made.

This is, however, not a paradox at all. The consequences of the theory of relativity—in this case, time dilation—can be applied only by observers in inertial reference frames. The earth is such a frame (or nearly so), whereas the spacecraft is not. The spacecraft accelerates at the start and end of its trip and, more important, when it turns around at the far point of its journey. During these acceleration periods, the spacecraft's predictions based on special relativity are not valid. The twin on earth is in an inertial frame and can make valid predictions. Thus there is no paradox. The traveling twin's point of view expressed above is not correct. The predictions of the earth twin are valid, and the prediction that the traveling twin returns having aged less is the correct one.†

24-6 ▌ Length contraction

Because time intervals are different in different reference frames, we might expect space intervals—lengths and distances—to be different as well. This is indeed the case, and we illustrate with a thought experiment.

An observer on earth watches a spacecraft traveling at speed v from earth to, say, Neptune (Figure 24-9*a*). The distance between the planets, as measured by an earth observer, is L_0. The time required for the trip, measured from earth, is $t = L_0/v$. In Figure 24-9*b* we see the point of view of an observer on the spacecraft. In this frame of reference the spaceship is at rest; the earth and Neptune move with speed v. (We assume that v is much greater than the relative speed of Neptune and earth, so the latter can be ignored.) But the time between the departure of earth and the arrival of Neptune is less for the spacecraft observer than for the earth observer, because of time dilation. From Equation 24-2, the time for the trip as viewed by the spacecraft is $t_0 = t\sqrt{1 - v^2/c^2}$. Since the spacecraft observer measures the same speed but less time between these two events, he must also measure the distance as less. If we let L be the distance between the planets as viewed by the spacecraft observer, then $L = vt_0$.

† Einstein's general theory of relativity, which deals with accelerating reference frames, confirms this result.

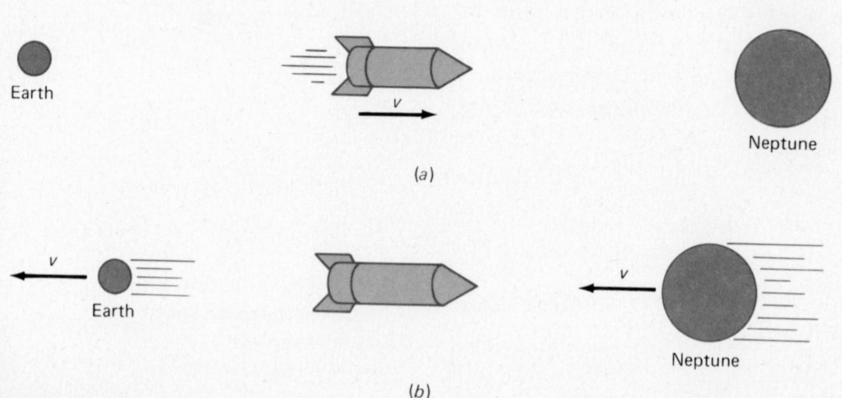

FIGURE 24-9
(*a*) A spaceship traveling at very high speed from earth to Neptune, as seen from earth's frame of reference. (*b*) As viewed by an observer on the spaceship, earth and Neptune are moving at the very high velocity v.

We have already seen that $t_0 = t\sqrt{1-v^2/c^2}$ and $t = L_0/v$, so we have $L = vt_0 = vt\sqrt{1-v^2/c^2} = L_0\sqrt{1-v^2/c^2}$. That is,

$$L = L_0\sqrt{1-v^2/c^2}. \tag{24-3}$$

This is a general result of the special theory of relativity and applies to lengths of objects as well as to distance. The result can be stated most simply in words as: *the length of an object is measured to be shorter when it is moving than when it is at rest.* This is called **length contraction**. The length L_0 in Equation 24-3 is called the *proper length.* It is the length of the object (or distance between two objects) as measured by an observer at rest with respect to it. Equation 24-3 gives the length that will be measured when the object travels by an observer at speed v. It is important to note, however, that length contraction occurs *only along the direction of motion.* For example, the moving spaceship in Figure 24-9a is shortened in length, but its height is the same as when it is at rest.

Length contraction, like time dilation, is not noticeable in everyday life because the factor $\sqrt{1-v^2/c^2}$ in Equation 24-3 differs from 1.00 significantly only when v is very large.

Equation 24-3 tells us what the length of an object will be *measured* to be when traveling at speed v. The *appearance* of the object is another matter. Suppose, for example, that you are traveling to the left past a tall building at a speed $v = 0.85c$. This is equivalent to the building moving past you to the right at speed v. The building will look narrower (and the same height), but you will also be able to see the side of the building even if you are directly in front of it. This is shown in Figure 24-10b (part a shows the building at rest). The fact that you see the side is not really a relativistic effect, but is due to the finite speed of light. To see how this occurs, we look at Figure 24-10c, which is a top view of the building, looking down. At the instant shown, the observer O is directly in front of the building (which is moving to the right with speed v). Light from points A and B reaches O at the same time. If the building were at rest, light from point C could never reach O. But the building is moving at very high speed ($v = 0.85c$) and does "get out of the way" so that light from C can reach O! Indeed, at the instant shown, light from point C when it was at an earlier location (C' on the diagram) can reach O because the building has moved. To reach the observer at the same time as light from A and B, light from C had to leave at an earlier time, since it must travel a greater distance. Thus it is light from C' that reaches the observer at the same time as light from A and B. This, then, is how an observer might see both the front and side of an object at the same time, even though he is directly in front of it.† It can be shown, by the same reasoning, that spherical objects will actually still have a circular outline even at high speeds. That is why the planets in Figure 24-9b are drawn round rather than contracted.

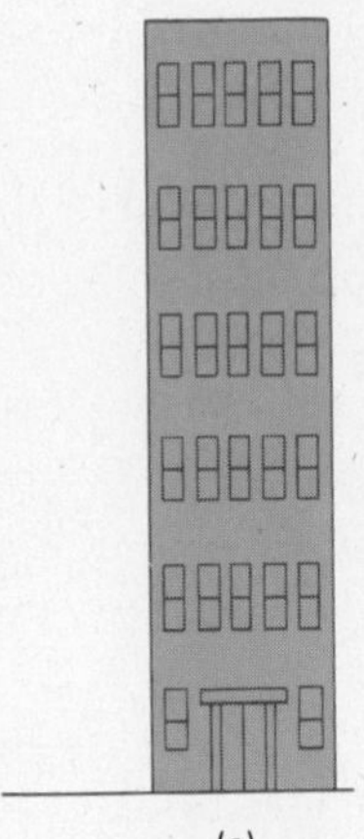

(a)

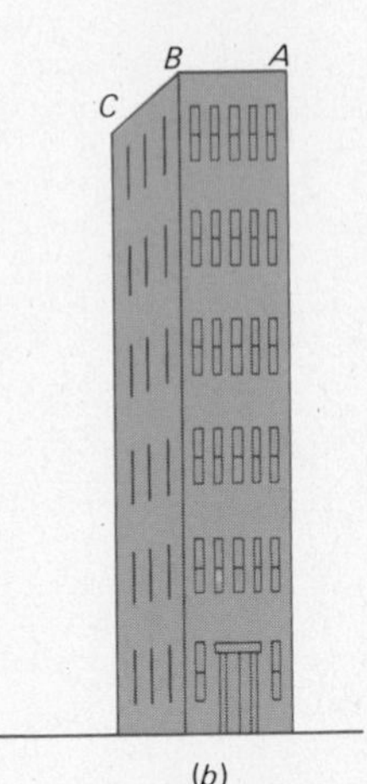

(b)

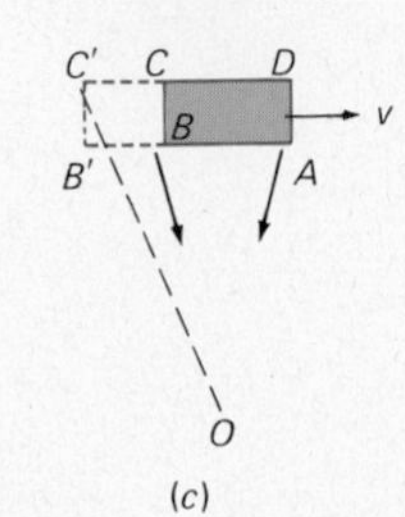

(c)

FIGURE 24-10
Building seen (a) at rest; (b) when moving past it at high speed. (c) Diagram explains why the side of the building is seen (see the text).

▌ EXAMPLE 24-2 A spaceship passes the earth at speed $v = 0.80c$. Describe the changes in length of a meter stick as it is slowly rotated from vertical to horizontal as viewed (a) by a person in the spaceship, and (b) by an observer on earth.

† It was thought, for a time, that the building in Figure 24-10b would look rotated. This is not correct, since in that case side A would look less high than side B. In fact, if the observer is directly in front, these sides appear equal in height. Thus the building looks contracted in its front face, but we also see the side, as described above.

SOLUTION (*a*) The meter looks 1.0 m long in all orientations since it is at rest. (*b*) The meter stick varies in length from 1.0 m (vertical) to $L = (1.0\text{ m})\sqrt{1-(0.80)^2} = 0.60$ m long in the horizontal direction (assuming that is the direction of motion). ▮▮

24-7 ▮ Four-dimensional space–time

Let us suppose that a person is on a train moving at a very high speed, say 0.65*c* (Figure 24-11). This person begins a meal at 7:00 and finishes at 7:15, according to a clock on the train. The two events, beginning and ending the meal, take place at the same point on the train. So the proper time between these two events is 15 min. To observers on earth, the meal will take longer—20 min according to Equation 24-2. Let us assume that the meal was served on a 20-cm-diameter plate. To observers on the earth, the plate is only 15 cm wide (length contraction). Thus to observers on the earth the food looks smaller but lasts longer.

In a sense, time dilation and length contraction balance each other. When viewed from the earth, what the food seems to lose in size it gains in the length of time it lasts. Space, or length, has been exchanged for time.

Such considerations led to the idea of four-dimensional "space–time"—space occupies three dimensions and time becomes a fourth dimension. Just as when we squeeze a balloon we make one dimension smaller and another becomes larger, so when we examine objects and events from different reference frames, a certain amount of space is exchanged for time, or vice versa. Space and time are intimately connected.

This idea of space–time is very satisfying from another point of view. Objects have temporal as well as spatial extent. They exist for a certain amount of time at a particular place, such as the food that existed for 15 min. Although the idea of four dimensions may seem strange, it refers to the fact that any object or event is specified by four quantities—three to describe where in space, one to describe when in time. The really unusual aspect of four-dimensional space–time is that space and time can intermix: a little of one can be exchanged for a little of the other when the reference frame is changed.

It may be difficult for many of us to understand this four-dimensional space–time. We may feel, just as physicists did before the advent of relativity, that space and time are separate entities. Yet, in our thought experiments we found that they are not completely separate. Our difficulty in accepting this is somewhat like the situation at the time of Galileo and Newton in the seventeenth century. Before Galileo, the vertical direction, that in which objects fall, was considered to be entirely different from the two horizontal dimensions. Galileo and Newton showed that the vertical dimension differs not at all; it simply happens to be the direction in which gravity acts. Otherwise, all three dimensions are equivalent, a fact that we all accept today. Now we are asked to accept one more dimension, time, which we had previously thought of as somehow being different. This is not to say that there is no distinction whatever between space and time. What relativity has shown is that space and time

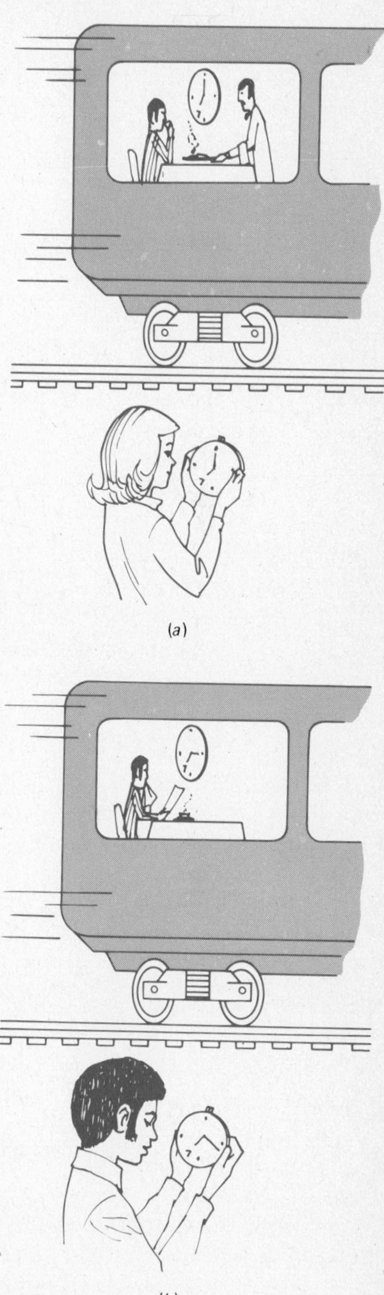
(*a*)
(*b*)

FIGURE 24-11
According to an accurate clock on a fast-moving train, a person begins dinner at 7:00 (*a*) and finishes at 7:15 (*b*). At the beginning of the meal, observers on the earth set their watches to correspond with the clock on the train. These observers measure the eating time as 20 min.

determinations are not independent of one another. Time cannot be thought of as being totally separate from the three spatial dimensions, any more than one of the spatial dimensions can be thought of as being totally separate from the other two.

24-8 ▮ Mass increase

The three basic mechanical quantities are length, time intervals, and mass. The first two have been shown to be relative—their value depends on the reference frame from which they are measured. We might expect that mass, too, is a relative quantity. Indeed, Einstein showed that *the mass of an object increases as its speed increases* according to the formula

$$m = \frac{m_0}{\sqrt{1 - v^2/c^2}}. \qquad (24\text{-}4)$$

In this **mass-increase** formula, m_0 is the *rest mass* of the object—the mass it has as measured in a reference frame in which it is at rest; and m is the mass it will be measured to have in a reference frame in which it moves at speed v (see Figure 24-12). Einstein's derivation of Equation 24-4 was based on the assumption that the law of conservation of momentum is valid.

Relativistic mass increase has been tested countless times on tiny elementary particles (such as muons) and the mass has been found to increase in accord with Equation 24-4.

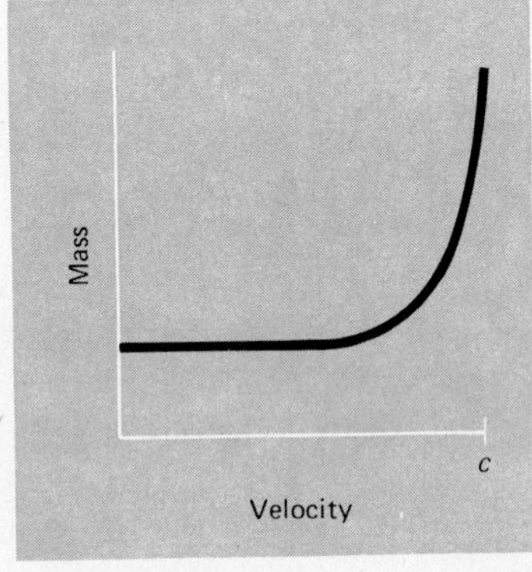

FIGURE 24-12
Increase in mass of an object as its velocity increases.

▮ EXAMPLE 24-3 An electron ($m_0 = 9.1 \times 10^{-31}$ kg) enters a magnetic field $B = 1.5$ T perpendicularly to the field lines with a speed $v = 0.50c = 1.50 \times 10^8$ m/s. What will be the radius of curvature r of its path?

SOLUTION We use $F = ma$ and obtain $evB = mv^2/r$ (see Chapter 17), where $e = 1.6 \times 10^{-19}$ C. Thus

$$r = \frac{mv}{eB} = \frac{m_0 v}{\sqrt{1 - v^2/c^2}\, eB} = 6.6 \times 10^{-4} \text{ m.}$$

▮▮

24-9 ▮ The ultimate speed

A basic result of the special theory of relativity is that no object can equal or exceed the speed of light. That the speed of light is a natural speed limit in the universe can be seen from any one of Equations 24-2, 24-3, and 24-4. It is perhaps easiest to see it from Equation 24-4, the mass-increase formula, $m = m_0/\sqrt{1 - v^2/c^2}$. As an object is accelerated to greater and greater speeds, its mass becomes larger and larger. Indeed, if v were to equal c, the denominator in this equation would be zero and the mass m would become infinite. To accelerate an object up to $v = c$ would thus require infinite energy, so is not possible. The fact that light itself travels at the speed c implies that its rest mass (if such a thing makes sense) must be zero; of course, light is never at rest.

If v were to exceed c, the factor $\sqrt{1-v^2/c^2}$ would be the square root of a negative number, which is imaginary. Thus ordinary objects cannot equal or exceed the speed of light. However, as was pointed out in the late 1960s, Einstein's equations do not rule out the possibility that objects exist whose speed is *always* greater than c. If such particles exist (the name "tachyon"—meaning "fast"—was proposed) the rest mass m_0 would have to be imaginary; in this way the mass m would be the ratio of two imaginary numbers for $v > c$, which is real. For such hypothetical particles, c would be a *lower* limit on their speed. In spite of extensive searches for tachyons, none has been found. Thus it seems that the speed of light *is* the ultimate speed in the universe.

24-10 ▮ $E = mc^2$; mass and energy

When a net force is applied to an object of rest mass m_0, the object increases in speed. Since the force is acting through a distance, work is done on the object and its energy increases. But as the speed of the object approaches c, the speed no longer increases as rapidly since it cannot exceed c. But the mass of the object also increases with increasing speed, and this is particularly noticeable as the speed of light is approached (Figure 24-12). Thus the work being done on the object goes into increasing its *mass*. Normally, the work done on an object increases its energy. This new twist from the theory of relativity led Einstein to the idea that mass is a form of energy.

To find the mathematical relationship between mass and energy, Einstein assumed that the work–energy theorem (Chapter 5) is still valid in relativity. That is, the net work done on a particle is equal to its change in kinetic energy (KE). Using this theorem, Einstein showed that at high speeds the formula $\text{KE} = \frac{1}{2}mv^2$ is not correct. You might think that using Equation 24-4 for m would give $\text{KE} = \frac{1}{2}m_0v^2/\sqrt{1-v^2/c^2}$, but this formula, too, is wrong. Instead, Einstein showed that the kinetic energy of a particle is given by

$$\text{KE} = mc^2 - m_0c^2, \tag{24-5}$$

where m is the mass of the particle traveling at speed v and m_0 is its rest mass.

But what does the second term in Equation 24-5—the m_0c^2—mean? Consistent with the idea that mass is a form of energy, Einstein called m_0c^2 the *rest energy* of the object. We can rearrange Equation 24-5 to get $mc^2 = m_0c^2 + \text{KE}$. We call mc^2 the *total energy* E of the particle (assuming no potential energy), and we see that the total energy equals the rest energy plus the kinetic energy:

$$\blacksquare\ E = mc^2$$
$$= m_0c^2 + \text{KE}. \tag{24-6}$$

Here we have Einstein's famous formula $E = mc^2$. The total energy of a particle at rest is $E_0 = m_0c^2$, which we have called its rest energy. This formula mathematically relates the concepts of energy and mass. But if the idea that mass is just one form of energy is to have any meaning from a practical point of view, then mass should be convertible to other forms of energy just as other types of energy are interconvertible. Einstein suggested that this might be possible, and indeed this has been experimentally confirmed countless times and forms the basis for many important processes.

The interconversion of mass and energy is most easily detected in nuclear and elementary particle physics. For example, an elementary particle called the π^0 (pi zero) meson is observed to decay into pure electromagnetic radiation (photons); the π^0 completely disappears in the process. The amount of electromagnetic energy produced is found to be exactly equal to that predicted by Einstein's formula, $E = m_0c^2$. The reverse process is also commonly observed by nuclear physicists: electromagnetic radiation under certain conditions can be converted into material particles such as electrons. On a larger scale, the energy produced in nuclear power plants is a result of the loss in mass of the uranium fuel as it undergoes a process called fission (Chapter 29). Indeed, even the great amount of radiant energy we receive from the sun is an instance of $E = mc^2$; the sun's mass is continually decreasing as it radiates energy outward.

The relation $E = mc^2$ is now believed to apply to all processes, although the changes are often too small to measure. That is, when the energy of a system changes by an amount ΔE, the mass of the system changes by an amount Δm given by

$$\Delta E = (\Delta m)(c^2).$$

In a chemical reaction where heat is gained or lost, the masses of the reactants and the products will be different. Even when water is heated on a stove, the mass of the water increases very slightly. This example is easy to see because as heat is added, the temperature and therefore the average speed of the molecules increases; and Equation 24-4 tells us that the mass also increases.

▮ EXAMPLE 24-4 A π^0 meson ($m_0 = 2.5 \times 10^{-28}$ kg) travels at a speed $v = 0.80c = 2.4 \times 10^8$ m/s. What is its kinetic energy? Compare to a classical calculation.

SOLUTION The mass of the π^0 at $v = 0.80c$ is

$$m = \frac{m_0}{\sqrt{1 - v^2/c^2}} = \frac{2.5 \times 10^{-28}\ \text{kg}}{\sqrt{1 - (0.80)^2}} = 4.1 \times 10^{-28}\ \text{kg}.$$

Thus its KE is

$$\begin{aligned} \text{KE} = (m - m_0)c^2 &= (4.1 \times 10^{-28}\ \text{kg} - 2.5 \times 10^{-28}\ \text{kg})(3.0 \times 10^8\ \text{m/s})^2 \\ &= 1.4 \times 10^{-11}\ \text{J}. \end{aligned}$$

Notice that the units of mc^2 are $\text{kg} \cdot \text{m}^2/\text{s}^2$, which is the joule. A classical calculation would give $\text{KE} = \frac{1}{2}m_0v^2 = \frac{1}{2}(2.5 \times 10^{-28}\ \text{kg})(2.4 \times 10^8\ \text{m/s})^2 = 7.3 \times 10^{-12}$ J, but this is not a correct result. ▮▮

▮ EXAMPLE 24-5 How much energy would be released if the π^0 meson in Example 24-4 were transformed by decay completely into electromagnetic radiation?

SOLUTION The rest energy of the π^0 is

$$E_0 = m_0c^2 = (2.5\times10^{-28}\text{ kg})(3.0\times10^8\text{ m/s})^2 = 2.3\times10^{-11}\text{ J}.$$

This is how much energy would be released if the π^0 decayed at rest. If it has $\text{KE} = 1.4\times10^{-11}$ J, the total energy released would be 3.7×10^{-11} J. ▌▌

Equation 24-5 for the kinetic energy can be written in terms of the speed v of the object with the help of Equation 24-4:

$$\text{KE} = m_0c^2\left(\frac{1}{\sqrt{1-v^2/c^2}}-1\right). \qquad (24\text{-}7)$$

At low speeds, $v \ll c$, we can expand the square root in Equation 24-7 using the binomial expansion (see the footnote in Section 24-2). Then we get

$$\text{KE} \approx m_0c^2\left(1+\frac{1}{2}\frac{v^2}{c^2}+\cdots-1\right)$$
$$\approx \tfrac{1}{2}m_0v^2,$$

where the dots in the first expression represent very small terms in the expansion which we have neglected since we assumed that $v \ll c$. Thus we see that at low speeds, the relativistic form for kinetic energy reduces to the classical form $\text{KE} = \frac{1}{2}m_0v^2$. This is, of course, what must be. Relativity would not be a valuable theory if it did not predict accurate results at low speed as well as at high. Indeed, the other equations of special relativity also reduce to their classical equivalents at ordinary speeds: length contraction, time dilation, and mass increase all disappear for $v \ll c$ since $\sqrt{1-v^2/c^2} \approx 1$.

A useful relation between the total energy E of a particle and its momentum p can also be derived. Since $E = mc^2$ and $p = mv$, where $m = m_0/\sqrt{1-v^2/c^2}$, we have

$$E^2 = m^2c^4 = m^2c^2(c^2+v^2-v^2)$$
$$= m^2c^2v^2 + m^2c^2(c^2-v^2)$$
$$= p^2c^2 + \frac{m_0^2c^4(1-v^2/c^2)}{1-v^2/c^2}$$

or

$$E^2 = p^2c^2 + m_0^2c^4. \qquad (24\text{-}8)$$

Thus the total energy can be written in terms of the momentum p or in terms of the kinetic energy (Equation 24-6).

24-11 ▌ Relativistic addition of velocities

FIGURE 24-13
Rocket 2 is fired from rocket 1 with speed $u = 0.60\ c$. What is the speed of rocket 2 with respect to the earth?

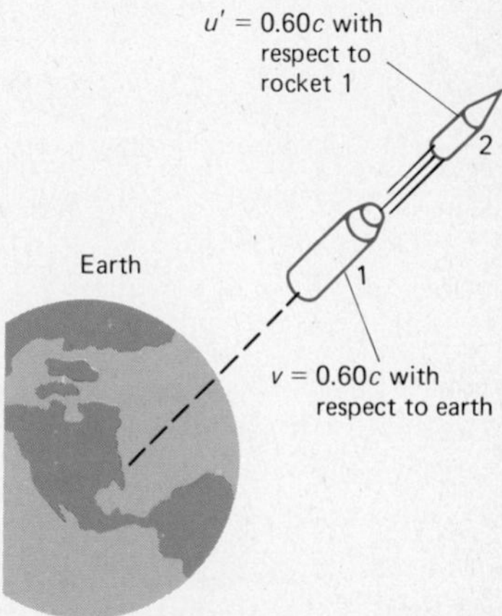

Consider a rocket ship that travels away from the earth with speed v, and assume that this rocket has fired off a second rocket that travels at speed u' with respect to the first (Figure 24-13). We might expect that the speed u of rocket 2 with respect to earth is $u = v + u'$, which in the case shown in the figure is $u = 0.60c + 0.60c = 1.20c$. But, as discussed in Section 24-9, no object can travel faster than the speed of light in any reference frame. Indeed, Einstein showed that since length and time are different in different reference frames, the old addition-of-velocities formula is no longer valid. Instead, the correct formula is

$$u = \frac{v + u'}{1 + vu'/c^2} \tag{24-9}$$

for a motion along a straight line. We derive this formula in the next section. If u' is in the opposite direction from v, then u' must have a minus sign and $u = (v - u')/(1 - vu'/c^2)$.

▌EXAMPLE 24-6 Calculate the speed of rocket 2 in Figure 24-13 with respect to earth.

SOLUTION Rocket 2 moves with speed $u' = 0.60c$ with respect to rocket 1. Rocket 1 has speed $v = 0.60c$ with respect to earth. The speed of rocket 2 with respect to earth is therefore

$$u = \frac{0.60c + 0.60c}{1 + \dfrac{(0.60c)(0.60c)}{c^2}} = \frac{1.20c}{1.36} = 0.88c.$$

▌▌

Notice that Equation 24-9 reduces to the classical form for velocities small compared to the speed of light since $1 + vu'/c^2 \approx 1$ for v and $u' \ll c$. Thus $u \approx v + u'$.

Let us test our formula in one more case, that of the speed of light. Suppose that rocket 1 in Figure 24-13 sends out a beam of light so that $u' = c$. Equation 24-9 tells us that the speed of this light with respect to earth is

$$u = \frac{0.60c + c}{1 + \dfrac{(0.60)(c)}{c^2}} = \frac{1.60c}{1.60} = c,$$

as it must if the second postulate of relativity is to hold.

*24-12 ▌ The Galilean and Lorentz transformations

In this section we examine in more detail the mathematics of relating quantities in one inertial reference frame to the equivalent quantities in another. In particular, we will see how positions and velocities *transform* (that is, change) from one frame to the other.

*Galilean transformation

We begin with the classical or Galilean viewpoint. Consider two reference frames S and S', each characterized by a set of coordinate axes (Figure 24-14). The axes x and y (z is not shown) refer to S and x' and y' to S'. The x' and x axes overlap one another, and we assume that frame S' moves to

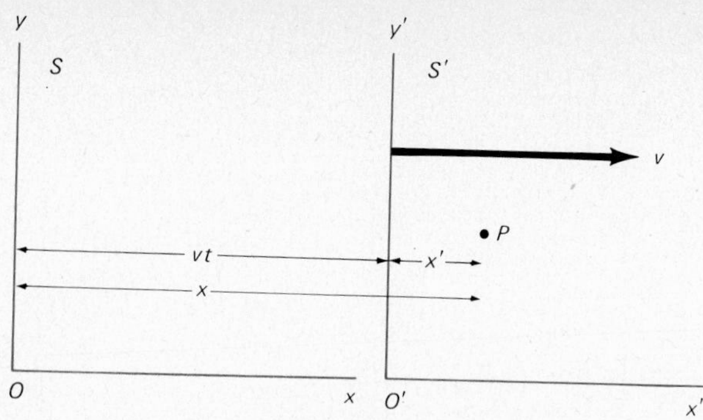

FIGURE·24-14
Inertial reference frame S' moves to the right at speed v with respect to frame S.

the right (in the x direction) at speed v with respect to S. For simplicity, let us assume that the origins O and O' of the two reference frames are superimposed at time $t = 0$.

Now consider an event that occurs at a point P (Figure 24-14) represented by the coordinates x', y', and z' in reference frame S' at the time t'. What will be the coordinates of P in S? Since S and S' overlap precisely initially, after a time t, S' will have moved a distance vt'. Therefore, at time t', $x = x' + vt'$. The y and z coordinates, on the other hand, are not altered by motion along the x axis; thus $y = y'$ and $z = z'$. Finally, since time is assumed to be absolute in Galilean–Newtonian physics, clocks in the two frames will agree with each other; so $t = t'$. We summarize these in the following *Galilean transformation* equations:

$$
\begin{aligned}
x &= x' + vt' \\
y &= y' \\
z &= z' \\
t &= t'.
\end{aligned}
\tag{24-10}
$$

These equations give the coordinate of an event in the S frame when those in the S' frame are known. If those in the S system are known, the S' coordinates are obtained from

$$x' = x - vt, \qquad y' = y, \qquad z' = z, \qquad t' = t.$$

These four equations are the "inverse" transformation and are very easily obtained from Equations 24-10. Notice that the effect is merely to exchange primed and unprimed quantities and replace v by $-v$. This makes sense because from the S' frame, S moves to the left (in the negative x direction).

Now suppose that the point P in Figure 24-14 represents an object that is moving. Let the components of its velocity vector in S' be u'_x, u'_y, and u'_z (we use u to distinguish it from the relative velocity of the two frames, v). Now $u'_x = \Delta x'/\Delta t'$, $u'_y = \Delta y'/\Delta t'$, and $u'_z = \Delta z'/\Delta t'$, by the definition of velocity [Chapter 1; the delta (Δ) means "small amount of"]. For example, if at time t'_1 the particle is at x'_1 and a short time later, t'_2, it is at x'_2, then

$$u'_x = \frac{x'_2 - x'_1}{t'_2 - t'_1} = \frac{\Delta x'}{\Delta t'}.$$

Now the velocity of P as seen from S will have components u_x, u_y, and u_z.

We can show how these are related to the velocity components in S' by using Equations 24-10. For example,

$$u_x = \frac{\Delta x}{\Delta t} = \frac{x_2 - x_1}{t_2 - t_1} = \frac{(x'_2 + vt'_2) - (x'_1 + vt'_1)}{t'_2 - t'_1}$$

$$= \frac{(x'_2 - x'_1) + v(t'_2 - t'_1)}{t'_2 - t'_1}$$

$$= \frac{\Delta x'}{\Delta t'} + v = u'_x + v.$$

For the other components, $u'_y = u_y$ and $u'_z = u_z$, so we have

$$\begin{aligned} u_x &= u'_x + v \\ u_y &= u'_y \\ u_z &= u'_z. \end{aligned} \qquad (24\text{-}11)$$

These are known as the *Galilean velocity transformation* equations. We see that the y and z components of velocity are unchanged, but the x components differ by v. This is just what we have used before when dealing with relative velocity. For example, if S' is a train and S the earth, and the train moves with speed v with respect to earth, a person walking toward the front of the train with speed u'_x will have speed $u_x = u'_x + v$ with respect to the earth.

*Lorentz transformation

The Galilean transformations, Equations 24-10 and 24-11, are valid only when the velocities involved are much less than c. We can see, for example, that the first of Equations 24-11 will not work for the speed of light; for light travelling in S' with speed $u'_x = c$ will have speed $c + v$ in S, whereas the theory of relativity insists it must also be c in S. Clearly, then, a new set of transformation equations is needed to deal with relativistic velocities.

We shall derive the required equations in a simple way, again looking at Figure 24-14. We assume the transformation is linear and of the form

$$x = \gamma(x' + vt'), \qquad y = y', \qquad z = z'.$$

That is, we modify the first of Equations 24-10 by multiplying by a yet-to-be-determined constant γ, but we assume the y and z equations are unchanged since there is no length contraction in these directions. We will not assume a form for t, but will derive it. The inverse equations must have the same form with v replaced by $-v$. (The principle of relativity demands it, since S' moving to the right with respect to S is equivalent to S moving to the left with respect to S'.) Therefore,

$$x' = \gamma(x - vt).$$

Now if a light pulse leaves the common origin of S and S' at time $t = t' = 0$, after a time t it will have traveled a distance $x = ct$ or $x' = ct'$ along the x axis. Therefore, from the equations for x and x' above,

$$ct = \gamma(ct' + vt') = \gamma(c + v)t'$$

$$ct' = \gamma(ct - vt) = \gamma(c - v)t.$$

We substitute t' from the second equation into the first and find $ct = \gamma(c+v)\gamma(c-v)(t/c) = \gamma^2(c^2-v^2)t/c$. We cancel out the t on each side and solve for γ to find

$$\gamma = \frac{1}{\sqrt{1-v^2/c^2}}.$$

Now that we have found γ, we need only find the relation between t and t'. To do so, we combine $x' = \gamma(x-vt)$ with $x = \gamma(x'+vt')$:

$$x' = \gamma(x-vt) = \gamma(\gamma[x'+vt'] - vt).$$

We solve for t and find that $t = \gamma(t'+vx'/c^2)$. In summary

$$\begin{aligned} x &= \frac{1}{\sqrt{1-v^2/c^2}}(x'+vt') \\ y &= y' \\ z &= z' \\ t &= \frac{1}{\sqrt{1-v^2/c^2}}(t'+vx'/c^2). \end{aligned} \qquad (24\text{-}12)$$

These are called the *Lorentz transformation equations.* They were first proposed by Lorentz in 1904 to explain the null result of the Michelson–Morley experiment. A year later Einstein derived them independently based on his theory of relativity. Notice that not only is the x equation modified as compared to the Galilean transformation, but so is the t equation; indeed, we see directly in this last equation how the space and time coordinates mix. The relativistically correct velocity equations are readily obtained. For example (we let $\gamma = 1/\sqrt{1-v^2/c^2}$),

$$\begin{aligned} u_x &= \frac{\Delta x}{\Delta t} = \frac{\gamma(\Delta x' + v\Delta t')}{\gamma(\Delta t' + v\Delta x'/c^2)} = \frac{(\Delta x'/\Delta t') + v}{1 + (v/c^2)(\Delta x'/\Delta t')} \\ &= \frac{u'_x + v}{1 + vu_x/c^2}. \end{aligned}$$

The others are obtained in the same way and we collect them here:

$$\begin{aligned} u_x &= \frac{u'_x + v}{1 + vu'_x/c^2} \\ u_y &= \frac{u'_y\sqrt{1-v^2/c^2}}{1 + vu'_x/c^2} \\ u_z &= \frac{u'_z\sqrt{1-v^2/c^2}}{1 + vu'_x/c^2}. \end{aligned} \qquad (24\text{-}13)$$

The first of these equations is just Equation 24-9 of the previous section. As we saw there, velocities do not add in our commonsense (Galilean) way, because of the denominator $(1 + vu'_x/c^2)$. We see, too, that the y and z components of velocity are also altered and that they depend on the x' component of velocity.

▮ EXAMPLE 24-7 Derive the length-contraction formula, Equation 24-3, from the Lorentz transformation equations.

SOLUTION Let an object of length L_0 be at rest on the x axis in S. The coordinates of its two end points are x_1 and x_2, so that $x_2 - x_1 = L_0$. At any instant in S', the end points will be at x'_1 and x'_2, as given by the Lorentz transformation equations. The length in S' is $L = x'_2 - x'_1$. An observer in S' measures this length by measuring x'_2 and x'_1 at the same time, so $t'_2 = t'_1$. Thus, from the first Equation 24-12,

$$L_0 = x_2 - x_1 = \frac{1}{\sqrt{1 - v^2/c^2}}(x'_2 + vt'_2 - x'_1 - vt'_1).$$

Since $t'_2 = t'_1$, we have

$$L_0 = \frac{1}{\sqrt{1 - v^2/c^2}}(x'_2 - x'_1) = \frac{L}{\sqrt{1 - v^2/c^2}},$$

or $L = L_0\sqrt{1 - v^2/c^2}$. ▮▮

▮ EXAMPLE 24-8 Derive the time-dilation formula.

SOLUTION The time T_0 between two events that occur at the same place ($x'_2 = x'_1$) in S' is measured to be $T_0 = t'_2 - t'_1$. Since $x'_2 = x'_1$, then from the last of Equations 24-12, the time T between the events as measured in S is

$$T = t_2 - t_1 = \frac{1}{\sqrt{1 - v^2/c^2}}\left(t'_2 + \frac{vx'_2}{c^2} - t'_1 - \frac{vx'_1}{c^2}\right)$$

$$= \frac{1}{\sqrt{1 - v^2/c^2}}(t'_2 - t'_1) = \frac{T_0}{\sqrt{1 - v^2/c_2}},$$

which is Equation 24-2. Notice that we chose S' to be the frame where the two events occur at the same place, so that $x'_1 = x'_2$ and these terms would cancel out. ▮▮

▮ Summary

An *inertial reference frame* is one in which Newton's law of inertia holds. Inertial reference frames can move at constant velocity relative to each other; accelerating reference frames are noninertial.

The special theory of relativity is based on two principles: the *relativity principle*, which states that the laws of physics are the same in all inertial reference frames, and the principle of the *constancy of the speed of light*, which states that the speed of light in empty space has the same value in all inertial reference frames.

One consequence of relativity theory is that two events that are simultaneous in one reference frame may not be simultaneous in another. Other effects are *time dilation*: moving clocks are measured to run slowly; *length contraction*: the length of an object is measured to be shorter when it is moving than when it is at rest; *mass increase*: the mass of a body increases with speed; and velocity addition must be done in a special way. All these effects are significant only at high speeds, close to the speed of light, which itself is the ultimate speed in the universe.

The theory of relativity has changed our notions of space and time, and of mass and energy. Space and time are seen to be intimately connected, with time being the fourth dimension in addition to space's three dimensions. Mass and energy are interconvertible; the equation $E = mc^2$ tells how much energy E is needed to create a mass m, or vice versa. The law of conservation of energy must include mass as a form of energy.

Questions

1 Give some examples of noninertial reference frames.

2 A woman stands on top of a moving railroad car. She throws a heavy ball straight up (it seems to her) in the air. Ignoring air resistance, will the ball land on the car or behind it?

3 Discuss a Michelson–Morley type of experiment done with sound waves. Under what conditions would you expect or not expect a null result?

4 Does the earth really go round the sun? Is it also valid to say that the sun goes around the earth? Discuss in view of the first principle of relativity (that there is no best reference frame).

5 If you were on a spaceship traveling at $0.5c$ away from a star, at what speed would the star light pass you?

6 Will two events that occur at the same place and same time for one observer be simultaneous to a second observer moving with respect to the first?

7 Explain why two events will be simultaneous to each of two observers moving with respect to each other only if the two events occur at the same point in each reference frame.

8 Analyze the thought experiment of Section 24-4 from O_1's point of view. (Make a diagram analogous to Figure 24-7.)

9 The time-dilation effect says that "moving clocks run slowly." Actually, this effect has nothing to do with motion affecting the functioning of clocks. What, then, does it deal with?

10 Does time dilation mean that time really passes more slowly in moving reference frames or that it simply *seems* to pass more slowly?

11 A young-looking woman astronaut has just arrived home from an extended trip. She greets an old gray-haired man and calls him her son. Is this possible?

12 If you were speeding away from earth at a speed of $0.5c$, would you notice a change in your waistline? Would your mass, height or heartbeat change? What would observers on earth using telescopes say?

13 The theory of relativity seems to contradict some of our commonsense ideas about the world. Make a detailed examination of these notions and determine if any measurable contradiction exists.

14 Do mass increase, time dilation, and length contraction occur at ordinary speeds, say, 90 km/h?

15 If the speed of light were infinite, what would happen to the relativistic predictions of length contraction, time dilation, and mass increase?

16 Explain how the length-contraction and time-dilation formulas might be used to indicate that c is the limiting speed in the universe.

17 Suppose a constant force is applied for an indefinite period of time to an object of mass m. Discuss how its velocity and mass change with time.

18 A white-hot iron rod is cooled to room temperature. Does its mass change?

19 Does the formula $E = mc^2$ conflict with the law of conservation of energy? Explain.

20 Does $E = mc^2$ apply to particles that travel at the speed of light? Does it apply only to them?

21 An electron is limited to travel at speeds less than c. Does this put an upper limit on the momentum of an electron? If so, what is this upper limit?

22 If mass is a form of energy, does this mean that a spring has more mass when compressed than when relaxed?

23 It is not correct to say that "matter can neither be created nor destroyed." What must we say instead?

24 A neutrino is an elementary particle with zero rest mass that travels at the speed of light. Could you ever catch up to a neutrino that passed you?

25 Discuss how our everyday lives would be different if the speed of light were only 25 m/s.

Problems

SECTIONS 24-5 AND 24-6

1 (I) A beam of a certain type of elementary particle travels at a speed of 2.4×10^8 m/s. At this speed, the average lifetime is measured to be 2.0×10^{-8} s. What is their lifetime at rest?

2 (I) What is the speed of a beam of pi mesons if their average lifetime is measured to be 3.5×10^{-8} s? At rest, their lifetime is 2.6×10^{-8} s.

3 (I) A rocketship passes you at a speed of $0.80c$. You measure its length to be 90 m. How long would it be when at rest?

4 (I) A space car travels by you at $0.50c$. People in the car claim it to be 100 m long. What length will you measure it to be?

5 (I) You are sitting in your sportscar when a really fast sportscar passes you at a speed of $0.10c$. A person in the car says his car is 6.00 m long and yours is 6.15 m long. What do you measure for these two lengths?

6 (I) Suppose that you decide to travel to a star 80 light-years away. How fast would you have to travel so the distance would be only 20 light-years?

7 (I) If you were to travel to a star 50 light-years from earth at a speed of 2.0×10^8 m/s, what would you measure this distance to be?

8 (II) A certain star is 20 light-years away. How long would it take a spaceship traveling $0.95c$ to reach that star from earth as measured by observers (*a*) on earth; (*b*) on the spaceship?

9 (II) A friend of yours travels by you in her rocket car at a speed of $0.60c$. It appears to be 5.0 m long and 1.5 m high. (*a*) What will be its length and height at rest? (*b*) How many seconds did you see elapse on your friend's watch when 10 s passed on yours? (*c*) How fast did you appear to be traveling according to your friend? (*d*) How many seconds did she see elapse on your watch when she saw 10 s pass on hers?

10 (III) How fast must a pi meson be traveling to travel 1.0 m before it decays? The average lifetime, at rest, is 2.6×10^{-8} s.

11 (III) The nearest star to us is Alpha Centauri, 4 light-years away. (*a*) At what constant velocity must a spacecraft travel from earth if it is to reach the star in 3 years time, as measured by travelers on the spacecraft? (*b*) How long does the trip take according to earth observers?

SECTION 24-8

12 (I) What is the mass of an electron traveling at $v=0.50c$?

13 (I) At what speed will an object's mass be twice its rest mass?

14 (I) At what speed v will the mass of an object be 1 percent greater than its rest mass?

15 (II) Derive a formula showing how the density of an object changes with speed v.

16 (II) Escape velocity from the earth is 40,000 km/h. What would be the increase in mass of a spacecraft 2.0×10^5 kg traveling at that speed?

17 (II) (*a*) What is the speed of an electron whose mass is 10^4 its rest mass? Such speeds are reached in the Stanford Linear Accelerator (SLAC). (*b*) If the electrons travel in the lab through a tube 3.0 km long (as at SLAC), how long is this tube in the electron's reference frame?

SECTION 24-10

18 (I) How many grams of mass is needed to produce 1 J of energy?

19 (I) How much energy can be obtained from conversion of 10 mg of mass? How much mass could this energy raise to a height of 100 m?

20 (I) What is the kinetic energy of an electron whose mass is four times its rest mass?

21 (II) Calculate the mass of a proton whose kinetic energy is 500 MeV. How fast is it traveling?

22 (II) A certain chemical reaction requires 120 kcal of energy input for it to go. What is the increase in mass of the products over the reactants?

23 (II) Suppose that a spacecraft of rest mass 10,000 kg is accelerated to $0.50c$. (*a*) How much kinetic energy would it have? (*b*) If this energy were provided by conversion of mass, how much mass would be needed?

24 (III) A rocket car traveling past you at $0.60c$ is accelerating at 10 m/s^2. What is the force exerted by the road on the car tires to cause this acceleration, assuming that the car's rest mass is 1500 kg?

25 (II) (*a*) By how much does the mass of the earth increase each year as a result of the sunlight reaching it? (*b*) How much mass does the sun lose per year? (Radiation from the sun reaches the earth at a rate of about 1400 W/m^2 of area perpendicular to the energy flow.)

26 (II) Calculate the rest energy of an electron in MeV ($1\text{ eV}=1.6\times10^{-19}$ J).

27 (II) What is the mass of an electron that has been accelerated by a voltage of 100 kV?

28 (II) Calculate the kinetic energy and momentum of a proton traveling 9.2×10^7 m/s.

29 (II) What is the momentum of a 500-MeV proton (that is, one with KE = 500 MeV)?

30 (II) What is the speed of a 200-MeV proton?

31 (II) What is the speed of an electron whose KE is 1.00 MeV?

32 (III) Show that the kinetic energy of a particle of rest mass m_0 is related to its momentum p by the equation $p = \sqrt{(\text{KE})^2 + 2(\text{KE})(m_0c^2)}/c$.

33 (III) A pi meson of rest mass m_π decays at rest into a muon (rest mass m_μ) and a neutrino of zero rest mass. Show that the kinetic energy of the muon is $\text{KE}_\mu = (m_\pi - m_\mu)^2c^2/2m_\pi$.

SECTION 24-11

34 (I) A person on a rocket traveling at $0.50c$ (with respect to the earth) observes a meteor come from behind and pass her at a speed she measures as $0.50c$. How fast is the meteor moving with respect to the earth?

35 (I) Two spaceships leave the earth in opposite directions, each with a speed of $0.50c$ with respect to the earth. (*a*) What is the velocity of spaceship 1 relative to spaceship 2? (*b*) What is the velocity of spaceship 2 relative to spaceship 1?

*SECTION 24-12

*36 (I) Suppose in Figure 24-14 that the origins of S and S' overlap at $t = t' = 0$ and that S' moves at speed $v = 30$ m/s with respect to S. In S', a person is resting at a point whose coordinates are $x' = 15$ m, $y' = 10$ m, and $z' = 0$. Calculate this person's coordinates in $S(x, y, z)$ at (*a*) $t = 5.0$ s, and (*b*) $t = 10.0$ s. Use the Galilean transformation.

*37 (I) Repeat Problem 36 using the Lorentz transformation and a relative speed $v = 1.0 \times 10^8$ m/s, but calculate at time t' equals (*a*) 5.0 μs and (*b*) 10.0 μs.

*38 (II) A spaceship traveling $0.50c$ away from earth fires a module with a speed of $0.70c$ at right angles to its own direction of travel (as seen by the first spaceship). What is the speed of this second spaceship and its direction of travel (relative to the first spaceship's direction) as seen by an observer on earth?

*39 (II) In Problem 36, suppose that the person moves with a velocity whose components are $u'_x = u'_y = 20$ m/s. What will be his velocity with respect to S? (Give magnitude and direction.)

*40 (II) In Problem 37, suppose that the person moves with a velocity (in a rocket car) whose components are $u'_x = u'_y = 1.2 \times 10^8$ m/s. What will be the person's velocity (magnitude and direction) with respect to S?

*41 (III) In the old West, a marshal riding on a train traveling 50 m/s sees a duel between two men standing on the earth 50 m apart parallel to the train. The marshal's instruments indicate that in his reference frame the two men fired simultaneously at the moment he was midway between them. (*a*) Which of the two men, the first one the train passes (A) or the second (B), should be arrested as firing the first shot? That is, in the gunfighters' frame of reference, who fired first? (*b*) How much earlier did he fire? (*c*) Who was struck first?

For additional reading

Casper, B. M., and R. J. Noer, *Revolutions in Physics*, W. W. Norton & Co., Inc., New York, 1972.

Clark, R. W., *Einstein: The Life and Times*, World Publishing Co., Inc., New York, 1971.

Einstein, A., *Ideas and Opinions*, Crown Publishers, Inc., New York, 1954.

Einstein, A., *Out of My Later Years*, Citadel Press, Secaucus, N.J., 1973.

Einstein, A., *Relativity: The Special and General Theory*, tr. R. W. Lawson, Crown Publishers, Inc., New York, 1961.

Gamow, G., *Mr. Tompkins in Wonderland*, Cambridge University Press, Cambridge, 1939.

Good, R. H., *Basic Concepts of Relativity*, Van Nostrand Rheinhold Company, New York, 1968.

Hume and Ivey, *Frames of Reference* (film).

Merman, N. D., *Space and Time in Special Relativity*, McGraw-Hill Book Company, New York, 1968.

Russell, B., *The ABC of Relativity*, The New American Library, Inc., New York, 1969.

Schilpp, P. A., ed., *Albert Einstein: Philosopher–Scientist*, Open Court Publishing Company, La Salle, Ill., 1973.

25 ▮ Early quantum theory

The second aspect of the revolution that shook the world of physics in the early part of the twentieth century (the first half was Einstein's theory of relativity) is the quantum theory. Unlike the special theory of relativity, whose basic tenets were put forth mainly by one person in a single year, the revolution of quantum theory required almost three decades to unfold, and many scientists contributed to its development. It began in 1900 with Planck's quantum hypothesis and culminated in the mid-1920s with the theory of quantum mechanics of Schrödinger and Heisenberg, which has been so effective in explaining the structure of atoms and molecules.

25-1 ▮ Planck's quantum hypothesis

One of the observations that was unexplained at the end of the nineteenth century was the spectrum of light emitted by hot objects. We saw in Chapter 9 that all objects emit radiation whose intensity is proportional to the fourth power of the Kelvin temperature (T^4). At normal temperatures we are not aware of this electromagnetic radiation because of its low intensity. At higher temperatures, there is sufficient infrared radiation that we can feel what we call heat if we are close to the body. At even higher temperatures, objects actually glow, such as the red-hot iron of an electric stove burner or the element in a toaster. At higher temperatures (above 2000 K), objects glow with a whitish color, such as white-hot iron and the filament of a light bulb. Thus, as the temperature increases, the electromagnetic radiation emitted by bodies emphasizes higher and higher frequencies.

The actual spectrum of light emitted by a hot dense object is shown in Figure 25-1. As can be seen, the spectrum contains a continuous range of frequencies; such a continuous spectrum is emitted by any heated solid or liquid, and even by dense gases. The 6000-K curve in Figure 25-1 corresponds to the temperature of the sun, and the visible part of the spectrum is seen to be nearly uniform with a peak in the center. For lower temperatures the total radiation is much less and the peak occurs at a higher wavelength. Hence the blue end of the visible spectrum (and the UV) is relatively weaker. It is found that the wavelength at the peak of the spectrum, λ_p, is related to the Kelvin temperature T by

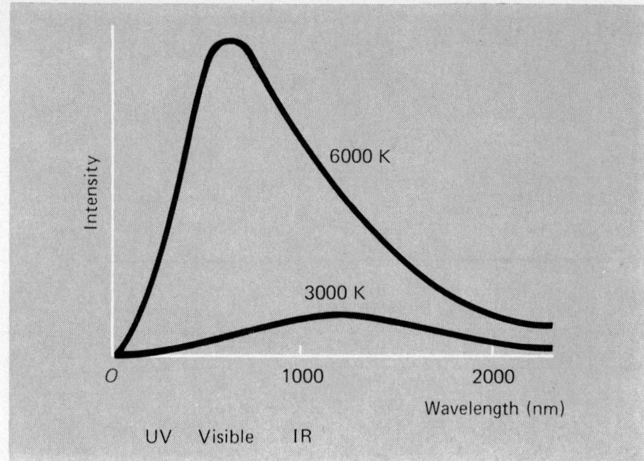

FIGURE 25-1
Spectrum of frequencies emitted by a blackbody.

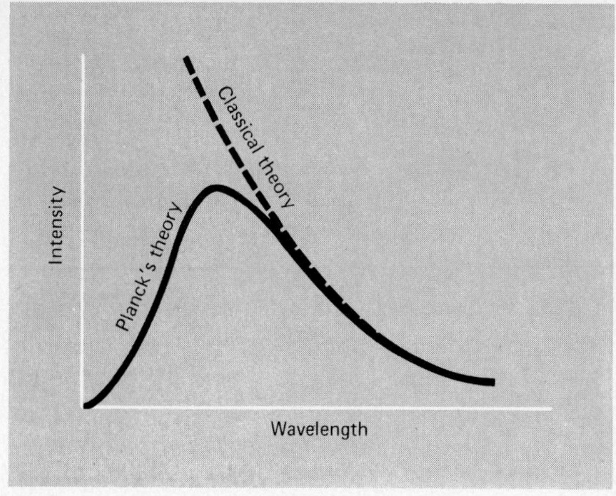

FIGURE 25-2
Comparison of classical theory with Planck's theory for spectrum of blackbody radiation.

$$\lambda_p T = 2.90 \times 10^{-3} \text{ m} \cdot \text{K}.$$

This is known as *Wien's law* and for the sun's temperature gives us $\lambda_p = (2.90 \times 10^{-3} \text{ m} \cdot \text{K})/(6.0 \times 10^3 \text{ K}) \approx 500 \text{ nm}$.

The experimental curves of Figure 25-1 are for a body that approximates an idealized nonreflecting object. Such a body would appear black (when not too hot) and the radiation it emits is the simplest to deal with. The problem that faced scientists in the 1890s was to explain this *black body radiation.* Maxwell's electromagnetic theory had shown that oscillating electric charges produce electromagnetic waves. The radiation emitted by a hot body was assumed to be due to the oscillations of electric charges in the molecules of the material. Although this explains where the radiation comes from, it did not correctly predict the observed spectrum of emitted light. Indeed, theory based on Maxwell's electromagnetic theory predicted that the intensity increases with frequency as shown by the dashed curve in Figure 25-2. Thus all objects would glow with a blue color, and in fact the curve goes to infinity in the UV region. This problem with classical theory was thus known as the "ultraviolet catastrophe." Clearly, classical theory was inconsistent with the experimental results.

The break came in 1900. In that year the German physicist Max Planck (1858–1947) proposed a new theory that was fully in accord with the experimental data (solid curve in Figure 25-2). But to achieve this accord, Planck had to make a radical assumption. He hypothesized that the vibrating molecules in a heated material can vibrate only with certain discrete amounts of energy. In particular, Planck assumed that the minimum energy of vibration E_{min} is proportional to the natural frequency of oscillation f:

$$E_{\text{min}} = hf;$$

here h is a constant, now called *Planck's constant,* whose value was estimated by Planck by fitting his formula for the blackbody radiation curve (h appears in this formula) to experiment. The value accepted today is

$$h = 6.626 \times 10^{-34} \text{ J} \cdot \text{s}.$$

Furthermore, Planck said, the energy of any molecular vibration could only be some whole-number multiple of this minimum energy:

$$E = nhf, \qquad n = 1, 2, 3, \ldots. \tag{25-1}$$

This, then, is **Planck's quantum hypothesis**. (*Quantum* means "fixed amount.") It says that there cannot be vibrations whose energy lies between the values given by Equation 25-1.† That is, energy is not a continuous quantity as had been believed for centuries. Rather, according to Planck, it is **quantized**—it exists only in discrete amounts. The smallest amount of energy possible (hf) is called the quantum of energy.

Planck's hypothesis was not readily accepted by his colleagues. Indeed, Planck himself was not comfortable with it and reported that he proposed it only to bring theory into accord with experiment; and he said he hoped that a better explanation would soon come forth. Alas, it did not, and Planck's quantum hypothesis has come to be accepted as a fact of nature.

The concept of quantization is not entirely new. The mass of a block of copper, for example, is a whole-number multiple of the mass of one copper atom. Nowhere will you find a copper block with a fractional number of atoms, say $936\frac{1}{2}$. Similarly, electric charge is quantized. The smallest charge found in nature is e, the charge on the electron. And the net charge on any object is a whole-number multiple of this quantum of charge. Thus a body may have a charge of $1e$, $2e$, $3e$, or $317e$, but never $317.2e$. Nonetheless, the idea that energy is quantized was not easy to accept. But in view of Einstein's epochal idea that mass and energy are equivalent ($E = mc^2$), since mass is quantized, why not energy?

25-2 ▮ Photon theory of light and the photoelectric effect

Einstein's photon theory

In 1905, the same year as he introduced the special theory of relativity, Einstein made a bold extension of the quantum idea by proposing a new theory of light. Planck had assumed that the vibrational energy of molecules in a radiating object is quantized with energy $E = nhf$. Einstein reasoned that if the energy of the molecular oscillators is thus quantized, then to conserve energy, the light must be emitted in pockets or quanta each with an energy

$$E = hf. \tag{25-2}$$

Again h is Planck's constant. Thus, since all light ultimately comes from a radiating source, Einstein proposed that *light is transmitted as tiny particles*, or **photons**, as they are now called, rather than as waves. This, too, was a radical departure from classical ideas. And Einstein proposed a simple test of his photon theory: careful experiments on the photoelectric effect.

† Another way of saying this is that not just any amplitude of vibration is possible. The possible values for the amplitude are related to the frequency f.

The photoelectric effect

The *photoelectric effect* refers to the fact that when light shines on a metal surface, electrons are observed to be emitted from the surface. (The photoelectric effect occurs in other materials but is most easily observed with metals.) This effect can be observed using the apparatus shown in Figure 25-3. A metal plate P along with a smaller electrode C are placed inside an evacuated glass tube (called a *photocell*). The two electrodes are connected to an ammeter and a battery as shown. When the photocell is in the dark, the ammeter reads zero. But when light of sufficiently high frequency is shone on the plate, the ammeter indicates a current flowing in the circuit. To complete the circuit in this case, electrons must be flowing across the tube from the plate to the "collector" C as shown in the diagram.

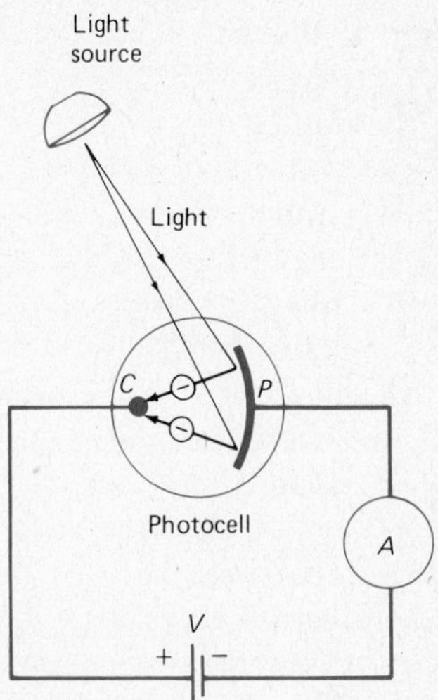

FIGURE 25-3
Photoelectric effect.

The fact that electrons should be emitted when light shines on a metal is consistent with the electromagnetic (EM) wave theory of light; for the electric field of the EM wave can exert a force on electrons in the metal and thrust some of them out. Einstein pointed out, however, that the wave theory and his new photon theory of light would give quite different predictions on the details of this effect. For example, one thing that can be measured with the apparatus of Figure 25-3 is the maximum kinetic energy (KE_{max}) of the emitted electrons. This can be done by using a variable voltage source and reversing the terminals so that electrode C is negative and P is positive. The electrons emitted from P will be repelled by the negative electrode, but if this reverse voltage is small enough, the fastest electrons will still reach C and there will be a current in the circuit. If the reversed voltage is increased, a point is reached where no electrons have sufficient kinetic energy to reach C. This point is called the "stopping potential," V_0, and from its measurement KE_{max} can be determined.

Now let us examine the details of the photoelectric effect in view of the wave theory versus Einstein's particle theory. First the wave theory. We assume that monochromatic light is being used. The two important properties of a light wave are its amplitude and its frequency (or wavelength). When these two quantities are varied, the wave theory makes the following predictions:

1 If the intensity is increased, the number of electrons ejected and their maximum KE is increased; this is because the higher intensity means a greater amplitude, and the greater electric field will thrust electrons out with higher speed.

2 The frequency of the light should not affect the KE of the ejected electrons. Thus only the intensity of the light affects KE_{max}.

The photon theory makes completely different predictions. First we note that in a monochromatic beam, all photons have the same energy ($=hf$). Increasing the intensity of the light beam means increasing the number of photons in the beam but does not affect the energy of each as long as the frequency is not changed. An electron is ejected from the metal by a collision with a single photon. In the process, all the photon energy is transferred to the electron and the photon ceases to exist. Since electrons are held in the metal by attractive forces, some minimum energy W_0 (called the work function, which is different for different materials) is required just to get an electron out of the surface. If the frequency f of the incoming light is so low that hf is less than W_0, the photons will not have enough energy to eject any electrons at all. If $hf > W_0$, electrons will be ejected and energy will be conserved in the process. That is, the input energy (of the photon), hf, will equal the outgoing KE of the electron plus the energy required to get it out of the metal W:

$$hf = \text{KE} + W. \tag{25-3}$$

For the least tightly held electrons, W is the work function W_0 and KE in this equation becomes KE_{max}. Electrons deeper in the metal will require more energy W to just get out of the metal, and thus the KE of such electrons will be less than the maximum.

From these considerations, the photon theory makes the following predictions:

1 An increase in intensity of the light beam means that more photons are incident, so more electrons will be ejected; but since the energy of each photon is not changed, the maximum KE of electrons is not changed.

2 If the frequency of the light is increased, the maximum KE of the electrons increases linearly, according to Equation 25-3. That is,

$$\text{KE}_{max} = hf - W_0.$$

This is plotted in Figure 25-4.

3 If the frequency f is less than the "cutoff" frequency f_0, where $hf_0 = W_0$, no electrons will be ejected.

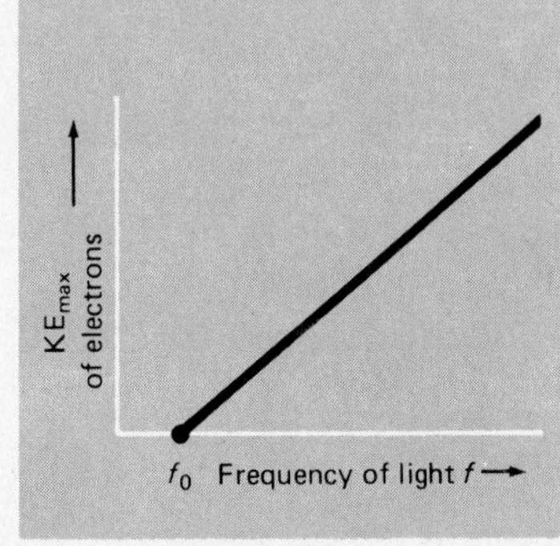

FIGURE 25-4
Photoelectric effect: maximum kinetic energy of ejected electrons increases linearly with frequency of incident light. No electrons are emitted if $f < f_0$.

These three predictions of the photon theory are clearly very different from the predictions of the wave theory. In particular, the EM wave theory predicts that the maximum KE of ejected electrons depends only on the light intensity, whereas the photon theory predicts that it depends only on the frequency of the light (and that this is a linear dependence). In 1913–1914 careful experiments were carried out by R. A. Millikan. The results were fully in agreement with Einstein's photon theory. Thus Einstein's brilliant theory was confirmed.

A number of other experiments were carried out in the early twentieth century which also supported Einstein's theory. One of these was the *Compton effect*, named after its discoverer, A. H. Compton (1892–1962). Compton scattered short-wavelength light (actually x rays) from various materials. He found that the scattered light was slightly less in frequency than the incident light, indicating a loss of energy. This, he showed, could only be explained on the basis of the photon theory as incident photons colliding with electrons of the material (Figure 25-5); he applied the laws of conservation of energy and momentum to such collisions and found that the predicted energy of the scattered photon was in accord with experimental results. Thus the photon theory of light rests on a firm experimental foundation.

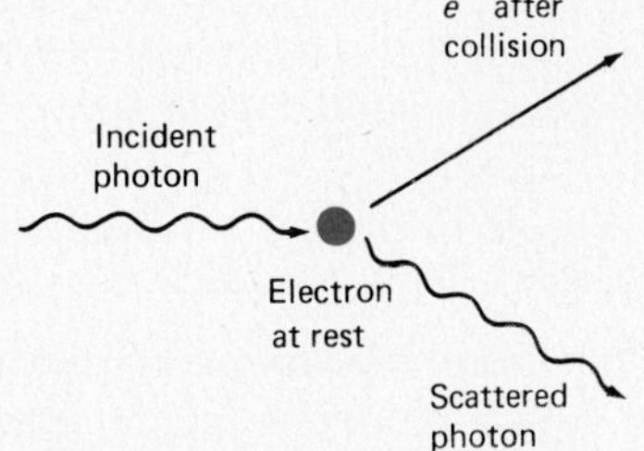

FIGURE 25-5
Compton effect.

▌ EXAMPLE 25-1 Calculate the energy of a photon of blue light, $\lambda = 450$ nm.

SOLUTION Since $f = c/\lambda$, we have

$$E = hf = \frac{hc}{\lambda} = \frac{(6.63 \times 10^{-34}\ \text{J} \cdot \text{s})(3.0 \times 10^8\ \text{m/s})}{4.5 \times 10^{-7}\ \text{m}} = 4.4 \times 10^{-19}\ \text{J},$$

or $(4.4 \times 10^{-19}\ \text{J})/(1.6 \times 10^{-19}\ \text{J/eV}) = 2.7$ eV. ▌▌

❚ EXAMPLE 25-2 What is the maximum kinetic energy and speed of an electron ejected from a sodium surface whose work function is $W_0 = 2.28$ eV when illuminated by light of wavelength (*a*) 410 nm, and (*b*) 550 nm?

SOLUTION (*a*) For $\lambda = 410$ nm, $hf = hc/\lambda = 4.85 \times 10^{-19}$ J or 3.03 eV. From Equation 25-3, with $W = W_0$, $KE_{max} = 3.03\text{ eV} - 2.28\text{ eV} = 0.75$ eV or 1.2×10^{-19} J. Since $KE = \frac{1}{2}mv^2$, where $m = 9.1 \times 10^{-31}$ kg,

$$v = \sqrt{2KE/m} = 5.1 \times 10^5 \text{ m/s}.$$

Notice that we used the nonrelativistic equation for KE; if v had turned out more than about $0.1c$, our calculation would have been sufficiently inaccurate that we would have had to redo it using the relativistic form (Equation 24-7). (*b*) For $\lambda = 550$ nm, $hf = 3.60 \times 10^{-19}$ J $= 2.25$ eV. Since this photon energy is less than the work function, no electrons are ejected. ❚❚

Applications of the photoelectric effect

The photoelectric effect, besides playing an important historical role in confirming the photon theory of light, also has many practical applications. Burglar alarms and automatic door openers often make use of the photocell circuit of Figure 25-3. When a person interrupts the beam of light, the sudden drop in current in the circuit activates a switch—often a solenoid—to operate a bell or open the door. UV or IR light is sometimes used in burglar alarms because of its invisibility. Many smoke detectors use the photoelectric effect to detect tiny amounts of smoke that interrupt the flow of light and so alter the electric current. Photographic lightmeters use this circuit as well. The brighter the light, the more current that flows in the circuit as indicated by the ammeter needle. Similar, but smaller devices are used in medical diagnosis such as to measure blood flow. Photocells are used in many other devices to measure light intensity, such as in absorption spectrophotometers. Finally, one type of film soundtrack is a variably shaded narrow section at the side of the film. Light passing through the film is thus "modulated" and the output electrical signal of the photocell detector follows the frequencies on the sound track.

For many applications today, the vacuum-tube photocell of Figure 25-3 has been replaced by a semiconductor device known as a *photodiode*. In these semiconductors, the absorption of a photon liberates a bound electron, which changes the conductivity of the material. Thus the current through a photodiode changes when light shines on it and so its operation is similar to a photocell.

Photon theory in biology

The photon theory of light is often used in biology and medicine, particularly where low light levels occur. For example, the sensitivity of the eye is specified in terms of photons. Research indicates that receptors (rods and cones) in the retina actually respond to single photons. However, the brain does not register "seeing" for a single photon. Instead several photons (at least 5 to 10 and usually more) are needed, either simultaneously at one receptor or at several adjacent receptors. This coincidence requirement is a valuable mechanism, for it means that our minds are not bothered by random signals that are meaningless. The photon concept is also used in the analysis of *photosynthesis*. Photosynthesis is the complex biochemical process by which pigments such as chlorophyll in plants use the energy of light from the sun to change CO_2 to useful carbohydrate with the release of molecular oxygen.

▮ EXAMPLE 25-3 It is believed that in photosynthesis about nine photons are needed to change one molecule of CO_2 to carbohydrate and O_2. Assuming light of wavelength $\lambda = 670$ nm (chlorophyll absorbs most strongly in the range 650 to 700 nm), how efficient is the photosynthetic process? The reverse chemical reaction has a heat of combustion of 4.9 eV/molecule of CO_2.

SOLUTION The energy of nine photons, each of energy $hf = hc/\lambda$, is $9(6.6\times10^{-34}\ \text{J}\cdot\text{s})(3.0\times10^{8}\ \text{m/s})/(6.7\times10^{-7}\ \text{m}) = 2.7\times10^{-18}$ J, or 17 eV. Thus the process is (4.9 eV)/(17 eV) = 29 percent efficient. ▮▮

25-3 ▮ Photon interactions; pair production

The photon is truly a relativistic particle, for it travels at the speed of light. (It *is* light!) Thus we must use relativistic formulas for dealing with its mass, energy, and momentum. The mass m of any particle is given by $m = m_0/\sqrt{1 - v^2/c^2}$. Since $v = c$ for a photon, the denominator is zero. Thus the rest mass, m_0, of a photon must be zero, for otherwise its mass m would be infinite, and so would its energy, $E = mc^2$. Of course, a photon never is at rest.

The momentum p of a photon, from Equation 24-8 with $m_0 = 0$, is $p = E/c$. The momentum of a photon, like the energy, can be written in terms of the wavelength or frequency since $p = E/c = hf/c$. Thus

$$p = \frac{hf}{c} = \frac{h}{\lambda}. \tag{25-4}$$

When a photon passes through matter, it interacts with the atoms and electrons. There are four important types of interactions that a photon can undergo. First, the photon can be scattered off an electron (or a nucleus) and in the process lose some energy; this is the Compton effect (Figure 25-5). But notice that the photon is not slowed down. It still travels with speed c, but its frequency will be less. A second type of interaction is the photoelectric effect; a photon may knock an electron out of an atom and in the process itself disappear. The third process is similar: the photon may knock an atomic electron to a higher-energy state in the atom if its energy is not sufficient to knock it out altogether. In this process the photon also disappears, and all its energy is given to the atom; such an atom is then said to be in an excited state, and we shall discuss this more later. Finally, a photon can actually create matter. The most common process is the production of an electron and a positron, Figure 25-6. (A positron has the same mass as an electron, but the opposite charge—that is,

FIGURE 25-6
Pair production: a photon disappears and produces an electron and a positron.

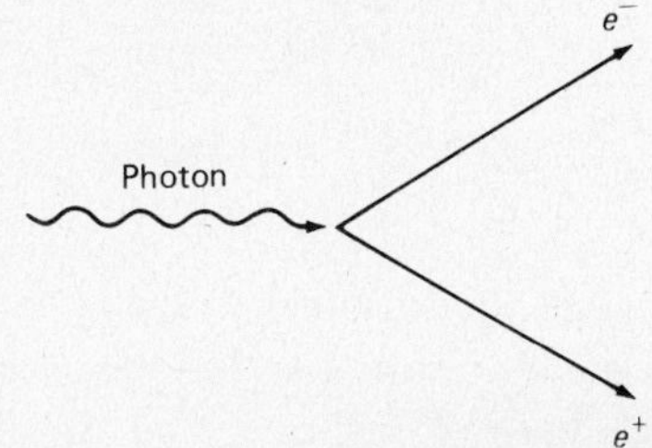

$+e$.)† This is called **pair production** and the photon disappears in the process. The photon, in a sense, *becomes* the electron–positron pair. This is an example of rest mass being created from pure energy, and it occurs in accord with Einstein's equation $E = mc^2$. Notice that a photon cannot create a single electron, since electric charge would not then be conserved.

▌EXAMPLE 25-4 What is the minimum energy of a photon, and its wavelength, that can produce an electron–positron pair?

SOLUTION Because $E = mc^2$ and two electron masses are being produced, the photon must have energy $E = 2(9.1 \times 10^{-31}\ \text{kg})(3.00 \times 10^8\ \text{m/s})^2 = 1.64 \times 10^{-13}\ \text{J} = 1.02 \times 10^6\ \text{eV}$, or $1.02\ \text{MeV}$. ($1\ \text{MeV} = 10^6\ \text{eV} = 1.6 \times 10^{-13}\ \text{J}$.) A photon with less energy cannot undergo pair production. Since $E = hf = hc/\lambda$, the wavelength of a 1.02-MeV photon is

$$\lambda = \frac{hc}{E} = \frac{(6.6 \times 10^{-34}\ \text{J} \cdot \text{s})(3.0 \times 10^8\ \text{m/s})}{1.64 \times 10^{-13}\ \text{J}} = 1.2 \times 10^{-12}\ \text{m},$$

which is 0.0012 nm. Thus the wavelength must be very short. Such photons are in the gamma-ray (or very short x-ray) region of the electromagnetic spectrum (Figure 20-8). ▌▌

Pair production cannot occur in empty space, for energy and momentum could not simultaneously be conserved. In the example above, for instance, energy is conserved; but for minimum energy, the electron–positron pair have no momentum to carry away the initial momentum of the photon. Indeed, it can be shown that an additional massive object, such as an atomic nucleus, must take part in the interaction by carrying off some of the momentum.

25-4 ▌X rays and x-ray diffraction

Discovery and nature of x rays

In 1895, W. C. Roentgen (1845–1923) discovered that, when electrons (then called cathode rays) were accelerated by a high voltage in a vacuum tube and allowed to strike a glass or metal surface, certain minerals placed nearby would fluoresce, and film placed nearby in a closed wrapper would become exposed. Roentgen attributed these effects to a new type of radiation, which he termed **x rays**, after the algebraic symbol x, meaning an unknown quantity. He showed that x rays could pass readily through cardboard and other opaque materials. It was soon found that x rays penetrated through some materials better than others and would expose film placed behind an object. For example, x rays pass readily through flesh and internal organs of the body but are absorbed to a large extent by bone. Indeed, within a few weeks of Roentgen's discovery, the first x-ray photographs of bones in the hand were made. The production of x rays today is done in a tube (Figure 25-7) not unlike Roentgen's, using typical voltages of 30 to 150 kV.

FIGURE 25-7
X-ray tube. Electrons emitted by a heated filament in a vacuum tube are accelerated by high voltage. When they strike the surface of the anode, the "target," x rays are emitted.

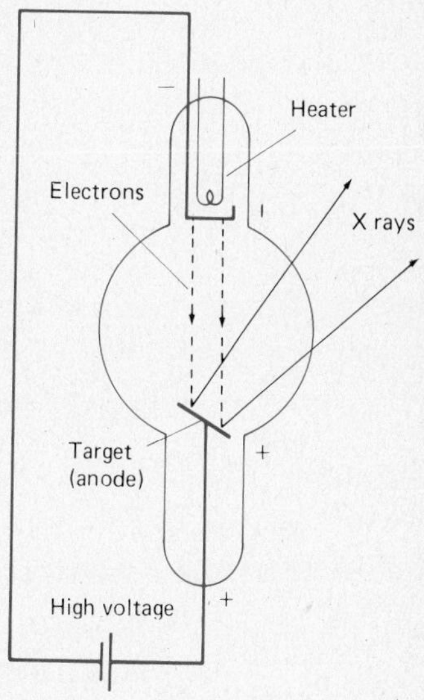

† Positrons do not last long in nature because, when they collide with an electron, they annihilate each other and produce two or more photons. This pair-annihilation process is also an example of $E = mc^2$.

The question naturally arose as to the nature of these x rays. Since they could not be deflected by electric or magnetic fields, it was clear that x rays were not charged particles such as electrons. It was thought, therefore, that they might be a form of invisible light. However, it was nearly 20 years before this hypothesis was tested. From Einstein's photon theory, it appeared that at the high voltages used, if x rays were indeed light, their wavelengths would be on the order of 0.1 nm. To detect light waves of such short wavelength would be very difficult; for as we saw in Chapter 22, properties of light such as interference and diffraction are noticeable only when the size of diffracting objects or slits is on the order of the wavelength of the light. However, it was suggested in 1913 that the atoms in a solid crystal might be arranged in a regular array (as in Figure 7-1*a*); since this spacing between atoms in a solid is on the order of 0.1 nm, such crystals could serve as a sort of diffraction grating for x rays. When the experiment was done, it was found that x rays scattered from a crystal did indeed show the peaks and valleys of a diffraction pattern (Figure 25-8). Thus it was shown, in one blow, that x rays have a wave nature and that atoms are arranged in a regular way in crystals. Indeed, x rays today are recognized as electromagnetic radiation with wavelengths in the range from roughly 10^{-2} nm to 10 nm, the range readily produced in an x-ray tube.

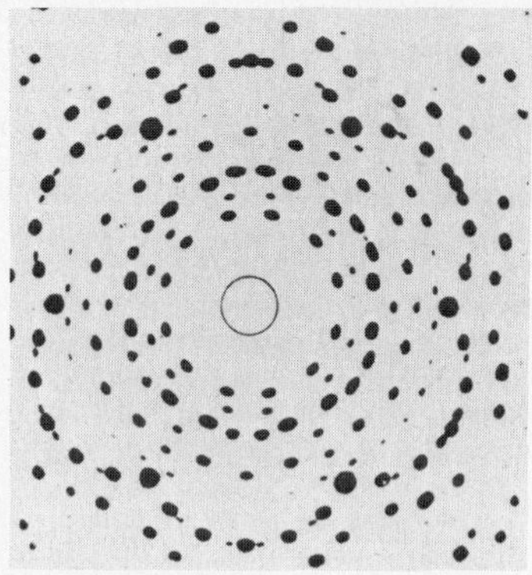
FIGURE 25-8
Diffraction pattern produced by x rays scattered from a beryl crystal. (Courtesy of Eastman Kodak.)

We are now in a position to explain how x rays are produced. First, we look at the spectrum of wavelengths emitted by an x-ray tube; a typical example is shown in Figure 25-9. The spectrum consists of two parts: a continuous spectrum with a cutoff at some λ_0 which depends only on the voltage across the tube; and peaks which are superimposed on top. The smooth curve and the cutoff wavelength λ_0 move to the left as the voltage across the tube increases. But the peaks remain at the same wavelength when the voltage is changed, although located at different places when different target materials are used. This suggests that the peaks are characteristic of the material used and, indeed, in Chapter 27 we will see they are due to electrons jumping from one energy level to another within the atom.

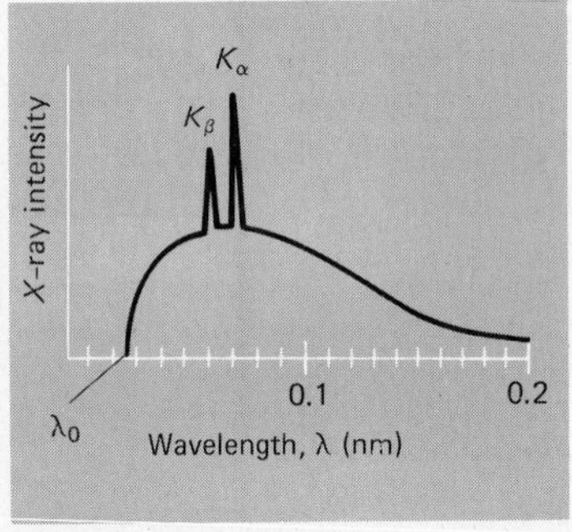

FIGURE 25-9
Spectrum of x rays emitted from a molybdenum target in an x-ray tube operated at 50 kV.

We now examine the continuous spectrum. When electrons strike the target, they collide with atoms of the material and give up most of their energy as heat (about 99 percent, so x-ray tubes must be cooled, usually with water). However, electrons can also give up energy by emitting a photon of light. As shown in Figure 25-10, an electron can be slowed down (decelerated) by the positive nucleus of an atom; and an accelerating charge can emit radiation (Chapter 20). This is called *bremsstrahlung* (German for "braking radiation"). Because energy is conserved, the energy of the emitted photon, *hf*, must equal the loss of kinetic energy of the electron, $\Delta KE = KE - KE'$, so

$$hf = \Delta KE.$$

An electron may lose all or a part of its energy in such a collision. The continuous x-ray spectrum (Figure 25-9) is due to such bremsstrahlung collisions in which varying amounts of energy are lost by the electrons. The shortest-wavelength x ray produced (the highest frequency) must be due to an electron that gives up all its kinetic energy to one photon in a single collision. Since the initial kinetic energy of the electrons is equal to the energy given it by the accelerating voltage, *V*, then $KE = eV$; in a

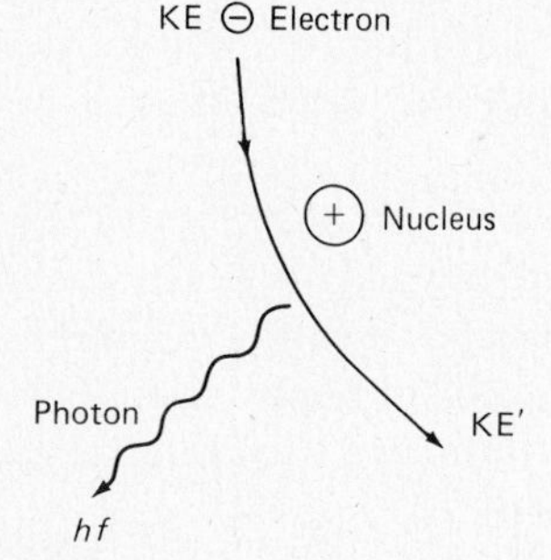

FIGURE 25-10
Bremsstrahlung photon produced by an electron that is decelerated by a positive nucleus.

collision in which the electron is brought to rest, we have

$$hf_0 = eV,$$

or

$$\lambda_0 = \frac{hc}{eV}, \qquad (25\text{-}5)$$

where $\lambda_0 = c/f_0$ is the cutoff wavelength (Figure 25-9). This prediction for λ_0 corresponds precisely with that observed experimentally. This is further evidence that x rays are a form of light† and that the photon theory of light is valid.

▌ EXAMPLE 25-5 What is the shortest-wavelength x-ray photon emitted in an x-ray tube subjected to 50 kV?

SOLUTION From Equation 25-5,

$$\lambda_0 = \frac{(6.6\times10^{-34}\ \mathrm{J\cdot s})(3.0\times10^{8}\ \mathrm{m/s})}{(1.6\times10^{-19}\ \mathrm{C})(5.0\times10^{4}\ \mathrm{V})} = 2.5\times10^{-11}\ \mathrm{m},$$

or 0.025 nm. This agrees well with experiment (Figure 25-9). ▌▌

X-ray diffraction

We saw in Chapter 23 that the shorter the wavelength of light used, the finer the resolution possible when magnifying a small object. X rays are much shorter than visible light; but, alas, no one has discovered an effective material to use as lenses for x rays. However, another technique has proved very effective in giving us information on the microscopic world of atoms and molecules, using the very short wavelengths of x rays. This is called *x-ray diffraction* or *crystallography.*

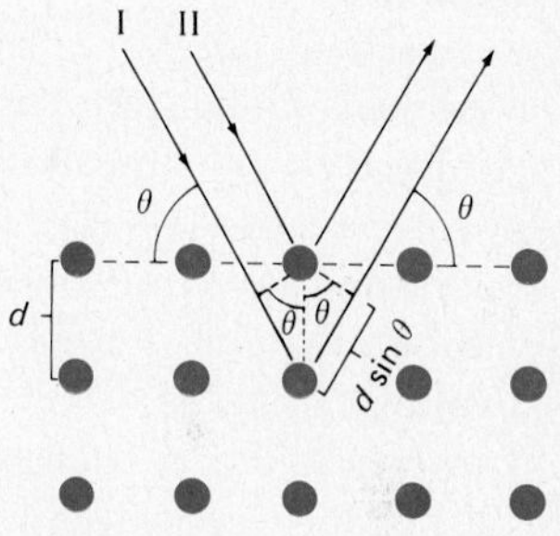

Let us consider a simple crystal, such as NaCl, in which the atoms are arranged in an orderly cubical fashion (Figure 25-11), with the atoms spaced a distance d apart. A beam of x rays is incident on the crystal at an angle θ to the surface, and we assume that the two rays shown are reflected from two subsequent planes of atoms as shown. The two beams, upon reflection from atoms in the two different planes of the crystal, will constructively interfere if the extra distance ray I travels is a whole number of wavelengths farther than the distance ray II travels. This extra distance is $2d \sin \theta$. Therefore, constructive interference will occur when

FIGURE 25-11
X-ray diffraction by a crystal.

$$m\lambda = 2d \sin \theta, \qquad m = 1, 2, 3, \ldots. \qquad (25\text{-}6)$$

where m can be any integer. This is called the *Bragg equation* after W. H. Bragg (1862–1942) and his son W. L. Bragg (1890–1971), who developed the theory of x-ray diffraction by crystals in 1913. (Notice that θ is *not* the angle with respect to the normal to the surface as has usually been the case in other situations.) Thus, if the x-ray wavelength is known and the angle θ at which constructive interference occurs is measured, d can be obtained. This is the basis for x-ray crystallography.

Actual x-ray diffraction patterns are quite complicated. First, a crystal is a three-dimensional object, and x rays can be diffracted from planes at various angles within the crystal, as shown in Figure 25-12. Although the

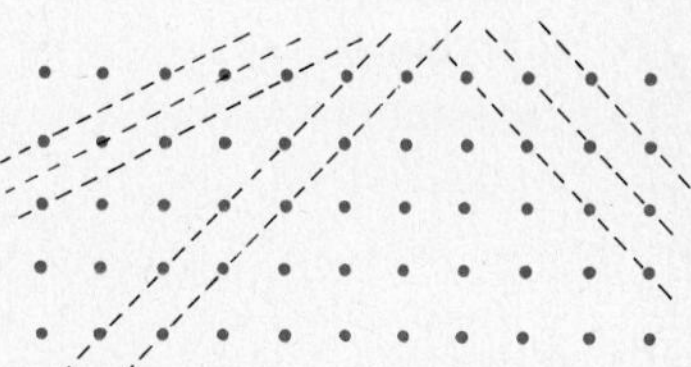

FIGURE 25-12
There are many possible planes existing within a crystal, from which x rays can be diffracted.

† If x rays were not photons but rather a neutral particle with rest mass m_0, Equation 25-5 would not hold.

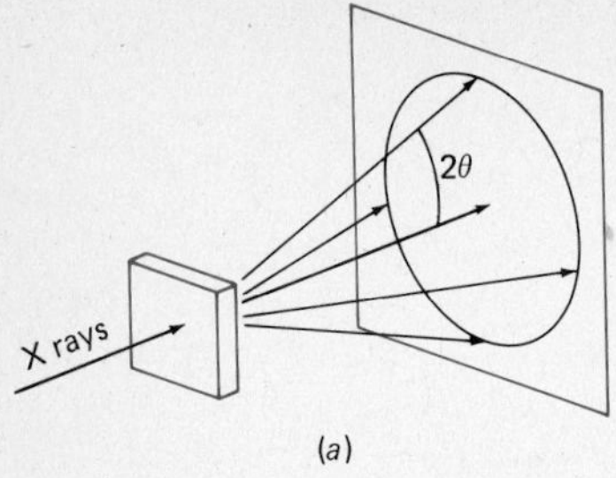

(a)

(b)

FIGURE 25-13
(a) Diffraction of x rays from a polycrystalline substance produces a set of circular rings as in (b) which is for polycrystalline sodium acetoacetate. (Part b Lester V. Bergman & Assoc., Inc.)

analysis is complex, a great deal can be learned about any substance that can be put in crystalline form. If the substance is not a single crystal but a mixture of many tiny crystals—as in a metal or a powder—then instead of a series of spots as in Figure 25-8, one obtains a series of circles (Figure 25-13); each circle corresponds to diffraction of a certain order m (Equation 25-6) from a particular set of parallel planes.

X-ray diffraction has been very useful in determining the structure of biologically important molecules. Often it is possible to make a crystal of such molecules. The analysis is complex, and it is usually necessary to make various guesses of the structure of the molecule. Predictions of the diffraction patterns for each guessed structure can then be compared to that actually obtained. For larger molecules, such as proteins and nucleic acids, an important innovation has been the "heavy atom technique." Since very large atoms scatter x rays more strongly than the ordinary C, N, O, and H atoms of biological molecules, heavy atoms can be used as "markers." The heavy atoms are chemically added to particular spots on the molecule (say, a protein)—hopefully without disturbing its structure significantly. Analysis of the *changes* in the resulting diffraction pattern gives helpful information.

Even when a good crystal cannot be obtained, if the molecule under study has a regularly repeating shape (such as many proteins and DNA have), x-ray diffraction can reveal it. In a sense, each molecule is then like a single crystal and a sample is a collection of such tiny crystals. Indeed, it was with the help of x-ray diffraction that the α structure of proteins (Section 13-4) was proposed. And in 1953 J. D. Watson and F. H. C. Crick worked out the double-helix structure of DNA with the aid of x-ray diffraction photographs.

Around 1960, the first detailed structure of a protein molecule was elucidated with the aid of x-ray diffraction; this was for myoglobin, a relative of the important constituent of blood, hemoglobin. Soon the structure of hemoglobin itself was worked out and since then the structure of a great many molecules have been determined with the help of x rays.

25-5 ▌ Wave–particle duality; the principle of complementarity

The photoelectric effect, the Compton effect, x rays, and other experiments have placed the particle theory of light on a firm experimental basis. But what about the classic experiments of Young and others

(Chapter 22), which clearly showed that light is a wave? These carefully performed experiments on interference and diffraction showed that the wave theory of light also rests on a firm experimental basis.

We seem to be in a dilemma. Some experiments indicate that light behaves like a wave; and other experiments indicate that it behaves like a stream of particles. These two theories seem to be incompatible but both have been shown to have validity. Physicists have finally come to the conclusion that this duality of light must be accepted as a fact of life. It is referred to as the **wave–particle duality**. Apparently, light is a more complex phenomenon than just a simple wave or a simple beam of particles.

To clarify the situation, the great Danish physicist Niels Bohr (1885–1962) proposed his famous **principle of complementarity**. It states that to understand any given experiment, we must use either the wave or the photon theory, but not both. Yet we must be aware of both the wave and particle aspects of light if we are to have a full understanding of light. Therefore, these two aspects of light complement one another—they are *complementary*.

It is not possible to "visualize" this duality. We cannot picture a combination of wave and particle. Instead, we must recognize that the two aspects of light are different "faces" that light shows to experimenters.

Part of the difficulty stems from the way we think. Visual pictures (or models) in our minds are based on what we see in the everyday world. We use the concepts of waves and particles because in the macroscopic world we see that energy is transferred from place to place by these two methods. We cannot see directly whether light is a wave or particle—so we do indirect experiments. And to explain the experiments we apply the concepts of waves or particles to the nature of light. But this is an abstraction of the human mind. When we try to conceive of what light really "is," we insist on a visual picture. Yet there is no reason why light should conform to these visual images from the macroscopic world. The "true" nature of light—if that means anything—is not possible to visualize. The best we can do is recognize that our knowledge is limited to the indirect experiments, and that in terms of everyday language and images, light reveals both wave and particle properties.

It is worth noting that Einstein's equation $E = hf$ itself links the particle and wave properties of a light beam. In this equation E refers to the energy of a particle; and on the other side of the equation we have the frequency f of the corresponding wave.

25-6 ▮ Wave nature of matter

In 1923, a Frenchman, Louis de Broglie (1892–), extended the idea of the wave–particle duality. He sensed deeply the symmetry in nature. Therefore, he argued, if light sometimes behaves like a wave and sometimes like a particle, then perhaps those things in nature thought to be particles–such as electrons and other material objects—might also have wave properties. DeBroglie proposed that the wavelength of a material particle would be related to its momentum in the same way as for a photon† (Equation 25-4), $p = h/\lambda$. That is, for a particle of mass m traveling with speed v, the wavelength λ is given by

$$\lambda = \frac{h}{mv}. \qquad (25\text{-}7)$$

This is sometimes called the *de Broglie wavelength* of a particle.

▮ EXAMPLE 25-6 Calculate the de Broglie wavelength of a 0.20-kg ball moving with a speed of 15 m/s.

SOLUTION $\lambda = h/mv = (6.6\times10^{-34}\ \text{J}\cdot\text{s})/(0.20\ \text{kg})(15\ \text{m/s})$

$= 2.2\times10^{-34}\ \text{m}.$ ▮▮

This is an incredibly small wavelength. Even if the speed were imperceptible, say 0.0001 m/s, the wavelength would be about 10^{-29} m. Indeed, the wavelength of any ordinary object is much too small to be measured and detected. The problem is that the properties of waves, such as interference and diffraction, are significant only when the size of objects or slits is not much larger than the wavelength. And there are no known objects or slits to diffract waves only 10^{-30} m long; so the wave properties of ordinary objects go undetected.

But tiny elementary particles such as electrons are another matter. Since the mass m appears in the denominator in Equation 25-7, a very small mass should give a much larger wavelength. Indeed, such is the case, as the following example shows.

▮ EXAMPLE 25-7 Determine the wavelength of an electron that has been accelerated through a potential difference of 100 V.

SOLUTION We assume that the speed of the electron will be much less than c, so we use nonrelativistic mechanics. (If this assumption comes out wrong—which it doesn't—we would have to use relativistic formulas.) The gain in KE will equal the loss in PE, so $\frac{1}{2}mv^2 = eV$ and $v = \sqrt{2eV/m} = \sqrt{(2)(1.6\times10^{-19}\ \text{C})(100\ \text{V})/(9.1\times10^{-31}\ \text{kg})} = 5.9\times10^6\ \text{m/s}$. Then

$$\lambda = \frac{h}{mv} = \frac{6.6\times10^{-34}\ \text{J}\cdot\text{s}}{(9.1\times10^{-31}\ \text{kg})(5.9\times10^6\ \text{m/s})} = 1.2\times10^{-10}\ \text{m},$$

or 0.12nm. ▮▮

From this example we see that electrons can have wavelengths on the order of 10^{-10}m. Although small, this wavelength can be dealt with. The spacing of atoms in a crystal is on the order of 10^{-10} m, and the orderly array of atoms in a crystal could be used as a sort of diffraction grating as had already been done for x rays (see Section 25-4). In 1927, C. J. Davisson and L. H. Germer performed the crucial experiment. They scattered electrons from the surface of a metal crystal and observed that the electrons came off in regular peaks (Figure 25-14). When they interpreted these peaks as a diffraction pattern, the wavelength of the diffracted electron wave was found to be just that predicted by de Broglie (Equation 25-7). Later experiments showed that protons, neutrons, and other particles also have wave properties.

FIGURE 25-14
Diffraction pattern of electrons scattered from aluminum foil, as recorded on film. (Courtesy of Education Development Center, Newton, Mass. 02160.)

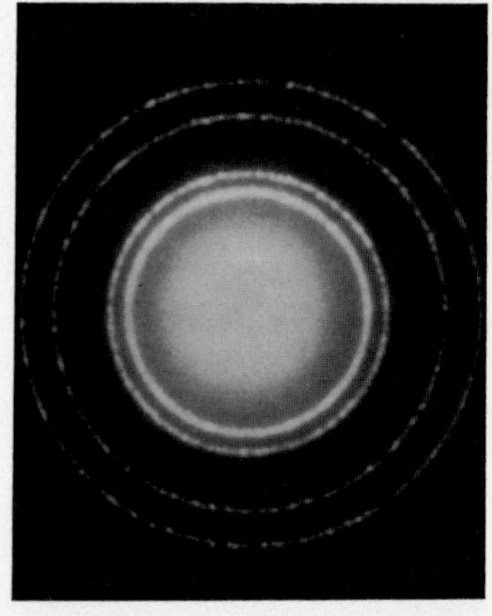

Thus it is clear that the wave–particle duality applies to material objects as well as to light. The principle of complementarity applies to matter as well. That is, we must be aware of both the particle and wave aspects in order to understand matter, including electrons. But again we must recognize that a visual picture of a "wave–particle" is not possible.

We might then ask ourselves: What is an electron? The early experiments of J. J. Thomson (see Section 17-4) indicated a glow in a tube that moved when a magnetic field was applied. These and other experiments were best interpreted as being caused by tiny negatively charged particles which we now call electrons. No one, however, has actually seen an electron directly. The drawings we sometimes make of electrons as tiny spheres with a negative charge on them are merely convenient pictures (now recognized to be inaccurate). Again, we must rely on experimental results, some of which are best interpreted using a particle picture and others using a wave picture. And again, these are mere pictures that we use to extrapolate from the macroscopic world to the tiny microscopic world of the atom. And there is no reason to expect that these models somehow reflect the reality of an electron. We thus use a wave or a particle model (whichever works best in a situation) so that we can talk about what is happening. But we should not be led to believe that an electron *is* a wave or a particle. Perhaps it is easiest to think of an electron as being something between a wave and a particle. But it may be more accurate to say that an electron is the set of its properties that we can measure. Bertrand Russell said it well when he remarked that an electron is "a logical construction."

*25-7 ▌ Electron microscopes

Shortly after the experimental confirmation of the wave properties of electrons, it was suggested that electrons could be used to magnify objects to much greater detail than with the ordinary light microscope. As we saw in Section 23-7 the resolution of a microscope is limited by the wavelength of the radiation used. For a light microscope the best that can be done is with an oil-immersion objective, and the maximum resolution is about half of the wavelength of light used, or 200 to 300 nm. This corresponds to a magnification of about 500×. A beam of electrons, on the other hand, can easily be obtained with a wavelength somewhat less than 1 nm (see Example 25-7). We therefore expect an electron microscope to increase our magnifying ability by a factor of 100 to 1000. The first electron microscopes were built in the 1930s.

*Transmission electron microscope

A transmission electron microscope is patterned after an ordinary light microscope except that glass lenses are replaced by "magnetic lenses" (Figure 25-15). Magnetic lenses are coils of current-carrying wire whose magnetic field exerts a force on the electrons. If carefully designed, the coils will bend the electron paths and bring them to a focus, just as a glass lens does for light. As shown in Figure 25-15, electrons emitted from a hot filament are accelerated by a high voltage of 50 to 100 kV or more. They

FIGURE 25-15

Electron microscope. The squares represent magnetic field coils for the "magnetic lenses."

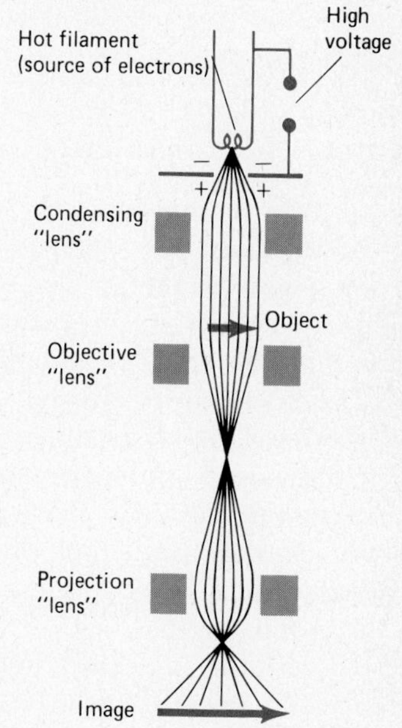

pass through a condensing "lens" that renders the beam parallel before it passes through the specimen to be viewed. The beam then passes through lenses corresponding to the objective and eyepiece of a light microscope. The latter is called the projection lens (rather than eyepiece) because the image cannot be viewed by the eye (our eyes are sensitive to light, not electrons) and the image is projected onto a fluorescent screen or onto film.

The whole electron path, and the specimen itself, must be kept under a high vacuum to avoid scattering of electrons off molecules of air and thus blurring the image. The specimen must be very thin (20 to 50 nm or so) to avoid slowing down the electrons; otherwise, the slower electrons would be bent differently by the magnetic field and focused at a different point and blur the image (the equivalent of chromatic aberration).

High voltages are used to produce very short de Broglie wavelengths. For example, 100-keV electrons have a wavelength of about 0.004 nm. You might therefore expect the resolution of an electron microscope to approach this value. This is not possible, however, for several reasons. The numerical aperture (see Equations 23-6 and 23-7), which for light microscopes is on the order of 1, can at best be about 0.01 for the electron microscope; this is because larger-aperture magnetic lenses distort the image greatly and thus reduce resolution drastically. Other aberrations are present, some of which are associated with the problem of maintaining a precisely constant current in the lens coils. (Variation in current means variation in focal length.) As a result, the best resolution obtained is in the vicinity of 0.2 to 0.5 nm. This is not to be scoffed at for it is 1000 times finer than that obtainable with a light microscope. This corresponds to a magnification of about 10^6 and is achieved only with the greatest difficulty. More typical magnifications are in the range 10,000 to 100,000×.

Finally, it should be mentioned that the powerful electron beams can be very damaging to the thin specimens. It is a constant worry of electron microscopists that the image they see is only of the charred remains of the specimen rather than of the specimen itself. Consequently, the shortest exposure times must be used.

*Scanning electron microscope

The ordinary transmission electron microscope produces a flat two-dimensional image. In recent years a good deal of work has gone into developing the scanning electron microscope, which is capable of producing remarkably three-dimensional images. Figure 25-16 shows examples of micrographs made with each type.

In the scanning electron microscope, a carefully directed electron beam is focused on the specimen (Figure 25-17). This tiny "flying spot" is made to move across the specimen by a "scanning" magnetic field. At each point, secondary electrons ejected from the material by collision with the incoming electron beam are collected as shown. Since the secondary electrons typically have energies of only a few electron volts, the collector held at about +200 V will attract nearly all of them. The image is displayed with the aid of a cathode ray tube (CRT) as in television. The scanning coils which move the beam across the specimen are connected to

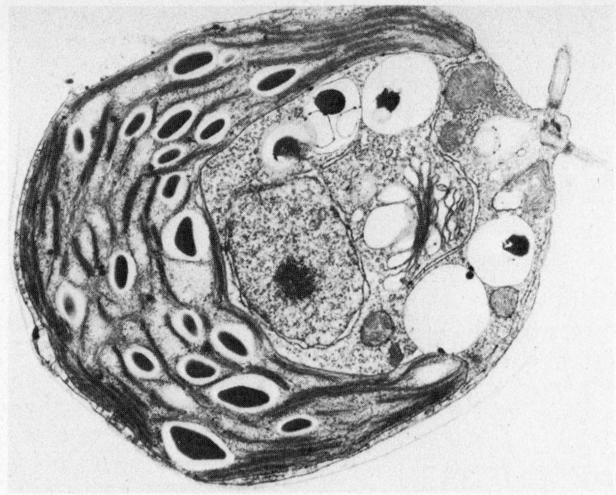

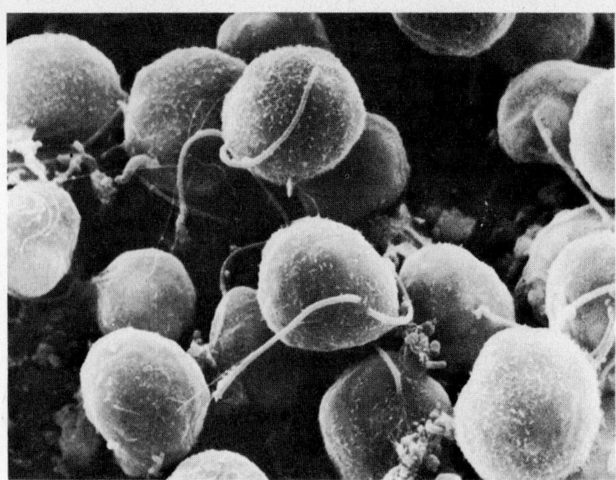

FIGURE 25-16
Micrographs of the green alga *Tetraspora* sp.: (*a*) made by the transmission electron microscope and (*b*) made by the scanning electron microscope. (Courtesy of J. D. Pickett-Heaps.)

the sweep of the CRT. Thus each point on the specimen corresponds to a point on the screen of the CRT. The electron collector is connected to the control grid of the CRT, so the number of secondary electrons arriving at the collector determines the brightness of the spot on the CRT screen. Contrast and the effect of depth results from the fact that the number of secondary electrons produced depends on the angle between the primary electron beam and the surface: many more secondary electrons are produced when the beam strikes the surface at a glancing angle than when it strikes the surface at right angles. Often the specimen is tilted at a 30° to 60° angle to improve picture quality. Two pictures taken at slightly different angles can be viewed in a stereoscopic viewer to produce a more truly three-dimensional image.

Besides collecting secondary electrons, it is also possible to detect primary electrons that have been "backscattered" with nearly their original energy. This produces a different image with additional information. A few x rays are also produced, and detection of these can reveal what elements are in the sample.

The resolution of a scanning electron microscope is less than for the transmission type, being at best on the order of 5 to 10 nm today. The magnification can be readily changed by simply limiting the extent of the scan. Useful magnifications vary from 15 to 50,000×. Since thin slices are not needed, the preparation of samples is much simpler for a scanning electron microscope, although the sample must still be placed in vacuum. Certain organisms, such as insects, can actually be viewed live if they are not kept too long in the chamber.

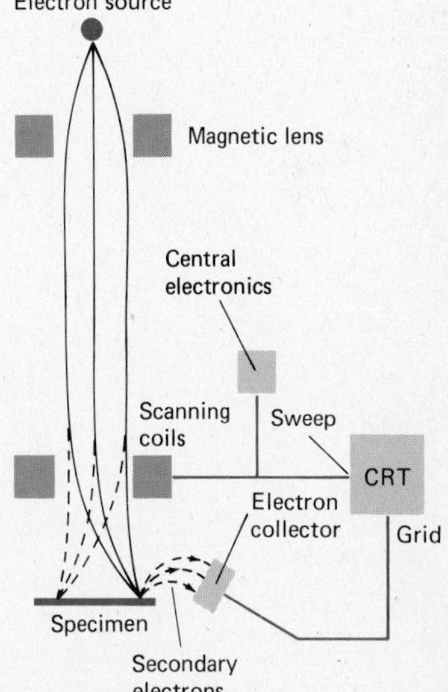

FIGURE 25-17
Scanning electron microscope. Scanning coils move an electron beam back and forth across the specimen. Secondary electrons produced when the beam strikes the specimen are collected and modulate intensity of the beam in the CRT to produce a picture.

*Scanning transmission electron microscope

A scanning electron microscope can be designed so that the *transmitted* beam of electrons is detected. Such a "scanning transmission electron microscope" combines advantages of the other two types. It is able to scan the specimen with a pinpoint beam of electrons and thus reduce scattering from other parts of the object that reduce contrast in the ordinary transmission electron microscope, yet can match the latter's resolution. Indeed, micrographs of uranium atoms have actually been obtained.

Summary

Quantum theory has its origins in *Planck's quantum hypothesis* that molecular oscillations are *quantized*: their energy E can only be integer (n) multiples of hf, where h is Planck's constant and f is the natural frequency of oscillation: $E = nhf$. This hypothesis explained the spectrum of radiation emitted by (black) bodies at high temperature.

Einstein proposed that light must be emitted as quanta (or particles) we now call *photons*, each with energy $E = hf$. He proposed the photoelectric effect as a test for the photon theory of light. In this effect, the photon theory says that each incident photon can strike an electron in a material and eject it if it has sufficient energy; the maximum energy of ejected electrons is then linearly related to the incident-light frequency. The photon theory is also supported by the Compton effect and the observation of electron–positron *pair production*.

X rays are high-energy photons. They are produced when high-speed electrons are decelerated upon striking a metal plate. Since x rays have short wavelengths, they can be used in the technique of x-ray diffraction to obtain information about crystal and molecule structure.

The *wave–particle duality* refers to the idea that light and matter (such as electrons) have both wave and particle properties. The wavelength of a material object is $\lambda = h/mv$, where mv is the momentum of the object. The *principle of complementarity* states that we must be aware of both the particle and wave properties of light and matter for a complete understanding of them.

Questions

1 What can be said about the relative temperature of whitish-yellow, reddish, and bluish stars?

2 If energy is radiated by all objects, why can't we see them in the dark? (See also Chapter 9.)

3 Does a light bulb at a temperature of 2500 K produce as white a light as the sun at 6000 K? Explain.

4 Why do jewellers often examine diamonds in daylight rather than with indoor light?

5 "Orthochromatic" film is not sensitive to red light. Give an explanation based on the photon theory of light.

6 If the threshold wavelength in the photoelectric effect increases when the emitting metal is changed, what can you say about the work functions of the two metals?

7 UV light causes sunburn, whereas visible light does not. Explain.

8 If an x-ray photon is scattered by an electron, does its wavelength change? If so, does it increase or decrease?

9 Why is it said that light has wave properties? Why is it said that light has particle properties?

10 Why do we say electrons have particle properties? Why do we say electrons have wave properties?

11 How do a photon and an electron differ? Be specific.

12 If an electron and a proton travel at the same speed, which has the shorter wavelength?

Problems

SECTION 25-1

1 (I) An HCl molecule vibrates with a natural frequency of 8.1×10^{13} Hz. What is the difference in energy (in joules and eV) between possible values of the oscillation energy?

2 (I) At what temperature will the peak of a blackbody spectrum be at 1.0 nm?

3 (I) What is the wavelength at the peak of a blackbody spectrum if the body is at a temperature of 800 K?

4 (II) A child's swing has a natural frequency of 0.40 Hz. What is the separation between possible energy values (in joules)? If the swing reaches a height of 30 cm above its lowest point and has a mass of 20 kg (including the child), what is the value of the quantum number n? Would quantization be measurable in this case?

SECTION 25-2

5 (I) What is the energy of a photon of wavelength (*a*) 400 nm, and (*b*) 700 nm?

6 (I) Calculate the wavelength of a 4.6-eV photon in nanometers.

7 (I) What is the energy of photons (in eV) emitted by a 100-MHz FM radio station?

8 (I) What minimum frequency of light is needed to eject electrons from a metal whose work function is 3.9×10^{-19} J?

9 (II) What is the longest wavelength of light that will emit electrons from a metal whose work function is 2.3 eV?

10 (II) Barium has a work function of 2.48 eV. What is the maximum kinetic energy of electrons if the metal is illuminated by light of wavelength 480 nm? What is their speed?

11 (II) What is the maximum kinetic energy of electrons ejected from barium ($W_0 = 2.48$ eV) when illuminated by white light, $\lambda = 400$ to 700 nm?

12 (II) Sunlight reaching the earth has an intensity of about 1400 W/m^2. How many photons cross a 1.0-cm^2 area perpendicular to the sun rays per second? Take the average wavelength to be 550 nm.

13 (II) When UV light of wavelength 280 nm falls on a metal surface, the maximum kinetic energy of emitted electrons is 0.85 eV. What is the work function of the metal?

14 (II) Can 550-nm light eject electrons from a surface whose work function is 2.42 eV? Why or why not?

15 (II) The threshold wavelength for emission of electrons from a given surface is 380 nm. What will be the maximum energy of ejected electrons when the wavelength is changed to (*a*) 480 nm, and (*b*) 280 nm?

16 (II) A certain type of film is sensitive only to light whose wavelength is less than 640 nm. What is the energy (kcal/mol) needed for the chemical reaction to occur which causes the film to change?

17 (III) If a 100-W light bulb emits 3 percent of the input energy as visible light (average wavelength 550 nm) uniformly in all directions, approximately how many photons per second will strike the pupil (4.0 mm diameter) of the eye of an observer, 10 km away?

18 (III) When 230-nm light falls on a metal, the current through a photoelectric circuit (Figure 25-3) is brought to zero at a reverse voltage of 1.64 V. What is the work function of the metal?

19 (III) In the Compton effect, a 0.100-nm photon strikes a free electron in a head-on collision and knocks it into the forward direction. The rebounding photon recoils directly backwards. Use conservation of energy and momentum to determine (*a*) the kinetic energy of the electron, and (*b*) the wavelength of the recoiling photon. Assume the electron's kinetic energy is given by the nonrelativistic formula.

SECTION 25-3

20 (I) How much total kinetic energy will an electron–positron pair have if produced by a 3.6-MeV photon?

21 (I) What is the momentum of a 450-nm photon of blue light?

22 (II) What is the momentum and the effective mass of a 0.10-nm x-ray photon?

23 (II) What is the minimum photon energy needed to produce a μ^+–μ^- pair? The mass of each μ is 207 times the mass of the electron. What is the wavelength of such a photon?

24 (II) An electron–positron pair, each with a kinetic energy of 350 keV, is produced from a photon. What was the energy and wavelength of the photon?

SECTION 25-4

25 (I) If the shortest-wavelength bremsstrahlung x rays emitted from an x-ray tube have $\lambda = 0.019$ nm, what is the voltage across the tube?

26 (I) What are the shortest-wavelength x rays emitted by electrons striking the face of a 30-kV television picture tube? What is the longest wavelength?

27 (II) An x-ray tube operates at 100 kV with a current of 20 mA, and nearly all the electron energy goes into heat. If the specific heat capacity of the 82-g plate is 0.11 kcal/kg · C°, what will be the temperature rise per minute if no cooling water is used?

28 (II) X rays of wavelength 0.13 nm fall on a crystal whose atoms are spaced 0.30 nm apart. At what angle is the first diffraction maximum observed?

SECTION 25-6

29 (I) Calculate the wavelength of a 0.10-kg ball traveling at 0.20 m/s.

30 (I) What is the wavelength of a neutron ($m_0 = 1.67 \times 10^{-27}$ kg) traveling 1.0×10^4 m/s?

31 (II) By what potential difference must a proton ($m_0 = 1.67 \times 10^{-27}$ kg) be accelerated to have a wavelength $\lambda = 0.0011$ nm?

32 (II) What is the wavelength of an electron of energy (*a*) 5.0 eV, and (*b*) 50 eV?

33 (II) Show that if an electron and a proton have the same kinetic energy, the proton has the shorter wavelength.

34 (II) Calculate the ratio of KE of an electron to that of a proton whose wavelengths are equal. Assume that the speeds are much less than the speed of light.

35 (II) What is the wavelength of an N_2 molecule in the air at room temperature? (*Hint*: See Chapter 7.)

36 (III) What is the wavelength of a proton whose KE is 1000 MeV?

37 (III) Show that for a particle of rest mass m_0, if $\lambda = h/mv$, it cannot be true that $E = hf$, where E is (*a*) kinetic energy, or (*b*) KE plus rest mass energy, and $v = f\lambda$ is the speed of the particle.

38 (III) What is the wavelength of an electron with energy (*a*) 50 keV, and (*b*) 1.0 MeV. Relativistic formulas must be used.

39 (III) What is the energy and wavelength of a "thermal" neutron (one that is in equilibrium at room temperature—see Chapter 7).

40 (III) Show that the wavelength of a particle of mass m_0 with kinetic energy KE is given by the relativistic formula $\lambda = hc/\sqrt{(\text{KE})^2 + 2m_0c^2(\text{KE})}$.

*SECTION 25-7

41 (II) What voltage is needed to produce electron wavelengths of 0.10 nm? (Assume that electrons are nonrelativistic.)

42 (II) What is the wavelength of electrons in a 40-kV electron microscope? (Assume that electrons are nonrelativistic.) What is the upper limit to the resolution achievable by such an EM?

43 (III) Electrons are accelerated by 1200 V in an electron microscope. To achieve a resolution of 4.0 nm, what numerical aperture is required?

44 (III) What is the theoretical limit of resolution for an electron microscope whose electrons are accelerated through 50 kV? (Relativistic formulas should be used.)

For additional reading

Cawley, J. M., and S. Iijima, "Electron Microscopy of Atoms in Crystals," *Physics Today*, **30**:3 (Mar., 1977), p. 32.

Cline, B., *Men Who Made a New Physics: Physicists and the Quantum Theory*, The New American Library, Inc., New York, 1969. Very readable account of the development of modern physics, particles, quantum theory.

Crewe, A. V., "A High-resolution Scanning Electron Microscope," *Sci. American*, **224**:4 (Apr., 1971), p. 26.

Everhart, T. E., and T. L. Hayes, "The Scanning Electron Microscope," *Sci. American*, **226**:1 (Jan., 1972), p. 54.

Gamow, G., *Thirty Years that Shook Physics*, Doubleday & Co., Inc., New York, 1966.

Goldhaber, A. S., and M. M. Nieto, "The Mass of the Photon," *Sci. American*, **234**:5 (May, 1976), p. 86.

Hawking, S. W., "The Quantum Mechanics of Black Holes," *Sci. American*, **236**:1 (Jan., 1977), p. 34.

Hoffman, B., *Strange Story of the Quantum*, Dover Publications, Inc., New York, 1959.

Kendrew, J. C., *The Thread of Life: An Introduction to Molecular Biology*, Harvard University Press, Cambridge, Mass., 1966. X-ray diffraction in molecular biology.

Lehninger, A. L., *Bioenergetics: The Molecular Basis of Biological Energy Transformations*, 2nd ed., W. A. Benjamin, Inc., Menlo Park, Calif., 1971. Photons in biology—photosynthesis.

Magie, W. F., *Source Book in Physics*, Harvard University Press, Cambridge, Mass., 1963. Photoelectric effect.

Medicus, H. A., "Fifty Years of Matter Waves," *Physics Today*, **27**:2 (Feb., 1974), p. 38. Discovery of matter waves.

Setlow, R. B., and E. C. Pollard, *Molecular Biophysics*, Addison-Wesley Publishing Co., Inc., Reading, Mass., 1962. Applications of x rays.

Stern, E. A., "Analysis of Materials by X-ray Absorption," *Sci. American*, **234**:4 (Apr., 1976), p. 96.

26 ▮ Models of the atom; quantum mechanics

The idea that matter is made up of atoms was accepted by most scientists by 1900. With the discovery of the electron in the 1890s, it became clear that the atom itself must have a structure and that electrons were part of that structure. However, as we shall see in this chapter, the structure of the atom could not be understood until the quantum theory and the wave–particle duality were applied to it. Indeed, the search for an understanding of atomic structure led to the further development of the quantum theory and the introduction of the far-reaching and powerful theory of *quantum mechanics.* We now trace the development of our modern understanding of the atom, and of the quantum theory with which it is intertwined.†

Positively charged material

FIGURE 26-1
Plum-pudding model of the atom.

26-1 ▮ Early models of the atom

Soon after his discovery of the electron, J. J. Thomson (1856–1940) proposed a model of the atom in which the negatively charged electrons were imagined to be embedded in a sphere of uniform positive charge (Figure 26-1), much like plums in a plum pudding.

Around 1911, Ernest Rutherford (1871–1937) and his colleagues performed experiments which indicated that the "plum-pudding model" of the atom could not be correct. In these experiments a beam of positively charged "alpha (α) particles" was directed at a thin sheet of metal foil, such as gold (Figure 26-2*a*). (These newly discovered α particles were emitted by certain radioactive materials and were soon shown to be ionized helium atoms—see Chapter 28). It was expected from the plum-pudding model that the alphas would not be deflected significantly to one side since they never approached any massive concentration of positive charge to strongly repel them, nor would collision with an electron cause significant deflection since the mass of the alphas was many thousands of times that of an electron.

The experimental results completely contradicted these predictions.

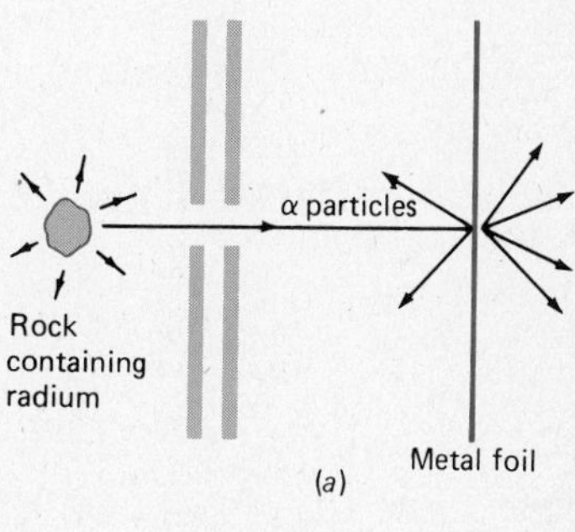

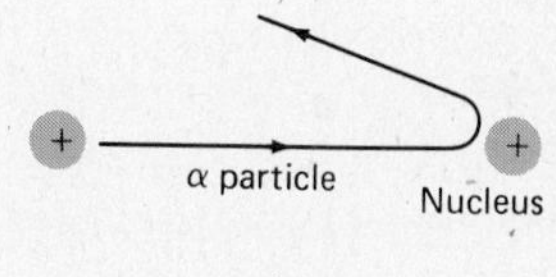

FIGURE 26-2
(*a*) Experimental setup for Rutherford's experiment: α particles emitted by radium strike metallic foil and some rebound backward; (*b*) backward rebound of α particles explained as repulsion from heavy positively charged nucleus.

† Some readers may prefer that we tell them the facts as we know them today, and not bother with the historical background and its "outmoded" theories. Alas, such an approach would not only ignore the creative aspect of science and thus give a false impression of how science develops, but it would not really be possible to understand today's view of the atom and the quantum without discussing the concepts that led to it.

It was found that most of the α particles passed through the foil unaffected, as if the foil were mostly empty space. Of those deflected, it was found that a few of them were deflected at very large angles; some even rebounded back in nearly the direction from which they had come. This could happen, Rutherford reasoned, only if the positively charged α particles were being repelled by a massive positive charge concentrated in a very small region of space (Figure 26-2*b*). He concluded that the atom must consist of a tiny but massive positively charged nucleus, containing over 99.9 percent of the mass of the atom, surrounded by electrons some distance away. The electrons must be moving in orbits about the nucleus—much like the planets moving around the sun—since if they were at rest, they would fall into the nucleus because of electrical attraction (Figure 26-3). From his experiments Rutherford concluded that the nucleus must have a radius of about 10^{-15} to 10^{-14} m. From kinetic theory, and especially Einstein's analysis of Brownian movement (see Section 7-1), the radius of atoms was estimated to be about 10^{-10} m. Thus the electrons must orbit the nucleus at a distance 10,000 to 100,000 times the radius of the nucleus itself, so an atom would be mostly empty space.

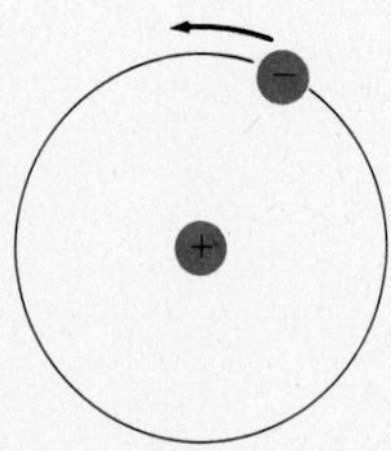

FIGURE 26-3
Rutherford's model of the atom, in which electrons orbit a tiny positive nucleus (not to scale). The atom is visualized as mostly empty space.

Rutherford's "planetary" model of the atom is roughly how we view the atom today. It was, however, not a complete model and presented some problems, as we shall now see.

26-2 ▮ Atomic spectra: key to the structure of the atom

At the beginning of Chapter 25 we saw that heated solids emit light with a continuous spectrum of wavelengths. (The same is true of liquids and dense gases.) This radiation is assumed to be due to oscillations of atoms and molecules, which are largely governed by the interaction of each atom or molecule with its neighbors.

It is also possible to excite rarefied gases to emit light. This is done by intense heating, or more commonly by applying a high voltage to a "discharge tube" containing the gas at low pressure (Figure 26-4). The radiation from excited gases had been observed as far back as the early nineteenth century, and it was found that the spectrum was not continuous, but *discrete*. That is, excited gases emit light of only certain wavelengths, and when this light is analyzed in a spectroscope or spectrometer, a *line spectrum* is seen. The line spectrum emitted by a number of elements in the visible region is shown in Figure 22-22 (color plate). We already discussed this briefly in Chapter 22 and we saw that the spectrum of each type of gas is different. Thus the *emission spectrum* is characteristic of the material and can serve as a type of "fingerprint" for identification purposes. We also saw that if a continuous spectrum passes through a gas, dark lines are observed in the spectrum, corresponding to lines normally emitted by the gas. This is called an *absorption spectrum*. It became clear that gases absorb light at the same frequencies at which they emit it. With the use of film sensitive to ultraviolet and to infrared light, it was found that gases emit and absorb discrete frequencies in these regions as well as in the visible.

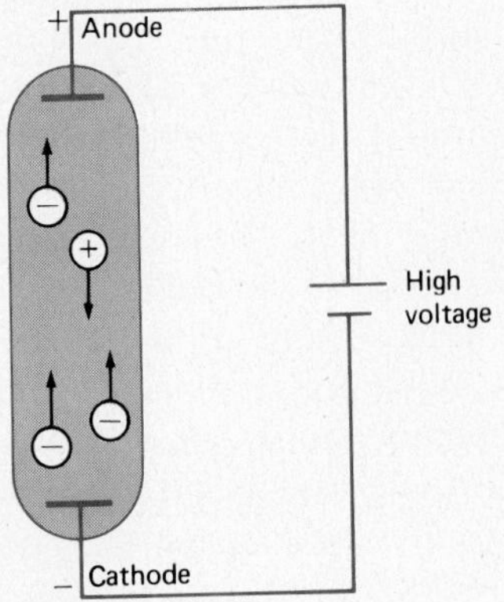

FIGURE 26-4
Gas discharge tube.

For our purposes here, the importance of the line spectra is that they are emitted (or absorbed) by gases that are not dense. In such gases the atoms are far apart on the average, and hence the light emitted or absorbed must be by individual atoms rather than through interactions between atoms as in a solid. Thus the line spectra serve as a key to the structure of the atom: any theory of atomic structure must be able to explain why atoms only emit light of discrete wavelengths, and it should be able to predict what these frequencies are.

Hydrogen is the simplest atom—it has only one electron orbiting its nucleus. And it also happens to have the simplest spectrum. The spectrum of most atoms shows little regularity; but for hydrogen, as seen in Figure 22-22, the spacing between lines decreases in a regular way. Indeed, in 1885, J. J. Balmer (1825–1898) showed that the four visible lines in the hydrogen spectrum (measured to be 656, 486, 434, and 410 nm) would fit a simple formula, which we write (slightly modified) as:

$$\frac{1}{\lambda}=R\left(\frac{1}{2^2}-\frac{1}{n^2}\right), \qquad n=3, 4, \ldots, \qquad (26\text{-}1)$$

where n takes on the values 3, 4, 5, 6 for the four lines and R (called the *Rydberg constant*) has the value $R=1.097\times10^7\ \mathrm{m}^{-1}$. Later it was found that this *Balmer series* of lines extended into the UV region, ending at $\lambda=365$ nm, as shown in Figure 26-5. Balmer's formula, Equation 26-1, worked for these lines as well with higher integer values of n. The lines near 365 nm became too close together to distinguish, but the limit of the series at 365 nm corresponds to $n=\infty$, so $1/n^2=0$ in the formula.

Later experiments on hydrogen showed that there were other series of lines in the UV and IR, and these additional series had a pattern just like the Balmer series, but at different wavelengths. Each of these series was found to fit a formula resembling Equation 26-1 but with the $1/2^2$ replaced by $1/1^2$, $1/3^2$, $1/4^2$, and so on. For example, the *Lyman series* contains wavelength 91 to 122 nm and fits the formula

$$\frac{1}{\lambda}=R\left(\frac{1}{1^2}-\frac{1}{n^2}\right), \qquad n=2, 3, \ldots.$$

The *Paschen series* fits

$$\frac{1}{\lambda}=R\left(\frac{1}{3^2}-\frac{1}{n^2}\right), \qquad n=4, 5, \ldots.$$

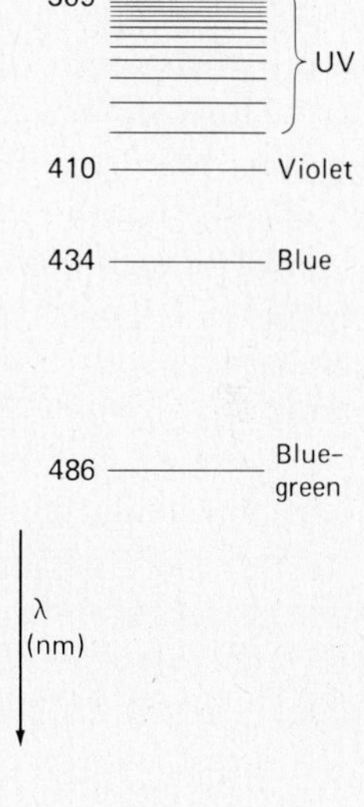

FIGURE 26-5
Balmer series of lines for hydrogen.

The Rutherford model, as it stood, was unable to explain why atoms emit line spectra. It had other difficulties as well. According to the Rutherford model, electrons orbit the nucleus; and since their paths are curved, the electrons are accelerating. As we saw in Chapter 20, accelerating electric charges give rise to electromagnetic waves. Thus, in the Rutherford model, electrons will give off light; and because energy is conserved, the electron's own energy must decrease to compensate. Hence electrons would be expected to spiral into the nucleus. As they spiraled inward, their frequency would increase gradually and so too would the frequency of the light emitted. Thus two main difficulties of the Rutherford model are these: (1) it predicts that light of a continuous range of frequencies will be emitted, whereas experiment shows line spectra; and (2) it predicts that atoms are unstable—electrons quickly spiral into the nucleus—but we know that atoms in general are stable, since the matter around us is stable.

Clearly, Rutherford's model was not sufficient. Some sort of modification was needed, and it was Niels Bohr who provided it by adding an essential idea—the quantum hypothesis.

26-3 ▮ The Bohr model

Bohr had studied in Rutherford's laboratory for several months in 1912 and was convinced that Rutherford's planetary model of the atom had validity. But to make it work he felt that the newly developing quantum theory would somehow have to be incorporated in it. The work of Planck and Einstein had shown that in heated solids, the energy of oscillating electric charges must change discontinuously—from one discrete energy state to another with the emission of a quantum of light. Perhaps, Bohr argued, the electrons in an atom also cannot lose energy continuously but must do so in quantum "jumps." In working out his theory during the next year, Bohr postulated that electrons move about the nucleus in circular orbits, but that only certain orbits are allowed. He further postulated that an electron in each orbit has a definite energy and moves in the orbit *without radiating energy* (even though this violated classical ideas); he thus called the possible orbits **stationary states**. Light is emitted, he hypothesized, only when an electron jumps from one orbit or stationary state to another of lower energy. When such a jump occurs, a single photon of light is emitted. Since energy must be conserved, the photon energy is given by

$$hf = E_u - E_l, \tag{26-2}$$

where E_u is the energy of the upper state and E_l the energy of the lower state.

His next problem was to determine what energies these orbits have, since then the spectrum of light emitted can be predicted from Equation 26-2. When he became aware of the Balmer formula in early 1913, he had the key he was looking for. Bohr quickly found that his theory would be in accord with the Balmer formula if he assumed the electron's angular momentum L was equal to an integer n times $h/2\pi$. We saw in Chapter 5 that the angular momentum of a particle of mass m moving in a circle of radius r with speed v is $L = mvr$. Bohr's "quantum condition" then is

$$L = mvr_n = n\frac{h}{2\pi}, \qquad n = 1, 2, 3, \ldots. \tag{26-3}$$

Here n is an integer and r_n is the radius of the nth possible orbit. The allowed orbits are numbered $1, 2, 3, \ldots$ according to the value of n, which is called the **quantum number** of the orbit.

Equation 26-3 was an ad hoc postulate. Bohr had searched for some "quantum condition," and such tries as $E = hf$ (where E represents the energy of the electron in an orbit) did not give results in accord with experiment. Bohr's reason for using Equation 26-3 was simply that it worked, as we shall now see.

An electron in a circular orbit of radius r_n has a centripetal acceleration v^2/r_n; this is produced by the electrical force of attraction between the negative electron and the positive nucleus according to Coulomb's law, $F = kq_1q_2/r^2$ (Figure 26-6). The charge on the electron is $q_1 = e$ and that on the nucleus is $q_2 = Ze$, where Z is the number of positive charges (or protons). For the hydrogen atom, $Z = +1$. [We include Z in our derivation so that we can treat other single-electron ("hydrogenlike") atoms such as the ions He^+ $(Z = 2)$ and Li^{2+} $(Z = 3)$. Helium in the neutral state has two electrons; when one is missing, the remaining He^+ ion consists of one electron revolving around a nucleus of charge $+2e$; similarly, doubly ionized lithium, Li^{2+}, also has a single electron and in this case $Z = 3$.] From Newton's second law ($F = ma$) we substitute $a = v^2/r_n$ and Coulomb's law for F and obtain

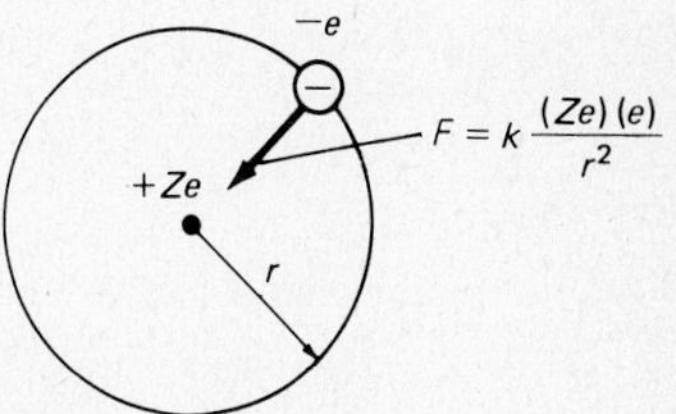

FIGURE 26-6
Electric force (Coulomb's law) keeps negative electron in orbit around a positively charged nucleus.

$$\frac{kZe^2}{r_n^2} = \frac{mv^2}{r_n}.$$

We solve this for r_n and substitute for v from Equation 26-3 (which says $v = nh/2\pi mr_n$):

$$r_n = \frac{kZe^2}{mv^2} = \frac{kZe^2 4\pi^2 mr_n^2}{n^2h^2}.$$

We solve for r_n (it appears on both sides, so we cancel one of them) and find

$$r_n = \frac{n^2h^2}{4\pi^2 mkZe^2}. \tag{26-4}$$

This equation gives the radii of the possible orbits. The smallest orbit is for $n = 1$ and for hydrogen ($Z = 1$) has the value

$$r_1 = \frac{(1)^2(6.6\times10^{-34}\ \text{J}\cdot\text{s})^2}{4(3.14)^2(9.1\times10^{-31}\ \text{kg})(9.0\times10^{9}\ \text{N}\cdot\text{m}^2/\text{C}^2)(1)(1.6\times10^{-19}\ \text{C})^2}$$

$$r_1 = 0.53\times10^{-10}\ \text{m}. \tag{26-5}$$

The radius of the smallest orbit in hydrogen, r_1, is sometimes called the *Bohr radius.* From Equation 26-4, we can see that the radii of the larger orbits increase as n^2, so

$$r_2 = 4r_1 = 2.1\times10^{-10}\ \text{m}$$

$$r_3 = 9r_1 = 4.8\times10^{-10}\ \text{m},$$

and so on. The first four are shown in Figure 26-7. Notice that, according to Bohr's model, an electron can exist only in the orbits given by Equation 26-4; it cannot be in between.

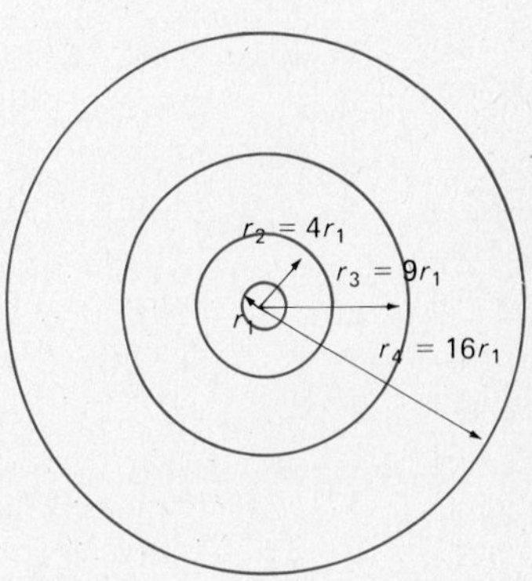

FIGURE 26-7
Possible orbits in the Bohr model of hydrogen; $r_1 = 0.53\times10^{-10}$ m.

In each of these orbits, the electron has a definite energy, as the following calculation shows. The total energy is the sum of the kinetic and potential energies. The potential energy (PE) of the electron, from Equation 14-1, is $\text{PE} = qV = -eV$, where V is the potential due to a point charge $+Ze$ as given by Equation 14-7: $V = kQ/r = kZe/r$. So $\text{PE} = -eV = -kZe^2/r$. The total energy E_n for an electron in the nth orbit of radius r_n is then

$$E_n = \text{KE} + \text{PE} = \tfrac{1}{2}mv^2 - \frac{kZe^2}{r_n}.$$

When we substitute v from Equation 26-3 and r_n from Equation 26-4 into this equation, we obtain

$$E_n = -\frac{2\pi^2 Z^2 e^4 mk^2}{h^2}\frac{1}{n^2}, \qquad n = 1, 2, 3, \ldots. \tag{26-6}$$

For hydrogen ($Z = 1$), the lowest energy level has $n = 1$, and when numbers are substituted into Equation 26-6, we find

$$E_1 = -2.17 \times 10^{-18}\ \text{J} = -13.6\ \text{eV},$$

where we have converted joules to electron volts, as is customary in atomic physics. Since n^2 appears in the denominator of Equation 26-6, the energies of the larger orbits are given by

$$E_n = \frac{-13.6\ \text{eV}}{n^2}$$

Thus

$$E_2 = \frac{-13.6\ \text{eV}}{4} = -3.4\ \text{eV},$$

$$E_3 = \frac{-13.6\ \text{eV}}{9} = -1.5\ \text{eV},$$

and so on. Thus we see that not only are the orbit radii quantized, but so is the energy. Notice that although the energy for the larger orbits has a smaller numerical value, all the energies are less than zero. Thus -3.4 eV is a greater energy than -13.6 eV. Hence the orbit (r_1) closest to the nucleus has the lowest energy. The reason the energies have negative values has to do with the way we defined the potential energy,† for, remember, we can choose the zero of PE wherever we wish. For two point charges, $\text{PE} = kq_1q_2/r$ corresponds to zero PE when the two are infinitely far apart. Thus an electron without KE that is free from the atom, which corresponds to $n = \infty$, has $E = 0$. Electrons that are part of an atom have $E < 0$. Also, we know that it requires energy to remove an electron from an atom—this is called the *binding energy* or *ionization energy*; thus, if $E = 0$ for a free electron, an electron bound to an atom must have $E < 0$. The ionization energy for hydrogen has been measured to be 13.6 eV, and this corresponds precisely to removing an electron in the lowest state, $E_1 = -13.6$ eV, up to $E = 0$, where it will be free.

It is useful to show the various possible energy values as horizontal lines on an energy-level diagram. This is shown for hydrogen in Figure 26-8. The quantum number n that labels the orbit radii also labels the energy levels. In fact, we often use the terms "orbit," "energy level," and "energy state" interchangeably. The lowest state, which is $n = 1$, has energy E_1 and is called the **ground state**. The higher states, E_2, E_3, and so on, are called **excited states**. The electron in a hydrogen atom can be in any one of these levels. But it can never be in between, say at -9.0 eV. At room temperature the electrons of nearly all H atoms will be in the ground state. At higher temperatures, or during an electric discharge when there are many collisions between free electrons and atoms, many

† This is not unlike the case of the gravitational potential energy of a stone, $\text{PE} = mgh$; if we measure h from the top of a table, say, then when the stone is on the floor below the table, it will have negative potential energy.

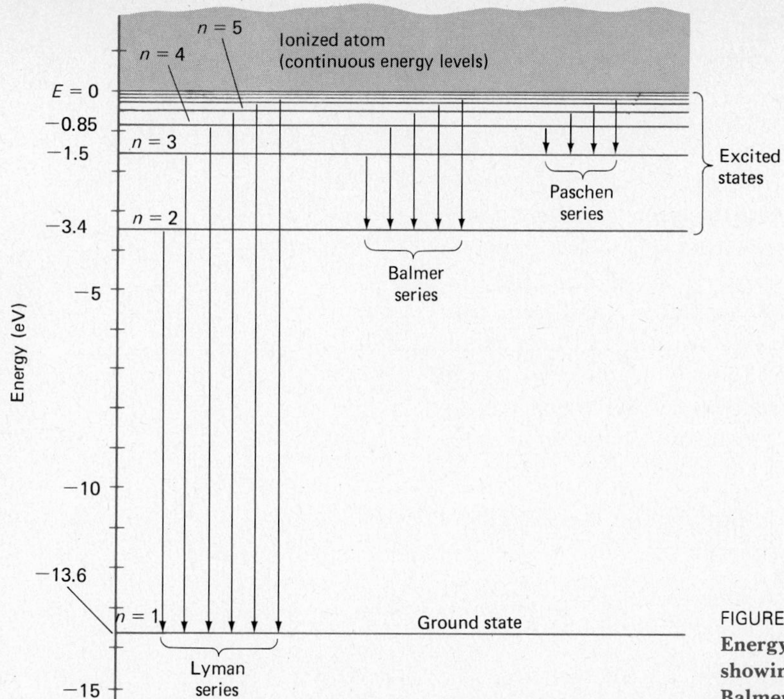

FIGURE 26-8
Energy-level diagram for the hydrogen atom, showing origin of spectral lines for the Lyman, Balmer, and Paschen series.

atoms can be in excited states. Once in an excited state, an electron can jump down to a lower state and give off a photon in the process. This is the origin of the line spectra emitted by excited gases. The vertical arrows in Figure 26-8 represent the transitions or jumps that correspond to the various observed spectral lines. For example, an electron jumping from the level $n=3$ to $n=2$ gives rise to the 656-nm line in the Balmer series, and the jump from $n=4$ to $n=2$ gives rise to the 486-nm line (see Figure 26-5). We can predict wavelengths of the spectral lines emitted by combining Equation 26-2 with Equation 26-6. Since $hf=hc/\lambda$, we have

$$\frac{1}{\lambda}=\frac{hf}{hc}=\frac{1}{hc}(E_n-E_{n'})$$

or

$$\frac{1}{\lambda}=\frac{2\pi^2Z^2e^4mk^2}{h^3c}\left(\frac{1}{n'^2}-\frac{1}{n^2}\right), \tag{26-7}$$

where n refers to the upper state and n' to the lower state. This theoretical formula has the same form as the experimental Balmer formula, Equation 26-1, with $n'=2$. Thus we see that the Balmer series of lines corresponds to transitions or "jumps" that bring the electron down to the second energy level. Similarly, $n'=1$ corresponds to the Lyman series and $n'=3$ to the Paschen series (see Figure 26-8). When the constant $(2\pi^2Z^2e^4mk^2/h^3c)$ in Equation 26-7 is evaluated with $Z=1$, it is found to have the measured value of the Rydberg constant (Equation 26-1), $R=1.0974\times10^7\ \text{m}^{-1}$, in accord with experiment.

The great success of Bohr's theory is that it explains why atoms emit line spectra and accurately predicts, for hydrogen, the wavelengths of emitted light. The Bohr theory also explains absorption spectra: photons of just the right wavelength can knock an electron from one energy level to a higher one. The photon must have just the right energy or such a process could not occur, because energy must be conserved. Thus we see why a continuous spectrum passing through a gas will have dark (absorption) lines (Figure 22-22) at the same frequencies as the emission lines.

The Bohr theory also explains the stability of atoms. The ground state is the lowest state an electron can be in. There is no lower energy level to which it can go and emit more energy. Hence the atom is stable. Finally, as we saw above, the Bohr theory correctly predicts how much energy is required to remove an electron from a hydrogen atom. This, the ionization energy, is measured for hydrogen to be 13.6 eV and is fully in accord with predictions of the Bohr theory for an upward jump of an electron from the ground state ($E_1 = -13.6$ eV) to where it is free of the atom ($E = 0$), Figure 26-8. (Note that an electron can have any energy above $E = 0$, for here it is free; thus there is a continuum of energy states above $E = 0$, as indicated in Figure 26-8.)

▌ EXAMPLE 26-1 Use Figure 26-8 to determine the wavelength of the first Lyman line, the transition from $n = 2$ to $n = 1$.

SOLUTION In this case, $hf = E_2 - E_1 = 13.6\text{ eV} - 3.4\text{ eV} = 10.2\text{ eV} = 1.63 \times 10^{-18}$ J. Since $\lambda = c/f$, we have

$$\lambda = \frac{c}{f} = \frac{hc}{E_2 - E_1} = \frac{(6.63 \times 10^{-34}\text{ J}\cdot\text{s})(3.00 \times 10^{8}\text{ m/s})}{1.63 \times 10^{-18}\text{ J}} = 1.22 \times 10^{-7}\text{ m},$$

or 122 nm, which is in the UV. ▌▌

▌ EXAMPLE 26-2 Determine the wavelength of light emitted when a hydrogen atom makes a transition from the $n = 6$ to the $n = 2$ energy level.

SOLUTION We can use Equation 26-7 or its equivalent, Equation 26-1, with $R = 1.097 \times 10^{7}\text{ m}^{-1}$. Thus

$$\frac{1}{\lambda} = (1.097 \times 10^{7}\text{ m}^{-1})(\tfrac{1}{4} - \tfrac{1}{36}) = 2.44 \times 10^{6}\text{ m}^{-1},$$

so $\lambda = 4.10 \times 10^{-7}$ m or 410 nm. This is the fourth line in the Balmer series (Figure 26-5) and is violet in color. ▌▌

▌ EXAMPLE 26-3 Determine the ionization energy of the ion He^{+}, which has a single electron. Also calculate the minimum wavelength a photon must have to cause ionization.

SOLUTION We want to determine the energy required to lift the electron from its ground state to the free state, $E = 0$. The ground-state energy of He^{+} is given by Equation 26-6 with $Z = 2$ and $n = 1$. Since all the symbols in Equation 26-6 are the same as for the calculation for hydrogen, except Z is 2 instead of 1, we see that E_1 will be $Z^2 = 2^2 = 4$ times larger than for hydrogen. That is,

$$E_1 = 4(-13.6\text{ eV}) = -54.4\text{ eV}.$$

Thus to ionize the He^{+} ion requires 54.4 eV. The minimum wavelength photon that can cause ionization will have energy $hf = 54.4$ eV and have wavelength $\lambda = c/f = hc/hf = (6.63 \times 10^{-34}\text{ J}\cdot\text{s})(3.00 \times 10^{8}\text{ m/s})/(54.4\text{ eV})(1.60 \times 10^{-19}\text{ J/eV}) = 22.8$ nm. If the atom absorbs a photon of greater energy (wavelength shorter than 22.8 nm), the atom will still be ionized and the freed electron will have kinetic energy of its own. ▌▌

In Example 26-3 we saw that E_1 for the He^+ ion is four times lower than that for hydrogen. Indeed, the energy-level diagram for He^+ will look just like that for hydrogen (Figure 26-8) except that the numerical values for each energy level will be four times larger. It is important to note, however, that we are talking here about the He^+ *ion.* Normal (neutral) helium has two electrons and its spectrum is entirely different.

Bohr's theory worked very well for hydrogen and for one-electron ions. It did not, however, prove as successful for multielectron atoms. We will discuss this problem shortly.

We should note that Bohr made some radical assumptions that were at variance with classical ideas. He assumed that electrons in fixed orbits do not radiate light even though they are accelerating, and he assumed that the angular momentum is quantized. Furthermore, he was not able to say how an electron moved when it made a transition from one energy level to another. On the other hand, there is no real reason to expect that in the tiny world of the atom electrons would behave as ordinary-sized objects do. Nonetheless, he felt that, where quantum theory overlaps with the macroscopic world, it should predict classical results. This he called the *principle of correspondence.* This principle does seem to work for his theory of the hydrogen atom. The orbit sizes and energies are quite different for $n = 1$ and $n = 2$, say. But orbits number 100,000,000 and 100,000,001 would be very close in size and energy (see Figure 26-8). Indeed, jumps between such large orbits, which would approach everyday sizes, would be imperceptible; such orbits would thus appear to be continuous, which is what we expect in the everyday world.

Bohr's theory was largely of an ad hoc nature. Assumptions were made so that theory would agree with experiment. But Bohr could give no reason why the orbits were quantized. Ten years later, a reason was found by de Broglie.

26-4 ▌ De Broglie's hypothesis

We saw in Chapter 25 that in 1923, Louis de Broglie proposed that material particles, such as electrons, have a wave nature. And this hypothesis was confirmed by experiment several years later.

One of de Broglie's original arguments in favor of the wave nature of electrons was that it provided an explanation for Bohr's theory of the hydrogen atom.

As discussed earlier, de Broglie hypothesized that a particle of mass m moving with speed v would have a wavelength

$$\lambda = \frac{h}{mv}.$$

Each electron orbit in an atom, he proposed, is actually a standing wave. We saw in Chapter 11 that when a violin or guitar string is plucked, a vast number of wavelengths are excited. But only certain ones—those that have nodes at the ends—are sustained. These are the *resonant* modes of the string. All other wavelengths interfere with themselves upon reflection and their amplitudes quickly drop to zero. Since electrons move in circles, according to Bohr's theory, de Broglie argued that the electron wave must be a *circular* standing wave that closes on itself (Figure 26-9). If the wavelength is such that it does not close on itself, as in Figure 26-10, destructive interference takes place as the wave travels around the loop and it quickly dies out. Thus the only waves that persist are those for which the circumference of the circular orbit contains a whole number of wavelengths (Figure 26-11). The circumference of an orbit of radius r_n is $2\pi r_n$, so we must have

$$2\pi r_n = n\lambda, \qquad n = 1, 2, 3, \ldots.$$

When we substitute $\lambda = h/mv$, we get

$$2\pi r_n = \frac{nh}{mv},$$

or

$$mvr_n = \frac{nh}{2\pi}.$$

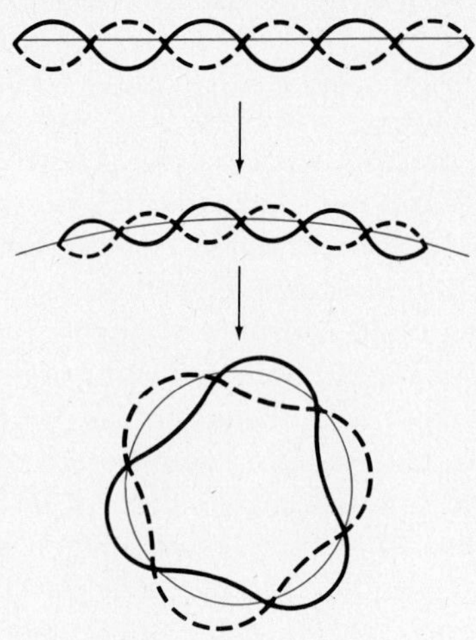

FIGURE 26-9
An ordinary standing wave compared to a circular standing wave.

This is just the quantum condition Bohr proposed on an ad hoc basis (Equation 26-3)—and it is from this equation that the discrete orbits and energy levels were derived. Thus we have the explanation for the quantized orbits and energy states in atoms: they are due to the wave nature of the electron and the fact that only resonant "standing" waves can persist. Thus we see that the wave–particle duality is at the root of atomic structure.

It should be noted in viewing the circular electron waves of Figure 26-11 that the electron does *not* follow the oscillating wave pattern. The electron considered as a particle moves in a simple circle. The circular wave represents the *amplitude* of the electron "matter wave"; and in Figure 26-11 the wave amplitude is shown superimposed on the circular path of the particle orbit for convenience.

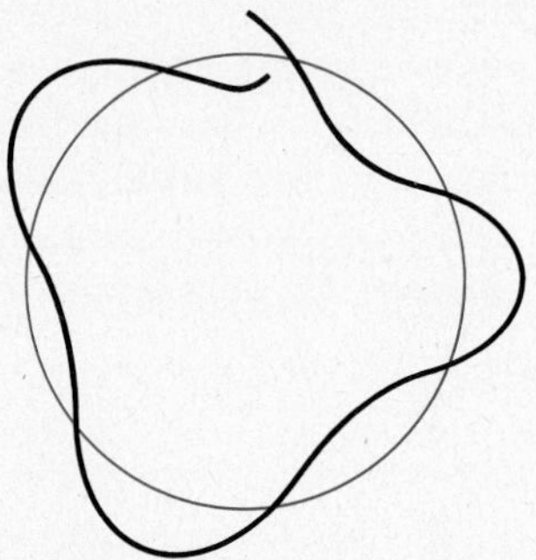

FIGURE 26-10
When a wave does not close (and hence interferes with itself), it rapidly dies out.

26-5 ▮ Quantum mechanics

A new theory

Bohr's model of the atom gave us a general picture of what an atom is like. It explained the emission and absorption of light by atoms at discrete wavelengths, as well as the stability of atoms. The wavelengths of the line spectra and ionization energy for hydrogen and other one-electron atoms (or ions) are calculated to be in excellent agreement with experiment. But the Bohr theory had limitations as well. It was not able to predict the line spectra of more complex atoms—not even for the neutral helium atom, which has only two electrons. Nor could it explain why the emission lines, when viewed with great precision, actually consist of two or more very closely spaced lines (called "fine structure"). The Bohr theory also did not explain why some spectral lines were brighter than others. And it couldn't explain the bonding of atoms in molecules or in solids and liquids.

From a theoretical point of view, too, the Bohr theory was not really satisfactory. For it was a strange mixture of classical and quantum ideas. Also, the wave–particle duality was still not really resolved.

We mention these limitations of the Bohr theory not to disparage it—for it was a landmark in the history of science and is still useful for visualizing certain aspects of atoms. Rather, they are mentioned to show why, in the early 1920s, it became increasingly evident that a new, more comprehensive theory was needed. It was not long in coming. Less than two years after de Broglie gave us his matter wave hypothesis, Erwin Schrödinger (1887–1961) and Werner Heisenberg (1901–1976) independently developed a new comprehensive theory.

The new theory, called **quantum mechanics,** unifies the wave–particle duality into a single consistent theory. Quantum mechanics as a theory has been fantastically successful. It has successfully dealt with the spectra emitted by complex atoms, even the fine details. It explains the relative brightness of spectral lines and how atoms form molecules. It is also a much more general theory that covers all quantum phenomena from blackbody radiation to atoms and molecules. It has explained an incredibly wide range of natural phenomena and from its predictions many new practical devices have become possible. Indeed, it has been so successful that it is accepted today by nearly all physicists as the fundamental theory underlying physical processes.

Quantum mechanics deals mainly with the microscopic world of atoms and light; but in our everyday macroscopic world, we do perceive light and we accept that ordinary objects are made up of atoms. This new theory must therefore also account for the verified results of classical physics; that is, when it is applied to macroscopic phenomena, quantum mechanics must be able to produce the old classical laws. This is the *principle of correspondence* mentioned earlier. This requirement is met fully by quantum mechanics. This does not mean that we throw away classical theories such as Newton's laws. In the everyday world they are far easier to apply and give an accurate description. But when we deal with high speeds, close to c, we must use the theory of relativity; and when we deal with the tiny world of the atom, we use quantum mechanics.

Although we will not go into the detailed mathematics of quantum mechanics, we will discuss the ideas involved and how they are applied to explain atomic structure and other applications.

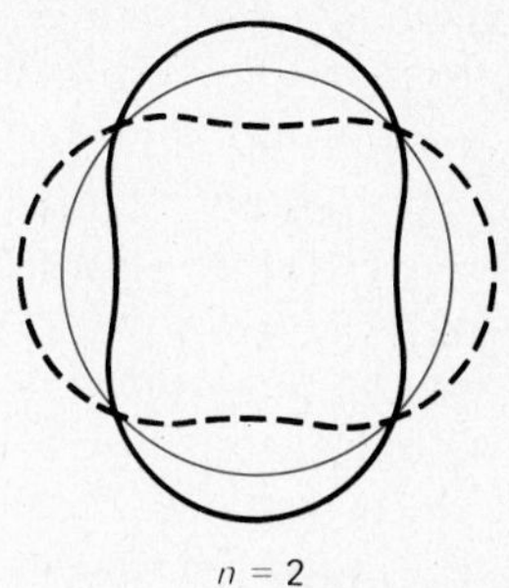

$n = 2$

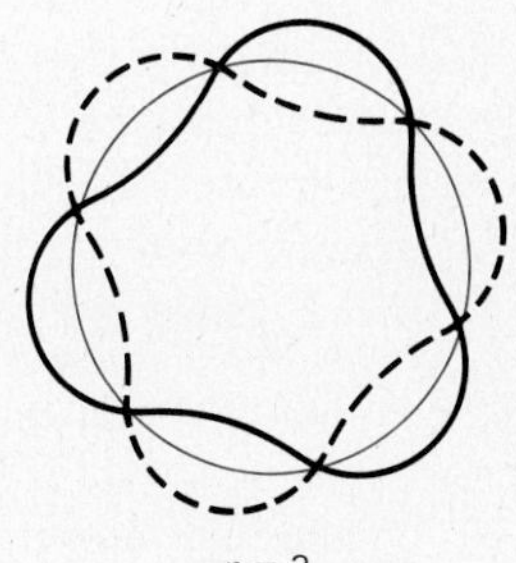

$n = 3$

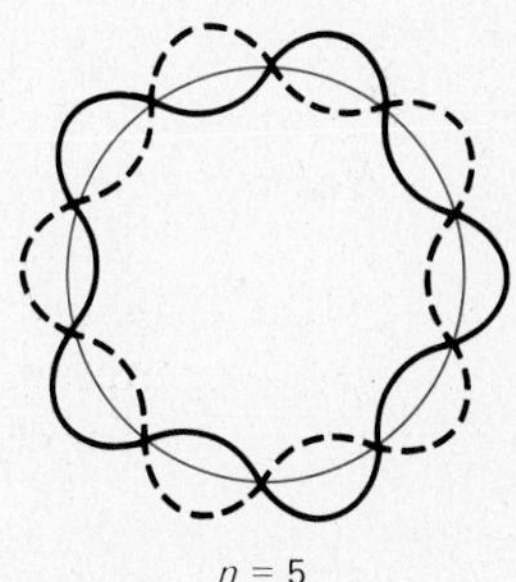

$n = 5$

FIGURE 26-11
Standing circular waves for two, three, and five wavelengths on the circumference; n, the number of wavelengths, is also the quantum number.

* The wave function and its interpretation

The important properties of any wave are its wavelength, frequency, and amplitude. For an electromagnetic wave, the wavelength determines whether the light is visible or not, and if so, what color it is. We also have seen that the wavelength (or frequency) is a measure of the energy of the corresponding photon ($E = hf$). The amplitude of an electromagnetic wave at any point is the strength of the electric (or magnetic) field at that point, and is related to the intensity of the wave (the brightness of the light).

For material particles such as electrons, quantum mechanics relates the wavelength to momentum according to de Broglie's formula, $\lambda = h/mv$. But what about the amplitude of a matter wave? In quantum mechanics the amplitude of, say, an electron wave is called the *wave function* and is given the symbol ψ (the Greek lowercase letter psi). Thus ψ represents the amplitude of a new kind of field, which we might call a "matter" field or a matter wave.

To calculate the wave function ψ in a given situation (say, for an electron in an atom) is one of the basic tasks of quantum mechanics. Indeed, the development of an equation to do so was Schrödinger's great contribution. The *Schrödinger wave equation,* as it is called, is a differential equation (so we will not deal with it here) and is considered to be a basic equation for the description of material particles.

The actual form of the Schrödinger equation, and how it is solved, need not concern us here. What will be useful are the solutions obtained and the meaning of the wave function, ψ, itself. One way to interpret ψ is simply as the amplitude at any point in space of a matter wave, just as E (the electric field vector) represents the amplitude of an EM wave. Another interpretation is possible, however, based on the wave–particle duality. To understand this, we make an analogy with light.

We saw in Chapter 11 (Equation 11-16) that the intensity I of any wave is proportional to the square of the amplitude. This holds true of light waves as well, as we saw in Chapter 20; that is,

$$I \propto E^2,$$

where E is the electric field strength. From the *particle* point of view, the intensity of a light beam is proportional to the number of photons, N, that pass through a given area per unit time. The more photons there are, the greater the intensity. Thus

$$I \propto E^2 \propto N.$$

This proportion can be turned around, so we have

$$N \propto E^2.$$

That is, the number of photons (striking a page of this book, say) is proportional to the square of the electric field strength.

If the light beam is very weak, only a few photons will be involved. Indeed, it is possible to "build up" a photograph in a camera using very weak light so that the effect of individual photons can be seen. If we are dealing with only one photon, the relationship above ($N \propto E^2$) can be interpreted in a slightly different way. At any point the square of the electric field strength, E^2, is a measure of the *probability* that a photon will be at that location. Where E^2 is large, there is a high probability the photon will be there, and vice versa.

We interpret matter waves in the same way. The wave function ψ may vary in magnitude from point to point in space and time. If ψ describes a collection of many electrons, then the value of ψ^2 at any point will be proportional to the number of electrons expected to be found at that point. When dealing with small numbers of electrons we cannot make very exact predictions, so ψ^2 takes on the character of a probability. If ψ, which depends on time and position, represents a single electron (say in an atom), ψ^2 is interpreted as follows: ψ^2 *at a certain point in space and time represents the probability of finding the electron at the given place and time.*

To understand this better, let us examine the familiar double-slit experiment, and consider it both for light and for electrons.

* Double-slit experiment

Consider two slits whose size and separation are on the order of the wavelength of whatever we direct at them, either light or electrons (Figure 26-12). We know very well what happens in this case for light, since this is just Young's famous double-slit experiment. An interference pattern will be seen on the wall behind. If light is replaced by electrons with wavelength comparable to the slit size, they, too, will produce an interference pattern. [This is basically a simplification of the Davisson–Germer experiment (Figure 25-14).] In the case of light, the pattern will be visible to the eye or can be recorded on film. For electrons, a fluorescent screen can be used (it glows where an electron strikes).

Now, if we reduce the flow of electrons (or photons) so that only one can pass through the slits at a time, we will see a glow each time an electron strikes the screen. At first, the flashes will seem random. Indeed, there is no way to predict just where any one electron will hit the screen. If we let the experiment run for a long time, however, and keep track of where each electron hits the screen, we will soon see a pattern emerging—the interference pattern predicted by the wave theory. Thus, although we cannot predict where a given electron (or photon) will strike the screen, we can predict probabilities. The probability, as mentioned before, is proportional to ψ^2. Thus, where ψ is zero, we get a minimum in the interference pattern; and where ψ is a maximum, we get a peak in the interference pattern.

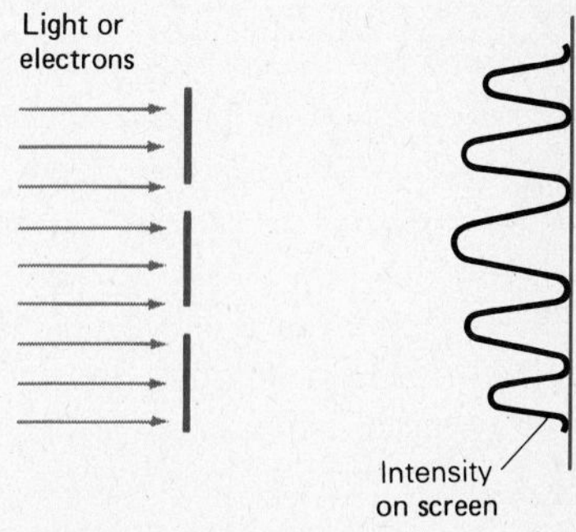

FIGURE 26-12
Parallel beam of light or electrons falls on two slits whose sizes are comparable to the wavelength. An interference pattern is observed.

Since the interference pattern occurs even when electrons (or photons) pass through the slit one at a time, it is clear that the interference pattern does not arise from the interaction of one electron with another. It is as if the electron passed through both slits at the same time! This is possible, for remember that an electron is not precisely a particle. It is as much a wave as it is a particle, and a wave can certainly travel through both slits at once. But what happens if we cover one of the slits so we know that the electron passed through the other one, and after a time we cover the second slit so that the electron has to have passed through the first? The result is that no interference pattern is seen. We merely see two bright spots on the screen behind each slit. This confirms our idea that when both slits are open, each electron passes through both, as if it were a wave. But it makes a tiny spot on the screen as if it were a particle.

The main point of this discussion is this. If we treat electrons (and other matter) as if they were particles, we must treat them on a *probabilistic* basis. The wave function ψ represents a probability, such that ψ^2 is the probability of finding a given electron at a given point. We cannot predict—or even follow—the path of a single electron precisely through space and time.

26-6 ▮ Quantum mechanical view of atoms

Quantum mechanics applied to atoms is far more complete than the old Bohr theory. It retains certain aspects of the older theory, such as that electrons in an atom exist only in discrete states of definite energy; and that a photon of light is emitted (or absorbed) when an electron makes a transition from one state to another. But quantum mechanics is not simply an extension of the Bohr theory. It is a far deeper theory, and has provided us with a very different view of the atom. According to quantum mechanics, electrons do not exist in the well-defined circular orbits of the Bohr theory. Rather, the electron is spread out in space, as a "cloud" of negative charge. The shape and size of the electron cloud can be calculated for a given state of an atom. The electron cloud for the ground state of hydrogen is spherically symmetric as shown in Figure 26-13. Clearly, an "orbit" is diffuse. The electron cloud roughly indicates the "size" of an atom; but just as a cloud does not have a distinct border, atoms do not have a precise boundary or a well-defined size. Not all electron clouds have a spherical shape, as we shall see in the next chapter.

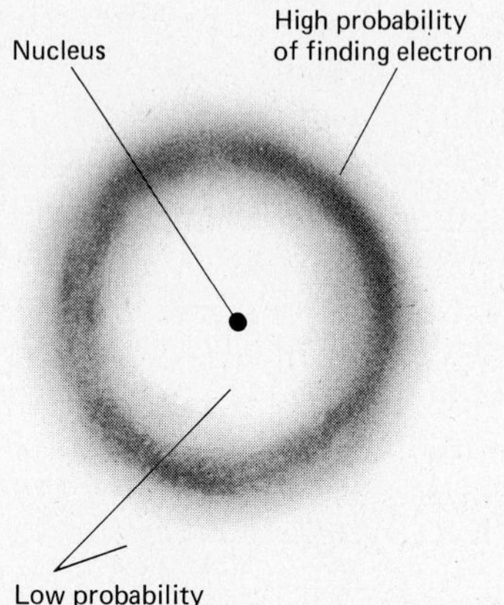

Electron cloud or "probability distribution" for the ground state of the hydrogen atom. The cloud is densest—indicating the highest probability—at a distance from the nucleus of 0.53×10^{-10} m, which is just what the Bohr model predicts for the radius of the first orbit. But unlike the Bohr model, quantum mechanics tells us the electron can be within or beyond that distance at any given time.

The electron cloud can be interpreted using either the particle or the wave viewpoint. Remember that by a particle we mean something that is *localized* in space—it has a definite position at any moment. But a wave is spread out in space. Thus the electron cloud, spread out in space as in Figure 26-13, is a result of the wave nature of electrons. Electron clouds can also be interpreted as *probability distributions* for a particle. If you were to make 500 different measurements of the position of an electron (thinking of it as a particle) the majority of the results would show the electron at points where the probability is high (dark area in Figure 26-13); only occasionally would the electron be found where the probability is low. We cannot predict the path an electron will follow. After one measurement of its position we cannot predict exactly where it will be at a later time. We can only calculate the probability that it will be found at different points. This is clearly different from classical Newtonian physics. Indeed, as Bohr later pointed out, since an electron is not simply a particle, it is meaningless to even ask how it gets from one state to another when the atom emits a photon of light.

The fact that we cannot predict the exact position of an electron in an atom, but can only give the probability distribution, is essentially a result of its wave properties. This fact has deep philosophic consequences, and we discuss it more fully in the next section.

In Chapter 27 we shall look in more detail at what quantum mechanics tells us about atomic structure.

26-7 ▮ The Heisenberg uncertainty principle and its implications

There is always some uncertainty or error when a measurement is made. For example, you cannot make an absolutely exact measurement of the length of a table. Even with a measuring stick that has markings 1 mm apart, there will be an inaccuracy of about $\frac{1}{2}$ mm or so. By using more precise instruments, more precise measurements can be made; but there is always some uncertainty involved in a measurement no matter how good the measuring device. However, we expect that, by using more precise instruments, the uncertainty in a measurement can be made indefinitely small.

But according to quantum mechanics, there is actually a limit to the accuracy of certain measurements. This limit is not a restriction on how well instruments can be made but, rather, it is inherent in nature. It is the result of two factors: the wave–particle duality, and the unavoidable interaction between the thing observed and the observing instrument. Let us look at this in more detail.

To make a measurement on an object without somehow disturbing it, at least a little, is not possible. Consider trying to locate a Ping-Pong ball in a completely dark room. You grope about trying to find its position; and just when you touch it with your finger, it bounces away! Whenever we measure the position of an object, whether it is a Ping-Pong ball or an electron, we always touch it with something else which gives us the information about its position. To locate a lost Ping-Pong ball in a dark room you could probe about with a stick or your hand; or you could shine a light and detect the light reflecting off the ball. When you search with a stick or your hand, you find the ball's position when you touch it. But when you touch the ball, you unavoidably bump it and give it some momentum; thus you will not know its *future* position. The same would be true, but to a much lesser extent, if you observe the Ping-Pong ball using light: in order to "see" the ball, at least one photon must bounce off it, and the reflected photon must enter your eye or some other detector. When a photon strikes an ordinary-sized object, it does not alter the motion or position of it very much. But when a photon strikes an electron, it can transfer much of its momentum to the electron and thus greatly change its motion and position in an unpredictable way. (It's much like your hand searching for the ping-pong ball in the dark.) The mere act of measuring the position of an object at one time makes our knowledge of its future position inaccurate.

Now let us see where the wave–particle duality comes in. Imagine a thought experiment in which we are trying to measure the position of an electron (or other tiny object) with photons, although the arguments would be similar if we were using an electron microscope. As we saw in Chapter 23, objects can be seen to an accuracy no greater than the wavelength of the radiation used. We must therefore use a short wavelength if we want an accurate position measurement. But a short wavelength corresponds to a high frequency and high energy (since $E = hf$), and the more energy the photons have, the more momentum they can give our electron when they collide. If photons of longer wavelength and corresponding lower energy are used, the electron's motion will not be affected as much, but its position will be less accurately known. Thus we see that the act of observing produces a significant uncertainty in either the *position* or the *momentum* of the electron. This is the essence of the *uncertainty principle* first enunciated by Heisenberg in 1927.

Quantitatively, we can make an approximate calculation of the magnitude of this effect. If we use light of wavelength λ, the position can be measured at best to an accuracy λ. That is, the uncertainty in the position measurement Δx, is

$$\Delta x \approx \lambda.$$

Suppose that the object can be detected by a single photon (this is the best case; normally many photons will be needed). The photon has a momentum $p = mv = h/\lambda$; and when it strikes our object it will give some or all of this momentum to the object. Therefore, the final momentum of our object will be uncertain in the amount

$$\Delta p \approx \frac{h}{\lambda}$$

since we cannot tell beforehand how much will be transferred. The product of these uncertainties is

$$(\Delta x)(\Delta p) \approx h.$$

Of course, the uncertainties could be worse than this, depending on the apparatus. In Heisenberg's more careful calculation, he found that at the very best,

$$(\Delta x)(\Delta p) \geq \frac{h}{2\pi}. \tag{26-8}$$

This is a mathematical statement of Heisenberg's **uncertainty principle**. It tells us that we cannot measure both the position and the momentum of an object precisely at the same time. The more accurately we try to measure the position, so that Δx is small, the greater will be the uncertainty in momentum, Δp. If we try to measure the momentum very precisely, the uncertainty in the position becomes large. The uncertainty principle does not forbid single exact measurements, however. For example, we could measure, in principle, the position of an object exactly. But then its momentum would be completely unknown. Thus, although we could know the position of the object exactly at one instant, we could have no idea where it would be a moment later.

Another useful form of the uncertainty principle is obtained as follows. The object to be detected has an uncertainty in position $\Delta x \approx \lambda$. Now the photon (or electron) used to detect it travels with speed v ($=c$ for a photon); and it takes a time $\Delta x/v \approx \lambda/v$ to pass through the distance of uncertainty. Hence, the precise time when our object is at a given position is uncertain by an amount

$$\Delta t \approx \frac{\lambda}{v}.$$

Since the photon can transfer some or all of its energy ($=hf = hv/\lambda$) to our object, the uncertainty in energy of our object as a result is

$$\Delta E \approx \frac{hv}{\lambda}.$$

The product of these two uncertainties is

$$(\Delta E)(\Delta t) \approx h.$$

Heisenberg's more careful calculation gives

$$(\Delta E)(\Delta t) \geq \frac{h}{2\pi}. \tag{26-9}$$

This form of the uncertainty principle tells us that the energy of an object can be uncertain, or may even be nonconserved, by an amount ΔE for a time $\Delta t \approx h/2\pi\Delta E$.

We have been speaking, in this discussion, of the position and velocity of an electron (or other object) as if it were a particle. But it is not a particle. Indeed, we have the uncertainty principle because an electron—and matter in general—is not purely particulate. What the uncertainty principle really tells us is that if we insist on thinking of the electron as a particle, then there are certain limitations on this simplified view—namely that the position and velocity cannot both be known precisely at the same time, and that the energy can be uncertain (or nonconserved), in the amount ΔE for a time $\Delta t \approx h/(2\pi\ \Delta E)$.

The uncertainties expressed in the uncertainty principle are usually negligible on the macroscopic level because Planck's constant h is so small. But at the level of the atom, the uncertainties are significant. Because ordinary-sized objects are made up of atoms containing nuclei and electrons, the uncertainty principle is relevant to our understanding of all of nature. The uncertainty principle expresses, perhaps most clearly, the probabilistic nature of quantum mechanics; it thus is often used as the basis for philosophic discussion.

Probability versus determinism

The classical Newtonian view of the world is a mechanistic one. One of its basic ideas is that once the position and velocity of an object are known at a particular time, its future position can be predicted if the forces on it are known. This implies that the future unfolding of the universe, assumed to be made up of particulate bodies, is completely determined.

But in the twentieth century, we find that the basic entities, such as electrons, cannot even be considered to be particles, since they have wave properties as well. And we therefore cannot know both the position and velocity of an object accurately. Indeed, the classical deterministic view of the physical world has been radically altered by quantum mechanics. Quantum mechanics only allows us to calculate the probability that the electron (when thought of as a particle) will be observed at various different places. Only approximate predictions are possible. There seems to be an inherent unpredictability in nature.

It is important to note that there is a difference in the probability imposed by quantum mechanics and that used in the nineteenth century to understand thermodynamics and the behavior of gases in terms of molecules (Chapters 7 and 10). In thermodynamics, probability is used because there are far too many molecules to be kept track of. But the molecules were still assumed to move and interact in a deterministic way according to Newton's laws. Probability in quantum mechanics is quite different, for it is seen as *inherent* in nature. It is not a limitation on our abilities to calculate, but is actually a part of nature itself.

Although a few physicists have not given up the deterministic view of nature and have refused to accept quantum mechanics as a complete theory—one was Einstein, who insisted that "God does not play dice"—nonetheless, the vast majority of physicists do accept quantum mechanics and the probabilistic view of nature. This view, which as presented here is the generally accepted one, is called the "Copenhagen interpretation" of quantum mechanics in honor of Niels Bohr's home, since it was largely developed there through discussions between Bohr and other prominent physicists.

Let us now investigate what quantum mechanics says about causality. If you drop a rock, it generally falls to the ground. But have you ever seen a violation of this phenomenon—for example, a stone suddenly jump upward? Probably not. Similarly, when it is cold enough, we generally observe that water freezes. Our experience tells us that it always does. This repeatability of similar events is what leads to the idea of causality. But causality does not hold at the atomic level. Electrons that are treated the same in a given experiment will not all end up in the same place; certain probabilities exist that an electron will arrive at different points, as we saw in Section 26-5 (the double-slit experiment).

Since matter is considered to be made up of atoms, even ordinary-sized objects are expected to be governed by chance and probability rather than by strict causality. For example, there is a finite probability that when you let go of a rock, it will fly upward instead of down; or that a lake will freeze on a hot summer day. Although the probability that such strange things will occur is extremely small, it is not zero. A quantum mechanical calculation might predict that such an unusual occurrence might happen once in a hundred trillion years. Quantum mechanics predicts with very high probability that ordinary objects will behave just as the classical laws of physics predict; but these predictions are probabilities, not certainties. The reason that macroscopic objects behave in accordance with classical laws with very high probability is because of the large number of molecules involved. Deviations from the average are negligible in any statistical situation. When large numbers of objects are present, it is the average arrangement of vast numbers of molecules that follows the "fixed laws" of classical physics with such high probability, and gives rise to an apparent "causality." Deviations from classical laws are readily observed when small numbers of molecules are dealt with.

Distinction between nature and our description of nature

Because electrons are not simply particles, they cannot be thought of as following particular paths in space and time. This suggests that a description of matter in space and time may not be completely correct. This deep and far-reaching conclusion has been a lively topic of discussion among philosophers. Perhaps the most important and influential philosopher of quantum mechanics was Niels Bohr. He argued that a space–time description of actual atoms and electrons is not possible. But, he pointed out, a description of experiments on atoms or electrons must be given in terms of space and time and other concepts familiar to ordinary experience, such as waves and particles. Yet we must not let our *descriptions* of experiments lead us into believing that atoms or electrons themselves actually exist in space and time as particles. This distinction between our interpretation of experiments and what is "really" happening in nature is crucial.

▮ Summary

Early models of the atom include Thomson's plum-pudding model and Rutherford's planetary (or nuclear) model. Rutherford's model, which was created to explain the back scattering of α particles from thin metal foils, assumes that an atom consists of a massive positively charged nucleus surrounded (at a relatively great distance) by electrons.

To explain the line spectra emitted by atoms, as well as their stability, Bohr proposed a theory that postulated (1) that electrons exist only in orbits with discrete radius and energy; (2) that an electron in such a *stationary state* emits no radiation; (3) that, if an electron jumps to a lower state, it emits a photon whose energy equals the difference in energy between the two states; (4) that the angular momentum L of atomic electrons is quantized by the rule $L = nh/2\pi$, where n is an integer called a *quantum number*. The $n = 1$ state in hydrogen is the *ground state*, which has an energy $E_1 = -13.6$ eV; higher values of n correspond to *excited states* and their energies are $E_n = -13.6\ \text{eV}/n^2$. Atoms are excited to these higher states by collisions with other atoms or electrons or by absorption of a photon of just the right frequency.

De Broglie's hypothesis that electrons (and other matter) have a wavelength $\lambda = h/mv$ gave an explanation for Bohr's quantized orbitals by bringing in the wave–particle duality: the orbits correspond to circular standing waves in which the circumference of the orbit equals a whole number of wavelengths.

In 1925, Schrödinger and Heisenberg worked out a new theory, *quantum mechanics*, which integrated the wave and particle aspects into a unified whole. Quantum mechanics is now considered to be the basic theory at the atomic level. It is a statistical theory rather than a deterministic one. According to quantum mechanics, the electrons in an atom do not have well-defined orbits but instead exist as a "cloud." Electron clouds can be interpreted as an electron wave spread out in space, or as a probability distribution for electrons as particles.

An important aspect of quantum mechanics is the Heisenberg *uncertainty principle*. It results from the wave–particle duality and the unavoidable interaction between the observed object and the observer. One form of the uncertainty principle states that the position and momentum of an object cannot both be measured precisely at the same time; the products of the uncertainties, $(\Delta x)(\Delta p)$, can be no less than $h/2\pi$. Another form states that the energy can be uncertain, or nonconserved, by an amount ΔE for a time $\Delta t \approx h/(2\pi\ \Delta E)$.

Questions

1 In Rutherford's planetary model of the atom, what keeps the electrons from flying off into space?

2 Which of the following can emit a line spectrum: (*a*) gases, (*b*) liquids, (*c*) solids? Which can emit a continuous spectrum?

3 Why doesn't the O_2 gas in the air around us give off light?

4 How can you tell if there is oxygen on the sun?

5 When a wide spectrum of light passes through hydrogen gas at room temperature, absorption lines are observed that correspond only to the Lyman series. Why don't we observe the other series?

6 Explain how the closely spaced energy levels for hydrogen near the top in Figure 26-8 correspond to the closely spaced spectral lines at the left of Figure 26-5.

7 Discuss the differences between Rutherford's and Bohr's theory of the atom.

8 In a helium atom, which contains two electrons, do you think that on the average the electrons are closer to the nucleus or farther away than they are in a hydrogen atom? Why?

9 How can the spectrum of hydrogen contain so many lines when hydrogen contains only one electron?

10 In an atom, why cannot an electron orbit exist whose circumference is equal to $3\frac{1}{2}$ wavelengths?

11 Explain why Bohr's theory of the atom is not compatible with quantum mechanics, particularly the uncertainty principle.

12 Explain why it is that the more massive an object is, the easier it becomes to predict its future position.

13 In view of the uncertainty principle, why does a baseball seem to have a well-defined position and speed, whereas an electron does not?

14 Discuss whether something analogous to the uncertainty principle operates when taking a public opinion survey. That is, do we alter what we are trying to measure when we take such a survey?

15 A cold thermometer is placed in a hot bowl of soup. Will the temperature reading of the thermometer be the same as the temperature of the hot soup before the measurement was made?

16 If Planck's constant were much larger than it is, how would this affect our everyday life?

17 In what ways is Newtonian mechanics contradicted by quantum mechanics?

Problems

SECTION 26-1

1 (II) In certain of Rutherford's experiments, the α particles (mass $= 6.68 \times 10^{-27}$ kg) had a kinetic energy of 4.8 MeV. How close could they get to a gold nucleus (charge $= +79e$)? Ignore the recoil motion of the nucleus.

SECTION 26-3

2 (I) How much energy is needed to ionize a hydrogen atom in the $n = 3$ state?

3 (I) For what maximum kinetic energy is a collision between an electron and a hydrogen atom in its ground state definitely elastic?

4 (I) The third longest wavelength in the Paschen series in hydrogen (Figure 26-8) corresponds to what transition?

5 (I) Sodium emits bright light of wavelength 589 nm. Calculate the energy difference in the atomic energy levels when such a transition occurs.

6 (I) Determine the wavelength of the second Balmer line ($n = 4$ to $n = 2$ transition) using Figure 26-8.

7 (I) Determine the wavelength of the first Lyman line ($n = 2$ to $n = 1$ transition) using Figure 26-8.

8 (I) Calculate the wavelength of the third Balmer line ($n = 5$ to $n = 2$).

9 (II) Construct the energy-level diagram for the He^+ ion.

10 (II) Construct the energy-level diagram for doubly ionized lithium (Li^{2+}) which has $Z = 3$.

11 (II) Calculate the ionization energy of doubly ionized lithium ($Z = 3$).

12 (II) What is the wavelength of light emitted when an $n = 50$ to $n = 30$ transition occurs in hydrogen? In what region of the spectrum does such a photon lie?

13 (II) Light of wavelength 388 nm is emitted by a hydrogen discharge tube. Between what energy levels is the atom jumping?

14 (II) By what fraction does the mass of an H atom decrease when it makes an $n = 2$ to $n = 1$ transition?

15 (II) If the moon's angular momentum about the earth were quantized according to Bohr's quantum condition (Equation 26-3), what would be the value of n?

16 (II) What is the potential energy and kinetic energy of an electron in the ground state of the hydrogen atom?

17 (II) An excited hydrogen atom could, in principle, have a radius of 1.0 mm. What would be the value of n for a Bohr orbit of this size?

18 (II) Calculate the ratio of the gravitational to electric force for the electron in a hydrogen atom. Can the gravitational force be safely ignored?

19 (II) Are nonrelativistic formulas justified in the Bohr atom? To check, calculate the electron's velocity, v, in terms of c for the ground state of hydrogen.

20 (II) What is the energy of the emitted photon when a He^+ ion makes a transition from $n = 3$ to $n = 1$?

21 (II) What is the longest-wavelength light capable of ionizing a hydrogen atom in the ground state?

22 (III) Electrons accelerated by a potential difference of 12.3 V pass through a gas of hydrogen atoms at room temperature. What wavelengths of light will be emitted?

23 (III) Atoms can be formed in which a mu meson (mass $= 207 \times$ mass of electron) replaces one of the electrons in an atom. Calculate, using Bohr theory, the energy of the photon emitted when a mu meson makes a transition from $n = 2$ to $n = 1$ in a mu-mesic $^{208}_{82}Pb$ (lead whose nucleus has a mass $208 \times 1.67 \times 10^{-27}$ kg and charge $+82e$).

24 (III) How much kinetic energy will an electron have if it is freed from a hydrogen atom in the ground state with a photon whose wavelength is 82.6 nm?

SECTION 26-4

25 (III) Suppose that a particle of mass m is confined to a one-dimensional box of width L. According to quantum theory, the particle's wave (with $\lambda = h/mv$) is a standing wave with nodes at the edges of the box. (*a*) Show the possible modes of vibration on a diagram. (*b*) Show that the kinetic energy of the particle has quantized energies given by $KE = n^2h^2/8mL^2$, where n is an integer. (*c*) Calculate the ground-state energy ($n = 1$) for an electron confined to a box of width 0.50×10^{-10} m. (*d*) What is the ground-state energy for a baseball ($m = 140$ g) in a box 0.50 m wide?

26 (III) An electron confined to a box has a ground-state energy of 10 eV. What is the width of the box? See Problem 25.

SECTION 26-7

27 (I) A proton is traveling with a speed of $(3.36 \pm 0.02) \times 10^5$ m/s. With what maximum accuracy can its position be ascertained?

28 (I) An electron remains in an excited state of an atom for typically 10^{-8} s. What is the minimum uncertainty in the energy of the state (in eV)? What is this uncertainty, in percent, of the first excited state in hydrogen?

29 (I) If an electron's position can be measured to an accuracy of 1.0×10^{-9} m, how accurately can its velocity be known?

30 (II) Estimate the least energy of a neutron contained in a typical nucleus of radius 10^{-15} m.

31 (II) A 10-g bullet leaves a rifle at a speed of 500 m/s. (*a*) What is the wavelength of this bullet? (*b*) If the position of the bullet is known to an accuracy of 0.50 cm (radius of the barrel), what is the minimum uncertainty in its momentum? (*c*) If the accuracy of the bullet were determined only by the uncertainty principle (an unreasonable assumption), by how much might the bullet miss a pinpoint target 300 m away?

32 (II) An electron and a 140-g baseball are each traveling 500 m/s measured to an accuracy of 0.05 percent. Calculate and compare the uncertainty in position of each.

33 (II) Use the uncertainty principle to show that if an electron were present in the nucleus ($r \approx 10^{-15}$ m), its energy would be hundreds of MeV. (Since such electron energies are not observed, we conclude that electrons are not present in the nucleus.)

34 (III) Use the uncertainty principle to estimate the position uncertainty for the electron in the ground state of the hydrogen atom. (*Hint*: Determine the momentum using the Bohr theory of Section 26-3.)

* 35 (III) In a double-slit experiment on electrons (or photons), suppose that we use indicators to determine which slit each electron went through (Section 26-5). These indicators must tell us the y coordinate to within $a/2$, where a is the distance between slits. Use the uncertainty principle to show that the interference pattern will be destroyed. (*Note*: First show that the angle θ between maxima and minima of the interference pattern is given by $\lambda/2a$, Figure 26-14.)

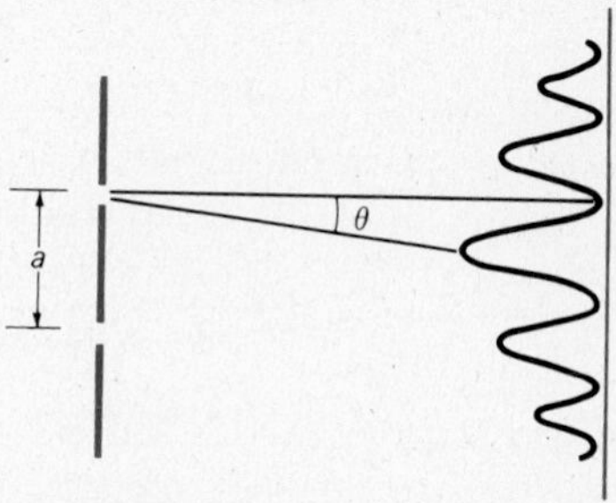

FIGURE 26-14

36 (III) How accurately can the position of a 2.0-keV electron be measured, assuming that its energy is known to 1 percent?

For additional reading

Baker, A., *Modern Physics and Anti-Physics*, Addison-Wesley Publishing Co., Inc., Reading, Mass., 1970.

Bohr, N. H., *Collected Scientific Works*, Elsevier North-Holland, Inc., New York, 1971.

Bohr, N. H., *Essays, 1958–1962, on Atomic Physics and Human Knowledge*, John Wiley & Sons, Inc., New York, 1966.

Cline, B., *Men Who Made a New Physics: Physicists and the Quantum Theory*, The New American Library, Inc., New York, 1969.

Eddington, A., *Nature of the Physical World*, Ann Arbor Books, Ann Arbor, Mich., 1958.

Gamow, G., *Mr. Tompkins Explores the Atom*, Cambridge University Press, New York, 1945.

Gamow, G., *Thirty Years that Shook Physics*, Doubleday & Co., Inc., New York, 1966.

Heisenberg, W., *Physics and Beyond: Encounters and Conversations*, tr. A. J. Pomeran, ed. R. N. Anshen, Harper & Row Publishers, Inc., New York, 1971.

Heisenberg, W., *Physics and Philosophy: The Revolution in Modern Science*, Harper & Row Publishers, Inc., New York, 1958.

Hoffman, B., *Strange Story of the Quantum*, Dover Publications, Inc., New York, 1959.

Jeans, J. H., *Physics and Philosophy*, Ann Arbor Books, Ann Arbor, Mich., 1958.

Magie, W. F., *Source Book in Physics*, Harvard University Press, Cambridge, Mass., 1963.

Russell, B., *Our Knowledge of the External World*, The New American Library, Inc., New York, 1960.

Setlow, R. B., and E. C. Pollard, *Molecular Biophysics*, Addison-Wesley Publishing Co., Reading, Mass., 1962. Application of spectroscopy.

27 ▮ Quantum theory of atoms and molecules

Since its development in the 1920s, quantum mechanics has had a profound influence on our lives, both intellectually and technologically. Even the way we view the world has changed, as we saw in Chapter 26. In the present chapter we will discuss how quantum mechanics has given us an understanding of the structure of atoms and molecules, and we will also discuss a number of applications; but because of their advanced nature, our discussion will mainly be qualitative.

27-1 ▮ Quantum mechanics of the hydrogen atom; quantum numbers

We first look more closely at what quantum mechanics tells us about the hydrogen atom. Much of what we say here also applies to more complex atoms, which are discussed in the next section.

Quantum mechanics predicts exactly the same energy levels for hydrogen as does the Bohr theory (Figure 26-8). That is,

$$E_n = -\frac{13.6\ \text{eV}}{n^2}, \qquad n = 1, 2, 3, \ldots,$$

where n is an integer. In the simple Bohr theory, we used only one quantum number, n. In quantum mechanics, it turns out that four different quantum numbers are needed to specify each state in the atom. We now discuss them, and for ease of visualization, we will sometimes speak of the electron as "orbiting" the nucleus, as in the Bohr model.

The quantum number n from Bohr theory is retained in quantum mechanics and is called the *principal quantum number*. It can have any integer value from 1 to ∞. The total energy of a state depends on n, as we saw above.

The *orbital quantum number*, l, is related to the angular momentum of the electron; l can take on integer values from 0 to $(n-1)$. For the ground state ($n = 1$) l can only be zero; but for $n = 3$, say, l can be 0, 1, or 2. The actual magnitude of the angular momentum L is related to the quantum number l by the formula $L = \sqrt{l(l+1)}(h/2\pi)$. The value of l does not affect the total energy in the hydrogen atom. But in atoms with two or more electrons, the energy does depend on l as well as on n, as we shall see in the next section.

The *magnetic quantum number*, m_l, is related to the direction of the electron's angular momentum and it can take on integer values ranging from $-l$ to $+l$; for example, if $l=2$, then m_l can be $-2, -1, 0, +1$, or $+2$. Since angular momentum is a vector, it is not surprising that both its magnitude and its direction would be quantized. For $l=2$, the five different directions allowed can be represented as in Figure 27-1. Usually, the direction is specified by giving the component of the angular momentum along the z axis. Then L_z is related to m_l by the equation $L_z = m_l(h/2\pi)$. The name for m_l derives not from theory (which relates it to L_z) but from experiment: it was found that when a gas discharge tube was placed in a magnetic field, the spectral lines were split into several very closely spaced lines. (This is called the Zeeman effect.) This implies that the energy levels must be split (Figure 27-2), and thus the energy of a state depends not only on n but also on m_l when a magnetic field is applied—hence the name "magnetic quantum number." Why the energy should depend on the direction of **L** has to do with the fact that a moving electron can be considered as an electric current which interacts with the magnetic field (see Chapter 17).

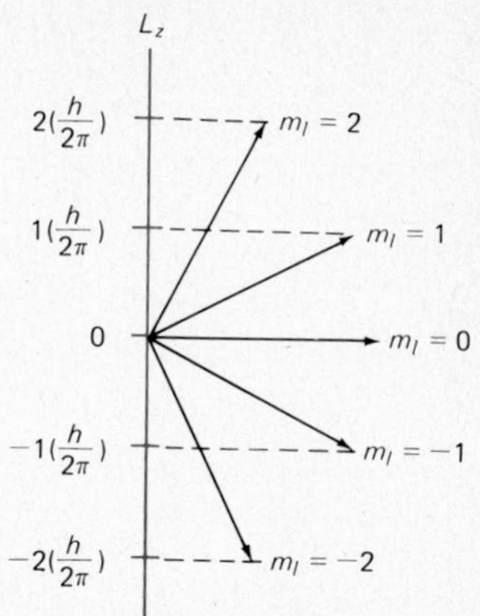

FIGURE 27-1
Quantization of angular momentum direction for $l=2$.

Finally, there is the *spin quantum number*, m_s, and it can have only two values, $m_s = +\frac{1}{2}$ and $m_s = -\frac{1}{2}$. The existence of this quantum number did not come out of Schrödinger's original theory, as did n, l, and m_l. Instead, a subsequent modification showed that it was present as a relativistic effect. The fact that m_s was needed, however, first came from experiment. A careful study of the spectral lines of hydrogen showed that each actually consisted of two (or more) very close lines. This was called *fine structure.* It was hypothesized that this tiny splitting of energy levels might be due to angular momentum associated with a spinning of the electron; that is, the electron might spin on its axis as well as orbit the nucleus, just as the earth spins on its axis and also orbits the sun. The interaction between the tiny current of the spinning electron could then interact with the magnetic field due to the orbiting charge and cause the small observed splitting of energy levels. This picture of the electron as spinning cannot be taken too seriously, however. For one thing, it is not correct to even view an electron as a localized object, much less a spinning one. What is important is that the electron can have two different states, owing to some intrinsic property, and we call this property "spin." The two possible values of m_s ($+\frac{1}{2}$ and $-\frac{1}{2}$) are said to be "spin up" and "spin down," referring to the two possible directions of the spin angular momentum.

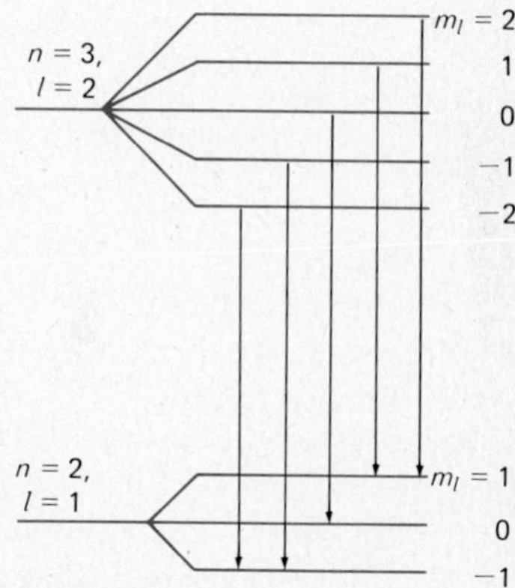

FIGURE 27-2
When a magnetic field is applied, an $n=3$, $l=2$ energy level is split into five separate levels, corresponding to the five values of m_l (2, 1, 0, −1, −2). An $n=2$, $l=1$ level is split into three levels ($m_l =$ 1, 0, −1). Transitions can occur between levels (not all are shown), with photons being given off with several frequencies, which are only slightly different.

The possible values of the four quantum numbers for an electron in the hydrogen atom are summarized in Table 27-1.

TABLE 27-1
Quantum numbers for an electron

Name	Symbol	Possible values
Principal	n	**1, 2, 3, . . . , ∞.**
Orbital	l	**For a given n, l can be $0, 1, 2, \ldots, n-1$.**
Magnetic	m_l	**For given n and l, m_l can be $l, l-1, \ldots, 0, \ldots, -l$**
Spin	m_s	**For each set of n, l, and m_l, m_s can be $+\frac{1}{2}$ or $-\frac{1}{2}$.**

▮ EXAMPLE 27-1 How many different states are possible for an electron whose principal quantum number is $n = 3$?

SOLUTION For $n = 3$, l can have the values $l = 2, 1, 0$. For $l = 2$, m_l can be $2, 1, 0, -1, -2$, which is five different possibilities; for each of these, m_s can be either up or down ($+\frac{1}{2}$ or $-\frac{1}{2}$); so for $l = 2$ there are $2 \times 5 = 10$ states. For $l = 1$, m_l can be $1, 0, -1$, and since m_s can be $+\frac{1}{2}$ or $-\frac{1}{2}$ for each of these, we have 6 more possible states. Finally, for $l = 0$, m_l can only be 0, and there are only 2 states corresponding to $m_s = +\frac{1}{2}$ and $-\frac{1}{2}$. The total number of states is $10 + 6 + 2 = 18$. The quantum numbers for each of these 18 states can be listed as follows, where each line represents one state:

n	l	m_l	m_s	n	l	m_l	m_s
3	2	2	$\frac{1}{2}$	3	2	−2	$-\frac{1}{2}$
3	2	2	$-\frac{1}{2}$	3	1	1	$\frac{1}{2}$
3	2	1	$\frac{1}{2}$	3	1	1	$-\frac{1}{2}$
3	2	1	$-\frac{1}{2}$	3	1	0	$\frac{1}{2}$
3	2	0	$\frac{1}{2}$	3	1	0	$-\frac{1}{2}$
3	2	0	$-\frac{1}{2}$	3	1	−1	$\frac{1}{2}$
3	2	−1	$\frac{1}{2}$	3	1	−1	$-\frac{1}{2}$
3	2	−1	$-\frac{1}{2}$	3	0	0	$\frac{1}{2}$
3	2	−2	$\frac{1}{2}$	3	0	0	$-\frac{1}{2}$

▮▮

Although l and m_l do not affect the energy levels in hydrogen (in the absence of a magnetic field), they do affect the electron probability distribution in space. For $n = 1$, l and m_l can only be zero and the electron cloud is as shown in Figure 26-13. For $n = 2$, l can be 0 or 1. The cloud for $n = 2$, $l = 0$ is shown in Figure 27-3*a*, and it is seen to differ from that for the ground state, although it is still spherically symmetric. For $n = 2$, $l = 1$, the cloud is not spherically symmetric, but is "barbell"-shaped, as shown in Figure 27-3*b*. The distribution shown is for one value of m_l, say $m_l = 1$; for $m_l = 0$ and $m_l = -1$, the distributions are also barbell-shaped but are centered around the other axes.

Although the spatial distribution of the electron can be calculated for the various states, it is difficult to measure them experimentally. Indeed, most of the experimental information about the atom has come from careful examination of the emission spectra under various conditions.

FIGURE 27-3
Electron cloud (or probability distribution) for $n = 2$ states in hydrogen.

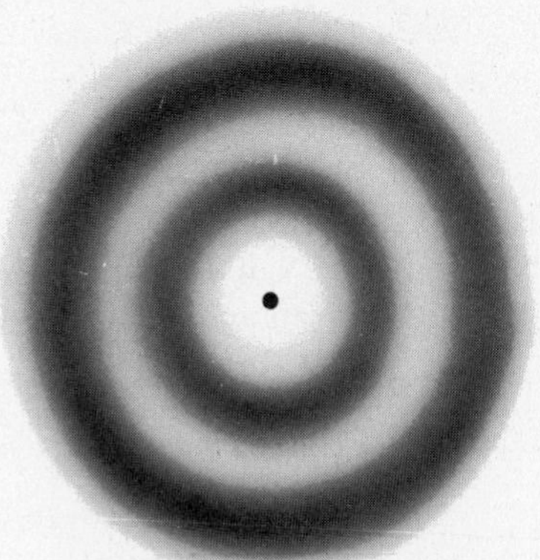

(*a*) $n = 2$, $l = 0$, $m_l = 0$

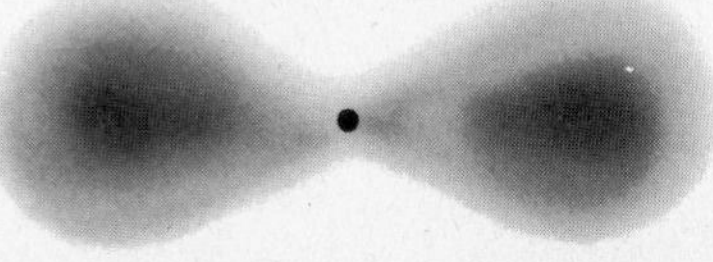

(*b*) $n = 2$, $l = 1$, $m_l = 1$

27-2 ▮ Complex atoms and the exclusion principle

We have discussed the hydrogen atom in detail because it is the simplest to deal with. Now we discuss more complex atoms, atoms that contain more than one electron. The number of electrons an atom contains in the neutral state is called its atomic number, Z; Z also equals the number of positive charges (protons) in the nucleus, and determines what kind of atom it is; that is, Z determines most of the properties of atoms that distinguish one kind from another.

The energy levels of complex atoms can be determined experimentally from an analysis of the emission spectra. Energy-level diagrams for each type of atom can then be constructed that resemble those for hydrogen (Figure 26-8) except that the energy values are different. It is found that the energy levels depend on both n and l; that is, a state with $n = 3$ and $l = 2$ has a different energy than one with $n = 3$ and $l = 0$, whereas for hydrogen these two states have the same energy.

Although the Bohr theory had been modified in an attempt to deal with complex atoms, the development of quantum mechanics in the years after 1925 proved far more successful. The mathematics becomes horrendous, however. This is because in multielectron atoms, each electron is not only attracted to the nucleus but is repelled by the other electrons.

The simplest approach has been to treat each electron in an atom as occupying a particular state characterized by the quantum numbers n, l, m_l, and m_s. But to understand the possible arrangements of electrons in an atom, a new principle was needed. It was introduced by Wolfgang Pauli (1900–1958) and is called the Pauli **exclusion principle**. It states:

> ▮ **No two electrons in an atom can occupy the same quantum state.**

Within an atom, this means that no two electrons can have exactly the same set of the quantum numbers n, l, m_l, and m_s. The Pauli exclusion principle forms the basis not only for understanding complex atoms, but also for understanding molecules and bonding, and other phenomena as well.

Let us now look at the structure of some of the simpler atoms when they are in the ground state. After hydrogen, the next simplest atom is helium, which has two electrons. Both electrons can have $n = 1$, since one can have spin up ($m_s = +\frac{1}{2}$) and the other spin down ($m_s = -\frac{1}{2}$), thus satisfying the exclusion principle. Of course, since $n = 1$, l and m_l must be zero (Table 27-1). Thus the two electrons have the quantum numbers indicated in the small table in the margin.

Lithium has three electrons, two of which can have $n = 1$; but the third cannot without violating the exclusion principle. Hence the third electron must have $n = 2$. Since it happens that the $n = 2$, $l = 0$ level has a lower energy than $n = 2$, $l = 1$, the electrons in the ground state have the quantum numbers indicated in the margin. Of course, the quantum numbers of the three electrons could also be $(1, 0, 0, \frac{1}{2})$, $(1, 0, 0, -\frac{1}{2})$, and $(3, 1, 0, \frac{1}{2})$, say; but the atom in this case would be in an excited state, since it would have greater energy. It would not be long before it jumped to the ground state with the emission of a photon. At room temperature, unless energy is put in (as in a discharge tube), the vast majority of atoms are in the ground state.

We can continue in this way to describe the quantum numbers of each electron in the ground state of larger and larger atoms. That for sodium, with its 11 electrons, is shown in the margin. The ground-state configuration for all atoms is given in the periodic table inside the cover, as explained in the next section.

Helium, $Z = 2$

n	l	m_l	m_s
1	0	0	$\frac{1}{2}$
1	0	0	$-\frac{1}{2}$

Lithium, $Z = 3$

n	l	m_l	m_s
1	0	0	$\frac{1}{2}$
1	0	0	$-\frac{1}{2}$
2	0	0	$\frac{1}{2}$

Sodium, $Z = 11$

n	l	m_l	m_s
1	0	0	$\frac{1}{2}$
1	0	0	$-\frac{1}{2}$
2	0	0	$\frac{1}{2}$
2	0	0	$-\frac{1}{2}$
2	1	1	$\frac{1}{2}$
2	1	1	$-\frac{1}{2}$
2	1	0	$\frac{1}{2}$
2	1	0	$-\frac{1}{2}$
2	1	−1	$\frac{1}{2}$
2	1	−1	$-\frac{1}{2}$
3	0	0	$\frac{1}{2}$

27-3 ▮ The periodic table of elements

A century ago, Dmitri Mendeleev (1834–1907) arranged the then known elements into what we now call the *periodic table* of the elements. The atoms were arranged according to increasing mass, but it was found that the table could also be arranged simultaneously so that atoms with similar chemical properties would fall in the same columns. For example, there are the alkali metals, which are very reactive and tend to give up one electron when they combine to form molecules. These form the first column of the periodic table. The second column contains metals that always give up two electrons when they combine. Further across there are the halogens (fluorine, chlorine, etc.) in the next-to-last column; they are very reactive, tend to acquire an electron when they react, and in the pure state are gases at room temperature formed of diatomic molecules. And the last column contains the inert gases (helium, neon, etc.), which react almost not at all.

Since Mendeleev's time, many new elements have been discovered and gaps in the table have been filled in. Today's version is shown on the inside front cover. Each box contains the atomic number Z, the symbol for the element, and the atomic mass (in atomic mass units: Chapter 28). Finally, in the lower left corner the electron configuration of the ground state of the atom is given. This requires some explanation. Electrons with the same value of n are referred to as being in the same *shell*. Electrons with $n = 1$ are in one shell (often called the K shell). Electrons with $n = 2$ are in a second shell (the L shell). Electrons with $n = 3$ are in a third shell (the M shell), and so on. Electrons with the same value of l are sometimes referred to as being in the same *subshell*. Letters are often used to specify the value of l as indicated in Table 27-2. That is, $l = 0$ is called the s subshell; $l = 1$ is the p subshell; $l = 2$ is the d subshell; beginning with $l = 3$, the letters follow the alphabet, f, g, h, i, and so on. (The first letters s, p, and d were originally abbreviations of "sharp," "principal," and "diffuse," and were experimental terms referring to the spectra.)

TABLE 27-2

Values of l

l value	Letter	Maximum number of electrons in subshell
0	s	2
1	p	6
2	d	10
3	f	14
4	g	18
5	h	22
⋮		

Also shown in Table 27-2 is the maximum number of electrons that can be in each subshell according to the Pauli exclusion principle. Since for any value of l there are $2l + 1$ different m_l values (m_l can be any integer from 1 to l, from -1 to $-l$, or zero), and two different m_s values, there can be at most $2(2l + 1)$ electrons in any l subshell. For example, for $l = 2$, five m_l values are possible $(2, 1, 0, -1, -2)$, and for each of these m_s can be $+\frac{1}{2}$ or $-\frac{1}{2}$, for a total of $2(5) = 10$ states.

Since the energy levels depend almost entirely on the values of n and l, it is customary to specify the electron configuration more simply than was done in the previous section by giving only the n and l values. This is done by giving the n value and the appropriate letter for l from Table 27-2. Thus, an electron in the $n = 2$, $l = 1$ state is said to be in the $2p$ state or the $2p$ subshell. Also, the number of electrons in each subshell is given as a superscript. The electron configuration of sodium, for example, is written as $1s^2 2s^2 2p^6 3s^1$, which means that there are two electrons in the $1s$ state ($n = 1$, $l = 0$), two in $2s$ ($n = 2$, $l = 0$), six in the $2p$ ($n = 2$, $l = 1$), and one in the $3s$ ($n = 3$, $l = 0$) state. In the periodic table on the inside front cover, this is simplified by specifying the configuration only of the outermost electrons and any other nonfilled subshells. Thus, since the $n=1$ and $n=2$ shells are completely filled up for sodium, sodium is simply said to be in the $3s^1$ configuration.

Now let us see how quantum mechanics explains the grouping of similar atoms into the same column of the periodic table. All the noble gases (in the last column of the table) have completely filled shells or subshells. That is, their outermost subshell is completely full, the electron distribution is spherically symmetric, and there is no opportunity for additional electrons. This is the reason they are nonreactive (more on this when we discuss molecules and bonding in Section 27-5). Column seven contains the halogens, which lack one electron from a filled shell. Because of the shapes of the orbits (see Section 27-5), an additional electron can be accepted from another atom, and hence these are quite reactive with a valence of -1 (meaning that when such an electron is accepted, the resulting ion has a net charge of $-1e$). At the left of the table, column I contains the alkali metals, all of which have a single outer s electron. This electron spends most of its time outside the inner closed shells and subshells which shield it from most of the nuclear charge; indeed, it is relatively far from the nucleus and is attracted to it by a net charge of only $+1e$, because of the shielding effect of the other electrons. Hence this outer electron is easily removed and can spend much of its time around another atom, forming a molecule; this is why the alkali metals have a valence of $+1$. The other columns of the table can be treated similarly.

The presence of the transition elements in the center of the table, as well as the lanthanides (rare earths) and actinides below, is a result of incomplete inner shells. For the lowest Z elements, the subshells are filled in a simple order: first $1s$, then $2s$, followed by $2p$, $3s$, and $3p$. You might expect that $3d$ ($n=3$, $l=2$) would be filled next, but it isn't. Instead, the $4s$ level actually has a slightly lower energy than the $3d$, so it fills first (K and Ca). Only then does the $3d$ shell start to fill up, beginning with Sc. (The $4s$ and $3d$ levels are close, so some elements have only one $4s$ electron, such as Cr.) Most of the chemical properties are governed by the relatively loosely held $4s$ electrons and hence these "transition elements" mainly have valences of $+1$ or $+2$. A similar effect is responsible for the rare earths, which are shown at the bottom of the periodic table for convenience; all have very similar chemical properties because their outer $6s$ or $7s$ electrons control their chemical properties, whereas the different numbers of electrons in the unfilled inner shells have little effect.

27-4 ▮ X rays and atomic number

The line spectra of atoms in the UV, visible, and IR regions of the EM spectrum are mainly due to transitions between states of the outer electrons. For these electrons, much of the charge of the nucleus is shielded from them by the negative charge on the inner electrons. But the innermost electrons in the $n=1$ shell see the full charge of the nucleus. Since the energy of a level is proportional to Z^2 (see Equation 26-6), for an atom with $Z=50$, we would expect wavelengths about $50^2=2500$ times shorter than those found in the Lyman series of hydrogen (around 100 nm), or 10^{-2} to 10^{-1} nm. These are in the x-ray region!

Normally, the inner shells are filled, so we would not expect to see such short-wavelength light. However, in Section 25-4, we saw that in an x-ray tube, there is not only a continuous spectrum but a line spectrum as well—see the peaks in Figure 25-9. Now we can see where these peaks come from. The electrons accelerated by the high voltage of the tube reach sufficient energies that when they collide with the atoms of the target, they can knock out one of the very tightly held inner electrons. The *characteristic* x rays (the peaks in Figure 25-9) are photons emitted when an electron in an upper state drops down to fill the vacated lower state.

Measurement of x-ray spectra has allowed a determination of the inner energy levels of atoms. It has also allowed the determination of Z values for many atoms, since (as we have seen) the wavelength of the shortest x rays emitted will be inversely proportional to Z^2. Actually, for an electron jumping from, say, the $n = 2$ to the $n = 1$ level, the wavelength is inversely proportional to $(Z-1)^2$ because the nucleus is shielded by the one electron that still remains in the $1s$ level. In 1914, H. G. J. Moseley found that a plot of $\sqrt{1/\lambda}$ versus Z produced a straight line (Figure 27-4). The Z values of a number of elements was determined by fitting them to such a "Moseley plot." The work of Moseley put the concept of atomic number on a firm experimental basis.

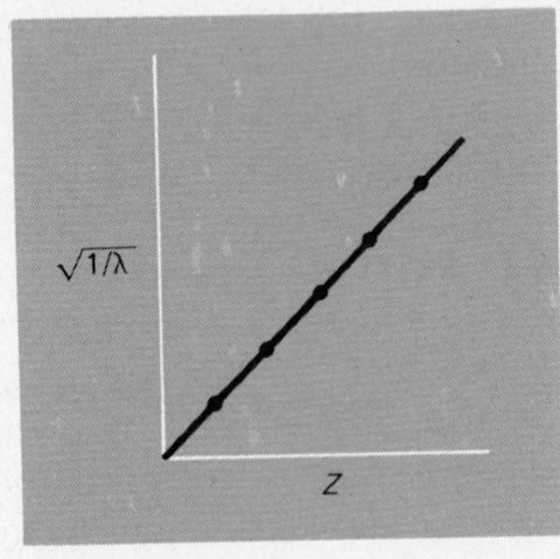

FIGURE 27-4
Plot of $\sqrt{1/\lambda}$ versus Z for K_α x-ray lines.

▮ EXAMPLE 27-2 Estimate the wavelength for an $n = 2$ to $n = 1$ transition in molybdenum ($Z = 42$).

SOLUTION We use the Bohr formula, Equation 26-7, with Z^2 replaced by $(Z-1)^2 = (41)^2$. Or, more simply, we can use the result of Example 26-1 for the $n = 2$ to $n = 1$ transition in hydrogen ($Z = 1$). Since $\lambda \propto 1/(Z-1)^2$, we will have

$$\lambda = (1.22 \times 10^{-7}\ \text{m})/(41)^2 = 0.073\ \text{nm}.$$

This is close to the measured value (Figure 25-9) of 0.071 nm. ▮▮

27-5 ▮ Bonding in molecules and solids

The study of molecules and their formation is usually considered the subject of chemistry. However, one of the great successes of quantum mechanics was to give scientists, at last, an understanding of the nature of chemical bonds. Since it is based in physics, and because this understanding is so important in many fields, we discuss it here.

By a molecule we mean a group of two or more atoms that are strongly held together so that it functions like a single unit. When atoms make such an attachment, we say that a chemical *bond* has been formed.

There are two main types of strong chemical bond: covalent and ionic. Many bonds are actually intermediate between these two types. We shall discuss each in turn.

Covalent bonds

To understand how covalent bonds are formed, we take the simplest case, that which holds two hydrogen atoms together to form the hydrogen molecule H_2. The mechanism is basically the same for other covalent bonds.

As two H atoms approach each other, the electron clouds begin to overlap, and the electrons from each atom can "orbit" both nuclei. (This is sometimes called "sharing" electrons.) If both electrons are in the ground state ($n = 1$) of their respective atoms, two possibilities can occur: they can have their spins parallel (both up or both down), or they can be opposite ($m_s = +\frac{1}{2}$ for one, $m_s = -\frac{1}{2}$ for the other). We shall now see that a bond is formed only when the spins are opposite. First suppose that the spins are the same; the two electrons cannot both be in the lowest energy state, for then they would have identical quantum numbers. The exclusion principle tells us that the two electrons can then not come close together. That is, since no two electrons can occupy the same quantum state, if they have the same quantum numbers, they must be different in some other way—namely, by being in different places in space. When the two atoms approach, the electrons will stay away from each other as shown by the probability distributions of Figure 27-5. The positively charged nuclei then repel each other, and no bond is formed.

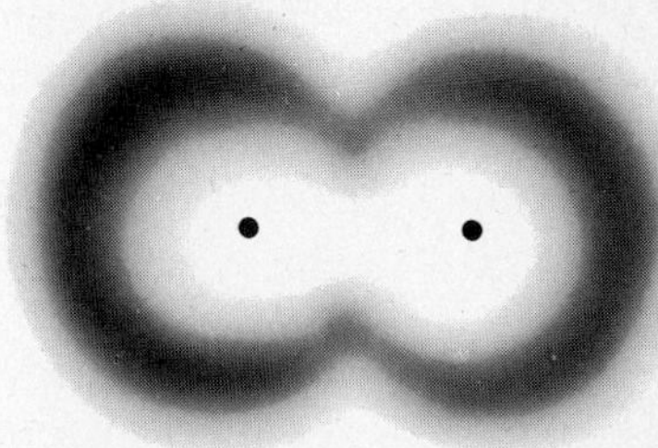

FIGURE 27-5
Electron probability distribution (electron cloud) for two H atoms when their spins are the same.

If the spins are opposite, the two electrons are automatically in different quantum states and then they can come close together. In this case, the probability distribution looks like Figure 27-6. As can be seen, the electrons spend much of their time between the two nuclei. The two positively charged nuclei are attracted to this negatively charged electron cloud between them and it is this attraction that holds the two atoms together to form a molecule. This is a *covalent bond.*

The probability distributions of Figures 27-5 and 27-6 can perhaps be better understood on the basis of waves. What the exclusion principle requires is that when the spins are the same, there is destructive interference of the electron waves in the region between the two atoms. But when the spins are opposite, constructive interference occurs in the region between the two atoms, resulting in a large amount of negative charge there. Thus a covalent bond can be said to be the result of constructive interference of electron waves in the space between the two atoms, and the electrostatic attraction of the two positive nuclei for the negative charge concentration between them.

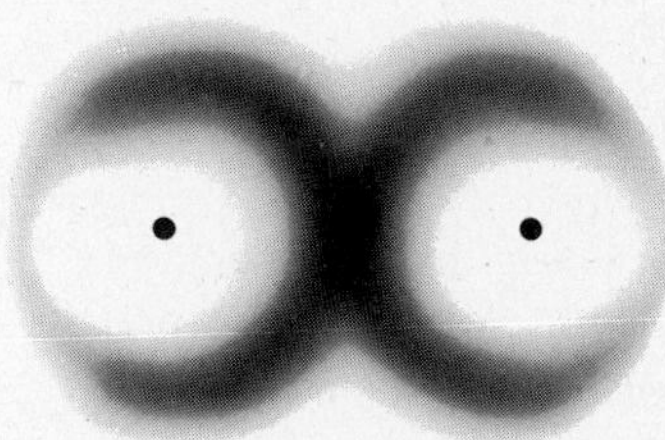

FIGURE 27-6
Electron probability distribution (cloud) around two H atoms when their spins are opposite. In this case a bond is formed because the positive nuclei are attracted to the concentration of negative charge between them. This is a hydrogen molecule, H_2.

Ionic bonds

An ionic bond is, in a sense, an extreme case of the covalent bond. Instead of the electrons being shared equally, they are shared unequally. For example, in sodium chloride (NaCl) the outer electron of the sodium spends nearly all its time around the chlorine (Figure 27-7). The chlorine atom has a net negative charge as a result of the extra electron, whereas the sodium is positive. The electrostatic attraction between these two charged atoms is what holds them together and forms the bond. This is called an *ionic bond* because it is the attraction between the two ions (Na^+ and Cl^-) that holds them together.

But to understand the ionic bond, we must understand why the extra electron from the sodium spends so much of its time around the chlorine. After all, the chlorine is neutral; why should it attract another electron? The answer lies in the probability distributions of the two neutral atoms.

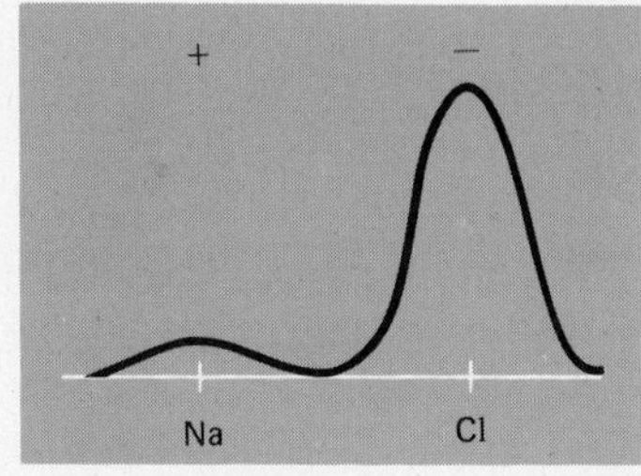

FIGURE 27-7
Probability distribution for last electron of Na in NaCl.

Sodium contains 11 electrons, 10 of which are in spherically symmetric closed shells (Figure 27-8). The last electron spends most of its time beyond these closed shells. Because the closed shells have a total charge of $-10e$ and the nucleus has $+11e$, the outermost electron in sodium feels a net attraction due to $+1e$. It is thus not held terribly strongly. On the other hand, 12 of chlorine's 17 electrons form closed shells or subshells (corresponding to $1s^2\,2s^2\,2p^6\,3s^2$). These 12 form a spherically symmetric shield around the nucleus. The other five electrons are in $3p$ orbits, which are barbell-shaped. Two of these $3p$ electrons are in the dumbbell-shaped orbit (or cloud) shown in Figure 27-9. Two more are in an orbit perpendicular to the page (not shown). The fifth is in an orbit perpendicular to these and is shown only in faint outline. The exclusion principle allows one more electron to be in this orbit (it will have spin opposite to that of the electron already there). If an extra electron—say from a Na atom—happens to be in the vicinity, it can be in this orbit, say at point x in Figure 27-9. It will experience an attraction due to $+5e$ because the $+17e$ of the nucleus is shielded at this point only by the 12 inner electrons. Clearly, the outer electron of a Na atom will be more strongly attracted by the $+5e$ of the chlorine than by the $+1e$ of its own atom.† This is why we get the charge distribution of Figure 27-7, and thus an ionic bond.

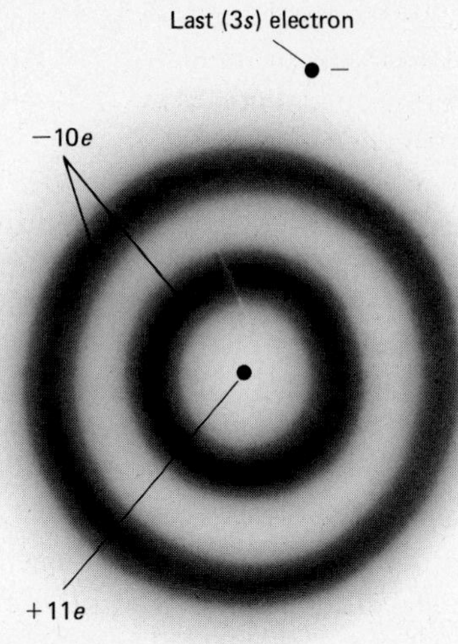

FIGURE 27-8
In a neutral sodium atom, the 10 inner electrons shield the nucleus; so the single outer electron is attracted by a net charge of $+1e$.

Polar molecules—partial ionic character of covalent bonds

A pure covalent bond in which the electrons are shared equally occurs mainly for symmetrical molecules like H_2, O_2, and Cl_2. When the atoms involved are different from each other, it is usual to find that the shared electrons are found more on one atom than the other. The extreme case is an ionic bond, but in intermediate cases the covalent bond is then said to have a *partial ionic character*. The molecules themselves are polar—that is, one part (or parts) of the molecule has a net positive charge and other parts a negative charge.

† Coulomb's law also involves distance between the two charges, but that is not greatly different in the two cases.

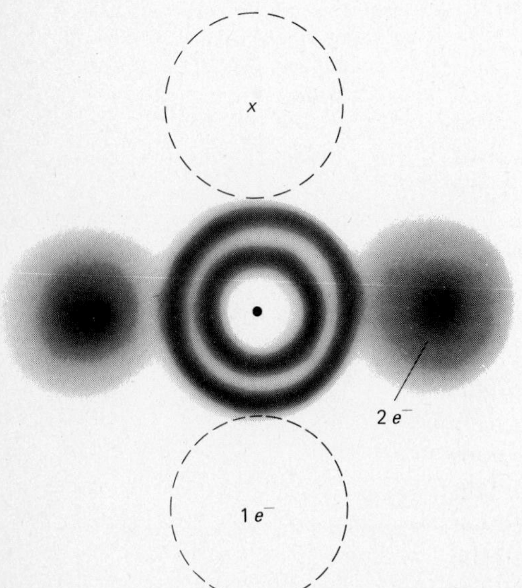

FIGURE 27-9
Neutral chlorine atom. The $+17e$ of the nucleus is shielded by the 12 electrons in the inner shells and subshells. Five electrons are in dumbbell-shaped clouds ($l = 1$). An extra electron at x will be attracted by a net charge of $+5e$.

An example is the water molecule, H_2O (Figure 27-10). The shared electrons gather around the oxygen atom more than around the two hydrogens. The reason is similar to that causing ionic bonds. Oxygen has eight electrons ($1s^2\,2s^2\,2p^4$), of which four form a spherically symmetric core and the other four are in barbell-shaped orbits. One possible barbell-shaped orbit is empty, so electrons from hydrogen atoms are attracted by a net charge of $+4e$. They are also attracted by the H nuclei, so they actually orbit the H atoms as well as the O atom. The net effect is that there is a net positive charge on each H atom (less than $+1e$), because the electron spends only part of its orbit there. And there is a net negative charge on the O atom.

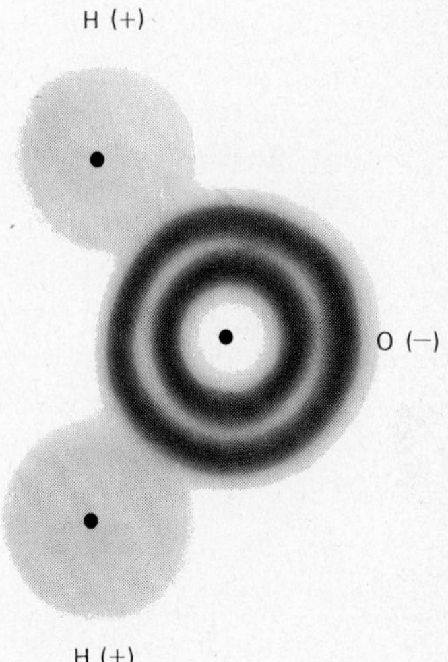

FIGURE 27-10
The water molecule is polar.

Binding energy; weak and strong bonds

Once a bond is made, energy must normally be put in to break the bond and separate the atoms. This energy is called the *bond energy* or *binding energy*. The binding energy for covalent and ionic bonds is typically 2 to 5 eV. These bonds, which hold atoms together to form molecules, are often called *strong bonds* to distinguish them from so-called *weak bonds*. The term "weak bond" as we use it here refers to an attachment between molecules because of simple electrostatic attraction—such as between polar molecules. The strength of the attachment is much less than for the strong bonds; binding energies are typically in the range 0.04 to 0.3 eV—hence their name "weak bonds."

Weak bonds are particularly important in molecular biology. The average kinetic energy of molecules in a cell is around 0.04 eV, about the magnitude of weak bonds. This means that a weak bond can readily be broken just by a molecular collision. Hence weak bonds are not very permanent—they are, instead, brief attachments. But because of this, they play an important role in the cell, as we saw in Chapters 13 and 14. On the other hand, strong bonds—those that hold molecules together—are almost never broken simply by molecular collision. Thus they are relatively permanent. They can be broken by chemical action (the making of even stronger bonds), and this usually happens in the cell with the aid of an enzyme.

Bonding in solids

The molecules of a solid are held together in a number of ways. The most common are by covalent bonding (such as between the carbon atoms of diamond) or ionic bonding (as in a NaCl crystal). Often the bonds are partially covalent and partially ionic.

Another type of bond occurs in metals. Metals have relatively free outer electrons, and present theories indicate that in a metallic solid, these outer electrons roam rather freely among all the metal atoms which, without their outer electrons, act like positive ions. The electrostatic attraction between the metal ions and the negative electron "gas" (as it is called) is what is believed to hold the solid together. This theory nicely accounts for the shinyness of smooth metal surfaces: the free electrons can vibrate at any frequency; so when light of almost any frequency falls on a metal, the electrons can vibrate in response and re-emit light of that same frequency. Hence the reflected light has the same frequency as the incident light. Compare this to ordinary materials that have a distinct color—the electrons exist only in certain energy states and thus can resonate only at certain frequencies.

The atoms or molecules of some materials, such as the inert gases, can only form *weak bonds* with each other. As we just saw, weak bonds have very low binding energies and would not be expected to hold atoms together as a liquid or solid at room temperature. Such materials condense only at very low temperatures, where the atomic kinetic energy is small and the bonds can then hold the atoms together.

* 27-6 ▌ Potential-energy diagrams

It is useful to analyze the interaction between two objects—say, between two atoms or molecules—with the use of a potential-energy diagram, which is simply a plot of the potential energy versus distance.

Let us first examine the simple case of one point charge, q_1, in the presence of a second point charge, q_2. As we saw in Chapter 14, the potential energy of charge q_2 in the presence of q_1 a distance r away is

$$\text{PE} = k\frac{q_1 q_2}{r},$$

where $k = 9.0 \times 10^9\ \text{N}\cdot\text{m}^2/\text{C}^2$.

If the two charges are the same sign, the PE is positive for all values of r; a graph of PE versus r in this case is shown in Figure 27-11*a*. The force is repulsive (the charges have the same sign) and we see that the curve rises as r decreases.

If the two charges are of opposite sign, the PE will be negative because the product q_1q_2 will be negative. The force is attractive in this case and $\text{PE} = k(q_1q_2/r)$ versus r looks like Figure 27-11*b*. The PE in this case becomes more *negative* as r decreases. Keep in mind this fact: for a repulsive force, the PE increases as r decreases; and for an attractive force, the PE decreases as r decreases.

Now let us look at the potential-energy diagram for the formation of a covalent bond, such as for the hydrogen molecule. The potential energy of one H atom in the presence of the other is plotted in Figure 27-12. The PE decreases as the atoms approach, because the electrons concentrate between the two nuclei (Figure 27-6), so attraction occurs. However, at very short distances, the electrons are "squeezed out"—there is no room for them between the two nuclei. Without the electrons between them, each nucleus feels a repulsive force† due to the other, so the curve rises as r decreases further. There is an optimum separation of the atoms, r_0, in Figure 27-12, where the energy is lowest. This is the point of greatest stability for the hydrogen molecule, and r_0 is the average separation of atoms in the H_2 molecule. The depth of this "well" is the binding energy, as shown. This is how much energy must be put into the system to separate the two atoms to infinity, where the $\text{PE} = 0$.

The potential-energy diagram for ionic bonds is similar in shape. In

FIGURE 27-11
Potential energy as a function of separation for two point charges of (*a*) like sign and (*b*) opposite sign.

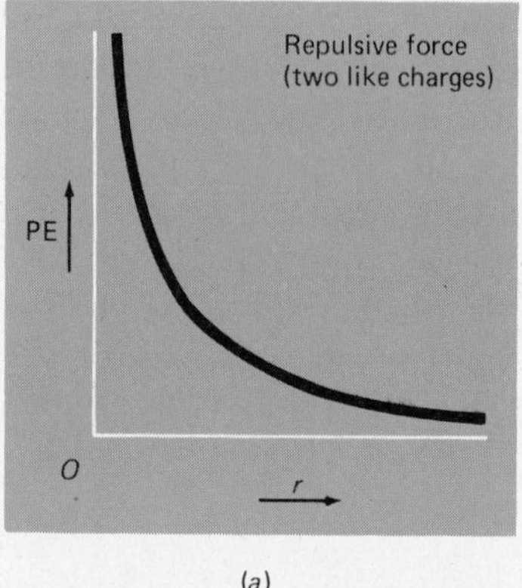

(*a*)

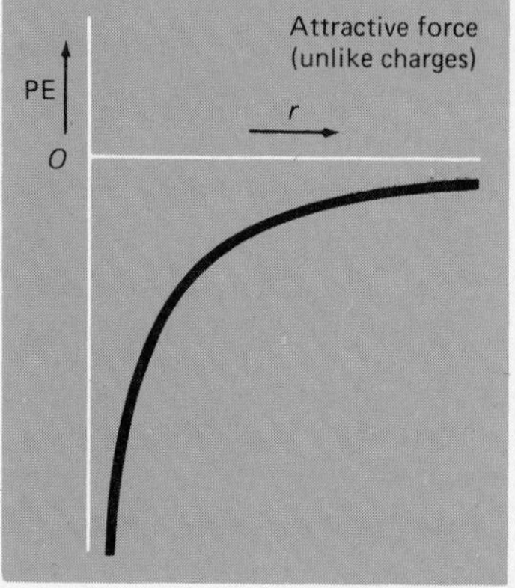

(*b*)

† In larger atoms, say oxygen or nitrogen, repulsion also occurs at short distances, because the closed inner electron shells begin to overlap and the exclusion principle forbids their coming too close.

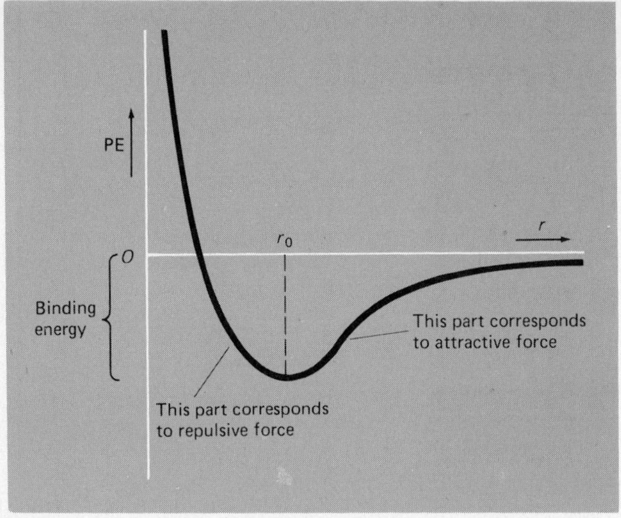

FIGURE 27-12
Potential-energy diagram for H_2 molecule; r is separation of the two H atoms.

NaCl, for example, the Na^+ and Cl^- ions attract each other at distances above some r_0, but at smaller distances the overlapping of inner electron shells gives rise to repulsion. The two atoms thus are most stable at some intermediate separation r_0.

*27-7 ▮ Activation energy, enzymes, and ATP

For some bonds, the potential-energy curve has the shape shown in Figure 27-13. There is still an optimum distance r_0 at which the molecule is stable. But when the atoms approach from a large distance, the force is initially repulsive rather than attractive. The atoms thus do not interact spontaneously. Instead, some additional energy must be injected into the system to get it over the "hump" (or barrier) in the potential-energy diagram. This required energy is called the *activation energy*.

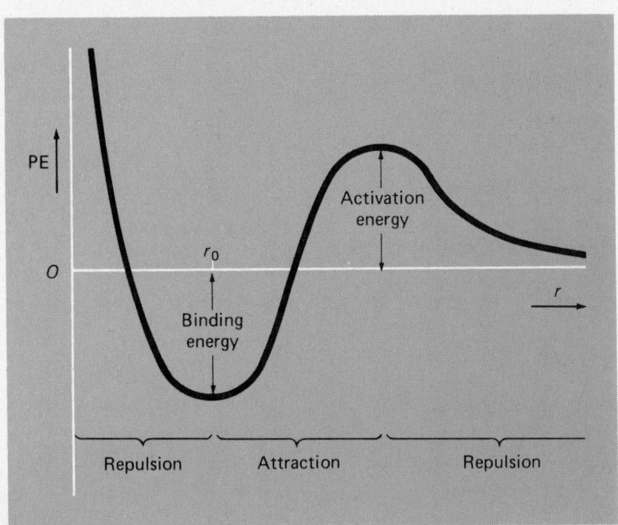

FIGURE 27-13
Potential-energy diagram for a bond requiring an activation energy.

The curve of Figure 27-13 is far more common than that of Figure 27-12. The activation energy is usually needed to break other bonds, before the one under discussion can be made. For example, to make water from O_2 and H_2, the H_2 and O_2 molecules must first be broken into H and O atoms; this is what the activation energy does. Then the H and O atoms can combine to form H_2O with the release of a great deal more energy than was put in initially. The initial activation energy can be provided by applying an electric spark to a mixture of H_2 and O_2, breaking them into atoms; the resulting explosive release of energy when the atoms combine to form H_2O quickly provides the activation energy for further reactions, so further H_2 and O_2 molecules are broken up and recombined to form H_2O.

In a living cell, many chemical reactions are occurring that involve the making and breaking of strong bonds. The activation energies are often on the order of several eV. Such energies are not easy to overcome in the cell. This is where enzymes come in. They act as catalysts, which means that they act to lower the activation energy so that reactions can occur that otherwise would not. Enzymes act by distorting the bonding electrons so that the initial bonds are easily broken.

Sometimes the potential energy of a bond looks like that of Figure 27-14. In this case the energy of the bonded molecule, at a separation r_0, is greater than when there is no bond ($r = \infty$). That is, an energy *input* is required to make the bond. (Hence the binding energy is negative.) And there is energy release when the bond is broken. Such a bond is stable only because there is the barrier of the activation energy. This type of bond is very important in biological cells, for it is by this means that energy can be stored efficiently in certain molecules, particularly ATP (adenosine triphosphate). The bond that connects the last phosphate group (designated Ⓟ in Figure 27-14) to the rest of the molecule (ADP, meaning adenosine diphosphate, since it contains only two phosphates) is of the form shown in Figure 27-14. Energy is actually stored in this bond.† When the bond is broken (ATP → ADP + Ⓟ) energy is released and this energy can be used to make other chemical reactions "go."

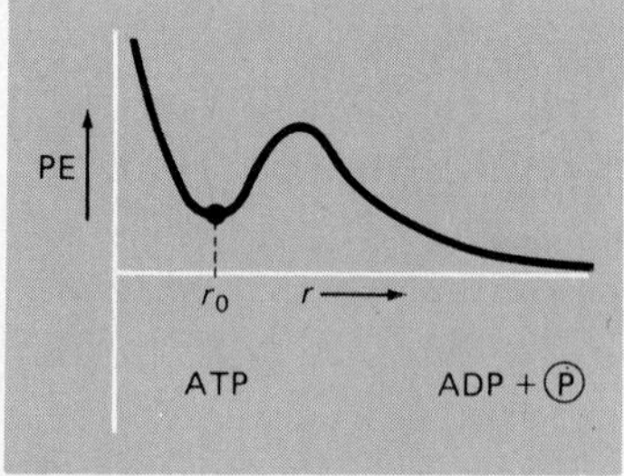

FIGURE 27-14
Potential-energy diagram for formation of ATP from ADP and phosphate (Ⓟ)

*27-8 ▮ Molecular spectra

When atoms combine to form molecules, the energy levels of the outer electrons are altered because they now interact with each other. Additional energy levels also become possible because the atoms can vibrate with respect to each other, and the molecule as a whole can rotate. The energy levels for both vibrational and rotational motion are quantized. However, they are very close together, particularly for the rotational levels (typically 10^{-3} eV apart). Each atomic energy level thus becomes a set of closely spaced levels corresponding to the vibrational and rotational motions (Figure 27-15). Transitions from one level to another appear as many very closely spaced lines. In fact, the lines are not always distinguishable, and such spectra are called *band spectra*. Each type of

† The second phosphate is also connected by such a "high-energy" bond, as it is sometimes called.

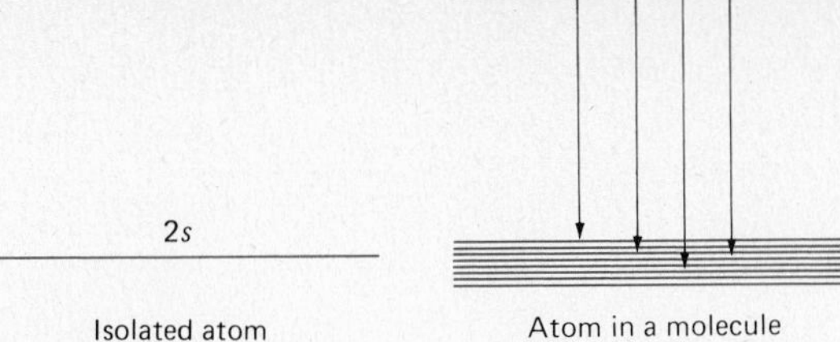

FIGURE 27-15
The simple energy levels of an isolated atom become bands of closely spaced levels in molecules as well as in solids and liquids.

molecule has its own characteristic spectrum, which can be used for identification and for determination of structure.

When atoms or molecules are in the solid (or liquid) state, their outer electrons also overlap. Hence their energy levels are changed somewhat, and because of the interaction, the energy levels are spread out into *energy bands* as in Figure 27-15. Because there are so many atoms present, the energy levels in such a band are extremely close together. They thus seem continuous, which is why the spectrum of heated solids, such as red-hot iron or a light-bulb filament, appears continuous. The energy bands in semiconductors play an important role for transistors and diodes, as we discussed in Chapter 19.

* 27-9 ▌ Fluorescence and phosphorescence

When an atom is excited from one energy state to a higher one by the absorption of a photon, it may return to the lower level in a series of two (or more) jumps if there is an energy level in between (Figure 27-16). The photons emitted will consequently have lower energy and frequency than the absorbed photon. This is called *fluorescence*; common fluorescent rocks and paints can emit visible light after absorbing UV light.

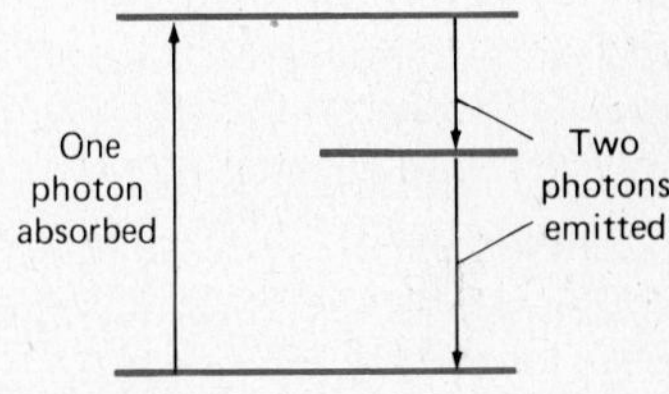

FIGURE 27-16
Fluorescence.

The wavelength for which fluorescence will occur depends on the energy levels of the particular atoms. Because the frequencies are different for different substances, and because many substances fluoresce readily, fluorescence is a powerful tool for identification of compounds. It is also used for assaying—determining how much of a substance is present—and for following substances along a natural pathway as in plants and animals. For detection of a given compound, the stimulating light must be monochromatic, and solvents or other materials present must not fluoresce in the same region of the spectrum. Often the observation of fluorescent light being emitted is sufficient; in other cases, spectrometers are used to measure the wavelengths and intensities of the light.

Fluorescent light bulbs work in a two-step process. The applied voltage accelerates electrons that strike atoms of the gas in the tube and cause them to be excited. When the excited atoms jump down to their normal levels, they emit UV photons which strike a fluorescent coating on the inside of the tube. The light we see is a result of this material fluorescing in response to the UV light striking it.

Materials such as those used for luminous watch dials are said to be *phosphorescent.* In a phosphorescent substance, atoms can be excited by absorption of a photon to an energy level said to be metastable. When an atom is raised to a normal excited state, it drops back down within about 10^{-8} s. *Metastable* states last much longer—even a few seconds or longer. In a collection of such atoms, many of the atoms will descend to the lower state fairly soon, but many will remain in the excited state for over an hour. Hence light will be emitted even after long periods. When you put your watch dial close to a bright lamp, it excites many atoms to metastable states, and you can see the glow a long time after.

* 27-10 ▌ Lasers and holography

A laser is a device that can produce a very narrow intense beam of monochromatic coherent light. By *coherent* light, we mean that the photons emitted by the source are in phase with one another; so the emitted beam is a nearly perfect plane wave. Ordinary light sources (other than lasers) emit light in all directions (so the intensity decreases rapidly with distance), and the emitted light is incoherent. It is incoherent (meaning that the photons in the beam are not in phase with each other) since the excited atoms that emit the light act independently; each emitted photon can be considered as a short wave train, typically 30 cm long lasting 10^{-8} s, and these wave trains bear no phase relation to one another.

The action of a laser is based on quantum theory. We have seen that a photon can be absorbed by an atom if (and only if) its energy hf corresponds to the energy difference between an occupied energy level of the atom and an available excited state (Figure 27-17*a*). This is, in a sense, a resonant situation. If the atom is already in the excited state, it may of course jump spontaneously to the lower state with the emission of a photon. However, if a photon with the same frequency strikes the excited atom, it can stimulate the atom to make the transition to the lower state (Figure 27-17*b*). This is called *stimulated emission,* and it can be seen that not only do we still have the original photon, but also a second one of the same frequency as a result of the atom's transition. An important fact is that these two photons are exactly *in phase.* This is how coherent light is produced in a laser—hence the name "laser," which is an acronym for *l*ight *a*mplification by *s*timulated *e*mission of *r*adiation.

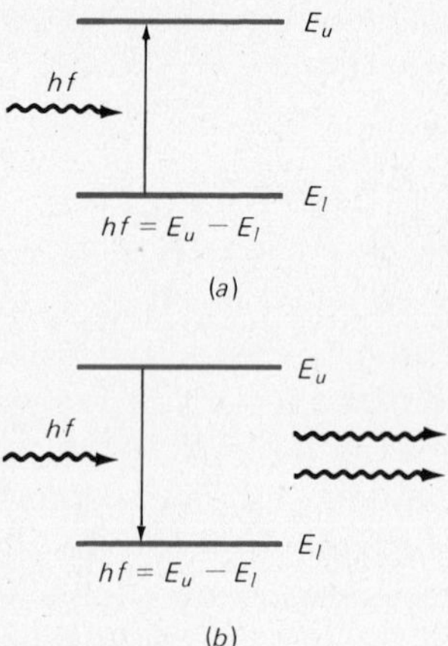

FIGURE 27-17
(*a*) Absorption of a photon. (*b*) Stimulated emission.

Normally, most atoms are in the lower state, so incident photons will mostly be absorbed. To obtain the coherent light from stimulated emission, two conditions must be satisfied. First, the atoms of the material used in a laser must be excited to the higher state. That is, an *inverted population* is needed so that *emission* of photons will dominate over absorption. Second, the higher state must be metastable—the electrons must remain in this state longer than normal so that the transition to the lower state occurs by stimulated emission rather than spontaneously. How these conditions are achieved for different lasers will be discussed shortly. For now, we assume that the atoms have been excited to an upper state. Figure 27-18 is a schematic diagram of a laser: the "lasing" material is placed in a long, narrow tube at the ends of which are two mirrors, one of which is

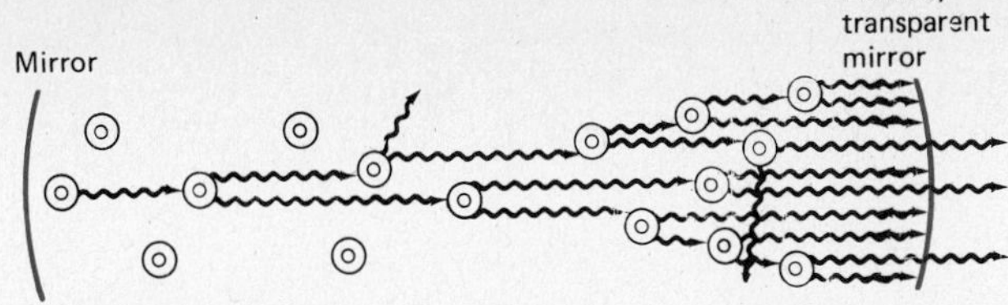

FIGURE 27-18
Laser diagram, showing excited atoms stimulated to emit light.

partially transparent (perhaps 1 or 2 percent). Some of the excited atoms drop down fairly soon after being excited. One of these is shown on the left in Figure 27-18. If the emitted photon strikes another atom in the excited state, it stimulates this atom to emit a photon of the same frequency and in phase with it. These two photons then move on to strike other atoms, causing more stimulated emission. As the process continues, the number of photons multiplies. When the photons strike the end mirrors, most are reflected back, and as they move in the opposite direction, they continue to stimulate more atoms to emit photons. As the photons move back and forth between the mirrors, a small percentage pass through the partially transparent mirror at one end. These photons make up the narrow coherent laser beam that can be used.

Some photons will not be emitted parallel to the axis of the tube, and these will merely go out the side and not contribute to the main beam. Thus the beam can be very narrow. In a well-designed laser, the spreading of the beam is limited only by diffraction, so the angular spread is $\approx \lambda/D$ (see Equation 22-2), where D is the diameter of the end mirror; the diffraction spreading can be incredibly small. The light energy, instead of spreading out in space as it does for an ordinary light source, is directed in a pencil-thin beam.

The excitation of the atoms in a laser can be done in several ways to produce an inverted population so that stimulated emission dominates over absorption. In a ruby laser, the lasing material is a ruby rod consisting of Al_2O_3, with a small percentage of Al atoms replaced by chromium (Cr) atoms. The Cr atoms are the ones involved in lasing. They are excited by strong flashes of light of wavelength 550 nm, which corresponds to a photon energy of 2.2 eV. As shown in Figure 27-19, the Cr atoms are excited from state E_0 to state E_2. This process is called optical pumping. The atoms quickly decay either back to E_0 or to the intermediate state E_1, which is metastable with a lifetime of about 3×10^{-3} s (compared to 10^{-8} s for ordinary levels). With strong pumping action, more atoms can be forced into the E_1 state than are in the E_0 state. Thus we have the inverted population needed for lasing. As soon as a few atoms in the E_1 state jump down to E_0, they produce stimulated emission of the other atoms and the lasing action begins. A ruby laser thus emits a beam whose photons have energy 1.8 eV (the energy difference between E_1 and E_0), which corresponds to a wavelength of 694.3 nm or "ruby-red" light.

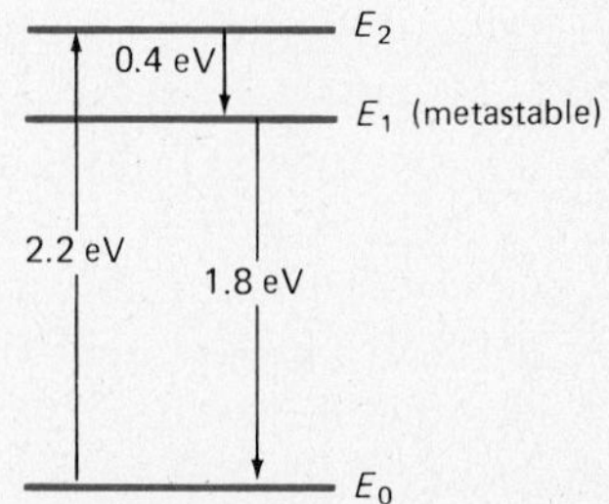

FIGURE 27-19
Energy levels of chromium in ruby crystal. Photons of energy 2.2 eV "pump" atoms from E_0 to E_2, which then decay to metastable state E_1. Lasing action occurs by stimulated emission of photons in transition from E_1 to E_0.

In a helium–neon (He–Ne) laser, the lasing material is a gas, a mixture of about 15 percent He and 85 percent Ne. This combination works as a lasing material because of certain compatible properties of the two gases. In this laser the atoms are excited by applying a high voltage to the tube so that an electric discharge takes place within the gas. In the

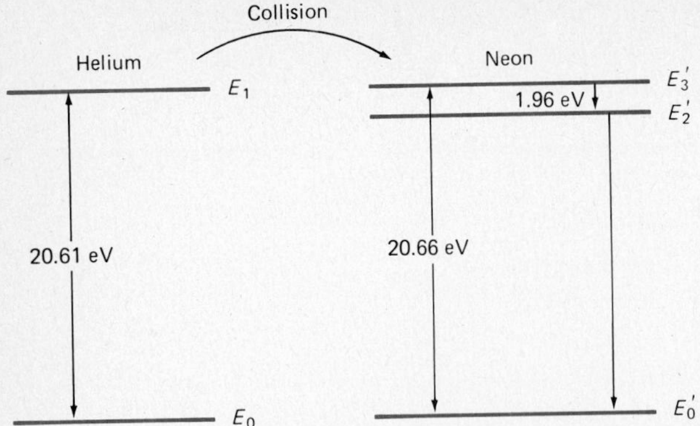

FIGURE 27-20
Energy levels for He and Ne. He is excited in electric discharge to the E_1 state. This energy is transferred to the E_3' level of Ne by collision. E_3' is metastable and decays to E_2' by stimulated emission.

process, some of the He atoms are raised to the metastable state E_1 shown in Figure 27-20, which corresponds to a jump of 20.61 eV.

Now Ne atoms have an excited state that is almost exactly the same energy above the ground state, 20.66 eV. The He atoms, since they cannot return readily to the ground state by spontaneous emission, instead give their excess energy to a Ne atom when they collide; in such a collision, the He drops to the ground state and the Ne atom is excited to the state E_3' (the prime refers to neon states). The slight difference in energy (0.05 eV) is supplied by the kinetic energy of the moving molecules. In this manner, the E_3' state in Ne—which is metastable—becomes more populated than the E_2' level. The inverted population between E_3' and E_2' is what is needed for lasing.

Another type of laser makes use of a *pn* junction in which the transitions occur between the bottom of the conduction band and the upper part of the valence band (Chapter 19).

The excitation of the atoms in a laser can be done continuously or in pulses. In a *pulsed laser*, the atoms are excited by periodic inputs of energy; the multiplication of photons continues until all the atoms have been stimulated to jump down to the lower state, and the process is repeated with each pulse. In a *continuous laser*, the energy input is continuous so that as atoms are stimulated to jump down to the lower level, they are soon excited back up to the upper level. In either case, of course, the laser is not a source of energy. Energy must be put in, and the laser converts a part of this into an intense narrow beam. Thus a laser, like an electronic amplifier, uses energy to amplify a certain type of signal.

The unique feature of light from a laser is, as mentioned before, that it is a coherent narrow beam of a single frequency (or several distinct frequencies). Because of this feature, the laser has found many applications. Lasers are a useful surgical tool. The narrow intense beam can be used to destroy tissue in a localized area; or, because of the heat produced, a laser beam can be used to "weld" broken tissue, such as a detached retina. Tiny organelles within a living cell have been destroyed by researchers using lasers to study how the absence of that organelle affects the behavior of the cell. Recently, lasers have been used in the treatment of cancer. The finely focused beam can destroy cancerous and precancerous cells; at the same time the heat seals off capillaries and lymph vessels, thus "cauterizing" the wound in the process to prevent spread of the disease. The intense heat produced in a small area by a laser beam is also used for

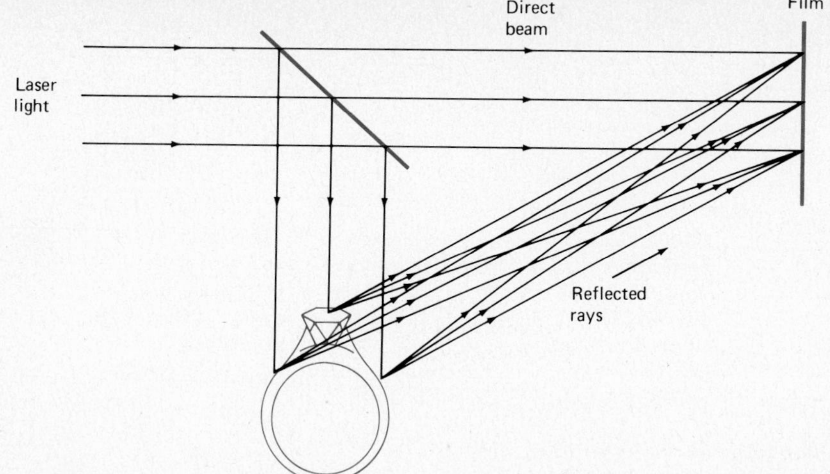

FIGURE 27-21
Making a hologram.

welding and machining metals and for drilling tiny holes in hard materials. The beam of a laser is narrow in itself (typically a few mm). But because the beam is coherent, monochromatic, and essentially parallel and narrow, lenses can be used to focus the light into incredibly small areas without the usual aberration problems; the limit thus becomes diffraction, and the energy crossing unit area per unit time can be very large.

The precise straightness of a laser beam is useful to surveyors for lining up equipment precisely, especially in inaccessible places. One example of its use was for alignment of the tunnels under San Francisco Bay for new rapid transit (BART) trains.

New uses for lasers are continually being thought of. One of the most interesting is the production of three-dimensional images called *holograms.* In an ordinary photograph, the film simply records the intensity of light reaching it at each point; when the photograph or transparency is viewed, light reflecting from it or passing through it gives us a two-dimensional picture. In the process of holography, the images are formed by interference, without lenses. When a laser hologram is made on film, a broadened laser beam is split into two parts by a half-silvered mirror (Figure 27-21). One part passes to the object to be photographed, from which it is reflected to the film. The other portion goes directly to the film, as shown. Light from every point on the object reaches each point on the film, and the interference of the two beams allows the film to record both the intensity and relative phase of the light at each point. After the film is developed, it is placed again in a laser beam and a three-dimensional image of the object is seen. You can walk around such an image and see it from all sides as if it were the original object. Yet, if you try to touch it with your hand, there will be nothing material there.

The details of how the image is formed are quite complicated. But we can get the basic idea by considering one single point on the object. In Figure 27-22*a* the rays *OA* and *OB* have reflected from one point on our object. The rays *CA* and *DB* come directly from the source and interfere with *OA* and *OB* at points *A* and *B* on the film. A set of interference fringes is produced as shown in Figure 27-22*b*. The spacing between the fringes changes from top to bottom as shown. Why this happens is

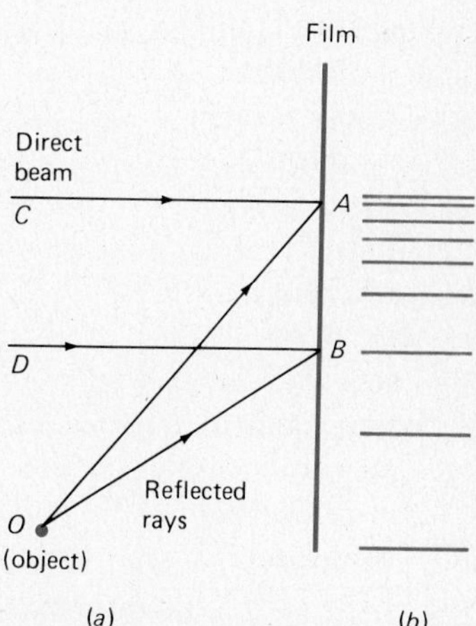

FIGURE 27-22
Light from point *O* on the object interferes with light of direct beam (rays *CA* and *DB*).

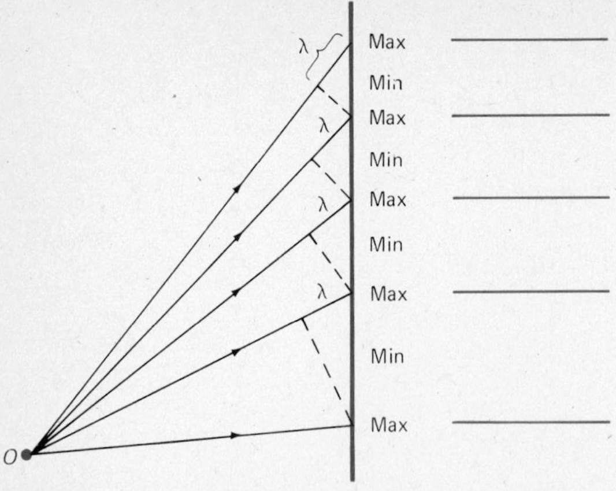

FIGURE 27-23
Each of the rays shown leaving point *O* is one wavelength shorter than the one above it. If the top ray is in phase with the direct beam (not shown), which has the same phase at all points on the screen, all the rays shown produce constructive interference. From this diagram it can be seen that the fringe spacing increases toward the bottom.

explained in Figure 27-23. Thus the hologram of a single point object would have the pattern shown in Figure 27-22*b*. The film in this case looks like a diffraction grating with variable spacing. Hence when coherent laser light is passed back through the developed film, the diffracted rays in the first-order maxima occur at slightly different angles because the spacing changes (remember Equation 22-2, $\sin\theta = \lambda/D$; so where the spacing D is greater, the angle θ is less). Hence the rays diffracted upward (in first order) seem to diverge from a single point (Figure 27-24). This is a virtual image of the original object, which can be seen with the eye. Rays diffracted *downward* in first order converge to make a real image which can be seen but can also be photographed. (Note that the straight-through undiffracted rays are of no interest.) Of course, real objects consist of many points, so a hologram will be a complex interference pattern which, when laser light shines on it, will reproduce an image of the object; each image point will be at the correct (three-dimensional) position with respect to other points, so the image accurately represents the original object; and it can be viewed from different angles as if viewing the original object. Holograms can now be made in which a viewer can walk entirely around the image (360°) and see all sides of it.

Normally, the light beam used to reconstruct the image is identical to that used to form the hologram in the first place. In this case the magnification is one (image size equals object size). However, it is possible to produce different magnifications by using spherical waves of different

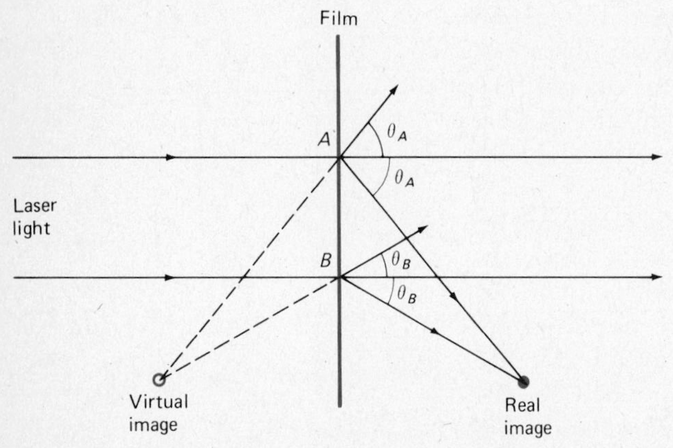

FIGURE 27-24
Laser beam strikes film that is like a diffraction grating of variable spacing. Rays corresponding to the first diffraction maxima are shown emerging. The angle $\theta_A > \theta_B$ because the spacing at *B* is greater than at *A* ($\sin\theta = \lambda/D$). Hence real and virtual images of the point are reproduced as shown.

radii of curvature instead of plane waves as discussed above; lenses are used to change the plane-wave laser beam into a spherical wave that seems to be emanating from a point.

A hologram records every plane in a specimen. It can thus be used to observe objects within translucent materials, such as particular cells within living tissue. Film can be placed at various positions and angles within the hologram image to obtain a great deal more information on life processes than is possible with ordinary microscopes. Such holographic "microscopes" are being developed and could be used for many biological purposes, from examining the interior of the eye or a complex capillary bed structure to studies of cell division and brain tissue.

Holography using nonvisible light is also possible. Lasers can be made that produce IR, UV, and microwave radiation. It may also prove feasible to develop instruments using x rays and electrons. These types of holography could have many uses. For example, UV holograms could not only produce better resolution (shorter wavelength) but could use the specific absorption of nucleic acids and proteins in the UV region to produce greater-contrast images. Acoustical holography (using coherent sound waves, particularly at ultrasonic frequencies) could have many uses. In medicine, acoustical holograms of the interior of a body would allow reconstruction of all the internal organs and their shape and placement. This would be of great value in diagnosis.

Finally, the principles of holography are being applied to moving pictures, and it may not be long before we have truly three-dimensional movies.

Summary

According to quantum mechanics, the state of an electron in an atom is specified by four quantum numbers: n, l, m_l, and m_s. The principal quantum number, n, can take on any integer values $(1, 2, 3, \cdots)$ and corresponds to the quantum number of the old Bohr theory; l can take on values from 0 up to $n-1$; m_l can take on integer values from $-l$ to $+l$; and m_s can be $+\frac{1}{2}$ or $-\frac{1}{2}$. The energy levels in the hydrogen atom depend on n, whereas in other atoms they depend on n and l; when an external magnetic field is applied, the spectral lines are split (the Zeeman effect), indicating that the energy depends also on m_l in this case. Even in the absence of a magnetic field, precise measurements of spectral lines show a tiny splitting of the lines called fine structure, whose explanation is that the energy depends very slightly on the spin quantum number m_s.

The arrangement of electrons in multielectron atoms is governed by the Pauli *exclusion principle*, which states that no two electrons can occupy the same quantum state—that is, they cannot have the same set of quantum numbers n, l, m_l, and m_s. Electrons, as a result, are grouped into shells (according to the value of n) and subshells (according to l). This shell structure gives rise to a periodicity in the properties of the elements.

When a transition occurs between inner energy states of large atoms, the energy difference is large, so the emitted photon is an x ray. The wavelengths of these x rays are characteristic of the type of atom; for a given transition, particularly for $n = 2$ to $n = 1$, the wavelength is inversely proportioned to $(Z-1)^2$. This fact helped in the determination of the atomic number Z for a number of atoms.

Quantum mechanics explains the bonding together of atoms to form molecules. In a covalent bond, the electron clouds of two or more atoms overlap because of constructive interference between the electron waves; the positive nuclei are attracted to this concentration of negative charge between them, which forms the bond. An ionic bond is an extreme case of a covalent bond in which one or more electrons from one atom spend much more time around the other atom than around their own; the atoms then act as oppositely charged ions which attract each other—which is the bond.

Questions

1 Discuss the differences between Bohr's view of the atom and the quantum mechanical view.

2 The 589-nm yellow line in sodium is actually two very closely spaced lines. It is due to an "internal" Zeeman effect. Can you explain this? (*Hint*: Put yourself in the reference frame of the electron.)

3 Which of the following electron configurations are forbidden? (*a*) $1s^2\,2s^2\,2p^6\,3s^3$; (*b*) $1s^2\,2s^2\,2p^4\,3s^2\,4p^2$; (*c*) $1s^2\,2s^2\,2p^8\,3s^1\,3p^6$.

4 Give the complete electron configuration for a uranium atom.

5 In what column of the periodic table would you expect to find the atom with each of the following configurations? (*a*) $1s^2\,2s^2\,2p^6\,3s^2$; (*b*) $1s^2\,2s^2\,2p^6\,3s^2\,3p^6$; (*c*) $1s^2\,2s^2\,2p^6\,3s^2\,3p^6\,4s^1$; (*d*) $1s^2\,2s^2\,2p^5$.

6 The ionization energy for neon ($Z = 10$) is 21.6 eV and that for sodium ($Z = 11$) is 5.1 eV. Explain the large difference.

7 Why do chlorine and iodine exhibit similar properties?

8 Explain why potassium and sodium exhibit similar properties.

9 Why are the chemical properties of the rare earths so similar?

*10 Why do we not expect perfect agreement between measured values of x-ray lines and those calculated using Bohr theory, as in Example 27-2?

*11 Certain dyes and other materials fluoresce by emitting visible light when UV light falls on them. Can infrared light produce fluorescence?

*12 Compare spontaneous emission to stimulated emission.

*13 How does laser light differ from ordinary light? How is it the same?

Problems

SECTIONS 27-1 to 27-3

1 (I) How many different states are possible for an electron whose principal quantum number is $n = 4$?

2 (I) List the quantum numbers for each electron in the ground state of boron ($Z = 5$).

3 (I) How many electrons can be in the $n = 5$, $l = 2$ subshell?

4 (I) For $n = 6$, what values can l have?

5 (I) For $n = 4$, $l = 3$, what are the possible values of m_l and m_s?

6 (I) List the quantum numbers for each electron in carbon.

7 (II) Calculate the magnitude of the angular momentum of an electron in the $n = 3$, $l = 2$ state of hydrogen.

8 (II) Using the Bohr formula for the radius of an electron orbit, estimate the average distance from the nucleus for an electron in the innermost ($n = 1$) orbit in uranium ($Z = 92$). Approximately how much energy would be required to remove this innermost electron?

9 (II) Estimate the binding energy of the third electron in lithium using Bohr theory. (*Hint*: This electron has $n = 2$ and "sees" a net charge of approximately $+1e$.) The measured value is 5.39 eV.

SECTION 27-4

10 (II) Estimate the wavelength for an $n = 2$ to $n = 1$ transition in iron ($Z = 26$).

11 (II) Use Bohr theory to estimate the wavelength for an $n = 3$ to $n = 1$ transition in molybdenum. The measured value is 0.063 nm. Why do we not expect perfect agreement?

12 (II) Use the result of Example 27-2 to estimate the x-ray wavelength emitted when Ti ($Z = 22$) atom jumps from $n = 2$ to $n = 1$.

*SECTION 27-10

*13 (II) A laser used to weld detached retinas puts out 25-ms-long pulses of 640-nm light, which average 0.50-W output during a pulse. How much energy can be deposited per pulse, and how many photons does each pulse contain?

*14 (III) Estimate the angular spread of a laser beam due to diffraction if the beam emerges through a 5.0-mm-diameter mirror. Assume that $\lambda = 694$ nm. How large would be the diameter of this beam if it struck a satellite 1000 km above the earth?

For additional reading

Ackerman, E., *Biophysical Science*, Prentice-Hall, Inc., Englewood Cliffs, N.J., 1962. Enzymes, activation energy, molecular spectroscopy.

Feleppa, E. J., "Biomedical Applications of Holography," *Physics Today*, **22**:7 (Jul., 1969), p. 25.

Herzenberg, L. A., L. A. Herzenberg, and R. G. Sweet, "Fluorescence-activated Cell Sorting," *Sci. American*, **234**:3 (Mar., 1976), p. 108.

Hinkle, P. C., and R. E. McCarty, "How Cells Make ATP," *Sci. American*, **238**:3 (Mar., 1978), p. 104.

Lehninger, A. L., *Bioenergetics: The Molecular Basis of Biological Transformation*, 2nd ed., W. A. Benjamin, Inc., Menlo Park, Calif., 1971. Activation energy and ATP.

Leith, E. N., "White-light Holograms," *Sci. American*, **235**:4 (Oct., 1976), p. 80.

Porter, A. G., and S. George, "An Elementary Introduction to Practical Holography," *Am. J. Phys.*, **43** (1975), p. 954.

Setlow, R. B., and E. C. Pollard, *Molecular Biophysics*, Addison-Wesley Publishing Co., Inc., Reading, Mass., 1962.

Snell, F. M., *et al.*, *Biophysical Principles of Structure and Function*, Addison-Wesley Publishing Co., Inc., Reading, Mass., 1965.

Strong, C. L., "How to Make Holograms . . . ," *Sci. American*, **216**:2 (Feb., 1967), p. 122.

28 ▌ The nucleus and radioactivity

In the early part of the twentieth century Rutherford's experiments led to the idea that at the center of an atom there is a tiny but massive nucleus. At the same time that the quantum theory was being developed and scientists were attempting to understand the structure of the atom and its electrons, investigations into the nucleus itself had also begun. In this chapter and the next, we study this tiny and mysterious world of **nuclear physics**.

28-1 ▌ Structure of the nucleus

An important question to physicists in the early part of this century was whether the nucleus had a structure, and what this structure might be. It turns out that the nucleus is a complicated entity and is not completely understood even today. However, by the early 1930s a model of the nucleus had been developed that still has validity today. According to this model, a nucleus is considered as an aggregate of two types of particles: protons and neutrons. (Of course, we must remember that these "particles" also have wave properties, but for ease of visualization and language we simply refer to them as "particles.") A **proton** is the same thing as the nucleus of the simplest atom, hydrogen. It has a positive charge ($=+e=+1.6\times10^{-19}$ C) and a mass $m_p=1.672\times10^{-27}$ kg. The **neutron**, whose existence was ascertained only in 1932 by the Englishman James Chadwick (1891–1974), is electrically neutral ($q=0$), as its name implies; its mass, which is almost identical to that of the proton, is 1.675×10^{-27} kg. These two constituents of a nucleus, neutrons and protons, are referred to collectively as **nucleons**.

Although the hydrogen nucleus consists of a single proton alone, the nuclei of other elements consist of both neutrons and protons. The different types of nuclei are often referred to as *nuclides*. The number of protons in a nucleus (or nuclide) is called the **atomic number** and is designated by the symbol Z. The total number of nucleons, neutrons plus protons, is designated by the symbol A and is called the **atomic mass number**. This name is used since the mass of a nucleus is very closely A times the mass of one nucleon. A nuclide with 7 protons and 8 neutrons thus has $Z=7$ and $A=15$. The *neutron number* N equals $A-Z$.

To specify a given nuclide, we need give only A and Z. A special symbol is used which takes the form

$$^{A}_{Z}X,$$

where X is the chemical symbol for the element (see the periodic table on inside front cover), A the atomic mass number, and Z the atomic number. For example, $^{15}_{7}N$ means a nitrogen nucleus containing 7 protons and 8 neutrons, for a total of 15 nucleons. Now in a neutral atom, the number of electrons orbiting the nucleus is equal to the atomic number Z (since the charge on an electron has the same magnitude but opposite sign to that of a proton). The main properties of an atom are determined by the number of electrons. Hence Z determines what kind of atom it is: carbon, oxygen, gold, or whatever. It is redundant to specify both the symbol of a nucleus and its atomic number Z, as described above. If the nucleus is nitrogen, for example, we know immediately that $Z=7$. The subscript Z is thus sometimes dropped and the $^{15}_{7}N$ nucleus is then written simply ^{15}N; in words, we say "nitrogen fifteen." The subscript Z is often kept, nonetheless, for convenience.

For a particular type of atom (say carbon), nuclei are found that contain different numbers of neutrons, although they all have the same number of protons. For example, carbon nuclei have 6 protons but may have 5, 6, 7, 8, 9, or 10 neutrons. Nuclei that contain the same number of protons but different numbers of neutrons are called **isotopes**. Thus $^{11}_{6}C$, $^{12}_{6}C$, $^{13}_{6}C$, $^{14}_{6}C$, $^{15}_{6}C$, and $^{16}_{6}C$ are all isotopes of carbon. Of course, the isotopes of a given element are not all equally common. For example, 98.9 percent of naturally occurring carbon (on earth) is the isotope $^{12}_{6}C$ and 1.1 percent is $^{13}_{6}C$. These percentages are referred to as the *natural abundances*.† Many isotopes that do not occur naturally can be produced in the laboratory by means of nuclear reactions (more on this later). Indeed, all elements beyond uranium ($Z>92$) do not occur naturally and are only produced artificially.

The approximate size of nuclei was determined originally by Rutherford by scattering of charged particles. The most accurate recent measurements have been done by scattering of high-speed electrons off nuclei. It is found that nuclei have a roughly spherical shape with a radius that increases with A according to the approximate formula

$$r \approx (1.2\times 10^{-15}\text{ m})(A^{1/3}). \tag{28-1}$$

Since the volume of a sphere is $V=\frac{4}{3}\pi r^3$, we see that the volume is proportional to the number of nucleons, $V\propto A$. This is what we would expect if nucleons were like impenetrable billiard balls: if you double the number of balls, you double the total volume. This result indicates that to form a nucleus, nucleons combine as if they were each impenetrable spheres, and all nuclei have nearly the same density.

The rest mass of nuclei can be determined by measuring the radius of curvature of fast-moving nuclei in a magnetic field using a mass spectrograph as discussed in Section 17-4. Indeed, as mentioned there, the existence of different isotopes of the same element was discovered using this device. Nuclear masses are specified in *unified atomic mass units* (u). On

† The mass values for the elements as given in the periodic table (inside the front cover) are averaged over the natural abundances.

TABLE 28-1

Rest masses in kilograms, atomic mass units, and MeV/c^2

Object	Mass: kg	u	MeV/c^2
Electron	9.11×10^{-31}	0.00055	0.511
Proton	1.672×10^{-27}	1.00728	938.3
$^1_1$H atom	1.673×10^{-27}	1.00783	938.8
Neutron	1.675×10^{-27}	1.00867	939.6

this scale, a neutral $^{12}_{6}$C atom is given the precise value 12.0000 u. A neutron then has a mass of 1.0087 u, a proton 1.0073 u, and a neutral hydrogen atom, $^1_1$H (proton plus electron), 1.0078 u. The masses of many nuclides are given in Appendix E. (It should be noted that the masses in this table, as is customary, are for the *neutral atom,* not for a bare nucleus.)

Masses are often specified using the electron-volt energy unit. This can be done because mass and energy are related, and the precise relationship is given by Einstein's famous equation $E = mc^2$ (Chapter 24). Since the mass of a neutral $^1_1$H atom is 1.673×10^{-27} kg or 1.0078 u, then $1.0000\ \text{u} = (1.0000/1.0078)\times(1.673\times10^{-27}\ \text{kg}) = 1.660\times10^{-27}$ kg; this is equivalent to an energy $E = mc^2 = (1.660\times10^{-27}\ \text{kg})(2.998\times10^8\ \text{m/s})^2/(1.602\times10^{-19}\ \text{J/eV}) = 931.5$ MeV. Thus

$$1\ \text{u} = 1.660\times10^{-27}\ \text{kg} = 931.5\ \text{MeV}/c^2.$$

(Note that, since $E = mc^2$, we state masses in MeV/c^2, although sometimes this is shortened to MeV.) The rest masses of some of the basic particles are given in Table 28-1.

28-2 ▮ Binding energy and nuclear forces

The total mass of a nucleus is always less than the sum of the masses of its constituent protons and neutrons as the following example shows.

▮ EXAMPLE 28-1 Compare the mass of a $^4_2$He nucleus to that of its constituent nucleons.

SOLUTION The mass of a neutral $^4_2$He atom, from Appendix E, is 4.0026 u. The mass of two neutrons and two protons (including the two electrons) is

$$2m_n = 2.0174\ \text{u}$$
$$2m_{^1_1\text{H}} = \underline{2.0157\ \text{u}}$$
$$4.0331\ \text{u}$$

(We almost always deal with masses of neutral atoms—i.e., nuclei with Z electrons—since this is how masses are measured. We must therefore be sure to balance out the electrons when we compare masses, which is why we used the mass of $^1_1$H in this example rather than that of the proton alone.) ▮▮

Thus the mass of $^4_2$He is measured to be 4.0331 u − 4.0026 u = 0.0305 u less than the masses of its constituents. How can this be? Where has this mass gone?

It has, in fact, gone into energy. In the case of $^4_2$He, this energy is (0.0305 u)(931.5 MeV/u) = 28.4 MeV. This is called the **total binding energy** of the nucleus. The total binding energy represents the amount of energy that must be put into a nucleus in order to break it apart into its constituent protons and neutrons. If the mass of, say, a $^4_2$He nucleus were exactly equal to the mass of two neutrons plus two protons, the nucleus would immediately fall apart without our having to put in any energy. To be stable, then, the mass *must* be less than that of its constituents.

This situation can be compared to the binding energy of electrons in an atom. We saw in Chapter 26 that the binding energy of the one electron in the hydrogen atom, for example, is 13.6 eV. Indeed, the mass of a $^1_1$H atom is less than that of a single proton plus a single electron by 13.6 eV. Compared to the total mass (938 MeV), this is incredibly small (1 part in 10^8), and for practical purposes the mass difference can be ignored. The binding energies of nuclei are on the order of 10^6 times greater than the binding energies of electrons and are therefore far more important.

The **average binding energy per nucleon** is defined as the total binding energy of a nucleus divided by A, the total number of nucleons. For $^4_2$He it is 28.4 MeV/4 = 7.1 MeV. Figure 28-1 shows the average binding energy per nucleon as a function of A for stable nuclei. The curve rises as A increases and reaches a plateau at about 8 MeV per nucleon above A = 15. Beyond about A = 60, the curve decreases slowly, indicating that larger nuclei (like the lightest ones) are held together a little less tightly than those in the middle of the periodic table. (We will see in Chapter 29 that this fact allows the release of nuclear energy in the processes of fission and fusion.)

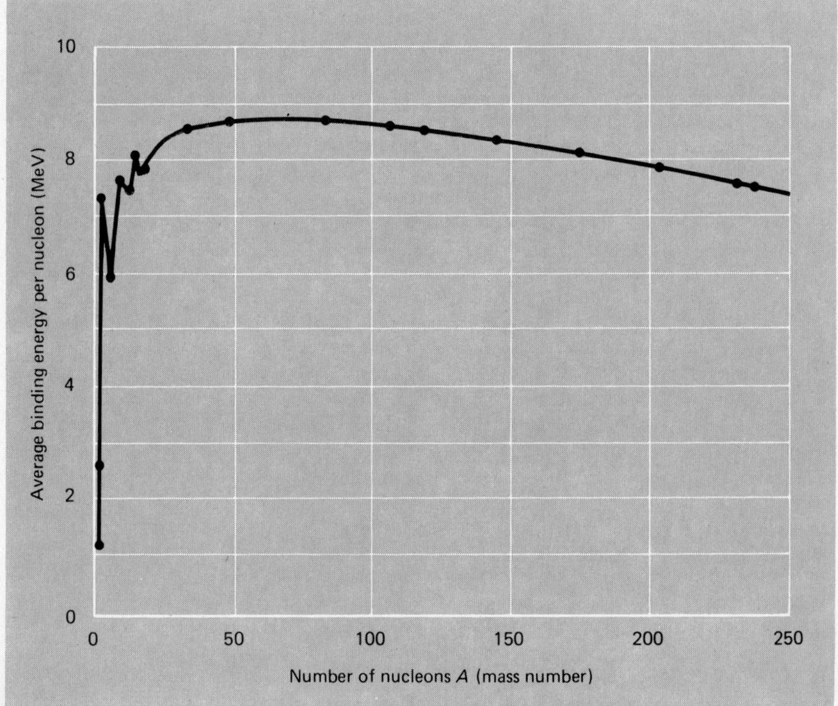

FIGURE 28-1
Average binding energy per nucleon as a function of mass number A for stable nuclei.

We can analyze nuclei not only from the energy point of view but also from the point of view of the forces that hold them together. We would not expect a collection of protons and neutrons to spontaneously come together since protons are all positively charged and they exert repulsive forces on each other. Indeed, the question arises as to how a nucleus stays together at all in view of the fact that the electric force between protons would tend to break it apart. Since stable nuclei *do* stay together, it is clear that another force must be acting. Because this new force is stronger than the electric force (which in turn is much stronger than gravity at the nuclear level), it is called the **strong nuclear force.** The strong nuclear force is an attractive force that acts between all nucleons, protons and neutrons alike. Thus protons attract each other via the nuclear force at the same time they repel each other via the electric force. Neutrons, since they are electrically neutral, only attract other neutrons or protons via the nuclear force.

The nuclear force turns out to be far more complicated than the gravitational and electromagnetic forces. A precise mathematical description is not yet possible. Nonetheless, a great deal of work has been done to try to understand the nuclear force. One important aspect of the strong nuclear force is that it is a *short-range* force: it acts only over a very short distance. It is very strong between two nucleons if they are less than about 10^{-15} m apart; but it is essentially zero if they are separated by a distance greater than this. Compare this to electric and gravitational forces, which can act over great distances and are therefore called *long-range* forces. The strong nuclear force has some strange quirks. For example, if a nuclide contains too many or too few neutrons relative to the number of protons, the nuclear force is weakened; nuclides that are too unbalanced in this regard are unstable. As shown in Figure 28-2, stable nuclei tend to have the same number of protons as neutrons ($N = Z$) up to about $A \approx 30$ or 40. Beyond this, stable nuclei contain more neutrons than protons. This reflects the fact that as Z increases, the electrical repulsion increases, so a greater number of neutrons—which only exert the attractive nuclear force—are required to maintain stability. For very large Z, no number of neutrons can overcome the greatly increased electric repulsion; indeed, there are no completely stable nuclides above $Z = 82$.

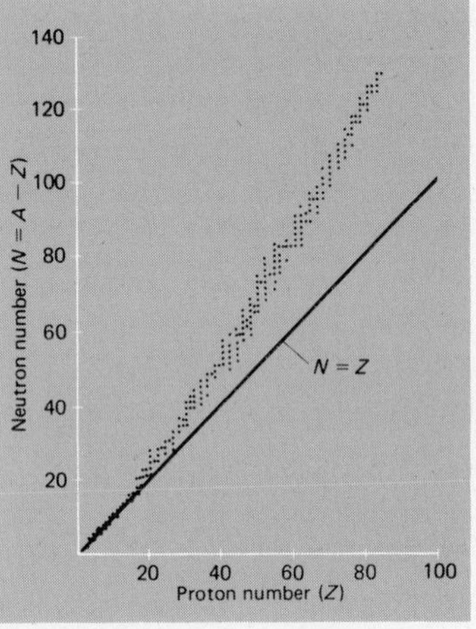

FIGURE 28-2
Number of neutrons versus number of protons for stable nuclides, which are represented by dots.

What we mean by a stable nucleus is one that stays together indefinitely. What then is an unstable nucleus? It is one that comes apart. This coming apart of an unstable nucleus is known as *radioactivity.*

Before we discuss the important subject of radioactivity (the next section), we note that there is a second type of nuclear force that is much weaker than the strong nuclear force. It is called the **weak nuclear force,** and we are aware of its existence only because it appears in certain types of radioactive decay. These two nuclear forces, the strong and the weak, together with the gravitational and electromagnetic forces, comprise the four known types of force in nature.

28-3 ▮ Discovery of radioactivity

Nuclear physics actually had its beginnings in 1896. For in that year, Antoine Henri Becquerel (1852–1908) made an important discovery. In his studies of phosphorescence, he found quite by accident that a mineral called "pitchblende" would darken a photographic plate even when the plate was wrapped to exclude light. It was clear that the pitchblende emitted some new kind of radiation which, unlike phosphorescence or x rays, occurred without any external stimulus. This new phenomenon eventually came to be called **radioactivity**.

Soon after Becquerel's discovery, Marie Curie (1867–1934) and her husband, Pierre Curie (1859–1906), isolated two previously unknown elements from pitchblende which were very highly radioactive. These were named radium and polonium. Other radioactive elements were soon discovered as well. The radioactivity (the emission of radiation that could, among other things, darken unexposed film) was found in every case to be unaffected by the strongest physical and chemical treatments: strong heating or cooling and the action of strong chemical reagents did not affect the radioactivity. It soon became clear that the source of radioactivity must be deep within the atom. And with Rutherford's theory of the nuclear atom, scientists realized that radioactivity must emanate from the nucleus. It soon became apparent that radioactivity is the result of the *disintegration* or *decay* of an unstable nucleus. Certain isotopes are not stable under the action of the nuclear force, and they decay with the emission of radiation or "rays."

Many unstable isotopes occur in nature, and such radioactivity is called "natural radioactivity." Other unstable isotopes can be produced in the laboratory by nuclear reactions (Chapter 29); these are said to be produced "artificially" and they are sometimes said to have "artificial radioactivity."

28-4 ▌ Radioactive decay modes: α, β, and γ

Rutherford and others began studying the nature of the rays emitted in radioactivity about 1898. They found that the rays could be classified into three distinct types according to their penetrating power. One type of radiation could barely penetrate a piece of paper. The second type could pass through as much as 3 mm of aluminum. The third was extremely penetrating: it could pass through several centimeters of lead and still be detected on the other side. They named these three types of radiation alpha (α), beta (β), and gamma (γ), respectively, after the first three letters of the Greek alphabet.

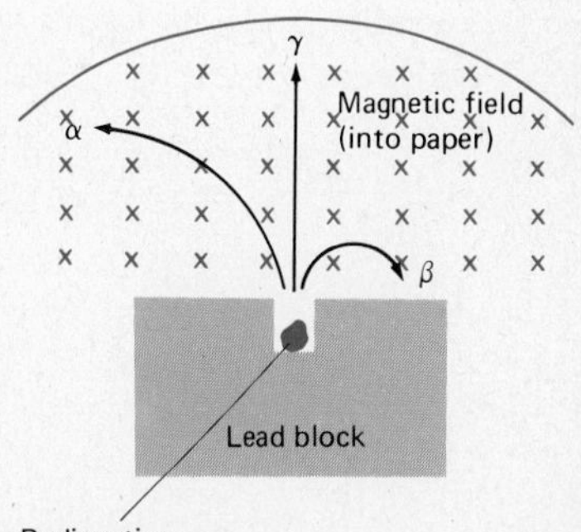

FIGURE 28-3
Alpha and beta rays are bent in opposite directions by a magnetic field, whereas gamma rays are not bent at all.

Each type of ray was found to have a different charge and hence is bent differently in a magnetic field (Figure 28-3); α rays are positively charged, β rays are negatively charged, and γ rays are neutral. It was soon found that all three types of radiation consisted of familiar kinds of particles. Gamma rays are very high energy photons whose energy (and frequency) is even higher than that of x rays. Beta rays are electrons, identical to those that orbit the nucleus (yet they have their origin within the nucleus itself). Alpha rays (or α particles) are simply the nuclei of helium atoms, ${}^4_2\text{He}$; that is, an α ray consists of two protons and two neutrons bound together.

We now discuss each of these three types of radioactive decay in more detail.

Alpha decay

When a nucleus emits an α particle ($^4_2$He), it is clear that the remaining nucleus will be different from the original: for it has lost two protons and two neutrons. Radium 226 ($^{226}_{88}$Ra), for example, is an α emitter. It decays to a nucleus with $Z = 88 - 2 = 86$ and $A = 226 - 4 = 222$. The nucleus with $Z = 86$ is radon (Rn). Thus the radium decays to radon with the emission of an α particle. This is written

$$^{226}_{88}\text{Ra} \rightarrow {}^{222}_{86}\text{Rn} + {}^4_2\text{He}.$$

It is clear that when α decay occurs, a new element is formed. The *daughter* nucleus ($^{222}_{86}$Rn in this case) is different from the *parent* nucleus ($^{226}_{88}$Ra in this case). This changing of one element into another is called **transmutation**.

Alpha decay occurs because the strong nuclear force is unable to hold very large nuclei together. Because the nuclear force is a short-range force, it acts only between neighboring nucleons. But the electric force can act clear across the nucleus. For very large nuclei, the large Z means the repulsive electric force becomes very large (Coulomb's law) and acts between all protons; the strong nuclear force, since it acts only between neighboring nucleons, is overpowered and is unable to hold the nucleus together.

We can express the instability in terms of binding energy. That is, the binding energy of a radioactive nucleus is too low for it to remain stable. More specifically, the mass of the parent nucleus must be greater than the mass of the daughter nucleus plus the mass of the α particle if radioactive decay is to occur. The mass difference appears as kinetic energy which is carried away mainly by the α particle. If the parent has *less* mass than the daughter plus α particle, the decay could not occur for the conservation-of-energy law would be violated.

▌ EXAMPLE 28-2 Calculate the kinetic energy of the α particle emitted when $^{232}_{92}$U (mass = 232.0372 u) decays to $^{228}_{90}$Th (228.0287 u).

SOLUTION Since the mass of a $^4_2$He atom (see Example 28-1) is 4.0026 u, the total mass in the final state is 228.0287 u + 4.0026 u = 232.0313 u. The mass lost when the $^{232}_{92}$U nucleus decays is 232.0372 u − 232.0313 u = 0.0059 u. This mass appears as kinetic energy; since 1 u = 931.5 MeV/c^2, the KE released is (0.0059 u)(931.5 MeV/u) = 5.4 MeV. (Actual measurements show that the α particle in this decay has 5.3 MeV. Thus the daughter nucleus—which recoils in the opposite direction from the emitted α particle because of momentum conservation—has about 0.1 MeV of kinetic energy.) ▌▌

Why, you may wonder, do nuclei emit this combination of four nucleons called an α particle? Why not just four nucleons, or even one? The answer lies in the fact that the α particle is very strongly bound and its mass is so much less than four separate nucleons. As we saw in Example 28-1, two protons and two neutrons separately have a total mass of 4.0331 u. A $^{228}_{90}$Th nucleus plus these four nucleons has a total mass of 232.0617 u, which is much greater than the mass of the parent nucleus. Such a decay could not occur because it would violate conservation of energy. It is also nearly always true that the emission of a single nucleon is also energetically impossible.

Beta decay

Transmutation of elements also occurs when a nucleus decays by β decay—that is, with the emission of an electron or β particle. The nucleus $^{14}_{6}C$, for example, decays as follows:

$$^{14}_{6}C \rightarrow {}^{14}_{7}N + {}^{0}_{-1}e.$$

(The symbol $^{0}_{-1}e$ stands for an electron whose charge corresponds to $Z = -1$ and, since it is not a nucleon and has very small mass, has $A = 0$.) No nucleons are lost when an electron is emitted, so the total number of nucleons, A, is the same in the daughter as in the parent. But because an electron has been emitted, the charge on the daughter is different from the parent. The parent had $Z = +6$. In the decay, the nucleus loses a charge of -1, so the nucleus remaining behind must have an extra + charge for a total of 7. Hence $Z = 7$ for the daughter, which is a nitrogen nucleus. Another way of seeing this is to remember that charge is conserved. The parent nucleus has a charge of $+6e$. Therefore, the total charge in the final state must be $+6e$; and it is because $+7e + (-1e) = +6e$.

It must be carefully noted that the electron emitted in β decay is *not* an orbital electron. Rather, the electron is created *within the nucleus itself*. It is as if one of the neutrons changes to a proton and in the process (to conserve charge) throws off an electron. Indeed, free neutrons actually do decay in this fashion: $n \rightarrow p + e^{-}$. Because of their origin in the nucleus, the electrons emitted in β decay are sometimes referred to as "β particles," rather than as electrons, to remind us of their origin.

▎EXAMPLE 28-3 How much energy is released when $^{14}_{6}C$ ($m =$ 14.00324 u) decays to $^{14}_{7}N$ (14.00307 u) by β emission? The masses given are, as is usual, those of the neutral atom.

SOLUTION Because the masses given are those of the neutral atoms, we have to keep track of the electrons involved. Assume that the parent nucleus has six orbiting electrons, so that it is neutral and its mass is as given above. The daughter, in this decay $^{14}_{7}N$, is not neutral, however, since it has the same six electrons circling it but the nucleus has a charge of $+7e$. However, the mass of this daughter with its six electrons, plus the mass of the emitted electron (which makes a total of seven electrons), is just the mass of a neutral nitrogen atom. That is, the mass in the final state is

$$\begin{aligned}&(\text{mass of } {}^{14}_{7}\text{N nucleus} + 6 \text{ electrons}) + (\text{mass of 1 electron})\\&\quad = \text{mass of neutral } {}^{14}_{7}\text{N (includes 7 electrons)}\\&\quad = 14.00307 \text{ u}.\end{aligned}$$

Hence the mass before decay is 14.00324 u and after decay is 14.00307 u, so the mass difference is 0.00017 u, which corresponds to 0.16 MeV, or 160 keV. ▮▮

According to this example, we would expect the emitted electron to have a kinetic energy of 0.16 MeV. (The daughter nucleus, because its mass is very much larger than that of the electron, recoils with very low velocity and hence gets very little of the kinetic energy.) Indeed, very careful measurements indicate that a few emitted β particles do have kinetic energy close to this calculated value; but the vast majority of emitted electrons have somewhat less energy. In fact, the energy of the emitted electron can be anywhere from zero up to the maximum value calculated above. This was found to be true not only in the decay of $^{14}_{6}C$ but for all β decays. It was as if the law of conservation of energy was being violated! Careful experiments also indicated that linear momentum and angular momentum also did not seem to be conserved. Physicists were very troubled at the prospect of having to give up these laws, which had worked so well in all other situations. In 1930, Wolfgang Pauli proposed an alternative solution; perhaps, he said, a new particle that was very difficult to detect was emitted during β decay in addition to the electron. This hypothesized particle could be carrying off the energy, momentum, and angular momentum required to maintain the conservation laws. This new particle was named the **neutrino**—meaning "little neutral one"—by the great Italian physicist Enrico Fermi (1901–1954), who in 1934 worked out a detailed theory of β decay. (It was Fermi, in this theory, who postulated the existence of the fourth force in nature, which we call the weak nuclear force.) The neutrino has zero charge and seems to have zero rest mass. Like the photon, it can have energy and, since its rest mass seems to be zero, must travel at the speed of light. However, it is very difficult to detect. In 1956, complex experiments produced further evidence for the existence of the neutrino, but by then most physicists had already accepted its existence.

The symbol for the neutrino is the lowercase Greek letter nu (ν). Now we write the decay of $^{14}_{6}C$ as

$$^{14}_{6}C \rightarrow {}^{14}_{7}N + {}^{0}_{-1}e + \bar{\nu}.$$

The bar over the neutrino symbol is to indicate that it is an "antineutrino." (More on antiparticles later; why this is called an antineutrino rather than simply a neutrino need not concern us now; it is discussed in Chapter 30.)

Many isotopes decay by electron emission. They are always isotopes that have too many neutrons compared to the number of protons. That is, they are isotopes that lie above the stable isotopes plotted in Figure 28-2. But what about unstable isotopes that have too few neutrons compared to their number of protons, those that fall below the stable isotopes of Figure 28-2? These, it turns out, decay by emitting a *positron* instead of an electron. A positron (sometimes called an e^+ or β^+ particle) has the same mass as the electron, but it has a positive charge of $+1e$. Because it is like an electron, except for its charge, it is called the *antiparticle* to the electron. An example of a β^+ decay is that of $^{19}_{10}Ne$:

$$^{19}_{10}Ne \rightarrow {}^{19}_{9}F + {}^{0}_{1}e + \nu,$$

where ${}^{0}_{1}e$ stands for a positron (e^+) and ν is a neutrino. [Note that the ν emitted here is a neutrino, whereas that emitted in β^- decay is called an antineutrino. Thus an antielectron (= positron) is emitted with a neutrino, whereas an antineutrino is emitted with an electron; by naming them this way, we keep a certain symmetry.]

Besides β^- and β^+ emission, there is a third related process. This is *electron capture* and occurs when a nucleus absorbs one of its orbiting electrons. An example is $^{7}_{4}Be$, which as a result becomes $^{7}_{3}Li$; the process is written

$$^{7}_{4}Be + {}^{0}_{-1}e \rightarrow {}^{7}_{3}Li + \nu.$$

Usually, it is an electron in the innermost (K) shell that is captured, and then it is called "K-capture." The electron disappears in the process and a

TABLE 28-2

The three types of radioactive decay†

α decay

$$N(A, Z) \rightarrow N(A-4, Z-2) + {}^{4}_{2}\mathrm{He}$$

β decay

β^-: $N(A, Z) \rightarrow N(A, Z+1) + {}^{0}_{-1}e + \bar{\nu}$

β^+: $N(A, Z) \rightarrow N(A, Z-1) + {}^{0}_{1}e + \nu$

Electron capture: $N(A, Z) + {}^{0}_{-1}e \rightarrow N(A, Z-1) + \nu$

γ decay

$$N^*(A, Z) \rightarrow N(A, Z) + \gamma$$

† $N(A, Z)$ means a nucleus with atomic number Z and mass number A; * indicates an excited state of a nucleus.

proton in the nucleus becomes a neutron; a neutrino is emitted as a result.

In β decay it is the weak nuclear force that plays the crucial role. The neutrino is unique in that it interacts with matter only via the weak force. This is why it is so hard to detect.

Gamma decay

Gamma rays are very-high-energy photons. And the decay of a nucleus by emission of a γ ray is much like emission of photons by excited atoms. Like an atom, a nucleus itself can be in an excited state. When it jumps down to a lower energy state, or to the ground state, it emits a photon. The possible energy levels of a nucleus are much farther apart in energy than are those of an atom: on the order of keV or MeV, as compared to a few eV for electrons in an atom. Hence the emitted photons have energies that can range from a few keV to several MeV. For a given decay, the γ ray always has the same energy. Since a γ ray carries no charge, there is no change in the element as a result of a γ decay.

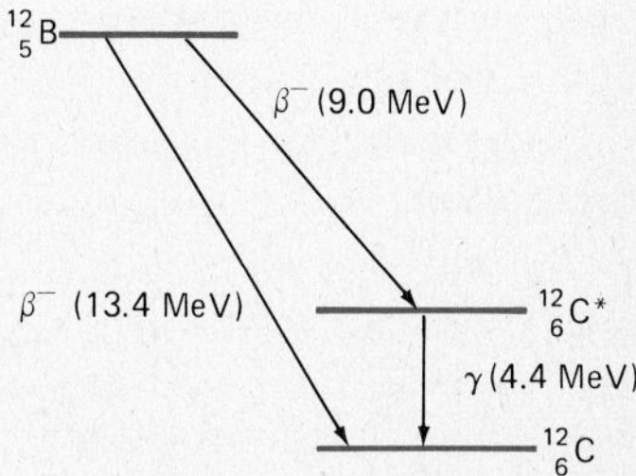

FIGURE 28-4
Energy-level diagram, showing how ${}^{12}_{5}$B can decay to the ground state of ${}^{12}_{6}$C by β decay (total energy released = 13.4 MeV) or can β-decay to an excited state of ${}^{12}_{6}$C (indicated by *), which subsequently decays to its ground state by emitting a 4.4-MeV γ ray.

How does a nucleus get into an excited state? It may occur because of a violent collision with another particle; or more commonly the nucleus remaining after a previous radioactive decay may be in an excited state. A typical example is shown in the energy-level diagram of Figure 28-4. ${}^{12}_{5}$B can decay by β decay directly to the ground state of ${}^{12}_{6}$C; or it can go by β decay to an excited state of ${}^{12}_{6}$C, which then decays by emission of a 4.4-MeV γ ray to the ground state.†

In some cases a nucleus may remain in an excited state for some time before it emits a γ ray as in Figure 28-4. The nucleus is then said to be in a *metastable state* and is called an *isomer*.

Table 28-2 gives a summary of α, β, and γ decays.

Conservation of nucleon number and other conservation laws

In all three types of radioactive decay the classical conservation laws hold. Energy, linear momentum, angular momentum, and electric charge are

† An excited nucleus can sometimes return to the ground state by a process known as *internal conversion.* In this process the excited nucleus interacts with one of the orbital electrons and ejects this electron from the atom with the same KE that an emitted γ ray would have had.

all conserved; these quantities are the same before the decay as after. But a new conservation law is also revealed. This is the law of **conservation of nucleon number**. According to this law, the total number of nucleons (A) remains constant in any process, although one type can change into the other type (protons into neutrons or vice versa). This law holds in all three types of decay (see Table 28-2).

28-5 ▮ Half-life and rate of decay

Decay constant and half-life

A macroscopic sample of any radioactive isotope consists of a vast number of radioactive nuclei. These nuclei do not all decay at one time. Rather, they decay one by one over a period of time. This is a random process: we cannot predict exactly when a given nucleus will decay. But we can determine, on a probabilistic basis, approximately how many nuclei in a sample will decay over a given time period.

The number of decays that occur in a short time interval Δt is found to be proportional to Δt and to the total number N of radioactive nuclei present:

$$\Delta N = -\lambda N \, \Delta t. \tag{28-2}$$

In this equation λ is a constant of proportionality called the *decay constant*, which is different for different isotopes. The greater λ is, the greater is the rate at which decays occur and the more radioactive that isotope is said to be. ΔN in Equation 28-2 represents the number of decays that occur in the short time interval Δt; we call this ΔN (meaning "change in N") since each decay that occurs corresponds to a decrease by one in the number N of nuclei present. That is, radioactive decay is a "one-shot" process (Figure 28-5): once a particular parent nucleus decays into its daughter, it cannot do it again. Hence, after (say) 10 nuclei have decayed, N has decreased by 10. The minus sign in Equation 28-2 is placed there to indicate that N is decreasing.

The number of parent nuclei in a sample decreases in time as shown in Figure 28-6*a*. Since the number of decays per second, $\Delta N/\Delta t$ (called the *activity* of the sample), is proportional to N (see Equation 28-2), it, too, decreases with time at the same rate (Figure 28-6*b*). Both of these curves are examples of "exponential" curves.†

The rate of decay of any isotope is often specified by giving its "half-life" rather than the decay constant. The **half-life** of an isotope is defined as the time it takes for half the original amount of isotope in a given sample to decay. For example, the half-life of ${}^{14}_{6}\mathrm{C}$ is about 5700 years. If at some time a piece of petrified wood contains, say, 1.00×10^{22} ${}^{14}_{6}\mathrm{C}$ nuclei, then 5700 years later it will contain only 0.50×10^{22} of these

(*a*)

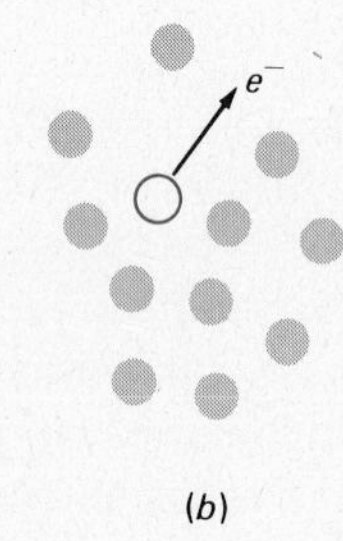

(*b*)

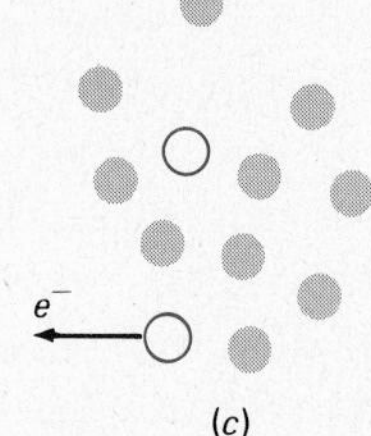

(*c*)

Legend
● ${}^{14}_{6}\mathrm{C}$ atom (parent)
○ ${}^{14}_{7}\mathrm{N}$ atom (daughter)

FIGURE 28-5
Radioactive nuclei decay one by one. Hence the number of parent nuclei in a sample is continually decreasing. When a ${}^{14}_{6}\mathrm{C}$ nucleus emits the electron, it becomes a ${}^{14}_{7}\mathrm{N}$ nucleus.

† The N versus t curve has the form $N = N_0 e^{-\lambda t}$, where N_0 is the number of parent nuclei present at time $t = 0$. The curve for rate of decay is written $\Delta N/\Delta t = (\Delta N/\Delta t)_0 e^{-\lambda t}$, where $(\Delta N/\Delta t)_0$ is the rate of decay at $t = 0$. The symbol e is a number called the "natural exponential" and has the value $e = 1 + \frac{1}{1} + \frac{1}{1 \cdot 2} + \frac{1}{1 \cdot 2 \cdot 3} + \cdots = 2.718 \ldots$. Why we use e instead of 2 or 10, say, to represent this exponential arises from calculus, and we will not deal with it here.

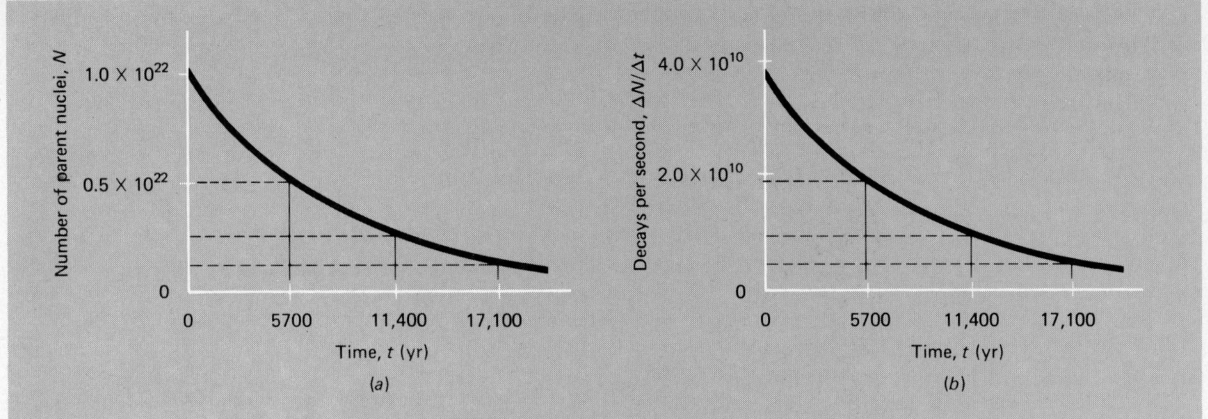

FIGURE 28-6
(a) The number N of parent nuclei in a given sample of $^{14}_{6}C$ decreases exponentially. (b) The number of decays per second also decreases exponentially. The half-life of $^{14}_{6}C$ is about 5700 yr, which means that the number of parent nuclei, N, and the rate of decay, $\Delta N/\Delta t$, decrease by half every 5700 yr.

nuclei. After another 5700 years it will contain 0.25×10^{22} nuclei, and so on. This is shown in Figure 28-6a. Since the rate of decay $\Delta N/\Delta t$ is proportional to N, it, too, decreases by a factor of 2 every half-life (Figure 28-6b).

The half-lives of known radioactive isotopes vary from about 10^{-22} s to 10^{28} s (about 10^{21} yr). The half-lives of many isotopes are given in Appendix E. It should be clear that the half-life (which we designate $T_{\frac{1}{2}}$) bears an inverse relationship to the decay constant. The longer the half-life of an isotope, the more slowly it decays, and hence λ is smaller; and very active isotopes (large λ) have very short half-lives. The precise relationship between half-life and decay constant is (calculus is needed to prove it)

$$T_{\frac{1}{2}} = \frac{0.693}{\lambda}. \tag{28-3}$$

EXAMPLE 28-4 The isotope $^{14}_{6}C$ has a half-life of 5700 yr. If at some time a sample contains 1.0×10^{22} carbon 14 nuclei, what is the activity of the sample?

SOLUTION First we calculate the decay constant λ from Equation 28-3 and obtain

$$\lambda = \frac{0.693}{T_{\frac{1}{2}}} = \frac{0.693}{(5700\ \text{yr})(3.15 \times 10^7\ \text{s/yr})} = 3.8 \times 10^{-12}\ \text{s}^{-1},$$

since there are $(60)(60)(24)(365) = 3.15 \times 10^7$ s in a year. From Equation 28-2, the activity or rate of decay (we ignore the minus sign) is

$$\frac{\Delta N}{\Delta t} = \lambda N = (3.8 \times 10^{-12}\ \text{s}^{-1})(1.0 \times 10^{22})$$
$$= 3.8 \times 10^{10}\ \text{decays/s}.$$

Note that the graph of Figure 28-6b starts at this value, corresponding to the original value of $N = 1.0 \times 10^{22}$ nuclei in Figure 28-6a.

EXAMPLE 28-5 A laboratory has 1.49 μg of pure $^{13}_{7}N$, which has a half-life of 10.0 min (600 s). (a) How many nuclei are present initially? (b) What is the activity initially? (c) What is the activity after 1.0 h? (d) After approximately how long will the activity drop to less than one per second?

SOLUTION (*a*) Since the atomic mass is 13, then 13 g will contain 6.02×10^{23} nuclei (Avogadro's number). Since we have only 1.49×10^{-6} g, the number of nuclei, N, that we have is given by the ratio

$$\frac{N}{6.02\times10^{23}}=\frac{1.49\times10^{-6}\text{ g}}{13\text{ g}},$$

so $N=6.9\times10^{16}$ nuclei. (*b*) From Equation 28-3, $\lambda=(0.693)/(600\text{ s})=1.16\times10^{-3}\text{ s}^{-1}$. Then

$$\frac{\Delta N}{\Delta t}=\lambda N=8.0\times10^{13}\text{ decays/s}.$$

(*c*) Since the half-life is 10 min, the decay rate decreases by half every 10 min. We can make the following table of activity after given periods of time:

Time (min)	Activity
0	8.0×10^{13}
10	4.0×10^{13}
20	2.0×10^{13}
30	1.0×10^{13}
40	0.50×10^{13}
50	0.25×10^{13}
60	0.13×10^{13}

Thus after 1.0 h, the activity is 1.3×10^{12} decays/s. A simpler way of seeing this is to note that 60 min is 6 half-lives, so the activity will decrease by $(\frac{1}{2})(\frac{1}{2})(\frac{1}{2})(\frac{1}{2})(\frac{1}{2})(\frac{1}{2})=(\frac{1}{2})^6=\frac{1}{64}$ of its original value, or $(8.0\times10^{13})/(64)=1.3\times10^{12}$ per second. (*d*) After each hour, the activity drops by a factor of 64, so our table is extended to read as follows:

Time (h)	Activity
0	8.0×10^{13}
1	1.3×10^{12}
2	2.0×10^{10}
3	3.1×10^{8}
4	4.9×10^{6}
5	7.7×10^{4}
6	1.2×10^{3}
7	$1.9\times10^{1}=19$
8	<1

So the activity drops to less than one per second within 8 h. ▌▌

Decay series

It is often the case that one radioactive isotope decays to another isotope that is also radioactive. Sometimes this daughter decays to a further isotope which also is radioactive. Such successive decays are said to be a *decay series.* An important example is illustrated in Figure 28–7. As can be

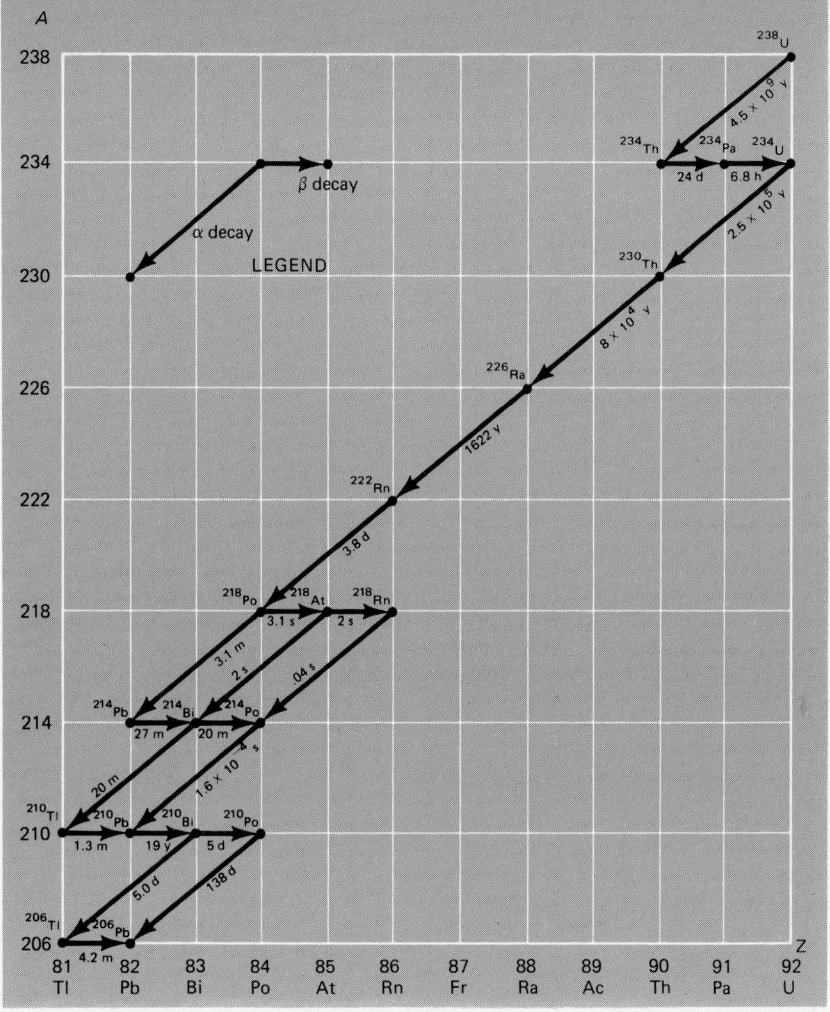

FIGURE 28-7
Decay series beginning with $^{238}_{92}U$. Nuclei in the series are specified by a dot representing *A* and *Z* values. Half-lives are given in seconds (s), minutes (m), hours (h), days (d), or years (y). Note that a horizontal arrow represents β decay (*A* does not change), whereas a diagonal line represents α decay (*A* changes by 4, *Z* changes by 2).

seen, $^{238}_{92}U$ decays by α emission to $^{234}_{90}Th$, which in turn decays by β decay to $^{234}_{91}Pa$. The series continues as shown, with several possible branches near the bottom. For example, $^{218}_{84}Po$ can decay either by α decay to $^{214}_{82}Pb$ or by β decay to $^{218}_{85}At$. The series ends at the stable lead isotope $^{206}_{82}Pb$. Other radioactive series also exist.

Because of such decay series, certain radioactive elements are found in nature that otherwise would not be. For when the solar system acquired its present form about 5 billion years ago, it is believed nearly all nuclides were formed (by the fusion process, Section 29-4). Many isotopes with short half-lives decayed quickly and no longer exist in nature today. But long-lived isotopes, such as $^{238}_{92}U$ with a half-life of 4.5×10^9 yr, still do exist in nature today; indeed, about half of the original $^{238}_{92}U$ still remains (assuming that the origin of the solar system was about 5×10^9 yr ago). We might expect, however, that radium ($^{226}_{88}Ra$), with a half-life of 1600 yr, would long since have disappeared from the earth. Indeed, the original $^{226}_{88}Ra$ nuclei must by now have all decayed. However, because $^{238}_{92}U$ decays (in several steps) to $^{226}_{88}Ra$, the supply of $^{226}_{88}Ra$ is continually replenished, which is why it still exists on earth today. The same can be said for many other radioactive nuclides.

* Stability and tunneling

We saw in Section 28-4 that radioactive decay occurs only when the mass of the parent nucleus is greater than the sum of the masses of the daughter nucleus and all particles emitted. For example, ${}^{238}_{92}U$ decays to ${}^{234}_{90}Th$, because the mass of ${}^{238}_{92}U$ is greater than the mass of the ${}^{234}_{90}Th$ plus the mass of the α particle. Since systems tend to go in the direction that reduces their internal or potential energy (a ball rolls downhill, a positive charge moves toward a negative charge), you may wonder why an unstable nucleus doesn't fall apart immediately. In other words, why do ${}^{238}_{92}U$ ($T_{\frac{1}{2}} = 4.5 \times 10^9$ yr) and other isotopes have such long half-lives?

The answer has to do with the nature of the forces involved. The simplest way to view the situation is with the aid of a potential-energy diagram, as in Figure 28-8. Here we consider the particular case of the decay ${}^{238}_{92}U \rightarrow {}^{234}_{90}Th + {}^{4}_{2}He$. The curved line represents the potential energy, including rest mass. Actually, it is simplest to imagine the α particle as a separate entity in the ${}^{238}_{92}U$ nucleus, and then the curve of Figure 28-8 represents the potential energy of the α particle within the uranium nucleus (A on the diagram) and when it is free of the uranium nucleus (C). In order to get to point C, the α particle has to get by the barrier shown. In classical physics, this could only be done by putting in an energy equal to the height of the barrier above point A. (A similar situation was discussed in Section 27-7, where the required input energy was called the "activation energy.") Nuclei decay spontaneously, however, without any input of energy. How, then, does the α particle get from point A to point C? It actually passes through the barrier in a process known as *tunneling*. Classically, this could not happen, because an α particle at point B (within the barrier) would be violating the conservation-of-energy principle. The uncertainty principle, however, tells us that energy conservation can be violated by an amount ΔE for a length of time Δt given by

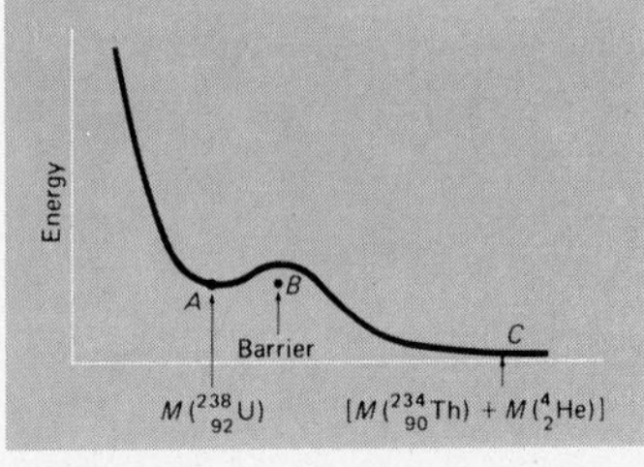

FIGURE 28-8
Potential-energy diagram for the decay ${}^{238}_{92}U \rightarrow {}^{234}_{90}Th + {}^{4}_{2}He$.

$$(\Delta E)(\Delta t) \geqslant \frac{h}{2\pi}.$$

We saw in Section 26-7 that this is a result of the wave–particle duality. Thus quantum mechanics allows conservation of energy to be violated for brief periods that may be long enough for an α particle to "tunnel" through the barrier. The higher and wider the barrier, the less time the α particle has to escape and the less likely it is to do so. It is therefore the height and width of this barrier that controls the rate of decay and half-life of an isotope.

28-6 ▮ Radioactive dating

Radioactive decay has many interesting applications. One is the technique of *radioactive dating*, in which the age of ancient materials can be determined.

The age of any object made from once-living matter, such as wood, can be determined using the natural radioactivity of ${}^{14}_{6}C$. All living plants absorb carbon dioxide (CO_2) from the air, utilizing the carbon and expelling oxygen. Although the vast majority of these carbon atoms are $^{12}_{6}C$, about 1 in 10^{12} are the radioactive isotope $^{14}_{6}C$. The ratio of $^{14}_{6}C$ to $^{12}_{6}C$ in the atmosphere will have remained roughly constant over many thousands of years even though $^{14}_{6}C$ decays with a half-life of about 5700 yr. The reason for this lies with neutrons in the cosmic radiation that comes from outer space and collides with atoms of the atmosphere. In particular, collisions with nitrogen nuclei produce the nuclear transformation $n + {}^{14}_{7}N \rightarrow {}^{14}_{6}C + p$. That is, a neutron strikes and is absorbed by a $^{14}_{7}N$ nucleus, and a proton is ejected in the process. What remains is $^{14}_{6}C$, and its continual production in the atmosphere roughly balances the loss of $^{14}_{6}C$ by radioactive decay.

As long as a plant or tree is alive, it continues to use the carbon from carbon dioxide in the air to form new tissue and to replace the old. Animals eat plants, so they, too, are continually receiving a fresh supply of carbon for their tissues. Since the ratio of $^{14}_{6}C$ to $^{12}_{6}C$ in the atmosphere is nearly constant, and organisms cannot distinguish $^{14}_{6}C$ from $^{12}_{6}C$, the ratio of the two isotopes within living organisms remains nearly constant as well. But when an organism dies, carbon dioxide is no longer absorbed and utilized; and because the $^{14}_{6}C$ decays radioactively, the ratio of $^{14}_{6}C$ to $^{12}_{6}C$ in a dead organism decreases with time. Since the half-life of $^{14}_{6}C$ is about 5700 yr, the $^{14}_{6}C/^{12}_{6}C$ ratio decreases by half every 5700 yr. Archeologists and others can utilize the $^{14}_{6}C/^{12}_{6}C$ ratio to determine the age of objects they find that are made of organic matter. For example, if the $^{14}_{6}C/^{12}_{6}C$ ratio of an ancient wooden tool is half of what it is in living trees, the object must have been made from a tree that was felled about 5700 yr ago.

Because the $^{14}_{6}C/^{12}_{6}C$ ratio in the atmosphere has not remained precisely constant over time, corrections must be made. The determination of what this ratio has been over the centuries has been done by comparing the expected ratio to the actual ratio for objects whose age is known (such as for very old trees whose annual rings can be counted).

Carbon dating is useful only for age determination of objects less than about 40,000 yr old, because the amount of ^{14}C remaining in older objects is usually too small to accurately measure. However, radioactive isotopes with longer half-lives can sometimes be used to measure the age of older objects. For example, the decay of $^{238}_{92}U$, because of its long half-life of 4.5×10^{9} yr, is useful in determining the age of rocks on a geologic time scale. When molten material solidifies into rock, the uranium present in the material becomes fixed in its position and the daughter nuclei that result from the decay of uranium will also be fixed in that position. Thus by measuring the amount of $^{238}_{92}U$ remaining in the material relative to the amount of daughter nuclei, a scientist can determine the time when the rock solidified.

Radioactive dating methods using $^{238}_{92}U$ and other isotopes have shown the age of the oldest earth rocks to be about 4×10^{9} yr. The age of rocks in which the oldest fossilized organisms are embedded indicates that life appeared at least 3 billion years ago. The earliest fossilized remains of mammals are found in rocks 200 million years old, and the first human-like creatures seem to have appeared about 2 million years ago. Radioactive dating has been indispensable for the reconstruction of earth's history and the evolution of its biological organisms.

* 28-7 ▍ Biological uses and dangers of radioactivity

The applications of radioactivity and radiation to human beings and other organisms is a vast field that has filled many books. This subject is often called *radiation biology*, and when applied in medicine it is called *nuclear medicine*. We now briefly examine some of these applications.

* Tracers

Radioactive isotopes are commonly used in biological and medical research as *tracers*. A given compound is artificially synthesized using a radioactive isotope such as $^{14}_{6}C$ or $^{3}_{1}H$. Such "tagged" molecules can then be traced as they move through an organism or as they undergo chemical reaction. The presence of these tagged molecules (or parts of them, if they undergo chemical change) can be detected by a Geiger or scintillation counter, which detects emitted radiation (see Section 28-9).

The details of how food molecules are digested, and to what parts of the body they are diverted, can be traced in this way. Radioactive tracers have been used to determine how amino acids and other essential compounds are synthesized by organisms.

The permeability of cell walls to various molecules and ions can be determined using radioactive isotopes: the tagged molecule or ion is injected into the extracellular fluid and the radioactivity present inside and outside the cells is measured as a function of time.

In each of these cases, the cells or chemicals involved are normally removed from the organism and placed in the Geiger or scintillation counter. To follow a process in time, different sets of organisms are treated for different periods of time before the cells or chemicals are extracted.

In a technique known as *autoradiography*, the position of the radioactive isotopes is detected on film. For example, the distribution of carbohydrates produced in the leaves of plants from absorbed CO_2 can be observed by keeping the plant in an atmosphere where the carbon atom in the CO_2 is $^{14}_{6}C$. After a time, a leaf is placed firmly on a photographic plate and the emitted radiation darkens the film most strongly where the isotope is most strongly concentrated (Figure 28-9). Autoradiography using labeled nucleotides (that form DNA) has revealed much about the details of DNA replication (Figure 28-10).

Nonradioactive isotopic tracers are also used sometimes. In this case the common isotope of an element is replaced by a heavier or lighter one. For example, compounds can be synthesized using $^{15}_{7}N$ in place of ordinary $^{14}_{7}N$. Such molecules will be slightly heavier and they can be separated from their lighter counterparts in an ultracentrifuge. We mentioned one example of this, the famous Meselson–Stahl experiment, at the end of Chapter 6.

FIGURE 28-9
Autoradiograph of a mature leaf of the squash plant *Cucurbita melopepo* exposed for 30 s to $^{14}CO_2$. The photosynthetic (green) tissue has become radioactive; the nonphotosynthetic tissue of the veins is free of ^{14}C and therefore does not blacken the x-ray sheet. This technique is very useful in following patterns of nutrient transport in plants. (Courtesy of R. Turgeon.)

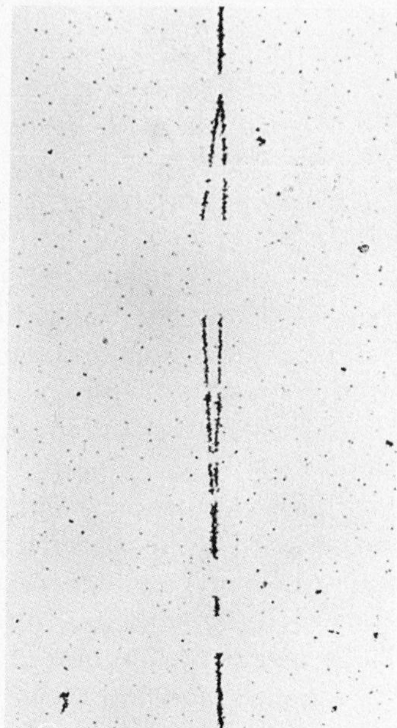

FIGURE 28-10
An autoradiogram of a fiber of chromosomal DNA isolated from the higher plant *Arabidopsis thaliana*. The dashed arrays of silver grains show the Y-shaped growing point of replicating DNA. (Courtesy of J. Van't Hof.)

* Passage of radiation through matter

When we speak of *radiation*, we include the α, β, or γ rays emitted by radioactive isotopes; but the term "radiation" also includes x rays—which, after all, are photons just as γ rays are—as well as protons, neutrons, and more recently discovered particles such as pi mesons (Chapter 30).

Because charged particles can ionize the atoms or molecules of any material they pass through, they are sometimes referred to as *ionizing radiation.* And because radiation produces ionization, it can cause considerable damage to materials, particularly to biological tissue.

Charged particles, such as α and β rays and protons, cause ionization because of the electric force. That is, when they pass through a material they can attract or repel electrons strongly enough to remove them from the atoms of the material. Since the α and β rays emitted by radioactive substances have energies on the order of 1 MeV (more precisely, from 10^4 to 10^7 eV), whereas ionization of atoms and molecules requires from about 10 to 50 eV, it is clear that a single α or β particle can cause thousands of ionizations.

Neutral particles also give rise to ionization when they pass through materials. For example, x-ray and γ-ray photons can ionize atoms by knocking out electrons by means of the photoelectric and Compton effects. Furthermore, if a γ ray has sufficient energy (greater than 1.02 MeV), it can undergo pair production: an electron and a positron are produced (Section 25-3). The charged particles produced in all three of these processes themselves can go on to produce further ionization. Neutrons, on the other hand, interact with matter mainly by collisions with nuclei, with which they interact strongly. Often the nucleus is broken apart by such a collision, which alters the molecule of which it was a part. And the fragments produced can move on to cause ionization.

* Radiation damage to cells

Radiation passing through matter can cause considerable damage. In metals and other structural materials, the strength of the material can be weakened if the radiation is very intense. This is a considerable problem in nuclear reactor power plants, and for space vehicles that must pass through areas of intense cosmic radiation.

Very important is the radiation damage produced in biological organisms. The damage is due primarily to the ionization produced in cells. Several related processes can occur. Ions or radicals are produced which are highly reactive and take part in chemical reactions that interfere with the normal operation of the cell. An important example is the production of the (neutral) free radicals H and OH from water, which is the major constituent of cells; these free radicals are very reactive and cause chemical reactions that may break chemical bonds in vital molecules such as proteins.

All forms of radiation can ionize atoms by knocking out electrons. If these are bonding electrons, the molecule may break apart; or the structure may be altered so that it does not perform its normal function or performs a harmful function. In the case of proteins, the loss of one molecule is not serious if there are other copies of that particular one in the cell, and additional ones can be made from its gene. However, large doses of radiation may damage too many molecules so that new copies cannot be made quickly enough, and the cell dies. Damage to the DNA is more serious, since a cell may have only one copy. Each alteration in the DNA affects a gene and can alter the molecule it codes for, so that needed proteins or other materials cannot be made at all. Again the cell may die.

The death of a single cell is not normally a problem, since the body can replace it with a new one. (There are exceptions, such as neurons, which are *not* replaceable, so their loss is serious.) But if many cells die, the organism may not be able to recover.

On the other hand, a cell may survive but be defective. It may go on dividing and produce many more defective cells to the detriment of the whole organism. Thus radiation can cause cancer—the rapid production of defective cells. And it can do so even at low levels of radiation.

Radiation damage to biological organisms is often separated into categories according to its location in the body: "somatic" and "genetic." Somatic damage refers to that in any part of the body except the reproductive organs. Somatic damage can seriously affect that particular organism, causing cancer and at high doses radiation sickness (characterized by nausea, fatigue, loss of body hair, and other symptoms) or even death. Genetic damage refers to damage to the reproductive apparatus and so affects an individual's offspring. Damage to the genes results in mutation, the majority of which are harmful; and mutations, if they occur in the reproductive organs, are transmitted to future generations. Radiation, including that from diagnostic use of x rays, is commonplace, and its effect on the future of the human race is a cause of great concern.

* Radiation therapy

Radiation can cause cancer. It can also be used to treat it. Rapidly growing cancer cells are more susceptible to destruction by radiation. Nonetheless, large doses are needed to kill the cancer cells, and some of the surrounding normal cells are inevitably killed as well. It is for this reason that cancer patients receiving radiation therapy often suffer side effects characteristic of radiation sickness. To minimize the destruction of normal cells, a narrow beam of γ or x rays is often used when the cancerous tumor is well localized. The beam is directed at the tumor and the source (or body) is rotated so that the beam passes through various parts of the body to keep the dose at any one place as low as possible—except at the tumor and its immediate surroundings, where the beam passes at all times. The radiation may be from a radioactive source such as $^{60}_{27}\mathrm{Co}$; or it may be from an x-ray machine that produces photons in the range 200 keV to 5 MeV. π mesons, which interact strongly with nuclei, are also used for cancer therapy.

In some cases a tiny radioactive source may be inserted directly inside a tumor; this will eventually kill the majority of the cells. A similar technique is used to treat cancer of the thyroid with the radioactive isotope $^{131}_{53}\mathrm{I}$. The thyroid gland tends to concentrate any iodine present in the bloodstream; so when $^{131}_{53}\mathrm{I}$ is injected into the blood, it becomes concentrated in the thyroid, particularly in any area where abnormal growth is taking place. The intense radioactivity emitted can then destroy the defective cells.

Although radiation can increase the lifespan of many patients, it is not completely effective. It is not always possible to kill all the diseased cancer cells, so a recurrence of the disease is possible. Many cases, especially when the cancerous cells are not well localized in one area, are difficult to treat at all without seriously damaging the rest of the organism.

Another application of radiation is for sterilization since bacteria and viruses can be killed or deactivated by it. Bandages, surgical equipment, and even packaged foods are sterilized in this fashion by large doses of radiation.

28-8 ▌ Measurement of radiation; dosimetry

It is important to be able to quantify the amount, or *dose*, of radiation that passes through a material. This is the subject of *dosimetry*—the measuring of doses. There are many different ways to measure the dose, and we discuss the important ones here.

When dealing with radioactive isotopes, we can quantify the strength of a source by specifying at a given time how many disintegrations occur per second; this, of course, is equal to the number of α, β, or γ rays emitted per second and is referred to as the *activity* of the source. A special unit is used, called the *curie* (Ci), defined as follows:

$$1\ \text{Ci} = 3.70 \times 10^{10}\ \text{disintegrations per second.}$$

(The curie was originally defined as the number of disintegrations per second from 1 g of radium, ${}^{226}_{88}\text{Ra}$; the early measurements were slightly off, for later precise measurements show that 1 g of ${}^{226}_{88}\text{Ra}$ produces 3.60×10^{10} disintegrations per second). The activity of tracers are often specified in curies, or more commonly in μCi (10^{-6} Ci). Of course commercial suppliers specify the activity at a given time, and the activity decreases as time goes on, particularly for short-lived isotopes.

▌ EXAMPLE 28-6 In a tracer experiment, 0.016 μCi of ${}^{32}_{15}\text{P}$ is injected into a medium containing a culture of bacteria. After 1 h the cells are washed and a scintillation counter that is 70 percent efficient (counts 70 percent of emitted β rays) records 720 counts per minute from all the cells. What percentage of the original ${}^{32}_{15}\text{P}$ was taken up by the cells?

SOLUTION The total number of disintegrations per second originally was $(0.016 \times 10^{-6})(3.7 \times 10^{10}) = 590$. The counter could be expected to count 70 percent of this, or 410 per second. Since it counted $720/60 = 12$ per second, then $12/410 = 0.029$, or 2.9 percent, was incorporated into the cells. ▌▌

The earliest unit of dosage was the *roentgen* (R), which was defined in terms of the amount of ionization produced by the radiation: *1 R is the amount of radiation that will produce* 2.1×10^9 *ion pairs/cm*^{3} *of air at STP.* (A single ionization collision produces one "ion pair": a positive ion and a negative ion or electron.) The roentgen can be defined in an equivalent way as that amount of x or γ radiation that deposits 0.83×10^{-2} J of energy per kilogram of air.

The roentgen has been used primarily for x and γ rays. It has been largely superseded by another unit of dosage applicable to any type of radiation. This new unit is the *rad* and is defined as follows: *1 rad is that amount of radiation which deposits* 10^{-2} *J of energy per kilogram of absorbing material.* (This is quite close to the roentgen for x and γ rays.) The radiation dose in terms of rads depends not only on the strength of a given radiation beam (number of particles per second) but also on the type of material that is absorbing the radiation. Since bone, for example, is denser than flesh and absorbs more of the radiation, the same beam passing through the body deposits a greater dose (in rads) in bone than in the flesh.

The rad is a physical unit of dose—it is the energy deposited per unit mass of material. It is, however, not the most meaningful unit for measuring the biological damage produced by radiation. This is because equal doses of different types of radiation cause differing amounts of damage. For example, 1 rad of α radiation does 10 to 20 times more damage than 1 rad of β or γ rays does. This is largely because α rays (and other heavy particles such as protons and neutrons) move much more slowly than equal energy β and γ rays because of their greater mass. Hence ionizing collisions occur closer together so more irreparable damage is done. The *relative biological effectiveness*† (RBE) of a given type of radiation is defined as the number of rads of x or γ radiation that produces the same biological damage as 1 rad of the given radiation. The RBE for several types of radiation is given in Table 28-3. The numbers are not exact, since they depend somewhat on the energy of the particles and on the type of damage that is used as the criterion.

TABLE 28-3

Relative biological effectiveness (RBE) of different kinds of radiation

Type	RBE
x and γ rays	**1.0**
β (electrons)	**1.0**
α particles	**10–20**
Slow neutrons	**3–5**
Fast neutrons and protons	**10**

The product of the dose in rads and the RBE gives a unit known as the *rem* (which stands for *rad equivalent man*):

$$\text{rem} = \text{rad} \times \text{RBE}.$$

By this definition, 1 rem of any type of radiation does approximately the same amount of biological damage. For example, 50 rem of fast neutrons does the same damage as 50 rem of γ rays. But note that 50 rem of fast neutrons is only 5 rads, whereas 50 rem of γ rays is 50 rad.

We are constantly exposed to low-level radiation from natural sources: cosmic rays, natural radioactivity in rocks and soil, and naturally occurring radioactive isotopes that occur in our food, such as $^{40}_{19}\text{K}$. This natural radioactivity supplies about 0.13 rem per year per person on the average. From medical x rays the average person receives about 0.07 rem. The U.S. government specifies the recommended upper limit of allowed radiation, exclusive of natural sources, for an individual in the general populace, at about 0.5 rem. However, since even low doses of radiation are believed to increase the chances of cancer or genetic defects, the attitude today is to keep the radiation dose as low as possible.

People who work around radiation—in hospitals, in power plants, in research—often are subjected to much higher doses than 0.5 rem/yr. The upper limit for such occupational exposures has been set somewhat higher, on the order of 5 rem/yr whole-body dose (presumably because such people know what they are getting into). To monitor radiation dose for such workers, some sort of measuring device must be carried on the body. The most common device is a *film badge*—a carefully wrapped piece of film kept from the light. When developed, the darkness of the film is a measure of the radiation dose received.

Large doses of radiation can cause reddening of the skin, drop in white-blood-cell count, and a large number of unpleasant symptoms such as nausea, fatigue, and loss of body hair. Such effects are some-

† Sometimes called the *quality factor* (QF).

times referred to as radiation sickness. Large doses can also be fatal, although the time span of the dose is important. A short dose of 1000 rem is nearly always fatal. A 400-rem dose in a short period of time is fatal in 50 percent of the cases. However, the body possesses remarkable repair processes so that a 400-rem dose spread over several weeks is not usually fatal. It will, nonetheless, cause considerable damage to the body.

▌EXAMPLE 28-7 What whole-body dose is received by a 70-kg patient exposed to a 1000-Ci $^{60}_{27}$Co source if 2 percent of the γ rays reach the patient. $^{60}_{27}$Co emits γ rays of energy 1.33 MeV and 1.17 MeV in equal amounts. Approximately 50 percent of the γ rays interact in the body and deposit all their energy. (The rest pass through.)

SOLUTION The average γ-ray energy is 1.25 MeV, so the total energy passing through the body is $(1000\text{ Ci})(3.7\times10^{10}$ decays/Ci · s$)(1.25\text{ MeV})(0.02)=9.3\times10^{11}$ MeV/s. (The factor 0.02 represents the 2 percent that reaches the body.) Only half this energy is deposited in the body, so the dose rate is $(\frac{1}{2})(9.3\times10^{11}\text{ MeV/s})(1.6\times10^{-13}\text{ J/MeV})=7.4\times10^{-2}$ J/s. Since 1 rad $=10^{-2}$ J/kg, the whole-body dose rate is $(7.4\times10^{-2}\text{ J/s})/(10^{-2}\text{ J/kg})(70\text{ kg})=0.11$ rad/s. ▌▌

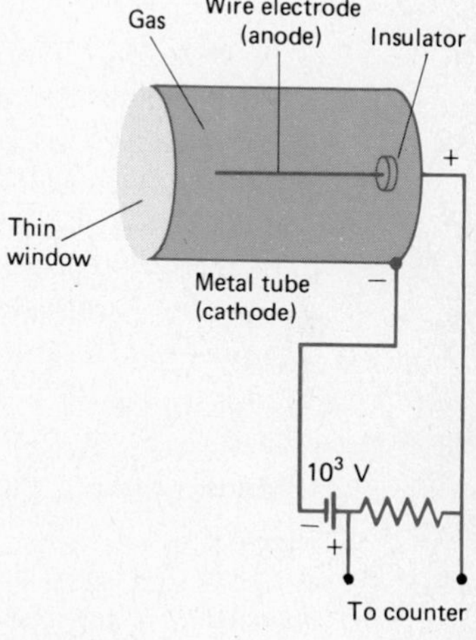

FIGURE 28-11
Diagram of a Geiger counter.

28-9 ▌Detection of radiation

Individual particles such as electrons, protons, α particles, neutrons, and γ rays are not detected directly by our senses. Consequently, a variety of instruments have been developed to detect them.

One of the most common is the *Geiger counter*. As shown in Figure 28-11, it consists of a cylindrical metal tube filled with a certain type of gas. A long wire runs down the center and is kept at a high positive voltage ($\sim10^3$ V) with respect to the outer cylinder. The voltage is just slightly less than that required to ionize the gas atoms. When a charged particle enters through a thin "window" at one end of the tube, it ionizes a few atoms of the gas. The freed electrons are attracted toward the positive wire and as they are accelerated they strike and ionize additional atoms. An "avalanche" of electrons is quickly produced, and when it reaches the wire anode it produces a voltage pulse. The pulse, after being amplified, can be sent to an electronic counter, which keeps track of how many particles have been detected. Or the pulses can be sent to a loudspeaker and each detection of a particle is heard as a "click."

A *scintillation counter* makes use of a solid, liquid, or gas known as a *scintillator* or *phosphor*. The atoms of a scintillator are easily excited when struck by an incoming particle and emit visible light when they return to their ground states. Typical scintillators are crystals of NaI and certain plastics. A solid scintillator is wrapped with opaque material to keep it lighttight—except for one face, which is cemented to a photomultiplier tube. The *photomultiplier (PM) tube* converts the energy of the scintillator-emitted photon(s) into an electric signal. A PM tube is a vacuum tube containing several electrodes (typically 8 to 14), called dynodes, which are maintained at successively higher voltages as shown in Figure 28-12. At its top surface is a photoelectric surface, called the photocathode, whose work function is low enough that electrons are easily released when struck

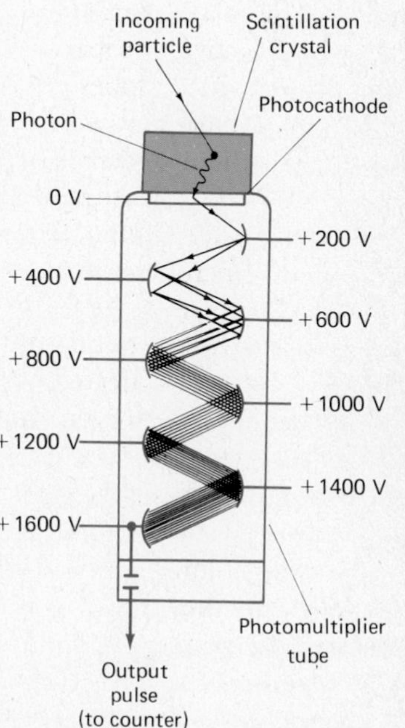

FIGURE 28-12
Scintillation counter with a photomultiplier tube.

by a photon from the scintillator. Such an electron is accelerated toward the first dynode. When it strikes the first dynode, the electron has acquired sufficient kinetic energy so that it can eject two to five more electrons. These, in turn, are accelerated to the second dynode, and a multiplication process begins. The number of electrons striking the last dynode may be 10^6 or more. Thus the passage of a particle through the scintillator results in an electric signal at the output of the PM tube that can be sent to an electronic counter just as for a Geiger tube. Because a scintillator crystal is much more dense than the gas of a Geiger counter, it is a more efficient detector—especially for γ rays, which interact less with matter than do β rays.

In tracer work and other biological experiments, liquid scintillators are often used. Radioactive samples taken at different times or from different parts of an organism are placed directly in small bottles containing the liquid scintillator. This is particularly convenient for detection of β rays from ${}^3_1H$ and ${}^{14}_6C$, which have very low energies and have difficulty passing through the outer covering of a crystal scintillator or Geiger tube. A PM tube is still used to produce the electric signal.

A recent development is the *semiconductor detector*. It consists of a reversed biased *pn* junction diode. A particle passing through the junction causes ionization, and the released charges produce a short electrical pulse that can be counted just as for Geiger and scintillation counters.

The three devices discussed so far are used for counting the number of particles (or decays of a radioactive isotope). Other devices have been developed that allow the track of a charged particle to be seen. The simplest is the *photographic emulsion*. A particle passing through a layer of photographic emulsion ionizes the atoms along its path. This results in a chemical change at these points; and when the emulsion is developed, the particle's path is revealed.

In a *cloud chamber*, a gas is cooled to a temperature slightly below its usual condensation point. (It is said to be "supercooled.") The gas molecules begin to condense on any ionized molecules present. Thus the ions produced when a charged particle passes through serve as centers on which tiny bubbles form (Figure 28-13). Light scatters more from these bubbles than from the gas background, so a photo of the cloud chamber at the right moment shows the track of the particle.

The *bubble chamber*, invented in 1952 by D. A. Glaser (b. 1926), makes use of a superheated liquid. The liquid is kept close to its boiling point and the bubbles characteristic of boiling form around the ions produced by the passage of a charged particle (Figure 28-14). A photograph of the chamber thus reveals the path of particles that recently passed through. Because the bubble chamber uses a liquid—often liquid hydrogen—the density of atoms is much greater than in a cloud chamber; hence it is a much more efficient device for observing particles and their interactions with the nuclei of the liquid. Usually, a magnetic field is applied across the chamber and the momentum of the moving particles can be determined from the radius of curvature of their paths.

A *spark chamber* consists of a set of closely spaced parallel plates; alternate plates are grounded and the ones in between are kept at very high voltage. When a charged particle passes through, the ions produced in the gas between the plates avalanche and a large current results which

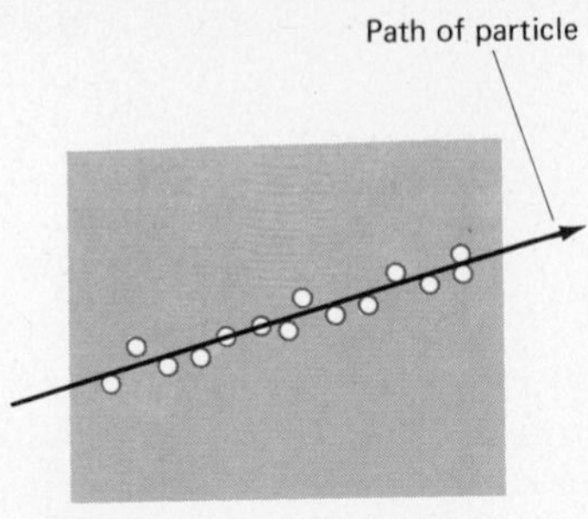

FIGURE 28-13
In a cloud or bubble chamber, bubbles are formed by the passage of a charged particle.

(*a*)

(*b*)

FIGURE 28-14
(*a*) Bubble chamber; (*b*) particle tracks in bubble chamber. (Courtesy of Fermilab.)

produces a visible spark. Thus the path of the particle is made visual. In a high-voltage *streamer chamber* there are only two plates, widely spaced, and the electric discharge closely follows the particle's path, which is seen as a glow or streamer. Spark and streamer chambers have the advantage that they recycle more quickly than bubble chambers (so more particles or events can be observed in a given time), although they can handle fewer tracks per photograph.

Summary

Nuclear physics is the study of atomic nuclei. Nuclei contain *protons* and *neutrons*, which are collectively known as *nucleons*. The total number of nucleons, A, is the *atomic mass number*. The number of protons, Z, is called the *atomic number*. The number of neutrons equals $A-Z$. *Isotopes* are nuclei with the same Z, but with different numbers of neutrons. For an element X, an isotope of given Z and A is represented by ${}^{A}_{Z}\mathrm{X}$. The nuclear radius is proportional to $A^{1/3}$, indicating that all nuclei have about the same density. Nuclear masses are specified in atomic mass units (u), where the mass of ${}^{12}_{6}\mathrm{C}$ is defined as exactly 12.000 u, or in terms of their energy equivalent (because $E=mc^2$), where $1\ \mathrm{u}=931.5\ \mathrm{MeV}/c^2$.

The mass of a nucleus is less than the sum of the masses of its constituent nucleons. The difference in mass (times c^2) is the *total binding energy*; it represents the energy needed to break the nucleus into its constituent nucleons. The *binding energy per nucleon* averages about 8 MeV per nucleon, and is lowest for very light and very heavy nuclei.

Unstable nuclei undergo *radioactive decay*; they change into other nuclei with the emission of an α, β, or γ particle. An α particle is a ${}^{4}_{2}\mathrm{He}$ nucleus; a β particle is an electron or positron; and a γ ray is a high-energy photon. In β decay, a *neutrino* is also emitted. The transformation of the parent into the daughter nucleus is called *transmutation* of the elements.

Nuclei are held together by the *strong nuclear force*; the *weak nuclear force* makes itself apparent in β-decay; these two forces, plus the gravitational and electromagnetic forces, are the four known types of force.

Radioactive decay is a statistical process. The number that decay (ΔN) in a time Δt is proportional to the number N of parent nuclei present: $\Delta N=-\lambda N\,\Delta t$; the proportionality constant, λ, is called the decay constant and is characteristic of the given nucleus. The *half-life*, $T_{\frac{1}{2}}$, is the time required for half the nuclei of a radioactive sample to decay. It is related to the decay constant by $T_{\frac{1}{2}}=0.693\lambda$. Electric charge, linear and angular momentum, mass–energy, and *nucleon number* are *conserved* in all decays. Radioactive decay occurs spontaneously only when the rest mass of the products is less than the mass of the parent nucleus; the loss in mass appears as kinetic energy of the products.

Radioactive nuclei are used for dating and for many biological purposes. Radiation can cause damage to biological tissue, and it is important to quantify amounts of radiation; this is the subject of dosimetry. The curie (Ci) is a unit that measures the activity or rate of decay of a sample: $1\ \mathrm{Ci}=3.70\times10^{10}$ disintegrations per second. The rad measures the amount of energy deposited per unit mass of absorbing material:

1 rad is the amount of radiation that deposits 10^{-2} J/kg of material. The *rem* = rad × RBE, where RBE is the relative biological effectiveness of a given type of radiation; 1 rem of any type of radiation does approximately the same amount of biological damage. The average dose received per person per year is about 0.20 rem.

Detectors of radiation include Geiger counters, scintillation counters, and semiconductor detectors. The paths of elementary particles can be seen in bubble chambers, cloud chambers, and spark chambers.

Questions

1 How do we know that there is such a thing as the strong nuclear force?

2 What do different isotopes of a given element have in common? How are they different?

3 What are the elements represented by the X in the following? (*a*) $^{232}_{92}X$; (*b*) $^{18}_{7}X$; (*c*) $^{1}_{1}X$; (*d*) $^{82}_{38}X$; (*e*) $^{247}_{97}X$.

4 How many protons and how many neutrons do each of the isotopes in Question 3 have?

5 Why are the atomic masses of many elements (Appendix E) not close to whole numbers?

6 What is the experimental evidence in favor of radioactivity being a nuclear process?

7 The isotope $^{64}_{29}Cu$ is unusual in that it can decay by γ, β^-, and β^+ emission. What is the resulting nuclide for each case?

8 A $^{238}_{92}U$ nucleus decays to a nucleus containing how many neutrons?

9 What is the mass of a gamma ray?

10 What is the mass of a neutrino?

11 Describe, in as many ways as possible, the differences between α, β, and γ rays.

12 What element is formed by the radioactive decay of (*a*) $^{24}_{11}Na$ (β^-); (*b*) $^{22}_{11}Na$ (β^+); (*c*) $^{210}_{84}Po$ (α)?

13 What element is formed by the decay of (*a*) $^{32}_{15}P$ (β^-); (*b*) $^{35}_{16}S$ (β^-); (*c*) $^{211}_{83}Bi$ (α)?

14 Fill in the missing particle or nucleus:

(*a*) $^{45}_{20}Ca \rightarrow ? + e^- + \bar{\nu}$

(*b*) $^{58}_{29}Cu \rightarrow ? + \gamma$

(*c*) $^{46}_{24}Cr \rightarrow ^{46}_{23}V + ?$

(*d*) $^{234}_{94}Pu \rightarrow ? + \alpha$

(*e*) $^{239}_{93}Np \rightarrow ^{239}_{92}U + ?$

15 Immediately after a $^{238}_{92}U$ nucleus decays to $^{234}_{90}Th + ^{4}_{2}He$, the daughter thorium nucleus still has 92 electrons circling it. Since thorium normally holds only 90 electrons, what do you suppose happens to the two extra ones?

16 Do isotopes that undergo electron capture generally lie above or below the line of stability in Figure 28-2?

17 Some ancient rocks are found to contain equal amounts of $^{238}_{92}U$ and its daughter nuclides. How old are the rocks?

18 Why are many artificially produced radioactive isotopes rare in nature?

19 Can hydrogen or deuterium emit an α particle?

20 Can $^{14}_{6}C$ dating be used to measure the age of stone walls and tablets of ancient civilizations?

21 Explain the absence of β^+ emitters in the radioactive decay series of Figure 28-7.

22 How might radioactive tracers be used to find a leak in a pipe?

23 Why is the recommended maximum radiation dose higher for women beyond the child-bearing age than for younger women?

Problems

SECTION 28-1

1 (I) Given the masses of the particles in Table 28-1 in atomic mass units, show that the masses in MeV/c^2 are correct as given there.

2 (I) What is the rest energy of an α particle in MeV?

3 (I) A pi meson has a mass of 139 MeV/c^2. What is this in atomic mass units?

4 (I) What is the approximate radius of a $^{64}_{29}Cu$ nucleus?

5 (I) Approximately what is the value of *A* for a nucleus whose radius is 3.6×10^{-15} m?

6 (II) Determine the density of nuclear matter in kg/m^3.

7 (II) What stable nucleus (approximately) has half the radius of a uranium nucleus? (*Hint*: Find A and use Appendix E to get Z.)

SECTION 28-2

8 (I) Estimate the total binding energy for $^{40}_{20}\text{Ca}$, using Figure 28-1.

9 (I) Use Figure 28-1 to estimate the total binding energy of $^{238}_{92}\text{U}$.

10 (II) Use Appendix E and calculate the binding energy of $^{2}_{1}\text{H}$.

11 (II) Calculate the total binding energy and the binding energy per nucleon for $^{6}_{3}\text{Li}$.

12 (II) Calculate the binding energy of the last neutron in a $^{12}_{6}\text{C}$ nucleus. (*Hint*: Compare the mass of $^{12}_{6}\text{C}$ and $^{11}_{6}\text{C}+^{1}_{0}n$; use Appendix E.)

13 (II) Calculate the binding energy per nucleon for a $^{12}_{6}\text{C}$ nucleus.

14 (II) (*a*) Show that the nucleus $^{8}_{4}\text{Be}$ (mass = 8.005305 u) is unstable to decay into two α particles. (*b*) Is $^{12}_{6}\text{C}$ stable against decay into three α particles? Show why or why not.

SECTION 28-4

15 (I) $^{60}_{27}\text{Co}$ in an excited state emits a 1.33-MeV γ ray as it jumps to the ground state. What is the atomic mass of the excited cobalt?

16 (II) Show that the decay $^{11}_{6}\text{C}\rightarrow ^{10}_{5}\text{B}+\text{p}$ is not possible because energy would not be conserved.

17 (II) A $^{232}_{92}\text{U}$ nucleus emits an α particle with KE = 5.32 MeV. What is the final nucleus and what is its approximate atomic mass (in u)?

18 (II) When $^{23}_{10}\text{Ne}$ (mass = 22.9945 u) decays to $^{23}_{11}\text{Na}$ (mass = 22.9898 u), what is the maximum kinetic energy of the emitted electron? What is its minimum energy? What is the energy of the neutrino in each case?

19 (II) The nuclide $^{32}_{15}\text{P}$ decays by emitting an electron whose maximum kinetic energy can be 1.71 MeV. What is the daughter nucleus, and what is its atomic mass (in u)?

20 (II) The isotope $^{218}_{84}\text{Po}$ can decay by either α or β^- emission. What is the energy release in each case? The mass of $^{218}_{84}\text{Po}$ is 218.008969 u.

21 (II) Use Figure 28-7 and calculate the relative decay rates for α and β decay of $^{218}_{84}\text{Po}$.

22 (II) What is the energy of the α particle emitted in the decay $^{210}_{84}\text{Po}\rightarrow ^{206}_{82}\text{Pb}+\alpha$?

23 (II) How much energy is released in electron capture by beryllium: $^{7}_{4}\text{Be}+^{0}_{-1}e\rightarrow ^{7}_{3}\text{Li}+\nu$?

24 (III) The nuclide $^{191}_{76}\text{Os}$ decays with β^- energy of 0.14 MeV accompanied by γ rays of energy 0.042 MeV and 0.129 MeV. (*a*) What is the daughter nucleus? (*b*) Draw an energy-level diagram showing the ground states of the parent and daughter and excited states of the daughter. To which of the daughter states does β decay of $^{191}_{76}\text{Os}$ occur?

25 (III) Show that when a nucleus decays by β^+ decay, the total energy released is equal to $(M_P - M_D - 2m_e)c^2$, where M_P and M_D are the masses of the parent and daughter atoms (neutral) and m_e is the mass of an electron or positron.

26 (III) Use the result of Problem 25 to determine the maximum kinetic energy of β^+ particles released when $^{11}_{6}\text{C}$ decays to $^{11}_{5}\text{B}$. What is the maximum energy the neutrino can have? What is its minimum energy?

27 (III) In α decay of, say, a $^{226}_{88}\text{Ra}$ nucleus, show that the nucleus carries away a fraction $1/(1+A_D/4)$ of the total energy available, where A_D is the mass number of the daughter nucleus. (*Hint*: Use conservation of momentum as well as conservation of energy.) Approximately what percentage of the energy available is thus carried off by the α particle in the case cited?

SECTIONS 28-5 AND 28-6

28 (I) A radioactive material registers 1280 counts per minute on a Geiger counter at one time, and 6 h later registers 320 counts per minute. What is its half-life?

29 (I) In a series of decays, the nuclide $^{235}_{92}\text{U}$ becomes $^{207}_{82}\text{Pb}$. How many α and β^- particles are emitted in this series?

30 (I) What is the decay constant of $^{238}_{92}\text{U}$ whose half-life is 4.5×10^9 yr?

31 (I) The decay constant of a given nucleus is $6.2\times 10^{-5}\ \text{s}^{-1}$. What is its half-life?

32 (I) What is the activity of a sample of $^{14}_{6}\text{C}$ that contains 2.0×10^{20} nuclei?

33 (I) How many nuclei of $^{238}_{92}\text{U}$ remain in a rock if the activity registers 1.8×10^4 decays per second?

34 (I) What fraction of a sample of $^{68}_{32}\text{Ge}$, whose half-life is about 9 months, will remain after 4.5 yr?

35 (II) $^{124}_{55}\text{Cs}$ has a half-life of 30.8 s. (*a*) If we have 5 μg initially, how many nuclei are present? (*b*) How many are present 2 min later? (*c*) What is the activity at this time? (*d*) After how much time will the activity drop to less than about 1 per second?

36 (II) Calculate the activity of a pure 3.5-μg sample of $^{32}_{15}\text{P}$ ($T_{\frac{1}{2}} = 1.24\times 10^6$ s).

37 (II) The activity of a sample of $^{35}_{16}\text{S}$ ($T_{\frac{1}{2}} = 7.5\times 10^6$ s) is 5.2×10^6 decays per second. What is the mass of sample present?

38 (II) A sample of $^{233}_{92}\text{U}$ ($T_{\frac{1}{2}} = 1.6\times 10^5$ yr) contains 2.0×10^{18} nuclei. (*a*) What is the decay constant? (*b*) Approximately how many disintegrations will occur per minute?

39 (II) A 50-g sample of pure carbon contains 1 part in 10^{14} of $^{14}_{6}\text{C}$. How many disintegrations occur per second?

40 (II) The $^{3}_{1}\text{H}$ isotope of hydrogen, which is called *tritium* (because it contains three nucleons), has a half-life of 12.3 yr. It can be used to measure the age of objects up to about 100 yr. It is produced in the

upper atmosphere by cosmic rays and is brought to earth by rain. As an application, determine approximately the age of a bottle of wine whose $^3_1$H radiation is about $\frac{1}{10}$ that present in new wine.

41 (II) The rubidium isotope $^{87}_{37}$Rb, a β emitter with a half-life of 4.9×10^{10} yr, is used to determine the age of rocks and fossils. Rocks containing fossils of early animals contain a ratio of $^{87}_{38}$Sr to $^{87}_{37}$Rb of 0.010. Assuming that there was no $^{87}_{38}$Sr present when the rocks were formed, calculate the age of these fossils. (*Hint*: Use Equation 28-2.)

42 (III) An ancient club is found that contains 240 g of carbon and has an activity of 5 decays per second. Determine the age of the club using the fact that in living trees about one nucleus of carbon in 10^{12} is $^{14}_{6}$C.

SECTION 28-8

43 (I) A 0.018-μCi sample of $^{32}_{15}$P is injected into an animal for tracer studies. If a Geiger counter intercepts 20 percent of the emitted β particles, and is 90 percent efficient in counting them, what will be the counting rate?

44 (I) An average adult body has about 0.1 μCi of $^{40}_{19}$K, which comes from food. How many decays occur per second?

45 (I) A dose of 500 rem of γ rays in a short period would be lethal to about half the people subjected to it. How many rads is this?

46 (I) Fifty rads of α-particle radiation is equivalent to how many rads of x rays, in terms of biological damage?

47 (I) How many rads of slow neutrons will do as much biological damage as 50 rad of fast neutrons?

48 (I) How much energy is deposited in the body of a 70-kg adult exposed to a 50-rad dose?

49 (II) A 1.0-mCi source of $^{32}_{15}$P (in $NaHPO_4$), a β^- emitter, is implanted in an organ where it is to administer 5000 rad. The half-life of $^{32}_{15}$P is 14.3 days and 1 mCi delivers about 1 rad/min. Approximately how long should the source remain implanted?

50 (II) About 35 eV is required to produce one ion pair in air. Calculate the dose in rads when air absorbs 1 R (roentgen) of x rays.

51 (II) $^{57}_{27}$Co emits 122-keV γ rays. If a 70-kg person swallowed 2.0 μCi of $^{57}_{27}$Co, what would be the dose rate (rad/day) averaged over the whole body? Assume that 50 percent of the γ-ray energy is deposited in the body. (*Hint*: Determine the rate of energy deposited in the body and use the definition of the rad.)

52 (II) What is the mass of a 1.0-μCi $^{14}_{6}$C source?

53 (II) In a certain town the average yearly background radiation consists of 25 mrad of x and γ rays plus 3.0 mrad of particles having an RBE of 10. How many rems will a person receive per year on the average?

54 (II) A shielded γ-ray source yields a dose rate of 0.050 rad/h at a distance of 1.0 m for an average-sized person. If workers are allowed a maximum dose rate of 5.0 rem/yr, how close to the source may they operate assuming a 40-h work week? Assume that the intensity of radiation falls off as the square of the distance. (It actually falls off more rapidly than $1/r^2$ because of absorption in the air, so the answer above will give a better-than-permissible value.)

For additional reading

Ackerman, E., *Biophysical Science*, Prentice-Hall, Inc., Englewood Cliffs, N.J., 1962. Tracers, effects of radiation.

Casorett, A. P., *Radiation Biology*, Prentice-Hall, Inc., Englewood Cliffs, N.J., 1968.

Davidovits, P., *Physics in Biology and Medicine*, Prentice-Hall, Inc., Englewood Cliffs, N.J., 1975.

Hendee, W. R., *Radioactive Isotopes in Biological Research*, John Wiley & Sons, Inc., New York, 1973.

Lapp, R. E., and H. L. Andrews, *Nuclear Radiation Physics*, 4th ed., Prentice-Hall, Inc., Englewood Cliffs, N.J., 1972.

Magie, W. F., *Source Book in Physics*, Harvard University Press, Cambridge, Mass., 1963. Original writings of Roentgen, Becquerel, and the Curies.

Shonle, J. I., *Environmental Applications of General Physics*, Addison-Wesley Publishing Co., Inc., Reading, Mass., 1974.

29 ▌ Nuclear reactions and nuclear energy

We continue our study of nuclear physics in this chapter, where we will emphasize the literally world-shaking investigations of nuclear energy. We begin with a discussion of nuclear reactions.

29-1 ▌ Nuclear reactions and the transmutation of elements

When a nucleus undergoes α or β decay, the daughter nucleus is that of a different element from the parent. The transformation of one element into another, called *transmutation*, also occurs by means of nuclear reactions. A **nuclear reaction** is said to occur when a given nucleus is struck by another nucleus, or by a simpler particle, such as a γ ray or neutron, so that an interaction takes place; usually the products are different than the original particles. Ernest Rutherford was the first to report seeing a nuclear reaction. In 1919 he observed α particles pass through nitrogen gas; some of the particles were absorbed and protons emitted. He concluded that nitrogen nuclei had been transformed into oxygen nuclei via the reaction

$$^{4}_{2}\text{He} + ^{14}_{7}\text{N} \rightarrow ^{17}_{8}\text{O} + ^{1}_{1}\text{H},$$

where $^{4}_{2}\text{He}$ is an α particle and $^{1}_{1}\text{H}$ is a proton.

Since then, a great many nuclear reactions have been observed. Indeed, many of the radioactive isotopes used in the laboratory are made by means of nuclear reactions. Nuclear reactions can be made to occur in the laboratory, but they also occur regularly in nature. In Chapter 28 we saw an example of this: $^{14}_{6}\text{C}$ is continually being made in the atmosphere via the reaction $n + ^{14}_{7}\text{N} \rightarrow ^{14}_{6}\text{C} + p$.

Nuclear reactions are sometimes written in a shortened form; for example, the reaction $n + ^{14}_{7}\text{N} \rightarrow ^{14}_{6}\text{C} + p$ is written $^{14}_{7}\text{N}(n, p)^{14}_{6}\text{C}$. The symbols outside the parentheses on the left and right represent the initial and final nuclei, respectively. The symbols inside the parentheses represent the bombarding particle (first) and the emitted small particle (second).

In any nuclear reaction, the total electric charge is conserved, and the total number of nucleons is conserved. These conservation laws are often useful, as the following example shows.

▌ EXAMPLE 29-1 A neutron is observed to strike an $^{16}_{8}$O nucleus and a deuteron is given off. (A deuteron, or deuterium, is the isotope of hydrogen containing one proton and one neutron, $^{2}_{1}$H.) What is the nucleus that results?

SOLUTION We have the reaction $^{1}_{0}n + ^{16}_{8}\text{O} \rightarrow\ ? + ^{2}_{1}\text{H}$. The total number of nucleons initially is $16+1=17$, and the total charge is $8+0=8$; the same totals apply to the right side of the reaction. Hence the product nucleus must have $Z=7$ and $A=15$. From the periodic table we find that it is nitrogen that has $Z=7$, so the nucleus produced is $^{15}_{7}$N. The reaction, then, can be written $^{16}_{8}\text{O}(n, d)^{15}_{7}\text{N}$, where d represents deuterium, $^{2}_{1}$H. ▌▌

Energy (as well as momentum) is conserved in nuclear reactions. This fact can be used to determine whether a given reaction can occur or not. For example, if the total mass of the products is less than the total mass of the initial particles, then energy will be released by the reaction—it will appear as kinetic energy of the outgoing particles. But if the total mass of the products is greater than the total mass of the initial reactants, the reaction requires energy; that is, the reaction will not occur at all unless the bombarding particle has sufficient kinetic energy. The minimum KE required of the projectile in such a reaction is called the *threshold energy*.

▌ EXAMPLE 29-2 Can the reaction $^{13}_{6}\text{C}(p, n)^{13}_{7}\text{N}$ occur when $^{13}_{6}$C is bombarded by 2.0-MeV protons?

SOLUTION We look up the masses of the nuclei in Appendix E. The total masses before and after the reaction are:

Before	After
$m(^{13}_{6}\text{C}) = 13.003355$	$m(^{13}_{7}\text{N}) = 13.005739$
$m(^{1}_{1}\text{H}) = 1.007825$	$m(^{1}_{0}n) = 1.008665$
total = 14.011180	total = 14.014404

(We must use the mass of the $^{1}_{1}$H atom rather than that of the bare proton, because the masses of $^{13}_{6}$C and $^{13}_{7}$N include the electrons, and we must include an equal number of electrons on each side of the equation.) The products have an excess mass of $0.003224\ \text{u} \times 931.5\ \text{MeV/u} = 3.00$ MeV. This reaction requires energy, and the 2.0-MeV protons do not have enough to make it go. Hence this reaction will not occur.† ▌▌

A number of other reactions were observed in the decade that followed Rutherford's first laboratory observation of a nuclear reaction in 1919. The artificial transmutation of elements took a great leap forward in the 1930s when Enrico Fermi realized that neutrons would be the most effective projectiles for causing nuclear reactions, and in particular for producing new elements. Because neutrons have no net electric charge, they are not repelled by positively charged nuclei as protons or α particles

† The proton would have to have somewhat more than 3.00 MeV; 3.00 MeV would be enough to conserve energy, but a proton of this energy would produce the $^{13}_{7}$N and $^{1}_{0}n$ with no KE, and hence no momentum; since the incident proton has momentum, conservation of momentum would be violated. To conserve both energy and momentum, the proton would have to have somewhat more than 3.00 MeV of energy. A more complicated calculation shows that the threshold energy is 3.23 MeV in this case.

are. Hence the probability of a neutron reaching the nucleus and causing a reaction is much greater than for charged projectiles,† particularly at low energies. Between 1934 and 1936, Fermi and his coworkers in Rome produced many previously unknown isotopes by bombarding different elements with neutrons. Fermi soon realized that if the heaviest known element, uranium, were bombarded with neutrons, it might be possible to produce new elements whose atomic numbers were greater than that of uranium! After several years of hard work, it was suspected that two new elements had been produced, neptunium ($Z = 93$) and plutonium ($Z = 94$). Confirmation that such "transuranic" elements could be produced came several years later at the University of California, Berkeley. The reactions are shown in Figure 29-1.

It was soon shown that what Fermi actually had observed was an even stranger process—one that was destined to play an extraordinary role in the world at large. But Fermi's work was interrupted at this critical point. In 1938 he was awarded the Nobel Prize and he used the trip to Sweden to escape the fascist regime of Mussolini. As we shall see, he was soon to reenter the developing story, this time in the United States.

(*a*) $n + {}^{238}_{92}\text{U} \rightarrow {}^{239}_{92}\text{U}$

Neutron captured by ${}^{238}_{92}\text{U}$

(*b*) ${}^{239}_{92}\text{U} \rightarrow {}^{239}_{93}\text{Np} + e^- + \bar{\nu}$

${}^{239}_{92}\text{U}$ decays by β decay to Neptunium-239.

(*c*) ${}^{239}_{93}\text{Np} \rightarrow {}^{239}_{94}\text{Pu} + e^- + \bar{\nu}$

${}^{239}_{93}\text{Np}$ itself decays by β decay to produce plutonium-239.

FIGURE 29-1
Neptunium and plutonium are produced in this series of reactions, after bombardment of ${}^{238}_{92}\text{U}$ by neutrons.

29-2 ▌ Nuclear fission

Discovery of fission

In 1939, the German scientists Otto Hahn and Fritz Strassmann made an amazing discovery. Following up on Fermi's work, they found that uranium bombarded by neutrons sometimes produced smaller nuclei, which were roughly half the size of the original uranium nucleus. Lise Meitner and Otto Frisch, two refugees from Nazi Germany working in Sweden, quickly realized what had happened: the uranium nucleus, after absorbing a neutron, actually had split into two roughly equal pieces. This was startling, for until then the only known nuclear reactions involved knocking out only a tiny fragment (for example, n, p, or α) from a nucleus.

This new phenomenon was named **nuclear fission** because of its resemblance to biological fission (cell division). It occurs much more readily for ${}^{235}_{92}\text{U}$ than for the more common ${}^{238}_{92}\text{U}$. The process can be visualized by imagining the uranium nucleus to be like a liquid drop. According to this *liquid-drop model*, when a neutron is absorbed by a ${}^{235}_{92}\text{U}$ nucleus, it gives the nucleus extra internal energy (like heating a drop of water). This intermediate state, or *compound nucleus*, is ${}^{236}_{92}\text{U}$ (because of the absorbed neutron). The extra energy of this nucleus (it is in an excited state) appears as increased motion of the individual nucleons, which causes the nucleus to take on abnormal elongated shapes (Figure 29-2). When the nucleus elongates into the shape shown in part *c*, it keeps going and splits in two. The two resulting nuclei, N_1 and N_2, are called *fission fragments*, and in the process a number of neutrons (typically 2 or 3) are also given off. The reaction can be written

$$ {}^{1}_{0}n + {}^{235}_{92}\text{U} \rightarrow {}^{236}_{92}\text{U} \rightarrow N_1 + N_2 + \text{neutrons}. \tag{29-1}$$

† That is, positively charged particles. Electrons rarely cause nuclear reactions, because they do not partake of the strong nuclear force.

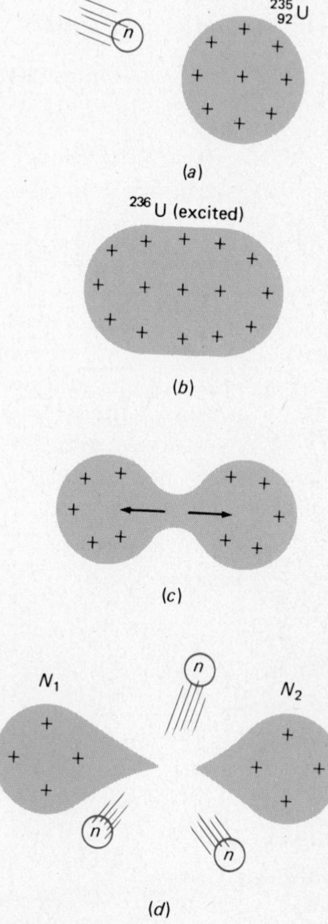

FIGURE 29-2
Fission of a ${}^{235}_{92}\text{U}$ nucleus after capture of a neutron.

The compound nucleus, $^{236}_{92}U$, exists for less than 10^{-12} s, so the process occurs very quickly. The two fission fragments have roughly half the mass of the uranium, although rarely are they exactly equal in mass. A typical fission reaction is

$$^{1}_{0}n + ^{235}_{92}U \rightarrow ^{141}_{56}Ba + ^{92}_{36}Kr + 3\,^{1}_{0}n, \tag{29-2}$$

although many others also occur.

Energy release and chain reactions: the first nuclear reactor

A tremendous amount of energy is released in a fission reaction. This is because the mass of $^{235}_{92}U$ is greater than that of the fission fragments. That this is so can be seen from the binding energy per nucleon curve of Figure 28-1. There we can see that the binding energy per nucleon for uranium is about 7.6 MeV/nucleon; but for fission fragments that have intermediate mass (in the center portion of the graph, $Z \approx 100$), the average binding energy per nucleon is about 8.5 MeV/nucleon. Since the fission fragments are more tightly bound, they have less mass. The difference in mass (or energy) between the original uranium nucleus and the fission fragments is $8.5 - 7.6 = 0.9$ MeV per nucleon. Since there are 236 nucleons involved in each fission, the total energy released per fission is

$$(0.9\ \text{MeV/nucleon})\ (236\ \text{nucleons}) \approx 200\ \text{MeV}.$$

This is an enormous amount of energy on the nuclear scale!

At a practical level, the energy from one fission is, of course, tiny. It was soon recognized, however, that a great deal of energy at the macroscopic level would be available if many such fissions could occur at once. Indeed, such a possibility exists, as Fermi recognized soon after his arrival in the United States in 1939: the neutrons released in each fission (Equation 29-1 or 29-2) could be used to create a **chain reaction**. That is, one neutron initially causes one fission of a uranium nucleus; the two or three neutrons released can go on to cause additional fissions, and the process can multiply, as shown schematically in Figure 29-3.

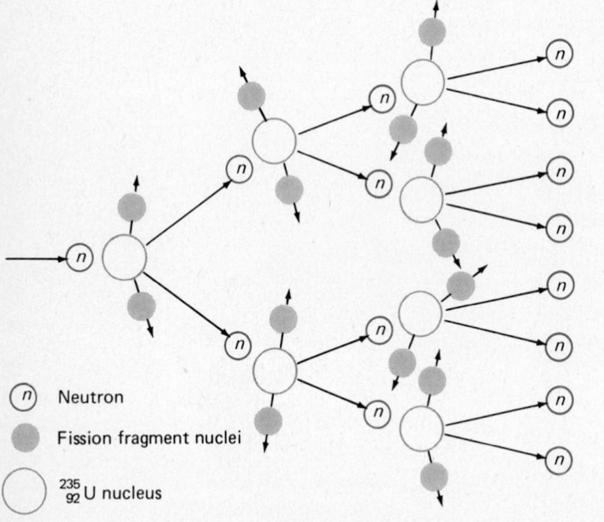

FIGURE 29-3
Chain reaction.

Thus the possibility of obtaining the enormous energy available in nuclear fission seemed at hand. The problem was to find out if a *self-sustaining* chain reaction was actually possible in practice. Fermi and his coworkers at the University of Chicago set out to build the first *nuclear reactor.*

Many problems had to be overcome. First, the probability that a $^{235}_{92}U$ nucleus will absorb a neutron is significant only for slow neutrons. This is because slowly moving neutrons have more time to interact with the nuclei. The problem is that the neutrons emitted during each fission, which are needed to sustain a chain reaction, are moving very fast. A substance referred to as a *moderator* must thus be used to slow down the neutrons. The most effective moderator will consist of atoms whose mass is as close as possible to that of the neutrons. (To see why this is true, recall that a billiard ball striking an equal mass at rest can itself be stopped in one collision; but a billiard ball striking a heavy object bounces off with nearly the same speed it had). The best moderator would thus contain ^1_1H atoms; unfortunately, ^1_1H tends to absorb neutrons; but the isotope of hydrogen ^2_1H (deuterium) does not absorb many neutrons and is thus the most ideal moderator. It is used in the form of *heavy water*, where the hydrogen atoms in the H_2O have been replaced by deuterium. Another common moderator is graphite, which consists of $^{12}_6C$ atoms.

A second problem is that the neutrons produced in one fission may be absorbed by other nuclei in the reactor and produce other nuclear reactions rather than produce further fissions. For example, $^{238}_{92}U$ absorbs neutrons to form $^{239}_{92}U$ in the reaction $n + ^{238}_{92}U \rightarrow ^{239}_{92}U + \gamma$. Naturally occurring uranium contains 99.3 percent of $^{238}_{92}U$ and only 0.7 percent of fissionable $^{235}_{92}U$. To increase the probability of fission of $^{235}_{92}U$ nuclei, natural uranium is often *enriched* to increase the percentage of $^{235}_{92}U$. Since the mass of $^{235}_{92}U$ is less than that of $^{238}_{92}U$, the percentage of the former can be increased by complicated processes involving diffusion or centrifugation.

To have a self-sustaining chain reaction, it is clear that on the average at least one neutron produced in each fission must go on to produce another fission. The average number of neutrons per each fission that do go on to produce further fissions is called the *multiplication factor f*. For a self-sustaining chain reaction, we must have $f \geqslant 1$. Now in the fission of $^{235}_{92}U$, an average of 2.5 neutrons is produced per fission. Of this number, some will be absorbed and produce useless nuclear reactions. Some will also escape through the surface of the uranium fuel before thay have a chance to cause further fissions (Figure 29-4). If the mass of uranium is too small, most of the neutrons will escape (or be absorbed) before they can cause another fission, and the chain reaction will not be sustained. The minimum mass of uranium needed for a self-sustaining chain reaction ($f \geqslant 1$) is called the **critical mass**. The value of the critical mass depends on the moderator used, what fuel is used (in some situations $^{239}_{94}Pu$ is used instead of $^{235}_{92}U$†), and how much the fuel is enriched, if at all. Typical values are on the order of a few kilograms (that is, not grams nor 1000s of kilograms).

† $^{238}_{92}U$ will fission, but only with fast neutrons ($^{238}_{92}U$ is more stable than $^{235}_{92}U$). The probability of absorbing a fast neutron and producing a fission is too low to produce a self-sustaining chain reaction.

FIGURE 29-4
If the amount of uranium exceeds the critical mass, as in (*b*), a sustained chain reaction is possible. If the mass is less than critical, as in (*a*), most neutrons escape before additional fissions occur, and the chain reaction is not sustained.

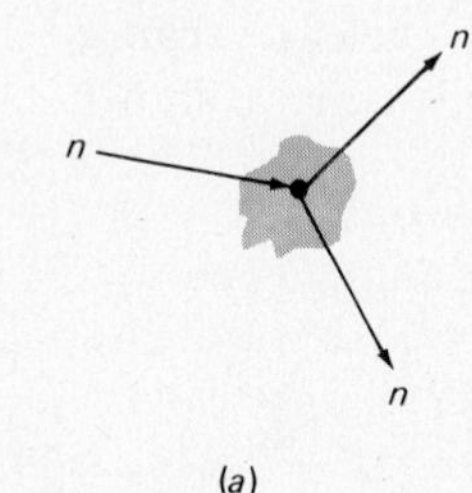

(*a*)

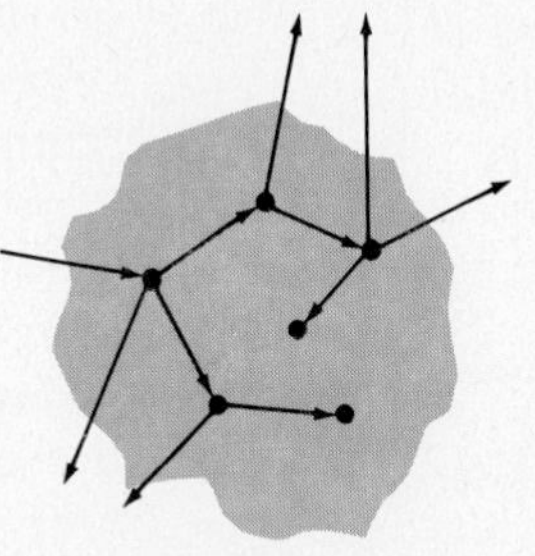
(*b*)

FIGURE 29-5
Sketch of the first nuclear reactor, built by Fermi under the grandstand of Stagg Field at the University of Chicago. (Courtesy of Argonne National Laboratory.)

The first nuclear reactor was built by Fermi in 1942 (Figure 29-5) and was ready for testing in December of that year. Natural uranium was used with graphite as the moderator. The reactor was equipped with cadmium control rods, whose function was to absorb neutrons and thus prevent the reactor from producing too much energy in a short time, which might lead to its destruction. With the rods in place, the cadmium—which strongly absorbs neutrons—would keep the reactor "subcritical" (that is, $f<1$ and no chain reaction). When the control rods were pulled out, fewer neutrons would be absorbed, so more would be available to cause fission. If all went well, it was hoped that a sustained chain reaction would occur. On December 2, 1942, the critical test was made: Fermi slowly withdrew the control rods—and the reactor went "critical." The first self-sustaining chain reaction had been produced.

For better or worse, the nuclear age had begun. The possibilities for the future seemed immense. It was clear that nuclear energy could be used to help mankind in peaceful endeavors; but it could also be used for destructive purposes, and—to the great misfortune of mankind—that was how it was first used.

The bomb

The building of the first reactor in 1942 was part of a plan to see if a bomb based on nuclear energy was feasible. In early 1940, with Europe already at war, Hitler banned the sale of uranium from the Czech mines he had recently taken over. Research into the fission process suddenly was enshrouded in secrecy by Germany and by the western powers. Physicists in the United States were alarmed. A group of them approached Einstein—a man whose name was a household word—to write a letter to President Roosevelt about the possibilities of using nuclear fission for a bomb far more powerful than any previously known, and inform him that Germany might already have begun development of such a bomb.

Roosevelt responded by authorizing the program known as the Manhattan Project, to see if a bomb could be built. Work began in earnest after Fermi's demonstration that a sustained chain reaction was possible. A new secret laboratory was developed on an isolated mesa in New Mexico known as Los Alamos. Under the direction of J. Robert Oppenheimer (1904–1967), it became the home of famous scientists from all over Europe and the U.S.; the roster of names of those who worked there during the next couple of years is like a who's who of science.

One of the problems was how to build a bomb that was subcritical during transport but which could be made critical (to produce a chain reaction) at just the right moment. This was achieved by using two pieces of uranium each of which was less than the critical mass but when placed together became greater than the critical mass. The two masses would be kept separate until the moment of detonation arrived; then a special kind of gun would force the two pieces together very quickly, a chain reaction of explosive proportions would occur, and a tremendous amount of energy would be released very suddenly. The first fission bomb (popularly known as an "atomic bomb") was tested in the New Mexico desert in July, 1945. It was successful. In early August, a fission bomb using uranium was dropped on Hiroshima and a second, using plutonium, was dropped on Nagasaki. World War II ended shortly thereafter, but the dropping of these bombs aroused much controversy.

Scientists were later criticized for working on the bomb, and many regretted having done so. Critical judgment, however, should perhaps be placed in the context of the times. In the early 1940s, Hitler's armies were overrunning Europe, and it was believed that German scientists were trying to build a bomb (they never did). Japan had entered the war and the world seemed in danger of being overwhelmed by oppressive powerful regimes. The scientists at Los Alamos were aware they were developing a dangerous weapon. But they had to make a choice. Stopping Hitler and his allies was uppermost in their minds.

Nuclear fallout

Besides its great destructive power, a fission bomb produces many highly radioactive isotopes. This happens because although uranium has about 50 percent more neutrons than protons, stable isotopes in the middle of the periodic table ($Z = 30$ to 60) have more nearly equal numbers of neutrons and protons (see Figure 28-2). Hence the fission fragments produced in fission have a large excess of neutrons—they lie above the line of stability in Figure 28-2. They are very unstable and decay mainly by β^- emission. When a fission bomb explodes, these radioactive isotopes are released into the atmosphere and are known as *radioactive fallout.* The effects of nuclear fallout were most critically observed in the survivors of Hiroshima and Nagasaki. Many of those who survived the initial blast later died or were crippled by the intense radiation from the fallout. Much of our knowledge of the effects of radiation has been obtained from studies of these victims.

Testing of nuclear bombs in the atmosphere after World War II was also a cause of concern, for the movement of air masses spread the fallout all over the globe. Radioactive fallout eventually settles to the earth, particularly in rainfall, and is absorbed by plants and grasses and enters the food chain. This is a far more serious problem than the same radioactivity on the exterior of our bodies, since α and β particles are largely absorbed by clothing and the outer (dead) layer of skin. But once inside our bodies via food, the isotopes are in direct contact with living cells. One particularly dangerous radioactive isotope is $^{90}_{38}\text{Sr}$, which is chemically much like calcium and becomes concentrated in bone, where it causes bone cancer and destruction of bone marrow. The 1963 treaty signed by over 100 nations that bans nuclear weapons testing in the atmosphere was motivated because of the hazards of fallout.

29-3 ▮ Nuclear reactors

Soon after World War II, peaceful uses of nuclear energy began. Nuclear reactors were built for use in research and to produce electric power. Fission produces many neutrons and a "research reactor" is basically an intense source of neutrons. These neutrons can be used as projectiles in nuclear reactions to produce nuclides not found in nature. Isotopes used as tracers and for medical therapy are generally produced in this way.

A "power reactor" is used to produce electricity for homes and industry. The energy released in the fission process appears as heat, which is used to boil water and produce steam to drive a turbine connected to an electric generator. Figure 29-6 shows a typical design.

The *core* of a nuclear reactor consists of the fuel and a moderator, usually graphite or heavy water. The fuel is usually uranium enriched so that it contains 2 to 4 percent $^{235}_{92}U$. The difference between a nuclear power reactor and a nuclear bomb is the rate of release of the energy: in a bomb, the neutron multiplication factor is somewhat greater than 1, so the chain reaction occurs swiftly and an explosion occurs; in a nuclear reactor, the multiplication of neutrons is held very close to 1.0 so that the reactor is kept just barely "critical" and the energy is released slowly. This is accomplished by careful and continuous adjustment of control rods that act to absorb neutrons. Water or other liquid (such as liquid sodium) is allowed to flow through the core. The thermal energy it absorbs is used to produce steam in the heat exchanger (Figure 29-6), so the fissionable fuel acts as the heat input for a heat engine (Section 10-5).

Many problems are associated with nuclear reactors, ranging from safety to environmental pollution. Although there seems little likelihood of a bomblike explosion, accidents or malfunctions can occur in which the highly radioactive fission fragments are released into the atmosphere; the effect on human beings† could be serious, depending on how much

† It should not be ignored that other sources of power also endanger human lives: for example, coal miners suffer serious diseases (from coal dust) and die in accidental cave-ins.

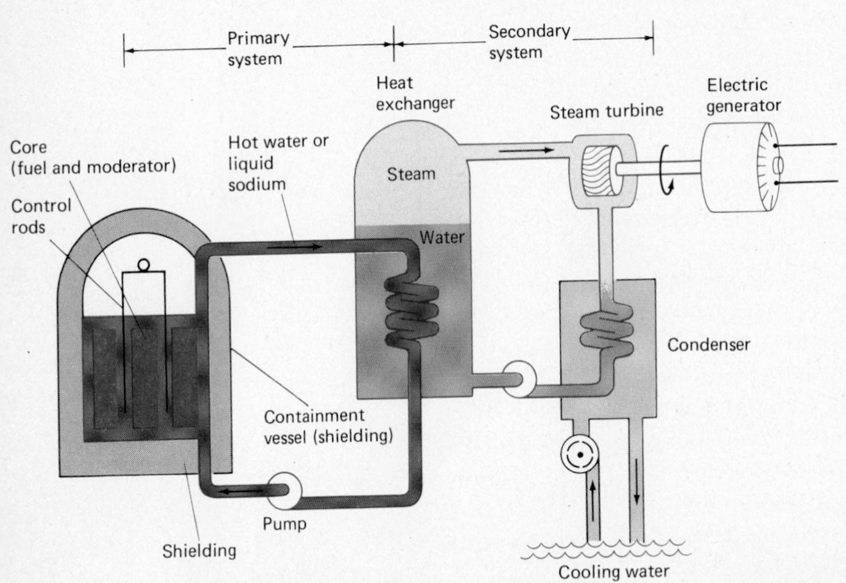

FIGURE 29-6
Nuclear reactor. The heat generated by the fission process in the fuel rods is carried off by hot water or liquid sodium and is used to boil water to steam in the heat exchanger. The steam drives a turbine to generate electricity and is then cooled in the condenser.

radioactivity is released (see Sections 28-7 and 28-8). The Three Mile Island accident (1979) casts doubt on present safety systems. Another serious problem is the disposal of the spent fuel, which contains highly radioactive fission fragments. Leakage of the radioactive wastes is possible and has in fact occurred. Indeed, a satisfactory method of disposal is still not at hand. The earth, because of its finite size, may not be able to safely hold all the nuclear waste products produced.

Other problems of nuclear power plants are that they require cooling water, which is given off at a higher than normal temperature, usually into the ocean, a river, or the air. Thermal pollution may upset the marine ecology nearby, or affect the weather in the case of air cooling towers.

Nuclear power offers the advantage over conventional fossil-fuel plants of normally producing little air pollution. And in the face of dwindling supplies of fossil fuels, it can offer an alternative source of energy. However, the supply of fissionable uranium is also limited. A *breeder reactor* alleviates this problem somewhat. A breeder reactor is one in which some of the neutrons produced in the fission of ${}^{235}_{92}U$ are absorbed by ${}^{238}_{92}U$, and ${}^{239}_{94}Pu$ is produced via the set of reactions shown in Figure 29-1. ${}^{239}_{94}Pu$ is fissionable with slow neutrons, so after separation it can be used as a fuel in a nuclear reactor. Thus a breeder reactor "breeds" new fuel† (${}^{239}_{94}Pu$) from otherwise useless ${}^{238}_{92}U$. Since natural uranium is 99.3 percent ${}^{238}_{92}U$, this means that the supply of fissionable fuel could be increased by more than a factor of 100.

But breeder reactors present additional problems. First, plutonium is a highly toxic substance that poses a serious danger to health, in large part because of its long half-life of 24,000 years. Also important is the fact that plutonium produced can readily be used in a bomb. Thus the use of a breeder reactor, even more than a conventional uranium reactor, presents the danger of nuclear proliferation: even poor nations might be able to produce nuclear bombs. And the possibility of theft of fuel by terrorists who could produce a bomb further increases the probability of nuclear holocaust.

It is clear that nuclear power presents many risks. But since other forms of power also have problems, nuclear power must be examined in the light of problems associated with the alternatives. Reduction of energy use must also be seriously considered.

29-4 ▮ Fusion

The mass of every stable nucleus is less than the sum of the masses of the protons and neutrons that compose it. The mass of the helium isotope ${}^{4}_{2}He$, for example, is less than the mass of two protons plus the mass of two neutrons. In other words, if two protons and two neutrons were to come together to form a helium nucleus, there would be a loss of mass, accompanied by the release of a large amount of energy. The process of building up nuclei by bringing together individual protons and neutrons, or building larger nuclei by combining small nuclei, is called **nuclear fusion**. A glance at Figure 28-1 shows how small nuclei can combine to form larger ones with the release of energy: it is because the binding

† A breeder reactor does *not* produce more fuel than it uses, for a nearly endless supply, as newspaper articles often imply.

energy per nucleon is smaller for light nuclei than it is for those of intermediate size ($A = 50$ to 100). It is believed that all the elements in the universe were originally formed through the process of fusion. Today, fusion is continually taking place within our sun and other stars, and the intense light energy they emit is obtained from these fusion reactions.

▌ EXAMPLE 29-3 One of the simplest fusion reactions involves the production of deuterium ${}^2_1H$ from a neutron and a proton: ${}^1_1H + {}^1_0n \rightarrow {}^2_1H + \gamma$. How much energy is released in this reaction?

SOLUTION From Appendix E the initial rest mass is 1.007825 u + 1.008665 u = 2.016490 u and after the reaction the mass is that of the ${}^2_1H$, namely 2.014102 u. The energy released is thus (0.002388 u) × (931.5 MeV/u) = 2.22 MeV, and is carried off by the ${}^2_1H$ nucleus and the γ ray. ▌▌

The possibility of utilizing the energy released in fusion to make a power reactor is very attractive, but so far a successful reactor has not been achieved. The fusion reactions most likely to succeed in a reactor involve the isotopes of hydrogen, ${}^2_1H$ (deuterium) and ${}^3_1H$ (tritium), and are as follows, with the energy released given in parentheses:

$${}^2_1H + {}^2_1H \rightarrow {}^3_1H + {}^1_1H \qquad (4.0\ \text{MeV}) \tag{29-3a}$$

$${}^2_1H + {}^2_1H \rightarrow {}^3_2He + {}^1_0n \qquad (3.3\ \text{MeV}) \tag{29-3b}$$

$${}^2_1H + {}^3_1H \rightarrow {}^4_2He + n \qquad (17.6\ \text{MeV}) \tag{29-3c}$$

$${}^3_1H + {}^3_1H \rightarrow {}^4_2He + 2{}^1_0n \qquad (11.3\ \text{MeV}). \tag{29-3d}$$

The energy released in fusion reactions is greater for a given mass of fuel than in fission. Furthermore, fusion presents much less of a radioactive waste problem, for almost no long-lived isotopes are produced. For fuel, a fusion reactor could use deuterium, which is very plentiful in the water of the oceans (one ${}^2_1H$ atom in every 3500 molecules of water, or 0.1g of deuterium per gallon). It is thought by many that the use of fusion as a source of power could be a great boon to humankind.

Unfortunately, many difficulties still exist, and a usable fusion reactor is not yet a reality. The major problems involve the fact that all nuclei have a positive charge and thus repel each other. However, if they can be brought close enough together to permit the short-range attractive nuclear force to operate, this force will pull the nuclei together and fusion will occur. For this to happen, the two nuclei must have very high speeds, and since high speed in atoms corresponds to high temperature, very high temperatures are required for fusion to occur. Because of this, fusion reactors are often referred to as *thermonuclear* devices. The sun and other stars are very hot, many millions of degrees, so the nuclei are moving fast enough for fusion to take place readily. The energy released by fusion reactions keeps the temperature high, so such reactions continue to occur. Indeed, the sun and the stars represent self-sustaining thermonuclear reactors, but on earth such high temperatures are not easily attained in a controlled manner.

It was realized after World War II that the temperature produced within a fission (or "atomic") bomb was close to 10^8 K; this suggested that a fission bomb could be used to ignite a fusion bomb (popularly known as a thermonuclear or hydrogen bomb) which uses the vast energy released in fusion.

Although it has been possible to produce an uncontrollable release of fusion energy in an H bomb, it has proved extremely difficult to realize usable energy from fusion at a slow and controlled rate. The high temperatures needed can now be produced by an infusion of concentrated energy, say from a high-powered laser. The real difficulty is to contain the nuclei long enough for sufficient reactions to occur that a usable amount of energy is obtained. At the temperatures needed for fusion, the atoms are ionized; and this collection of ions and electrons is referred to as a *plasma.* Ordinary materials vaporize at a few thousand degrees at best, and hence could not be used to contain a high-temperature plasma. One technique used to try to contain a hot plasma is a "magnetic bottle," as shown in Figure 29-7. The paths of the charged particles in the plasma are bent by the magnetic field (Chapter 17) and where the lines are close together, the force on the particles is such that they are reflected back toward the center. Unfortunately, this magnetic bottle develops "leaks" and the charged particles leak out before sufficient fusion takes place. More complex arrangements of magnetic fields are being tried, and confinement times are slowly being increased.

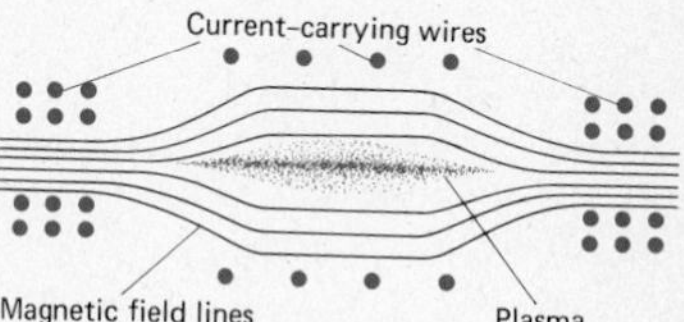

FIGURE 29-7
"Magnetic bottle" used to confine a plasma.

Another containment technique is to form solid pellets of fuel which are quickly heated by an intense laser or electron beam. This technique also shows promise, but controlled fusion still seems to be some time off in the future.

One attractive aspect of fusion is that it may offer the possibility of generating electric power without using the inefficient heat engine cycle (Chapter 10) as an intermediate stage. Such direct conversion of energy would greatly increase efficiency; one possibility would be the use of magnetohydrodynamics (MHD), which we discussed in Section 18-6.

▮ Summary

A *nuclear reaction* occurs when two nuclei collide and two or more other nuclei (or particles) are produced. In this process, as in radioactivity, transmutation (change) of elements occurs.

In *fission* a heavy nucleus such as uranium splits into two intermediate-sized nuclei, usually after being struck by a neutron. $^{235}_{92}\mathrm{U}$ is fissionable by slow neutrons, whereas some fissionable nuclei require fast neutrons. Much energy is released in fission because the binding energy per nucleon is lower for heavy than for intermediate-sized nuclei, so the mass of a heavy nucleus is greater than the total mass of its fission products. The fission process releases neutrons so that a *chain reaction* is possible. The *critical mass* is the minimum mass of fuel needed to sustain a chain reaction.

In a nuclear reactor or nuclear bomb, a moderator is needed to slow down the released neutrons. A breeder reactor not only produces energy but changes $^{238}_{92}\mathrm{U}$ into the isotope $^{239}_{94}\mathrm{Pu}$, which is readily fissionable in a nuclear reactor.

The *fusion* process, in which small nuclei combine to form larger ones, also releases energy. It has not yet been possible to build a fusion reactor for power generation, partly because of the difficulty in containing the fuel at the high temperature required.

Questions

1 Fill in the missing particles or nuclei: (*a*) $^{137}_{56}Ba(n, \gamma)$?; (*b*) $^{137}_{56}Ba(n, ?)^{137}_{55}Cs$; (*c*) $^{2}_{1}H(d, ?)^{4}_{2}He$; (*d*) $^{197}_{79}Au(\alpha; d)$?

2 The isotope $^{32}_{15}P$ is produced by an (n, p) reaction. What must be the target nucleus?

3 When $^{22}_{11}Na$ is bombarded by deuterons ($^{2}_{1}H$) an α particle is emitted. What is the resulting nuclide?

4 Why are neutrons such good projectiles for producing nuclear reactions?

5 A proton strikes a $^{20}_{10}Ne$ nucleus, and an α particle is observed to come out. What is the residual nucleus? Write down the reaction equation.

6 Are fission fragments β^+ or β^- emitters?

7 If $^{235}_{92}U$ released only 1.5 neutrons per fission on the average, would a chain reaction be possible? What would be different?

8 $^{235}_{92}U$ releases an average of 2.5 neutrons per fission compared to 2.7 for $^{239}_{94}Pu$. Pure samples of which of these two nuclei do you think would have the smaller critical mass?

9 Discuss how the course of history might have been changed if, during World War II, scientists had refused to work on developing a nuclear bomb. Do you think it would have been possible to delay the building of a bomb indefinitely?

10 Research in molecular biology is moving toward the ability to perform genetic manipulations on human beings. The moral implications of future discoveries along these lines has led to a warning that this may be the molecular biologists' "Hiroshima." Discuss.

11 The energy produced by a nuclear reactor appears in what form?

12 Discuss the relative merits and disadvantages, including pollution and safety, of power generation by fossil fuels, nuclear fission, and nuclear fusion.

13 What breeds what in a breeder reactor?

14 The energy from nuclear fission appears in the form of thermal energy—but the thermal energy of what?

15 A reactor that uses highly enriched uranium can use ordinary water (instead of heavy water) as a moderator and still have a self-sustaining chain reaction. Explain.

16 Why must the fission process release neutrons if it is to be useful?

17 Why should a porous block of uranium be more likely to explode if kept under water rather than if kept in air?

18 In a nuclear reactor, why do you suppose that high-temperature water heated directly by the fissioning fuel is not used to drive a turbine directly? That is, what is the reason for the secondary system (Figure 29-6)?

19 What is the basic difference between fission and fusion?

20 Light energy emitted by the sun and stars comes from the fusion process. What conditions in the interior of stars makes this possible?

Problems

SECTION 29-1

1 (I) Determine whether the reaction $^{2}_{1}H(d, n)^{3}_{2}He$ requires a threshold energy.

2 (I) Is the reaction $^{238}_{92}U(n, \gamma)^{239}_{92}U$ possible with slow neutrons?

3 (I) Does the reaction $^{7}_{3}Li(p, \alpha)^{4}_{2}He$ require energy or does it release energy?

4 (II) (*a*) Can the reaction $^{24}_{12}Mg(n, d)^{23}_{11}Na$ occur if the bombarding particles have 1.00 MeV of KE? (*b*) If so, how much energy is released?

5 (II) (*a*) Can the reaction $^{7}_{3}Li(p, \alpha)^{4}_{2}He$ occur if the incident proton has KE = 500 keV? (*b*) If so, what is the total kinetic energy of the products?

6 (II) In the reaction $^{14}_{7}N(\alpha, p)^{17}_{8}O$, the incident α particles have 7.68 MeV of kinetic energy. Can this reaction occur? If so, what is the total kinetic energy of the products? The mass of $^{17}_{8}O$ is 16.999131 u.

7 (III) Use conservation of energy and momentum to show that a bombarding proton must have an energy of 3.23 MeV to make the reaction $^{13}_{6}C(p, n)^{13}_{7}N$ occur. (See Example 29-2.)

8 (III) Show, using the laws of conservation of energy and momentum, that for a nuclear reaction requiring energy the minimum kinetic energy of the bombarding particle (the "threshold energy") is equal to $[Qm_{pr}/(m_{pr} - m_b)]$, where Q is the energy required (the difference in total mass between products and reactants), m_b the mass of the bombarding particle, and m_{pr} the total rest mass of the products.

SECTION 29-2

9 (I) Calculate the energy released in the fission reaction $^{1}_{0}n + ^{235}_{92}U \rightarrow ^{88}_{38}Sr + ^{136}_{54}Xe + 12\,^{1}_{0}n$. Use Appendix E and assume that the initial KE of the neutron is very small.

10 (I) What is the energy released in the fission reaction of Equation 29-2? (The masses of $^{141}_{56}Ba$ and $^{92}_{36}Kr$ are 140.9141 and 91.9250, respectively.)

11 (II) One means of enriching uranium is by diffusion of the gas UF_6. Calculate the ratio of the speeds of molecules of this gas containing $^{235}_{92}U$ and $^{238}_{92}U$, on which this process depends.

12 (II) What was the mass loss of $^{235}_{92}U$ actually fissioned in the first atomic bomb, whose energy was the equivalent of 20 kilotons of TNT (1 kiloton of TNT releases 5×10^{12} J)?

SECTION 29-3

13 (II) Suppose that the average power consumption, day and night, of an average house is 300 W. What mass of $^{235}_{92}U$ would be needed to supply the electrical needs of such a house for a year? (Assume that 200 MeV is released per fission.)

14 (II) What mass of $^{235}_{92}U$ is required to operate a 500-MW reactor for 1 yr?

15 (II) How many fissions take place per second in a 25-MW reactor? Assume that 200 MeV is released per fission.

16 (III) Suppose that the neutron multiplication factor is 1.0004. If the time between fissions is 10^{-5} s, by what factor will the reaction rate increase in 1 s?

17 (III) Consider a system of nuclear power plants that produce 4000 MW. (*a*) What mass of $^{235}_{92}U$ would be required to operate these plants for 1 yr (3.16×10^7 s), assuming that 200 MeV is released per fission? (*b*) Typically 3 percent of the $^{235}_{92}U$ mass is converted to $^{90}_{38}Sr$, a β^--emitter with a half-life of 29 years. What is the total radioactivity of the $^{90}_{38}Sr$, in curies, after its production. (Neglect the fact that some of it decays during the 1-yr period.) (*Hint*: See Section 28-5.)

SECTION 29-4

18 (I) What is the average kinetic energy of protons at the center of a star where the temperature is 10^7 K?

19 (II) If an average house requires 300 W of electric power on average, how much deuterium will be needed to supply these electrical needs for 1 yr? Assume the reaction of Equation 29-3*b*.

20 (II) Show that the energy released in the fusion reaction $^2_1H + ^3_1H \rightarrow ^4_2He + n$ is 17.6 MeV.

21 (II) Show that the energy released when two deuterium nuclei fuse to form ^3_2He with the release of a neutron is 3.3 MeV.

22 (II) Calculate approximately the energy release per gram of fuel for the reaction of Equation 29-3*a*. Compare to the energy release per gram of uranium in the fission process.

23 (II) In the "carbon cycle" that occurs in the sun, ^4_2He is built from four protons starting with $^{12}_6C$. First, $^{12}_6C$ absorbs a proton to form nucleus X_1. X_1 decays by β^+ emission to X_2. X_2 absorbs a proton to become X_3, which itself absorbs a proton to become X_4. X_4 decays to X_5 by β^+ decay and X_5 reacts via $X_5(p, \alpha)X_6$. Determine the intermediate nuclei and write out each step in detail. Show that X_6 is again $^{12}_6C$, which is thus not used up in the process.

24 (III) How much energy (in joules) is contained in 1.0 kg of water if natural deuterium is used in the fusion reaction of Equation 29-3*a*? Compare this with the energy obtained from the burning of 1.0 kg of gasoline, about 5×10^7 J.

For additional reading

APS Study Group, "The Nuclear Fuel Cycle: An Appraisal," *Physics Today*, **30**:10 (Oct., 1977), pp. 32–35. Management of nuclear wastes.

Bebbington, W. P., "Reprocessing of Nuclear Fuels," *Sci. American*, **235**:6 (Dec., 1976), pp. 30–41.

Bronowski, J., *The Ascent of Man*, Little, Brown & Company, Boston, 1974. Chapter 11.

Coppi, B., and J. Rem, "Tokamak Approach in Fusion Research," *Sci. American*, **227**:1 (Jul., 1972), pp. 65–75.

Cowan, G. A., "A Natural Fission Reactor," *Sci. American*, **235**:1 (Jul., 1976), pp. 36–47.

Davis, N. P., *Lawrence and Oppenheimer*, Simon & Schuster, Inc., New York, 1968.

Emmett, J. L., J. Nuckolls, and L. Wood, "Fusion Power by Laser Implosion," *Sci. American*, **230**:6 (Jun., 1974), pp. 24–37.

Engleman, D. M., and P. B. Moore, "Neutron-Scattering Studies of the Ribosome," *Sci. American*, **235**:4 (Oct., 1976), p. 44.

Fermi, L., *Atoms in the Family: My Life with Enrico Fermi*, University of Chicago Press, Chicago, 1954.

Inglis, D. R., *Nuclear Energy: Its Physics and Its Social Challenge*, Addison-Wesley Publishing Co., Inc., Reading, Mass., 1973.

Jungk, R., *Brighter than a Thousand Suns*, Harcourt Brace Jovanovich, Inc., New York, 1958.

Kipphardt, H., *In the Matter of J. Robert Oppenheimer*, R. Speirs, tr., Hill & Wang, New York, 1969.

Seaborg, G. T., and J. L. Bloom, "Fast Breeder Reactors," *Sci. American*, **223**:5 (Nov., 1970), p. 13.

Vendryes, G. A., "Superphénix: A Full-scale Breeder Reactor," *Sci. American*, **236**:3 (Mar., 1977), pp. 26–35.

The World of Enrico Fermi (film). Interesting and human film about the life and achievements of Fermi, including the development of fission and atomic energy.

30 ▮ Elementary particles

In this, the final chapter, we discuss the exciting subject of elementary particles. This might be called the last frontier, for it represents the human endeavor to understand the basic building blocks of all matter.

In the years after World War II, it was found that if the incoming particle in a nuclear reaction has sufficient energy, new types of particles can be produced. In order to produce high-energy particles for nuclear reactions, physicists constructed various types of particle accelerators or "atom smashers." Most commonly, they accelerate protons or electrons, although heavy ions such as helium and oxygen can also be accelerated. They have been used to probe the nucleus more deeply, and to produce and study new particles. We begin this chapter with a discussion of particle accelerators, and then turn directly to the exciting field of elementary particle physics.

30-1 ▮ Particle accelerators

Particle accelerators have been a principal means of investigating the structure of nuclei. The accelerated particles are projectiles that probe the interior of the nuclei they strike. An important factor is that faster-moving projectiles can reveal more detail about nuclei. The wavelength of particles such as electrons or protons is given by de Broglie's wavelength formula,

$$\lambda = \frac{h}{mv}.$$

From this relation we can see that the greater the momentum of the bombarding particle, the shorter the wavelength and the more detail that can be obtained. (As discussed in Chapter 23, resolution of details is limited by the wavelength.) This is one reason that particle accelerators of higher and higher energy have been built in recent years.

FIGURE 30-1
Distribution of electric charge for bismuth nucleus, determined using 1.3-GeV electrons.

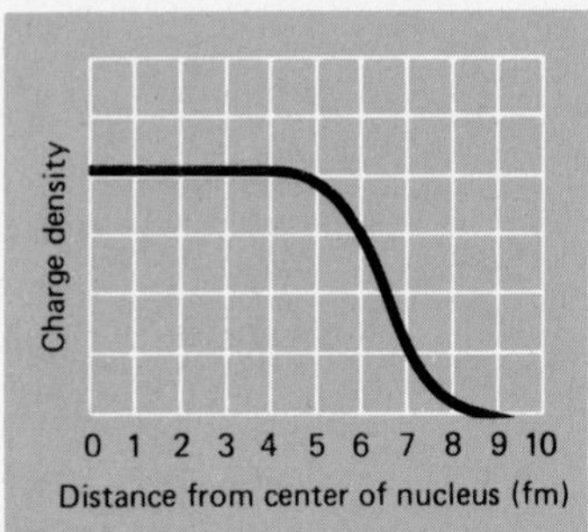

▮ EXAMPLE 30-1 To explore the distribution of charge within nuclei, very-high-energy electrons are used. (Electrons are used rather than protons because they do not partake in the strong nuclear force, so only the electric charge is investigated.) Experiments at the Stanford linear accelerator using electrons with 1.3 GeV of kinetic energy obtained the charge distribution for the bismuth nucleus shown in Figure 30-1. What is the expected resolution—that is, the size of the smallest details that can be detected?

SOLUTION 1.3 GeV = 1300 MeV, which is about 2500 times the mass of the electron. We are clearly dealing with relativistic speeds here, and it is easily shown that the speed of the electron is only very slightly less than $c = 3.0 \times 10^8$ m/s. Therefore,

$$\lambda = \frac{h}{mv} \approx \frac{h}{mc} = \frac{hc}{mc^2},$$

where $mc^2 = 1.3$ GeV. Hence

$$\lambda = \frac{(6.6 \times 10^{-34}\ \text{J}\cdot\text{s})(3.0 \times 10^8\ \text{m/s})}{(1.3 \times 10^9\ \text{eV})(1.6 \times 10^{-19}\ \text{J/eV})} = 0.96 \times 10^{-15}\ \text{m},$$

or 0.96 fm (1 fm $= 10^{-15}$ m = 1 femtometer or 1 fermi—in honor of Enrico Fermi). This is somewhat less than the size of nuclei. Notice in Figure 30-1 that nuclei have a nearly uniform charge distribution on the interior, but their boundary is not distinct. Studies with even higher energy electrons on protons and neutrons show structural details which may indicate that nucleons consist of still smaller particles. ▮▮

High-energy particles produced in accelerators are used not only to investigate nuclear structure, but also to produce many new types of particles, as we shall soon see. Now we discuss the operation of several types of particle accelerators.

Van de Graaff accelerator

A Van de Graaff generator, invented in 1931, is diagrammed in Figure 30-2. A nearly spherical large hollow conductor is supported by an insulating column above the base. The sphere is charged to a high potential by a nonconducting moving belt in the following way. A voltage of typically 50,000 V is applied to a pointed conductor *A*, which "sprays" positive charge onto the moving belt. (Actually, electrons are pulled off the belt onto the electrode *A*.) The belt carries the positive charge into the interior of the sphere, where it is "wiped off" the belt at *B* and races to the outer surface of the spherical conductor— remember that charge collects on the outer surface of any conductor, since the charges repel each other and try to get as far from each other as possible. As more and more charge is brought upward, the sphere becomes more highly charged and reaches greater voltages. The process requires energy, of course, since the upward-moving charged belt is repelled by the charged sphere; the energy is supplied by the motor driving the belt. Very high potential differences can be obtained in this way, often as high as 30 million volts when two or more generators are used in tandem. Connected to the Van de Graaff generator is an evacuated tube which serves as the particle accelerator. A source of H or He ions (p or α) is located inside the tube and the large positive voltage repels them so they are accelerated toward the grounded target at the far end of the tube. A 30-MV generator can produce 30-MeV protons.

FIGURE 30-2
Van de Graaff generator and accelerator.

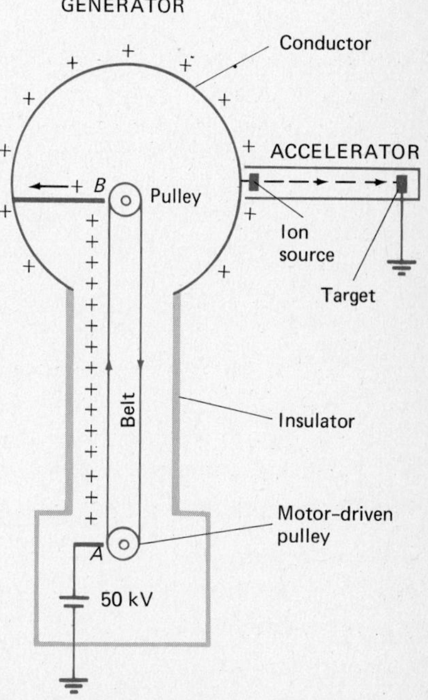

The cyclotron

The cyclotron was developed in 1930 by E. O. Lawrence (1901–1958), at the University of California at Berkeley. It uses a magnetic field to maintain the ions—usually protons—in nearly circular paths (Chapter 17). The protons move within two D-shaped cavities, as shown in Figure 30-3. Each time they pass into the gap between the "dees," however, a voltage is applied that accelerates them. This increases their speed and also increases the radius of curvature of their path. After many revolutions, the protons acquire high energy and reach the outer edge of the cyclotron. They then either strike a target placed inside the cyclotron, or leave the cyclotron with the help of a carefully placed "bending magnet" and are directed to an external target.

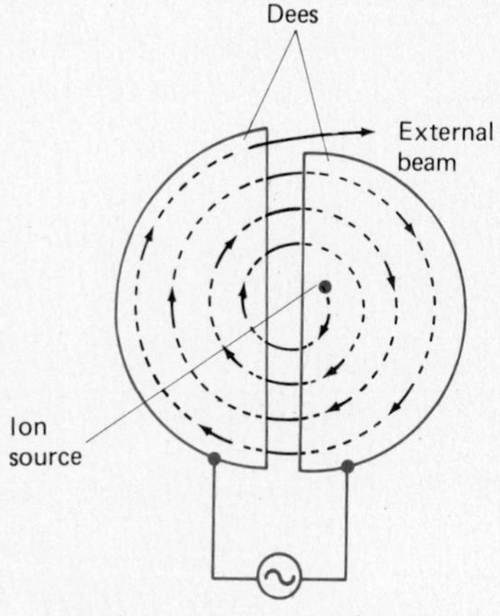

FIGURE 30-3
Diagram of a cyclotron. The magnetic field, applied by a large electromagnet, points out of the paper.

The voltage applied to the dees to produce the acceleration must be alternating. When the protons are moving to the right across the gap in Figure 30-3, the right dee must be negative and the left one positive. A half-cycle later, the protons are moving to the left, so the left dee must be negative in order to accelerate them. The frequency of the applied voltage must be equal to that of the circulating protons, and this can be determined as follows. When the ions are circulating within the dees, the net force F on each is simply that due to the magnetic field B, so $F = qvB$, where q is the charge on the ion and v its speed at a given moment (Equation 17-4). (We ignore the electric force due to the applied voltage since it acts only when the protons are between the dees.) Since the protons move in circles, the acceleration is centripetal and equals v^2/r, where r is the radius of the ion's path at a given moment. We use Newton's second law, $F = ma$, and find that

$$F = ma$$

$$qvB = \frac{mv^2}{r}$$

or

$$v = \frac{qBr}{m}.$$

The time required for a complete revolution is the period T and is equal to

$$T = \frac{\text{distance}}{\text{speed}} = \frac{2\pi r}{qBr/m} = \frac{2\pi m}{qB}.$$

Hence the frequency of revolution f is

$$f = \frac{1}{T} = \frac{qB}{2\pi m}. \qquad (30\text{-}1)$$

This is known as the *cyclotron frequency*.

▮ EXAMPLE 30-2 A small cyclotron of maximum radius $R = 0.50$ m accelerates protons in a 1.7-T magnetic field. Calculate (a) what frequency is needed for the applied alternating voltage, and (b) the kinetic energy of protons when they leave the cyclotron.

SOLUTION (*a*) From Equation 30-1

$$f=\frac{qB}{2\pi m}=\frac{(1.6\times10^{-19}\text{ C})(1.7\text{ T})}{(6.28)(1.67\times10^{-27}\text{ kg})}=2.6\times10^{7}\text{ Hz}.$$

(*b*) The protons leave the cyclotron at $r=R=0.50$ m. Then, since $v=qBr/m$,

$$\text{KE}=\tfrac{1}{2}mv^2=\tfrac{1}{2}m\frac{q^2B^2R^2}{m^2}=\frac{q^2B^2R^2}{2m}$$

$$=\frac{(1.6\times10^{-19}\text{ C})^2(1.7\text{ T})^2(0.50\text{ m})^2}{(2)(1.67\times10^{-27}\text{ kg})}=5.5\times10^{-12}\text{ J},$$

or 34 MeV. Note that the magnitude of the voltage applied to the dees does not affect the final energy; but the higher this voltage, the fewer revolutions are required to bring the protons to full energy. ▌▌

An interesting aspect of the cyclotron is that the frequency of the applied voltage, as given by Equation 30-1, does not depend on the radius r. That is, the frequency does not have to be changed as the ions start from the source and are accelerated to paths of larger and larger radii. Unfortunately, this is only true at nonrelativistic energies. For at higher speeds, the mass of the ions will increase according to Einstein's formula, $m=m_0/\sqrt{1-v^2/c^2}$, where m_0 is the rest mass. This is already a problem for the 34-MeV protons in the example above, for the KE is about 3 percent of the mass. As can be seen from Equation 30-1, as the mass increases, the frequency of the applied voltage must be reduced. To achieve large energies, complex electronics is needed to decrease the frequency as a packet of protons increases in speed and reaches larger orbits. Such a modified cyclotron is called a *synchrocyclotron*.

FIGURE 30-4

(*a*) Aerial view of Fermilab at Batavia, Illinois; the accelerator is a circular ring 2.0 km in diameter. (Courtesy of Fermilab.) (*b*) Interior photograph, showing magnets that keep ions moving in a circle. (Courtesy of Fermilab.)

(*a*)

(*b*)

Synchrotron

Another way to deal with the increase in mass with speed is to increase the magnetic field B as the particles speed up. Such a device is called a *synchrotron*. The largest circular accelerators today fall into this category, and they are enormous. The Fermi National Accelerator Laboratory (Fermilab) at Batavia, Illinois, has a radius of 1.0 km and that at CERN (European Council for Nuclear Research) is 1.1 km in radius. They can accelerate protons to about 500 GeV and that at Fermilab is expected to soon reach 1000 GeV using superconducting magnets. These large synchrotrons do not use enormous magnets 1 km in radius. Instead, a narrow ring of magnets is used (see Figure 30-4) which are all placed at the same radius from the center of the circle. The magnets are interrupted by gaps where high voltage accelerates the ions. After the ions are injected, they must then be maintained in motion in a circle of constant radius. This is done by giving them considerable energy initially in a much smaller accelerator, and then slowly increasing the magnetic field as they speed up in the large synchrotron.

One problem of any accelerator is that accelerating electric charges radiate electromagnetic energy (see Chapter 20). Since ions or electrons are accelerated in any accelerator, we can expect considerable energy to

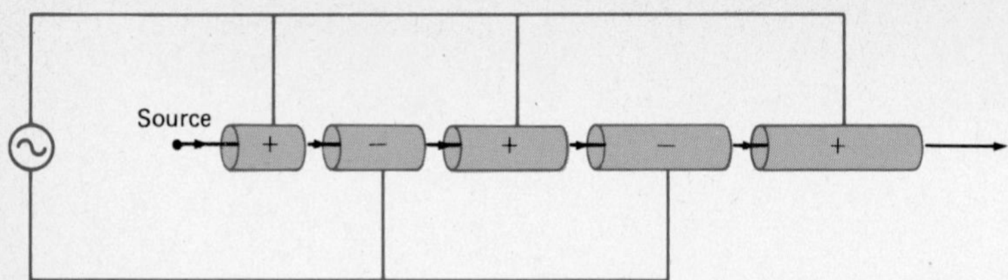

FIGURE 30-5
Simple linear accelerator.

be lost by radiation. The effect increases with speed and is especially important in circular machines where centripetal acceleration is present. It is mainly important in synchrotrons and hence is called *synchrotron radiation.*

Synchrotron radiation can actually be useful, however. Since it consists of photons of a wide range of frequencies (typically UV and x rays), it can be used in experiments—or for treatment—where intense beams of photons are needed.

Linear accelerators

A Van de Graaff is essentially a linear accelerator since the ions move in a linear path. But the name *linear accelerator* is usually reserved for a more complex arrangement in which particles are accelerated many, many times along a straight-line path. Figure 30-5 is a diagram of a simple "linac." The ions pass through a series of tubular conductors. The voltage applied to the tubes must be alternating so that when the ions reach a gap, the tube in front of them is negative and the one they just left is positive. This assures that they are accelerated at each gap.

As the ions increase in speed, they cover more distance in the same amount of time. Consequently, the tubes must be longer the farther they are from the source. The Heavy Ion Linear Accelerator (Hilac) at Berkeley, California, can accelerate ions from hydrogen through uranium, imparting to them as much as 8.5 MeV per nucleon.

Linear accelerators are of particular importance for accelerating electrons. Because of their small mass, electrons reach high speeds very quickly. (Indeed, an electron linac such as the one shown in Figure 30-5 would have tubes nearly equal in length, since the electrons would be traveling close to $c = 3.0 \times 10^8$ m/s for almost the entire distance.)

The largest electron linear accelerator is that at Stanford, called SLAC (Stanford Linear Accelerator Center). It is over 3 km (2 miles) long and can accelerate electrons to over 20 GeV.

Colliding beams

The usual arrangement in *high-energy-physics* experiments (as this field in which high-energy accelerators are used is called) is to allow the beam of particles from an accelerator to strike a stationary target. An important way to increase the energy of a collision is the recent development of *colliding beams*—that is, the target particles are moving as well as the projectile particles. This can be accomplished with only one accelerator through the use of *storage rings.* The accelerator accelerates one type of particle (say electrons or protons) to a maximum energy and then magnets are used to steer these particles into one circular storage ring where the particles can continue to circulate for many hours. It then accelerates a second type of particle (say positrons, or it could be the same as the first type, such as protons), and these are sent to a second storage ring. The two storage rings overlap at several places and the two beams, circulating in opposite directions, collide head on with each other at these points of intersection.

Storage rings for colliding-beam experiments are in use at a number of facilities around the world and have played an important role in recent advances in elementary particle physics.

30-2 ▮ Beginnings of elementary particle physics—the Yukawa particle

By the mid 1930s, it was recognized that all atoms can be considered to be made up of neutrons, protons, and electrons. Thus the basic constituents of the universe were no longer considered to be atoms but rather the proton, neutron, and electron. Besides these three *elementary particles*, as they could be called, several others were also known: the positron (a negative electron), the neutrino, and the γ particle (or photon), for a total of six elementary particles. (We discussed these in Chapter 28.)

Things seemed fairly simple in 1935. But in the decades that followed, hundreds of other elementary particles were discovered. The properties and interactions of these particles, and which ones should be considered as basic or "elementary," is the substance of the exciting and active field of *elementary particle physics*.

This field, as it exists today, can be said to have begun in 1935 when the Japanese physicist Hideki Yukawa (1907–) predicted the existence of a new particle that would in some way mediate the strong nuclear force. To understand Yukawa's theoretical breakthrough, let us first look at the electromagnetic force.

When we first discussed electricity we saw that the electric force acts over a distance, without contact. To better perceive how a force can act over a distance, we saw that Faraday introduced the idea of a field. The force that one charged particle exerts on a second can be said to be due to the electric field set up by the first. Similarly, the magnetic field can be said to carry the magnetic force. Later (Chapter 20) we saw that electromagnetic fields can travel through space as waves. Finally, in Chapter 25, we saw that an electromagnetic wave (light) can also be considered as a collection of particles called photons (or γ particles); this was one example of the wave–particle duality.

Now, because of the wave–particle duality, it is possible to imagine that the electromagnetic force between charged particles is due (1) to the electromagnetic field set up by one and felt by the other, or (2) to an exchange of γ particles between them. We want to concentrate on (2) here; and an example of how an exchange of particles could give rise to a force is illustrated in the crude example of Figure 30-6. In part *a*, two children on rollerskates start throwing pillows at each other; each catch results in the child being moved backward by the impulse. This is the equivalent of a repulsive force. On the other hand, if the two children

FIGURE 30-6
Forces equivalent to particle exchange. (*a*) Repulsive force (children throwing pillows). (*b*) Attractive force (children grabbing pillows from each other's hand).

(*a*) Repulsive force (children throwing pillows)

(*b*) Attractive force (children grabbing pillows from other's hand)

exchange pillows by grabbing them out of the other person's hand, they will be pulled toward each other, as when an attractive force acts.

In the case of the electromagnetic force between two charged particles, it is photons that are exchanged between the two particles that give rise to the force. A simple diagram of how this might work is shown in Figure 30-7. Such a diagram is called a *Feynman diagram.* The simplest case is shown in which a single photon is exchanged. One of the charged particles emits the photon and recoils somewhat as a result; and the second particle absorbs the photon. In any such collision or *interaction*, energy and momentum is transferred from one particle to the other, and it is carried by the photon. Because the photon is absorbed by the second particle very shortly after it is emitted by the first, and hence is not observable, it is referred to as being a *virtual* photon, as compared to one that is free and can be detected by instruments.

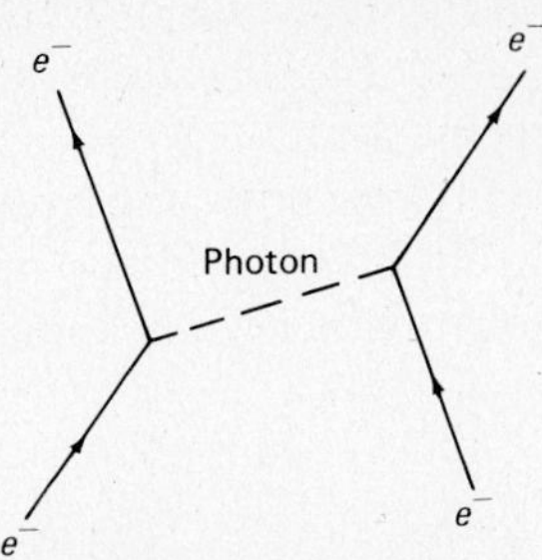

FIGURE 30-7
Feynman diagram, showing how a photon acts as carrier of electromagnetic force between two electrons.

Now to Yukawa's prediction. In analogy to photon exchange to mediate the electromagnetic force, Yukawa argued that there ought to be a particle that mediated the strong nuclear force—the force that holds nucleons together in the nucleus. Just as the photon is called the quantum of the electromagnetic field or force, so the Yukawa particle would represent the quantum of the strong nuclear force.

Yukawa predicted that this new particle would have a mass intermediate between that of the electron and the proton. Hence it was called a **meson**, meaning "in the middle," and Figure 30-8 is a Feynman diagram of meson exchange. We can make a rough approximation of the mass of the meson as follows. Suppose that the proton on the left in Figure 30-8 is at rest. For it to emit a meson would require energy (to make the mass), which would have to come from nowhere; such a process would violate conservation of energy. But the uncertainty principle allows nonconservation of energy of an amount ΔE if it occurs only for a time Δt given by $(\Delta E)(\Delta t) \approx h/2\pi$. We set ΔE equal to the energy needed to create the mass m of the meson: $\Delta E = mc^2$. Now conservation of energy is violated only as long as the meson exists, which is the time Δt required for the meson to pass from one nucleon to the other. If we assume that the meson travels at relativistic speeds, close to the speed of light c, then Δt will be at most about $\Delta t = d/c$, where d is the maximum distance that can separate the interacting nucleons ($d \approx 1.5 \times 10^{-15}$ m). Thus we have

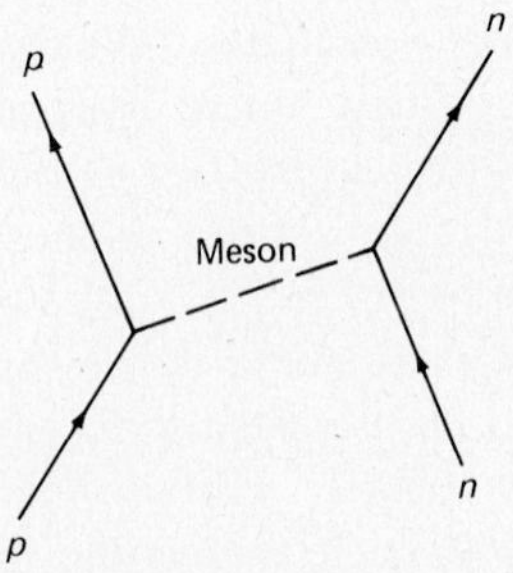

FIGURE 30-8
Meson exchange when proton and neutron interact via strong nuclear force.

$$\Delta E\ \Delta t \approx \frac{h}{2\pi}$$

$$mc^2\left(\frac{d}{c}\right) \approx \frac{h}{2\pi},$$

or

$$mc^2 \approx \frac{hc}{2\pi d} \tag{30-2}$$

$$\approx \frac{(6.6 \times 10^{-34}\ \text{J}\cdot\text{s})(3.0 \times 10^{8}\ \text{m/s})}{(6.28)(1.5 \times 10^{-15}\ \text{m})} \approx 2.2 \times 10^{-11}\ \text{J} = 130\ \text{MeV}.$$

The mass of the predicted meson is thus very roughly 130 MeV/c^2 or about 250 times the electron mass (0.51 MeV/c^2). (Note, incidentally, that for the electromagnetic force, which has infinite range, $d = \infty$, Equation

30-2 tells us that the exchanged particle will have zero rest mass, which is indeed the case for a photon.)

Just as photons can be observed as free particles, as well as acting in an exchange, so it was expected that mesons might be observed directly. Such a meson was searched for in the cosmic radiation that enters the earth's atmosphere from the sun and other sources in the universe. In 1937 a new particle was discovered whose mass was 106 MeV (207 times the electron mass). This is quite close to the mass predicted, but this new particle, called the *muon* (or *mu meson*), did not interact strongly with matter. It could hardly mediate the strong nuclear force if it did not interact by means of the strong nuclear force. Thus the muon, which can have either a + or − charge and seems to be nothing more than a very massive electron, is not the Yukawa particle.

The particle predicted by Yukawa was finally found in 1947. It is called the "π" or pi meson, or simply the *pion.* It comes in three charge states, +, −, or 0. The π^+ and π^- have mass of 140 MeV and the π^0 a mass of 135 MeV. All three interact strongly with matter. Soon after their discovery in cosmic rays, pions were produced in the laboratory using a particle accelerator. Reactions observed included

$$\begin{aligned} p+p &\rightarrow p+p+\pi^0 \\ p+p &\rightarrow p+n+\pi^+. \end{aligned} \tag{30-3}$$

The incident proton from the accelerator must have sufficient energy to produce the additional mass of the pion.

Since the discovery of the pion, a number of other mesons have been discovered. These also mediate the strong nuclear force, and more will be said of them later.

So far we have discussed the particles that mediate the electromagnetic and strong nuclear forces. But there are four known types of force—or interaction—in nature. What about the other two: the weak nuclear force, and gravity? Theorists believe that these are also mediated by particles. The quantum of the weak force is referred to as the *W particle.* The quantum of the gravitational force is called the *graviton.* Despite extensive searching, neither the W nor the graviton has been positively identified. Hence their existence remains in doubt. A comparison of the four forces is given in Table 30-1, where they are listed according to their relative strengths. Notice that although gravity may be the most obvious force in daily life (because of the huge mass of the earth), on a nuclear scale it is much the weakest of the four forces and its effect at this level can almost always be ignored.

TABLE 30-1

The four forces in nature

Type	Relative strength	Field particle
Strong nuclear	**1**	**Mesons**
Electromagnetic	10^{-2}	**Photon**
Weak nuclear	10^{-13}	***W* particle (?)**
Gravitational	10^{-40}	**Graviton (?)**

30-3 ▌ Particles and antiparticles

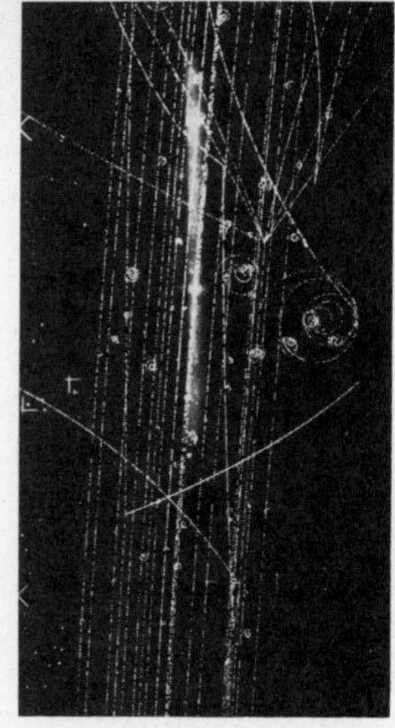

The positron, as we saw earlier, is basically a positive electron. That is, many of its properties are the same as for the electron, such as mass, but it has the opposite charge. The positron is said to be the **antiparticle** to the electron. The positron was discovered in 1932. It was later predicted that other particles should also have antiparticles. However, it was not until 1955 that another antiparticle was found. This was the antiparticle to the proton, called the *antiproton* ($\bar{p}$), Figure 30-9. (The bar over the p indicates it is the antiparticle of the proton.) A large amount of energy was needed to produce this massive particle, and its discovery (by E. Segrè and O. Chamberlain) was made only after the completion of the large accelerator (the Bevatron) at the University of California at Berkeley. Soon after, the antineutron ($\bar{n}$) was found. Most other particles also have antiparticles. But the photon, the π^0, and a few others do not have distinct antiparticles—or we say that they are their own antiparticles.

Antiparticles are produced in nuclear reactions when there is sufficient energy available. Antiparticles do not live very long in the presence of matter. For example, when a positron encounters an electron, the two annihilate each other. The energy of their vanished mass, plus any kinetic energy they initially possessed, is converted to the energy of γ rays or of other particles. Annihilation occurs for all other particle–antiparticle pairs. Far out in space there may be *antimatter*—matter whose atoms are made up of antiparticles. Nuclei would consist of antineutrons and negatively charged antiprotons, and they would be encircled by positrons instead of electrons. These "antiatoms" would be much like ordinary atoms—they could combine chemically, emit photons, and so on. But if an antiworld made of antimatter approached our world of ordinary matter, annihilation would take place and a vast explosion would result.

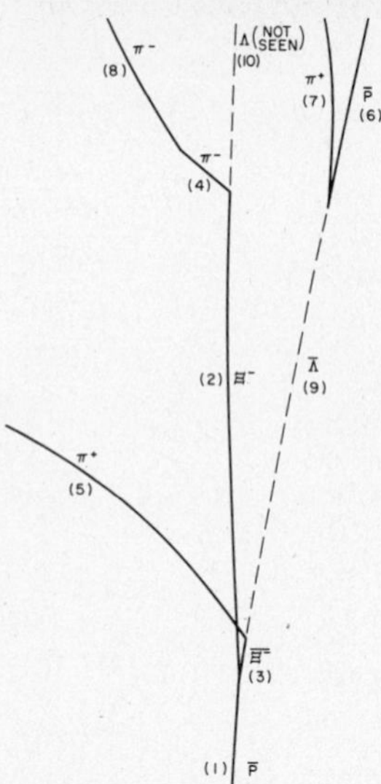

FIGURE 30-9
Liquid-hydrogen bubble-chamber photograph of an antiproton ($\bar{p}$) colliding with a proton, producing a hyperon pair ($\bar{p}+p \rightarrow \Xi^- + \bar{\Xi}^-$), which subsequently decays into other particles. The key indicates the assignment of a particle to each track. Neutral-particle paths are shown by dashed lines since neutral particles produce no bubbles. (Courtesy of Brookhaven National Laboratory.)

30-4 ▌ Particle interactions and conservation laws

One of the important uses of high-energy accelerators is to study the interactions of elementary particles with each other. As a means of understanding this subnuclear world, the conservation laws are indispensable. The laws of conservation of energy, of momentum, of angular momentum, and of electric charge are found to hold precisely in all particle interactions. (Although the uncertainty principle allows nonconservation of energy for times $\Delta t \approx h/2\pi\,\Delta E$, as we have seen, the times involved are much too short for experimental observation and violation has never been detected.)

A study of particle interactions has revealed a number of new conservation laws, some of which we now discuss. These new conservation laws (like the old ones) are ordering principles: they help to explain why some reactions occur and others do not. For example, the following reaction has never been found to occur:

$$p + n \nrightarrow p + p + \bar{p},$$

even though charge, energy, and so on, are conserved ($\bar{p}$ indicates an antiproton and $\nrightarrow$ that the reaction does not occur). To understand why it does not occur, physicists hypothesized a new conservation law, the **conservation of baryon number**. (Baryon number is the same as nucleon number, which we saw earlier is conserved in nuclear reactions.) An important addition to this law is the proposal that whereas all nucleons have baryon number $B=+1$, all antinucleons (antiprotons, antineutrons) have $B=-1$. The reaction above does not conserve baryon number, since on the left side we have $B=(+1)+(+1)=+2$ and on the right $B=(+1)+(+1)+(-1)=+1$. On the other hand, the following reaction *does* occur if the incoming proton has sufficient energy:

$$p+n \rightarrow p+n+\bar{p}+p$$

$$B=+1+1=+1+1-1+1.$$

As indicated, $B=+2$ on both sides of this equation. From these and other reactions, the conservation of baryon number has been established as a basic law of physics.

Two other "number" laws are also useful. They are associated with weak interactions, mainly decays. Two classes of decay occur: ordinary β decay, in which an electron or positron is emitted along with a neutrino or antineutrino; and a similar type of decay, in which a muon is emitted instead of an electron. The neutrino that accompanies an emitted electron is found to be different from the neutrino that accompanies an emitted muon. Hence there are two types of neutrinos, the electron neutrino (ν_e) and the muon neutrino (ν_μ); they each have an antiparticle ($\bar{\nu}_e$ and $\bar{\nu}_\mu$). In ordinary β decay we have, for example,

$$n \rightarrow p+e^-+\bar{\nu}_e,$$

but never $n \rightarrow p+e^-+\bar{\nu}_\mu$ or $n \rightarrow p+e^-+\bar{\nu}_e+\nu_e$. To explain why these do not occur, the concept of electron lepton number, L_e, was invented. If the electron (e^-) and the electron neutrino (ν_e) are given $L_e=+1$, and e^+ and $\bar{\nu}_e$ are given $L_e=-1$, whereas all other particles have $L_e=0$, then all observed decays conserve L_e. For example, in $n \rightarrow p+e^-+\bar{\nu}_e$, $L_e=0$ initially, and $L_e=0+(+1)+(-1)=0$ after the decay. Those decays that do not occur, even though other conservation laws would not be violated, would not conserve L_e; hence it is believed that L_e is conserved in all interactions.

In a decay involving muons, such as

$$\pi^+ \rightarrow \mu^+ + \nu_\mu,$$

a second quantum number, muon lepton number (L_μ), is conserved. The μ^- and ν_μ are assigned $L_\mu=+1$ and μ^+ and $\bar{\nu}_\mu$ have $L_\mu=-1$, whereas other particles have $L_\mu=0$. It is found that L_μ is also conserved in all interactions or decays.

❚ EXAMPLE 30-3 Which of the following decay schemes is possible for muon decay? (*a*) $\mu^- \rightarrow e^- + \bar{\nu}_e$; (*b*) $\mu^- \rightarrow e^- + \bar{\nu}_e + \nu_\mu$; (*c*) $\mu^- \rightarrow e^- + \nu_e$.

SOLUTION A μ^- has $L_\mu=+1$ and $L_e=0$. This is the initial state, and the final state (after decay) must also have $L_\mu=+1$, $L_e=0$. In (*a*) the final state has $L_\mu=0+0=0$ and $L_e=+1-1=0$; L_μ would not be conserved, and indeed this decay is not observed to occur. The final state of (*b*) has

$L_\mu = 0+0+1 = +1$ and $L_e = +1-1+0 = 0$, so both L_μ and L_e are conserved. The μ^- does, in fact, decay in this way. Finally, (*c*) does not occur because $L_e(= +2$ in the final state) is not conserved. Note that all other conservation laws, such as for energy and charge, would hold for all three examples. ▮▮

Besides conservation of baryon number (B), it is now believed that laws of **conservation of lepton numbers** (L_e, L_μ) are also valid.

Note, finally, that antiparticles have not only the opposite charge but also opposite B, L_e, and L_μ from their particles.

30-5 | Particle classification

In the decades following the discovery of the π meson in the late 1940s, a great many other subnuclear particles were discovered; today they number in the hundreds. Much theoretical and experimental work has been done to try to understand this multitude of particles. One important aid to understanding is to arrange the particles in categories according to their properties. The common way of doing this today is according to their interactions. Since not all particles take part in all four of the forces known in nature, this fact is used as a classification scheme. Table 30-2 lists many of the known particles classified in this way and shows many of the properties of the particles. The **photon** takes part only in the electromagnetic force, and it is in a class by itself. The **leptons** are those particles that do not interact by way of the strong force but do interact via the weak nuclear force (as well as the much weaker gravitational force); and those that carry electric charge also interact via the electromagnetic force. The four known leptons are the electron, the muon, and two types of neutrino: the electron neutrino, ν_e, and the muon neutrino, ν_μ. They each have antiparticles as indicated in Table 30-2.

The third category of particle is the **hadron**. Hadrons are those particles that can interact via the strong nuclear force. Hence they are said to be *strongly interacting particles*. They also interact via the other forces, but the strong force predominates at short distances. The hadrons include nucleons, pions, and a large number of other particles. They are divided into two subgroups†: **baryons**, which are those particles that have baryon number $+1$ (or -1 in the case of their antiparticles); and **mesons**, which have baryon number $= 0$.

Notice that the baryons Λ, Σ, Ξ, and Ω all decay to lighter mass baryons, and eventually to a proton or neutron. All these processes conserve baryon number. But since there is no lighter particle than the proton with $B = +1$, the proton itself cannot decay and is stable.

† Originally, particles were divided according to their mass into leptons (meaning light particles), baryons (meaning "heavy"), and those of intermediate mass, the mesons (meaning "middle"). The newer classification according to their interactions is almost consistent with this. The muon, however, is now called a lepton (it does not interact strongly), although it was once called the mu meson because of its mass. Other exceptions also exist; for example, the newly found ψ/J particles are very heavy but have $B = 0$ so are classified as mesons (Section 30-8).

TABLE 30-2

Elementary particles (stable or "long-lived")

Category	Particle name	Symbol	Antiparticle	Rest mass (MeV)	B	L_e	L_μ	S	Lifetime (s)	Principal decay modes
Photon	**Photon**	γ	**Self**	**0**	**0**	**0**	**0**	**0**	**Stable**	
Leptons	**Neutrino (electron)**	ν_e	$\bar{\nu}_e$	**0**	**0**	**+1**	**0**	**0**	**Stable**	
	Neutrino (muon)	ν_μ	$\bar{\nu}_\mu$	**0**	**0**	**0**	**+1**	**0**	**Stable**	
	Electron	e^-	e^+	**0.51**	**0**	**+1**	**0**	**0**	**Stable**	
	Muon	μ^-	μ^+	**105.7**	**0**	**0**	**+1**	**0**	2.2×10^{-6}	$\mu^- \to e^- + \bar{\nu}_e + \nu_\mu$
Hadrons										
Mesons	**Pion**	π^+	π^-	**139.6**	**0**	**0**	**0**	**0**	2.6×10^{-8}	$\pi^+ \to \mu^+ + \nu_\mu$
		π^0	**Self**	**135.0**	**0**	**0**	**0**	**0**	0.8×10^{-16}	$\pi^0 \to \gamma + \gamma$
	Kaon	K^+	K^-	**493.8**	**0**	**0**	**0**	**+1**	1.2×10^{-8}	$K^+ \to \mu^+ + \nu_\mu$ $\to \pi^+ + \pi^0$
		K_S^0	$\overline{K_S^0}$	**497.8**	**0**	**0**	**0**	**+1**	0.9×10^{-10}	$K_S^0 \to \pi^+ + \pi^-$ $\to \pi^0 + \pi^0$
		K_L^0	$\overline{K_L^0}$	**497.8**	**0**	**0**	**0**	**+1**	5.7×10^{-8}	$K_L^0 \to \pi^\pm + e^\mp + \overset{(-)}{\nu}_e$ $\to \pi^\pm + \mu^\mp + \overset{(-)}{\nu}_\mu$ $\to \pi^0 + \pi^0 + \pi^0$
	Eta	η^0	**Self**	**548.8**	**0**	**0**	**0**	**0**	$<10^{-16}$	$\eta^0 \to \gamma + \gamma$
Baryons	**Proton**	p	$\bar{p}$	**938.3**	**+1**	**0**	**0**	**0**	**Stable**	
	Neutron	n	$\bar{n}$	**939.6**	**+1**	**0**	**0**	**0**	10^3	$n \to p + e^- + \bar{\nu}_e$
	Lambda	Λ^0	$\overline{\Lambda^0}$	**1116**	**+1**	**0**	**0**	**−1**	2.5×10^{-10}	$\Lambda^0 \to p + \pi^-$ $\to n + \pi^0$
	Sigma	Σ^+	$\bar{\Sigma}^-$	**1189**	**+1**	**0**	**0**	**−1**	0.8×10^{-10}	$\Sigma^+ \to p + \pi^0, n + \pi^+$
		Σ^0	$\bar{\Sigma}^0$	**1192**	**+1**	**0**	**0**	**−1**	$<10^{-14}$	$\Sigma^0 \to \Lambda^0 + \gamma$
		Σ^-	$\bar{\Sigma}^+$	**1197**	**+1**	**0**	**0**	**−1**	1.5×10^{-10}	$\Sigma^- \to n\pi^-$
	Xi	Ξ^0	$\overline{\Xi^0}$	**1315**	**+1**	**0**	**0**	**−2**	3×10^{-10}	$\Xi^0 \to \Lambda^0 + \pi^0$
		Ξ^-	$\bar{\Xi}^+$	**1321**	**+1**	**0**	**0**	**−2**	1.7×10^{-10}	$\Xi^+ \to \Lambda^0 + \pi^+$
	Omega	Ω^-	Ω^+	**1672**	**+1**	**0**	**0**	**−3**	1.3×10^{-10}	$\Omega^- \to \Xi^0 + \pi^0$, $\Lambda^0 + K^-$

30-6 ▌ Particle stability and resonances

The particles listed in Table 30-2 are those that are either stable or are rather long-lived (lifetime $> 10^{-18}$ s). The lifetime of an unstable particle depends on which force is most active in causing the decay. When we say that the strong nuclear force is stronger than the electromagnetic, we mean that two particles will interact more quickly and more frequently if this force is acting. When a stronger force influences a decay, the decay occurs more quickly. Decays caused by the weak force have lifetimes of 10^{-10} s or longer. Particles that decay via the electromagnetic force have much shorter lifetimes, typically about 10^{-16} s. All the unstable particles listed in Table 30-2 decay either via the weak or the electromagnetic interaction; those that involve a γ (photon) are electromagnetic; the others shown decay via the weak interaction (note the lifetimes).

A great many particles have been found that can decay via the strong interaction (these are not listed in Table 30-2). Such particles decay into other strongly interacting particles (say, n, p, π, but never involving γ, e, etc.) and their lifetimes are incredibly short, typically 10^{-23} s. In fact their lifetimes are so short that they do not go far enough before decaying to be seen in a bubble chamber or spark chamber. Their decay products can be detected, however, and it is from them that the existence of such short-lived particles is inferred. To see how this is done, let us consider the first such particle discovered (by Fermi, Figure 30-10). Fermi used a beam of π^+ directed through a hydrogen target (protons) with varying amounts of energy. A graph of the number of interactions (π^+ scattered to one side) versus the pion's initial kinetic energy is shown in Figure 30-11. The large peak around 200 MeV was much higher than expected and certainly much higher than the number of interactions at other energies. This led Fermi to conclude that the π^+ and proton combined momentarily with high probability to form a short-lived particle before coming apart again, or at least resonated back and forth. Indeed, the large peak in Figure 30-11 resembles a resonance curve (see Figures 11-11 and 19-8) and this new "particle"—now called the Δ—is referred to as a *resonance*. Hundreds of other resonances have been found in a similar way. Many resonances are regarded as excited states of other particles, such as the nucleon.

FIGURE 30-10
Enrico Fermi (1901–1954). (Wide World Photo.)

The width of a resonance—in Figure 30-11 the width of the Δ peak is about 100 MeV—is an interesting application of the uncertainty principle. If a particle lives only 10^{-23} s, then its mass (that is, its rest energy) will be uncertain by an amount $\Delta E \approx h/2\pi\,\Delta t \approx (6.6 \times 10^{-34}\ \mathrm{J \cdot s})/(6)(10^{-23}\ \mathrm{s}) \approx 10^{-11}\ \mathrm{J} \approx 100$ MeV, which is what is observed. Actually, the lifetimes of $\approx 10^{-23}$ s for such resonances is inferred by the reverse process: from the measured width being ≈ 100 MeV.

30-7 ▮ Strange particles

In the early 1950s, certain of the newly found particles, namely the K, Λ, and Σ, were found to behave rather strangely in two ways. First, they were always produced in pairs. For example, the reaction

$$\pi^- + p \rightarrow K^0 + \Lambda^0$$

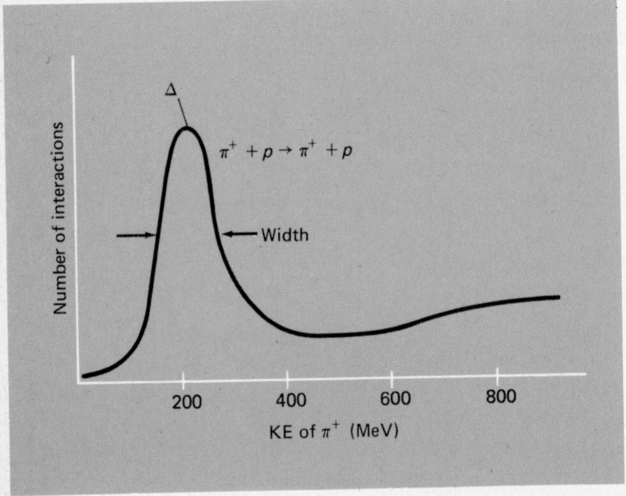

FIGURE 30-11
Number of π^+ being scattered by a proton target as a function of the incident π^+ kinetic energy.

occurred with high probability, but the reaction $\pi^- + p \nrightarrow K^0 + n$ was never observed to occur; this seemed strange since no known conservation law would have been violated and there was plenty of energy available. The second feature of these *strange particles* (as they came to be called) was that although they were clearly produced via the strong interaction (that is, at a high rate), they did not decay at a rate characteristic of the strong interaction; this, in spite of the fact that they decayed only into strongly interacting particles (for example, $K \rightarrow 2\pi$, $\Sigma^+ \rightarrow p + \pi^0$). Instead of lifetimes of 10^{-23} s as expected for strongly interacting particles, strange particles have lifetimes of 10^{-10} to 10^{-8} s, which are characteristic of the weak interaction. (See Table 30-2.)

To explain these observations, a new quantum number, **strangeness** (S), and a new conservation law, *conservation of strangeness*, were introduced. By assigning the strangeness numbers indicated in Table 30-2, the production of strange particles in pairs was readily explained; antiparticles were assigned opposite strangeness from their particles: one of each pair had $S = +1$ and the other $S = -1$. For example, in the reaction $\pi^- + p \rightarrow K^0 + \Lambda^0$, the initial state has strangeness $S = 0 + 0 = 0$, and the final state has $S = +1 - 1 = 0$, so strangeness is conserved. But for $\pi^- + p \nrightarrow K^0 + n$, the final state has $S = +1 + 0 = +1$, so strangeness would not be conserved; hence the reaction is not observed.

To explain the decay of strange particles, it is assumed that strangeness is conserved in the strong interaction but is *not* conserved in the weak interaction. Thus, although the strange particles were forbidden by strangeness conservation to decay to lower-mass nonstrange particles via the strong interaction, they could undergo such decay by means of the weak interaction. This would occur much more slowly, of course, which accounts for their longer lifetimes of 10^{-10} to 10^{-8} s.

The conservation of strangeness is the first example of a "partially conserved" quantity. In this case, the quantity strangeness is conserved by strong interactions but not by weak.

30-8 ▮ Quarks: how physics became charming and found truth and beauty

Nearly all observed particles fall into one of two groups: leptons or hadrons. The principal difference between these two groups is that the hadrons interact via the strong interaction whereas the leptons do not. Another important difference that physicists have had to deal with in the 1960s and 1970s is that there were only four known leptons (e^-, μ^-, ν_e, ν_μ), but well over a hundred hadrons.

The four leptons are considered to be truly elementary particles since they do not seem to break down into smaller entities, do not show any internal structure, and have no measurable size.

The hadrons, on the other hand, are more complex. Experiments indicate that they do have an internal structure. And the fact that there are so many of them suggests that they cannot all be elementary. To deal with this problem, Murray Gell-Mann and George Zweig in 1963 independently proposed that none of the hadrons so far observed was elementary. They proposed instead that hadrons are made up of combinations of three, more fundamental, pointlike entities called **quarks**.† Quarks, then, would be considered truly elementary particles, like leptons. The three quarks were given the names *up*, *down*, and *strange*. They were assumed to have fractional charge ($\frac{1}{3}$ or $\frac{2}{3}$ the charge on the electron—that is, less than the previously thought smallest charge); other properties of quarks and antiquarks are indicated in Table 30-3. All hadrons known at the time could be constructed in theory from these three quarks. Mesons would consist of a quark–antiquark pair. For example, a π^+ meson is considered a $u\bar{d}$ pair (note that for the $u\bar{d}$ pair, $Q = \frac{2}{3}e + \frac{1}{3}e = +1e$, $B = \frac{1}{3} - \frac{1}{3} = 0$, etc., as it must for a π^+). On the other hand a $K^+ = u\bar{s}$ with $Q = +1$, $B = 0$, $S = +1$. Baryons, on the other hand, consist of three quarks; for example, a neutron is $n = ddu$, whereas an antiproton is $\bar{p} = \bar{u}\bar{u}\bar{d}$.

Soon after the quark theory was proposed, physicists began looking for these fractionally charged particles. Although there is indirect experimental evidence in favor of their existence, none has yet been detected directly. In fact some theoretical physicists believe that quarks are so tightly bound together that they may not ever exist in the free state.

In 1964, several physicists proposed that there ought to be a fourth quark. Their argument was based on the expectation that there exists a deep symmetry in nature, including a connection between quarks and leptons. If there are four leptons, symmetry in nature would suggest there should also be four quarks. This fourth quark was said to be *charmed*; its charge would be $+\frac{2}{3}e$ and it would have another property to distinguish it from the other three quarks. This new property, or quantum number, was called **charm** (see Table 30-3). Charm was assumed to be like strangeness: it would be conserved in strong and electromagnetic interactions but would not be conserved by the weak. The new charmed quark would have charm $C = +1$ and its antiquark $C = -1$.

Experimentally, however, there seemed to be no need for a charmed quark. Before 1974, all known hadrons could be explained as combinations of the three original quarks. In fact, hadrons corresponding to all of the three quark, and quark–antiquark, combinations had been found. Indeed, the Ω^- baryon had been predicted by the three-quark theory ($\Omega^- = sss$) and was discovered soon after (Figure 30-12). But in 1974, a new heavy meson was discovered simultaneously by two different groups of experimenters. (The leaders of the groups, Burton Richter and Samuel Ting, received the Nobel Prize in 1976.) This new meson, called the ψ/J or simply the ψ, did not fit the old three-quark scheme. All the places were filled, and the new ψ was left over. The ψ, whose mass was 3095 MeV, far higher than other known mesons, could also not be an excited state of a smaller mass meson (into which it would decay) because its lifetime would have to be about 10^{-23} s. In fact, its lifetime was found to be 1000 times greater than this, about 10^{-20} s. It soon became clear that the existence of the ψ could be accounted for on the basis of the charmed quark: a ψ would be a combination of a charmed quark and its antiquark ($c\bar{c}$). The charm of the ψ itself is zero ($C = +1 - 1 = 0$), so it can decay strongly into hadrons (usually several pions, $p\bar{p}$, $\Lambda\bar{\Lambda}$, etc.), which is observed. The question of why

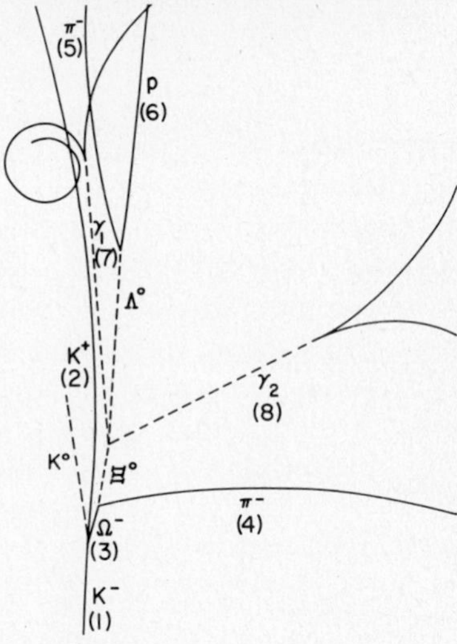

FIGURE 30-12
Liquid-hydrogen bubble-chamber photograph showing the production of a negatively charged omega meson (Ω^-) by the interaction of a negative *K* meson (K^-) with a proton (a hydrogen nucleus in the bubble chamber). (Courtesy of Brookhaven National Laboratory.)

† Gell-Mann chose the word from the phrase "Three quarks for Muster Mark" in James Joyce's *Finnegan's Wake*.

TABLE 30-3

Properties of quarks and antiquarks

	Quarks					Antiquarks				
Name	Symbol	Charge	Baryon number	Strange-ness	Charm	Symbol	Charge	Baryon number	Strange-ness	Charm
Up	u	$+\frac{2}{3}e$	$\frac{1}{3}$	0	0	$\bar{u}$	$-\frac{2}{3}e$	$-\frac{1}{3}$	0	0
Down	d	$-\frac{1}{3}e$	$\frac{1}{3}$	0	0	$\bar{d}$	$\frac{1}{3}e$	$-\frac{1}{3}$	0	0
Strange	s	$-\frac{1}{3}e$	$\frac{1}{3}$	−1	0	$\bar{s}$	$\frac{1}{3}e$	$-\frac{1}{3}$	+1	0
Charmed	c	$+\frac{2}{3}e$	$\frac{1}{3}$	0	1	$\bar{c}$	$-\frac{2}{3}e$	$-\frac{1}{3}$	0	−1

it lives 1000 times longer than other strongly decaying particles apparently has to do with the fact that the c and $\bar{c}$ quarks that make it up must each be converted into noncharmed quarks that become the hadrons seen in the decay. This process would inhibit the speed at which the decay could occur, so the charmed quark model yielded a useful explanation.

Soon after the ψ was discovered, a ψ' meson (mass = 3684 MeV) was also found. More than half a dozen other ψ-like mesons were also found, and these are all believed to be bound states of a $c\bar{c}$ pair.

Although the ψ meson and its relatives have no net charm themselves, it stands to reason that in the decay of one of the larger mass $c\bar{c}$ combinations, there ought to be mesons that do have charm. That is, the c and $\bar{c}$ quarks ought to appear in separate particles and lend a charm of +1 or −1 to them. In such a decay process, other quarks would also be produced. For example, the decay

$$\psi \rightarrow D^+ + D^-$$

might be written

$$\psi = c\bar{c} \rightarrow c\bar{c}d\bar{d} = c\bar{d} + \bar{c}d,$$

where the $d\bar{d}$ quarks were produced from energy. The $c\bar{d}$ combination, which has $Q = +1$, $B = 0$, $S = 0$, and $C = +1$, has been dubbed the D^+ meson; and $\bar{c}d$ is its antiparticle, D^-. This meson, along with its neutral sister, D^0, was searched for and believed found in 1977 with a mass around 1870 MeV. Experiments at Fermilab hint that charmed baryons may also exist.

The discovery of charmed particles does not end the story, however. Recent experimental evidence indicates that there may be another lepton, the tau (τ), with a mass of about 2000 MeV. This lepton, like the electron and muon, presumably has a neutrino associated with it. Thus the family of leptons may have to be expanded from four to six. This would upset the balance between leptons and quarks, the presumed basic building blocks of matter, unless two new quarks also exist. Indeed, theoretical physicists have postulated the existence of a fifth and sixth quark. These have been given the symbols t and b. Agreement on their names, however, is lacking. Some physicists call them *top* and *bottom* quarks (since they resemble the "up" and "down" quarks). Others prefer the names *truth* and *beauty* for these two new quarks, and for the new properties (quantum numbers) that distinguish them from the old quarks. In 1977 a new meson was detected, the heaviest yet found at 9400 MeV, which is believed to be a combination of a "beauty" quark and its antiquark ($b\bar{b}$).

The world of elementary particles has apparently just opened into new vistas. What happens in the near future is bound to be exciting.

▮ Summary

Particle accelerators are used to accelerate charged particles such as electrons and protons to very high energy. High-energy particles have short wavelength and so can be used to probe in great detail; high kinetic energy also allows the creation of new particles through collision (via $E = mc^2$). Van de Graaff and linear accelerators use high voltage to accelerate particles along a line. Cyclotrons and synchrotrons use a magnetic field to keep the particles in a circular path and they are accelerated at intervals by high voltage.

Just as the electromagnetic force can be said to be due to an exchange of photons, the strong nuclear force is thought to be carried by *mesons* that have rest mass.

An *antiparticle* has the same mass as a particle but opposite charge. Certain other properties may also be opposite: for example, the antiproton has *baryon number* (nucleon number) opposite to that for the proton. In all nuclear and particle reactions, the following *conservation laws* hold: momentum, mass energy, angular momentum, electric charge, *baryon number*, and the two *lepton numbers*. Certain particles have a property, called *strangeness*, which is conserved by the strong nuclear force (or interaction) but not by the weak force.

The elementary particles are classified as *leptons* and *hadrons*, plus the *photon*. Leptons participate in the weak and electromagnetic interactions. Hadrons participate in the strong interaction as well. The hadrons are subdivided into *mesons*, with baryon number zero, and *baryons* with nonzero baryon number.

Of the hundreds of subatomic particles known, only the photon, electron, neutrino, and proton are stable. All others decay with half-lives varying from 10^{-23} s to 10^3 s. The half-life depends on which force is predominant in the decay. Weak decays have half-lives greater than about 10^{-10} s. Electromagnetic decays have half-lives on the order of 10^{-16} s. The shortest-lived particles, called resonances, decay via the strong interaction and live typically for only about 10^{-23} s.

The latest theories of elementary particle physics postulate the existence of *quarks* as the basic building blocks of the hadrons. Initially, three quarks were proposed. Recent evidence suggests that a fourth, *charmed*, quark is needed and perhaps also a fifth and a sixth.

Questions

1 What limits the maximum energy attainable for protons in an ordinary cyclotron? How is this overcome in a synchrotron?

2 A proton in a synchrotron has a speed of $0.99c$. What must be done to increase its energy?

3 Give a reaction between two nucleons that could produce a π^-, similar to Equation 30-3.

4 Why is it that a neutron decays via the weak interaction even though the neutron and one of its decay products (proton) are strongly interacting?

5 Which of the four interactions (strong, electromagnetic, weak, gravitational) does an electron take part in? A neutrino? A proton?

6 Check that charge and baryon number are conserved in each of the decays in Table 30-2.

7 Which of the particle decays in Table 30-2 occur by the electromagnetic interaction?

8 Which of the particle decays in Table 30-2 occur by the weak interaction?

9 Which of the following decays are possible? For those that are forbidden, explain which laws are violated.

(*a*) $\Xi^0 \rightarrow \Sigma^+ + \pi^-$
(*b*) $\Omega^- \rightarrow \Sigma^0 + \pi^- + \nu$
(*c*) $\Sigma \rightarrow \Lambda + \gamma + \gamma$

10 Which of the following reactions are possible, and by what interaction could they occur? For those forbidden, explain why.

(*a*) $\pi^- + p \rightarrow K^+ + \Sigma^-$
(*b*) $\pi^+ + p \rightarrow K^+ + \Sigma^+$
(*c*) $\pi^- + p \rightarrow \Lambda^0 + K^0 + \pi^0$
(*d*) $\pi^+ + p \rightarrow \Sigma^0 + \pi^0$
(*e*) $\pi^- + p \rightarrow \underline{p} + e^- + \bar{\nu}_e$
(*f*) $\pi^- + p \rightarrow \overline{K^0} + p + \pi^0$
(*g*) $K^- + p \rightarrow \Lambda^0 + \pi^0$
(*h*) $K^+ + n \rightarrow \Sigma^+ + \pi^0 + \gamma$
(*i*) $K^+ \rightarrow \pi^0 + \pi^0 + \pi^+$
(*j*) $\pi^+ \rightarrow e^+ + \nu_e$

11 What are the quark combinations that can form (*a*) a neutron, (*b*) an antineutron, (*c*) Λ^0, and (*d*) $\overline{\Xi^0}$?

12 What particles do the following quark combinations produce? (*a*) uud; (*b*) $\bar{u}\bar{u}\bar{s}$; (*c*) $\bar{u}s$; (*d*) $d\bar{u}$?

13 What is the quark combination needed to produce a D^0 meson ($Q = B = S = 0$, $C = +1$)?

14 An F^+ meson has been postulated that has $Q = S = C = +1$, $B = 0$. What quark combination would produce it?

Problems

SECTION 30-1

1 (I) What is the total energy of a proton whose kinetic energy is 15 GeV?

2 (I) Calculate the wavelength of 20-GeV electrons.

3 (I) What strength of magnetic field is used in a cyclotron in which protons make 1.9×10^7 revolutions per second?

4 (I) If α particles are accelerated by the cyclotron of Example 30-2, what must be the frequency of voltage applied to the dees?

5 (I) If the cyclotron of Example 30-2 accelerated α particles, what maximum energy could they attain? What would their speed be? Repeat for deuterons ($^2_1$H).

6 (I) What is the time for one complete revolution for a very high energy proton in the 2.0-km-diameter Fermilab accelerator?

7 (I) The voltage across the dees of a cyclotron is 50 kV. How many revolutions do protons make to reach a kinetic energy of 10 MeV?

8 (II) Show that the speed of the electron in Example 30-1 is nearly c.

9 (II) What is the wavelength, and maximum resolving power attainable, using 400-GeV protons at Fermilab?

10 (II) Protons are injected into the 1.0-km-radius Fermilab synchrotron with an energy of 8.0 GeV. If they are accelerated by 2.5 MV each revolution, how far do they travel and approximately how long does it take for them to reach 400 GeV?

11 (III) What magnetic field intensity is needed at the 1.0-km-radius Fermilab synchrotron for 400-GeV protons? Use the relativistic mass.

12 (III) Show that the energy of a particle (charge e) in a synchrotron, in the relativistic limit ($v \approx c$), is given by E (in eV) $= Brc$, where B is magnetic field strength and r the radius of the orbit (SI units).

SECTION 30-2

13 (I) Two protons are heading toward each other with equal speeds. What minimum energy must each have if a π^0 meson is to be created in the process? (See Table 30-2.)

14 (I) Could a π^+ meson be produced if a 100-MeV proton struck a proton at rest?

15 (III) The mass of a π^0 can be measured by observing the reaction $\pi^- + p \rightarrow \pi^0 + n$ at very low incident π^- kinetic energy (assume it is zero). The neutron is observed to be emitted with a KE of 0.60 MeV. Use conservation of energy and momentum to determine the π^0 mass.

SECTION 30-3

16 (I) (*a*) How much energy is released when an electron and a positron annihilate each other? (*b*) How much energy is released when a proton and an antiproton annihilate each other?

17 (I) How much energy is required to produce a neutron–antineutron pair?

18 (II) What are the wavelengths of the two photons when a proton–antiproton pair at rest annihilate?

19 (II) What would be the wavelengths of the two photons produced when an electron–positron pair, each with 300 keV of KE, annihilate?

SECTION 30-5

20 (I) How much energy is released in the decay $\pi^+ \rightarrow \mu^+ + \nu_\mu$?

21 (I) About how much energy is released when a Λ^0 decays? (See Table 30-2.)

22 (I) What minimum kinetic energy must a neutron and proton each have if they are traveling at the same speed toward each other and produce a K^+K^- pair? (See Table 30-2.)

23 (III) In the decay $\pi^+ \rightarrow \mu^+ + \nu_\mu$, the μ always has a kinetic energy of 4.5 MeV. In the rare decay $\pi^+ \rightarrow e^+ + \nu_e$, what is the kinetic energy of the positron?

SECTIONS 30-6 AND 30-8

24 (I) The measured width of the ψ meson is about 70 keV. Estimate its lifetime.

25 (I) The measured width of the ψ' meson is about 230 keV. Estimate its lifetime.

For additional reading

Cline, D. B., A. K. Mann, and C. Rubbia, "The Search for New Families of Elementary Particles," *Sci. American*, **234**:1 (Jan., 1976), pp. 44–54.

Davis, N. P., *Lawrence and Oppenheimer*, Simon & Schuster, Inc., New York, 1968.

Feinberg, G., *What Is the World Made Of?: The Achievements of Twentieth Century Physics*, Doubleday & Co., Inc., New York, 1977.

Fermi, L., *Atoms in the Family: My Life with Enrico Fermi*, University of Chicago Press, Chicago, 1954.

Freedman, D. Z., and P. van Nieuwenhuizen, "Supergravity and the Unification of the Laws of Physics," *Sci. American*, **238**:2 (Feb., 1978), p. 126. New theory for unifying the four forces.

Glashow, S. L., "Quarks with Color and Flavor," *Sci. American*, **233**:4 (Oct., 1975), pp. 38–50.

Gourian, R., *Particles and Accelerators*, McGraw-Hill Book Company, New York, 1967.

Livingston, M. S., *Particle Physics: The High Energy Frontier*, McGraw-Hill Book Company, 1968.

Nambu, Y., "The Confinement of Quarks," *Sci. American*, **235**:5 (Nov., 1976), pp. 48–60.

Perl, M. L., and W. T. Kirk, "Heavy Leptons," *Sci. American*, **238**:3 (Mar., 1978), pp. 50–7.

Rowe, E. M., and J. H. Weaver, "The Uses of Synchrotron Radiation," *Sci. American*, **236**:6 (Jun., 1977), pp. 32–41.

Schwitters, R., "Fundamental Particles with Charm," *Sci. American*, **237**:4 (Oct., 1977), pp. 56–70.

Weinberg, S., "Unified Theories of Elementary-particle Interaction," *Sci. American*, **231**:1 (Jul., 1974), pp. 50–59.

Wilson, R. R., "The Batavia Accelerator," *Sci. American*, **230**:2 (Feb., 1974), pp. 72–83. Fermilab.

Wilson, R. R., "The Tevatron," *Physics Today*, **30**:10 (Oct., 1977), p. 23.

Appendix A Mathematical review

A-1 ▌ Relationships, proportionality, and equations

One of the important aspects of physics is the search for relationships between different quantities—that is, determining how one quantity affects another. For example, how does temperature affect the air pressure in a tire? Or how does the net force on an object affect its acceleration? Sometimes a given quantity is affected by two or more quantities; for instance, the acceleration of an object is related to both its mass and the applied force. If it is suspected that a relationship exists between two or more quantities, one can try to determine the precise nature of this relationship. This is done by varying one of the quantities and measuring how the other varies as a result. If it is likely that a particular quantity will be affected by more than one factor or quantity, only one of the latter is varied at a time, while the others are held constant.†

As a simple example, the ancients found that if one circle has twice the diameter of a second circle, the first also has twice the circumference. If the diameter is three times as large, the circumference is also three times as large. In other words, an increase in the diameter results in a proportional increase in the circumference. We say that the circumference is directly proportional to the diameter. This can be written in symbols as $C \propto D$, where "$\propto$" means "is proportional to" and C and D refer to the circumference and diameter of a circle, respectively. The next step is to change this proportionality to an equation, which will make it possible to link the two quantities numerically. This merely entails inserting a proportionality constant, which in many cases is determined by measurement. (In some cases it can be chosen arbitrarily, as in Section 2-4.) The ancients found that the ratio of the circumference to the diameter of any circle was 3.1416 (to keep only the first few decimal places). This number is designated by the Greek letter π. It is the constant of proportionality for the relationship $C \propto D$, and to obtain an equation we merely insert it into the proportion and change the $\propto$ to $=$. Thus $C = \pi D$.

Other kinds of proportionality occur as well. For example, the area of a circle is proportional to the *square* of its radius. That is, if the radius is doubled, the area becomes four times as large; and so on. In this case we

† When one quantity affects another, we often use the expression "is a function of" to indicate this dependence; for example, we say that the pressure in a tire is a function of the temperature.

can write $A \propto r^2$, where A stands for the area and r for the radius of the circle.

Sometimes two quantities are related in such a way that an increase in one leads to a proportional decrease in the other. This is called *inverse proportion.* For example, the time required to travel a given distance is inversely proportional to the speed of travel. The greater the speed, the less time it takes. We can write this inverse proportion as: time $\propto 1/\text{speed}$. The larger the denominator of a fraction, the smaller is the fraction as a whole. For example, $\frac{1}{4}$ is smaller than $\frac{1}{2}$. Thus if the speed is doubled, the time is halved, which is what we want to express by this inverse proportionality relationship.

Whatever kind of proportion is found to hold, it can be changed to an equality by insertion of the proper proportionality constant. Quantitative statements or predictions about the physical world can then be made with the equation.

A-2 ▮ Accuracy and significant figures

Accurate measurements are an important part of physics. But no measurement is absolutely precise; there is an uncertainty associated with every measurement. Uncertainty arises from different sources; among the most important, other than blunders, are the limited accuracy of every measuring instrument and the inability to read an instrument beyond some fraction of the smallest division shown. For example, if you were to use a centimeter ruler to measure the width of a spice box, the result could be claimed to be accurate to about 0.1 cm, the smallest division on the ruler (although half of this value might be a valid claim as well). The reason for this is that it is difficult for the observer to interpolate between the smallest divisions, and the ruler itself has probably not been manufactured to an accuracy much better than this.

When giving the result of a measurement, it is good practice to state the precision, or assumed *uncertainty,* in the measurement. For example, the width of a spice box would be written as 3.2 ± 0.1 cm. The ± 0.1 cm ("plus or minus 0.1 cm") represents the assumed uncertainty in the measurement, so that the actual length most likely lies between 3.1 and 3.3 cm. If you use vernier calipers, the uncertainty can be reduced to perhaps 0.02 cm.

Sometimes the *percent uncertainty* (sometimes called "percent error" or "expected error") in a measurement is used which is defined as follows†:

$$\text{percent uncertainty} = \frac{\text{uncertainty}}{\text{value}} \times 100.$$

† You might want to compare your own measurement of a particular quantity with the accepted value for that measurement. In this case you specify the percent difference or percent error; this is defined as the difference in the measured and accepted values, divided by the accepted value, and multiplied by 100. For example, the distance between two cities may be given by the highway department as 276 km, but your car's odometer measures it to be 287 km; then the percent error is

$$\frac{287-276}{276} \times 100 = 4\%.$$

Thus, the percent uncertainty in the length measurement 3.2 ± 0.1 cm is

$$\frac{0.1\text{ cm}}{3.2\text{ cm}} \times 100 = 3\%.$$

Often the uncertainty in a measured value is not specified explicitly; in this case it is generally accepted that the uncertainty is approximately one or two units in the last digit specified. Although this is not as precise as actually specifying the uncertainty, it is often adequate. For example, if a length is given as 3.2 cm, the uncertainty is assumed to be about 0.1 cm. If this is true, it is important that you not write 3.20 cm, for this implies an uncertainty of 0.01 cm; it assumes that the length is probably between 3.19 cm and 3.21 cm, when actually you only believe it is between 3.1 and 3.3 cm.

The number of reliably known digits in a number is called the number of *significant figures.* Thus there are three significant figures in the number 3.21 and two in the number 0.062 cm.

When making measurements, or when doing calculations, one should never keep more digits in the final answer than the number of significant figures. For example, to calculate the area of a rectangle 11.3 by 6.8 cm, the result of multiplication would be 76.84 cm^2, but this answer is clearly not accurate to 0.01 cm^2, since (using the outer limits of the assumed uncertainty) the result could be between $11.2 \times 6.7 = 75.04$ cm^2 and $11.4 \times 6.9 = 78.46$ cm^2. At best, we can quote the answer as 77 cm^2, which implies an uncertainty of about 1 or 2 cm^2. The other two digits must be dropped, since they are not significant. As a general rule, *the final result of a multiplication or division should have only as many digits as the least significant number used in the calculation.* In our example, 6.8 cm was the least significant number and it had two significant figures; thus the 76.84 cm^2 is rounded off to 77 cm^2.

Another way to indicate significant figures is shown in Section A-4.

A-3 ▮ Exponents

When we write 10^4 we mean that you multiply 10 times itself four times: $10^4 = 10 \times 10 \times 10 \times 10 = 10{,}000$. The superscript 4 is called an *exponent,* and 10 is said to be raised to the fourth power. Any number or symbol can be raised to a power; special names are used when the exponent is 2 (a^2 is "a squared") or 3 (a^3 is "a cubed"). For any other power we say a^n is "a to the nth power." If the exponent is 1, it is usually dropped: $a^1 = a$, since no multiplication is involved.

The rules for multiplying numbers expressed as powers are as follows:

$$(a^n)(a^m) = a^{n+m}. \tag{A-1}$$

That is, the exponents are added. To see why, consider the result of the multiplication of 3^3 by 3^4:

$$(3^3)(3^4) = (3)(3)(3) \times (3)(3)(3)(3) = (3)^7.$$

Here the sum of the exponents is $3 + 4 = 7$, so rule A-1 works. Notice that this rule works only if the base numbers (a in Equation A-1) are the same.

Thus we *cannot* use the rule of summing exponents for $(6^3)(5^2)$; these numbers would have to be written out. However, if the base numbers are different but the exponents are the same, we can write a second rule:

$$(a^n)(b^n) = (ab)^n. \tag{A-2}$$

For example, $(5^3)(6^3) = (30)^3$ since

$$(5)(5)(5)(6)(6)(6) = (30)(30)(30).$$

The third rule involves a power raised to another power: $(a^3)^2$ means $(a^3)(a^3)$, which is equal to $a^{3+3} = a^6$. The general rule is then

$$(a^n)^m = a^{nm}. \tag{A-3}$$

In this case, the exponents are multiplied.

Negative exponents are used for reciprocals. Thus

$$\frac{1}{a} = a^{-1}, \qquad \frac{1}{a^3} = a^{-3},$$

and so on. The reason for using negative exponents is that it allows us to use the multiplication rules given above. For example, $(a^5)(a^{-3})$ means

$$\frac{(a)(a)(a)(a)(a)}{(a)(a)(a)} = a^2.$$

Rule A-1 gives us the same result:

$$(a^5)(a^{-3}) = a^{5-3} = a^2.$$

What does an exponent of zero mean? That is, what is a^0? Any number raised to the zeroth power is defined as being equal to 1:

$$a^0 = 1.$$

This definition is used because it follows from the rules for adding exponents. For example,

$$a^3a^{-3} = a^{3-3} = a^0 = 1.$$

But *does* a^3a^{-3} actually equal 1? Yes, because

$$a^3a^{-3} = \frac{a^3}{a^3} = 1.$$

Fractional exponents are used to represent *roots.* For example, $a^{1/2}$ means the square root of a; that is $a^{1/2} = \sqrt{a}$. Similarly, $a^{1/3}$ means the cube root of a, and so on. The fourth root of a means that if you multiply the fourth root of a by itself four times, you again get a:

$$(a^{1/4})^4 = a.$$

This is consistent with rule A-3 since $(a^{1/4})^4 = a^{4/4} = a^1 = a$.

A-4 ▌ Powers of 10, or exponential notation

Writing out very large and very small numbers such as the distance of Neptune from the sun, 4,500,000,000 km, or the diameter of a typical atom, 0.00000001 cm, is inconvenient and prone to error. It also leaves in question the number of significant figures. (How many of the zeros are significant in the number 4,500,000,000 km?) We therefore make use of the "powers of 10," or exponential notation. The distance from Neptune to the sun is then expressed as 4.50×10^9 km (assuming that the value is significant to three digits) and the diameter of an atom 1.0×10^{-8} cm. This way of writing numbers is based on the use of exponents, where a^n signifies a multiplied by itself n times. For example, $10^4 = 10 \times 10 \times 10 \times 10 = 10{,}000$. Thus $4.50 \times 10^9 = 4.50 \times 1{,}000{,}000{,}000 = 4{,}500{,}000{,}000$. Notice that the exponent (9 in this case) is just the number of places the decimal point is moved to the right to obtain the fully written out number (4.500,000,000.).

When two numbers are multiplied (or divided), you first multiply (divide) the simple parts and then the powers of 10. Thus 2.0×10^3 multiplied by 5.5×10^4 equals $(2.0 \times 5.5) \times (10^3 \times 10^4) = 11 \times 10^7$, where we have used the rule for adding exponents (Section A-3). Similarly, 8.2×10^5 divided by 2.0×10^2 equals

$$\frac{8.2 \times 10^5}{2.0 \times 10^2} = \frac{8.2}{2.0} \times \frac{10^5}{10^2} = 4.1 \times 10^3.$$

For numbers less than 1, say 0.01, the exponent or power of 10 is written with a negative sign: $0.01 = 1/100 = 1/10^2 = 1 \times 10^{-2}$. Similarly, $0.002 = 2 \times 10^{-3}$. The decimal point has again been moved the number of places expressed in the exponent. The negative exponent allows calculations to come out correctly. Thus $0.020 \times 3600 = 72$; in exponential notation $(2.0 \times 10^{-2}) \times (3.6 \times 10^3) = 7.2 \times 10^1 = 72$.

Notice also that $10^1 \times 10^{-1} = 10 \times 0.1 = 1$, and by the law of exponents, $10^1 \times 10^{-1} = 10^0$. Therefore, $10^0 = 1$.

When writing a number in exponential notation, it is usual to make the simple number be between 1 and 10. Thus it is conventional to write 4.5×10^9 rather than 45×10^8, although they are the same number.† This notation also allows the number of *significant figures* to be clearly expressed. We write 4.50×10^9 if this value is accurate to three significant figures, but 4.5×10^9 if it is accurate to only two.

A-5 ▮ Algebra

Physical relationships between quantities can be represented as equations involving symbols (usually letters of the alphabet) that represent the quantities. The manipulation of such equations is the field of algebra, and is used a great deal in physics. An equation involves an equals sign, which tells us that the quantities on either side of the equals sign have the same value. Examples of equations are

$$3 + 8 = 11$$

$$2x + 7 = 15$$

$$a^2 b + c = 6.$$

The first equation involves only numbers, so is called an arithmetic

† Another convention used, particularly with computers, is that the simple number be between 0.1 and 1. Thus we would write 4,500,000,000 as 0.450×10^{10}.

equation. The other two equations are algebraic since they involve symbols. In the third equation, the quantity a^2b means the product of a times a times b: $a^2b = a \times a \times b$.

Solving for an unknown

Often we wish to solve for one (or more) symbols, and we treat it as an *unknown.* For example, in the equation $2x + 7 = 15$, x is the unknown; this equation is true, however, only when $x = 4$. Determining what value (or values) the unknown(s) can have to satisfy the equation(s) is called *solving the equation.* To solve an equation, the following rule can be used:

An equation will remain true if any operation performed on one side is also performed on the other side: for example, (*a*) addition or subtraction of a number or symbol; (*b*) multiplication or division by a number or symbol; (*c*) raising each side of the equation to the same power, or taking the same root (such as square root).

▌ EXAMPLE A-1 Solve for x in the equation

$$2x + 7 = 15.$$

SOLUTION We first subtract 7 from both sides:

$$2x + 7 - 7 = 15 - 7$$

$$2x = 8.$$

Then we divide both sides by 2 to get

$$\frac{2x}{2} = \frac{8}{2}$$

$$x = 4,$$

and this solves the equation. ▌▌

▌ EXAMPLE A-2 (*a*) Solve the equation

$$a^2b + c = 24$$

for the unknown a in terms of b and c. (*b*) Solve for a assuming that $b = 2$ and $c = 6$.

SOLUTION (*a*) We are trying to solve for a, so we first subtract c from both sides

$$a^2b = 24 - c,$$

then divide by b:

$$a^2 = \frac{24 - c}{b},$$

and finally take square roots,

$$a = \sqrt{\frac{24 - c}{b}}.$$

(*b*) If we are given that $b = 2$ and $c = 6$, then

$$a = \sqrt{\frac{24 - 6}{2}} = 3.$$ ▌▌

To check a solution, we put it back into the original equation (this is really a check that we did all the manipulations correctly). In the equation

$$a^2b+c=24$$

we put in $a=3$, $b=2$, $c=6$ and find

$$(3)^2(2)+(6)\stackrel{?}{=}24$$

$$24=24,$$

which checks.

Two or more unknowns

If we have two or more unknowns, one equation is not sufficient to find them. In general, if there are n unknowns, n independent equations are needed. For example, if there are two unknowns, we need two equations. If the unknowns are called x and y, a typical procedure is to solve one equation for x in terms of y, and substitute this into the second equation.

▮ EXAMPLE A-3 Solve the following pair of equations for x and y:

$$3x-2y=19$$

$$x+4y=-3.$$

SOLUTION We solve the second equation for x in terms of y by subtracting $4y$ from both sides:

$$x=-3-4y.$$

We substitute this expression for x into the first equation, and simplify

$$3(-3-4y)-2y=19$$

$$-9-12y-2y=19 \qquad \text{(carried out the multiplication by 3)}$$

$$-14y=28 \qquad \text{(added 9 to both sides)}$$

$$y=-2 \qquad \text{(divided both sides by } -14\text{)}.$$

Now that we know $y=-2$, we substitute this into the expression for x:

$$x=-3-4y$$

$$=-3-4(-2)=-3+8=5.$$

Our solution is $x=5$, $y=-2$. We check this solution by putting these values back into the original equations:

$$3x-2y\stackrel{?}{=}19$$

$$3(5)-2(-2)\stackrel{?}{=}19$$

$$15+4\stackrel{?}{=}19$$

$$19=19 \qquad \text{(it checks)}$$

and

$$x+4y\stackrel{?}{=}-3$$

$$5+4(-2)\stackrel{?}{=}-3$$

$$-3=-3 \qquad \text{(it checks).}$$

▮▮

Other methods for solving two or more equations, such as the method of determinants, can be found in an algebra textbook.

The quadratic formula

We sometimes encounter equations that involve an unknown, say x, that appears not only to the first power, but squared as well. Such a *quadratic equation* can be written in the form

$$ax^2+bx+c=0.$$

The quantities a, b, and c are typically given numbers or constants.† The general solutions to such an equation are given by the *quadratic formula*:

$$x=\frac{-b\pm\sqrt{b^2-4ac}}{2a}.$$

The ± sign indicates that there are two solutions for x: one where the plus sign is used, the other where the minus sign is used.‡

▮ EXAMPLE A-4 Find the solutions for x in the equation

$$3x^2-5x=2.$$

SOLUTION First we write this equation in the standard form $ax^2+bx+c=0$ by subtracting 2 from both sides:

$$3x^2-5x-2=0.$$

In this case, a, b, and c in the standard formula take the values $a=3$, $b=-5$, and $c=-2$. The two solutions for x are

$$x=\frac{+5+\sqrt{25-(4)(3)(-2)}}{(2)(3)}=\frac{5+7}{6}=2$$

and

$$x=\frac{+5-\sqrt{25-(4)(3)(-2)}}{(2)(3)}=\frac{5-7}{6}=-\frac{1}{3}.$$ ▮▮

In this example the two solutions are $x=2$ and $x=-\frac{1}{3}$. In physics problems, however, only one of these solutions may correspond to a real-life solution; in this case the other solution is discarded. In other cases both solutions may correspond to physical reality.

Notice, incidentally, that b^2 must be greater than $4ac$, so that $\sqrt{b^2-4ac}$ yields a real number. If (b^2-4ac) is less than zero (negative), there is no real solution. The square root of a negative number is called *imaginary*.

A-6 ▮ Plane geometry

We review here a number of theorems involving angles and triangles that are useful in physics.

† Or one or more of them could be variables, in which case additional equations are needed.
‡ A second-order equation—one in which the highest power of x is 2—has two solutions; a third-order equation—involving x^3—has three solutions, and so on.

1 *Equal angles.* Two angles are equal if any of the following conditions are true:
(*a*) They are vertical angles (Figure A-1); *or*
(*b*) the left side of one is parallel to the left side of the other, and the right side of one is parallel to the right side of the other (the left and right sides are as seen from the vertex, where the two sides meet; Figure A-2); *or*
(*c*) the left side of one is perpendicular to the left side of the other, and the right sides are likewise perpendicular (Figure A-3).

2 *The sum of the angles* in any plane triangle is 180°.

3 *Similar triangles.* Two triangles are said to be similar if all three of their angles are equal (in Figure A-4 $\theta_1 = \phi_1$, $\theta_2 = \phi_2$, $\theta_3 = \phi_3$). Similar triangles thus have the same basic shape but may be different sizes and have different orientations. Two useful theorems about similar triangles are:
(*a*) Two triangles are similar if any two of their angles are equal. (This follows because the third angles must also be equal since their sum is 180°.)
(*b*) The ratio of corresponding sides of two similar triangles are equal. That is (Figure A-4)

$$\frac{a_1}{b_1} = \frac{a_2}{b_2} = \frac{a_3}{b_3}.$$

4 *Congruent triangles.* Two triangles are congruent if one can be placed precisely on top of the other. That is, they are similar triangles and they have the same size. Two triangles are congruent if any of the following holds:
(*a*) The three corresponding sides are equal.
(*b*) Two sides and the enclosed angle are equal ("side–angle–side").
(*c*) Two angles and the enclosed side are equal ("angle–side–angle").

5 *Right triangles.* A right triangle has one angle that is 90° (a *right angle*); that is, the two sides that meet at the right angle are perpendicular. The two other (acute) angles in the right triangle add up to 90°.

6 *Pythagorean theorem.* In any right triangle, the square of the length of the hypotenuse (the side opposite the right angle) is equal to the sum of the squares of the lengths of the other two sides. From Figure A-5,

$$c^2 = a^2 + b^2.$$

FIGURE A-1

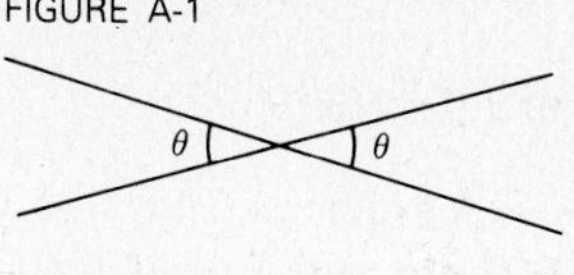

FIGURE A-2

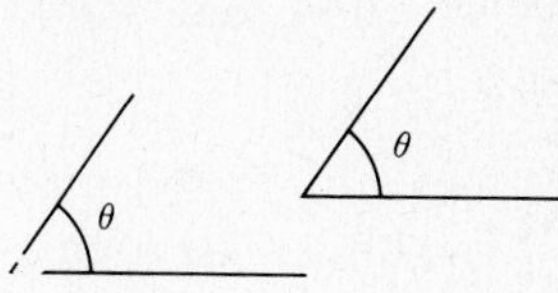

FIGURE A-3

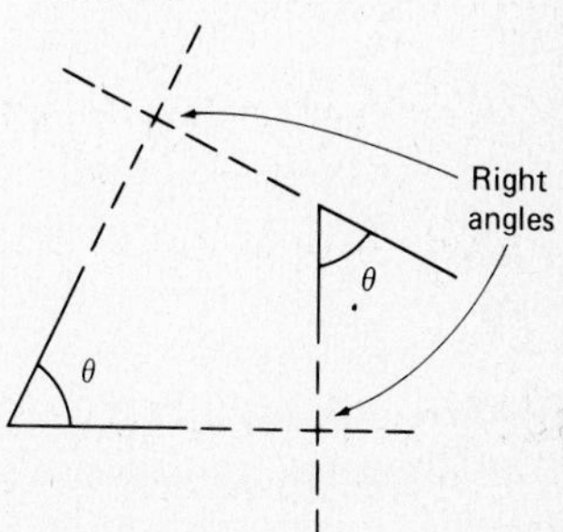

FIGURE A-4

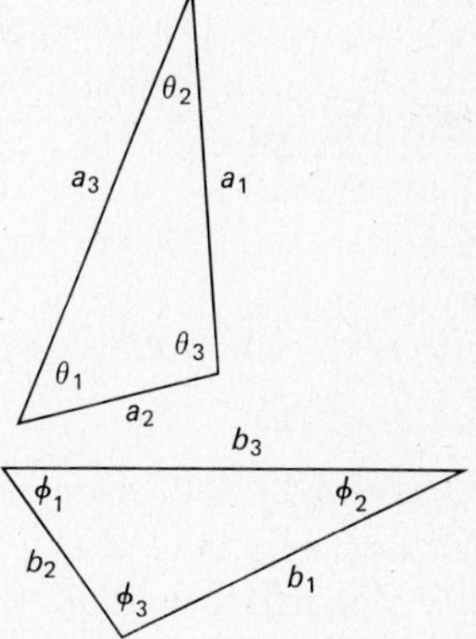

FIGURE A-5

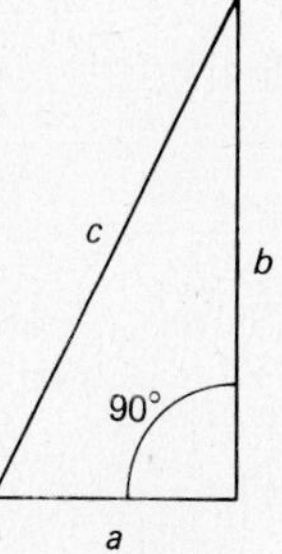

A-7 ▮ Logarithms

Logarithms are defined in the following way:

$$\text{if} \quad y = A^x, \quad \text{then} \quad x = \log_A y.$$

That is, the logarithm of a number y to the base A is that number which, as the exponent of A, gives back the number y. For *common logarithms*, the base is 10, so

$$\text{if} \quad y = 10^x, \quad \text{then} \quad x = \log y.$$

The subscript 10 on $\log_{10}$ is usually omitted when dealing with common logs. Another base sometimes used is the natural number $e = 2.718\ldots$.†

† The natural number e can be written as an infinite series:

$$e = 1 + \frac{1}{1} + \frac{1}{1\cdot 2} + \frac{1}{1\cdot 2\cdot 3} + \frac{1}{1\cdot 2\cdot 3\cdot 4} + \cdots.$$

Such logarithms are called *natural logarithms* and are written ln. Thus

$$\text{if} \quad y = e^x, \quad \text{then} \quad x = \ln y.$$

For any number y, the two types of logarithm are related by

$$\ln y = 2.3026 \log y.$$

Some simple rules for logarithms are as follows:

$$\log (ab) = \log a + \log b. \tag{A-4}$$

This is true because if $a = 10^n$ and $b = 10^m$, then $ab = 10^{n+m}$. From the definition of logarithm, $\log a = n$, $\log b = m$, and $\log (ab) = n + m$; hence $\log (ab) = n + m = \log a + \log b$. In a similar way, we can show that

$$\log \left(\frac{a}{b}\right) = \log a - \log b \tag{A-5}$$

and

$$\log a^n = n \log a. \tag{A-6}$$

These three rules apply not only to common logs but to natural or any other kind of logarithm.

Logs were once used as a technique for simplifying certain types of calculation. Because of the advent of electronic calculators and computers, they are not often used any more for this purpose. However, logs do appear in certain physical equations, so it is helpful to know how to deal with them. If you do not have a calculator that calculates logs, you can easily use a *log table*, such as the small one (Table A-1) below. The number N is given to two digits (some tables give N to three or more digits); the first digit is in the vertical column to the left, the second digit is in the horizontal row across the top. For example, the table tells us that $\log 1.0 = 0.000$, $\log 1.1 = 0.041$, and $\log 4.1 = 0.613$; note that the table does not include the decimal point—it is understood. The table gives logs for numbers between 1.0 and 9.9; for larger or smaller numbers we use rule A-4:

$$\log (ab) = \log a + \log b.$$

TABLE A-1

Short table of common logarithms

N	0.0	0.1	0.2	0.3	0.4	0.5	0.6	0.7	0.8	0.9
1	**000**	**041**	**079**	**114**	**146**	**176**	**204**	**230**	**255**	**279**
2	**301**	**322**	**342**	**362**	**380**	**398**	**415**	**431**	**447**	**462**
3	**477**	**491**	**505**	**519**	**531**	**544**	**556**	**568**	**580**	**591**
4	**602**	**613**	**624**	**633**	**643**	**653**	**663**	**672**	**681**	**690**
5	**699**	**708**	**716**	**724**	**732**	**740**	**748**	**756**	**763**	**771**
6	**778**	**785**	**792**	**799**	**806**	**813**	**820**	**826**	**833**	**839**
7	**845**	**851**	**857**	**863**	**869**	**875**	**881**	**887**	**892**	**898**
8	**903**	**908**	**914**	**919**	**924**	**929**	**935**	**940**	**944**	**949**
9	**954**	**959**	**964**	**968**	**973**	**978**	**982**	**987**	**991**	**996**

For example,

$$\log(380) = \log(3.8 \times 10^2) = \log(3.8) + \log(10^2).$$

From the table, $\log 3.8 = 0.580$; and from rule A-6, $\log(10^2) = 2\log(10) = 2$ since $\log(10) = 1$. [This follows from the definition of the logarithm: if $10 = 10^1$, then $1 = \log(10)$.] Thus

$$\begin{aligned}\log(380) &= \log(3.8) + \log(10^2)\\ &= 0.580 + 2\\ &= 2.580.\end{aligned}$$

Similarly,

$$\begin{aligned}\log(0.081) &= \log(8.1) + \log(10^{-2})\\ &= 0.908 - 2 = -1.092.\end{aligned}$$

Sometimes we need to do the reverse process: find the number N whose log is, say, 2.670. This is called "taking the antilogarithm." To do so, we separate our number 2.670 into two parts, making the separation at the decimal point:

$$\begin{aligned}\log N &= 2.670 = 2 + 0.670\\ &= \log 10^2 + 0.670.\end{aligned}$$

We now look in the table to see what number has its log equal to 0.670; none does, so we must *interpolate*: we see that $\log 4.6 = 0.663$ and $\log 4.7 = 0.672$. So the number we want is between 4.6 and 4.7, and closer to the latter by 7/9. Approximately we can say that $\log 4.68 = 0.670$. Thus

$$\begin{aligned}\log N &= 2 + 0.670\\ &= \log(10^2) + \log(4.68) = \log(4.68 \times 10^2),\end{aligned}$$

so $N = 4.68 \times 10^2 = 468$. If the given logarithm is negative, say, -2.180, we proceed as follows:

$$\begin{aligned}\log N &= -2.180 = -3 + 0.820\\ &= \log 10^{-3} + \log 6.6 = \log 6.6 \times 10^{-3},\end{aligned}$$

so $N = 6.6 \times 10^{-3}$. Notice that what we did was to add to our given logarithm the next largest integer (3 in this case) so that we have an integer, plus a decimal number between 0 and 1.0 whose antilogarithm can be looked up in the table.

Appendix B Order of magnitude: rapid estimating

We are sometimes interested only in an approximate value for a quantity. This might be because an accurate calculation would take more time than it is worth or would require additional data that are not available. In other cases, we may want to make a rough estimate in order to check an accurate calculation made on a calculator to make sure that no blunders were made when entering the numbers. Also, the correct power of 10 may be lost on a calculator or slide rule, and a rough estimate can be used to obtain it.

In general, a rough estimate is made by rounding off all numbers to one significant figure plus its power of 10, and after the calculation is made, again only one significant figure is kept. Such an estimate is called an *order-of-magnitude* estimate and can be assumed to be accurate within a factor of 10, and usually better; in fact, the phrase "order of magnitude" is sometimes used to refer simply to the power of 10.

As an example, suppose that a person wants to find out how much water there is in a particular lake, which is roughly circular, about 1 km across, and has an average depth of 10 m. To find the volume, we simply multiply the average depth of the lake times its surface area (as if it were a cylinder). We approximate the surface to be a circle, so the area is πr^2, which is approximately $3 \times (5 \times 10^2 \text{ m})^2 \approx 8 \times 10^5 \text{ m}^2$, where the radius r is 500 m and π was rounded off to 3 ($\approx$ means "approximately equal to"). Then the volume is about $(8 \times 10^5 \text{ m}^2) \times (10 \text{ m}) = 8 \times 10^6 \text{ m}^3$, which is on the order of 10^7 m^3. Because of all the estimates that went into this calculation, the order-of-magnitude estimate (10^7) is probably better to quote than the 8×10^6 figure.

Appendix C Dimensional analysis

When we speak of the *dimensions* of a quantity, we are referring to the type of units that must be used. The dimensions of area, for example, are always length squared (abbreviated $[L^2]$, using square brackets) and the units can be square meters, square feet, and so on. Velocity, on the other hand, can be measured in units of km/h, m/s, and mi/h, but the dimensions are always a length $[L]$ divided by a time $[T]$, that is, $[L/T]$. The formula for a quantity may be different in different cases, but the dimensions remain the same. For example, the area of a triangle of base b and height h is $A = \frac{1}{2}bh$, whereas the area of a circle of radius r is $A = \pi r^2$. The formulas are different in the two cases, but the dimensions in both cases are the same: $[L^2]$.

When we specify the dimensions of a quantity, we usually do so in terms of basic quantities, not derived quantities (see Section 2-4). The four common basic quantities are length $[L]$, time $[T]$, mass $[M]$, and electric current $[I]$. (There are also three other, less used, basic quantities—see Section 2-4.) Thus, force, which by Newton's second law has the same units as mass $[M]$ times acceleration $[L/T^2]$, has dimensions of $[ML/T^2]$.

Dimensions can be used as a help in working out relationships, and such a procedure is referred to as *dimensional analysis.* One useful technique is the use of dimensions to check a relationship for correctness. Two simple rules apply here. First, we can add or subtract quantities only if they have the same dimensions (we do not add centimeters and pounds); second, the quantities on each side of an equals sign must have the same dimensions.

For example, suppose that you derived the equation $v = v_0 + \frac{1}{2}at^2$, where v is the speed of an object after a time t, when it starts with an initial speed v_0 and undergoes an acceleration a. Let us do a dimensional check to see if this equation is correct. We write a dimensional equation as follows, remembering that the dimensions of speed are $[L/T]$ and of acceleration are $[L/T^2]$:

$$\left[\frac{L}{T}\right] \stackrel{?}{=} \left[\frac{L}{T}\right] + \left[\frac{L}{T^2}\right][T^2]$$

$$\stackrel{?}{=} \left[\frac{L}{T}\right] + [L].$$

The dimensions are incorrect: on the right side, we have the sum of quantities whose dimensions are not the same. Thus we conclude that an error was made in the derivation of the original equation.

If such a dimensional check does come out correct, it does not prove that the equation is correct; for example, a dimensionless numerical factor (such as $\frac{1}{2}$ or 2π) could be wrong. Thus a dimensional check can only tell you when a relationship is wrong; it cannot tell you if it is completely right.

Another use of dimensional analysis is for a quick check on an equation you are not sure about. For example, suppose that you cannot remember whether the equation for the period T of a mass m on the end of a spring with spring constant k is $T=2\pi\sqrt{k/m}$ or is $T=2\pi\sqrt{m/k}$. A dimensional check (with the dimensions of k being $[M/T^2]$, since from Hooke's law $k=$ force/distance) shows that the latter is correct:

$$[T]=\sqrt{\frac{[M]}{[M/T^2]}},$$

whereas the former is not,

$$[T]\neq\sqrt{\frac{[M/T^2]}{[M]}}.$$

Finally, an important use of dimensional analysis, but one with which much care must be taken, is to obtain the *form* of an equation. That is, we may want to determine how one quantity depends on others. To take a concrete example, let us try to find an expression for the period T of a simple pendulum. First we try to figure out what T could depend on, and make a list of these variables. It might depend on its length l, on the mass m of the bob, on the angle of swing θ, and on the acceleration due to gravity, g. It might also depend on air resistance (we would use the viscosity of air), the gravitational pull of the moon, and so on; but everyday experience suggests that the earth's gravity is the major force involved, so we ignore the others. So let us assume that T is a function of l, m, θ, and g, and that each of these factors is present to some power:

$$T=Cl^w m^x \theta^y g^z.$$

C is a dimensionless constant, and w, x, y, and z are exponents we want to solve for. We now write down the dimensional equation for this relationship:

$$[T]=[L]^w[M]^x[L/T^2]^z;$$

because θ has no dimensions (a radian is a length over a length—see Equation 3-1), it does not appear. We simplify and obtain

$$[T]=[L]^{w+z}[M]^x[T]^{-2z}.$$

To have dimensional consistency, we must have

$$1=-2z$$

$$0=w+z$$

$$0=x.$$

We solve these equations and find that $z=-\frac{1}{2}$, $w=\frac{1}{2}$, and $x=0$. Thus our desired equation must be

$$T=C\sqrt{l/g}\,f(\theta), \tag{C-1}$$

where $f(\theta)$ is some function of θ that we cannot determine using this technique. Nor can we determine in this way the dimensionless constant C. (Of course, to obtain C and f, we would have to do an analysis such as that in Chapter 11 using Newton's laws, which reveals that $C=2\pi$ and $f\approx 1$ for small θ.) But look what we *have* found, using only dimensional consistency. We obtained the form of the expression that relates the period of a simple pendulum to the major variables of the situation, l and g (see Equation 11-8) and saw that it indeed does not depend on the mass m.

How did we do it? And how useful is this technique? Basically, we had to use our intuition as to what variables were important and which were not. This is not always easy, and often requires a lot of insight. As to usefulness, the final result in our example could have been obtained from Newton's laws, as in Chapter 11. But in many physical situations, such a derivation from other laws cannot be done. In those situations dimensional analysis can be a powerful tool.

In the end, any expression derived by the use of dimensional analysis (or by any other means, for that matter) must be checked against experiment. For example, in our derivation of Equation C-1, we can compare the periods of two pendula of different lengths, l_1 and l_2, whose amplitudes (θ) are the same. For, using Equation C-1, we would have

$$\frac{T_1}{T_2}=\frac{C\sqrt{l_1/g}\,f(\theta)}{C\sqrt{l_2/g}\,f(\theta)}=\sqrt{\frac{l_1}{l_2}}.$$

Because C and $f(\theta)$ are the same for both pendula, they cancel out, so we can experimentally determine if the ratio of the periods varies as the ratio of the square roots of the lengths. This comparison to experiment checks our derivation, at least in part. C and $f(\theta)$ could be determined by further experiment.

Appendix D Conversion factors

Length

1 in = 2.54 cm
1 cm = 0.394 in
1 ft = 30.5 cm
1 m = 39.4 in = 3.28 ft
1 mi = 5280 ft = 1.61 km
1 km = 0.621 mi
1 nautical mile = 6080 ft = 1.85 km
1 fermi = 1 femtometer (fm) = 10^{-15} m
1 angstrom (Å) = 10^{-10} m
1 light-year = 9.46×10^{15} m

Time

1 day = 8.64×10^4 s
1 year = 3.15×10^7 s

Speed

1 mi/h = 1.47 ft/s = 1.61 km/h
1 km/h = 0.278 m/s = 0.621 mi/h
1 ft/s = 0.305 m/s = 0.682 mi/h
1 m/s = 3.28 ft/s = 3.60 km/h

Angle

1 radian (rad) = 57.30° = 57°18′
1° = 0.01745 rad
1 rev/min (rpm) = 0.1047 rad/s

Mass

1 atomic mass unit (u) = 1.660×10^{-27} kg
1 slug = 14.6 kg
1 kg = 0.0685 slug
[1 kg has a weight of 2.21 lb where g = 9.80 m/s^2.]

Force

1 lb = 4.45 N
1 N = 0.225 lb

Energy and work

1 J = 0.738 ft · lb = 10^7 ergs
1 ft · lb = 1.36 J = 1.29×10^{-3} Btu = 3.25×10^{-4} kcal
1 kcal = 4.18×10^3 J = 3.97 Btu
1 Btu = 252 cal = 778 ft · lb = 1054 J
1 eV = 1.60×10^{-19} J
1 kWh = 3.80×10^6 J = 860 kcal

Power

1 W = 1 J/s = 0.738 ft · lb/s
1 hp (U.S.) = 550 ft · lb/s = 746 W
1 hp (metric) = 750 W

Pressure

1 atm = 1.013 bar = 1.013×10^5 N/m^2
= 14.7 lb/in^2 = 760 torr
1 lb/in^2 = 6.90×10^3 N/m^2
1 Pa = 1 N/m^2 = 1.45×10^{-4} lb/in^2

Appendix E
Selected isotopes†

(1) Atomic number Z	(2) Element	(3) Symbol	(4) Mass number, A	(5) Atomic mass‡	(6) Percent abundance, or decay mode if radioactive	(7) Half-life (if radioactive)
0	(Neutron)	n	1	1.008665	β^-	10.6 min
1	Hydrogen	H	1	1.007825	99.985	
	Deuterium	D	2	2.014102	0.015	
	Tritium	T	3	3.016049	β^-	12.33 yr
2	Helium	He	3	3.016029	0.00014	
			4	4.002603	≈ 100	
3	Lithium	Li	6	6.015123	7.5	
			7	7.016005	92.5	
4	Beryllium	Be	7	7.016930	EC, γ	53.3 days
			9	9.012183	100	
5	Boron	B	10	10.012938	19.8	
			11	11.009305	80.2	
6	Carbon	C	11	11.011433	β^+, EC	20.4 min
			12	12.000000	98.89	
			13	13.003355	1.11	
			14	14.003242	β^-	5730 yr
7	Nitrogen	N	13	13.005739	β^+	9.96 min
			14	14.003074	99.63	
			15	15.000109	0.37	
8	Oxygen	O	15	15.003065	β^+, EC	122 s
			16	15.994915	99.76	
			18	17.999159	0.204	
9	Fluorine	F	19	18.998403	100	
10	Neon	Ne	20	19.992439	90.51	
			22	21.991384	9.22	
11	Sodium	Na	22	21.994435	β^+, EC, γ	2.602 yr
			23	22.989770	100	
			24	23.990964	β^-, γ	15.0 h
12	Magnesium	Mg	24	23.985045	78.99	
13	Aluminum	Al	27	26.981541	100	
14	Silicon	Si	28	27.976928	92.23	
			31	30.975364	β^-, γ	2.62 h
15	Phosphorus	P	31	30.973763	100	
			32	31.973908	β^-	14.28 days
16	Sulfur	S	32	31.972072	95.0	
			35	34.969033	β^-	87.4 days
17	Chlorine	Cl	35	34.968853	75.77	
			37	36.965903	24.23	
18	Argon	Ar	40	39.962383	99.60	

† Data are taken from *Chart of the Nuclides*, 12th ed., 1977, and from C. M. Lederer and V. S. Shirley, eds., *Table of Isotopes*, 7th ed., John Wiley & Sons, Inc., New York, 1978.

‡ The masses given in column (5) are those for the neutral atom, including the Z electrons.

(1) Atomic number Z	(2) Element	(3) Symbol	(4) Mass number, A	(5) Atomic mass	(6) Percent abundance, or decay mode if radioactive	(7) Half-life (if radioactive)
19	Potassium	K	39	38.963708	93.26	
			40	39.964000	β^-, EC, γ, β^+	1.28×10^9 yr
20	Calcium	Ca	40	39.962591	96.94	
21	Scandium	Sc	45	44.955914	100	
22	Titanium	Ti	48	47.947947	73.7	
23	Vanadium	V	51	50.943963	99.75	
24	Chromium	Cr	52	51.940510	83.79	
25	Manganese	Mn	55	54.938046	100	
26	Iron	Fe	56	55.934939	91.8	
27	Cobalt	Co	59	58.933198	100	
			60	59.933820	β^-, γ	5.271 yr
28	Nickel	Ni	58	57.935347	68.3	
			60	59.930789	26.1	
29	Copper	Cu	63	62.929599	69.2	
			65	64.927792	30.8	
30	Zinc	Zn	64	63.929145	48.6	
			66	65.926035	27.9	
31	Gallium	Ga	69	68.925581	60.1	
32	Germanium	Ge	72	71.922080	27.4	
			74	73.921179	36.5	
33	Arsenic	As	75	74.921596	100	
34	Selenium	Se	80	79.916521	49.8	
35	Bromine	Br	79	78.918336	50.69	
36	Krypton	Kr	84	83.911506	57.0	
37	Rubidium	Rb	85	84.911800	72.17	
38	Strontium	Sr	86	85.909273	9.8	
			88	87.905625	82.6	
			90	89.907746	β^-	28.8 yr
39	Yttrium	Y	89	88.905856	100	
40	Zirconium	Zr	90	89.904708	51.5	
41	Niobium	Nb	93	92.906378	100	
42	Molybdenum	Mo	98	97.905405	24.1	
43	Technetium	Tc	98	97.907210	β^-, γ	4.2×10^6 yr
44	Ruthenium	Ru	102	101.904348	31.6	
45	Rhodium	Rh	103	102.90550	100	
46	Palladium	Pd	106	105.90348	27.3	
47	Silver	Ag	107	106.905095	51.83	
			109	108.904754	48.17	
48	Cadmium	Cd	114	113.903361	28.7	
49	Indium	In	115	114.90388	95.7; β^-	5.1×10^{14} yr
50	Tin	Sn	120	119.902199	32.4	
51	Antimony	Sb	121	120.903824	57.3	
52	Tellurium	Te	130	129.90623	34.5; β^-	2×10^{21} yr
53	Iodine	I	127	126.904477	100	
			131	130.906118	β^-, γ	8.04 days
54	Xenon	Xe	132	131.90415	26.9	
			136	135.90722	8.9	
55	Cesium	Cs	133	132.90543	100	
56	Barium	Ba	137	136.90582	11.2	
			138	137.90524	71.7	
57	Lanthanum	La	139	138.90636	99.911	
58	Cerium	Ce	140	139.90544	88.5	
59	Praseodymium	Pr	141	140.90766	100	
60	Neodymium	Nd	142	141.90773	27.2	
61	Promethium	Pm	145	144.91275	EC, α, γ	17.7 yr
62	Samarium	Sm	152	151.91974	26.6	
63	Europium	Eu	153	152.92124	52.1	
64	Gadolinium	Gd	158	157.92411	24.8	
65	Terbium	Tb	159	158.92535	100	

(1) Atomic number Z	(2) Element	(3) Symbol	(4) Mass number, A	(5) Atomic mass	(6) Percent abundance, or decay mode if radioactive	(7) Half-life (if radioactive)
66	Dysprosium	Dy	164	163.92918	28.1	
67	Holmium	Ho	165	164.93033	100	
68	Erbium	Er	166	165.93031	33.4	
69	Thulium	Tm	169	168.93423	100	
70	Ytterbium	Yb	174	173.93887	31.6	
71	Lutecium	Lu	175	174.94079	97.39	
72	Hafnium	Hf	180	179.94656	35.2	
73	Tantalum	Ta	181	180.94801	99.988	
74	Tungsten (wolfram)	W	184	183.95095	30.7	
75	Rhenium	Re	187	186.95577	62.60, β^-	4×10^{10} yr
76	Osmium	Os	191	190.96094	β^-, γ	15.4 days
			192	191.96149	41.0	
77	Iridium	Ir	191	190.96060	37.3	
			193	192.96294	62.7	
78	Platinum	Pt	195	194.96479	33.8	
79	Gold	Au	197	196.96656	100	
80	Mercury	Hg	202	201.97063	29.8	
81	Thallium	Tl	205	204.97441	70.5	
82	Lead	Pb	206	205.97446	24.1	
			207	206.97589	22.1	
			208	207.97664	52.3	
			210	209.98418	α, β^-, γ	22.3 yr
			211	210.98874	β^-, γ	36.1 min
			212	211.99188	β^-, γ	10.64 h
			214	213.99980	β^-, γ	26.8 min
83	Bismuth	Bi	209	208.98039	100	
			211	210.98726	α, β^-, γ	2.15 min
84	Polonium	Po	210	209.98286	α, γ	138.38 days
			214	213.99519	α, γ	164 μs
85	Astatine	At	218	218.00870	α, β^-	≈ 2 s
86	Radon	Rn	222	222.017574	α, γ	3.8235 days
87	Francium	Fr	223	223.019734	α, β^-, γ	21.8 min
88	Radium	Ra	226	226.025406	α, γ	1.60×10^3 yr
89	Actinium	Ac	227	227.027751	α, β^-, γ	21.773 yr
90	Thorium	Th	228	228.02873	α, γ	1.9131 yr
			232	232.038054	100, α, γ	1.41×10^{10} yr
91	Protactinium	Pa	231	231.035881	α, γ	3.28×10^4 yr
92	Uranium	U	232	232.03714	α, γ	72 yr
			233	233.039629	α, γ	1.592×10^5 yr
			235	235.043925	0.72; α, γ	7.038×10^8 yr
			236	236.045563	α, γ	2.342×10^7 yr
			238	238.050786	99.275; α, γ	4.468×10^9 yr
			239	239.054291	β^-, γ	23.5 min
93	Neptunium	Np	239	239.052932	β^-, γ	2.35 days
94	Plutonium	Pu	239	239.052158	α, γ	2.41×10^4 yr
95	Americium	Am	243	243.061374	α, γ	7.37×10^3 yr
96	Curium	Cm	245	245.065487	α, γ	8.5×10^3 yr
97	Berkelium	Bk	247	247.07003	α, γ	1.4×10^3 yr
98	Californium	Cf	249	249.074849	α, γ	351 yr
99	Einsteinium	Es	254	254.08802	α, γ, β^-	276 days
100	Fermium	Fm	253	253.08518	EC, α, γ	3.0 days
101	Mendelevium	Md	255	255.0911	EC, α	27 min
102	Nobelium	No	255	255.0933	EC, α	3.1 min
103	Lawrencium	Lr	257	257.0998	α	≈ 35 s
104	Rutherfordium (?)	Rf	261	261.1087	α	1.1 min
105	Hahnium (?)	Ha	262	262.1138	α	0.7 min
106			263	263.1184	α	0.9 s
107			261	261	α	1–2 ms

Answers to odd-numbered Problems

Chapter 1

1 85 km/h
3 7.8 h
5 70, 45, 55, 55, 60, 70 km/h
7 0.32 m/s
9 Total time = 4.5 h, 910 km/h
11 3800 cars/h
13 3.9 m/s^2
17 90 m
19 3.5 m/s^2
7 m/s
21 (*a*) 160 m, (*b*) 15 s, (*c*) 21.3 m, 18.3 m
23 (*a*) 69.3 m, (*b*) 39.1 m
25 (*a*) $t_R - (v_0/2a) + (d_1/v_0)$, (*b*) 3.1 s
29 23.5g
31 (*a*) 29 m, (*b*) 4.9 s
33 7 m/s
35 28 m
37 5.8 s
41 1.6 m
43 d = 41 m
45 9.1 m/s
49 6.3 blocks, 18° south of east
51 $-V_1$
V_{wrong} V_2

53 0.06 h
55 (*a*) 337 km/h, θ = 6° east of north; (*b*) 17.5 km
57 (*a*) 89 m, (*b*) θ = 26°
59 16 km/h
61 2 min, 2 km, 0.007 h, 0.4 km
63 6.4 s
65 (*a*) 0.3 m/s, (*b*) 1.4 m/s, (*c*) 0.3 m/s, (*d*) 1.8 m/s, (*e*) −0.9 m/s
67 (*a*) At about 50 s; (*b*) from 90 to 105 s; (*c*) from 0 to about 30 s, from 60 to about 80 s, and from 90 to 105 s; (*d*) from about 60 to 80 s

Chapter 2

1 850 N
3 5×10^{-4} N
5 2.44 kg
7 3500 N
9 1100 N
11 1.9×10^4 N
13 250 N
15 The child must accelerate downward at 2.5 m/s^2.
17 (*a*) 9.9 m/s , (*b*) 4000 N
19 $T_2 = 2ma = 2T_1$
21 Horizontal = 1500 N, vertical = 1300 N
23 Southwest
25 F = 110 N, 0
27 98 N
29 8 m/s^2
31 v = 26 m/s
33 3.8 s
35 5.4 m/s^2
37 16 m/s
39 (*a*) 1.1 m/s^2, (*b*) 420 N
41 160 N
43 (*a*) 33 N · s/m, (*b*) 130 N
45 10°
47 19.6 m, 7 m
49 8.7 m
51 6.4 s
53 0.8 m
55 10 m/s
57 (*a*) $x = (v_0^2 \sin 2\theta)/g$, (*b*) θ = 45°, (*c*) due to air resistance
59 45°

Chapter 3

1 (*a*) 0.52 rad, (*b*) 1.57 rad, (*c*) 7.33 rad
3 6.8 km
5 210 rad/s
7 −8.4 rad/s^2
11 (*a*) 2×10^{-7} rad/s, (*b*) 7×10^{-5} rad/s
13 v = 6.4 m/s
15 8.5g
17 3.6×10^{22} N
19 (*a*) 1.1 N, (*b*) 8.0 N
21 0.34
23 (*a*) 10 m, (*b*) 40 m
25 Yes, 5500 N down the bank
27 $\tan \theta = w^2 r/g$
29 1.3×10^{-8} N
31 1.7 m/s^2
33 6500 m/s
35 1.9 times greater
37 3.5×10^7 m
39 3.4×10^8 m
41 1.4 h
43 (*b*) 9.5×10^{26} kg
45 225 days
47 $v_N/v_F = d_F/d_N$
49 106 N · m
51 0.15 kg · m^2
53 (*a*) 0.003 kg · m^2, (*b*) 0.11 N · m
55 17.4 kg
57 5.4 m/s^2
59 (*a*) 8.4 N · m, (*b*) 340 N
61 (*a*) 3.7 rad/s^2, (*b*) 7.4 m/s^2, (*c*) 370 m/s^2, (*d*) 2700 N, (*e*) θ = 1.1°
63 18.9% of body height from shoulder joint
65 3.14 m
67 9.4% of full body height below median line
69 θ = 26.6°
71 9570 m

Chapter 4

1 1.44 N
3 2100 N · m
5 710 N
7 F_T = 200 N, F_w = 200 N at 30°
9 $F_2 = 1.5 \times 10^4$ N, $F_1 = -2.9 \times 10^3$ N
11 53 N
13 F_2 = 2700 N, F_1 = 1800 N
15 F_2 = 4400 N, F_1 = 4600 N
17 490 N
19 2100 N, no
21 6.3 m up
23 50 cm
25 72 N
27 F = 450 N
29 F = 990 N
31 F_M = 1800 N
33 1700 N
35 6w
37 F_V = 2700 N
39 F_M = 1500 N, F_J = 2400 N
41 The equilibrium is stable, 7 m off-center.
43 9.8×10^4 N/m^2, 2.0×10^{-6}
45 1.7 mm, 0.58 mm

47 0.13%, 0.15%
49 90×10^6 N/m^2, 900 atm
51 (*a*) No, (*b*) 0.11 cm
53 5.1×10^4 N
55 7.7 mm
57 1.7 cm
59 2.5×10^6 N

Chapter 5

1 0.12 kg · m/s
3 (*a*) 450 kg · m/s, (*b*) 450 kg · m/s, (*c*) 450 N
5 6×10^7 N
7 23 m/s
9 (*a*) 1.2×10^9 N/m^2. (*b*) The bone will break.
11 1.89 m
13 (*a*) $v = 0.17$ m/s, (*b*) 1900 N
15 3.3×10^{-2} m
17 7.2 m/s
19 (*a*)

500
400
300
200
100
1 2 3 4
t (ms)

(*b*) 0.72 N · s, (*c*) 2.3×10^{-3} kg
21 1.9 kg · m^2/s
23 (*a*) *I* increases; so ω decreases. (*b*) 1.5
25 0.38 rev/s
27 If they jump on at the edge, ω = 0.46 rad/s. If they are jumping *off*, not enough information to determine.
29 2500 J
31 9.4×10^6 J
33 17.5
35 25 J
37 6 ropes on the pulleys supporting the weight
39 (*a*) 3.9, (*b*) 1.55
41 5400 J
43 0.25×10^{-6} J
45 3.9×10^5 J
47 (*a*) $\frac{1}{2}k(x^2 - x_0^2)$, (*b*) 0
49 4.3×10^4 J
51 3.3 m, no
53 $v_0 = 6.6$ m/s
55 15.3 m/s
57 $\frac{1}{2}kx_0^2 = \frac{1}{2}kx^2 + \frac{1}{2}mv^2$
59 2.1×10^6 J
61 (*a*) 20 m/s, (*b*) 1.9 N
63 3.0 m/s, −2.0 m/s
65 18 m/s
67 (*a*) 7.8×10^4 J, (*b*) 960 N
71 510 N
73 0.5%
75 410 W
77 (*a*) 7.5×10^5 J, (*b*) 52 W, 0.07 hp, (*c*) 5×10^6 J
79 1.7 W
81 θ = 24°
83 6.6 m/s, or 24 km/h

Chapter 6

1 7.5×10^{-3} m^3
3 100 kg
5 0.8563
7 (*a*) 120 torr, (*b*) 1.6×10^4 N/m^2, (*c*) 0.16 atm, (*d*) 2.3 lb/in^2
9 $P = 1.2 \times 10^5$ N/m^2, $F = 1.4 \times 10^7$ N. The pressure on the side of the pool near the bottom will be near 1.2×10^5 N/m^2.
11 (*a*) 0.81 m, (*b*) 0.60 m, (*c*) 0.24 m
13 (*a*) 1.2×10^{-3} N, (*b*) 0.19 N
15 0.63 N
17 13 cm
19 5800 kg
23 0.98 W
25 1.005×10^3 kg/m^3
27 1.01 g/cm^3
29 0.57
31 (*a*) $\rho_0 = 1004$ kg/m^3, (*b*) $w' = 0$, it floats
33 1.8×10^7 kg
35 20 people
37 1.0005 kg
39 1.36 kg
41 3.4×10^{-2} N/m
43 No.
45 2.3×10^{-3} N down (at 20°C)
47 3 times more surface energy
51 1.0×10^5 N
53 4.2 m/s
55 1.1×10^{-3} m^3/s
57 2.0 m/s, 0.87 atm
59 (*b*) 6.9×10^{-3} m/s
61 250 N/m^2
63 A factor of 7.7
65 0.96 N/m^2
67 5.5 m
69 4.0 cm
71 (*b*) 4.8 m/s

Chapter 7

1 1.8 kg
3 20°C
5 75°
7 2.2×10^{-2} m
9 93°
11 3.7×10^6 N. The beam will fracture.
13 230°C above room temperature
15 −49°C
19 (*a*) 29°C, (*b*) 8500 N
21 (*a*) 310 K, (*b*) 305 K, (*c*) 77 K
23 3.34×10^{25} molecules
25 2.7×10^{25} molecules/m^3
27 1.42 kg/m^3
29 (*a*) 38 liters, (*b*) 316 K
31 810 atm
33 2.7 kg
35 3.20 atm
37 3.1 cm^3
39 4.6%
41 5.65×10^{-21} J
43 (*a*) 3.3, (*b*) 3.9
45 $v_{1rms}/v_{2rms} = \sqrt{M_2/M_1}$
47 2.7×10^4 molecules/cm^3
49 1.1
51 1.9×10^6 J
53 (*a*) 480 m/s, (*b*) 14 per second
55 1.004

Chapter 8

1 15°C
3 0.47 atm
5 1.5 atm
7 1.2 kg
9 29%
11 (*a*) Solid and vapor, (*b*) −56.6°C to 31°C; 5.11 atm to 73 atm
13 Vapor
15 0.16 s; 2×10^{-7} of its thermal speed
17 (*a*) $Q \propto 1/\sqrt{m}$. (*b*) Nitrogen diffuses faster by 7%.
19 4.9×10^5 N/m^2

Chapter 9

1 400 kcal
3 11.1°C
7 62.5 J, or 1.5×10^{-2} kcal
9 7.9×10^3 kg
11 600 kcal
13 211°C
15 1 BTU/lb · °F
17 0.35 C°
19 0.60 kcal/kg · C°
21 3.4°C
23 0.085 kcal/kg · C°
25 52.5 kcal
27 1440 kcal
29 65 kcal/kg
31 (*a*) 1.0 h, (*b*) 6.1 h additional
33 1.9×10^{-3} kg/min
35 $T = 100$°C
37 2.4×10^{-3} kg
39 18.4 J/s
41 6.3 kW
43 0.68 mm
45 20 kcal/h
47 101°C
49 1.9×10^6 J/h = 460 kcal/h (assuming 1000 W/m^2 reaches the earth's surface)
51 (*a*) 23.5 kcal/s, (*b*) $Q = 3.5 \times 10^4$ kcal, (*c*) \$190/month
53 (*b*) $\frac{Q}{t} = \frac{A(T_n - T_0)}{l_n/k_n + \cdots + l_1/k_1}$, where T_n and T_0 are the inside and outside temperatures
55 8.8×10^4 s

57 (*a*) 2.4 C°/s, (*b*) 400 K. (Assume 1000 W/m^2 reaches the earth's surface.) (*c*) Evaporation, convection, and conduction through the stem
59 $A = 73\ m^2$
61 (*a*) 1.03×10^4 kg/day, (*b*) supplies only about $\frac{1}{2}$ (pessimistic because of temperature gradient near walls)

Chapter 10

1 Yes, 3700 J
3 −700 J
5 (*a*) -3.3×10^5 J, (*b*) $U = 0$
7 (*a*) 1.2×10^5 J, (*b*) 1.6×10^6 J
9 3800 kcal
11 5.1 W
13 120 W
15 12 K/h
17 0.04%
19 $C_V = 4.97$ cal/mol · K, or 2.49 kcal/kg · °C; $C_P = 6.96$ cal/mol · K, or 3.48 kcal/kg · °C
21 54 K
23 21%
25 2.2×10^{10} J
27 690°C
29 63 kg/s
31 $8.8 \times 10^9\ m^3$/day
33 0.3 kcal/K
35 −2.9 kcal/K
37 0.26 cal/K · s
39 (*a*) $\frac{5}{16}$, (*b*) $\frac{1}{64}$

Chapter 11

1 35 N/m
3 1.2 Hz
5 $k = 820$ N/m, $f = 3.1$ Hz
9 3.4 Hz
11 0.25 s
13 (*a*) 8.7 cm, (*b*) 5.0 cm
15 (*a*) $x =$ (0.080 m) sin 29 t, (*b*) $t =$ 0.055 s, $t = 0.16$ s
17 416 m/s
19 3.2 times as large
23 1.3 Hz
25 (*a*) 1.25 s, (*b*) 0.80 Hz
27 (*a*) 1.4 s. (*b*) There would be no restoring force; so T would be infinite.
29 (*a*) 1.0 Hz, (*b*) 0.54 m/s
31 (*b*) 1.7 s
33 $\frac{1}{3}$
35 3.3 m/s
37 Between 190 and 550 m; between 2.8 and 3.4 m
39 4.4 N
41 (*a*) 1.4×10^3 m/s, (*b*) 4.1×10^3 m/s
43 0.36 m
45 (*a*) 1.4×10^6 m. (*b*) You need more than one.
47 The high-frequency wave has 3.2 times as much energy.
49 (*a*) The intensity decreases by 4. (*b*) The amplitude decreases by 2.
53 8.1 km/s
55 9.0°
57 261 Hz
59 30 Hz, 60 Hz, 90 Hz, etc.
61 50 Hz
63 $f_1 = 310$ Hz, $f_2 = 620$ Hz, $f_3 = 930$ Hz
65 $f = \dfrac{n}{2L}\sqrt{F_T/(m/L)}$

Chapter 12

1 (*a*) 5.76×10^{-3} m, (*b*) 1.32×10^{-3} m
3 0.35 m
5 40 dB
7 3.2×10^5
9 (*a*) 9, (*b*) 9.5 dB
11 130 dB
13 40 dB
15 19 W
17 45 dB
19 310 W
21 120 dB
23 10 dB
25 Estimating from Fig. 12-5, we see the lowest frequency is about 300 Hz and the highest is about 15 kHz.
27 0.53 m
29 0.58 m from mouthpiece
31 (*a*) about 120 overtones, (*b*) about 120 overtones
33 3300 Hz, yes
35 660 Hz (at 20°C)
37 F″ = 1408 Hz
39 $\frac{3}{4}$ tone off
41 0.5 Hz
43 (*a*) 128 or 136 Hz, (*b*) 6%
45 4.4 Hz
47 780 Hz (at 20°C)
49 (*a*) 11.9 Hz, (*b*) 57.5 m
51 (*a*) 3.4 s, (*b*) 3.8 s
53 (*a*) 850 Hz, (*b*) 950 Hz
55 19 m/s
57 $f(1 \pm v_0/v)/(1 \pm v_s/v)$
59 1.6×10^3 Hz
61 90/min
65 2.8 m/s

Chapter 13

1 6×10^{12} electrons
3 68 N
5 1.5×10^{-8} m
7 $F_e = 8.2 \times 10^{-8}$ N, $F_g = 3.6 \times 10^{-47}$ N. Therefore the gravitational force is insignificant compared to the electrical force.
9 14 N away from the center
11 1.9×10^5 N away from the center
13 2.1×10^{-10} m
15 $x = 0.35$ m away from −1.3 μC
17 1.3×10^6 N/C, upward
19 9.6×10^{-17} N
21 At A: $E = 6.4 \times 10^6$ N/C in the vertical direction; at B: $E = 17 \times 10^6$ N/C at 56° from the vertical; yes
23 -4.5×10^5 C
25 5.7×10^7 N/C
27 $r = 38{,}000$ km from the moon
29 5.0×10^{-10} N
31 8.7 turns/s

Chapter 14

1 0.09 m
3 18 V
5 4.8×10^{-17} J
7 The speed would double.
9 3.5×10^{-2} eV
11 200 V
13 2.9° from horizontal
15 5 μF
17 35 pF
19 7500 V/m
21 2.2×10^{-7} C
23 65 V on each, $Q_2' = 6.4 \times 10^{-4}$ C, $Q_1' = 2.6 \times 10^{-4}$ C
25 4.43×10^{-8} J/m^3
27 Energy is also doubled.
29 14 V
31 5.4×10^{-2} J
33 (*a*) 0.70 cm from the negative charge, (*b*) 0.13 cm from the negative charge toward the positive charge
35 8.0×10^{-30} C · m
37 (*b*) 2.0×10^{-20} J
39 0.48 eV

Chapter 15

1 1.1×10^5 C
3 3600 C
5 4.5 A
7 28 Ω
9 $1.9 \times 10^{18}\ s^{-1}$
11 0.1 siemens
13 3.8×10^{-4} m
15 The tungsten wire must have a diameter of 1.8 mm for the resistance to be the same.
17 1800°C
19 $R_{ON} = 560\ \Omega$, $R_{OC} = 440\ \Omega$
21 19 Ω
23 3.6 W
25 1800 W
27 $0.13
29 $I_t = 12$ h
31 2.6×10^6 J
33 2800 kg/year
35 160 W vs. 2500 W
37 (*a*) 2.2 hp, (*b*) 220 km, (*c*) yes
39 930 W
41 5.7 A
43 0.42 A
45 5800 W, 11,500 W, 0
47 (*a*) 19 Hz, (*b*) 1.8 A, (*c*) 125 sin 120t
49 69 mol/m^3

51 It is not in equilibrium. There must be a K^+ pump; $C_I = 95\ mol/m^3$.
53 Energy = 5×10^{-11} J, $P = 5 \times 10^{-5}$ W

Chapter 16

1 In series: 1200 Ω; in parallel: $R = 33$ Ω
3 5.7 V
5 The terminal voltage of 3 mercury cells is greater; also, internal power loss is greater in the dry cells.
9 1.6×10^3 Ω
11 $R = 1200$ Ω; $P_L/P_T = 1.4 \times 10^{-2}$, or 1.4%
13 (a) 0.1 A, (b) 0.1 A, (c) $I_p = 0.04$ A
15 $R = 2000$ Ω
17 In series, then separately; 288 Ω, 144 Ω
19 $V_{cord} = 4$ V, $V_{app} = 116$ V
21 43 V
23 $r = 0.1$ Ω, $\varepsilon = 51$ V
25 $V_2 = 43$ V, $V_1 = 77$ V
27 30 Ω: 0.1 A; 20 Ω: 0.15 A; 10 Ω: 0.05 A; 15 Ω: 0.18 A; 25 Ω: 0.13 A
29 $I_1 = 0.76$ A, $I_2 = 0.71$ A, $I_3 = 0.05$ A, $V = 6.0$ V
31 1.0×10^6 Ω
33 (a) 1.4×10^{-4} Ω in parallel, (b) 71×10^6 Ω in series
35 1000 Ω in series with the ammeter; S = 100 Ω/V
37 $I_A = 1.0 \times 10^{-3}$ A, $V = 3.9$ V
39 2.4×10^5 Ω
41 9.0 V
43 $R_{sh} = 45$ Ω, $R_{ser} = 30$ kΩ
45 0.301 V
47 4.473 V
49 13.6 Ω
51 100 m
53 0.036°C
55 1.5×10^{-3} mm
57 a 1-μF capacitor in parallel
59 1.0×10^6 Ω
61 (a) $V_1 = 3.0$ V, $V_2 = 6.0$ V; (b) 0.60×10^{-6} C; (c) $Q_1 = 1.8 \times 10^{-6}$ C, $Q_2 = 0.9 \times 10^{-6}$ C, $V = 9$ V
63 $C = 0.12$ F
65 1.2×10^5 Ω

Chapter 17

1 (a) 0.50 N/m, (b) 0.35 N/m
3 12 A
5 8.2×10^{-14} N due north
7 1.6 N
9 17 T westward
11 8.3×10^{-2} m
13 7.0×10^{-2} T
15 1.0×10^{-8} m
17 70, 72, 73, 74
19 2.0 m
21 1.5×10^6 m/s, 3.4 mm
23 6 electrons
27 44 μA
29 10%
33 (a) $d = 5.6 \times 10^{-4}$ m, (b) 0.11 Ω
35 1.8×10^4 A
37 2.0×10^{-4} m
39 4.5×10^{-18} N
41 40 T, east
43 20 A downward
47 (a) 94 turns, (b) 1.2×10^{-3} T. (c) B decreases as N increases.
49 48 A, parallel

Chapter 18

1 3.9×10^{-2} V
3 (a) 6.3×10^{-2} V, (b) clockwise
5 (a) 0.080 V, (b) 0.40 V/m
7 2.8×10^{-5} J
11 2.3 C
13 110 V
17 10 rev/s
19 110 V
21 (b) 31 A, (c) 4.4 A
23 4 turns
25 $\frac{1}{4}I_p$
27 $N_s/N_p = 0.05$
29 (a) stepdown, (b) $V_s = \frac{1}{3}V_p$
31 (a) 61 kV, (b) 0.02
33 5600 W
35 -45×10^{-3} V
37 1.0 H
39 0.8 J
41 1.13 A
43 Rough estimate, 0.28 H; careful calculation, 0.18 H
45 2.3×10^{-5} H
47 210 m^2, yes
49 4.4×10^{13} J/day

Chapter 19

1 370 Hz
3 at $f = 10$ Hz, $X_C = 1.6 \times 10^4$ Ω; at f = 1000 Hz, $X_C = 1.6 \times 10^2$ Ω
5 $X_L = 1.2 \times 10^3$ Ω, $I_{rms} = 0.32$ A
7 (a) 4.0×10^4 Ω, (b) 3.6×10^{-2} A
9 (a) 1.9%, (b) 95%
11 (a) 440 pF, (b) 0.1 m^2, no
13 (a) 10×10^3 Ω, (b) 10×10^3 Ω
15 (a) 4.5 mA, (b) $\phi = -80°$, (c) 9.3×10^{-2} W, (d) $V_R = 21$ V; $V_C = 120$ V
17 3.2×10^{-3} W
19 (a) 2.7 A, (b) $\phi = 70°$, (c) 15 W
21 $Z = 4.0 \times 10^4$ Ω, $\phi = -88°$, I_{rms} = 7.5 mA
25 1.5×10^6 Hz
27 1.2 pF
29 1.9×10^{-2} A
31 1.2×10^3 loops
33 4 mA/V
35 3.4×10^{-13} N
37 $+6.7 \times 10^4$ to -6.7×10^4 V/m
39 2.7 V
41 (a) 2.3×10^{-2} A, (b) 1.6×10^{-2} A
43 12 mA
45 (a) 25 mA, (b) 50 mA
47 $I_B = 7.3$ μA
49 6.7 kΩ
51 7.2×10^{-4} A
53 (a) 200, (b) 9.5
55 (a) 500, (b) 5% change
57 26 turns
59 4-Ω terminals

Chapter 20

1 8.85×10^{-10} A
3 4.0×10^{-8} T
5 2.76×10^{-2} m
7 5.0×10^{-13} T
9 The person by the radio hears it 0.14 s sooner.
11 9.5×10^{-2} W/m^2
13 4.5×10^{-12} J
15 200 m
17 180 pF
19 L must range from 2.4×10^{-9} to 3.6×10^{-9} H.
21 (a) 2.0×10^{-5} W/m^2, (b) 8.7×10^{-2} V/m
23 1.01×10^{-3} V/m; 1.35×10^{-9} W/m^2
25 (a) in parallel, (b) 9.8 to 88 pF, (c) 9.5×10^{-4} H

Chapter 21

1 2.21×10^8 m/s
3 6.7×10^8 mi/h, 1.08×10^9 km/h
5 500 s
7 4.0 m
13 −7.5 cm
15 It is a virtual image 14 cm behind the mirror. No, you do not see yourself clearly.
21 $f = -1.7$ m
23 −6.0 m
25 $\theta_{air} = 42°$
27 $\theta_1 = 53°$
29 $\theta_g = 24°$, $\theta_w = 27°$; $\theta_w = 27°$
33 33°
35 1.2
39 $d_i = -7.2$ cm; virtual image
41 (a) $f = -13$ cm; a diverging lens and a virtual image
(b) $f = 93$ cm; a converging lens and a virtual image
43 −0.12 m; a diverging lens
45 (a) $d_o = 6.7$ cm; (b) $d_o = 3.3$ cm
47 4.7×10^{-4} m
49 (a) $d_i = 0.14$ m, 0.13 cm high, real image, (b) $d_i = -0.13$ m, −0.12 cm high, virtual image
51 $f = 0.14$ m
55 3.5 m
57 $d_i = 6.3$ cm behind the diverging lens
59 $f = -11.8$ cm
61 $R = 8.3$ cm
63 $d_i = 480$ cm; $M = -4$

Chapter 22

1 5.4×10^{-6} m
3 1.1×10^{-4} m
5 6.0 cm
7 3.0 mm
9 0.16°
11 0.4 cm
13 570 nm
15 3×10^{-5} m
17 1.1×10^{-4} m
19 $\theta = 12°$
21 Two complete spectral orders plus part of a third
23 1.3×10^{4} lines/cm
25 8×10^{-7} m
27 518 nm, 593 nm, 737 nm
29 0.4 cm
31 90 nm
33 1.0×10^{-5} m
35 Bright: 135 nm; dark: 0 or 270 nm
37 11 m; 22 m
39 $\lambda = 581$ nm
41 2.6×10^{-4} m
43 Moving the mirror changes the phase between the two interfering waves; 0.87 mm.
45 $\theta_p = 50°$
47 $\frac{1}{4} I_0$
49 37°
51 (*a*) No light is transmitted; (*b*) $\frac{1}{8} I_0$; (*c*) nothing gets through the last two.
53 490 kg/m^3
55 0.24 wavelength

Chapter 23

1 36 mm
3 2.3
5 $\frac{1}{50}$ s
7 3.8 m to infinity
11 (*a*) near-sighted, (*b*) 14 cm
13 (*a*) −2.9 diopters, (*b*) 26 cm
15 −0.4 diopters, 1.7 diopters
17 (*a*) $f = 2.0$ cm; (*b*) $f = 1.8$ cm
19 10 cm
21 (*a*) 5.2×, (*b*) 4.2×
23 2.3×; 4.1×
25 4.2 cm
27 (*a*) 4.8×, (*b*) 1.8×
29 −25×, 62.4 cm
31 83×
33 15×
35 $f_e = 1.6$ cm, $f_o = 82.4$ cm
37 1.2 cm
39 (*a*) 1.1 cm, (*b*) 190×
41 (*a*) 16.1 cm, (*b*) 160×
43 $f = 5$ cm
45 −2.3 diopters
47 57.5°, 2.3×10^{-7} m
49 (*a*) 1.4, (*b*) 2.4×10^{-7} m
51 0.84
53 0.13 m, 2.3 km

Chapter 24

1 1.2×10^{-8} s
3 150 m
5 6.18 m is the length of your car and 5.97 m is the length of his.
7 37 light years
9 (*a*) 6.25 m, 1.5 m, (*b*) 8.0 s, (*c*) 0.60c, (*d*) 8.0 s
11 (*a*) 0.8c, (*b*) 5 years
13 0.87c
15 $\rho = \dfrac{\rho_0}{1 - v^2/c^2}$
17 (*a*) $(1 - 0.5 \times 10^{-8})c$, (*b*) 0.30 m
19 9×10^{11} J, 9.2×10^{8} kg
21 2.56×10^{-27} kg, 0.76c
23 (*a*) 1.4×10^{20} J, (*b*) 1600 kg
25 (*a*) 6.3×10^{7} kg/year, (*b*) 1.39×10^{17} kg/year
27 1.09×10^{-30} kg
29 5.8×10^{-19} kg · m/s
31 2.82×10^{8} m/s
35 (*a*) 0.80c, (*b*) 0.80c
37 (*a*) $x = 550$ m, $y = 10$ m, $z = 0$ m; (*b*) $x = 1100$ m, $y = 10$ m, $z = 0$ m
39 54 m/s, 22° (refer to Problem 36)
41 (*a*) *A* fired first; (*b*) 2.8×10^{-14} s; (*c*) *B* is struck first.

Chapter 25

1 5.4×10^{-20} J, or 0.33 eV
3 3.6×10^{-6} m
5 (*a*) 5.0×10^{-19} J, (*b*) 2.8×10^{-19} J
7 4.1×10^{-7} eV
9 540 nm
11 0.63 eV
13 3.6 eV
15 (*a*) The electrons will not be ejected. (*b*) 1.9×10^{-19} J = 1.2 eV
17 8.3×10^{4} photons/s
19 (*a*) 570 eV, (*b*) 0.105 nm
21 1.5×10^{-27} kg · m/s
23 3.4×10^{-11} J, 5.9×10^{-15} m
25 65 kV
27 3200 C°/min
29 3.3×10^{-32} m
31 680 V
35 2.8×10^{-11} m
39 KE = 0.040 eV, $\lambda = 1.5 \times 10^{-10}$ m
41 150 V
43 5.4×10^{-3}

Chapter 26

1 4.7×10^{-14} m
3 10.2 eV
5 2.11 eV
7 120 nm
9 The lowest six energy levels are −54.4 eV, −13.6 eV, −6.0 eV, −3.4 eV, −2.2 eV, and −1.5 eV.
11 122.4 eV
13 from $n = 8$ to $n = 2$
15 2.8×10^{68}
17 4300
19 Yes, 2.2×10^{6} m/s
21 91 nm
23 14 MeV
25 (*c*) 2.4×10^{-17} J = 150 eV, (*d*) 1.6×10^{-66} J
27 $\pm 3.2 \times 10^{-11}$ m
29 $\pm 1.2 \times 10^{5}$ m/s
31 (*a*) 1.3×10^{-34} m, (*b*) 2.1×10^{-32} kg · m/s, (*c*) 1.3×10^{-30} m

Chapter 27

1 32 states
3 10 electrons
5 $m_l = 3, 2, 1, 0, -1, -2, -3$; $m_s = \pm\frac{1}{2}$
7 2.6×10^{-34} J · s
9 3.4 eV
11 0.061 nm
13 1.25×10^{-2} J, 4×10^{16} photons/pulse

Chapter 28

3 0.149 u
5 27
7 $^{30}_{14}$Si
9 1800 MeV
11 31.994 MeV, 5.332 MeV/nucleon
13 7.680 MeV
15 59.93525 u
17 228.0288 u, $^{228}_{90}$Th
19 $^{32}_{16}$S, 31.97207 u
21 β decay is 60 times as fast as α decay.
23 0.862 MeV
27 98.2%
29 7 α particles and 4 β^- particles
31 1.1×10^{4} s
33 3.7×10^{21} nuclei
35 (*a*) 2.4×10^{16} nuclei, (*b*) 1.5×10^{15} nuclei, (*c*) 3.4×10^{13} decays/s, (*d*) about 23 min
37 3.3×10^{-12} kg
39 0.082 decay/s
41 7.1×10^{8} years
43 120 counts/s
45 500 rad
47 100–150 rad
49 about 4 days
51 8.9×10^{-5} rad/day
53 55×10^{-3} rem

Chapter 29

1 No threshold energy is required.
3 The reaction releases energy.

5 (*a*) The reaction could go. (*b*) 17.8 MeV
9 126.5 MeV
11 The $^{235}_{92}UF_6$ has 1.004 the speed of the $^{238}_{92}UF_6$.
13 1.2×10^{-4} kg/year
15 7.8×10^{17} fissions/s
17 (*a*) 1500 kg/year, (*b*) 6.1×10^6 Ci
19 0.12 g/year
23 X_1 is $^{13}_{7}N$; X_2 is $^{13}_{6}C$; X_3 is $^{14}_{7}N$; X_4 is $^{15}_{8}O$; X_5 is $^{15}_{7}N$; X_6 is $^{12}_{6}C$.

Chapter 30

1 16 GeV
3 1.2 T
5 For the α particle: KE = 34 MeV, $v = 4.1 \times 10^7$ m/s; for the deuterons: KE = 17 MeV, $v = 4.1 \times 10^7$ m/s
7 100 rev
9 $\lambda = 3.1 \times 10^{-18}$ m; the resolving power is about 3×10^{-18} m.
11 $B = 1.3$ T
13 68 MeV
15 133.5 MeV
17 1879.2 MeV
19 1.5×10^{-12} m
21 38 MeV
23 69.4 MeV
25 2.9×10^{-21} s

Index

The symbol *fn* refers to a footnote.

Aberration:
  chromatic, 604–5
  of eye, 606
  of lenses 604–5, 606
  spherical, 543, 604
Aberration of starlight, 625
Absolute temperature scale, 192
Absolute zero, 192
Absorption coefficient, 329
Absorption lines, 575, 668
Absorption spectra, 668, 673–4
Acceleration, 12–21, 27–30, 39, 61–69
  angular, 61–64
  average, 12
  centripetal, 64–69
  of gravity, 18, 44, 73–4, 282
  instantaneous, 13
  linear, 12–21, 27–30, 39, 62, 64–68
  related to force, 39–40
  tangential, 62
  uniform, 14–17
Accelerators (*see* Particle accelerators)
Acceptor, 503
Accommodation of eye, 596
Achromatic lens, 605
Acoustics, 329–30
  (*see also* Sound)
Action at a distance, 349
Action potential, 401–3
Action-reaction (Newton's third law), 40–42
Activation energy, 202, 699–700
Active transport, 220
Activity of nuclear radiation, 720, 729–31
Addition of velocities:
  classical, 23–27
  relativistic, 639–40
Adhesion, 162
Adiabatic process, 252
Air columns, vibration of, 317–20
Air conditioners, 262–3
Airfoil, 168
Airplane wing, 168
Airy disc, 607
Algebra, review of, 773–6
Alpha decay, 715–6, 719–20
  and tunneling, 724
Alpha particles, 715, 716
Alternating current (ac), 395–7, 489–97, 505–6
Alternators, 470–1
Ammeter, 416–7, 418–20
Amorphous solids, 215
Ampere (unit), 388, 455–6
  operational definition of, 455–6
Ampère, André, 388, 453, 520
Ampère's law, 452–5, 520–3
Amplification, electronic, 500, 501, 507–11
Amplifiers, 510–12
Amplitude:
  pressure, 312
  of vibration, 275
  of wave, 287–8
Amplitude modulation (AM), 533
AM radio, 532–4
Anechoic chamber, 330
Aneroid barometer, 154
Aneroid gauge, 153–4
Angle:
  Brewster's, 583
  critical, 549
  of incidence, 294, 296, 539, 547
  phase, 496
  radian measure of, 60
  of reflection, 294–5, 539
  of refraction, 296, 547
Angstrom (unit), 568*fn*
Angular acceleration, 61–64
Angular magnification, 599
Angular momentum, 121–3
  in atoms, 670, 688–9
  law of conservation of, 122–3
Angular quantities, 60–64
Angular velocity, 61–64
Anode, 387, 499–500
Antenna, 523, 532, 533
Antielectron (*see* Positron)
Antimatter, 758
Antineutrino, 718, 759
Antineutron, 758
Antinodes, 300–3, 318–20
Antiparticles, 718, 758, 760–1
Antiproton, 758, 759
Apparent weight, 75–6, 158
Apparent weightlessness, 76
Approximate calculations, 55, 780
Arago, F., 570
Arches, 107–8
Archimedes, 157–8
Archimedes' principle, 156–9
Area under a curve or graph, 29–30
Aristotle, 1, 18, 37–38
Armature, 450, 469
Astigmatism:
  of eye, 597
  of spherical lens, 604
Astronomical telescope, 600–2
Atmosphere (unit of pressure), 151
Atmosphere, scattering of light by, 586–7
Atmospheric pressure, 151
Atomic bomb, 742–3, 746
Atomic mass, 182
Atomic mass number, 710–11
Atomic mass unit, 182, 711–2
Atomic number, 690–2, 710–11
  determined from x-ray analysis, 693–4
Atomic spectra, 668–70, 673, 676–7, 690, 691
  fine structure of, 676, 689
Atomic structure, 667–694
  Bohr model of, 670–6
  of complex atoms, 690–4
  of hydrogen, 670–5, 679–80, 688–90
  planetary model of, 342–3, 668
  plum pudding model of, 667
  quantum mechanics of, 679–80, 688–94
Atomic theory, 181–3, 196–204
  (*see also* Atomic structure, Atoms, Kinetic theory)
Atoms, 181–3, 196–204
  complex, 690–3
  electric charge in, 342–3
  energy levels in, 670–6, 679–80, 688–94
  excited states of, 672
  ground state of, 672, 691
  probability distributions in, 679–80, 690
  quantum theory of, 670–694
  stationary states in, 670–6, 679–80
  (*see also* Atomic structure, Kinetic theory)
ATP, 359–60, 700
Atwood's machine, 57
Audible range, 309
Audio amplifier, 510
Autoradiography, 726
Average binding energy per nucleon, 713

Avogadro, Amedeo, 196
Avogadro's hypothesis, 196
Avogadro's number, 196

Back, forces in human, 99–100
Back emf, 471–2
Back torque, 472
Balance, 100–101
Ballistic pendulum, 146
Balmer, J. J., 669
Balmer formula, 669, 673
Balmer series, 669, 673
Band spectra, 700
Banking of curves, 68
Bar (unit), 153
Barometer, 153, 154–5
Baryon number and its conservation, 759
Baryons, 760–1
  quark theory of, 763–4
Basal metabolic rate, 254
Base (of transistor), 506–9
Base and derived quantities and units, 43–4
Basic amplifier, 510
Battery, 384–7
Beats, 327–8
Becquerel, A. H., 714–5
Bernoulli, Daniel, 166
Bernoulli's equation, 166–9
Bernoulli's principle, 166, 168–9
Beta decay, 715, 717–20
Beta particle, 715, 717
  (*see also* Electron)
Bias voltage, 500, 504–5, 506
Bicycle, 59, 88, 122, 144, 145, 146
Binding energy:
  in atoms, 672–5
  of molecules, 697–700
  of nuclei, 712–3
  related to radioactivity, 716
  (*see also* Interaction energy)
Binoculars, 549, 601–2
Birefringence, 585–6
Blackbody radiation, 649
Blood:
  diffusion of gases in, 219–20
  electromagnetic measurement of flow, 468
  flow, 156, 165–6, 172–3
  gas exchange in, 197
  pressure, 154
Bohr, Niels, 659, 670–5, 683–4
Bohr model of the atom, 670–6
Bohr radius, 671
Boiling (*see* Phase, changes of)
Boltzmann, Ludwig, 267
Boltzmann's constant, 196, 269
Bond energy, 697–700
  (*see also* Interaction energy)
Bonds:
  covalent, 694–5
  dipole-dipole, 377
  ionic, 695
  partial ionic character of covalent, 696–7
  in solids, 697–8
  weak, 376–8, 697–8

Bourdon gauge, 153–4
Boyle, Robert, 191, 308
Boyle's law, 191, 193
Bragg, W. H., 657
Bragg, W. L., 657
Bragg equation, 657
Brahe, Tycho, 77
Brake, hydraulic, 152
Braking distances, 16–17, 32
Breeder reactor, 745
Bremsstrahlung, 656
Brewster's angle, 583
Brown, Robert, 182
Brownian movement, 182–3
Brunelleschi, 109
Btu (British thermal unit), 227
Bubble chamber, 732
Bulk modulus, 103, 105
Buoyancy, 156–9
Buoyant force, 156–9, 174

Caloric theory, 226
Calorie (unit), 227
  relation to the joule, 228, 251
Calorimeter, 231–2
Calorimetry, 231–2
Camera, 592–5
Candela (unit), 44
Cantilever, 97
Capacitance, 369–70
Capacitor microphone, 424
Capacitors, 369–72, 426–7, 492–7
  energy stored in, 371–2
  impedance of, 492–3
  reactance of, 492–3
  in series and parallel, 424–6
Capacity (*see* Capacitance)
Capillarity, 162–3
Carbon dating, 724–5
Carnot, S., 260–2
Carrier frequency, 532–3
Cartridge, phonograph, 468
Cassegrainian focus, 601
Cathode, 387, 498–500
Cathode rays, 446, 500 (*see also* Electron)
Cathode ray tube, 500–1, 534
Causality, 78–9, 683–4
Cavendish, Henry, 73
Cell, electric 386–7
Celsius temperature scale, 184–5
Center of gravity, 83–6
Center of mass, 84
Centigrade (*see* Celsius temperature scale)
Centipoise (unit), 170
Centrifugal (pseudo) force, 66–67, 70
Centrifuge, 63, 68–69, 70, 175–6
Centripetal acceleration, 64–69
CERN, 753
Cgs system of units, 43
Chadwick, James, 710
Chain reaction, 740–1
  self-sustaining, 741
Chamberlain, O., 758
Characteristic curves of transistor, 507–8
Charge (*see* Electric charge)
Charles, Jacques, 192
Charles' law, 192–3

Charm, and charmed quark, 764–5
Chart recorder, 450, 510
Chemical bonds (*see* Bonds)
Choke coil, 477
Chromatic aberration, 604–5
Chromosome (*see* DNA)
Circle of confusion, 593, 604
Circuit (*see* Electric circuit)
Circuit breaker, 394–5
Circular motion, 60–69
  nonuniform, 69
  uniform, 64–8
Citizens' band (CB) receiver, 534
Clausius, R. J. E., 259, 265
Clock paradox (*see* Twin paradox)
Cloud chamber, 732
Coating of lenses, 578–9
Coherent light, 702
Coherent source, 568
Cohesion, 162
  of water, 163–4
Coil (*see* Inductor)
Collector (transistor), 506–9
Colliding beams of elementary particles, 754–5
Collision, 119–21, 137–9
  elastic, 137
  inelastic, 138
Color, 568–9
Coma, 604
Commutator, 451, 470
Compass, magnetic, 436–8
Complementarity, principle of, 658–9, 661
Components of vector, 24–26
Compound, 181
Compound microscope, 602–3
Compound nucleus, 739
Compression, 104–110
Compressive stress, 104–110
Compton, A. H., 652
Compton effect, 652, 654
Condenser lens, 603
Conduction:
  electric, 343–4
  heat, 236–8
Conduction band, 503
Conductivity, thermal, 236–7
Conductors:
  electric, 343–4, 391, 503
  electric field and, 352–3
  heat, 237
Conservation laws:
  of angular momentum, 122–3
  apparent violation of in beta decay, 718
  of baryon number, 759
  of electric charge, 342
  in elementary particle interactions, 758–60
  of energy, 116, 133–8, 250–1
  of lepton numbers, 759–60
  of momentum, 116, 119–21, 137–8
  in nuclear decays and reactions, 719–20, 737–8
  of nucleon number, 719–20
  of strangeness, 763
Conservative force, 131–2

Contact lenses, 598
Contact potential, 424
Continuity, equation of, 164-6
Contrast in microscope images, 610-3
Control, feedback and, 511-3
Control grid, 499-501
Control rods in nuclear reactor, 742, 744
Convection, 238-40
Conversion factors, table of, 784
Coordinate systems, 8-9
Copenhagen interpretation of quantum mechanics, 683
Copernicus, 2
Core, of nuclear reactor, 744
Correspondence principle, 675, 677
Coulomb (unit), 345, 456
  operational definition of, 456
Coulomb, Charles, 345
Coulomb's law, 345-8, 520
Counter emf, 471-2
Counter torque, 472
Covalent bond, 694-5
  in molecules, 694-5
  partial ionic character of, 696-7
  in solids, 697
Creativity in science, 1-3
Crick, F. H. C., 658
Critical angle, 549
Critical damping, 283-4
Critical mass, 741
Critical point, 213
Critical temperature, 213
CRT, 500-1, 534
Crystallography, 657-8
Curie (unit), 729
Curie, Marie and Pierre, 715
Curie temperature, 439
Current (*see* Electric current)
Current sensitivity, 416, 419-20
Curvature of field, 604
Cybernetics, 513
Cyclotron, 752-3
Cyclotron frequency, 752

Dalton, John, 182, 197
Dalton's law of partial pressures, 197
Damping, 283-4, 293
Dark-field microscopy, 611
Dating, radioactive, 724-5
Daughter nucleus, 716
Davisson, C. J., 660
Dc amplifier, 510
De Broglie, Louis, 659-60, 675-6
De Broglie wavelength, 660, 675, 750
Decay, nuclear (*see* Radioactivity)
Decay constant, 720-1
Deceleration, 13
Decibel (unit of sound intensity), 310
Declination, magnetic, 438
Dees, 752
Defects of the eye, 596-8, 606
Definite proportions, law of, 182
De Forest, Lee, 499
Degrees of freedom, 257-9
Delbrück, Max, 3
Density, 147-9
Depth of field, 594
Descartes, R., 564
Detection of nuclear radiation, 731-3
  bubble chamber, 732
  cloud chamber, 732
  geiger counter, 731
  photographic emulsion, 732
  scintillation counter, 731
  semiconductor detector, 732
  spark chamber, 732-3
  streamer chamber, 733
Determinism, 79, 683-4
  versus probability, 683-4
Deuterium, 741, 746
Dew point, 211
Dextrorotatory, 584
Diamagnetism, 457*fn*
Diamond, 569
Dichroism, 586-7
Dielectric constant, 348, 370
Dielectrics, 370-1
Differential amplifier, 510
Diffraction, 299, 564, 569-72
  of electrons, 660
  Fraunhofer, 570
  Fresnel, 570
  of light, 564, 569-72
  of matter, 660
  by single slit, 569
  around solid object, 570
  of water waves, 299
  x-ray, 657-8
Diffraction grating, 572-6
Diffraction limit of lens resolution, 607-8
Diffraction pattern, 570
Diffusion, 216-20
  Fick's law of, 216-7, 220
  in living organisms, 217-20
Diffusion constant, 216
Dimensional analysis, 781-3
Dimensions, 781
Diodes, 498, 504-6
  junction, 504-6
  Zener, 505
Diopter, 551
Dipole, electric, 374-5
Dipole-dipole bond, 377
Dipole moment, 375
Direct conversion, 483-4, 747
Direct current (dc) (*see* Electric current)
Discharge tube, 668
Disintegration, nuclear (*see* Radioactivity)
Disorder and order, 264-70
Dispersion, 569
Displacement, 9-10, 21-22, 27-30
  of a spring, 275
Displacement current, 522
Display, electronic, 509
Dissipative force, 136
Dissolving of salt, 348
Dissonance, 328
Distortion, by lenses, 604
Diver, 122
DNA and its replication, 176, 353-5, 376-8, 658
Domains, magnetic, 438-9
Domes, 108-110
Donor, 503
Doping of semiconductors, 502
Doppler effect, 332-4
Dosimetry, 729-31
Double refraction, 580, 585-6
Double slit experiment:
  for electrons, 679
  for light, 566-8
Drag force, 173
Dry cell, 386
Duality, wave-particle, 658-9, 661
Dynamics, 7, 36-54
  of rotation, 79-83
Dynamo (*see* Electric generator)
Dyne (unit), 43
Dynodes, 731-2

$E = mc^2$, 637-9
Ear, 313-6
Earth:
  magnetic field of, 437-8
  mass of, 73
  shape of, 71
Earthquake waves, 291-2, 297
ECG, 378-80, 510-11
Eddy, 164
Eddy currents, 472-3
Edison, Thomas, 498
Efficiency of heat engine, 261-2
Einstein, A., 3, 182, 258, 618, 622, 625-44, 650-2, 683, 742
Elasticity, 101-6
Elastic limit, 102
Elastic moduli, 102-5, 189, 291, 309
Elastic region, 102
Electric battery, 384-7
Electric car, 486-7
Electric cell, 386-7
Electric charge, 341-5
  in atom, 342-3
  conservation of, 342
  on electron, 346, 447-8
  induced, 344
  motion of, in magnetic field, 444
  "test," 349
Electric circuits, 387
  ac, 489-97, 505-6
  containing capacitors, 424-6
  dc, 408-30
  feedback in, 511-13
  grounding, 428-9
  household, 394-5
  impedance matching of, 513-4
  integrated, 509
  and Kirchhoff's rules, 412-15
  LR, 479
  LRC, 494-7
  RC, 426-7
  rectifier, 505-6
  containing resistors, 408-10
  resonant, 497
  time constants of, 427, 479
Electric current, 384-430, 489-97
  alternating, 395-7, 489-97, 505-6
  conventional, 388
  direct, defined, 395
  displacement, 522
  hazards of, 427-9, 493-4
  leakage, 493-4

produces magnetic field, 439
measurement of, 416-7, 418-20
and Ohm's law, 389
Electric dipole, 374-5
Electric energy, 364-9
in molecular biology, 376-8
production, 479-84
stored in capacitor, 371-2
stored in electric field, 372-3
Electric field, 348-53
and conductors, 352-3
in EM wave, 523-5
energy stored in, 372-3
intensity, defined, 349
produced by magnetic field, 466-8, 523
produces magnetic field, 520, 523
quantum of, 755-6
relation to electric potential, 366-8
Electric field lines, 351-2
Electric flux, 523
Electric force, 341, 345-8
Coulomb's law for, 345-8
in molecular biology, 353-60
quantum of, 755-6
Electric generators, 469-72, 480-3
Electric motor, 450-1, 486
counter emf in, 471-2
Electric potential, 364-8
of dipole, 374
relation to electric field, 366-8
of single point charge, 373-4
(*see also* Potential difference)
Electric power, 393-5, 396
in ac circuits, 396, 489, 490, 492, 496, 497
and impedance matching, 513-4
production of, 480-4
transmission of, 474-6
Electrocardiogram (ECG), 378-80, 510-11
Electrode, 386, 397, 498-501
Electrolysis, 397-8
Electromagnet, 440-1
Electromagnetic blood-flow measurement, 468
Electromagnetic force, 341, 755-7
quantum of, 755-7
Electromagnetic induction, 463-79
Electromagnetic pumping, 445
Electromagnetic spectrum, 528-9
Electromagnetic waves, 520, 523-34, 563, 620-7
Electrometer, 345
Electromotive force, 387, 411
back, 471-2
counter, 471-2
of generator, 469-71
Hall, 445
induced, 463-8
series and parallel, 415
Electron:
in atoms (*see* Atomic structure)
charge on, 346
discovery of, 445-7
mass of, 448, 712
measurement of charge on, 447-8
measurement of e/m, 445-7
in molecules, 694-8
in pair production, 654-5
wavelength of, 660
wave-particle nature of, 660-1
Electron capture, 718-9
Electron "cloud," 680
(*see also* Probability distributions of electrons)
Electron gun, 500
Electronics, 498-514
Electron lepton number, 759-60
Electron microscope, 661-3
scanning, 662-3
scanning-transmission, 663
transmission, 661-2
Electron neutrino, 759, 761
Electron-positron pair production, 655
Electron spin, 689
Electron volt (unit), 368-9, 712
Electrophoresis, 398
Electroplating, 398
Electroscope, 344-5
Electrostatics, 341-80
Elementary charge, 346
Elementary particles, 750-66
accelerators of, 750-5
classification of, 760-1
colliding beams of, 754-5
and conservation laws, 758-60
interactions of, 758-60
J/ψ, 760*fn*, 764-5
short-lived (resonances), 762
stability of, 761-2
strange, 762-3
table of, 761
wavelength of, 750
Yukawa, 755-7
Elements, 181
Emf (*see* Electromotive force)
Emission spectra, 668, 673, 676-7, 690, 691
Emissivity, 241
Emitter, 506-9
Emulsion, photographic, for nuclear radiation detection, 732
Energy, 116, 123
activation, 202, 699-700
in EM waves, 524, 530-2
equipartition of, 258-9
equivalence to mass, 637-9
and first law of thermodynamics, 250-1
heat as, 227-9
ionization, 672
kinetic, 127-9
mechanical, 134
stored, 131
thermal, 132, 136, 137, 228-9
transformation of, 132-3, 139
vibrational, 258-9, 276-7
of waves, 286, 292-3, 310, 530-2
relation to work, 128, 131-3
(*see also* Binding energy, Conservation laws, Internal energy, Nuclear energy, Potential energy, Solar energy)
Energy bands:
in semiconductors, 503
in solids and liquids, 701
Energy level (*see* Energy states)
Energy level diagram:
atomic, 672-3, 689
for fluorescence, 701
for lasers, 702-4
in molecules, liquids, and solids, 700-1
nuclear, 719
Energy states:
in atoms, 670-6, 679-80, 688-94
in molecules, 700-1
in nuclei, 719
Engine, heat (*see* Heat engine)
Enrichment of uranium, 741, 744
Entropy, 265-70
in life processes, 266-7
Environmental pollution:
air, 203-4, 481
noise, 323-4
from power plants, 481, 744-5
thermal, 263-4, 481, 745
Equation of state, 191
for an ideal gas, 194
Equilibrium, 93-101, 107-110
Equipartition of energy, 258-9
Equipotential lines, 367-8
Equipotential surface, 367-8
Erg (unit), 123, 128
Ether, 621-6
Evaporation, 208-10, 235
Evolution and entropy, 266-7
Exchange of particles, as carriers of force, 755-7
Excited state:
of atom, 672
of nucleus, 719
Exclusion principle, 691, 695
Expansion, thermal (*see* Thermal expansion)
Exponential notation, 772-3
Exponents (review), 771-2
Extraordinary ray, 585
Eye, 595-8
aberrations of, 606
defects of, 596-8
resolution of, 610
structure and function of, 595-6
Eyeglass lenses, 596-8
Eyepiece, 600, 602, 622

Fahrenheit temperature scale, 184-5
Falling bodies, 17-21
Fallout, radioactive, 743
Farad (unit of capacitance), 369
Faraday, Michael, 349, 463-5, 520
Faraday's law, 465-8, 520
Far point of eye, 596
Farsighted eye, 597-8
Feedback, 186-7, 511-13
in electronics, 511-13
in humans, 186-7, 513
negative, 186, 512
positive, 186, 512
in thermostat, 186-7
Feedback amplifier, 511-12
Fermi, Enrico, 718, 738-42, 751, 762
Fermilab, 753, 765

Ferromagnetism and ferromagnetic materials, 436–7, 440, 457–8
Feynman diagrams, 756
Fiber optics, 549–50
Fick, Adolf, 216
Fick's law of diffusion, 216–7, 220
Field, 349
(*see also* Electric field, Gravitational field, Magnetic field)
Film badge, 730
Fine structure, 676, 689
Fission (*see* Nuclear fission)
Fission fragments, 739, 743
Fitzgerald, G. F., 625
Fleming, J., 498
Floating, 158
Florence, cathedral of, 109
Flow of fluids, 164–76
laminar, 164, 170–2, 173–4
streamline, 164
in tubes, 170–3
turbulent, 164, 172–3, 174
Flow rate, 164–6, 170–3
Fluid, 147–76
object moving in, 173–6
(*see also* Blood, Gases, Liquids)
Fluorescence, 701
Fluorescent light bulbs, 701
Flute, 319–20
Flux:
electric, 573
magnetic, 464–6
FM radio, 532–4
Focal length:
of lens, 551, 605–6
of spherical mirror, 542
Focal plane, 551
Focal point, 542, 550–1
Focusing of camera lens, 593–4
Force, 36–7, 38–47, 757
buoyant, 156–9, 174
conservative, 131–2
dissipative, 136
drag, 173
in equilibrium, 93–101, 107–110
of friction, 37–8, 47–51, 131, 136–7
exerted by inanimate object, 42
lines of, 351–2
long-range, 714
in muscles and bones, 98–100, 118
net, 39, 45
in Newton's laws, 39–42
normal, 47
short-range, 714
units of, 42–44
(*see also* Electric force, Gravitational force, Magnetic force, Nuclear force)
Fossil-fuel plants, 480–1
Foucault, J., 565
Four-dimensional space-time, 635–6
Fourier's theorem, 284
Fracture, 101, 105–7, 189–90
Frames of reference (*see* Reference frames)
Franklin, Benjamin, 342
Fraunhofer diffraction, 570
Free-body diagram, 46, 95
Freezing (*see* Phase, changes of)
Frequency, 62, 275, 288
of audible sound, 309
carrier, 532–3
fundamental, 300, 316, 318
resonant, 285, 300–2, 497
of rotation, 62
of vibration, 275
of wave, 288
Frequency modulation (FM), 533
Fresnel, A., 569–70
Fresnel diffraction, 570
Friction, 37–8, 47–51, 131, 136–7
coefficients of, 47–8
Fringes, 566–8
Frisch, O., 739
f-stop, 592–3
Fundamental frequency, 300, 316, 318
Fuse, 394–5
Fusion, heat of, 233–4
Fusion, nuclear, 745–7

Gain of amplifier, 508, 511–2
in dB, 338
of feedback amplifier, 511–2
Galilean telescope, 601
Galilean transformation, 640–2
Galileo, 1–2, 7, 18, 37–38, 51, 78, 108, 184, 538, 601, 618–20
Galvani, Luigi, 384–6
Galvanometer, 416, 449–50
Gamma decay, 715, 719–20
Gamma ray, 715, 719
as elementary particle (photon), 755–6, 760–1
Gas constant, 194
Gases, 147, 183, 191–204
change of phase, 213–5, 233–6
diatomic, 257
heat capacities of, 256–8
ideal, 191–204
light emitted by, 575, 668–70
monatomic, 257
real, 212–15
triatomic, 257
work done by, 252–3
Gas laws, 191–6
Gauge pressure, 151
Gauges, pressure, 153–5
Gauss (unit), 442
Gay-Lussac, Joseph, 193
Gay-Lussac's law, 193
Gedanken experiment, 627
Geiger counter, 731
Gell-Mann, M., 763
General motion, 86, 136
Generator, electric, 469–72, 480–3
Geometric optics, 537–58, 592–606
Geometry (review of), 776–7
Geothermal power plants, 481
Germanium, 502
Germer, L. H., 660
Glaser, D. A., 732
Glasses, eye, 596–8
Graphical analysis, 27–30, 119, 125
Grating, diffraction, 572–6
Gravitation, universal law of, 71–4, 76–9
Gravitational field, 352
quantum of, 757
Gravitational force, 36, 39, 44, 71–9
quantum of, 757
Gravitational potential energy, 129–30
Graviton, 757
Gravity, 17–21, 71–9
acceleration of, 18, 44, 73–4, 282
center of, 83–6
free fall under, 17–21
Grid (of a tube), 499–501
Grimaldi, F., 564
Grounding, electrical, 428–9
Ground state, in atom, 672, 691
Gyroscope, 123

Hadrons, 760–1
quark theory of, 763–6
Hahn, O., 739
Half-life, 720–2
and radioactive dating, 724–5
table of, for elementary particles, 761
table of, for nuclei, 785–7
Hall effect, 444–5
Harmonic motion:
non-, 283–4
simple, 274–83
Harmonics, 301–2, 318
Harmony, musical, 326, 328
Hazards of electricity, 427–9
Hazards of radioactivity, 727–31, 743
Heart, 156, 378–80
ECG, 378–80
electromagnetic pumping for, 445
pacemakers for, 429–30
as pump, 156
Heat, 226–70
compared to work, 250
distinguished from energy and temperature, 228–9
in first law of thermodynamics, 250–1
as flow of energy, 227–9
in human body, 253–5
mechanical equivalent of, 228
Heat capacity, 229–30, 255–9
Heat death, 267
Heat engine, 259–64
efficiency of, 261–2
internal combustion, 260–1, 263
in nuclear reactor, 744
steam, 260–2
stratified charge, 263
Heat of fusion, 233–4
Heat of vaporization, 233–4
Heat reservoir, 251
Heat transfer, 236–43
Heavy water, 741
Heisenberg, W., 677
Heisenberg uncertainty principle (*see* Uncertainty principle)
Henry (unit), 476–7
Henry, Joseph, 463
Hertz (unit of frequency), 275
Hertz, Heinrich, 528
High-energy physics, 754 (and all of Chapter 30)
High jump, 86
Holes (in semiconductor), 502–3

Holograms, 705–7
Holography, 705–7
Homeostasis, 186–7
Hooke's law, 101–2, 130, 276, 278
Horsepower, 139
Humidity, 210–11
Huygens, C., 563
Huygens' principle, 563–4, 565
Hydraulic brakes, 152
Hydraulic lift, 152
Hydraulic pressure, 152
Hydroelectric power plants, 482
Hydrogen:
  Bohr theory of, 670–5
  quantum mechanics of, 679–80, 688–90
  spectrum of, 669–70
Hydrogen bomb, 746–7
Hydrometer, 158–9
Hypothesis, 1
Hysteresis, 456–8

Ideal gas law, 191, 193–6
Ignition system of automobile, 474–5
Image:
  formed by plane mirror, 540–1
  formed by spherical mirror, 543
  real, 541, 552
  virtual, 541, 552
Image distance, 540, 544, 552–3
Impedance, 477, 491–7, 513–4
Impedance matching, 513–4
Impulse, 118–9
Inclines, motion on, 49–51
Index of refraction, 539, 569
  in birefringent materials, 585–6
  dependence on wavelength, 569
  in Snell's law, 548
Induced electric charge, 344
Induced emf, 463–79
  in electric generator, 469–71
  in transformer, 473–6
Inductance:
  mutual, 476–7
  self, 477–9
Inductor, 477, 489–91, 494–7
Inertia, 38
  law of, 38
  rotational, 81–3
Inertial mass, 74
Inertial reference frame, 70, 619–20
Infrared radiation, 529, 568–9
Infrasonic waves, 309
Instrumentation, electronic, 509–11
Instruments:
  electronic, 509–11
  optical, 592–613
  (*see also* Ammeter, Musical instruments, Voltmeter, etc.)
Insulators:
  electrical, 343–4, 391, 503
  thermal, 237
Integrated circuits, 509
Intensity:
  electric field, 349
  of sound, 310–3
  of waves, 292–3
Intensity level, 310–3
Interaction energy, 376–7
Interference, 297–8
  of light waves, 566–8
  of sound waves, 326–8
  by thin films, 576–9
  of water waves, 298
  of waves on a string, 297
Interference fringes, 566–8
Interference microscope, 611–2
Interferometers, 579, 622–5
Internal energy, 228, 229, 250–8
Internal reflection, 548–50
Internal resistance, 411–2
Interval on musical scale, 324
Intrinsic semiconductor, 503
Inverted population, 702
Ionic bond, 695–7
Ionization energy, 672–5
Ionizing radiation, 727
Irreversible process, 262
Isobaric process, 252
Isomer, 719
Isothermal process, 251–2
Isotopes, 448, 711
  table of, 785–7
  (*see also* Nucleus, Radioactivity)

J/ψ particles, 760*fn*, 764–5
Joule (unit of work and energy), 123, 128
  relation to the calorie, 228, 251
Joule, James Prescott, 227–8
Junction diode, 504–6
Junction transistor, 506–9

K-capture, 718
Kelvin (unit), 192
Kelvin temperature scale, 192
Kepler, Johannes, 76–8
Kepler's laws, 76–9
Kilocalorie (unit), 227
Kilogram (unit), 42
Kilowatt-hour (unit), 394
Kinematics:
  of rotational motion, 60–5
  of translational motion, 7–30, 52
Kinetic energy, 127
  in collisions, 137–9
  molecular, relation to temperature, 197–203
  relativistic, 637–9
  rotational, 128–9
  translational, 127
Kinetic friction, 47–8
Kinetic theory, 181–3, 196–204, 208–23, 235–6, 256–9, 268
Kirchhoff's rules, 412–5, 521, 522*fn*

Laminar flow, 164, 170–2, 173–4
Land, Edwin, 580
Lasers, 702–7
  continuous, 704
  helium-neon, 703–4
  and holography, 705–7
  pulsed, 704
  ruby, 703
  semiconductor, 704
Latent heats, 233–6
Lawrence, E. O., 752
Laws, 5, 40
  (*see also* specific name of law)
LC current, 497, 533
Leakage current, 493–4
Length contraction, 633–5, 643–4
Lens:
  achromatic, 605
  coating of, 578–9
  compound, 605–6
  condenser, 603
  contact, 598
  converging, 551
  cylindrical, 597
  diverging, 551
  of eye, 595–8
  eyeglass, 596–8
  eyepiece, 600, 602
  focal length of, 551, 605–6
  magnetic, 661–2
  magnification of, 554
  objective, 600, 602
  power of (diopters), 551
  resolution of, 607–10
  telephoto, 594
  thin, 550–8
  wide-angle, 594
  zoom, 594
Lens aberrations, 604–5, 606
Lens combinations, 555–6, 600–6
Lens equation, 552–4
Lens-maker's equation, 556–8
Lenz's law, 465
Leonardo da Vinci, 3, 47, 108
Lepton numbers and their conservation, 759–60
Leptons, 760–1, 763, 765
Lever, 96, 125–6
Lever arm, 79–80
Levorotatory, 584
Lift, 168
Light, 537–613
  as electromagnetic wave, 528, 563
  infrared, 529
  monochromatic, 576
  photon theory of, 650–5
  polarized, 580–3
  scattering of, 586–7
  speed of, 528, 538–9, 621–6, 636–7
  ultraviolet, 529, 568–9
  visible, 528–9
  wave theory of, 563–87
  (*see also* Diffraction, Interference, Reflection, Refraction, Spectra)
Lightning, 388
Light pipe, 549
Linear accelerator, 754
Linear momentum, 116–21
Lines of force, 351–2
Line spectrum, 575, 668–70, 673, 676–7
Liquid-drop model, 739
Liquids, 147, 183
Liquefaction, 213
Logarithms, 777–9
Longitudinal wave, 289–92
Lorentz, H. A., 625, 643
Lorentz transformation, 642–4

Loudness, 314-6
(*see also* Intensity)
Loudspeakers, 451
LR circuit, 479
Lucretius, 182-3
Lungs, human, 161-2, 219-20
Lyman series, 669, 673

Machines, simple, 125-7, 152
Mach number, 334-5
Macroscopic, 183
Macrostate, 267
Magnetic declination, 438
Magnetic domains, 438-9
Magnetic field, 437
of circular loop, 440, 462
of earth, 437-8
in EM wave, 523-5
energy stored in, 478-9
path of charged particle in, 444
produced by electric field, 520, 523
produces electric field, 466-8, 523
of solenoid, 454
of straight wire, 439, 452
Magnetic field strength, 437, 442, 451-5
Magnetic flux, 464-6
Magnetic flux density, 437
(*see also* Magnetic field strength)
Magnetic force:
on electric current, 441
on moving electric charge, 443-4
Magnetic induction, 437
(*see also* Magnetic field strength)
Magnetic lenses, 661-2
Magnetic moment, 449
Magnetic permeability, 452, 455, 457
Magnetic poles, 436
Magnetic quantum number, 689
Magnetism, 436-79
produced by electric currents, 439-40
Magnetohydrodynamics, 483-4, 747
Magnets, 436-9, 457-8
domains of, 438-9
Magnification:
angular, 599
of lens, 554
of magnifying glass, 599
of microscope, 602
of mirror, 545
of telescope, 601
useful, 610
Magnifier, simple, 598-600
Magnifying glass, 598-600
Magnifying power, 599, 601
(*see also* Magnification)
Manhattan Project, 742-3
Manometer, 153
Marconi, Guglielmo, 532
Mass, 38-40, 42
gravitational vs. inertial, 74
Mass-energy transformation, 637-9
Mass increase in relativity, 636
Mass number, 710-11
Mass spectrograph (spectrometer), 448-9
Mathematical review, 769-79
Matter waves, 659-61, 675-6
Maxwell, James Clerk, 201, 520-5, 528, 620-6
Maxwell distribution of molecular speeds, 201-3, 268
Maxwell's equations, 520-3, 621
Mechanical advantage, 96, 126-7
Mechanical energy, 134
Mechanical equivalent of heat, 228
Meitner, L., 739
Membranes, 220, 221-3
Mendeleev, D., 692
Mercury, mm of (unit of pressure), 153
Meson, 756-7, 760-1
quark theory of, 763-4
Meson exchange, 756-7
Metabolism, 253-5
Metals, bonding in, 697-8
Metastable state, 702, 719
Meter (unit), 11
determining the standard, 579
Meters:
correction for resistance of, 419-20
(*see also* Ammeter, Galvanometer, Voltmeter, etc.)
Metric system, 11, 42-4
MeV (million electron volts) (*see* Electron volt)
MHD, 483-4, 747
Michelson, A. A., 538-9, 579, 621-5
Michelson interferometer, 579, 622-5
Michelson-Morley experiment, 621-6
Microphones:
capacitor, 424
magnetic, 468
ribbon, 468
Microscope:
compound, 602-3
dark field, 611
interference, 611-2
magnification of, 602
phase contrast, 612-3
polarizing, 613
resolving power of, 608-10
UV, 611
Microstate, 267
Microwaves, 529
Millikan, R. A., 447-8, 652
Millikan oil-drop experiment, 447-8, 461
Mirror equation, 543-7
Mirrors:
focal length of, 542
parabolic, 543
plane, 540-1
spherical, 541-7
used in telescope, 601
MKS system of units, 43*fn*
(*see also* SI units)
Models, 4-5
Moderator in nuclear reactor, 741, 744
Modulation, 532-4
Moduli of elasticity, 102-5, 189, 291, 309
Molar heat capacity, 256-9
of gases, 256-8
of solids, 258-9
Mole, 193
volume of, for a gas, 194
Molecular biology, 3, 4, 353-60, 376-8, 699-700
Molecular mass, 182
Molecular spectra, 700-1
Molecular speeds, 197-203
Molecule, 181
(*see also* Atoms, Kinetic theory)
Moment of a force, 79
Moment of inertia, 81-3, 122, 129, 282
Momentum, 116-21
angular, 121-3
in collisions, 119-21, 137-8
conservation of angular, 122-3
conservation of linear, 116, 119-21, 137-8
Monitoring of nuclear radiation, 730
Monochromatic light, 576
Moon, 71-2
Morley, E. W., 621-5
Morse code, 532
Moseley, H. G. J., 694
Motion, 7-87, 116-41, 618-44
circular, 60-9
description of (kinematics), 7-30, 52, 60-5
general, 86, 136
graphical analysis of, 27-30
under gravity (free fall), 17-21
horizontal, 1-2, 37-8
linear, 7-21, 27-30
Newton's laws of, 38-42, 70, 79-83, 117
nonharmonic, 283-4
nonuniform circular, 69
projectile, 51-4
relative, 23-7, 618-44
rotational, 60-71, 79-83, 86, 121-3
simple harmonic, 274-83
translational, 7-55, 86
uniform circular, 64-8
uniformly accelerated, 14-7
uniform rotational, 63-4
of waves (*see* Waves)
Motor, electric, 450-1, 486
counter emf in, 471-2
Multimeter, 418-20
Multiplication factor, in nuclear reactor, 741
Mu meson, 757, 761
Muon, 757, 761
Muon lepton number, 759-60
Muon neutrino, 759, 761
Muscle contraction, 358-60
Muscles and joints, 80-1, 98-100, 118
Music, 316-23, 324-6, 328
Musical instruments, 302, 316-20
tuning of, 328
Musical scales, 324-6
Mutual inductance, 476-7
Myopic eye, 596-8

Natural abundances, 711, 785-7
Natural frequency (*see* Resonant frequency)
Near point, of eye, 596
Nearsighted eye, 596-8
Negative electric charge, 342
Negative feedback, 186, 512

Negative pressure, 163–4
Neptune, 78
Nernst equation, 400–1
Nerves and nerve conduction, 398–403
Neutrino, 718–9, 759, 761
  electron, 759, 761
  muon, 759, 761
Neutron, 710, 712, 761
  produced by reactor, 744
  used to produce new elements, 738–9
  role in fission, 739–41, 744
Neutron number, 710
Newton (unit), 42, 44
Newton, Isaac, 7, 38, 71–3, 76–9, 564–5, 576, 618
Newtonian focus, 601
Newton's law of cooling, 249
Newton's law of universal gravitation, 71–4, 76–9
Newton's laws of motion, 38–42, 70, 117
  applications of, 45–51
  for rotational motion, 79–83
Newton's rings, 576–7
Newton's synthesis, 76–9
Nodes, 300–3, 318–20
Noise, 322–4
Nonharmonic motion, 283–4
Nonlinear device, 505
Normal force, 47
npn transistor, 506–7
n-type semiconductor, 502–3
Nuclear binding energy, 712–3
Nuclear energy:
  levels of, 719
  released in fission, 740
  released in fusion, 746–7
Nuclear fallout, 743
Nuclear fission, 739–45
  and bomb, 742–3
  and chain reaction, 740–1
  discovery of, 739
  energy release in, 740
  in nuclear reactors, 741–2, 744–5
Nuclear force:
  and binding energy, 712–3
  quanta of, 755–7
  role in radioactive decay, 714, 724
  short range of, 714
  strong, 714, 724, 755–7
  weak, 714, 718–9, 757
Nuclear fusion, 745–7
Nuclear masses, 711–2, 785–7
Nuclear physics, 710–66
Nuclear power, 481, 744–5, 747
Nuclear radiation, 715, 726–7
  activity of, 720, 729
  alpha, beta, gamma, 715
  damage to biological organisms, 727–8, 730–1, 743
  detection of, 731–3
  dosimetry for, 729–31
  medical uses of, 726, 728–9
  passage through matter, 726–7
Nuclear reactions, 737–9
Nuclear reactors, 741–2, 744–5
  breeder, 745
  "critical," 742, 744
  first, 741–2
  fusion, 747
  for power, 744–5
  problems of, 481, 744–5
  for research, 744
  "subcritical," 742
Nuclear testing, 743
Nuclear transmutation, 716–7, 737–9
Nuclei, 343, 710–20, 737–47
  compound, 739
  daughter, 716
  half-lives of, 720–2
  masses of, 711–2, 785–7
  parent, 716
  radioactive decay of, 714–24
  size of, 711
  stability of, 714, 715, 716, 718
  structure of, 710–11
  table of, 785–7
Nucleon, 710
Nuclide, 710
  masses of, table of, 785–7
Numerical aperture, 609

Object distance, 540, 544, 552–3
Objective lens, 600, 602, 662
  oil-immersion, 609
Octave, 324
Ocular lens (*see* Eyepiece)
Oersted, H. C., 439, 441, 520
Ohm (unit of resistance), 389
Ohm, G. S., 389
Ohmmeter, 433
Ohm's law, 389
Oil-drop experiment, 447–8, 461
Oil-immersion objective lens, 609
Operational definitions, 456
Oppenheimer, J. Robert, 742
Optical activity, 583–4
Optical illusions, 547–8
Optical instruments, 592–613
Optic axis, 585
Orbital quantum number, 688–9
Order and disorder, 264–70
Order of interference pattern, 567, 572
Order of magnitude and rapid estimating, 55, 780
Ordinary ray, 585
Organ pipes, 317–20
Orthodonture, 93–4
Oscillations (*see* Vibrations)
Oscillator, 511
Oscilloscope, 500–1, 510
Osmosis, 221–3
  in life processes, 222
  reverse, 222–3
Osmotic pressure, 221–2
Overtone, 301–2, 318

Pacemakers, 429–30
Pair production, 654–5
Pantheon, 108–9
Paramagnetism, 457*fn*
Parent nucleus, 716
Partial ionic character, 696–7
Partial pressure, 197, 210, 217
  Dalton's law of, 197
Particle accelerators, 750–5
  cyclotron, 752–3
  linear accelerators, 754
  synchrocyclotron, 753
  synchrotron, 753–4
  Van de Graaf, 751
Particle exchange as carrier of force, 755–7
Pascal (unit of pressure), 149
Pascal's principle, 151–3
Pauli, W., 691, 718
Pauli exclusion principle, 691, 695
Pendulum:
  physical, 282–3
  simple, 280–2
Period:
  of planet, 77–8
  of vibration, 275, 277–8
Periodic table, 692–3, inside front cover
Periodic wave, 287
Permeability:
  magnetic, 452, 455, 457
  of membrane, 220
Permittivity, 347, 348, 370
Phase:
  in ac circuits, 489–97
  changes of, 208–15, 233–6
  of waves, 298
  (*see also* States of matter)
Phase angle, 496
Phase-contrast microscope, 612–3
Phase diagram, 213
Phasor diagram, 494–7
Phon (unit), 316
Phonograph cartridge, 468
Phosphor, 501, 731
Phosphorescence, 701–2
Photocathode, 731
Photocell, 651
Photodiode, 653
Photoelectric effect, 651–3
  applications of, 653
Photographic emulsion, 732
Photomultiplier (PM) tube, 731–2
Photon as elementary particle, 755–6, 760–1
  virtual, 756
Photon exchange, 755–6
Photon theory of light, 650–5
  and atoms, 670–4, 679–80
  in biology, 653–4
  (*see also* Gamma rays, X rays)
Photosynthesis, 654
Photovoltaic cells, 483
Piezoelectric effect, 423
Pi meson, 757, 761
Pion, 757, 761
Pipe, vibrating air columns in, 317–20
Pitch of a sound, 309
Planck, Max, 648–50
Planck's constant, 649, 650
Planck's quantum hypothesis, 648–50
Plane-polarized light, 580
Planets and planetary motion, 72, 76–8
Plane waves, 526
Plasma, 147, 483, 747
Plum-pudding model of the atom, 667
Pluto, 78
Plutonium, 739, 741, 743, 745
pn junction diode, 504–6

pnp transistor, 507
Poise (unit), 170
Poiseuille, J. L., 170–1
Poiseuille's law, 170–2
Poisson, S., 570
Polarization, 580–7
Polarized light, 580–7
Polarizing angle, 582–3
Polarizing microscope, 613
Polar molecules, 343, 371, 696–7
Polaroid, 580–2
Poles of magnet, 436
Pole vault, 135
Pollution (*see* Environmental pollution)
Positive electric charge, 342
Positive feedback, 186, 512
Positive holes, 502–3
Positivism, 40
Positron, 654–5, 718, 758
Post-and-beam, 107
Potential difference, electric, 365–7, 369–71, 389
Potential energy, 129–32
  chemical, 132
  elastic, 130–1
  electric, 364–9, 371–3
  gravitational, 129–30
  vibrational, 258–9, 276–7
  (*see also* Nuclear energy)
Potential-energy diagrams, 698–700, 724
Potentiometer, 420–1
Pound (unit), 42
Power, 139–40
  (*see also* Electric power, Nuclear power)
Power amplifiers, 510
Power factor (ac circuit), 496
Power generation, 479–84
Power of a lens, 551, 554, 597–8, 606
Power plants, 479–84
Powers of ten, 772–3
Power supply, 511
Poynting vector, 531
Precipitator, thermal gradient, 203–4
Pressure, 149–56
  atmospheric, 151, 153
  blood, 154, 172
  of gas, 191–200
  gauge, 151
  measurement of, 153–5
  negative, 163–4
  osmotic, 221–2
  partial, 197, 210, 217
  units for and conversions, 153
  vapor, 209–11
Pressure amplitude, 312
Pressure transducers, 422–3
Pressure waves, 312
Prestressed concrete, 106–7, 110
Primary, of transformer, 473
Principal axis, 542, 550
Principal planes, 605
Principal quantum number, 688–9
Principle of complementarity, 658–9, 661
Principle of correspondence, 675, 677
Prism, 569, 575
Prism binoculars, 549, 601–2
Probability:
  in nuclear decay, 720
  in quantum mechanics, 678–80, 683–4
  and second law of thermodynamics, 267–70
  determinism, 683–4
Probability distributions of electrons, 679–80, 690, 695–7
Problem solving techniques, 15–6, 55, 95
Projectile motion, 51–4
Proportionality (review), 769–70
Proteins:
  synthesis and structure of, 355–7
  x-ray diffraction of, 658
Proton, 710, 712, 761
Pseudoforce, 70
Psi/*J* particle, 760*fn,* 764–5
Ptolemy, 2
p-type semiconductor, 502–3
Pulleys, 125–7
Pulse amplifier, 510
Pumps, 155–6
PV diagram, 212–3
Pyrex glass, 189

Quality factor (QF) of nuclear radiation, 730
Quality of a sound, 321–2
Quantities, base and derived, 43–4
Quantum, 650
Quantum hypothesis, Planck's, 648–50
Quantum mechanics, 676–707
  of atoms, 679–80, 688–94
  Copenhagen interpretation of, 683
  of molecules, 694–8
Quantum number, 670
  magnetic, 689
  orbital, 688–9
  principal, 688–9
  spin, 689
Quantum theory, 648–707
  of atoms, 670–94
  of blackbody radiation, 648–50
  of light, 650–9
  (*see also* Quantum mechanics)
Quarks, 763–6
Q factor of resonant circuit, 517

Rad (unit for nuclear radiation), 729–30
Rad equivalent mass (rem), 730
Radian measure for angles, 60
Radiation (electromagnetic), 523–34
  blackbody, 649
  gamma, 715, 719
  from hot objects, 241–3, 648–50
  from humans, 242
  infrared, 241, 529
  microwave, 529
  synchrotron, 754
  ultraviolet, 529, 568–9
  x (*see* X rays)
  (*see also* Nuclear radiation)
Radiation damage, 727–8, 743
Radiation therapy, 728–9
Radio, 532–4
Radioactive dating, 724–5
Radioactive decay constant, 720–1
Radioactive decay modes, 715–20
Radioactive fallout, 743
Radioactive series, 722–3
Radioactive tracers, 726, 729
Radioactivity, 714–31
  in fallout, 743
  hazards of, 727–8, 730–1
  medical uses of, 726, 728–9
  probabilistic nature of, 720
  (*see also* Nuclear radiation)
Radio waves, 528–9, 534
Radius of gyration, 82–3
Rainbow, 569
Ray, 294, 537
  extraordinary, 585
  ordinary, 585
Rayleigh criterion, 608
Ray model of light, 537–8
  (*see also* Geometric optics)
Ray tracing, 551–2
RBE, 730
RC circuit, 426–7, 501
Reactance, 491–3
Reactions, nuclear, 737–9
Reactor (*see* Nuclear reactor)
Real image, 541, 552
Receivers, radio and television, 533–4
Recoil, 121
Rectification, 498, 505–6
Rectifier, 498, 505–6, 516
Reference frames, 8–9, 70, 619
  inertial, 70, 619–20
  noninertial, 70
  rotating, 70–1
Reflecting telescope, 601
Reflection, 293–5, 539–47
  diffuse, 540
  law of, 539
  of light, 539–47
  phase changes during, 577
  from thin films, 576–9
  total internal, 548–50
  of water waves, 293–5, 299
  of waves on a string, 294
Reflection grating, 574
Refraction, 295–7, 547–8, 565
  double, 585–6
  index of, 539, 585–6
  law of, 548, 565
  of light, 547–8, 565
  and Snell's law, 548
  by thin lenses, 550–8
  of water waves, 295–7
Refrigerator, 262–3
Relative biological effectiveness (RBE), 730
Relative humidity, 210–11
Relative motion, 23–7, 618–44
Relative velocity, 23–7, 639–40
Relativity, Galilean-Newtonian, 618–22
Relativity, special theory of, 618–44
  four-dimensional space-time, 635–6
  of length, 633–5
  and Lorentz transformation, 642–4
  of mass, 636
  mass-energy relation in, 637–9
  postulates of, 626
  simultaneity in, 627–9

of time, 629–33
Relativity principle, 619, 626
Relay, magnetic, 459
Rem (unit), 730
Resistance, 389
internal, 411–2
Resistance thermometer, 392
Resistivity, 390–2, 503
temperature coefficient of, 391–2
Resistors, 389–90
in ac circuit, 489
in parallel, 408–10
in series, 409–10
shunt, 416
Resolution:
of electron microscope, 661–3
of eye, 610
of lens, 607–8
of light microscope, 608–10
of telescope, 608–10
Resolving power, 608–10
(*see also* Resolution)
Resonance, 284–5
in ac circuit, 497
Resonances, elementary particle, 762
Resonant frequency, 285, 300–2, 497
Resonator, 317
Resting potential, 400–1
Rest mass, 636
Retentivity (magnetic), 458
Retina, 595
Reverberation, 329–30
Reverse current, 504
Reversible process, 261–2
Review of mathematics, 769–79
Reynolds' number, 172, 173–4
Ribosomes, 175
Richter, B., 764
Right-hand rules, 439, 442, 443
RNA, 356–7
Rockets, 41, 121
Roemer, Olaf, 538
Roentgen (unit), 729
Roentgen, W. C., 655
Rotational inertia, 81–3
Rotational kinetic energy, 128–9, 136
molecular, 257
Rotation and rotational motion, 60–71, 79–83, 86, 121–3, 128–9, 136
Rotatory power, optical, 584
Rotor, 68, 450
Rumford, Count, 227, 229
Russell, Bertrand, 661
Rutherford, Ernest, 667–8, 715, 737–8
Rydberg constant, 669, 673

Sailboat and Bernoulli's principle, 168
Satellites, 74–6
Saturation (magnetic), 457
Scalar, 10
Scales, musical, 324–6
Scattering of light, 586–7
Schrödinger, Erwin, 3, 677–8
Science:
creativity in, 1–3
hypotheses in, 1–2
laws and principles in, 5, 40
models in, 4–5
theories in, 2, 4–5
Scintillation counter, 731–2
Scintillator, 731–2
Second (unit), 11
Secondary of transformer, 473
Sedimentation, 68, 174–6
Sedimentation velocity, 174–5
Segrè, E., 758
Self-inductance, 477–8
Semiconductor, 343–4, 501–9
in diodes, transistors, integrated circuits, 504–9
energy bands in, 503
intrinsic, 503
n and *p* types, 502–3
resistivity of, 343–4
Semiconductor detector, 732
Sensitivity:
current, 416
of meter, 419–20
Servomechanism, 513
Shear modulus, 103, 104–5
Shear stress, 104–7
Shells, atomic, 692–3
SHM (*see* Simple harmonic motion)
Shock waves, 334–5
Shunt resistor, 416
Shutter speed, 592
Significant figures, 770–1, 773
Silicon, 502
Simple harmonic motion, 274–80
applied to pendulum, 280–3
Simple harmonic oscillator, 276
Simultaneity, 627–9
Single-lens reflex camera (SLR), 594
Sinusoidal curves, 279–80
SI units (*Système International*), 42–4
Sky, color of, 586–7
Slope, of a curve or graph, 28
Slug (unit), 42
Snell, W., 548
Snell's law, 548
Soap-bubble colors due to interference, 576–8
Solar cells, 483
Solar constant, 242
Solar energy, 243–4, 482–4
Solenoid, 440–1
Solids:
bonding in, 697–8
energy levels in, 700
(*see also* Phase, changes of, Specific heat capacity)
Solid-state detector (*see* Semiconductor detector)
Solid-state devices, 501–9
Sonar, 330–1
Sonic boom, 334–6
Sound "barrier," 335
Sounding board, 317
Sound waves, 308–36
detection of, 308, 313–6
Doppler shift of, 332–4
intensity of (and dB), 310–3
interference of, 326–8
source of, 308, 316–20
speed of, 309
Space heating, 243–4
Space-time, 635–6
Space travel, 632
Spark chamber, 732–3
Sparks, 388
Special theory of relativity (*see* Relativity)
Specific gravity, 149
Specific heat capacity, 229–30, 255–8
Spectrograph, mass, 448
Spectrometer:
light, 575
mass, 448
Spectrophotometer, 575
Spectroscope, 574–6
Spectrum:
absorption, 668, 673–4
band, 700
continuous, 575, 648–9
electromagnetic, 528–9
emitted by atoms, 668–70, 673, 676–7, 690, 691
emitted by hot object, 648–9
line, 575, 668–70, 673, 676–7
molecular, 700–1
visible light, 568–9
Speed, 7–9
molecular, 197–203
(*see also* Velocity)
Speed of light, 528, 538–9, 621–6
constancy of, 626
as ultimate speed, 636–7
Spherical aberration, 543, 604
Spin quantum number, 689
Spring, vibration of (*see* Vibrations)
Spyglass, 601
Stability, 100–1
Standard conditions (STP), 194
Standards and units, 10–2, 42–4
Standing waves, 300–2, 316–20
State:
changes of, 213–5, 233–6
energy (*see* Energy states)
of matter, 147, 183
as physical condition of system, 191
Static electricity, 341
Static friction, 47–8
Statics, 93–101, 107–10
Stationary states in atom, 670–6, 679–80
(*see also* Energy states)
Statistics and entropy, 267–70
Stator, 471
Steam engine, 186–7, 260–2
Stefan-Boltzmann constant, 241
Stimulated emission, 702
Stokes' law, 174
Stopping distances for car, 16–7, 32
Storage rings, 754
Strain, 102–4
Strain gauge, 422–3, 434
Strangeness, 763
Strange particles, 762–3
Strassmann, F., 739
Streamer chamber, 733
Streamline flow (*see* Laminar flow)
Stress, 102–7
compressive, 104–10
shear, 104–7
tensile, 104, 106–7
thermal, 188–90

Strings, vibrating, 300–3, 316–7
Strong nuclear force, 714
  and elementary particles, 755–7, 760–6
  quantum of, 755–7
  role in alpha-decay tunneling, 724
Sublimation, 214
Subshell, atomic, 692–3
Sunsets, 586–7
Superconductivity, 392–3
Superfluidity, 215
Superposition, principle of, 297
Supersonic speed, 334–6
Surface energy, 160
Surface tension, 159–62
Surfactant, 161
Svedberg (unit), 175
Swim bladder, 159
Synchrocyclotron, 753
Synchrotron, 753–4
Synchrotron radiation, 754
Systems:
  closed, 250
  isolated, 250
  open, 250
  as set of objects, 250
  of units, 11, 42–4

Tape-recorder head, 468–9
Telephoto lens, 594
Telescopes, 600–2
  astronomical, 600–2
  Galilean, 601
  magnification of, 601
  reflecting, 601
  resolution of, 608–10
  terrestrial, 601–2
Television, 500–1, 532–4
Temperature, 183–5
  distinguished from heat and energy, 228–9
  effect on chemical reactions and life processes, 202–3
  relation to molecular velocities, 197–203
  operating (of heat engine), 260
  scales of, 184–5
Tensile stress, 104, 106–7
Tension, 104, 106–7
Terminal velocity, 19*fn,* 174
Terminal voltage, 411–2
Terrestrial telescope, 601–2
Tesla (unit of magnetic field strength), 442
Test charge, 349
Testing of hypothesis, 1–2
Theory of relativity (*see* Relativity)
Thermal conductivity, 236
Thermal energy, 132, 136, 137, 228–9
  distinguished from heat and temperature, 228–9
Thermal expansion 187–91
  coefficients of, 187–90
  of water, 190–1
Thermal gradient precipitator, 203–4
Thermal pollution, 263–4, 481, 745
Thermal stress, 188–90
Thermal transmission coefficient, 249
Thermionic emission, 498–500
Thermistor, 392
Thermocouple, 423–4
Thermodynamics, 250–70
  first law of, 250–7
  second law of, 259–70
Thermography, 243
Thermometers, 183–5
Thermonuclear devices, 746
Thermostat, 186–7
Thin-film interference, 576–9
Thin lenses, 550–8
Thompson, Benjamin (*see* Rumford, Count)
Thomson, J. J., 446–7, 661, 667
Threshold energy, 738, 748
Thyratron, 427
Tidal power, 482
Timbre, 321–2
Time constant, 427, 479
Time dilation, 629–33, 644
"Time's arrow," 267
Ting, S., 764
Top, spinning, 122
Torque, 79–83, 95–100, 108, 122
  counter, 472
  on current loop, 449
Torr (unit), 153
Torricelli, Evangelista, 153, 167
Torricelli's theorem, 167
Total binding energy of nucleus, 713
Total internal reflection, 548–50
Tracers, 726, 729
Transconductance of triode, 517
Transducers, 422–4, 468–9, 509
Transformations:
  Galilean, 640–2
  Lorentz, 642–4
Transformer, 473–6, 514
Transition temperature, 392
Transistors, 506–9
Translational motion defined, 7
Transmission grating, 574
Transmission lines, 529–30
Transmutation, 716–7, 737–9
Transport processes, 223
Transverse waves, 289–92
Trees, fluid flow in, 163, 164
Triangles, geometric theorems of, 776–7
Trigonometric functions, 25
  table of, inside back cover
Triode, 499–500
Triple point, 214–5
Tritium, 746
Tropical seas, power from, 482
Tube, vibrating column of air in, 317–20
Tubes, vacuum, 498–501
Tuning circuit, 497, 533
Tunneling, 724
Turbulent flow, 164, 172–3, 174
Twin paradox, 632–3

Ultimate strength, 102, 105–7, 118
Ultrasonic waves, 309, 331
Ultrasound, 331
Ultraviolet "catastrophe," 649
Ultraviolet (UV) light, 529, 568–9
Uncertainty principle, 680–3, 724
  and lifetime of particle resonance, 762
  and mass of exchange particle, 756
  philosophic implications of, 683–4
  and tunneling in nuclei, 724
Unified atomic mass unit, 711–2
Uniform circular motion, 64–8
Uniformly accelerated motion, 14–7
Uniformly accelerated rotational motion, 63–4
Units of measurement, 10–2, 42–4
Universal gas constant, 194
Universal law of gravitation, 71–4, 76–9
Unpolarized light, 580
Uranium:
  in bomb, 742–3
  used in dating, 725
  enrichment of, 741, 744
  fission of, 739–42
  radioactivity of, 723–4
  in reactors, 741–2, 744–5
Uranus, 78
Useful magnification, 610
UV microscope, 611

Vacuum tubes, 498–501
Valence band, 503
Valve (*see* Vacuum tubes)
Van de Graaff generator and accelerator, 751
Vapor, 213
  (*see also* Gases)
Vaporization, latent heat of, 233–6
Vapor pressure, 209–11
Vectors, 10
  addition of, 21–7
  components of, 24–6
  resolution of, 24
  resultant, 21–3
  subtraction of, 23
Velocity, 9–30
  addition of, 23–7, 639–40
  angular, 61–4
  average, 9, 14
  of EM waves, 525–30
  instantaneous, 10
  of light, 528, 538–9, 621–6, 636–7
  linear, 9–30, 61
  molecular and relation to temperature, 197–203
  relative, 23–7, 639–40
  rms (root mean square), 200
  of sound, 309
  supersonic, 334–6
  terminal, 19*fn,* 174
  of waves, 288, 291
Venturi meter, 169, 180
Venturi tube, 168–9
Vibrational energy, 276–7
  molecular, 258–9
Vibrations, 274–85
  of air columns, 317–20
  of atoms and molecules, 183, 258–9
  forced, 284–5
  as source of wave, 287
  of strings, 300–3, 316–7
Virtual image, 541, 552
Virtual photon, 756

Viscosity, 164, 170–4
  coefficient of, 170, 171
Visible light, 528–9, 568–9
Visible spectrum, 568–9
Vision, 595–8
Vocal chords, 317
Volt (unit), 365
Volta, Alessandro, 365, 384–6
Voltage:
  bias, 500
  terminal, 411–2
  (*see also* Potential difference, electric)
Voltage amplifier, 510
Voltage drop, 413
Voltmeter, 417–20, 510

Water:
  desalination of, 222–3
  heavy, 741
  latent heats of, 233–4
  polar nature of, 343
  thermal expansion of, 190–1
Water equivalent, 232
Watson, J. D., 658
Watt (unit), 139
Watt, James, 139*fn*
Wave:
  continuous, 287
  earthquake, 291–2, 297
  electromagnetic, 520, 523–34, 563, 620–7
  energy in, 286, 292–3, 310, 530–2
  infrasonic, 309
  longitudinal, 289–92
  matter, 659–61
  periodic, 287
  plane, 526
  pressure, 312
  radio, 528–9, 534
  spherical, 293
  standing, 300–2, 316–20
  surface, 292
  transverse, 289–92
  ultrasonic, 309
  velocity of, 288, 291
  water, 286, 292–9
  (*see also* Diffraction, Interference, Light, Reflection, Refraction, Sound)
Wave front, 294
Wave function, 677–9
Wave intensity, 292–3
Wavelength, 287–8
  de Broglie, 660, 675–6
  as limit to resolution, 609–10, 661–2, 750
  of material particles, 659–61, 675–6, 750
  of visible light, 528–9, 568–9
Wave motion, 286–302
Wave nature of matter, 659–61, 675–6
Wave–particle duality, 658–9, 661, 677
Wave pulse, 286
Wave theory of light, 563–87
Wave velocity, 288, 291
Weak bonds, 376–8, 697–8
Weak nuclear force, 714
  and elementary particles, 757, 760–1, 763
  quantum of, 757
  role in beta decay, 714, 718–9
Weber (unit), 442, 465
Weight, 39, 44, 73, 75–6
Weightlessness, 74–6
Wheatstone bridge, 421–2
Wide-angle lens, 594
Wiener, Norbert, 513
Wien's law, 649
Wind:
  acting on building, 97–8
  as convection, 238
  of Yosemite (convection on hiking trail), 239–40
Wind power, 482
Wing of airplane, lift on, 168
Work, 123–8, 139
  compared to heat, 250
  done by a gas, 252–3
  in first law of thermodynamics, 250–1
  from heat engines, 261–3
  relation to energy, 128, 131–3
Work-energy theorem, 128, 131–2, 167
W particle, 757

X-ray diffraction, 657–8
X rays, 529, 655–8
  and atomic number, 693–4
  characteristic, 656, 694
  effects on matter, 727, 730
  in electromagnetic spectrum, 529

Yosemite, 239–40
Young, Thomas, 566–8
Young's double-slit experiment, 566–8
Young's modulus, 102–4, 189
Yukawa, H., 755–7
Yukawa particle, 755–7

Zeeman effect, 689
Zener diode, 505
Zero, absolute, 192
Zoom lens, 594
Zweig, G., 763